Advances in Renewable Energy and Energy Storage

Advances in Renewable Energy and Energy Storage

Editors

Luis Hernández-Callejo
Jesús Armando Aguilar Jiménez
Carlos Meza Benavides

MDPI

Basel • Beijing • Wuhan • Barcelona • Belgrade • Novi Sad • Cluj • Manchester

Editors

Luis Hernández-Callejo
Department of Agricultural and Forestry Engineering, University of Valladolid, Campus Duques de Soria
Soria, Spain

Jesús Armando Aguilar Jiménez
Facultad de Ingeniería, Universidad Autónoma de Baja California
Mexicali, Mexico

Carlos Meza Benavides
Department of Electrical, Mechanical and Industrial Engineering, Anhalt University of Applied Sciences
Köthen, Germany

Editorial Office
MDPI
St. Alban-Anlage 66
4052 Basel, Switzerland

This is a reprint of articles from the Topic published online in the open access journals *Solar* (ISSN 2673-9941), *Energies* (ISSN 1996-1073), *Batteries* (ISSN 2313-0105), *Electronics* (ISSN 2079-9292), *Sustainability* (ISSN 2071-1050), and *Applied Sciences* (ISSN 2076-3417) (available at: https://www.mdpi.com/topics/Energy_Storage).

For citation purposes, cite each article independently as indicated on the article page online and as indicated below:

Lastname, A.A.; Lastname, B.B. Article Title. *Journal Name* **Year**, *Volume Number*, Page Range.

ISBN 978-3-0365-9552-8 (Hbk)
ISBN 978-3-0365-9553-5 (PDF)
doi.org/10.3390/books978-3-0365-9553-5

Cover image courtesy of Luis González Uribe and Jesús Armando Aguilar Jiménez
Solar Photovoltaic Parking Lot of the Engineering Faculty of the Autonomous University of Baja California.

Contents

About the Editors

Luis Hernández-Callejo

Prof. Dr. Luis Hernández-Callejo holds degrees as an Electrical Engineer at Universidad Nacional de Educación a Distancia (UNED, Spain), a Computer Engineer at UNED, and a PhD at Universidad de Valladolid (Spain). A professor and researcher at the Universidad de Valladolid, his areas of interest are renewable energy, microgrids, photovoltaic energy, wind energy, smart cities, artificial intelligence.

Prof. Dr. Hernandez Callejo is an editor in numerous scientific journals, and is a guest editor in many Special Issues. He has directed four doctoral theses, and at the moment, he is directing six doctoral theses. He is a professor in wind energy, solar energy, and microgrids, and he collaborates with many universities in Spain and in the rest of the world.

Jesús Armando Aguilar Jiménez

Dr. Jesús Armando Aguilar Jiménez holds a PhD in Engineering from the Institute of Engineering at the Autonomous University of Baja California (UABC). He is a faculty member at the Faculty of Engineering, Mexicali Campus, UABC, and is recognized as a Level 1 member of the National System of Researchers. His research interests encompass renewable energies, energy simulation, energy conservation, and efficient energy use, as well as energy innovation. Currently engaged in energy projects for both the public and private sectors, he actively participates in academic committees for the evaluation of scientific research. Dr. Aguilar-Jiménez also contributes to the development of human capital through teaching at both undergraduate and postgraduate levels, including the supervision of postgraduate theses.

Carlos Meza Benavides

Dr. Carlos Meza Benavides is an electronics engineer with a Master's degree in Embedded System Design from the Università della Svizzera italiana (USI), Lugano, Switzerland; an Advanced Studies Diploma in Dynamic Systems; and a PhD in Advanced Automation and Robotics from the Universitat Politècnica de Catalunya—BarcelonaTech (UPC), Spain. He has held various research positions, including as a visiting researcher at the University of Groningen, Netherlands; a postdoctoral researcher at the Institute of Applied Sustainability for the Built Environment in Cannobio, Switzerland; and a senior postdoctoral researcher at UNESCO's International Centre for Theoretical Physics in Trieste, Italy. Dr. Meza has served as a consultant on energy issues in Latin America for European projects and the Organization of American States. He was the founder and first president of the Costa Rican Solar Energy Association (Acesolar). Currently, he is a professor in the Master's program in Photovoltaic Engineering at Anhalt University of Applied Sciences in Germany and teaches and advises students at the Costa Rica Institute of Technology (TEC) and the Valencia International University. His research interests include photovoltaic system analysis, embedded system design for energy systems, and power electronics for renewable energy and storage systems ....

Preface

"Advances in Renewable Energy and Energy Storage" constitutes a thorough examination of the crucial transition towards a global energy matrix based on renewable sources. This reprint focuses on cutting-edge technologies, highlighting enhanced efficiency, superior energy integration, and cost-effectiveness to foster the widespread adoption of renewable energies. The objectives encompass guiding collective efforts towards sustainable development, mitigating environmental pollution, and addressing the challenges of climate change.

Motivated by the urgent need for sustainable energy solutions, this reprint aims to inform and inspire diverse stakeholders, including academics, industry professionals, and policymakers. The authors, distinguished experts in the field, have collaborated to distill innovative research, providing a comprehensive resource for decision makers in both the public and private sectors. We extend our sincere gratitude to those who contributed to this effort, recognizing the collective commitment needed to steer our world towards a resilient and environmentally conscious energy future.

Luis Hernández-Callejo, Jesús Armando Aguilar Jiménez, and Carlos Meza Benavides
Editors

solar

MDPI

Article

Assessing the Potential of Qatari House Roofs for Solar Panel Installations: A Feasibility Survey

Ayed Banibaqash, Ziad Hunaiti and Maysam Abbod *

Department of Electronic and Electrical Engineering, College of Engineering, Design and Physical Sciences, Brunel University London, Uxbridge, London UB8 3PH, UK
* Correspondence: maysam.abbod@brunel.ac.uk

Abstract: Qatar's ambitious Vision 2030 includes a major shift towards clean energy, and residential solar PV installation can be an obvious option, given its abundant sunlight and high power for residential cooling. Despite significant solar panel farm investment, there has been limited progress in deploying solar panels on home roofs, and further research is needed to identify the potential for such an initiative and its impact on the country's move towards clean energy. This field survey assesses the potential for residential rooftop solar panel installation across Qatar, considering space availability, currently utilized space, remaining space, shading, and roof type. It also provided indications of potential obstacles and shading that might affect panel sunlight exposure. The results showed that there is significant potential for installing solar panels on Qatari homes, which could contribute to a considerable portion of the energy consumed by households during peak usage periods, particularly in the summer months. Moreover, excess energy generated could be exported to other countries with high demand during periods of low demand in Qatar. The study's findings complement previous research efforts and provide insights for policymakers and stakeholders to develop strategies that endorse the vision for 2030 and promote the transition towards clean energy in Qatar.

Keywords: renewable energy; solar panels; Qatar; roof solar panels; roof survey; installation feasibility; solar panel installation drivers

Citation: Banibaqash, A.; Hunaiti, Z.; Abbod, M. Assessing the Potential of Qatari House Roofs for Solar Panel Installations: A Feasibility Survey. *Solar* **2023**, *3*, 650–662. https://doi.org/10.3390/solar3040035

Academic Editor: Jürgen Heinz Werner

Received: 22 July 2023
Revised: 25 September 2023
Accepted: 26 October 2023
Published: 10 November 2023

1. Introduction

Qatar is committed to achieving its Vision 2030, which aims to create a sustainable and diversified economy as well as a healthy and secure society. One of the key strategies to achieve this vision is through a major shift towards clean energy [1]. Given the country's geographic location, Qatar is well-suited to harnessing the power of the sun to achieve its sustainable electricity goals. The country has abundant sunlight throughout the year, making it an ideal location for the installation of solar PV arrays [2]. Qatar has already taken significant steps towards achieving its vision of clean energy and has invested heavily in solar panel farms, such as the 800 MW Al-Kharsaah Solar Power Plant, which is expected to power around 10% of the country's energy needs [3]. These farms are a critical component of Qatar's clean energy strategy, but more can be undertaken. The majority of Qatar's energy load is domestic use, especially for cooling during the summer months, as noted by the Annual Statistics Report 2021 of Qatar General Electricity & Water Corporation (KAHRAMAA), accounting for 38,284,270 MWh out of the total amount of electricity generated nationally per year (Figure 1) [4]. It is clear from the figure that domestic users account for nearly four times the number of industrial consumers. This fact supports the objective of this study, indicating that residential buildings have the potential to play a significant role in the transition toward clean energy and achieve milestones toward the national net-zero target. At the time of this study, there is still limited progress in deploying solar panels on home roofs, despite the country's high levels of sunshine and the potential benefits that such installations can bring [2].

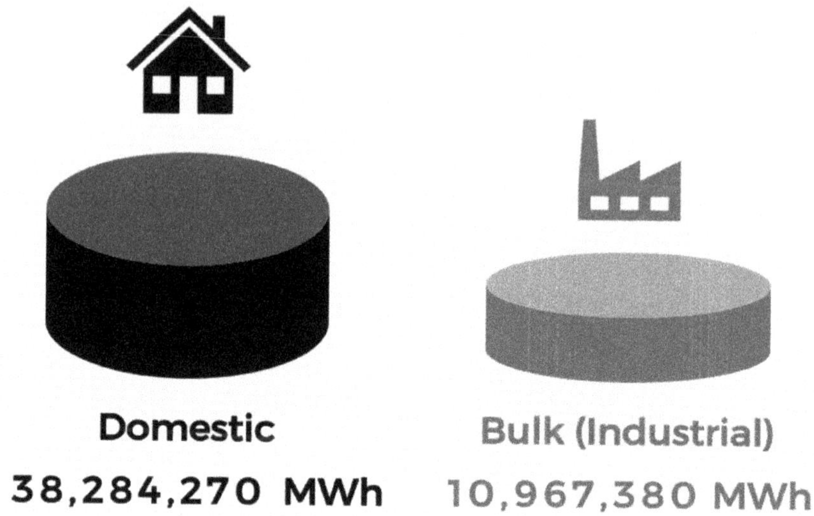

Figure 1. Domestic energy consumption [4].

Deploying solar panels on home roofs can contribute to the country's clean energy goals by creating a distributed network of clean energy production. This can help reduce the country's dependence on fossil fuels (as well as the cannibalization of its primary national exports), lower carbon emissions, and create a more resilient energy system. Given that the country is one of the main energy exporters and experiences a high number of sunny days, it can enable significant clean energy generation that can be exported [2]. However, further research is needed to identify the potential for this initiative and its impact on the country's move towards clean energy. This includes examining factors such as the availability of suitable roof space, the orientation of the roofs, and the presence of any shading or obstructions that might limit the effectiveness of the solar panels. By conducting such research, stakeholders can better understand the potential of solar panels on residential roofs and develop strategies to accelerate and optimize their deployment.

2. Drivers for Roof Solar Panels in Qatar

Qatar has made progress in large-scale solar farm investments, but residential solar panel deployment is still needed to achieve the nation's clean energy vision by 2030 [1]. Previous research has identified key challenges, as summarized in Figure 2, that must be addressed before scaling up solar panel deployment in Qatar [5]. These challenges include a lack of interest in renewable energy, competition from other energy sources, uncompetitive pricing due to subsidized conventional electricity, limited awareness of renewable energy, insufficient government initiatives, and a lack of environmental concern. Similar challenges exist in other Gulf Cooperation Council (GCC) countries. Thus, addressing these issues is necessary before considering residential and commercial solar panel deployment in Qatar. A national roadmap for sustainable energy is crucial for achieving this long-term goal.

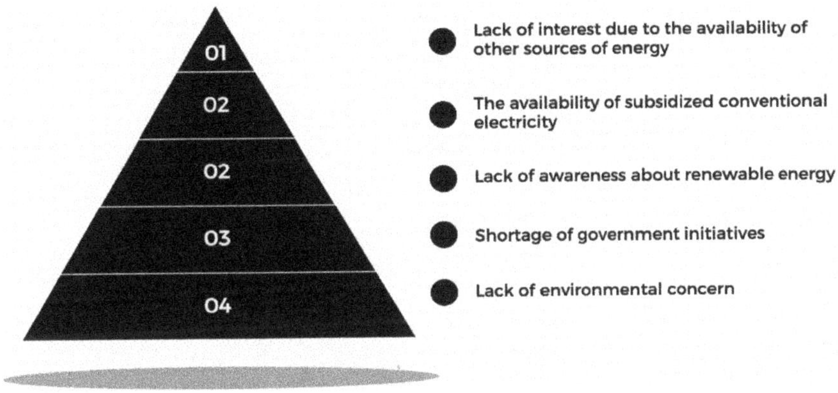

Figure 2. Five key challenges to residential solar PV adoption [5].

By considering the six drivers for promoting solar panel adoption, namely potential, awareness, the net-zero pathway, energy efficiency, lowering subsidies, and sustainability, stakeholders can develop effective strategies to encourage the installation of solar panels and achieve the vision of a sustainable future (Figure 3).

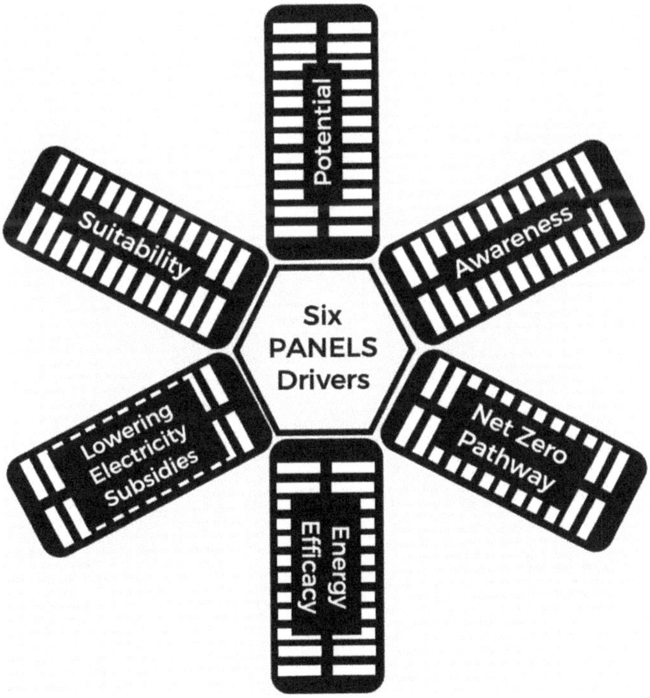

Figure 3. Six PANELS drivers [authors].

2.1. Potential

Qatar has vast potential for solar energy investment, positioning itself as a leader in clean energy exports. Factors such as high sun hours, market experience, and financial capacity make Qatar an ideal location for solar energy ventures [6]. Qatar is well-suited for solar energy investments, with an average of 3600 h of direct solar radiation annually [7].

The country is committed to reducing its carbon footprint and aims to generate 20% of its energy from renewables by 2030, with solar energy playing a significant role [8].

A recent feasibility study explored various scenarios of solar panel deployment in Qatar's residential and commercial buildings, considering panel sizes, efficiency, and daily sun exposure. Results indicated substantial potential for rooftop solar panels, generating significant energy during peak usage, particularly in hot summer months. Surplus energy could be exported to countries with high energy demands, positioning Qatar as a major contributor to solving global energy crises [2].

Qatar's extensive experience in the energy market, particularly in the oil and gas industry, can be leveraged to develop the solar energy sector and gain a competitive advantage in the global clean energy market [9]. The country's financial capacity and resources enable large-scale solar energy investments [10]. Notably, Qatar has already invested in significant solar energy projects, including the Al-Kharsaah Solar Power Plant, one of the world's largest [11]. With its potential for solar energy investment and clean energy exports, Qatar can contribute to its own energy needs, reduce carbon emissions, and address global energy demands [12]. Its high sun hours, market experience, and financial capacity make it an ideal destination for solar energy investment [2].

2.2. *Awareness*

Awareness plays a vital role in promoting the adoption of renewable energy resources. In Qatar, a lack of awareness has been identified as a significant challenge hindering the deployment of renewable energy [13]. To address this challenge, the government should implement projects that allow households to actively participate and experience the benefits of installing solar panels. This can effectively raise awareness throughout the nation [14]. Given Qatar's socially connected society, visible solar panels on homes would initiate conversations and spark interest in renewable energy, leading to a snowball effect of adoption [15]. The installation of solar panels serves as a strategic tool to raise awareness and drive the transition toward clean energy use [16]. Not only does it offer tangible benefits like reduced energy bills, but it also visually demonstrates individuals', businesses', and nations' commitment to reducing carbon emissions and contributing to global climate change mitigation efforts [17]. This visual representation can generate public interest and awareness of renewable energy, inspiring more individuals and businesses to embrace clean energy sources [18].

Moreover, increasing public awareness about renewable energy can garner support for policies and initiatives promoting clean energy adoption [17]. This creates a favorable environment for renewable energy investment and deployment in Qatar, encouraging businesses to invest in renewable energy projects. Therefore, enhancing public understanding of the nature and significance of clean and renewable energy sources is a crucial step in establishing them as a normative and primary component of the national energy mix. In Qatar, the installation of solar panels on homes acts as a catalyst for raising awareness and driving the transition to clean energy. By generating public interest and support for renewable energy, Qatar can contribute to global climate change mitigation efforts and foster a more sustainable future for all.

2.3. *Net-Zero Building Pathway*

To achieve more sustainable and energy-efficient buildings, implementing a net-zero strategy is crucial. Net-zero buildings have the capability to generate as much energy as they consume, resulting in a neutral or net-zero energy balance [19]. Currently, homes in Qatar heavily rely on external energy sources, but the installation of solar panels on rooftops can pave the way towards net-zero buildings [20]. By incorporating solar panels, homes in Qatar can reduce their dependence on conventional energy sources and promote the use of renewable energy. The region's abundant sunshine makes solar energy an ideal solution for meeting residential energy needs [13]. Local energy production allows homes to decrease their reliance on grid-supplied electricity, which is often costly and contributes to carbon emissions [2].

Apart from the environmental advantages, transitioning to net-zero buildings can lead to significant cost savings for homeowners. By generating their own energy, homeowners can reduce their electricity bills and even potentially sell surplus energy back to the grid [19]. This introduces a new revenue stream and helps offset the upfront costs of solar panel installation [2]. Additionally, adopting a net-zero strategy improves overall energy efficiency in buildings, resulting in reduced energy consumption and lower carbon emissions. These efforts align with global initiatives to combat climate change [21]. Therefore, installing solar panels on homes in Qatar serves as a strategic step towards achieving net-zero buildings. Through localized energy production, homeowners can decrease their reliance on conventional energy sources, lower energy expenses, and contribute to global climate change mitigation [20]. Thus, implementing a net-zero strategy is essential for creating a more sustainable and energy-efficient built environment in Qatar.

2.4. Energy Efficiency

Energy efficiency is crucial for a sustainable energy strategy in Qatar, where all homes currently rely on fossil fuel-generated energy. Promoting energy efficiency is vital to reduce consumption, lower carbon emissions, and create a sustainable built environment [22]. By installing solar panels and implementing a net-zero strategy, homeowners are expected to shift towards using energy-efficient appliances and monitoring their energy usage [23]. Solar panel installation on Qatar's homes can transform the way homeowners consume energy. Producing their own energy raises awareness about consumption and environmental impact, fostering a greater appreciation for energy efficiency and a willingness to invest in low-energy appliances [23].

Furthermore, a net-zero strategy incentivizes energy efficiency through a feedback loop between production and consumption [23]. As homeowners become adept at generating their own energy, they become conscious of consumption patterns and take steps to reduce energy usage. This includes adopting energy-efficient habits, investing in efficient appliances, and monitoring consumption for potential savings. Therefore, promoting energy efficiency is vital for a sustainable energy strategy, especially in Qatar, where homes solely rely on fossil fuel-generated energy. Through solar panel installation and a net-zero approach, homeowners can enhance awareness, invest in efficient technologies, and contribute to a sustainable built environment [24]. Energy efficiency is essential for achieving a sustainable and energy-efficient future in Qatar.

2.5. Lowering Electricity Subsidies

Qatar can transition to a sustainable energy future by deploying solar panels on homes. Energy subsidies pose a challenge as the government provides free energy to citizens, hindering solar panel installation. However, the government can support panel installation to reduce non-clean energy use and subsidies [10,13]. This empowers homeowners to generate their own energy, reducing their reliance on non-clean sources [25]. It shifts the energy production burden from the government to individuals, lowering subsidies [10]. Homeowners can contribute to the grid and export excess energy, generating revenue [2]. Government support for solar panels demonstrates a commitment to sustainability and emission reduction, fosters environmental consciousness, and encourages investment in sustainable energy [25,26].

2.6. Suitability

The deployment of solar panels in Qatar can contribute significantly to achieving sustainability across the three main pillars: the economy, society, and the environment. Solar energy is a clean and renewable source of energy that can support the economy by providing a new source of energy and sustaining the other available sources for future generations [10]. By investing in solar energy, the country can reduce its reliance on non-clean sources of energy, which can reduce the costs associated with importing and transporting non-renewable fuels. This can, in turn, support economic growth and development [27]. Moreover, solar energy can benefit society by minimizing pollution, particularly carbon emissions. The use of solar panels can help reduce the amount of carbon particles in the air, which can lead to a healthier and more sustainable society. This can contribute to the overall well-being of the population, as reducing pollution can lead to a reduction in respiratory illnesses and other health issues [28]. By reducing the generation of CO_2 from clean sources of energy, solar panels can contribute to a more sustainable environment that can be kept for future generations. This can also help reduce the negative impact of climate change on the environment and support the preservation of natural resources [29].

In order to effectively facilitate the implementation of the six drivers for solar panel deployment in Qatar, it is essential to carry out a field survey aimed at evaluating the viability of installing solar panels on different types of residential buildings. This survey will provide valuable insights to homeowners and decision-makers, enabling them to develop suitable scenarios for the installation and effective utilization of solar panels on domestic roofs in Qatar [30].

3. Materials and Methods

To evaluate the feasibility of installing solar panels on homes in Qatar, a survey was conducted using a manual approach with a structured interview sheet with closed-ended questions as the data collection method, based on the previous literature (as explained in Table 1). The reason for opting for manual surveying instead of remote sensing methods, such as the use of satellite images [31], is primarily due to cost, accessibility, and issues related to legal and ethical approvals. Additionally, the use of drones has been considered; however, as of now, drone surveying is not yet allowed in Qatar [32].

The survey aimed to gather data on the availability of space for solar panel installation, currently utilized space, remaining space, shading, and roof type. The use of a structured interview sheet with closed-ended questions provided a standardized approach to data collection and made the process more efficient. Closed-ended questions were used to collect quantitative data, which could be easily analyzed statistically to identify patterns and relationships in the data [33]. As shown in Figure 4, the research design consisted of four stages to complete the research project. The first stage was designing and testing the data collection instrument, which was the most important part of the research design. It involved identifying the main aspects to be included within the data collection instruments, and an interview sheet was established. Each item included had a justification, as shown in Table 1.

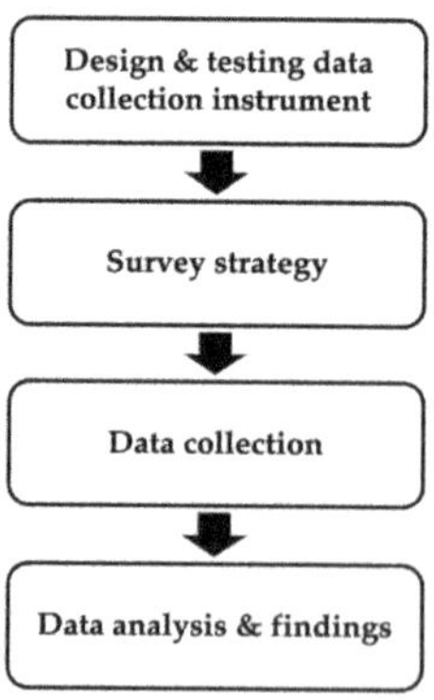

Figure 4. Research design [authors].

Table 1. Collected data and justification.

Interview Question	Rationale
House location	The location of a house in the field survey is essential to obtaining accurate geographical representation [3,34].
Number of similar houses in the street or compound	Knowing the number of similar houses in the same area is essential for a wider and more representative study. It helps with solar panel installation planning, as a community solar setup can be considered [34].
Number of bedrooms in the house	The number of bedrooms in a house is crucial for solar panel installation. It helps estimate energy consumption and the correct number of panels needed. Also, analyzing energy usage patterns identifies areas for conservation measures, reducing consumption, and enhancing solar panel effectiveness [35].
Items on the roof	Before installing solar panels, it is essential to identify existing objects that may hinder the process or limit available space. These can include AC units, chimneys, skylights, and other structures. Assessing their current use and placement on the roof helps determine the best solar panel installation approach [3].
Approximate total roof size in square meters	Determining the approximate size of a roof (in square meters) is essential when considering the installation of solar panels. This information can be used to calculate the amount of viable free space available for solar panel installation [3].
Average remaining empty space on the roof	To install solar panels on a roof effectively, assess the available space for optimal energy generation. Utilize the average remaining empty space, avoiding obstructions like vents or chimneys, to maximize panels and energy output. Plan and design carefully for an aesthetically pleasing integration with the roof's architecture [3].
Recreational use of the roof	Before installing solar panels on a recreational roof, assess their potential impact on leisure activities. The size and orientation of the panels might obstruct or limit recreational use [3]. However, with careful planning, solar panels can serve a dual purpose by providing shade and protection for leisure activities while generating renewable energy for the home [34].
Roof type	To install solar panels on a house, assess the roof type, condition, orientation, and angle for sufficient sunlight. Evaluating the roof's suitability ensures the maximum benefits of renewable energy for the home [3].
House height	When assessing the feasibility of solar panels for a house, consider the building height compared to surrounding structures. If taller neighboring buildings block direct sunlight from the east or west, optimal sun exposure may be challenging [3]. Also, evaluate the roof angle and orientation. Proper assessment ensures efficient solar panel installation, maximizing the benefits of solar energy for the home [34].
Solar shading	Solar shading is the process of identifying potential obstructions, like nearby buildings or tall trees, that may block sunlight from reaching a rooftop or solar panel installation. By understanding shading risks, effective measures can be taken to optimize sunlight exposure, maximize energy generation, and reduce the need for additional energy sources [3].

The second stage of the research design was the survey strategy. Since the majority of the country's population is based in Doha, the study selected areas to be included in the survey. Moreover, due to the fact that homes in similar neighborhoods typically have many similarities, the interview included asking the homeowners if nearby homes are similar to their own. This enabled achieving geographical representation as well as statistical representation to represent the whole number of homes in Qatar, which is expected to be around 365,000 domestic customers, according to KAHRAMAA's Annual Statistics Report 2021 [4].

The third stage of the research design involved data collection. To facilitate this process, the interview questions were digitized into an electronic form that allowed for convenient and efficient data collection on the spot [36]. This eliminated the need for further processing and streamlined the data collection process. Additionally, using an electronic form allowed for anonymous data storage, thereby ensuring the confidentiality of the participants [37]. Electronic data collection methods have become increasingly popular in recent years as they offer a range of advantages over traditional paper-based methods. For example, electronic forms can reduce the risk of errors and inconsistencies as well as improve the speed and efficiency of data collection. Furthermore, electronic forms can be easily stored and accessed, making it easier to analyze and interpret the data.

The final stage of the research design involved data analysis and the presentation of the findings. To achieve this task, Excel was used to plot graphs and perform the necessary statistical analyses to convert the collected data into meaningful information. Excel is a widely used spreadsheet software that can be used for data analysis and has many built-in statistical functions that can aid in the analysis process [38]. The use of Excel in data analysis has several advantages, such as the ability to handle large amounts of data, perform calculations efficiently, and present data in an easily understood way using graphs and charts [39]. Furthermore, the use of objective data in presenting research findings helps draw meaningful conclusions.

4. Results

Given the homogeneity of homes in Qatar [40], the use of a cluster sampling plan [41] enabled the surveying of 10 different areas in Doha, with a total of 50 homes physically surveyed from these areas, resulting in a total sample of 1068 similar homes in the same street or compound, as shown in Table 2. This sample size is considered larger than the typically required sample of 384 from the total population of 365,000, ensuring a statistically representative sample [42].

Table 2. Areas and number of houses surveyed.

Surveyed House Area	Number	Similar Houses in the Street or Compound
Al-Kheesa	20	700
Al Waab	10	240
The Pearl-Qatar	4	25
Al Waab	3	20
Musheireb	3	20
Al Sadd	3	20
Abu Hamour	2	13
Ain Khaled	2	12
Lusail	1	8
Al-Hilal	2	10
Total	50	1068

Table 3 summarizes the survey results.

Table 3. Survey results.

Survey Item	Outcome				
Average number of bedrooms per house	4				
Average total roof area [m^2]	244				
Average remaining total roof area [m^2]	144				
Average house height [m]	9				
Use of roof for recreational purposes	Yes 10%		No 90%		
Possibility for shading	Yes 30%		No 70%		
Type of house roof	Flat roof 100%	Gable roof 0%		Bonnet roof 0%	
Items on the roof	External AC units 80%	Satellite dishes 80%	Water tank 100%	Storage shed (box) 10%	Solar panel for water heating 20%

The survey findings indicate that the majority of homes in the area are relatively spacious, with 40% being five-bedroom homes and 20% being four-bedroom homes. The remaining 40% comprise three- and two-bedroom homes. The average total size of the homes is 244 square meters, which is quite substantial.

Regarding the features on the roofs of these homes, all houses have water tanks, and a significant majority (80%) have external AC units and satellite dishes installed. However, storage sheds are less common, present on only 10% of houses, and solar panels for water heating are found on 20% of homes.

Approximately 59% of the roof space remains empty after accounting for the mentioned items, suggesting that over 50% of houses have enough available space to potentially accommodate solar panel installations.

Furthermore, it appears that roofs are rarely used for recreational purposes, with 90% of respondents confirming that their roofs are not utilized for any other activities, likely due to the hot climate and modern indoor lifestyles. This means that if solar panels were installed on the roofs, they would not interfere with daily activities in the homes.

An important consideration for solar panel installation is the roof orientation. Fortunately, all the roofs in the area are flat, making them suitable for solar panel placement to maximize energy generation.

Moreover, the majority of houses have a height ranging from 7 to 10 m, with an average height of 9 m. This height distribution indicates that there is a low likelihood of shading or obstruction of sunlight, which is favorable for solar energy generation.

Finally, the survey revealed that 70% of surveyed houses do not have nearby buildings, trees, or other objects that may cause shading, while 30% reported potential shading issues. Overall, these results demonstrate a high potential for solar PV installation in the area to generate an optimal amount of clean energy.

5. Discussion

The presented results indicate that Qatari homes are highly germane to the installation of solar panels due to their relatively large area and spaciousness. The study found that 50% or more of the roof space is available for solar panel installation without interfering with the daily activities of the homeowners. This is in line with a previous study, which reported that most Qatari homes have spacious rooftops with significant potential for solar panel installation [43]. Moreover, the study found that there are limited problems caused by shading from trees or tall neighboring structures. This is because most Qatari homes are of similar height in wholly residential neighborhoods; thus, they are not typically in proximity to obstructions that could otherwise block sunlight, as reported by the homeowners. However, if there is latent shading, solar panels can be installed in locations that provide a minimum degree of shading and the best sun view. This is consistent with the findings of a study [43], which indicated that shading can be minimized by choosing the best location for solar panel installation. Additionally, solar PV arrays can themselves be used as shading devices, thus reducing direct solar heat gain by homes (while having unimpeded exposure to sunlight for power generation) and thus reducing the latent energy demand required for cooling [44].

The study also discovered that the prevalence of flat roofs on Qatari homes offers several benefits for solar panel installation, as depicted in Figure 5. These flat roofs offer easy access for technicians during installation and allow for the adjustment of panel orientation to maximize energy generation, a crucial factor impacting performance. Moreover, flat roofs support the use of automatic platforms for sun tracking, significantly enhancing energy generation—unlike angled roofs, which have limitations in this regard. Tracking systems can boost energy production by up to 25%, a valuable advantage, particularly in regions with high solar irradiance like Qatar [45,46].

Furthermore, the high similarities between homes in Qatar [40], as shown in Figure 6, enable the establishment of an easy strategy for technical teams to replicate the installation design in similar homes, which can reduce time costs and achieve the intended deployments within a realistic timescale. The similarities between homes can make it easier for homeowners to make the decision to install solar panels, as they can see examples of installations on homes similar to theirs, and installers will become highly proficient in installing required arrays and equipment in similar types of structures. This is in line with previous research on the benefits of standardization in building design for renewable energy adoption. For example, a study [47] found that standardizing building design can facilitate the integration of renewable energy systems, particularly in residential buildings. Another study [47] found that standardizing building design can reduce the cost of renewable energy systems by streamlining the installation process.

Figure 5. Example of a surveyed home roof.

Figure 6. Example of structural homogeneity of Qatari homes.

The outcomes from this study complement previous research conducted by the authors, aimed at establishing the analytical feasibility of deploying solar panels on Qatar's houses [2]. The primary objective was to calculate various solar panel deployment scenarios using analytical methods, considering different panel sizes, efficiency, and daily sun exposure. These scenarios were then compared to estimate the generated energy and its comparison with actual consumption over a twelve month period. The findings revealed the most viable scenarios that enable homes in Qatar to sufficiently generate solar energy to cover their consumption. Moreover, there is a high possibility that surplus energy generated during the low season can be exported to areas experiencing high energy demands, such as the Western world. This potential not only benefits Qatar but also supports global efforts towards transitioning to clean energy, providing a valuable solution for energy crises [48].

Hence, the findings of this study support the potential for solar panel installation on Qatari homes. With their spaciousness, limited shading, and flat roofs, Qatari homes are ideally suited for solar panel installation. This aligns with the Qatari government's ambitious goal of generating 200 MW of solar energy by 2022 [49]. By encouraging and facilitating the installation of solar panels on homes, Qatar can move towards a more sustainable future. This should begin with pilot projects involving the installation of solar panels on real homes to gather actual information on their performance, taking into account factors such as solar radiation, temperature, and dust that might reduce generation potential. This is essential to ensure optimal and sustainable performance [50].

6. Conclusions

The study has successfully achieved its primary objective of assessing the suitability of Qatari homes for solar panel installation. The obtained results are highly promising and offer significant value to various stakeholders involved in decision-making regarding the deployment of solar panels in residential areas. Furthermore, these findings hold invaluable importance for the parties responsible for the deployment, as they can utilize them to develop a comprehensive national strategy for replicating the deployment plan in different regions, considering the high similarities observed among homes in Qatar. These findings also complement and align with past research efforts conducted using analytical means.

However, to draw a definitive conclusion, it is recommended to conduct real pilot projects involving the installation of solar panels on actual homes and gather additional information on their performance in situ. This should include aspects like solar radiation, temperature, air quality (as dust can reduce generation potential), connectivity with the main grid, storage, and all relevant variables. Such an approach will yield crucial insights into potential challenges that may arise during installation and usage, thereby enabling the formulation of practical solutions based on user experiences and operational evidence.

Despite the potential difficulties that may arise, the positive outcomes of this study indicate a promising future for solar panel installation on Qatari homes. By implementing such initiatives, Qatari households can effectively reduce their carbon footprint and contribute to the creation of a greener and more sustainable environment for future generations. This will undoubtedly play a significant role in Qatar's commitment to clean energy and global efforts towards combating climate change.

Author Contributions: Conceptualization, A.B. and Z.H.; methodology, A.B. and Z.H.; investigation, A.B.; writing—original draft preparation, A.B.; writing—review and editing, Z.H.; supervision, M.A. All authors have read and agreed to the published version of the manuscript.

Funding: This research received no external funding.

Institutional Review Board Statement: Not applicable.

Informed Consent Statement: Not applicable.

Data Availability Statement: The data can be shared upon request.

Acknowledgments: The authors would like to thank all homeowners who participated in this study

Conflicts of Interest: The authors declare no conflict of interest.

References

1. Al-Hababi, R. The evolvement of Qatar's environmental sustainability policy: The strategies, regulations, and institutions. In *Sustainable Qatar: Social, Political and Environmental Perspectives*; Cochrane, L., Al-Hababi, R., Eds.; Springer Nature: Singapore 2023; pp. 17–35. [CrossRef]
2. Banibaqash, A.; Hunaiti, Z.; Abbod, M. An analytical feasibility study for solar panel installation in Qatar based on generated to consumed electrical energy indicator. *Energies* **2022**, *15*, 9270. [CrossRef]
3. Torcellini, P.A.; Crawley, D.B. Understanding zero-energy buildings. *ASHRAE J.* **2006**, *48*, 62–69.
4. Qatar General Electricity & Water Corporation "KAHRAMAA". *Annual Statistics Report 2021*; KAHRAMAA: Doha, Qatar, 2022; pp. 26–46.
5. Scharfenort, N. Urban development and social change in Qatar: The Qatar National Vision 2030 and the 2022 FIFA World Cup. *J. Arab. Stud.* **2012**, *2*, 209–230. [CrossRef]
6. Omer, A.M. Energy, environment and sustainable development. *Renew. Sustain. Energy Rev.* **2008**, *12*, 2265–2300. [CrossRef]
7. Darwish, S.; Abdo, H.; Alshuwaiee, W.M. Opportunities, challenges and risks of transition into renewable energy: The case of the Arab Gulf Cooperation Council. *Int. Energy J.* **2018**, *18*, 391–400.
8. World Data: Sunrise and Sunset in Qatar. Available online: https://www.worlddata.info/asia/qatar/sunset.php (accessed on 13 March 2023).
9. Bayram, I.S.; Saffouri, F.; Koc, M. Generation, analysis, and applications of high resolution electricity load profiles in Qatar. *J. Clean. Prod.* **2018**, *183*, 527–543. [CrossRef]
10. Bohra, M.; Shah, N. Optimising Qatar's energy transition through model-based analysis. *Energy Transit* **2020**, *4*, 11–29. [CrossRef]
11. Oxford Analytica. Qatar's Green Strategy Will Focus on Greening of LNG. *Emerald Expert Briefings*, 26 October 2022. [CrossRef]
12. Global Investors Move into Renewable Infrastructure: Reviewing the World's Top Renewable Energy Financiers. Available online: https://ieefa.org/sites/default/files/resources/Global-Investors-Move-Into-Renewable-Infrastructure_July-2021.pdf (accessed on 13 March 2023).
13. Obaideen, K.; AlMallahi, M.N.; Alami, A.H.; Ramadan, M.; Abdelkareem, M.A.; Shehata, N.; Olabi, A.G. On the contribution of solar energy to sustainable developments goals: Case study on Mohammed bin Rashid Al Maktoum Solar Park. *Int. J. Thermofluids* **2021**, *12*, 100123. [CrossRef]
14. Bergman, N.; Eyre, N. What role for microgeneration in a shift to a low carbon domestic energy sector in the UK? *Energy Effic.* **2011**, *4*, 335–353. [CrossRef]
15. Al-Ammari, B.; Romanowski, M.H. The impact of globalisation on society and culture in Qatar. *Pertanika J. Soc. Sci. Humanit.* **2016**, *24*, 1535–1556.
16. International Renewable Energy Agency. *Community Renewable Energy Deployment: The Role of Community Energy in Accelerating Renewable Energy Deployment*, IRENA: Masdar City, United Arab Emirates, *in press*.
17. Seyfang, G.; Haxeltine, A. Growing grassroots innovations: Exploring the role of community-based initiatives in governing sustainable energy transitions. *Environ. Plan. C Gov. Policy* **2012**, *30*, 381–400. [CrossRef]
18. Almulhim, A.I. Understanding public awareness and attitudes toward renewable energy resources in Saudi Arabia. *Renew. Energy* **2022**, *192*, 572–582. [CrossRef]
19. Voss, K.; Musall, E. *Net Zero Energy Buildings: International Projects of Carbon Neutrality in Buildings*; Walter de Gruyter: Berlin, Germany, 2012.
20. Tsalikis, G.; Martinopoulos, G. Solar energy systems potential for nearly net zero energy residential buildings. *Sol. Energy* **2015**, *115*, 743–756. [CrossRef]
21. Cao, X.; Dai, X.; Liu, J. Building energy-consumption status worldwide and the state-of-the-art technologies for zero-energy buildings during the past decade. *Energy Build.* **2016**, *128*, 198–213. [CrossRef]
22. Jaber, S.; Saidur, R. The role of energy efficiency in the sustainable development of Qatar: A review. *Renew. Sustain. Energy Rev.* **2020**, *120*, 109675.
23. Torres, A.R.; Herold, M. Energy consumption behavior in households with photovoltaic panels. *Energy Build.* **2019**, *195*, 68–79.
24. Saidur, R.; Jaber, S.; Mohamed, E.A. Opportunities and challenges for renewable energy in the State of Qatar. *Renew. Energy* **2021**, *177*, 583–599.
25. Residential Solar Panels and Their Impact on the Reduction of Carbon Emissions. Available online: https://nature.berkeley.edu/classes/es196/projects/2013final/ArifM_2013.pdf (accessed on 13 March 2023).
26. Karakaya, E.; Hidalgo, A.; Nuur, C. Motivators for adoption of photovoltaic systems at grid parity: A case study from Southern Germany. *Renew. Sustain. Energy Rev.* **2015**, *43*, 1090–1098. [CrossRef]
27. Bryan, H.; Rallapalli, H.; Ho, J.J. Designing a solar ready roof: Establishing the conditions for a high-performing solar installation. In Proceedings of the 39th ASES National Solar Conference, Phoenix, AZ, USA, 17–22 May 2010; Volume 5, pp. 4081–4110.
28. Shahsavari, A.; Akbari, M. Potential of solar energy in developing countries for reducing energy-related emissions. *Renew. Sustain. Energy Rev.* **2018**, *90*, 275–291. [CrossRef]

29. Monna, S.; Abdallah, R.; Juaidi, A.; Albatayneh, A.; Zapata-Sierra, A.J.; Manzano-Agugliaro, F. Potential Electricity Production by Installing Photovoltaic Systems on the Rooftops of Residential Buildings in Jordan: An Approach to Climate Change Mitigation. *Energies* **2022**, *15*, 496. [CrossRef]
30. Hussain, H.M.; Rahi, K.; Al Tarawneh, M.; Preece, C. Developing applicable scenarios to install and utilize solar panels in the houses of Abu Dhabi city. *Sustainability* **2022**, *14*, 15361. [CrossRef]
31. Starková, L. Toward a High-Definition Remote Sensing Approach to the Study of Deserted Medieval Cities in the Near East. *Geosciences* **2020**, *10*, 369. [CrossRef]
32. AL-Dosari, K.; Hunaiti, Z.; Balachandran, W. Civilian UAV Deployment Framework in Qatar. *Drones* **2023**, *7*, 46. [CrossRef]
33. Fink, A. *How to Conduct Surveys: A Step-by-Step Guide*; Sage Publications Ltd.: London, UK, 2019.
34. Khan, M.M.A.; Asif, M.; Stach, E. Rooftop PV Potential in the Residential Sector of the Kingdom of Saudi Arabia. *Buildings* **2017**, *7*, 46. [CrossRef]
35. Bekele, M.T.; Atakara, C. Residential Building Energy Conservation in Mediterranean Climate Zone by Integrating Passive Solar and Energy Efficiency Design Strategies. *Buildings* **2023**, *13*, 1073. [CrossRef]
36. Salgado, C.D.; Segura, O.R.; León, I.G. Electronic data collection methods in research. *Rev. Med. Inst. Mex. Seguro Soc.* **2017**, *55*, 236–242.
37. Creswell, J.W.; Creswell, J.D. *Research Design: Qualitative, Quantitative, and Mixed Methods Approaches*; Sage Publications Ltd.: London, UK, 2018.
38. Khan, J. Data analysis using Excel. *J. Appl. Res. High. Educ.* **2017**, *9*, 307–315.
39. Hulland, J. Use of partial least squares (PLS) in strategic management research: A review of four recent studies. *Strateg. Manag. J.* **1999**, *20*, 195–204. [CrossRef]
40. Scheller, F.; Doser, I.; Schulte, E.; Johanning, S.; McKenna, R.; Bruckner, T. Stakeholder dynamics in residential solar energy adoption: Findings from focus group discussions in Germany. *Energy Res. Soc. Sci.* **2021**, *76*, 102065. [CrossRef]
41. Thompson, S.K. Adaptive Cluster Sampling: Designs with Primary and Secondary Units. *Biometrics* **1991**, *47*, 1103–1115. [CrossRef]
42. Bujang, M.A. An Elaboration on Sample Size Planning for Performing a One-Sample Sensitivity and Specificity Analysis by Basing on Calculations on a Specified 95% Confidence Interval Width. *Diagnostics* **2023**, *13*, 1390. [CrossRef] [PubMed]
43. Morphology of Urban Qatari Homes. Available online: https://sites.northwestern.edu/buildingdoha/morphology-of-urba (accessed on 25 October 2023).
44. Mohammed, A.; Tariq, M.A.U.R.; Ng, A.W.M.; Zaheer, Z.; Sadeq, S.; Mohammed, M.; Mehdizadeh-Rad, H. Reducing the Cooling Loads of Buildings Using Shading Devices: A Case Study in Darwin. *Sustainability* **2022**, *14*, 3775. [CrossRef]
45. Ayoub, J.; Al-Jibouri, S. Analysis of the impact of solar panel orientation on the electricity generation using solar energy. *Int. J. Energy Environ. Eng.* **2021**, *12*, 63–73.
46. Gupta, R.; Nayak, J.K. Solar tracking systems: Technologies and trackers. In *Handbook of Research on Energy Systems and Sustainable Energy*; Sharma, D.K., Reddy, K.S., Eds.; IGI Global: Hershey, PA, USA, 2019; pp. 379–411. [CrossRef]
47. Murshed, S.; Shafie, S.; Saidur, R. Review on solar energy storage and thermal energy storage for sustainable energy houses. *Renew. Sustain. Energy Rev.* **2018**, *82*, 1518–1542.
48. Environmental Sustainability Initiatives in Qatar. Available online: https://www.qf.org.qa/research/environmental-sustainability (accessed on 25 October 2023).
49. Al-Sulaiman, F.A.; Zubair, S.M. Solar energy potential and policies in Qatar: Opportunities and challenges. *Renew. Sustain. Energy Rev.* **2019**, *103*, 325–333. [CrossRef]
50. Olorunfemi, B.O.; Ogbolumani, O.A.; Nwulu, N. Solar Panels Dirt Monitoring and Cleaning for Performance Improvement: A Systematic Review on Smart Systems. *Sustainability* **2022**, *14*, 10920. [CrossRef]

energies

MDPI

Article

Optimal Design and Analysis of a Hybrid Hydrogen Energy Storage System for an Island-Based Renewable Energy Community

Robert Garner and Zahir Dehouche *

College of Engineering, Design and Physical Sciences, Brunel University London, London UB8 3PH, UK; robert.garner2@brunel.ac.uk
* Correspondence: zahir.dehouche@brunel.ac.uk

Abstract: Installations of decentralised renewable energy systems (RES) are becoming increasing popular as governments introduce ambitious energy policies to curb emissions and slow surging energy costs. This work presents a novel model for optimal sizing for a decentralised renewable generation and hybrid storage system to create a renewable energy community (REC), developed in Python. The model implements photovoltaic (PV) solar and wind turbines combined with a hybrid battery and regenerative hydrogen fuel cell (RHFC). The electrical service demand was derived using real usage data from a rural island case study location. Cost remuneration was managed with an REC virtual trading layer, ensuring fair distribution among actors in accordance with the European RED(III) policy. A multi-objective genetic algorithm (GA) stochastically determines the system capacities such that the inherent trade-off relationship between project cost and decarbonisation can be observed. The optimal design resulted in a levelized cost of electricity (LCOE) of 0.15 EUR/kWh, reducing costs by over 50% compared with typical EU grid power, with a project internal rate of return (IRR) of 10.8%, simple return of 9.6%/year, and return on investment (ROI) of 9 years. The emissions output from grid-only use was reduced by 72% to 69 gCO_2e/kWh. Further research of lifetime economics and additional revenue streams in combination with this work could provide a useful tool for users to quickly design and prototype future decentralised REC systems.

Keywords: decentralised energy systems; renewable energy community; hydrogen energy storage system; decarbonisation; techno-economic assessment; multi-objective optimisation

Citation: Garner, R.; Dehouche, Z. Optimal Design and Analysis of a Hybrid Hydrogen Energy Storage System for an Island-Based Renewable Energy Community. *Energies* **2023**, *16*, 7363. https://doi.org/10.3390/en16217363

Academic Editors: Luis Hernández-Callejo, Jesús Armando Aguilar Jiménez and Carlos Meza Benavides

Received: 2 August 2023
Revised: 26 October 2023
Accepted: 27 October 2023
Published: 31 October 2023

1. Introduction

The current state of the energy generation landscape is undergoing a significant change as concerns are raised over climate change, energy cost, and energy security. The aim as stipulated in the Paris Agreement [1] of keeping the average surface temperature increase below 2 °C by 2050 is unlikely given global trends [2], and will be impossible without an ambitious sustainable energy development and technological innovation [3]. Recent events on the global stage have also caused nations in Europe and around the world to reconsider their energy security strategies [4,5]. The adoption of renewable energy at scale should include measures to increase the effectiveness whilst providing cost reductions [6].

The introduction of renewable energy systems (RES), including photovoltaic (PV) solar panels and wind turbines, have been the key driving force in removing global dependence on fossil fuels from the energy sector [7]. These types of generation assets are known as non-dispatchable as they are completely dependent on weather conditions [8] and so cannot be precisely controlled. This is a problem for transmission service operators (TSOs) as ensuring the voltage and frequency are balanced at a grid level becomes challenging [9]. The increasing volume of decentralised RES installed at the demand side is also problem for grid operators [10], as they can induce bidirectional grid flows and put additional strain on the network.

One solution to this problem is the use of energy storage systems (ESS) to store excess energy and increase the share of the total RES production directly through self consumption [11,12]. Electro-chemical storage such as batteries have been deployed in many cases for use as grid-level storage [13–15], as they have the advantage of fast response to demand and can be installed in most global climates. A number of battery technologies including high-performance solid state chemistries are a promising solution due to their long-term stability and high capacity retention [16,17]. Most grid storage applications deploy LiFePO4 variants as they are widely available and have a relatively low cost [18] Crucially for the research methods used in this work, the retrieval of reliable cost and environmental data is vital for an accurate result, which for LiFePO4 is widely available within the literature. Hydrogen has often been considered for long-term seasonal storage [19], due in part to the mentioned capacity retention challenges of battery storage. Hydrogen is also a flexible energy vector for many other uses, such as heating and industrial processes [20]. A hybrid battery and hydrogen ESS has a great potential to increase the share of renewables within the energy mix [21], thus decreasing the reliance on traditional power stations. The advent of widely available ESS has meant that it is now possible to emphasise the self-consumption of energy at a local level to reduce the problems of grid stress and planning. A 'prosumer' (an end user that is able to both consume and produce energy [22]) or group of prosumers could install decentralised RES coupled with storage technology, and self-consume the power generated at a local level.

To address these challenges, this work presents and evaluates the application of decentralised renewable energy communities (RECs). RECs in practice have many advantages and solve the most common issues associated with increased decentralised generation, while also promoting the further self-consumption of electricity. In a REC configuration, consumers and prosumers are no longer restricted to buying and selling energy from their utility company and can virtually 'share' the excess energy between actors within the energy community itself. This is mutually beneficial for both the network operator, as they no longer need to manage unpredictable grid flows, and for the REC participants as they receive direct renumeration and a reduction in carbon emissions.

The REC considered in this work was based on the policy recommendations recently implemented by a number of EU countries outlined in the Renewable Energy Directive (RED-II) (EU) 2018/2001 [23]. The directive defines a REC as "a legal entity that is based on open and voluntary participation, it is autonomous and controlled by shareholders or members located in the proximity of renewable energy plants belonging to the community itself. The members may be physical persons, companies or local authorities...". While the directive has been transposed into several other national laws and decrees, including Austria [24], France [25], Germany [26], and the Netherlands [27], the REC modelled in this work most closely resembles the framework practiced in Italy as discussed by Trevisan et al. [28]. Although the study was in Spanish territory, it was chosen to follow the Italian implementation as there are more example cases available and, as of 2021, further improvements to the 2019 Spanish REC policy are currently in progress [29].

As laid out in decree-law 199/2021 [30], a group of self-consuming members within the REC must be located within the same low-voltage (LV) network downstream of the same LV/MV substation. Energy is shared in the existing physical network using a virtual network model. The difference between the energy consumed and energy produced by the REC is resolved over each one-hour period to determine the capacity available to be shared [31]. The model created in this work uses the principles of the relevant regulation to design the virtual REC.

A number of studies including operational renewable energy communities have investigated the use of ESS within a REC to further improve the economic performance of centralised renewables. Trevisan et al. presented an optimised energy model considering PV solar and ESS to provide renewable power to a port REC, showing a decrease in energy bills of 28% compared with the business-as-usual case [32]. Bartolini et al. investigated how to size a mixed RES to fully self-consume all generation at a community level, as well as meeting the heat energy needs, and showed that using hydrogen generation and storage is an economically viable alternative to battery systems [33]. Although less explored in the literature, there have also been studies focused on the environmental and emissions reductions possible with such a community-based system. Wang et al. proposed a community-based virtual power plant solution in Japan with PV and battery ESS with the ability to reduce carbon emissions by 16.26% [34].

Several different modelling and optimisation software tools have also emerged to assist in model-based design and assessment. An in-depth review by Cuesta et al. presented popular renewable energy modelling tools, including the ability to model different renewable assets and output different technical, economic, environmental, and social key performance indicators [35]. Software such as HOMER (version 3.16), TRNSYS (version 18), and MATLAB/Simulink (2022b) are most often used due to their ease of use and available documentation. However, they can be restrictive for some REC cases due to their proprietary nature. Creating the model in Python will provide the flexibility of an open-source platform and a scalable product suitable for deployment as a lightweight software or web applications.

A number of optimisation procedures have been addressed and utilised in the literature to determine the optimal design of hybrid RES and ESS. Most cases vary the design capacities to achieve one or more competing criteria such as economics, grid independence, and environmental impact. Niveditha and Rajan Singaravel consider a multi-objective design criteria for achieving near zero energy buildings (NZEB), using the functions of cost, loss of load probability (LLP), and total energy transfer (TET) to determine the best sizing arrangement for the PV-wind-battery storage system [36]. Zhang et al. presents a capacity configuration for both an on-grid and off-grid mixed renewable system with hydrogen and batteries [37]. The NSGA-II algorithm was used to determine the trade-off relationship between system cost, renewable curtailment, and loss of load probability (LLP), which can be considered analogous to grid independence for grid-connected configurations. Xu et al. considers the design of an off-grid PV-wind-hydrogen storage system using the multi-objective criteria of LCOE, LLP, and power abandonment rate (PAR). The pareto optimal solution produces an LCOE of 0.226 USD/kWh at acceptable LLP and PAR values [38]. Studying the emissions associated with the grid independence would more accurately determine the positive environmental impact, which was of particular focus in this work. Results from the literature also do not consider the implementation of such an optimization procedure for RECs, and the impact of trading arrangements between members. Other algorithms including multi-objective particle swarm optimisation (MOPSO) [39] and multi-objective evolutionary algorithm with decision-making (MOEA-DM) [40] have also been applied to ESS design; however, NSGA-II remains very popular and has proven robustness in energy flow optimisation problems [41].

2. Contribution

In this study, a techno-economic and carbon emissions assessment was conducted for a decentralised REC. The case study location was chosen as Formentera; a largely rural Balearic Island located in the Mediterranean Sea as illustrated in Figure 1. Emphasis is put on the isolated nature of the energy grid, which naturally increases the energy cost and embedded carbon of electricity usage, making it an ideal location for the study. A comparison of the base case scenario was used to compare the improvements made with the implementation of the REC.

The community has shared usage of PV solar and wind power to produce energy, and a hybrid battery and regenerative hydrogen fuel cell to store excess production. The combination of battery and hydrogen minimises the potential shortcomings of decentralised storage. A virtual trading scheme based on the EU decree-law 199/2021 for REC implementation was used to evaluate the energy shared between community members, without considering incentives or feed in tariffs. Through the implementation of key economic and environmental parameters, the multi-objective optimisation determines the best design topology within the defined REC boundary conditions.

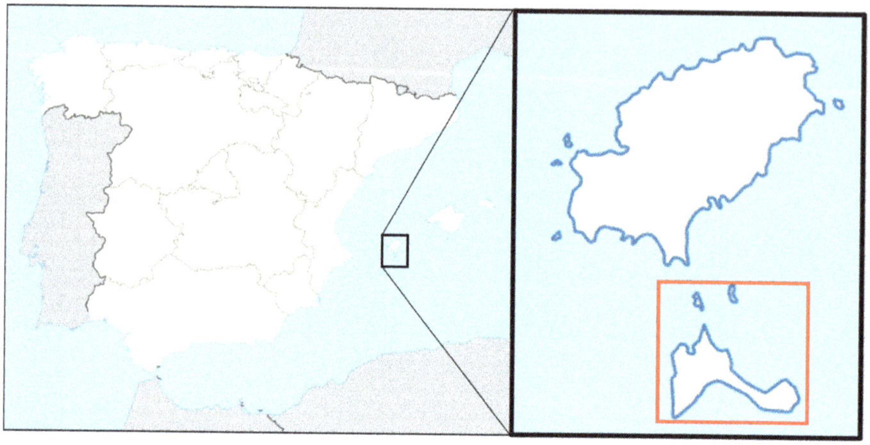

Figure 1. Formentera Island, highlighted in red, is located east of the Spanish mainland in the Mediterranean [42].

The multi-objective results reveal an inherent trade-off relationship between low-cost energy and the ability to decarbonise supply, and that this approaches a critical limit at both extremes of the pareto front. This work shows that across the pareto optimal sets, the hybridisation of energy storage provides a better overall performance than a battery-only or hydrogen-only case. Additional constraints can be applied to the objective domain to assist in design decision-making.

The implementation of the model in Python allows for the creation of a scalable product, which following digitisation trends in model-based design could provide a vital tool for communities and policymakers to determine the best method for assisting communities to reach net-zero emissions. To summarise, the novelty of this work is summarised as follows:

- The modelling of a hybridised battery and RHFC system for a remote renewable energy community application using real-world power consumption data from a rural island location;
- the use of multi-objective optimisation to evaluate the system pareto front based on economic and environmental performance;
- the inclusion of a virtual trading layer based on the latest RED(III) REC policies;
- the formulation of a scalable and modular renewable energy community modelling and simulation platform.

3. Materials and Methods

For the purposes of simplification, the simulation model was discretised into one hour time steps using kWh as the function unit for all energy flows within the system. The case study input assumptions including building load and meteorological datasets are defined first. The meteorological data at the chosen coordinate location were obtained from the National Aeronautics and Space Administration (NASA) Langley Research Center (LaRC) Prediction of Worldwide Energy Resource Project funded through the NASA Earth Science/Applied Science Program [43]. A combination of hourly and monthly energy consumption collected from the case study location was used to recreate typical annual load profiles for each of the seven buildings within the virtual REC. A selection of 24 industrial, commercial, and residential load profiles produced by Farhad et al. (2020) were used to augment the profiles where required [44].

3.1. Renewable Energy Community Implementation

It was assumed that the community members will have a shared capital investment in the generation and storage assets. PV solar and wind assets can either be installed in the low-voltage energy grid within the same secondary substation of the REC, or spread out between the members, installing in open areas such as rooftops. The stationary ESS consisting of both a lithium-ion battery and a RHFC was installed with the REC boundary conditions in accordance with the EU decree-law 199/2021, with the capacity to accept but also release energy to the physical energy grid. A simple diagram of the system architecture is shown in Figure 2.

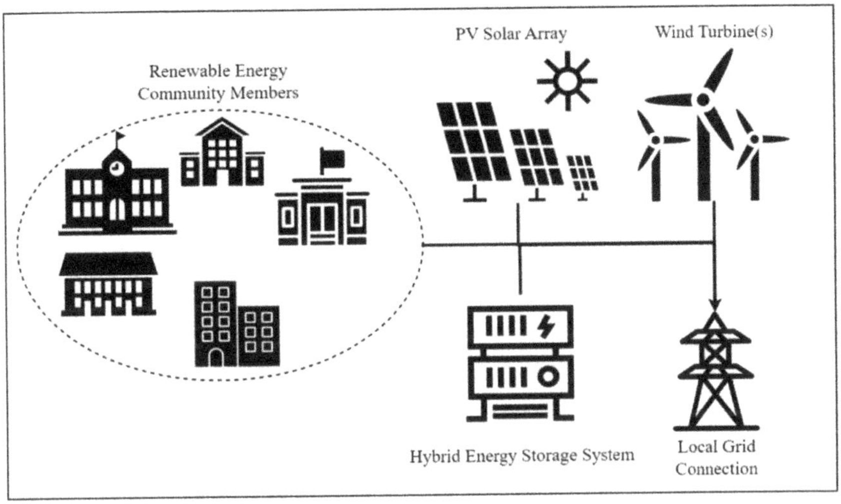

Figure 2. Renewable energy community system architecture. The community buildings grouped on the left are connected virtually to the distributed generation and storage assets, which are also able to export to the local power grid.

The control strategy consists of a load-following authority, but with additional considerations for the hybrid ESS. Since batteries have an improved performance as short-term storage, these are allowed to discharge first to cover the load of the REC. Once the battery depth-of-discharge (DOD) limit is reached, the hydrogen system is then activated to cover the remaining demand. During the charging phase, this control scheme occurs in reverse. By evaluating the excess energy available between the total REC consumption and production over each one-hour increment in line with decree-law 162/19 for community implementation in Italy, the total virtual energy flows between community members were derived. This case does not consider incentives to reduce financial strain and instead evaluates through a techno-economic assessment over a 20-year project period whether the hybrid system was able to provide net-positive economic and environmental performance over the business-as-usual case.

The electrical load profiles form the foundation of the assessment of economic and environmental improvements to the REC. The community consists of seven member buildings; a community centre, a small school, a large school, local government offices, and three typical residential units. For the community centre, two schools, and offices, sample daily load profiles, as well as the monthly average energy consumption, were collected directly from the test site. For the residential units, a combination of the annual heating, cooling, and appliances usage of 80.7 kWh/m^2 was used to evaluation the typical characteristics of a residence in Spain [45], where the buildings were assumed to be 50 m^2 in area. The monthly and yearly consumption was used to create a spline, over which the daily load profile was interpolated and repeated to create the one-year load profiles for each building. The total yearly consumption for each member is included in Table 1, with the monthly and daily load profiles shown in Figure 3. The three mixed homes have been combined to represent a mixed family building and to improve visibility within the analysis.

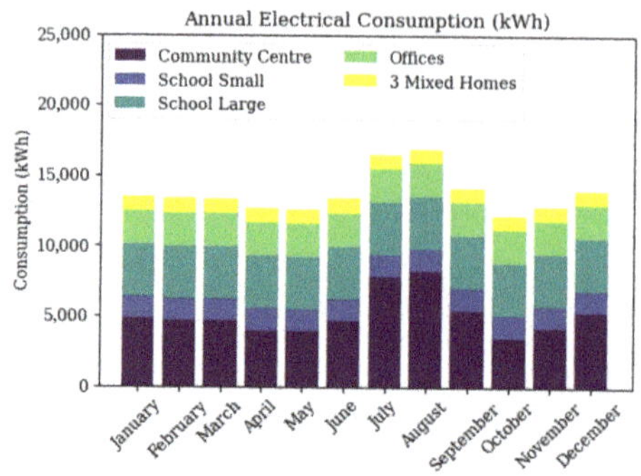

(**a**) Monthly electricity demand

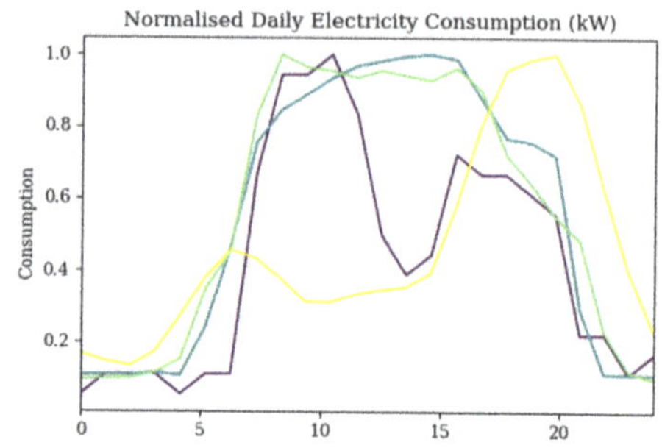

(**b**) Normalised daily electricity demand

Figure 3. Building electrical energy service demand model based on the requirements of the renewable energy community. Figure 3b displays the average daily power demand curves of the different building types within the model.

Table 1. Total annual electrical consumption for each member of the renewable energy community.

	Annual Consumption
Community centre	66,500
Elementary school	19,200
High school	46,200
Government offices	28,900
3× Residential units	12,000

3.2. *Weather and Environment Data*

The weather data were collected for the year 2022 at the coordinate location of the chosen REC case study site. The model requires accurate measurements of ambient temperature, wind speed, and global horizontal irradiance (GHI) solar conditions to evaluate the hour-by-hour power output of the renewable generation technologies. Figure 4 shows the hourly mean temperature and GHI for each month over one year. A higher GHI is observed in the summer period as expected in the northern hemisphere.

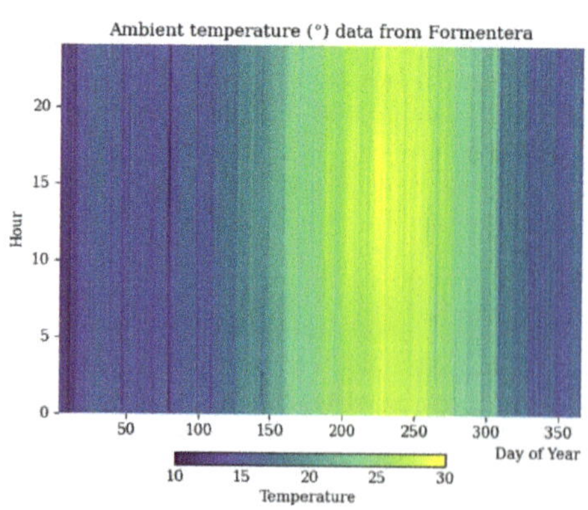

(**a**) Ambient temperature (°C)

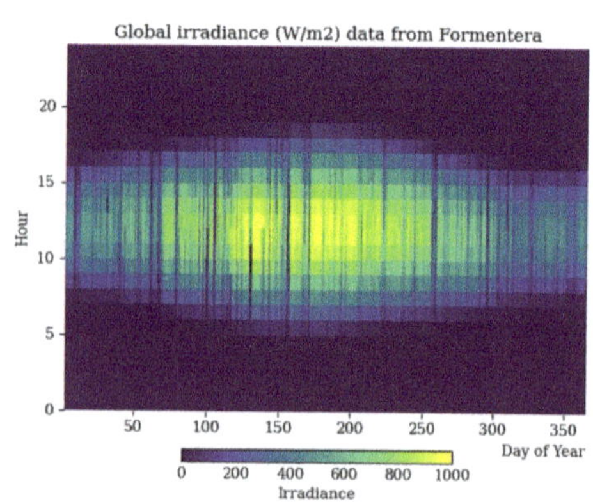

(**b**) Global horizontal irradiance (W/m^2)

Figure 4. Temperature and solar conditions over a one-year period at the island location. The weather conditions are assumed to remain constant year-on-year through the lifetime of the system.

3.3. System Design and Characteristics

3.3.1. PV Solar Array Model

The Mediterranean region's warm and dry climate promotes the use of PV solar systems to generate clean energy. For the purposes of the study, the solar array was assumed to be installed at 180° directly to the south, and at an optimal tilt angle of 38.7°. The power output of the solar panels P_{PV} was modelled using the following governing equation [46]:

$$P_{PV} = C_{PV} d_f \left(\frac{G(t)_{module}}{G_{STC}}\right)[1 + \alpha_P (T_c - T_{c,STC})] \quad (1)$$

where C_{PV} is the generation capacity (kW) of the solar installation under standard conditions, d_f is the derate factor, $G(t)_{module}$ is the direct solar irradiance in W/m^2, G_{STC} is the direct solar irradiance under standard test conditions (1000 W/m^2), α_P is the thermal power coefficient (%/°C), and $T_{c,STC}$ is the PV cell temperature under standard test conditions (25 °C). T_c is the PV cell temperature and is calculated by considering the measured nominal operating cell temperature (NOCT). NOCT is the cell measured temperature at a solar irradiance G_{NOCT} of 800 W/m^2, an ambient temperature $T_{a,NOCT}$ of 20 °C, and a wind speed of 1 m/s [44]. This known thermal characteristic can then be used to adjust the cell temperature and find the corrected power output using the following equation [47]:

$$T_c = T(t)_a + (T_{c,NOCT} - T_{a,NOCT})\left(\frac{G(t)_{module}}{G_{NOCT}}\right)\left(\frac{1 - \eta_{mp}}{\tau\alpha}\right) \quad (2)$$

where $T(t)_a$ is the ambient temperature at timestep t and η_{mp} is the cell efficiency. The constants $\tau\alpha$ can be assumed to be 0.9 for most cases. Since η_{mp} is not known, the efficiency under standard conditions $\eta_{mp,STC}$ is substituted into the cell temperature equation above and the result yields the following:

$$T_c = \frac{T(t)_a + (T_{c,NOCT} - T_{a,NOCT})(G(t)_{module}/G_{NOCT})[1 - (\eta_{mp,STC}(1 - \alpha_P T_{c,STC}))/\tau\alpha])}{(1 + (T_{c,NOCT} - T_{a,NOCT})(G(t)_{module}/G_{NOCT})[(\alpha_P \eta_{mp,STC})/\tau\alpha]} \quad (3)$$

The GHI input data need to be adjusted based on the local latitude φ and module tilt β to find the module irradiance $G(t, module)$ for the time of day and year. This is found with the following equations [48]:

$$G(t)_{module} = \frac{G(t)_{horizontal} \sin(\alpha + \beta)}{\sin \alpha} \quad (4)$$

$$\alpha = 90° - \varphi + \delta \quad (5)$$

$$\delta = 23.45° \cdot \sin[360/365(284 + d)] \quad (6)$$

where $G(t)_{module}$ is the module irradiance, $G(t)_{horizontal}$ is the GHI data, α is the elevation angle, and δ is the declination angle which deviates from the earth's tilt of 23.45° depending on the day of the year d.

3.3.2. Wind Turbine Model

A generic dynamic wind turbine model was used to calculate the expected power output in the selected location using the following [49]:

$$P(t) = \frac{1}{2} C_p \rho(t) A V^3(t)(\eta_m \cdot \eta_e) \quad (7)$$

where C_p is the power coefficient, $\rho(t)$ is the air density at the hub height, A is the selected swept area in m^2, $V(t)$ is the wind speed in m/s at the time step t, and η_m and η_e are the mechanical and electrical efficiencies. The wind speed is usually measured at a different height compared to the hub height, Z_{hub}. Therefore, the model uses the logarithmic law to derive the hourly wind speed at the hub height as follows [49]:

$$V_{hub} = V_{anem}(\frac{\text{In}(Z_{hub}/Z_0)}{\text{In}(Z_{anem}/Z_0)}) \quad (8)$$

where Z_0 is the surface roughness length (m), Z_{anem} is the anemometer height (m), V_{hub} is the wind speed at the required hub height (m/s), and the V_{anem} is the measured wind speed at the anemometer height (10 m for the dataset used). For simplicity, C_p is evaluated by way of a 2D-look up table based on the four classes of wind turbines described in IEC 61400 standard [50]. The average wind speed and distribution was evaluated and the most appropriate characteristic was chosen from the four available classes ranked from low to high wind speeds [51]. The normalised power range for each class of "Offshore", "IEC-1", "IEC-2", and "IEC-3" are shown in Figure 5a. The energy output over the course of one year can also be determined analytically by assessing the wind speed distribution. The Rayleigh distribution, shown in Figure 5b, has been overlaid to show that the wind speed distribution data follow this statistical law, which indicates that the normalised power curves will operate effectively for the model.

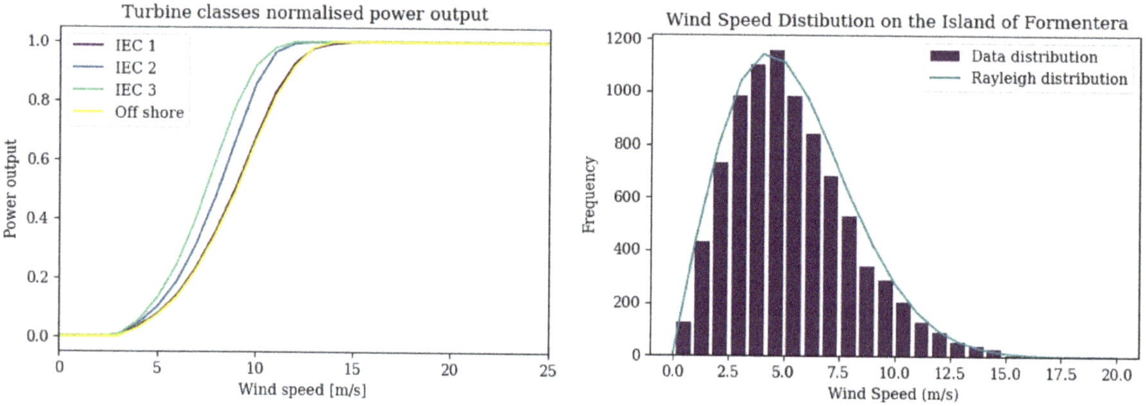

(**a**) Turbine normalised power output curve

(**b**) Wind speed distribution at Formentera

Figure 5. Wind model input assumptions are primarily a combination of standardised wind turbine power coefficients and the load wind speed data measured at 10 m above sea level.

3.3.3. Lithium Ion Battery Model

The battery model utilises a simplified version of the Shepard battery model [52], replacing internal and other resistive losses with a total charge $\eta charge$ discharge $\eta discharge$ efficiency for the hourly discharge case. The simplification allows for less information to be known about the chemistry and dynamics of the specific battery to perform calculations for the current capacity and state of charge (SOC). The battery system contains two parts: a charge model and a discharge model. The models take the power requirement from the battery and output the resulting SOC for the end of the timestep. These parts are defined as follows:

$$\begin{cases} \text{SOC}(t+1)_{batt} = \dfrac{Q(t)_{batt} + \int P(t)_{batt}\eta_{charge}, dt)}{Q(t_0)_{batt}} \cdot 100 & \text{charging} \\ \text{SOC}(t+1)_{batt} = \dfrac{Q(t)_{batt} - \int P(t)_{batt}\eta_{discharge}, dt)}{Q(t_0)_{batt}} \cdot 100 & \text{discharging} \end{cases} \quad (9)$$

where $\text{SOC}_{t+1,batt}$ is the next timestep battery SOC, $Q_{t,batt}$ is the battery state of charge at timestep t, $Q_{t0,batt}$ is the initial SOC, $P_{t,charge}$ is the average charge power draw, and $P_{t,discharge}$ is the discharge power draw. These outputs are subject to the minimum and maximum SOC limits SOC_{min} and SOC_{max}. The model includes degradation in the battery capacity linearly as a function of charge cycles, as shown below:

$$Q(l,t)_{batt} = Q(t_0)_{batt} - \alpha l \quad (10)$$

where $Q(l, t)_{batt}$ is the dynamic capacity in kWh as a function of cycles the cycles l, and α is the ageing factor (kWh/cycle).

3.3.4. Regenerative Hydrogen Fuel Cell

The RHFC model provides an alternative energy storage facility to the electrochemical battery. The model consists of a PEM fuel cell and PEM electrolyser capable of consuming and producing hydrogen, respectively. The system also considers a hydrogen storage module with its own rated capacity and efficiency. The overall equations are like that of the simplified battery model in that the electrolyser and fuel cell analogously represent the charge and discharge elements. The system can therefore be shown as the following:

$$\left\{ \begin{array}{ll} Q(t+1)_{H2} = Q(t)_{H2} - \int P(t)_{fc}\eta_{fc}, dt & \text{Fuel cell} \\ Q(t+1)_{H2} = Q(t)_{H2} + \int P(t)_{el}\eta_{el}, dt & \text{Electrolyser} \end{array} \right. \quad (11)$$

where $Q(t+1)_{H2}$ is the next timestep hydrogen energy stored (kWh), $Q(t)_{H2}$ is the current timestep hydrogen energy stored (kWh) $P(t)_{fc}$ is the average fuel cell power production [kW] in the current one-hour timestep t, and $P(t)_{el}$ is the average electrolyser power consumption [kW]. ηfc and ηel are the average lifetime fuel cell and electrolyser efficiencies [%], respectively. Like the battery, these energy values are also subject to $Q_{H2,min}$ and $Q_{H2,max}$ limits.

3.3.5. Model Input Assumptions

Table 2 contains the necessary input assumptions for the energy models, including efficiencies and other system dynamics that determine the output power generated or stored. The PV panel characteristics are based on the Sunpower Maxeon panel series, while the wind turbine is an approximation of common small-scale turbine systems on the market. The roughness length assumption of 0.05 is defined as rural, farmland area with low crops and without many trees [53]. The hydrogen system efficiency values are based on industry knowledge gathered from leading European fuel cell and electrolyser manufacturers.

Table 2. Hybrid renewable energy system design input assumptions across the different included technologies.

PV Solar	
Panel Power (W)	400
Panel Area (m^2)	2
Thermal Coefficient (%/°C)	−0.3
NOCT (°C)	42
Lifetime (years)	20
Wind turbine	
Hub Height (m)	20
Roughness Height (m)	0.05
Lifetime (years)	20
Lithium battery	
Total Efficiency (%)	95
Maximum Cycles	8000
Maximum Age (years)	10
Regenerative hydrogen fuel cell	
Fuel Cell Efficiency (%)	46
Electrolyser Efficiency (%)	68
Lifetime (years)	20

3.4. Energy Management Strategy

The energy management strategy for the hybrid storage system is shown in Figure 6. When generation supply is available in excess of demand, the battery charges first, followed by the larger capacity hydrogen storage via the electrolyser. When the demand outgrows the supply, the battery discharges first, followed by the activation of the fuel cell. In practical terms, the battery is actually being charged by the fuel cell while active, as the fuel cell cannot module its output without incurring performance losses. The charge and discharge states are is shown in Figure 6.

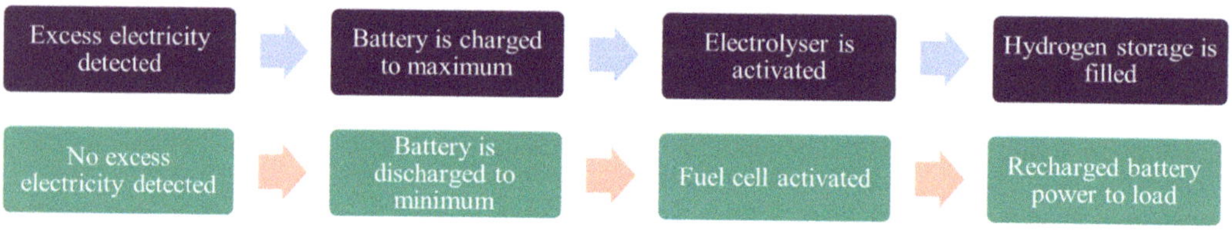

Figure 6. Energy management strategy of the hybrid storage system.

Virtual trading was used to fairly satisfy the community members based on the shared energy available. The excess energy available is shared equally, satisfying each load in ascending order of magnitude. This means that it is more likely that a member's electricity demand will be fully satisfied if smaller. It should be noted that this algorithm can be modified to suit any location-specific REC policy.

3.5. Economic and Environmental Indicators

A selection of three different system configurations: the best economic outcome, best environmental outcome, and a midpoint configuration between the two would be assessed in the model. It is important from a financial perspective to understand the investment requirements and expected returns for prospective REC members. The net present value (NPV) is commonly employed to determine economic feasibility, as well as the internal rate of return (IRR), simple return [%], payback period [years], and levelized cost of electricity (LCOE) for energy specific cases. Generally, if the NPV is positive compared to the base scenario, the investment is worthwhile [54].

$$NPV = \sum_{n=1}^{N} \frac{C_{O\&M,n} + C_{f,n}}{(1+R)^n} - C_0 \tag{12}$$

where $C_{O\&M,t}$ is the operation and maintenance cashflow for year t, $C_{f\,,t}$ is the fuel input cashflow, R is the discount rate, and C_0 is the initial capital investment. It was assumed that any grid consumption is included in $C_{f\,,t}$ in units of EUR/year. The capital requirement and operating cashflows are summed for each generation and storage asset to solve for the system NPV. The IRR evaluates the rate of return if the NPV is set to zero, at which point the project breaks even.

$$NPV = \sum_{n=1}^{N} \frac{C_n}{(1+IRRR)^n} \tag{13}$$

Calculating the LCOE is beneficial when assessing the economic feasibility of different technologies. The LCOE was evaluated against the grid cost to assess the cost savings per unit of electricity which could be expected by the community members. LCOE is defined as the total cost or lifetime cost of the asset divided by the total electricity delivered to the consumer [55].

$$LCOE(/kWh) = \frac{\sum_{n=1}^{N} C_{0,n} + C_{O\&M,n} + C_{f,n}}{\sum_{n=1}^{N} E_n} \tag{14}$$

where $C_{0,n}$ is the capital cost of the asset, and E_n is the lifetime energy delivered. A range of different assessments exist for the economics of renewable assets, as it is highly dependent on the capital requirement, location, delivery and installation cost, and available labour among other factors. The resulting CAPEX, OPEX, and lifetime parameters are shown in Table 2. The costs include the balance of plant (BOP), such as DC-AC inverters and IoT control equipment. The project has an assumed discount rate R of 5% and an estimated inflation rate of 2% per year, as well as a year-one electricity grid unit cost of 0.30 EUR/kWh for each building. Where the asset lifetime is less than 20 years, the asset is retired and the cost of a new equivalent system was included in the NPV assessment in that given replacement year. This method assumes the BOP cost is relatively low.

The environmental impact was estimated through the global warming potential (GWP) of the assets, which when summed together and divided by the total energy delivered over the system lifetime derives the emissions intensity, measured in $extgCO_2eext/kWh$. The values are then compared with the grid emissions intensity for the island, for which the total decarbonisation potential was evaluated. The grid emissions were found using generation data gathered from the national TSO (Red Electrica de Espana) for the year 2021 and found to have an average of 325 gCO_2e/kWh.

$$EI_{total} = \frac{\sum_{j=1}^{m}(EI_j \cdot E_j)}{\sum_{j=1}^{m}(E_j)} \tag{15}$$

EI_j is the emissions intensity and E_j is the energy output for m number of generators and energy storage systems. This calculation was performed for each timestep of the simulation to find the dynamic emissions value depending on the instantaneous energy mix of the REC. The emissions intensity found within the literature can vary due to the range of manufacturing techniques and factors considered when performing the life cycle assessment (LCA). For this reason, some values such as those used for the hydrogen system are taken as an educated estimation of the emissions impact based on a variety of sources. The GWP embedded during manufacturing and installation for the assets and technology costs are shown in Table 3.

Table 3. Hybrid renewable energy system economic and climate impact assumptions for the different modelled technologies.

Technology	CAPEX	OPEX	Lifetime	Emissions Embedded
PV Solar Array [55–57]	2500 EUR/kW	30 EUR/kW/year	20 years	1826 kgCO2e/kW
Wind Turbine [58,59]	2850 EUR/kW	32 EUR/kW/year	20 years	520 kgCO2e/kW
Lithium-Ion LFP [18,60]	328 EUR/kWh	5 EUR/kWh/year	10 years or 8000 cycles	254 kgCO2e/kWh
PEM Fuel Cell [18,61]	1200 EUR/kW	13 EUR/kW/year	20 years	73.3 kgCO2e/kWh
AEM Electrolyser [18,62,63]	1500 EUR/kW	14 EUR/kW/year	20 years or 35,000 h	239 kgCO2e/kWh
Hydrogen Storage Vessel [64–66]	30 EUR/kWh	-	20+ years	5.1 kgCO2e/kWh

3.6. Multi-Objective Optimisation Procedure

Designing and configuring the optimal system sizing for a hybrid decentralized energy system is a complex process. There are a number of non-linear phenomena being simulated, as well as many potential design objectives and constraints. The chosen objective functions considering both cost and carbon reduction are the NPV and the equivalent GWP. The objective functions rely on varying the capacities of the PV solar, wind, battery, and RHFC installations at the site.

The NSGA-II uses a heuristic evolutionary learning algorithm with a population of potential design solutions within the defined constraints. It then ranks the population based on a non-dominated sorting, producing a pareto front of optimal solutions by minimizing both objective functions [36]. Each individual in the population was determined based on the simulation of the model of a one-year period and evaluating the two objectives. The best-performing individuals are passed to the next generation, whereas a combination of mutations and created offspring (crossover) determines the remaining individuals. NSGA-II provides several advantages including the use of elitism and reduced computational complexity [67]. The solving process for NSGA-II implementation is shown in Figure 7. The algorithm also requires inputs, including the population size, number of offspring, stopping conditions, and variable constraints, as shown in Table 4. The lower limit for all system assets was set to zero, while the upper limit was set to 200 kW in line with the adopted REC regulation for this study. The *pymoo* module created and maintained by Blank et al. [68] was used to implement the NSGA-II algorithm in Python.

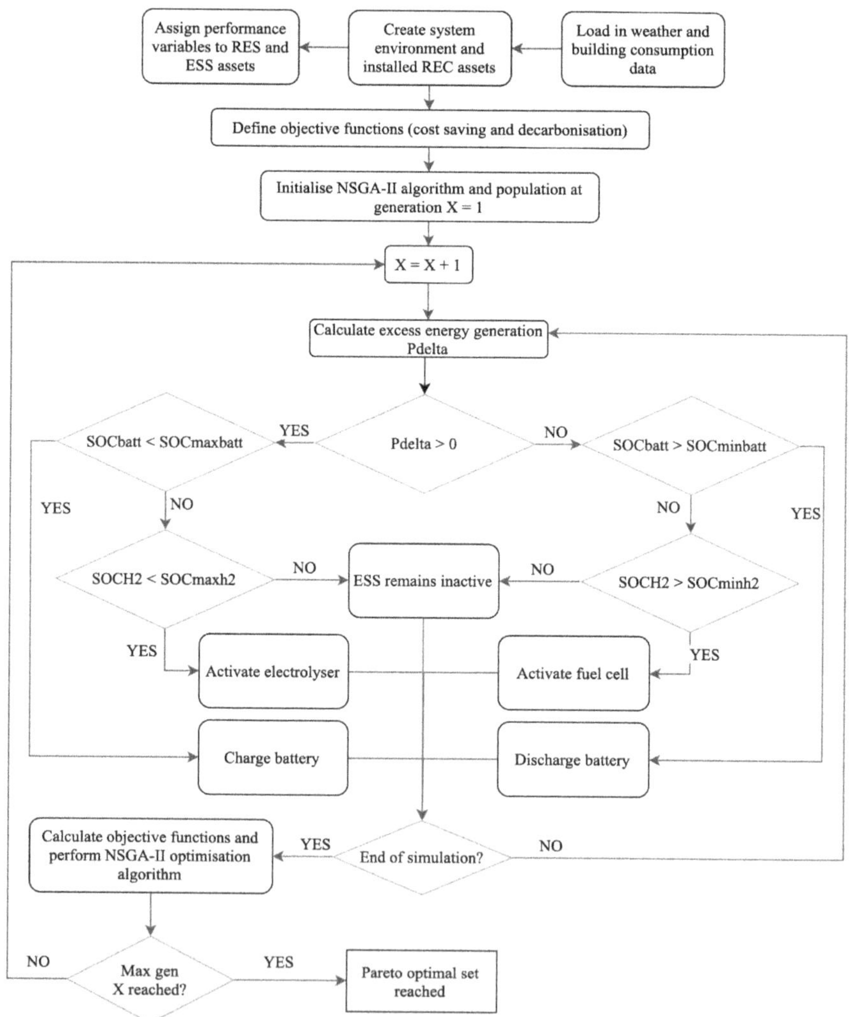

Figure 7. Hybrid energy system model approach with multi-objective optimisation algorithm NSGA-II solving process.

Table 4. Initialisation parameters and constraints of the NSGA-II optimisation algorithm.

Parameter	Value
Population Size	72
No. of Offspring	24
Max No. of Generations	400
Lower Bounds (all assets)	0 kW/kWh
Upper Bounds (all assets)	200 kW/kWh

The input parameters were set into the simulation model with the selected objective functions and run within the NSGA-II algorithm. The optimisation ran to the maximum allowed generations before terminating. Due to the bound nature of the problem, the component capacity variables start as a random distribution, from which the non-dominated solutions on the pareto front are derived. Well-performing individuals are moved forward to the next generation, as well as a selection of offspring and individuals that have experienced random mutation. As the generations progress, the population steadily converges on a large set of non-dominated solutions that align with the pareto front between the best system economics and decarbonisation performance, denoted by the objective functions of cost savings and emissions intensity. The graph in Figure 8 shows the convergence of the objective function products during the progression through the first 200 generations of the hybrid system optimisation, which will converge towards a single value.

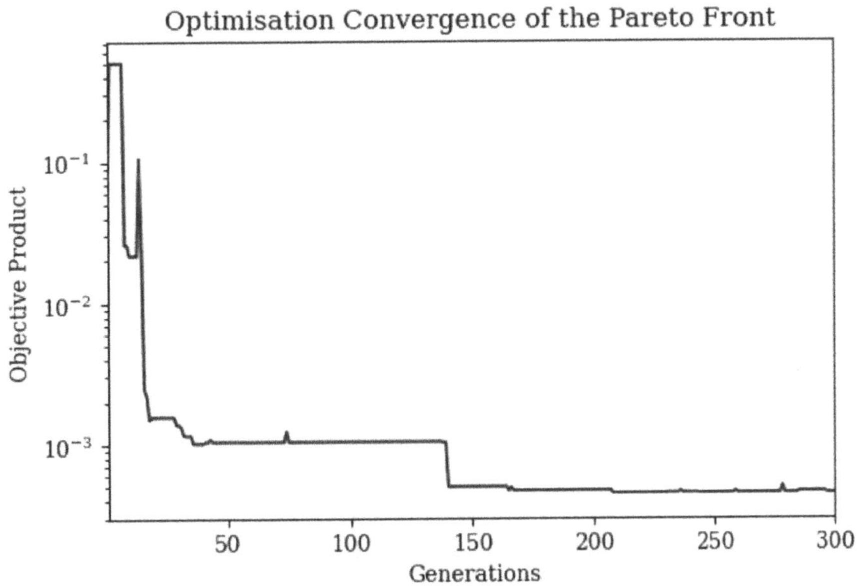

Figure 8. Convergence of the optimisation pareto front as shown by an aggregated scalar objective function minimising towards a single value.

4. Results and Discussion

4.1. Optimisation Results of the Hybrid Energy Generation and Storage Renewable Energy Community

The primary case studied was the hybrid architecture consisting of both a lithium battery and an RHFC. Within the resulting pareto front in Figure 9a, each point on the graph represents a different combination of design capacities ranging from the configuration able to achieve the highest economic returns to the system able to deliver the lowest net carbon output. The lifetime cost savings potential ranges from approximately EUR 130k to EUR 186k, while the emission intensity ranges from 82 to 140 gCO_2e/kWh. It is interesting to note that the savings do not start at zero, implying that below EUR 130k returns, the configuration was able to increase in economic performance as well as decarbonisation before reaching an inflection point. At this point, it is clear that the net savings were able to continue increasing, while the net emissions reached its minimum and began to climb again. At the other end of the front, the gradient began to increase as both the returns increase but also the emissions intensity. This continued up to the point where the system can no longer provide additional savings without an exponential increase in embedded emissions and therefore environmental impact.

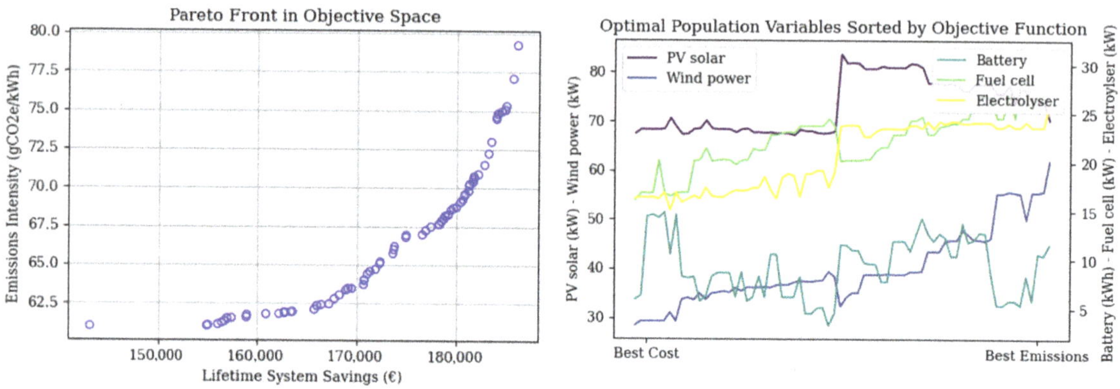

(**a**) Pareto front in objective space

(**b**) Optimal variables sorted by objective

Figure 9. Key outputs from the multi-objective optimisation process, indicating relationship between cost reduction and climate impact of the system design.

The resulting pareto front presents several crucial outcomes and challenges for providing a low cost and net-zero energy system. Firstly, an inherent trade-off relationship was observed between the ability to decarbonise and ensure net profitability. Secondly, the REC architecture, within the context and constraints of the study, can reduce carbon emissions by over 75% compared to local grid usage. However, this is a hard limit due to the capacity factors of the components and the embedded carbon within the system during manufacturing. Additionally, trying to decrease the carbon emissions further only incurs a financial penalty, which would be hard to incentivise to the REC members.

The graph in Figure 9b displays the capacities of PV solar, wind power, battery, fuel cell, and electrolyser systems with the final population arranged by the two objective functions. The best economic outcome is on the left, while the best environmental outcome is on the right. It can be observed that all systems generally tend towards an increase in capacity as the emissions improve. This is most likely because a larger total off-grid capacity has a higher self-consumption rate, and therefore is relying less on the grid which has a high emissions intensity of 325 gCO_2e/kWh. The REC was consequently able to reduce emissions to a greater extent. This, of course, negatively impacts the economics of the REC as more capital has to be invested into a more substantial design. It appears from the graph that the wind power, as well as the fuel cell and electrolyser which make up the RHFC are most sensitive to changes in the objective functions. The following section explores the chosen optimal design, and details why the capacities affect the objective functions in this way.

4.2. *Best Hybrid System Design for the Renewable Energy Community*

The pareto front provides a range of potential non-dominated solutions in which neither objective function is favoured over the other. There are several methods that can be used to choose a nominally 'best' system from the population to perform further analysis. Based on the research conducted by Wang and Rangaiah [69], it was chosen to use simple additive weighting (SAW). SAW normalises both objective function values, where zero is the worst possible result and one is the most improved. The values are then summed for each member of the population to find the best overall solution.

$$\begin{cases} F_{ij} = \frac{f_{ij}}{f_j^+} & \text{for a maximisation criterion, where } f_{i^+} = \max_{i \in m} f_{ij} \\ F_{ij} = \frac{f_{ij}}{f_j^-} & \text{for a maximisation criterion, where } f_{i^-} = \min_{i \in m} f_{ij} \end{cases} \quad (16)$$

$$A_i = \sum_{j=1}^{n} F_{ij} \quad (17)$$

F_{ij} is the normalised set of objective functions j for the pareto population i and f_{ij} is the initial set. f_{i^+} and f_{i^-} are the maximum and minimum criteria of the set, respectively. A_i then provides the best set of design variables to use in the hybrid REC, given in Table 5. The system was then simulated to perform analysis of all performance indicators.

Table 5. Optimal installed capacities of the energy system assets.

REC Asset	Optimal Values
PV Solar	71 kW
Wind Turbine(s)	32 kW
Lithium Battery	14 kWh
PEM Fuel Cell	20 kW
AEM Electrolyser	18 kW

Figure 10 contains two one-week sample periods obtained from the simulation, displaying the balance of each asset and their contribution to balancing the total REC load. Typical summer and winter periods are used to observe the seasonal variation in the system response. The REC load was higher on average during the summer period, leading to increased reliance and leading the energy grid to fill gaps in the consumption requirement when the ESS was unavailable. The winter period, by contrast, was able to satisfy the load requirement with the exception of some short periods. This shows that although the REC can operate largely off-grid, it is still beneficial from both an economic and emissions perspective to remain grid-connected from the short period when the REC generation and hybrid storage cannot fully balance the consumption. The hydrogen system requires a maximum storage of 1835k, which was evaluated from the simulation as the storage required to avoid any state-of-charge limits. The value therefore is a worst-case scenario for the system, as it is likely that a smaller storage would be chosen in accordance with the installation space available within the REC. Given the lower heating value (LHV) of hydrogen and the average fuel cell efficiency of 46%, the system would require approximately 14 Nm^2 of hydrogen stored at 35bar to supply the required quantity of a one-year period.

Table 6 below shows a full breakdown of the economic and environmental performance of each grouped asset. The solar array was able to deliver the most energy to the REC due to the high capacity of 71 kW, but also the higher solar potential on the island of Formentera of 4.7 kWh/m^2, compared to London, UK, of 2.9 kWh/m^2. Energy generated from wind provides the next greatest portion of over 24%, the benefit of which is that energy is generated during the night period as well as the day to charge the battery and a steady quantity of hydrogen. The battery itself was relatively small compared to the other components at 14 kWh and responds only when the energy generated is no longer available in excess of supply. The fuel cell and electrolyser were sized at 20 kW and 18 kW, respectively. It is interesting to note that the electrolyser was smaller in power input capacity than the fuel cell, even though the efficiencies would dictate the fuel cell would need approximately half the rated power of the electrolyser to achieve the same capacity factor. The increased generation from wind power over more of the simulation may allow the electrolyser to run for longer periods and make up the fuel cell's lower efficiency.

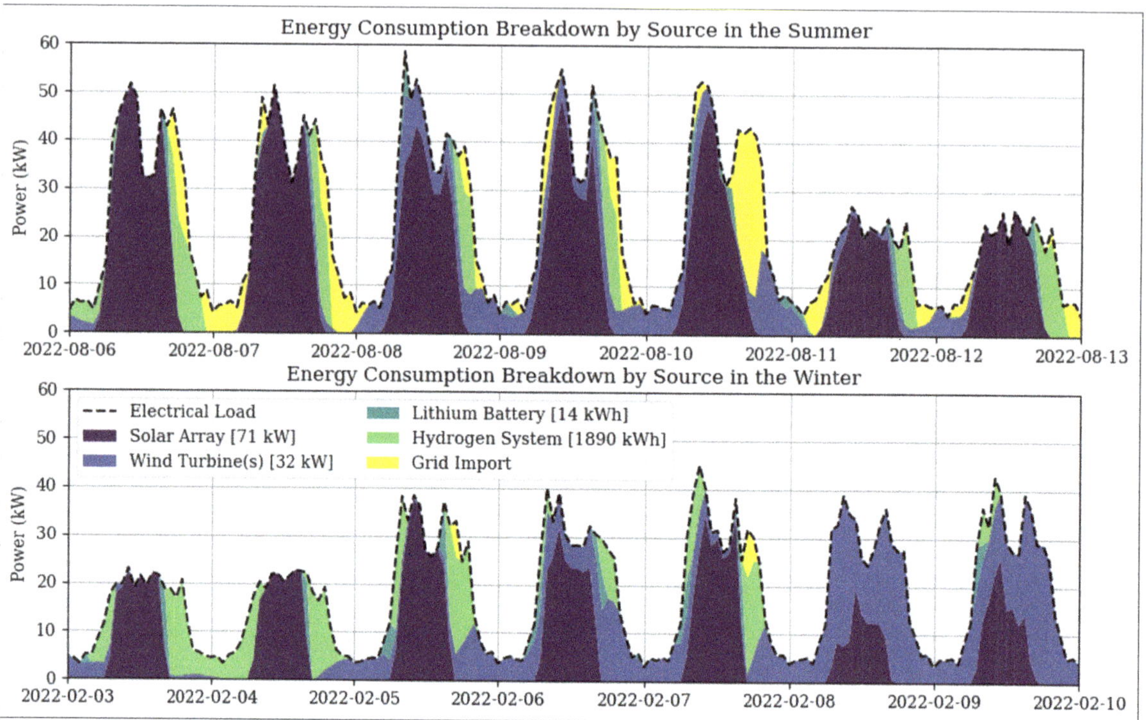

Figure 10. Energy generation hour-by-hour breakdown by source. Example shown includes typical summer and winter weeks.

Table 6. The economic and environmental performance of the different REC assets in the optimal design configuration.

Technology Asset	Energy-Delivered (kWh)	Capacity-Factor (%)	CAPEX (EUR)	OPEX (EUR/year)	LCOE (EUR/kWh)	Emissions (gCO_2e/kWh)
PV Solar [71 kW]	141,184	19	177,500	2130	0.07	40.7
Wind Turbine(s) [32 kW]	60,517	22	91,200	1024	0.09	15.9
Lithium Battery [14 kWh]	4910	8	5446	308	0.08	72.4
Hydrogen System [1836 kWh]	26,360	15	106,000	570	0.17	18.6

4.2.1. Techno-Economic Assessment

It is important to analyse each component on an individual basis to fully understand their contribution to the economic and emissions performance within the system. This would not only help to confirm the results seen in the pareto optimality, but also from a practical perspective assist a potential system designer to identify the most important assets, any particularly sensitive parameters, and assess the risks associated with each.

The solar array has the largest capital and operational costs compared with the wind power alternative due to the higher unit costs per kW. Despite this, the PV solar was able to achieve a greater capacity factor, which is the measure of energy output as a ratio of the total potential output of the same period. PV solar is naturally limited by the hours of solar available, while wind power is limited by the average wind speed and distribution.

The higher capacity is the main mechanism which produced a lower LCOE for the PV solar of 0.07 compared with wind power's 0.09 EUR/kWh, despite the higher CAPEX and OPEX costs. This result also implies that although wind power is possible on with the REC, it may be beneficial from a to study a PV solar generation only option due to the potential impracticalities of local wind turbines. The inverse was then observed for the environmental impact, in that the PV solar has considerably higher embedded emissions of 40.7 gCO_2e/kWh compared with the 15.9 gCO_2e/kWh expected from equivalent wind energy. These results show a good agreement with the reported embedded emissions from the IPCC AR5 report [70] of 45 gCO_2e/kWh and 13 gCO_2e/kWh for solar and wind, respectively.

At EUR 106k, the RHFC CAPEX was a factor of twenty higher than the battery. This trend carries over into the LCOE results, where an approximate doubling of the levelized cost was observed for the hydrogen system compared with lithium batteries. These outcomes are in line with similar hydrogen system results found in the literature [55,71]. It is widely known that hydrogen technology is a less financially viable alternative for many applications, so this result was somewhat expected. This could change in the near future as the costs of hydrogen technology reduce.

The emissions output from the battery per kWh delivery was far higher than the hydrogen solution at 72.4 and 18.6 gCO_2e/kWh, respectively. The trend is also supported by the population variables in Figure 9b, in which it is noted that as the hydrogen assets increased in capacity, the emissions result improved, while the net savings deteriorated.

Figure 11 contains the present value curve of the grid-only case, that is when electricity cost is paid to the utility company over the project period. The curve starts at zero as there is no capital cost associated with grid usage, but the operational cost per year is high. By contrast, the modelled REC requires an initial investment of EUR 380k. However, the lower year-on-year cost means that the system can pay off the investment cost, described as the payback term, in 9 years. The project ends with a final total savings of EUR 178k when the inflation and discounts rates of 2% and 5%, respectively, are considered. The result produces an IRR of 10.9%, year's returns of 9.6%, and an average system LCOE of 0.16 EUR/kWh. This assessment was based on the cost of equipment and installation since 2020.

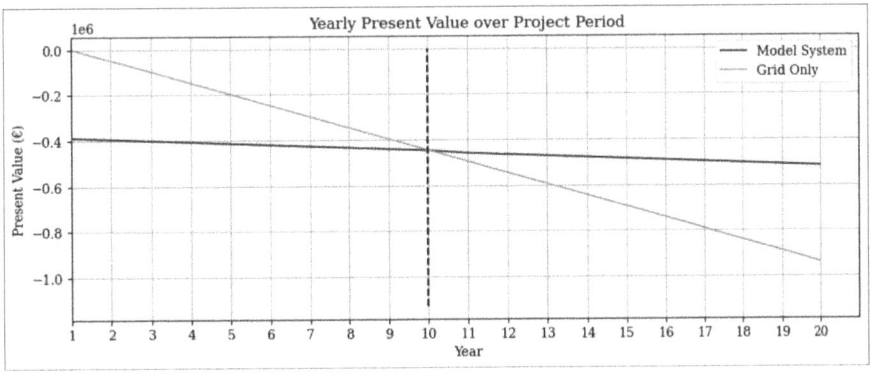

Figure 11. The present value over the project period. The system was primarily compared with a 'business-as-usage' grid-only scenario.

Considering the historical and currently observed trends in renewable generation and storage equipment cost, it is projected that by 2030 and beyond there will continue to be a substantial decrease in the financial requirements for this type of system. The results shown here are therefore towards the upper bounds in terms of uncertainty about the future cost of an REC implementation.

4.2.2. REC Members' Net Savings and Environmental Impacts

The model not only provides a global view of the potential impact of an REC configuration but is also able to analyse the reduction in cost and emissions on a per load basis. There were seven discretised loads within the model, with each being able to mutually accept and trade energy with the decentralised assets. Table 7 below shows the average LCOE and emissions intensity for each building and the percentage decrease in emissions. The REC provided a considerable degree of self-consumption, ranging from 91.1% for the largest load to over 98% for the smallest. In terms of the impact on the energy cost, the new LCOE ranged between 0.16 and 0.17 EUR/kWh compared to 0.30 EUR/kWh for grid-only. The decarbonisation of energy usage was also seen to be in the range of 75–77% in the first year of installation.

Table 7. Quantity of energy delivered to the REC compared with the quantity of energy delivered from the grid.

REC Member	REC Delivered (kWh)	Grid Delivered (kWh)
Community centre	61,078	5461
Elementary school	18,807	348
High school	44,743	1437
Council offices	28,244	683
Residential units	11,975	514

4.3. *Best Case and Extremes Comparison*

During the study, it was vital to understand not only the characteristics of the system at the 'best' pareto result, but also the performance at the extremes of the multi-objective optimisation. The result gives an indication as to how sensitive the result was to changing parameters. Table 8 contains the results of the three chosen REC configurations in terms of hybrid generation and storage capacities.

Table 8. Comparison REC configurations for extreme cases for net savings and decarbonisation potential compared with the chosen nominal case.

	Best Net Savings	Nominal	Best Emissions Savings
REC Delivered (kWh)	151,493	156,536	158,823
Self-Consumption (%)	91.0	94.5	96.2
LCOE (EUR/kWh)	0.15	0.16	0.19
Net Savings (EUR)	187,080	178,229	139,647
Savings (%)	51	47	36
IRR (%)	12.6	10.9	7.1
Simple Payback (%)	10.1	9.6	8.0
Payback Term (years)	8.3	9.0	11.5
Emissions (gCO2e/kwh)	79	69	61
Decarbonisation (%)	75.6	78.8	81.2

4.4. *Pareto Front Comparison of Energy Storage System Technologies*

The hybrid ESS comprised of a lithium battery and RHFC system produces differing performance outcomes based on the relative capacities of the technologies. Therefore, a comparison of the multi-objective optimisation for the same REC set-up with an additional battery-only ESS and RHFC-only ESS are required to ensure that the hybrid design was able to provide the best performance in terms of environmental impact and cost savings for the REC members. Figure 12 compares these pareto fronts, from which the combination of the technologies was able to produce a significant improvement over the technologies working independently. This was likely due to the fact that the battery is better at providing a short-term response but suffers from increased degradation if used frequently for charge and discharge, and similarly the hydrogen system requires a high capital cost and is best suited to the long-term storage of grid energy. The battery alone also has increased embedded carbon, which limited its ability to decarbonise. The hydrogen-only system suffered from the limitation that the electrolyser only runs at rated power, limiting flexibility.

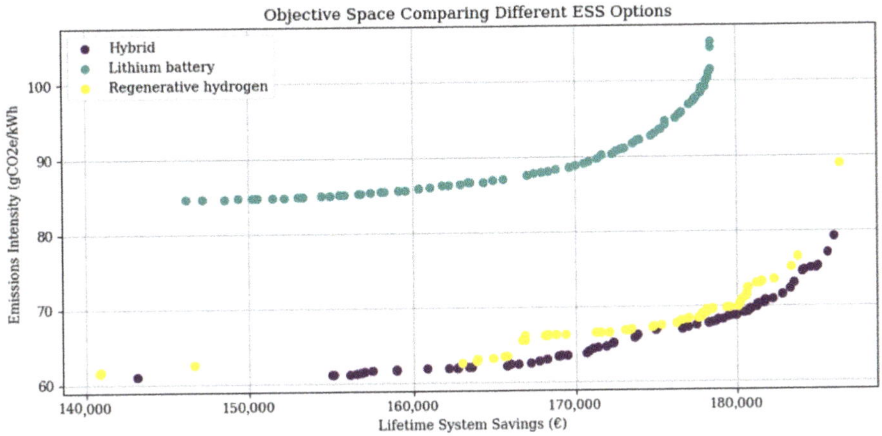

Figure 12. The present value over project period compared with the 'business-as-usual' grid-only scenario.

Table 9 below summarises a range of LCOE results from the literature with similar smart grid and renewable energy architectures, incorporating hydrogen storage where available. It should be noted that specific parameters such as renewable resource availability, local technology costs, and system sizing may induce uncertainty in the resulting total system LCOE. For example, locations with a higher solar potential are naturally able to achieve a lower PV solar LCOE, and conversely, remote areas with high delivery and installation costs would experience higher project costs. The resulting average is 0.13 EUR/kWh, which is in good agreement with the central economic and environmental trade-off case produced in this work.

Table 9. Comparison of the levelized cost of electricity results of similar hybrid hydrogen smart grid systems within the literature.

Reference	Smart Grid Architecture	Assets	Location	Scale	LCOE (EUR/kWh)
This work	Energy community	Solar, wind, battery/RHFC	Formentera, Spain	<100 kW	0.15
[55]	DC microgrid	Solar, battery/RHFC	Sub-Saharan	<100 kW	0.16
[72]	AC microgrid	Solar/wind, genset/RHFC	Morocco	<1 MW	0.07
[73]	AC microgrid	Solar/wind, hydrogen	India	<1 MW	0.08
[74]	Energy community	Solar/wind, battery/RHFC	Ghana	<100 kW	0.26
[75]	Energy community	Solar/wind, hydrogen	Canada	>1 MW	0.08

5. Conclusions

This work presents a novel decentralised hybrid generation and ESS implementing both battery and hydrogen technology for use in a geographically isolated rural renewable energy community. A review of the existing state of the art was presented and highlighted the gaps in knowledge for such a system, particularly when considering both economic feasibility and a dynamic calculation of the environmental impact. This discussion comes at an interesting time for Europe and around the world as policymakers work to facilitate the potential benefits of aggregating decentralised renewables, one such method being the REC. The steady cost reduction in PV solar and wind power as seen over the past years has also accelerated growth in the decentralised energy sector. The rise of cost-effective hydrogen technology is set to make a considerable impact on how energy is stored and transported as a vector.

The results from this study show that there is an inherent trade-off relationship between cost reduction and the ability to decarbonise the energy system. By using a model built in Python, several different economic and environmental scenarios can be assessed. The implementation of a multi-objective algorithm gives potential system designers and policymakers a range of possible solutions. In this case, the optimal design results in an LCOE of 0.15W EUR/kWh, a project IRR of 10.8%, and an ROI of 9 years. Greenhouse gas emissions were reduced by 72% in the first year of installation to 69 gCO_2e/kWh.

In further studies, emphasis should be placed on performing a sensitivity analysis and understanding where uncertainty may arise in the energy model. In particular, varying the component input assumptions such as capital cost and embedded emissions in line with reported ranges from the literature would further account for future uncertainty in performance. Additionally, further improvements to the system sizing optimisation method, including additional objective functions such as the loss of load probability from the literature, quantifying social impacts, and applying local space constraints would capture other potential strengths and weaknesses of the hybrid storage technology. Finally, research into the implementation of REC architectures in other locations around the world beyond the case study in this work would not only provide additional validation but also give a global perspective of how effective an REC could be in keeping energy costs stable and curbing the impacts of climate change.

Author Contributions: R.G.: Conceptualization, methodology, software, validation, formal analysis, investigation, resources, data curation, writing—original draft preparation, writing—review and editing, visualization. Z.D.: Supervision, project administration, funding acquisition, writing—review and editing. All authors have read and agreed to the published version of the manuscript.

Funding: This research is sponsored by the EU Horizon 2020 research and innovation program under the grant agreement No 957852: Virtual Power Plant for Interoperable and Smart isLANDS—VPP4ISLANDS. More information is available at https://cordis.europa.eu/project/id/957852 (accessed on 30 August 2023).

Data Availability Statement: Data will be made available on request.

Acknowledgments: Special thanks are extended to the Consel de Formentera for allowing access to energy consumption data, from which the energy community model was built.

Conflicts of Interest: The authors declare no conflict of interest.

Abbreviations

The following abbreviations are used in this manuscript:

CAPEX	Capital Expenditure
DOD	Depth of Discharge
ESS	Energy Storage System
GA	Genetic Algorithm
GHI	Global Horizontal Irradiance
GWP	Global Warming Potential
IEC	International Electrotechnical Commission
IRR	Internal Rate of Return
LCOE	Levelised Cost of Electricity
NPV	Net Present Value
NSGA	Non-dominated Sorting Genetic Algorithm
OPEX	Operational Expenditure

PEM	Proton Exchange Membrane
PV	Photovoltaic
REC	Renewable Energy Community
RED	Renewable Energy Directive
RES	Renewable Energy System
RHFC	Regenerative Hydrogen Fuel Cell
ROI	Return on Investment
SOC	State of Charge
TSO	Transmissions System Operator

References

1. United Nations. *The Paris Agreement*; United Nations: Paris, France, 2015.
2. Intergovernmental Panel on Climate Change. *Synthesis Report of the IPCC Sixth Assessment Report (AR6)*; IPCC Secretariat: Geneva, Switzerland, 2022.
3. Wang, R.; Usman, M.; Radulescu, M.; Cifuentes Faura, J.; Lorente, D. Achieving ecological sustainability through technological, innovations, financial development, FDI and energy consumption in developing European countries. *Gondwana Res.* **2023**, *119*, 138–152. [CrossRef]
4. Department for Business, Energy & Industrial Strategy. *British Energy Security Strategy*; HM Goverment UK: London, UK, 2022.
5. Boie, I.; Fernandes, C.; Frías, P.; Klobasa, M. Efficient strategies for the integration of renewable energy into future energy, infrastructures in Europe—An analysis based on transnational modeling and case studies for nine European regions. *Energy Policy* **2014**, *67*, 170–185. [CrossRef]
6. Kabeyi, M.J.B.; Olanrewaju, O.A. Sustainable Energy Transition for Renewable and Low Carbon Grid Electricity Generation and Supply. *Front. Energy Res.* **2022**, *9*, 743114. [CrossRef]
7. Gielen, D.; Boshell, F.; Saygin, D.; Bazilian, M.D.; Wagner, N.; Gorini, R. The role of renewable energy in the global energy transformation. *Energy Strategy Rev.* **2019**, *24*, 38–50. [CrossRef]
8. Saboori, H.; Mohammadi, M.; Taghe, R. Virtual Power Plant (VPP), Definition, Concept, Components and Types. In Proceedings of the 2011 Asia-Pacific Power and Energy Engineering Conference, Washington, DC, USA, 25–28 March 2011; pp. 1–4. [CrossRef]
9. Hirth, L.; Ziegenhagen, I. Balancing power and variable renewables: Three links. *Renew. Sustain. Energy Rev.* **2015**, *50*, 1035–1051. [CrossRef]
10. Mohd Azmi, K.H.; Mohamed Radzi, N.A.; Azhar, N.A.; Samidi, F.S.; Thaqifah Zulkifli, I.; Zainal, A.M. Active Electric Distribution Network: Applications, Challenges, and Opportunities. *IEEE Access* **2022**, *10*, 134655–134689. [CrossRef]
11. Rezaeimozafar, M.; Monaghan, R.F.; Barrett, E.; Duffy, M. A review of behind-the-meter energy storage systems in smart grids. *Renew. Sustain. Energy Rev.* **2022**, *164*, 112573. [CrossRef]
12. Aneke, M.; Wang, M. Energy storage technologies and real life applications—A state of the art review. *Appl. Energy* **2016**, *179*, 350–377. [CrossRef]
13. Hesse, H.C.; Schimpe, M.; Kucevic, D.; Jossen, A. Lithium-Ion Battery Storage for the Grid—A Review of Stationary Battery Storage System Design Tailored for Applications in Modern Power Grids. *Energies* **2017**, *10*, 2107. [CrossRef]
14. Wali, S.; Hannan, M.; Ker, P.J.; Rahman, M.A.; Mansor, M.; Muttaqi, K.; Mahlia, T.; Begum, R. Grid-connected lithium-ion battery energy storage system: A bibliometric analysis for emerging future directions. *J. Clean. Prod.* **2022**, *334*, 130272. [CrossRef]
15. Vykhodtsev, A.V.; Jang, D.; Wang, Q.; Rosehart, W.; Zareipour, H. A review of modelling approaches to characterize lithium-ion battery energy storage systems in techno-economic analyses of power systems. *Renew. Sustain. Energy Rev.* **2022**, *166*, 112584. [CrossRef]
16. Jamal, H.; Khan, F.; Si, H.R.; Kim, J.H. Enhanced compatibility of a polymer-based electrolyte with Li-metal for stable and dendrite-free all-solid-state Li-metal batteries. *J. Mater. Chem. A* **2021**, *9*, 27304–27319. [CrossRef]
17. Jamal, H.; Khan, F.; Si, H.R.; Kim, J.H. Enhancement of the ionic conductivity of a composite polymer electrolyte via surface functionalization of SSZ-13 zeolite for all-solid-state Li-metal batteries. *J. Mater. Chem. A* **2021**, *9*, 4126–4137. [CrossRef]
18. Mongird, K. *2020 Grid Energy Storage Performance Assessment*; US Department of Energy: Richland, WA, USA, 2020.
19. Zhang, W.; Maleki, A.; Rosen, M.A.; Liu, J. Optimization with a simulated annealing algorithm of a hybrid system for renewable energy including battery and hydrogen storage. *Energy* **2018**, *163*, 191–207. [CrossRef]
20. French, S. The Role of Zero and Low Carbon Hydrogen in Enabling the Energy Transition and the Path to Net Zero Greenhouse Gas Emissions: With global policies and demonstration projects hydrogen can play a role in a net zero future. *Johns. Matthey Technol. Rev.* **2020**, *64*, 357–370.
21. Coppitters, D.; De Paepe, W.; Contino, F. Robust design optimization and stochastic performance analysis of a grid-connected photovoltaic system with battery storage and hydrogen storage. *Energy* **2020**, *213*, 118798. [CrossRef]
22. Haji Bashi, M.; De Tommasi, L.; Le Cam, A.; Relaño, L.S.; Lyons, P.; Mundó, J.; Pandelieva-Dimova, I.; Schapp, H.; Loth-Babut, K.; Egger, C.; et al. A review and mapping exercise of energy community regulatory challenges in European member states based on a survey of collective energy actors. *Renew. Sustain. Energy Rev.* **2023**, *172*, 113055. [CrossRef]
23. European Parliament. *Directive (EU) 2018/2001 of the European Parliament and of the Council of 11 December 2018 on the Promotion of the Use of Energy from Renewable Sources*; European Parliament: Brussels, Belgium, 2018.

24. Fina, B.; Monsberger, C. Legislation for renewable energy communities and citizen energy communities in Austria: Changes from the legislative draft to the finally enacted law. *J. World Energy Law Bus.* **2022**, *15*, 237–244. [CrossRef]
25. Sebi, C.; Vernay, A.L. Community renewable energy in France: The state of development and the way forward. *Energy Policy* **2020**, *147*, 111874. [CrossRef]
26. Broska, L.H.; Vögele, S.; Shamon, H.; Wittenberg, I. On the Future(s) of Energy Communities in the German Energy Transition A Derivation of Transformation Pathways. *Sustainability* **2022**, *14*, 3169. [CrossRef]
27. Swens, J.; Diestelmeier, L. 4—Developing a legal framework for energy communities beyond energy law. In *Energy Communities*, Löbbe, S., Sioshansi, F., Robinson, D., Eds.; Academic Press: Cambridge, MA, USA, 2022; pp. 59–71. [CrossRef]
28. Trevisan, R.; Ghiani, E.; Pilo, F. Renewable Energy Communities in Positive Energy Districts: A Governance and Realisation Framework in Compliance with the Italian Regulation. *Smart Cities* **2023**, *6*, 563–585. [CrossRef]
29. Gallego-Castillo, C.; Heleno, M.; Victoria, M. Self-consumption for energy communities in Spain: A regional analysis under the new legal framework. *Energy Policy* **2021**, *150*, 112144. [CrossRef]
30. Official Gazette. *Legislative Decree no. 192/2021*; Official Gazette: Rome, Italy, 2021.
31. Di Silvestre, M.L.; Ippolito, M.G.; Sanseverino, E.R.; Sciumè, G.; Vasile, A. Energy self-consumers and renewable energy communities in Italy: New actors of the electric power systems. *Renew. Sustain. Energy Rev.* **2021**, *151*, 111565. [CrossRef]
32. Trevisan, R.; Ghiani, E.; Ruggeri, S.; Mocci, S.; Pisano, G.; Pilo, F. Optimal sizing of PV and Storage for a Port Renewable Energy Community. In Proceedings of the 2022 2nd International Conference on Energy Transition in the Mediterranean Area (SyNERGY MED), Thessaloniki, Greece, 17–19 October 2022; pp. 1–5. [CrossRef]
33. Bartolini, A.; Carducci, F.; Muñoz, C.B.; Comodi, G. Energy storage and multi energy systems in local energy communities with high renewable energy penetration. *Renew. Energy* **2020**, *159*, 595–609. [CrossRef]
34. Wang, Y.; Gao, W.; Li, Y.; Qian, F.; Yao, W. Techno-economic analysis of the transition toward the energy self-sufficiency community based on virtual power plant. *Front. Energy Res.* **2023**, *11*, 1010846. [CrossRef]
35. Cuesta, M.; Castillo-Calzadilla, T.; Borges, C. A critical analysis on hybrid renewable energy modeling tools: An emerging opportunity to include social indicators to optimise systems in small communities. *Renew. Sustain. Energy Rev.* **2020**, *122*, 109691. [CrossRef]
36. Niveditha, N.; Rajan Singaravel, M. Optimal sizing of hybrid PV–Wind–Battery storage system for Net Zero Energy Buildings to reduce grid burden. *Appl. Energy* **2022**, *324*, 119713. [CrossRef]
37. Zhang, Y.; Sun, H.; Tan, J.; Li, Z.; Hou, W.; Guo, Y. Capacity configuration optimization of multi-energy system integrating wind turbine/photovoltaic/hydrogen/battery. *Energy* **2022**, *252*, 124046. [CrossRef]
38. Xu, C.; Ke, Y.; Li, Y.; Chu, H.; Wu, Y. Data-driven configuration optimization of an off-grid wind/PV/hydrogen system based on modified NSGA-II and CRITIC-TOPSIS. *Energy Convers. Manag.* **2020**, *215*, 112892. [CrossRef]
39. Xuan, J.; Chen, Z.; Zheng, J.; Zhang, Z.; Shi, Y. Optimal planning of hybrid electric-hydrogen energy storage systems via multi-objective particle swarm optimization. *Front. Energy Res.* **2023**, *10*, 1034985. [CrossRef]
40. He, Y.; Guo, S.; Zhou, J.; Ye, J.; Huang, J.; Zheng, K.; Du, X. Multi-objective planning-operation co-optimization of renewable energy system with hybrid energy storages. *Renew. Energy* **2022**, *184*, 776–790. [CrossRef]
41. Wang, H.; Xie, Z.; Pu, L.; Ren, Z.; Zhang, Y.; Tan, Z. Energy management strategy of hybrid energy storage based on Pareto optimality. *Appl. Energy* **2022**, *327*, 120095. [CrossRef]
42. ©NordNordWest. CC-BY-SA-3.0. Spain Location Map. 2008. Available online: https://en.wikipedia.org/wiki/File:Spain_location_map.svg (accessed on 30 August 2023).
43. NASA. *LARC Power Data Access Viewer*; NASA: Houston, TX, USA, 2023.
44. Farhad, A.; Ali, G.; Mohsen, J. *Dataset on Hourly Load Profiles for a Set of 24 Facilities from Industrial, Commercial, and Residential Enduse Sectors*; Mendeley Data; Rutgers The State University of New Jersey: New Brunswick, NJ, USA, 2020; p. 1.
45. Pinto, E.S.; Serra, L.M.; Lázaro, A. Energy communities approach applied to optimize polygeneration systems in residential buildings: Case study in Zaragoza, Spain. *Sustain. Cities Soc.* **2022**, *82*, 103885. [CrossRef]
46. Chung, M.; Shin, K.Y.; Jeoune, D.S.; Park, S.Y.; Lee, W.J.; Im, Y.H. Economic Evaluation of Renewable Energy Systems for the Optimal Planning and Design in Korea—A Case Study. *J. Sustain. Dev. Energy Water Environ. Syst.* **2018**, *6*, 725–741. [CrossRef]
47. Duffie, J.; Beckman, W. *Solar Engineering of Thermal Processes*; Wiley: Hoboken, NJ, USA, 1991.
48. Reda, I.; Andreas, A. *Solar Position Algorithm for Solar Radiation Applications*; National Renewable Energy Laboratory: Golden, CO, USA, 2008.
49. Corke, T.; Nelson, R. *Wind Energy Design*; Taylor and Francis Group: Miami, FL, USA, 2018.
50. Wood, D. *Improvements to the Simplified Loads Methodology in IEC 61400-2*; National Renewable Energy Laboratory: Golden, CO, USA, 2021.
51. Draxl, C.; Clifton, A.; Hodge, B.M.; McCaa, J. The Wind Integration National Dataset (WIND) Toolkit. *Appl. Energy* **2015**, *151*, 355–366. [CrossRef]
52. Shepherd, C.M. Design of Primary and Secondary Cells: II. An Equation Describing Battery Discharge. *J. Electrochem. Soc.* **1965**, *112*, 657. [CrossRef]
53. Burton, T.; Sharp, D.; Jenkins, N.; Bossanyi, E. *Wind Energy Handbook*; John Wiley and Sons: Chichester, UK, 2001.
54. Warner, S.; Hussain, S. *The Finance Book*; Pearson Education Ltd.: Harlow, UK, 2017.

55. Jansen, G.; Dehouche, Z.; Corrigan, H. Cost-effective sizing of a hybrid Regenerative Hydrogen Fuel Cell energy storage system for remote and off-grid telecom towers. *Int. J. Hydrog. Energy* **2021**, *46*, 18153–18166. [CrossRef]
56. Feldman, D.; Dummit, K.; Zuboy, J.; Margolis, R. *Fall 2022 Solar Industry Update*; National Renewable Energy Laboratory: Golden, CO, USA, 2022.
57. Carvalho, M.; Menezes, V.L.; Gomes, K.C.; Pinheiro, R. Carbon footprint associated with a mono-Si cell photovoltaic ceramic roof tile system. *Environ. Prog. Sustain. Energy* **2019**, *38*, 13120. [CrossRef]
58. Stehly, T.; Beiter, P.; Duffy, P. *2019 Cost of Wind Energy Review*; National Renewable Energy Laboratory: Golden, CO, USA, 2019.
59. Intergovernmental Panel on Climate Change. *Climate Change 2021: The Physical Science Basis Contribution of Working Group I to the Sixth Assessment Report of the Intergovernmental Panel on Climate Change*; Cambridge University Press: Cambridge, UK, 2021.
60. Romare, M.; Dahllöf, L. *The Life Cycle Energy Consumption and Greenhouse Gas Emissions from Lithium-Ion Batteries*; IVL Swedish Environmental Research Institute: Stockholm, Sweden, 2017.
61. Ballard. *Fuel Cell Life Cycle Assessment, Burnaby*; Ballard: Burnaby, BC, Canada, 2018.
62. Department for Business, E.; Strategy, I. *Hydrogen Production Costs*; HM Government UK: London, UK, 2021.
63. Enapter. *A Small Carbon Footprint for Big Climate Impact*; Enapter: Saerbeck, Germany, 2022.
64. Reddi, K.; Elgowainy, A.; Rustagi, N.; Gupta, E. Techno-economic analysis of conventional and advanced high-pressure tube trailer configurations for compressed hydrogen gas transportation and refueling. *Int. J. Hydrog. Energy* **2018**, *43*, 4428–4438. [CrossRef]
65. Agostini, A.; Belmonte, N.; Masala, A.; Hu, J.; Rizzi, P.; Fichtner, M.; Moretto, P.; Luetto, C.; Sgroi, M.; Baricco, M. Role of hydrogen tanks in the life cycle assessment of fuel cell-based auxiliary power units. *Appl. Energy* **2018**, *215*, 1–12. [CrossRef]
66. Salameh, T.; Al-Othman, A.; Olabi, A.G.; Issa, S.; Tawalbeh, M.; Alami, A.H. Comparative life cycle assessment for PEMFC stack including fuel storage materials in UAE. In Proceedings of the 2020 Advances in Science and Engineering Technology International Conferences (ASET), Dubai, United Arab Emirates, 4 February–9 April 2020; pp. 1–5. [CrossRef]
67. Deb, K.; Pratap, A.; Agarwal, S.; Meyarivan, T. A fast and elitist multiobjective genetic algorithm: NSGA-II. *IEEE Trans. Evol. Comput.* **2002**, *6*, 182–197. [CrossRef]
68. Blank, J.; Deb, K. Pymoo: Multi-Objective Optimization in Python. *IEEE Access* **2020**, *8*, 89497–89509. [CrossRef]
69. Wang, Z.; Rangaiah, G.P. Application and Analysis of Methods for Selecting an Optimal Solution from the Pareto-Optimal Front obtained by Multiobjective Optimization. *Ind. Eng. Chem. Res.* **2017**, *56*, 560–574. [CrossRef]
70. Intergovernmental Panel on Climate Change. *Climate Change 2014: Synthesis Report*; IPCC Secretariat: Geneva, Switzerland, 2014.
71. Hasan, T.; Emami, K.; Shah, R.; Hassan, N.; Belokoskov, V.; Ly, M. Techno-economic Assessment of a Hydrogen-based Islanded Microgrid in North-east. *Energy Rep.* **2023**, *9*, 3380–3396. [CrossRef]
72. El Hassani, S.; Oueslati, F.; Horma, O.; Santana, D.; Moussaoui, M.A.; Mezrhab, A. Techno-economic feasibility and performance analysis of an islanded hybrid renewable energy system with hydrogen storage in Morocco. *J. Energy Storage* **2023**, *68*, 107853. [CrossRef]
73. Alluraiah, N.C.; Vijayapriya, P. Optimization, Design, and Feasibility Analysis of a Grid-Integrated Hybrid AC/DC Microgrid System for Rural Electrification. *IEEE Access* **2023**, *11*, 67013–67029. [CrossRef]
74. Pai, P.F.; Acakpovi, A.; Adjei, P.; Nwulu, N.; Asabere, N.Y. Optimal Hybrid Renewable Energy System: A Comparative Study of Wind/Hydrogen/Fuel-Cell and Wind/Battery Storage. *J. Electr. Comput. Eng.* **2020**, *2020*, 1756503. [CrossRef]
75. Temiz, M.; Dincer, I. Development of solar and wind based hydrogen energy systems for sustainable communities. *Energy Convers. Manag.* **2022**, *269*, 116090. [CrossRef]

batteries

MDPI

Article

Battery Degradation Impact on Long-Term Benefits for Hybrid Farms in Overlapping Markets

Pedro Luis Camuñas García-Miguel [1,*], Jaime Alonso-Martinez [1], Santiago Arnaltes Gómez [1,*], Manuel García Plaza [2] and Andrés Peña Asensio [2]

[1] Electrical Engineering Department, Carlos III University of Madrid, Avda. Universidad, 30, 28911 Leganés, Madrid, Spain

[2] Siemens Gamesa Renewable Energy, C. de Ramírez de Arellano, 37, 28043 Madrid, Spain; manuel.garcia.p@siemensgamesa.com (M.G.P.); andres.pena@siemensgamesa.com (A.P.A.)

* Correspondence: pcamunas@pa.uc3m.es (P.L.C.G.-M.); arnalte@ing.uc3m.es (S.A.G.)

Abstract: Participation in the electricity market requires making commitments without knowing the real generation or electricity prices. This is problematic for renewable generators due to their fluctuating output. Battery energy storage systems (BESSs) integrated with renewable sources in a hybrid farm (HF) can alleviate imbalances and increase power system flexibility. However, the impact of battery degradation on long-term profitability must be taken into account when choosing the correct market participation strategy. This study evaluates the state-of-the-art on energy management systems (EMS) for HFs participating in day-ahead and intraday markets, incorporating both BESSs' calendar and cycling degradation. Results suggest that efforts to attain additional profits in intraday markets can be detrimental, especially when the degradation effect is considered in the analysis. A new market participation strategy is proposed that aims to address the limitations of market overlapping and forecasting errors. The results demonstrate that the proposed method can enhance long-term benefits while also reducing battery degradation.

Keywords: BESS; optimization; degradation; markets

Citation: Camuñas García-Miguel, P.L.; Alonso-Martínez, J.; Arnaltes Gómez, S.; García Plaza, M.; Peña Asensio, A. Battery Degradation Impact on Long-Term Benefits for Hybrid Farms in Overlapping Markets. *Batteries* **2023**, *9*, 483. https://doi.org/10.3390/batteries9100483

Academic Editors: Pascal Venet, Luis Hernández-Callejo, Jesús Armando Aguilar Jiménez and Carlos Meza Benavides

Received: 27 July 2023
Revised: 13 September 2023
Accepted: 20 September 2023
Published: 22 September 2023

1. Introduction

Renewable energy systems face uncertainty in resource availability, which can create challenges in participating in electricity markets that require prior commitments. Energy storage systems (ESS) can mitigate this uncertainty by storing energy for later use [1].

The energy commitment for the day-ahead market is made by submitting offers to the market operator the day before delivery. Forecasting techniques are applied to predict electricity prices and resource availability [2], using techniques such as the SARIMA model, which can capture seasonal correlations in historical data. The authors in [3] demonstrate how a SARIMA model can outperform deep-learning techniques. In this work, SARIMA models are used to forecast both electricity prices and wind speeds, and have been shown to outperform deep-learning techniques in previous studies.

An offering strategy for energy commitment is typically formulated as a constrained optimization problem [4]. The decision vector includes offers for each hour of the market, and the objective is to maximize revenue during the session. Constraints include physical parameters of the plant and market rules, which heavily influence the feasible solution space.

Multi-market participation, which includes day-ahead and intraday market sessions, cannot be formulated as a single optimization problem due to the different timeframes of each market. A progressive optimization approach, as proposed by the authors in [5], is used in this work.

Intraday markets can be utilized to increase profits through revenue stacking, which typically involves combining energy and power services. Studies such as [6] demonstrate that combining frequency response and arbitrage can increase revenue by up to 25%. Similar results are found in studies such as [7], which consider multiple spot market participation. This is also proposed by the authors in [7]. These approaches are compared in this work.

The aforementioned studies do not model the forecast generation and market offering processes. Generated forecasts at different times of the day can contradict each other and lead to issues when various markets overlap. This work addresses this issue.

Balancing mechanisms in electricity markets, such as penalties for deviations, can be used to address deviations caused by forecasting errors. Intraday markets can be used to correct errors [8], but require prior commitments. An alternative solution is to use energy storage systems for real-time corrections through a service called capacity firming (CF) [9].

Capacity firming has received increasing attention in recent literature as a service provided in real-time as opposed to arbitrage. Studies such as [10] propose energy conservation methods for control schemes of BESSs integrated with a PV system. Other works, such as [11], show how a simple control algorithm can achieve capacity firming in a BESS combined with a wind farm, although the storage system is only used for this service.

The Iberian electricity markets are used as an example in this work since, like many other systems, they have day-ahead and intraday spot markets. Market operator rules are incorporated into the optimization algorithm and the calculation of net benefits. Participation in intraday markets and the use of the CF service are compared, and the revenue stacking of different services is evaluated. This work also considers the effect of degradation on long-term profits, an approach that has not been considered in the previous literature.

A HF model consisting of a Gamesa G128 Wind Turbine Generator (WTG) and a BESS is presented. Unlike a virtual power plant (VPP), the system components are not distributed; thus, they share a point of common coupling (PCC). The SARIMA forecasting model uses wind historical data from the Sotavento experimental wind farm [12] and electricity prices from the Iberian market as inputs for the EMS.

The results show that participating in all markets may be counter-productive due to market overlap. The best results in terms of profits per degradation are obtained by allowing the BESS to participate only in the day-ahead market and performing capacity firming in real time.

The work proposes a new service called SOC Emptying (SE), which involves dividing the BESS into two virtual energy storage systems (VESS). One VESS provides regular services, while the other is used to empty the BESS whenever the combined state of charge (SOC) exceeds a specific threshold. This service aims to reduce upward deviations and give the BESS more maneuverability. The inclusion of this service further improves the results, resulting in increased profits per percentage of capacity loss and higher net present values when extrapolating the results for the entire project.

The contributions of this work are outlined as follows:

- A formulation for a progressive optimization algorithm for hybrid farm multi-market participation that takes into account degradation effects caused by operation.
- The implementation of a real-time BESS degradation model that assesses capacity and power losses in a simulation framework to evaluate different market participation strategies.
- Evidence shows that using the BESS in all markets does not necessarily result in a better income after comparing different simulation scenarios.
- A proposed service that enhances the BESS utilization efficiency and increases the project's net present value (NPV).

The paper is structured as follows: Section 2 provides an overview of the Iberian electricity market rules. Section 3 describes the HF model used in the study. The day-ahead and intraday market offering optimization models are analyzed in Section 4. Section 5 presents the simulation use cases and results. Conclusions and future research directions are discussed in Section 6.

2. Iberian Market's Rules

Before each market session opening time, the EMS must have forecasted data of prices and wind power generation. It is therefore necessary to know when each market session takes place and which hourly delivery periods are negotiated, both for forecasting and optimization problem definition. To obtain real benefits, this study follows the regulations of the Iberian markets, which take into account deviation costs.

This section introduces the rules of the Iberian markets, starting with the day-ahead market, followed by the intraday markets, and finally, the deviation rules are described, which consider four different deviation costs.

2.1. Day-Ahead Market

The majority of energy traded in the Iberian wholesale markets is conducted through the day-ahead market. In 2020, it accounted for 74% of the total energy traded [13]. Therefore, it is the most important market for arbitrage operations. The Iberian day-ahead market session takes place every day of the year at 12:00 CET.

The price and volume of energy is determined for each hour of the following day by the intersection of demand and supply. Market agents submit their offers through the market operator OMIE [14]. As a result, the EMS has to submit 24-hourly offers for the following day, using price and generation forecasts generated 12 to 36 hours prior to delivery time.

2.2. Intraday Markets

After the day-ahead market, intraday markets accounted for 14% of the energy traded in 2020 [13]. Since intraday spot markets had six times more energy than continuous intraday markets, this work only considers the former. In Table 1, the closing times and delivery hours of the market sessions are shown.

Table 1. Intraday market sessions in 2018.

Session	1	2	3	4	5	6
Closing time	18:50	21:50	1:50	4:50	8:50	12:50
Delivery hours	22(D-1)–23	0–23	4–23	7–23	11–23	15–23

The closing times in Table 1 are the deadlines for submitting offers to the market operator. Decisions must be made before this time. The delivery hours in the table are the hours during which the energy negotiated in each intraday market session will be delivered on day D.

Since Sessions 1 and 2 cover the same hours, Session 1 is neglected as Session 2 has a closer opening time to the delivery. Intraday markets allow agents to correct their schedules in the day-ahead market. This can be performed by purchasing energy during hours when a deviation from the day-ahead program is expected. Arbitrage can also be performed to gain additional liquidity. Both options are considered in the simulation cases.

2.3. Adjustment Mechanism

Four different deviation costs need to be considered in the Iberian market:

- Positive deviation costs for upward deviations that have a negative impact on the system;
- Positive deviation costs for upward deviations that have a positive impact on the system;
- Negative deviation costs for downward deviations that have a negative impact on the system;
- Negative deviation costs for downward deviations that have a positive impact on the system.

A deviation during an hourly period h is calculated as follows:

$$\lambda(h) = |E_d(h) - E_c(h)|, \quad (1)$$

where the following apply:

- $\lambda(h)$: Deviation during hour h (MWh).
- $E_d(h)$: Delivery during hour h (MWh).
- $E_c(h)$: Commitment during hour h (MWh).

If the deviation is upwards and in favor of the system, the additional energy is remunerated at the day-ahead price during hour h; therefore, the bonus is calculated as:

$$\beta(h) = \lambda(h) * \Pi_{dM}(h), \tag{2}$$

where the following apply:

- $\beta(h)$: Bonus obtained during hour h (EUR).
- $\Pi_{dM}(h)$: Day-ahead price during hour h (EUR/MWh).

If the deviation is upwards and against the system, the energy excess is remunerated at less than the day-ahead price during hour h. This bonus is calculated as:

$$\beta(h) = \lambda(h) * \Pi_{dM}(h) * (1 - \lambda_{coef}(h)), \tag{3}$$

where the following applies:

- $\lambda_{coef}(h)$: Coefficient for deviations against the system during hour h.

If the deviation is downwards and in favor of the system, the energy deficit is charged at the same price as the day-ahead price during hour h; therefore, the penalty is calculated as:

$$\rho(h) = \lambda(h) * \Pi_{dM}(h), \tag{4}$$

where the following applies:

- $\rho(h)$: Penalty during hour h (EUR).

If the deviation is downwards and against the system, the energy deficit is charged at a rate surpassing the day-ahead price during hour h; therefore, the penalty is calculated as:

$$\rho(h) = \lambda(h) * \Pi_{dM}(h) * (1 + \lambda_{coef}(h)). \tag{5}$$

The total deviation cost is formulated as follows. As can be seen, it can be negative when more energy is available than committed:

$$\lambda_{cost}(h) = \rho(h) - \beta(h) \tag{6}$$

where the following applies:

- $\lambda_{cost}(h)$: Deviation costs during hour h (EUR).

The coefficient $\lambda_{coef}(h)$ represents the system's vulnerability to deviations against it. A higher coefficient means that a higher penalty will be paid. As seen in (3), if the coefficient is greater than 1, the bonus for upward deviations can be negative, which implies a penalty. During the same hourly period, if downward deviations are against the system, upward deviations are in favor of the system, and vice versa. The deviation coefficient is determined by the system operator and is the same for both types of deviations.

3. Hybrid Plant Model Overview

The model consists of the physical systems and their control architecture. In this section, the model inputs' generation is described for both wind speed and market prices. Afterwards, the plant model and its control architecture are introduced. Lastly, the daily earning calculation is formulated.

3.1. Model Inputs

3.1.1. Wind Power

Wind power is obtained through a two-stage approach, as in [15]. First, hourly wind speed is forecasted using a SARIMA; then, forecasted data are fed into a function that expresses a Gamesa G128 wind turbine power curve. The optimization and energy management system (EMS) models directly receive the forecasted and real wind power hourly values.

Wind speed historical data are obtained from Sotavento experimental park in Galicia, Spain [12]. This source was chosen due to its publicly available data and its location within the Iberian market region. The data have a resolution of one hour, and the measured wind speed for the year 2018 is presented in Figure 1.

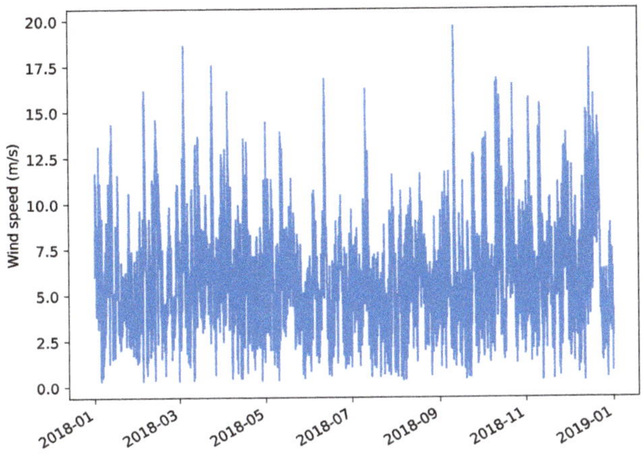

Figure 1. Hourly wind speed historical data for the year 2018.

Hourly wind speed forecasts are required during the opening hours of market sessions. The first forecast is at 12:00 h, when the day-ahead market commences, and subsequent forecasts are generated during the opening hours of intraday market sessions. For real-time operations, actual measured wind speed data are utilized. The SARIMA model utilized in this work has an order of $(2,0,3)(2,1,3)_{12}$, obtained from [16]. The configuration process for the model is not covered in this work. Figure 2 illustrates the 2018 average mean absolute percentage error (MAPE) of the wind speed forecasts generated during different market sessions. It is observed that the prediction error tends to decrease.

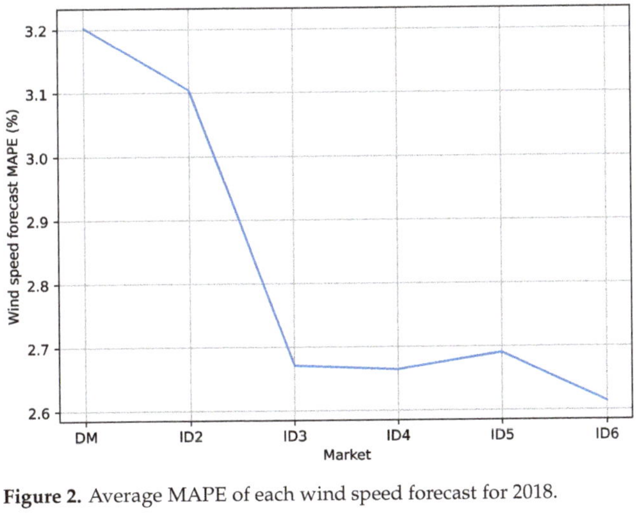

Figure 2. Average MAPE of each wind speed forecast for 2018.

3.1.2. Electricity Price Forecasts

The 24-hourly prices for the next day are predicted at 12:00 h on the previous day. A SARIMA model of order $(2,1,3)(1,0,1)_{24}$ was obtained from [16]. In Figure 3, the predicted and real day-ahead prices for April the 16th are shown. Intraday prices are considered as known beforehand for simplicity.

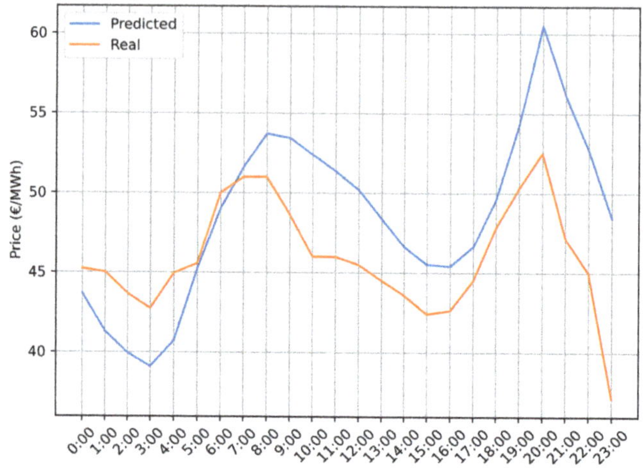

Figure 3. Day-ahead market price forecast for April the 16th.

3.1.3. Deviation Prices

The hourly deviation coefficients derived from 2018 historical data are used to calculate the final deviation costs. However, a deviation coefficient is required for the optimization model of intraday market offerings. Since deviation coefficients are only known after the delivery period has ended, a forecasting technique for predicting the direction (favorable or unfavorable) of deviation is required, but it is outside the scope of this paper. Therefore, a deviation coefficient of 21%, the average of 2017, is considered when participating in the intraday market.

3.2. *Plant Model Components*

The HF model consists of a wind turbine generator and a BESS.

3.2.1. Wind Turbine Generator

The generation system is a single Gamesa G128 WTG with a nominal power of 4.5 MW. The power curve is taken from [17]. Only a WTG is considered for convenience. Generator converter efficiency is considered part of the power curve characteristic.

3.2.2. Battery Energy Storage System

The storage system is composed of a 10 MWh/2.5 MW lithium iron phosphate (LFP) battery, whose round-trip efficiency is assumed to be constant at 90%, as in [18]. A cycling degradation model, based on depth of discharge (DOD) is implemented, sourced from a finalization of the degradation curve presented in [19]. The model calculates degradation by using rainflow counting of the cycles performed during the day and determines the State of Health (SOH) lost based on the DOD of these cycles, in accordance with the curve shown in Figure 4.

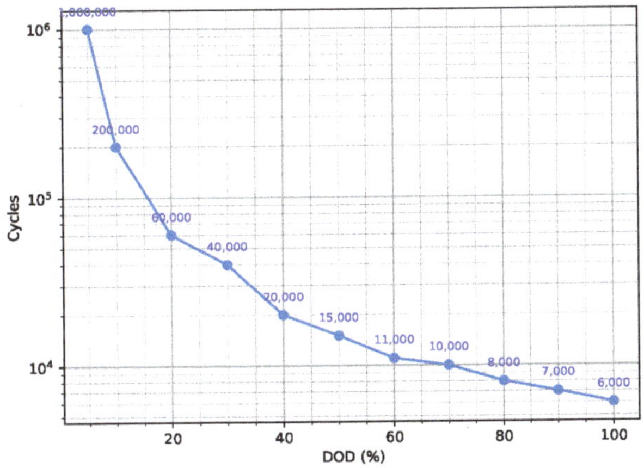

Figure 4. Cycling degradation model.

A calendar degradation model, sourced from [20], is incorporated into the plant model. This model is based on tests conducted on different LFP batteries over 899 days, where they were maintained at various SOCs and temperatures. The test results for 25 °C were selected and are presented in Table 2.

Table 2. Calendar degradation results at 25 °C [20].

SOC (%)	Capacity Lost (%)
0	0.002
50	0.0055
100	0.012

As depicted in Table 2, daily calendar degradation can be represented as a linear function of SOC. Since it is also considered linear with respect to time, the following expression has been obtained from the table:

$$Deg_{cal}(h) = SOC(h) \times \frac{1.2 \times 10^{-4}}{24} \tag{7}$$

where the following apply:

- $Deg_{cal}(h)$: Calendar degradation during hour h (%).
- SOC(h): SOC at the end of hour h (%)

3.3. Control Architecture

The EMS operates on three levels. The tertiary level operates in advance, and is responsible for formulating strategies for offerings in various energy markets. It receives as inputs the predicted prices, generation, and the real-time SOC for the intraday market and SE optimization models, which are executed at this control level. It then sends market commitments to the secondary level.

The secondary control level operates in real-time and generates the setpoints for the BESS based on the difference between generation and commitments generated on the third level. The BESS setpoints are calculated as follows:

$$P_{ESS}(h) = P_{Com}(h) - P_{Gen}(h), \tag{8}$$

where the following apply:

- $P_{ESS}(h)$: BESS setpoint during hour h (MW).
- $P_{Gen}(h)$: Generation during hour h (MW).

- $P_{Com}(h)$: Commitment during hour h (MW).

As shown, when an upward deviation is anticipated, the BESS will receive a setpoint to charge the excess. Conversely, when a downward deviation is expected, a discharging setpoint is generated.

The primary control level manages the energy storage system; it receives charge–discharge setpoints from the secondary level and adjusts them based on the real-time SOC. If the BESS does not have enough energy to cover the missing energy, a downward deviation will occur during that hour. If the BESS is unable to store the excess energy, then an upward deviation will occur.

A sketch of the EMS control architecture is illustrated in Figure 5.

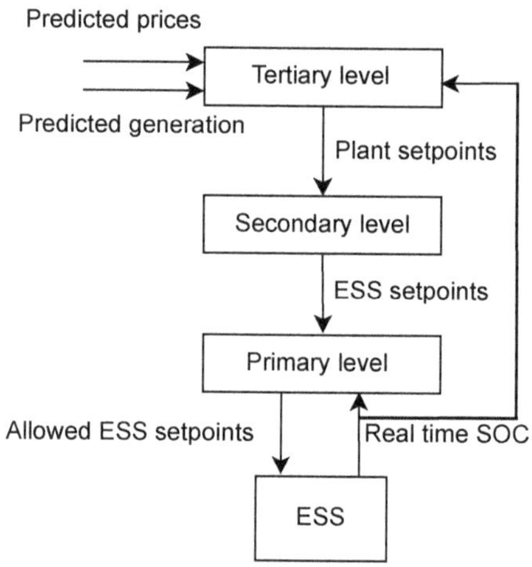

Figure 5. EMS control levels.

3.4. Daily Profits Calculation

The daily profits are calculated as the sum of the hourly profits. In Equation (9), the calculation of daily earnings is depicted. The daily losses are formulated as in Equation (10).

$$Earnings = \sum_{h=1}^{24} \left(E_{DM,C}(h) \cdot \Pi_{DM}(h) + E_{ID,C}(h) \cdot \Pi_{ID}(h) \right). \tag{9}$$

$$Losses = \sum_{h=1}^{24} \left(P_{\varphi}(h) \cdot \Pi_{ID}(h) + \lambda_{cost}(h) \right). \tag{10}$$

where the following apply:

- *Earnings*: Daily earnings (EUR).
- $E_{DM,C}(h)$: Energy commitment in day-ahead market during hour h (MWh).
- $\Pi_{DM}(h)$: Day ahead market real price during hour h (EUR/MWh).
- $E_{ID,C}(h)$: Energy commitment in intraday markets during hour h (MWh).
- $\Pi_{ID}(h)$: Intraday market real price during hour h (EUR/MWh).
- *Losses*: Daily losses (EUR).
- $P_{\varphi}(h)$: Power purchased in intraday market during hour h (MW).

Downward deviations can be corrected in intraday markets in two ways:

- By purchasing the expected energy deficit in the intraday market ($P_{\varphi}(h)$).
- By using stored energy to cover the imbalance.

The optimization algorithm chooses how to correct expected deviations depending on the intraday market prices and deviation costs at each hour.

4. Optimization Problems for Market Participation

In this work, market scheduling strategies are formulated as mathematical optimization problems. The progressive optimization approach, similar to the one described in [5], is employed. The day-ahead market scheduling problem is first solved to generate an hourly power schedule vector P_{Sch}. This vector is updated and sent to the secondary control level in real-time throughout the day. The process is illustrated in Figure 6.

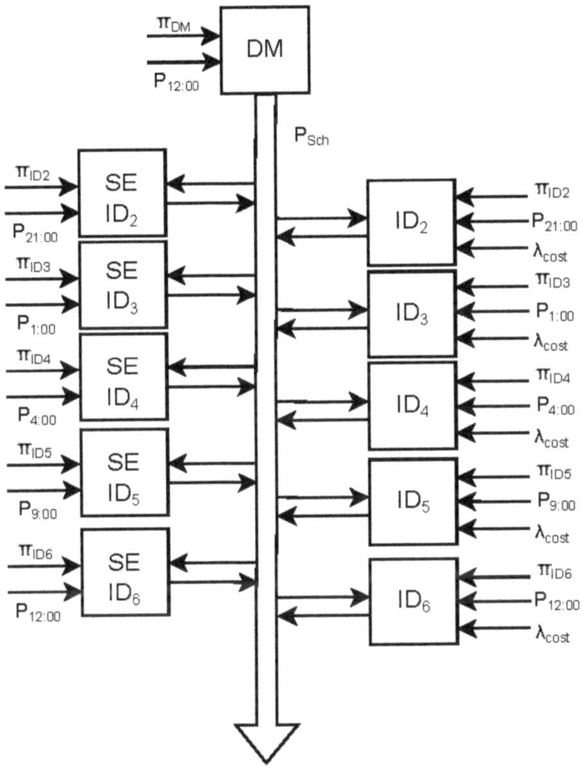

Figure 6. Daily optimization process.

This section presents the formulation of the optimization problems. The day-ahead market offering is explained first, followed by the intraday market offering, and finally the proposed SE service optimization is described as a separate problem.

4.1. Day-Ahead Market Offering

The inputs for the day-ahead market offering include the price and power generation forecasts for the next day generated at 12:00 PM and the plant assets' parameters. The optimization problem is subject to the following constraints:

1. The charge power cannot be higher than the nominal value:

$$P_{ch}(h) \leq P_{nom,BESS} \cdot \phi_d(h), \tag{11}$$

where the following apply:

- $P_{ch}(h)$: Charging power during hour h (MW).
- P_{nom}: Nominal power (MW).
- $\phi_d(h)$: A binary variable that equals 1 when the battery is not discharging during hour h.

2. The discharge power cannot be higher than the nominal value:

$$P_{dis}(h) \leq P_{nom,ESS} \cdot \phi_c(h), \tag{12}$$

where the following apply:

- $P_{dis}(h)$: Discharging power during hour h (MW).
- P_{nom}: Nominal power (MW)
- $\phi_c(h)$: A binary variable that equals 1 when the battery is not charging during hour h.

3. The charging power is always positive:

$$P_{ch}(h) \geq 0. \tag{13}$$

4. For optimization algorithm simplicity, the discharge power is also always positive:

$$P_{dis}(h) \geq 0. \tag{14}$$

5. The simultaneous charge and discharge is not possible:

$$\phi_c(h) + \phi_d(h) \leq 1. \tag{15}$$

6. The energy stored at the end of each period is calculated as follows:

$$E(h) = E(h-1) + \left(P_{ch}(h) \cdot \xi - \frac{P_{dis}(h)}{\xi}\right), \tag{16}$$

where the following apply:

- E(h): Energy stored in the battery at the end of hour h (MWh).
- ξ: Efficiency (%)
- E(h − 1): Energy stored on the battery at the beginning of hour h, which is considered to be 0 at the first hour.

7. The stored energy cannot have a negative value:

$$E(h) \geq 0. \tag{17}$$

8. The BESS cannot discharge if its participation in the day-ahead market is disabled; this constraint is activated depending on the case of study:

$$P_{dis}(h) = 0. \tag{18}$$

9. The power flow of the plant is defined as follows:

$$P_{gen}(h) = P_s(h) + P_{ch}(h), \tag{19}$$

where the following apply:

- $P_{gen}(h)$: Total generation power during hour h (MW).
- $P_s(h)$: Generation power sent directly to the grid during hour h (MW).
- $P_{ch}(h)$: Generation power sent to the storage system during hour h (MW).

The following objective function seeks to maximize the income:

$$Max\left\{\sum_{h=1}^{24} \Pi_{DM}(h) \cdot \left(P_s(h) + P_{dis}(h)\right)\right\}. \tag{20}$$

The objective of this function is to maximize profits through the generation of the optimal schedule based on forecasted energy and prices. Deviations are not accounted for in this optimization problem and are addressed in the intraday market optimization. The output of this offering strategy is the hourly schedules for the following day, which serve as input for the next optimization problems.

As it can be seen, the BESS does not purchase energy from the market. As per the regulations of the Renewable Energy Economic Regime [21], a BESS is not allowed to purchase energy from electricity markets when operating in hybrid plants.

4.2. Intraday Market Participation

The intraday market participation aims to adjust the hourly schedule in the case of expected deviations. It does so through two means: selling expected excess energy if enabled and purchasing energy in case of expected downward deviation. The optimization function takes the following inputs: committed hourly schedules, generated power prediction, deviation costs, intraday market prices, and the expected state of charge at the beginning of delivery.

Each intraday market session occurs three hours before delivery, as shown in Table 1. The expected SOC at the start of delivery is communicated by the tertiary control level. The constraints for this optimization are the same as those for the day-ahead market scheduling problem, with the addition of the following constraints:

1. The upward deviations are

$$\lambda_{\uparrow}(h) = P_{PCC}(h) - P_{sch}(h), \tag{21}$$

where the following apply:

- $\lambda_{\uparrow}(h)$: Upward deviation during hour h (MW).
- $P_{PCC}(h)$: PCC power during hour h (MW).
- $P_{sch}(h)$: Scheduled power during hour h (MW).

2. The downward deviations are

$$\lambda_{\downarrow}(h) = P_{sch}(h) - P_{PCC}(h), \tag{22}$$

where the following applies:

- $\lambda_{\downarrow}(h)$: Downward deviation during hour h (MW).

3. The downward deviations are considered always positive, for simplicity purposes:

$$\lambda_{\downarrow}(h) \geq 0. \tag{23}$$

4. The same happens with upward deviations:

$$\lambda_{\uparrow}(h) \geq 0. \tag{24}$$

As can be seen, when one type of deviation takes place, the other is equal to zero.

5. The hourly deviation costs are

$$\lambda_{cost}(h) = \lambda_{\downarrow}(h) \cdot \rho(h) - \lambda_{\uparrow}(h) \cdot \beta(h), \tag{25}$$

The deviation penalty and bonuses are calculated from expressions (3) to (5). The intraday optimization uses real day-ahead market prices, available at the time of the intraday market. The deviation coefficient is set at 21% and the average is from 2017, and it is used only as an assumption at this stage, while historical deviation coefficients are used later to calculate real benefits.

6. The internal power flow constraint depicted in (19) is modified:

$$P_{gen}(h) = P_s(h) + P_{ch}(h) + P_{del}(h) + P_{curt}(h), \tag{26}$$

where the following apply:

- $P_{del}(h)$: Generated power used to cover deviations during hour h (MW).
- $P_{cur}(h)$: Generation power curtailed during hour h (MW).

The generated power is allocated to either cover expected deviations or to maximize profits in the intraday market.

7. The constraint regulating stored energy is a modification of (16) as follows:

$$E(h) = E(t-1) + \left(P_{ch}(h) \cdot \xi - \frac{P_{dis}(h) + P_{dis,s}(h)}{\xi}\right), \tag{27}$$

where the following applies:

- $P_{dis,s}(h)$: BESS power sold during hour h (MW).

This division of discharged power into two parts—one used to cover deviations and one used for arbitrage—is similar to the division of generated power.

8. If arbitrage in the intraday market is disabled, the following constraints are applied:

$$P_{dis,s}(h) = 0. \tag{28}$$

$$P_s(h) = 0. \tag{29}$$

9. The PCC output power is computed as follows:

$$P_{PCC}(h) = P_{del}(h) + P_{dis}(h) + P_{\varphi}(h), \tag{30}$$

The intraday market purchased power is not physically received by the plant and serves to fulfill commitments in the day-ahead market in case of deviations. It is therefore not included in the PCC output power constraint, which measures expected deviations. The exchanged power is part of the scheduled power vector input for the subsequent intraday market optimization, as shown in the constraints represented by expressions (21) and (22).

The objective function is defined in (31) with two goals: minimizing expected deviations and maximizing profits through energy trading.

$$Max\left\{\sum_{h=1}^{ID_{len}} \Pi_{ID}(h) \cdot \left(P_s(h) + P_{dis,s}(h) - P_{\varphi}(h)\right) - \lambda_{cost}(h)\right\}, \tag{31}$$

where the following applies:

- ID_{len}: Intraday market length.

The hourly commitment vector is updated using the outputs of the intraday market optimization problem:

$$P_{Sch,new}(h) = P_{Sch,prev}(h) - P_{\varphi}(h) + (P_s(h) + P_{dis,s}(h)) \tag{32}$$

where the following apply:

- $P_{Sch,new}(h)$: New hourly schedule (MW).
- $P_{Sch,prev}(h)$: Previous hourly schedule (MW).

4.3. State of Charge Emptying

As previously discussed, the secondary control level sends generated energy to the grid when the BESS is full and hourly commitments are fulfilled, leading to an upward deviation. The market operator only pays for excess energy at the day-ahead market price when the deviation is in favor of the system, as described in Equations (2) and (3). This can result in a missed opportunity to sell energy at higher prices when the deviation is against the system.

Moreover, the highest calendar degradation occurs when the BESS is full, as shown in Table 2. A new operating mode is proposed that involves selling part of the stored energy on the nearest intraday market when the BESS SOC exceeds a set threshold. This service differs from intraday market arbitrage in the following ways:

- It operates simultaneously with the intraday market optimization process.
- It only manages energy above a certain threshold, not the entire BESS.
- It is limited to offering energy only in the first hours of each intraday market, to prevent overlap with the next one.

In Figure 7, the operating hours of the proposed service are shown, marked in blue, over the delivery hours of the intraday market, which are represented as white bars.

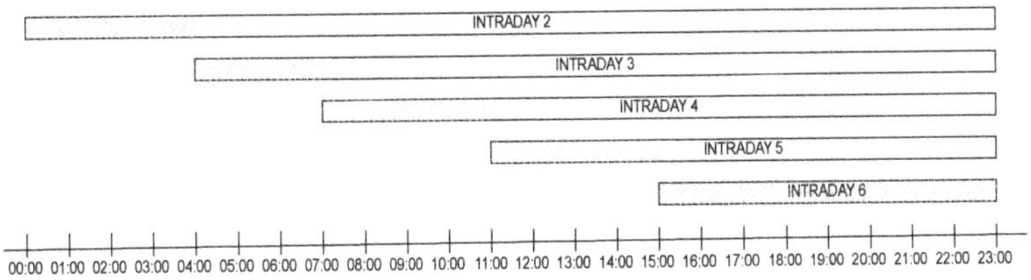

Figure 7. SE service operating hours.

This service splits the BESS into two virtual energy storage systems (VESS), one for profit generation and the other for deviation reduction. The stored energy is divided with a threshold of 75% set for the SE service. This value was determined through testing various values in the study and was found to be the optimal balance between profits and deviation reduction.

$$SOC_i = SOC_{SE} - SOC_{thr}, \tag{33}$$

where the following apply:

- SOC_{SE}: SOC used for SE service (%).
- SOC_{thr}: SOC threshold (%).

The SE service is managed as an optimization problem identical to the one for intraday market participation but limited to the first hours of the next intraday market. The variables used for deviation coverage are disabled as the objective of this service is solely profit generation. The objective function is as follows:

$$Max\left\{\sum_{h=1}^{SE_{len}} \Pi_{ID} \cdot P_{SE}(h)\right)\right\}, \tag{34}$$

where the following apply:

- SE_{len}: SE length for the next intraday market .
- $P_{SE}(h)$: Power sold during hour h (MW).

The goal is to sell available energy at the most expensive hours. The scheduled power vector is updated as in (32).

$$P_{Sch,new}(h) = P_{Sch,prev}(h) + P_{SE}(h) \tag{35}$$

5. Simulations and Results

In this section, the simulation scenarios are presented, each showcasing a different approach to using the BESS on the HF. Each scenario involves simulating the HF with the respective approach operating during 2018 in the Iberian electricity markets. The aim is to compare and assess whether revenue-stacking is more efficient than focusing on individual services, and the performance of the proposed service.

At the end of each day, the accumulated degradation and SOC serve as initial values for the following day's operation. The average daily profits and degradation under each scenario are used in a full project extrapolation for a comprehensive view of the different cases.

5.1. Simulation Cases

The simulation cases used are the following:

- Ideal: Perfect foresight of prices and power generation, which operates only in the day-ahead market.
- DM: Energy sales in the day-ahead and intraday markets to cover deviations.
- DM + SE: Same as DM, but with additional SOC emptying service.

- DM + ID: Energy sales in day-ahead market and intraday markets to sell excess and cover deviations.
- ID: Only participates in intraday markets for energy sales and covering deviations.
- SE: Only participates in intraday markets to cover deviations and performs SE service
- CF: No market participation; covers deviations through capacity firming services (already provided in previous cases).

5.2. Simulation Results

The simulation was conducted using the same input data in each case. Expected profits were calculated by summing the earnings from day-ahead and intraday market commitments and real prices. Real profits were calculated as the difference between daily earnings (Equation (9)) and daily losses (Equation (10)) using historical data of electricity prices and deviation coefficients. Figure 8 displays the accumulated profits and costs for each scenario, and Table 3 shows the numerical results.

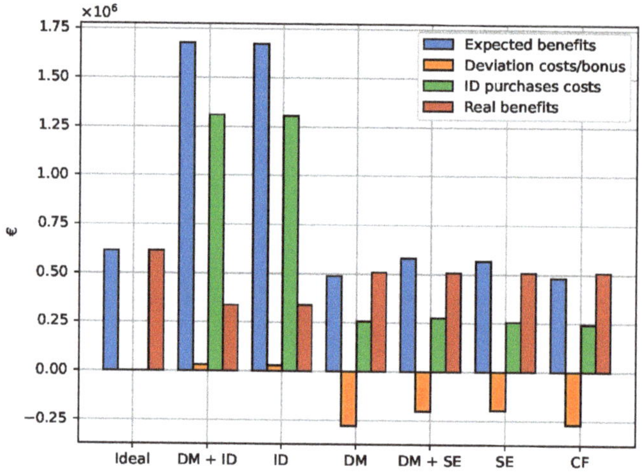

Figure 8. Simulations results.

Table 3. Numerical results.

Case	Expected Profits (EUR M)	Deviation Costs (EUR M)	Purchases Costs (EUR M)	Real Profits (EUR M)
Ideal	0.614	0	0	0.614
DM + ID	1.677	0.031	1.309	0.336
ID	1.671	0.029	1.305	0.338
DM	0.488	−0.277	0.256	0.509
DM + SE	0.581	−0.201	0.276	0.507
SE	0.568	−0.195	0.255	0.508
CF	0.482	−0.27	0.244	0.509

The results indicate that expected profits increase with intraday market arbitrage, as anticipated. These profits are calculated based on the delivery of all committed energy to the market, resulting in higher profits in scenarios with intraday market participation as all expected energy excess can be sold. The worst expected outcomes occur in scenarios without intraday market participation, and similar results are observed when the BESS performs capacity firming.

The deviation costs are negative in almost all scenarios, indicating when upward deviations are more frequent. Purchasing energy in the intraday market effectively avoids downward deviations. The fewer services the BESS provides, the greater the negative deviation costs, suggesting that the BESS tends to be fully charged most of the time, leading to upward deviations.

Purchases in the intraday market are the primary cause of profitability losses. These costs are significantly higher with intraday market participation. The fact that intraday market purchases occur when the BESS cannot cover expected deviations highlights that increased BESS involvement in markets can have negative effects.

Committing more energy to various markets increases the risk of having to make corrections by purchasing energy in the intraday market. The best results seem to be achieved by letting the BESS operate solely in the day-ahead market or providing capacity firming services. As shown in Figure 9, intraday market participation has resulted in the need to cover up to 40% of committed energy through intraday market purchases.

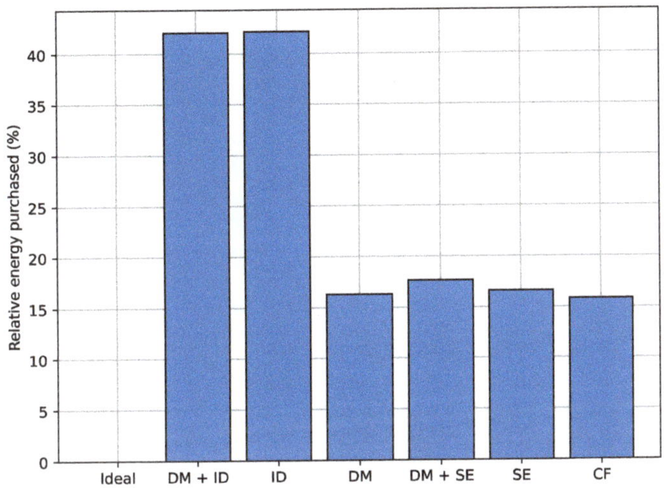

Figure 9. Committed energy covered with purchases in intraday markets.

Figure 10 displays the accumulated degradation. In scenarios where the BESS provides the most services, it tends to be emptier, which reduces calendar degradation and results in lower overall aging of the BESS.

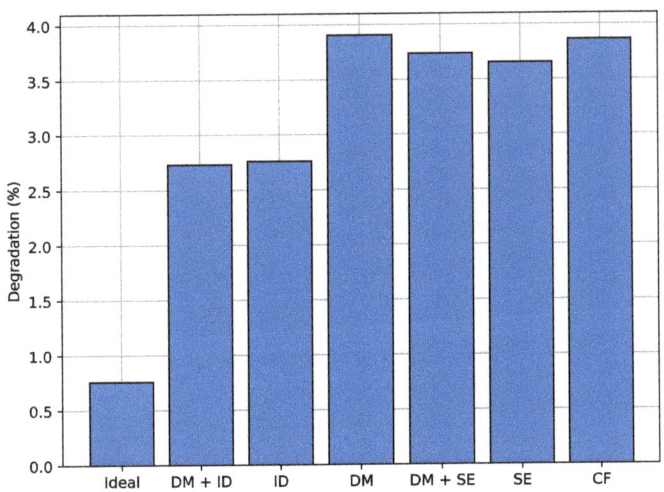

Figure 10. Degradation in each case.

Figure 11 shows the comparison between real profits and degradation. It reveals that reserving the BESS for SOC emptying and capacity firming yields the best outcome when profits are compared to capacity loss, which indicates a more efficient usage of the BESS.

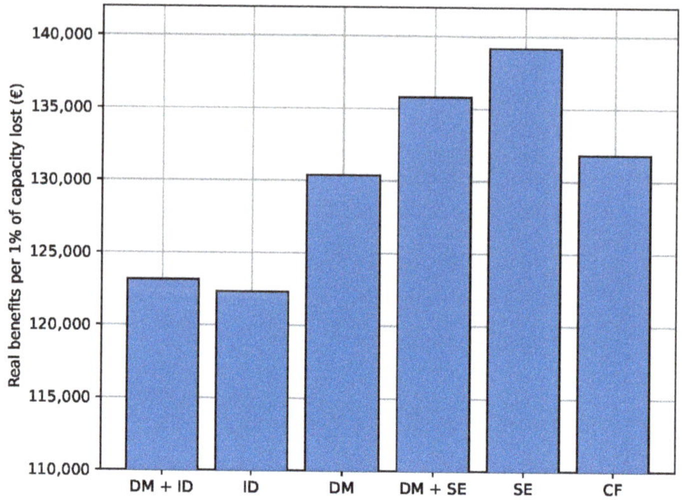

Figure 11. Real profits per 1% of capacity lost.

5.3. Full Project Extrapolation Results

The trade-off between maximizing short-term profits and stretching the life of the energy storage system is a crucial factor to consider. The NPV is a useful metric that takes into account future cash flows and discount rates to determine a project's overall profitability.

In the table, the NPV of each case is calculated, discounting the daily cash flow and the purchase cost of the BESS. The results show that the best NPV is achieved when the BESS is reserved for SE and capacity firming. When only participating in the day-ahead market, the NPV is not very different, but participating in the intraday market significantly lowers the NPV, indicating lower project profitability.

The results of each simulation scenario are extrapolated for the entire project life. The average daily cash flow is determined from the average daily results of 2018, and daily degradation is similarly calculated to estimate project completion. The project ends when the accumulated degradation of the BESS reaches 20%, a commonly used value in the relevant literature (e.g., [22,23]). The average values and estimated project lifetimes for each case are presented in Table 4.

Table 4. Project extrapolation under each case.

Case	Average Daily Capacity Loss (%)	Average Daily Profits (EUR)	Lifetime (Years)
Ideal	2.08×10^{-3}	1682.28	26.31
DM + ID	7.48×10^{-3}	921.44	7.32
ID	7.56×10^{-3}	925.11	7.25
DM	1.06×10^{-2}	1394.37	5.12
DM + SE	1.02×10^{-2}	1389.89	5.36
SE	1×10^{-2}	1392.7	5.48
CF	1.05×10^{-2}	1394.74	5.18

It is worth noting that the BESS is not operated in the ideal scenario. In this case, the BESS is only utilized for allocating generated energy when prices are the highest. However, due to the degradation costs and the BESS efficiency, this operation is not profitable.

The SE has lower daily profits but a higher NPV compared to the case with CF. The NPV of each project is calculated using a 7.5% discount rate and is presented in Table 5. The results show that using the BESS for the SE service and capacity firming provides the highest NPV.

Table 5. NPV under each case.

Case	NPV (EUR M)
Ideal	7.226
DM + intraday market	1.912
Intraday market	1.905
DM	2.179
DM + SE	2.253
SE	2.299
CF	2.2

6. Conclusions

A three-level EMS for a HF to participate in the Iberian electricity market has been presented. The day-ahead and intraday market offering strategies are formulated as linear programming problems, with generation and price forecasts at the start of each market session as inputs. The model has incorporated adjustments based on Iberian market regulations to account for committed power correction impacts. It also includes a real-time degradation model of the BESS to assess capacity and power losses.

A simulation framework has been proposed to assess various BESS market participation strategies. The results indicate that participating in both day-ahead and intraday wholesale markets generates the highest expected profits, but actual profits are lower compared to other strategies. The proposed optimization algorithm and results of the study can be applied to any electrical system that operates with concurrent day-ahead and intraday spot markets, which is a common practice in many countries. This can be used as a guide for agents trying to optimize their plants.

The results show that using the BESS in all markets does not necessarily result in a higher income, challenging the common belief that utilizing the BESS in all markets leads to the best outcome. The proposed SE service enables participation in intraday markets without the drawback of market overlap, enhancing the BESS utilization efficiency and increasing the NPV of the project. Moreover, it highlights the crucial role of adjustment mechanisms on actual income and the importance of modeling them thoroughly. Lastly, it is worth remarking that the plant proposed in this work is considered a price-taker; the effect of a plant or a number of plants large enough to impact electricity market prices is proposed as a future work.

Author Contributions: The authors contributed equally to this work. All authors have read and agreed to the published version of the manuscript.

Funding: This research received no external funding.

Data Availability Statement: Data is available by request.

Conflicts of Interest: The authors declare no conflict of interest.

Abbreviations

The following abbreviations are used in this manuscript:

BESS	Battery energy storage system
HyF	Hybrid farm
EMS	Energy management system
DOD	Depth of discharge
SOC	State of charge
SOH	State of Health
NPV	Net present value
EOL	End of life
MILP	Mixed-integer linear program

References

1. Günter, N.; Marinopoulos, A. Energy storage for grid services and applications: Classification, market review, metrics, and methodology for evaluation of deployment cases. *J. Energy Storage* **2016**, *8*, 226–234. [CrossRef]
2. Sweeney, C.; Bessa, R.J.; Browell, J.; Pinson, P. The future of forecasting for renewable energy. *WIREs Energy Environ.* **2020**, *9*, e365. [CrossRef]
3. Liu, X.; Lin, Z.; Feng, Z. Short-term offshore wind speed forecast by seasonal ARIMA—A comparison against GRU and LSTM. *Energy* **2021**, *227*, 120492. [CrossRef]
4. Legrand, M.; Labajo-Hurtado, R.; Rodríguez-Antón, L.M.; Doce, Y. Price arbitrage optimization of a photovoltaic power plant with Liquid Air Energy Storage. implementation to the Spanish case. *Energy* **2022**, *239*, 121957. [CrossRef]
5. Wang, Y.; Zhao, H.; Li, P. Optimal offering and operating strategies for wind-storage system participating in spot electricity markets with progressive stochastic-robust hybrid optimization Model Series. *Math. Probl. Eng.* **2019**, *2019*, 2142050. [CrossRef]
6. Pusceddu, E.; Zakeri, B.; Castagneto Gissey, G. Synergies between energy arbitrage and fast frequency response for battery energy storage systems. *Appl. Energy* **2021**, *283*, 116274. [CrossRef]
7. Englberger, S.; Jossen, A.; Hesse, H. Unlocking the Potential of Battery Storage with the Dynamic Stacking of Multiple Applications. *Cell Rep. Phys. Sci.* **2020**, *1*, 100238. [CrossRef]
8. Shinde, P.; Amelin, M. A Literature Review of Intraday Electricity Markets and Prices. In Proceedings of the 2019 IEEE Milan PowerTech, Milan, Italy, 23–27 June 2019; pp. 1–6. [CrossRef]
9. Pantos, M.; Riaz, S.; Chapman, A.C.; Verbič, G. Capacity firming of intermittent generation by dispersed energy storage. In Proceedings of the 2017 Australasian Universities Power Engineering Conference (AUPEC), Melbourne, VIC, Australia, 19–22 November 2017; pp. 1–6. [CrossRef]
10. Abdelrazek, S.A.; Kamalasadan, S. Integrated PV Capacity Firming and Energy Time Shift Battery Energy Storage Management Using Energy-Oriented Optimization. *IEEE Trans. Ind. Appl.* **2016**, *52*, 2607–2617. [CrossRef]
11. Robles, M.; Kim, J.S.; Song, H. Capacity firming for wind generation using one-step model predictive control and battery energy storage system. *J. Electr. Eng. Technol.* **2017**, *12*, 2043–2050. [CrossRef]
12. Sotavento Experimental Wind Park. Available online: https://www.sotaventogalicia.com/area-tecnica/datos-tiempo-real/instantaneos-parque/ (accessed on 30 May 2022).
13. CNMC . *Informe de Supervisión del Mercado Peninsular Mayorista al Contado de Electricidad*; CNMC: Madrid, Spain, 2020 .
14. Day-Ahead-Market-Operation. Available online: https://www.omie.es/sites/default/files/inline-files/day_ahead_market.pdf. (accessed on 30 May 2022).
15. Zheng, D.; Shi, M.; Wang, Y.; Eseye, A.T.; Zhang, J. Day-Ahead Wind Power Forecasting Using a Two-Stage Hybrid Modeling Approach Based on SCADA and Meteorological Information, and Evaluating the Impact of Input-Data Dependency on Forecasting Accuracy. *Energies* **2017**, *10*, 1988 . [CrossRef]
16. McHugh, C.; Coleman, S.; Kerr, D.; McGlynn, D. Forecasting Day-ahead Electricity Prices with A SARIMAX Model. In Proceedings of the 2019 IEEE Symposium Series on Computational Intelligence (SSCI), Xiamen, China, 6–9 December 2019; pp. 1523–1529. [CrossRef]
17. The Wind Power. Available online: https://www.thewindpower.net/turbines_manufacturers_es.php. (accessed on 30 May 2022).
18. Kim, R.K.; Glick, M.; Olson, K.; Kim, Y.S. MILP-PSO Combined Optimization Algorithm for an Islanded Microgrid Scheduling with Detailed Battery ESS Efficiency Model and Policy Considerations. *Energies* **2020**, *13*, 1898. . [CrossRef]
19. Cheng, S.; Liu, Y.H.; Hesse, H.; Naumann, M.; Truong, C.; Jossen, A. A PSO-Optimized Fuzzy Logic Control-Based Charging Method for Individual Household Battery Storage Systems within a Community. *Energies* **2018**, *11*, 469. [CrossRef]
20. Naumann, M.; Schimpe, M.; Keil, P.; Hesse, H.C.; Jossen, A. Analysis and modeling of calendar aging of a commercial LiFePO4/graphite cell. *J. Energy Storage* **2018**, *17*, 153–169. [CrossRef]
21. BOE-A-2020-13591, 2020. Available online: https://www.boe.es/buscar/doc.php?lang=en&id=BOE-A-2020-13591 (accessed on 14 April 2023).

22. Smith, K.; Saxon, A.; Keyser, M.; Lundstrom, B.; Ziwei Cao.; Roc, A. Life prediction model for grid-connected Li-ion battery energy storage system. In Proceedings of the 2017 American Control Conference (ACC), Seattle, WA, USA, 24–26 May 2017; pp. 4062–4068. [CrossRef]
23. Maheshwari, A.; Paterakis, N.G.; Santarelli, M.; Gibescu, M. Optimizing the operation of energy storage using a non-linear lithium-ion battery degradation model. *Appl. Energy* **2020**, *261*, 114360. [CrossRef]

electronics

Article

Research on a High-Precision State-of-Charge Estimation Method Based on Forgetting Factor Recursive Least Squares and Adaptive Extended Kalman Filter Applied to LiFePO4 Battery

Yihui Xia, Zhihao Ye *, Liming Huang, Lucheng Sun and Yunxiang Jiang

School of Electrical Engineering, Naval University of Engineering, Wuhan 430033, China; xiayihui2005@163.com (Y.X.); huangliming1998@163.com (L.H.); d23380807@nue.edu.cn (L.S.); jiangyunxaing@163.com (Y.J.)
* Correspondence: yxyx928@126.com

Abstract: The state-of-charge (SOC) estimation accuracy is closely associated with the estimation method and the battery parameter identification performance. The battery parameter identification method based on forgetting factor recursive least squares (FFRLS) has the advantages of high parameter identification accuracy and fast dynamic response speed. On this basis, the performance of two SOC estimation methods, the extended Kalman filter (EKF) and adaptive extended Kalman filter (AEKF) are compared and studied. The results show that AEKF has better steady-state and dynamic SOC estimation performance, but the estimation accuracy and dynamic response performance are still not objective. To further improve the performance of SOC estimation, a joint SOC estimation method based on FFRLS-AEKF is proposed, and the SOC estimation experimental results with FFRLS-AEKF and AEKF are conducted. The experimental results show that the proposed joint SOC estimation method based on FFRLS-AEKF has a better steady-state and dynamic performance of SOC estimation. The maximum absolute error of the proposed algorithm is 4.97%. As the battery working time increases, the SOC estimation accuracy continues to converge to the true value, and the average absolute error is reduced to 2.5%. The proposed method and theoretical analysis are proven to be correct and feasible.

Keywords: LiFePO4 battery; high precision; FFRLS; AEKF; SOC estimation

Citation: Xia, Y.; Ye, Z.; Huang, L.; Sun, L.; Jiang, Y. Research on a High-Precision State-of-Charge Estimation Method Based on Forgetting Factor Recursive Least Squares and Adaptive Extended Kalman Filter Applied to LiFePO4 Battery. *Electronics* **2023**, *12*, 3670. https://doi.org/10.3390/electronics12173670

Academic Editors: Luis Hernández-Callejo, Jesús Armando Aguilar Jiménez and Carlos Meza Benavides

Received: 2 July 2023
Revised: 3 August 2023
Accepted: 4 August 2023
Published: 31 August 2023

1. Introduction

The LiFePO4 battery has advantages such as good safety, long service life, and high power density, and so on [1,2], which is widely used in electric vehicles and energy storage systems. To prevent battery failures such as overcharging and over-discharging, it is necessary to monitor the battery operating status in real time, especially its state of charge (SOC).

The methods of accurate SOC estimation can generally be categorized into the following types: open-circuit voltage, ampere-hour integral-based estimation method, data-driven-based estimation method, and filtering algorithms based on equivalent circuit models [3–5]. There is a monotonic one-to-one correspondence between the open-circuit voltage and the SOC value of the battery. The corresponding table of open-circuit voltage and SOC can be established, and the SOC value can be determined by looking up the table [6]. However, this method requires a long time after the battery is charged and discharged before measuring the open-circuit voltage. Reference [7] proposed a fast open-circuit voltage prediction method. The battery only needs to be placed within 15 min to accurately predict the open-circuit voltage. However, the calculation process is complex, and the requirements for the processor are high. The ampere-hour integral method tracks the SOC by the change in the current integral value. The measurement error will accumulate in the integral, resulting in a large error of SOC. Reference [8] analyzed the error sources of the ampere-hour analysis method and proposed a method to reduce the cumulative error of time and the proportional error of the SOC. However, it is more dependent on the current accuracy, and the accuracy of the current sensor is higher.

If the data-driven algorithm is trained properly, it can predict the SOC of any type of battery without knowing its inherent characteristics in advance. Reference [9] proposed a method of predicting SOC via artificial neural network machine learning. The root-mean-squared error (RMSE) of the LiFePO4 battery used in this paper is 0.33% during the test cycle. The author of Reference [10] proposed a recurrent neural network (RNN) with long short-term memory (LSTM) to accurately estimate the SOC of lithium-ion batteries, and it can achieve high accuracy at temperatures from 10 °C to 25 °C. Reference [11] used the support vector machine (SVM) method to estimate the SOC of the battery from the experimental data set, which has high accuracy. In the method of machine learning or deep learning, a data-driven algorithm collects terminal voltage, working current, temperature, and other data in the work of the battery, and it uses neural network and machine learning to train a large amount of data to obtain the current SOC value. This method needs to collect a large amount of data for calculation, which requires a powerful processor and is sensitive to the type of battery. If the internal parameters of the battery change, the data-driven model cannot follow the internal changes well, resulting in a larger SOC estimation error.

The filtering algorithms based on equivalent circuit models mainly use derivative algorithms of Kalman filtering (KF) and particle filtering algorithms for SOC estimation [12–17]. As the most popular model-based method, the KF-based algorithm has enhanced robustness to measurement and process noise and has a high estimation accuracy [18]. Based on the extended Kalman filter (EKF) algorithm, an adaptive dual EKF algorithm is proposed in Ref. [12]; the EKF algorithm is used to improve the nonlinear battery model for SOC estimation in Refs. [17,19] for the LiFePO4 battery. A strong tracking volume EKF algorithm (STEKF) is proposed in Ref. [20] to provide accurate SOC prediction and a faster computing time. A multi-rate strong tracking EKF algorithm (MRSTEKF) is proposed by introducing multi-rate control strategy and enhancement technology in Ref. [21], and the SOC tracking stability and estimation accuracy is improved from 55.34% and 49.51% with STEKF to 52.66% and 33.88% with MRSTEKF, respectively. An online adaptive EKF algorithm based on the Davidson model is used to estimate SOC [22], and the SOC estimation error can be reduced to 2%. The sigma points Kalman filter (SPKF) is another nonlinear system filtering algorithm that can be used to estimate SOC, which uses numerical approximations instead of analytical approximations of EKF. A joint battery model and SOC estimation algorithm based on SPKF is proposed in Ref. [23], which has the same computational complexity as EKF but has higher accuracy. In Ref. [24], a combination of the volumetric Kalman filtering algorithm and the forgetting factor recursive least squares (FFRLS) algorithm is used to estimate SOC, and the maximum estimation error of SOC under high-rate pulses is reduced to be less than 1%. On the other hand, the method of approximating a probability density function based on a particle filter is used to estimate the SOC. In Ref. [25], a volumetric particle filter for accurate SOC estimation is proposed. In Ref. [26], dual-scale, dual-particle, and dual-scale adaptive particle filters for SOC estimation are proposed. However, compared with SOC estimation algorithms based on the Kalman filter, the SOC estimation algorithms based on the particle filter have a larger computational burden [27]. In Ref. [28], an analytical mathematical formulation of storage and its SOC is presented. In Ref. [29], a power allocation strategy based on cluster switching to relieve the stated problem in two levels is proposed, which is used to eliminate the imbalanced SOC and decrease battery energy loss. To address the power allocation challenges, a novel optimized state-of-charge (SOC) feedback-based energy management strategy in Ref. [30] is proposed for HESS in IPS to restrain the DC bus voltage fluctuation in this paper.

In this paper, a FFRLS-AEKF joint estimation algorithm based on time-varying parameter model is proposed, which considers the characteristics that the FFRLS algorithm can update the battery parameters in real time and the AEKF algorithm can correct the initial value error of SOC and follows the parameters of the battery. Firstly, the working principle of FFRLS for battery parameter identification is analyzed. Secondly, the second-order equivalent circuit state discretization equation based on the AEKF algorithm is established, and the steps of the battery SOC estimation method based on AEKF are discussed. Thirdly, a joint SOC estimation method based on FFRLS-AEKF is proposed. Fourthly, the SOC estimation performance of three methods as EKF, AEKF, and FFRLS-AEKF is compared by experiments. The experimental results show that the FFRLS-AEKF algorithm can have a higher SOC estimation and faster accuracy convergence speed. Finally, the thesis is summarized.

2. The Basic Principle of Parameter Identification Based on the FFRLS

2.1. Mathematical Modeling of a Second-Order RC Circuit

The second-order RC equivalent circuit model of the LiFePO4 battery is shown in Figure 1. The two RC circuits, respectively, describe the electrochemical polarization process with a small time constant and the concentration polarization process with a large time constant. Among them, U_{OC} is the open-circuit voltage of the battery, R_0 is the internal resistance, and R_{p1} and C_{p1} are the electrochemical polarization internal resistance and electrochemical polarization capacitance, respectively. R_{p2} and C_{p2} are the concentration difference polarization internal resistance and concentration difference polarization capacitance, respectively. U_L and I_L are load voltage and load current, respectively.

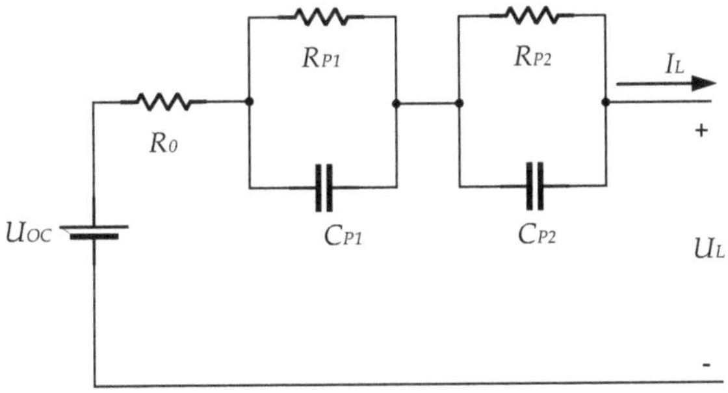

Figure 1. Second-order RC model.

According to Figure 1, the state space equation expression of the second-order RC model circuit can be listed as follows:

$$\begin{gathered}\begin{pmatrix}\frac{dU_{p1}}{dt}\\ \frac{dU_{p2}}{dt}\end{pmatrix}=\begin{pmatrix}-\frac{1}{R_{p1}C_{p1}} & 0\\ 0 & -\frac{1}{R_{p2}C_{p2}}\end{pmatrix}\begin{pmatrix}U_{p1}\\ U_{p2}\end{pmatrix}+\begin{pmatrix}\frac{1}{C_{p1}}\\ \frac{1}{C_{p2}}\end{pmatrix}I_L\\ U_{OC}=U_{p1}+U_{p2}+R_0I_L+U_L\end{gathered} \tag{1}$$

where $\frac{dU_{p1}}{dt}$ and $\frac{dU_{p2}}{dt}$ represent the electrochemical polarization voltage and the concentration difference polarization voltage, respectively.

2.2. The Basic Principle of the FFRLS

The traditional recursive least squares (RLS) method is widely used in the field of system identification. For linear systems, its mathematical description can be expressed as follows:

$$\begin{gathered}Y(k)=\varphi(k)\theta(k)^T+e(k)\\ \varphi(k)=[-Y(k-1)\cdots-Y(k-n)\;u(k)\cdots u(k-n)]\end{gathered} \tag{2}$$

where $Y(k)$ is the output signal of the system at k time, $u(k)$ is the input signal of the system at k time, $\theta(k)$ represents the parameter vector to be estimated, $\varphi(k)$ is the input matrix of the system at time k, n is the parameter to be estimated, and $e(k)$ is the noise of the system. The recursive formula of the FFRLS method is as follows:

$$\begin{aligned} \hat{\theta}(k+1) &= \hat{\theta}(k) + K(k+1)[Y(k+1) - \varphi^T(k+1)\theta(k)] \\ K(k+1) &= P(k+1)\varphi(k+1) \\ P(k+1) &= P(k) - \frac{P(k)\varphi(k+1)\varphi^T(k+1)}{1+\varphi^T(k+1)P(k)\varphi(k+1)} \end{aligned} \tag{3}$$

In Equation (3), $P(k)$ is the covariance matrix, and $K(k)$ is the gain matrix. During the recursive process, RLS uses the difference between the estimated value and the measured value of the system output, as well as the gain matrix $K(k)$, to make adjustments to the parameter vector $\hat{\theta}(k)$. During initialization, the initial value of $\hat{\theta}(k)$ and $K(k)$ can be any value, $P(0) = \alpha I$, α is as large as possible a constant, and I is the identity matrix.

To reduce the influence of past data and give greater weight to new data, a forgetting factor is introduced by setting it to $\lambda (0 < \lambda < 1)$, leading to the development of FFRLS [31,32]. Equation (3) is modified as follows:

$$\begin{aligned} \hat{\theta}(k+1) &= \hat{\theta}(k) + K(k+1)[y(k+1) - \varphi^T(k+1)\hat{\theta}(k)] \\ K(k+1) &= \frac{P(k)\varphi(k+1)}{\lambda+\varphi^T(k+1)P(k)\varphi(k+1)} \\ P(k+1) &= \frac{1}{\lambda}[1 - K(k+1)\varphi^T(k+1)]P(k) \end{aligned} \tag{4}$$

Equation (4) is a recursive formula based on the forgetting factor recursive least squares method, where the forgetting factor λ is generally taken between 0.95 and 1. The larger the value of λ, the smaller the weight of new data. When $\lambda = 1$, it degenerates into the ordinary RLS method.

2.3. Online Identification of Parameters for a Second-Order RC Circuit

From Figure 1, the KVL (Kirchhoff's voltage law) equation of the circuit can be obtained:

$$U_{OC} = U_{p1} + U_{p2} + R_0 I_L + U_L \tag{5}$$

According to Equation (1), taking the Laplace transform on both sides:

$$U_{OC}(s) = \left(\frac{R_{p1}}{R_{p1}C_{p1}s+1} + \frac{R_{p2}}{R_{p2}C_{p2}s+1} + R_0\right)I(s) + U_L(s) \tag{6}$$

Let the time constants be $\tau_{p1} = R_{p1}C_{p1}$ and $\tau_{p2} = R_{p2}C_{p2}$, and let $a = \tau_{p1}\tau_{p2}$, $b = \tau_{p1} + \tau_{p2}$, $c = R_{p1} + R_{p2} + R_0$; then, Equation (6) can be simplified as:

$$aU_{OC}s^2 + bU_{OC}s + U_{OC} = aR_0 I_s^2 + dI_s + cI + aU_s^2 + U_L \tag{7}$$

Equation (8) can be obtained via discretion using Equation (7) as follows:

$$U(k) = \frac{-bT-2a}{T^2+bT+a}U(k-1) + \frac{a}{T^2+bT+a}U(k-2) + \frac{cT^2+dT+aR_0}{T^2+bT+a}I(k) + \frac{-dT-2aR_0}{T^2+bT+a}I(k-1) + \frac{aR_0}{T^2+bT+a}I \tag{8}$$

After simplification, the expression (8) becomes:

$$U(k) = k_1U(k-1) + k_2U(k-2) + k_3I(k) + k_4I(k-1) + k_5I(k-2) \tag{9}$$

Substituting Equation (9) into FFRLS, the value of parameter $\theta(k) = (k_1, k_2, k_3, k_4, k_5)$ can be calculated, and then the circuit model parameter R_0, R_{p1}, R_{p2}, C_{p1}, C_{p2} from the identification results can be deduced as follows:

$$\begin{aligned} \tau_{p1} &= \frac{T}{2(k_1+k_2+1)}\left[\sqrt{k_1^2-4k_2} - k_1 - 2k_2\right] \\ \tau_{p2} &= -\frac{T}{2(k_1+k_2+1)}\left[\sqrt{k_1^2-4k_2} + k_1 + 2k_2\right] \end{aligned} \tag{10}$$

Combining Equations (4), (8) and (10), Equation (11) can be obtained as follows:

$$\begin{cases} R_0 = \frac{k_5}{k_2} \\ R_{p1} = \frac{\tau_{p1}\frac{k_3+k_4+k_5}{k_1+k_2+1}+\tau_{p2}R_0+\frac{T(k_4+2k_5)}{k_1+k_2+1}}{\tau_{p1}-\tau_{p2}} \\ R_{p2} = \frac{k_3+k_4+k_5}{k_1+k_2+1} - R_{p1} - R_0 \\ C_{p1} = \frac{\tau_{p1}}{R_{p1}} \\ C_{p2} = \frac{\tau_{p2}}{R_{p2}} \end{cases} \tag{11}$$

3. SOC Estimation Method Based on AEKF

3.1. Adaptive Extended Kalman Filter State Equation

The Kalman filter algorithm applies to linear systems. If it is used in a nonlinear system, the Taylor expansion formula needs to be used to locally approximate the nonlinear state equation as a linear equation. The method used for nonlinear systems is called the extended Kalman filter algorithm. Nonlinear systems can be represented by Formula (12):

$$\begin{cases} x_{k+1} = f(x_k, u_k) + w_k \\ y_k = g(x_k, u_k) + v_k \end{cases} \tag{12}$$

where x_k is the state variable, y_k is the observation variable, $f(x_k, u_k)$ is the nonlinear state function, $g(x_k, u_k)$ is the nonlinear observation function, and w_k and v_k are gaussian white noise with zero means and covariances of Q_k and R_k, respectively. The EKF algorithm takes advantage of the local linearity property of nonlinear functions by locally linearizing both nonlinear functions. As described in the previous section, $\hat{x}_{k/k}$ this is the optimal estimated value of the state variable at k time. By performing a first-order Taylor expansion of the nonlinear state function $f(x_k, u_k)$ around $\hat{x}_{k/k}$, Equation (13) can be obtained as follows:

$$f(x_k, u_k) = f(\hat{x}_{k/k}, u_k) + \frac{\partial f}{\partial x_k}\Big|_{x_k=\hat{x}_{k/k}} (x_k - \hat{x}_{k/k}) + o(x_k - \hat{x}_{k/k}) \tag{13}$$

In Equation (13), ignoring the high-order term $o(x_k - \hat{x}_{k/k})$ and letting $\frac{\partial f}{\partial x_k}\Big|_{x_k=\hat{x}_{k/k}} = F_k$, Equation (2) can be simplified as follows:

$$x_{k+1} = f(\hat{x}_{k/k}, u_k) + F_k(x_k - \hat{x}_{k/k}) + w_k \tag{14}$$

Expanding the nonlinear observation function $g(x_k, u_k)$ around the prior state estimate $\hat{x}_{k/k-1}$ at k time by a first-order Taylor series, Equation (15) can be obtained as follows:

$$g(x_k, u_k) = g(\hat{x}_{k/k-1}, u_k) + \frac{\partial g}{\partial x_k}\Big|_{x_k=\hat{x}_{k/k-1}} (x_k - \hat{x}_{k/k-1}) + o(x_k - \hat{x}_{k/k-1}) \tag{15}$$

Ignoring higher-order terms in $o(x_k - \hat{x}_{k/k-1})$ and setting $\frac{\partial g}{\partial x_k}\Big|_{x_k=\hat{x}_{k/k-1}} = G_k$, Equation (14) can be simplified as follows:

$$y_k = g(\hat{x}_{k/k-1}, u_k) + G_k(x_k - \hat{x}_{k/k-1}) + v_k \tag{16}$$

If both the nonlinear state equation and the observation equation are linear, then Equation (11) can be rewritten as follows:

$$\begin{cases} x_{k+1} = F_k x_k + f(\hat{x}_{k/k}, u_k) - F_k\hat{x}_{k/k} + w_k \\ y_k = G_k x_k + g(\hat{x}_{k/k-1}, u_k) - G_k\hat{x}_{k/k-1} + v_k \end{cases} \tag{17}$$

Here, matrices F_k and G_k can be obtained by calculating the Jacobian matrices of f and g. If the state variable x is n-dimensional, i.e., $x = [x_1, x_2, \ldots, x_n]^T$, then the solution for matrices F_k and G_k is as follows.

$$F_k = \frac{\partial f}{\partial x} = \begin{bmatrix} \frac{\partial f_1}{\partial x_1} & \frac{\partial f_1}{\partial x_2} & \cdots & \frac{\partial f_1}{\partial x_n} \\ \frac{\partial f_2}{\partial x_1} & \frac{\partial f_2}{\partial x_2} & \cdots & \frac{\partial f_2}{\partial x_n} \\ \vdots & \vdots & \cdots & \vdots \\ \frac{\partial f_n}{\partial x_1} & \frac{\partial f_n}{\partial x_2} & \cdots & \frac{\partial f_n}{\partial x_n} \end{bmatrix} \tag{18}$$

$$G_k = \frac{\partial g}{\partial x} = \begin{bmatrix} \frac{\partial g_1}{\partial x_1} & \frac{\partial g_1}{\partial x_2} & \cdots & \frac{\partial g_1}{\partial x_n} \\ \frac{\partial g_2}{\partial x_1} & \frac{\partial g_2}{\partial x_2} & \cdots & \frac{\partial g_2}{\partial x_n} \\ \vdots & \vdots & \cdots & \vdots \\ \frac{\partial g_n}{\partial x_1} & \frac{\partial g_n}{\partial x_2} & \cdots & \frac{\partial g_n}{\partial x_n} \end{bmatrix} \tag{19}$$

The EKF algorithm requires advanced calibration of the covariance matrices for observation noise and process noise, which is often calculated by experience, and the covariance matrices Q_k and R_k are fixed values. However, in high-rate conditions of lithium iron phosphate batteries, the noise often changes due to the internal chemical reaction and resulting temperature variation and is no longer a fixed value. To improve the accuracy of SOC estimation, the AEKF algorithm is introduced. Based on the EKF algorithm, the AEKF adds the Sage–Husa adaptive filtering algorithm, enabling the observation noise covariance matrix and the process noise covariance matrix in the EKF algorithm to be adaptively updated, thus improving the accuracy of the SOC estimation. The steps of the AEKF algorithm are listed as follows:

Step 1. Initialization

Set $\hat{x}_0 = x_0, y_0, P_0, Q_0, R_0$ when $k = 0$, where $\hat{x}_0$ is the initial estimate of the state variables, y_0 is the initial observation value, P_0 is the initial value of the error covariance matrix, and Q_0 and R_0 are the initial values of the process covariance matrix and observation noise covariance matrix, respectively.

Step 2. State prediction

$$\hat{x}_{k/k-1} = f(\hat{x}_{k-1/k-1}, u_{k-1}) \tag{20}$$

Step 3. Prediction of error covariance

$$P_{k/k-1} = F_{k-1}P_{k-1/k-1}F_{k-1}{}^T + Q_{k-1} \tag{21}$$

Step 4. Calculate the Kalman gain

$$K_k = P_{k/k-1}G_k^T\left[G_kP_{k/k-1}G_k^T + R_{k-1}\right]^{-1} \tag{22}$$

Step 5. State estimation

$$\hat{x}_{k/k} = \hat{x}_{k/k-1} + K_k[y_k - g(\hat{x}_{k/k-1}, u_k)] \tag{23}$$

Step 6. Update the noise covariance matrix

$$e_k = y_k - g(\hat{x}_{k/k-1}, u_k) \tag{24}$$

$$\begin{cases} E_k = E(e_k e_k^T) \\ d = e_k^T E_k^{-1} e_k \end{cases} \tag{25}$$

In Equations (24) and (25), e_k is called the innovation matrix, and d is the estimated residual value expressed using the adaptive window factor in the windowing function, which is used to calculate the observation dimension M. According to the principle of the windowing function, if M is small, the computational burden of the adaptive algorithm is reduced, but the accuracy of the algorithm is lower. Conversely, if M is large, the accuracy of the algorithm is significantly improved, but the computational burden is too high, which is not suitable for recursive estimation algorithms. Therefore, M needs to be adjusted according to the convergence time, as shown in Equation (26).

$$\begin{cases} M = 1, d = 1 \\ M = k, d = 0 \\ M = k \times \eta^d, 0 < d < 1 \end{cases} \tag{26}$$

In Equation (26), η is the window convergence rate, which has a range of $0 < \eta < 1$. After calculating M, the noise covariance matrix can be updated as shown in Equations (27) and (28):

$$H_k = \frac{1}{M} \sum_{i=k-M+1}^{i=k} e_k e_k^T \tag{27}$$

$$\begin{cases} R_k = H_k - G_k P_{k/k-1} G_k^T \\ Q_k = K_k H_k K_k^T \end{cases} \tag{28}$$

Step 7. Estimation of error covariance.

$$P_{k/k} = (I - K_k G_k) P_{k/k-1} \tag{29}$$

Subsequently, repeat step 2~step 7 for recursive estimation to obtain the optimal estimate of the state variables.

3.2. Discretization of State Space Equation

Taking the operating current I_L as the input variable, battery SOC, battery electrochemical polarization voltage U_{p1}, and concentration difference polarization voltage U_{p2} as state variables, and U_{OC} as the observation variable, Equation (30) can be obtained by combining Equations (8) and (9):

$$\begin{pmatrix} SOC(t) \\ \frac{dU_{p1}(t)}{dt} \\ \frac{dU_{p2}(t)}{dt} \end{pmatrix} = \begin{pmatrix} 1 & 0 & 0 \\ 0 & -\frac{1}{R_{p1}(t)C_{p1}(t)} & 0 \\ 0 & 0 & -\frac{1}{R_{p2}(t)C_{p2}(t)} \end{pmatrix} \begin{pmatrix} SOC_{t0} \\ U_{p1}(t) \\ U_{p2}(t) \end{pmatrix} + \begin{pmatrix} -\frac{\int_{t_0}^{t} \eta dt}{Q_0} \\ \frac{1}{C_{p1}(t)} \\ \frac{1}{C_{p2}(t)} \end{pmatrix} I(t) \tag{30}$$

$$U_L(t) = U_{OC}[SOC(t)] - I(t)R_0(t) - U_{p1}(t) - U_{p2}(t) \tag{31}$$

Equations (30) and (31) are the continuous state equation and continuous observation equation, respectively. Here, $U_{OC}[SOC(t)]$ is the open-circuit voltage at time t obtained using the OCV-SOC relation function.

Setting the Coulomb efficiency η to be 1 and defining the sampling period as T, Equations (30) and (31) are discretized as follows:

$$\begin{pmatrix} SOC(k+1) \\ U_{p1}(k+1) \\ U_{p2}(k+1) \end{pmatrix} = \begin{pmatrix} 1 & 0 & 0 \\ 0 & e^{\frac{-T}{\tau_{p1}(k)}} & 0 \\ 0 & 0 & e^{\frac{-T}{\tau_{p2}(k)}} \end{pmatrix} \begin{pmatrix} SOC(k) \\ U_{p1}(k) \\ U_{p2}(k) \end{pmatrix} + \begin{pmatrix} -\frac{T}{Q_0} \\ R_{p1}(k)(1 - e^{\frac{-T}{\tau_{p1}(k)}}) \\ R_{p2}(k)(1 - e^{\frac{-T}{\tau_{p2}(k)}}) \end{pmatrix} I(k) \tag{32}$$

$$U_L(k) = U_{OC}[SOC(k)] - U_{p1}(k) - U_{p2}(k) - I(k)R_0(k) \tag{33}$$

The vector composed of $SOC(k)$, $U_{p1}(k)$, and $U_{p2}(k)$ in the above equation is the state vector x_k of the system at k time in Equation (11), while $U_L(k)$ is the observation vector y_k of the system at k time. Equation (32) is a linear equation, while the nonlinearity of the battery system state space equation is reflected in the $U_{OC}[SOC(k)]$ part of Equation (33). Therefore, the Kalman filtering algorithm for SOC estimation is carried out using both EKF and AEKF, and the results are compared with the coulomb counting method for SOC estimation.

3.3. SOC Estimation Process for LiFePO4 Battery

Step 1. Data input. Input the test data of the LiFePO4 battery, including the working current sampling data $I_L(k)$, terminal voltage $U_L(k)$, battery capacity of 13 Ah, parameter model identification results, and sampling period T = 0.1 s.

Step 2. Algorithm initialization. Initialize the battery state vector $x_0 = [SOC(0), 0, 0]$, estimation error covariance matrix $P_0 = \begin{bmatrix} 10^{-6} & 0 & 0 \\ 0 & 10^{-6} & 0 \\ 0 & 0 & 10^{-6} \end{bmatrix}$, process noise covariance matrix $Q_0 = \begin{bmatrix} 10^{-6} & 0 & 0 \\ 0 & 10^{-6} & 0 \\ 0 & 0 & 10^{-6} \end{bmatrix}$, and observation noise covariance matrix $R_0 = 0.05$.

Step 3. Calculate the prior estimate of the state variable and the predicted covariance matrix. Firstly, according to Equation (10), calculate the state transition matrix $A_k = diag[1, e^{\frac{-T}{\tau_{p1}(k)}}, e^{\frac{-T}{\tau_{p2}(k)}}]$ and the system input matrix $B_k = [-\frac{T}{Q_0}, R_{p1}(1 - e^{\frac{-T}{\tau_{p1}(k)}}), R_{p2}(1 - e^{\frac{-T}{\tau_{p2}(k)}})]^T$ using the model parameters at time k. Secondly, substitute the state vector $x_k = [SOC(k)$, $U_{p1}(k)$, $U_{p2}(k)]$ and the operating current $I(k)$ at k time into Equation (20) to calculate the prior estimate value $\hat{x}_{k+1/k}$ of the state variable. Finally, calculate the predicted error covariance matrix $P_{k+1/k}$ according to Equation (21).

Step 4. Calculating the Kalman gain matrix. Firstly, the nonlinear observation equation is linearized based on Equation (8), and the Jacobian matrix G_k of the observation equation is calculated, as given in equation (33). Secondly, the Jacobian matrix G_k, the predicted error covariance matrix $P_{k+1/k}$, and the observation noise matrix R_k are substituted into Equation (22) to calculate the Kalman gain matrix K_k.

$$G_k = [\frac{\partial U_{OC}[SOC] - IR_0}{\partial SOC}|_k, \frac{\partial - U_{p1}}{U_{p1}}|_k, \frac{\partial - U_{p2}}{U_{p2}}|_k] = [\frac{\partial U_{OC}[SOC]}{\partial SOC}|_k, -1, -1] \quad (34)$$

Step 5. Calculating the optimal estimated value of the state vector and updating the error covariance matrix. Based on Equation (34), the observation value, i.e., the estimated value $U_L(k)$ of the battery terminal voltage at k time, is calculated. Then, Equation (23) is used to calculate the optimal estimated value $\hat{x}_{k+1/k+1} = [SOC(k+1), U_{p1}(k+1), U_{p2}(k+1)]^T$ of the state vector, where $SOC(k + 1)$ is the SOC value estimated by AEKF at $(k + 1)$ time. Finally, Equation (13) is used to update the error covariance matrix $P_{k+1/k}$.

Step 6. Updating the noise covariance matrix. The process noise covariance matrix Q_k and the observation noise covariance matrix $R(k)$ are calculated based on Equations (24) and (25). Steps 2 to 6 are repeated recursively to estimate the SOC value at each moment.

3.4. Comparison and Analysis of SOC Estimation Results between EKF and AEKF Algorithms

To compare and analyze the performance of EKF and AEKF algorithms for SOC estimation of lithium iron phosphate batteries, a comparative study is conducted on SOC estimation based on the second-order equivalent circuit parameter model. Through battery charge–discharge testing, the corresponding relationship between battery OCV and SOC can be obtained through fitting. A 13 Ah-rated capacity and 3.2 V-rated voltage LiFePO4 battery is selected for HPPC testing to obtain its SOC and OCV data. The testing procedure involved discharging at 1 C for 10 s, resting for 40 s, and then charging at 0.75 C for 10 s. After resting for 45 min for every 10% drop in SOC, the OCV is measured, followed by another cycle of testing until the SOC of the battery reached 0.06, which is considered as the end of the discharge. Under the global discharge, the battery terminal voltage U_L, the output current I_L, and the real values of SOC are shown in Figures 2 and 3, respectively.

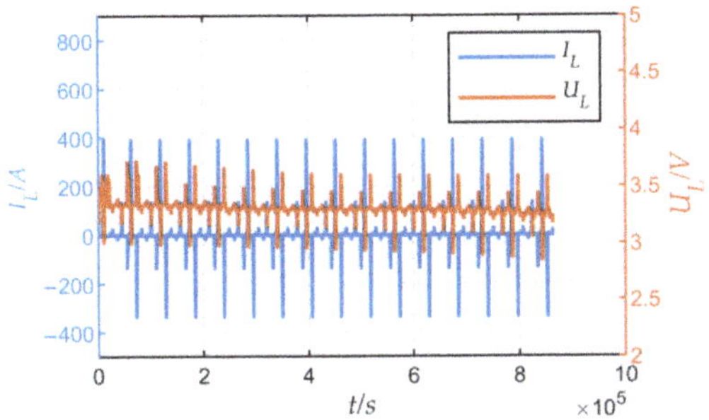

Figure 2. The waveform of terminal voltage U_L and the output current I_L.

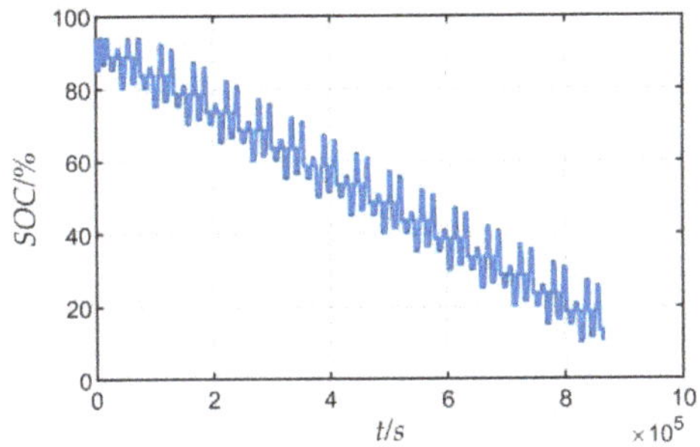

Figure 3. The waveform of real values of SOC.

Figures 4 and 5, respectively, show the SOC estimation results and estimation errors under the EKF and AEKF algorithms. From Figures 4 and 5, it can be seen that compared with the EKF algorithm, the AEKF algorithm can converge to the vicinity of the true value more quickly. The convergence time of the EKF algorithm and the AEKF algorithm is 532.8 s and 124.5 s, respectively. The SOC estimation accuracy of the AEKF algorithm is higher than that of the EKF algorithm. Experimental results show that the AEKF algorithm has a better SOC estimation convergence speed and accuracy.

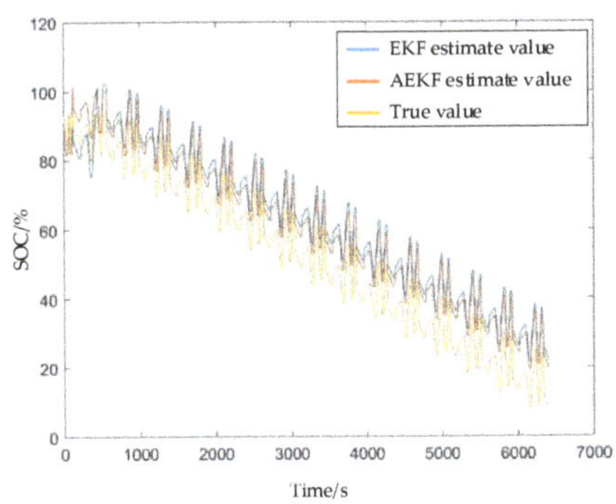

Figure 4. The SOC estimating results with EKF and AEKF.

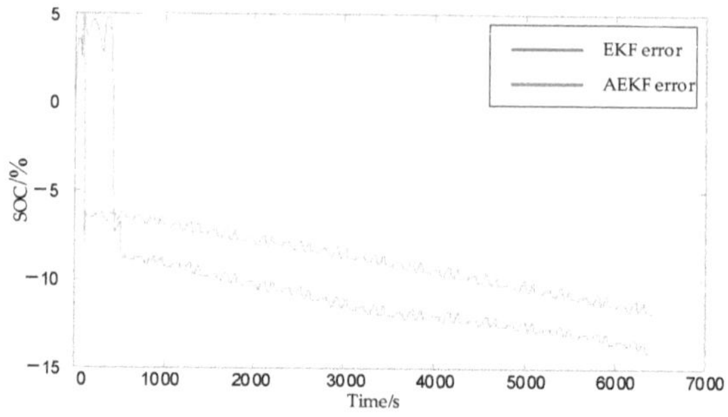

Figure 5. The SOC estimating errors with EKF and AEKF.

The specific statistical characteristics of SOC estimation errors under EKF and AEKF algorithms are shown in Table 1 (absolute error, AE; relative error, RE; mean absolute error, MAE; mean relative error, MRE), and the relevant features are calculated based on the converged results. From Table 1, it can be seen that the SOC estimation accuracy of the AEKF algorithm is better than that of the EKF algorithm, with a maximum absolute error (MaxAE) of 11.9% and a maximum relative error (MaxRE) of 152.6%, which occurs at the end of discharge. And the MAE of the SOC estimation is 8.8%, and the MRE of the SOC estimation is 25.4% with the AEKF algorithm, which is lower than that with the EKF. It can also be seen from Table 1 that the errors of both EKF and AEKF algorithms gradually increase. Therefore, the adaptive noise covariance matrix updating method of the AEKF algorithm can reduce SOC estimation errors. However, the SOC estimation precision is still large, which cannot meet the actual use needs of the battery energy management system. And, further optimization of the SOC estimation algorithm is needed.

Table 1. SOC estimation error under the EKF algorithm and AEKF algorithm.

Estimation Algorithm	MaxAE	MaxRE	MAE	MRE
EKF	14.1%	180.3%	10.9%	31.2%
AEKF	11.9%	152.6%	8.8%	25.4%

4. SOC Estimation Method Based on FFRLS-AFKF

The FFRLS-AEKF joint estimation algorithm uses the SOC value estimated using the AEKF algorithm to replace the estimated value of the ampere-hour integration method in the FFRLS algorithm to improve the battery equivalent circuit parameters identification accuracy. The identified model parameters are substituted into the AEKF algorithm to recursively estimate the SOC and improve the accuracy of the AEKF algorithm. Then, the estimated SOC value is fed back to FFRLS, and through positive feedback, the SOC estimation accuracy is finally improved. The specific calculation process is shown in Figure 6.

Step 1. Initialize the FFRLS algorithm and the AEKF algorithm, where SOC(0) is obtained by the open-circuit voltage method.

Step 2. Obtain the model parameters at k time through the FFRLS algorithm, substitute them into the AEKF algorithm, construct the corresponding transfer matrix and input matrix, and obtain the SOC at (k + 1) time.

Step 3. Substitute the SOC at (k + 1) time into the OCV-SOC relationship formula, and then use the FFRLS algorithm to obtain the model parameters at (k + 1) time.

Step 4. Update the transfer matrix and input matrix in the AEKF algorithm to obtain the SOC of the next time step; repeat steps 2 to step 4 and recursively obtain the SOC at each time step. The specific calculation process is shown in the diagram.

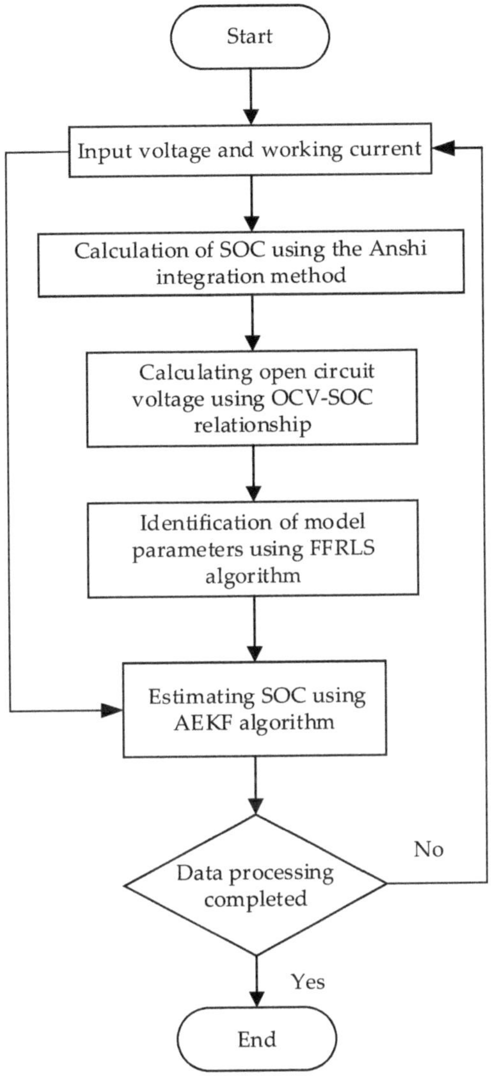

Figure 6. FFRLS + AEKF processes.

5. Experimental Results

The experimental parameters setting is the same as in Section 3.2. The structure of the battery pack used in the experiment is first parallel and then series. The batteries used are all new, and the working temperature is 25 °C. According to the method in Section 2.3, the battery parameters R_0, R_{p1}, C_{p1}, R_{p2}, and C_{p2} can be identified at 25 °C, and the identification results are in Table 2. Four single cells form a parallel unit, and ten parallel units, a total of forty battery cells form a series battery pack. The parallel unit is composed of ten parallel units, and a total of forty battery cells are composed of series battery packs. The model of the battery is IFP9380, the rated capacity is 15 Ah, the nominal voltage is 3.2 V, the operating voltage range is 2.0–3.65 V, and the maximum discharge current is 2.00 A. The main control chip of the switching circuit selects STM32F103VB8T6 battery voltage measurement using the battery management chip LTC6811. The measurement error is less than 1 mV, and the sampling frequency is greater than 3 kHz. The sampling of the circuit current uses a shunt, and the sampling error is less than 0.1%. The experimental platform is shown in Figure 7.

Table 2. The identification results.

R_0/Ω	R_{p1}/Ω	C_{p1}/F	R_{p2}/Ω	C_{p2}/F
0.0005206917	0.0001007158	1256.806552	0.0024315296	26355.45136

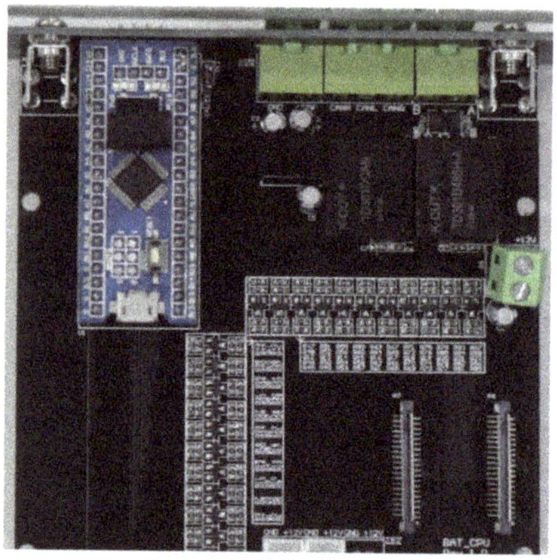

(**a**)

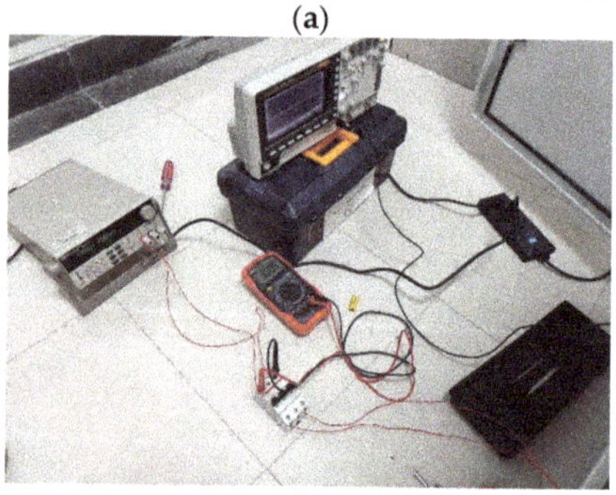

(**b**)

(**c**)

Figure 7. Experimental platform of battery charge and discharge. (**a**) Control board physical diagram. (**b**) DC power supply charging. (**c**) Battery pack interface diagram.

Figures 8 and 9 show the SOC estimation results and errors using the AEKF and FFRLS-AEKF algorithms, respectively. From Figures 6 and 7, it can be seen that the estimation results with the FFRLS-AEKF algorithm are closer to the true values, and the SOC estimation errors with the FFRLS-AEKF algorithm are smaller than with the AEKF algorithm. The estimation errors of the AEKF and FFRLS-AEKF algorithms are shown in Table 3. It can be seen from Table 3 that the maximum absolute error of the joint estimation algorithm is 4.97%, and the error is controlled within 5%. As the battery operating time increases, the SOC estimation accuracy continuously converges to the vicinity of the true value, with the average absolute error decreasing to 2.5%. The experimental results show that the FFRLS-AEKF joint estimation algorithm has good convergence performance and high estimation accuracy, verifying that the proposed method is correct and feasible. Therefore, the SOC estimation performances with EKF, AEKF, and FFRLS-AEKF are shown in Table 4.

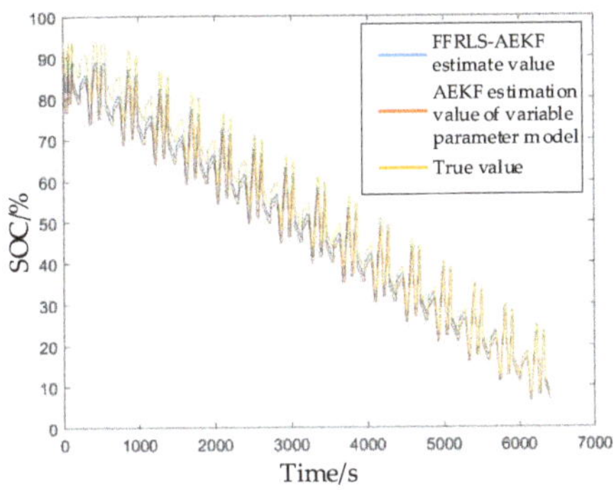

Figure 8. The SOC estimating results of the FFRLS-AEKF and AEKF.

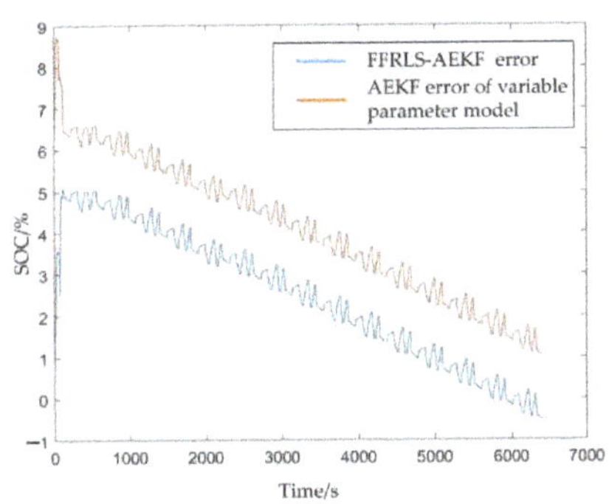

Figure 9. The SOC estimating errors of the FFRLS-AEKF and AEKF.

Table 3. Compared FFRLS-AEKF and AEKF algorithm of SOC errors.

Estimation Algorithm	MaxAE	MaxRE	MAE	MRE
AEKF	6.6%	15.7%	4.0%	8.1%
FFRLS-AEKF	4.97%	6.8%	2.5%	4.3%

Table 4. The SOC estimation performances with EKF, AEKF, and FFRLS-AEKF.

Estimation Algorithm	EKF	AEKF	FFRLS-AEKF
Steady-precision	low	lower	lowest
Convergence speed	fast	faster	fastest

6. Conclusions

The working principle of FFRLS for battery parameter identification is analyzed. The second-order equivalent circuit state discretization equation based on the AEKF algorithm is established, and the steps of the battery SOC estimation method based on AEKF are discussed and simulated. The simulation results show that it has the disadvantages of low steady-state accuracy and slow convergence rate. To improve SOC estimation precision, combining the advantages of high precision and adaptability of FFRLS and AEKF, a joint SOC estimation method based on FFRLS-AEKF is proposed and experimented with. The experimental results show that the FFRLS-AEKF algorithm can have higher SOC estimation and faster accuracy convergence speed, verifying that the proposed method is correct and feasible.

In the actual use of the battery pack, it is often necessary to combine the battery cells in series and parallel to provide sufficient capacity and voltage level. However, in the process of series and parallel grouping of the battery pack, there will be problems such as more complex model parameters and difficult data sampling. Therefore, the next step in this research field needs to consider the influence of physical parameters such as the small number of sampling points, incomplete sampling data, and contact resistance on the accuracy of the battery model. The establishment of a perfect equivalent model of the battery pack is the focus of the next step in this research.

Author Contributions: Conceptualization, Y.X.; methodology, Z.Y.; software, L.H.; validation, Y.X., Z.Y. and L.S.; formal analysis, L.H.; investigation, L.H.; resources, Z.Y.; data curation, Y.X.; writing—original draft preparation, L.H.; writing—review and editing, Y.X.; visualization, Z.Y.; supervision, L.H.; project administration, L.S.; funding acquisition, Y.J. All authors have read and agreed to the published version of the manuscript.

Funding: This research was funded by the National Natural Science Foundation of China (Grant No. 51507183 and 51877212).

Data Availability Statement: The data that support the findings of this study are available from the corresponding author upon reasonable request.

Conflicts of Interest: The authors declare no conflict of interest.

References

1. Marc, D.; Thomas, F.F.; John, N. Modeling of Galvanostatic Charge and Discharge of the Lithium/Polymer/Insertion Cell. *J. Electrochem. Soc.* **2019**, *140*, 13–21.
2. Hu, X.S.; Tang, X.L. Review of modeling techniques for lithium-ion traction batteries in electric vehicles. *J. Mech. Eng.* **2017**, *16*, 20–31. [CrossRef]
3. Qays, M.O.; Buswig, Y.; Hossain, M.L.; Abu-Siada, A. Recent progress and future trends on the state of charge estimation methods to improve battery-storage efficiency: A review. *CSEE J. Power Energy Syst.* **2022**, *8*, 105–114.
4. Xiong, R.; Cao, J.; Yu, Q.; He, H.; Sun, F. Critical Review on the Battery State of Charge Estimation Methods for Electric Vehicles. *IEEE Access* **2017**, *6*, 1832–1843. [CrossRef]
5. Naguib, M.; Kollmeyer, P.; Emadi, A. Lithium-Ion Battery Pack Robust State of Charge Estimation, Cell Inconsistency, and Balancing: Review. *IEEE Access* **2021**, *9*, 50570–50582. [CrossRef]
6. He, H.; Zhang, X.; Xiong, R.; Xu, Y.; Guo, H. Online model-based estimation of state-of-charge and open-circuit voltage of lithium-ion batteries in electric vehicles. *Energy* **2012**, *39*, 310–318. [CrossRef]
7. Meng, J.; Stroe, D.-I.; Ricco, M.; Luo, G.; Swierczynski, M.; Teodorescu, R. A Novel Multiple Correction Approach for Fast Open Circuit Voltage Prediction of Lithium-Ion Battery. *IEEE Trans. Energy Convers.* **2018**, *34*, 1115–1123. [CrossRef]
8. Movassagh, K.; Raihan, A.; Balasingam, B.; Pattipati, K. A Critical Look at Coulomb Counting Approach for State of Charge Estimation in Batteries. *Energies* **2021**, *14*, 4074. [CrossRef]
9. Vidal, C.; Hausmann, M.; Barroso, D.; Shamsabadi, P.M.; Biswas, A.; Chemali, E.; Ahmed, R.; Emadi, A. Hybrid Energy Storage System State-Of-Charge Estimation Using Artificial Neural Network for Micro-Hybrid Applications. In Proceedings of the 2018 IEEE Transportation Electrification Conference and Expo (ITEC), Long Beach, CA, USA, 13–15 June 2018.
10. Chemali, E.; Kollmeyer, P.J.; Preindl, M.; Ahmed, R.; Emadi, A.; Kollmeyer, P. Long Short-Term Memory Networks for Accurate State-of-Charge Estimation of Li-ion Batteries. *IEEE Trans. Ind. Electron.* **2018**, *65*, 6730–6739. [CrossRef]
11. Antón, J.C.; Nieto, P.J.G.; Viejo, C.B.; Vilán, J.A.V. Support Vector Machines Used to Estimate the Battery State of Charge. *IEEE Trans. Power Electron.* **2013**, *28*, 5919–5926. [CrossRef]
12. Hou, J.; Yang, Y.; He, H.; Gao, T. Adaptive Dual Extended Kalman Filter Based on Variational Bayesian Approximation for Joint Estimation of Lithium-Ion Battery State of Charge and Model Parameters. *Appl. Sci.* **2019**, *9*, 1726. [CrossRef]
13. Hu, X.; Li, S.; Peng, H. A comparative study of equivalent circuit models for Li-ion batteries. *J. Power Sources* **2012**, *198*, 359–367. [CrossRef]

14. Wan, E.A.; Van Der Merwe, R. The unscented Kalman filter for nonlinear estimation. In Proceedings of the IEEE 2000 Adaptive Systems for Signal Processing, Communications, and Control Symposium (Cat. No.00EX373), Lake Louise, AB, Canada, 4 October 2000.
15. Plett, G.L. Extended Kalman filtering for battery management systems of LiPB-based HEV battery packs—Part 3. State and parameter estimation. *Power Sources* **2004**, *134*, 277–292. [CrossRef]
16. Bhangu, B.; Bentley, P.; Stone, D.; Bingham, C. Nonlinear Observers for Predicting State-of-Charge and State-of-Health of Lead-Acid Batteries for Hybrid-Electric Vehicles. *IEEE Trans. Veh. Technol.* **2005**, *54*, 783–794. [CrossRef]
17. Pavkovic, D.; Smetko, V.; Hrgetic, M.; Komljenovic, A. Dual Kalman filter-based SOC/SoH estimator for an ultracapacitor module. In Proceedings of the 2014 IEEE Conference on Control Applications (CCA), Juan Les Antibes, France, 8–10 October 2014.
18. Lai, X.; Zheng, Y.; Sun, T. A comparative study of different equivalent circuit models for estimating state-of-charge of lithium-ion batteries. *Electrochim. Acta* **2018**, *259*, 566–577. [CrossRef]
19. Zhou, D.; Yin, H.; Xie, W.; Fu, P.; Lu, W. Research on Online Capacity Estimation of Power Battery Based on EKF-GPR Model. *J. Chem.* **2019**, *2019*, 1–9. [CrossRef]
20. Gao, Z.C.; Chin, C.S.; Toh, W.D.; Chiew, J.; Jia, J. State-of-Charge Estimation and Active Cell Pack Balancing Design of Lithium Battery Power System for Smart Electric Vehicle. *J. Adv. Transp.* **2017**, *2017*, 1–14. [CrossRef]
21. Jia, J.; Lin, P.; Chin, C.S.; Toh, W.D.; Gao, Z.; Lyu, H.; Cham, Y.T.; Mesbahi, E. Multirate strong tracking extended Kalman filter and its implementation on lithium iron phosphate (LiFePO4) battery system. In Proceedings of the 11th IEEE International Conference on Power Electronics and Drive Systems, Sydney, NSW, Australia, 9–12 June 2015.
22. Ali, M.U.; Khan, H.F.; Masood, H.; Kallu, K.D.; Ibrahim, M.M.; Zafar, A.; Oh, S.; Kim, S. An adaptive state of charge estimator for lithium-ion batteries. *Energy Sci. Eng.* **2022**, *10*, 2333–2347. [CrossRef]
23. Ge, D.; Zhang, Z.; Kong, X.; Wan, Z. Online SoC Estimation of Lithium-Ion Batteries Using a New Sigma Points Kalman Filter. *Appl. Sci.* **2021**, *11*, 11797. [CrossRef]
24. Zhang, Z.Y.; Wang, G.S.; Nie, S.X.; Xing, P.X. State of Charge Estimation of LiFePO4 battery under Pulse High Rate Discharge. *J. Electr. Eng. Technol.* **2019**, *8*, 1769–1779.
25. Xia, B.; Sun, Z.; Zhang, R.; Lao, Z. A Cubature Particle Filter Algorithm to Estimate the State of the Charge of Lithium-Ion Batteries Based on a Second-Order Equivalent Circuit Model. *Energies* **2017**, *10*, 457. [CrossRef]
26. Ye, M.; Guo, H.; Xiong, R.; Yu, Q. A double-scale and adaptive particle filter-based online parameter and state of charge estimation method for lithium-ion batteries. *Energy* **2017**, *144*, 789–799. [CrossRef]
27. Schwunk, S.; Armbruster, N.; Straub, S.; Kehl, J.; Vetter, M. Particle filter for state of charge and state of health estimation for lithium–iron phosphate batteries. *J. Power Sources* **2013**, *239*, 705–710. [CrossRef]
28. Giannelos, S.; Borozan, S.; Aunedi, M.; Zhang, X.; Ameli, H.; Pudjianto, D.; Konstantelos, I.; Strbac, G. Modelling Smart Grid Technologies in Optimisation Problems for Electricity Grids. *Energies* **2023**, *16*, 5088. [CrossRef]
29. Li, X.; Lyu, L.; Geng, G.; Jiang, Q.; Zhao, Y.; Ma, F.; Jin, M. Power Allocation Strategy for Battery Energy Storage System Based on Cluster Switching. *IEEE Trans. Ind. Electron.* **2021**, *69*, 3700–3710. [CrossRef]
30. Gao, X.; Fu, L. SOC Optimization Based Energy Management Strategy for Hybrid Energy Storage System in Vessel Integrated Power System. *IEEE Access* **2020**, *8*, 54611–54619. [CrossRef]
31. Song, Q.; Mi, Y.; Lai, W. A Novel Variable Forgetting Factor Recursive Least Square Algorithm to Improve the Anti-Interference Ability of Battery Model Parameters Identification. *IEEE Access* **2019**, *7*, 61548–61557. [CrossRef]
32. Badoni, M.; Singh, A.; Singh, B. Variable Forgetting Factor Recursive Least Square Control Algorithm for DSTATCOM. *IEEE Trans. Power Deliv.* **2015**, *30*, 2353–2361. [CrossRef]

batteries

MDPI

Article

Optimal Capacity Configuration of Wind–Solar Hydrogen Storage Microgrid Based on IDW-PSO

Ge He [1], Zhijie Wang [1,*], Hengke Ma [1] and Xianli Zhou [2]

[1] Department of Electricity, Shanghai Dianji University, Shanghai 201306, China
[2] Department of Electricity, Shanghai University of Electric Power, Shanghai 201306, China
* Correspondence: wzjsdstu@163.com; Tel.: +86-18964586826

Abstract: Because the new energy is intermittent and uncertain, it has an influence on the system's output power stability. A hydrogen energy storage system is added to the system to create a wind, light, and hydrogen integrated energy system, which increases the utilization rate of renewable energy while encouraging the consumption of renewable energy and lowering the rate of abandoning wind and light. Considering the system's comprehensive operation cost economy, power fluctuation, and power shortage as the goal, considering the relationship between power generation and load, assigning charging and discharging commands to storage batteries and hydrogen energy storage, and constructing a model for optimal capacity allocation of wind–hydrogen microgrid system. The optimal configuration model of the wind, solar, and hydrogen microgrid system capacity is constructed. A particle swarm optimization with dynamic adjustment of inertial weight (IDW-PSO) is proposed to solve the optimal allocation scheme of the model in order to achieve the optimal allocation of energy storage capacity in a wind–hydrogen storage microgrid. Finally, a microgrid system in Beijing is taken as an example for simulation and solution, and the results demonstrate that the proposed approach has the characteristics to optimize the economy and improve the capacity of renewable energy consumption, realize the inhibition of the fluctuations of power, reduce system power shortage, and accelerate the convergence speed.

Keywords: independent microgrid system; wind and solar complementary power generation; hydrogen energy storage; IDW-PSO; capacity configuration

Citation: He, G.; Wang, Z.; Ma, H.; Zhou, X. Optimal Capacity Configuration of Wind–Solar Hydrogen Storage Microgrid Based on IDW-PSO. *Batteries* **2023**, *9*, 410. https://doi.org/10.3390/batteries9080410

Academic Editors: Luis Hernández-Callejo, Jesús Armando Aguilar Jiménez and Carlos Meza Benavides

Received: 24 May 2023
Revised: 21 July 2023
Accepted: 26 July 2023
Published: 6 August 2023

1. Introduction

In recent years, wind and photovoltaic power generation have been essential for new power systems mainly based on new energy sources. With the promotion of carbon neutrality and the increasingly prominent problem of energy shortage, the large-scale application of new energy generation has become the trend of power system development. Because wind and sunlight are the primary energy sources of new energy generation, they are randomly influenced by the environment, temperature, geographical location, and other aspects, quickly leading to the imbalance of power conversion and the instability of power generation time, affecting the system's stability. Given the intermittent and uncertain influence of new energy, with the popularization of new energy, the proportion is increasing, which will significantly impact the stability, safety, and system operating rate for new energy sources [1]. The application of energy storage technology in new energy systems helps to improve the utilization rate of power generation in new energy power systems, keep the system stable in power supply during peak power generation and low power generation due to external influence, and improve the utilization efficiency of power generation systems [2]. Aiming at the intermittency and instability nature of new energy power generation, the energy storage is utilized in order to store and release the electric energy, and power supplementation is carried out so as to improve the energy utilization rate and the stability of the power supply of the power system.

Because of the uncertainty and fluctuation of scenery, large-scale access to clean energy will also contain more uncertain factors, which will cause the phenomenon of abandoning wind and light and affect power stability [3]. The incorporation of hydrogen energy storage in the system has the advantages of being pollution-free, sustainable, and energy-saving, and is a green and clean energy storage system. The use of a hydrogen energy storage system allows for the storage of excess electricity from wind and solar energy abandonment, realizing the use of clean energy in the form of integrated energy of electricity–hydrogen–electricity, and improving the efficiency of the available renewable energy sources. Hydrogen energy storage has high energy density, a low cost of running and sustaining compared with other energy storage, long-term storage and non-pollution, applicable to both instantaneous power supply and long time power supply, applicable to different situations. Compared with the traditional battery, the fuel cell in hydrogen energy storage can be between the hydrogen chemical energy converted to electrical energy without the mutual conversion of other energy forms, which significantly reduces the loss and increases the power generation efficiency, and has the advantage of high energy conversion, environmental protection and the like.

With the gradual increase in the occupation of new energy sources, the percentage of wind and solar farms has been increasing, and consequently the research on power fluctuation problems has been gradually deepened. There have been many studies at home and abroad on the problem of optimal allocation of integrated energy system capacity. Literature [4] uses variational mode decomposition (VMD) to analyze the unbalanced power in a wind–solar hybrid microgrid. It establishes a model for the optimal allocation of hybrid energy capacity for the storage of batteries and supercapacitors. Literature [5] established a capacity optimization allocation method to reduce grid-connected photovoltaic power for photovoltaic power plants as well as hybrid energy storage systems. According to [6], the megawatt isolated microgrid consisting of photovoltaic/wind turbines, energy storage, diesel, and gas turbines is optimized in capacity allocation to solve electricity supply problems for powerless remote locations. Literature [7] proposes a quantitative optimal configuration method for a wind and solar complementary power supply system. Literature [8] puts forward an optimization strategy of double-layer hybrid energy storage capacity for a distribution grid by accounting for the distribution network loss and the total cost of the stored capacity system. In literature [9], a configuration model of a multi-source microgrid is constructed considering three aspects: installation location, unit arrangement and combination, extraordinary load of electric vehicles, and the dynamic matching problem between power generation and power consumption is analyzed. In reference [10], a multi-objective optimal allocation method of energy storage systems is proposed to deal with the issue of energy storage allocation under one-party investment and multiple benefits. These studies all show that the adoption with hybrid energy storage is crucial in rational distribution of microgrids, both consider the issues of reducing power loss and system investment cost or maximum economy, but do not consider the utilization of clean energy. In literature [11], the coupling of renewable energy power generation and hydrogen energy storage is shown to be a powerful means of achieving clean and carbon-neutral energy consumption, which has excellent potential for development. Literature [12] explains the application value system of hydrogen energy storage in the "source-network load" of the new power system. Hydrogen energy storage has the advantages of cross-season, cross-regional and large-scale storage. It has a specific rapid response ability, which has substantial application value in all aspects of the "source-network load" of new power systems. Literature [13] mainly considers how to construct a multi-energy combined storage and supply model in the integrated energy system with hydrogen storage as the conversion hub of multiple energy forms in low carbon parks, so as to achieve the optimized allocation method with low carbon emission and low cost; literature [14] uses the hydrogen produced by the hydrogen storage system to work together with a heat storage system and air source heat pump to reduce the use cost of electricity, heat and cold energy in the park, and establishes a capacity optimal allocation model taking integrated energy cost minimization as the optimization target. The research on hydrogen energy storage systems mainly focuses on using hydrogen without considering the conversion of hydrogen energy storage as electric energy. Literature [15] builds a typical wind and solar hydrogen storage capacity configuration model based on wind energy, solar photovoltaic, electric energy storage, and hydrogen production equipment, Then establishes a demand response model of day-ahead segmented electricity price load to reduce the total cost of running the system. The application of hydrogen energy storage focuses on the recycling of hydrogen. Increasing hydrogen energy conversion into electric energy is conducive to improving the rate of abandoning wind and light and transforming into a green and low-carbon environment [16].

Based on the issues described above, a wind–solar hydrogen storage microgrid system with a wind turbine, photovoltaic generator, hydrogen storage system, and battery system as subsystems is constructed in the paper, and the particle swarm algorithm for improving the dynamic adjustment of inertia weights is applied to the system's capacity configuration, and the optimal configuration proposal on system capacity is obtained, which makes the highest system economy, achieves power stabilization, reduces the rate of abandoning wind and solar power, as well as the reduction of the system's shortage of power.

2. Wind–Solar Hydrogen Microgrid System

The distributed new energy system has wind power, photovoltaic, geothermal, tidal, and other subsystems with random output. The integrated development of a comprehensive energy system enhances the combined application of energy [17]. This paper's wind, light, and hydrogen microgrid system consists of wind turbines, photovoltaic generators, hydrogen production units, hydrogen storage units, fuel cells, and other auxiliary equipment. Hydrogen energy storage is used to realize the interconnection of electricity–hydrogen–electricity, to suppress the power fluctuation of distributed new power sources, which can contribute to the sustainability of energy [18]. The electricity–hydrogen system architecture of the park is schematically displayed in Figure 1. By matching and coupling various forms of energy storage, the energy utilization rate can be improved, achieving peak shaving and valley filling, stabilizing power fluctuation, and leading to certain economic benefits.

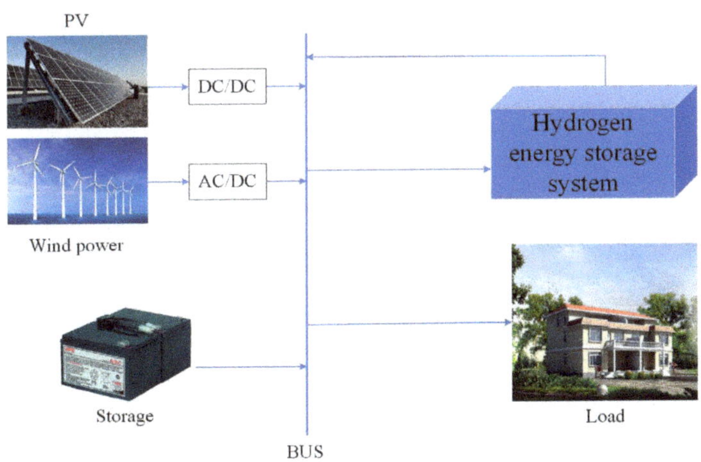

Figure 1. Schematic diagram of electric–hydrogen integrated energy structure.

Figure 1 is a schematic structural diagram of an electric–hydrogen system, mainly consisting of a generation unit, a capacity storage unit and load. The electricity generation unit comprises of wind power generation and photovoltaic power generation, which takes renewable energy as the main energy source of the system, while the energy storage unit comprises a storage battery and hydrogen storage system, which has a role of balancing the load of the system, and absorbs excess energy through the energy storage system when there is sufficient wind and light energy; and supplements it through the energy storage system when there is insufficient wind and light energy.

2.1. Modeling of Generator Set

2.1.1. Photovoltaic Generation System

Physical photovoltaic (PV) power generation is the direct transformation of solar energy through PV panels into electrical energy. Photovoltaic power generation is affected by solar radiation, the radiation angle, sunshine hours, and environmental temperature. The specific photovoltaic power generation output model is shown in Formula (1) [19]:

$$P_{\mathrm{pv}}(t) = P_{\mathrm{pvN}} f_{\mathrm{pv}} \frac{G(t)}{G_{\mathrm{ref}}} [1 + \alpha(T_c(t) - T_{\mathrm{ref}})] \tag{1}$$

where: $P_{\mathrm{pv}}(t)$ represents the actual electricity generation power of the photovoltaic panel; P_{pvN} is the nominal capacity of the photovoltaic panel to generate electricity; f_{pv} is the operation efficiency of photovoltaic system; $G(t)$ is the actual light intensity at time t; G_{ref} is the reference standard light intensity; α is the temperature coefficient; $T_c(t)$ is the operating temperature of the photovoltaic panel at time t; T_{ref} is the reference standard ambient temperature.

Photovoltaic panel operation temperature transformation is caused by the ambient temperature, and solar light intensity is caused by many aspects of the influence of the relationship shown in Formula (2):

$$T_C(t) = T(t) + \frac{T_{\mathrm{rated}}}{800} G(t) \tag{2}$$

where: $T(t)$ is the actual ambient temperature; and T_{rated} is the standard reference temperature of a photovoltaic panel.

2.1.2. Wind Power Generation Model

Wind turbine power output is affected by various aspects such as the speed of wind, blades, ambient temperature, air pressure, etc. [20]. Wind power generation means to convert the kinetic energy to mechanical energy and utilize a turbine to convert the mechanical energy into electrical power. The wind speed changes randomly due to external factors, leading to intermittent and fluctuating wind power generation. The power delivery by wind turbine is most affected by the wind speed, and approximated relationship with the wind speed is shown in Formula (3):

$$P_{wt}(t) = \begin{cases} 0, \ v(t) < v_{\mathrm{in}} \ v(t) > v_{\mathrm{o}} \\ P_{\mathrm{r}} \frac{v(t) - v_{\mathrm{in}}}{v_{\mathrm{r}} - v_{\mathrm{in}}}, v_{\mathrm{in}} \le v(t) \le v_{\mathrm{r}} \\ P_{\mathrm{r}}, \ v_{\mathrm{r}} < v(t) \le v_{\mathrm{o}} \end{cases} \tag{3}$$

where: $P_{\mathrm{wt}}(t)$ is the active output power of the wind farm in time t, P_{r} is the rated output power of the wind turbine, $v(t)$ is the actual wind speed of the wind turbine in time t, v_{r} is the rated wind speed, v_{in} is the cut-in wind speed, and v_{o} is the cut-out wind speed.

2.2. Modeling of Hydrogen Energy Storage System

The hydrogen energy storage system is an integral part for the energy storage system in an independent microgrid system. The hydrogen energy storage system mainly comprises electrolytic cells, fuel cells, and hydrogen storage equipment. Its structural schematic diagram is shown in Figure 2.

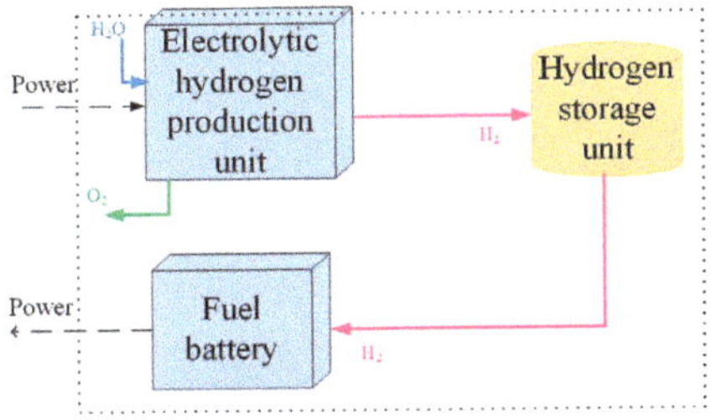

Figure 2. Structure diagram of hydrogen energy storage system.

2.2.1. Mathematical Modeling of Electrolytic Cell

Electrolyzed water is a widely used primary hydrogen production method in industrial hydrogen production [21]. The electrolyzer electrolyzes water into hydrogen and oxygen. There are alkaline electrolyzers, proton exchange membrane electrolyzers, and solid oxide electrolyzers. Compared with other electrolyzers, alkaline electrolyzers have higher efficiency and the best hydrogen production capacity. The system life is twice as long as that of proton exchange membrane electrolyzers, and the cold start-up time is shorter than that of solid oxide electrolyzers. An alkaline electrolyzer is the safest, most mature, and most widely used. The hydrogen production rate of an alkaline electrolyzer is [22]

$$\begin{cases} V_{H_2} = \eta_F \frac{N_C}{2F} I \\ \eta_F = 96.5e^{\left(\frac{0.09}{I} - \frac{75.5}{I^2}\right)} \end{cases} \tag{4}$$

where: V_{H2} is the hydrogen production rate; η_F is Faraday efficiency; N_C is the number of electrolyzers; F is Faraday constant (C/mol); I is the current in the electrolytic cell.

In practice, the electrolytic cell cannot be electrolyzed entirely, and its conversion efficiency is represented by

$$P_{H_2}(t) = \eta_{EL} P_e(t) \tag{5}$$

where: $P_{H2}(t)$ is the power generated by hydrogen production in the electrolytic cell; $P_e(t)$ is the electricity consumption of the electrolytic cell; and η_{EL} is the conversion efficiency of electricity and hydrogen in the electrolyzer.

2.2.2. Mathematical Modeling of Fuel Cell

A mathematical example is provided for the fuel cell:

$$Q_{H_2fc} = \frac{N_S P_{fc}}{U_{fc}(2F)} \tag{6}$$

where: Q_{H2fc} is the hydrogen consumption of the fuel cell; N_S is the number of batteries connected in series; P_{fc} is the output power of the fuel cell; U_{fc} is the battery voltage; and F is the Faraday constant (C/mol).

2.2.3. Mathematical Modeling of Hydrogen Storage Device

Most hydrogen storage devices use hydrogen storage tanks, which can store the hydrogen produced by electrolytic cells and provide hydrogen for fuel cells. The hydrogen storage tank device has the characteristics of low cost, high safety, and fast charging and discharging speed. A hydrogen storage tank is characterized by a mathematical model:

$$Pa_{H_2} = \frac{RT_a}{V} \int_{t_1}^{t_2} (V_{H_2} - Q_{H_2fc}) dt \tag{7}$$

where: Pa_{H2} is the pressure of the hydrogen storage tank; R is a gas constant; T_a is the thermodynamic temperature of the gas; V is the total capacity of the hydrogen storage tank;

t_1 and t_2 are the start time for starting hydrogen production and the end time for stopping hydrogen production, respectively.

2.3. Battery Modeling

A storage battery is a kind of galvanic energy storage, while chemical energy storage is a relatively stable and high-grade energy storage method. This article selected a lithium battery as a storage battery, and the running state of the storage system is marked by the state of charge (*SOC*) of the lithium battery. When $SOC = 1$, the battery capacity reaches a maximum value [23]. The battery output model is

Charging status:

$$SOC(t) = SOC(t-1)(1-\sigma) + P_c(t)\eta_c \frac{\Delta t}{E_{\max}} \tag{8}$$

Discharge state:

$$SOC(t) = SOC(t-1)(1-\sigma) - P_f(t)\frac{\Delta t}{\eta_f E_{\max}} \tag{9}$$

where: σ is the charge and discharge rate of the storage battery; P_c and P_f are the charging and discharging power of the battery in t time; η_c and η_f are charge and discharge efficiency; and $E_{\max}$ is the maximum capacity of the battery.

3. Capacity Optimal Allocation Model

Based on the microgrid system of wind–solar hydrogen storage, this paper not only considers the economy of the independent microgrid of wind–solar hydrogen storage; but also to consider the power fluctuations on the wind generated by the wind and light abandonment, so as to make the wind utilization rate to reach the highest, and put forward the corresponding optimization scheme.

3.1. Objective Function

In this paper, the most economical price is used as the objective function in the independent wind, solar, and hydrogen storage microgrid.

Objective function 1 includes: In the hydrogen-containing composite energy storage system, investment and recovery costs are considered to achieve the best economic benefits. In this paper, the objective function of minimizing the integrated operating cost of the system can be expressed as:

C_T is the system's total operation cost, composed of each piece of equipment's investment and operation cost. Each piece of equipment is comprised of wind turbines, photovoltaics, batteries, and hydrogen storage, and their installation cost, replacement cost, in addition to operation and maintenance cost together constitute the operating cost of the investment [24].

$$C_T = \min\{C_{IN} + C_{RE} + C_{OM}\} \tag{10}$$

where: C_T is the total operating cost of the system; C_{IN} is the installed cost of the equipment; C_{RE} is the replacement cost of the equipment; and C_{OM} is the operation and maintenance cost of the equipment.

C_{IN} represents the installed cost for the equipment, which can be expressed as

$$C_{IN} = QP_{IN}\frac{r(1+r)^m}{(1+r)^m - 1} \tag{11}$$

where: C_{IN} is the installed cost of the equipment; Q is the rated capacity of the equipment; P_{IN} is the unit installed cost of the equipment; r is the discount rate of equipment; and m is the service life of the equipment.

C_{RE} as replacement cost for devices, which can be expressed as

$$C_{RE} = QP_{RE}\frac{r(1+r)^m}{(1+r)^m - 1} \quad (12)$$

where: C_{RE} is the replacement cost of the equipment; Q is the rated capacity of the equipment; P_{RE} is the unit replacement cost of equipment; r is the discount rate of equipment; and m is the service life of the equipment.

C_{OM} is the operation and maintenance cost for equipment, which can be expressed as

$$C_{OM} = \lambda QP \quad (13)$$

where: C_{OM} is the operation and maintenance cost of the equipment; Q is the rated capacity of the equipment; P is the unit cost of equipment; and λ is the ratio of the operation and maintenance cost of each equipment to the total cost of each equipment.

Objective function 2 can restrain the power fluctuation and build an optimal configuration model of a microgrid, including the wind–hydrogen storage as well as the energy storage system formed by the battery, which can be expressed as follows:

$$E_A = E_{\mathrm{PV}} + E_{\mathrm{W}} + E_{\mathrm{Li}} \quad (14)$$

where: E_{A} is the sum capacity for energy storage of wind and solar batteries. Because the battery has the function of charging and discharging, E_{A} has a maximum and minimum value.

$$\min E_A \le E_A \le \max E_A \quad (15)$$

Hydrogen storage system discharge:

$$E_A + E_{\mathrm{H_2}} = E_{\mathrm{L}} \quad (16)$$

Charging (hydrogen storage) of hydrogen energy storage system:

$$E_A = E_{\mathrm{H_2}} + E_{\mathrm{L}} \quad (17)$$

where: E_{H2} is the energy storage capacity of the hydrogen energy storage system; and E_{L} is the total capacity of the load.

Objective function 3 for the wind–solar hydrogen storage-independent microgrid, due to many external influences, after increasing the energy storage system, cannot completely guarantee the reliability of the system power supply [25]. The critical index of load shortage can refer to the power fluctuation problem. When the load is short of electricity, the system is more stable and reliable, and vice versa. Figure 3 shows the load power shortage operation flow in the wind–solar hydrogen storage microgrid system. E_{lps} shows a power shortage of the load, which can be expressed as

$$f_{\mathrm{L}} = \sum_{k=1}^{n} E_{lps}(k) / \sum_{k=1}^{n} E_L(k) \quad (18)$$

When power generation from wind and solar meets the load requirements, when $\Delta E > 0$, power shortage $E_{lps} = 0$, and the ΔE is judged, and different charging combination devices are selected. When ΔE is greater than or equal to the max range between the battery and hydrogen storage capacity, the maximum value of the battery and hydrogen storage is used for charging; When ΔE is greater than the maximum range of the battery capacity, the maximum value of the battery is used to charge, and the remaining electricity is used to charge the hydrogen storage. When ΔE exceeds the upper range of the hydrogen storage capacity, the maximum amount in hydrogen storage is used for charging. When ΔE becomes less than the maximum range of capacity of hydrogen storage, hydrogen storage is used for charging.

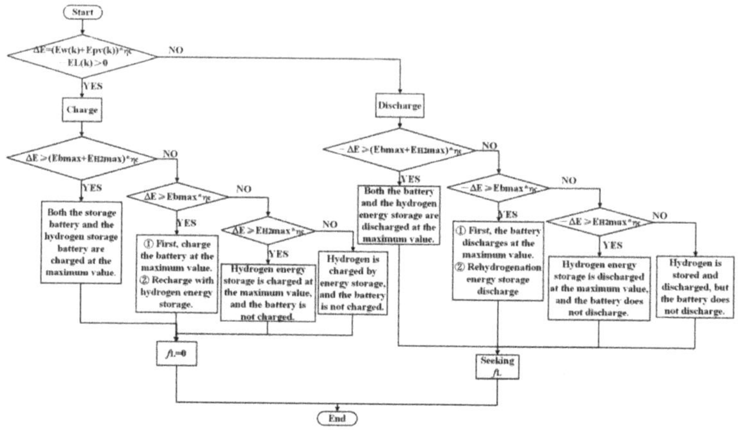

Figure 3. Operation flow chart of load power shortage.

When the amount of wind–solar power generation satisfies the load requirement, that is, $\Delta E < 0$, the wind–solar power generation is insufficient, so it is necessary to discharge the stored power to supplement the shortage of the system, judge ΔE, and choose different discharge combination distribution methods. When $-\Delta E$ is greater than or equal to the maximum range of the storage battery and hydrogen storage capacity, the maximum value of storage battery and hydrogen storage capacity is used for discharge. When $-\Delta E$ is greater than the maximum range of the battery capacity, the maximum value of the battery is employed for discharge, and the balance of the power is used to discharge on hydrogen energy storage. When $-\Delta E$ is greater than the maximum range of the hydrogen storage capacity, the maximum value of hydrogen storage is used for discharge. When $-\Delta E$ is less than the maximum range of hydrogen storage capacity, hydrogen storage is used for discharge.

3.2. Constraints

(1) Capacity constraints of hydrogen storage equipment:

$$V_{\min} \leq V \leq V_{\max} \tag{19}$$

where: $V_{\min}$ and $V_{\max}$ are the minimum and maximum capacities of the hydrogen storage tank.

(2) Capacity constraints of hydrogen storage equipment:

$$SOC \in [0, 1] \tag{20}$$

where: SOC is the operating state of charge in the energy storage system.

(3) Power load constraint:

$$P_{\mathrm{L}} = P_{\mathrm{W}} + P_{\mathrm{PV}} + P_{\mathrm{H_2}} + P_{\mathrm{Li}} \tag{21}$$

where: P_L is the power load of the system, kW.

(4) Energy waste rate constraint:

Excess power is generated in the system when the power generated in the system is greater than the load demand in the system and the maximum energy stored in the system, resulting in wasted energy. This is because the capacity allocation is unreasonable, and the power generation unit is configured with excessive output. When the system energy waste rate is less, the system capacity allocation is more reasonable, which can significantly reduce the waste of resources.

$$f_{\mathrm{L}} < f_{set} \tag{22}$$

where: f_{set} is the control range of the energy waste efficiency of the system for a setting of 0.5.

To solve the multi-objective function problem, fuzzy mathematics is used in this paper. In most cases, the objective functions may affect each other, so it is not easy to achieve simultaneous optimization. The sub-objective optimization and multi-objective are solved by fuzzy mathematics. Under the condition of satisfying all constraints, each objective function is fuzzified, and the objective function is solved based on fuzzy statistics by taking the maximum value according to the affiliation function, and the optimal solution is obtained.

In this paper, the membership function of the distributed function of half Γ decline is

$$u_k(t) = \begin{cases} 1, & F_k(t) \leq F_{k\min}(t) \\ \exp\left(\frac{F_{k\min}(t) - F_k(t)}{F_{k\min}(t)}\right), & F_k(t) > F_{k\min}(t) \end{cases} \quad (23)$$

where: $F_{kmin}(t)$ is the minimum value of the single objective function $F_k(t)$ under the constraint conditions.

Under the principle of maximum affiliation, the fuzzy multi-objective optimization problem is converted into a nonlinear but targeted optimization issue, and the multi-objective function solving model becomes

$$\max u(t) = \begin{cases} u(t) \leq u_1(t) \\ u(t) \leq u_2(t) \\ u(t) \leq u_3(t) \end{cases} \quad (24)$$

where: $u(t)$, $u_1(t)$, $u_2(t)$, and $u_3(t)$ are, respectively, the satisfaction of fuzzy optimization, the satisfaction of system total operation cost, the satisfaction of power fluctuation, and the satisfaction of load power shortage.

4. Improved Particle Swarm Optimization Algorithm

4.1. Particle Swarm Optimization

The particle swarm optimization (PSO) algorithm was inspired by the study of artificial life and was proposed as a global stochastic search algorithm in the simulation of bird flock foraging [26]. The particle swarm optimization algorithm has strong anti-interference ability, good results, fast speed, and memory function, and is also a multi-agent optimization system. Each particle can dynamically adjust according to the surrounding state to find its optimal and overall solution [27].

In the optimization process, after initializing the target population in N-dimensional space, the particles constantly update their positions and velocities, generating a new position each time and solving the next time to attain the optimal result. The iterative expression for the update rate as well as the position in the process of particle solution is

$$\begin{cases} v_i^{k+1} = \omega v_i^k + c_1 r_1 (P_M - x_i^k) + c_2 r_2 (G_M - x_i^k) \\ x_i^{k+1} = v_i^{k+1} + x_i^k \end{cases} \quad (25)$$

where: k is the number of iterations; ω is the inertia weight coefficient; v_i^k and x_i^k are the velocity and position of particles; c_1 and c_2 are acceleration factors; r_1 and r_2 are random numbers between [0,1]; P_M is the particle individual extreme value; and G_M is the extreme value of the population.

4.2. Improved Particle Swarm Optimization

For the particle swarm optimization algorithm, it is susceptible to problems such as falling into the local extremum and poor local search ability. To solve these problems, the particle swarm optimization algorithm is improved. A particle swarm optimization with dynamic adjustment of intrinsic weight is used, and exponential function and random function of beta distribution [28] are used to improve it, achieve the algorithm's global search ability, and reduce the possibility of falling into local extremum.

The inertia weight ω is an essential variable among the particle swarm, determining influence degree of particle velocity on the velocity in the next iteration. The capability of global search is strong when ω is large, the capability of local search is weak, and it can easily fall into the local optimization state. With the increase in iteration times, the exponential function $e^{\frac{-k}{k_{\max}}}$ is adopted to increase the global search ability and the later search accuracy.

The improved expression for the inertia weights is

$$\omega = \omega_{\min} + (\omega_{\max} - \omega_{\min})e^{\frac{-k}{k_{\max}}} + \sigma \times \text{betarnd}(p, q) \tag{26}$$

where: $k_{\max}$ is the maximum number of iterations; σ is the inertia adjustment factor; $\omega_{\max}$ is the initial inertia weight; $\omega_{\min}$ is the inertia weight of the maximum number of iterations.

In particle swarm optimization, the acceleration factor determines the influence of individual particle experience information and other particle experience information on the optimization trajectory. For the acceleration factor, c_1 is the global acceleration factor and c_2 is the local acceleration factor [29]. To converge faster, the acceleration factor is improved so that c_1 gradually increases and c_2 gradually decreases, thus strengthening the convergence ability of particles to the global optimum.

$$\begin{aligned} c_1 &= c_0 \sin^2\left[\frac{\pi}{2}\left(1 - \frac{k}{k_{\max}}\right)\right] \\ c_2 &= c_0 \sin^2\left(\frac{\pi}{2}\frac{k}{k_{\max}}\right) \end{aligned} \tag{27}$$

where: c_0 is the initial value.

4.3. Solution Steps

In solving the problem of optimal system capacity configuration, the paper applies the improved IDW-PSO problem solving and the flow is shown in Figure 4.

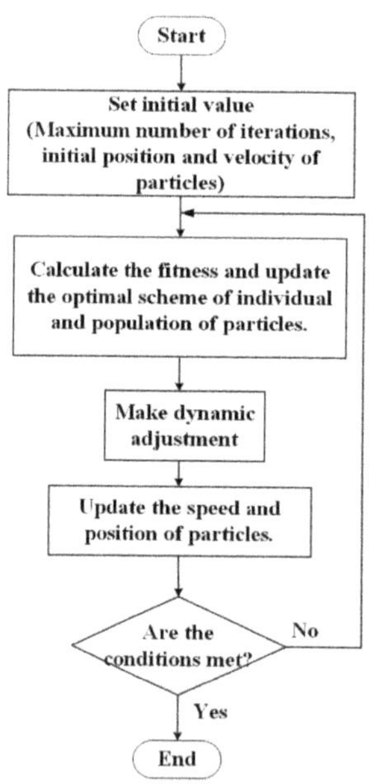

Figure 4. Solution flow.

5. Model Solving

This paper uses the improved particle swarm optimization algorithm with dynamically adjusted inertia weight to optimize the configuration of each device's hydrogen-containing hybrid energy storage microgrid capacity. Firstly, the fundamental parameter models of each device unit, including illumination intensity and load parameters, are determined—secondly, input parameters, including capacity range, conversion rate, etc. Then, the system's capacity is optimized according to the system's total operating cost. The core of the system optimal analysis method based on IDW-PSO is to determine the optimal capacity allocation under the condition of minimum total system cost, to reduce the loss of power supply probability, reduce the fluctuation of power, preventing instability of the microgrid. A flow diagram of the operation strategy is illustrated in Figure 5:

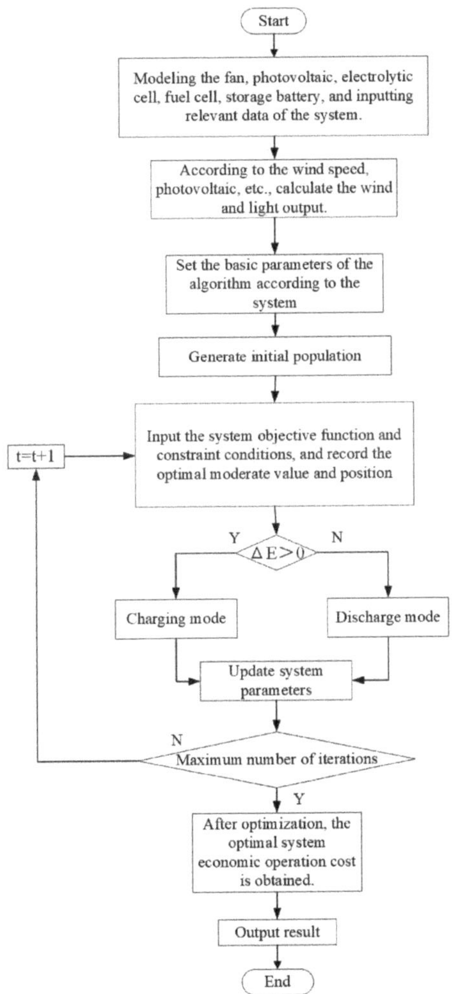

Figure 5. Operation strategy flow chart.

Step (1): Modeling each unit of the microgrid, inputting data such as wind speed and load, and inputting relevant parameters;
Step (2): Initialize the calculated output of photovoltaic and wind power generation;
Step (3): Input system constraints and objective functions;

Step (4): Calculate whether the wind and solar power generation meet the load demand ΔE, execute an objective function, and select different operation processes;
Step (5): After executing the objective function, check whether the microgrid is within the constraint range and update the relevant parameters of the system;
Step (6): Whether the maximum number of iterations has been reached. Without satisfying the condition, the program will continue to run;
Step (7): Obtain optimal economic operating cost and related data.

6. Example Analysis

Taking a wind, solar, and hydrogen microgrid system in Beijing as an example, the capacity of centralized photovoltaic units is 200 KW, and that of centralized wind turbines is 350 KW. Figure 6 shows the data on wind power, photovoltaic power generation, and load consumption on a specific day in this area. Relevant parameters of battery and hydrogen storage of the system are presented in Table 1. The optimization model of the capacity optimization of wind and hydrogen storage system constructed in this paper is solved by MATLAB.

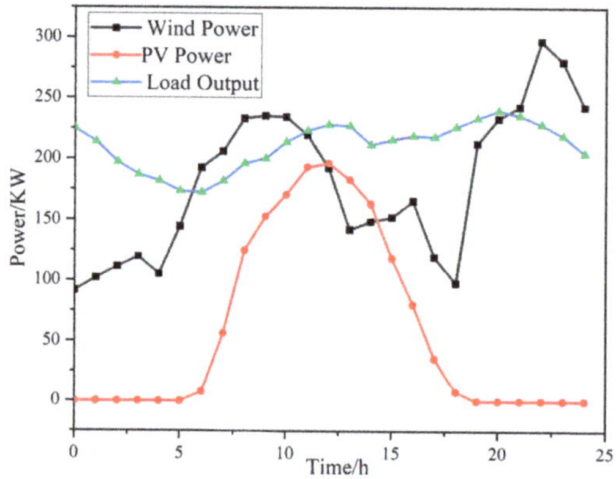

Figure 6. Wind–solar load curve.

Table 1. Example parameters.

Parameter	Value
The daily investment cost of storage battery is RMB/set	397
Battery charging efficiency/%	0.75
Battery discharge efficiency/%	0.85
The daily investment cost of hydrogen energy storage is RMB/set	534
Hydrogen storage charging efficiency/%	0.75
Hydrogen energy storage discharge efficiency/%	0.6
Efficiency of inverter/%	0.95
Type	**Service life/year**
Wind generator [30]	20
PV	25
Storage battery	15
Electrolytic bath	15
Hydrogen storage tank	25
Fuel battery	10
Inverter	20

In the whole microgrid system, the equipment at the power generation end includes a wind turbine, photovoltaic equipment, storage battery, and hydrogen energy storage system. According to a defined objective value function and various parameters, wind turbine, photovoltaic unit, and hydrogen energy storage jointly bear the load consumption of the system and keep the power balance in real time. At 5~18 h, due to the sun's rising, the light amplitude appeared and reached the maximum peak at 12 h at noon, when the light incident angle reached the maximum. Because wind turbines are built in places with abundant wind resources, they can generate electricity 24 h a day.

Simulation Results Analysis

To comprehensively analyze the economic advantages of the energy storage operation of the system according to the improved IDW-PSO algorithm, the effects of system load shortage and power balance on the configuration results are considered, this paper sets the following four schemes for comparative analysis. Scheme 1: Choose the energy storage configuration scheme that is currently widely used. Scheme 2: The battery and hydrogen energy storage is selected as the energy storage schemes of the system for optimal configuration, and the compression factor particle swarm optimization algorithm is used. Scheme 3: Select battery and hydrogen energy storage as the system energy storage scheme for optimal configuration, and use the IDW-PSO algorithm. Scheme 4: Select battery and hydrogen energy storage as the system energy storage scheme to optimize the structure and use the improved IDW-PSO algorithm. The simulation establishes the population scale of 200, and iteration number of 200, acceleration factor initial value is set to 1.65, the battery SOC lower limit is 0.1, the SOC upper limit is 0.9, while a battery SOC initial value is set to 0.5, and a simulation results are as follows:

Upon calculation, the results of the optimized configuration under each scenario are obtained as shown in Figure 7 and Table 2. Scheme 2, Scheme 3, and Scheme 4, respectively, show the differences in iteration times, average time, and total cost caused by different operation results. The above three schemes can all be applied to the capacity optimization arrangement of hydrogen-containing composite power storage system. From a comparative analysis of Figure 7, the improved IDW-PSO algorithm can reduce the number of iterations, speed up the calculation time, and calculate the optimal system operating cost more effectively. The convergence speed and accuracy of the improved IDW-PSO are different from those of IDW-PSO and compressed factor particle swarm optimization, which reduces the local optimal solution, slow divergence speed and efficiency. The algorithms in the present paper have fewer iteration times and operation times. Compared with other algorithms, this algorithm is superior to different algorithms with the capacity optimization configuration scheme, which improves the system economy.

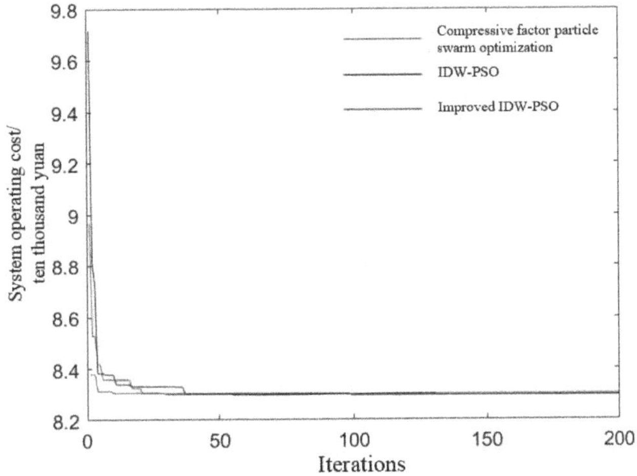

Figure 7. Comparison of convergence curves of various algorithms.

Table 2. Comparison of economic results of each algorithm.

Algorithm Name	Minimum Operating Cost/Ten Thousand Yuan	Average Number of Iterations	Average Time/s
Compressive factor particle swarm optimization [31]	8.359	150	16.12
IDW-PSO	8.328	144	15.15
Improved IDW-PSO	8.326	124	15.05

At this time, see Table 3 for the system capacity optimization configuration scheme through Scheme 1, Scheme 3, and Scheme 4.

Table 3. Optimization result.

Application Scheme	Storage Battery/Unit	Supercapacitor/Unit	Hydrogen Energy Storage/Unit	LPSP	Minimum Cost/Ten Thousand Yuan
Scheme 1	2034	28,956	—	0.0321	8.321
Scheme 3 (Before improvement)	1975	—	127	0.0297	8.359
Scheme 4 (After improvement)	1989	—	106	0.014	8.323

Table 3 shows that the optimal configuration for the microgrid system in the hybrid energy storage of supercapacitors and storage batteries in Scheme 1 is 2034 storage batteries and 28,956 supercapacitors. In this case, the system loss of power supply probability is 0.0321, and the system's total operating cost is 83,210 yuan. The optimal configuration of the Scheme 3 microgrid system before improvement is 1975 batteries and 127 hydrogen storage batteries; now the system loss of power supply probability is 0.0297, and the comprehensive operating of the system costs 83,590 yuan. The optimal configuration of the improved Scheme 4 microgrid system is 1989 batteries and 106 hydrogen storage batteries. Currently, the system loss of power supply probability is 0.014, and the system's total operating cost is 83,230 yuan. From the comparative analysis of Table 3, in the hybrid energy storage of battery and supercapacitor, the minimum price for Scheme 3 is higher than that of Scheme 1 because the cost of hydrogen energy storage is much higher than that of the supercapacitor, and the operating cost of Scheme 4 is unchanged. The scheme proposed in this paper (Scheme 4) reduces the power shortage by 56.4% in Scheme 1 and 52.9% in Scheme 3 while maintaining the running cost unchanged. Therefore, the scheme used in this paper is superior to other projects, which improves the power shortage problem caused by the system's unbalanced configuration.

From the perspective of solving the problem of the power shortage rate, the scheme in this paper has increased by 1.81% and 1.57%, respectively, compared with Scheme 1 and 3. In this paper, Beijing's average daily electricity consumption in 2022 is 210 degrees. According to this ratio, the scheme in this paper can solve the power shortage of about 3.5 degrees by storing energy. Thermal power plants need about 320 g standard coal for the first generation of electricity, saving 1 kg of typical coal = reducing emissions by 2.493 kg "carbon dioxide" = reducing emission by 0.68 kg "carbon." Then, this scheme can reduce carbon dioxide and carbon emissions by about 1019.13 kg a year in this area. Therefore, the scheme proposed in this paper has great practical significance for promoting carbon neutrality.

For wind–solar hybrid electricity generation, both wind turbines and photovoltaic units have limited capacities, and the adjustment range is relatively small. Hydrogen storage has excellent advantages for power generation because hydrogen storage can perform charging and discharging functions and has a wide range of power adjustments. As can be seen from Figure 8, from 0:00 to 6:00, since the load output is higher than the wind and light energy export power, batteries and hydrogen energy storage are discharged; At 6:00–16:00, because the load output is less than the wind and light output power, the battery and hydrogen energy storage are charged. From 16:00 to 22:00, the battery and hydrogen energy storage discharge because the load output is greater than the wind and light output power. From 22:00 to 24:00, the battery and hydrogen energy storage charge because the load output is less than the wind and light output power.

Based on the economic and system loss of power supply probability, the system optimizes the capacity allocation scheme for wind, light, and hydrogen storage systems, thus achieving the purpose of shaving peaks and filling valleys and restraining power fluctuation. When the generating power of wind and light is greater than the load output, the hydrogen storage is optimized by the algorithm to realize "peak clipping". When the generating power of wind and light is less than the load output, the hydrogen storage system is discharged through algorithm optimization to realize "valley filling" of the system power.

Figure 8 is a power comparison chart before and after system optimization, which is, respectively, the power comparison in the whole system before and after the optimization of the hydrogen energy storage system. Before the optimal configuration of the hydrogen energy storage system, a variance of the output power of the whole system was 9171.78 kW^2. After the optimal configuration, the variance of the whole system's output power is 6582.22 kW^2, with an obvious decrease in the fluctuation of the output power. The 0-A region represents the supplementary power region where the fuel cell of the hydrogen energy storage system discharges to supplement wind power and photovoltaic power, thus achieving the function of "valley filling" for the system power. The area A-B region represents the electrolyze in the hydrogen storage system for hydrogen storage, to absorb the force of wind power and photovoltaic, and thus achieve the "peak clipping" effect on the system power. The B—C region is the same as the 0—A region, and the C—D region is the same as the A—B region. Meanwhile, the waste power of the hydrogen energy storage system before configuration is 3.7435 kW, while after configuration, it is 1.8263 kW, which significantly reduces the waste air volume.

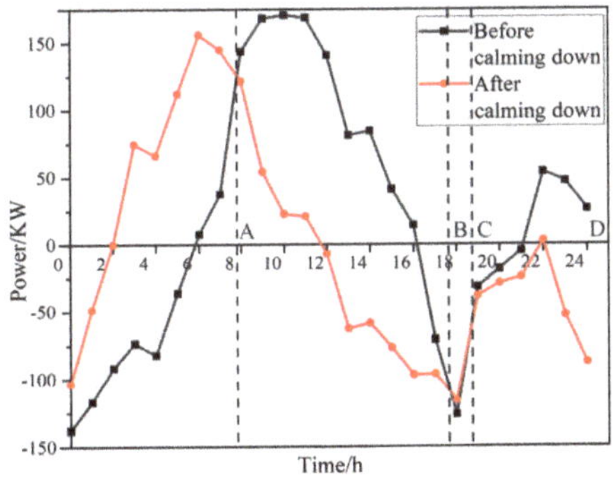

Figure 8. Power comparison diagram before and after optimization.

7. Conclusions

In this paper, the influence of wind turbines, photovoltaic systems, hydrogen storage systems, and battery output ratios on the operation of independent microgrid systems is considered, which is experimentally verified to improve the degree of renewable energy consumption.

In this paper, a particle swarm algorithm is proposed to improve the particle swarm algorithm for dynamically adjusting the inertia weights, establish a capacity optimization model, and obtain a capacity optimization scheme with the optimal system operating cost, system loss of power supply probability and system power. The integrated operation cost of the wind–solar hydrogen storage microgrid system is reduced by 0.431%, and the variance of the whole system after hydrogen storage configuration is 28.23% of that before configuration. The power fluctuation and the rate of abandoning wind and light are reduced.

The improved IDW-PSO algorithm will improve the optimization accuracy, reduce the risk of falling into the optimal local solution, and accelerate the iteration speed so that the system can get the optimal solution faster and better. The improved algorithm is improved from 142 to 124 iterations, and the accuracy of the results is improved by 0.024%.

This scheme can reduce carbon dioxide and carbon emissions by about 1019.13 km annually in this area. When building the optimal configuration of wind and solar hydrogen storage systems, considering the economic indicators and power shortage load, the complementary relationship between hydrogen energy storage and storage battery is used to combine them and realize the suppression of power fluctuations, to achieve the purpose of peak shaving peaks and filling in valleys, to eliminate the wind and light rejection rate better, increase system economic efficiency, to reduce the amount of power shortage in the system, and strive to make the system more environmentally friendly by realizing carbon neutrality.

This paper in the microgrid capacity optimization configuration model allows for a wind-optical-hydrogen-storage microgrid system to provide some support. Starting from the energy demand side, this model can not only configure the capacity of wind, light, storage battery, and hydrogen storage, but also add more forms of energy—for example, flywheel energy storage, biological energy storage, pumped storage, and so on. The system is more environmentally friendly for the application of hydrogen energy storage, and the interconnection of electricity and hydrogen is used to realize carbon emission reduction. With the development of time and the application of hydrogen energy, the cost of hydrogen production is reduced, the difficulty of hydrogen production is reduced, and the efficiency of hydrogen production is improved. Hydrogen energy storage can optimize power and energy simultaneously in electricity storage and power generation and will be further studied. High-density, pollution-free, and sustainable utilization of hydrogen is a significant trend and the significance of efficient production. In future research, the load-side model or the model of demand-side response and load-side interaction will be further considered when optimizing the system to be closer to the actual demand.

Author Contributions: Conceptualization, G.H. and Z.W.; methodology, G.H.; software, G.H.; H.M.; validation, G.H.; formal analysis, G.H.; investigation, G.H.; H.M.; X.Z.; resources, G.H.; H.M.; data curation, G.H.; Z.W.; writing—original draft preparation, G.H.; writing—review and editing, G.H.; Z.W.; visualization, G.H.; Z.W.; supervision, Z.W.; project administration, G.H.; Z.W.; funding acquisition, none. All authors have read and agreed to the published version of the manuscript.

Funding: This research received no external funding.

Institutional Review Board Statement: Not applicable.

Informed Consent Statement: Not applicable.

Data Availability Statement: Data sharing not applicable.

Conflicts of Interest: The authors declare no conflict of interest.

References

1. Xia, D. Overview of new energy power system stability. *Sci. Technol. Innov. Appl.* **2021**, *331*, 69–72.
2. Lian, Z. Research on Capacity Allocation Optimization of Hybrid Energy Storage System of Wind Power Heat Pump Based on Particle Swarm Optimization. Master's Thesis, Shenyang University of Technology, Shenyang, China, 2021.
3. Xiong, Y.; Chen, L.; Zheng, T.; Si, Y.; Mei, S. Optimal configuration of hydrogen energy storage in low-carbon park integrated energy system considering electricity-heat-gas coupling characteristics. *Electr. Power Autom. Equip.* **2021**, *41*, 31–38.
4. Lei, S.; He, Y.; Zhang, J.; Deng, K. Optimal Configuration of Hybrid Energy Storage Capacity in a Microgrid Based on Variational Mode Decomposition. *Energies* **2023**, *16*, 4307. [CrossRef]
5. Li, W.; Jin, R.; Ma, X.; Zhang, G. Capacity Optimal Allocation Method and Frequency Division Energy Management for Hybrid Energy Storage System Considering Grid-Connected Requirements in Photovoltaic System. *Energies* **2023**, *16*, 4154. [CrossRef]
6. Xie, H.; Zheng, S.; Ni, M. Microgrid Development in China: A method for renewable energy and energy storage capacity configuration in a megawatt-level isolated microgrid. *IEEE Electrif. Mag.* **2017**, *5*, 28–35. [CrossRef]

7. Xiu, X. Research on Optimal Allocation of Energy Storage System Capacity and Life Cycle Economic Evaluation Method. Master's Thesis, China Agricultural University, Beijing, China, 2018.
8. Sun, H.; Jie, C.; Liang, S.; Hu, J.; Zang, J.; Li, Y. Optimization of double-layer energy storage capacity of wind, solar and electricity grid-connected based on improved slime mold algorithm. *Power Grid Clean Energy* **2023**, *39*, 128–136.
9. Sun, S. Study on Optimal Allocation and Economic Operation Model of Multi-Energy Microgrid. Master's Thesis, Hefei University of Technology, Hefei, China, 2012.
10. Ye, C.; Huang, M.; Wang, Y.; Sun, F.; Zhong, Y. Optimal configuration of wind-solar hybrid power supply system based on discrete probability model. *Power Syst. Autom.* **2013**, *37*, 48–54.
11. Zhou, J.L.; Wu, Y.N.; Zhong, Z.M. Modeling and configuration optimization of the natural gas-wind-photovoltaic-hydrogen integrated energy system: A novel deviation satisfaction strategy. *Energy Convers. Manag.* **2021**, *243*, 114340. [CrossRef]
12. Xu, C.; Liu, J. Application value, challenge and prospect of hydrogen energy storage in China's new power system. *China Eng. Sci.* **2022**, *24*, 89–99. [CrossRef]
13. Zhang, J.; Cheng, H.; Hu, Z.; Ma, Z.; Zhang, J.; Yao, L. Power System Probabilistic Production Simulation Including Wind Farms. *Proc. CSEE* **2009**, *29*, 34–39.
14. Yu, X. Study on Optimal Allocation and Operation of Comprehensive Energy System in the Park with Hydrogen Storage and Heat Storage. Master's Thesis, North China Electric Power University, Beijing, China, 2020.
15. Hong, F.; Xu, F.; Liu, G. Optimal capacity allocation of photovoltaic hydrogen storage system considering load demand response. *Power Supply* **2023**, *40*, 45–51. [CrossRef]
16. Zhao, B.; Zhao, P.; Hu, J.; Niu, M.; Xiao, Y.; Liu, F. Overview of Hydrogen Energy Storage Technology in Large Scale Intermittent Renewable Energy Integration Application. *Dianqi Yu Nengxiao Guanli Jishu* **2018**, *16*, 1–7.
17. Peng, S.; Zhu, L.; Dou, Z.; Liu, D.; Yang, R.; Pecht, M. Method of Site Selection and Capacity Setting for Battery Energy Storage System in Distribution Networks with Renewable Energy Sources. *Energies* **2023**, *16*, 3899. [CrossRef]
18. Liu, H.; Wang, S.; Liu, G.; Zhang, J.; Wen, S. SARAP Algorithm of Multi-Objective Optimal Capacity Configuration for WT-PV-DE-BES Stand-Alone Microgrid. *IEEE Access* **2020**, *8*, 126825–126838. [CrossRef]
19. Zhu, Z.; Guo, J.; Yu, G.; Xu, M.; Hu, Z. Bi-Layer Optimization Model for Energy Storage Systems Under Wind and PV Access. *Acta Energiae Solaris Sin.* **2022**, *43*, 443–451.
20. Du, G.; Zhao, D.; Liu, X. Research review on optimal scheduling considering wind power uncertainty. *Proc. CSEE* **2022**, *10*, 1–21.
21. Xie, X.; Wang, H.; Tian, S.; Liu, Y. Optimal capacity configuration of hybrid energy storage for an isolated microgrid based on QPSO algorithm. In Proceedings of the International Conference on Electric Utility Deregulation & Restructuring & Power Technologies, Changsha, China, 26–29 November 2015. [CrossRef]
22. Li, Q.; Zhao, S.; Pu, Y.; Chen, W.; Yu, J. Capacity Optimization of Hybrid Energy Storage Microgrid Considering Electricity-Hydrogen Coupling. *Trans. China Electrotech. Soc.* **2021**, *36*, 486–495.
23. Wang, C.; Meng, J.; Wang, Y.; Li, C. Multi-source Coordinated Control Strategy Considering Battery's SOC for Islanded DC Microgrid. *High Volt. Eng.* **2018**, *44*, 160–168.
24. Pan, X.; Liu, K.; Wang, J.; Hu, Y.; Zhao, J. Capacity Allocation Method Based on Historical Data-Driven Search Algorithm for Integrated PV and Energy Storage Charging Station. *Sustainability* **2023**, *15*, 5480. [CrossRef]
25. Xu, L.; Ruan, X.; Zhang, B.; Mao, C. Improved optimal allocation method of wind-solar hybrid power generation system capacity. *J. China Electr. Eng.* **2012**, *32*, 11.
26. Dang, X.; He, B.; Sun, J.; Kong, L.; Meng, F. Photovoltaic maximum power point tracking based on improved particle swarm optimization. *Shandong Electr. Power Technol.* **2022**, *49*, 36–43.
27. Shi, Y.H.; Eberhart, R.C. Empirical study of particle swarm optimization. In Proceedings of the Congress on Evolutionary Computation, Washington, DC, USA, 6–9 July 1999. [CrossRef]
28. Dong, H.; Li, D.; Zhang, X. A Particle Swarm Optimization Algorithm with Dynamic Adjustment of Inertia Weight. *Comput. Sci.* **2018**, *45*, 98–102+139.
29. Mao, K.F.; Bao, G.Q.; Xu, C. Particle Swarm Optimization Algorithm Based on Non-symmetric Learning Factor Adjusting. *Comput. Eng.* **2010**, *36*, 182–184.
30. Li, Y.; Guo, X.; Dong, H.; Gao, Z. Optimal Capacity Configuration of Wind/PV/Storage Hybrid EnergyStorage System in Microgrid. *Power Syst. Its Acta Autom. Sin.* **2020**, *32*, 6. [CrossRef]
31. Kennedy, J.; Eberhart, R.C. A discrete binary version of the particle swarm algorithm. In Proceedings of the 1997 IEEE International Conference on Systems, Man, and Cybernetics. Computational Cybernetics and Simulation, Orlando, FL, USA, 12–15 October 1997. [CrossRef]

electronics MDPI

Article

Optimal Configuration of Battery Energy Storage for AC/DC Hybrid System Based on Improved Power Flow Exceeding Risk Index

Yanming Tu [1], Libo Jiang [2], Bo Zhou [1], Xinwei Sun [1], Tianwen Zheng [2,*], Yunyang Xu [1] and Shengwei Mei [2]

1 State Grid Sichuan Electric Power Research Institute, Chengdu 610041, China; zbv_s@126.com (B.Z.); sunxiw09@126.com (X.S.)
2 Sichuan Energy Internet Research Institute, Tsinghua University, Chengdu 610200, China; jliboc3@163.com (L.J.); Msw9821@126.com (S.M.)
* Correspondence: tianwenscu@163.com

Abstract: After the fault disturbance (DC bi-polar blocking) in the AC/DC hybrid system, when the battery energy storage system (BESS) near the fault location is used to eliminate the power transfer, some sensitive and vulnerable transmission lines still have the problem of power flow exceeding the limit value. Therefore, an optimal configuration of BESS for AC/DC hybrid systems based on power flow exceeding risk index is proposed, which is used to eliminate the impact of power transfer on transmission lines. Firstly, considering the line outage distribution factor, the power flow exceeding risk index is established, which is used to judge the sensitive and vulnerable transmission lines on the shortest path power flow after the fault in the AC/DC hybrid system. The shortest path power flow is found by using the Dijkstra algorithm; the transmission lines nodes of the shortest path power flow are selected as candidate nodes for BESS configuration. Secondly, considering the safe and stable operation capability of the transmission lines, a multi-objective optimal mathematical model of BESS configuration for the AC/DC hybrid system is established, which minimizes the annual investment cost of BESS and maximizes the sum of the power flow exceeding risk index. Finally, the CEPRI36V7 power grid model in Power System Analysis Software Package (PSASP) is used for simulation analysis to verify the effectiveness of the proposed method.

Keywords: AC/DC hybrid system; battery energy storage system (BESS); improved power flow exceeding risk index; fault disturbance; optimal configuration; line outage distribution factor

Citation: Tu, Y.; Jiang, L.; Zhou, B.; Sun, X.; Zheng, T.; Xu, Y.; Mei, S. Optimal Configuration of Battery Energy Storage for AC/DC Hybrid System Based on Improved Power Flow Exceeding Risk Index. *Electronics* **2023**, *12*, 3169. https://doi.org/10.3390/electronics12143169

Academic Editors: Luis Hernández-Callejo, Jesús Armando Aguilar Jiménez and Carlos Meza Benavides

Received: 20 June 2023
Revised: 15 July 2023
Accepted: 18 July 2023
Published: 21 July 2023

1. Introduction

With the emergence of new phenomena, such as the widespread interconnection of power grids and the high penetration of renewable energy, in the past decades, cascading failures of power systems have caused several large-scale power outages worldwide, such as the power outages in California [1], and the disconnection accident in the power grid of Europe [2], which have caused huge economic losses and threatened the stable operation of power grids.

Due to its good technical and economic benefits in large capacity, long-distance, and flexible transmission, DC transmission technology has been widely used in long-distance power transmission, power grid interconnection, and other aspects. DC transmission technology has improved the ability of friendly large-scale renewable energy and effectively solved the imbalance between regional supply and demand of electricity [3]. With the continuous development of HVDC transmission technology and the application of many DC projects, China has built a large-scale complex AC/DC hybrid power grid [4,5]. However, the DC blocking fault in the DC transmission converter station will inevitably transfer the power from the DC transmission line to the AC transmission line, which will cause the power flow of the AC transmission line to exceed the thermal stability limit power and the cascading failure. And then blackout accidents of power systems will happen, which seriously influence the local society and economy [6,7]. Therefore, to ensure the safe and stable operation of the AC/DC hybrid system, it is of great practical significance to eliminate the out-of-limit power flow caused by DC power transfer in AC transmission lines.

The energy storage has good dynamic active and reactive power regulation capabilities, and it can adapt to operational control requirements of different time scales. To reduce the load shedding after the system failure, improve the system operation flexibility and stability, and ensures the safe, reliable, and efficient operation of the AC/DC hybrid system, the DC power transfer of the AC/DC hybrid system is eliminated by using the energy storage [8]. However, due to the different impact of DC power transfer on other lines, the energy storage near the fault location is used to eliminate the power transfer after the fault disturbance (DC bi-polar blocking) in the AC/DC hybrid system, some sensitive and vulnerable transmission lines still have the problem of power flow exceeding the limit value. It is necessary to quickly identify the sensitive line set that has a great impact on the transmission power, and configure the energy storage in this node line, which can quickly eliminate the power limit, improve the system stability, and prevent the occurrence of major accidents.

At this stage, the method of identifying the vulnerability of the power system based on the dynamic characteristics of the power grid has been widely used [9]. In [10], the risk theory assessment method is used to identify key lines by simulating the hidden fault model in the chain fault, but it requires many simulation results to determine the probability of line disconnection through tests, which increases the workload and is difficult to achieve online application and reduces the practicality of the project. In [11], a vulnerability assessment method of power grid cascading fault propagation elements based on power flow entropy is proposed, which can distinguish the vulnerability of branches from impact and consequence. This method accurately models the physical characteristics of the power grid and can improve the simulation speed by reducing the fault search space. However, there is still a contradiction between sampling times and simulation accuracy, which is difficult to achieve online application. In [12], a comprehensive index is proposed to identify the vulnerable lines, which applies the impact vulnerability to represent its impact-resistance ability and the transfer vulnerability to represent the damage caused by its removal from the system. However, this method does not consider the margin of power flow out-of-limit capacity of other lines after the disconnected transmission line. The method of identifying the sensitive and vulnerable transmission lines based on power flow exceeding risk index is proposed in [13]. The method would not have to repeatedly calculate the impedance matrix of the line disconnection and connection. But it is difficult to apply to AC/DC hybrid systems, weak power grids, and other power systems. The scope of the application is limited. The improved power flow exceeding risk index is used for AC/DC hybrid systems and other various power systems. And this method preserved the advantages of traditional methods.

The optimal configuration of BESS is mainly to determine its optimal access location and capacity, to better play its performance, and to improve the absorption rate of renewable energy. In [14,15], the proposed coordinated operational planning for wind farms with BESS is that it can reduce the impacts of wind power forecast errors. Considering the uncertainty and curtailment rate constraint of wind power, reference [16] focuses on the BESS configuration method in wind farms. In [17], the capacity allocation of BESS is used to smooth wind power fluctuations, and the BESS capacity size at different confidence levels is studied. In [18], this paper proposes a bi-level optimal energy storage system (ESS) siting and sizing algorithm to mitigate the voltage deviation in distribution networks. A capacity allocation method of BESS in secondary frequency regulation with the goal of maximum net benefit is proposed in [19]. The literature [14–19] focuses on a single application scenario, such as reducing prediction error, improving new energy consumption, and ensuring power grid stability to achieve BESS configuration, which has significant limitations. They do not fully explore their advantages in coordinated operation or multiple application scenarios. Large BESS capacity needs to be configured, and the utilization rate of BESS is low.

References [20–22] propose an optimized configuration method for the coordinated operation of BESS and renewable energy. Collaborative configuration of distributed generation and BESS in microgrids considering the state of health is studied in [23]. In [24], considering the uncertainty of the net load, this study provides an approach to analyzing the BESS demand capacity for peak shaving and frequency regulation. In reference [25], the feasibility and compatibility of using such idle capacity and power of BESS to participate in the electricity energy market and reserve ancillary service market are explored, and a coordinated operation strategy for the three application scenarios of BESS is proposed to improve its utilization. When the power system is in a steady state, to achieve economic efficiency, ensure grid stability, and improve the utilization rate of BESS, the optimization configuration of BESS for the collaborative operation of BESS and renewable energy and multiple application scenarios of BESS services is studied by domestic and foreign scholars. But the frequent occurrence of extreme weather would seriously affect the safe and stable operation ability of the power grid. It is necessary to study the optimal configuration of BESS considering the influence of extreme weather on the power grid. This can enhance the safe and stable operation capacity of the power grid.

To cope with the impact of extreme weather, such as typhoons and freezing rain, on the power grid, BESS has been configured to improve the reliability and flexibility of the power grid in recent years. Literature [26] takes the load-shedding cost of the system under extreme events as the toughness index and studies the optimal allocation of BESS considering the toughness of the distribution network. In [27], a distributed energy storage planning model for the distribution network considering the influence of typhoon weather is established, and a decomposition collaborative solution method based on the Benders decomposition is proposed. In [28], a new quantitative index of toughness and formulates of a method of BESS planning were proposed to enhance the seismic capacity of the distribution network. Literature [29] proposed a distribution network BESS planning method considering toughness and established a two-stage robust optimization model, which can effectively ensure the uninterrupted power supply of important loads. The above literature configures BESS to improve the flexibility or toughness of the power grid by ensuring a continuous power supply of important loads in extreme weather. However, with the increase in the penetration rate of new energy, the probability of power grid failures has increased, such as exceeding the power limit of transmission lines and cascading faults in the power grid. It is very necessary to allocate BESS reasonably after a power grid failure, such as to quickly eliminate the over-limit of AC line power and improve the stability of the system; how to reasonably configure BESS after the DC locking fault occurs in the AC/DC hybrid system.

And then, a large amount of research has been conducted domestically and internationally on solution methods for BESS optimization configuration. Intelligent optimization algorithms, such as genetic algorithm [30] and particle swarm optimization (PSO) [31], have been widely applied. The optimization configuration method proposed provides a good reference for the solution in this article.

Aiming at the advantages and disadvantages of the existing research, considering the millisecond level active dynamic response capability of the BESS system, an optimized configuration of BESS in the AC/DC hybrid system based on the improved power flow exceeding risk index is proposed. Firstly, the improved power flow exceeding risk index is established to evaluate the sensitivity and vulnerability of other lines to the transferred power flow after the branch is disconnected. Secondly, the Dijkstra algorithm is used to find out the shortest path of the closed loop formed by the breaking line, and the key nodes are selected as the candidate sites for BESS by calculating the improved power flow exceeding risk index of the shortest path. Finally, a multi-objective function with the maximum sum of improved power flow exceeding risk index and the minimum annual investment cost of BESS is established, and particle swarm optimization (PSO) is used to obtain the optimal configuration scheme of BESS in the AC/DC hybrid system. Meanwhile, the BESS system adopted the active power control strategy, including plant-level control and local control, which quickly eliminates the power exceeding the limit of the AC line and suppresses the power fluctuation of the power grid.

This paper is organized as follows: The identification of sensitive and vulnerable lines is talked about in Section 2. The mathematical model for the optimal allocation of BESS is proposed in Section 3. The Model-solving method is given in Section 4. In Section 5, the effectiveness and feasibility performance of the proposed method are examined on the CEPRI36V7 grid model. Section 6 is the conclusion.

2. Identification of Sensitive and Vulnerable Lines

2.1. Line Outage Distribution Factor

If line A of the AC/DC hybrid system is faulty, and it causes the line disconnection (DC line causes blocking fault, etc.), which causes the transfer of active power flow in the system, that is, the active power flow of other lines is changed. The relationship between the change of normal line power flow and the original power flow of the disconnected line can be expressed by the Line Outage Distribution Factor (LODF) [32]:

$$D_{R-A} = \frac{\Delta P_{R-A}}{P_A} \tag{1}$$

where ΔP_{R-A} is the change of line R's active power flow after line A is disconnected; D_{R-A} is the LODF that causes the change of line R's active power flow after line A is disconnected; P_A is the steady-state initial active power of line A.

Assuming that the nodes at both ends of line A are i and j, and the injected active power remains unchanged before and after disconnection, the change of node active power flow caused by line A disconnection is

$$\Delta P = [0 \cdots 1 \cdots -1 \cdots 0]^T P_A = M_A P_A \tag{2}$$

where M_A is the node-branch associated $n \times 1$ order column vector of branch A, and the row corresponds to the node number.

The $n \times 1$ order change $\Delta\theta$ of node voltage phase angle caused by line A disconnection. $\Delta\theta$ can be expressed as

$$\Delta\theta = \left(B - M_A x_A^{-1} M_A^T\right)^{-1} M_A P_A \tag{3}$$

where B is the $n \times n$ order admittance matrix; the admittance matrix is sparse type; x_A is the reactance of line A.

Then, the change of active power flow of branch R ($R \neq A$) caused by line A disconnection is

$$\Delta P_{R-A} = \frac{M_R^T \Delta\theta}{x_R} = D_{R-A} P_A = \frac{M_R^T \left(B - M_A x_A^{-1} M_A^T\right)^{-1} M_A P_A}{x_R} \tag{4}$$

where M_R is the node-branch associated $n \times 1$ order column vector of branch R; x_R is the 1 order reactance of line R.

Let $B^{-1} = X$, after simplification, the expression of D_{R-A} is:

$$D_{R-A} = \frac{X_{R-A}/x_R}{1 - X_{A-A}/x_R} \tag{5}$$

Among them,

$$X_{R-A} = M_R^T X M_A \tag{6}$$

$$X_{A-A} = M_A^T X M_A \tag{7}$$

where X is the $n \times n$ order impedance matrix; X_{A-A}, X_{R-A} is the 1 order self-impedance and mutual impedance between nodes of port R and port A, respectively.

After the DC blocking fault occurs in the AC/DC hybrid system, the influence of DC power transfer on the AC line can be measured by calculating the LODF of the AC line.

2.2. The Improved Power Flow Exceeding Risk Index

The improved power flow exceeding risk index takes into account the impact of power transfer on other lines and the margin of the out-of-limit capacity of line power flow. After considering the margin of the out-of-limit capacity of line power flow, it is not necessary to consider the problem of reverse power flow of other lines caused by power flow transfer separately, which reduces unnecessary calculations, and can better reflect the sensitivity and vulnerability of other lines, and identify the sensitive and vulnerable lines in the system. Under the change of power flow of line R caused by the disconnection of line A, the margin of the out-of-limit capacity of the power flow of line R can be expressed as follows:

When $D_{R-A} < 0$, this is true, the expression of line power flow out-of-limit capacity margin is as follows:

$$\Delta P' = \begin{cases} |P_{R,\max} + P_R| & P_R \geq 0 \\ |-P_{R,\max} - P_R| & P_R < 0 \end{cases} \tag{8}$$

When $D_{R-A} > 0$, this is true, the expression of line power flow out-of-limit capacity margin is as follows:

$$\Delta P' = \begin{cases} |-P_{R,\max} + P_R| & P_R \geq 0 \\ |P_{R,\max} - P_R| & P_R < 0 \end{cases} \tag{9}$$

Combined with the LODF, the improved power flow exceeding risk index is given to evaluate the sensitivity and vulnerability of other lines to the transferred power flow after the branch break, as follows:

$$\Phi_{R-A} = \frac{\Delta P'}{D_{R-A} \cdot P_{\mathrm{A,max}}} \tag{10}$$

where $P_{\mathrm{A,max}}$ is the thermal stability limit value of the breaking line.

After a DC blocking fault occurs in the AC/DC hybrid system, the sensitivity and vulnerability of each AC line can be effectively evaluated by calculating the improved power flow exceeding risk index of each AC line. The smallest the improved power flow exceeding risk index of each AC line, the lower its ability to receive the transferred power flow, and the higher the improved power flow exceeding risk index. This paper selects the sensitive vulnerability line when the absolute value of the improved power flow exceeding risk index is less than 0.5. Secondly, after the DC blocking fault occurs in the AC/DC hybrid system, the DC power flow is mainly transferred to the shortest path that forms a closed loop with the DC line, so the shortest path set of DC power transfer needs to be searched.

2.3. The Shortest Path Search Based on the Dijkstra Algorithm

Using the knowledge of graph theory, the AC/DC hybrid system is simplified and abstracted into a graph G (V, E), where V represents the bus set in the grid, E represents the line set between buses, and the line side weight value is the line reactance. Then, the shortest path algorithm related to graph theory is adopted.

The shortest path search algorithms in graph theory include the Dijkstra algorithm, Floyd algorithm, etc. The Dijkstra algorithm has small time complexity and is easy to expand; the Floyd algorithm has high time complexity and space complexity, which increases the calculation amount. Its advantage is that it can be used to search the shortest path of the line with negative weight. Because of the scalability of the Dijkstra algorithm and the fact that there are no branches with negative weights in the graph, and to meet the requirements of fast calculation, the shortest path search based on the Dijkstra algorithm is adopted.

After the DC blocking fault occurs in the AC/DC hybrid system, the shortest path forming a closed loop with the DC line can be searched by using the Dijkstra algorithm, and then the sensitive AC line can be identified by combining the improved power flow exceeding risk index, and the power input node of the sensitive, vulnerable line is used as the candidate location for configuring BESS.

3. The Mathematical Model for Optimal Allocation of BESS

3.1. Objective Function

After the DC locking fault occurs in the AC/DC hybrid system, when the system optimizes the configuration of BESS, it should also have a certain economy while eliminating the power flow over-limit on the sensitive and vulnerable lines. Therefore, the multi-objective function of the optimal configuration of BESS in the AC/DC hybrid system is as follows:

$$\max\Gamma = \sum_{k=1}^{K}\left|\frac{\Delta P'}{D_{k-A}\cdot P_{\text{dc}-\max}}\right| \tag{11}$$

$$\min G_{inv} = \frac{r(1+r)^{y}}{(1+r)^{y}-1}\cdot(c_1\cdot P_e + c_2\cdot E_e) \tag{12}$$

where Γ is the sum of the improved power flow exceeding risk index of sensitive vulnerable lines after DC blocking fault; D_{k-A} is the LODF; $P_{\text{dc}-\max}$ is the thermal stability limit value of DC line; G_{inv} is the annual investment cost of BESS; c_1, c_2 are the unit power cost and capacity cost of BESS; P_e, E_e are the rated power and rated capacity of the BESS, respectively; r is the annual interest rate of the fund; Y is the life cycle of BESS; K is the sensitive vulnerability line.

3.2. Constraint

(1) Power balance constraints

$$\sum_{m=1}^{M} P_{G,m} + \sum_{n=1}^{N} P_{L,n} + \sum_{s=1}^{S} P_{B,s} = 0 \tag{13}$$

where $P_{G,m}$, $P_{L,n}$, $P_{B,s}$ are, respectively, the output of generator m, the required power of load n, and the charging and discharging power of BESS s; S is the quantity of configured BESS in the system; M and N are the number of generators and the number of loads in the system.

(2) Line loss constraint

$$P_R' U_R^2 \geq R_R\cdot({P_R'}^2 + {Q_R'}^2) \tag{14}$$

where P_R', Q_R' are the active power and reactive power transmitted by the receiving end of the R line, respectively; R_R is the resistance of the R line; U_R is the voltage amplitude of the receiving terminal node of the Rth line.

(3) Generator power constraint

$$\overline{P}_{G,m} \leq P_{G,m} \leq \underline{P}_{G,m} \tag{15}$$

where $\overline{P}_{G,m}$, $\underline{P}_{G,m}$ are the upper and lower limits of generator output, respectively.

(4) Line power constraint

$$\overline{P}_{L,m} \leq P_{L,m} \leq \underline{P}_{L,m} \tag{16}$$

where $\overline{P}_{L,m}$, $\underline{P}_{L,m}$ are the upper and lower limits of the active power of the transmission line.

(5) Capacity constraints of BESS

$$S_{\min} \leq S_r \leq S_{\max} \tag{17}$$

where S_{min}, S_{max} are the minimum and maximum capacities of BESS, respectively.

(6) Power constraint of BESS charge and discharge

$$\begin{array}{c} -P_{r,\max} \leq P_r^c \leq 0 \\ 0 \leq P_r^d \leq P_{r,\max} \end{array} \tag{18}$$

where $P_{r,\max}$ is the maximum value of BESS discharge power; P_r^c, P_r^d are the charge power and discharge power of the BESS system, respectively.

4. Model-Solving Method

The sharing of information among the entire population is beneficial for the population towards a better position in genetic algorithms. Only the best individual's information is shared in PSO, and the entire search process is tracking the optimal solution. So, the PSO algorithm has faster convergence than the genetic algorithm. And due to its advantages of high accuracy and fast convergence, the PSO algorithm is widely used in BESS capacity configuration [33,34]. Therefore, this article chooses the PSO algorithm for solving BESS capacity configuration.

To reasonably obtain the location of BESS, the PSO algorithm is used to solve for the optimal capacity of BESS, and the optimal location of BESS is selected from candidate nodes. Meanwhile, BESS adopts the active power control strategy, including plant-level control and local control, which quickly eliminates the power exceeding the limit of the AC line, suppresses the power fluctuation of the power grid, and ensures the safe and stable operation ability of the power grid.

The specific process is shown in Figure 1, and the solution steps are as follows:

(1) The graph obtained by abstracting the power grid is G0, and the reactance value of each line in the power system is taken as the weight value of each side;
(2) After DC blocking occurs in the converter station at the sending end, the Dijkstra algorithm is used to find the shortest path between the converter station and the designated node;
(3) The branch set contained in the target source point and destination point are combined to obtain the branch set of power flow transfer;
(4) Calculate the improved power flow exceeding risk index of all branches in the power flow transfer branch set, and select the branches whose absolute value of the improved power flow exceeding risk index is less than 0.5 to form the main branch set;
(5) For the lines in the main branch set, if there is a reverse flow and the branch flow meets $|P_R'| \leq |P_R|$, it will be removed from the main branch set (P_R' is the branch flow after the line is disconnected);
(6) Input parameters. Input PSO controlling variables of the original parameters. Set PSO algorithm parameters: the maximum iteration number is 300, and the population size is 200;
(7) Initialize the population. According to Equation (19), the N solutions are generated, such as the energy storage power and capacity, and they also are guaranteed to satisfy the condition. The objective function value is calculated for all the scenes using Equations (11) and (12).

$$x_{ij} = x_j^{\min} + rand() \cdot (x_j^{\max} - x_j^{\min}) \tag{19}$$

where i = 1, 2, ..., N is a D-dimension vector; j = 1, 2, ..., d; $rand()$ represents random numbers between 1 and 0; $x_j^{\max}$, $x_j^{\min}$ are the maximum and minimum values of particles, respectively;

(8) Calculate the fitness value for particles by using (20). It is updated local optimal position and global optimal position by using Equations (21) and (22).

$$p_i = \frac{F_i}{\sum\limits_{k=1}^{N} F_k} \tag{20}$$

$$v_{ij}(t+1) = w \times v_{ij}(t) + c_1 \times rand() \times (p_{ij}(t) - x_{ij}(t)) + c_2 \times rand() \times (p_{gj}(t) - x_{ij}(t)) \tag{21}$$

$$x_{ij}(t+1) = x_{ij}(t) + v_{ij}(t+1) \tag{22}$$

where F_i is the corresponding fitness value for particles I; $v_{ij}(t+1)$, $v_{ij}(t)$ are the velocity of the ith particle at t + 1, t times, respectively; w is the inertia factor; $w = 0.8$, c_1 and c_2 are the learning rate; $c_1 = 0.9, c_2 = 0.9$. $p_{ij}(t)$, $p_{gj}(t)$ respectively represent the individual optimal value and the global optimal value of particles;

(9) Output optimal solution. If the iteration number is greater than the set value, then output the Parote optimal. Otherwise, return to step (7).

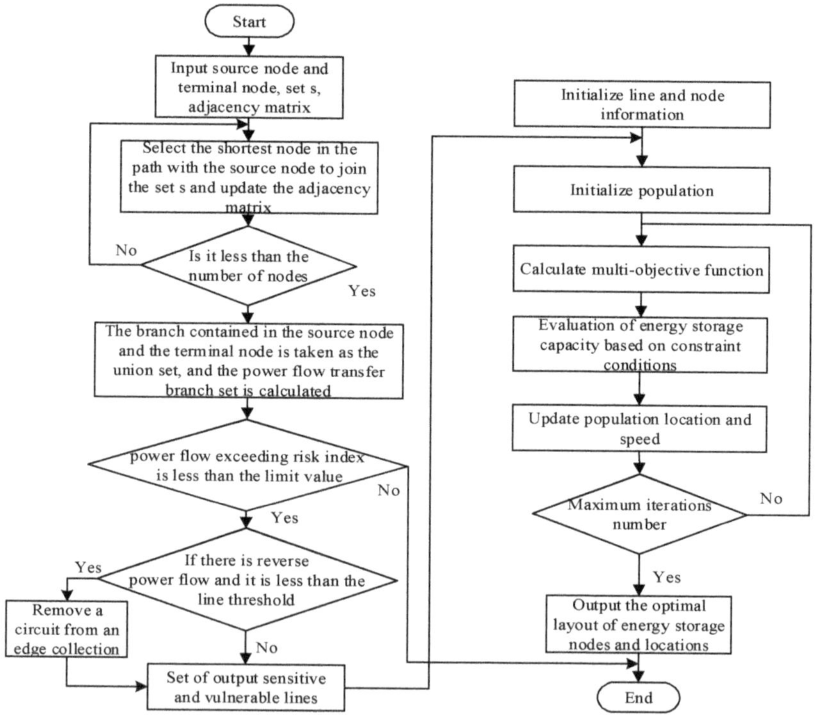

Figure 1. Solution flow of BESS optimization configuration.

5. Simulation Analysis

5.1. Parameter Design

The CEPRI36V7 power grid model in the power system analysis comprehensive program (PSASP) was used for simulation analysis to verify the effectiveness of the proposed BESS configuration strategy. The parameters of the CEPRI36V7 model are referred to in reference [35]. The topology of the CEPRI36V7 model is shown in Figure 2. The parameters of the CEPRI36V7 model are referred to in reference [35]. The node parameters, generator parameters, and branch parameters of the CEPRI36V7 model are shown in Tables 1–3, respectively.

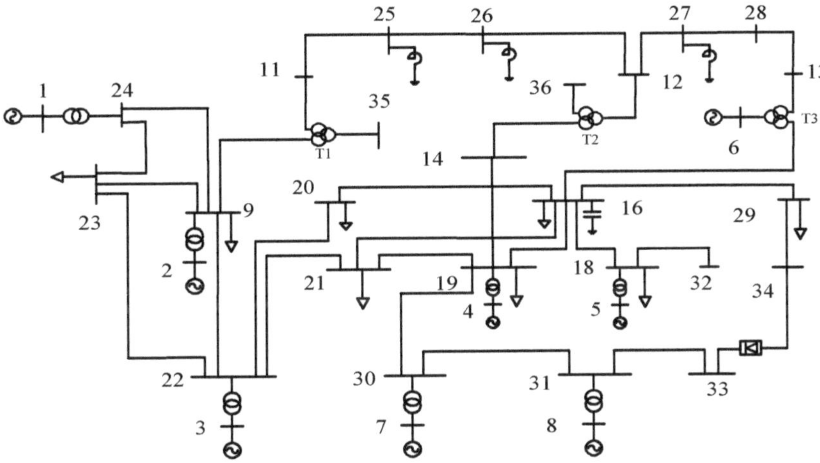

Figure 2. The power grid structure of CEPRI36V7.

Table 1. The node parameters of the CEPRI36V7 model.

Bus_i	Type	P_d/MW	Q_d/Mvar	Base/kV	Bus_i	Type	P_d/MW	Q_d/Mvar	Base/kV
1	3	0	0	10.5	19	1	86.4	66.2	220
2	1	0	0	20	20	1	71.9	47.4	220
3	2	0	0	10.5	21	1	70	50	220
4	1	0	0	15.7	22	1	226.5	169	220
5	1	0	0	10.5	23	1	287	144	220
6	2	0	0	10.5	24	1	0	0	220
7	2	0	0	10.5	25	1	0	0	500
8	2	0	0	10.5	26	1	0	0	500
9	1	376	221	220	27	1	0	0	500
10	1	0	0	20	28	1	0	0	500
11	1	0	0	500	29	1	520	10	220
12	1	0	0	500	30	1	0	0	220
13	1	0	0	500	31	1	0	0	220
14	1	0	0	220	32	1	0	0	220
15	1	0	0	20	33	1	0	0	220
16	1	500	230	220	34	1	0	0	220
17	1	0	0	20	35	1	0	0	0
18	1	430	220	220	36	1	0	0	0

Table 2. The generator parameters of the CEPRI36V7 model.

Bus	P_g/MW	Q_g/Mvar	V_g/p.u.
1	0	0	1
2	600	360	1
3	310	0	1
4	160	70	1
5	430	334	1
6	−1	0	1
7	225	0	1
8	306	0	1

Table 3. The branch parameters of the CEPRI36V7 model.

Fbus	Tbus	r	x	b	Ratio	Fbus	Tbus	r	x	b
11	25	0	0.0001	0	0	31	32	0	0.0001	0
12	26	0	0.0001	0	0	9	22	0.0559	0.218	0.3908
12	27	0	0.0001	0	0	9	23	0.0034	0.0131	0
13	28	0	0.0001	0	0	9	24	0.0147	0.104	0
14	19	0.0034	0.02	0	0	24	1	0	0.015	0
16	18	0.0033	0.0333	0	0	9	2	0	0.0217	0
16	19	0.0578	0.218	0.3774	0	22	3	0	0.0124	0
16	20	0.0165	0.0662	0.4706	0	19	4	0	0.064	0
16	21	0.0374	0.178	0.328	0	18	5	0	0.0375	0
16	29	0	0.0001	0	0	30	7	0	0.0438	0
18	34	0	0.001	0	0	31	8	0	0.0328	0
19	21	0.0114	0.037	0	0	12	15	0	0.018	0
19	30	0.0196	0.0854	0.162	0	6	17	0	0.0337	0
20	22	0.0214	0.0859	0.6016	0	9	10	0	−0.002	0
21	22	0.015	0.0607	0.4396	0	14	15	0	−0.002	0
22	23	0.0537	0.19	0.3306	0	13	17	0	0.01	0
23	24	0.0106	0.074	0	0	11	10	0	0.018	0
25	26	0.0033	0.0343	3.7594	0	36	15	0	0.0001	0
27	28	0.00245	0.0255	2.79	0	16	17	0	0.001	0
29	33	0	0.0001	0	0	35	10	0	0.001	0
30	31	0	0.0001	0	0					

The capacity of the CEPRI36V7 power grid is 2600 MW, and nodes 33 to 34 are DC transmission lines, with a DC transmission capacity of 2 × 200 MW; DC power flows from 33 nodes to 34 nodes, and nodes 10, 15, and 17 in the system are the central nodes of three-winding transformers T1, T2, and T3.

This article uses lithium BESS. The maximum rated power and capacity of the total installed BESS are 800 MW and 1600 MW. h respectively, and the cost coefficients are 1500 yuan/kW and 2000 yuan/kWh. The charge and discharge rate of BESS is 0.5 C.

5.2. Sensitive Line Identification

The simplified structure of the power grid is shown in Figure 3. The red line represents the DC bipolar locking fault line. The blue line and black line represent the non-shortest path AC line. The green line represents the shortest path AC line. The converter station of node 33 has a DC bipolar locking fault, and the Dijkstra algorithm is used to obtain the shortest path composed of node 33 and node 34, namely: 33, 31, 30, 19, 14, 15, 12, 27, 28, 13, 17, 16, 29, and 34, a total of 13 AC lines. The power change curve of the AC line is shown in Figures 4 and 5 after the converter station of node 33 has a dual-machine locking fault at 5 s.

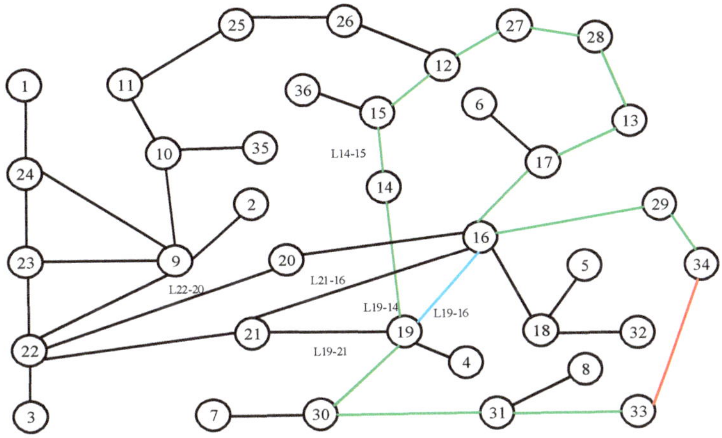

Figure 3. The simplified grid structure of CEPRI36V7.

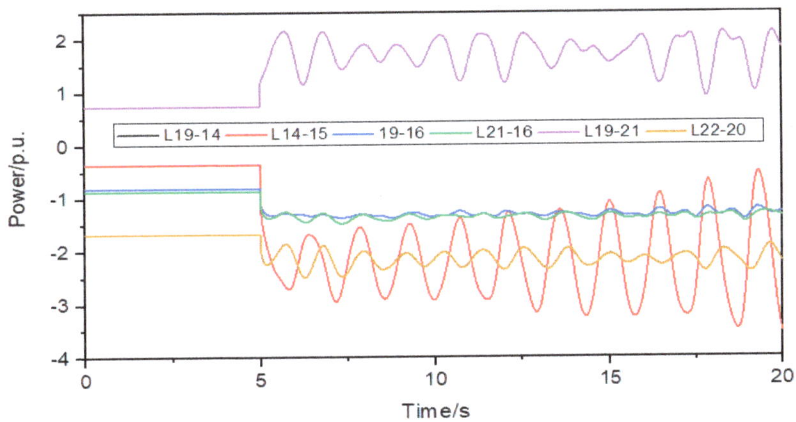

Figure 4. Change curve of AC line after DC blocking.

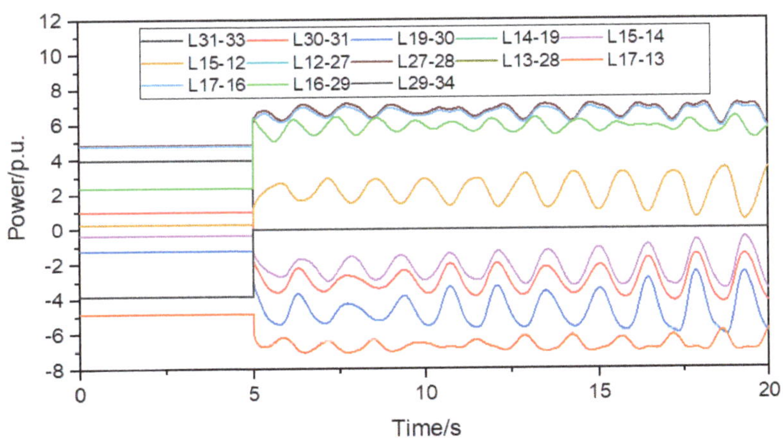

Figure 5. Power change curve of the shortest path line after DC blocking.

It can be seen from Figure 4 that the power variation of lines L19-14 and L14-15 is the largest, with a power variation greater than 2 p.u., and the power variation of other lines is less than 1.3 p.u. It can be seen from Figure 5 that the power variation of the AC line on the shortest path is greater than 1.5 p.u. It can be seen that after the DC bipolar locking fault, the power flow is mainly in the shortest path composed of node 33 and node 34.

The improved power flow exceeding risk index of each line on the shortest path is shown in Table 4 after the bipolar locking fault occurs at the converter station of node 33 at 5 s. From Table 4, it can be seen that the absolute value of the improved power flow exceeding risk index of lines L30-31, L19-30, L14-19, L15-14, L15-12, L17-13, L17-16, and L16-29 is less than 0.5, so the above AC lines are sensitive and vulnerable. Since node 15 and node 17 are the central nodes of the three-winding transformer, and BESS is configured at the sending end of the AC line, the candidate nodes for BESS are 31, 19, 14, and 16.

Table 4. The improved power flow exceeding risk index of the shortest path.

AC Line	Initial Power/p.u.	Power after Fault/p.u.	LODF	The Improved Power Flow Exceeding Risk Index
L31-33	4.00	0.00	−1.00	1.67
L30-31	1.02	−2.98	−1.00	0.41
L19-L30	−1.20	−5.17	−1.00	0.70
L14-19	−0.34	−2.94	−0.82	0.42
L15-14	−0.34	−2.94	−0.65	0.32
L15-12	0.34	2.94	0.65	0.14
L12-27	4.86	7.18	0.58	1.29
L27-28	4.86	7.18	0.58	1.29
L13-28	−4.61	−7.04	−0.61	3.70
L17-13	−4.80	−7.04	−0.56	3.22
L17-16	4.79	7.01	0.56	0.36
L16-29	2.38	6.20	1.02	0.32
L29-L34	−3.82	0.00	−0.96	2.15

5.3. Optimization Configuration Results of Single BESS

To verify that the BESS is configured in the sensitive and vulnerable line, the BESS to improve the improved power flow exceeding risk index is the best. Firstly, the PSO algorithm is used to obtain a set of Pareto solutions of the BESS configuration capacity and location, as shown in Table 5.

Table 5. The single BESS configuration capacity and location.

ESS Power/MW	ESS Capacity/MW·h	ESS Location	Annual Investment Cost/Million Yuan	the Sum of the Improved Power Flow Exceeding Risk Index
541.15	541.15 × 2	31	443.47	24.18
276.31	276.31 × 2	16	226.44	18.29
391.88	391.88 × 2	16	321.15	19.47

From Table 5, there are two sets for the BESS configuration at node 16. There are two sets for the BESS configuration at node 16. The annual investment cost of BESS is 226.44 million yuan and 321.15 million yuan. Similarly, the BESS configuration also obtains the optimal solution at 31 nodes; the annual investment cost of BESS is 226.44 million yuan and 443.47 million yuan. Therefore, we obtained the lowest annual investment costs in Pareto solutions in Table 2.

To verify that the PSO algorithm has obtained the optimal solution, the iterative convergence curve comparison between PSO and GA (Genetic algorithm) is shown in Figure 6. According to the iterative convergence curve of Figure 6, it can conclude that the PSO has fewer convergence times, more effectively avoid local convergence, and has better stability than the GA. We know that the PSO algorithm can obtain the optimal solution.

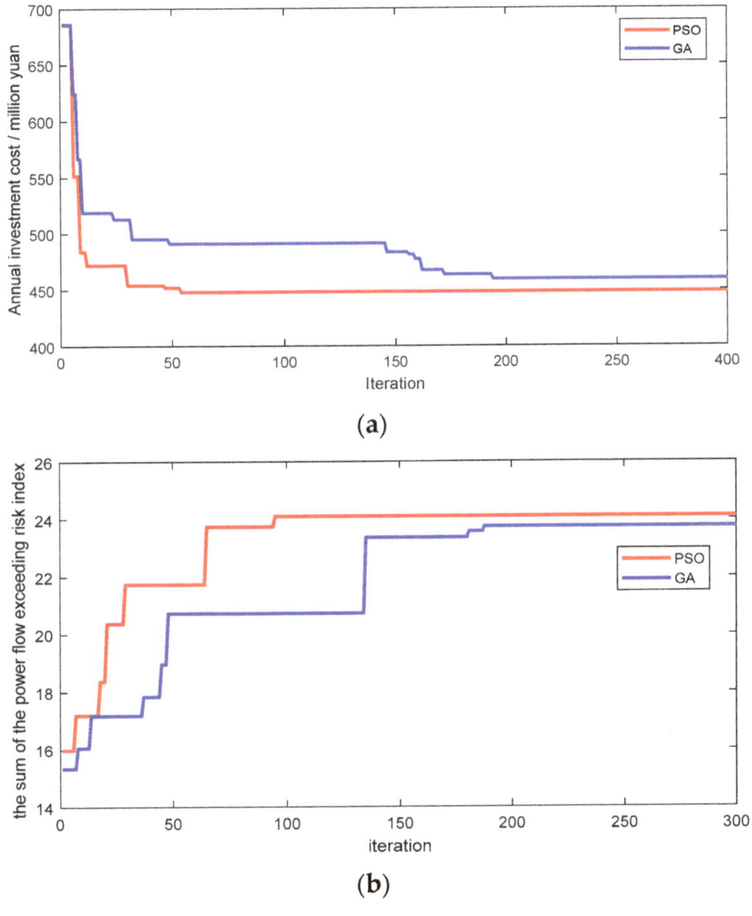

Figure 6. The iterative convergence curve. (**a**) Iterative convergence curve of energy storage annual investment cost. (**b**) Iterative convergence curve of the sum of improved power flow exceeding the risk index.

Then, the following three scenarios are compared and analyzed: (1) the BESS is configured in a node in the sensitive and vulnerable line, and 31 nodes are selected in this paper; (2) the BESS is configured in the nodes in the shortest path except for the sensitive and vulnerable lines. This paper selects 27 nodes; (3) 21 nodes are selected for other nodes with BESS configured outside the shortest path. Under the three conditions, each node is connected to 541.15 MW of BESS, and the improved power flow exceeding risk index of sensitive and vulnerable AC lines is shown in Tables 6–8.

Table 6. The improved power flow exceeding risk index of the AC line with vulnerability after node 31 is connected to BESS.

AC Line	Initial Power/p.u.	Power after Fault/p.u.	LODF	The Improved Power Flow Exceeding Risk Index
L30-31	1.02	−1.68	−0.68	−0.71
L14-19	−0.34	−1.3	−0.41	−0.83
L15-14	−0.34	−1.3	−0.24	−0.87
L15-12	0.34	1.3	0.24	0.39
L17-16	4.79	6.37	0.39	0.51
L16-29	2.38	6.79	1.1	0.29

Table 7. The improved power flow exceeding risk index of the AC line with vulnerability after node 27 is connected to BESS.

AC Line	Initial Power/p.u.	Power after Fault/p.u.	LODF	The Improved Power Flow Exceeding Risk Index
L30-31	1.02	−2.98	−1	−0.48
L14-19	−0.34	−4.17	−1.13	−0.3
L15-14	−0.34	−4.17	−0.96	−0.22
L15-12	0.34	4.17	0.96	0.1
L17-16	4.79	6.36	0.39	0.51
L16-29	2.38	6.69	1.08	0.3

Table 8. The improved power flow exceeding risk index of the AC line with vulnerability after node 21 is connected to BESS.

AC Line	Initial Power/p.u.	Power after Fault/p.u.	LODF	The Improved Power Flow Exceeding Risk Index
L30-31	1.02	−2.98	−1	−0.48
L14-19	−0.34	−1.69	−0.51	−0.67
L15-14	−0.34	−1.69	−0.34	−0.61
L15-12	0.34	1.69	0.34	0.27
L17-16	4.79	6.9	0.53	0.38
L16-29	2.38	7.03	1.16	0.28

It can be seen from Tables 6–8 that after the BESS is arranged at the 31 nodes of the sensitive and vulnerable line, the risk index of tidal current out-of-limit of lines L31-30, 19-14, L14-15 and L17-16 exceeds 0.5, and only the risk index of tidal current out-of-limit of lines L15-12 and L16-29 is lower than 0.5; After the BESS is arranged at 27 nodes, only the improve power flow exceeding risk index of line L17-16 exceeds 0.5, and the improve power flow exceeding risk index of other lines does not exceed 0.5; After the BESS is configured at node 21, the improve power flow exceeding risk index of lines 19-14 and L14-15 exceeds 0.5, and the improve power flow exceeding risk index of other lines does not exceed 0.5; It can be seen that the BESS configuration on the sensitive and vulnerable lines has greatly improved the safe operation ability of AC lines.

It can be seen from Figures 7–9 that after the BESS is incorporated into 31 nodes, it is helpful to suppress the oscillation of line power, and the change of line power is lower than 1.5 p.u; after the BESS is incorporated into 27 nodes, the oscillation of line power is increased, and the power variation of some lines is greater than 1.5 p.u; after the BESS is connected to 21 nodes, the power of AC lines on the shortest path changes irregularly, and the power of some lines increases gradually after 17 s, greatly reducing the system stability. It can be seen that only when the BESS is connected to the sensitive and vulnerable lines that the safe and stable operation capacity of the AC lines can be effectively improved.

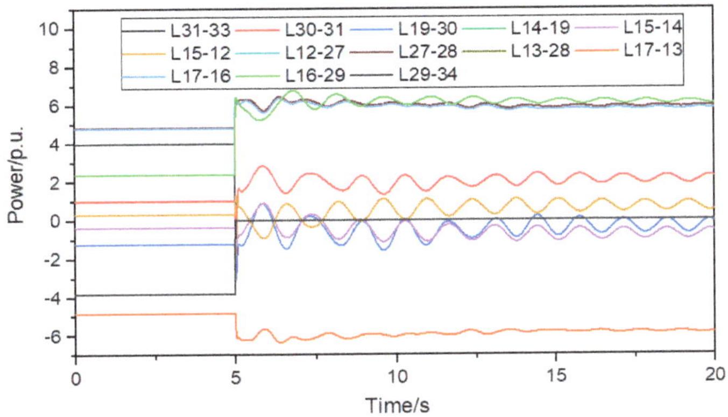

Figure 7. The shortest path power change curve after BESS is connected to node 31.

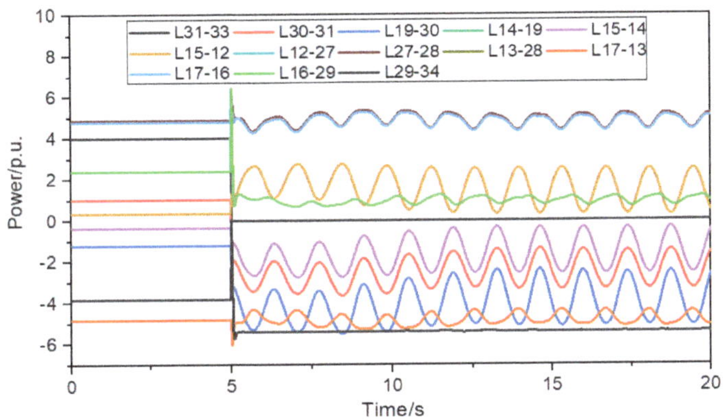

Figure 8. Shortest path power change curve after BESS access node 27.

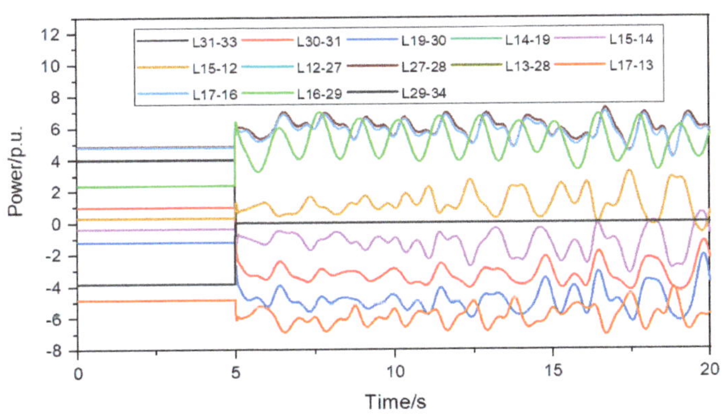

Figure 9. The shortest path power change curve after BESS is connected to node 21.

5.4. Optimization Configuration Results of Multi-BESS

When the system configures two and three BESS, a set of Pareto solutions is shown in Table 9. The improved power flow exceeding risk index of sensitive and vulnerable AC lines after configurations two and three BESS is shown in Tables 10 and 11. The shortest path power change curve after BESS is shown in Figures 10 and 11.

Table 9. The multi-BESS configuration capacity and location.

Nodes	BESS Power/MW	ESS Capacity/MW·h	BESS Location	Investment and Construction Costs/Million Yuan	the Sum of the Improved Power Flow Exceeding Risk Index
2 nodes	88.09	88.09 × 2	29	248.89	17.78
	299.25	299.25 × 2	31		
3 nodes	88.09	88.09 × 2	29	248.89	17.78
	11.87	11.87 × 2	19		
	287.38	287.38 × 2	31		

Table 10. The improved power flow exceeding risk index of sensitive and vulnerable AC lines after the configuration of two BESS.

AC Line	Initial Power/p.u.	Power after Fault/p.u.	LODF	The Improved Power Flow Exceeding Risk Index
L30-31	1.02	−2.98	−1	−1.38
L14-19	−0.34	−4.17	−1.13	−1.48
L15-14	−0.34	−4.17	−0.96	−0.5
L15-12	0.34	4.17	0.96	0.5
L17-16	4.79	6.36	0.39	0.58
L16-29	2.38	6.69	1.08	0.5

Table 11. The improved power flow exceeding risk index of sensitive and vulnerable AC lines after configuration three BESS.

AC Line	Initial Power/p.u.	Power after Fault/p.u.	LODF	The Improved Power Flow Exceeding Risk Index
L30-31	1.02	−2.98	−1	−1.24
L14-19	−0.34	−1.69	−0.51	−1.48
L15-14	−0.34	−1.69	−0.34	−0.5
L15-12	0.34	1.69	0.34	0.5
L17-16	4.79	6.9	0.53	0.58
L16-29	2.38	7.03	1.16	0.5

It can be seen from Table 6 that when two and three BESS are configured on sensitive lines, the total power of the BESS is the same, and the annual investment cost and the sum of improved power flow exceeding risk index are also the same. Moreover, the sum of the power of 19 nodes and 31 nodes, when three BESS are configured on sensitive lines, is equal to the sum power of configured two BESS. It can be seen from Tables 7 and 8 that the improved power flow exceeding risk index for sensitive and vulnerable AC lines is greater than 0.5, and when three BESS are configured, the impact on the improved power flow exceeding risk index is relatively small. It can be seen from Figures 9 and 10 that when two and three BESS are configured for sensitive lines, the power variation of the line is much lower than configuring one BESS for sensitive lines. And suppressing power oscillation is greatly improved. We know that configuring BESS for multiple nodes eliminating the impact of DC power transfer on AC lines is better than the BESS configured for a single node. Configuring BESS for multiple nodes to improve the safe and stable operation ability of AC lines is better than the BESS configured for a single node. Secondly, when the configured BESS quantity is greater than 2, it has a small impact on the sum of improved power flow exceeding risk index and annual investment cost.

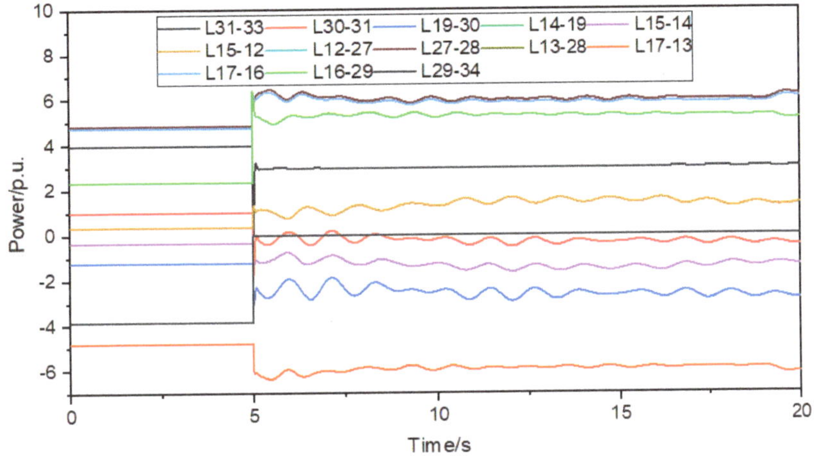

Figure 10. Shortest path power change curve after configuring 2 BESS.

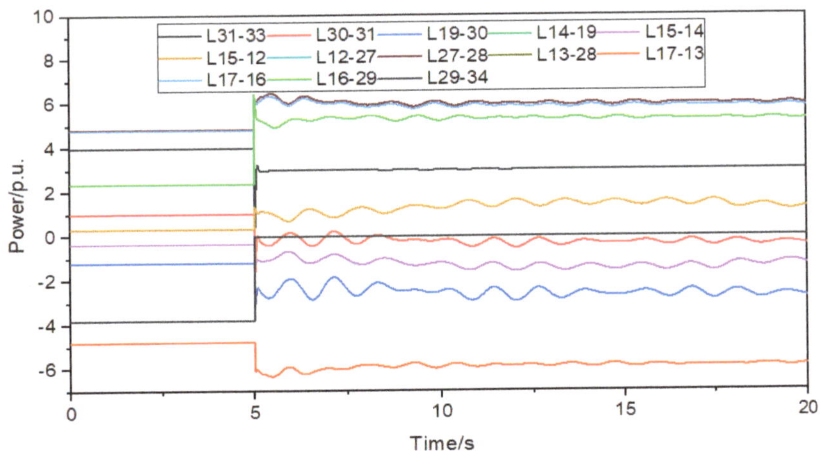

Figure 11. Shortest path power change curve after configuring 3 BESS.

6. Conclusions

The method of identifying the sensitive and vulnerable transmission lines based on improved power flow exceeding risk index is proposed, and it can apply to AC/DC hybrid systems, weak power grids, and other power systems. And the proposed method can quickly identify the sensitive and vulnerable transmission lines in the shortest path and simplify the method of calculating the branch disconnection coefficient. The method would not have to repeatedly calculate the impedance matrix of the line disconnection and connection.

The optimal energy storage configuration can be obtained by the multi-objective optimal mathematical model, including minimizing the annual investment cost of BESS and maximizing the sum of the improved power flow exceeding the risk index of the sensitive and vulnerable transmission lines. The BESS for sensitive and fragile lines can quickly eliminate the impact of DC power transfer on AC lines, and the BESS can also suppress power fluctuations and greatly improve the transient stability operation ability of the power grid.

The proposed energy storage configuration method does not only improve the transient stability of the power grid. Simultaneously, during the stability power grid, the configured energy storage can actively supports voltage, frequency, etc. The configuration method proposed in this article enriches the application scenarios of energy storage. The safe and stable operation ability of the system is greatly improved.

In the future, we will study the HIL implementation of the proposed system along with the proposed optimal allocation method. After this, the fault disturbance, power system planning BESS resources will be explored through the robust optimization-based model.

Author Contributions: Conceptualization, L.J. and T.Z.; investigation, Y.T., L.J., B.Z., X.S. and Y.X.; resources, B.Z., X.S. and Y.X.; data curation, L.J. and T.Z.; writing—original draft preparation, L.J. and T.Z.; writing—review and editing, L.J., T.Z. and S.M.; supervision, T.Z. and S.M.; project administration, Y.T.; funding acquisition, Y.T. All authors have read and agreed to the published version of the manuscript.

Funding: This research was funded by the Technology Project of State Grid Sichuan Electric Power Company (No. B1199721009N).

Data Availability Statement: The data presented in this study are available on request from the corresponding author.

Conflicts of Interest: The authors declare no conflict of interest.

References

1. ENTSO-E. *System Separation in the Continental Europe Synchronous Area on 8 January 2021—Interim Report[R/OL]*; ENTSO-E: Geneva, Switzerland, 26 February 2021. Available online: https://www.entsoe.eu/news/2021/02/26/system-separation-in-the-contnetal-europe-synchronous-area-on-8-january-2021-interim-report/ (accessed on 14 June 2023).
2. California ISO. *AWE Grid History Report[R/OL]*; California ISO: Folsom, CA, USA, 9 September 2020. Available online: http://www.caiso.com/Documents/AWE-Grid-History-Report-1998-Present.pdf (accessed on 14 June 2023).
3. Wang, S.; Zhou, L.; Wang, T.; Chen, T.; Wang, Y. Fast protection strategy for DC transmission lines of MMC-based MT-HVDC grid. *Chin. J. Electr. Eng.* **2021**, *7*, 83–92. [CrossRef]
4. Liu, Y.; Shen, C. Review and research on construction methods of the energy function for hybrid AC-DC power system. *Proc. CSEE* **2022**, *42*, 2842–2853.
5. Xin, B.; Guo, M.; Wang, S.; Li, X. Friendly HVDC transmission technologies for large-scale renewable energy and their engineering practice. *Autom. Electr. Power Syst.* **2022**, *45*, 1–8.
6. Zheng, C.; Ma, S.; Shen, X.; Liu, D. Definition, connotation and form of strong HVDC and weak AC and countermeasures for stable operation of the hybrid power grid. *Power Syst. Technol.* **2017**, *14*, 2491–2498.
7. Hines, P.; Balasubramaniam, C.; Sanchez, C. Cascading failures in power grids. *IEEE Potentials* **2009**, *28*, 24–30. [CrossRef]
8. Li, J.; Wang, J.; Ge, L.; Meng, G.J.; Yuan, X.D.; Zhou, J.H. Review on application technology of electrochemical energy storage power station group in ultrahigh voltage AC/DC hybrid receiver power grid. *Proc. CSEE* **2020**, *46*, 51–61.
9. Ma, Z.; Liu, F.; Shen, C.; Zhang, S.; Tian, B. Rapid identification of vulnerable lines in power grid using modified PageRank algorithm—Part I: Theoretical foundation. *Proc. CSEE* **2016**, *36*, 6363–6370+6601.
10. Lijie, D.; Meijun, L.; Yijia, C. Power system key-lines identification based on hidden failure model risk theory. *Autom. Electr. Power Syst.* **2007**, *31*, 1–6.
11. Li, Y.; Liu, J.; Liu, X.; Jiang, L.; Wei, Z.; Xu, W. Vulnerability assessment in power grid cascading failures based on the entropy of power flow. *Autom. Electr. Power Syst.* **2012**, *36*, 11–16.
12. Ren, J.; Wei, J. Vulnerableline identification and its transmission section search for power system. *Electr. Power Autom. Equip.* **2015**, *35*, 37–42.
13. Ruo-fa, C.; Xin, F. Fast search of transmission section in power flow transferring resulting from simultaneous multi-branch removal. *Adv. Technol. Electr. Eng. Energy* **2020**, *39*, 62–71.
14. Naziri Moghaddam, I.; Chowdhury, B.; Mohajeryami, S. Predictive operation and optimal sizing of battery energy storage with high wind energy penetration. *IEEE Trans. Ind. Electron.* **2017**, *65*, 6686–6695.

15. Luo, F.; Meng, K.; Dong, Z.Y.; Zheng, Y.; Chen, Y.; Wong, K.P. Coordinated operational planning for wind farm with battery energy storage system. *IEEE Trans. Sustain. Energy* **2015**, *6*, 253–262. [CrossRef]
16. Xiaowei, D.; Guiping, Z.; Liangzhong, Y. Two-Stage optimization of battery energy storage capacity to decrease wind power curtailment in grid-connected wind farms. *IEEE Trans. Power Syst.* **2018**, *33*, 3296–3305.
17. Sun, Y.S.; Tang, X.S.; Sun, X.Z.; Jia, D.Q.; Zhang, G.W.; Wang, P. Research on energy storage capacity allocation method for smoothing wind power fluctuations. *Proc. CSEE* **2017**, *37*, 88–97.
18. Jiang, T.; Li, X.; Chen, H.; Zhang, R.; Li, G.; Kou, X.; Wang, M.; Li, F. Optimal energy storage siting and sizing to mitigate voltage deviation in distribution networks. In Proceedings of the 2019 IEEE Power & Energy Society General Meeting (PESGM), Atlanta, GA, USA, 4–8 August 2019; IEEE: Piscateville, NJ, USA, 2020; pp. 1–5.
19. Tang, J.; Li, X.R.; Huang, J.Y.; Xu, P.; He, C. Capacity allocation of BESS in secondary frequency regulation with the goal of maximum net benefit. *Trans. China Electrotech. Soc.* **2019**, *34*, 963–972.
20. Zheng, L.; Hu, W.; Lu, Q.; Min, Y. Optimal energy storage system allocation and operation for improving wind power penetration. *IET Gener. Transm. Distrib.* **2015**, *9*, 2672–2678. [CrossRef]
21. Hou, L.; Chen, Y.; Su, K.; Wang, Y.; Zhang, Z. Research on bi-level planning method of active distribution network considering flexible interconnection of source network load and storage. In Proceedings of the 2019 IEEE Sustainable Power and Energy Conference, Beijing, China, 21–23 November 2019; IEEE: Piscateville, NJ, USA, 2020.
22. Zhang, Y.; Dong, Z.; Luo, F.; Zheng, Y.; Meng, K.; Wong, K.P. Optimal allocation of battery energy storage systems in distribution networks with high wind power penetration. *IET Renew. Power Gener.* **2016**, *10*, 1105–1113. [CrossRef]
23. Xiu, X.; Tang, W.; Li, J.; Tian, C. Collaborative Configuration of distributed generation, energy storage and load in microgrid considering state of health. *High Volt. Eng.* **2017**, *43*, 3118–3126.
24. Wang, S.; Li, F.; Zhang, G.; Yin, C. Analysis of energy storage demand for peak shaving and frequency regulation of power systems with high penetration of renewable energy. *Energy* **2023**, *267*, 126586.
25. Song, T.; Li, T.H.X. Coordinated operation strategy of energy storage system participating in multiple application scenarios. *Autom. Electr. Power Syst.* **2021**, *45*, 43–51.
26. Yuan, W.; Wang, J.; Qiu, F.; Chen, C.; Kang, C.; Zeng, B. Robust optimization-based resilient distribution network planning against natural disasters. *IEEE Trans. Smart Grid* **2016**, *7*, 2817–2826. [CrossRef]
27. Shen, Z.; Sun, F.; Dan, Y.; Dai, P.; Song, K.; Wang, L. Optimal Planning of Distributed Energy Storage in Distribution Network Considering the Influence of Typhoon-like Extreme Weather. In Proceedings of the 2021 3rd International Academic Exchange Conference on Science and Technology Innovation (IAECST), Guangzhou, China, 10–12 December 2021; IEEE: Piscateville, NJ, USA, 2021; pp. 1882–1886.
28. Nazemi, M.; Moeini-Aghtaie, M.; Fotuhi-Firuzabad, M.; Dehghanian, P. Energy storage planning for enhanced resilience of power distribution networks against earthquakes. *IEEE Trans. Sustain. Energy* **2020**, *11*, 795–806. [CrossRef]
29. Haibo, Z.; Shentong, M.; Xin, C. Energy storage planning method of distribution network to ensure uninterrupted power supply of important loads. *Power Syst. Technol.* **2021**, *45*, 259–268.
30. Liu, Z.; Tan, Q.; Zhou, Y.; Xu, H. Syncretic application of IBAS-BP algorithm for monitoring equipment online in power system. *IEEE Access* **2021**, *9*, 2169–3536. [CrossRef]
31. Al Ahmad, A.K.; Sirjani, R. Optimal allocation of energy storage system in transmission system considering wind power. In Proceedings of the 2020 7th International Conference on Electrical and Electronics Engineering, Antalya, Turkey, 14–16 April 2020; IEEE: Piscateville, NJ, USA, 2020; pp. 181–187.
32. Shu, H.; Shen, X.; Xu, L.; Guo, Q.; Sun, H. A validity test method for transmission betweens and transmission sections based on chain attack analysis and line outage distribution factors. In Proceedings of the 2018 2nd IEEE Conference on Energy Internet and Energy System Integration (EI2), Beijing, China, 20–22 October 2018; pp. 1–6.
33. Clerc, M.; Kennedy, J. The particle swarm-explosion, stability, and convergence in a multidimensional complex space. *IEEE Trans Evol. Comput.* **2002**, *6*, 58–73. [CrossRef]
34. Ozcan, E.; Mohan, C. Particle swarm optimization: Surfing the waves. In Proceedings of the Congress on Evolutionary Computation, Piscataway, NJ, USA, 6–9 July 1999; Volume 1, pp. 1939–1944.
35. Li, Y.; Shuai, Z.; Liu, X.; Hong, Y.; Wu, X.; Shen, Z.J. Study on damping low frequency oscillation of an AC/DC power grid based on UPFC & DC-SDC. *Power Syst. Prot. Control* **2020**, *48*, 73–79.

sustainability

MDPI

Article

A Causal Relationship between the New-Type Urbanization and Energy Consumption in China: A Panel VAR Approach

Cheng Chen [1,*], Yajie Gao [1] and Yidong Qin [2]

[1] School of Business, Hunan University of Science and Technology, Xiangtan 411201, China; 21301503001@mail.hnust.edu.cn
[2] School of Economics, Wuhan University of Technology, Wuhan 430070, China; qyd@whut.edu.cn
* Correspondence: 20301503001@mail.hnust.edu.cn

Abstract: The accelerated urbanization process has been considered to be the root cause of increasingly severe energy consumption growth in China. However, energy is still an essential factor for the urbanization process, so arbitrarily mitigating energy use currently will unquestionably slow down the urbanization process. The principal contribution of this paper is to comprehensively analyze the dynamic interaction mechanism between the new-type urbanization and energy consumption, and further put forward a new idea of comparing the benefit of an increase in the level of new-type urbanization resulting from energy consumption and the negative externality of environmental damage related with energy consumption. This paper conducts an empirical study on the causal relationship between new-type urbanization and energy consumption using Chinese provincial administrative units from 1999 to 2020. And we find that new-type urbanization leads to energy consumption negatively and energy consumption leads to new-type urbanization positively for provinces in the eastern region. There is only a one-way effect of energy consumption on new-type urbanization for provinces in the central and northeastern regions, and there is negative feedback causality for provinces in the western region. Additionally, the benefit of an increase in the level of new-type urbanization resulting from energy consumption is larger than the negative externality of environmental damage related to energy consumption for provinces in the eastern, central, and northeastern regions, yet it is totally opposite for provinces in the western region. Finally, we propose some fruitful policy recommendations to construct new-type urbanization under the background of clear reduction targets for energy consumption in China.

Keywords: new-type urbanization; energy consumption; environmental Kuznets curve; PVAR approach

Citation: Chen, C.; Gao, Y.; Qin, Y. A Causal Relationship between the New-Type Urbanization and Energy Consumption in China: A Panel VAR Approach. *Sustainability* **2023**, *15*, 11117. https://doi.org/10.3390/su151411117

Academic Editors: Luis Hernández-Callejo, Jesús Armando Aguilar Jiménez, Carlos Meza Benavides and Sergio Nardini

Received: 9 May 2023
Revised: 4 July 2023
Accepted: 10 July 2023
Published: 17 July 2023

1. Introduction

When the "Reform and Opening-up" policy was put into practice in the year of 1978, China's urbanization experienced a steady upward trend [1,2]. According to the statistics from the National Bureau of Statistics (NBS), China's urbanization rate increased from 17.9% in 1978 to 65.2% in 2022. Nevertheless, much of China remains to be urbanized, especially the inland regions. As China has a larger population, millions of rural residents will move into urban areas every year during to-be urbanization process. Although China's urbanization greatly raises people's living standards, it has also been deemed to be the root cause of increasingly severe energy consumption growth; for example, the heavy use of fossil energy, which undoubtedly brings about a series of environmental pollution [3]. The total energy consumption in China even made up 23.6% of the global total in the year 2018 [4]. Specifically, urban areas solely account for 75.15% of total energy consumption in China [5]. As a responsible country, China has been determined to set clear reduction targets for energy consumption. Because the urbanization process in China is still accelerating at present, energy is still an essential factor for the urbanization process. Arbitrarily mitigating energy use currently will unquestionably slow down the urbanization process. Therefore, there are two kinds of opposite effects simultaneously, which are the benefit of an increase in the level of urbanization resulting from energy consumption and the negative externality of environmental damage related to energy consumption respectively. Could the advantages sufficiently offset the disadvantages? Investigating these two kinds of opposite effects has vital theoretical and practical significance for promoting the urbanization process under the background of clear reduction targets for energy consumption in China. If the advantages of energy consumption outweigh its disadvantages, it pays to promote the urbanization process by increasing energy consumption. However, if energy consumption does not promote or even adversely affect the urbanization process, an energy conservation policy should be adopted to offset the negative externality of environmental damage related to energy consumption.

Previous studies have extensively investigated this subject, yet the majority emphasized the one-way effect of urbanization on energy consumption in the beginning. Many scholars found that the urbanization process leads to energy consumption growth [6–9]. In contrast, some scholars argued that urbanization may lead to energy consumption negatively [10–16]. Based on Northam [17], the third strand of research further examined the nonlinear relationship between the two due to the mixed findings. Most scholars confirmed that the nexus of the two is indeed nonlinear [2,18–20]. Another branch of the literature used city size as a proxy variable for urbanization rate, which indicated that there also was a nonlinear relationship between city size and energy consumption [21,22]. There are two kinds of explanations for the nonlinear relationship between the two. On the one hand, the essence of urbanization can be ascribed to the agglomeration effect, scale effect, and spatial spillover effect, which are conducive to reducing the energy consumption of residents or increasing energy efficiency [5,14,23–25]. On the other hand, the urbanization process reduces energy consumption through industrial structure upgrading and technical innovation [19,26–28].

The level of urbanization is mainly measured by the single index method in the above studies. These simple indicators can only mirror the level of population-oriented urbanization rather than the improvement of production and lifestyle. Especially after China issued the "National New-type Urbanization Plan" in 2014, human-centered urbanization has been put into practice. Therefore, investigating the relationship between new-type urbanization and energy consumption has much more practical significance currently. Recently, a few scholars used the comprehensive index method to measure its connotation and further examine its effect on energy consumption. Liu et al. [14] used a spatial econometric model for China's regions on this subject and found that new-type urbanization leads energy consumption negatively, yet its effect on adjacent areas or the spatial spillover effect is positive. Lin and Zhu [4] examined the effect of new-type urbanization on energy saving and its transmission channels based on Chinese cities and found that it can bring about an energy-saving effect. Yu [29] examined the ecological effect of new-type urbanization and found that China's new-type urbanization can improve energy efficiency. Feng et al. [30] examined the effect of new-type urbanization on energy efficiency based on Chinese cities and found that it has a double-threshold effect. Shao and Wang [31] examined the effect of new-type urbanization on green total factor energy efficiency and found that it has a heterogeneous effect for different cities. Not surprisingly, the relationship between the two is much more complicated compared to traditional urbanization.

Apart from that, as an essential factor of economic development, energy consumption is also conducive to promoting the level of urbanization. Ghosh and Kanjilal [32] investigated the cointegration relationship between the two for India and found that there is causality running from energy consumption to urbanization. Wu et al. [2] estimated the direct effects of various energy consumption patterns in China and found that energy consumption leads to urbanization positively, and the positive effect is dependent on energy consumption intensity, energy consumption scale, and energy consumption structure. Xu and Wang [33] examined the threshold effect of energy consumption on new-type urbanization in China and found that there was a significant threshold effect. Some scholars further explored how energy consumption affected the urbanization process. These studies found that the carbon emission reduction effect [34], agglomeration economy effect and economies of scale [32,35,36], and the industrial structure effect [36–38] are primary transmission channels through which energy consumption affects urbanization.

The above studies in this field indicate that there should be a bi-directional causality between the two. So far, a wealth of studies primarily investigated uni-directional causality on this subject, yet the bi-directional causality between the two is still scarce. Comparatively, a large body of studies proved that economic growth and energy consumption present a bi-directional causal relationship [39–44]. As urbanization is widely considered a symbol of economic development [37–39], an abundance of support can be indirectly found for the bi-directional causality relationship on our subject. To the best of our knowledge, only Tang et al. [45] explored the two-way correlation mechanism between new-type urbanization and clean energy consumption based on Chinese provincial data and found that there is a significant two-way promoting effect between the two. To sum up, the extant literature in this field actually denotes that energy consumption probably affects urbanization by means of its effects on economic development.

Despite the existing studies in this field having explored extensively the relationship between the two, there are still a few drawbacks on this subject. Firstly, the bulk of empirical studies simply examined how urbanization affects energy consumption, or whether energy consumption promoted the level of urbanization, and these empirical results did not compare the benefit of an increase in the level of urbanization resulting from energy consumption and the negative externality of environmental damage related to energy consumption, so they cannot provide corresponding policy implications for promoting the urbanization process. Secondly, there should be a bi-directional causality between the two, and the existing studies mainly examined the one-way effect of urbanization on energy consumption. Thirdly, the empirical studies on this subject are generally conducted based on the linear relationship hypothesis, and the estimated results are always inconsistent. The nonlinear relationship hypothesis may be more realistic, especially for the effect of new-type urbanization, which remains to be further examined empirically. In this regard, our study makes the following contributions. Firstly, this paper comprehensively analyzes the dynamic interaction mechanism between new-type urbanization and energy consumption, aiming to reveal the bi-directional causality between the two and extend the depth and breadth of this subject. Secondly, this paper put forward a new idea of comparing the benefit of an increase in the level of new-type urbanization resulting from energy consumption and the negative externality of environmental damage related to energy consumption, which can provide targeted policy recommendations. Thirdly, this paper creatively explains the estimated results with the concept of the well-known Environmental Kuznets Curve (EKC) hypothesis based on some energy-related data, which not only adds new empirical evidence for the EKC relation but also provides a robustness test for our regression results given that new-type urbanization and energy consumption factually present the two-way causal relationship.

The remainder of this article is arranged as follows. The following section introduces the measurement methods and analysis of the measurement results; Section 3 outlines the econometric specification and presents empirical results; and Section 4 concludes.

2. Measurement Methods and Analysis of Measurement Results

2.1. Measurement Methods

The most critical things are methods for calculating the level of new-type urbanization and energy consumption in this paper. According to Ma et al. [46], the amount of energy consumption per capital is applied to measure the level of energy consumption. Similar to the existing studies [4,47], the composite index method is used to fully measure the level of new-type urbanization. The index system is made up of two levels, which include a total of 19 computable indices, as is shown in Table 1. To overcome some shortcomings of the subjective weighting method, the entropy method is employed to calculate the constructed comprehensive urbanization index. The "+" and "−" of Index Attributes in Table 1 signify the influence of 19 computable indices on the comprehensive index, and "+" indicates an increase in the metric of indices would promote the comprehensive index, and "−" indicates an increase in the metric of indices would decrease the comprehensive index.

Table 1. China's comprehensive urbanization index system. Reprinted from [48]. Copyright 5584080645167 (2023) with permission from Elsevier.

Index I	Index II	Index Attribute
Population	Urban population density	+
	Full-time equivalent of R&D personnel	+
	Urban population ratio	+
	Number of college degrees or above per ten thousand people	+
	Proportion of employed persons in the tertiary industry	+
Economy	Gross domestic products per capita	+
	Consumption proportion of urban to rural residents	−
	Disposable income of urban household per capita	+
	Tertiary industry as a percentage of regional GDP	+
Living environment	Urban wastewater treatment ability per day	−
	Green covered area as a percentage of completed area	+
	Greenery area of per capital park	+
	Area under a cleaning program per square kilometer of built-up area	+
Living conditions	Urban gas access rate	+
	Urban water access rate	+
	Number of public toilets per ten thousand people	+
	Number of public transportation vehicles per ten thousand people	+
	Urban per capita area of paved roads	+
	Number of patent grants per ten thousand people	+

The sample data of our empirical research spans 22 years, from 1999 to 2020, and includes all provincial administrative units in China. The original data for energy consumption are obtained from the China Energy Statistical Yearbook and the China Statistical Yearbook. All relevant data for calculating the comprehensive urbanization index are obtained from the China Statistical Yearbook, China Energy Statistical Yearbook, China City Statistical Yearbook, China Statistical Yearbook on Science and Technology, each provincial statistical yearbook, and so on. Considering the data unavailability, Tibet, Hong Kong, Macao, and Taiwan are deleted from the sample in our empirical study. Consequently, the research sample finally consists of 30 provincial administrative units. And these 30 provincial administrative units can be classified into four categories according to the National Bureau of Statistics: east, northeast, central, and west. So as to cancel the impact of the price level in different years, all data related to nominal GDP are revised to a constant price based on the 1999 price index in the process of computation.

2.2. *Analysis of Measurement Results*

According to the specific calculation method introduced above, the provincial level of energy consumption in China from 1999 to 2020 is calculated firstly. To visually describe the characteristics of the temporal–spatial evolution in the provincial level of energy consumption, the measurement results are reported in the form of topographic maps. Figure 1 shows the concrete results. Due to space constraints, the results from 1999, 2007, 2013, and 2020 are only displayed. It can be clearly seen in Figure 1 that the dark blue areas represent the highest level of energy consumption in 30 provincial administrative units. Obviously, the number of dark blue regions increased from zero in 1999 to ten in 2020. For the convenience of discussions, all provincial administrative units are called "province". Therefore, we can conclude from these results that the provincial level of energy consumption in China is on the rise over the study period [43,49]. And in 2020, the provincial level of energy consumption demonstrated visible spatial differentiation. Among the provinces with the highest level of energy consumption, four provinces lie in the western region: Xinjiang, Qinghai, Gansu, and Inner Mongolia. Four provinces, Beijing, Shanghai, Hebei, and Jiangsu, are in the eastern region. Only one province, Shaanxi, lies in the central region, and only one province, Liaoning, lies in the northeastern region. Generally, the western region has the highest level of energy consumption among the four types of regions, followed by the eastern region. And the rest of the two regions have comparatively lower levels of energy consumption. The provincial level of energy consumption from the eastern to the central, northeastern, and western regions is similar to the U-shaped curve [49]. The spatial characteristics of the provincial level of energy consumption imply preliminarily that it is necessary to give thought to regional heterogeneity on this subject.

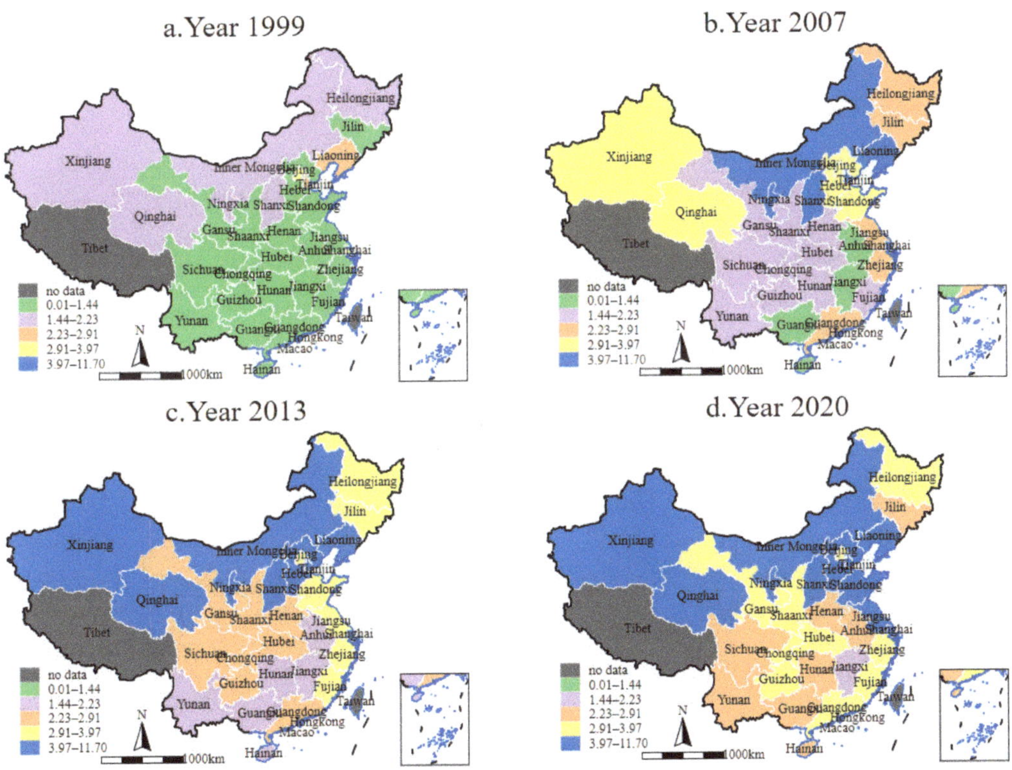

Figure 1. Chinese provincial level of energy consumption for some years (million tons).

Then, the comprehensive urbanization index is calculated by the widely used entropy method. To visually describe the characteristics of the temporal–spatial evolution of this index, the measurement results are reported similarly in the same way. Figure 2 shows the concrete results. Due to space constraints, the results from 1999, 2007, 2013, and 2020 are only displayed. It can be clearly seen in Figure 2 that the dark blue areas represent the highest level of new-type urbanization in 30 provincial administrative units. Obviously, the number of dark blue regions is zero in 1999, yet nearly all provinces in the eastern and central regions belonged to the highest group in 2020. Therefore, we can conclude from these results that the provincial level of new-type urbanization has also increased greatly over the study period, especially after the year 2013 [2,4,49]. On the one hand, the main reason for these time series characteristics can be probably ascribed to the "National New-type Urbanization Plan", which makes the economy, society, and ecology balanced in the process of urbanization. On the other hand, the increase in the level of energy consumption may also be conducive to speeding up the process of urbanization. And in 2020, the provincial level of new-type urbanization has also displayed significant spatial differentiation. Specifically, the majority of provinces have achieved the highest status in 2020. And only a few provinces have not reached the highest level, these are Xinjiang, Jilin, Shanxi, Gansu, Guizhou, Qinghai, Yunnan, and Guangxi. Additionally, these provinces mainly lie in the western and central regions. The average comprehensive urbanization index of the provinces in the eastern, central, northeastern, and western regions is calculated to be 0.299, 0.169, 0.188, and 0.159, respectively, which verifies that the provincial level of new-type urbanization looks like the inverted S-shaped curve, which is slightly different with the provincial level of energy consumption [46,47]. Therefore, the spatial characteristics of the provincial level of new-type urbanization further imply that it is necessary to give thought to regional heterogeneity on this subject.

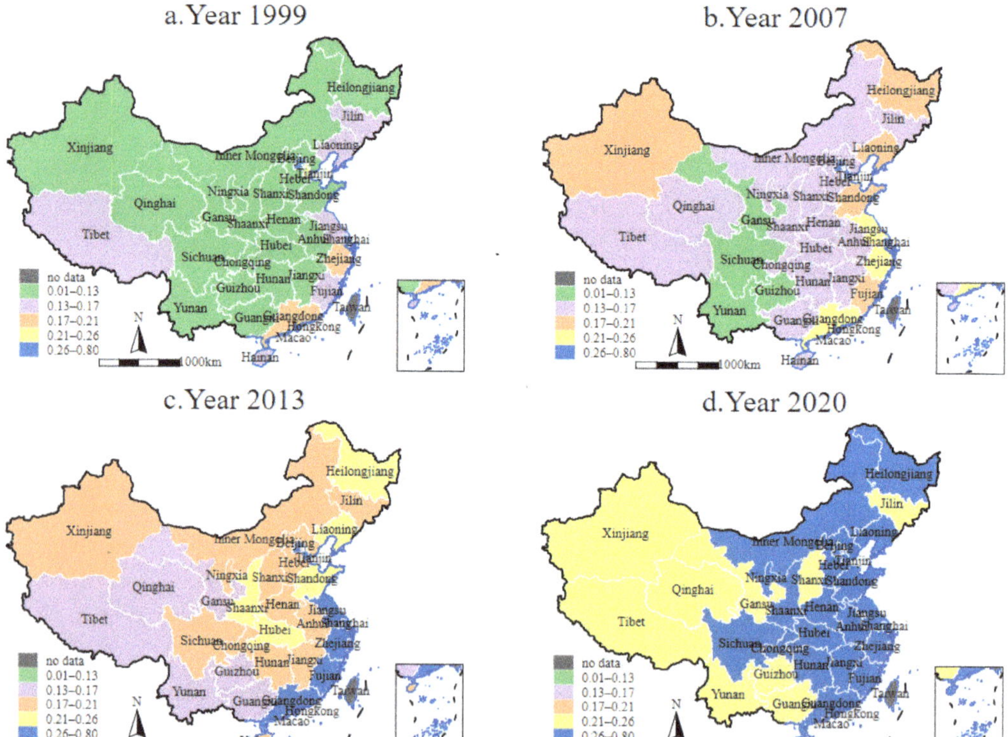

Figure 2. Chinese provincial level of new-type urbanization for some years.

To display the tendency intuitively, the growth rate of the above measured variables is depicted with a scatter diagram in Figure 3 for the four groups, eastern (Figure 3a), central (Figure 3b), northeastern (Figure 3c), and western regions (Figure 3d), respectively. In Figure 3, the relationship between comprehensive urbanization index growth and energy consumption growth is significantly different in the four groups. Specifically, there seems to be a strong relationship between the northeast and western regions, yet no such relationship emerges in the eastern and central regions. These results further prove that we cannot neglect the "one size for all" homogeneity issue among provinces when investigating the causal relationship between the two, and it is reasonable to classify the data into four groups: east, northeast, central, and west. Secondly, the weak relationship between the two in the eastern and central regions denotes there may be a nonlinear relationship between comprehensive urbanization index growth and energy consumption growth, and the traditional linear model is not suitable for this subject. Nevertheless, it should be noted that these kinds of scatter charts roughly reflect the possible correlation relationship between the two, and the exact relationship between the two remains to be confirmed rigorously by employing reasonable econometric models.

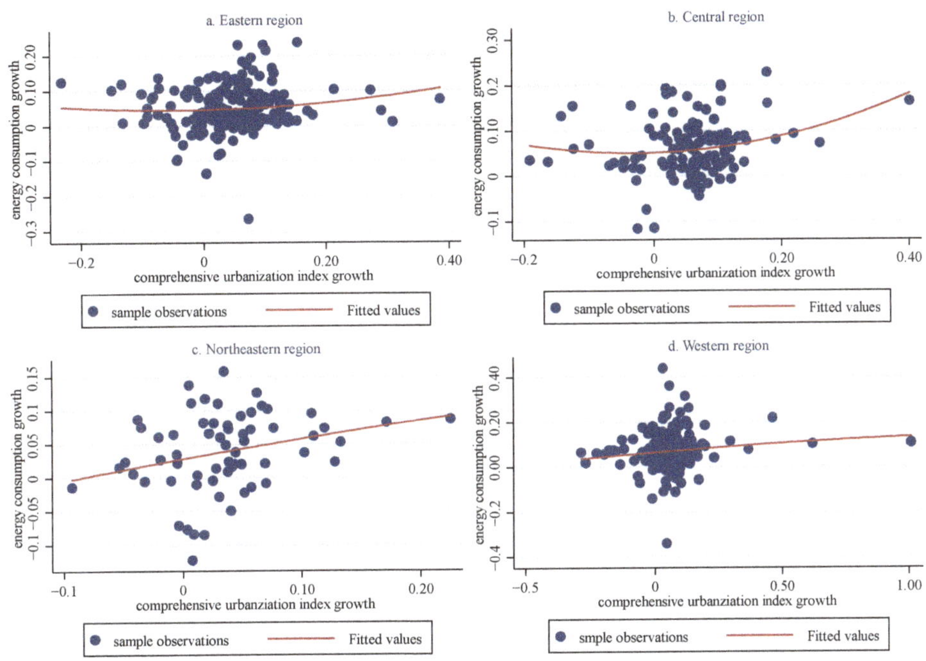

Figure 3. Comprehensive urbanization index growth vs. energy consumption growth.

3. PVAR Model Regression Results and Analysis

3.1. Model Specification

So as to fully reveal the causal on this subject, the panel vector autoregression (PVAR) model is employed to conduct empirical research. Compared with the widely used vector autoregression (VAR) model, the PVAR model has the advantage of dealing with long-term panel data and endogenous causality in the traditional linear regression model. The corresponding PVAR model is constructed as follows:

$$Y_{it} = \Gamma_0 + \sum_{p=1}^{n} \Gamma_p Y_{it\text{-}p} + \delta_i + f_t + \varepsilon_{it}$$

where the subscripts i and t denote province and year, respectively; Y_{it} is a multi-dimensional variable, which is *NU* and *EC*, representing the level of new-type urbanization and energy consumption, respectively, measured by the methods introduced hereinabove; $Y_{it\text{-}p}$ is a p-period lag term of Y_{it}; and δ_i and f_t indicate individual province fixed effects and year fixed effects, respectively. ε_{it} is termed the random disturbance.

As the PVAR model includes the lag term of dependent variables and individual province fixed effects, it can be considered a typical dynamic panel data model. Thus, the

OLS estimation will be biased and inconsistent. Therefore, the system GMM proposed by Blundell and Bond [49] is used to estimate the PVAR model. The PVAR model is commonly conducted by the following steps [50,51]. Firstly, the stationarity of the panel data and the causal relationship between *NU* and *EC* have to be tested. Secondly, the optimal lag order needs to be selected, and the PVAR model can be estimated logically. Thirdly, the impulse response graph will be displayed based on the PVAR model regression. The last step is the variance decomposition.

3.2. *PVAR Model Regression Analysis*

3.2.1. Stationarity Tests

First of all, it is necessary to inspect the stationarity of time series variables. The IPS, Fisher-ADF, and Fisher-PP are comprehensively used to implement unit root tests, and the results are reported in Table 2. In Table 2, the *p*-values of all tests for the first-difference series of *NU* and *EC* are all 0.000, which rejects the null hypothesis at the 1% significance level. Hence, we can conclude that the first-difference series of *NU* and *EC* are stationary. In other words, these original data are integrated processes of order one.

Table 2. Unit root test results.

Variables	IPS		Fisher-ADF		Fisher-PP	
	Statistics	***p*-Value**	**Statistics**	***p*-Value**	**Statistics**	***p*-Value**
NU	−3.272	0.001	107.964	0.000	39.533	0.988
NU_{-1}	−13.091	0.000	290.177	0.000	615.121	0.000
EC	−1.569	0.363	141.268	0.000	45.468	0.943
EC_{-1}	−10.008	0.000	219.605	0.000	290.094	0.000

Notes: The null hypothesis is that the time series variables have a unit root process.

3.2.2. Benchmark Regression Results

Before estimating the PVAR model, an optimal lag *p*-period of time series variables remains to be explored. Based on the standard procedure in empirical studies, the AIC, BIC, and HQIC are used to select the optimal lag order. The selection of two lag periods is reasonable. Therefore, 2 is the optimal lag order in this paper. The estimation results of the PVAR model are reported in Table 3. The *EC* equation reflects the effects of *EC* and *UN* on *EC*, and the *UN* equation reflects the effects of *UN* and *EC* on *UN*. The estimated results for all provinces as a whole are reported in the top line of Table 3. As shown in Table 3, in the *EC* equation, the first-period lag of *EC* has a significant positive effect on *EC*, and the second-period lag of *EC* has a significant negative effect on *EC*, which indicates that energy consumption presents the characteristics of path-dependent inertia in the short run, yet it tends to converge in the long run [4,47]. The first-period lag of *NU* has a significant negative effect on *EC* and the second-period lag of *NU* has a significant positive effect on *EC*, which indicates that new-type urbanization leads to energy consumption negatively in the short run [10–16], yet the new-type urbanization leads to energy consumption positively in the long run [6–9]. In the *NU* equation, the first-period lag of *NU* has a significant positive effect on *NU*, and the second-period lag of *NU* does not have any significant effect on *NU*, which indicates that new-type urbanization also presents the characteristics of path-dependent inertia in the short run. The first-period lag of *EC* has a significant positive effect on *NU* [33], and the second-period lag of *EC* has a significant negative effect on *NU* [45], which indicates that an increase in energy consumption brings about a further increase in the level of new-type urbanization in the short run, yet it is detrimental to new-type urbanization in the long run.

Table 3. Estimation results of the PVAR model.

Sample	Type	Variable	Coefficient	Variable	Coefficient
Countrywide	*EC* equation	EC_{-1}	1.553 *** (14.04)	NU_{-1}	−1.279 ** (2.00)
		EC_{-2}	−0.513 *** (−8.32)	NU_{-2}	1.138 * (1.81)
	NU equation	NU_{-1}	1.342 *** (4.84)	EC_{-1}	0.021 ** (2.37)
		NU_{-2}	−0.257 (−0.84)	EC_{-2}	−0.022 *** (−2.58)
Eastern	*EC* equation	EC_{-1}	1.492 *** (15.46)	NU_{-1}	−0.879 * (−1.80)
		EC_{-2}	−0.465 *** (−5.45)	NU_{-2}	0.638 (1.18)
	NU equation	NU_{-1}	1.083 *** (5.09)	EC_{-1}	0.042 *** (4.42)
		NU_{-2}	0.008 (0.04)	EC_{-2}	−0.035 *** (−4.80)
Central	*EC* equation	EC_{-1}	1.451 *** (11.31)	NU_{-1}	−1.188 (−1.40)
		EC_{-2}	−0.495 *** (−4.21)	NU_{-2}	1.321 * (1.69)
	NU equation	NU_{-1}	1.507 *** (7.37)	EC_{-1}	0.031 ** (2.35)
		NU_{-2}	−0.448 ** (−2.02)	EC_{-2}	−0.027 ** (−2.02)
Northeastern	*EC* equation	EC_{-1}	1.689 *** (10.52)	NU_{-1}	−10.142 (−1.35)
		EC_{-2}	−0.408 *** (−3.43)	NU_{-2}	5.158 * (1.87)
	NU equation	NU_{-1}	0.891 *** (3.59)	EC_{-1}	0.029 *** (2.81)
		NU_{-2}	0.189 (0.94)	EC_{-2}	−0.026 *** (−2.93)
Western	*EC* equation	EC_{-1}	1.508 *** (18.47)	NU_{-1}	−3.508 *** (−3.05)
		EC_{-2}	−0.415 *** (−5.21)	NU_{-2}	1.462 * (1.95)
	NU equation	NU_{-1}	1.339 *** (4.57)	EC_{-1}	0.010 (2.37)
		NU_{-2}	−0.284 (−0.97)	EC_{-2}	−0.008 * (−1.66)

Notes: ***, **, and * show significant levels at 1%, 5%, and 10%, respectively. The value given in parentheses is *t* statistics. The subscripts $_{-1}$ and $_{-2}$ represent the first-period lag and the second-period lag, respectively.

The estimated results for all provinces as a whole neglect regional heterogeneity. To explore this issue, groups of provinces are classified into four types of regions according to NBS. The estimated results for provinces in the eastern region are reported in the top second line of Table 3. As shown in Table 3, in the *EC* equation, the first-period lag of *EC* and the second-period lag of *EC* have a similar effect on *EC* compared to the whole country. The first-period lag of *NU* has a significant negative effect on *EC*, and the second-period lag of *NU* does not have any effect on *EC*, which indicates that new-type urbanization leads to energy consumption negatively in the short run [4,30], yet this inhibitory effect gradually disappears over time. In the *NU* equation, both *NU* and *EC* have a similar effect on *UN* compared to the whole country.

The estimated results for provinces in the central region are reported in the top third line of Table 3. As shown in Table 3, in the *EC* equation, the first-period lag of *EC* and the second-period lag of *EC* have a similar effect on *EC* compared to the whole country. Both the first-period lag of *NU* and the second-period lag of *NU* do not have any significant effect on *EC*, which indicates that an increase in the level of new-type urbanization does not bring about energy consumption [52]. In the *NU* equation, the first-period lag of *NU* has a significant positive effect on *NU*, and the second-period lag of *NU* has a significant negative effect on *NU*. The first-period lag of *EC* and the second-period lag of *EC* have a similar effect on *UN* compared to the whole country.

The estimated results for provinces in the northeastern region are reported in the fourth line of Table 3. As shown in Table 3, in the *EC* equation, the first-period lag of *EC* and the second-period lag of *EC* have a similar effect on *EC* compared to the whole country. Both the first-period lag of *NU* and the second-period lag of *NU* do not have any significant effect on *EC*, which also indicates that an increase in the level of new-type urbanization does not bring about energy consumption [52]. In the *NU* equation, both *NU* and *EC* have a similar effect on *UN* compared to the whole country.

The estimated results for provinces in the western region are reported in the fifth line of Table 3. As shown in Table 3, in the *EC* equation, the first-period lag of *EC* and the second-period lag of *EC* have a similar effect on *EC* compared to the whole country. The first-period lag of *NU* and the second-period lag of *NU* have a similar effect on *EC* as the eastern region, which indicates that new-type urbanization leads to energy consumption negatively in the short run [4,30], yet this inhibitory effect gradually disappears over time. In the *NU* equation, the first-period lag of *NU* and the second-period lag of *NU* have a similar effect on *NU* compared to the whole country. The first-period lag of *EC* does not have any effect on *UN*, and the second-period lag of *EC* has a significantly negative effect on *UN*, which indicates that an increase in energy consumption is detrimental to new-type urbanization.

3.2.3. Discussions with Concept of EKC

According to existing studies, energy consumption may lead to economic development and environmental pollution simultaneously [53–55]. Since new-type urbanization is widely considered a symbol of economic development [38], the critical issue is whether energy consumption can bring about larger benefits with respect to its cost. This basic benefit–cost tradeoff can be inferred from the causal relationship between the two. When energy consumption is conducive to promoting the level of new-type urbanization, it may indicate that the benefit of an increase in the level of new-type urbanization resulting from energy consumption is larger than the negative externality of environmental damage related to energy consumption. On the contrary, if the new-type urbanization leads to energy consumption positively, it may indicate that the advantage of energy consumption exceeds its disadvantage. From the estimated results for all provinces as a whole, we can conclude that the advantage of energy consumption is larger than its disadvantage in the short run, yet the relationship is opposite over time. From the estimated results for provinces in the eastern region, we can conclude that the advantage of energy consumption is always larger than its disadvantage over time. This again proves that those provinces in the eastern region may have started to cope with the possible environmental damage related to energy consumption. From the estimated results for provinces in the central, northeastern, and western regions, the advantages of energy consumption and its disadvantages are similar to the whole country, where the negative externality of environmental damage exceeds its benefit over time.

The PVAR model regression results can also be interpreted with the concept of the EKC, which assumes that economic growth and environmental pollution present an "inverted U" relationship [56]. Initially, as the level of new-type urbanization is relatively low, there are not too many industrial activities that lead to environmental pollution. Therefore, an increase in the level of new-type urbanization is conducive to reducing energy consumption for all samples in the short run. As the pace of new-type urbanization accelerates, there will be more and more high-pollution industries, and environmental pollution may gradually increase. As the estimation results indicated, an increase in the level of new-type urbanization may enhance energy consumption in the long run for the central, northeast, and western regions. Even so, for provinces in the eastern region, new-type urbanization is not conducive to increasing energy consumption over time. As the economy improves, these provinces may start to focus on the possible environmental damage related to energy consumption and attempt to take some remedial actions [57]. Generally speaking, as long as new-type urbanization reaches a high level as the eastern region, more resources may be dedicated to environmental protection. Consequently, an increase in the level of new-type urbanization will be conducive to reducing environmental pollution. Our main target is not to explore the EKC, yet the implications of the estimated results are compatible with the EKC prediction. To sum up, as the pace of new-type urbanization accelerates, a negative externality, such as environmental pollution related to energy consumption, gradually increases. Once a province reaches a high level of new-type urbanization, it may conversely reduce the negative externality related to energy consumption, as indicated by the EKC assumption.

3.3. Impulse Response Analysis

So as to further investigate the dynamic relationship between the variables in the PVAR model, the impulse response function is computed. Specifically, we set up Monte Carlo simulations for four types of subsamples and finally obtained a 4 × 4 impulse response graph, and the corresponding results are displayed in Figure 4 for the different groups: eastern (Figure 4a), central (Figure 4b), northeastern (Figure 4c), and western regions (Figure 4d), respectively.

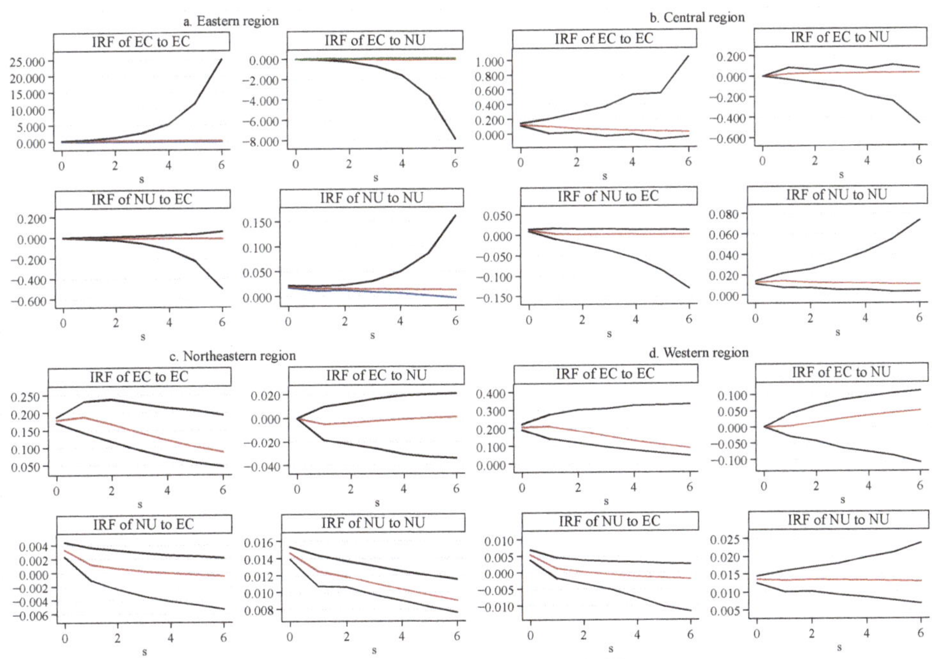

Note: Black lines represent a band of two standard deviations from mean value. Red line represents the IRF.

Figure 4. Variable impulse responses.

For provinces in the eastern regions, only *NU* shows a positive response to a standard deviation in its own unit. These results further verify that provinces in the eastern region may have been committed to environmental protection, hence there is no dynamic relationship. For provinces in the central region, the impulse response of *EC* to *NU* is positive, which is consistent with the PVAR model regression analysis. The implication is that these provinces should pay much attention to the high-pollution industries while the pace of new-type urbanization accelerates. The impulse response of *NU* to *EC* is positive in the current period, and then it decreases in the first period and tends to be weak after the second period, which is also consistent with the PVAR model regression analysis. The implication is that the negative externality of environmental damage related to energy consumption may offset the benefit of an increase in the level of new-type urbanization from energy consumption over time. While energy consumption is not conducive to promoting the level of new-type urbanization, these provinces should adopt some conservation policies. For provinces in the northeastern region, the impulse response of *EC* to *NU* is negative, which quickly reaches the highest value and then converges to zero. The implication is that the negative externality of environmental damage related to energy consumption may gradually increase over time. The impulse response of *NU* to *EC* is positive in the current period, and then it decreases in the first period and tends to be weak after the fourth period, which is the same as the provinces in the central region. For the provinces in the western region, the impulse response of *EC* to *NU* and the impulse response of *NU* to *EC* are virtually the same as the provinces in the central region, hence it is not necessary to interpret these results. In contrast to other regions, it is noteworthy that the impulse response of EC to NU does not diminish over time. This further denotes that the urbanization process in the western region relies predominantly on high energy consumption and highly pollutant-intensive inputs, and the legacy of conventional economic development approaches remains an arduous obstacle to overcome.

3.4. Variance Decomposition

In order to compare the relative contribution degree of unit standard deviation in *EC* and *NU*, we further adopt the variance decomposition. The variance decomposition results of period 5, period 10, and period 20 are displayed in Table 3 for different groups: (a) eastern, (b) central, (c) northeastern, and (d) western regions. As shown in Table 4, for provinces in the eastern, central, and western regions, the variance decomposition of *EC* is dominated by its own shock, and *NU* has a small effect. However, for provinces in the northeastern region, the variance decomposition of *EC* is dominated by the effect of *NU*, and its own shock has a small effect. For provinces in the eastern region, the variance decomposition of *NU* to its own shock is almost equal to the effect of *EC*. However, for provinces in central, northeastern, and western regions, the variance decomposition of *NU* is dominated by its own shock, and *EC* has a small effect. Combined with the variance decomposition of both variables, the effect of *EC* on *NU* is much larger than the effect of *NU* on *EC* for provinces in the eastern region, yet it is the opposite for provinces in central, northeastern, and western regions. These variance decomposition results also further confirm that energy consumption may bring about greater advantages relative to its disadvantage for provinces in the eastern region, yet it is imperative to cope with the possible cost of environmental pollution related to energy consumption for the provinces in central, northeastern, and western regions.

Table 4. Variance decomposition.

Response Variable	**s**	**Eastern**		**Central**		**Northeastern**		**Western**	
		EC	*NU*	*EC*	*NU*	*EC*	*NU*	*EC*	*NU*
EC	5	0.979	0.021	0.916	0.084	0.443	0.557	0.986	0.014
NU	5	0.057	0.943	0.177	0.823	0.233	0.767	0.036	0.964
EC	10	0.950	0.050	0.818	0.182	0.255	0.745	0.913	0.087
NU	10	0.194	0.806	0.124	0.876	0.118	0.882	0.032	0.968
EC	15	0.924	0.076	0.756	0.244	0.160	0.840	0.820	0.180
NU	15	0.360	0.640	0.106	0.894	0.069	0.931	0.040	0.960
EC	20	0.903	0.097	0.724	0.276	0.103	0.897	0.744	0.256
NU	20	0.507	0.493	0.099	0.901	0.043	0.957	0.047	0.953

So as to visually display the relative contribution degree of unit standard deviation in *EC* and *NU*, we further draw the variance decomposition results of *EC* and *NU* in the form of a coordinate axis. Figure 5 shows the variance decomposition of *EC* to *NU* for different groups: variance decomposition of *EC* to *NU* (Figure 5a) and variance decomposition of *NU* to *EC* (Figure 5b), respectively. As shown in Figure 5a, *NU* has a remarkable influence on *EC* for the provinces in the northeastern region, and the contribution rate is the largest among the four groups. Additionally, *NU* has an increasing contribution rate to unit standard deviation in *EC* for all groups. As explained above, these results imply that it is imperative to cope with the possible cost of environmental pollution related to energy consumption. As shown in Figure 5b, *EC* has decreasing contribution rate to unit standard deviation in *NU* for provinces in the central, northeastern, and western regions, yet *EC* has an increasing contribution rate to unit standard deviation in *NU* for provinces in the eastern region, which became the largest after the tenth forecast period. These results also imply that environmental pollution related to energy consumption may gradually increase as the pace of new-type urbanization accelerates for provinces in the central, northeastern, and western regions, which are consistent with the estimation results of the PVAR model.

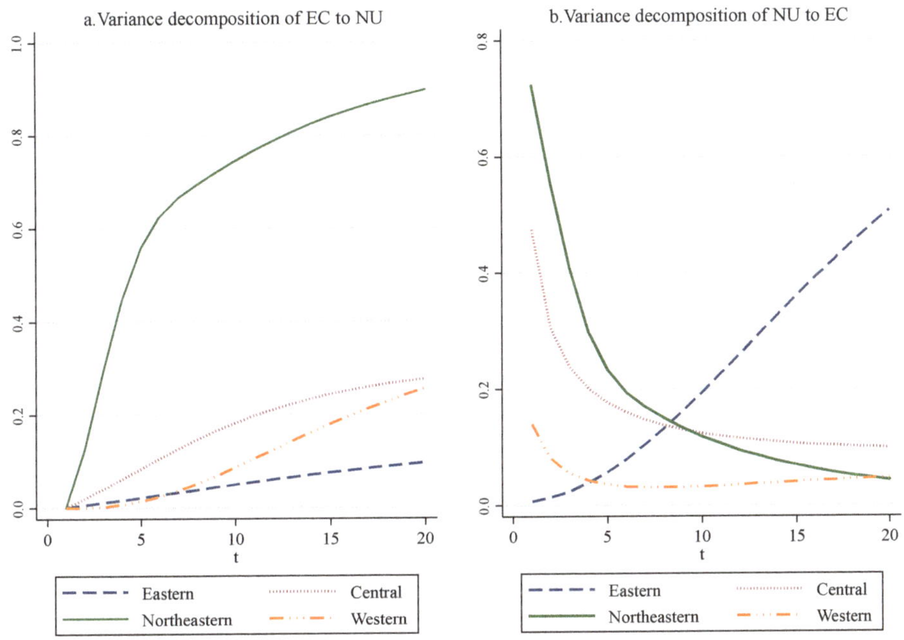

Figure 5. Variance decomposition of *EC* and *NU*.

4. In-Depth Analysis of Energy-Related Data

Finally, the pollution-related calculations are included for an in-depth analysis. We collect supplementary data, such as carbon emissions per real CNY 10,000, energy used per real CNY 10,000 (energy consumption intensity), the ratio of output value in the tertiary industry to real GDP, and the average per capita real GDP. All the relevant data are from each year's China Statistical Yearbooks, China Energy Statistical Yearbooks, each provincial statistical yearbook, and so on. As original data for the carbon emissions can be obtained from 2003 to 2019, the sample data span from 2003 to 2019. Figure 6 shows these calculations for different groups: carbon emissions (Figure 6a), energy consumption intensity (Figure 6b), industrial structure (Figure 6c), and per capita real GDP (Figure 6d), respectively.

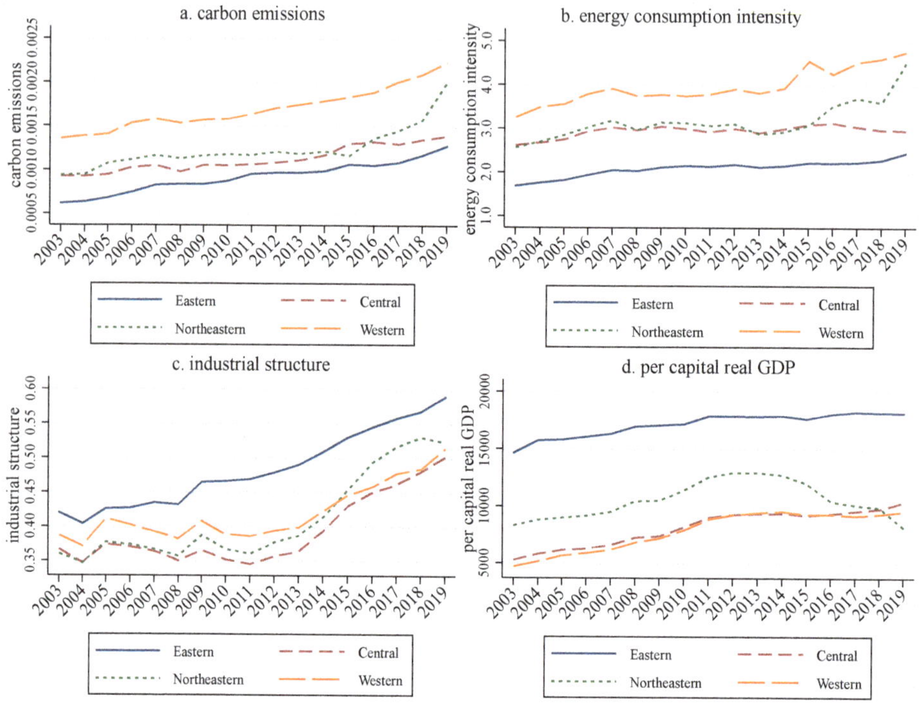

Figure 6. Some energy-related data.

As shown in Figure 6a, the largest carbon emitter appears in the western region, followed by the northeastern and central regions, and the least carbon emitter happens to be in the eastern region. As the pace of the new-type urbanization accelerates, provinces in the eastern region may make great strides to reduce the pollution of carbon emissions related to energy consumption. These results undoubtedly denote that there is an EKC relation. As shown in Figure 6b, the western region has the highest energy consumption intensity, followed by the northeastern and central regions, and finally the eastern region. As energy consumption intensity represents the efficiency of energy use, the characteristics of this variable in the four different groups denote that the energy is the most efficiently used in the eastern regions, followed by central and northeastern regions, and finally the western region. As the pace of new-type urbanization accelerates, provinces in the eastern region may successfully change their energy mix, and a decrease in carbon emissions is also can be anticipated. For example, pollution-free electrical energy can be massively put to use. To sum up, these two variables both confirm the causal relationship that new-type urbanization leads to energy consumption negatively for provinces in the eastern region, and new-type urbanization leads to energy consumption positively for provinces in the central, northeastern, and western regions over time, which seems to be in agreement with the estimation results of the PVAR model.

Figure 6c,d further provides the economic reality for the EKC relation and the causal relationship between the two. As shown in Figure 6c, the ratio of output value in the tertiary industry to real GDP in the eastern region is the highest among the four groups, followed by the northeastern region, the western region, and the central region. As shown in Figure 6d, per capita real GDP in the four groups is the same as Figure 6c. As the economy develops in the eastern region, there may be a trend toward the tertiary industry, which produces low-pollution products. Therefore, new-type urbanization leads to negative energy consumption. Provinces in the central, northeastern, and western regions are generally eager to raise the per capita real GDP level, so standards are lax concerning relevant environmental regulations. Even some high-pollution industries are encouraged in these regions, and spontaneously their ratio of output value in the tertiary industry to real GDP is behind the eastern region. Hence, the economic reality is consistent with the EKC relation and the causal relationship between the two.

5. Concluding Remarks

The panel data of Chinese provincial administrative units are classified into four categories according to the widely adopted standard in this paper. In order to explore the causal relationship between new-type urbanization and energy consumption, we employ the PVAR model to investigate this issue and obtain the following conclusions. (1) For provinces in the eastern region, new-type urbanization leads to energy consumption negatively, and energy consumption leads to new-type urbanization positively, which becomes negative over time. For the provinces in the central and northeastern regions, an increase in the level of new-type urbanization does not bring about energy consumption, while the uni-directional causal relationship running from energy consumption to new-type urbanization is similar to provinces in the eastern region. For provinces in the western region, there is negative feedback causality between the two. (2) For provinces in the eastern region, the advantage of energy consumption is larger than its disadvantage in the short run, yet the relationship is the opposite in the long run. For provinces in the central and northeastern regions, the advantages of energy consumption and its disadvantages are similar to provinces in the eastern region. However, for provinces in the western region, the disadvantage of energy consumption may exceed its advantage. These findings are also consistent with the EKC relation. Once a province reaches a high level of new-type urbanization, it may conversely reduce the negative externality of environmental damage related to energy consumption. (3) The impulse response analysis further presents the dynamic relationship between the two. The variance decomposition demonstrates that the effect of energy consumption on new-type is much larger than the effect of new-type urbanization on energy consumption for the provinces in the eastern region, while it is totally the opposite for provinces in the central, northeastern, and western regions. (4) The largest carbon emitter appears in the western region, followed by the northeastern and central regions, and the smallest carbon emitter happens to be in the eastern region. The western region has the highest energy consumption intensity, followed by the northeastern and central regions, and finally the eastern region. The ratio of output value in tertiary industry to real GDP in the eastern region is the highest among the four groups, followed by the northeastern region, the western region, and the central region, and the per capita real GDP in the four groups is the same as the ratio of output value in the tertiary industry to real GDP.

On the basis of our conclusions above, several straightforward policy implications can be put forward. As the causal relationship presents regional heterogeneity, a one-size-for-all energy policy will not work effectively. Our government should take into consideration the different levels of new-type urbanization while implementing an energy consumption policy. Specifically, for provinces in the eastern region, they should spare no effort to promote the level of new-type urbanization and mitigate energy use in the construction of new-type urbanization. For provinces in the central region, energy consumption is not conducive to promoting the level of new-type urbanization, hence they should adopt some conservation policies to avoid the negative externality of environmental damage related to energy consumption. For provinces in the northeastern and western regions, the negative externality of environmental damage related to energy consumption may gradually increase, and they should especially pay attention to cope with the possible environmental damage and take remedial actions. Finally, according to the in-depth analysis of energy-related data, for the provinces in the northeastern and western regions, they should try to raise the efficiency of energy use, so as to reduce carbon emissions. For example, our Chinese government can encourage enterprises to carry out energy-saving technologies in the production of goods and services. Additionally, for provinces in the northeastern and western regions, some high-pollution industries may have been encouraged, hence standards have to be strict concerning environmental protection and relevant environmental regulations.

The used methodology in this paper has some shortcomings, which calls for further research. Firstly, the constructed comprehensive urbanization index does not take into consideration the aspect of income inequality due to data availability. A comprehensive urbanization index should be constructed to reflect more about social welfare. Secondly, the empirical study in this paper does not consider the spatial spillover effect when exploring the causal relationship between energy consumption and urbanization. As we all know, both the level of energy consumption and urbanization probably present a spatial autocorrelation relationship, so future studies should employ spatial econometric techniques in this topic to deal with the possible estimation bias of the non-spatial econometric models. Thirdly, the empirical study in this paper is still based on Chinese provincial-level data. As the urban areas are the primary body to promote the urbanization process currently, it is of great value to conduct in-depth investigations at the city level.

Author Contributions: Conceptualization, Y.Q.; methodology, Y.Q.; formal analysis, C.C. and Y.G.; data curation, Y.G.; writing—original draft preparation, C.C.; writing—review and editing, C.C. All authors have read and agreed to the published version of the manuscript.

Funding: This research received no external funding.

Institutional Review Board Statement: Not applicable.

Informed Consent Statement: Not applicable.

Data Availability Statement: The data presented in this study are available on request from the corresponding author.

Conflicts of Interest: The authors declare no conflict of interest.

References

1. He, S.; Li, S.M.; Chan, K.W. Migration, communities, and segregation in Chinese cities: Introducing the special issue. *Eurasia. Geogr. Econ.* **2015**, *56*, 223–230. [CrossRef]
2. Wu, H.T.; Hao, Y.; Weng, J.H. How does energy consumption affect China's urbanization? New evidence from dynamic threshold panel models. *Energy Policy* **2019**, *127*, 24–38. [CrossRef]
3. Lv, Y.L.; Chen, W.; Cheng, J.Q. Effects of urbanization on energy efficiency in China: New evidence from short run and long run efficiency models. *Energy Policy* **2020**, *147*, 111858. [CrossRef]
4. Lin, B.Q.; Zhu, J.P. Impact of China's new-type urbanization on energy intensity: A city-level analysis. *Energy Econ.* **2021**, *99*, 105292. [CrossRef]
5. Xu, J.J.; Wang, J.C.; Li, R.; Yang, X.J. Spatio-temporal effects of urbanization on CO_2 emissions: Evidences from 268 Chinese cities. *Energy Policy* **2023**, *177*, 113569. [CrossRef]
6. Ma, B. Does urbanization affect energy intensities across provinces in China? Long-run elasticities estimation using dynamic panels with heterogeneous slopes. *Energy Econ.* **2015**, *49*, 390–401. [CrossRef]
7. Han, F.; Xie, R.; Lu, Y.; Fang, J.; Liu, Y. The effects of urban agglomeration economies on carbon emissions: Evidence from Chinese cities. *J. Clean. Prod.* **2018**, *172*, 1096–1110. [CrossRef]
8. Zhou, N.; Khanna, N.; Feng, W.; Ke, J.; Levine, M. Scenarios of energy efficiency and CO_2 emissions reduction potential in the buildings sector in China to year 2050. *Nat. Energy* **2018**, *3*, 978–984. [CrossRef]
9. Nathaniel, S.P. Modelling urbanization, trade flow, economic growth and energy consumption with regards to the environment in Nigeria. *GeoJournal* **2020**, *85*, 1499–1513. [CrossRef]
10. Chen, H.; Jia, B.; Lau, S.S.Y. Sustainable urban form for Chinese compact cities: Challenges of a rapid urbanized economy. *Habitat Int.* **2008**, *32*, 28–40. [CrossRef]
11. Mishra, V.; Sharma, S.; Smyth, R. Are fluctuations in energy consumption per capital transitory? Evidence from a panel of Pacific Island countries. *Energy Policy* **2009**, *37*, 2318–2326. [CrossRef]
12. Sun, C.; Ouyang, X.; Cai, H.; Luo, Z.; Li, A. Household pathway selection of energy consumption during urbanization process in China. *Energy Convers. Manag.* **2014**, *84*, 295–304. [CrossRef]
13. Wang, Q. Effects of urbanization on energy consumption in China. *Energy Policy* **2014**, *65*, 332–339. [CrossRef]
14. Liu, Y.; Xiao, H.; Lv, Y.; Zhang, N. The effect of new-type urbanization on energy consumption in China: A spatial econometric analysis. *J. Clean. Prod.* **2017**, *163*, s299–s305. [CrossRef]

15. Fang, J.C.; Gozgor, G.; Mahalik, M.K.; Mallick, H.; Padhan, H. Does urbanization induce renewable energy consumption in emerging economies? The role of education in energy switching policies. *Energy Econ.* **2022**, *111*, 106081. [CrossRef]
16. Ding, R.; Zhou, T.; Yin, J.; Zhang, Y.L.; Shen, S.W.; Fu, J.; Du, L.Y.; Du, Y.M.; Chen, S.H. Does the urban agglomeration policy reduce energy intensity? Evidence from China. *Int. J. Environ. Res. Public Health* **2022**, *19*, 14764. [CrossRef]
17. Northam, R.M. Vacant urban land in the American city. *Lan Econ.* **1971**, *47*, 345–355. [CrossRef]
18. Lin, B.Q.; Zhu, J.P. Energy and carbon intensity in China during the urbanization and industrialization process: A panel VAR approach. *J. Clean. Prod.* **2017**, *168*, 780–790. [CrossRef]
19. Wang, Q.; Yang, X. Urbanization impact on residential energy consumption in China: The roles of income, urbanization level, and urban density. *Environ. Sci. Pollut. Res.* **2019**, *26*, 3542–3555. [CrossRef]
20. Chen, Z.; Zhou, M. Urbanization and energy intensity: Evidence from the institutional threshold effect. *Environ. Sci. Pollut. Res.* **2021**, *28*, 11142–11157. [CrossRef]
21. Hu, W.; Fan, Y. City size and energy conservation: Do large cities in China consume more energy. *Energy Econ.* **2020**, *92*, 104943. [CrossRef]
22. Zarco-Soto, I.M.; Zarco-Periñán, P.J.; Sánchez-Durán, R. Influence of cities population size on their energy consumption and CO_2 emissions: The case of Spain. *Enviro. Sci. Pollut. Res.* **2021**, *28*, 28146–28167. [CrossRef]
23. Parikh, J.; Shukla, V. Urbanization, energy use and greenhouse effects in economic development: Results from a cross-national study of developing countries. *Glob. Environ. Chang.* **1995**, *5*, 87–103. [CrossRef]
24. Lenze, M.; Wier, M.; Cohen, C.; Hayami, H.; Pachauri, S.; Schaeffer, R. A comparative multivariate analysis of household energy requirement in Australia, Brazil, Denmark, India and Japan. *Energy* **2006**, *31*, 181–207. [CrossRef]
25. Ostuka, A. Population Agglomeration and Residential Energy Consumption: Evidence from Japan. *Sustainability* **2018**, *10*, 469.
26. Shen, L.; Cheng, S.; Gunson, A.J.; Wan, H. Urbanization, sustainability and the utilization of energy and mineral resources in China. *Cities* **2005**, *22*, 287–302. [CrossRef]
27. Sadorsky, P. Do urbanization and industrialization affect energy intensity in developing countries. *Energy Econ.* **2013**, *37*, 52–59. [CrossRef]
28. Cheng, Z.H.; Wang, L. Can new urbanization improve urban total-factor energy efficiency in China. *Energy* **2023**, *266*, 126494. [CrossRef]
29. Yu, B.B. Ecological effects of new-type urbanization in China. *Renew. Sustain. Energy Rev.* **2021**, *135*, 110239. [CrossRef]
30. Feng, Y.D.; Yuan, H.X.; Liu, Y.B. The energy-saving effect in the new transformation of urbanization. *Econ. Anal. Policy* **2023**, *78*, 41–59. [CrossRef]
31. Shao, J.; Wang, L.H. Can new-type urbanization improve the green total factor energy efficiency? Evidence from China. *Energy* **2023**, *262*, 125499. [CrossRef]
32. Ghosh, S.; Kanjilal, K. Long-term Relationship between urbanization, energy consumption and economic activity: Empirical evidence from India. *Energy* **2014**, *66*, 324–331. [CrossRef]
33. Xu, Y.Z.; Wang, Q.T. The impact of energy consumption on new urbanization—An analysis based on threshold effect. *East China Econ. Manag.* **2018**, *32*, 5–13.
34. Shafiei, S.; Salim, R. Non-renewable and renewable energy consumption and CO emissions in OECD countries: A comparative analysis. *Energy Policy* **2014**, *66*, 547–556. [CrossRef]
35. Bakirtas, T.; Akpolat, A.G. The relationship between energy consumption, urbanization, and economic growth in new emerging-market countries. *Energy* **2018**, *147*, 110–121. [CrossRef]
36. Moomaw, R.L.; Shatter, A.M. Urbanization and economic development: A bias toward large cities. *J. Urban Econ.* **1996**, *40*, 13–37. [CrossRef]
37. Davis, J.C.; Henderson, J.V. Evidence on the political economy of the urbanization process. *J. Urban Econ.* **2003**, *53*, 98–125. [CrossRef]
38. Zhao, J.; Tang, J. Industrial structure change and economic growth: A China-Russia comparison. *China Econ. Rev.* **2018**, *47*, 219–233. [CrossRef]
39. Lee, C.C.; Chang, C.P. Energy consumption and GDP revisited: A panel analysis of developed and developing countries. *Energy Econ.* **2007**, *29*, 1206–1223. [CrossRef]
40. Huang, B.N.; Hwang, M.J.; Yang, C.W. Causal relationship between energy consumption and GDP growth revisited: A dynamic panel data approach. *Ecol. Econ.* **2008**, *67*, 41–54. [CrossRef]
41. Shahbaz, M.; Lean, H.H. The dynamics of electricity consumption and economic growth: A revisit study of their causality in Pakistan. *Energy* **2012**, *39*, 146–153. [CrossRef]
42. Ben-Salha, M.; Sebri, O. On the causal dynamics between economic growth, renewable energy consumption, CO_2 emissions and trade openness: Fresh evidence from BRICS countries. *Renew. Sustain. Energy Rev.* **2014**, *39*, 14–23.
43. Zheng, W.; Walsh, P.P. Economic growth, urbanization and energy consumption-A provincial level analysis of China. *Energy Econ.* **2019**, *80*, 153–162. [CrossRef]
44. Ha, N.M.; Ngoc, B.H. Revisiting the relationship between energy consumption and economic growth nexus in Vietnam: New evidence by asymmetric ARDL cointegration. *Appl. Econ. Lett.* **2021**, *28*, 978–984. [CrossRef]
45. Tang, L.Z.; Shi, J.X.; Li, Y.J. Two-way correlation mechanism and measurement between new urbanization and clean energy consumption. *Soc. Sci. Nanjing* **2022**, *18*, 27–36.

46. Ma, M.Y.; Zheng, J.W.; Ma, T. Spatiotemporal characteristics of the impact of new urbanization on China's carbon dioxide emissions from a multi-dimensional perspective. *Acta Sci. Circumstantiae* **2021**, *41*, 2474–2486.
47. Hao, Y.; Peng, H. On the convergence in China's provincial per capital energy consumption: New evidence from a spatial econometric analysis. *Energy Econ.* **2017**, *68*, 31–43. [CrossRef]
48. Chen, C.; Qin, Y.; Gao, Y. Does new urbanization affect CO_2 emissions in China: A spatial econometric analysis. *Sustain. Cities Soc.* **2023**, *96*, 104687. [CrossRef]
49. Blundell, R.; Bond, S. Initial conditions and moment restrictions in dynamic panel data models. *J. Econ.* **1998**, *87*, 115–143. [CrossRef]
50. Jawadi, F.; Mallick, S.K.; Sousa, R.M. Fiscal and monetary policies in the BRICS: A panel VAR approach. *Econ. Model.* **2016**, *58*, 535–542. [CrossRef]
51. Hou, J.; Lu, X.; Wu, S.M.; Ke, S.G.; Li, J.; Fu, J. Analysis of the dynamic relationship between green economy efficiency and urban land development intensity in China. *Int. J. Environ. Res. Public Health* **2022**, *19*, 7960. [CrossRef]
52. Feng, Y.D.; Liu, Y.B.; Yuan, H.X. The spatial threshold effect and its regional boundary of new-type urbanization on energy efficiency. *Energy Policy* **2022**, *164*, 112866. [CrossRef]
53. Omri, A. CO_2 emissions, energy consumption and economic growth nexus in MENA countries: Evidence from simultaneous equations models. *Energy Econ.* **2013**, *40*, 657–664. [CrossRef]
54. Ozcan, B.; Tzeremes, P.; Tzeremes, N. Energy consumption, economic growth and environmental degradation in OECD countries. *Econ. Model.* **2020**, *84*, 203–213. [CrossRef]
55. Xiong, J.H.; Xu, D.Y. Relationship between energy consumption, economic growth and environmental pollution in China. *Environ. Res.* **2021**, *194*, 110718. [CrossRef] [PubMed]
56. Grossman, G.M.; Krueger, A.B. Economic growth and the environment. *Q. J. Econ.* **1995**, *110*, 353–377. [CrossRef]
57. Sun, Y.P.; Li, H.N.; Andlib, Z.; Genie, M.G. How do renewable energy and urbanization cause carbon emission? Evidence from advanced panel estimation techniques. *Renew. Energy* **2022**, *185*, 996–1005. [CrossRef]

energies

MDPI

Article

A Health-Aware Energy Storage Sharing Mechanism for a Renewable Energy Base

Chong Shao [1], Bolin Zhang [1], Bo Wei [1], Wenfei Liu [2], Yong Yang [2] and Zhaoyuan Wu [3,*]

1 State Grid Gansu Electric Power Company, Lanzhou 730030, China; shaoch_dkzx@gs.sgcc.com.cn (C.S.); zhangblgsepc@163.com (B.Z.); edisonjoke@163.com (B.W.)
2 State Grid Gansu Electric Power Research Institute, Lanzhou 730070, China; liuwenfeitgh@163.com (W.L.); yy8801@163.com (Y.Y.)
3 School of Electrical and Electronic Engineering, North China Electric Power University, Beijing 102206, China
* Correspondence: wuzy@ncepu.edu.cn

Abstract: With the increasing global demand for renewable energy (RE), the growing share of new energy sources has become an inevitable trend. However, due to the uncertainty and fluctuation of renewable energy generation, this poses challenges to the stability of the power system. To mitigate the volatility of wind power output, ensure reliable power supply, and improve energy storage utilization, shared energy storage (SES) can be deployed in renewable energy bases (REBs) to alleviate the pressure on the power supply. However, electrochemical energy storage (EES) faces issues such as lifespan degradation and maintenance cost allocation. In this regard, this paper establishes an EES characterization model considering the dynamic degradation characteristics of batteries and analyzes the coupled relationship between lifespan degradation laws and key parameters in SES operation. Additionally, to assess the impact of electrochemical energy storage's dynamic degradation characteristics on energy capacity allocation and operational strategies, an optimization model for SES in REBs is developed. Building upon this, a cost allocation mechanism is designed based on the marginal contribution in both the day-ahead and the real-time markets to address the differing demands for SES among different units within the REBs. Case studies are conducted to validate the rationality of the proposed optimization model for SES in REBs and the adaptability of the cost allocation mechanism. The results provide valuable insights for practical applications.

Keywords: shared energy storage; renewable energy base; dynamic degradation characteristics; two-stage market optimization; cost allocation mechanism

Citation: Shao, C.; Zhang, B.; Wei, B.; Liu, W.; Yang, Y.; Wu, Z. A Health-Aware Energy Storage Sharing Mechanism for a Renewable Energy Base. *Energies* **2023**, *16*, 5356. https://doi.org/10.3390/en16145356

Academic Editors: Luis Hernández-Callejo, Jesús Armando Aguilar Jiménez and Carlos Meza Benavides

Received: 14 June 2023
Revised: 8 July 2023
Accepted: 11 July 2023
Published: 13 July 2023

1. Introduction

With the implementation of the dual-carbon target, it has become clear that large-scale renewable energy generation, specifically through wind and photovoltaic power, is the direction and necessary choice for new power systems in the future [1–3]. To achieve the strategic goals of building a new power system, China has proposed to further build REBs to facilitate the high-quality and rapid development of RE. The "14th Five-Year Plan for China's Economic and Social Development and the Long-Range Objectives through the Year 2035" released in March 2021 proposes to focus on developing nine clean energy bases and four offshore wind power bases during the duration of the 14th Five-Year Plan. In June 2022, the National Development and Reform Commission and nine other departments issued the "14th Five-Year Plan for the Development of RE", which explicitly proposes active steps to promote the development of wind and solar power generation facilities, and expedite the construction of large-scale REB projects, with a particular emphasis on desert, Gobi, and other barren regions [4–6]. However, the "anti-peak" characteristic of wind power and the weather impact on photovoltaic power generation have increased volatility of the net load curve, which imposes higher requirements on flexible resources for the new power system. Several provinces have implemented policies mandating that RE plants install a specific percentage of ES to reduce the influence of RE integration on the safety and stability of the power grid. The integration of energy storage systems with renewable energy sources addresses the mismatch between renewable energy generation and load demand and reduces the uncertainty of renewable energy output, thereby enhancing the overall operational efficiency of the grid, lowering power supply costs, improving system stability, and enhancing power quality [7–9].

Due to the instability of RE, ES is needed to improve the total efficiency and stability of the power grid, reduce electricity supply costs, and enhance the utilization rate of RE. Currently, EES is the main ES technology, and its application has become increasingly widespread as its technology continues to develop and costs continue to decline [10]. Nevertheless, when ES is solely coupled with REBs, its usage rate is comparably low, making it challenging to recoup the expenses of ES, particularly in the present scenario where raw material costs are surging while RE project prices continue to fall [11,12]. The utilization rate of ES will further decrease when coupled solely with REBs, thereby hindering the promotion of the development of RE. To address the low utilization rate and poor economics of ES paired only with REBs, SES can provide an effective solution. By using SES, ES expenses can be reduced and utilization rates can be increased, thereby better supporting the development of RE [13–15].

A large-scale REB is composed of multiple wind and photovoltaic units, as well as their collection and transmission networks. Essentially, configuring SES for REBs means meeting the ES needs of various RE units within the base. Therefore, we need to consider the differences in ES needs of different RE units within the base to achieve fair distribution of ES costs. The core of the SES mechanism for REBs is how to quantify the contribution of SES to the ES capacity requirements of various RE units and then share the cost of SES accordingly [7]. Several studies have been conducted on the modes of operation for SES and cost allocation among RE units. Some researchers have investigated the impact of performance quality and prediction errors of renewable energy units on the demand for energy storage capacity and the allocation of energy storage costs [16]. Ref. [17] proposed a wind power cluster and SES coordination optimization mechanism, and allocated the benefits of each member of the alliance to demonstrate that the coordination optimization mechanism is conducive to reducing operating costs of each member and ensuring fairness of benefit distribution. In [18], a novel non-cooperative game mechanism is proposed, which optimally regulates the operation of distributed generation and flexibility resources by considering economic factors and electric power quality. Ref. [19] introduced a new two-stage credit-based model for SES which, considering time accumulation effects, developed an SES pricing strategy and a capacity planning scheme, and demonstrated the advantages of the proposed novel shared model in the field of economic efficiency and the utilization rate of ES. The research results indicate that SES has enormous potential value, not only for configuring SES in REBs but also for applying SES to various links in the power supply chain. Ref. [20] aimed to examine the real advantages of implementing SES in residential neighborhoods, established optimization operation models for independent and SES, compared and analyzed the optimal energy operations of the two, and developed an efficient control strategy suitable for the use of SES to demonstrate the advantages of SES in saving electricity costs and improving the ES utilization rate. Some studies have also proposed transactional operation mechanisms for energy storage systems based on non-cooperative game theory [21,22]. In [22], an interactive energy management scheme is defined for multiple SES systems and users to achieve information sharing. These studies have provided certain theoretical support and a decision-making basis for the formulation of SES modes among a considerable quantity of RE units within a large REB. Existing research has primarily focused on optimal capacity allocation and economic benefits of SES among renewable energy units while neglecting the impact of cost allocation mechanisms on the sustainable operation of SES. Therefore, this paper formulates a fair cost allocation mechanism considering the differential ES demands of each renewable energy unit within the base, aiming to achieve equitable distribution of ES costs.

EES, mainly consisting of energy storage batteries, is one of the most economically advantageous ES technologies among existing RE storage technologies. However, it should be noted that EES has the characteristic of dynamic degradation of its lifespan. Although SES operation modes can improve ES utilization rate, they can accelerate the degradation of the lifespan of EES. However, in the research on SES for renewable energy integration currently, the influence of shared operation on energy storage battery lifespan degradation is often overlooked or simplified. For instance, some studies consider the working efficiency of energy storage batteries as a constant value, disregarding the dynamic changes in charging and discharging efficiencies [23,24]. However, in actual operation, energy storage batteries experience dynamic changes in charging and discharging efficiencies due to power losses generated during operation to meet load demand [25,26]. Additionally, to simplify the complex degradation variations in practical operation, certain studies assume the same degree of lifespan degradation under different operating conditions, without accounting for the nuanced degradation of ES capacity under different operational scenarios [27]. However, it is clear that the lifespan of EES will gradually shorten with changes in the state of charge (SOC) and the number of charging-and-discharging cycles. To more accurately estimate the degradation of the lifespan of energy storage batteries, equivalent circuit models, empirical models, and aging mechanism models are currently mainly used [28]. Ref. [29] proposed a method to detect the decay of the available capacity of energy storage batteries using the discharge curve and capacity data based on a first-order Thevenin equivalent circuit model. Ref. [30] established a calendar aging model of energy storage batteries based on experimental data and quantified the impact of SOC, temperature, and battery operating time on the degree of battery life decay. Ref. [31] studied the dynamic performance changes of energy storage batteries under different environmental conditions in a residential photovoltaic energy storage battery system, and analyzed the impact of charging-and-discharging curves on battery aging. Moreover, existing research tends to focus on the lifespan degradation characteristics of distributed independent energy storage systems, lacking investigations on the impact of dynamic degradation characteristics of SES on system operation. This paper, however, conducts a refined analysis of the dynamic degradation characteristics in the actual operation of EES.

Moreover, due to the diverse output characteristics of different renewable energy units, there are variations in the capacity requirements for SES. Existing research has primarily focused on optimal capacity allocation and economic benefits of SES among renewable energy units while neglecting the impact of cost allocation mechanisms on the sustainable operation of SES. To analyze the influence of dynamic degradation characteristics on the operational strategies and capacity allocation of electrochemical shared energy storage in REBs, and to address the issue of uneven cost allocation resulting from differences in ES capacity requirements, this paper presents a refined modeling of SES lifespan degradation. Building upon the health status of EES, known as the state of health (SoH), this research investigates the optimization of operational strategies and cost allocation mechanisms for SES in REBs by considering dynamic degradation characteristics. The main innovations can be summarized in the following three aspects:

- This paper refines the coupling relationship between the degradation laws and key parameters in the operation process of shared energy storage, and establishes a refined degradation model for the operation of electrochemical energy storage sharing. This model can better reflect the changes in performance parameters such as shared energy storage charging and discharging efficiencies and state of health (SoH), thus quantifying the degree of degradation in the lifespan of shared energy storage. It also provides important theoretical support for the practical application of shared energy storage.
- A renewable energy base–shared energy storage operation framework that considers dynamic lifespan degradation is designed. This framework fully utilizes the advantages of shared energy storage and enhances the profitability of various units within high renewable energy bases in the day-ahead market through "peak shaving and valley filling". At the same time, it mitigates the uncertainty of wind power output

and reduces the assessment cost of real-time balancing markets. The design of this framework can better promote the sustainable development of renewable energy generation.

- A shared energy storage cost allocation mechanism is proposed for renewable energy bases based on the marginal contribution in both the day-ahead and the real-time market. This mechanism can meet the energy storage demands of different renewable energy generators and incentivize compatibility. The numerical results demonstrate a positive correlation between the shared energy storage costs allocated to different renewable energy generators and their corresponding energy storage demands. The implementation of this mechanism can better promote the coordinated optimization of renewable energy and shared energy storage operations, achieving a win-win situation.

The rest of this paper is structured in the following manner. The SES operation framework for a REB is proposed in Section 2. The refined model of dynamic life decay of EES is introduced in Section 3. In Section 4, an optimized operational approach for SES in a REB is presented, taking into account the dynamic degradation characteristics of EES. The SES cost allocation mechanism based on the marginal contribution in both the day-ahead and real-time markets is introduced in Section 5. Section 6 contains the conclusions and future prospects of this study.

2. Framework of Energy Storage Sharing

A two-stage optimal collaborative operation strategy for a REB and SES is proposed by combining day-ahead optimization and real-time optimization. This strategy includes two stages: during the first stage, the optimization of day-ahead scheduling is carried out, and each unit in the REB optimizes its day-ahead operation strategy based on the day-ahead output prediction data. In the second stage, the charging and discharging operation statuses of SES vary according to real-time electricity prices and the uncertainty of wind power output, then embedded it into the day-ahead optimization model of the first stage. Meanwhile, the dynamic degradation characteristics of the SES's lifespan are taken into account, and the influence of battery health status changes on ES capacity allocation is considered. Considering the differential SES capacity demands of different units within the REB, this paper measures the contribution of each member to the overall alliance and allocates the investment and operational costs of SES among various renewable energy units within the base. This allocation is performed in a manner that reflects the varying needs of different units and ensures a fair distribution of the investment and operational costs associated with SES. This strategy aims to optimize the overall operation of the system, fully consider the uncertainty and fluctuations of wind power output in actual operation, and improve the efficiency and economic viability of RE generation. The Operation framework for energy storage sharing in a renewable energy base is shown in Figure 1.

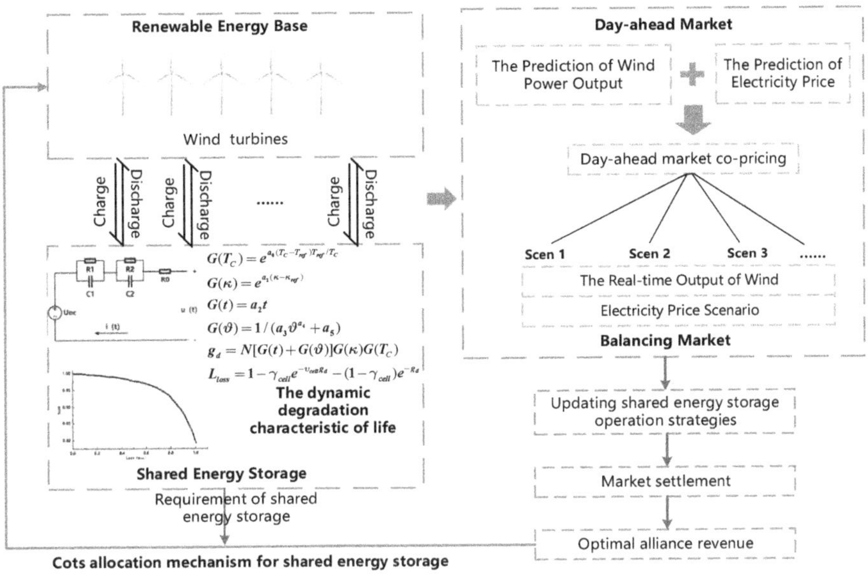

Figure 1. Operation framework for energy storage sharing in a renewable energy base.

3. Dynamic Degradation Model in Battery Energy Storage Sharing

To alleviate the impact of large-scale RE, such as wind and solar power, it may be necessary to frequently adjust the charge and discharge states of RE batteries and the power flow in and out of the grid. Therefore, the capacity degradation caused by the changes in charge and discharge behavior of RE batteries in a brief span of time cannot be ignored. It is necessary to consider the influence of changes in charge and discharge power on the performance and lifespan of ES devices. Therefore, the fine-grained dynamic degradation characteristics of EES are of great significance. In this section, the dynamic degradation characteristics of EES will be finely modeled to provide more theoretical support for the subsequent research on operation and cost allocation mechanisms of SES in REB.

3.1. Health-Aware Perception Model

The previous literature has used the SoH of an electrochemical battery to characterize its degree of life degradation. Within this segment, a health-aware perception model utilizing the battery's equivalent circuit is established. Since the dual-polarization (DP) equivalent circuit model is superior to the Thevenin model in the field of balance estimation precision and calculative speed, the DP model has been chosen to characterize the SOH of the battery. The DP model is essentially a series circuit, consisting of a power source, internal resistance, and a second-order RC parallel circuit, shown in Figure 2 [32].

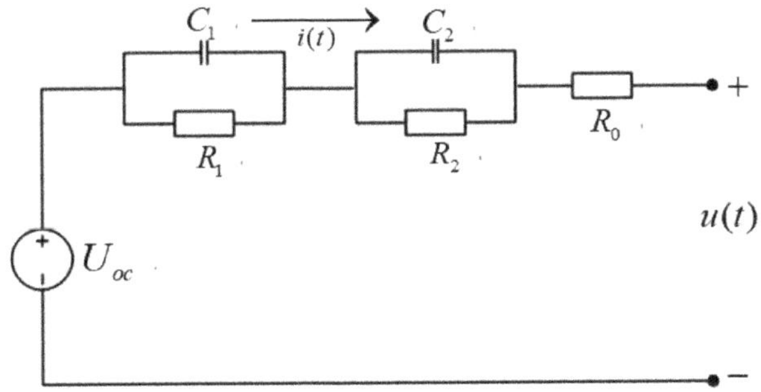

Figure 2. DP equivalent circuit.

The spatial state equations for the DP equivalent circuit can be derived as follows:

$$\begin{cases} u_t = U_{oc} - u_{c1,t} - u_{c2,t} - i_t R_0 \\ i_t = \frac{u_{c1,t}}{R_1} + C_1 \frac{du_{c1,t}}{dt} = \frac{u_{c2,t}}{R_2} + C_2 \frac{du_{c2,t}}{dt} \end{cases} \tag{1}$$

where U_{oc} is the open circuit voltage, R_0 is the internal resistance of the battery, $i(t)$ is the internal current of the battery, and $u_{c1,t}$ and $u_{c2,t}$ represent voltage across the two RC circuits. By measuring the open circuit voltage and voltage across each RC circuit, the circuit current can be calculated using the spatial state equations. Battery capacity can then be calculated using Equation (2).

$$E_{SES,t} = E_{SES,t_0} - \eta_{SES} \int_{t_0}^{t} i_t dt \tag{2}$$

where η_{SES} is the Coulombic efficiency of the battery, E_{SES,t_0} represents the rated capacity of the battery at initial state, $E_{SES,t}$ represents the capacity of the battery at time t, and t_0 represents the initial operating time of the battery. Based on the stored energy capacity, the SoH of the ES battery can be calculated.

The SoH of the ES battery at time t SOH_t is expressed as the present available capacity divided by the rated capacity at initial state, as follows:

$$SOH_t = \frac{E_{SES,t}}{E_{SES,t_0}} \times 100\% \tag{3}$$

As energy storage batteries undergo continuous charging-and-discharging cycles, internal aging occurs, resulting in increased internal resistance and decayed capacity. For a brand-new RE storage battery, its initial SoH is 1. When SoH_t is lower than a certain value δ or the internal resistance of the battery increases to more than twice the initial resistance, the energy storage battery should be dismantled and recycled. Therefore, the final state of health for the battery is δ, which is set to 0.8 in this paper [33–35]. Thus, the SoH_t and internal resistance of energy storage batteries are subject to the following constraints:

$$\begin{cases} SOH_t \geq \delta \\ Z_t \geq 2Z_{start} \end{cases} \tag{4}$$

where Z_t is the internal resistance of the battery at time t. The relationship between SoH and battery internal resistance can be further established using DP equivalent circuits, which facilitates the determination of whether the battery is still suitable for ES based on its internal resistance, as shown below:

$$Z_t = Z_{start} + \frac{SOH_{t_0} - SOH_t}{SOH_{t_0} - SOH_{t_N}} (Z_{end} - Z_{start}) \tag{5}$$

where SOH_{t_0} and SOH_{t_N}, respectively, represent the initial and final health status of the battery during its lifespan, Z_{start} and Z_{end} are the battery's resistance at the beginning and end of the entire lifespan, respectively, and t_N is the end time of the battery's lifespan.

In addition, the relationship between the maximum power output P_t of a battery at a certain moment and its SoH can be derived from DP equivalent circuits, as shown in Equation (6):

$$P_t = \frac{Z_{start}}{Z_t} P_0 \tag{6}$$

where P_0 refers to the maximum power output that the battery can produce at the initial state.

3.2. Energy Storage Battery Life Degradation Model

The life degradation of ES batteries is influenced by various external stress factors, such as temperature and operating time, and is also affected by the battery's life status. Therefore, its degradation can be considered as a nonlinear process that is the result of the combined effect of external stress and time by establishing a nonlinear life degradation model that can be decomposed into multiple stress factor models.

The life degradation caused by calendar aging L_{cal}, can be considered as a function of the average state of charge $Soc(\overline{\kappa})$, battery average temperature $\overline{T_c}$, and time t.

$$L_{cal} = g_t(\overline{\kappa}, t, \overline{T_C}) \tag{7}$$

Each charge and discharge cycle of the battery results in life degradation. The cumulative life degradation is obtained by adding up life degradation of each cycle.

$$L_{cyc} = \sum_{t=1}^{N} \omega_t g_c(\kappa_t, \vartheta_t, T_c) \tag{8}$$

where $L_{cyc,loss}$ refers to the life degradation caused by cycle aging; ϑ_t represents the depth of discharge of the battery in the t-th cycle; and ω_t is a 0–1 variable that characterizes the operating state of the battery in the cycle t, with a value of 1 indicating that the battery is in a cyclic charging-and-discharging state and 0 indicating that the battery is not undergoing charging and discharging, during which the life degradation of the battery only includes the life degradation caused by calendar aging. The life degradation of the entire lifetime of the battery can be represented as the function g_d of $\kappa, t, T_c, \vartheta$:

$$g_d(\kappa, t, T_C, \vartheta) = g_t(\overline{\kappa}, t, \overline{T_C}) + \sum_{t=1}^{N} \omega_t g_c(\kappa_t, \vartheta_t, T_c) \tag{9}$$

If cycles are identical, in a single cycle, the average temperature and SOC equals the overall averages of the battery lifetime, thus $\overline{T_c} = T_c, \overline{\kappa} = \kappa$. Equation (8) can be simplified as shown below:

$$g_d(\kappa, t, T_C, \vartheta, N) = N g_d(\kappa, t, T_C, \vartheta, 1) = N g_{d,1} \tag{10}$$

where $g_d(\kappa, t, T_c, \vartheta, 1)$ is denoted as $g_{d,1}$, which represents the life degradation during a single cycle, and N is the number of charge and discharge cycles.

The life degradation caused by calendar aging and cycle aging can be expressed in a product form of multiple linear stress factor models as follows:

$$g_{d,1} = [G(t) + G(\vartheta)]G(\kappa)G(T_C) \tag{11}$$

(1) The temperature stress model:

$$G(T_C) = e^{a_0(T_C - T_{ref})\frac{T_{ref}}{T_C}} \tag{12}$$

(2) The state-of-charge stress model:

$$G(\kappa) = e^{a_1(\kappa - \kappa_{ref})} \tag{13}$$

(3) The time stress model:

$$G(t) = a_2 t \tag{14}$$

(4) The depth-of-discharge stress model:

$$\begin{aligned} G(\vartheta) &= \frac{1}{a_3 \vartheta^{a_4} + a_5} \\ \vartheta_t &= \frac{P_t^{ch} + P_t^{dis}}{2E_{c,t}} \end{aligned} \tag{15}$$

where a_0 represents the temperature stress coefficient; T_{ref} refers the reference temperature in Kelvin (K); a_1 is the SOC stress coefficient; κ_{ref} is the reference SOC, which can be generally taken as 0.4–0.5; a_2 is the time stress coefficient, indicating that after excluding factors such as temperature and life dependence, the degradation rate has a linear relationship with time; $P^{ch}(t)$ and $P^{dis}(t)$ represent the charging and discharging powers of the battery during the t-th cycle, respectively; and a_3, a_4, and a_5 are stress coefficients about DOD.

According to the empirical formula [30], the life degradation of an energy storage battery can be calculated as follows:

$$L_{loss} = 1 - \gamma_{cell} e^{-v_{cell} g_d} - (1 - \gamma_{cell}) e^{-g_d} \tag{16}$$

where L_{loss} represents the life degradation of the battery over its entire life cycle (pu), and γ_{cell} and v_{cell} are parameters related to the formation process of SEI film.

3.3. *Dynamic Efficiency Model for Energy Storage Batteries*

3.3.1. Segmented Linearization of Power for EES Batteries

At any given time, the operating power of the battery $P(t)$ satisfies the following equation:

$$P_{c,t} = P_t^{ch} + P_t^{dis} \tag{17}$$

However, it is impossible for one battery to charge and discharge simultaneously at any given time, and one of P_t^{ch} or P_t^{dis} must be zero. The dynamic behavior of the battery is simulated using the aforementioned DP equivalent circuit model, and the relationship between the battery SoH and internal resistance is calculated. The charge–discharge cycles of the battery will cause dynamic changes in its SoH and internal resistance, resulting in dynamic changes in the operational efficiency during different periods. To simplify the model, the charging and discharging power of the battery are segmented and linearized separately. If the charging power is divided into M_1 segments, it satisfies:

$$P_t^{ch} = \sum_{m=1}^{M_1} P_{cm,t} \tag{18}$$

$$P_{cm}^{\min} \omega_{c,m,t} \le P_{cm,t} \le P_{cm}^{\max} \omega_{c,m,t}, m = 1, 2, \ldots, M_1 \tag{19}$$

where P_{cm}^{max} and P_{cm}^{min} are the maximum and minimum of the charging power for the segment, respectively, and $\omega_{c,m,t}$ is a 0–1 state variable that characterizes the charging state at time t, where a value of 1 indicates that the charging power is within the mth segment.

$$\sum_{m=1}^{M_1} \omega_{c,m,t} = 1 \tag{20}$$

The above equation indicates that the battery charging power can only be within one power segment at any given moment. Similarly, for the discharging power, it is divided into M_2 segments, which satisfies:

$$P_t^{dis} = \sum_{j=1}^{M_2} P'_{cj,t} \tag{21}$$

$$P_{cj,t}^{\min\prime} \pi_{c,j,t} \le P'_{cj,t} \le P_{cj,t}^{\max\prime} \pi_{c,j,t}, j = 1, 2, \ldots, M_2 \tag{22}$$

$$\sum_{j=1}^{M_2} \pi_{c,j,t} = 1 \tag{23}$$

where $P_{cj,t}^{max\prime}$ and $P_{cj,t}^{min\prime}$ are the maximum and minimum of the discharging power for the j-th segment of the battery, respectively, and $\pi_{c,j,t}$ is a 0–1 state variable that characterizes the discharging power at t, where a value of 1 indicates that the discharging power is within the j-th segment.

3.3.2. Storage Capacity of EES Batteries

After the time Δt, the incremental capacity of the battery is as follows:

$$\Delta E_{c,t} = [\sum_{m=1}^{M_1} P_{cm,t}\eta_{cm}^{ch} + \frac{1}{\eta_{cj}^{dis}}\sum_{j=1}^{M_2} P'_{cj,t}]\Delta t \tag{24}$$

$$\Delta E_{c,t} = E_{c,t-1} - E_{c,t} \tag{25}$$

$$E_c^{\min} \leq E_{c,t} \leq E_c^{\max} \tag{26}$$

where η_{cj}^{ch} and η_{cj}^{dis} refer to the charging and discharging efficiencies of the m-th segment, respectively, and E_c^{min} and E_c^{max} are the minimal and maximal battery capacity, respectively.

4. Shared Energy Storage Operation Model

In this section, we investigate the optimal strategy for the joint operation of RE units containing multiple wind turbines and SES, considering the dynamic degradation characteristics.

4.1. Objective Function

The objective function is to maximize profit, which comprises the revenue of the day-ahead electricity market and real-time electricity market, and the total cost of SES.

$$\max \sum_{t\in\phi T} (R_t^{DA} + R_t^{BA} - C_t^{sum}) \tag{27}$$

where R_t^{DA} denotes the day-ahead market revenue at time t, R_t^{BA} denotes the real-time market revenue at time t, and C_t^{sum} denotes the total cost of SES at time t.

The day-ahead energy market (DEM) revenue R_t^{DA} and real-time energy market (REM) revenue R_t^{BA} can be expressed as follows:

$$R_t^{DA} = \lambda_t^{DA} P_t^{DA} \tag{28}$$

$$R_t^{BA} = \lambda_t^{down} P_t^{down} - \lambda_t^{up} P_t^{up} \tag{29}$$

$$\lambda_t^{down} = \phi^{down}\lambda_t^{DA}, \lambda_t^{up} = \phi^{up}\lambda_t^{DA} \tag{30}$$

where λ_t^{DA} is the DEM price at time t; P_t^{DA} is the total power in the DEM at time t; P_t^{down} and P_t^{up} represent the positive and negative power imbalance at time t respectively; λ_t^{down} and λ_t^{up} are the settlement prices for the positive and negative imbalances of electricity quantities in the balancing market, respectively; and ϕ^{down} and ϕ^{up} are the penalty factors corresponding to positive and negative imbalances of electricity quantities, respectively.

The total cost of SES C_t^{sum} is expressed as Equation (33), and C_t^{sum} includes the investment cost of SES and its degradation cost, as follows:

$$C_t^{at} = \frac{C_t^{inv} L_{loss,t}}{24 \times (1 - 40\%)} \tag{31}$$

$$C_t^{inv} = \frac{\gamma(1+\gamma)^y}{(1+\gamma)^y - 1} \times \frac{c_P^{SES} P_{\max}^{SES} + c_E^{SES} E_{SES}}{T \times 365} \tag{32}$$

$$C_t^{sum} = C_t^{inv} + C_t^{at} \tag{33}$$

where T is all the dispatch cycle number of SES within one day, and $L_{loss,t}$ is the lifespan degradation rate of EES batteries during period t. Typically, when the health status of a lithium-ion battery drops below 80%, the battery's utilization rate cannot meet the ES requirements and it should be recycled. C_t^{inv} is the SES investment cost, γ refers to the annual percentage rate of funds, y is the lifespan of the SES device, c_P^{SES} and c_E^{SES} are the unit cost prices of shared energy per unit power and unit capacity, respectively, P_{max}^{SES} is the maximum power of the SES, and E_{SES} is the capacity of SES.

The above utilizes the concept of unitized cost, which converts ES replacement cost into cost per unit charge–discharge capacity, in order to obtain the ES degradation cost at each moment.

4.2. Constraints

In this paper, the wind turbine units and the collaborative entity of the SES are selected to participate in the day-ahead energy DEM. Therefore, the total power in the DEM is equal to the sum of the day-ahead power outputs from the wind turbines and the energy storage unit. Similarly, when they participate in the REM, the total power in the REM is equal to the sum of the real-time power outputs from the wind turbines and the energy storage unit, as follows:

$$P_t^{DA} = P_{wind,t}^{DA} + P_{SES,t}^{DA} \tag{34}$$

$$P_\delta^{BA} = P_{wind,\delta}^{BA} + P_{SES,\delta}^{BA} \tag{35}$$

where $P_{wind,t}^{DA}$ is the wind power of the DEM at time t, $P_{SES,t}^{DA}$ is the ES power of the DEM at time t, P_δ^{BA} is the total power in the REM at time t, $P_{wind,\delta}^{BA}$ is the wind power of the REM at time δ, and $P_{SES,\delta}^{BA}$ is the ES power of the REM at time δ. The participation of the energy storage unit in the DEM is determined by both its charging power and discharging power. It is the difference between the discharging power and the charging power. Since the energy storage unit can only be in either a charging or discharging state at a given time t, when it is in the discharging state, it generates positive revenue from participating in the DEM. On the other hand, when it is in the charging state, it incurs negative revenue from participating in the DEM. Therefore, the day-ahead power of the energy storage unit participating in the market is the difference between the discharging power and the charging power. The same principle applies to the power of the energy storage unit participating in REM.

$$P_{SES,t}^{DA} = P_{dis,t}^{DA} - P_{ch,t}^{DA} \tag{36}$$

$$P_{SES,\delta}^{BA} = P_{dis,\delta}^{BA} - P_{ch,\delta}^{BA} \tag{37}$$

where $P_{ch,t}^{DA}$ is charging power of the DEM at time t, $P_{dis,t}^{DA}$ is the ES discharging power of the DEM at time t, $P_{ch,t}^{BA}$ is the ES charging power of the REM at time t, and $P_{dis,t}^{BA}$ is the ES discharging power of the REM at time t. All four of these variables take non-negative values.

Constraints on the system are as follows:

(1) the constraints of wind power output are

$$0 \le P_{wind,t}^{DA} \le P_{sum,t}^{DA} \tag{38}$$

$$0 \leq P^{BA}_{wind,\delta} \leq P^{BA}_{sum,t} \tag{39}$$

where $P^{DA}_{sum,t}$ represents the day-ahead forecasted total power for multiple wind turbines at time t, and $P^{BA}_{sum,t}$ represents the real-time forecasted total power for multiple wind turbines at time t.

(2) Energy storage capacity and power constraints: Equation (40) represents the constraint on the ES capacity. Equation (41) represents the minimal and maximal of the charging and discharging powers. Equation (42) restricts the device from charging and discharging energy simultaneously. Equation (43) denotes the energy balance constraint of the ES; it means that the charging-and-discharging capacity during the 24 h regulation process must maintain balance with the initial energy level.

$$0 \leq E^{DA}_{SES,t} \leq E_{SES}, 0 \leq E^{BA}_{SES,\delta} \leq E_{SES} \tag{40}$$

$$\begin{cases} 0 \leq P^{DA}_{ch,t} \leq \mu^{ch}_t P^{SES}_{\max} \\ 0 \leq P^{DA}_{dis,t} \leq \mu^{dis}_t P^{SES}_{\max} \end{cases}, \begin{cases} 0 \leq P^{BA}_{ch,\delta} \leq \mu^{ch}_t P^{SES}_{\max} \\ 0 \leq P^{BA}_{dis,\delta} \leq \mu^{dis}_t P^{SES}_{\max} \end{cases} \tag{41}$$

$$\mu^{char}_t + \mu^{dis}_t \leq 1 \tag{42}$$

$$E^{DA}_{SES,0} = E^{DA}_{SES,T}, E^{BA}_{SES,0} = E^{BA}_{SES,T} \tag{43}$$

where $E^{DA}_{SES,t}$ and $E^{BA}_{SES,t}$ represent the day-ahead and REM storage capacities at different times, P^{SES}_{max} represents the maximum operational power of the SES, μ^{ch}_t and μ^{dis}_t are 0–1 variables that represent the operation status of the ES at time t, and $E^{DA}_{SES,T}$ and $E^{BA}_{SES,T}$ represent the DEM and REM storage capacities at the last time t of the day.

Additionally, the energy iteration relationship of the ES unit in the DEM and REM is shown in Equations (44) and (45):

$$E^{DA}_{SES,t+1} = E^{DA}_{SES,t} + P^{DA}_{ch,t} \cdot \eta^{SES}_t - P^{DA}_{dis,t} / \eta^{SES}_t, E^{DA}_{SES,0} = 60\% E_{SES} \tag{44}$$

$$E^{BA}_{SES,\delta+1} = E^{BA}_{SES,\delta} + (P^{BA}_{ch,\delta} \cdot \eta^{SES}_t - P^{BA}_{dis,\delta} / \eta^{SES}_t)\Delta t, \delta \in [t, t+1], E^{BA}_{SES,0} = 60\% E_{SES} \tag{45}$$

where η^{SES}_t represents the working efficiency of the ES at time t.

Taking into account the dynamic degradation characteristics of EES devices and using SoH as a medium, the changes in the performance parameters of ES are incorporated into the above constraints to reflect the influence of the degradation of ES life on the operation and benefits. The analysis in Section 3 reveals a nonlinear dependence between ES device parameters and SoH. To facilitate the optimization calculation, this nonlinear relationship is first linearized.

By using the idea of piecewise model linearization, the linear relationship between battery internal resistance Z_t and SoH can be fitted as follows:

$$Z_t = \alpha SOH_t + \beta \tag{46}$$

Substituting Equation (46) into the expression for ES charging and discharging powers yields.

$$P^{SES}_{t,\max} = \frac{Z_{start}}{\alpha SOH_t + \beta} \cdot P^{SES}_{0,\max} \tag{47}$$

Similarly, the expression for different segments of ES capacity can be uniformly linearized.

$$E_{SES,t} = aSOH_t + b \tag{48}$$

We can substitute the ES parameter expressions obtained from Equations (47) and (48) into the constraints for ES power and capacity and update them on an hourly basis to account for the dynamic degradation characteristics of the ES device.

(3) Power balance constraints:

$$P_{wind,\delta}^{BA} + g_{SES,t} - d_{SES,t} - P_{wind,t}^{DA} = P_t^{down} - P_t^{up} \tag{49}$$

$$0 \leq P_t^{up} \leq M_3(1 - z_t) \tag{50}$$

$$0 \leq P_t^{down} \leq M_4 z_t \tag{51}$$

where $g_{SES,t}$ and $d_{SES,t}$ represent the charging and discharging quantities respectively, of the ES system, z_t is a binary variable indicating the power imbalance status, and M_3 and M_4 are sufficiently large positive numbers.

Equations (27)–(51) are used to establish the coupling relationship between the DEM and REM of the alliance which is composed of REBs and SES.

5. Cost Allocation Mechanism of Shared Energy Storage

To ensure equitable distribution of investment costs for SES, this paper introduces the concept of a "revenue increase rate" as a measure to quantify the demand level of a REB for SES. This metric evaluates the number of occupied ES resources by the REB and the potential revenue that can be obtained. The analysis in this paper considers the market revenue of a REB, including electricity value and system flexibility, in both day-ahead and real-time balancing markets, and compares them with the benefits obtained when the REB participates individually in the market. Finally, the costs of the SES are allocated to each REB based on comprehensive revenue-increase-rate metrics for REB.

5.1. Electricity Value in the Day-Ahead Market

Before allocating the SES costs, it is essential to calculate the revenue obtained by each alliance member from selling electricity in the day-ahead market. The calculation method of this revenue is obtained by multiplying the declared power with the day-ahead market clearance price. To better solve the cost allocation problem, this paper uses the Vickrey–Clarke–Groves (VCG) mechanism, which is a widely used incentive-compatible mechanism for allocating social welfare [36]. In addition, this mechanism can help define the substitute value of SES for other electricity users. As multiple renewable energy base–energy storage systems jointly quoting can only obtain the alliance's overall quotation curve, this paper analyzes the demand for SES by different REBs through the substitute value method to obtain individual quotation curves for each entity. This method indirectly obtains the individual quotation curves of each entity by comparing the changes in the alliance's overall quotation curve, thereby better solving the cost allocation problem, as in Equation (52).

$$R_{DA,i,t}^{\mathrm{WPP}} = \sum_{t \in \Gamma} \lambda_t^{DA} \cdot (P_{DA,t}^{C*} - P_{DA,-i,t}^{C*}), \forall i,j,t \tag{52}$$

where $R_{DA,i,t}^{WPP}$ represents the revenue obtained by a certain REB in the DEM, $P_{DA,t}^{C*}$ represents the optimal day-ahead declared power obtained by the optimization model, and $P_{DA,-i,t}^{C*}$ represents the optimal declared power corresponding to removing a specific REB.

5.2. Flexibility Value in the Real-Time Balancing Market

The revenue earned by the alliance members' flexibility during real-time operations is known as the flexibility value in the REM, which is obtained by multiplying the respective imbalanced electricity quantity with the corresponding settlement price. When comparing the individual deviation direction with the system's overall deviation direction, if they are opposite, this indicates that the member has alleviated the extent of system deviation, reduced the system's demand for flexibility, and increased overall revenue. Then, the member's revenue is positive. If the directions are the same, it indicates that the member has intensified the degree of system deviation, further increased the system's demand for flexibility, and decreased overall revenue, and the member's revenue is negative. Similar to measuring the energy value of each member in the DEM, we can extract the flexibility value of a member in the real-time market and represent the revenue of REB members in the real-time balancing market through the deviation that appears in the alliance as a whole, as shown in Equation (53):

$$R_{BA,i,t}^{WPP} = \sum_{t\in\Gamma}\sum_{s\in\Omega} \delta_s(\lambda_{s,t}^{+}(P_{s,t}^{C+*} - P_{s,-i,t}^{C+*}) + \lambda_{s,t}^{-}(P_{s,t}^{C-*} - P_{s,-i,t}^{C-*})), \forall s,i,t \tag{53}$$

where δ_s is the possibility of the scenario, $R_{BA,i,t}^{WPP}$ refers to the revenue obtained by the REB in the real-time balancing market, $P_{s,t}^{C+*}$ and $P_{s,t}^{C-*}$ represent the most effective methods for making bids of positive and negative imbalance power of the alliance, respectively, obtained by solving the above optimization model, and $P_{s,-i,t}^{C+*}$ and $P_{s,-i,t}^{C-*}$ represent the positive and negative imbalance power of the alliance, respectively, corresponding to removing a specific wind power merchant.

Similarly, by measuring the revenue obtained by a REB's individual participation in the market from both the day-ahead and real-time dimensions and comparing them with the revenue obtained after forming the alliance, the revenue increment of each REB in the alliance can be obtained, as shown in Equation (54).

$$\Delta R_{i,t}^{WPP} = R_{DA,i,t}^{WPP} + R_{BA,i,t}^{WPP} - R_{i,t,D}^{WPP}, \forall i,t \tag{54}$$

where $\Delta R_{i,t}^{WPP}$ represents the revenue increase in a certain REB in the alliance, and $R_{i,t,D}^{WPP}$ represents the revenue obtained by this REB's individual participation in the market.

Therefore, the revenue increase rate of each REB in the alliance's joint bidding is the ratio of the total revenue increase in this member in the DEM and REM and the total revenue increase in all members in the alliance, as shown in Equation (55).

$$\tau_i^{WPP} = \frac{\sum\limits_{t\in\Gamma} T_{i,t}^{WPP}}{\sum\limits_{i\in M}\sum\limits_{t\in\Gamma} T_{i,t}^{WPP}}, \forall i,t \tag{55}$$

The main revenue improvement rate characterizes the proportion of the revenue that alliance members can obtain in the DEM and REM from two dimensions: the value of electricity energy and the value of flexibility. The coupling connection between the DEM and REM is established by the concept of imbalance power. Through the determination of penalty prices, a REB can be guided to overproduce or underproduce a certain amount of power in the DEM. SES can exert a controlling function over smoothing the power imbalance that arises as a result, thus achieving the goal of maximizing overall revenue. On this basis, the investment and depreciation costs of SES can be allocated according to the aforementioned revenue improvement rate, as shown in Equation (56).

$$C_i^{WPP} = C_{SES} \cdot \tau_i^{WPP}, C_{SES} = C^{inv} + C^{at} \tag{56}$$

where C_i^{WPP} represents the SES cost that must be allocated to a certain REB, and C_{SES} represents the overall cost of SES.

6. Case Study

6.1. Data Description

In this case study, the alliance consists of 10 wind turbines and 1 shared energy storage, and the operational parameters of the SES are shown in Table 1. To model the wind power output uncertainty, this paper adjusted the real electricity generated by a wind power plant in a certain northwest region proportionally as the wind power forecast value, and used Monte Carlo sampling to generate multiple wind power output scenarios for each sub-wind-power plant based on the statistical error of the forecast value. Then, this paper used the scenario reduction method to limit the number of scenarios to 10. The market prices were obtained from publicly available data of the PJM electricity market, and the price penalty factors φ^{down} and φ^{up} in REM were taken as 0.8 and 1.2, respectively.

Table 1. Model parameters.

Parameter	Value	Parameter	Value
T_{ref}	20 °C	μ_{ref}	0.5
a_0	0.0693	a_1	1.04
a_2	4.14×10^{-10}/s	a_3	1.40×10^5
a_4	−0.501	a_5	-1.23×10^5
γ_{cell}	0.0575	ν_{cell}	121
E_{SES}	100 MW·h	P_{max}^{SES}	30 MW
SOH_0	100%	c_P^{SES}	300 USD/kW
c_E^{SES}	1200 USD/kWh	γ	8%

6.2. Results and Discussion

Based on the existing REB, the reasonable allocation of EES battery capacity and power is critical to SES system planning. If the ES capacity allocated to wind turbines in the REB is too small, it will be difficult to effectively absorb wind power. If the ES capacity is too large, the investment and operation costs will be too high, which could significantly affect the financial advantages of SES. This paper studies the effect of SES capacity allocation on alliance revenue. Figure 3 demonstrates that alliance members experience an increase in market revenue as the energy storage capacity rises, reaching a peak at a certain point when the capacity is relatively low. However, surpassing the optimal energy storage capacity linked to maximum profit leads to a subsequent decline in the members' profitability. This decline can be attributed to excessive energy storage capacity, which introduces redundancy in energy storage resources. The resulting high investment and operational costs associated with this surplus capacity contribute to diminished profits for alliance members. In this case study, for most wind turbines, the ES capacity ratio corresponding to the maximum revenue is mostly between 14% and 21%.

Additionally, Figure 4 shows a significant increase in revenue for wind turbines 7–10, indicating that these four wind turbines need to bear a high SES investment cost. To balance the economic profits of the alliance and ES investment costs, the most suitable SES capacity ratio is within the 17–20% interval, which means that the SES capacity allocation for the REB is most suitable within the range of 90–110 MW. This paper chooses 100 MW as the optimal SES capacity configuration for the REB, and studies the optimized operation and cost allocation of the 100 MW.

The scheduling and operation status of SES with or without considering the cost of dynamic degradation of ES are presented in Figure 5.

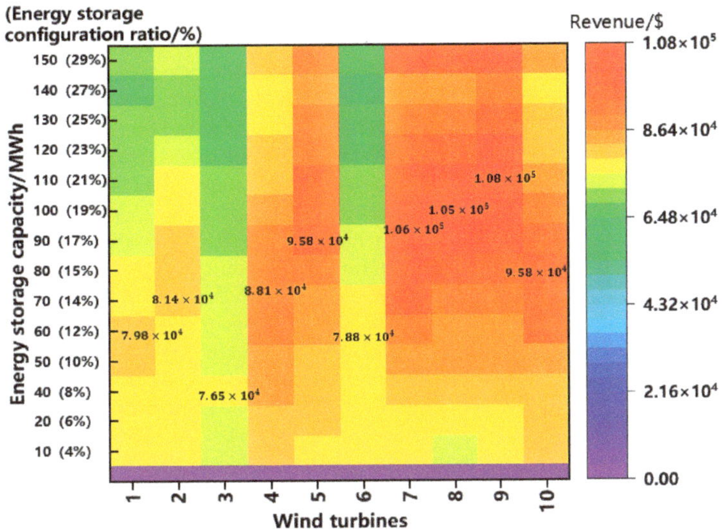

Figure 3. Revenue of alliance members.

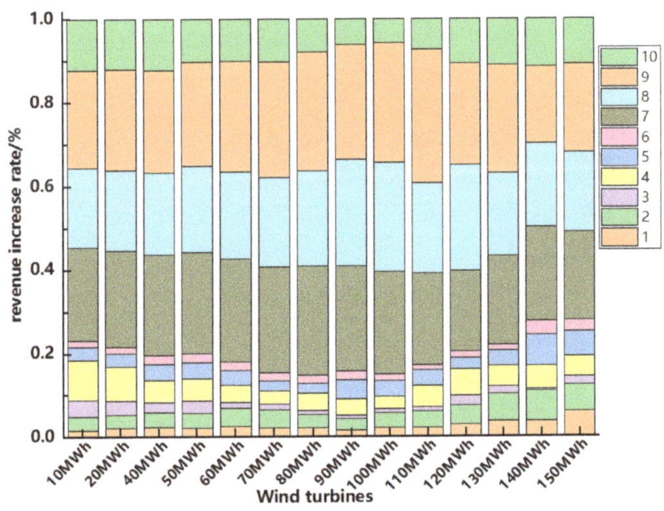

Figure 4. The revenue increase rate of alliance members.

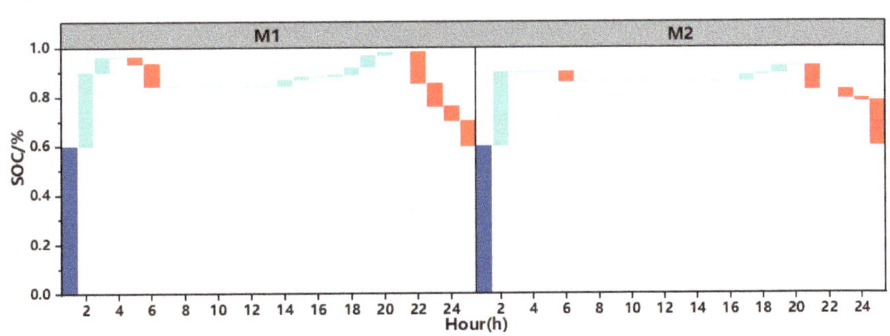

Figure 5. The SOC of energy storage.

In Figure 5, M1 represents the SoC of SES when dynamic attenuation characteristics are not taken into consideration, and M2 represents the SoC of SES when dynamic attenuation characteristics are taken into consideration. And dark blue represents the initial SoC of the ES, green represents an increase in the value of the SoC, and red represents a decrease in SoC. By comparing the SOC of SES with and without considering the dynamic attenuation, it can be observed that when there is a significant difference between the real-time output of the wind turbine and the previously declared power, the SOC of SES changes extensively. This indicates that the REM power of the ES is also substantial, and the storage device can adjust the overall clear power through its charging-and-discharging behavior to maintain the power balance. Consequently, the storage device can minimize the effects of unpredictability in wind power output on system operation, improving the overall alliance revenue. Moreover, compared to the scenario without considering dynamic attenuation, when dynamic attenuation is considered, the SOC change frequency and amplitude of SES are more conservative. This results in a reduced number of charging-and-discharging cycles and a reduced frequency of deep charging or discharging, leading to a smoother change in energy capacity. This prolongs the service life of SES. When dynamic attenuation is not considered, the ES device tends to increase the revenue of various entities in the energy market through frequent charging-and-discharging behavior to maximize overall revenue. However, when dynamic attenuation is considered, the utilization rate of the storage device is significantly reduced to achieve higher total alliance revenue, leading to a lower clear power in the real-time balance market and a lowered frequency of charging-and-discharging behavior.

By analyzing the cross-sections of different stages of SES, it is shown in Figure 6 that when the ES capacity drops below 20%, the optimal scheduling requirement under the initial state has deviated. This will reduce its ability to improve the anti-peak properties of wind power and reduce the imbalance settlement cost of wind power. This further indicates the necessity of considering the dynamic degradation of ES to avoid excessive use of SES.

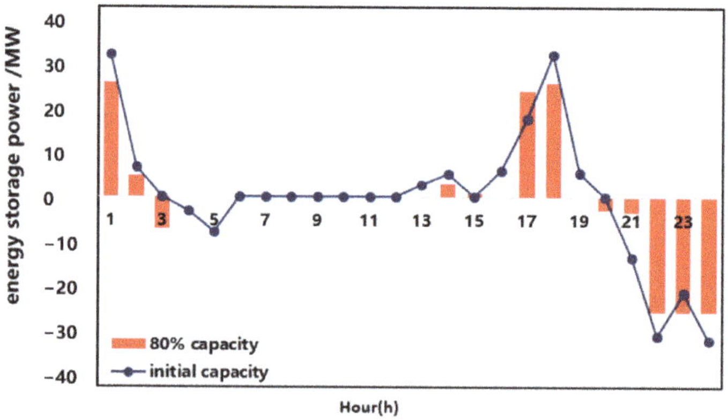

Figure 6. Operation of energy storage before and after degradation.

If fixed decay cost for payment of SES is applied without considering its dynamic characteristic changes, and if ES state parameters are not updated during its usage, and if the charging and discharging costs remain constant and do not change with the changes in charging and discharging capacities, then the health status of the ES cannot be adequately represented, and the overuse of ES cannot be avoided in the sharing mode. This paper proposes that updating the characteristic parameters of ES in a timely manner according to its operating conditions and accounting for the decay cost of ES during different usage stages based on charging and discharging quantities can more reasonably improve the system benefits on the basis of reducing the loss of ES life.

Based on the revenue obtained by wind turbines and SES forming alliances and wind turbines participating in the market individually, the revenue increase brought by forming alliances can be derived. The results are shown in Table 2. Comparing the final revenue situation, it can be seen that after multiple wind turbines share ES, their revenue has increased, but the increase proportion is different, mainly due to the different prediction accuracies of each wind turbine. Wind turbine 9 achieved a 28.61% increase in revenue compared to participating in the market individually, while wind turbine 3, which has the least revenue increase, has also achieved 1.01 times the revenue when participating in the market individually after constructing SES. It is evident that the model proposed in this paper, which involves multiple wind turbines jointly sharing ES to participate in market operation, can take into account the interests of all parties and improve the overall revenue.

Table 2. Revenue increase rate of each wind turbine.

Wind Turbine Number	1	2	3	4	5
Revenue increase rate	2.24%	3.47%	1.00%	2.99%	3.81%
Wind turbine number	6	7	8	9	10
Revenue increase rate	1.51%	24.44%	26.22%	28.61%	5.72%

Comparing the different profits of wind power with the same installed capacity in Figure 7, it can be observed that the distinct profitability of wind power producers with the same capacity are due to different prediction errors. The larger the deviation between the real and reported power data of a wind turbine, the higher the corresponding imbalance costs it incurs, which results in lower actual profits of the RE station than expected. Meanwhile, the profit of a single wind turbine, when forming alliances with other wind turbines and ES, may not increase significantly even if its profitability is high, mainly because the wind turbine itself is already highly matched with the load, and the complementary effects between multiple wind turbines and the regulating function of ES do not significantly reduce their output deviation.

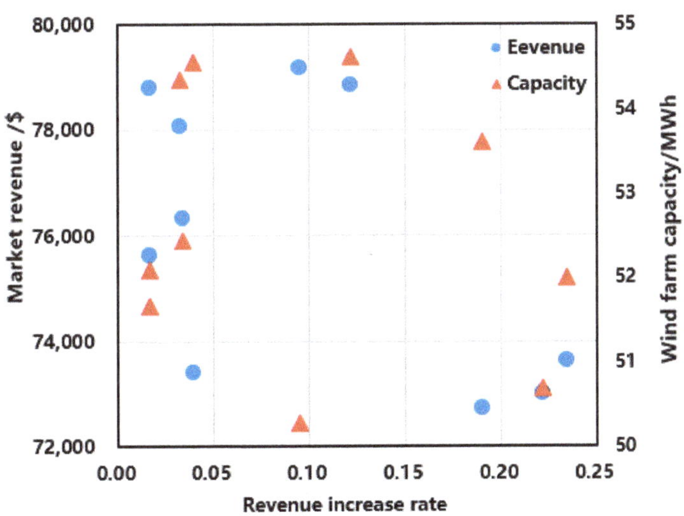

Figure 7. Revenue increase rate of each wind farm.

According to the proportion of revenue improvement from each wind turbine to the total revenue increase, the investment and degradation costs of SES are allocated, and the higher the revenue increase rate of a wind turbine, the more SES cost it needs to bear.

This can ensure enthusiasm for cooperation of all alliance members and the stability of the alliance.

As shown in Figure 8, by comparing the day-ahead and real-time revenues of ten wind farms, it can be seen that the capacity and prediction error of wind turbines both affect their share of investment in SES. In an RE field, wind turbines with larger errors between actual capacity and output and reported values need to bear higher initial investment costs of SES.

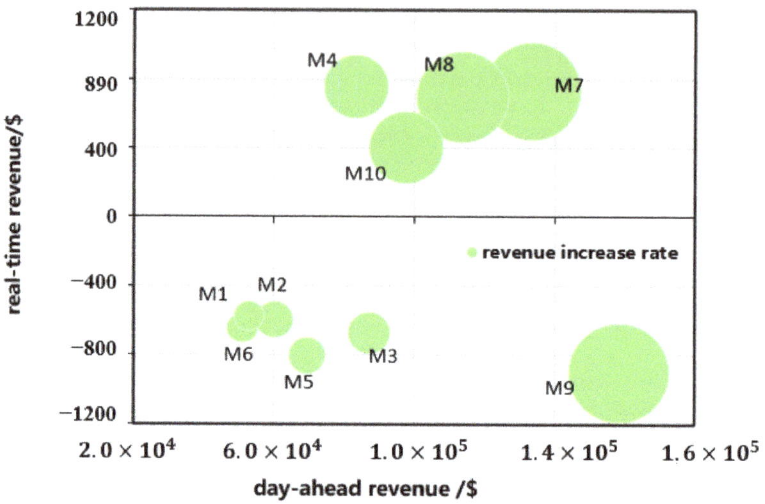

Figure 8. Revenue increase rate of different market members.

The day-ahead revenues of wind farm 3 and wind farm 4 are basically the same, but the investment expense of ES required varies greatly. This is mainly because wind farm 3 has better matching with the load, which can better meet the load demand in each period with its own output characteristics, and therefore has less demand for peak-shaving and filling of ES and flexibility value, resulting in a smaller proportion of corresponding shared investment. The real-time revenue of wind farm 4 and wind farm 7 is basically the same, but the cost of SES is different, indicating that wind turbines with different capacities still need to bear a larger proportion of initial investment costs of ES even if their prediction accuracy is similar.

The impact of wind turbine output prediction accuracy on the cost allocation of an SES alliance is analyzed below, and the results are presented in Figure 9.

As seen in Figures 8 and 9, it can be observed that wind turbines 7, 8, 9, and 10 have higher improvement rates in revenue and higher SES costs. As prediction accuracy improves, the SES cost of each wind turbine unit, especially those with higher revenue increase rates, decreases to varying degrees. This is because the improvement in prediction accuracy reduces the deviation between the pre-bid power and real-time output of each unit, leading to decreased demand for ES capacity and subsequently lowering the SES investment cost. Moreover, the improvement in prediction accuracy reduces the frequency and power of charging and discharging, resulting in a lower degradation cost due to the ES cycle. Therefore, each member of the alliance will also see a reduction in their corresponding SES costs. Additionally, due to the different prediction accuracy of each unit, their demand for ES when participating in the spot market is also different, leading to varying changes in the cost sharing. For turbine 9, which has the highest improvement rate in revenue, when its prediction accuracy improves by 5%, the shared cost can be reduced by about 6%, which is conducive to incentivizing alliance members to actively participate in the spot market and enhancing their market competitiveness.

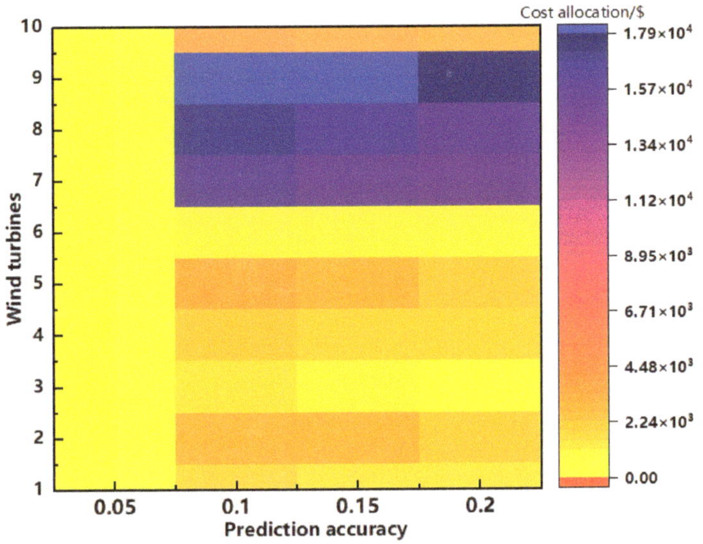

Figure 9. The allocation of ES costs for alliance members under different prediction accuracies.

As shown in Figure 10, the capacity demand of SES in each wind turbine unit in the REB also changes when the electricity price penalty coefficient in the REM of the alliance changes. Specifically, when the overgeneration price penalty coefficient ϕ^{down} decreases or the undergeneration price penalty coefficient ϕ^{up} increases, the optimal SES allocation capacity in the REB will gradually increase. This is because the change in the electricity price penalty coefficient will increase the capacity demand of SES of each wind turbine unit, which will increase the investment expense of SES, and the charging and discharging powers will also increase, resulting in more frequent charging-and-discharging behaviors, and the decay cost will also increase, leading to a continuous increase in the SES cost borne by each unit. However, since the capacity demand for ES of each turbine is different when the electricity price penalty coefficient changes, the increment of their ES allocation cost is also different.

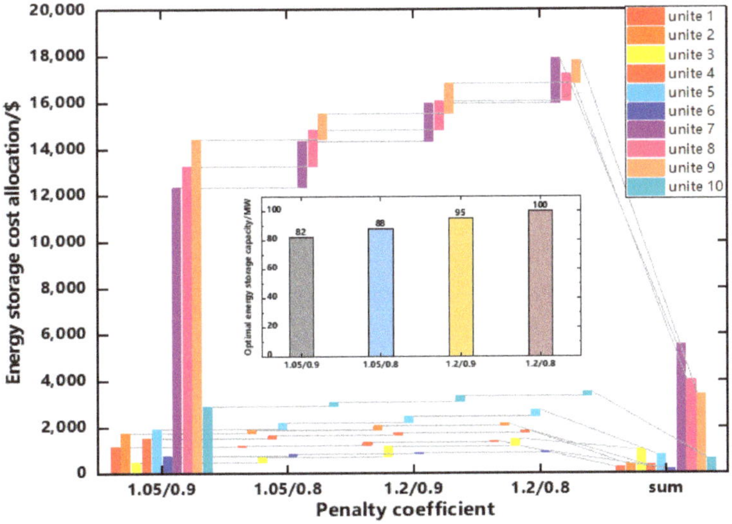

Figure 10. The allocation of ES costs for alliance members under different penalty coefficients.

The output prediction accuracy of each unit is set as shown in Table 3. Combined with Table 3 and Figure 10, for unit 7, 8, and 9, when the electricity price penalty coefficient changes, their increment of ES allocation cost is relatively large. This is mainly because their output prediction accuracy is low, and the day-ahead forecast error is large, which will face higher imbalance penalty fees in the REM. To improve their economic benefits, they will increase the demand for SES to compensate for the fluctuation between their awarded electricity volume and real-time power output, thus reducing the imbalance settlement cost.

Table 3. Prediction accuracy of wind turbine.

Wind Turbine	**1**	**2**	**3**	**4**	**5**
Prediction accuracy	15%	10.5%	6%	20%	13%
Wind turbine	6	7	8	9	10
Prediction accuracy	15%	5%	8%	8.3%	17%

For units with higher prediction accuracy, when the electricity price penalty coefficient changes, their increment of ES allocation cost increases more conservatively. When ϕ^{down} decreases from 0.9 to 0.8, the ES allocation cost for unit 9 increases by USD 1095, which is about 10 times the increment of the ES allocation cost for unit 4. In addition, for wind turbine units with similar installed capacity, such as units 3, 5, and 10, their increments of ES allocation cost are different due to different prediction accuracies. Obviously, for unit 10 with higher prediction accuracy, the increment in ES demand is smaller, and the ES allocation cost increases more conservatively. For unit 3 with lower prediction accuracy, its increment of ES allocation cost is 1.11 times higher than that of unit 10. In fact, according to the cost allocation mechanism proposed in this article, wind power producers must make a trade-off between implementing higher-cost yet more effective prediction technologies and bearing increased shared energy storage investment costs to maximize their own utility.

7. Conclusions

This paper provides a detailed modeling of the degradation of shared energy storage lifespan, and analyzes the impact of dynamic degradation characteristics on the operational strategies and capacity allocation schemes of shared energy storage in renewable energy bases. It establishes an optimization model for the optimal operation of shared energy storage in renewable energy bases, taking into account the dynamic degradation characteristics. Furthermore, a cost allocation mechanism is designed to address the diversity in shared energy storage demands. The following results have been confirmed:

(1) The most suitable capacity ratio for SES in a REB is in the range of 17% to 20%, which can balance the economic benefits of alliances and ES investment costs well. When wind turbines form an alliance with SES, their profits increase compared to when they participate in the market alone. Considering the dynamic decay of ES, the utilization rate of ES is considerably reduced to increase the overall profits of the alliance, and the charging-and-discharging frequency and quantity in the REM are also reduced.

(2) The capacity and prediction errors of wind turbines will affect their share of SES costs. When the prediction accuracy is similar, wind turbines with larger installed capacities need to bear a higher proportion of ES costs, and as the prediction accuracy improves, the SES costs borne by each wind turbine decrease to varying degrees.

(3) The penalty factor of the REM price in the alliance also affects the optimal SES capacity configuration and the SES costs of each wind turbine in the REB. When the penalty factor changes, the capacity demand for SES in each wind turbine in the REB increases, and the SES costs increase. Moreover, wind turbines with lower output prediction accuracy and similar installed capacity have larger increases in SES costs.

In conclusion, the SES optimization model for RE stations, taking into account dynamic decay of EES, is more objective and reasonable. The calculation of profits for different units in the REB has guidance significance for designing SES cost allocation mechanisms. The proposed model is also applicable to photovoltaic power stations. However, there are still some limitations in this paper that can be improved in the future. On the one hand, this paper does not consider network constraints and power flow constraints. In-depth research can be conducted on the capacity configuration of energy storage in renewable energy bases, taking into account power supply security, reliability, and power quality. On the other hand, this study primarily focuses on multiple renewable energy bases and a single shared energy storage system. Future research can investigate the capacity allocation problem for multiple shared energy storage stations.

Author Contributions: Conceptualization and investigation, C.S.; conceptualization and formal analysis, B.Z.; resources and methodology, B.W.; data curation and software, W.L.; writing—original draft preparation, Y.Y.; supervision, formal analysis and writing—review & editing, Z.W. All authors have read and agreed to the published version of the manuscript.

Funding: This research was funded by the State Grid Gansu Electric Power Company Science and Technology Project "Research on Energy Storage Participation in Marketization Trading Mechanism Based on New Power Systems" (Grant number: 52272222000H).

Data Availability Statement: Datasets are available upon reasonable request.

Acknowledgments: The authors are grateful to the editor and anonymous reviewers for their work.

Conflicts of Interest: The authors declare no conflict of interest.

References

1. Wu, Z.; Zhou, M.; Zhang, Z.; Zhao, H.; Wang, J.; Xu, J.; Li, G. An incentive profit-sharing mechanism for welfare transfer in balancing market integration. *Renew. Sustain. Energy Rev.* **2022**, *168*, 112762. [CrossRef]
2. IRENA. *Global Energy Transformation: A Roadmap to 2050*, 2019th ed.; International Renewable Energy Agency: Abu Dhabi, United Arab Emirates, 2019.
3. Wu, Z.; Chen, L.; Wang, J.; Zhou, M.; Li, G.; Xia, Q. Incentivizing the Spatiotemporal Flexibility of Data Centers Toward Power System Coordination. *IEEE Trans. Netw. Sci. Eng.* **2023**, *10*, 1766–1778. [CrossRef]
4. Wu, Z.; Wang, J.; Zhong, H.; Gao, F.; Pu, T.; Tan, C.-W.; Chen, X.; Li, G.; Zhao, H.; Zhou, M.; et al. Sharing Economy in Local Energy Markets. *J. Mod. Power Syst. Clean Energy* **2022**, *11*, 714–726. [CrossRef]
5. Chen, Y.; Yang, Y.; Xu, X. Towards transactive energy: An analysis of information-related practical issues. *Energy Convers. Econ.* **2022**, *3*, 112–121. [CrossRef]
6. Malka, L.; Daci, A.; Kuriqi, A.; Bartocci, P.; Rrapaj, E. Energy Storage Benefits Assessment Using Multiple-Choice Criteria: The Case of Drini River Cascade, Albania. *Energies* **2022**, *15*, 4032. [CrossRef]
7. Wu, Z.; Wang, J.; Zhou, M.; Xia, Q.; Tan, C.-W.; Li, G. Incentivizing Frequency Provision of Power-to-Hydrogen toward Grid Resiliency Enhancement. *IEEE Trans. Ind. Inform.* **2022**, 1–10. [CrossRef]
8. Liu, D.; Cao, J.; Liu, M. Joint Optimization of Energy Storage Sharing and Demand Response in Microgrid Considering Multiple Uncertainties. *Energies* **2022**, *15*, 3067. [CrossRef]
9. Yao, M.; Molzahn, D.K.; Mathieu, J.L. An Optimal Power-Flow Approach to Improve Power System Voltage Stability Using Demand Response. *IEEE Trans. Control Netw. Syst.* **2019**, *6*, 1015–1025. [CrossRef]
10. Chen, L.; Wang, J.; Wu, Z.; Li, G.; Zhou, M.; Li, P.; Zhang, Y. Communication Reliability-Restricted Energy Sharing Strategy in Active Distribution Networks. *Appl. Energy* **2021**, *282*, 116238. [CrossRef]
11. Atawi, I.E.; Al-Shetwi, A.Q.; Magableh, A.M.; Albalawi, O.H. Recent Advances in Hybrid Energy Storage System Integrated Renewable Power Generation: Configuration, Control, Applications, and Future Directions. *Batteries* **2023**, *9*, 29. [CrossRef]
12. Parag, Y.; Sovacool, B.K. Electricity market design for the prosumer era. *Nat. Energy* **2016**, *1*, 16032. [CrossRef]
13. Sousa, T.; Soares, T.; Pinson, P.; Moret, F.; Baroche, T.; Sorin, E. Peer-to-peer and community-based markets: A comprehensive review. *Renew. Sustain. Energy Rev.* **2019**, *104*, 367–378. [CrossRef]
14. Zhu, K.; Victoria, M.; Andresen, G.B.; Greiner, M. Impact of climatic, technical and economic uncertainties on the optimal design of a coupled fossil-free electricity, heating and cooling system in Europe. *Appl. Energy* **2020**, *262*, 114500. [CrossRef]
15. Zhang, C.; Wu, J.; Long, C.; Cheng, M. Review of existing peer-to-peer energy trading projects. *Energy Procedia* **2017**, *105*, 2563–2568. [CrossRef]
16. Chomać-Pierzecka, E.; Sobczak, A.; Soboń, D. Wind Energy Market in Poland in the Background of the Baltic Sea Bordering Countries in the Era of the COVID-19 Pandemic. *Energies* **2022**, *15*, 2470. [CrossRef]
17. Jiang, Y.; Zheng, C. Two-stage Operation Optimization for Grid-connected Wind Farm Cluster with Shared Energy Storage. *Power Syst. Technol.* **2022**, *46*, 3426–3439.

18. Scarabaggio, P.; Carli, R.; Dotoli, M. Noncooperative Equilibrium-Seeking in Distributed Energy Systems Under AC Power Flow Nonlinear Constraints. *IEEE Trans. Control Netw. Syst.* **2022**, *9*, 1731–1742. [CrossRef]
19. Lai, S.; Qiu, J.; Tao, Y. Credit-Based Pricing and Planning Strategies for Hydrogen and Electricity Energy Storage Sharing. *IEEE Trans. Sustain. Energy* **2022**, *13*, 67–80. [CrossRef]
20. Awnalisa, W.; Soongeol, K. Analysis on impact of shared energy storage in residential community: Individual versus shared energy storage. *Appl. Energy* **2021**, *282*, 116172.
21. Karunakaran, V.; Uma, G. Optimal power flow control of hybrid renewable energy system with energy storage: A WOANN strategy. *J. Renew. Sustain. Energy* **2019**, *11*, 015501.
22. Nicola, M.; Paolo, S.; Raffaele, C.; Dotoli, M. Control frameworks for transactive energy storage services in energy communities. *Control Eng. Pract.* **2023**, *130*, 105364.
23. Niu, J.; Tian, Z.; Lu, Y.; Zhao, H. Flexible dispatch of a building energy system using building thermal storage and battery energy storage. *Appl. Energy* **2019**, *243*, 274–287. [CrossRef]
24. Zafar, R.; Ravishankar, J.; Fletcher, J.E.; Pota, H.R. Multi-Timescale Model Predictive Control of Battery Energy Storage System Using Conic Relaxation in Smart Distribution Grids. *IEEE Trans. Power Syst.* **2018**, *33*, 7152–7161. [CrossRef]
25. Wu, J.; Lu, Z.; Qiao, Y.; Yang, H. Wind-Storage Power Station Operation Optimization Considering Dynamic Efficiency Characteristics of Energy Storage Charging and Discharging. *Autom. Electr. Power Syst.* **2018**, *42*, 41–47+101.
26. Sakti, A.; Gallagher, K.G.; Sepulveda, N.; Uckun, C.; Vergara, C.; de Sisternes, F.J.; Dees, D.W.; Botterud, A. Enhanced representations of lithium-ion batteries in power systems models and their effect on the valuation of energy arbitrage applications. *J. Power Sources* **2017**, *342*, 279–291. [CrossRef]
27. Musallam, M.; Johnson, C.M. An Efficient Implementation of the Rainflow Counting Algorithm for Life Consumption Estimation. *IEEE Trans. Reliab.* **2012**, *61*, 978–986. [CrossRef]
28. You, H.W.; Bae, J.I.; Cho, S.J.; Lee, J.M.; Kim, S.-H. Analysis of equivalent circuit models in lithium-ion batteries. *AIP Adv.* **2018**, *8*, 125101. [CrossRef]
29. Roscher, M.A.; Assfalg, J.; Bohlen, O.S. Detection of Utilizable Capacity Deterioration in Battery Systems. *IEEE Trans. Veh. Technol.* **2011**, *60*, 98–103. [CrossRef]
30. Xu, B.; Oudalov, A.; Ulbig, A.; Andersson, G.; Kirschen, D.S. Modeling of Lithium-Ion Battery Degradation for Cell Life Assessment. *IEEE Trans. Smart Grid* **2018**, *9*, 1131–1140. [CrossRef]
31. Förstl, M.; Azuatalam, D.; Chapman, A.; Verbič, G.; Jossen, A.; Hesse, H. Assessment of residential battery storage systems and operation strategies considering battery aging. *Int. J. Energy Res.* **2020**, *44*, 718–731. [CrossRef]
32. Bašić, M.; Vukadinović, D.; Višnjić, V.; Rakić, I. Dynamic Equivalent Circuit Models of Lead-Acid Batteries—A Performance Comparison. *IFAC-PapersOnLine* **2022**, *55*, 189–194. [CrossRef]
33. Lin, C.; Tang, A.; Wang, W. A Review of SOH Estimation Methods in Lithium-ion Batteries for Electric Vehicle Applications. *Energy Procedia* **2015**, *75*, 1920–1925. [CrossRef]
34. USABC. *Electric Vehicle Battery Test Procedures Manual*; Revision 2; Idaho National Engineering Laboratory: Idaho Falls, ID, USA, 1996.
35. Groenewald, J.; Grandjean, T.; Marco, J. Accelerated energy capacity measurement of lithium-ion cells to support future circular economy strategies for electric vehicles. *Renew. Sustain. Energy Rev.* **2017**, *69*, 98–111. [CrossRef]
36. Sessa, P.G.; Walton, N.; Kamgarpour, M. Exploring the Vickrey-Clarke-Groves Mechanism for Electricity Markets. *IFAC-PapersOnLine* **2017**, *50*, 189–194. [CrossRef]

energies

MDPI

Article

Research on Low-Frequency Stability under Emergency Power Supply Scheme of Photovoltaic and Battery Access Railway Traction Power Supply System

Ying Wang [1,2], Yueyang Xin [1,*], Ziyun Xie [1], Xiuqing Mu [1] and Xiaoqiang Chen [1,2]

[1] School of Automation and Electrical Engineering, Lanzhou Jiaotong University, Lanzhou 730070, China; wangying01@lzjtu.edu.cn (Y.W.)

[2] Key Laboratory of Opt-Electronic Technology and Intelligent Control Ministry of Education, Lanzhou 730070, China

* Correspondence: xyylzjtdx@163.com

Abstract: Photovoltaics and batteries can be connected to a traction power supply system through a railway power conditioner (RPC) to switch between different control strategies. This can address power quality issues or provide emergency traction for locomotives that unexpectedly lose power and even break through traditional energy barriers in the railway field, achieving a low-carbon power supply for railway energy, and a mutual backup with substations. However, methods to coordinate the control strategies of PV and the battery locomotive traction have not been clearly revealed, nor has the actual stability of the system. In this study, to address the above issues, an emergency power supply scheme is proposed for the first time that utilizes a dual-mode RPC inverter combined with a coordinated control strategy for the PV and battery, achieving the traction of locomotives. In addition, a one-dimensional impedance model was established for the PV system, battery system, locomotive (CRH3), and RPC projected onto the dq coordinate system, and the critical amplitude margin (CAM) was defined to quantitatively analyze the sensitivity and laws of different parameters concerning the low-frequency stability of the system. At the same time, impedance ratios and passive criteria were used to reveal the stability mechanism, and parameter adjustment criteria and design suggestions were put forward. Finally, the feasibility of the emergency power supply scheme of the "PV–battery locomotive network" coupling system and the correctness of the low-frequency stability study were verified using the Starsim semi-physical experiment platform.

Keywords: emergency power supply; PV–battery locomotive network; railway power conditioner; low-frequency stability; critical amplitude margin; verification and testing

Citation: Wang, Y.; Xin, Y.; Xie, Z.; Mu, X.; Chen, X. Research on Low-Frequency Stability under Emergency Power Supply Scheme of Photovoltaic and Battery Access Railway Traction Power Supply System. *Energies* **2023**, *16*, 4814. https://doi.org/10.3390/en16124814

Academic Editors: Luis Hernández-Callejo, Jesús Armando Aguilar Jiménez and Carlos Meza Benavides

Received: 16 May 2023
Revised: 18 June 2023
Accepted: 19 June 2023
Published: 20 June 2023

1. Introduction

The "2022 Climate Services Status" report released by the World Meteorological Organization (WMO), a specialized agency of the United Nations, states that the clean energy power supply must be doubled by 2030 to limit the global temperature rise. Railway transportation is a significant contributor to energy consumption and carbon emissions in the transportation sector due to its massive electricity usage. With the depletion of fossil fuels and the consequences of global warming, there is mounting pressure on the railway industry to take action and reduce its carbon emissions. To achieve the green transformation of railways, progress has been made internationally in the adoption of new energy sources for railways: JR-East has installed 453 kW solar panels at Tokyo Station, serving locomotives on the Tokaido line 3 [1]; the subway operator in Santiago, Chile, built two solar photovoltaic power stations in 2017, supplying 60% of the subway's electricity and achieving a renewable energy utilization rate of 76% [2]; 100% of the primary energy used by the Dutch railway is provided by wind [3]; a 2.2 MW rooftop photovoltaic system has been built at Wuhan Railway Station in China; a 10 MW solar power generation device has been installed on the roof of Hangzhou East Station [4]; the total installed capacity of the photovoltaic system of the Xiong'an high-speed railway in China is 6 MW [5]. The photovoltaic (PV) resources along China's AC-electrified railways are abundant, and their high proportion of utilization can promote China's achievement of the "dual carbon" goal and increase the national independent contribution under the global energy conservation and emission reduction goal [6].

Currently, access methods of PV and battery devices are roughly divided into two categories: one is to access the battery device from the side of the locomotive to recover braking energy and control the voltage fluctuation of the traction network [7], and the other is to access the traction network side. For example, China's electrified railway mainly considers access to PV and battery devices from the high-voltage side (110 kV) and the traction side (27.5 kV). There are two types of PV and battery inverters: one is a three-phase inverter combined with a three-phase/two-phase transformer, and the other is a back-to-back single-phase inverter [8]. The high-voltage side access method is suitable for small-capacity PV and battery devices, and the large capacity required for three-phase inverters leads to higher costs. At present, the railway power conditioners (RPCs) that utilize single-phase inverters to connect multiple source devices from the traction side have received frequent attention due to their excellent comprehensive performance [9]. The authors of [10–14] studied the unbalanced power compensation and harmonic control of the traction power supply systems with different RPC topologies. In [15], the train braking energy was recycled and reused through a RPC in conjunction with battery devices. In [16], a coordinated control scheme based on a RPC for PV and battery access to a traction power supply system was proposed, utilizing new energy to provide additional active power. Most of the above studies considered using PV or battery compensation power to reduce the consumption of fossil fuels but electricity is still mainly provided by thermal power generation. In [17], an overview of an emergency traction scheme for locomotive and substation coordination based on battery devices was presented and this paper aims to solve the problem of sudden accidents such as substation failures leading to locomotive power loss. In this state, the locomotive is completely powered by the battery. However, the research on the combined application of PV and battery for emergency traction in locomotives is not yet widely explored, and more research is needed to explore its potential and feasibility, if the PV power is connected to the battery for charging and collaborates with the battery to traction the locomotive, almost all of the electrical energy comes from renewable energy. However, to design the emergency power supply scheme for PV and battery systems and identify whether the collaborative integration of PV and battery systems can further change the railway energy supply system to achieve the long-term stable full-power traction of locomotives, the above-mentioned problems are worth deep research and solving. Due to the wide coverage of Chinese railways and superior energy storage conditions, as well as the long routes and abundant solar resources along such lines, solar power for AC-electrified railways has broad application prospects and extensive benefits. Therefore, this article takes the AC power supply system of Chinese railways with a rated voltage of 27.5 kV as an example and proposes the use of RPC access to PV and battery devices to achieve the emergency or long-term traction of locomotives, thus, expanding the function of RPCs and promoting the consumption of PV resources. In addition, due to differences in operational modes and design parameters between DC railways and AC railways, this article does not explore issues related to DC railways.

However, it is also important to determine whether the system remains stable when using the PV and battery locomotive traction and to reveal the main factors and laws that affect the stable operation of the system; all of these issues need to be addressed prior to the implementation of PV and battery traction locomotives in practice. At present, there are multiple frequency-scale instability issues in the traction power supply system of railways, such as low-frequency oscillation (LFO), harmonic resonance, and harmonic instability. Among them, LFO has been commonly reported in the electric railway domain around the world, thus, attracting widespread research attention [18,19]. The first reported occurrence of the LFO phenomenon dates back to 1996 in Norway when a rotating converter was adopted in a traction substation [20]. Since then, in Germany, Sweden, and Norway, rotating converters have been used to interface the traction network, transforming three-phase utility power into a single phase for the catenary network [21]. The main cause of LFO has been attributed to the electromechanical characteristics of rotating converters [22]. However, for other electric railway systems [23,24], the mechanism of LFO remains unclear. Recently, from 2008 to 2016, there has been an increasing frequency of LFOs observed in Chinese railways [25,26]. Many studies have revealed that the occurrence of LFO in China is caused by the introduction of more converters and impedance mismatch [27–29]. Concurrently, locomotives have a LC resonant onboard filter that may oscillate, in addition, such a filter combined with the line can be triggered in oscillation by electric arcs, which is quite a commonplace phenomenon caused by the sliding contact mechanism [30]. In PV and battery locomotive traction, more converters may be introduced, which may further induce LFO. Research on the low-frequency stability of multi-source connected traction power supply systems is not yet complete. In [31], the integration of PV power into the traction network through a RPC was considered and an impedance model of a "PV-locomotive-traction network" was established. It was found that unreasonable parameter settings during the integration of the PV systems did indeed induce LFO; at the same time, the parallel connection of the PV modules led to multiple increases in the converters, which can also lead to LFO in the system. However, the instability mechanism of the system and methods to improve the system stability were not specifically revealed. The variation in the source impedance caused by the parallel connection of the subsystem modules and the parameter adjustment of the converter controller may lead to artificial active enhancement or the weakening of the low-frequency stability of the system. Therefore, to further improve the system stability, it is necessary to specifically reveal the sensitivity and law of the influence of controller parameters on the system's stability.

Regarding the above issues, this article proposes an emergency power supply scheme based on RPC access to PV and battery devices in Section 2.1. Through coordinated control strategies, PV and batteries can be used independently for the day and night emergency traction of locomotives. They may achieve "low-carbon" locomotive operation and also serve as a backup power source for the long-term traction of locomotives. On this basis, an impedance model of the "PV–battery locomotive network" coupling system under this scheme is established on the RPC DC bus side in Section 2.2. Then, the critical amplitude margin is defined for the first time based on the impedance ratio criterion; the influence of parameters on the low-frequency stability of the system is quantitatively evaluated and passive evidence is introduced to reveal the mechanism of the influence of the parallel number of PV and battery modules on the stability. Furthermore, parameter adjustment criteria and main circuit impedance reshaping governance are proposed to prevent LFO in Section 2.3. Finally, the feasibility of the emergency power supply scheme and the correctness of the stability study are verified through testing on the Starsim platform in Section 3.

2. Materials and Methods

2.1. Proposal of Emergency Power Supply Scheme

2.1.1. System Topology

The topology of the "PV–battery locomotive network" system is shown in Figure 1, which includes a traction network, a high-speed locomotive, a railway power conditioner, and a new energy power supply system composed of PV and a battery.

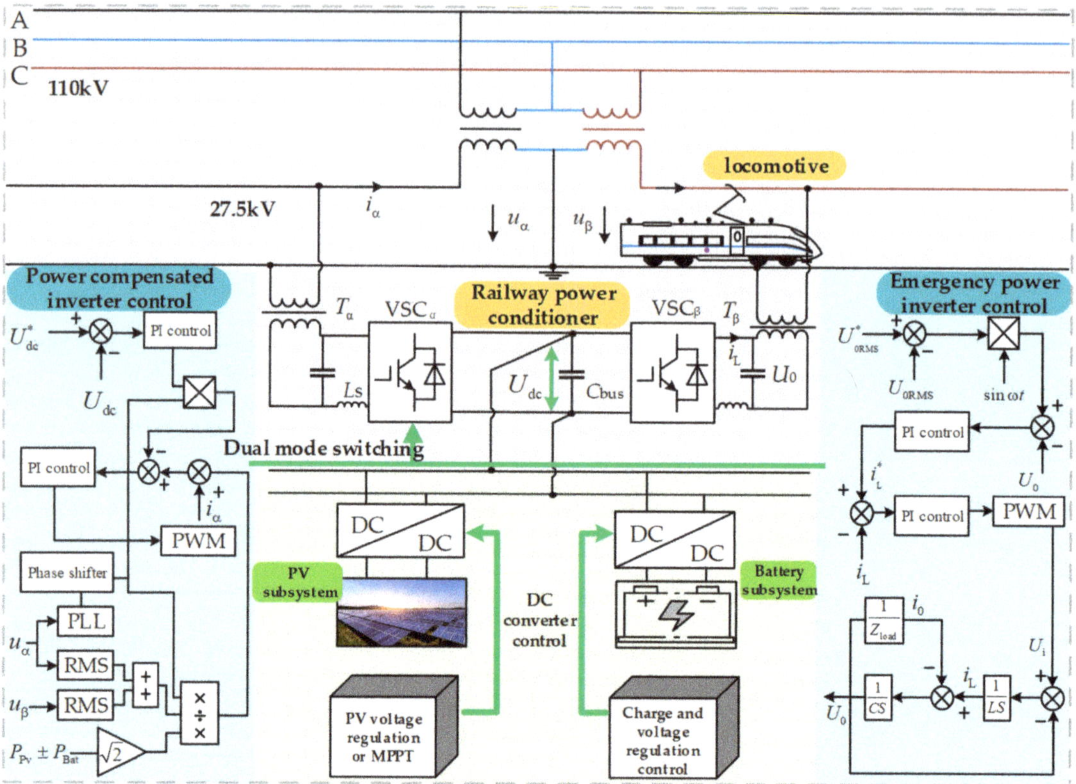

Figure 1. System topology and control block diagram.

The three-phase electricity from the grid is converted from the traction substation to a 27.5 kV single-phase AC current, which is sent to the α and β power supply arms. The PV and battery are integrated into a 2 kV DC bus via their respective DC/DC converters, and the RPC-side single-phase inverter and LC-type filtering circuit are used to convert the power into 1.5 kV single-phase AC power through step-up transformers T_α and T_β, which is then sent to the α and β power supply arms. The U_{dc} is the DC voltage on the input side of the RPC; U_0 is the single-phase AC voltage on the output side of the filtering circuit; U_i is the unfiltered voltage of the RPC; C_{bus} is the capacitor on the DC side of the RPC; VSC_α and VSC_β are single-phase inverters in RPC; i_α, u_α, and u_β are the corresponding voltage and current of the supply arm; RMS means root-mean-square extraction; P_{PV} and P_{Bat}, respectively, represent the output power of the PV and battery; PLL stands for phase-locked loop; U_{0RMS} is the root-mean-square value of U_0; i_L is the output current of the RPC inverter; i_0 is the output current after filtering; and $\sin\omega t$ is the reference sine wave and Z_{load} is the equivalent load in Section 2.2.2. DC converters in PV systems mainly include voltage regulation control and maximum power point tracking control (MPPT), while in battery systems they mainly include charging and voltage regulation control. Under the emergency power supply scheme, the PV and battery systems provide the locomotive's traction but it is still necessary to consider the overall plan for the connection of the PV and battery systems to the traction power supply system to formulate the control strategy.

The RPC converter is a dual-mode single-phase inverter. Its control strategy is shown in Figure 1. When it is applied to power compensation, the voltage U_0 is controlled to track the phase frequency of the traction network voltages u_α and u_β, and the output signal is compared with the harmonic and negative sequence reference signal, followed by the PWM input signal from the PI controller. At this time, the PV and battery are used as auxiliary power supplies, and the traction net is used as the main power source to control the single-phase inverter to support the voltage amplitude and frequency of both the DC and AC sides of the RPC. In the emergency power supply scheme, the PV and battery provide the DC-side voltage, the U_{dc} amplitude, and the AC-side U_0 frequency support through a DC converter, and the inverter independently outputs the sinusoidal AC power. At this time, the PV and battery serve as the main power supply. The DC converter coordinates with the RPC inverter according to the actual working conditions to achieve traction network voltage management and locomotive traction under different power supply schemes.

2.1.2. Division of Multiple Working Modes

The sum of the power inputs and outputs of the PV, battery, and RPC ports will affect the DC bus voltage U_{dc}, as shown in Figure 1. It is necessary to balance the power of each port through the division of the working modes. The flowchart for the system's power control is shown in Figure 2.

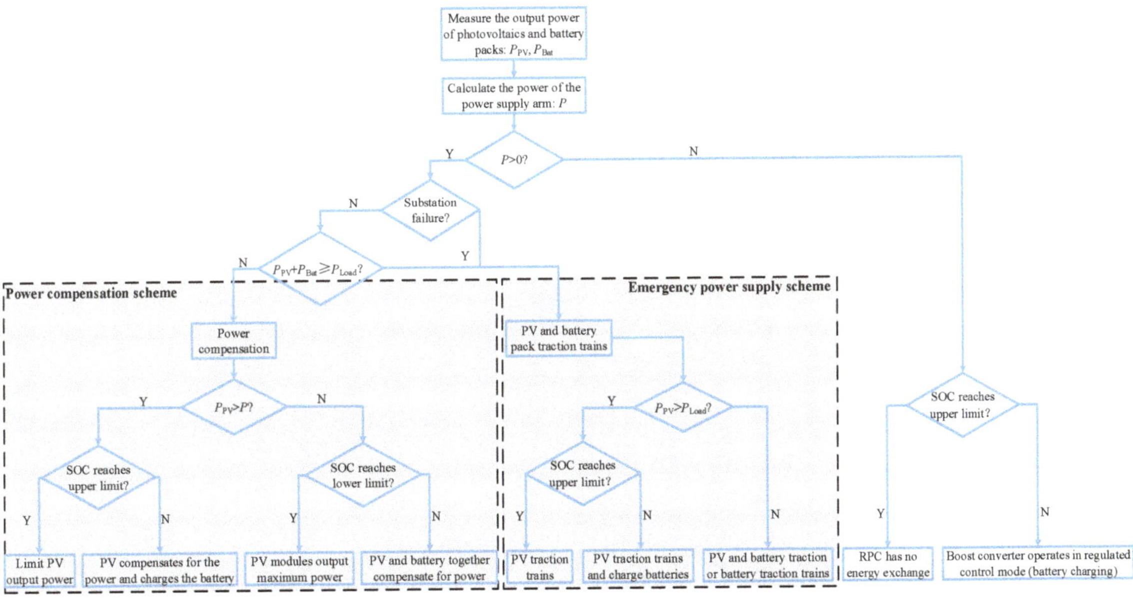

Figure 2. Flowchart for the traction power supply system's power control.

In the above figure, P_{PV} is the output power of the PV system, P_{Bat} is the output power of the battery system, P_{Load} is the maximum operating power of locomotives, and P is the power of the power supply arm calculated by measuring the voltage and current of the power supply arm. Based on the positive, negative, and zero values of P, the locomotive's operating conditions can be distinguished, which are namely, traction, braking, and no-load. In cases where $P \leq 0$, if there is remaining capacity in the battery, it can be used to absorb the energy generated by the braking of the locomotive, as well as the energy provided by the PV system; In cases where $P > 0$ but the total power generated from the PV and battery systems is still lower than P_{load}, the RPC system will operate under the power compensation scheme. However, if the traction substation experiences a fault or the output power generated from the PV and battery systems goes beyond P_{load}, the RPC system will automatically switch to the emergency power supply scheme in order to utilize the PV and battery systems for the locomotive traction. Furthermore, the state-of-charge (SOC) threshold value is considered to prevent over-charging and over-discharging from affecting the battery life.

This article focuses on the study of using PV and battery systems for emergency locomotive traction. For different working conditions, such as different levels of irradiance and state-of-charge (SOC), the proposed emergency power supply scheme includes different working modes as follows:

Mode One: When the PV power generation is much greater than the sum of the maximum operating power of the load and the lithium battery charging power, i.e., $P_{PV} \gg P_{Load} + P_{Bat}$, and the SOC is less than 90%, the PV system can be used to provide the locomotive's traction and simultaneously charge the battery. When $P_{PV} \gg P_{Load} + P_{Bat}$ but the SOC is above 95%, in order to prevent the battery from over-charging, the PV system will be responsible for the locomotive's traction while the battery will be on standby.

Mode Two: When the PV power generation is slightly greater than the maximum operating power of the load but there is no additional power to charge the battery, that is, $P_{Load} < P_{PV} \leq P_{Load} + P_{Bat}$, and the SOC is less than 90%, the PV system will only supply power to the locomotive, and the battery will be on standby.

Mode Three: When the PV power generation is less than the maximum operating power of the load, that is, $P_{PV} < P_{Load} < P_{PV} + P_{Bat}$, and the battery's SOC is higher than 5%, the PV and battery will work in voltage control mode to jointly provide the locomotive's traction.

Mode Four: When the PV power generation approaches zero, that is, $P_{PV} < P_{Load} < P_{PV} + P_{Bat}$, and the battery's SOC is higher than 5%, the PV system will be on standby, and the battery will provide the locomotive's traction.

The system topologies under the above four modes are different, meaning that it will be time-consuming and complex to analyze the system stability from the RPC AC side by using the impedance method. However, the operating conditions of the PV and battery combined locomotive traction under mode three are relatively common. Meanwhile, when carrying out RPC DC bus segmentation modeling, since the control method of each mode's DC converter is similar to mode three, the parallel connection or split of the DC converter output impedance is equivalent to the switching of modes, and the impedance resolution of the other modes becomes a high-proportion intersection or a subset of mode three, which is convenient to quickly reveal the factors and regularity affecting the stability under multiple working modes. The next section builds a mathematical model for the "PV–battery locomotive network" coupling system using the specific control method of mode three.

2.2. Impedance Modeling of Proposed Emergency Power Supply Scheme

2.2.1. PV and Battery Power Generation Systems

Due to how the topology and control of PV and battery systems are similar, this section will describe and establish the impedance model of the power generation system simultaneously. As shown in Figure 3, in the Boost circuit, u_{PV} is the output voltage of the PV side; i_{PV} is the output current of the PV side; i_{LPV} is the inductor current of the PV side; u_{Bus} and i_{Bus} are the DC bus voltage and current; L_{PV} is the energy storage inductor of the PV side; C_{PV} is the support capacitor of the PV side; C_{Bus1} is the DC-side capacitor of the RPC; d_{PV} is the duty cycle of the Boost converter's switching tube, $d^*_{PV} = 1 - d_{PV}$; and u^*_{Bus} and i^*_{LPV} are the reference signals for their corresponding variables. In the DC/DC circuit, u_{Bat} is the output voltage of the lithium battery side, i_{Bat} is the output current of the lithium battery, L_{Bat} is the energy storage inductor, C_{Bus2} is the DC-side capacitor of the RPC, d_{Bat} is the duty cycle of the T_1 switching tube, and the T_2 switching tube is complementary to the T_1 switching tube. In addition, u^*_{Bus} and i^*_{Bat} are the reference signals for their corresponding variables.

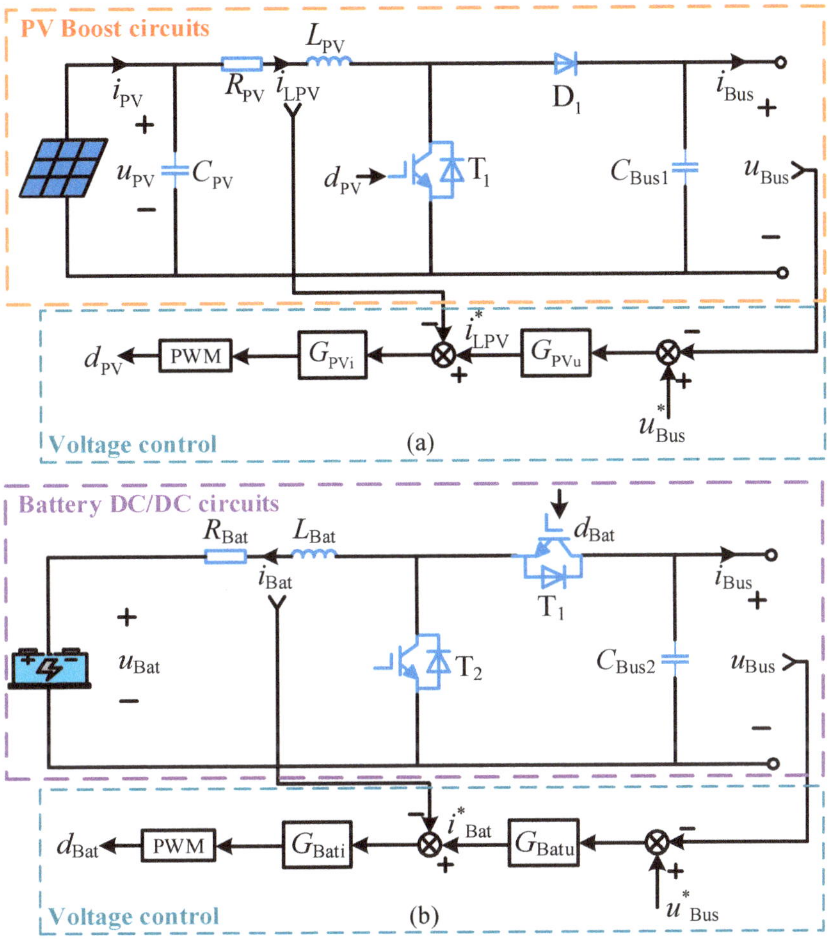

Figure 3. Topology and control of DC converters under mode three. (**a**) PV boost converter; (**b**) battery DC/DC converter.

The state equations for the PV and battery systems are given by Equations (1) and (2), respectively.

$$\begin{cases} L_{PV}\frac{di_{LPV}}{dt} + R_{PV}\cdot i_{LPV} = u_{PV} - d^*_{PV}\cdot u_{Bus} \\ C_{Bus1}\frac{du_{Bus}}{dt} = d^*_{PV}\cdot i_{LPV} - i_{Bus} \\ i_{PV} = i_{LPV} + C_{PV}\frac{du_{PV}}{dt} \end{cases} \tag{1}$$

$$\begin{cases} L_{Bat}\frac{di_{Bat}}{dt} + R_{Bat}\cdot i_{Bat} = u_{Bat} - d_{Bat}u_{Bus} \\ C_{Bus2}\frac{du_{Bus}}{dt} = d_{Bat}i_{Bat} - i_{Bus} \end{cases} \tag{2}$$

With small signal linearizing Equations (1) and (2), we obtain Equations (3) and (4):

$$\begin{cases} (sL_{PV} + R_{PV})\cdot\Delta i_{LPV} = -D^*_{PV}\cdot\Delta u_{Bus} + U_{Bus}\cdot\Delta d_{PV} + \Delta u_{pv} \\ sC_{Bus1}\cdot\Delta u_{Bus} = D^*_{PV}\cdot\Delta i_{LPV} - I_{LPV}\cdot\Delta d_{PV} - \Delta i_{Bus} \\ \Delta i_{PV} = \Delta i_{LPV} + sC_{PV}\cdot\Delta u_{PV} \end{cases} \tag{3}$$

$$\begin{cases} (sL_{Bat} + R_{Bat})\Delta i_{Bat} = -D_{Bat} \cdot \Delta u_{Bus} - U_{Bus} \cdot \Delta d_{Bat} + \Delta u_{Bat} \\ sC_{Bus2}\Delta u_{Bus} = D_{Bat} \cdot \Delta i_{Bat} + I_{Bat} \cdot \Delta d_{Bat} - \Delta i_{Bus} \end{cases} \quad (4)$$

According to the superposition theorem, the transfer functions between each variable of the PV and battery systems can be obtained from Equations (3) and (4), and the specific expressions are shown in Appendix A, where Equation (A1) includes the transfer function expansion of the PV system, and Equation (A2) includes the transfer function expansion of the battery system. Then, based on the control strategies of the PV and battery systems, draw the closed-loop block diagrams of the small signal transfer functions as shown in Figure 4.

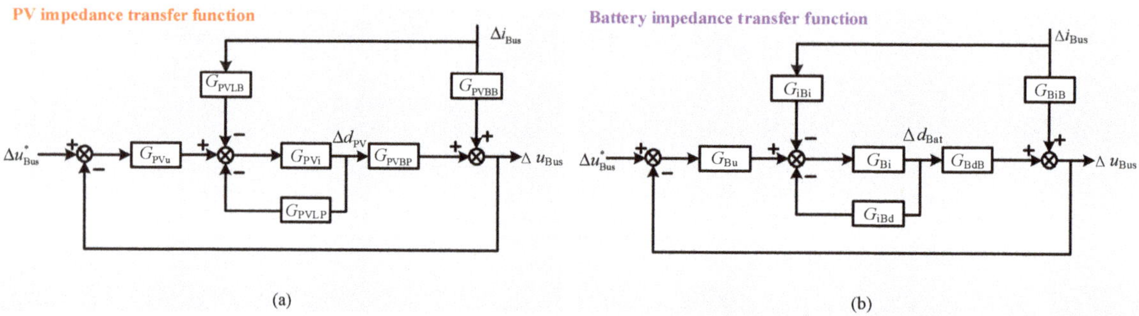

Figure 4. Small-signal closed-loop control block diagram. (**a**) PV transfer function; (**b**) battery transfer function.

Using the Mason's formula, the output impedances of the PV and battery systems are, respectively:

$$Z_{PV}(s) = \frac{\Delta u_{Bus}}{\Delta i_{Bus}} = \frac{G_{PVBB} + G_{PVBB}G_{PVi}G_{PVLP} - G_{PVLB}G_{PVi}G_{PVBP}}{1 + G_{PVi}G_{PVLP} + G_{PVu}G_{PVi}G_{PVBP}} \quad (5)$$

$$Z_{Bat}(s) = \frac{\Delta u_{Bus}}{\Delta i_{Bus}} = \frac{G_{BiB} + G_{BiB}G_{Bi}G_{iBd} - G_{iBi}G_{BdB}G_{Bi}}{1 + G_{Bi}G_{iBd} + G_{Bu}G_{Bi}G_{BdB}} \quad (6)$$

2.2.2. Locomotive Network and Railway Power Conditioner

In the study of the low-frequency stability of locomotive network coupling systems, two-level locomotives are often considered. The China railway high-speed 3 (CRH3) locomotive with four power units is the research object of this paper. In [32], the impedance expression of a single power unit is presented. After converting the impedance of the China Railway High-speed 3 (CRH3) locomotive to the RPC AC side, the expression is:

$$Z_{CRH3}(s) = \frac{{k_a}^2{k_b}^2U_{s2}}{4I_{s2}} = \frac{{k_a}^2{k_b}^2[1 + G_K G_{PWM} G_{Ln}(G_{CRH3up}G_{CRH3ui}G_{switch}G_{circuit} + G_{CRH3ui})]}{4G_{Ln}(1 - G_{delay}G_{PWM})} \quad (7)$$

where the transformation ratio of the onboard transformer is k_a and the transformation ratio of the RPC-side step-up transformer is k_b.

At present, the equivalent circuit model is commonly used in low-frequency stability analysis of traction networks [31] and its expression corresponds to the RPC AC side:

$$Z_s(s) = k_b^2 \begin{bmatrix} R_s + sL_s & -\omega_0 L_s \\ \omega_0 L_s & R_s + sL_s \end{bmatrix} \quad (8)$$

The equivalent impedance of the locomotive network converted to the RPC AC side is recorded as $Z_{load}(s)$, where $Z_{load}(s) = Z_s(s) + Z_{CRH3}(s)$.

RPC impedance modeling is conducted on the DC port of the inverter, that is, the input impedance of the RPC is studied. Considering that the back-to-back inverter in the RPC has a symmetrical structure and the impedance calculation and modeling method are similar, the β power supply arm is selected as the research object. For the emergency power supply scheme, the specific control method and topology of the inverter are shown in Figure 5.

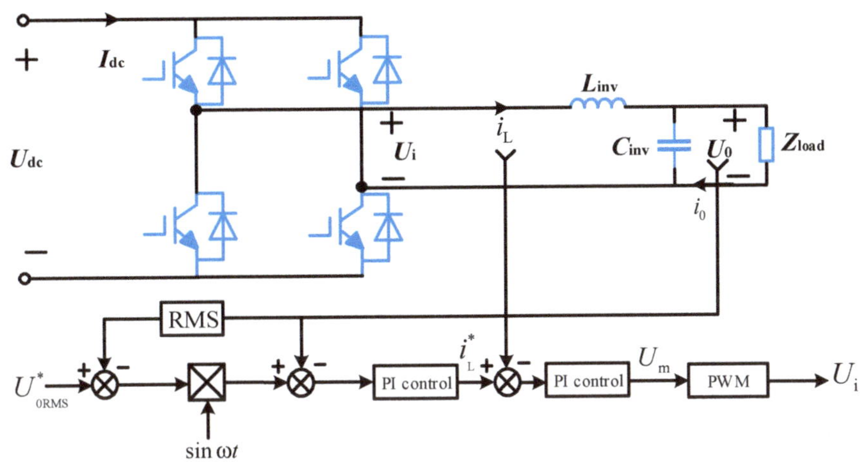

Figure 5. RPC inverter main circuit topology and control.

In the above figure, Z_{load} is the equivalent impedance of the locomotive network converted to the RPC AC side; U_{m} is the amplitude of the modulated wave signal; and $\sin\omega t$ is the sine wave frequency reference signal. Linearization is performed on the DC side and the variables are decomposed into a closed-loop impedance model in the dq coordinate system. According to the main circuit, the state equation under dq decoupling can be derived by:

$$\begin{cases} L_{\text{inv}}\frac{di_{Ld}}{dt} = U_{\text{id}} - U_{\text{0d}} + \omega L_{\text{inv}} I_{\text{Lq}} \\ L_{\text{inv}}\frac{di_{Lq}}{dt} = U_{\text{iq}} - U_{\text{0q}} - \omega L_{\text{inv}} I_{\text{Ld}} \\ C_{\text{inv}}\frac{du_{0d}}{dt} = I_{\text{Ld}} - I_{\text{0d}} + \omega C_{\text{inv}} U_{\text{0q}} \\ C_{\text{inv}}\frac{du_{0q}}{dt} = I_{\text{Lq}} - I_{\text{0q}} - \omega C_{\text{inv}} U_{\text{0d}} \\ U_{\text{0d}} = Z_{\text{load}} \cdot I_{\text{0d}} \\ U_{\text{0q}} = Z_{\text{load}} \cdot I_{\text{0q}} \end{cases} \quad (9)$$

Based on the basic derivation process of SPWM, the relationship between the output voltage U_{i} before filtering and the modulation wave U_{m} can be obtained using the switch average period method:

$$\begin{cases} U_{\text{id}} = 0.5U_{\text{dc}}U_{\text{md}}/U_{\text{tri}} \\ U_{\text{iq}} = 0.5U_{\text{dc}}U_{\text{mq}}/U_{\text{tri}} \end{cases} \quad (10)$$

In Equation (10), U_{tri} is the amplitude of the triangular carrier wave, and the remaining symbols are the system variables under dq decoupling. In the dual-loop control strategy, the response speed of the inner loop is much faster than that of the outer loop. This article assumes that the response of the current inner loop is approximately the same within the average time of a single switching cycle. Due to the PI controller used in the voltage outer loop, the input–output relationship of the controller is:

$$\begin{cases} (k_{\text{invp}} + k_{\text{invi}}/s) \cdot (U_{\text{vd}}^* - U_{\text{0d}}) = U_{\text{md}} \\ (k_{\text{invp}} + k_{\text{invi}}/s) \cdot (U_{\text{vq}}^* - U_{\text{0q}}) = U_{\text{mq}} \end{cases} \quad (11)$$

where U^*_{vd} and U^*_{vq} are the product of the effective value of the voltage outer-loop output and the reference sine signal; k_{invp} and k_{invi} are proportional integral parameters of the voltage outer-loop PI controller.

According to the input power and output power of the RPC that are equal:

$$U_{dc}I_{dc} = 1.5(U_{0d}I_{0d} + U_{0q}I_{0q}) \tag{12}$$

After processing the small signal in Equation (12), the following is obtained:

$$\begin{bmatrix} I_{dc} \\ U_{dc} \end{bmatrix}^T \begin{bmatrix} \Delta u_{dc} \\ \Delta i_{dc} \end{bmatrix} = 1.5\left(\begin{bmatrix} U_{0d} \\ U_{0q} \end{bmatrix}^T \begin{bmatrix} \Delta i_{0d} \\ \Delta i_{0q} \end{bmatrix} + \begin{bmatrix} I_{0d} \\ I_{0q} \end{bmatrix}^T \begin{bmatrix} \Delta u_{0d} \\ \Delta u_{0q} \end{bmatrix}\right) \tag{13}$$

Equations (9) to (11) can be substituted into Equation (13) to obtain the input admittance on the DC side of the β arm inverter:

$$Y_{inv} = \frac{\Delta i_{dc}}{\Delta u_{dc}} = \frac{3}{2U_{dc}}\left(\begin{bmatrix} U_{0d} \\ U_{0q} \end{bmatrix}^T (G_{Zload} + G_{Zload}^T) \begin{bmatrix} \frac{\Delta u_{0d}}{\Delta u_{dc}} \\ \frac{\Delta u_{0q}}{\Delta u_{dc}} \end{bmatrix}\right) - \frac{I_{dc}}{U_{dc}} \tag{14}$$

As output impedance is the reciprocal of admittance, it can be inferred that $Z_{inv} = 1/Y_{inv}$ and the variables in Equation (14) are as follows:

$$\begin{cases} \begin{bmatrix} \frac{\Delta u_{0d}}{\Delta u_{dc}} \\ \frac{\Delta u_{0q}}{\Delta u_{dc}} \end{bmatrix} = 0.5[G_{Linv}(G_{Cinv} + G_{Zload}) + E - 0.5U_{dc}G_{PI}]^{-1} \begin{bmatrix} U_{md} \\ U_{mq} \end{bmatrix} \\ G_{PI} = \begin{bmatrix} s & -\omega \\ \omega & s \end{bmatrix}^{-1} \begin{bmatrix} -(sK_{invp} + K_{invi}) & \omega K_{invp} \\ -\omega K_{invp} & -(sK_{invp} + K_{invi}) \end{bmatrix} \\ G_{Linv} = \begin{bmatrix} sL_{inv} & -\omega L_{inv} \\ \omega L_{inv} & sL_{inv} \end{bmatrix} \\ G_{Cinv} = \begin{bmatrix} sC_{inv} & -\omega C_{inv} \\ \omega C_{inv} & sC_{inv} \end{bmatrix} \\ G_{Zload} = \begin{bmatrix} Z_{load} & 0 \\ 0 & Z_{load} \end{bmatrix}^{-1} \end{cases} \tag{15}$$

2.2.3. Verification of Established Impedance Model

To verify the established one-dimensional impedance model, the disturbance injection method was used for the simulation measurements [33]. By injecting disturbance signals with specific frequencies during the steady-state state of the system and using the Fourier transform to process these signals, the impedances of the three ports of the "PV–battery locomotive network" can be simultaneously measured at specific frequencies, and it should be noted that the impedances of the three ports of the "PV-battery locomotive network" refer to the output impedances Z_{PV} towards the PV boost circuit port, Z_{Bat} towards the battery DC/DC circuit port, and Z_{inv} towards the inverter DC port. We will validate whether the theoretical deduced values of these impedances in the frequency domain match the simulation results.

Validation results are shown in Figure 6, the first row shows the corresponding impedance function's Bode plot, and the second row is the corresponding phase plot, with the blue curves representing the theoretical values of the model and the red circles representing the actual measured values. It can be seen that the modeling of the PV and battery system's output impedance and RPC's input impedance is consistent with the simulation test results.

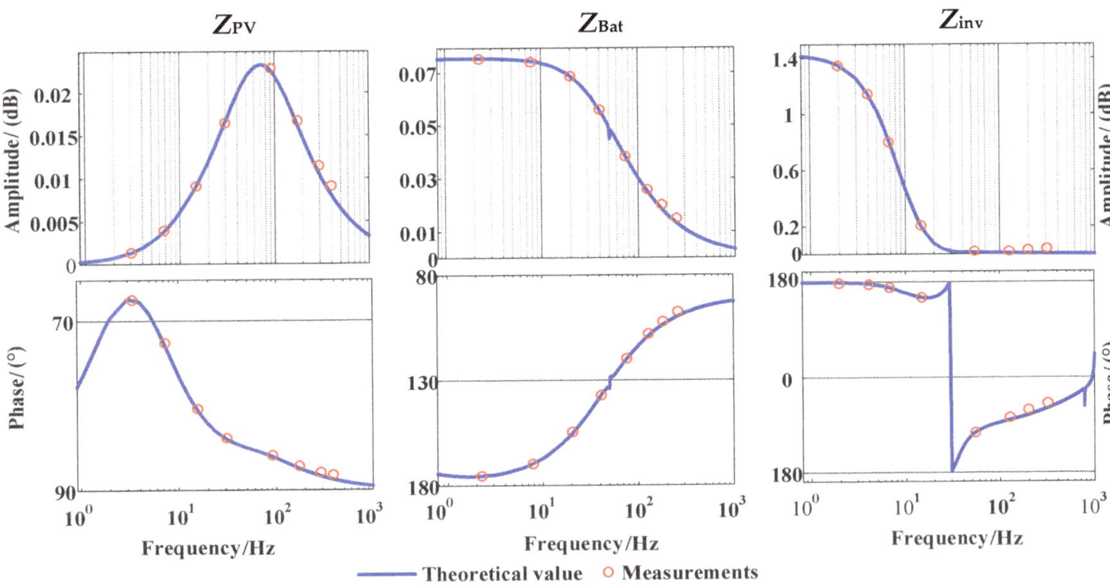

Figure 6. Three-port impedance verification.

2.3. Low-Frequency Stability Analysis of Proposed Emergency Power Supply System

2.3.1. Stability Judgment Based on Generalized Nyquist Criterion

According to the mathematical model established earlier, the impedance equivalent circuit of the PV and battery locomotive traction is shown in Figure 7a, where Z_0 is equal to the parallel value of Z_{PV} and Z_{Bat}, and Z_g is the impedance of the traditional locomotive network system source side. Due to the PI controller parameters of the converters in PV and battery systems that affect the output impedance of the system, Z_0, Z_{PV}, and Z_{Bat} are different from Z_g because the value of Z_g depends on factors such as the material of the transmission line, its length, and the capacity of transformers, which are difficult to adjust in practical operation. In comparison, the values of Z_0, Z_{PV}, and Z_{Bat} are related to the parameters of the PI controller, and adjusting the control parameters of the PI controller is less costly as well as more simple and fast. The introduction of a controllable equivalent impedance may exacerbate the LFO of the system or prevent system instability caused by impedance mismatch; this characteristic makes it particularly important to explore the impact of the control parameters and number of PV and battery parallel connections on the system's low-frequency stability. This section uses stability judgment methods based on impedance ratio criteria to conduct the research [27,28].

Firstly, it must be ensured that there are no RHP poles for Z_0, Z_{PV}, Z_{Bat}, and Y_{inv} and that each subsystem can operate independently and stably in the simulation. Under different traction modes, applying the one-dimensional impedance model established earlier, the Nyquisit curve based on the impedance ratio of Z_0/Z_{inv}, Z_{PV}/Z_{inv}, and Z_{Bat}/Z_{inv} can be obtained. Taking Z_0/Z_{inv} as an example, the expression for the system's impedance ratio can be represented by a transfer function of $G(s) = Z_0/Z_{inv}$. The stability of the system can be measured by plotting the Nyquist curve of $G(s)$ and determining whether it encircles $(-1, j0)$. Additionally, the variations in parameter values based on the impedance model discussed earlier can affect Z_0, thus, modifying the Nyquist curve of $G(s)$. Therefore, the parameters of each variable may have some effect on the stability of the system. Meanwhile, due to the lack of a clear concept of the stability margin in the impedance ratio criterion for MIMO equivalent systems, and to showcase the analysis process concisely and clearly, the critical amplitude margin (CAM) is defined to quantitatively analyze the system stability in order to reveal sensitive parameters and their influence laws. If CAM ≥ 0, the system is stable; if CAM < 0, the system is unstable. The Nyquist plot of the impedance ratio transfer function for this system may change due to the effect of different parameters, leading to two different scenarios, as shown in Figure 7b.

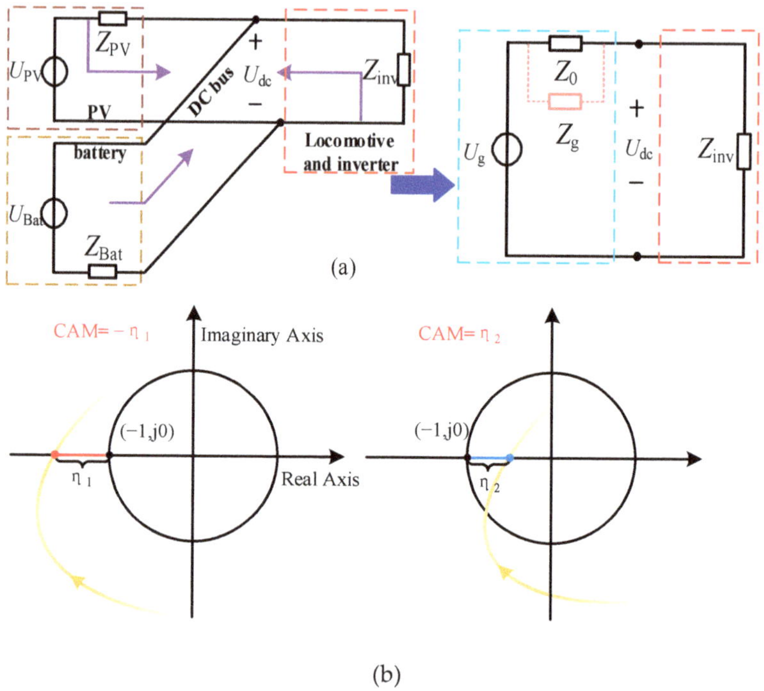

Figure 7. (**a**) System impedance equivalent circuit; (**b**) CAM value scenario diagram.

Scenario 1: When the curve surrounds (−1, j0), the distance from the intersection point of the real axis of the curve to (−1, j0) is expressed as η_1, CAM = $-\eta_1$;

Scenario 2: When the curve does not surround (−1, j0), the corresponding distance is expressed as η_2, CAM = η_2.

In summary, changes in system variables can affect the Nyquist curve of $G(s)$, thus, impacting the stability of the system. Such an impact can be translated into changes in CAM values to quantify the effect of a particular variable on the stability of the system. To further reveal the main controller parameters that affect the system stability, this article defines the sensitivity of the parameters as:

$$\varepsilon = \frac{\mathrm{CAM_n}}{(k_s - k_y)/k_y} \times 100\% \quad (16)$$

where k_y is a certain original parameter in Table 1, k_s is the value of the critical instability parameter with the smallest change from k_y, $(k_s - k_y)/k_y$ is the multiple of the minimum change in the original parameters when the system is in critical instability, and CAMn is the critical amplitude margin corresponding to the original parameter k_y under different operating conditions. The introduction of CAMn is aimed at correcting the sensitivity under different traction conditions (PV traction, battery traction, PV, and battery co-traction). This expression can be simply understood as the sensitivity to the electrical instability of the nearest value of the original parameter.

Table 1. Original simulation system parameters.

Subsystem	Parameter	Description	Value
Traction net	R_s	Grid-side equivalent resistance	0.2 Ω
	L_s	Grid-side equivalent inductance	6 mH
	e_s	Traction net voltage RMS	27.5 kV
	ω_0	Net-side angle frequency	314 rad·s^{-1}
Railway power conditioner	L_{inv}	Filter inductance	0.04 mH
	C_{inv}	Filter capacitors	6 mF
	C_{bus}	DC-side capacitance	300 mF
	u^*_{bus}	DC-side voltage reference value	2000 V
	ω	Reference angle frequency	314 rad·s^{-1}
	k_{invp}	Voltage-loop proportional gain	40
	k_{invi}	Voltage-loop integral gain	20
	k_b	Step-up transformer ratio	1500/27,500
PV subsystem	R_{PV}	PV-side resistance	1 mΩ
	L_{PV}	PV-side energy storage inductance	0.3 mH
	C_{PV}	PV-side support capacitor	2.2 mF
	k_{PVup}	Voltage-loop scaling factor	4
	k_{PVui}	Voltage-loop integration factor	20
	k_{PVip}	Current-loop scaling factor	0.4
	k_{PVii}	Current-loop integration factor	5
Battery subsystem	R_{Bat}	Energy-storage-side resistance	1 mΩ
	L_{Bat}	Energy storage inductance	1 mH
	k_{Bup}	Voltage-loop scaling factor	15
	k_{Bui}	Voltage-loop integration factor	2
	k_{Bip}	Current-loop scaling factor	0.02
	k_{Bii}	Current-loop integration factor	1
CRH3	k_a	On-board transformer ratio	27,500/1500
	L_n	leakage inductance of transformer	2 mH
	L_{CRH3}	DC-side filter inductance	0.84 mH
	C_{CRH3}	DC-side filter capacitors	3 mF
	C_d	DC side supports the capacitor	6 mF
	k_{PWM}	PWM equivalent gain	1
	k_{CRH3up}	Voltage-loop scaling factor	0.1
	k_{CRH3ui}	Voltage-loop integration factor	10
	k_{CRH3ip}	Current-loop scaling factor	1

2.3.2. Influence of DC Converter and RPC Parameters on the Low-Frequency Stability of the Proposed System

Based on the method proposed in Section 2.3.1, Figure 8a shows the electrical sensitivity bar chart of the control parameters of the DC converter and RPC inverter under three different traction modes when the ratio between the PV module, battery module, and CRH3 locomotive is 1:1:1. Due to the sensitivity of the inner-loop control parameters of the system under this scheme being less than 1%, the Nyquist curve hardly changes with the changes in the parameters, and the corresponding test shows that the system remains stable even after a large number of changes in the inner-loop control parameters; therefore, it is not included in the research object. The bar chart shows that the proportional gain of the outer ring is more sensitive to the system stability and the combined traction of the PV and battery will reduce the parameter sensitivity.

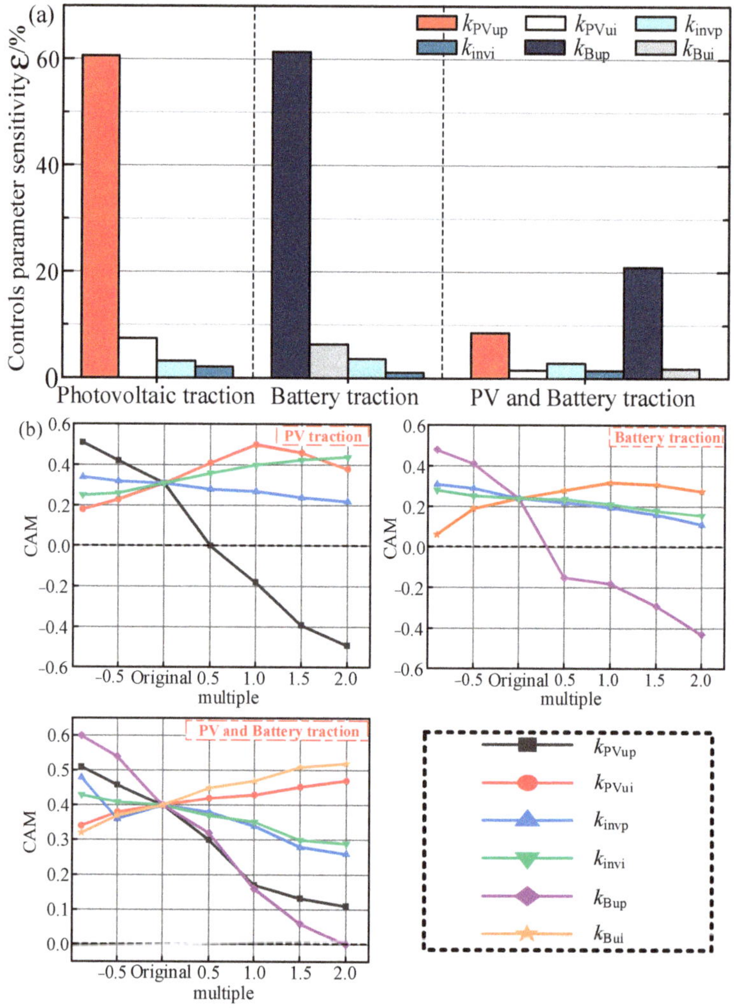

Figure 8. (**a**) Parameter sensitivity of the RPC and DC converters; (**b**) CAM change law of the control parameters.

The corresponding parameter influence law is shown in Figure 8b. Within the range of −0.9 to 2 times the original parameter value, the proportional gain of the outer ring of each converter exhibits monotonic, identical laws under different operating conditions, and the sensitivity (i.e., slope) is relatively high. However, the sensitivity of the integral gain of the outer ring is low, and the impact of the integral gain on the stability of the system exhibits diversification and non-monotonic characteristics under different traction conditions. Therefore, the outer-loop proportional gain of the system can be more easily used to control or prevent system impedance mismatch instability. The specific parameter adjustment criterion and case verification are expounded in Section 2.3.3.

In addition, LFO in locomotive network systems always occurs when the Nyquist curve of the impedance ratio (Z_g/Z_{CRH3}) passes through $(-1, j0)$. To investigate whether the occurrence of LFO under the topology and control scheme proposed in this paper is consistent with the theoretical analysis of locomotive network systems when the locomotive is separately tractioned by the PV or battery, the impedance ratio expression for the PV traction locomotive is Z_{PV}/Z_{inv}, and the expression for battery traction locomotive is Z_{Bat}/Z_{inv}, where Z_{PV}, Z_{Bat}, and Z_{inv} come from Equations (5), (6), and (14), and these impedance ratio expressions are used to plot Nyquist curves under different traction conditions. Then adjust the high sensitivity parameters k_{PVup} and k_{Bup} so that the Nyquist curve just surrounds $(-1, j0)$, thereby obtaining the Nyquist curve shown in Figure 9 and the test waveform shown in Figure 10. The results show that when the Nyquist curve of the impedance ratio for the system described in this paper precisely intersects the $(-1, j0)$ point, LFO at around 9.6 Hz also occurs. The principle and testing results are consistent under other parameters in Figure 8 but due to the large number of parameters, they are not listed in detail.

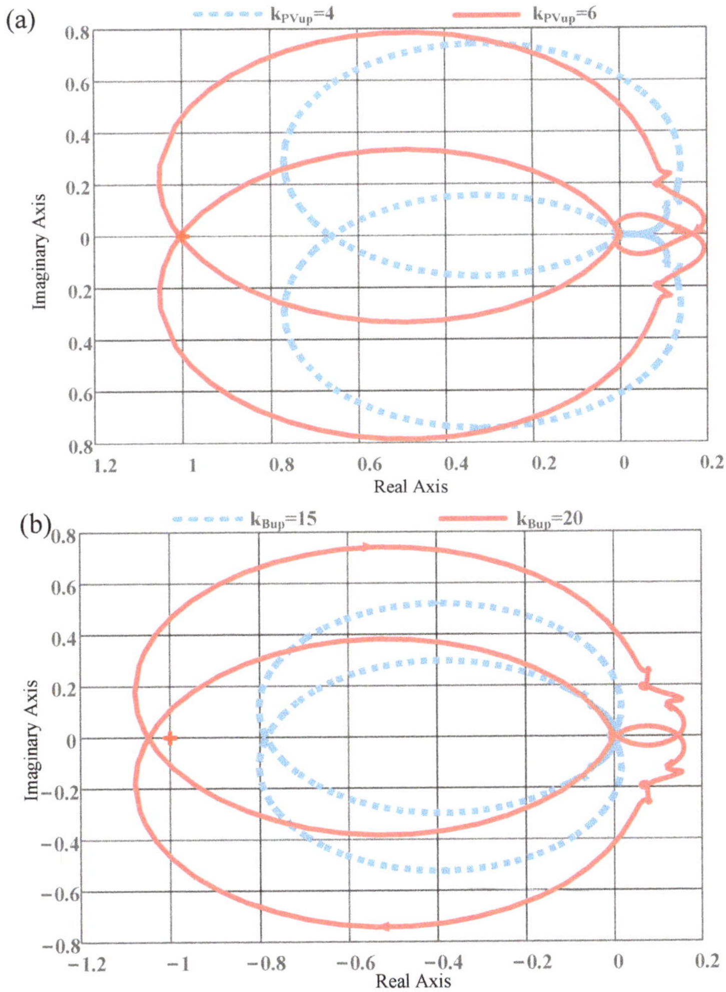

Figure 9. Nyquist curve before and after changes in high-sensitivity parameters: (**a**) working conditions of PV locomotive traction; (**b**) working conditions of battery locomotive traction.

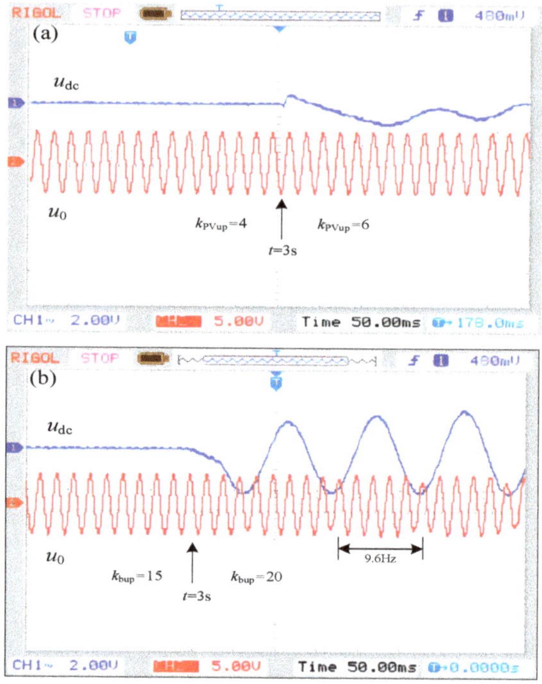

Figure 10. (**a**) Experimental waveform of the critical instability of photovoltaic locomotive traction; (**b**) experimental waveform of the critical instability of battery pack locomotive traction.

2.3.3. Parameter Adjustment Criterion and Case

Usually, an increase in the transformer capacity and the addition of other types of equipment such as static synchronous reactive power compensators on the source side are considered in locomotive network systems to address LFO caused by impedance mismatch. Such problems also exist in systems under emergency power supply schemes. However, when using PV and battery locomotive traction, based on the stability law of the control parameters revealed in Section 2.3.2, the goal can be achieved by adjusting the converter parameters accordingly. The specific parameter adjustment criteria are as follows: Firstly, if the system experiences instability, consider lowering the PV boost converter k_{PVup}, battery DC/DC converter k_{Bup}, and RPC single-phase inverter k_{invp}. Secondly, consider slightly lowering the k_{invi} during PV and battery co-traction or battery traction, and slightly increasing the k_{invi} during PV traction. It is worth noting that the RPC outer-loop proportional gain seriously affects the control accuracy (i.e., voltage amplitude); therefore, prioritize adjusting the proportional gain of the PV and battery DC converters unless faced with a situation that has special requirements. Taking the ratio of PV modules to battery modules as 1:1 as an example, the following explains the methods of parameter adjustment and governance in three cases. The Nyquist curve of Case 1 is shown in Figure 11a. When using the PV and battery locomotive co-traction, the battery system's energy storage inductance deteriorates from 1 mH to 1.23 mH, and the curve surrounds $(-1, j0)$; at this time, the adjustment k_{Bup} is reduced from 15 to 5, and the Nyquist criterion shows that the system will return to stability. The theoretical curve of Case 2 is shown in Figure 11b. When the k_{PVup} of the PV locomotive traction deteriorates, it causes a critical oscillation in the system. The k_{Bup} is adjusted and connected to the battery system; at this time, the PV and battery combined locomotive traction are stable. The test validation waveforms of the parameter adjustment criteria and related conclusions described in this section will be presented in Section 3.2.1.

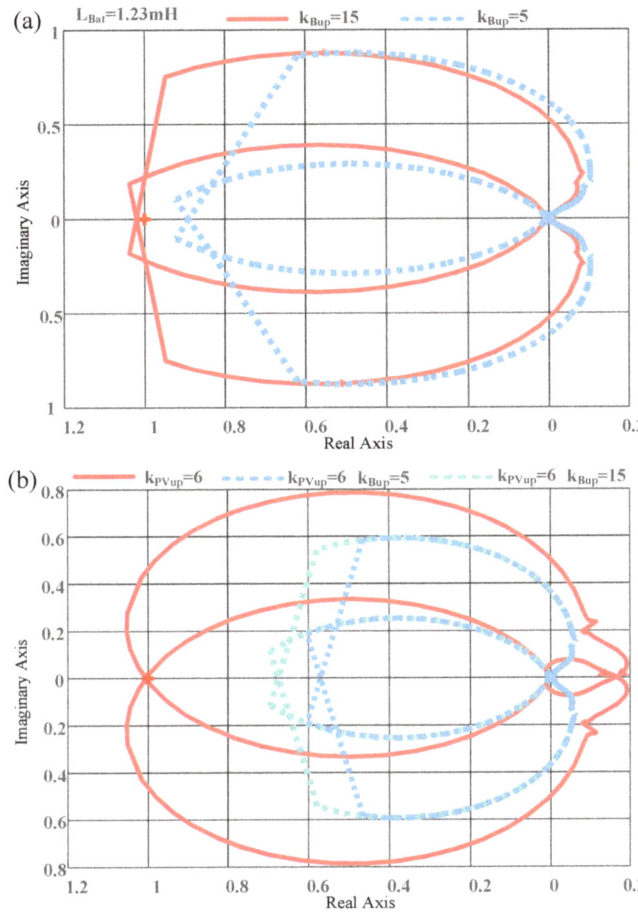

Figure 11. (**a**) Nyquist curve for Case 1; (**b**) Nyquist curve for Case 2.

2.3.4. The Influence of Mixed Proportional Parallel Numbers of PV and Battery Modules on System Stability

The number of PV and battery hybrid proportional parallel connections will also change the source-side impedance model to weaken or improve the stability of the system. Therefore, the original parameters were substituted into the hybrid parallel connection system. Based on the generalized Nyquist criterion, the bar chart shown in Figure 12a was obtained. The results show that the system operated stably under different PV and battery module proportions. To reveal its stability mechanism, the number of PV and battery modules were introduced as system variables, and because all the modules are connected in parallel, thus, expanding the impedance ratio expression Z_0/Z_{inv} in Section 2.3.1, we achieve the following formula:

$$L_{\text{hui}} = \left(\frac{\frac{Z_{\text{pv}} Z_{\text{Bat}}}{n_{\text{pv}} n_{\text{bat}}}}{\frac{Z_{\text{pv}}}{n_{\text{pv}}} + \frac{Z_{\text{bat}}}{n_{\text{bat}}}} \right) / Z_{\text{inv}} \tag{17}$$

where n_{pv} represents the number of PV modules, and n_{bat} represents the number of battery modules.

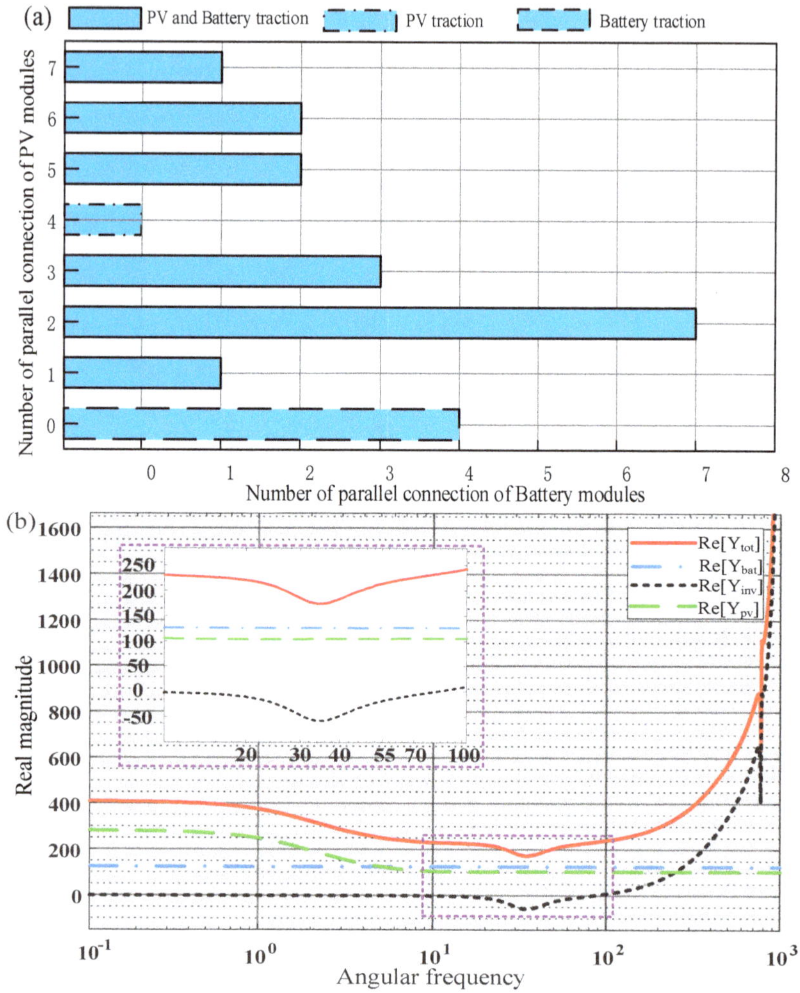

Figure 12. (**a**) The proportion of stable system operation when the PV and battery modules were connected in parallel; (**b**) analysis results of the passive criterion when the ratio of modules was 1:1.

If $n_{pv} = n_{bat} = n$, the expression is rewritten as $L_{hui} = (Z_0/Z_{inv})/n$, which is equivalent to proportionally reducing the Nyquist function curve (L_{hui}) by n times. Systems that do not originally include (−1, j0) will be more stable. In the previous work, it was proven and verified that the system was stable when the PV traction module ratio was 1:0, the battery traction module ratio was 0:1, and the PV and battery co-traction module ratio was 1:1 under the original parameters; therefore, the stability of the system with the PV and battery module ratios of 4:0, 0:4, and 3:3 in Figure 12a is explained. However, it is complex and difficult to give an answer for the stability mechanism of non 1:1, 1:0 systems using the impedance ratio criterion, since it changes a variable to directly judge the stability of the overall system. Therefore, this paper introduces a passive criterion applicable to the stability analysis of variables, subsystems, and overall systems. The judgment criterion is as follows: if the sum of the real parts of the total admittance of all parallel systems in the bus is guaranteed to be constantly greater than 0, all subsystems can operate stably at the same time. This passive criterion was proposed and verified by Riccobono A. in 2012 [34] but compared to the impedance ratio criterion, it is less used. Based on the "PV–battery locomotive network" coupling system, this article additionally provides the proof process of the passive criterion from the perspective of dissipative system stability, as shown in Appendix B. The criterion combines Equations (5), (6), and (14) and Figure 7a, resulting in the following expression:

$$\begin{aligned} \mathrm{Re}[Y_{\mathrm{tot}}(s)] &= \mathrm{Re}\left[n_{\mathrm{pv}} Y_{\mathrm{PV}}(s) + n_{\mathrm{bat}} Y_{\mathrm{Bat}}(s) + Y_{\mathrm{inv}}(s)\right] \geq 0 \\ &= \mathrm{Re}\left[\frac{n_{\mathrm{pv}}}{Z_{\mathrm{PV}}(s)} + \frac{n_{\mathrm{bat}}}{Z_{\mathrm{Bat}}(s)} + \frac{1}{Z_{\mathrm{inv}}(s)}\right] \geq 0 \end{aligned} \tag{18}$$

Applying Equation (18), the function curve is plotted under the original parameters when the locomotive traction has a PV and battery module ratio of 1:1, as shown in Figure 12b. At this time, the real parts of the admittance of the PV and battery modules are both greater than 0 in the entire frequency band. If the subsystem is paralleled in any proportion, it will increase the distance between Re[$Y_{\mathrm{tot}}(s)$] and the horizontal axis, and the system will obtain an additional stability margin to maintain stability. This reveals the reason why the system always maintains stability after the PV and battery module mixing ratio is paralleled. Based on the above analysis, the problem of LFO in PV and battery locomotive co-traction under specific working conditions can also be solved by using the law of the influence of the number of module parallel connections on stability to reshape the system impedance from the main circuit structure through the passive criterion. In Section 3.2.2 Case 1, we designed experiments and demonstrated the validity of the theoretical analysis.

Increasing the number of PV battery parallel connections in a targeted manner can address instability caused by multiple locomotives running together or unreasonable RPC parameter settings but it is difficult to use this method to solve instability caused by the deterioration of the system. The instability process of this phenomenon can be revealed using passive criteria, as shown in Figure 13, which shows the criterion curve after the 1:1 parameters of the PV and battery modules deteriorate. Increasing n_{pv} and n_{bat} will cause Re[$Y_{\mathrm{PV}}(s)$], Re[$Y_{\mathrm{bat}}(s)$], and Re[$Y_{\mathrm{tot}}(s)$] to extend towards negative infinity, thereby exacerbating the system instability, and this was validated in Section 3.2.2 Case 2.

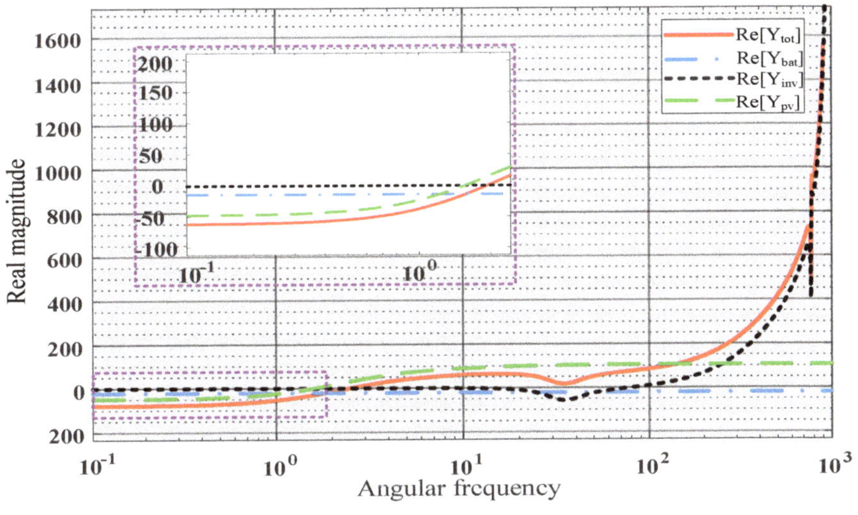

Figure 13. Passive criterion analysis results after deterioration of the system parameters.

Based on the research in this section, it is recommended to ensure that the real part of the admittance of the PV and battery subsystem is always greater than 0 during the design.

2.3.5. Passive Criterion Is Used to Reveal the Influence Law of Parameters on Stability

As shown in Section 2.3.4, the passive criterion based on the DC bus has clear boundary conditions and judgment criteria, which makes it easier to explore the law of the influence of parameters on the stability of the system, in comparison with the Nyquist criterion. This is reflected in the bivariate function curve, which directly shows the influence of a parameter of the system on its stability. This means that the law of stability near the original parameter will no longer be experimentally searched as before, and the three-dimensional plotting of the function can immediately reveal the influence of arbitrary values of a parameter on the stability of the original system. In order to verify the above analysis, passive criteria are used to reveal the stability influence laws of k_{PVup} and k_{Bup}, which have been verified before. By taking k_{PVup} and k_{Bup} as variables, we can obtain functions similar to Equation (19) but because the expressions are verbose, they are presented in the form of graphs, as shown in Figures 14 and 15, where the z axis is $\mathrm{Re}[Y_{tot}(s)]$.

$$\mathrm{Re}[Y_{tot}(s)] = F(k, \omega) \tag{19}$$

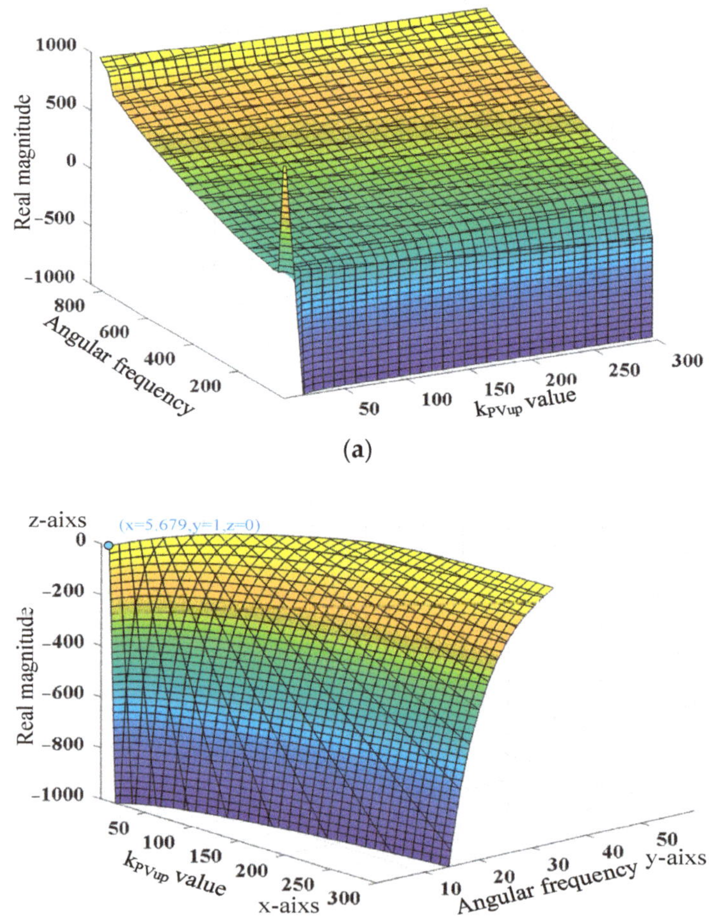

Figure 14. The influence of k_{PVup} on the system stability: (**a**) global diagram; (**b**) local diagram of the system instability.

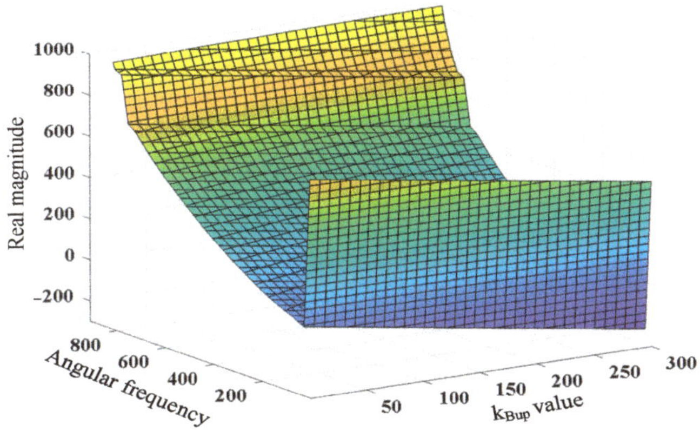

(a)

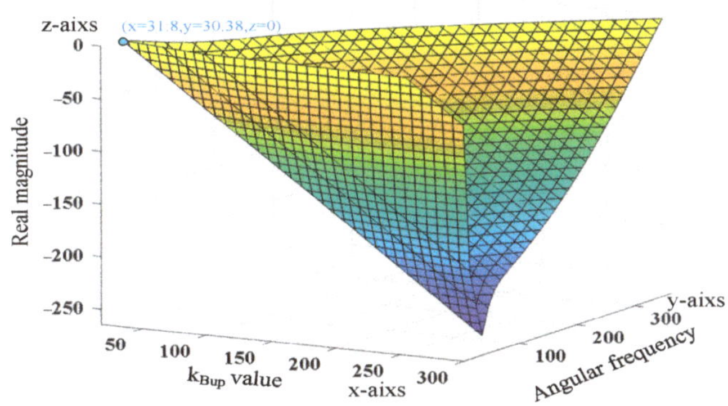

(b)

Figure 15. The influence of k_{Bup} on the system stability: (**a**) global diagram; (**b**) local diagram of the system instability.

Figures 14a and 15a show that with the increase in k_{PVup} and k_{Bup}, the surface gradually extends to the negative plane; in other words, the system gradually becomes unstable, which is consistent with the results obtained using Nyquist's criterion and in the semi-physical simulation, however, we are more concerned about the LFO phenomenon that often occurs in actual railway systems, that is, the critical instability of the system. Accordingly, the local amplification when the corresponding parameter is unstable is shown in Figures 14b and 15b, and the values of the LFO parameters given by the criterion are k_{PVup} = 5.679 or k_{Bup} = 31.8, which are obviously lower than the values given by the generalized Nyquist criterion, as shown in Figure 10b. The system ran stably in the simulation experiment under these parameters. It is also necessary to continue to increase the parameter values to cause LFO. This is caused by the large conservatism of the passive criterion. Assuming that the law revealed by the criterion is used to control the LFO, if the stability of a parameter affecting the law is not monotonic and there are lots of extreme points, this situation may lead to a counterproductive parameter adjustment. Here, the law revealed by the criterion is more suitable when parametric influences are monotonic over a large scale. However, in the system design, the conservativeness of the criterion is favorable, and it will leave a certain stability margin.

In addition, by applying passive criteria to the AC bus, the designed virtual impedance compensator successfully suppresses LFO [17] but the specific proof process in the AC system and its conservative improvement are the next research foci concerning this criterion.

3. Results and Discussion

3.1. Emergency Power Supply Scheme Testing

To verify the feasibility of the proposed emergency power supply scheme and the accuracy of the stability analysis, a simulation model of the PV and battery power generation systems, RPC, and CRH3 locomotive, as shown in Figure 1, was built based on Starsim. Figure 16 shows Shanghai Yuankuan Energy's Starsim HIL real-time simulation software 5.0 and HIL real-time simulator (Modeling Tech, Shanghai, China). This system can perform small-step real-time testing on power electronic models based on state equations, switch averaging, and modeling of large and small resistors. In this paper, the model was verified using a real-time test system (MT 6020) with a 5 μs step size and 20 kHz sampling frequency.

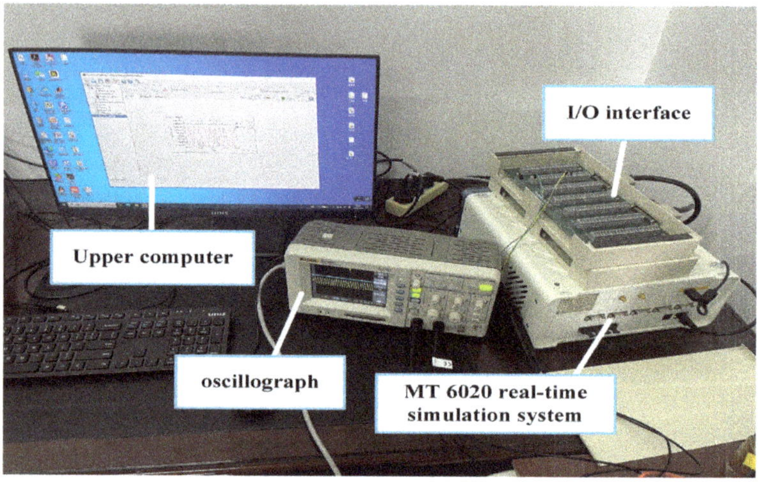

Figure 16. Hardware-in-the-loop real-time testing platform.

Here are the model parameters in the above-mentioned testing system, as shown in Table 1. In this paper, the parameters in Table 1 are referred to as the original parameters. Unless otherwise specified, analysis or testing will default to using the original parameters.

3.1.1. Emergency Traction Test based on Locomotive Power Failure

The upper computer calculation display is shown in Figure 17a, where P_{RPC} is the output power of the PV and battery, P_{net} is the output power of the substation, and P_{CRH3} is the absorbed power of one locomotive. At 1 s, the traction network unexpectedly lost power to the locomotive; after a short response time, the RPC identified the operation status of the PV and battery devices and switched to the corresponding mode to achieve the emergency power supply for the locomotive. At 2 s, the power supply to the substation was restored; to ensure passenger safety, the RPC delayed the power outage. Finally, at 2.5 s, the RPC exited the power supply or returned to the power compensation scheme. During this power outage process, the locomotive remained running, as shown in Figure 17b. If emergency-response-type locomotive traction is started in actual engineering, it can provide on-site measured data support and technical guidance for future new energy self-sufficient traction. For example, the working conditions of delayed power outages are equivalent to the actual test of changing to the substation supply when the new energy self-sufficient traction encounters an accident.

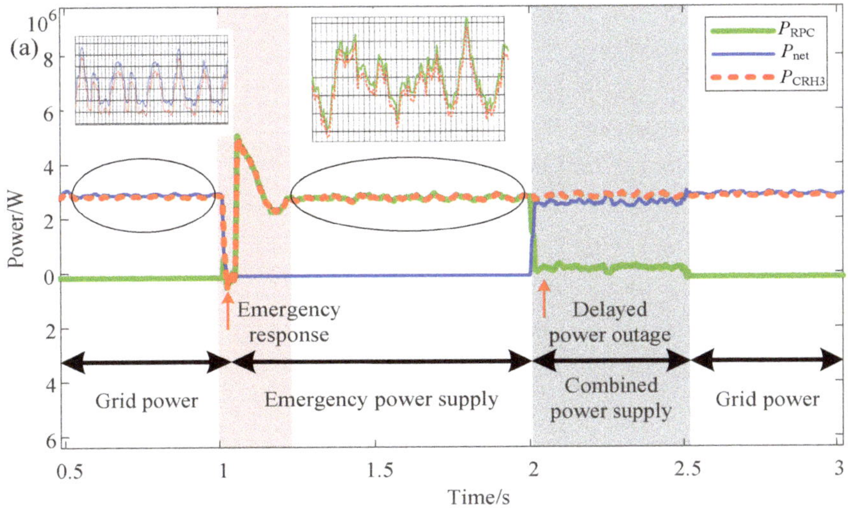

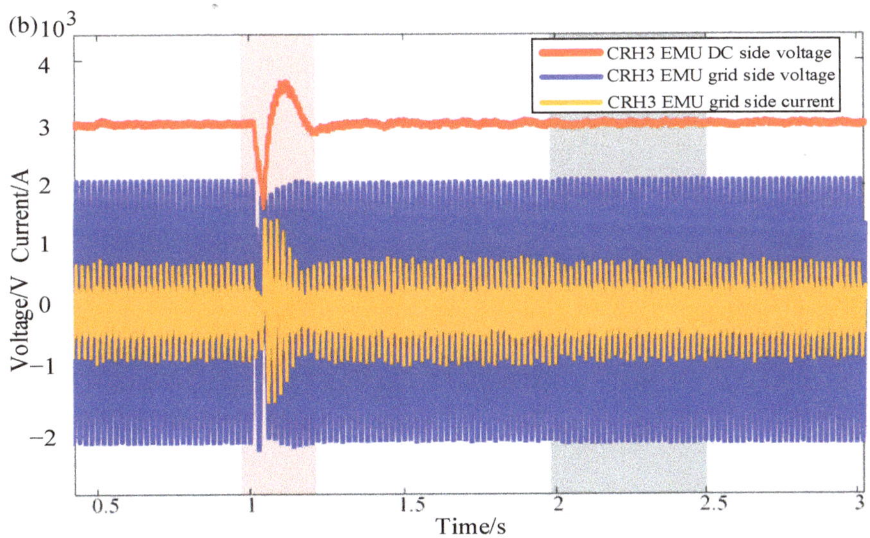

Figure 17. Emergency response traction power supply: (**a**) system energy flow; (**b**) locomotive voltage and current.

3.1.2. Coordinated Traction Test

The test setup of the operating power P_{load} of the high-speed locomotive is set to 3.6 MW, and the light intensity will continue to rise, as shown in Figure 18 obtained from the upper computer. Among the modes, mode four to mode one correspond to battery traction, PV and battery co-traction, PV traction, and PV locomotive traction and battery charging. By switching operation modes under different working conditions, the locomotive traction power can be stabilized while charging the battery. Under this scheme, the renewable energy continuous closed-loop power supply allows the locomotive to remain "low-carbon" for a long time, and a substation can be used for backup power in case of accidents.

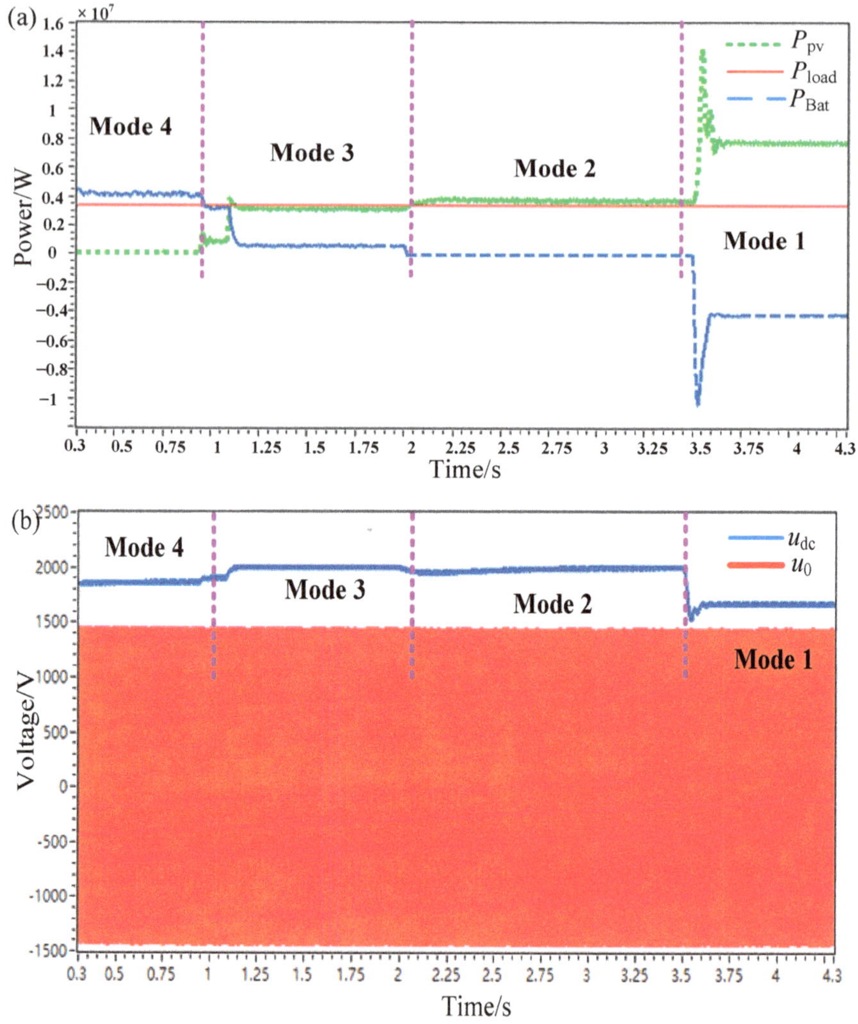

Figure 18. Energy self-sufficient traction power supply: (**a**) multi-mode power flow conditions; (**b**) voltage waveform on both sides of the RPC.

3.2. Verification of Low-Frequency Stability Analysis

The main purpose of this section is to verify the results of the low-frequency stability analysis in Section 2.3 to ensure the correctness of the analysis. The following subsections will present corresponding test results in the order of the analysis in Section 2.3.

3.2.1. Verification of Parameter Adjustment Criterion

1. **Case 1:** Adjusting control parameters to address instability caused by deteriorating circuit parameters. Under the original parameters, the PV and battery stably provided co-traction for the locomotive, which is verified in Figure 18. However, at this time, the parameters of the energy storage inductor L_{bat} deteriorated. The waveform of the traction-network-side voltage U_{ac} and current I_{ac} is shown in Figure 19b. The system experienced a low-frequency constant-amplitude oscillation of 5.7 Hz after 2 s, and after k_{Bup} decreased after 2 s, the system was stabilized and restored. This case

demonstrates that numerically adjustable control parameters can govern LFO caused by the deterioration of topology circuit parameters.

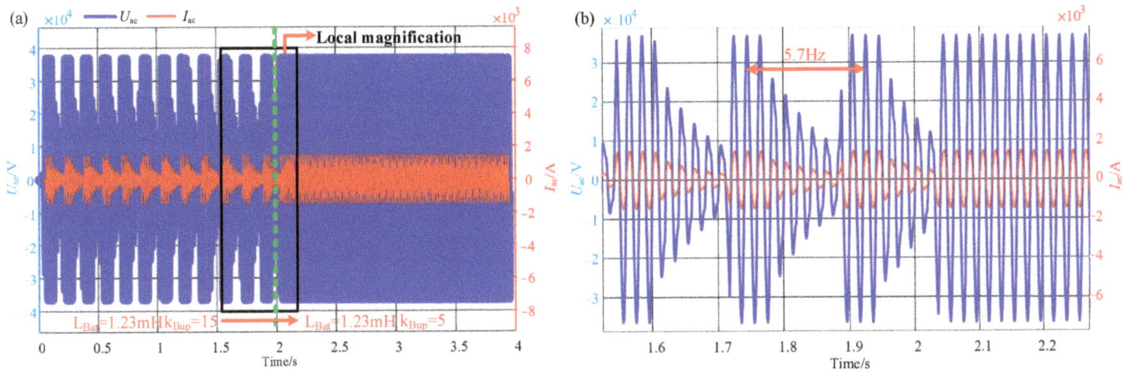

Figure 19. Waveform for parameter adjustment after the deterioration of the main circuit parameters: (**a**) global diagram; (**b**) local diagram.

2. **Case 2:** Connecting a parameter-adjusted other-source subsystem to address single-source traction instability. Due to the unreasonable setting of k_{PVup} during the PV locomotive traction, the system became unstable. Finally, a parameter-adjusted battery system was connected to control the constant-amplitude oscillation. The experimental waveform is shown in Figure 20, where u_{dc} and u_0 are the output voltages of the RPC DC and AC sides, respectively. By comparing and observing the oscilloscope, it was found that a lower value of k_{Bup} made the bus voltage fluctuation smaller and smoother when the battery system was connected. The reason for the successful governance of the parameter adjustment, in this case, is reflected in the addition of the circuit topology of the battery subsystem and the parameter setting under the parameter adjustment criteria, which improved the impedance matching relationship. At the same time, the relatively lower electrical sensitivity under the combined traction of the PV and battery provided the control parameters with a wider range of choices in the stable domain.

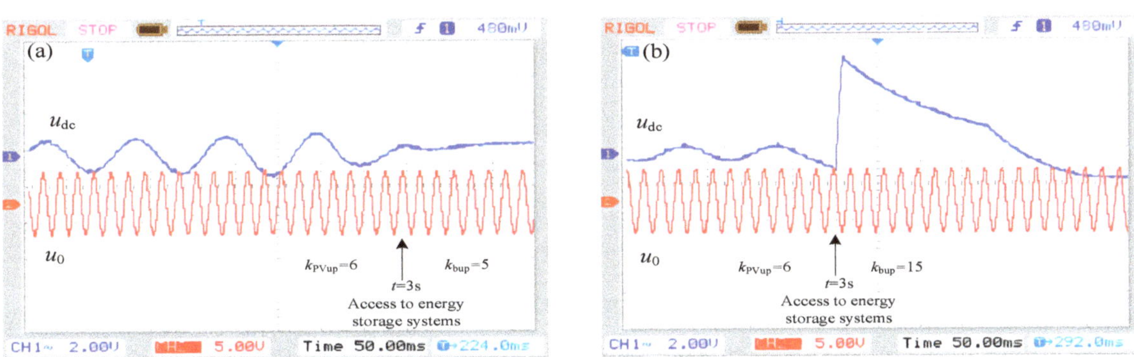

Figure 20. When the PV locomotive traction is unstable, it can be corrected by connecting a parameter-adjusted battery subsystem: (**a**) experimental waveform after the battery was connected when $k_{\text{Bup}} = 5$; (**b**) experimental waveform after the battery was connected when $k_{\text{Bup}} = 15$.

3. **Case 3:** Adjusting the control parameters to govern LFO caused by multi-locomotive operation (heavy load). The impedance ratio criterion mechanism shows that the smaller the load-side impedance, the more unstable the system will be. Therefore,

many studies on locomotive network systems have found that multi-locomotive operation or mixed operation is an important factor inducing LFO. The problem of heavy loads reducing the stability of traction power supply systems also exists in the study of PV and battery locomotive traction but the parameter-adjustable "artificially controllable traction network impedance" has a certain adaptability. The CRH3 high-speed locomotive was used as the research object, and the simulation waveform is shown in Figure 21, which includes the locomotive DC-side voltage U_{CRH3dc}, locomotive-network-side voltage U_{CRH3ac}, and locomotive-network-side current I_{CRH3ac}. After 2.5 s, the addition of three high-speed locomotives caused a 4Hz low-frequency constant-amplitude oscillation among the high-speed locomotives. According to the parameter adjustment criteria, k_{PVup} and k_{Bup} were decreased after 2.5 s, and then the locomotives were stabilized.

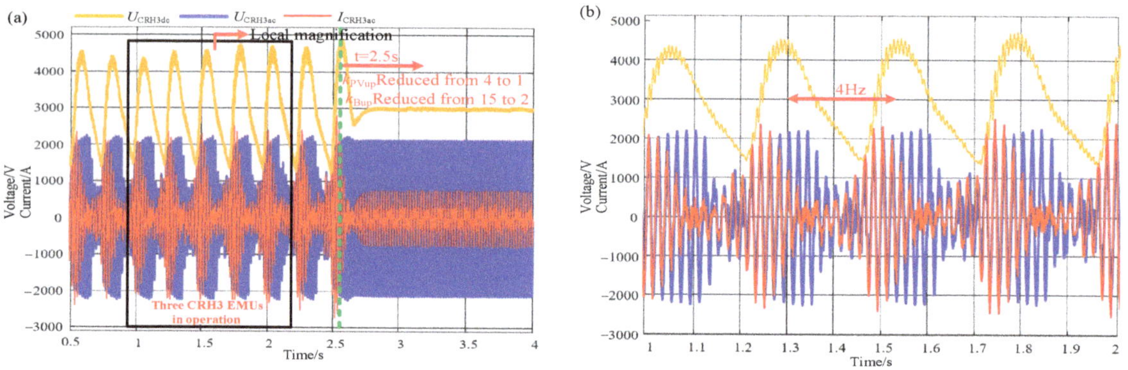

Figure 21. Adjustment of the parameters to control system instability caused by multi-locomotive parallel connection: (**a**) global diagram; (**b**) local diagram.

3.2.2. The Impact of Parallel Connections of PV and Battery Modules

1. **Case 1:** Worsening RPC parameters and overloading lead to LFO and then an increase in the number of parallel connections of the PV and battery modules. According to the passive criterion, overloading and deteriorating RPC parameters will cause Re[$Y_{inv}(s)$] to extend towards the negative plane until the system experiences critical instability. According to the impedance ratio criterion, overloading is equivalent to proportionally amplifying the Nyquist curve until it envelops (−1, j0). Under this operating condition, it will cause low-frequency divergent oscillations, as shown in Figure 22a, where U_{dc} and U_0 are the voltages on both the RPC DC and AC sides, respectively. Increasing the number of parallel PV battery modules and reshaping the impedance to provide an additional stability margin can compensate for this, as shown in Figure 22b. In this case, the system can regain stability after 1.6 s.
2. **Case 2:** Increasing the number of parallel modules after the module parameters deteriorate. As shown in Figure 23, multiple increases in the number of module parallel connections after the PV and energy storage inductance parameters and control parameters deteriorate will actually exacerbate the system instability.

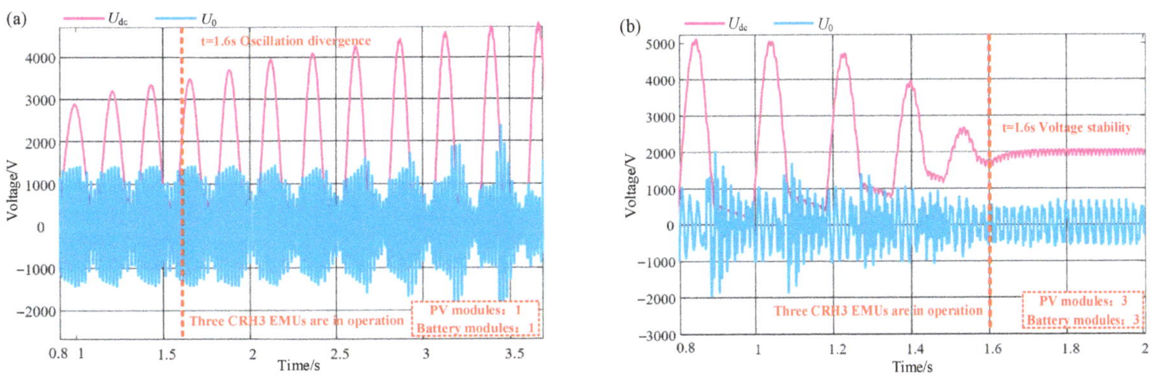

Figure 22. (**a**) Overloading and worsening RPC parameters lead to system instability: (**b**) the ratio of PV and battery module parallel connections can multiply and increase to reshape the system impedance.

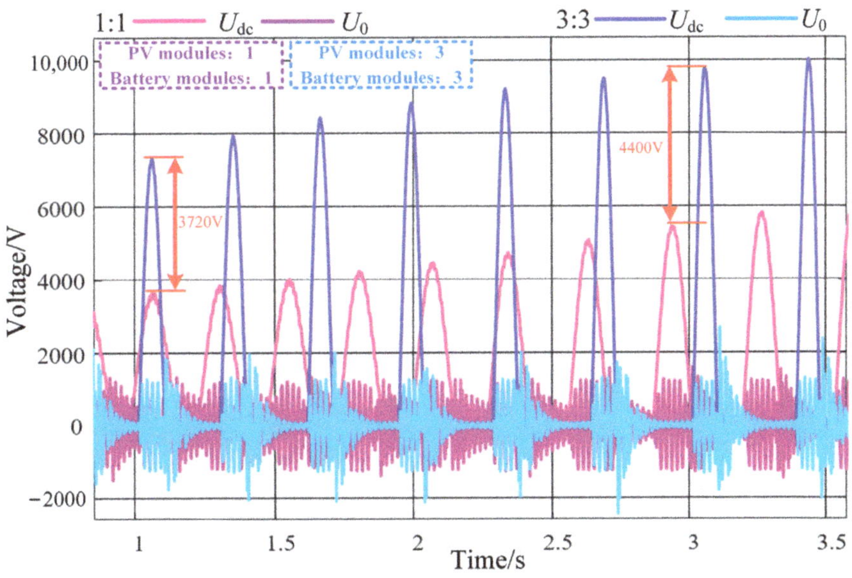

Figure 23. Deterioration of the system parameters leads to instability and multiple increases in the ratio of modules will aggravate the voltage oscillation.

4. Discussion

This paper mainly proposes an emergency power supply scheme to solve the problem of interruption in train power supply caused by unexpected faults in traction substations. The solution coordinates with the PV and battery systems to achieve emergency traction of locomotives, thereby expanding the functions of the railway power conditioner (RPC). Meanwhile, this paper proves through theoretical modeling and verification tests that the PV and battery traction locomotive additionally have the problem of low-frequency oscillation. To curb low-frequency oscillation occurring during the emergency power supply, this paper quantitatively analyzes the influence law of PI control parameters and topological structures on the low-frequency stability of the system, proposes a design method for the impedance real part greater than 0 of the PV and battery systems module, and provides parameter adjustment criteria to suppress or even prevent low-frequency oscillations.

Most studies on the integration of PV and battery into electrified railways focus on RPC grid compensation of traction power and harmonic governance. However, it is necessary to make RPC multifunctional, and few of the literature studies the coordinated control scheme of PV and battery for emergency train traction. If the majority of current RPC research's control strategies, such as the ones described in references [16,31], are used to independently traction the locomotive, the lack of phase information obtained by the phase-locked loop will cause the frequency and current of the traction network to be incorrect, forcing the locomotive to stop running. Therefore, this paper proposes the use of a dual-mode RPC with an independent power supply function, which can not only compensate for the power of the traction power supply system but also provide emergency traction for the locomotive.

Meanwhile, current research of the literature applies various criteria to analyze the low-frequency stability of a system. However, many research methods and processes in the literature are rather similar, in that they require graph redrawing every time a variable is modified to display the corresponding result [18,27,28]. This process is not intuitive enough and does not facilitate a thorough investigation of the impact of parameters on the stability and sensitivity of the system, simultaneously considering that there is little research on the low-frequency stability of the PV–battery locomotive network coupling system. Therefore, this paper defines CAM to quantify system stability and provides a method to calculate the sensitivity ε of parameters, which can weigh the importance of different parameters under different working conditions and explore the main influencing parameters and their influence laws. This paper also proposes the use of a passive criterion to reveal the influence mechanism of the number of PV and battery modules in parallel on system stability. Moreover, it finds that it has great potential for exploring parameter influence laws, it can intuitively give the law that a variable affects system stability within any value range through functions or three-dimensional drawings, however, the passive criterion still needs some improvement to achieve analysis accuracy similar to that of the Nyquist criterion.

Furthermore, PV and battery co-traction locomotives are environmentally friendly, because almost all of the electricity comes from renewable energy sources. Based on the research results of this article, further optimization and improvements in the power supply scheme of emergency traction locomotives, and exploring the impact of more variables on the low-frequency stability of the system, such as different types of locomotives running together under different working conditions, and conducting comprehensive research on the simultaneous impact of multiple parameters on the system, all have a profound significance and great value. Meanwhile, issues such as whether the emergency power supply scheme proposed in this article be used for long-term traction locomotives persist. How to configure the capacity of the PV and battery, whether the PV resources can be recycled by the power grid while the supply arm is idle and the locomotive is running, etc., are also the focus of future research. Compared to the traditional AC-electrified railway, the fact that the output of both PV and energy storage is DC suggests that there may be broader application prospects for PV and battery traction locomotives in DC railways and urban rail transit systems, and how to further explore its specific scheme and system stability are yet to be discussed.

5. Conclusions

This work demonstrates that it is feasible to independently drive locomotives using PV and battery equipment under different working conditions during the day and night, and it established a one-dimensional impedance model for a PV–battery locomotive network system from the DC side. At the same time, this study suggests that low-frequency stability analysis of the system of the PV–battery locomotive network may be very important. On this basis, the stability of the system was studied using the generalized Nyquist criterion and a passive criterion. The critical amplitude margin (CAM) and the sensitivity of the controller parameters were proposed to quantitatively evaluate the influence law of the source-side DC converter control parameters and RPC control parameters on this system stability under different emergency traction modes, and corresponding parameter tuning criteria and design suggestions were provided to improve the stability of the system. Finally, the findings were verified through time-domain simulation models and semi-physical testing. The main research conclusions are as follows:

(1) Proposal of a coordinated control strategy, achieving locomotives emergency traction by using PV and battery, thus, increasing the function of the RPC.
(2) When the locomotive is co-tractioned by the PV and battery, the sensitivity of the control parameters is reduced compared to when the locomotive is powered by a single source. Moreover, the parameter selection in the stable region is more extensive, and the control accuracy is higher. In addition, the influence of the proportional gain parameters of the outer loop of the PV and battery DC converters on the system stability is consistent under different traction conditions. Compared to the RPC control parameters and the integral gain sensitivity of the outer loop of the PV and battery DC converter controller, it is more reasonable to improve the system stability by adjusting the proportional gain of the PV and battery DC converters. Priority should be given to reducing the proportional coefficient k_{PVup} and k_{Bup} to improve the stability of the system.
(3) The numerical value of the connection ratio between the PV modules and battery modules has a significant impact on the low-frequency stability of the system. If the output admittance of the PV and battery modules is less than 0, increasing the number of parallel connections will weaken the system stability; if it is greater than 0, the stability margin will be increased to a certain extent. By utilizing this feature, the impedance can be reshaped during system design and operation.

Author Contributions: Conceptualization, Y.W.; methodology, Y.X.; software, Z.X.; validation, Y.W. and Y.X.; formal analysis, X.M.; data curation, X.C.; writing—original draft preparation, Y.X.; writing—review and editing, Y.W.; visualization, X.C. All authors have read and agreed to the published version of the manuscript.

Funding: This research was funded by National Natural Science Foundation of China (52067013); the Natural Science Key Foundation of Science and Technology Department of Gansu Province (22JR5RA318, 21JR7RA280, 22JR5RA353).

Data Availability Statement: Data are available on request from the authors. The data that support the findings of this study are available from the corresponding author upon reasonable request.

Conflicts of Interest: The authors declare no conflict of interest.

Appendix A

$$\begin{cases} G_{\mathrm{PVBB}} = \frac{\Delta u_{\mathrm{Bus}}}{\Delta i_{\mathrm{Bus}}} = \frac{s^2 L_{\mathrm{PV}} C_{\mathrm{PV}} + s R_{\mathrm{PV}} C_{\mathrm{PV}} + 1}{s^3 L_{\mathrm{PV}} C_{\mathrm{PV}} C_{\mathrm{Bus1}} + s^2 R_{\mathrm{PV}} C_{\mathrm{PV}} C_{\mathrm{Bus1}} + s\left(C_{\mathrm{Bus1}} + C_{\mathrm{PV}} D_{\mathrm{PV}}^{*2}\right)} \\ G_{\mathrm{PVBP}} = \frac{\Delta u_{\mathrm{Bus}}}{\Delta d_{\mathrm{PV}}} = \frac{-s^2 R_{\mathrm{PV}} C_{\mathrm{PV}} C_{\mathrm{Bus1}} + s\left(C_{\mathrm{PV}} D_{\mathrm{PV}}^{*} U_{\mathrm{Bus1}} - R_{\mathrm{PV}} L_{\mathrm{PV}} C_{\mathrm{PV}}\right) - I_{\mathrm{LPV}}}{s^3 L_{\mathrm{PV}} C_{\mathrm{PV}} C_{\mathrm{Bus1}} + s^2 R_{\mathrm{PV}} C_{\mathrm{PV}} C_{\mathrm{Bus1}} + s\left(C_{\mathrm{Bus1}} + C_{\mathrm{PV}} D_{\mathrm{PV}}^{*2}\right)} \\ G_{\mathrm{PVLB}} = \frac{\Delta i_{\mathrm{LPV}}}{\Delta i_{\mathrm{Bus}}} = \frac{s C_{\mathrm{PV}} D_{\mathrm{PV}}^{*}}{s^3 L_{\mathrm{PV}} C_{\mathrm{PV}} C_{\mathrm{Bus1}} + s^2 R_{\mathrm{PV}} C_{\mathrm{PV}} C_{\mathrm{Bus1}} + s\left(C_{\mathrm{Bus1}} + C_{\mathrm{PV}} D_{\mathrm{PV}}^{*2}\right)} \\ G_{\mathrm{PVLP}} = \frac{\Delta i_{\mathrm{LPV}}}{\Delta d_{\mathrm{PV}}} = \frac{s^2 C_{\mathrm{PV}} C_{\mathrm{Bus1}} U_{\mathrm{Bus}} + s L_{\mathrm{PV}} C_{\mathrm{PV}} D_{\mathrm{PV}}^{*}}{s^3 L_{\mathrm{PV}} C_{\mathrm{PV}} C_{\mathrm{Bus1}} + s^2 R_{\mathrm{PV}} C_{\mathrm{PV}} C_{\mathrm{Bus1}} + s\left(C_{\mathrm{Bus1}} + C_{\mathrm{PV}} D_{\mathrm{PV}}^{*2}\right)} \\ G_{\mathrm{PVu}} = k_{\mathrm{PVup}} + \frac{k_{\mathrm{PVui}}}{s} \\ G_{\mathrm{PVi}} = k_{\mathrm{PVip}} + \frac{k_{\mathrm{PVii}}}{s} \end{cases} \quad \text{(A1)}$$

$$\begin{cases} G_{BiB} = \frac{\Delta u_{Bus}}{\Delta i_{Bus}} = \frac{sL_{Bat}+R_{Bat}}{s^2L_{Bat}C_{Bus2}+sR_{Bat}C_{Bus2}+D_{Bat}^2} \\ G_{BdB} = \frac{\Delta u_{Bus}}{\Delta d_{Bat}} = \frac{(sL_{Bat}+R_{Bat})I_{Bat}-D_{Bat}U_{Bus}}{s^2L_{Bat}C_{Bus2}+sR_{Bat}C_{Bus2}+D_{Bat}^2} \\ G_{iBi} = \frac{\Delta i_{Bat}}{\Delta i_{Bus}} = \frac{D_{Bat}}{s^2L_{Bat}C_{Bus2}+sR_{Bat}C_{Bus2}+D_{Bat}^2} \\ G_{iBd} = \frac{\Delta i_{Bat}}{\Delta d_{Bat}} = \frac{sU_{Bus}C_{Bus2}+D_{Bat}L_{Bat}}{s^2L_{Bat}C_{Bus2}+sR_{Bat}C_{Bus2}+D_{Bat}^2} \\ G_{Bu} = k_{Bup} + \frac{k_{Bui}}{s} \\ G_{Bi} = k_{Bip} + \frac{k_{Bii}}{s} \end{cases} \quad \text{(A2)}$$

Appendix B

The passive criterion is additionally proven from the perspective of the dissipative system: Based on the DC bus, the equivalent impedance circuit of "PV-Battery-locomotive network" is obtained by using Thevenin's and Norton theorems, as shown in Figure A1, where $u(t) = U_{dc}$.

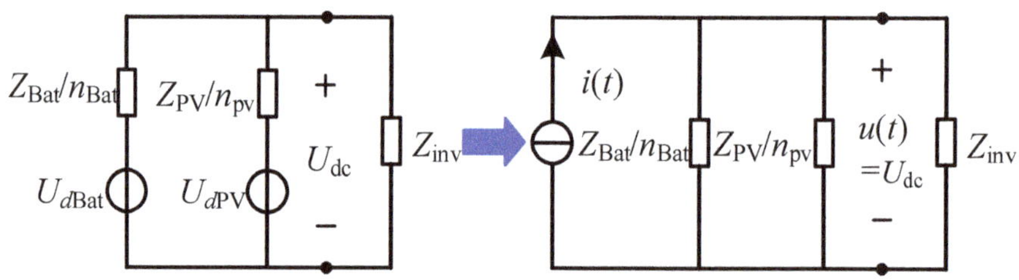

Figure A1. "PV-Battery-locomotive network" equivalent impedance circuit.

The energy function is listed using the output current and voltage, as shown in the following equation.

$$\rho(t) = \int_{-\infty}^{+\infty} u(t)i(t)dt \quad \text{(A3)}$$

When the above equation is always greater than zero, the system is a dissipative system and remains stable at all times. According to Passerval's theorem, the conversion to the frequency domain yields:

$$\begin{aligned} \int_{-\infty}^{+\infty} u(t)i(t)dt &= \frac{1}{2\pi}\int_0^{+\infty} (u^*(\omega)i(\omega) + u(\omega)i^*(\omega))d\omega \\ &= \frac{1}{2\pi}\int_0^{+\infty} (u(\omega))^2(Y_{tot}(\omega) + Y_{tot}{}^*(\omega))d\omega \end{aligned} \quad \text{(A4)}$$

Among them, * represents conjugation, Y_{tot} represents the sum of the admittance realities of the optical storage network, and ensuring that the above equation is greater than 0 is ensuring that the admittance Y_{tot} real part is greater than 0.

References

1. JR East Group. *CSR Report 2017: Aiming for a Sustainable Society*; JR East Group: Shibuya City, Japan, 2017.
2. Metro de Santiago. *Sustainability Report 2017*; Metro de Santiago: Santiago, Chile, 2018.
3. Zhong, J.; Bollen, M.; Rönnberg, S. Towards a 100% renewable energy electricity generation system in Sweden. *Renew. Energy* **2021**, *171*, 812–824. [CrossRef]
4. Zhou, X. A Study on Potential for Using New Energy and Renewable Energy Sources in Railways. *Int. J. Energy Power Eng.* **2019**, *8*, 45. [CrossRef]
5. Zhang, G.; Xue, H.; Wang, Y. Systematic study and innovation practice of new concept for the construction of Xiongan station. *China Railw.* **2021**, 50–56. [CrossRef]

6. Wang, J.; Yang, X.; Kumari, S. Investigating the Spatial Spillover Effect of Transportation Infrastructure on Green Total Factor Productivity. *Energies* **2023**, *16*, 2733. [CrossRef]
7. Chen, J.; Wang, Y.; Chen, X. An on board energy storage solution considering voltage fluctuations and harmonic control of traction networks under mountain area conditions. *J. Railw. Sci. Eng.* **2021**, *18*, 1582–1594.
8. Chen, W.; Wang, X.; Li, Q. Review on the development status of pv power station accessing to traction power supply system for rail transit. *Power Syst. Technol.* **2019**, *43*, 3663–3670.
9. Tanta, M.; Pinto, J.G.; Monteiro, V.; Martins, A.P.; Carvalho, A.S.; Afonso, J.L. Topologies and Operation Modes of Rail Power Conditioners in AC Traction Grids: Review and Comprehensive Comparison. *Energies* **2020**, *13*, 2151. [CrossRef]
10. Guo, Y.; Chen, X.; Wang, Y. Negative sequence optimization compensation strategy for energy storage type railway power conditioner considering grid voltage imbalance. *High Volt. Eng.* **2023**, *49*, 772–780.
11. Tanta, M.; Cunha, J.; Barros, L.A.M.; Monteiro, V.; Pinto, J.G.O.; Martins, A.P.; Afonso, J.L. Experimental Validation of a Reduced-Scale Rail Power Conditioner Based on Modular Multilevel Converter for AC Railway Power Grids. *Energies* **2021**, *14*, 484. [CrossRef]
12. Zhang, D.; Zhang, Z.; Wang, W.; Yang, Y. Negative Sequence Current Optimizing Control Based on Railway Static Power Conditioner in V/v Traction Power Supply System. *IEEE Trans. Power Electron.* **2015**, *31*, 200–212. [CrossRef]
13. Li, T.; Shi, Y. Power Quality Management Strategy for High-Speed Railway Traction Power Supply System Based on MMC-RPC. *Energies* **2022**, *15*, 5205. [CrossRef]
14. Barros, L.A.M.; Martins, A.P.; Pinto, J.G. Balancing the Active Power of a Railway Traction Power Substation with an sp-RPC. *Energies* **2023**, *16*, 3074. [CrossRef]
15. Lu, Q.; Gao, Z.; He, B.; Che, C.; Wang, C. Centralized-Decentralized Control for Regenerative Braking Energy Utilization and Power Quality Improvement in Modified AC-Fed Railways. *Energies* **2020**, *13*, 2582. [CrossRef]
16. Chen, W.; Wang, X.; Han, Y. Research on coordinated control method of pv and battery access traction power supply system based on RPC. In Proceedings of the 2020 IEEE Sustainable Power and Energy Conference (iSPEC), Chengdu, China, 23–25 November 2020.
17. Hu, H.; GE, Y.; Huang, Y. A novel "source-network-train-storage" integrated power supply system for electric railways. In Proceedings of the CSEE 2022, Lisbon, Portugal, 10–12 April 2022; Volume 42, pp. 4374–4391.
18. Liao, Y.; Liu, Z.; Zhang, H.; Wen, B. Low-Frequency Stability Analysis of Single-Phase System with *dq*-Frame Impedance Approach—Part II: Stability and Frequency Analysis. *IEEE Trans. Ind. Appl.* **2018**, *54*, 5012–5024. [CrossRef]
19. Hu, H.; Zhou, Y.; Li, X.; Lei, K. Low-Frequency Oscillation in Electric Railway Depot: A Comprehensive Review. *IEEE Trans. Power Electron.* **2020**, *36*, 295–314. [CrossRef]
20. Carlos, S.B. Direct Generation of Low Frequency Single Phase AC for the Railway in Norway and Sweden. Ph.D. Thesis, Royal Institute of Technology, Stockholm, Sweden, 2009.
21. Danielsen, S.; Molinas, M.; Toftevaag, T.; Fosso, O.B. Constant power load characteristic's influence on the low-frequency interaction between advanced electrical rail vehicle and railway traction power supply with rotary converters. In Proceedings of the 9th International Scientific Conference "Modern Electric Traction", Gdansk, Poland, 24–26 September 2009; pp. 1–6.
22. Danielsen, S. Electric Traction Power System Stability. Ph.D. Thesis, Norwegian University of Science and Technology, Trondheim, Norway, 2010.
23. Song, Y.; Liu, Z.; Ronnquist, A.; Navik, P.; Liu, Z. Contact wire irregularity stochastics and effect on high-speed railway pantograph-catenary interactions. *IEEE Trans. Instrum. Meas.* **2020**, *69*, 8196–8206. [CrossRef]
24. Song, Y.; Wang, Z.; Liu, Z.; Wang, R. A spatial coupling model to study dynamic performance of pantograph-catenary with vehicle-track excitation. *Mech. Syst. Signal Process.* **2020**, *151*, 107336. [CrossRef]
25. Frugier, D.; Ladoux, P. Voltage disturbances on 25 kV–50 Hz railway lines Modelling method and analysis. In Proceedings of the SPEEDAM 2010, Pisa, Italy, 14–16 June 2010; pp. 1080–1085. [CrossRef]
26. Jiang, X.; Hu, H.; He, Z.; Tao, H.; Qian, Q. Study on low-frequency voltage fluctuation of traction power supply system introduced by multiple modern trains. *Electr. Power Syst. Res.* **2017**, *146*, 246–257. [CrossRef]
27. Mu, X.; Wang, Y.; Chen, S. Stability research on high-speed railway vehicle network electric coupling system based on improved sum-norm criterion. *Trans. China Electrotech. Soc.* **2019**, *34*, 3253–3264.
28. Xia, W.; Wang, H.; Meng, X. Electrical stability research on vehicle-grid coupling system of high-speed railways for mixed passenger and freight traffic. *Power Syst. Technol.* **2021**, *45*, 4837–4848.
29. Zhang, X.; Wang, L.; Dunford, W.; Chen, J.; Liu, Z. Integrated Full-Frequency Impedance Modeling and Stability Analysis of the Train-Network Power Supply System for High-Speed Railways. *Energies* **2018**, *11*, 1714. [CrossRef]
30. Fan, F.; Wank, A.; Seferi, Y.; Stewart, B.G. Pantograph Arc Location Estimation Using Resonant Frequencies in DC Railway Power Systems. *IEEE Trans. Transp. Electrif.* **2021**, *7*, 3083–3095. [CrossRef]
31. Zhao, X.; Xie, Z.; Wang, Y. Low frequency stability analysis of photovoltaic connected traction power supply system based on extended forbidden region-based criterion. *High Volt. Eng.* **2023**, 1–12. [CrossRef]
32. Zhou, Y. *Research on Low Frequency Oscillation Suppression of Emu Based on On-Board Energy Storage and Virtual Synchronous Machine Technology*; Beijing Jiaotong University: Beijing, China, 2021.

33. Pedra, J.; Sainz, L.; Monjo, L. Three-Port Small Signal Admittance-Based Model of VSCs for Studies of Multi-Terminal HVDC Hybrid AC/DC Transmission Grids. *IEEE Trans. Power Syst.* **2020**, *36*, 732–743. [CrossRef]
34. Riccobono, A.; Santi, E. A novel Passivity-Based Stability Criterion (PBSC) for switching converter DC distribution systems. In Proceedings of the 2012 Twenty-Seventh Annual IEEE Applied Power Electronics Conference and Exposition, Orlando, FL, USA, 5–9 February 2012; pp. 2560–2567. [CrossRef]

electronics

MDPI

Article

Harmonic Distortion and Hosting Capacity in Electrical Distribution Systems with High Photovoltaic Penetration: The Impact of Electric Vehicles

Miguel Dávila-Sacoto [1], L. G. González [2], Luis Hernández-Callejo [1,*], Óscar Duque-Perez [3], Ángel L. Zorita-Lamadrid [3], Víctor Alonso-Gómez [4] and J. L. Espinoza [2]

1 Department of Agricultural and Forestry Engineering, University of Valladolid, Duques de Soria University Campus, 42004 Soria, Spain; miguelalberto.davila@estudiantes.uva.es
2 Department of Electrical Engineering, University of Cuenca, Central Campus, Cuenca 010104, Ecuador; luis.gonzalez@ucuenca.edu.ec (L.G.G.); juan.espinoza@ucuenca.edu.ec (J.L.E.)
3 Department of Electrical Engineering, School of Industrial Engineering, University of Valladolid, 47011 Valladolid, Spain; oscar.duque@eii.uva.es (Ó.D.-P.); zorita@eii.uva.es (Á.L.Z.-L.)
4 Department of Physics, University of Valladolid, Duques de Soria University Campus, 42004 Soria, Spain; victor.alonso.gomez@uva.es
* Correspondence: luis.hernandez.callejo@uva.es

Abstract: Electric vehicles and the charging stations and their operation require a thorough examination to evaluate the effects on the electrical network. This becomes particularly challenging in the case of high photovoltaic penetration, due to the variability of the solar resource and vehicle connection patterns, which cater to individual user preferences. The current study investigates the impact of harmonics generated by charging stations and electric vehicles on different photovoltaic penetration scenarios within an electrical distribution system. DC and AC charging stations are analyzed. The findings reveal a third harmonic magnitude increase exceeding 300% compared to other cases. Furthermore, this study demonstrates the effects of current and voltage variations on end-users and substation transformers. The impact of harmonics on the hosting capacity of the network is also analyzed, resulting in a 37.5% reduction in the number of vehicles.

Keywords: electric vehicle; charging stations; photovoltaic generation; distribution system; harmonics

Citation: Dávila-Sacoto, M.; González, L.G.; Hernández-Callejo, L.; Duque-Perez, Ó.; Zorita-Lamadrid, Á.L.; Alonso-Gómez, V.; Espinoza, J.L. Harmonic Distortion and Hosting Capacity in Electrical Distribution Systems with High Photovoltaic Penetration: The Impact of Electric Vehicles. *Electronics* **2023**, *12*, 2415. https://doi.org/10.3390/electronics12112415

Academic Editors: Jesús Armando Aguilar Jiménez and Carlos Meza Benavides

Received: 3 May 2023
Revised: 22 May 2023
Accepted: 24 May 2023
Published: 26 May 2023

1. Introduction

Over the past 5 years, there has been significant growth in the adoption of electric vehicles (EVs). However, integrating EVs into electrical distribution systems involves a thorough examination of their impact on the electrical network. The connection of EVs to the power grid may cause an elevation in voltage and current harmonic distortion, as well as a decrease in the lifespan of distribution transformers due to elevated temperatures [1]. In addition, harmonic distortion can also arise from the inverters of grid-connected photovoltaic (PV) systems, with some harmonics reaching high levels [2]. The integration of EVs with a high penetration of PV systems can create challenges for electrical distribution grids. Additionally, a large quantity of electric vehicles may result in power quality issues, such as service interruptions, fluctuations in voltage and current, and harmonic distortion [3]. Hence, it is crucial to carefully plan the integration of electric vehicles into PV-dependent systems to prevent any undesirable disruptions to the distribution system.

Several studies have investigated the influence of EVs on electrical distribution systems, analyzing both EV penetration and PV systems [3–7]. However, there are only a few case studies that investigate the integration of these technologies in electrical grids with a high concentration of PV systems. It has been demonstrated that the presence of EVs in the electrical grid has an impact on total harmonic distortion (THD). In [8], the effect of level two fast chargers on the grid was investigated. As the penetration of EVs on a feeder rises, the degree of harmonic distortion also rises, as observed. Furthermore, the study investigated the impact of the electric vehicle supply equipment (EVSE)'s location on the feeder. The outcomes indicated that the highest THD value was at the end of the feeder, and the voltage at the user's terminals decreased when the EVSE was located further from the feeder's origin. These findings have significant implications for the deployment of EV charging infrastructure in the power grid, as careful EVSE placement can mitigate the negative effects of EVs on harmonic distortion. Moreover, these findings emphasize the necessity for additional research on the influence of EVs on the electrical grid and the development of strategies to guarantee the dependable and effective integration of electric vehicles into the grid.

It has also been observed that the individual harmonic distortion caused by EVs limits the capacity of the electrical grid. In a study conducted by the authors in [9], a sensitivity analysis of models with uncertainty in medium voltage networks was assessed. The results showed that individual harmonics cause a decrease in the hosting capacity due to the probability of exceeding technical limits by approximately 67%. These findings underscore the need for careful consideration of the impact of EVs on the electrical grid and the potential limitations that may arise. It is essential to ensure the reliability and efficiency of the electrical grid as EV adoption continues to increase. Further research on the impact of EVs on the electrical grid's hosting capacity and the development of strategies to mitigate the negative effects of harmonic distortion will be crucial in facilitating the integration of EVs into the electrical grid.

This study focuses on investigating the harmonic effects of EV charging stations in an electrical distribution grid with high PV penetration, considering both peak and valley demand. To determine the actual harmonic values, the power quality of fast-charging stations employing direct current (DC) and slow-charging stations using alternating current (AC) for KIA- and BYD-brand vehicles was measured. Subsequently, the system was simulated utilizing OpenDSS, and validation was carried out using the IEEE European low-voltage test feeder case.

The current study is an extension of the article titled "Harmonic Impact of Electric Vehicles in Electrical Distribution Systems with High Photovoltaic Penetration", presented at the V Ibero-American Congress of Smart Cities.

This article's key contributions include (a) a comprehensive analysis and assessment of the impact and distinct levels of harmonics on the behavior of distribution systems with high PV penetration and EV loads, proposing that a high penetration of photovoltaic generators and EVs has a direct relationship with amplified harmonics and (b) indicating that THD effects on voltage values at loads are significant and must be considered.

2. Methodology

The electrical parameters for this study were acquired by measuring two EV brands and charging stations under real charging conditions. Furthermore, the charging stations and electric vehicles were modeled as both storage components and sources of harmonic generation. The specific modeling of EVs as storage components was deemed necessary due to the standard model available in OpenDSS, which permits the configuration of various efficiency parameters, including state of charge or discharge and power control. However, as there is no specific model designed for EVs, MATLAB must be employed to control their operations. In contrast, the standard OpenDSS model was deemed sufficient for modeling PV systems, enabling the incorporation of a wide range of parameters, including irradiance, efficiency, and temperature effects. Lastly, a standard electrical IEEE standard case study was utilized to conduct the simulations.

2.1. Charging Stations and Electric Vehicles

Tests were carried out at the University of Cuenca's laboratory [10], using a Fluke 435 II power quality analyzer, to gather real data on current harmonics. The measurements were conducted on a BYD T3 and Kia Soul EVs, as well as on a 7.2 kW AeroVironment

RS 25 AC electric vehicle supply equipment (EVSE) (sourced from California, USA) and a 50 kW Circontrol Raption 50 model (sourced from Barcelona, Spain) fast DC EVSE (refer to Figure 1).

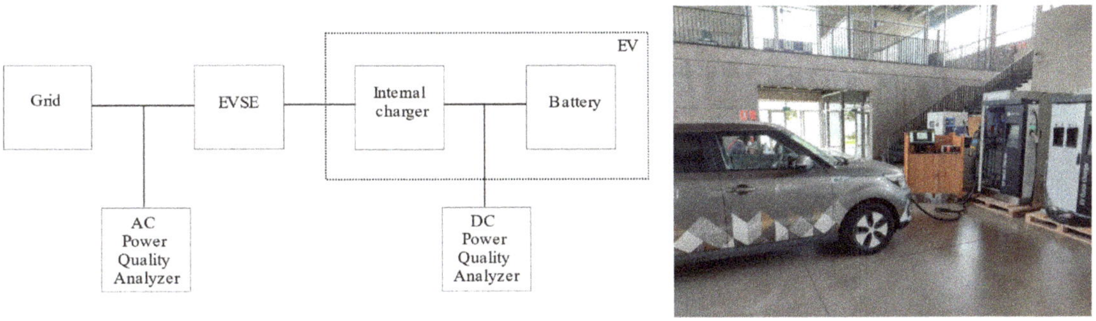

(a) (b)

Figure 1. Measurement equipment: (**a**) connection diagram; (**b**) EV and EVSE.

The laboratory measurements of voltage harmonics, encompassing both magnitude and angle, yielded the results that are shown in Table 1. Additionally, the values obtained from prior studies available in the literature involving other models of EVs have also been included. Table 1 shows the results of the laboratory measurements up to the eleventh harmonic for the parameters (param) in magnitude and phase angle of voltage. It presents the harmonics generated by an electric vehicle charger (EV charger), a Nissan Leaf vehicle, a BYD-brand vehicle with an alternating current charging station (BYD AC) and direct current (BYD DC), as well as a Kia-brand vehicle, likewise with an alternating current charging station (Kia AC) and a direct current one (Kia DC).

Table 1. Harmonics magnitude and angles obtained from measurements and other studies.

No.	Param	EV Charger [6]	Nissan Leaf [5]	BYD AC	Kia AC	BYD DC	Kia DC
1	Mag.	100	100	100	100	100	100
	Angle	−55	−26	136	143	8.6	28.6
3	Mag.	9.20	25.00	9.91	6.31	2.40	6.22
	Angle	120.00	−94.00	98.00	102.00	6.50	32.00
5	Mag.	62.20	17.00	8.47	1.15	1.73	14.55
	Angle	255.00	−96.00	120.00	131.00	7.20	36.00
7	Mag.	41.80	14.20	8.11	0.83	2.14	5.61
	Angle	−28.00	−72.00	146.00	153.00	6.80	31.00
9	Mag.	1.48	9.69	7.30	0.49	1.13	4.34
	Angle	−3.00	−68.00	152.00	161.00	5.70	34.00
11	Mag.	7.08	5.04	6.92	0.64	2.45	5.06
	Angle	−2.00	−49.00	168.00	182.00	5.60	35.00
Charge Efficiency		-	-	92.90%	89.80%	94.60%	91.70%

Additionally, measurements were carried out to evaluate the efficiency of both the charging station and the internal converter of the vehicle. The performance of an EVSE was assessed using instrumentation to ensure accurate measurements. An AC power quality analyzer Fluke 435 II was utilized at the input of the station to monitor critical electrical parameters, such as the current, voltage, and harmonics in both magnitude and angle.

Furthermore, a Hioki PW3337 DC (sourced from Nagano, Japan) power quality analyzer was connected to the internal charger circuit and the EV battery to obtain the charging efficiency. The charging processes of EVs were tested from a state of charge (SOC) of 10% to 95% to ensure a comprehensive analysis of charging performance. The experiment was conducted in the University of Cuenca's laboratory. The efficiency of EVSEs was examined in both AC and DC charging modes, with the highest efficiency observed during DC charging at 94.6%, while the lowest efficiency was recorded during AC charging at 89.8%.

2.2. Simulation Characteristics

OpenDSS was utilized for simulation purposes, wherein EVs were represented as storage elements and were dispersed across different phases of the system, while also acting as a source of harmonics. The simulation process involved the use of two distinct models, one for the EV and another for the PV. The EV was accurately modeled as a battery with a charge controller, customizable efficiency settings, and a current source to generate harmonic waves. The model employed consists of a battery storage element connected to each load bus of the system. It utilizes a two-phase model, with each phase operating at 230 VAC and a power factor of 1. The specific spectrum and load curve for the model can be found in the study's repository, specifically in the Input_data/circuit/autos folder. Similarly, the PV system was designed to operate with a DC-to-AC converter, which served as the control signal for a harmonic current source. Standard values were employed for efficiency, temperature, and irradiation to ensure consistency throughout the simulation process (see Figure 2). The PV system in OpenDSS is defined as a PVSystem element, connected to the system buses according to each scenario. It is configured as a single-phase system operating at 230 VAC and an operating temperature of 25 °C. Additionally, the irradiation, efficiency, harmonic spectrum, and power–temperature adjustment curve are defined in the repository files, specifically in the Input_data/circuit/fotovoltaico folder.

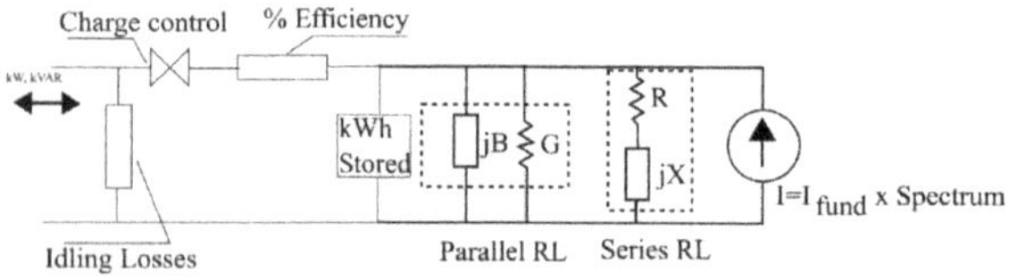

(a)

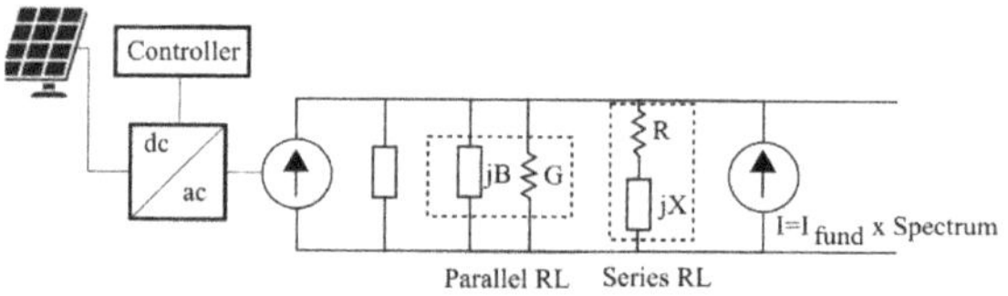

(b)

Figure 2. Simulation models. (**a**) Electric Vehicle. (**b**) PV generator.

The models employed for the EV and PV generator incorporate series and parallel impedances, specifically series RL and parallel RL. To accurately align with the power profile, OpenDSS computes the values of conductance (G), susceptance (jB), and impedance (R+jX) for both the vehicle's power consumption and the generator's power generation requirements according to its power delivery requirements. Regarding the harmonic spectrum, the harmonic values obtained from Table 1 are applied to EVs, whereas default values from OpenDSS are utilized for PV generators. Furthermore, the efficiencies (% efficiency) indicated in Table 1 are incorporated into the EV model, and ideal idling losses are assumed.

This approach enables the accurate modeling and analysis of harmonics impact of EVs and PV systems. A critical consideration during the simulation process was the connection time of EVs, which could significantly impact the results of the analysis. To precisely reflect the behavior of EV connections, a multimodal probability distribution function (PDF) was employed. Such a PDF accounts for multiple modes or peaks that demonstrate the connection of electric vehicles at various times of the day. For residential EVSEs, it is worth noting that, while the connection can be established at any point during the day, the maximum modes transpire at 08:00 and 18:00. These timeframes coincide with when users typically return home with their EVs and proceed to connect them for charging purposes [11]. The function used was designed with modes at 02:00, 08:00, and 19:00, and standard deviations of 2 h, to generate a realistic representation of EV connection behaviors (see Figure 3). In the simulation, EVs maintain a charging regimen lasting between 6 and 8 h, reaching a SOC ranging from 80% to 100%. This encompasses up to 55 distinct charging behaviors, depending on the specific scenario being analyzed. The OpenDSS file utilized to describe the charging duration, as well as the connection and disconnection times, can be downloaded from the repository associated with this study. It is located within the Input_data/circuit/autos folder.

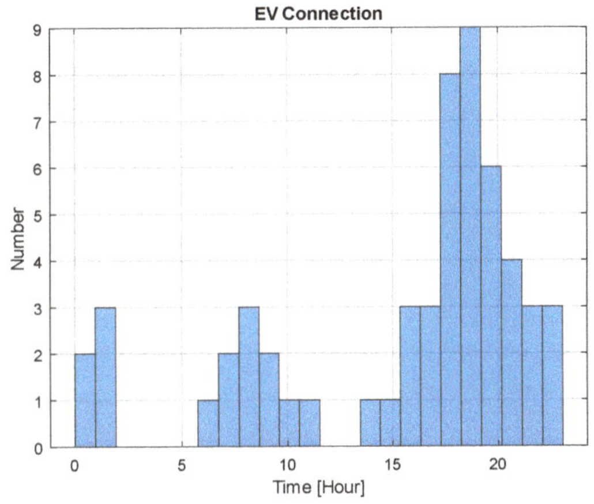

Figure 3. Electric vehicle connection probability distribution function used in simulation.

To automate the simulation process in OpenDSS, MATLAB was employed as a tool for controlling the timing of both customer load profiles and EV connections. The integration of MATLAB with OpenDSS allowed for a streamlined and efficient simulation process, which ultimately facilitated the accurate analysis of the performance of EVSEs.

2.3. *Application Scenarios*

To evaluate the impact of harmonics, a simulation was conducted utilizing the IEEE European low-voltage test feeder model. Three scenarios were analyzed based on the PV generation location: (1) 14 distributed PV systems with a capacity of 5 kW each and 14 EVSEs with a capacity of 7.2 kW each; (2) a central PV system with a capacity of 70 kW and 14 EVSEs with a capacity of 7.2 kW each, as illustrated in Figure 4; and (3) PV and EV distributed across all loads. The harmonic spectrum from Table 1 was assigned randomly to each scenario, while for the PV generators, the harmonic spectrum from [12] was employed. A clear sky irradiance profile was applied to the PV systems. The study examined the following scenarios:

- The initial test case solely examines residential loads within the system, devoid of EVs or PVs (Case 1).
- EVs connected without PV generation and without harmonic effects (Case 2).
- EVs connected with distributed PV generation and without harmonic effects (Case 3).
- EVs connected with single PV generation and without harmonic effects (Case 4).
- EVs and PV generation distributed across the loads without harmonic effects (Case 5).
- EVs connected without distributed PV generation and with harmonic effects (Case 6).
- EVs connected with distributed PV generation and with harmonic effects (Case 7).
- EVs connected with single PV generation and with harmonic effects (Case 8).
- EVs and PV generation distributed across the loads with harmonic effects (Case 9).

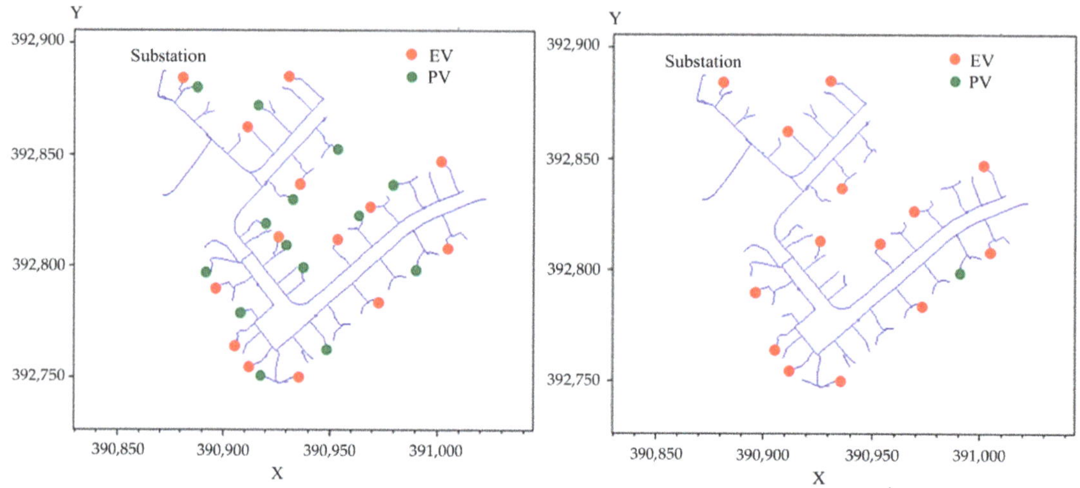

(a)
(b)

Figure 4. Test cases analyzed: (**a**) distributed EV and distributed PV systems; (**b**) distributed EV and single PV system.

The analysis focused on observing the impact on the substation response, the maximum total harmonic distortion (THD), and the voltage and current at the client terminals.

3. Results

During the simulation, the power rating of the transformer in the substation, as well as the highest and lowest voltages recorded across the loads, were taken into account. Moreover, the maximum THD in the loads was also considered.

3.1. Power of the Substation Transformer

Initially, the power of the substation transformer was analyzed for the scenarios outlined in Section 2.3. Figure 5 illustrates that the power of the substation transformer fluctuates as EVs are introduced into the network and the impact of harmonics caused by the EVs is taken into consideration. Case 2 without harmonics exhibits the lowest power value, whereas the highest power value among all scenarios is observed when harmonics are introduced into the network. The high current resulting from harmonics saturates the system's conductors, which diminishes the effect of power injection by the PV generation. Additionally, the situation deteriorates when there is only one PV generation in the distribution system. Likewise, an analysis of the network's performance in Cases 5 and 9, which include distributed PV and EV systems, reveals a distinct impact of harmonic distortions. Specifically, the presence of harmonics causes a reduction in the current output of the PV systems.

The power with harmonics is internally calculated by OpenDSS through multiple simultaneous simulations. First, the power of the fundamental component is computed, followed by repeated simulations for each of the harmonic components configured in the spectrum file. Consequently, OpenDSS provides separate power results for each harmonic, and in MATLAB, the contribution of each harmonic is summed. Equation (1) illustrates the outcome of this calculation.

$$P = \sum_{n=1}^{m} \frac{V_n I_n}{2} \cos(\theta_n - \varphi_n) \quad (1)$$

where:

P is the active power;

n is the harmonic;
V_n is the voltage of nth harmonic;
I_n is the current of nth harmonic;
θ_n is the voltage angle of nth harmonic;
φ_n is the current angle of nth harmonic.

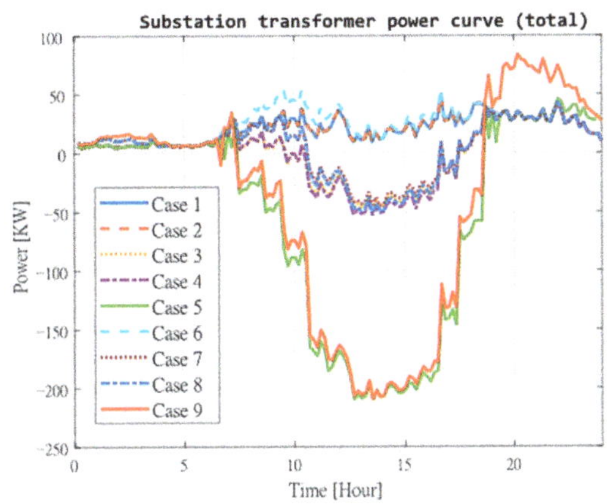

Figure 5. Substation transformer power curve for each case.

3.2. Load Voltages

This study examined the behavior of maximum and minimum voltages across the loads (customers) in the network to ensure that they meet the limit values prescribed by the ANSI C84.1-2006 standard for each simulation scenario. A base RMS voltage of 240 VAC was assumed for the network, and the standard limits of 0.9 pu and 1.05 pu were used, which correspond to 216 VAC and 252 VAC, respectively. The simulations showed that Cases 5 and 9 had lower RMS voltage values than case 8, which had the highest voltage at the customer terminal connected to the PV generator. It was observed that high voltage values coincided across different simulation cases (see Figure 6). Notably, minimum voltage values were found to be within the standard limits across all simulation scenarios.

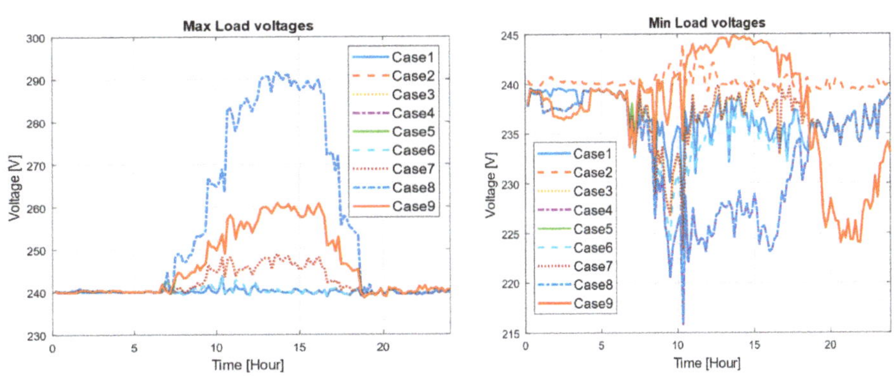

(**a**) (**b**)

Figure 6. Voltages at customer bars. (**a**) Maximum values on all customers by selected case. (**b**) Minimum values on all customers by selected case.

Table 2 displays the highest and lowest power values of the substation transformer, along with the voltage and current at the customer's terminals. It is evident that in Case 2 with harmonics, the transformer requires the most power, both in terms of maximum and minimum power during the 24 h period. With regard to voltage, scenarios with harmonics do not seem to have a significant impact. However, the current remains at a high level during the minimum of Case 2 with harmonics.

Table 2. Power, current, and voltage obtained from the analyzed scenarios.

Case		Transformer Power (kW)	Load Voltage (V)
(1) Base	Max	48.95	243.67
	Min	4.81	225.50
(2) Only EVs	Max	48.95	243.99
	Min	4.81	238.89
(3) EV and distributed PV	Max	43.29	249.01
	Min	−47.06	223.07
(4) EV and single PV	Max	43.29	291.93
	Min	−52.75	215.76
(5) Distributed EV and distributed PV	Max	45.76	260.91
	Min	−209.20	224.01
(6) Case 2 plus harmonics	Max	54.48	244.02
	Min	5.79	221.02
(7) Case 3 plus harmonics	Max	43.29	249.02
	Min	−44.94	223.23
(8) Case 4 plus harmonics	Max	43.29	291.97
	Min	−50.28	215.95
(9) Case 5 plus harmonics	Max	83.81	260.94
	Min	−209.20	224.04

3.3. Load Harmonics

Through a comprehensive analysis of the system's response to specific injected harmonics throughout a 24 h duration in each case, notable observations were made. The findings revealed that the third harmonic held a dominant presence, aligning with the measurements conducted in controlled laboratory settings. The obtained results exhibit distinct harmonic profiles for Case 6 and Case 7, showcasing interesting trends in the magnitudes of various harmonics. In Case 6, the analysis unveiled a maximum value of 2.5% for the third harmonic, indicating its prominent influence on the system. On the other hand, the seventh harmonic exhibited the lowest magnitude, with a minimum value of 0.3%. Case 7 demonstrated a similar pattern, with the third harmonic maintaining its position as the most significant harmonic in terms of magnitude, while the seventh harmonic retained its status as the least influential. Moreover, it is worth noting that the minimum values of the fifth harmonic displayed a noticeable increase of approximately 30% compared to the other harmonics. This observation suggests a distinct behavior and potential sources of harmonics within the system, warranting further investigation into the underlying causes and implications of this phenomenon.

These findings shed light on the harmonic characteristics of the system under investigation, highlighting the prevalence of the third harmonic and the varying magnitudes across different harmonics. This analysis provides valuable insights into the harmonic composition and distribution within the network, aiding in the identification of potential sources and facilitating the development of appropriate mitigation strategies.

The behavior in Case 8 was comparable; however, the minimums of the third, fifth, and seventh harmonics (H3, H5, and H7) increased by up to 60% in comparison to previous cases. Finally, Case 9 demonstrated the most significant harmonic behavior, with the dominance continuing in the third harmonic and its peaks reaching a magnitude of 4.5%. Moreover, the increase in the minimum values was substantial, reaching up to 300% compared to other cases (refer to Figure 7a). The analysis of the maximum THD behavior in loads, as depicted in Figure 7b, revealed that Cases 6 and 7 exhibited higher harmonic levels compared to other cases. Conversely, during the connections of EVs, Case 9 displayed a lower THD value.

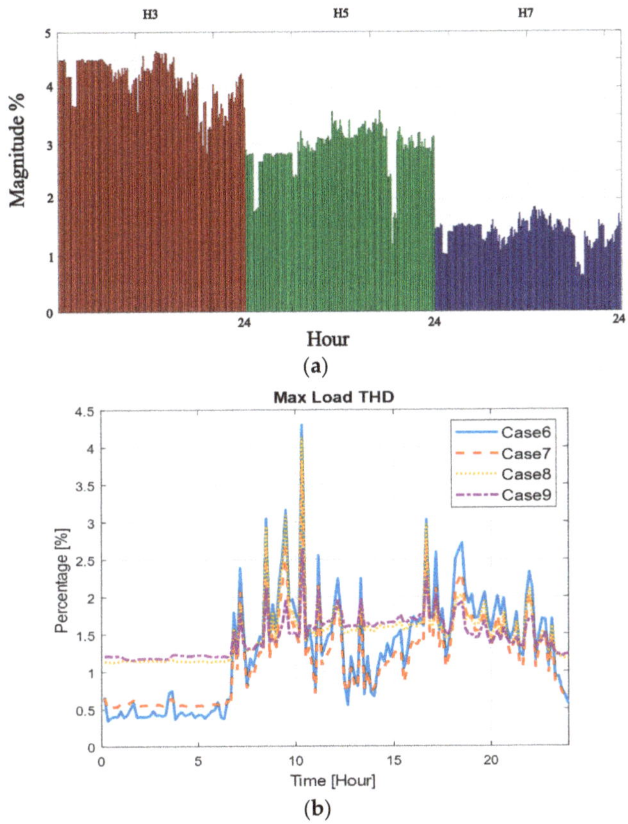

Figure 7. Harmonic behavior. (**a**) Case 9; (**b**) THD.

3.4. Hosting Capacity Limited by Harmonics

One crucial parameter in an electric distribution network with distributed generators and loads such as electric vehicles is the determination of hosting capacity. Traditionally, the analysis of this parameter relies on the limits of power supply quality and the thermal limits of network components. However, there is a limited number of studies addressing hosting capacity specifically considering harmonics.

Within the scope of this study, a hosting capacity analysis was undertaken to assess the influence of harmonic values recorded at customers' premises on the power distribution network. The analysis focused on a specific scenario, namely, Case 9, which encompasses a diverse distribution of both PV system EVs across the network's loads. By examining this particular case, this study aimed to examine the potential repercussions of harmonics on the network's overall performance and functionality. A base case was established with 55 EVs, each with a nominal power rating of 7.2 kW and an average energy consumption of 50.4 kWh, and 55 PV systems, each with a nominal power rating of 5 kW. Through MATLAB simulations, the number of EVs in the system was increased in multiples of their nominal power ratings while ensuring that the maximum THD and individual harmonic distortions remained in compliance with the IEEE 519-2022 standard (see Table 3). The results indicate that while the system can accommodate the connection of up to 104 EVs, the limitation on the third harmonic restricts the number of vehicles that can be connected to 65. The analysis also considered the probability function of connecting vehicles at different times.

Table 3. Maximum Load Total Harmonic Distortion.

EVs Connected	Max THD (%)	Max Harmonic (%)				
		3th	5th	7th	9th	11th
55	2.64	1.61	1.66	0.71	1.18	0.71
65	5.44	4.73	3.51	1.93	1.48	0.77
104	8.08	7.05	4.95	3.33	1.65	0.82

4. Conclusions and Future Work

This study aimed to investigate the impact of harmonics produced by EV charging stations on an electric distribution network with a high level of photovoltaic generation. Two distinct scenarios were examined, one with distributed photovoltaic power generation and the other with centralized generation. The results indicate that the scenario with single photovoltaic generation and EVs distributed throughout the system had the poorest performance. Overvoltages were observed at the terminals of nearby users, while undervoltages were observed at nodes further away from the system. When the harmonic analysis was added, it was found that undervoltages improved in this case. However, overvoltages were still observed at the terminals, and it was also observed that THD was higher at all nodes in the system. These findings highlight the importance of considering multiple factors when evaluating the performance of an electrical system with distributed electric vehicles and photovoltaic generation. The analysis of both voltage and harmonic distortion is crucial in identifying potential issues and developing strategies to address them. Further research is necessary to investigate the interaction between electric vehicles, photovoltaic generation, and other components of the electrical grid, such as energy storage systems, and to develop more comprehensive approaches for evaluating the performance of these systems. This research reveals that harmonics resulted in considerable variations in the voltage, current, and power of the substation transformer, with the magnitude of the third harmonic exceeding 300% compared to other cases. This study also collected information on the magnitude and angle values of harmonics for EVs and charging stations by directly measuring them at the battery terminals, enabling a comprehensive analysis of the entire energy conversion system of both the station and the EV. Upon reviewing the hosting capacity of Case 9, considering harmonics, it was observed that the high values obtained from the third harmonic decrease the total number of electric vehicles that the network can accommodate by 37.5% in comparison with a grid without PV and harmonics. Expanding the harmonic analysis with a greater number of electric vehicle brands is required, and it is also important to examine the effect of charging station efficiency. Additionally, it is necessary to conduct an analysis involving a wider variety of load types to verify their interaction with EVs, such as heat pumps, motors, and thermal generators, among others.

Author Contributions: Conceptualization and methodology, M.D.-S. and L.G.G.; investigation, M.D.-S. and L.H.-C.; writing—review, and editing, M.D.-S., Ó.D.-P., V.A.-G. and Á.L.Z.-L.; supervision and project administration, Á.L.Z.-L., V.A.-G. and J.L.E. All authors have read and agreed to the published version of the manuscript.

Funding: This research received no external funding.

Data Availability Statement: The data and simulation files used in this study are available at https://github.com/davilamds/harmonicsEVPV_2 (accessed on 15 May 2023).

Acknowledgments: The authors thank Universidad de Cuenca and Universidad de Valladolid, who, through a cooperation framework agreement and the specific agreement to regulate their collaboration in research in the field of electrical microgrids and renewable energies, made this work possible. The authors thank Universidad de Cuenca for easing access to the facilities of the Microgrid Laboratory of the Centro Científico Tecnológico y de Investigación Balzay (CCTI-B), for allowing the use of its equipment, and for authorizing its staff the provision of technical support necessary to carry out the experiments described in this article.

Conflicts of Interest: The authors declare no conflict of interest.

References

1. Alame, D.; Azzouz, M.; Kar, N. Assessing and mitigating impacts of electric vehicle harmonic currents on distribution systems. *Energies* **2020**, *13*, 3257. [CrossRef]
2. Phannil, N.; Jettanasen, C.; Ngaopitakkul, A. Power quality analysis of grid connected solar power inverter. In Proceedings of the 2017 IEEE 3rd International Future Energy Electronics Conference and ECCE Asia (IFEEC 2017—ECCE Asia), Kaohsiung, Taiwan, 3–7 June 2017; pp. 1508–1513. [CrossRef]
3. Wang, X.; He, H.; Sun, F.; Sun, X.; Tang, H. Comparative Study on Different Energy Management Strategies for Plug-In Hybrid Electric Vehicles. *Energies* **2013**, *6*, 5656–5675. [CrossRef]
4. Gusai, A. Harmonic Analysis in PV Connected Power System. *Int. J. Eng. Res.* **2020**, *V9*, 474–477. [CrossRef]
5. Aljanad, A.; Mohamed, A. Harmonic Impact of Plug-In Hybrid Electric Vehicle on Electric Distribution System. *Model. Simul. Eng.* **2016**, *2016*, 5968943. [CrossRef]
6. Masoum, M.A.S.; Moses, P.S.; Deilami, S. Load management in smart grids considering harmonic distortion and transformer derating. In Proceedings of the 2010 Innovative Smart Grid Technologies (ISGT), Gaithersburg, MD, USA, 19–21 January 2010; pp. 1–7. [CrossRef]
7. Staats, P.T.; Grady, W.M.; Arapostathis, A.; Thallam, R.S. A statistical method for predicting the net harmonic currents generated by a concentration of electric vehicle battery chargers. *IEEE Trans. Power Deliv.* **1997**, *12*, 1258–1264. [CrossRef]
8. Zarbil, M.S.; Vahedi, A. Power Quality of Electric Vehicle Charging Stations and Optimal Placement in the Distribution Network. *J. Oper. Autom. Power Eng.* **2023**, *11*, 193–202. [CrossRef]
9. Nakhodchi, N.; Bollen, M.H.J. Impact of modelling of MV network and remote loads on estimated harmonic hosting capacity for an EV fast charging station. *Int. J. Electr. Power Energy Syst.* **2023**, *147*, 108847. [CrossRef]
10. Espinoza, J.L.; Gonzalez, L.G.; Sempertegui, R. Micro grid laboratory as a tool for research on non-conventional energy sources in Ecuador. In Proceedings of the 2017 IEEE International Autumn Meeting on Power, Electronics and Computing (ROPEC), Ixtapa, Mexico, 8–10 November 2017; Volume 2018, pp. 1–7. [CrossRef]
11. Richardson, P.; Moran, M.; Taylor, J.; Maitra, A.; Keane, A. Impact of electric vehicle charging on residential distribution networks: An Irish demonstration initiative. In Proceedings of the 22nd International Conference and Exhibition on Electricity Distribution (CIRED 2013), Stockholm, Sweden, 10–13 June 2013; pp. 10–13. [CrossRef]
12. Pereira, J.L.M.; Leal, A.F.R.; de Almeida, G.O.; Tostes, M.E.d.L. Harmonic Effects Due to the High Penetration of Photovoltaic Generation into a Distribution System. *Energies* **2021**, *14*, 4021. [CrossRef]

energies

MDPI

Article

Balancing Electricity Supply and Demand in a Carbon-Neutral Northern Europe

Lisa Göransson

Department of Space, Earth and Environment, Chalmers University of Technology, 412 96 Göteborg, Sweden; lisa.goransson@chalmers.se

Abstract: This work investigates how to balance the electricity supply and demand in a carbon-neutral northern Europe. Applying a cost-minimizing electricity system model including options to invest in eleven different flexibility measures, and cost-efficient combinations of strategies to manage variations were identified. The results of the model were post-processed using a novel method to map the net load before and after flexibility measures were applied to reveal the contribution of each flexibility measure. The net load was mapped in the space spanned by the amplitude, duration and number of occurrences. The mapping shows that, depending on cost structure, flexibility measures contribute to reduce the net load in three different ways; (1) by reducing variations with a long duration but low amplitude, (2) by reducing variations with a high amplitude but short duration and low occurrence or (3) by reducing variations with a high amplitude, short duration and high occurrence. It was found that cost-efficient variation management was achieved by combining wind and solar power and by combining strategies (1–3) to manage the variations. The cost-efficient combination of strategies depends on electricity system context where electricity trade, flexible hydrogen and heat production (1) manage the majority of the variations in regions with good conditions for wind power while stationary batteries (3) were the main contributors in regions with good conditions for solar power.

Keywords: flexibility measures; variation management; VRE; electricity system modeling; wind power integration; solar power integration; sector coupling

Citation: Göransson, L. Balancing Electricity Supply and Demand in a Carbon-Neutral Northern Europe. *Energies* **2023**, *16*, 3548. https://doi.org/10.3390/en16083548

Academic Editors: Luis Hernández-Callejo, Jesús Armando Aguilar Jiménez and Carlos Meza Benavides

Received: 13 March 2023
Revised: 16 April 2023
Accepted: 18 April 2023
Published: 19 April 2023

1. Introduction

The aim of this work is to visualize how variations in a future northern European electricity system can be cost-efficiently managed. The work first examines climate neutral energy systems with extensive electrification of the industry, transport and heating sectors. Outlooks for Nordic [1], northern European [2,3] and European [4] electricity systems have been presented in recent years. The common features of these outlooks are a substantial increase in electricity demand which, to a large extent, is supplied by wind and solar power. Given the varying generation of wind and solar power, the ability of the electricity system to manage these variations have become a question of high priority [5].

There exists a wide range of technologies and strategies to provide flexibility, ranging from shifting the electricity demand for heat pumps in time in single-family dwellings by exploiting the thermal inertia of buildings, to using large-scale underground storage of hydrogen produced for industrial applications. Lund and Lindgren [6] have compiled a comprehensive overview of the different approaches for increasing energy system flexibility. Flexibility measures serve two main purposes: to assure the reliability or increase the cost efficiency of a given electricity system. This work concerns the latter subset, i.e., flexibility measures with the purpose to manage variability such that the social cost of electricity is reduced. This subset is referred to as variation management strategies (VMSs) [7,8]. The potential of flexibility measures to reduce the social cost of meeting the demand for electricity is well documented. Previous work investigating the impact of flexibility measures on the operation of a given electricity system reported reduced costs because of reduced curtailment [7,9] and because of reduced operation of peak generation units in favor of varying renewable electricity (VRE) and base load generation [10]. Studies on the impact of flexibility measures on the cost-optimal system composition reported reduced costs as investments in varying renewable electricity generation capacity replace investments in more expensive electricity generation options [8,11]. Johansson et al. [8] investigated the impact of a range of variation management strategies on the cost-optimal system composition of electricity systems with different preconditions for VRE. It was found that different strategies influenced the operation of the electricity system and investment in electricity generation capacity in different ways. In addition, it was found that the cost-optimal investments in variation management strategies depends on the preconditions for VRE generation. These findings indicate that there are different types of variability which need to be matched with the most cost-efficient strategies for each type. It is reasonable to assume that all types of variations are present in all electricity systems, albeit to different degrees, and what we look for is a balanced combination of strategies adequate for the given electricity system.

Few outlooks for the northern European electricity system address in detail how variations in the electricity system can be met cost efficiently. Pursiheimo and Kiviluoma [3] analyzed electrification scenarios for northern Europe with a focus on Finland. They applied the model of Backbone [12], an investment model with a high time resolution and consequentially with the ability to balance options to cost-efficiently manage variations. Variation management was addressed by illustrating the dispatch of one winter week and one summer week for the Finnish electricity system. The illustrations of the dispatch of electricity systems provided a detailed understanding of how the demand for electricity was met every hour, including how variations were managed when plotted for a limited time span such as a couple of weeks. However, to give an overview of how variations are managed more generally across one or several years, other options for visualization are needed. Several efforts have been made to measure and graphically represent flexibility provision in the electricity system. IEA [13] proposed a flexibility assessment tool (FAST) to estimate if there is sufficient flexibility in a given electricity system to accommodate additional wind and solar power, accounting for flexibility on the demand and supply side as well as the storage and interconnection capacity. Yasuda and Carlini [14] introduced flexibility charts which map the capacity for flexible generation and interconnections relative to the peak capacity of a given electricity system. The tools by IEA and Yasuda et al. rely on statistics which are easy to access to give a first idea of the ability of a given electricity system to manage variations. Using time series from actual or modelled electricity system operations, the contribution of each flexibility supplier can be assessed in more detail. Heggarty and Bourmaud [15] introduce two graphical tools, the flexibility solution modulation stacks and the flexibility solution contribution distribution, to visualize the contribution of flexibility suppliers on annual, weekly and daily timescales.

This paper adds to the previous outlooks for future northern European electricity systems by focusing on how variations in the electricity system are managed using a novel approach of visualizing the net load variability and the contribution of different strategies to reduce this net load. A cost-efficient combination of strategies to manage the net load was derived using a cost-minimizing electricity system model with high temporal resolution representing 11 strategies to manage variations on the demand side and the generation side, as well as pure storage technologies. The contribution of each strategy was visualized using a novel graphical representation in the space spanned by the amplitude, duration and the number of occurrences of the net load variations. Based on the results, the functionality of different strategies to manage variations can be generalized and connected to a functionality-based framework for strategies to manage variations which facilitate the choice of a strategy for a given electricity system context.

2. Method

To find cost-efficient strategies to meet the demand for electricity, heat and hydrogen in a carbon-neutral northern Europe, the cost-minimizing electricity system model ENODE was applied. Given a time-resolved demand for electricity, heat and hydrogen, together with costs and technical limitations of generation and storage technologies, ENODE generates investments in and operations of technologies for generating and storing electricity, heat and hydrogen. From these results, net loads (i.e., electricity demand which cannot be moved in time, reduced by wind and solar power generation) for each region with and without accounting for strategies to manage variations were produced to visualize the role of each strategy in a carbon-neutral northern Europe.

2.1. Electricity System Investment Model

In this work, cost-efficient deployment of variation management strategies was identified using the cost-minimizing electricity system investment model ENODE. The objective of the ENODE model is to minimize the cost of investments and operation while meeting a given demand for electricity, heat and electricity-based hydrogen. The ENODE model was first presented in [16] and further developed in [8,17] to include more options to manage variations and to better represent the heat sector. The model has a high temporal resolution and a detailed technology description, while the temporal scope is limited to 1–2 years. As such, the model results provide a good understanding of the interplay between electricity generation technologies and variation management strategies on the timescale of hours to seasons. In this work, the model was run with a 3 h time resolution for two years (with investment costs represented as annualized costs) representing the year 2050 in terms of technology costs and carbon emission limitation, i.e., only technologies without net carbon dioxide emissions were included as investment options. Existing hydropower was assumed to remain in place together with the new nuclear power in Finland and the nuclear power under construction in the UK.

In this work, the ENODE model was applied to northern Europe subdivided into 14 regions as given by the map in Appendix A Figure A1. The demand for electricity, heat and hydrogen must be met in each region at every timestep. Electricity could be traded between regions. The existing transmission grid was assumed to remain in place and additional investments in transmission capacity of up to 5 GW per connection was allowed.

On the supply side, nine technology options were included including thermal base load generation such as nuclear power, biomass-based combined heat and power and bio-blended coal with carbon capture and storage as well as thermal peak generation in terms of biogas open-cycle and combined-cycle gas turbines. A total list of the technologies together with their cost properties are given in Appendix B. On- and offshore wind power and solar PV were represented using time-resolved wind and solar power production potential for the investigated regions. The power production potential was derived using [18] which relies on ECMWF ERA5 [19] and the Global Wind Atlas [20] for the historical years 1991–1992. These two years were chosen since they represent one year with a lower hydropower inflow in the Nordic countries (1991) and one year with a higher hydropower inflow in the Nordic countries (1992). The potential for wind and solar PV investments per region together with their respective full load hours are given in Appendix B. The electricity demand corresponding to temperature variations in the historical years 1991–1992 was also derived using [18].

This work investigated a future northern European electricity system assuming a carbon-neutral energy system which was targeted to be attained by the European Union by around 2050 [4]. By 2050, it is expected that we will have already experienced a significant increase in temperature due to global warming [9]. Thus, in this work, we assumed an increase in global mean temperature of 2 degrees. In northern Europe, a warmer climate is expected to mainly impact the energy system through altered hydropower inflow, reduced demand for heating and increased number of incidents induced by extreme weather [21]. In this work, we accounted for the two former factors based on [22] which showed an increased annual average hydropower production distributed more evenly across the year in the Nordic countries together with reduced demand for district heating and electricity for heating purposes in the wintertime.

In this work, an electrification of the transport sector and the industry sector was assumed together with a shift from natural gas for heating purposes to individual heat pumps in Germany, the Netherlands, Poland and the UK. The electricity demand for the heating of single-family dwellings was calculated according to Nyholm [23], accounting for energy efficiency improvements. In addition, there was a heat demand corresponding to the demand for district heating in the investigated regions [24]. The heat demand could be met by heat-only boilers, but also combined heat and power plants (CHP), heat pumps and electric boilers. The inclusion of thermal energy storage enabled a temporal decoupling between heat production and electricity production/consumption and facilitated an adapted heat production which offered flexibility to the electricity system.

The electrification of industry includes electrification of steel, cement and ammonia. The total production of steel, cement and ammonia in the investigated regions was assumed to remain at today's levels. In cement production, we assumed plasma burners were used for high temperature heating with a continuous demand for electricity over time. There was an exogenous demand for electricity-based hydrogen (i.e., hydrogen produced from electricity) corresponding to the electricity demand from the steel industry and the ammonia industry. The hydrogen demand was a simplified representation of an electrification of industry and was distributed evenly across all hours of the year, assuming a continuous operation of industrial processes. By overinvesting in the electrolyzer capacity and hydrogen storage, an adapted hydrogen production, which offers flexibility to the electricity system, can be achieved.

A full electrification of the road transport was assumed. The majority of vehicles were assumed to charge as they are parked. However, 30% of the cars charged flexibly over time and could also discharge back to the grid. Vehicle charging was implemented as described by Taljegard [25]. It should be noted that flexible cars were also constrained to meet the demand for transportation as defined by the vehicle movement data collected by Kullingsjö and Karlsson [26].

Furthermore, the model included the option to invest in stationary batteries to manage variations as well as fuel cells to regenerate electricity stored as hydrogen. Thus, the strategies to manage the variations included in this work encompassed supply side, demand-side, and pure storage options. Table 1 list the strategies to manage variations included in this work together with the technologies they involve.

Table 1. Strategies to manage variations included in this work together with the technologies involved and their key properties. CCGT = Combined Cycle Gas turbine, CHP = Combined Heat and Power Plant.

Strategy	Technology	Investment Cost [kEUR/MW(h)]	Efficiency [%]	Fixed O&M Costs [kEUR/MW(h), year]	Lifetime [year]
Charging or discharging of electric vehicles		-	-	-	-
Charging or discharging Li-Ion batteries	Charge/discharge	80	1	1	25
	Storage capacity	70			25
Charging or discharging hydrogen storage system	Fuel cell	500	50	55	10
	Electrolyzer	390	79	18	20
	Hydrogen storage	10	100	-	40
Adapted hydrogen production	Electrolyzer	390	79	18	20
	Hydrogen storage	11	100	-	50
Adapted heat production	Heat pump	1000	300	8	25
	Tank storage	3	80	-	25
Opportunistic heat production	Electric boiler	50	100	-	20
Opportunistic biogas open-cycle electricity production	Gas turbine	450	42	17	30
Adapted biogas combined-cycle electricity production	CCGT	900	61	15	30
Adapted biomass combined heat and power production	CHP	3260	30	105	40
	Tank storage	3	80	-	25
Adapted biomass electricity production	Steam turbine	1980	35	52	40
Adapted nuclear power production	Steam turbine	3980	33	123	60

2.2. Visualizing the Net Load

The method to visualize the net load suggested in this work is based on the proposition that variability can be defined by its amplitude, duration and number of occurrences. Here, we refer to number of occurrences as how often variations with a certain amplitude and duration takes place. According to time series analyses, any time series in a linear system can be described in two dimensions: amplitude and frequency. Weather systems, wind speeds and thus the net load in electricity systems supplied to a large extent by wind power, are composed of a combination of linear and non-linear phenomena. For this reason, we find it practical to subdivide the frequency into duration and number of occurrences and we will refer to the three-dimensional space spanned by the amplitude, duration and number of occurrences as the variation space.

Algorithm 1 describes how to map the net load of any electricity system into the variation space in six steps. The resulting plot illustrates to which extent the net load variations to be managed are variations with a high amplitude, long duration, high occurrence or any combination of these. We refer to the map of the net load in the variation space as the variation profile of the electricity system. An overview of the variation profile is a first step in identifying cost-efficient strategies to manage the net load variations.

To understand the role of different strategies to manage variations in an existing electricity system, or a future system proposed by electricity system models, the remaining net load after the strategy is applied is calculated and plotted. By comparing the original variability profile and the variability profile after a strategy has been applied, it is possible to identify to which extent the strategy reduces variations with a high amplitude, long duration or high number of occurrences.

Algorithm 1. Creating the Variability Profile.

1: Calculate the net load $n_t \in N$ as the total electricity demand fixed in time and reduced by non-dispatchable generation for every timestep $t \in T$.

2: Define a list L:= (min N:max N, s) where the step length s is small enough to give a high resolution of the amplitude in the plots. Here, we use s = 0.1 GW.

3: Make a vector $c_{l,t}$ which counts up as long as the net load n is above amplitude $l \in L$ (or as long as net load n is below amplitude l for negative net loads).
 For $l \in L$:
  For $n_t \in N$:
   if $l > 0$ **and** $n_t < l$:
    $c_{l,t} = c_{l,t-1} + 1$
  else if $l < 0$ **and** $n_t > l$:
    $c_{l,t} = c_{l,t-1} + 1$
  $n_t = n_{t+1}$
 $l = l + s$

4: Find the end of each variation interval and obtain the duration $d_{l,t}$.
 For $l \in L$:
  For $n_t \in N$:
   if $l > 0$ **and** $n_{t-1} > l$ and $n_t < l$:
    $d_{l,t} = c_{l,t}$
   else if $l < 0$ **and** $n_{t-1} < l$ and $n_t > l$:
    $d_{l,t} = c_{l,t}$
  $n_t = n_{t+1}$
 $l = l + s$

5: Count the number of occurrences $o_{l,d}$ for a certain combination of duration and amplitude.
 For $d \in T$:
  $o_{l,d} = \sum_t d_{l,t}$
  $d = d + 1$

6: Plot occurrences of variations with amplitude l on the x-axis, duration d on the y-axis and occurrences o on the z-axis.

3. Results

Figure 1 shows the annual electricity demand for the investigated regions. The historic electricity demand (grey) represents the electricity demand prior to electrification. The direct electricity demand from the transport (yellow) and industry sectors (green) and the electricity for hydrogen production (blue) are a consequence of the assumptions and input data presented in the Section 2. The electricity demand for heating (orange) corresponds to electricity for individual heat pumps replacing natural gas, which is given exogenously, but also electricity demand for district heating, which can be met by combined heat and power plants and heat-only boilers as well as heat pumps and electric boilers. The total electricity demand for heating was thus a result of the optimization. For the case investigated, the increase in electricity demand from electrification corresponded to an increase in the annual electricity demand of around 80% in northern Europe. As illustrated in Figure 1, the electricity demand following the electrification of the transport sector was particularly high in the regions with a high population density such as southern Germany (DE_S), southern Poland (PO_S) and southern UK (UK_S). The electricity for hydrogen production was located in regions with an extensive industrial sector (Sweden, Germany, Poland, the BENELUX region and the UK).

Figure 2 shows the annual electricity production in the investigated regions calculated by the electricity system model presented in the Section 2. Applying the costs presented in the Appendix, it was found to be cost-efficient to meet 2/3 of the annual electricity demand in northern Europe with wind power (blue). As illustrated in Figure 2, there were two major clusters of offshore wind power in northern Europe: one off the coast of northern Germany, the Netherlands and Denmark and one off the coast of the UK. With very good conditions for wind power but a limited electricity-intensive industry, Denmark was the single largest exporter of electricity (DK exports 90 TWh/year). The opposite was true for southern Germany which imported 165 TWh/year for the investigated years. Solar PV energy supplied 1/5 of the electricity demand in northern Europe and was mainly located in the southmost regions (southern Germany and southern Poland) from which other regions imported solar power. The southern UK also had large solar PV investments to complement wind power since import capacity from continental Europe was limited. The model was limited to northern Europe and trade outside the modeled scope was omitted. In reality, there is significant transmission capacity between northern Europe and continental Europe and the results for regions on the border with central Europe (southern Germany and southern Poland) should be viewed with this simplification in mind.

Figure 3 shows the installed capacity of energy storage for the investigated regions. The total energy storage capacity in northern Europe corresponded to 11 TWh. The volume of energy storage capacity reflected the investment cost of energy storage capacity. Tank heat storages were the cheapest energy storage option but were only relevant in regions with district heating (district heating is very limited in Norway and the UK). Hydrogen storage was the second cheapest option for energy storage, and was relevant to regions with a demand for hydrogen. Batteries were significantly more expensive and battery investments were mainly located in regions with substantial electricity supplies from solar PV energy. To investigate the role of energy storage in the different regions and how electricity demand and supply was balanced every hour, four regions with different conditions for wind and solar power and different access to flexibility measures were further investigated.

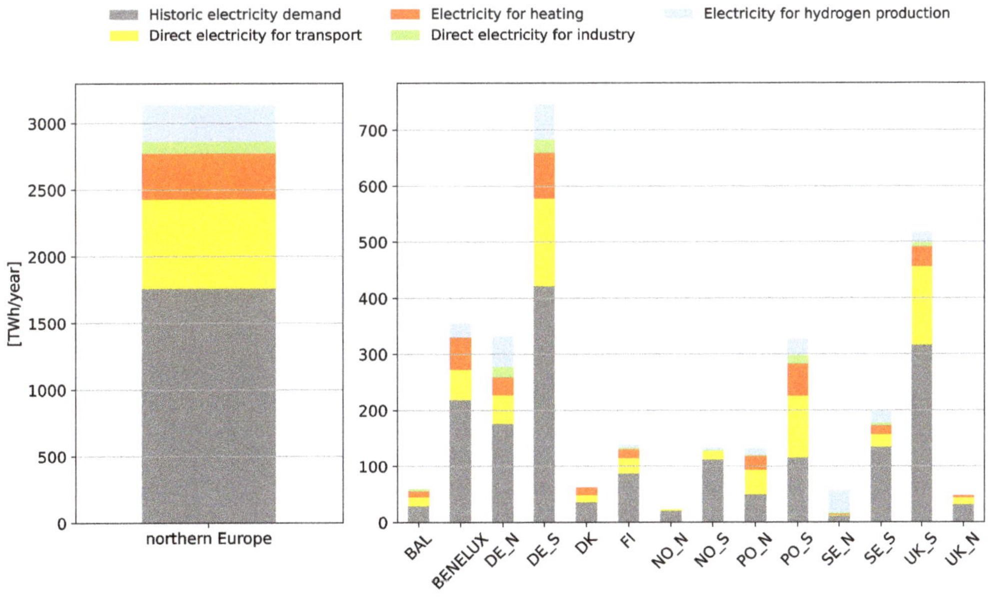

Figure 1. Annual electricity demand for northern Europe as a whole (**left**), together with electricity demand by region investigated in this work (**right**), subdivided into electricity demand of today (grey), electricity demand for heat pumps in district heating systems (orange), for new individual heat pumps replacing natural gas (yellow), for hydrogen production (blue), for light vehicles (light green) and for trucks and buses (dark green).

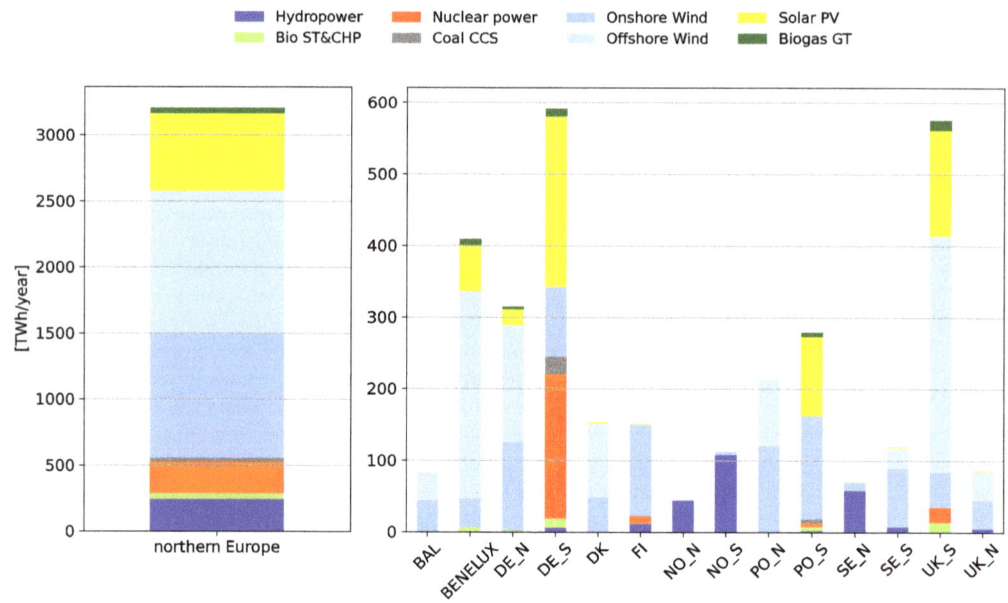

Figure 2. Annual electricity production in northern Europe (**left**) and by region (**right**). ST&CHP = steam turbine and combined heat and power, CCS = carbon capture and storage, GT = gas turbine (open cycle and combined cycle).

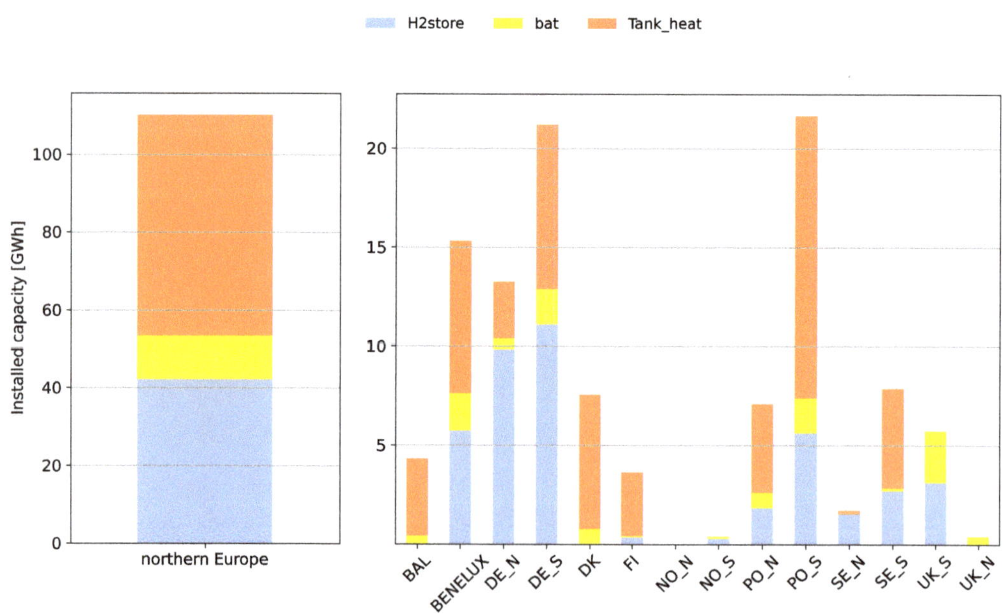

Figure 3. Installed energy storage capacity for northern Europe (**left**) and by investigated region (**right**), subdivided into underground lined-rock cavern hydrogen storage (blue), stationary lithium-ion batteries (yellow) and hot water tank storage in district heating systems (orange).

Figures 4a, 5a, 6a and 7a show the variation profile for DE_S, DK, UK_S and SE_S as defined in the Section 2. These four regions represent four different types of regions observed in the model of northern Europe; regions with extensive solar PV capacity (DE_S and PO_S), regions with extensive wind power for domestic use and export (DK, BAL, PO_N and UK_N), regions with a combination of wind and solar power (UK_S, BENELUX and DE_N) and regions with access to hydropower (SE_S, SE_N, FI, NO_N and NO_S). The color scale was limited at 50 occurrences, implying that the number of occurrences may be higher in the deep red fields. The y-axis was limited at 168 h (one week). Some variations with a longer duration occurred and plots with a y-axis of 300 h for all regions are shown in Appendix C (Figures A2–A15).

As Figures 4a, 5a, 6a and 7a show, net load variations were very different between the investigated regions. Net load variations in Denmark (Figure 5a), with a supply side dominated by wind power, had a low amplitude and number of occurrences but long duration. This was contrasted by net load variations of southern Germany (Figure 4a) with a high amplitude and high number of occurrences but few variations with a long duration. The pattern of the variations in solar power dominated Germany and represented the nature of solar PV energy with a high occurrence in positive net load at a duration of 12–15 h corresponding to night, high number of occurrences in positive net load with high amplitude and a duration of a few hours corresponding to afternoon peaks and high number of occurrences of negative net load with a duration of 2–10 h and varying amplitude corresponding to daytime solar PV production. Variations with a longer duration reoccurred with 24 h intervals and corresponded to cloudy days. The variation profile of southern UK (UK_S) was a combination of the variation profile of southern Germany and Denmark, with the typical pattern of solar PV variations, but with lower amplitudes and lower number of occurrences than those in southern Germany due to a lower total installed capacity, but there were also positive and negative variations with a longer duration which are characteristic of wind power. The variation profile of southern Sweden (SE_S) resembles the one of Denmark, indicating a wind-dominated net load. However, the net load was oriented towards positive amplitudes corresponding to a need for additional generation capacity to complement the wind power to meet the load.

Figures 4–7 illustrate how strategies to manage variations reduced the net load variability in the four example regions. The 11 strategies in Table 1 were aggregated into six groups of strategies to facilitate an overview. Here, the category Thermal includes all thermal electricity generation options except open-cycle gas turbines (adapted biogas combined-cycle electricity production, adapted biomass combined heat and power production, adapted biomass electricity production, adapted nuclear power production). Based on the electricity generation supplied by different sources (see Figure 2), nuclear power dominated this group of strategies. Batteries included stationary batteries and charging of electric vehicles since previous work showed that these strategies act on the same timescales in the electricity system [2]. This implies that we added a load, which was fixed to 70% and flexible to 30% according to assumptions, together with the charging and discharging of stationary batteries. Finally, here, Heat includes both heat production with heat pumps and electric boilers connected to the district heating system (adapted heat production and opportunistic heat production in Table 1). However, this group mainly represents variation management by heat pumps since electric boilers played a very minor role in all investigated regions.

Figure 4b–g shows the net load reduction by six different strategies for southern Germany. Electricity was imported to solar PV-dominated southern Germany, reducing the positive nighttime net load. The positive net load was further reduced by base load generation (Figure 4d). At the same time, base load generation increased the negative net load with long durations. Due to the limited flexibility of nuclear power, production was not only added when needed but also during hours of low net load. The negative net load with a long duration caused by base load generation was reduced by adapted production of hydrogen and heat (Figure 4f–g). However, in the solar PV-dominated southern Germany, the strategy with the largest impact on the net load was batteries (Figure 4e). Stationary batteries matched the charging of electric vehicles (in this work, 70% of the vehicle charging was fixed in time) to the solar PV generation and thereby reduced the net load variations. Batteries have a relatively low cost of charging and discharging power and high efficiency which makes them well suited for the solar PV variations with a high amplitude and high number of occurrences. In general, it can be observed that investments in stationary batteries were particularly substantial in regions with a large demand for electricity for transport (cf. Figures 1 and 3) and that solar PV generation was high in the same regions (Figure 2). The net load variations with a low amplitude, duration and a low number of occurrences which remained in Figure 4f was supplied by biogas open-cycle gas turbines.

Figure 5b–g shows the gradual net load reduction by the six strategies to manage the variations in Denmark. Denmark has some of the best conditions for offshore wind power in northern Europe and together with its central location between the Nordic countries and continental Europe it becomes an ideal exporter of electricity. The majority of this electricity was exported to northern Germany and onwards to load centers in southern Germany. The extensive export results in a strong reduction of the wind-dominated net load variations (Figure 5b). Trade with Germany also introduced solar variations in Denmark, as seen in Figure 5b. These variations were efficiently reduced by batteries. The remaining net load variations had a low amplitude but sometimes had a long duration. In Denmark, these variations were primarily reduced by heat pumps in the district heating system (Figure 5g).

Figure 6b–f shows the net load reduction by strategies to manage variations in the southern UK. The system composition here was a combination of wind and solar power, as illustrated in Figure 2, and the net load had both variations with a long duration and variations with a high amplitude and a high number of occurrences. It is important to note that by combining wind and solar power in the electricity mix, net load variations were, to some extent, already reduced from the start. In particular, solar PV production reduced the duration of low wind events and wind power reduced the positive nighttime net load of solar power. These complementary effects can be visualized by comparing the net loads in Figure 6a to those in Figures 4a and 5a. Trade, primarily with northern UK, further reduced net load variations with a 20–24 h duration. However, it was by combining wind and solar power together with the use of batteries for variations with a high amplitude and high number of occurrences that the majority of the variations in this region were managed. Some variations with a low duration and low number of occurrences were managed by adapted production of hydrogen. The net load which remained in Figure 5f was managed by biogas open-cycle gas turbines.

Figure 7b–f illustrates how the net load was reduced by five different strategies to manage variations in southern Sweden. Figure 7b shows the net load reduced by trade (except for northern Sweden). As the figure shows, trade shifted the net load to the left, indicating that southern Sweden imported electricity. However, since the electricity was mainly imported from Denmark, which is wind-dominated like southern Sweden, the variability of the net load curve was similar after accounting for trade. Some solar variations with a high number of occurrences and high amplitude but short duration were introduced with the trade. Figure 7c shows the net load reduced by trade and hydropower, including trade with northern Sweden. Even though electricity demand in northern Sweden increased substantially, hydropower from the north was exported to southern Sweden during low wind events since the demand in the north was predominantly composed of electricity for hydrogen production which was avoided during low wind events. As the figure shows, the positive net load was drastically reduced by hydropower and hydropower managed positive net loads of any duration. Batteries and the charging of electric vehicles reduced the amplitude of the remaining negative net load (Figure 4e). The remaining net load had a low amplitude and any duration which was managed by adapted production of hydrogen (Figure 4f) and heat (Figure 4g). With the high cost of electricity consumption capacity (electrolyzer and heat pump) but low cost of energy storage (underground hydrogen storage and storage of hot water), variations with a low amplitude but long duration matched the cost structure of these strategies. It is important to note that adding electricity consumption adapted to the net load, i.e., adapted production of heat and hydrogen, does not only imply that "excess" electricity was consumed. These flexible electricity demands stimulated investments in wind power production without increasing consumption during low wind events. Since there was also some wind power production during low wind events, adapted consumption of hydrogen and heat reduced the net load during these events. The little remaining net load visible in Figure 7f was met by open-cycle gas turbines.

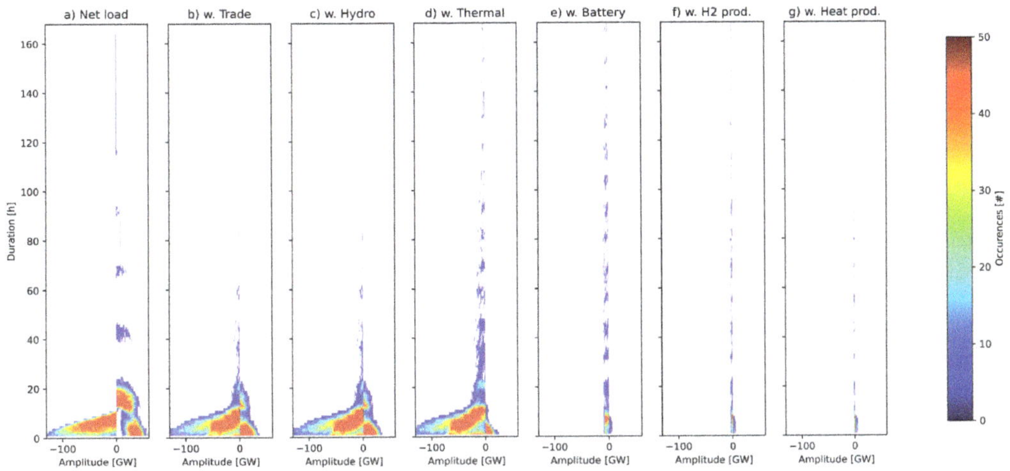

Figure 4. Net load of southern Germany (DE_S) (**a**) reduced by (**b**) trade, (**c**) trade and hydropower, (**d**) trade, hydropower and base load, (**e**) trade, hydropower, base load and batteries, (**f**) trade, hydropower, base load, batteries and adapted hydrogen production and (**g**) trade, hydropower, base load, batteries, adapted hydrogen production and adapted heat production. The color scale indicates number of (#) occurrences and was limited to 50 occurrences, implying that the number of occurrences may be higher in the deep red fields. The y-axis was limited at 168 h (one week).

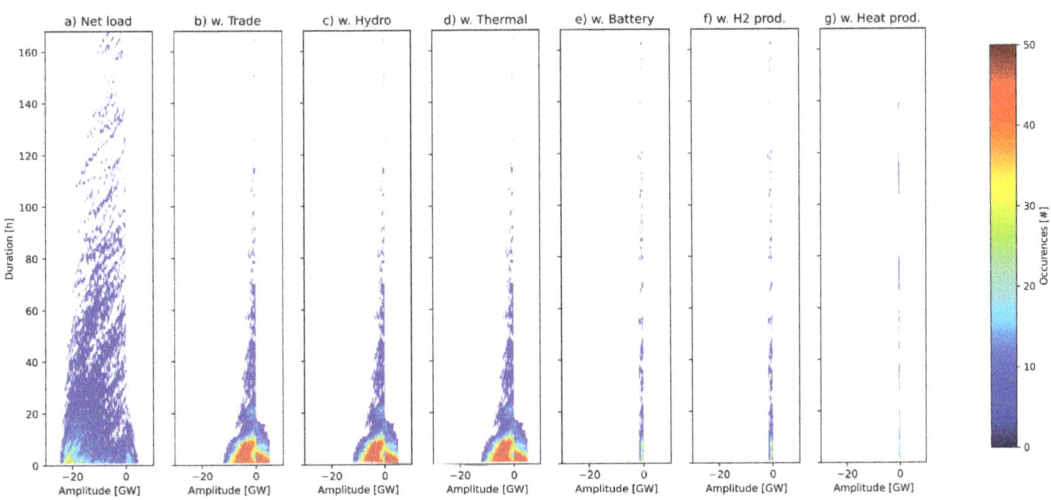

Figure 5. Net load of Denmark (DK_T) (**a**) reduced by (**b**) trade, (**c**) trade and hydropower, (**d**) trade, hydropower and base load, (**e**) trade, hydropower, base load and batteries, (**f**) trade, hydropower, base load, batteries and adapted hydrogen production and (**g**) trade, hydropower, base load, batteries, adapted hydrogen production and adapted heat production. The color scale indicates number of (#) occurrences and was limited to 50 occurrences, implying that the number of occurrences may be higher in the deep red fields. The y-axis was limited at 168 h (one week).

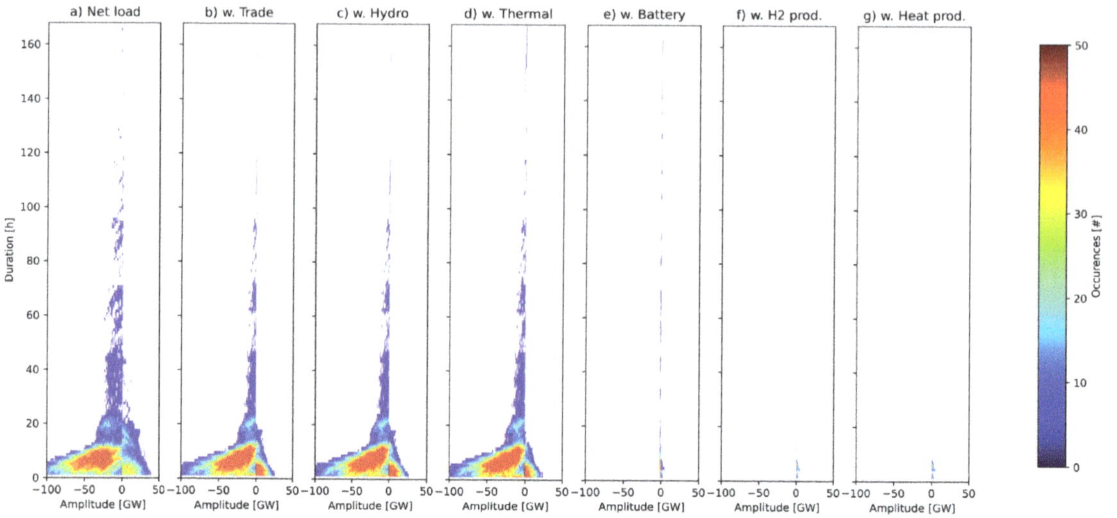

Figure 6. Net load in southern UK (UK_S) (**a**) reduced by (**b**) trade, (**c**) trade and hydropower, (**d**) trade, hydropower and base load, (**e**) trade, hydropower, base load and batteries, (**f**) trade, hydropower, base load, batteries and adapted hydrogen production and (**g**) trade, hydropower, base load, batteries, adapted hydrogen production and adapted heat production. The color scale indicates number of (#) occurrences and was limited to 50 occurrences, implying that the number of occurrences may be higher in the deep red fields. The y-axis was limited at 168 h (one week).

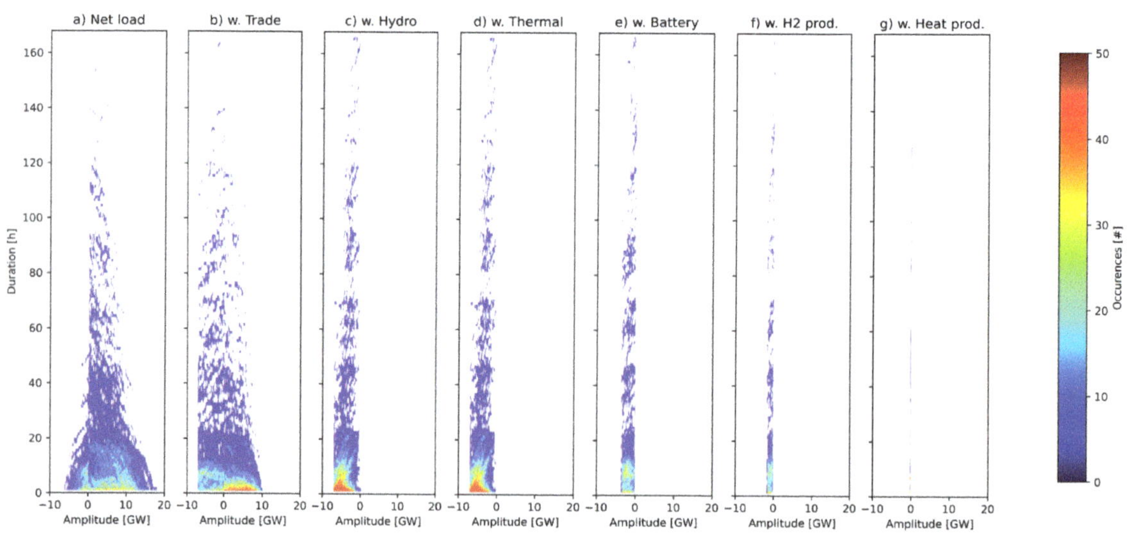

Figure 7. Net load in southern Sweden (SE_S) (**a**) reduced by (**b**) trade, (**c**) trade and hydropower, (**d**) trade, hydropower and base load, (**e**) trade, hydropower, base load and batteries, (**f**) trade, hydropower, base load, batteries and adapted hydrogen production and (**g**) trade, hydropower, base load, batteries, adapted hydrogen production and adapted heat production. The color scale indicates number of (#) occurrences and was limited to 50 occurrences, implying that the number of occurrences may be higher in the deep red fields. The y-axis was limited at 168 h (one week).

4. Categorizing the Cost-Efficient Contribution of Variation Management Strategies

The results confirm that different variation management strategies contributed with different functionalities in the electricity system. At the same time, the contribution from each strategy, in terms of reduction in net load amplitude and number of occurrences on different timescales, was qualitatively similar in the four investigated regions. However, the size of the investments in different strategies and their quantitative contribution was different in the different system contexts.

As the objective of the model applied in this work was to minimize the total cost of the electricity system, the investments in, and operation of, the variation management strategies as given in this work was cost efficient given the available options. That is, the functionality provided was motivated by the cost structure and technical limitations of each strategy. High reductions in the amplitude of recurring variations were cost-efficiently supplied by strategies with a low cost of charging and/or discharging power, such as batteries. To manage variations over longer timescales, a low cost of energy storage is required. In the examples given in this work, the cost of hydrogen storage represents the cost of storing hydrogen in large-scale underground caverns, which is low compared to energy storage in batteries. The cost of storing heat in tank storages is even lower. The cost of storing fuel for thermal generation was omitted in the model, which allows combined-cycle gas turbines and combined heat and power plants to manage the variations with a long duration. Variations with a low number of occurrences were managed most efficiently using strategies with a low investment cost, such as open-cycle gas turbines. With a low number of occurrences a high running cost is acceptable.

The roles identified by these measures can form the basis for a functionality-based categorization of strategies to manage variations. Whereas previous categorizations of strategies to manage variations were based on the technical properties of the strategies, a categorization supporting the understanding of the functionality of the strategies to manage variations in the electricity system should rely on the cost structure. Based on the above reasoning and the work by [7,8], the following categories are proposed:

- Peaking strategies, which manage variations with a low number of occurrences. Similar to peak production plants, peaking variation management strategies have low investment costs in terms of charging, discharging and storage capacity. However, the low investment cost is typically accompanied by a high operational cost, i.e., a high cost of electricity discharged or low value of electricity charged.
- Shifting strategies, which reduce recurring variability on shorter timescales. These strategies are associated with a low cost of charging and discharging capacity and a low cost of operation, since variability on shorter timescales have high amplitude and high number of occurrences. However, a low cost of charging and discharging capacity is typically associated with a high cost or a limited capacity to store energy. Thus, these strategies are mainly applied for shifting loads or production during shorter time intervals.
- Complementing strategies, which manage recurring variations on longer timescales. These strategies are associated with a low cost of energy storage. However, since the amplitude of variations on longer timescales is lower, a higher cost associated with the charging and discharging power is acceptable. A low to medium cost of operation is required from these strategies since the variations being managed have a long duration and medium recurrence. This category could be further subdivided to differentiate between complementing strategies managing variations on weekly and seasonal timescales.

A categorization based on the cost structure has the advantage that the category reveals the functionality of the strategy. This categorization thus facilitates the choice of variation management strategies for different system services and contexts.

5. Discussion and Limitations

In this work, net load variations were plotted in the space spanned by amplitude, duration and number of occurrences (i.e., the variation space) to better understand how different strategies to manage variations complement each other. For this analysis, the cost-efficient combination of a large set of strategies provided by the electricity system investment model was required. However, plotting a gradually reduced net load in the variation space can also be useful to understand how variations are managed in an existing system, using electricity production data.

Applying the method proposed to map the contribution of flexibility measures to reduce the net load, functionality-based categories of variation management strategies can be identified. The vast majority of previous categorizations have focused on the technical properties of the flexibility measures. For example, Fuchs and Lunz [27] and Zhao and Wu [28] subdivided electricity storage systems into electrical, mechanical, chemical and thermal storage units, while Sauer [29] differentiated between electricity-to-electricity storage, electricity-to-anything storage and anything-to-electricity storage systems. Palizban and Kauhaniemi [30] have mapped energy storage systems with respect to those applications for which they are suitable and unsuitable for, using a color-coded matrix. Other categorizations have focused on the motive for the investment, such as single-use/double-use [31] which distinguished between investments that are actively made and are dedicated to provide flexibility in the electricity system (single-use) and those cases in which assets that are already integrated in the electricity system for other purposes are used (double-use). While all these categorizations support the understanding of flexibility measures, measures with the same functionality in the electricity system are distributed between the categories in the above-mentioned frameworks. A previous work by the author [7] proposed instead a functionality-based framework, organizing variation management strategies according to the service they provide in the electricity system. This work further developed this framework.

The strategies applied in the regions modeled in this work are cost efficient from an energy-only perspective. However, without the perfect foresight of the modelled world, flexibility markets may be needed to stimulate investments prior to large electricity price variations. In the design of flexibility markets, a market for each of the above-mentioned categories would assure competition between strategies providing the same function to the electricity system and stimulate investments in a balanced set of strategies. The size of the different markets should be defined by the electricity system context, i.e., the need for the different functionalities based on the variability profile of the present and future electricity systems.

This work investigates strategies to manage variations on the timescale of hours to a couple of years. Variations within the hour as well as variations with very a low number of occurrences (e.g., once every ten or thirty years) is outside the scope of this work. Meng and Zafar [32] provided an overview on how variations on very short timescales, to recover the frequency after faults, can be managed in electricity systems with a high share of wind and solar power. They concluded that batteries with grid-forming converters, with the ability to provide active power very rapidly, are able to replace synchronous generators while maintaining the ability to recover the frequency after a fault. However, if synchronous generators can be replaced completely remains to be investigated. Ullmark [33] showed that if batteries are allowed to meet the demand for variations within the hour, these variations can be managed at a low cost (<1% of total system cost).

Ruhnau and Qvist [34] investigate the impact of variations in the German electricity system over a 30-year period and found that there were low wind events with a long duration (60 days with short periods of interruption) which occurred very rarely. As shown in this work, the choice of strategy for a certain type of variation depends on the cost structure. Variations with a long duration and low number of occurrences are cost-efficiently managed by strategies with a low cost of storage and low investment cost. A cost-efficient strategy to manage the variations identified by [34] would thus be biogas

gas-turbines with a sufficiently large storage of biogas. Sufficient biogas turbine capacity is likely to already be available to the electricity system to manage variations with a shorter duration occurring on a yearly basis. The fuel costs during the 60-day period would be significant but, since these events occur very rarely, this fuel cost would have a low impact on the total system cost and the overall electricity system composition.

6. Conclusions

In this work, a novel tool to map variations in electricity systems was applied to visualize how variations can be cost-efficiently managed in a future northern Europe. The results show that cost-efficient variation management depends on the system context. More specifically, it depends on the nature of the variations present in the electricity system as well as the context-specific prerequisites for variation management, such as extensive demand for hydrogen from industry, availability of district heating grids or access to hydropower. In particular, the access to strategies suitable to manage variations with a long duration varied between the investigated regions. As a consequence, the dominant strategy to manage variations with a long duration varied between, for example, Denmark, where adapted heat production has an important role, and southern Sweden, which relies more on hydropower and adapted hydrogen production to manage these variations.

Electricity systems with good conditions for solar power and extensive electricity demand for transportation were subject to variations with a high amplitude and high number of occurrences which can be cost-efficiently managed using batteries with a low cost of charging and discharging power and low losses per cycle. Examples of such regions in northern Europe were southern Germany and southern Poland, with inland locations and high population densities. In these regions, there may also be potential for some base load generation.

Electricity systems with good conditions for wind power, such as Demark, the Netherlands and northern Germany, managed variations through trade, including import of solar PV energy and adapted production of hydrogen and/or heat. Extensive industrial hydrogen demand or district heating systems facilitated the integration of wind power since wind variations have long durations and hydrogen and heat can be stored at a relatively low cost.

In regions with good conditions for wind power but limited trade, such as the UK, it was cost efficient to combine wind power capacity with solar PV capacity to reduce the duration of low wind events. It was often cost efficient to combine hydropower with batteries, where the former managed the positive net loads with short and long durations but the latter managed variations with a high amplitude including negative net load. In general, cost-efficient variation management was achieved by combining wind and solar power, and with strategies to manage variations with different cost structures.

Funding: This research was funded by MISTRA, grant number 2019-0026, and the Swedish Energy Agency, grant number 49341-1.

Data Availability Statement: The key data applied in this work are provided in the manuscript, in the appendix or is referenced. The data can be made available on request.

Acknowledgments: The author would like to thank Viktor Walter and Filip Johnsson for valuable discussions, as well as Tom Brown and Juha Kiviluoma for feedback on early work.

Conflicts of Interest: The author declares no conflict of interest.

Appendix A

Figure A1. Northern Europe subdivided into 14 regions as applied in this work.

Appendix B

Table A1 shows the investment costs together with the operation and maintenance costs (O&M) for electricity and heat generation applied in this work. The costs are based on IEA [35] and the Danish Energy Agency [36], except for nuclear power and wind power with low specific power (SP). The investment costs for nuclear power were estimated after dialogues with experts. It corresponds to the average investment cost in the case where several units are invested in and is lower compared to levels given for Europe by the IEA. The cost to manage waste (today 4 EUR/MWh in Sweden) was expressed as a small addition to the variable cost (0.5 EUR/MWh) to manage fuel waste and a larger fixed cost to manage the plant at the time of decommissioning (corresponding to 3.5 EUR/MWh after 60 years with 90% utilization) were allocated to fixed O&M. Nuclear energy was assumed to be able to vary its output between 70–100% of rated power. The majority of wind turbines today have a specific power around 300 W/m^2. However, turbines with a lower specific power (generator capacity over swept rotor area) are advantageous for sites with lower average wind speeds and offshore locations. Hodel and Göransson [37] assessed cost-efficient turbine designs in different system contexts using a cost model to assess the cost of a range of wind turbine designs. Investment costs for onshore wind turbines with a specific power of 100 W/m^2 and tower height of 150 m and for offshore wind turbines with 200 W/m^2 and 150 m tower height were based on their work.

Table A2 shows the cost and properties of the energy storage options. These costs were based on the Danish Energy Agency [36]. The cost of transmission capacity was assumed to 2k EUR/MW and km, where the distance is taken as the distance between a point in each region with extensive grid capacity. The calculations applied annual investment costs which were derived using a 5% interest rate and technical lifetimes as given in Table A1. Table A3 shows the fuel costs applied in this work.

Table A1. Costs and properties of electricity and heat generation units. CHP = Combined Heat and Power, CCS = Carbon Capture and Storage, CCGT = Combined Cycle Gas Turbine, OCGT = Open Cycle Gas Turbine.

Technology	Investment Cost [MEUR/MW(h)]	Variable O&M Cost [EUR/MWh]	Fixed O&M Cost [kEUR/MW, year]	Technical Lifetime [year]	Efficiency [%]
Biomass steam	2.0	2.1	52	40	35
Biomass CHP	3.3	2.1	105	40	30
Coal w. CCS [1]	3.5	2.1	107	40	40
Biogas CCGT	0.90	0.8	17	30	61
Biogas OCGT	0.45	0.4	15	30	42
Nuclear power	4.0	7.1	123	60	33
Solar PV power	0.3	0.5	7	40	100
Onshore wind power (100 W/m^2 150 m)	1.65	1.1	13	30	100
Onshore wind power (300 W/m^2 100 m)	1.0	1.1	13	30	100
Offshore wind power (200 W/m^2 150 m)	1.75	1.1	36	30	100
Heat pump	0.9	2.2	2	25	3
Electric boiler	0.1	1	1	20	1

[1] Coal was assumed to be mixed with biomass in order to compensate for emissions which are not captured.

Table A2. Costs and properties of storage technologies. Investment costs for batteries (power), electrolysis and fuel cells are given in MEUR/MW while costs for batteries (energy) hydrogen storage and heat storage are given in MEUR/MWh.

Technology	Investment Cost [MEUR/MW(h)]	Efficiency (ch/disch) [%]	Fixed O&M Cost [kEUR/MW(h), year]	Technical Lifetime [year]
Battery, Li-ion (energy)	0.08	96/96	-	25
Battery, Li-ion (power)	0.07	100	0.5	25
Electrolysis	0.4	70	18	20
Fuel cell	0.5	50	55	10
Hydrogen storage	0.011	100	-	40
Heat storage	0.003	100 [1]	0.009	25

[1] Heat storages have a continuous loss corresponding to 0.023% per unit of time and energy stored.

Table A3. Cost of fuel applied in the calculations.

Fuel	Fuel Cost [EUR/MWhth]
Biomass	40
Biogas	77
Uranium	1.65
Coal w. biomass blend	7.5

On- and offshore wind power and solar PV power were represented using time-resolved wind and solar power production potential for the investigated regions derived using [18] which rely on ECMWF ERA5 [19] and the Global Wind Atlas [20] for the historical years 1991–1992. These two years were chosen since they represent one year with a lower hydropower inflow in the Nordic countries (1991) and one year with a higher hydropower inflow in the Nordic countries (1992). The potential for wind and solar PV investments per region together with their respective full load hours are given in Tables A4 and A5. Electricity demand corresponding to temperature variations in the historical years 1991–1992 was also derived using [18]. After the removal of unsuitable and protected areas, the potentials for onshore wind power, offshore wind power and solar power were derived applying an assumed factor of social acceptability. For offshore wind power and solar PV power, 33% and 5% of the share of the remaining land was assumed to be available in all regions, respectively. For onshore wind power, 8% of the remaining land was assumed to be available except for Norway (1%), Sweden, the UK and Ireland (4%) where acceptance for wind power investments has proven to be low. The potential area for wind power was subdivided into five wind classes for each region representing different wind conditions. Wind turbine technologies were chosen to match the conditions of their respective area, with wind turbines with 100 SP (Specific Power, i.e., generator capacity relative to swept rotor area in W/m^2) and 150 m hub height for onshore sites with low average wind speeds (i.e., wind class 1–3), 300 SP and 100 m hub height for areas with high average wind speeds (i.e., wind class 4–5) and 200 SP and 150 m hub height for offshore wind turbines.

Table A4. Potential for electricity production (GW) for the five classes of onshore wind power (WON1-WON5), the five classes of offshore wind power (WOFF1–WOFF5) and solar PV parks (PV) for the regions considered in this work. SP = Specific Power, i.e., generator capacity relative to swept rotor area in W/m^2.

Region	WON1 100SP	WON2 100SP	WON3 100SP	WON4 300SP	WON5 300SP	WOFF1 200SP	WOFF2 200SP	WOFF3 200SP	WOFF4 200SP	WOFF5 200SP	PV
SE_N	0.7	5.1	5.3	3.4	2.1	0.0	0.0	1.7	16.3	2.8	86.3
SE_S	0.2	1.5	7.0	8.2	0.9	0.0	0.0	0.1	22.9	47.7	44.1
DE_N	0.0	0.0	2.9	18.7	9.6	0.0	0.0	0.0	0.0	29.4	161.4
DE_S	1.0	5.1	6.7	10.1	0.6	0.0	0.0	0.0	0.0	0.0	180.2
BAL	0.0	0.0	7.3	34.9	0.5	0.0	0.0	0.0	10.5	41.9	193.5
PO_S	0.1	0.9	8.4	27.3	0.1	0.0	0.0	0.0	0.0	0.0	256.3
NO_N	0.1	0.3	0.5	1.8	2.0	0.1	0.7	1.3	4.9	7.9	194.5
DK_T	0.0	0.0	0.0	1.1	10.8	0.0	0.0	0.0	0.1	77.4	53.2
BENELUX	0.0	0.0	0.4	6.2	4.2	0.0	0.0	0.0	0.0	83.5	51.3
FI	0.0	7.1	26.1	7.7	0.9	0.0	0.0	1.4	27.3	31.6	43.7
NO1	0.4	0.6	0.6	1.1	1.0	0.0	0.0	0.1	2.2	18.3	125.0
PO3	0.0	0.1	4.2	27.4	0.4	0.0	0.0	0.0	1.0	17.5	160.2
UK1	0.0	0.0	0.1	5.4	12.2	0.0	0.0	0.0	0.8	134.5	181.6
UK2	0.0	0.0	0.2	2.9	9.5	0.0	0.0	0.0	0.3	46.7	85.6

Table A5. Full load hours before curtailment for the five classes of onshore wind power (WON1-WON5), the five classes of offshore wind power (WOFF1–WOFF2) and solar PV parks (PV) for the regions considered in this work. SP = Specific Power, i.e., generator capacity relative to swept rotor area in W/m^2.

Region	WON1 100SP	WON2 100SP	WON3 100SP	WON4 300SP	WON5 300SP	WOFF4 200SP	WOFF5 200SP	PV
SE_N	4127	5062	-	-	4244	-	-	1006
SE_S	4162	5178	5823	3451	4131	5159	5348	1233
DE_N	-	4932	5734	3556	4265	5338	5577	1263
DE_S	3313	4560	5190	3417	3987	-	-	1320
BAL	-	5387	5799	3407	4076	5256	-	-
PO_S	3762	4808	5644	3364	4006	-	-	1288
NO_N	3614	-	-	-	4393	-	-	-
DK_T	-	-	-	3823	4270	-	5555	1299
BENELUX	-	5020	5475	3483	4106	-	5198	1267
FI	4607	5470	5852	-	4340	-	-	998
NO1	3319	-	-	3429	4251	-	4869	1156
PO3	-	5424	5838	3490	4086	5215	5390	1271
UK1	0	0	5303	0	4254	5026	5237	1237
UK2	0	4933	5345	3624	4558	0	5501	1087

Appendix C

Appendix C.1 Solar PV Dominated Regions (DE_S, PO_S)

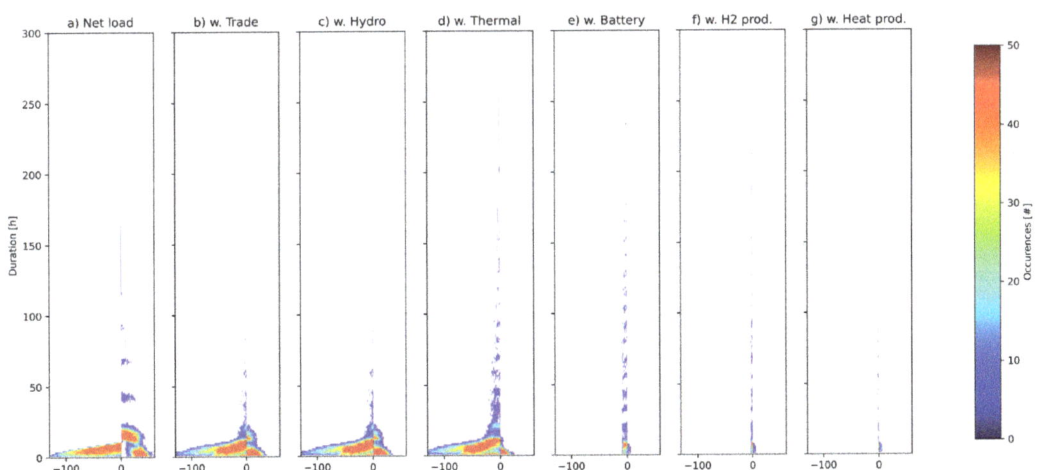

Figure A2. Net load in southern Germany (DE_S) (**a**) reduced by (**b**) trade, (**c**) trade and hydropower, (**d**) trade, hydropower and base load, (**e**) trade, hydropower, base load and batteries, (**f**) trade, hydropower, base load, batteries and adapted hydrogen production and (**g**) trade, hydropower, base load, batteries, adapted hydrogen production and adapted heat production. The color scale indicates number of (#) occurrences and was limited to 50 occurrences, implying that the number of occurrences may be higher in the deep red fields.

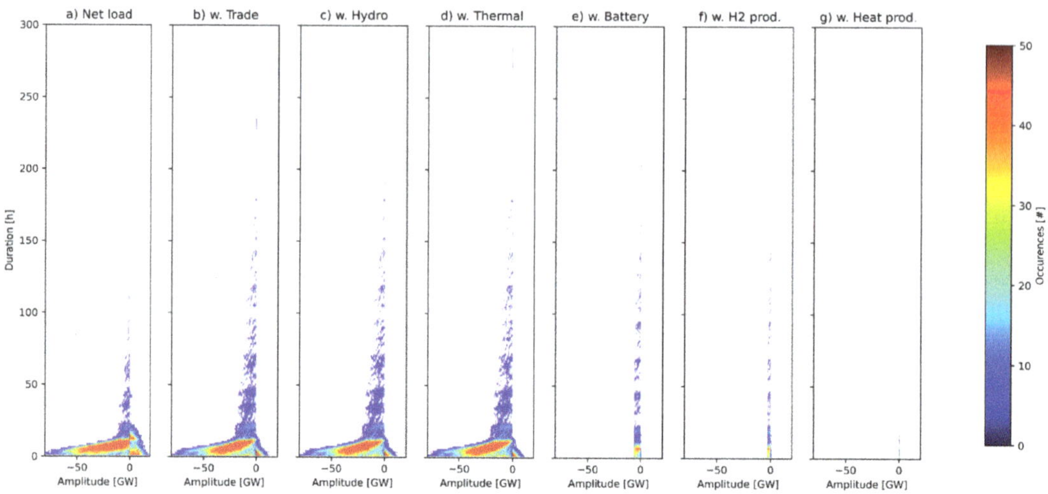

Figure A3. Net load in southern Poland (PO_S) (**a**) reduced by (**b**) trade, (**c**) trade and hydropower, (**d**) trade, hydropower and base load, (**e**) trade, hydropower, base load and batteries, (**f**) trade, hydropower, base load, batteries and adapted hydrogen production and (**g**) trade, hydropower, base load, batteries, adapted hydrogen production and adapted heat production. The color scale indicates number of (#) occurrences and was limited to 50 occurrences, implying that the number of occurrences may be higher in the deep red fields.

Appendix C.2 Wind Dominated Exporting Regions (DK_T, BAL, PO3, UK_N)

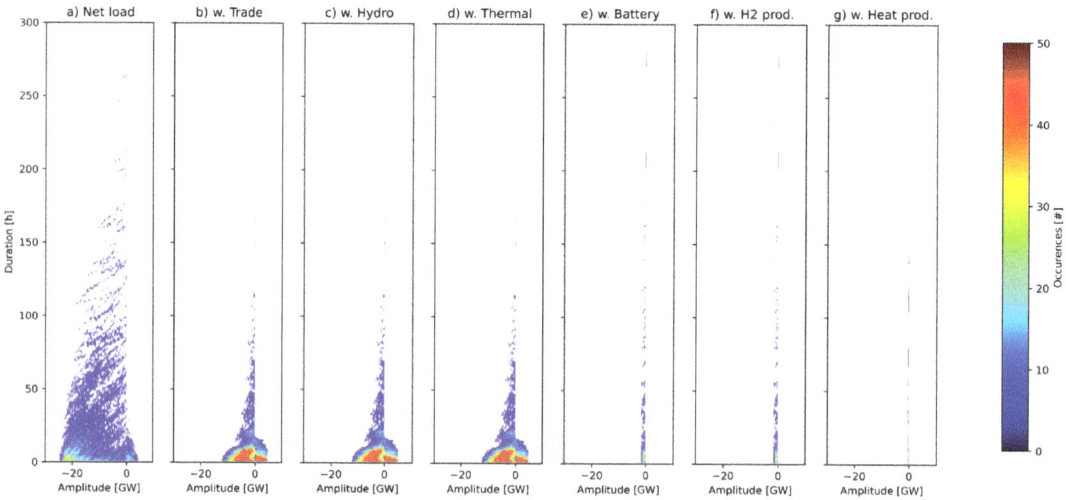

Figure A4. Net load in Denmark (DK_T) (**a**) reduced by (**b**) trade, (**c**) trade and hydropower, (**d**) trade, hydropower and base load, (**e**) trade, hydropower, base load and batteries, (**f**) trade, hydropower, base load, batteries and adapted hydrogen production and (**g**) trade, hydropower, base load, batteries, adapted hydrogen production and adapted heat production. The color scale indicates number of (#) occurrences and was limited to 50 occurrences, implying that the number of occurrences may be higher in the deep red fields.

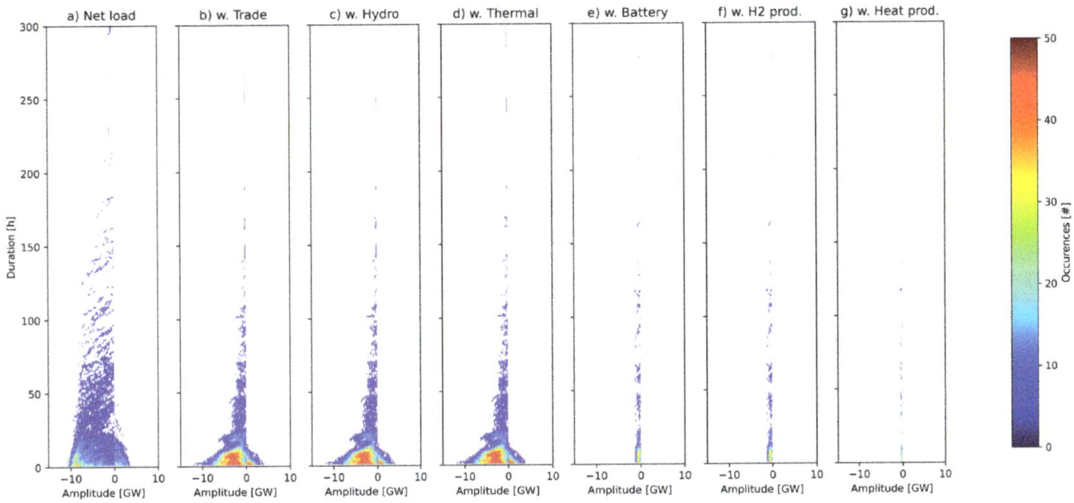

Figure A5. Net load in Estonia, Latvia and Lithuania (BAL) (**a**) reduced by (**b**) trade, (**c**) trade and hydropower, (**d**) trade, hydropower and base load, (**e**) trade, hydropower, base load and batteries, (**f**) trade, hydropower, base load, batteries and adapted hydrogen production and (**g**) trade, hydropower, base load, batteries, adapted hydrogen production and adapted heat production. The color scale indicates number of (#) occurrences and was limited to 50 occurrences, implying that the number of occurrences may be higher in the deep red fields.

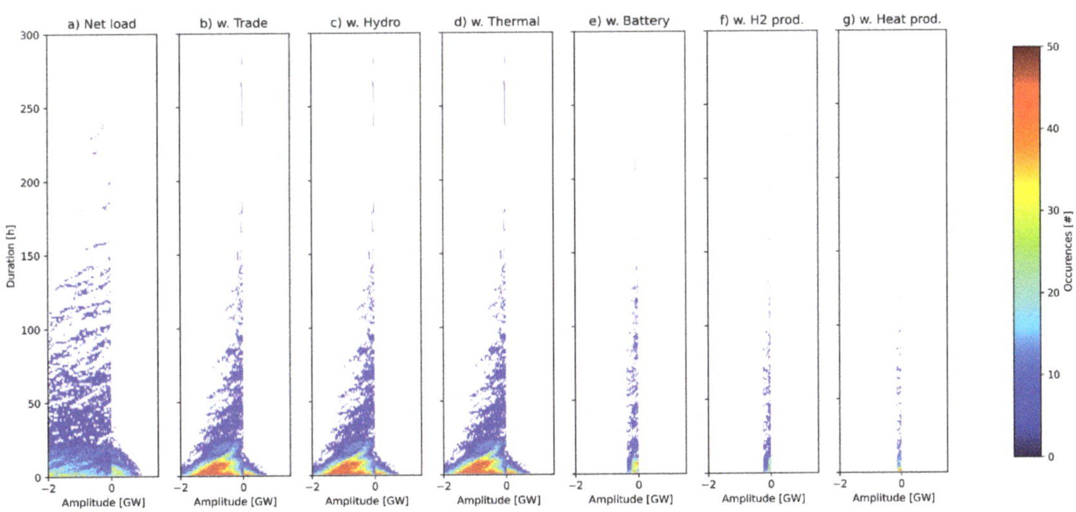

Figure A6. Net load in northern Poland (PO_N) (**a**) reduced by (**b**) trade, (**c**) trade and hydropower, (**d**) trade, hydropower and base load, (**e**) trade, hydropower, base load and batteries, (**f**) trade, hydropower, base load, batteries and adapted hydrogen production and (**g**) trade, hydropower, base load, batteries, adapted hydrogen production and adapted heat production. The color scale indicates number of (#) occurrences and was limited to 50 occurrences, implying that the number of occurrences may be higher in the deep red fields.

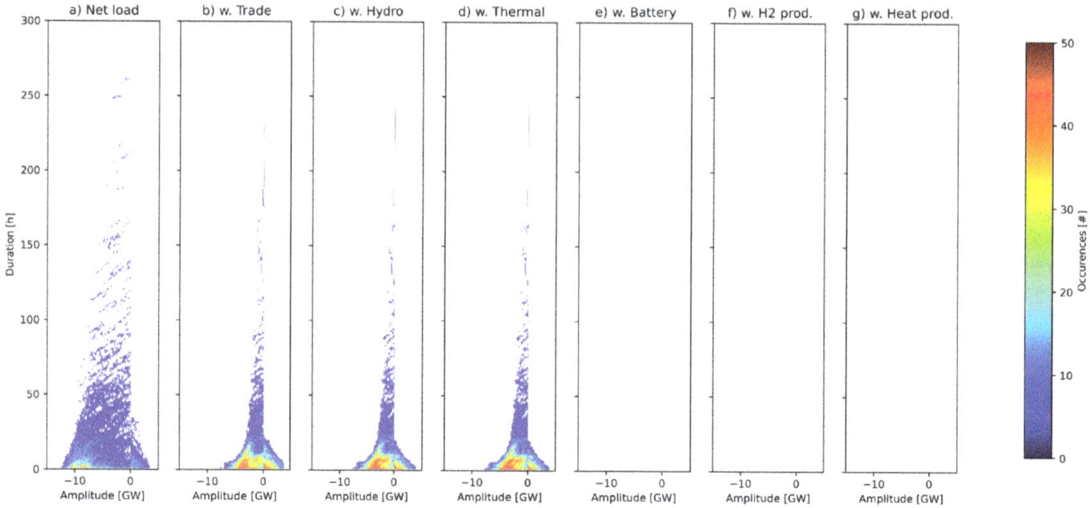

Figure A7. Net load in northern UK (UK_N) (**a**) reduced by (**b**) trade, (**c**) trade and hydropower, (**d**) trade, hydropower and base load, (**e**) trade, hydropower, base load and batteries, (**f**) trade, hydropower, base load, batteries and adapted hydrogen production and (**g**) trade, hydropower, base load, batteries, adapted hydrogen production and adapted heat production. The color scale indicates number of (#) occurrences and was limited to 50 occurrences, implying that the number of occurrences may be higher in the deep red fields.

Appendix C.3 Regions with Wind and Solar Power (UK_S, DE_N, BENELUX)

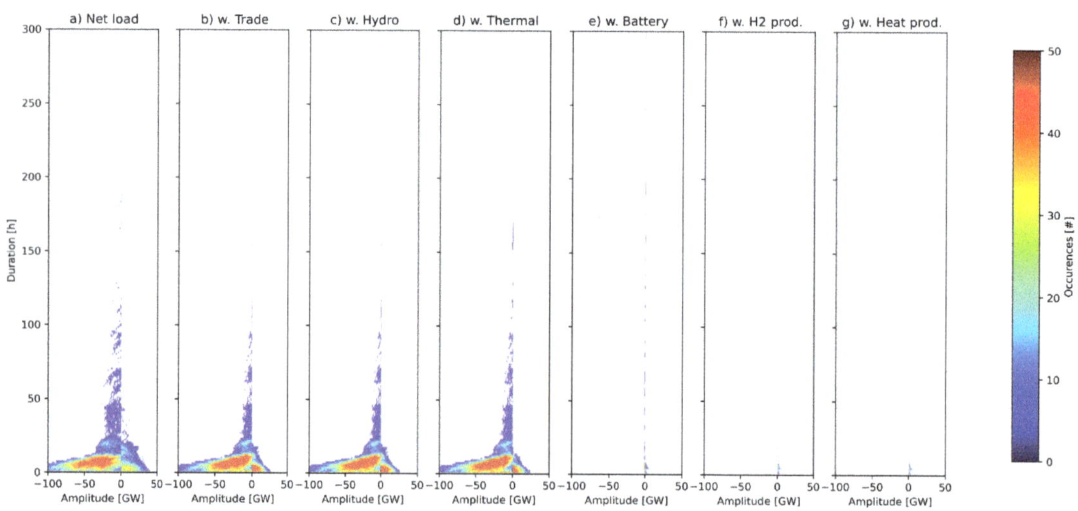

Figure A8. Net load in southern UK (UK_S) (**a**) reduced by (**b**) trade, (**c**) trade and hydropower, (**d**) trade, hydropower and base load, (**e**) trade, hydropower, base load and batteries, (**f**) trade, hydropower, base load, batteries and adapted hydrogen production and (**g**) trade, hydropower, base load, batteries, adapted hydrogen production and adapted heat production. The color scale indicates number of (#) occurrences and was limited to 50 occurrences, implying that the number of occurrences may be higher in the deep red fields.

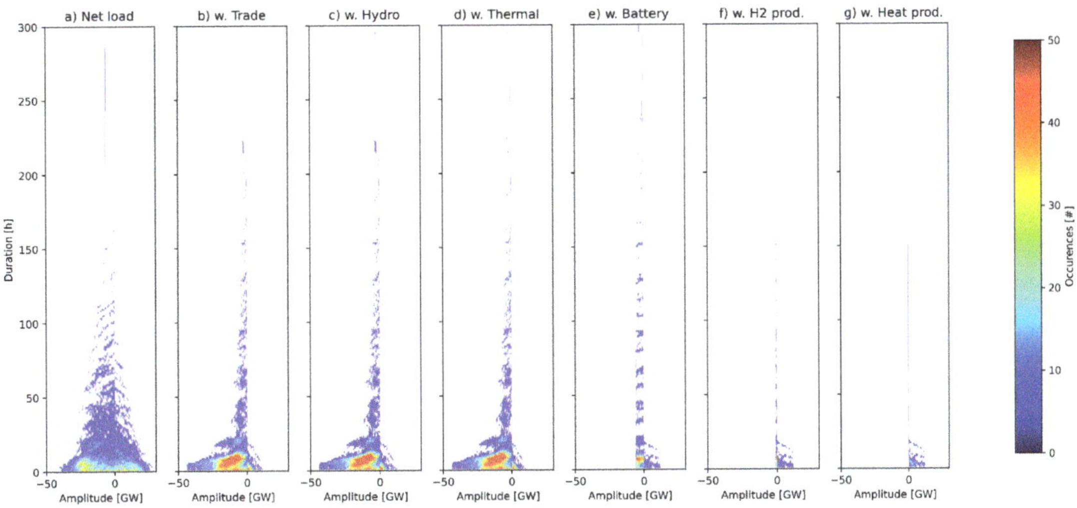

Figure A9. Net load in northern Germany (DE_N) (**a**) reduced by (**b**) trade, (**c**) trade and hydropower, (**d**) trade, hydropower and base load, (**e**) trade, hydropower, base load and batteries, (**f**) trade, hydropower, base load, batteries and adapted hydrogen production and (**g**) trade, hydropower, base load, batteries, adapted hydrogen production and adapted heat production. The color scale indicates number of (#) occurrences and was limited to 50 occurrences, implying that the number of occurrences may be higher in the deep red fields.

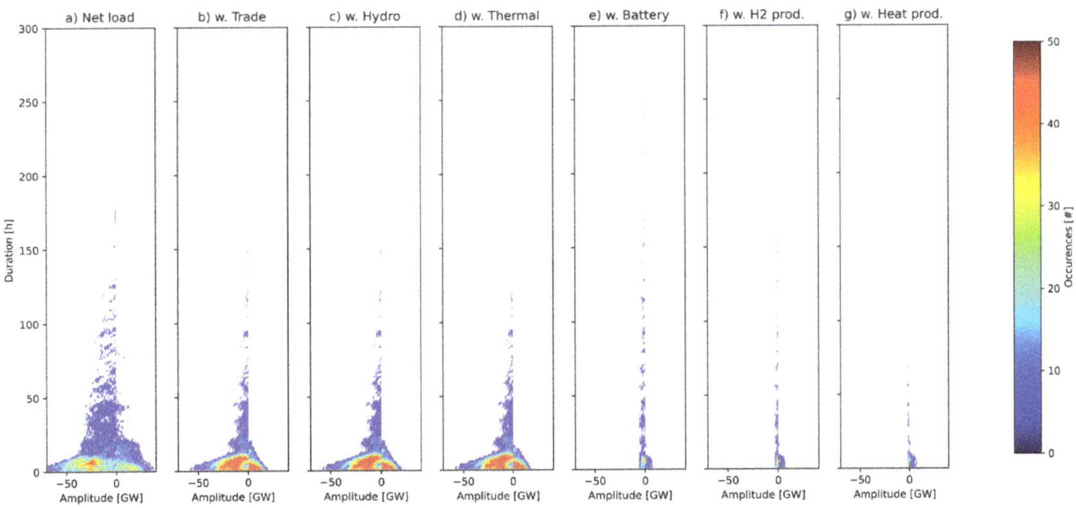

Figure A10. Net load in Belgium, the Netherlands and Luxemburg (BENELUX) (**a**) reduced by (**b**) trade, (**c**) trade and hydropower, (**d**) trade, hydropower and base load, (**e**) trade, hydropower, base load and batteries, (**f**) trade, hydropower, base load, batteries and adapted hydrogen production and (**g**) trade, hydropower, base load, batteries, adapted hydrogen production and adapted heat production. The color scale indicates number of (#) occurrences and was limited to 50 occurrences, implying that the number of occurrences may be higher in the deep red fields.

Appendix C.4 Regions with Access to Hydropower (SE_S, FI, NO1, SE_N, NO_N)

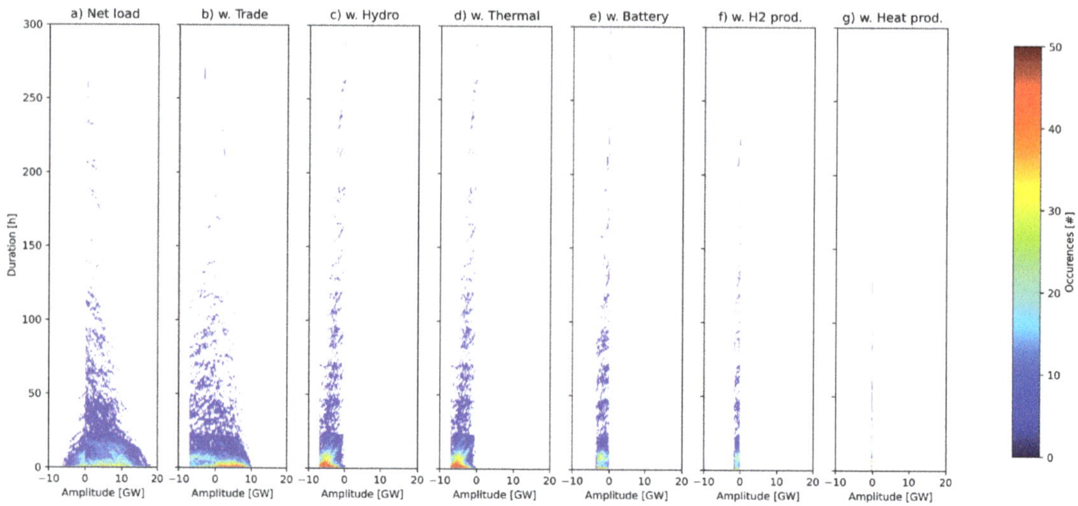

Figure A11. Net load in southern Sweden (SE_S) (**a**) reduced by (**b**) trade, (**c**) trade and hydropower, (**d**) trade, hydropower and base load, (**e**) trade, hydropower, base load and batteries, (**f**) trade, hydropower, base load, batteries and adapted hydrogen production and (**g**) trade, hydropower, base load, batteries, adapted hydrogen production and adapted heat production. The color scale indicates number of (#) occurrences and was limited to 50 occurrences, implying that the number of occurrences may be higher in the deep red fields.

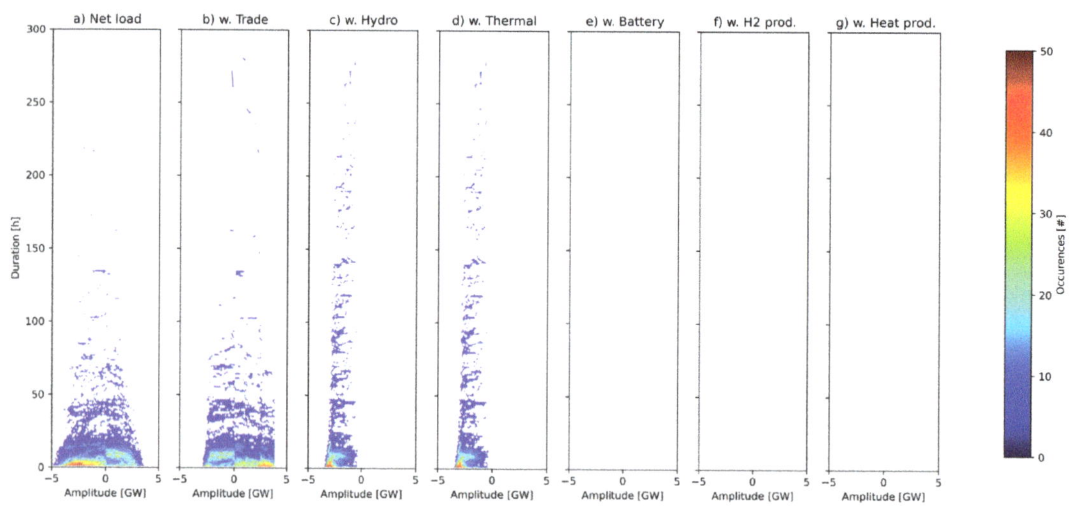

Figure A12. Net load in northern Sweden (SE_N) (**a**) reduced by (**b**) trade, (**c**) trade and hydropower, (**d**) trade, hydropower and base load, (**e**) trade, hydropower, base load and batteries, (**f**) trade, hydropower, base load, batteries and adapted hydrogen production and (**g**) trade, hydropower, base load, batteries, adapted hydrogen production and adapted heat production. The color scale indicates number of (#) occurrences and was limited to 50 occurrences, implying that the number of occurrences may be higher in the deep red fields.

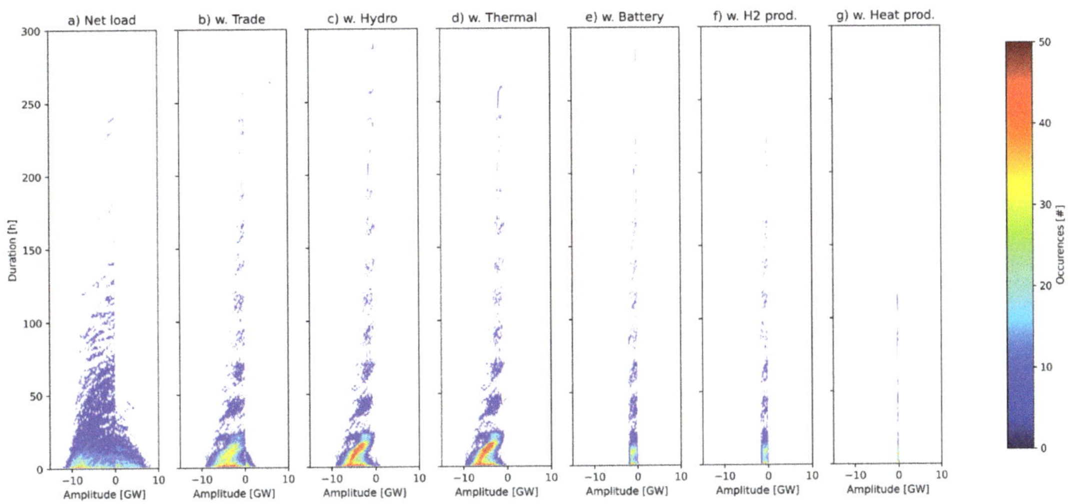

Figure A13. Net load in Finland (FI) (**a**) reduced by (**b**) trade, (**c**) trade and hydropower, (**d**) trade, hydropower and base load, (**e**) trade, hydropower, base load and batteries, (**f**) trade, hydropower, base load, batteries and adapted hydrogen production and (**g**) trade, hydropower, base load, batteries, adapted hydrogen production and adapted heat production. The color scale indicates number of (#) occurrences and was limited to 50 occurrences, implying that the number of occurrences may be higher in the deep red fields.

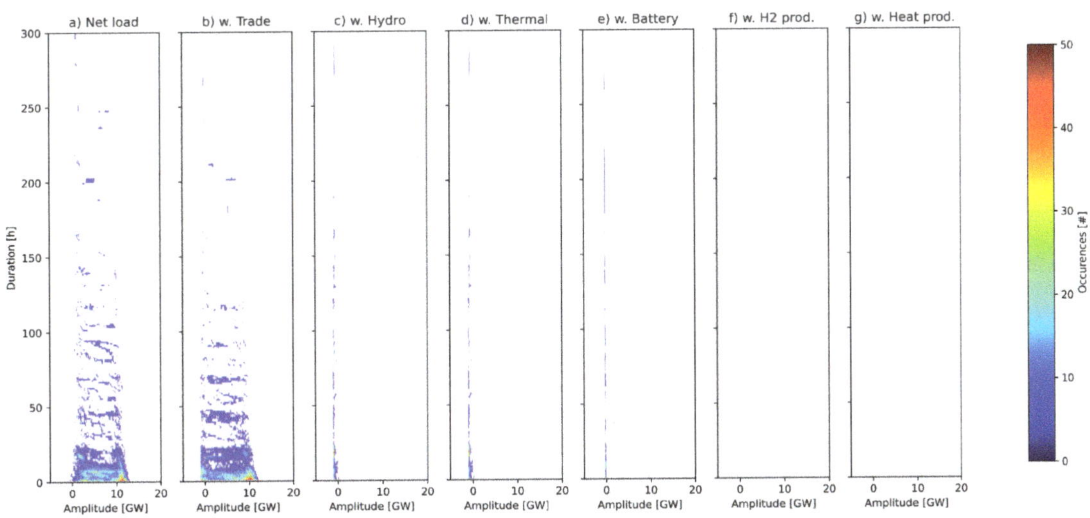

Figure A14. Net load in southern Norway (NO_S) (**a**) reduced by (**b**) trade, (**c**) trade and hydropower, (**d**) trade, hydropower and base load, (**e**) trade, hydropower, base load and batteries, (**f**) trade, hydropower, base load, batteries and adapted hydrogen production and (**g**) trade, hydropower, base load, batteries, adapted hydrogen production and adapted heat production. The color scale indicates number of (#) occurrences and was limited to 50 occurrences, implying that the number of occurrences may be higher in the deep red fields.

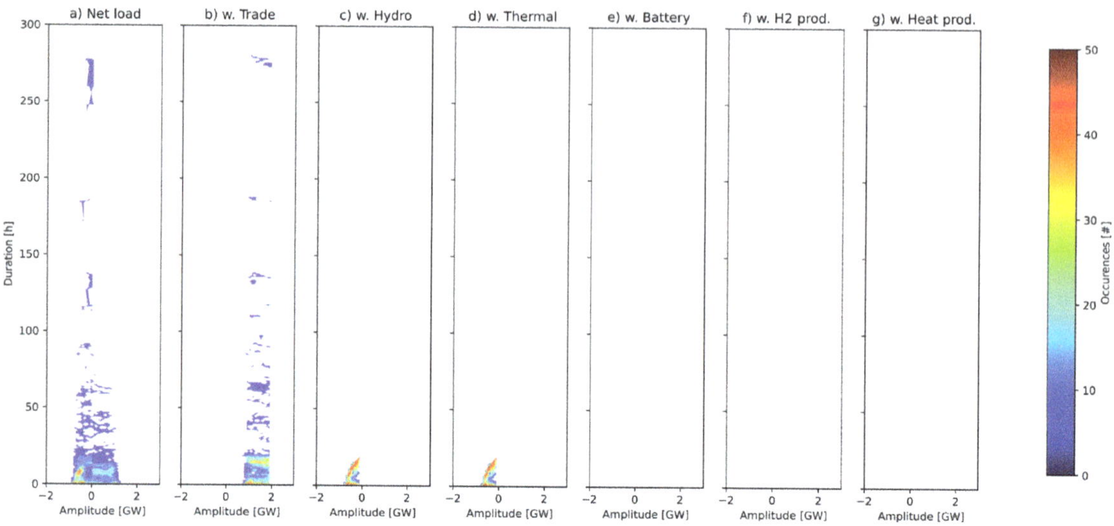

Figure A15. Net load in northern Norway (NO_N) (**a**) reduced by (**b**) trade, (**c**) trade and hydropower, (**d**) trade, hydropower and base load, (**e**) trade, hydropower, base load and batteries, (**f**) trade, hydropower, base load, batteries and adapted hydrogen production and (**g**) trade, hydropower, base load, batteries, adapted hydrogen production and adapted heat production. The color scale indicates number of (#) occurrences and was limited to 50 occurrences, implying that the number of occurrences may be higher in the deep red fields.

References

1. Wråke, M.; Karlsson, K.; Kofoed-Wiuff, A.; Folsland Bolkesjø, T.; Lindroos, T.J.; Hagberg, M.; Bosack Simonsen, M.; Unger, T.; Tennbakk, B.; Ogner Jåstad, E.; et al. *Nordic Clean Energy Scenarios*; Nordic Energy Research: Oslo, Norway, 2021.
2. Göransson, L.; Lehtveer, M.; Nyholm, E.; Taljegard, M.; Walter, V. The Benefit of Collaboration in the North European Electricity System Transition-System and Sector Perspectives. *Energies* **2019**, *12*, 24. [CrossRef]
3. Pursiheimo, E.; Kiviluoma, J. Chapter 12—Analyzing electrification scenarios for the northern European energy system. In *Electrification*; Aalto, P., Ed.; Academic Press: Cambridge, MA, USA, 2021; pp. 271–288.
4. European Commission. *A Clean Planet for All—A European Adapted Long-Term Vision for a Prosperous, Modern, Competitive and Climate Neutral Economy*; European Commission: Brussels, Belgium, 2018.
5. IEA. *Status of Power System Transformation*; International Energy Agency: Paris, France, 2018.
6. Lund, P.D.; Lindgren, J.; Mikkola, J.; Salpakari, J. Review of energy system flexibility measures to enable high levels of variable renewable electricity. *Renew. Sustain. Energy Rev.* **2015**, *45*, 785–807. [CrossRef]
7. Göransson, L.; Johnsson, F. A comparison of variation management strategies for wind power integration in different electricity system contexts. *Wind. Energy* **2018**, *21*, 837–854. [CrossRef]
8. Johansson, V.; Göransson, L. Impacts of variation management on cost-optimal investments in wind power and solar photovoltaics. *Renew. Energy Focus* **2020**, *32*, 10–22. [CrossRef]
9. Denholm, P.; Mai, T. Timescales of energy storage needed for reducing renewable energy curtailment. *Renew. Energy* **2019**, *130*, 388–399. [CrossRef]
10. Nyamdash, B.; Denny, E.; O'Malley, M. The viability of balancing wind generation with large scale energy storage. *Energy Policy* **2010**, *38*, 7200–7208. [CrossRef]
11. Kiviluoma, J.; Rinne, E.; Helistö, N. Comparison of flexibility options to improve the value of variable power generation. *Int. J. Sustain. Energy* **2018**, *37*, 761–781. [CrossRef]
12. Helistö, N.; Kiviluoma, J.; Ikäheimo, J.; Rasku, T.; Rinne, E.; O'Dwyer, C.; Li, R.; Flynn, D. Backbone—An Adaptable Energy Systems Modelling Framework. *Energies* **2019**, *12*, 3388. [CrossRef]
13. IEA. *Harnessing Variable Renewables—A Guide to the Balancing Challenge*; International Energy Agency: Paris, France, 2011.
14. Yasuda, Y.; Carlini, E.M.; Estanqueiro, A.; Eriksen, P.B.; Flynn, D.; Herre, L.F.; Hodge, B.-M.; Holttinen, H.; Koivisto, M.J.; Gómez-Lózaro, E.; et al. Flexibility chart 2.0: An accessible visual tool to evaluate flexibility resources in power systems. *Renew. Sustain. Energy Rev.* **2023**, *174*, 113116. [CrossRef]
15. Heggarty, T.; Bourmaud, J.-Y.; Girard, R.; Kariniotakis, G. Quantifying power system flexibility provision. *Appl. Energy* **2020**, *279*, 115852. [CrossRef]

16. Göransson, L.; Goop, J.; Odenberger, M.; Johnsson, F. Impact of thermal plant cycling on the cost-optimal composition of a regional electricity generation system. *Appl. Energy* **2017**, *197*, 230–240. [CrossRef]
17. Holmér, P.; Ullmark, J.; Göransson, L.; Walter, V.; Johnsson, F. Impacts of thermal energy storage on the management of variable demand and production in electricity and district heating systems: A Swedish case study. *Int. J. Sustain. Energy* **2020**, *39*, 446–464. [CrossRef]
18. Mattsson, N.; Verendel, V.; Hedenus, F.; Reichenberg, L. An autopilot for energy models—Automatic generation of renewable supply curves, hourly capacity factors and hourly synthetic electricity demand for arbitrary world regions. *Energy Strategy Rev.* **2021**, *33*, 100606. [CrossRef]
19. ECMWF. *ERA-Interim u- and v-Components of Horizontal Wind, Surface Solar Radiation Downward, Skin Temperature*; ECMWF: Reading, UK, 2010.
20. DTU. *Global Wind Atlas*; DTU Wind and Energy Systems Group, Ed.; DTU: Lyngby, Denmark, 2018.
21. Gode, J. *Klimatförändringarnas Inverkan På Energisystemet*; Energiforsk: Stockholm, Sweden, 2021.
22. Scharff, R. *Klimatförändringarnas Inverkan På Vatten- Kraftens Produktions- Och Reglerförmåga*; Energiforsk: Stockholm, Sweden, 2023.
23. Nyholm, E. The Role of Swedish Single-Family Dwellings in the Electricity System—The Importance and Impacts of Solar Photovoltaics, Demand Response, and Energy Storage. In *Energy and Environment*; Chalmers University of Technology: Gotheburg, Sweden, 2016.
24. Werner, S. *ECOHEATCOOL Work Package 1*; Euroheat and Power: Brussels, Belgium, 2005.
25. Taljegard, M. Electrification of Road Transport-Implications for the Electricity System; Chalmers: Gothenburg, Sweden. In *Space, Earth and Environment*; Chalmers University of Technology: Gotheburg, Sweden, 2019.
26. Kullingsjö, L.-H.; Karlsson, S. The Swedish Car Movement Data Project. In Proceedings of the European Electric Vehicle Congress, Brussels, Belgium, 19–22 November 2012.
27. Fuchs, G.; Lunz, B.; Leuthold, M.; Sauer, D.U. Overview of Nonelectrochemical Storage Technologies. In *Electrochemical Energy Storage for Renewable Sources and Grid Balancing*; Elsevier: Amsterdam, The Netherlands, 2015; pp. 89–102.
28. Zhao, H.; Wu, Q.; Hu, S.; Xu, H.; Rasmussen, C.N. Review of energy storage system for wind power integration support. *Appl. Energy* **2015**, *137*, 545–553. [CrossRef]
29. Sauer, D.U. Classification of storage systems. In *Electrochemical Energy Storage for Renewable Sources and Grid Balancing*; Elsevier: Amsterdam, The Netherlands, 2015; pp. 13–21.
30. Palizban, O.; Kauhaniemi, K. Energy storage systems in modern grids—Matrix of technologies and applications. *J. Energy Storage* **2016**, *6*, 248–259. [CrossRef]
31. Magnor, D.; Lunz, B.; Sauer, D.U. 'Double Use' of storage systems. In *Electrochemical Energy Storage for Renewable Sources and Grid Balancing*; Elsevier: Amsterdam, The Netherlands, 2015; pp. 453–463.
32. Meng, L.; Zafar, J.; Khadem, S.K.; Collinson, A.; Murchie, K.C.; Coffele, F.; Burt, G.M. Fast Frequency Response from Energy Storage Systems—A Review of Grid Standards, Projects and Technical Issues. *IEEE Trans. Smart Grid* **2020**, *11*, 1566–1581. [CrossRef]
33. Ullmark, J.; Göransson, L.; Chen, P.; Bongiorno, M.; Johnsson, F. Inclusion of frequency control constraints in energy system investment modeling. *Renew. Energy* **2021**, *173*, 249–262. [CrossRef]
34. Ruhnau, O.; Qvist, S. Storage requirements in a 100% renewable electricity system: Extreme events and inter-annual variability. *Environ. Res. Lett.* **2022**, *17*, 044018. [CrossRef]
35. IEA. *Global Energy and Climate Model Documentation 2022*; International Energy Agency: Paris, France, 2022.
36. ENS. *Technology Data—Energy Plants for Electricity and District Heating Generation*; Danish Energy Agency and Energinet: København, Danish, 2016.
37. Hodel, H.; Göransson, L.; Chen, P.; Carlsson, O. Cost Efficient Design of Wind Turbines from An Electricity System's Perspective. 2023, *submitted for publication*.

solar

MDPI

Article

Sheet Metal Design Approach for 3D Shaped Facade Elements with Integrated Solar Thermal Functionality

Peter Scholz [1,*], Dieter Weise [1], Linda Schmidt [2], Martin Dembski [2], Alexander Stahr [2], Martin Dix [1], Florin Duminica [3], Sebastien Le Craz [3] and Jiri Koziorek [4]

1 Department of Sheet Metal Forming, Fraunhofer Institute for Machine Tools and Forming Technology, 09126 Chemnitz, Germany
2 Hochschule für Technik, Wirtschaft und Kultur Leipzig HTWK, 04277 Leipzig, Germany
3 Department Advanced Surface and Smart Solutions, CRM Group, 4000 Liege, Belgium
4 Department of Cybernetics and Biomedical Engineering, VSB—Technical University of Ostrava, 70800 Ostrava, Czech Republic
* Correspondence: peter.scholz@iwu.fraunhofer.de

Abstract: This paper provides an overview of the development of a 3D formed and metal-based facade element that combines a custom design and solar thermal functionality. To achieve this, a novel simplified solar thermal collector structure was developed using formed sheet metal half-shells with an integrated channel structure on the inside and a special absorber coating on the outside. The sheet metal half-shells were manufactured by highly innovative incremental sheet forming (ISF), which allows seamless integration into existing facades. As a part of this paper, the initial test results on thermal efficiency and the energy accumulation of the new collector type are presented.

Keywords: absorbance coating; facade elements; sheet metal design; solar thermal collector

Citation: Scholz, P.; Weise, D.; Schmidt, L.; Dembski, M.; Stahr, A.; Dix, M.; Duminica, F.; Le Craz, S.; Koziorek, J. Sheet Metal Design Approach for 3D Shaped Facade Elements with Integrated Solar Thermal Functionality. *Solar* **2023**, *3*, 213–228. https://doi.org/10.3390/solar3020014

Academic Editors: Luis Hernández-Callejo and Jürgen Heinz Werner

Received: 13 January 2023
Revised: 3 April 2023
Accepted: 6 April 2023
Published: 13 April 2023

1. Introduction

The reduction of harmful greenhouse gases is one of the biggest challenges of our time. The use of unlimited available solar thermal power as renewable energy can help to support efforts to reduce CO_2 emissions worldwide. More than 40% of the total energy consumption in Europe can be allocated to heating, hot water, and illumination [1]. This corresponds to almost 20% of the total CO_2 emissions. Especially in the building industry, reducing CO_2 emissions is one of the main drivers. By using modern building envelopes, the required energy for heating and cooling can be minimized [2]. The major goal is the realization of affordable zero-energy buildings. One effective way of achieving these objectives is the use of solar thermal energy. The solar radiation at our latitudes is perfectly suited for thermal use.

In central Europe, solar radiation produces, on a 1 m^2 area, 1000 kWh/year to 1100 kWh/year [3], which is equivalent to 100 L of fuel oil or 100 m^3 of natural gas. For this reason, the building sector has a strongly growing interest in facade-integrated solar thermal absorbers, which shall be presented in this paper.

A broad range of solar thermal collectors have been available on the market for decades worldwide. These products are credible and generally made to a high technical standard, especially in Europe with a homogeneous market. However, there is a shortage in the field of solar thermal products that are suited for building envelope integration to create high-quality architecture [4,5]. Functional and constructive aspects, together with aesthetics, have to be considered. State-of-the-art solar thermal collectors are not flexible in shape and size due to the hydraulic fluid circuits fixed to solar absorbers. Architectural requests for design freedom require the hydraulic system concept to be redesigned, which is generally difficult and expensive for conventional means of production. Conventional production methods for absorbers limit the flexibility in collector design [6] and such a lack of flexibility significantly reduces the potential application of solar thermal systems [7]. The production of solar collectors in sheet metal half-shell design by means of hydroforming was investigated in the European Union (EU) project with the acronym "BIONICOL" [8] and in "Industrielle Gemeinschaftsforschung" (IGF) project No. 339 ZN [9]. However, curved surfaces cannot be obtained with the shown technologies and the channel structures can only be manufactured on both sides, which has a disadvantageous effect on the design freedom of the facade. The integration of flat-plate solar thermal collectors is only possible for opaque envelopes (roofs and facades) so far [10]. The structure of evacuated tubes allows mounting on transparent envelopes as sun shading, but this kind of application is rare. The application potential of incremental sheet forming (ISF) in solar absorbers was demonstrated in [11] by the production of absorber lamellae using incremental forming technology.

The thickness of solar collectors affects their integration, especially for facades. Thick solar thermal collectors are difficult to implement as functional elements or as sun shading. The appearance of solar collectors is affected by the glass (glazed collector) and absorber surface treatment. State-of-art collectors apply highly transparent, low-iron glass for glazing and black or dark-blue spectrally selective coatings with high absorptance (0.95) and low emittance (0.05). Solar collectors integrated into facades are more conspicuous than collectors installed on rooftops. Many studies and surveys have shown that architects prefer a large variety of absorber colors [12], and they even regard the possibility of a custom color choice as essential. Manufacturers can meet the demand for a variety of absorber colors by means of solar paint coatings, but such collectors show considerably reduced thermal performance compared with quality selective coatings in the usual solar collectors commercially available in Europe [13].

This paper presents a new design of a solar thermal collector. The focus of this paper is fabrication with ISF, concepts for integration into facade systems, possible design arrangements, and the development of the absorber coating. Furthermore, initial results on the thermal efficiency and performance of the new collector type are presented and evaluated.

2. Materials and Methods

2.1. Principal Collector Design

Figure 1 shows the layout of the developed collector design, which consists of two layers of cold-rolled deep-drawing steel DC06 (t = 0.8 mm) joined together in a fluid-tight manner by laser welding.

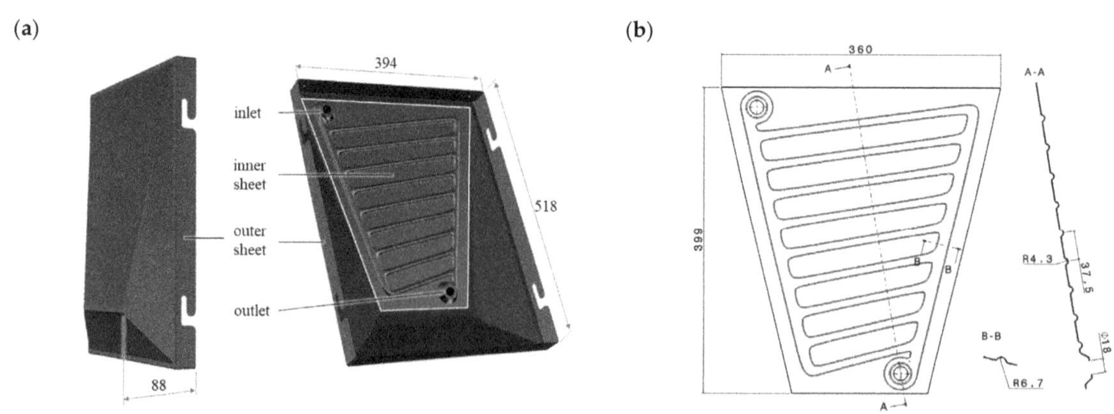

Figure 1. Principal collector design: (**a**) CAD model of the assembled collector and (**b**) drawing of the inner sheet with the channel structure.

The three-dimensional shaped outer sheet is designed as a closed cassette with integrated notches for hooking into the façade's substructure. To integrate the solar thermal functionality, the flat inner sheet has a two-dimensional channel structure and is attached to the outer sheet. The use of this simple geometry for the inner sheet ensures a good fit between the two components, which is essential for the joining process. The channel structure chosen was a harp structure, which has also been used in commercially available flat plate collectors. With this structure, the area to be cooled can be efficiently passed through with low pressure loss at the same time. For hydraulic connection, parallel planar surfaces were proposed in the inlet and outlet areas for the easy integration of conventional fittings.

The representative design of the geometry of the exterior sheet was determined by the results of the solar gain analysis, the architectural design boundaries, and the constraints of the forming process, which are described in the following sections.

2.2. Architectural Aspects and Facade Integration

2.2.1. Necessity of Research into Solar Thermal Facade Elements in Architecture

Endeavoring to be part of the solution, the publicly funded research project "FutureFacade" strives to do its part for climate protection in the construction industry by utilizing the inextricable connection to resource and energy efficiency. The integration of solar thermal energy into facades holds great potential for generating solar yields that have, so far, remained largely untapped. Although a study by the Leibniz Institute for Ecological and Regional Development (IÖR) shows that a total of 12,000 km^2 of facade area in Germany can be used for solar energy, no architecturally sophisticated solar thermal facade elements exist yet [14,15]. The combination of solar thermal energy with inner-city building surfaces can be used for domestic hot water, heating support, and cooling. Regarding unforeseeable developments in the cost of all fuels, solar thermal energy offers a sustainable and cost-effective alternative.

2.2.2. Architectural Innovation

While conventional collectors are visually recognizable by the glass and the dark-blue absorber surface, more complex appearances can be created using mono-material panels made of metal. Hence, through innovating solar modules, a novel structure was created in this project. Linking architectural aspiration and sustainability was paramount. The functional integration, which was not visible on the outside, was achieved by two combined sheet metal panels invisibly welded together.

2.2.3. Design Criteria for the Collector Geometry based on Solar Thermal Aspects

Optimizing the geometry of the outer panel was most important in order to adapt the modules as efficiently as possible to the irradiation conditions using site-specific weather data while also taking into account the manufacturing constraints (e.g., maximum wall angle and minimal radius). The adjustment of the module geometry and, thus, the solar-active area and the irradiation angle was regulated by a digital parametric design tool. This allowed the yields to be controlled and optimized, which is a pioneering development in sustainable building construction.

The solar gain analysis showed that the chosen shape could provide 792 kWh/m^2 per year for a southern orientation using the Chemnitz, Germany, region as an example (Figure 2). Different rectangular-based shapes were tested throughout the project. The analysis showed that flat collectors had the lowest potential for solar gain. Plenty of design studies have been conducted. Round, triangular, rectangular, and hexagonal formats were also tested as irregular freeform shapes. In the end, the final polygonal shape lent itself best to the chosen forming technique and the solar gains that were to be achieved. The rectangular base allowed standardized substructures to be used. The rather simple basic shape contained a lot of freedom in the arrangement and variation of the geometry as a pattern, ensuring that design freedom could be kept (Figure 3). Compared with the chosen polygonal shape, there would be a loss of 165 kWh/m^2 per year. Round double-curved shapes had the best solar-gain effect (809 kWh/m^2 per year), but were excluded due to the manufacturing technique after some forming tests, as there was a large spring-back effect after forming. Therefore, the polygonal shape was optimal from all tested design studies (Figure 2).

(a)

(b)

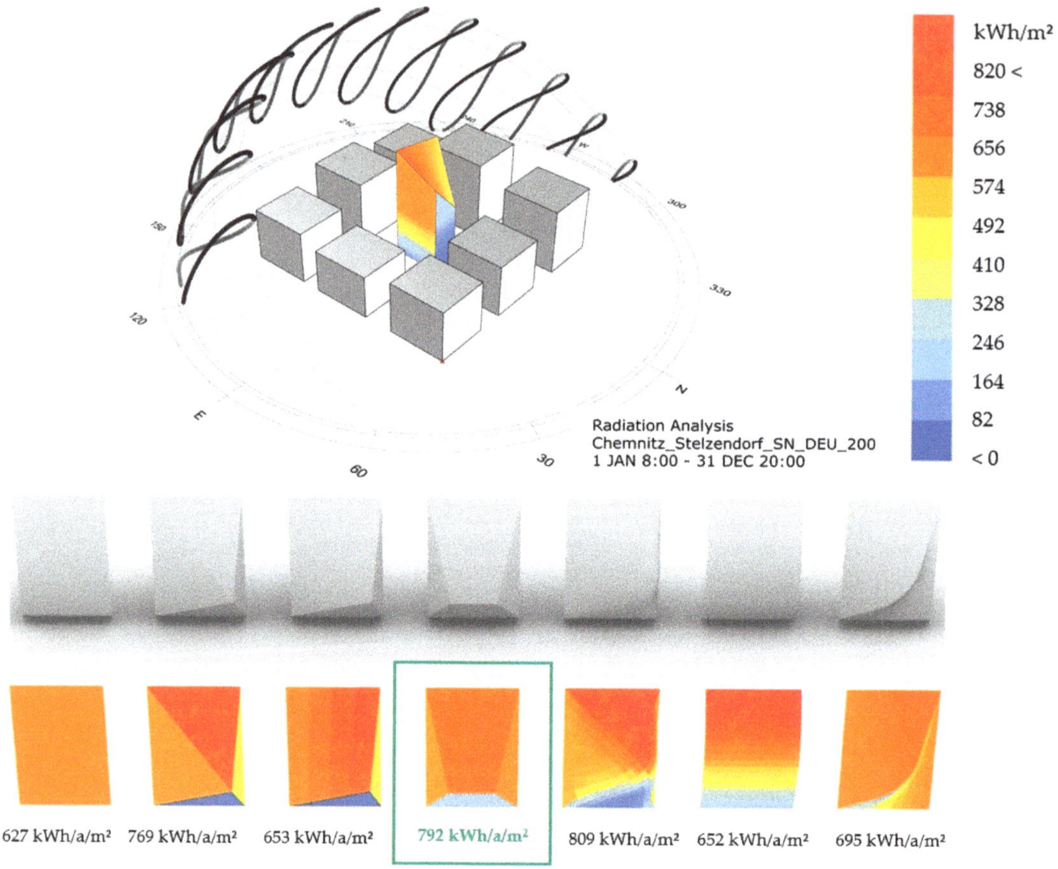

Figure 2. Solar gain analysis (**a**) of different shapes (**b**) with a rectangular base performed with the energy-simulation engine EnergyPlus™ in combination with the Grasshopper plugin Ladybug for Rhinoceros® 3D.

Figure 3. Variations and patterns of polygonal panels.

2.2.4. Facade Construction and Hydraulic System

The facade panel was designed as a simple standardized cassette and could thus be hooked into the bolt suspension. (Figure 4b). Owing to the standardized system used, it was possible to use facade (under-) constructions from various manufacturers. (Figure 4a). By using a standardized substructure, the effort required for assembly and maintenance can be decreased and the economic efficiency can be increased.

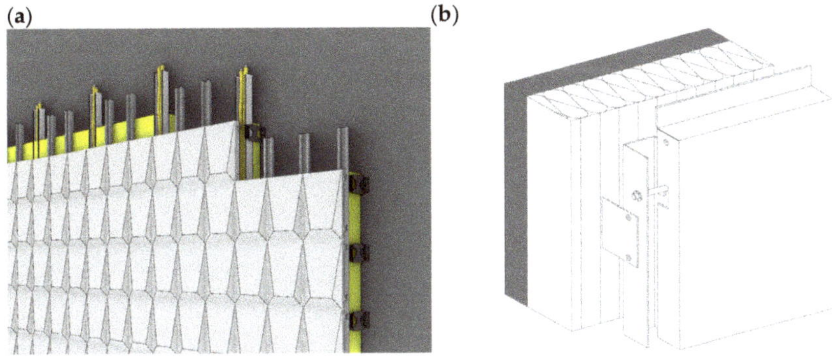

Figure 4. Visualization of facade construction (**a**) and under-construction detail (**b**).

The facade panel was a mono-material system. There was an outer panel and an inner panel welded together. The inner panel represented the channel structure, which was designed as a common harp shape. (Figure 5a) With regard to the most energy-efficient design of the fluid channels within the facade elements, computational fluid dynamics (CFD) simulations were carried out to achieve optimized performance for the facade element.

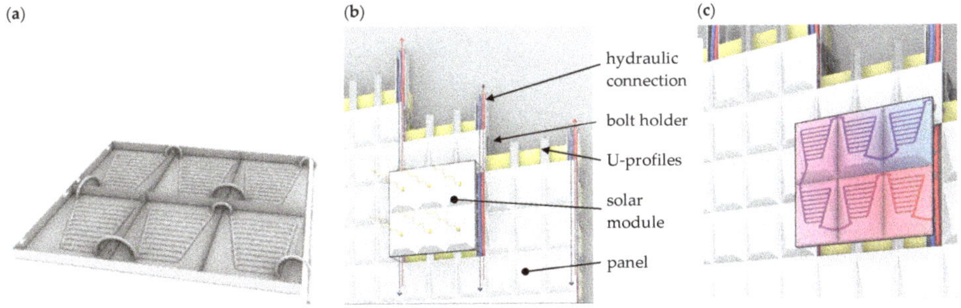

Figure 5. Schematic illustration of the addition of the panels (**a**), schematic illustration of facade construction (**b**), and schematic illustration of panel function (**c**).

To build a facade system that can operate as a power plant, the panels were connected in series and formed one module when connected in a mirrored position. (Figure 5c). One module at a time was then connected to the vertical cable routing. (Figure 5b). A pump ensured the circulation of the fluid in the facade. A heat exchanger provided the energy to the heating units, so the facade system became its own sustainable power plant to harvest renewable heat.

2.2.5. Architectural Results

The designed panel holds the potential to create new appearances for urban spaces and positively influence the use of renewable energy. (Figure 6) The panel can be used for new construction and existing projects. It can be used on large-scale installations in rural areas, as well as on urban facades, and thus has multiple applications. Therefore, the project is a huge step in Mono-Material Design, facade design, architecture, and the building industry and renewable energies in general.

Figure 6. Visualization of a sample façade.

2.3. Demonstrator Manufacturing

2.3.1. Forming

For the manufacturing of the solar collector sheets, incremental sheet metal forming (ISF) in combination with subsequently performed bending operations (only for the outer sheet) was used. With ISF, the final part of the geometry is successively formed by the CNC-controlled movement of a forming stylus over a counter die (Figure 7a). In comparison with conventional forming methods (e.g., deep drawing), this reduces tool costs and the time required to achieve the first finished part.

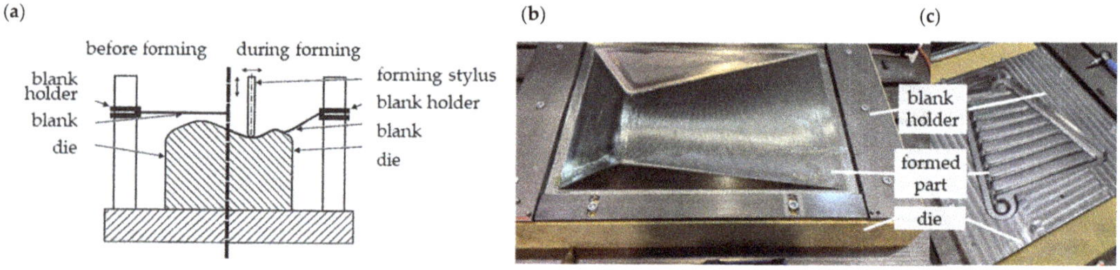

Figure 7. Incremental sheet forming: (**a**) process principle, (**b**) clamped outer sheet after forming, and (**c**) clamped inner sheet after forming.

For the forming of the test components, a polyurethane counter-die (material: obomodulan® 1600 sand) was used, in which the cut-to-size sheets were clamped by means of a blank holder (Figure 7b,c). As a lubricant, Raziol CLF 125F was used for all forming trials.

To avoid wrinkling during forming, the outer sheet was formed in two steps. In the first step, a stylus diameter of 100 mm with a z-increment of 0.2 mm was used. In the second step, the final shape was formed with a stylus diameter of 6 mm and an increment of 0.2 mm. For both steps, a z-constant tool path in helix design was chosen, with a total forming time of approx. 2 h. After unclamping, the edges of the outer sheet were folded and mechanically joined at the corners to create the closed cassette design shown in Figure 1.

The shape of the inner sheet could be formed in one step with a forming stylus diameter of 6 mm, a z-increment of 0.2 mm, and a forming time of approx. 45 min, also by using a z-constant tool path in helix design. With these parameters, the inner and outer sheets could be successfully formed without failures. For both geometries, the global shape deviation was approx. 1 mm compared with the original CAD geometry which is within the common tolerances for facade panels.

2.3.2. Joining

A welding Nd:YAG laser of 3 kW mounted on a robot was used for the assembly of the inner and outer formed sheets. The experimental set-up is presented in Figure 8a. Specific tools were used to optimize the contact between the welded surfaces and to prevent gaps from forming between the steel sheets; otherwise, the upper sheet could be pierced by the laser beam. The first tool (Figure 8b) was developed to allow the welding of the outer contour of deformed panels, and the second one (Figure 8c) was designed for the welding of lines between the channels, allowing the passage of fluid through the drawn network and avoiding the swelling of the panels. Finally, the fluid connectors were fixed using manual tungsten inert gas (TIG) welding (Figure 8d).

(a) (b) (c) (d)

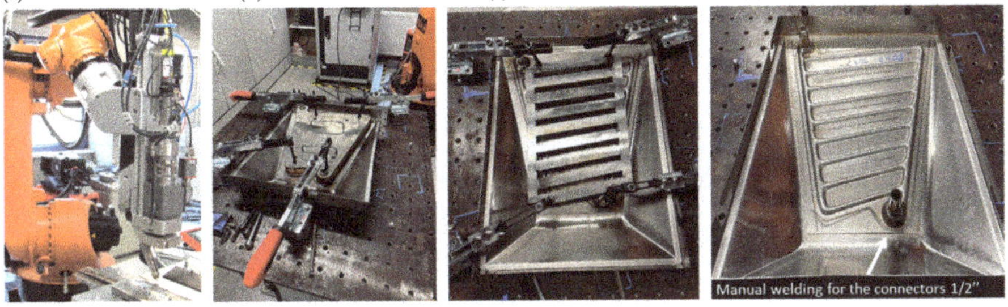

Figure 8. Laser welding setup: (**a**) laser head mounted on a robot, (**b**) clamped tool for outer contour, (**c**) clamped tool for inner channels; (**d**) TIG welding for fluid connectors.

Laser welding requires a compromise between the thermal load and the welding speed in order to limit damage to the outer sheet's steel surface. The laser beam must cross the top sheet and penetrate the bottom one without crossing it. Figure 9a shows a cross-section micrograph of the welding of two steel sheets. The focal spot is in the range of 0.8 mm to 1 mm for the following welding parameters: laser head speed of 3 m/min, steel sheet thickness of 0.8 mm, and laser power of 1.4 kW. Various parameters, such as the accuracy of the robot trajectory, the variation in the steel thickness generated by the incremental forming, as well as the distance between the steel sheets, can influence the welding beam penetration, leading to marks on the back side of the panel. Figure 9b (left) shows the impact of the welding beam on the back side of the panel. However, additional surface treatment, such as sandblasting, can remove the oxides generated by the laser beam overheating and can also considerably decrease the thermal stress near the welding lines. Figure 9b (right) shows the same surface as Figure 9b (left) after sandblasting.

A potential optimization requires an increase in the thickness of the back metal sheet to decrease the thermal effect and to increase the welding process parameters. However, for this project, the choice to simplify the assembly system drove the use of the same material and same thickness for both the inner and outer shaped steel sheets.

(a) (b)

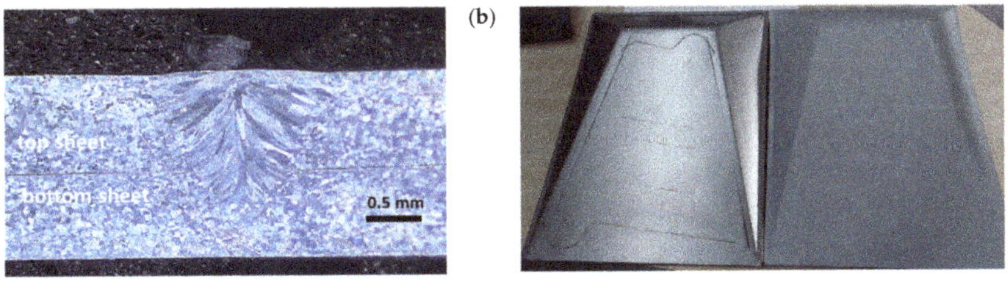

Figure 9. (**a**) Cross-section micrograph of the welding of two steel sheets and (**b**) welding lines on the front side of the panel: as initially welded (left) and after sand-basting (right).

2.3.3. Strategies and Concepts for Absorbing Coatings

The design of a non-glazed solar absorber for practical application requires the development of a robust coating applied by low-cost techniques. This coating must exhibit high absorption efficiency in the ultraviolet and visible regions of the solar radiation spectrum. Therefore, the reflectance should be minimized in these regions of the solar spectrum. Additionally, the loss of heat to the surroundings via convection or conduction should be reduced as much as possible. Thus, there are conflicting requirements for strong absorption in the solar spectrum, with minimum emission in the infrared range. The use of deep-drawing steel DC06 as a substrate requires additional corrosion protection. The outer formed steel sheet was protected by an 80 μm-thick aluminum coating deposited by the twin arc-wire thermal spray technique. The role of this additional coating is to prevent steel corrosion in case of surface scratches during the manipulation of the panels. Thermally sprayed aluminum, in combination with organic and ceramic coatings, is a common method used for corrosion protection for bridges, ships, and oil and gas installations. These systems are supposed to provide a long lifetime (>20 years) and, with that, be both cost-effective and environmentally friendly [16]. The sand-blasting used to remove the welding defects on the outer steel formed surface is a convenient surface treatment for the thermal spray arc-wire coating of aluminum. The twin arc-wire system was the PERFECT Spray from SMS Group, Düsseldorf, Germany. The typical conditions for arc spraying were arc parameters of 28 V and 90 A, and a wire speed of 2 m/min. The diameter of the aluminum wire was 1.6 mm. The coating was sprayed uniformly over all of the surface of the outer formed steel sheet.

A black absorbing coating was applied to the outer formed steel sheet by spraying, followed by curing at 300 °C. The benefits of such surfaces are their comparative simplicity to produce and the potential for cost-effectiveness. This novel coating chemistry combines high solar-absorbing performance, high thermal resistance to withstand excessive heat from the metal surface, and good resistance to UV and humid conditions to ensure a long service life. The coating thickness can be controlled in the range of a few micrometers to several tens of microns by using several steps of spraying and drying. The reflectance spectrum of the coating deposited as a solar absorber is presented in Figure 10. The solar absorptance in the UV-VIS-NIR spectra is 95% and can slightly change depending on the surface roughness and the coating thickness. The emissivity of the coating was relatively high, as the solar absorbance was low in the NIR region of the spectrum, leading to the solar absorber exhibiting non-selective behavior.

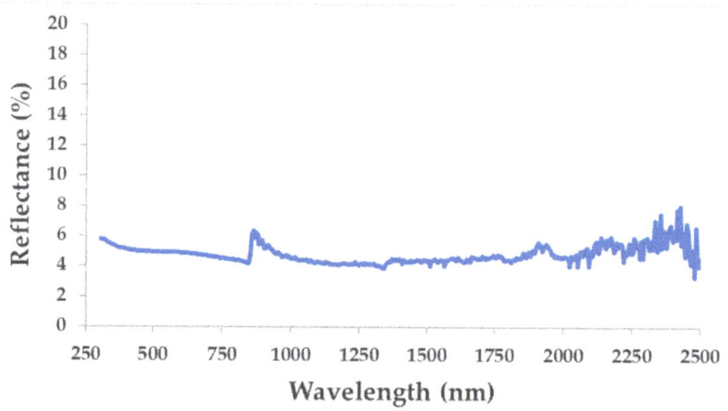

Figure 10. Reflectance spectrum of the silica–carbon sol-gel coating applied as the solar absorber.

2.4. Collector Testing Setup

A solar thermal collector integrating incremental formed panels coated with sol-gel absorbing layers was manufactured for thermal testing. The collector contained eight panels to generate a surface of 1.6 m^2, comparable to solar thermal collectors available on the market. Figure 11a shows a view of the rear side of the solar collector as manufactured, while Figure 11b shows the same view with insulation. The insulation material installed on the back of the solar panels was PIR foam boards laminated with aluminum foil, 4 cm thick, from IKO Enertherm (thermal conductivity 0.022 W/m·K). The flexible pipes were insulated with polyethylene slotted-pipe lagging (inner diameter 15 mm, wall thickness 13 mm, and thermal conductivity 0.036 W/m·K). Leakage tests were performed to confirm the quality of the welding and the connections of the flexible fluid pipes.

(**a**) (**b**)

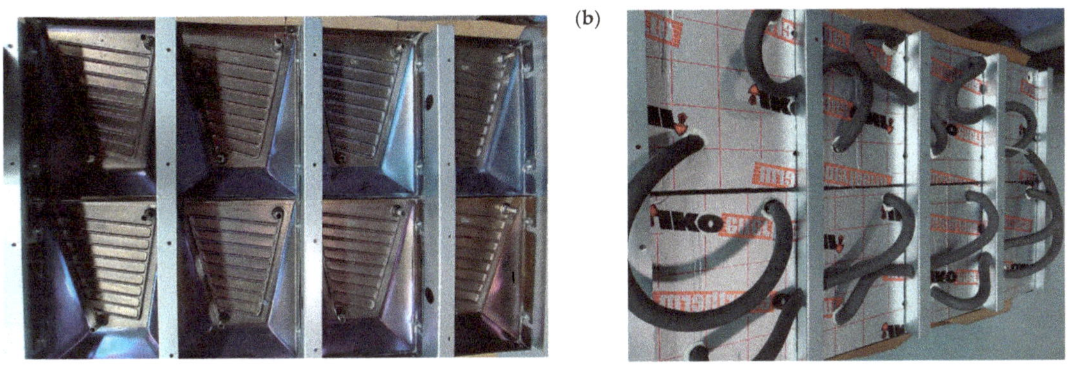

Figure 11. Pictures of the solar collector as manufactured without insulation (**a**) and with rear insulation (**b**).

3. Collector Characterization

A schematic diagram of the experimental set-up is presented in Figure 12. This system provides the fluid to the thermal panels via insulated pipes equipped with measurement probes and a circulator. The probes are placed directly at the inlet and outlet of the panel in order to minimize the losses and errors of measurements related to the pipes. The flow rate is adjusted via a manual valve in series on the sample supply circuit. The temperatures are measured at the inlet and outlet of the panel via thermocouples. In order to achieve very good stability for the inlet temperature, the measuring system has a large buffer tank and an internal temperature regulation system. This system is cooled via an external water cooling unit (chiller). The storage tank is equipped with a water-mixing unit to guarantee the homogeneity of the temperature. The heat output of the panel is determined by considering the characteristics of the fluid, the temperature variation, and the flow rate. The thermal conversion efficiencies of the panel are determined by controlling the flux density of the radiation source and knowing the active surface of the panel.

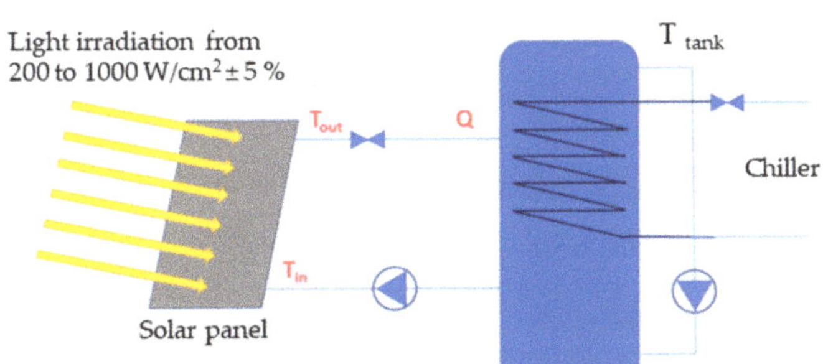

Figure 12. Schematic diagram of experimental test for thermal characterization.

The thermal performance tests were carried out in collaboration with ELIOSYS SA, Liege, Belgium. The investigated heat transfer fluid was water without glycol. As the thermal characteristics of the thermal solar collector were multifactorial, we decided to fix certain parameters. The sample inlet temperature was fixed at room temperature to avoid losses between the pipes and the environment. The measurement of the ambient temperature was carried out in a ventilated enclosure and by obscuring solar irradiance. The measurement time was at least 15 min in steady state. The irradiance measurement was carried out during all tests using a calibrated pyranometer and the thermal environment was controlled by a pyrgeometer. The stability of the irradiation was 0.1% for the duration of the test. Figure 13a shows the experimental setup. Knowing the characteristics of the heat transfer fluid, temperature variations, and flow rate, it was possible to determine the heat output of the system. By controlling the stability of the surface power of the radiation source and knowing the active surface of the sample, it was possible to calculate the thermal conversion efficiencies of the sample. The source of the irradiation was a continuous solar simulator capable of creating irradiation of 200 W/m^2 to 1100 W/m^2 with a maximum variation of less than 1% over 1 h. The spectrum of this simulator was classified B for the wavelength from 800 nm to 900 nm and A for all other wavelengths according to IEC 60904-9 [17].

(a)

(b)

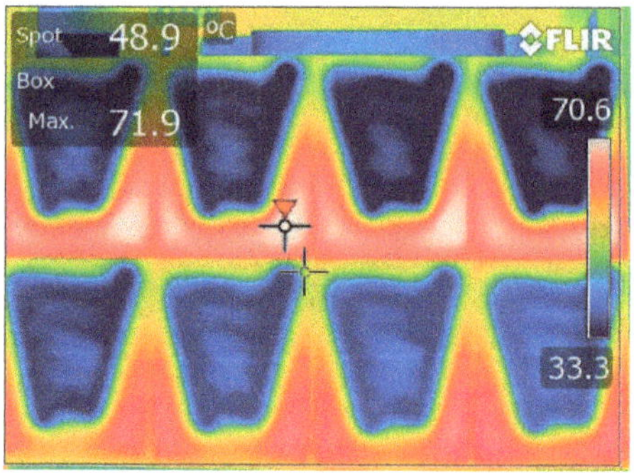

Figure 13. Experimental setup, including the solar collector and irradiance lamps (**a**) and infrared camera image of the solar collector in operation (**b**).

Figure 13b shows an example of thermal imaging, helping to visualize the interconnection of panels. The first panel (top right) was cooler than the last panel in the chain (bottom right). The irradiation power was fixed at 1000 W/m^2 and the water temperature input was fixed at 30 °C +/− 0.2 °C in the panel located at the top-right and exits at the temperature given at the bottom-right. Outside the cooled areas of the panels, a significant increase in temperature was detected, contributed by thermal conduction to the increase in fluid temperature.

3.1. Thermal Power

The thermal power of the collector was determined for various irradiance values (Figure 14a). For a typical irradiance of 1000 W/m^2 and a cooling water flow of 3.7 L/min, the total output power of the collector was 745 W/m^2. This graph also shows that the thermal conversion was better at low irradiance than at high irradiance. This phenomenon was slight and mainly originated from first- and second-order thermal losses, which increased with the temperature difference between the collector and the environment. The tilt angle of the collector significantly influenced the thermal power, as presented in Figure 14b). Indeed, the installation of the collector in a vertical orientation decreased the output power to around 17% compared with the horizontal orientation.

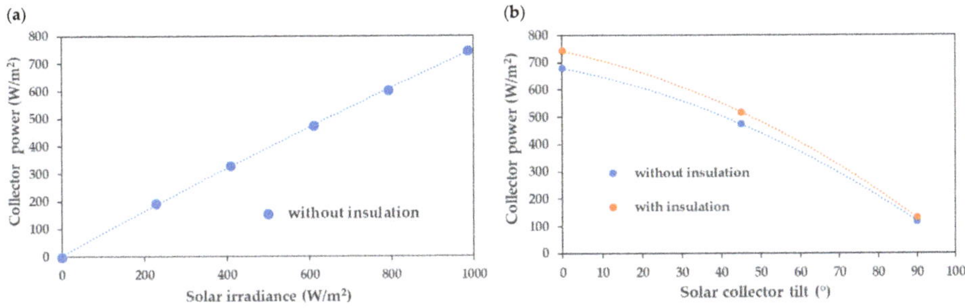

Figure 14. Variation in the thermal power of the collector versus the irradiance for a constant water flow rate of 3.7 L/min (**a**) and variation in the thermal power versus the irradiance angle (**b**).

3.2. Thermal Efficiency

The European standard EN 12975 [18] defines the efficiency η of a thermal collector on the basis of four parameters according to Equations (1) and (2):

$$\eta = \eta_0\left(\frac{a_1 \cdot \Delta T}{E_0}\right) - \left(\frac{a_2 \cdot \Delta T^2}{E_0}\right) \tag{1}$$

$$\Delta T = T_{water\ out} - T_{water\ in} \tag{2}$$

where:

η—collector efficiency;
η_0—optical efficiency;
a_1 and a_2—heat loss coefficients;
E_0—solar radiation;
ΔT—Temperature difference of the solar fluid between the inlet and outlet of the collector.

The overall efficiency η is, therefore, not a single value, but a characteristic curve. The optical efficiency η_0 represents the maximum efficiency of the thermal collector when the temperature of the fluid is at ambient temperature (no thermal losses). Measured under standardized test conditions (AM1.5 spectrum, solar irradiation E_0 = 1000 W/m^2, perpendicular to the sensor), the efficiency of a thermal collector depends on the properties of the glazing and/or the selectivity of the absorber. Figure 15 shows the variation in the collector efficiency as a function of the flow rate of the cooling fluid. The thermal insulation of the collector slightly influenced its efficiency. On the characteristic curve, the stagnation temperature of the collector was defined as the temperature difference at which the solar gains could not compensate for the thermal losses.

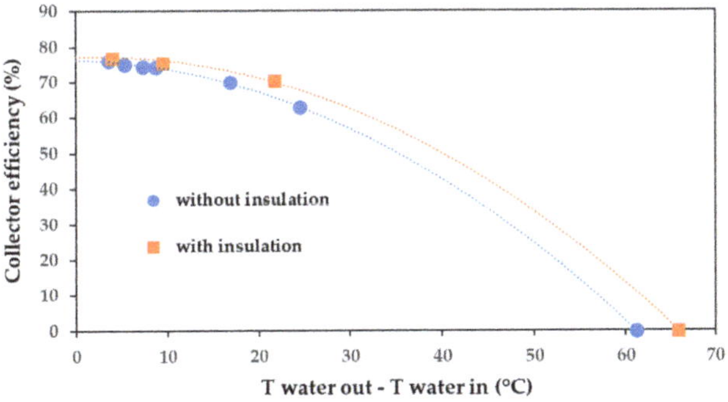

Figure 15. Collector efficiency with and without back insulation for E_0 = 1000 W/m^2.

4. Economic Analysis

An economic feasibility study was carried out to estimate the manufacturing costs of the new type of collector. To show the influence of the forming technology used on the total manufacturing costs, ISF was compared with a conventional deep drawing for quantities n of 1 and 1000 pieces.

The starting point for the cost calculation was the scaled collector geometry shown in Figure 1 and the consideration of the process chain described in Figure 16.

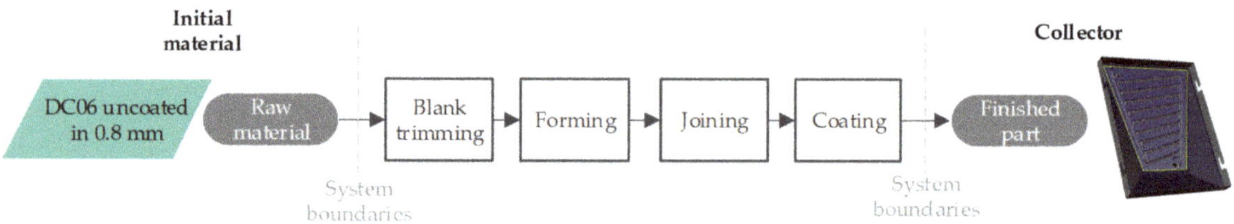

Figure 16. Simplified process chain for the manufacture of the collector.

In this simplified representation of the manufacturing process, the steps of the single-part production of the inner and outer sheets are summarized in the first three superordinate processes of blank trimming, forming, and trimming.

In addition to the main forming costs, the one-time costs for the formation of simulation (deep drawing), as well as the path generation (ISF), are included in the forming process step. All trimming and joining operations were performed using a laser. The coating process itself was divided into the sub-processes of corrosion protection and solar coating.

The assumed boundary conditions and the determined costs for each (sub-)process step are given in Table 1.

Table 1. Boundary conditions and detailed cost distribution by process step for collector-manufacturing depending on the formation technology and quantity n.

Process	Sub Process	ISF		Deep Drawing	
		$n = 1$	$n = 1000$	$n = 1$	$n = 1000$
Raw material (inner + outer sheet)	-	≈16 €		≈16 €	
Forming (inner + outer sheet)	Simulation (one-time)	-		≈800 €	
	Tool-path generation (one time)	≈200 €		-	
	Tooling costs (one time)	≈1500 €		≈20,000 €	
	Manufacturing time per collector	≈2.9 h		≈0.09 h	
	Forming cost per collector	≈292 €		≈22 €	
	Total forming cost per collector	≈2192 €	≈294 €	≈21,622 €	≈43 €
Trimming (inner + outer sheet)	Jig costs (one-time)	≈250 €		≈250 €	
	Laser trimming cost per collector	≈4.20 €		≈4.20 €	
	Total trimming cost per collector	≈254 €	≈5 €	≈254 €	≈5 €
Joining	Total joining cost per collector	≈270 €	≈25 €	≈270 €	≈25 €
Coating	Costs of corrosion protection (Al-Zn)	≈245 €	≈20 €	≈245 €	≈20 €
	Costs of solar coating (sol-gel)	≈37 €	≈17 €	≈37 €	≈17 €
	Total coating cost per collector	≈282 €	≈37 €	≈282 €	≈37 €
Total manufacturing costs per collector		≈3014 €	≈376 €	22,444 €	126 €

The results of the process chain analysis show that the formation and associated tooling investment costs were the drivers for all process variants investigated. As expected, for small quantities, ISF can take advantage of its universal forming stylus and low overall tooling costs. With total manufacturing costs per collector of approx. 3014 € for a quantity of one, the costs of ISF were 87% lower compared with the use of conventional deep-drawing technology. In contrast, deep drawing has advantages for larger quantities due to the significantly short process times. When considering a number of 1000 pieces, the unit price per collector of 126 € for the deep drawing variant was 33% lower than that of the ISF variant. For large quantities, conventional forming is clearly preferable. In the case of prototypes or small quantities, however, ISF is an interesting alternative. In this specific case, the break-even would be at 73 collectors. From this point on, the manufacturing costs for deep drawing are lower than those for ISF.

In view of the large-scale production capability of the collector type presented, the graphical representation of the cost distribution in Table 2 shows that, in addition to formation, coating and joining account for the largest shares in terms of total costs. In the deep drawing scenario at a quantity of n = 1000, the share of formation is approx. 34%, followed by coating (≈29%) and joining (≈20%).

Table 2. Overview of the estimated manufacturing costs per collector depending on the forming technology and quantity *n*.

<table>
<tr><th></th><th></th><th>n = 1</th><th>n = 1000</th></tr>
<tr><td rowspan="2">Forming technology</td><td>ISF</td><td>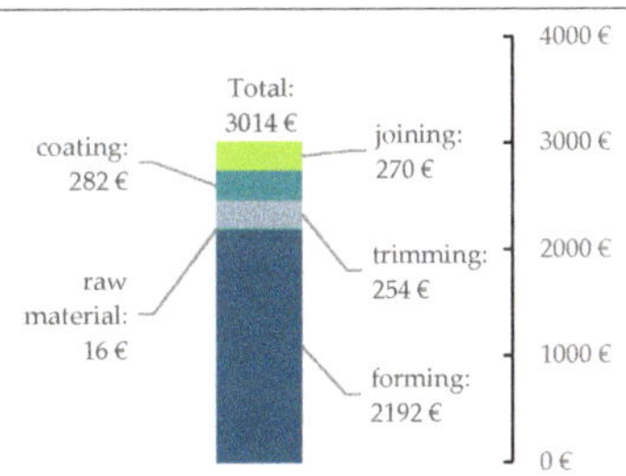
</td><td>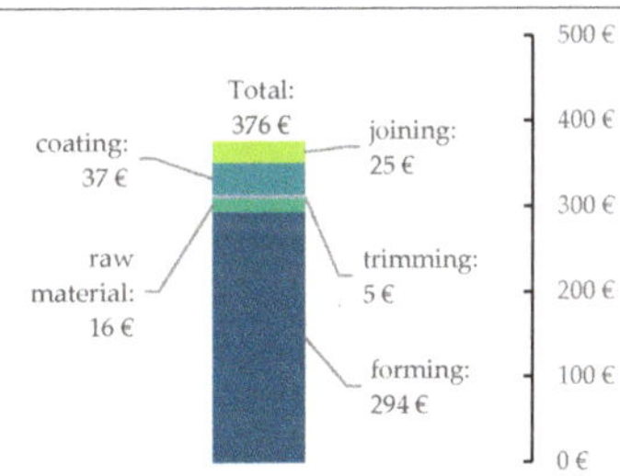
</td></tr>
<tr><td>Deep drawing</td><td>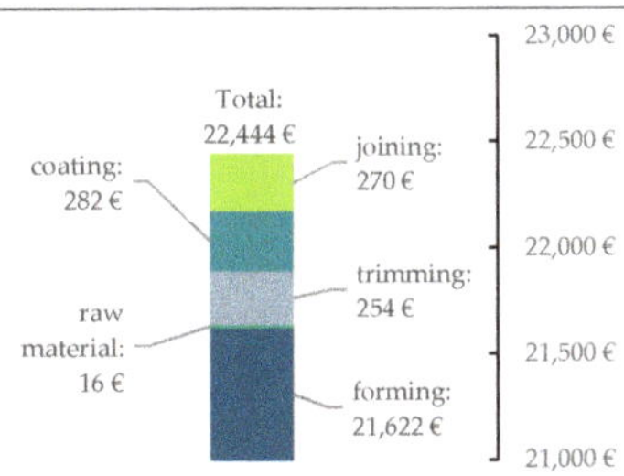
</td><td>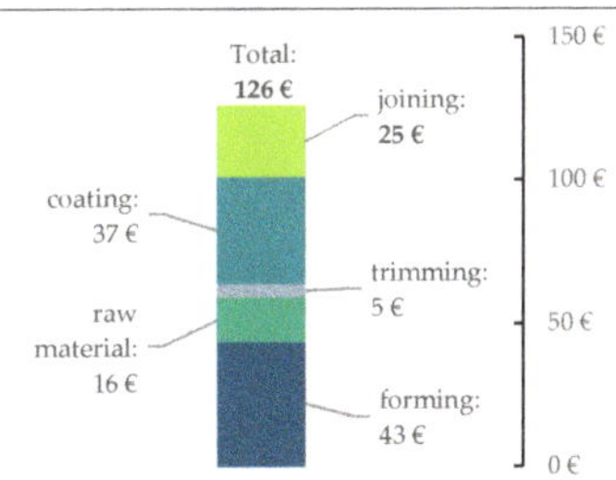
</td></tr>
<tr><td></td><td></td><td colspan="2">■ forming ■ raw material ■ trimming ■ coating ■ joining</td></tr>
</table>

If the costs are related to the collector area, the costs for the new collector type were approx. 625 €/m^2 under the assumptions made. In combination with the degree of design freedom made possible for the first time, the type of collector presented represents an interesting addition and extension to existing systems.

5. Discussion

As shown above, there is good potential for all-sheet metal solar collectors, as the thermal performance and cost were comparable to those of modern flat plate collectors, with the added benefit of three-dimensional design freedom for seamless integration into facade systems.

It was demonstrated that the manufacture of complex collector geometry with the chosen parameters without cracks was possible by ISF. However, as expected, the cost analysis showed that the area of application for this manufacturing technology was only limited to small quantities (e.g., prototypes or individualized elements) due to the long manufacturing time. For large quantities with recurring geometries, it would make more economic sense to use conventional forming technologies with high output rates. The high quantities amortize the initial investment costs in the forming tools and enable low marketable unit costs.

Laser welding was proven to be an efficient method for the assembly of an all-sheet metal collector, despite the 3D complex shape. However, the increase in the metal thickness could considerably improve the welding robustness influenced by sheet formation.

The use of sol-gel painting spray was an efficient and low-cost way to deposit the solar-absorbing layer on a complex-shaped surface. The high solar absorptance and high emissivity of the solar layer could generate a high thermal efficiency of the solar collector at a low temperature. The overheating effect of the present thermal panels was significantly reduced, as the maximum heating temperature was under 100 °C. Indeed, the overheating effect observed in the case of fluid stagnation (e.g., power cut, failure of the primary pump, or when the heat demand is low) for conventional glazed thermal panels using selective absorbing layers generates a high increase in the fluid temperature (>150 °C), leading to premature component degradation. The large surface available for the implementation of these new solar facades, combined with their high efficiency at low temperature, can generate an economical approach to saving energy. However, the coating design could be improved to decrease the surface emissivity and, consequently, to increase the efficiency at higher temperatures; however, this can be achieved only by using complementary antireflective coatings.

6. Conclusions

This project has shown that, within the research project "FutureFacade", new potential for renewable energies in cities can be created. Solar thermal-active facades are particularly suitable for large buildings with large facade areas, as these are the first places where it becomes economically viable.

The approach of using standardized facade substructures was proven successful. However, this project still has a lot of development potential. With regard to the manufacturing process, it is advisable to switch to the more economical manufacturing process for subsequent projects. This would also make the production of smaller facades and quantities more economical. The material could also be changed to aluminum instead of steel in order to minimize the weight of the facade construction. Changes in material and forming technology also allow new design freedom in the geometry of further developed facade panels, which can be further extended by adapting the coating. In terms of appearance, it would be desirable to be able to create more variance in color with the same or similar efficiency. In terms of efficiency, it would also be advisable to further develop the channel structure professionally with the help of meaningful CFD simulations. When considering the efficiency, its ratio with the absorber surface must always be taken into account. With regard to the panel size, adjustments could be made in the future. As long as the absorber surface area is 2 m^2, the collector remains comparable to conventional absorbers.

All in all, the "FutureFacade" approach has brought many new insights and provides new stimuli for further research on solar thermal facades.

Author Contributions: Conceptualization, P.S.; collector design, P.S., D.W., L.S. and M.D. (Martin Dembski); architectural aspects, L.S., M.D. (Martin Dembski) and A.S.; facade construction and integration, L.S. and M.D. (Martin Dembski); incremental forming tests, P.S. and D.W.; joining and coating experiments, F.D. and S.L.C.; collector characterization, F.D. and S.L.C.; economic analysis, P.S. and F.D; resources, J.K., M.D. (Martin Dix) and A.S.; writing—review and editing, P.S., D.W., L.S., M.D. (Martin Dembski), A.S., J.K. and M.D. (Martin Dix). All authors have read and agreed to the published version of the manuscript.

Funding: This research is supported by the Federal Ministry for Economic Affairs and Climate Action (BMWK) on the basis of a decision by the German Bundestag. This article is based on the project "Future Facade—Combination of individualised design and solar heat functionality by application of a new forming technology for metal based facade elements" coordinated by the German Federation of Industrial Research Associations (AiF) via Forschungsvereinigung Stahlanwendung e.V. (FOSTA) under Grant No. 269 EBR.

Institutional Review Board Statement: Not applicable.

Informed Consent Statement: Not applicable.

Data Availability Statement: Not applicable.

Acknowledgments: This paper was completed in association with the European Union's Horizon 2020 research and innovation program under grant agreement N°856670. The authors would like to acknowledge ELIOSYS SA, Liege, Belgium for their help in the thermal collector's characterization.

Conflicts of Interest: The authors declare no conflict of interest. The funders had no role in the design of the study; in the collection, analyses, or interpretation of data; in the writing of the manuscript; or in the decision to publish the results.

References

1. Bundesregierung. CO_2-Gebäudesanierung. Available online: https://www.bundesregierung.de/breg-de/aktuelles/co2-kohlenstoffdioxid-oder-kohlendioxid-gebaeudesanierung-614754 (accessed on 29 November 2022).
2. International Energy Agency. Technology Roadmap—Energy Efficient Building Envelopes. Available online: https://www.iea.org/reports/technology-roadmap-energy-efficient-building-envelopes (accessed on 29 November 2022).
3. Rynska, E. Review of PV Solar Energy Development 2011–2021 in Central European Countries. *Energies* **2022**, *15*, 8307. [CrossRef]
4. Kaufmann, D. Multifunktionales energieeffizientes Dach- und Fassaden-Solarelement—DAFASOL: Teilvorhaben: Entwicklung des Bauelementes und Anpassung an Einsatzfälle im Stahlbau: Sachbericht: Laufzeit des Vorhabens: 01.12.2014–30.11.2017. Available online: https://www.tib.eu/de/suchen?tx_tibsearch_search%5Baction%5D=download&tx_tibsearch_search%5Bcontroller%5D=Download&tx_tibsearch_search%5Bdocid%5D=TIBKAT%3A1017755027&cHash=ef067e89bb2d1593251944ea0064ceb0#download-mark (accessed on 29 November 2022).
5. Munari-Probst, M.; Roecker, C.; Schueler, A. Architectural integration of solar thermal collectors: Results of a European survey. In Proceedings of the ISES 2005 Solar World Congress, Orlando, FL, USA, 6–12 August 2005.
6. Munari-Probst, M.; Cristina, M.; Roecker, C. *Architectural Integration and Design of Solar Thermal Systems*, 2nd ed.; EPFL Press: Lausanne, France, 2011.
7. Buker, M.S.; Riffat, S.B. Building integrated solar thermal collectors—A review. *Renew. Sustain. Energy Rev.* **2015**, *51*, 327–346. [CrossRef]
8. Hermann, M.; Karin, L.; Hillerns, F. BIONICOL—Entwicklung eines bionischen Solarkollektors mit Aluminium-Rollbond-Absorber. In Proceedings of the 20. Symposium Thermische Solarenergie, Bad Staffelstein, Germany, 5–7 May 2010; pp. 74–79.
9. Tekkaya, E.; Rainer, S. Entwicklung von Solarabsorbern in Stahlbauweise auf Basis partiell plattierter Hybridhalbzeuge. In *Forschung für die Praxis P820*; Verl. und Vertriebsges. mbH: Düsseldorf, Germany, 2015.
10. Maurer, C.; Cappel, C.; Kuhn, T.E. Progress in building integrated solar thermal systems. *Sol. Energy* **2017**, *154*, 158–186. [CrossRef]
11. Schreiber, R.G.; Schaeffer, L. Manufacture of absorber fins for solar collector using incremental sheet forming. *J. Mater. Res. Technol.* **2019**, *8*, 1132–1140. [CrossRef]
12. Weiss, W.; Stadler, I. Facade integration—A new promising opportunity for thermal solar collectors. In *Proceedings of the Industry Workshop of the IEA Solar Heating and Cooling Programme*; Task 26; International Energy Agency: Delft, The Netherlands, 2001.
13. Matuška, T.; Sourek, B. Aspects of solar collector integration into building facade. In Proceedings of the 6th EuroSun, Glasgow, UK, 27–30 June 2006.
14. Behnisch, M.; Münzinger, M.; Poglitsch, H. Die vertikale Stadt als solare Energiequelle? Theoretische Flächenpotenziale für bauwerksintegrierte Photovoltaik und Abschätzung der solaren Einstrahlung. *Transform. Cities* **2020**, *4*, 58–62.
15. Behnisch, M.; Münzinger, M.; Poglitsch, H.; Willenborg, B.; Kolbe, T.H. Anwendungsszenarien von Geomassendaten zur Modellierung von Grünvolumen und Solarflächenpotenzial. In *IÖR Schriften Band 78*; Rhombos Verlag: Berlin, Germany, 2020; pp. 251–261.
16. Knudsen, O.Ø. *Coatings Systems for Long Lifetime: Thermally Sprayed Duplex Systems*; Final Report; SINTEF Rapport: Trondheim, Norway, 2010.

17. *IEC 60904-9:2020*; Photovoltaic Devices—Part 9: Classification of Solar Simulator Characteristics, 3rd ed. IEC Central Office: Geneva, Switzerland, 2020.
18. *DIN EN 12975:2022-06*; Solar Collectors—General Requirements. Beuth Verlag GmbH: Berlin, Germany, 2022. [CrossRef]

energies

MDPI

Article

Thermal-Hydraulic-Mechanical (THM) Modelling of Short-Term Gas Storage in a Depleted Gas Reservoir—A Case Study from South Germany †

Muhammad Zain-Ul-Abedin * and Andreas Henk *

Institute of Applied Geosciences, Technical University of Darmstadt, 64287 Darmstadt, Germany
* Correspondence: abedin@geo.tu-darmstadt.de (M.Z.-U.A.); henk@geo.tu-darmstadt.de (A.H.); Tel.: +49-6151-1622-347 (M.Z.-U.-A.); +49-6151-1622-344 (A.H.)

† This paper is an extended version of my PhD thesis with title "Coupled Thermal-Hydraulic-Mechanical (THM) Modelling of Underground Gas Storage—A Case from the Molasse Basin, South Germany" submitted in 2022 Technical university of Darmstadt, Darmstadt, Germany, March 2022; p. 103.

Abstract: This study addresses the use of former gas storage facilities as short-term storage for renewable energy through power-to-gas (PtG) technology in Germany. Three test cases with coupled thermal-hydromechanical (THM) modelling were conducted to evaluate short-term injection and production schedules. The operating rates were controlled by the upper and lower limits of the wellbore pressure. The maximum difference in pore pressure and effective stress was 0.6 MPa in all cases. Fault reactivation analysis was performed on the THM models to estimate fault stability. The critical pore pressure for safe reservoir operation was determined to be 1.25 times the original pore pressure, corresponding to a WBHP value of 20.25 MPa. The upper limit of the gas injection rate for safe storage operation was estimated to be between 100,000 and 150,000 m^3/day. The thermal stresses were found to be negligible for short-term cases. The storage capacity of PtG technology was reported to be up to 1,322,400 kWh/d of renewable electricity, which can contribute to Germany becoming a greenhouse gas neutral country by 2050. The workflows and results of the study are applicable to all gas storage in a porous medium, including methane, CO_2, and hydrogen.

Keywords: THM modelling; short term underground gas storage; dynamic modelling; "battery" for power-to-gas

Citation: Zain-Ul-Abedin, M.; Henk, A. Thermal-Hydraulic-Mechanical (THM) Modelling of Short-Term Gas Storage in a Depleted Gas Reservoir—A Case Study from South Germany. *Energies* **2023**, *16*, 3389. https://doi.org/10.3390/en16083389

Academic Editors: Luis Hernández-Callejo, Jesús Armando Aguilar Jiménez and Carlos Meza Benavides

Received: 27 February 2023
Revised: 5 April 2023
Accepted: 10 April 2023
Published: 12 April 2023

1. Introduction

High energy demand has intensified research into the "underground gas storage" (UGS) discipline [1]. As a result, both long-term (seasonal) and short-term (weekly) gas storage projects have gained the attention of UGS researchers [1]. An idea in this context is to store methane produced by "power-to-gas (PtG)" technology from excessive electricity produced by renewable sources, such as solar and wind. The stored gas can then be reused for power generation whenever needed. Thus, UGS is used as a kind of "battery" for surplus green energy.

Short-term storage cycles may cause some geomechanical issues in the porous reservoir. The pore fluid pressure in the reservoir fluctuates due to intensive gas injection and withdrawal phases. These fluctuations in fluid pressure change the effective stresses in the reservoir and may also change the in-situ stress state outside the reservoir area [2]. These changes have implications for geomechanical phenomena related to fault stability, caprock integrity, and surface deformation. Numerical modelling, e.g., 3D geomechanical modelling, provides a platform to integrate lithological and mechanical heterogeneities and investigate stress state changes during injection-production cycles of high frequency.

This study concentrates on a geomechanical assessment of a former gas field in the Bavarian Molasse Basin east of Munich (Germany) for which a hypothetical transformation into an UGS site is investigated. Various scenarios with variable short-term (weekly)

schedules to test cases for gas storage and withdrawal are considered to evaluate stresses (e.g., effective stress) and deformation due to pressure changes with high-frequency injection/production cycles. German data for excess electricity from renewable energy sources (such as solar and wind) throughout the calendar year 2017 are also considered in two cases to address the issue of irregular schedules in gas supply (via PtG) and energy demand.

The modelling results provide information about the stress state within and around the reservoir because of the production and injection of each selected time step, and they are compared with the stress states at the depletion and replenishment stages. The production history of the reservoir (porous media) is categorically useful to determine stress paths within and around the reservoir and wellbore periphery, as well as caprock integrity. Further, the modelling results provide information about ground surface subsidence during peak depletion and replenishment time steps, which can be useful to minimize geomechanical risks to any gas storage facility, not only for methane or CO_2 but also for hydrogen. Finally, fault reactivation analyses are also incorporated to obtain a safe gas injection rate for safe storage capacity.

The most popular gas storage method is underground gas storage among others, such as liquefied natural gas (LNG), storage tanks, and pipeline storage [3,4]. Two important issues are associated with UGS. First, it relies on gas imports due to increased demand for power generation (gas-to-power) and other domestic usages. Second, any damage to infrastructure could lead to higher gas prices or disruption of supply, with unpredictable, costly consequences for customers [1]. To address these challenges, various research has suggested using depleted gas/oil porous reservoirs or aquifers as UGS, with which significant volumes can be strategically stored [1,5].

One of the main advantages of depleted gas reservoirs is that they allow convenient and cost-effective gas storage, as they have suitable permeability properties and pore connections [6,7]. Some vital data, such as geological and geophysical characteristics, petrophysical properties, storage capacity, pressure, and the production history of depleted gas reservoirs, have been thoroughly studied and well recorded during the development phase of the reservoir field, enabling numerical modelling to simulate the injection-production process of underground gas reservoirs [3,6,8].

Water encroachment occurs in the porous reservoir vertically and laterally during the development phase of the gas reservoir. This water invasion causes fewer porous spaces in the reservoir, and different fluid distribution areas are formed in the longitudinal and traverse planes of the reservoir. During the injection and production operations of gas storage, the gas-water interface moves downward when gas is injected and moves upward when gas is produced [9]. Furthermore, as the foreign gas is injected into the reservoir, which has a different temperature than the reservoir's actual temperature, these thermal changes in the reservoir also cause some thermal stress changes in the reservoir, which is not the case in the reservoir exploitation phase. Therefore, it is necessary to conduct a systematic modelling study of the changes in the thermal-hydraulic-mechanical properties of UGS due to repeated water intrusion and multiphase seepage rules during high-speed injection production operations. The state-of-the-art thermal-hydro-mechanical (THM) modelling provides a platform to analyse and investigate all these issues related to porous UGS systems discussed above. THM models are typically derived from a wide range of geological, geophysical, and engineering data, including field measurements, core tests, well logs, drilling, and production data. After calibration, these THM models can be used for ground surface subsidence, thermal stress changes, maximum safe storage capacity, and maximum threshold pressure, avoiding fault reactivation and maintaining caprock integrity [10–12].

This modelling case study is a THM assessment of a former depleted gas reservoir field from the Molasse Basin in South Germany and has been presented in [10]. This dynamic modelling study addresses the following issues related to short-term and long-term operations of UGS: (1) the storage capacity of the reservoir; (2) thermal analysis with changing temperatures of the reservoir due to the injection of gas; (3) potential fault reactivation analysis; and (4) stress path changes due to changes in pore pressure because of short-term production/injection cycles. These types of THM models, particularly short-term scheduled models, have implications for PtG technology in which simulations help us to understand that excess power from renewable resources can be stored in underground gas storage reservoirs and vice versa.

2. Methodology

2.1. Hydraulic Model

The flow simulation accounts for multiphase fluid flow in porous media. A hydraulic model is usually conducted by reservoir simulation, i.e., a form of numerical modelling in which physical phenomena are quantified and interpreted throughout the history of a reservoir and beyond, with the ability to extend this model to future performances. Reservoir simulation is a proven and effective method for dealing with uncertainties during exploration and production [13]. It is also helpful to determine the amount of gas that can be stored in each underground gas (CO_2, hydrogen, or methane) storage reservoir [13]. The physical phenomenon behind (fluid flow) reservoir simulations are based mainly Darcy's law and mass material balance [10].

The composition of the fluid can be treated in different ways in reservoir simulations. Black oil simulators assume oil and gas phases to be one component through space and time. The properties of this component can change with pressure and temperature, while the composition does not change [13]. Thereby, the behaviour of the multiphase system can be described by complex PVT (pressure, volume, temperature) and SCAL (special core analysis) relations [13].

As a general solution method, the reservoir is divided into several cells with provided petrophysical properties, such as porosity and permeability. Then, the wells are placed within cells, and production rates are provided with different time steps. Last, the equations are solved to determine the pressure, temperature, and saturation for each cell. Each cell is solved simultaneously; therefore, the number of cells in the reservoir simulation is directly related to the time required to solve a time step [13].

2.2. Thermal Model

The thermal-flow-stress simulation model considers the proportional heat transfer in porous media when considering the multi-fluid flow concept in a THM simulation. The thermal flow model is usually performed by numerical modelling, in which the thermal hydraulic aspects are quantified together with the geomechanical simulation and interpreted throughout the history of a reservoir and storage operations in underground gas storage facilities. The main governing law in the thermal modelling is Fourier's law, also known as the law of heat conduction. The law states that the rate of heat transfer through a material is proportional to the negative temperature gradient and to the area at right angles to that gradient through which the heat flows. The governing equation of Fourier's law describes that the local heat flux density (q_h) is equal to the product of the thermal conductivity (k_t) and the negative local temperature gradient ($-\nabla T$) [14]:

$$q_h = -k_t \nabla T, \tag{1}$$

where q_h is in (W/m^2), k_t is in (W/m. K), and ∇T is in (K/m). Fourier's law can also be used in uni-dimensional form in any direction, i, j, k; for that reason, the equation becomes:

$$q_h = -k_t \frac{dT}{d(i,j,k)}, \tag{2}$$

2.3. Coupled Thermal-Hydraulic-Mechanical (THM) Modelling

The coupling of a reservoir simulator with a geomechanics module is an integral component for analysing hydrocarbon reservoirs in petroleum, underground gas storage, and the geothermal industry. A conventional geomechanical simulator provides surface subsidence, which is often estimated using a simple mechanical formula without knowing the full geomechanical response. The only geomechanical parameter considered may be the pore compressibility, which is not sufficient to reproduce the changes in pore volume caused by complex pressure and temperature variations [15,16]. For some problems, such as primary production and linear elastic responses of reservoirs, subsidence calculated by a reservoir simulator alone can produce results comparable to coupled solutions [16,17].

In a coupled simulator, flow can be strongly influenced by the stress and strain distributions that lead to changes in porosity and permeability, but effective stress changes are ignored in conventional simulation methods. Such approaches cannot provide adequate predictions when considering a stress-sensitive reservoir (e.g., underground gas storage reservoir) [15,18]. There are two main components of coupling: volume coupling and fluid flow coupling according to [15,17].

In volume coupling, the changes in pore volume occur in response to variations in stress, pressure, and temperature. For convergence, the calculated pore volume changes should be the same in both the fluid flow model and the geomechanical model. The pore volume changes of the geomechanical model are usually more accurate than those of the fluid flow model because they are calculated by volumetric strain via a more realistic complex material constitutive model. This method is well suited to shear and plastic deformation, which involve large changes in pore volume or porosity. These problems are common in unconsolidated heavy oils and oil sands, North Sea chalk, Californian diatomite, and possibly some other materials [15,17].

In case of fluid flow coupling, the changes in permeability and relative permeability are related to the changes in stress, shear stress, and compaction. Material parameters, such as permeability, relative permeability, compressibility, and others, change when conditions encounter a shear fracture. This fact is important in some reservoirs where the rock compressibility does not play a significant role in volumetric behaviour, such as gas reservoirs where volume coupling is not important. Another example is cold water/fluid injection, which leads to a thermally-induced decrease in horizontal stress until the injection pressure increases the minimum horizontal stress [19].

To achieve the stress states of the reservoir and surrounding formations throughout history, as well as during future gas storage operations, the fluid flow and the geomechanical simulation must be coupled. The pore pressure controls the effective stresses and, hence, deformation, in turn changing rock porosity and permeability, which again affect fluid flow.

2.3.1. Effective Stress and Poroelasticity

The effective stress for incompressible rock and the concept of one-dimensional consolidation are the fundamentals of poroelasticity formulated by Terzaghi in 1923 [17]. Successively, using the basic principles of continuum mechanics and applying the concept of the coupling of stress and pore pressure in a porous medium, Biot developed a comprehensive three-dimensional theory of consolidation [17,20]. Biot's theory and the papers that he published are more aligned towards geomechanics than flow models, due to which they are rarely compatible with the coupling of geomechanics and flow models. By introducing the so-called Skempton pore pressure parameters (A and B), Skempton in 1954 procured a relationship between the total stress and the pore pressure under undrained initial loading [17,21]. Later, however, the relationships among pore pressure, stress, and volume and the concept of compressibility in a porous medium were better clarified by Geerstma in 1957 [17,22]. Later, Van der Knaap (1959) extended Geerstma's work to nonlinear elastic geomaterials only for dense and uncemented sands [17,23]. By applying Biot's theory, Geerstma in 1966 examined subsidence problems in oil fields and published prototype geomechanical modelling, which is probably the first-ever coupled analysis of fluid flow [17,20]. Nur and Byerlee (1971) demonstrated that the effective stress law proposed by Biot is far more general and precise than that proposed by Terzaghi [17,24]. Nevertheless, there are certain limitations (one-dimensional analysis, neglect of the compressibility of fluids and rocks, etc.) that Terzaghi recognized in the assumptions that he made in the 1920s to solve problems of applied rock mechanics in clay consolidation [17]. Later in the 1970s, there were further developments on coupled flow stress issues; e.g., fluid compressibility was introduced into the classical soil mechanical consolidation theory of Ghaboussi and Wilson [17,25]. Rice and Cleary (1976) showed how poroelasticity problems could be solved using pore pressure and stress as primary variables, instead of the displacements used by Biot [17,26].

2.3.2. Simulation Concept and Governing Equations

The numerical modelling of an underground gas reservoir can contribute to the understanding of the interaction mechanisms between the injected gas and the deformation of the reservoir. The injection of cold foreign gas into the reservoir leads to thermal and mechanical disequilibrium in the reservoir by altering the transport properties, including porosity and permeability. A three-dimensional THM coupling model of a reservoir is created by incorporating the mechanical equilibrium equation, the fluid flow or seepage equation (Darcy equation), the heat transfer equation (Fourier equation) of the formation's rock matrix, and the THM stress equation. These equations are based on the porosity, permeability, thermal diffusivity, and other physical and mechanical parameters of the sandstone formation [27].

The mechanical equilibrium equation can be expressed as follows [27]:

$$S_{i,j,j} + f_i = 0, \tag{3}$$

where, $S_{i,j,j}$ is the total stress tensor (N/m^2), and f_i is the body force (N/m^3). The equation of the continuity of the fluid flow in the rock can be written as [27]:

$$\frac{\partial \rho_l}{\partial t} + \frac{\partial(\rho_l r v_r)}{r\partial r} + \frac{\partial(\rho_l v_\theta)}{r\partial \theta} + \frac{\partial(\rho_l w)}{\partial_z} = 0, \tag{4}$$

In this equation v_r, v_θ, and w are the Darcy velocities (m/s) along the radial, hoop, and well-depth directions, respectively, in the porous reservoir. The relationship between stress and porosity/permeability changes in porous rock can be described with the following equations [27,28],

$$\phi = \phi_r + (\phi_0 - \phi_r)exp(e.S_M), \tag{5}$$

$$k = k_0\, exp(c.(\frac{\phi}{\phi_0} - 1)), \tag{6}$$

In the above equations, S_M denotes mean effective stress; ϕ_0 and k_0 are the porosity and permeability at zero stress, respectively; ϕ_r represents the residual porosity at high stress; and the exponents e and c are determined experimentally.

The heat transfer process and the total energy conservation can be express by rewriting the Fourier equation [29]:

$$(\rho c)_t \frac{\partial T}{\partial t} - \frac{1}{r}\frac{\partial}{\partial r}(k_t r \frac{\partial T}{\partial t}) - \frac{1}{r^2}\frac{\partial}{\partial \theta}(k_t \frac{\partial T}{\partial \theta}) - \frac{\partial}{\partial z}(k_t \frac{\partial T}{\partial z}) - q_t r = 0, \tag{7}$$

$$(\rho c)_t = (1-\phi)c_s\rho_s + \phi c_f \rho_f, \tag{8}$$

$$k_{total} = \phi k_f + (1-\phi)k_s, \tag{9}$$

In the above equations, the total heat capacity of the solid and fluid phases is denoted by $(pc)_t$ in units ($J/(m^3.°C)$); k_{total} is the total thermal conductivity ($J/(m.s.°C)$); q_t is the intensity of the internal heat source ($J/(m^3.s)$); c_s and c_f are the specific heat capacities of the formation and the fluid, respectively (J/kg.K); ρ_s and ρ_f are the density of the formation and the fluid (Kg/m^3), respectively; and k_s and k_f are the thermal conductivity of the formation and the fluid (W/(m·K)), respectively [29].

Finally, the governing equation involving all important thermal-hydro-mechanical parameters is as follows [30]:

$$2\alpha\frac{(1-2\nu)}{1+\nu}\nabla^2 p + 6\beta K_B\frac{(1-2\nu)}{1+\nu}\nabla^2 T - \nabla.f - 3\frac{(1-\nu)}{1+\nu}\nabla^2 S_{m(total)} = 0, \quad (10)$$

where $S_{m(total)}$ is the mean total stress (MPa), ν is the Poisson's ratio of the rock mass, α is the Biot coefficient, β is the coefficient of linear thermal expansion (1/°C), and K_B is the bulk modulus of the rock (MPa). The term $(2\alpha\frac{(1-2\nu)}{1+\nu}\nabla^2 p)$ describes the effect of poroelastic stress; $(6\beta K_B\frac{(1-2\nu)}{1+\nu}\nabla^2 T)$ represents thermalelastic stress; and the term $(\nabla.f - 3\frac{(1-\nu)}{1+\nu}\nabla^2 S_{m(total)})$ shows the body force [30].

2.4. ECLIPSE™_VISAGE™ THM Modelling

The THM modelling is performed using two commercial software packages, referred to here as the ECLIPSE™-VISAGE™ coupling. ECLIPSE™ is a conventional reservoir simulator used as the flow simulator for the multiphase flow processes. In the following study, the ECLIPSE™ 100 finite difference black oil simulator is used for the flow calculations. VISAGE™ is one of the most advanced and comprehensive stress analysis simulators commercially available. It is designed for applications in which nonlinear mechanics play a greater role, for example, in disciplines such as rock mechanics and geomechanics. It is flexible and can also be used in many other scientific disciplines, e.g., fluid mechanics, heat transfer, materials science, etc. The system provides high computational power and sophisticated modelling for many analytical situations, which currently include mining, civil engineering, reservoir engineering, and geothermal energy.

The advanced and comprehensive finite element code (VISAGE™) has been coupled with the ECLIPSE™ reservoir simulator to include geomechanical processes in this modelling study. Fluid flow is first calculated in ECLIPSE™, and the results are then transferred from ECLIPSE™ to VISAGE™ via the ECL2VIS interface for specified time steps. Updates of porosity and permeability can be calculated via constitutive relations, e.g., Kozeny–Carman.

The coupled THM models are based on three fundamental laws, namely Hooke's law of elasticity, Fourier's law of heat conduction, and Darcy's law of fluid flow in porous media. The governing equations are discussed above.

One-way coupling is usually sufficient for THM modelling of gas reservoirs, as gas compressibility dominates the bulk rock compressibility, and the mass balance is mainly controlled by gas pressure rather than by the stresses of solid rock [31]. The THM-coupled model presented in this paper is a one-way coupled model. With this approach, pressure data are transferred from the reservoir simulator (fluid flow simulator) ECLIPSE™ to the mechanical finite element simulator (geomechanical simulator) VISAGE™ at regular and/or critical times. The fluid pressure drives the geomechanics, but mechanically induced changes in porosity and permeability are not fed back into the dynamic reservoir simulation. It is therefore assumed that permeability and porosity are the same for each time step in the modelling and are not influenced by changes in stresses or ageing effects of the reservoir. This type of approach is feasible in this case study due to the high permeability of up to 80 mD, which is characteristic of the reservoir.

3. Case Study

The presented modelling study is a working example of how to set up and populate a 3D coupled thermal-hydraulic-mechanical model of an underground gas storage site. The case study reservoir is a depleted gas reservoir located about 65 km east of Munich in the Central Molasse Basin. It is an anticlinal structural trap bounded by a normal fault. The reservoir formation is mainly Early Cenozoic Chattian sand, with 85-m thick, three gas-bearing layers, found at a depth of 1770 m (1200 m below sea level). The initial gas-water contact is at 1239 m below sea level (BSL). The reservoir has produced 528 million m^3 of gas over 18 years from 1958 till 1976; replenishment started in 1978 and has continued to the present, and the reservoir has not been in operation to date [10]. The modelling study uses pore pressure development during the production history and subsequent shut-in phase to calibrate the dynamic reservoir fluid model.

The 3D MEM model is built using a hydraulic model (green area in Figure 1), which comprises a high-resolution reservoir section and regions of lower resolution away from the reservoir section, called the sideburden, overburden, and underburden sections. Topography is extracted from the elevation maps of the ground level to include the top surface of the model. The horizons bounding the reservoir are used to make overburden layers and underburden layers. The basal unit of the model comprises crystalline basement rocks at a depth of about 5 km. However, as none of the wells has reached this depth, this information is inferred from regional geological knowledge. The final 3D THM model consists of 12 horizons and 11 lithostratigraphic units with dimensions of about $30 \times 24 \times 5$ km^3 in the X, Y, and Z directions, respectively. The grid of the THM with the reservoir model embedded is shown in Figure 1. The higher resolution in the area of interest (reservoir) and the lower resolution outside make up a grid that creates balance between simulation precision and computational demand. The initial pore pressures and elastic properties are upscaled and interpolated from the 1D MEM's. The calculated and calibrated log-derived properties, including pore pressure, Young's modulus, Poisson's ratio, and density, are upscaled from the well locations to the entire model domain. The Kriging interpolation method is used to populate the 3D geomechanical model. The precision is of course decreasing away from the wells, but the model fits well with overall trends.

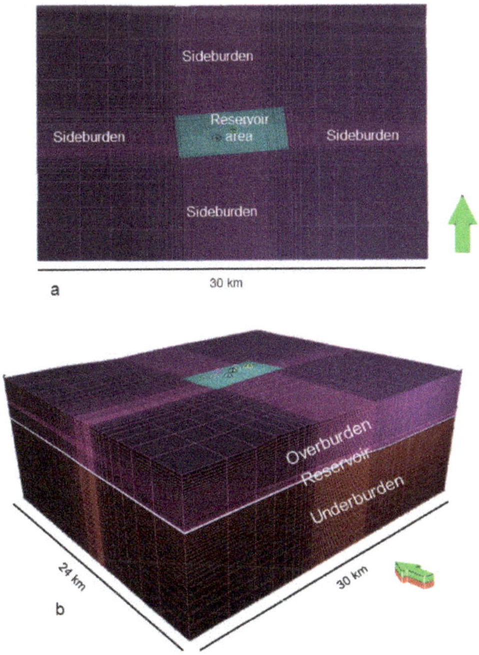

Figure 1. Reservoir model proper embedded in 3D geomechanical model with reservoir, overburden, underburden, and sideburden zones: (**a**) is the top view, and (**b**) is the oblique view, the arrows represent north direction [10].

Further details regarding the model setup, population of the model from 1D MEMs, history match, etc., are explained in our previous publication [10]. The starting point of the further modelling study in the following sections is the state after the replenishment phase, achieved by a history match of the production and pressure data from the production and subsequent shut-in phases, respectively.

4. Modelling

4.1. Modelling Scenarios

The following section describes the dynamic fluid flow models setup for future scenario testing cases designed for short-term (weekly) gas storage operations. The pressure profile of different future testing scenarios can be coupled with and incorporated into the THM model. The concept for these scenario tests is to evaluate geomechanical stresses on the reservoir due to pressure changes with intensive injection/production cycles. There are various models with different short-term cases that have been considered. German data for excess electricity throughout 2017 have also been considered in one case to address the issues of renewable energy aspects. This case implies that the excess of power energy (electricity) in Germany can be stored in underground gas storage with the power-to-gas (PtG) concept, and then the stored gas can be reused for power generation (gas-to-power) when needed. All the cases are summarized in Table 1. The starting pressure point of these future scenario testing cases is the end point pressure of the replenishment phase, i.e., ~15.8 MPa.

Table 1. All modelling scenarios with input parameters. WBHP is well bottom hole pressure, WGIR is well gas injection rate, and WGPR is well gas production rate.

Modelling Scenarios	Subdivisions		Input Parameters			
			WBHP upper limit (MPa)	WBHP lower limit (MPa)	WGIR (m^3/day)	WGPR (m^3/day)
Short-term (weekly) cases	Case A (with water-cut 5 m^3/day)	With three wells (two vertical wells, one horizontal well)	18.8	13.8	100,000	100,000
	Case B (without limited water-cut)	With three wells (two vertical wells and one horizontal well)	18.8	13.8	100,000	100,000
Real-world cases	Case C (with water-cut 5 m^3/day)	With one well	18.8	13.8	100,000	100,000
	Case D (without limited water-cut)	With one well	18.8	13.8	100,000	100,000

Short-term scenario cases represent weekly storage operations. The scenario scheme has been designed to compensate for the excess power produced in a season and to store power-to-gas energy into same underground gas storage. A short-term cycle consists of phases of one week of injection, one week of shut in, two weeks of production, and one week of shut (1wkInj-1wkShut-2wkProd-1wkShut) for one year. During the injection week, gas is injected into the reservoir, which builds up field pressure (but again limited by 18.8 MPa, the upper limit of WBHP); then, one week of shut in maintains the pressure, followed by a two-week production phase to withdraw gas, which drops the field pressure (lower limit constraint to 13.8 MPa), and again a shut in phase to maintain the well bore pressure.

4.2. Case A

There are three wells considered for this scenario: two wells are vertical wells (X2 and X6), and one is a horizontal well (H1). All the wells have been considered to have the same short-term weekly schedule as discussed above. This case comprises the same schedule as discussed above, but the well water production rate (WWPR) is limited to 5 m^3/day to take into consideration the economic aspects of operating the gas storage. The commercial storage industry limits the water production rates to minimize operating costs and enhance economic returns. Therefore, this aspect has also been considered in this case study. The gas rate for injection and production both is 100,000 m^3/day. Bottom hole pressure is constrained by an upper limit of 18.8 MPa and a lower limit of 13.8 MPa in cases of injection and production, respectively. The water cut-off is again 5 m^3/day in the production phase to limit the production of water from each well.

The FPR profile for this case is shown in Figure 2. The progressive oscillation cycles of FPR are injection (upward) and production (downward) phases. The overall upward trend of FPR from its initial pressure of ~15.39 MPa represents the buildup pressure with each passing schedule cycle.

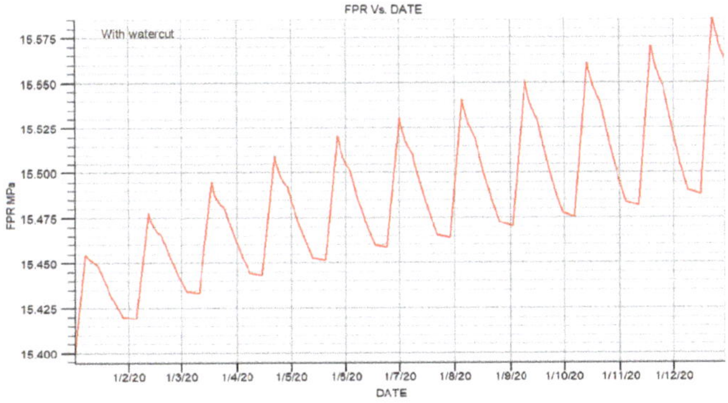

Figure 2. Field pressure (FPR) profiles of all three wells (X2, X6, and H1) with schedule 1wkInj-1wkShut-2wkProd-1wkShut for one year.

The upward trend of FPR throughout the schedule year is due to the well water production rate (WWPR) (Figure 3d), and as a result, the field pressure is not stabilized in such a short time span. To maintain material balance in this scenario, the injection rate should be lowered to the actual production rates of each cycle.

The comparison of the ptoperties of all three wells is summarized in Figure 3. The WBHP of X6 reaches the maximum limit of 18.8 MPa at the end of each injection cycle and drops back to ~15 MPa at the end of the production phase. The WBHP of X2 also shows similar behaviour but at a lower pressure; e.g., it varies between 16 MPa and 17 MPa at the end of each injection phase and drops back to the same level of ~15 MPa. The WBHP behaviour of H1 is however different from that of both X2 and X6. It reaches a maximum value of 16 MPa at the peak injection time and drops to the lowest level of ~14 MPa (Figure 3a). The WGPR behaviour of all three wells is similar with respect to linear increases with each increasing cycle. However, the rates are completely different for each well. The WGPR of well X2 ranges within ~10,000–11,000 m^3/day during the initial cycles but reaches up to 18,000 m^3/day at the end of the schedule. Contrarily, the WGPR of well X6 ranges within ~5000–6000 m^3/day during the initial cycles but reaches up to ~9000 m^3/day at the last cycle. An entirely different behaviour of WGPR is exhibited by well H1, showing ~50 m^3/day during the initial cycles but reaching up to ~1100 m^3/day at the end of the schedule year (Figure 3b).

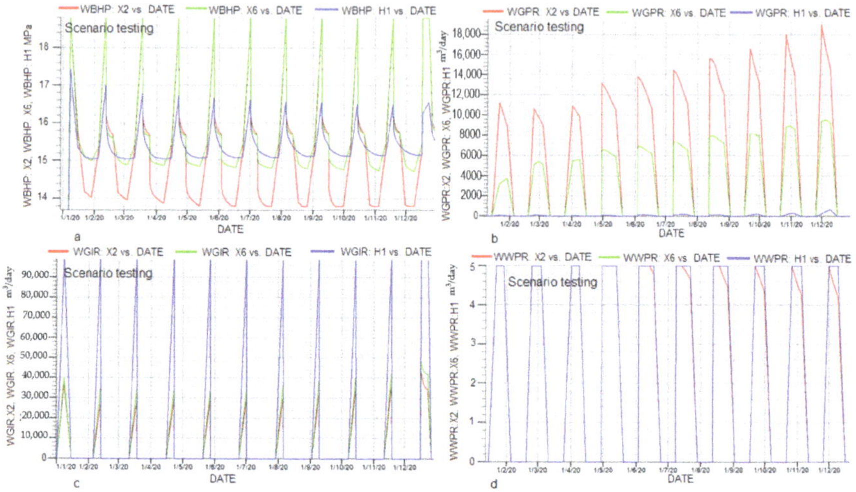

Figure 3. Well properties comparison of the three-well case scenario with the short-term schedule 1wkInj-1wkShut-2wkProd-1wkShut for one year. X2 and X6 are vertical wells, and H1 is the horizontal well: (**a**) well bottom hole pressure (WBHP) of all wells; (**b**) production gas rates (WGPRs) for all three wells; (**c**) injection gas rates (WGIRs) for all three wells and; (**d**) well water production rate (WWPR) for all the wells.

The WGIR profile of each well is completely different from the profile of the WGPR; the WGIR of well H1 shows the highest WGIR rates among other vertical wells (X2 and X6). The WGIR of H1 reaches to the maximum rate of 100,000 m^3/day. The WGIR of well X2 and X6 reach a maximum of 40,000 m^3/day and 50,000 m^3/day, respectively (Figure 3c).

The comparison shows that the horizontal well allows more gas injection and less gas production than the vertical wells and vice versa.

Results

The modelling results of this case are presented in the form of pore pressure and effective stress changes of the top layer of the reservoir. Two-time steps have been selected for the analyses of changes in pore pressure and effective stresses acting on the reservoir. Time step t1 (16 December 2020) represents the lowest pressure during the production phase of the schedule cycle, and t2 (23 December 2020) indicates the maximum injection pressure. The fluctuations in pore pressure and effective stress on the reservoir during t1 and t2 are the main results of this model.

Figure 4 shows the locations of three wells, which are denoted by H1, X2, and X6. The pore pressures at t1 and t2 for the well H1 are ~15.0 MPa and ~15.3 MPa, respectively, whereas the effective stress values are ~29.3 MPa at t1 and about 29.0 MPa at t2. There is a difference of about ~0.3 MPa for both pore pressure and effective stress from t1 to t2.

The vertical wells X2 and X6 are close to each other; therefore, the differences in the change in pore pressure and effective stress at these well locations are negligible. The values of pore pressure at both well locations at t1 and t2 are about 15.4 MPa and ~15.7 MPa, respectively. The effective stresses at t1 and t2 are ~28.5 and ~28.2 MPa, respectively, at both well locations. There is an increase in pore pressure of ~0.3 MPa from t1 to t2 and a decrease in effective stress of about 0.3 MPa.

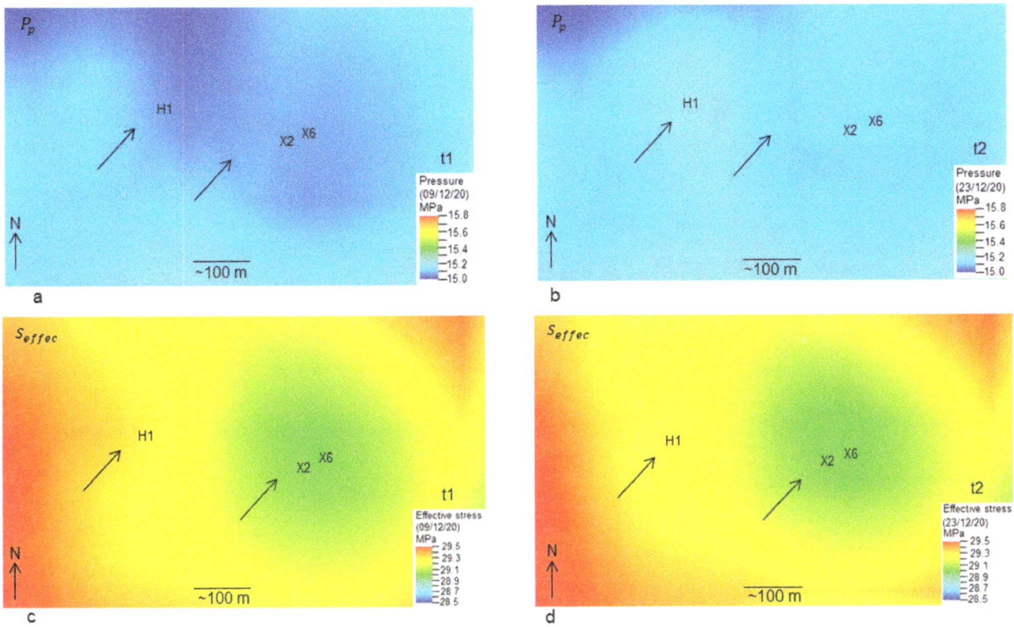

Figure 4. Pore pressure (P_p) and effective stress (S_{effec}) changes from t1 (16 December 2020) to t2 (23 December 2020) in the short-term case with three wells (X2, X6, and H1) with a water cut-off rate of 5 m^3/day. The arrows show the location of maximum observed fluctuations in P_p and S_{effec} from t1 to t2. The color scale is in MPa. (**a**) is pore pressure at time t1; (**b**) is pore pressure at time t2; (**c**) is effective stress at time t1; (**d**) is effective stress at time t2.

4.3. Case B

Three wells are considered for this scenario, two of which are vertical wells (X2 and X6), and one is a horizontal well (H1). All wells are assumed to have the same short-term weekly schedule as described above. WGIR and WGPR both have same value of 100,000 m^3/day; however, they are constrained by the upper limit of BHP of 18.8 MPa and the lower limit of BHP of 13.8 MPa in the cases of injection and production, respectively.

The FPR profile for this case is shown in Figure 5. The progressive oscillation cycles of the FPR are injection (upward) and production (downward) phases. The general trend of the FPR remains within the limits of ~15.85 MPa and ~15.375 MPa during the injection and production phases, respectively.

The comparison of the properties of all three wells is summarized in Figure 6. The WBHP of X6 reaches the maximum limit of 18.8 MPa at the end of each injection cycle and falls back to ~13.8 MPa at the end of the production phase. The WBHP of X2 also shows almost similar behaviour. However, the WBHP behaviour of H1 is different from both X2 and X6. It reaches a maximum value of 17.4 MPa at peak injection time and drops to the lowest level of ~13.8 MPa (Figure 6a). The WGPR behaviour of all three wells is similar in terms of linear increase with each increasing cycle. However, the rates are completely different for each well. The WGPR of well X2 is ~12,000–16,000 m^3/day during the initial cycles but reaches up to 24,000 m^3/day at the end of the schedule. In contrast, the WGPR of well X6 varies between 14,000 and 15,000 m^3/day during the initial cycles and reaches up to 22,000 m^3/day at the end of the last schedule cycle. The WGPR of well H1 is entirely different from the other two wells because it shows ~400–500 m^3/day during the initial cycles but reaches up to ~70,000 m^3/day at the end of the schedule year (Figure 6b).

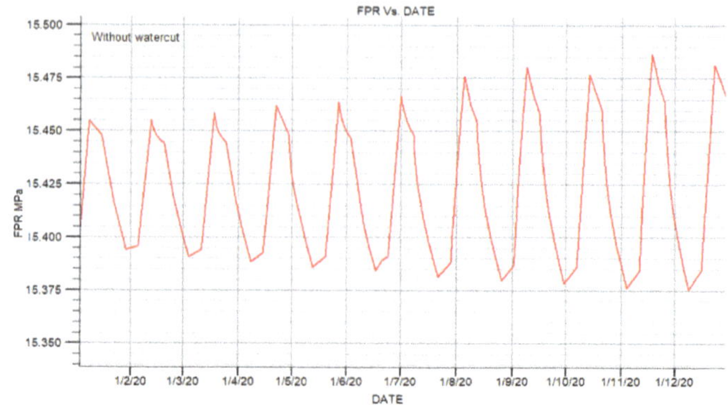

Figure 5. Field pressure (FPR) profile of all three wells (X2, X6, and H1) with schedule 1wkInj-1wkShut-2wkProd-1wkShut for 1 year.

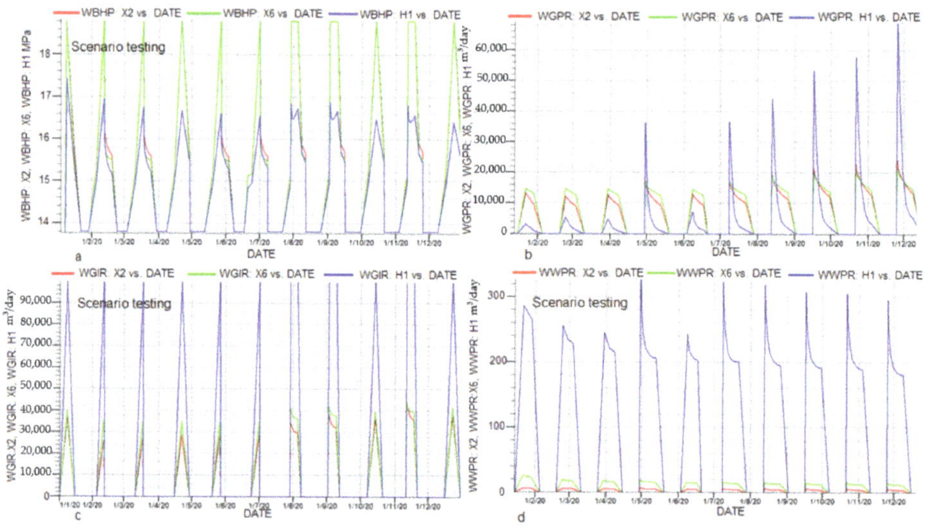

Figure 6. Well properties comparison of the three-well case scenario with short-term schedule 1wkInj-1wkShut-2wkProd-1wkShut 1 year. X2 and X6 are vertical wells, and H1 is the horizontal well: (**a**) well bottom hole pressure (WBHP) of all wells; (**b**) production gas rate (WGPR) for all three wells; (**c**) injection gas rate (WGIR) for all three wells and; (**d**) well water production rate (WWPR) for all the wells.

The WGIR profile of each well is completely different from the WGPR profile; the WGIR of well H1 shows the highest WGIR among the other vertical wells (X2 and X6). The WGIR of well H1 reaches the maximum value of 100,000 m^3/day. The WGIRs of wells X2 and X6 reach maximum values of 44,000 m^3/day and 45,000 m^3/day, respectively (Figure 6c). The comparison shows that the horizontal wells allow for higher WGIR and WGPR than the vertical wells under the same WBHP conditions.

The WWPR of well X2 allows a maximum rate of 5 m^3/day throughout schedule year, and the WWPR of well X6 also remain constant over the schedule year with a rate of 10 m^3/day, while the WWPR of well H1 remains higher, within the range of 250 m^3/day to 310 m^3/day (Figure 6d).

4.3.1. Results

The modelling results of this case are presented in the form of pore pressure and effective stress changes of the top layer of the reservoir. Two time steps with the greatest fluctuations in pore pressure have been selected. Time step t1 (9 December 2020) corresponds to the lowest pore pressure point, and t2 (23 December 2020) corresponds to the highest pore pressure point of the schedule cycle. The fluctuation in pore pressure and the effective stress on the reservoir during time t1 and t2 are the main results of this model.

Figure 7 shows the locations of three wells, which are denoted by H1, X2, and X6. These three wells are the operating wells for this scenario. The main changes in pore pressure, along with the effective stress, occur around these wells. The pore pressures at t1 and t2 for well H1 are ~15.5 MPa and ~15.4 MPa, respectively, whereas the effective stress values are ~29.1 MPa at t1 and about ~28.7 MPa at t2. There is a difference of about ~0.4 MPa for both pore pressure and effective stress at t1 and t2.

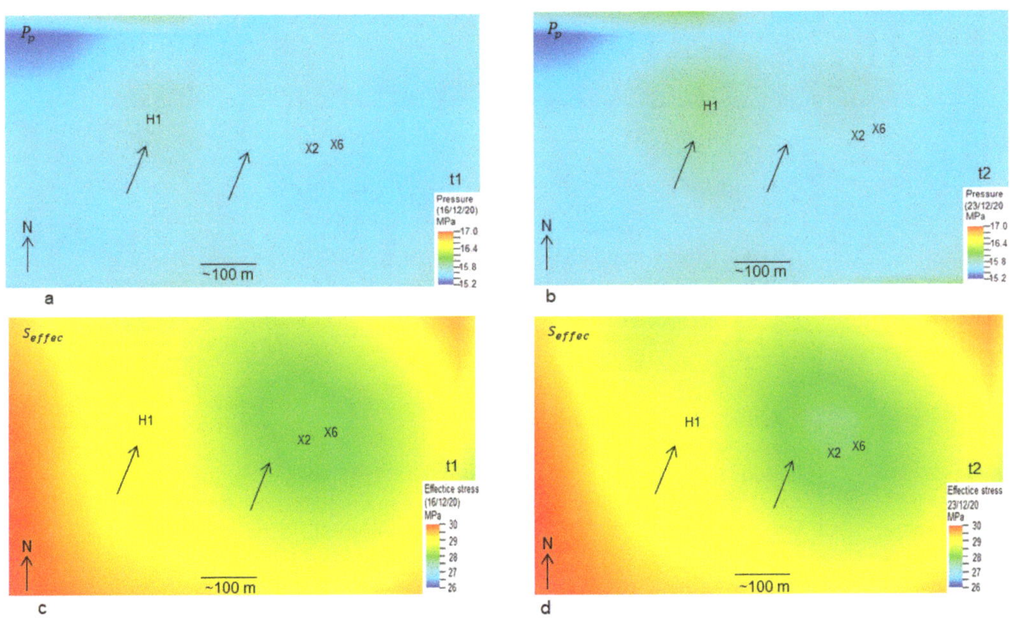

Figure 7. Pore pressure (P_p) and effective stress (S_{effec}) changes from t1 (9 December 2020) to t2 (23 December 2020) in the short-term case with three wells (X2, X6, and H1) without a water cut-off rate of 5 m^3/day. The arrows show the location of maximum observed changes in P_p and S_{effec} from t1 to t2. The color scale is in MPa. (**a**) is pore pressure at time t1; (**b**) is pore pressure at time t2; (**c**) is effective stress at time t1; (**d**) is effective stress at time t2.

4.3.2. Real World Cases

The excess of electricity produced [32,33] in Germany could be stored in underground gas storage by converting the power energy into gas (Power-to-Gas). PtG is a process of generation of a gas with high energy density through the electrolysis of water. The first intermediate product is power-to-hydrogen, which can be converted into synthetic methane gas power-to-methane in a subsequent methanation process that requires injection of CO_2. In this way, the same seasonal underground gas storage can also be used as a battery for excess energy in a calendar year. Figure 8 shows Germany's data on excess electricity produced in calendar year 2017. It can be seen from the data that, during the summertime (from March till August), electricity produced from renewable sources, such as wind and solar, increases enormously. The combined wind energy (onshore and offshore) shows high variation during the first and fourth quarters of the year, meaning this high variation of electricity production from renewable sources can be stored (power-to-gas) and reused (gas-to-power) in cases of excesses and shortages of electricity, respectively.

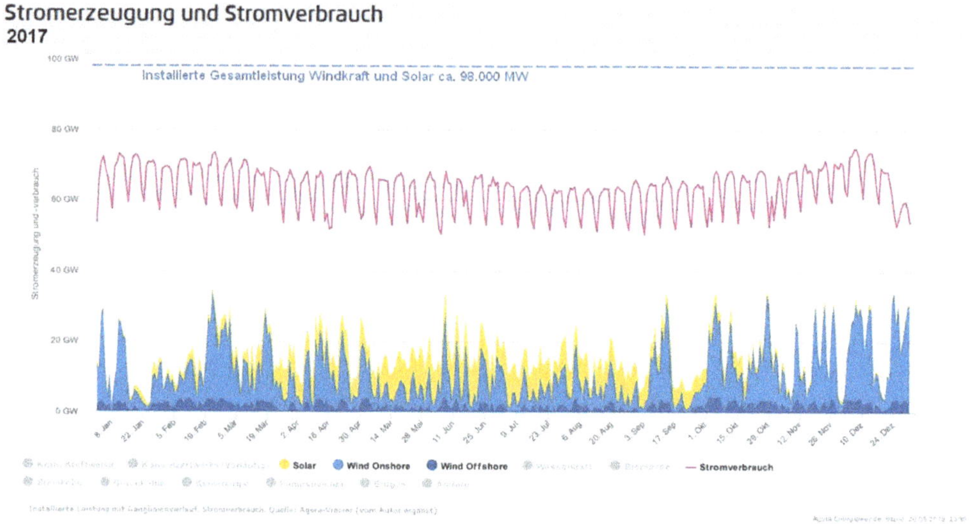

Figure 8. Data show the consumption of total electricity in Germany, along with electricity produced from renewable energy sources (such as wind, either offshore or onshore, and solar) in calendar year 2017 [32,33].

These scenarios have been performed on well X6. The schedule is based on the data shown in Figure 8. The baseline of 10 GW is the shut-in period. Greater than the 10-GW baseline is considered to have an excess of electricity that can be used as injection periods for power-to-gas storage, and less than the limit of 10 GW is a shortage of electricity. These periods have been considered for the production of gas for gas-to-power conversion. These data [32,33] on excess electricity from Germany have been used to conduct two short-term real world case schedule cases: one with limited water cut-off and one without limited water cut-off.

4.4. Case C

This scenario has been performed on vertical well X6. The schedule timeline of one year has been adopted from Figure 8. The well water production rate (WWPR) is restricted to 5 m^3/day (Figure 9c). The FPR eventually increases in this case to up to 15.72 MPa with slight variations during the gas production cycles. Gas injection succeeds to maintain the pressure, and it increases from ~15.39 MPa and reaches approximately ~15.72 MPa at the end of the one-year period (Figure 9a). The variability in WBHP is directly proportional to the gas production cycles. As shown in Figure 9a, as the gas production increases, the well bottom-hole pressure decreases even as the gas injection continues. The maximum WBHP reaches a value of 18.8 MPa, the minimum WBHP reaches a value of 14.4 MPa, and these maximum and minimum pressure values represent injection and production cycles (Figure 9b). The well gas injection rate (WGIR) varies during the year, having a minimum injection rate of approximately 32,000 m^3/day and a maximum rate of about 66,000 m^3/day. In contrast, the well gas production rate (WGPR) has minimum and maximum values of 4300 m^3/day and 19,000 m^3/day, respectively (Figure 9d). The lower WGPR compared to the WGIR is due to the limited WWPR, which does not allow the well to produce at a higher WGPR.

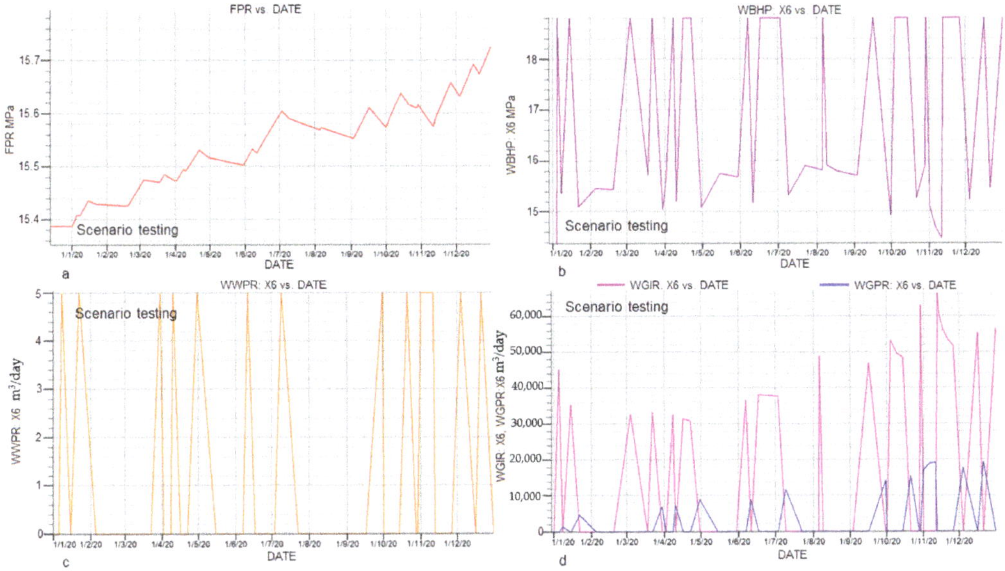

Figure 9. The fluctuation in electricity produced in Germany from renewable sources in 2017 is modelled into a future scenario testing case with limited water cut-off (5 m^3/day). The excess of energy can be stored in UGS and can be used when needed. This schedule is helpful to understand in which month of the year energy can be stored as gas in UGS and in which month of the year this energy can be utilized when shortage occurs: (**a**) field pressure (FPR) profile for these cycles; (**b**) well bottom hole pressure (WBHP) of well X6; (**c**) well water production rate (WWPR) for well X6; and (**d**) well gas injection rate (WGIR) and well gas production rate (WGPR) for well X6.

Results

The modelling results of this case are presented in the form of the pore pressure and effective stress changes in the top layer of the reservoir. Two time steps have been selected for the conclusion of the results for this model. Time step t1 is the starting point of the schedule case, i.e., 1 January 2020, and t2 is the end schedule point (31 December 2020). The pore pressure at the well X6 location is about 15.2 MPa at t1 and increases to about 15.8 MPa at t2; simultaneously, the effective stresses at well X6 is about 28.6 MPa, and it decreases to about 28.0 MPa at time steps t1 and t2 (Figure 10). There is an increase of 0.6 MPa in pore pressure and decrease of 0.6 MPa in effective stress at the top surface of the reservoir layer near well X6 from t1 to t2.

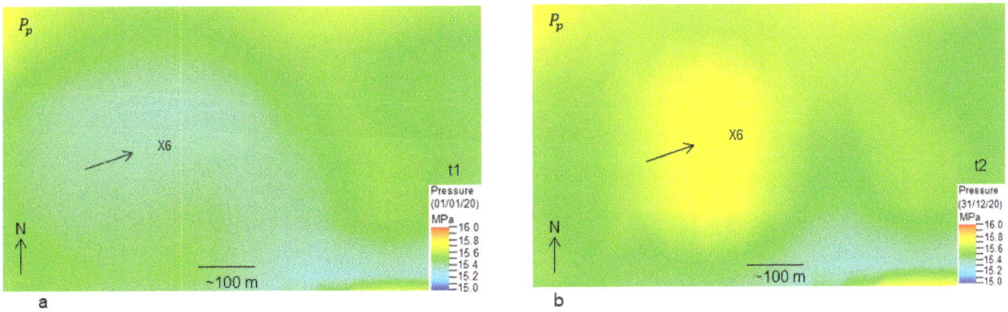

Figure 10. *Cont.*

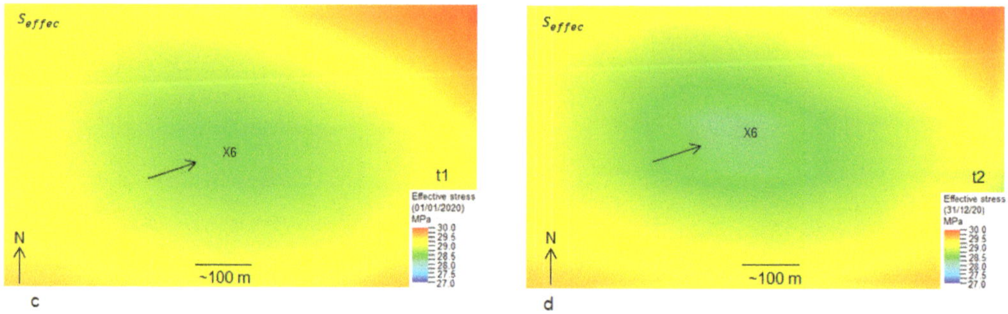

Figure 10. Pore pressure (P_p) and effective stress (S_{effec}) changes from t1 (1 January 2020) to t2 (31 December 2020) in the short-term case with one well (X6) with a water cut-off rate of 5 m^3/day and a random schedule. The arrows show the location of the maximum observed fluctuation in P_p and S_{effec} from t1 to t2. The color scale is in MPa. (**a**) is pore pressure at time t1; (**b**) is pore pressure at time t2; (**c**) is effective stress at time t1; (**d**) is effective stress at time t2.

4.5. Case D

This scenario has been performed on vertical well X6. The schedule timeline of one year has been adopted from Figure 8. There was no water cut-off rate limit set in this scenario; hence, the maximum water production rate (WWPR) increases up to 43 m^3/day and remains less than 20 m^3/day throughout the production and injection period of one year, respectively (Figure 11c). The FPR is sustained in this case by gas injection and increases to up to a maximum value of 15.62 MPa with slight variation during the gas production cycles in the one-year period (Figure 11a). The alteration in WBHP is directly proportional to the gas production cycles without a water cut-off limit. As shown in Figure 11b, the WBHP reaches a maximum value of 18.8 MPa and a minimum value of approximately 13.8 MPa. The well gas injection rate (WGIR) varies during the year, having a minimum injection rate of approximately 32,000 m^3/day and a maximum rate of about 60,000 m^3/day, whereas the well gas production rate (WGPR) has minimum and maximum values of 16,000 m^3/day and 40,000 m^3/day, respectively (Figure 11d).

Results

The modelling results of this case are presented in the form of pore pressure and effective stress changes of the top layer of the reservoir. Two time steps have been selected for the conclusion of the results for this model. Time step t1 is the starting point of the schedule case, i.e., 1 January 20, and t2 is the end schedule point (31 December 2020). The pore pressure at the well X6 location is about 15.1 MPa at t1 and increases to about 15.7 MPa at t2, whereas the effective stress at well X6 is about 28.7 MPa, and it decreases to about 28.1 MPa at time steps t1 and t2. There is an increase of 0.6 MPa in pore pressure and a decrease of 0.6 MPa in effective stress at the top layer of the reservoir around well X6 from t1 to t2 (Figure 12).

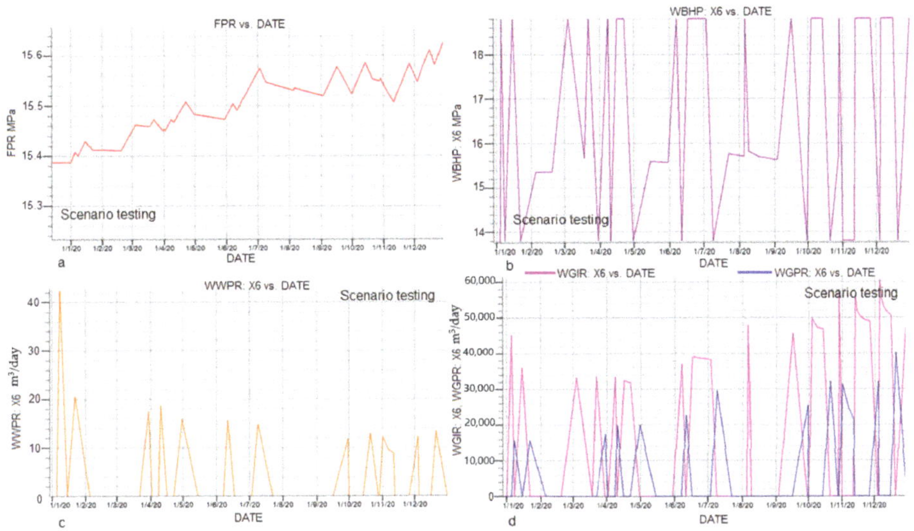

Figure 11. The fluctuation of electricity produced in German from renewable sources in 2017 is modelled into a future scenario testing case without limited water cut-off. The excess of energy can be stored in UGS reservoirs and can be used when needed. This schedule is helpful to understand which month of the year's energy can be stored as gas in UGS and in which month of the year this energy can be utilized when shortages occur: (**a**) field pressure (FPR) profile for these cycles; (**b**) well bottom hole pressure (WBHP) of well X6; (**c**) well water production rate (WWPR) for well X6; and (**d**) injection (WGIR) and production (WGPR) gas rates for well X6.

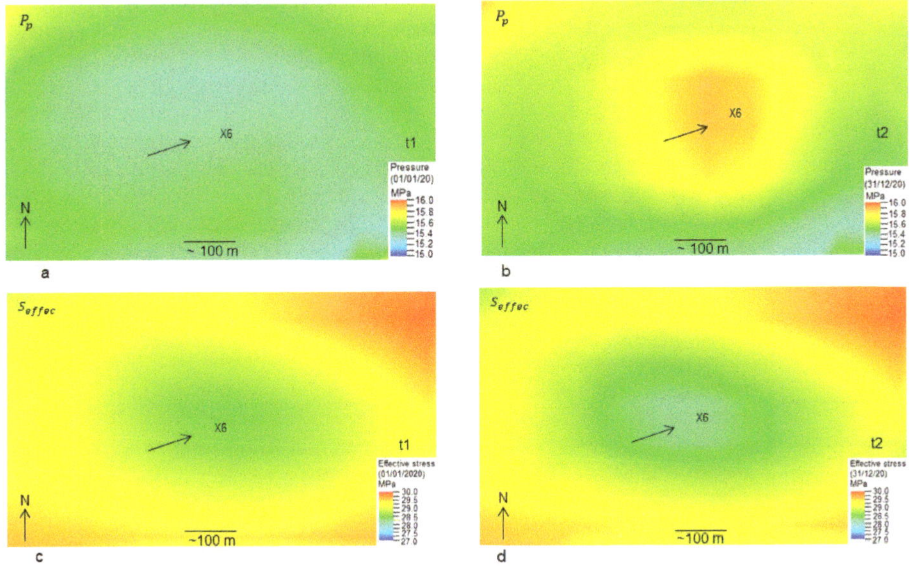

Figure 12. Pore pressure (P_p) and effective stress (S_{effec}) changes from t1 (1 January 2020) to t2 (31 December 2020) in short-term case with one well (X6) without a water cut-off rate of 5 m^3/day with a random schedule. The arrows show the location of the maximum observed fluctuation in P_p and S_{effec} from t1 to t2. The color scale is in MPa. (**a**) is pore pressure at time t1; (**b**) is pore pressure at time t2; (**c**) is effective stress at time t1; (**d**) is effective stress at time t2.

The summary of all the results is compiled in Table 2 to have better understanding of pore pressure and effective stress changes of all future testing cases with time (t1 to t2).

Table 2. Summary of results for all future test scenarios. The sign + in the pore pressure changes indicates a positive change or an increase in pore pressure from time step t1 to t2, while the sign—in the changes in effective stresses denotes a decrease in magnitudes of effective stresses for time step t1 to t2. These two quantities are inversely proportional to each other and are expressed in MPa and KPa for a better understanding of the changes.

Modelling Scenarios	**Subdivisions**		**Results**	
			Pore pressure changes ΔP_p	Effective stress changes ΔS_{effec}
Short-term (weekly) cases	Case A	With three wells (two vertical wells, one horizontal well)	+0.3 MPa +300 KPa	−0.3 MPa −300 KPa
	Case B	With three wells (two vertical wells, one horizontal well)	+0.4 MPa +400 KPa	−0.4 MPa −400 KPa
Real-world cases	Case C	With one well	+0.6 MPa +600 KPa	−0.6 MPa −600 KPa
	Case D	With one well	+0.6 MPa +600 KPa	−0.6 MPa −600 KPa

5. Thermal Analysis

The same dynamic model has been used for thermal analyses. Since long-term injection would impact the thermal changes in the reservoir significantly, a long-term seasonal case is used to analyse the temperature changes within the reservoir if a foreign gas is injected into it. Therefore, six months of gas injection and six months of gas withdrawal are considered in this modelling case. The initial reservoir temperature is ~45 °C, and the foreign gas temperature is 25 °C. Two cycles have been considered to analyse the temperature changes during these injection/production operations. Gas is injected into the reservoir for the first half year and produced in the second half of the year. Two wells, X2 and X6, are considered to analyse the temperature changes around the well bore vicinity. The bottom hole pressure (WBHP) for both wells is set to an upper limit of 18.8 MPa and a lower limit of 13.8 MPa in case of the injection and production phases, respectively. These pressure limits are set in place to avoid fault reactivation or fracture-inducing phenomena during the injection phase, as well as to avoid sand production or contraction of the reservoir during the production phase. The WGIR and WGPR are set to 100,000 m^3/day for both wells.

Thermal stresses are the stresses that occur due to the change in temperature in the system, i.e., original temperature minus final temperature. If foreign gas is injected into the underground gas reservoir, the temperature in the reservoir changes, which causes thermal-related stress changes in the reservoir. The relationship of temperature changes and thermal stress is expressed by the following equation [34]:

$$S_t = E * \alpha_t (T_f - T_0) = E * \alpha_t (\Delta T), \quad (11)$$

In the above equation, S_t is thermal stress, E is the Young's modulus, α_t is the thermal coefficient, T_f is the final temperature of the reservoir, T_0 is the original temperature of the reservoir, and ΔT is the temperature difference in the reservoir. Less ΔT causes less thermal stress in the reservoir and vice versa.

The following section describes the results, i.e., temperature changes with injection of colder foreign gas (25 °C) into the reservoir (i.e., about 45 °C) through space and time with seasonal cyclic injection/production phases. The top view of the reservoir surface is shown in the figure at different time steps (Figure 13). Two injection/production cycles with four-time steps have been selected to show thermal changes in the reservoir with injection and production phases. Time step t1 (1 January 2020) is the pre-operational history temperature of the reservoir at wells X2 and X6 (Figure 13a). Time step t2 (1 July 2020) represents the end of the injection time of colder foreign gas (25 °C), which is injected for the first half of 2020 (Figure 13b). Time step t3 (1 January 2021) is the end of the production period of the cycle (Figure 13c), and t4 (1 July 2021) is again the end of injection phase of the second cycle (Figure 13d).

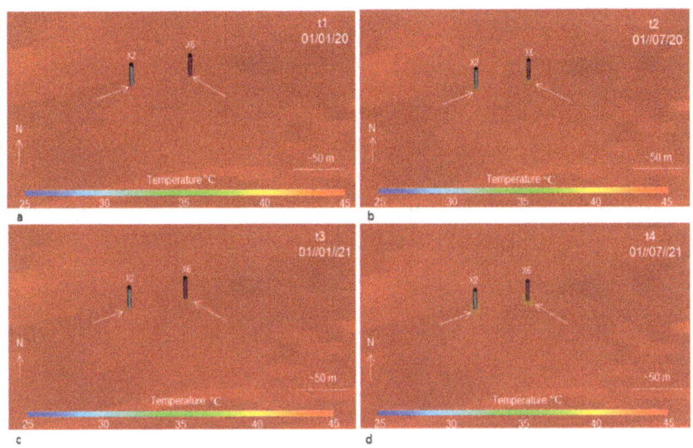

Figure 13. Temperature changes around wells X2 and X6 by injecting colder foreign gas at different time steps (t1, t2, t3, and t4). Time steps t1, t2, t3, and t4 correspond to 1 January 2020, 1 July 2020, 1 January 2021, and 1 July 2021, respectively. The arrows show the exact location of significant temperature differences during injection/production phases. The colour scale is in °C, whereas the arrows with N indicate a northward direction. (**a**) is temperature at time t1; (**b**) is temperature at time t2; (**c**) is temperature at time t3; (**d**) is temperature at time t4.

The reservoir temperature is 45 °C at time step t1, which is the pre-operational temperature. The temperature decreases to about 43 °C at well X6 and 42.5 °C at well X2 at time step t2 after constant injection of colder gas (25 °C), with a well gas injection rate (WGIR) of 100,000 m^3/day for six months. The temperature increases to about 43.5 °C and ~43 °C at wells X6 and X2, respectively, at time step t3. There is only about a 0.5 °C increase in temperature from t2 to t3. Temperatures at well X6 and X2 decrease to about 41.5 °C and 42 °C, respectively, again in the second cycle of injection at t4. Thermal changes are minor and occur only at the vicinity and around the well locations. The thermal effects on the stress are not significant in the reservoir even after injection of 100,000 m^3/day of colder gas for about a half year. This outcome shows that the thermal changes in the short-term cases are negligible for analysing the geomechanical stresses on the reservoir in storage operations.

6. Potential Fault Reactivation Analyses

6.1. Model Setup

Fault reactivation is the possibility of failure in geomechanical assessment of the reservoir, which can risk operational safety, cause micro seismicity within and around reservoirs, and provide a leakage path for gas to escape. Fault reactivation occurs when the shear stress acting on the fault planes exceeds the shear strength of the fault. The

Mohr–Coulomb failure criterion relationship of pore pressure and the principal stresses of this case study reservoir (lies in normal stress regime) are expressed by this equation [35]:

$$P_p = \frac{1}{\alpha}[\frac{1}{2}(S_v + S_{hmin}) + \frac{1}{2}(S_v - S_{hmin})\cos 2\theta - \frac{1}{2}(S_v - S_{hmin})\frac{\sin 2\theta}{\mu}], \quad (12)$$

where α is the Biot coefficient (assumed 1), S_v is the vertical and maximum principal stress, S_{hmin} is the minimum horizontal stress, θ is the angle between the dip line of the fault and the S_{hmin} direction, P_p is the critical pore pressure, and μ is the coefficient of friction.

These analyses include the calculation of the critical pore pressure with the aim of observing possible differences in pore pressure required for fault reactivation. The pore pressure derived from the history matching scenario is multiplied by a fixed factor controlled by gas rates until fault reactivation occurs. The upper limit of the BHP is removed to obtain a higher pore pressure. The factors used for this operation are 1.15, 1.25, and 1.5 (Figure 14). It is then possible to evaluate pressure changes in the reservoir required to reactivate the fault, as well as the safe storage capacity of the reservoir.

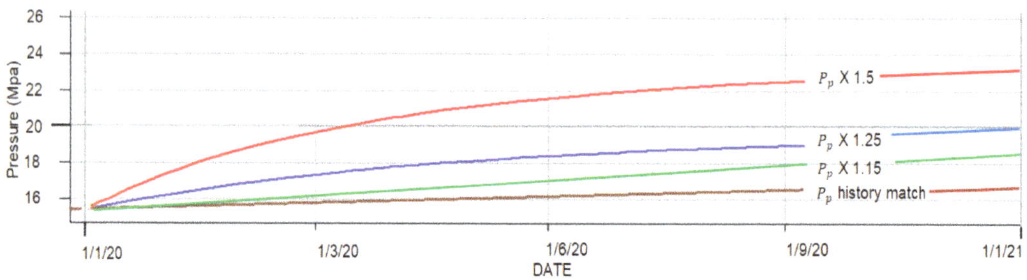

Figure 14. Pressure profiles for the history matching scenario and scenarios multiplied by fixed factors (P_p, P_p X 1.15, P_p X 1.25, and P_p X 1.5) controlled by gas injection rates until fault reactivation occurs.

Figure 14 shows different pressure profiles based on distinct gas injection scenarios. The P_p (history match pressure) curve exhibits a maximum value of ~16.2 MPa with gas injected at a rate of 100,000 m^3/day for one year, whereas to reach a pressure value 1.15 times the actual history match pressure case, 175,000 m^3/day of gas are injected for 1 year. The pressure increased up to 18.6 MPa. In case P_p X 1.25, 240,000 m^3/day is injected to reach a pressure of about 20.25 MPa. Similarly, in case of P_p X 1.5, a gas volume of 560,000 m^3/day is injected (for one year), which increases the pressure up to 24.3 MPa.

6.2. Results

The results of this section are based on the model setup explained in the above section. The methodology follows the different cases in which the initial reservoir pore pressure is multiplied by a factor of 1, 1.15, 1.25, and 1.5, corresponding to pore pressure values of ~16.2 MPa, ~18.6 MPa, ~20.25 MPa and ~24.3 MPa, respectively. The results display the oblique view of the topmost layer of the reservoir (Figure 15). The Mohr circles correspond to the well X6 location near the main fault (Figure 15). Two time steps have been considered for all the cases: t1 (1 January 2020) is the starting time step, and t2 (1 January 2021) is the end time step of the schedule year.

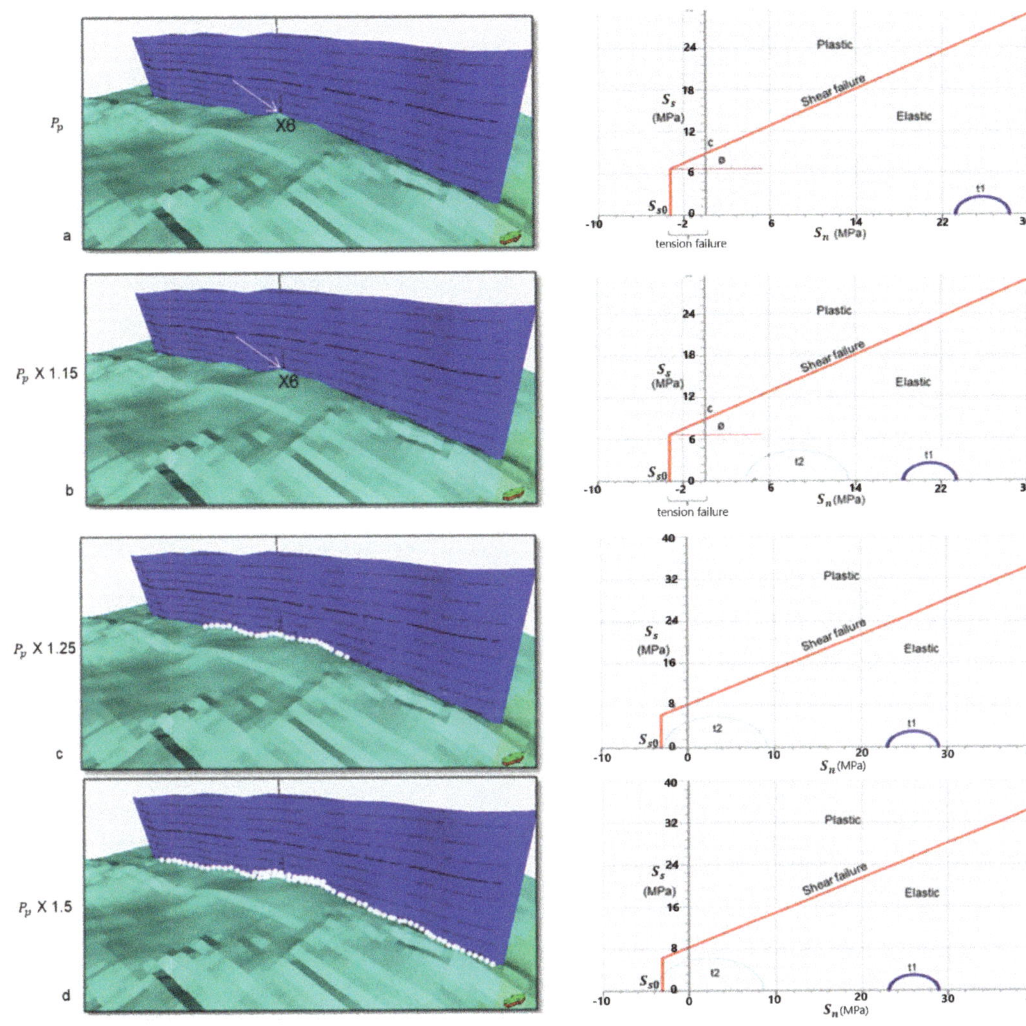

Figure 15. The computation of fault reactivation for initial pore pressure multiplied by a factor of 1 (**a**), 1.15 (**b**), 1.25 (**c**), and 1.5 (**d**) is shown on the left side. White boxes show the cells with stress states exceeding the failure criterion (arrows show northward direction). On the right side of the figure are the shear stress S_s vs. normal stress S_n diagrams, showing the Mohr–Coulomb failure criterion at the well X6 location (which is nearest the main fault of the reservoir). Time steps t1 and t2 correspond to the starting (1 January 2020) and final (1 January 2021) time steps of the tested cases, respectively. The increase in pore pressure leads to a decrease in effective stresses, causing the corresponding Mohr circle to shift to the left. If the failure line is finally touched, plastic straining and—in case of a fault zone—fault reactivation occur.

Fault reactivation is observed already at P_p X 1.25 (~20.25 MPa) (Figure 15b), at which the corresponding Mohr circle has just touched the tensile failure line. The amount of failure in the cells becomes more prominent as pore pressure increases to greater than the pore pressure factor of 1.25 (i.e., P_p X 1.5), and it causes the corresponding Mohr circles to move further left (Figure 15c,d). The increase in pore pressure causes decreases in effective stress, causing the Mohr circles to move to the tensile failure line. Thereafter, the material enters the plastic regime; therefore, in case of failure, fault reactivation occurs.

6.3. Safe Injection Rate for Safe Storage Capacity

The pore pressures for fault reactivation for the different scenarios calculated in the numerical modelling analysis provide an estimate of the pressure at different injection rates. The injection rate (in terms of volume rate) to achieve a perfect history match (16.2 MPa) is about 100,000 m^3/day, and fault reactivation is already observed at P_p times 1.25, i.e., about 20.25 MPa at an injection rate of 240,000 m^3/day. With the determination of the critical pore pressure, it is also possible to derive an upper limit for the injection rate to be selected for injection processes. By considering the highest safety margin, the injection rate between 100,000 m^3/day and 150,000 m^3/day would be the considered safe injection rate for safe storage for the case study reservoir. A gas injection rate greater than this threshold value can have a significant impact on the risk management and operational setup of underground gas storage.

6.4. Storage Capacity of Power-to-Gas and Gas-to-Power

Regarding the storage capacity of power-to-gas technologies, the case study reservoir can store 881,600 kWh/d up to maximum of 1,322,400 kWh/d of power from renewable or other resources with respect to the conversion of a natural gas volume of 100,000 m^3/day to a maximum of 150,000 m^3/day, respectively. Power-to-gas and gas-to-power convertible units are summarized in Table 3.

Table 3. Power-to-gas and gas-to-power convertible units.

m^3 Natural Gas	kWh Power
1	8.816
0.113	1

7. Discussions

7.1. Hydraulic Model

The results of the hydraulic model show variations in gas saturation at different phases during short-term injection-production cycles. Figure 16 shows that the injection-production cycle increases the gas saturation during the later cycles, as more gas is produced with increasing cycles until convergence is reached. This outcome indicates the adoption of high-speed multicycle injection-production in UGS with time. This characteristic of UGS indicates the increase in gas-containing pore space with time and explains encroachment of gas flow continuously into the ground water zone for production in the pore throat development area, as the injection-production cycle increases. Meanwhile, under the wetting function, the water film is concentrated in the small pore space compared to the larger pore space; the storage space occupied by the formation of water and dead gas zones is released by the gas execution. Further analysis of the data shows notable findings on the correlation between the time and the amplitude of the gas saturation increase, as shown in Figure 16.

During injection and production cycles, gas saturation increases from ~35% to ~50% from the first cycle to the sixth cycle. Therefore, it appears that, during multi-cycle injection and production in the pre-UGS gas zone, the pore gas storage space in reservoirs with low permeability generally increases sharply, as evident from the correlation between low permeability and high initial water saturation (according to [6]). Moreover, the higher the relative quantity of liquid phase that is continuously transported and dried during high velocity development, the greater the increase in gas saturation.

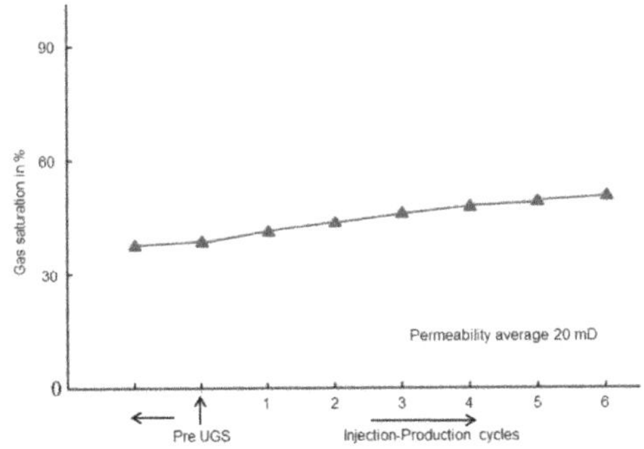

Figure 16. Gas saturation (%) during injection-production cycles.

The gas injection production capacity tends to improve in the UGS gas zone when gas production is increased during the injection production cycle. Further investigation of the UGS gas zone shows that the degree of gas production also increases with the injection-production cycle, but the degree of gas production varies according to the different physical properties and connectivity of the pore spaces [6]. Strong seepage is observed in reservoirs with high permeability, and the gas production and degree of gas recovery are also high under the same injection-production conditions [6].

The gas saturation first decreases in the pre UGS phase and then increases significantly during multi-cycle injection and production, showing the decrease in gas saturation and gas storage space when the gas-drive water zone (gas-water zone, where gas saturation is more dominant) is converted into gas storage. Due to the intrusion of lateral and bottom water, part of the gas storage space is occupied by water during the slow exploitation phase, and a capillary trap for the gas-containing space is also created. Further data show that higher permeability causes water to penetrate, leading to an even greater reduction in gas saturation. During the multi-cycle injection and production phase in UGS, there is no water encroachment, or the part where water encroachment occurs is displaced in the initial gas injection phase of gas storage into the gas-drive water zone. In contrast, under the influence of high velocity injection and production cycle extraction, residual water is produced that is like that in the gas zone before UGS [6]. The experimental analysis shows that the gas-drive water zone is the main expansion zone of the UGS, as the storage space and storage capacity are greatly increased in this gas-drive water zone [6].

The variation of gas saturation in the gas-drive water zone shows the same trends compared to both the trend in the single core experiment and the trends of average gas saturation in the parallel multicore experiment. However, in the multicore parallel experiment, the reduction in gas saturation caused by water encroachment is relatively small [6]. The distinctive feature of reservoir heterogeneity, a formation with high permeability, has a great impact on the water body as it promotes water intrusion, while formations with medium and low permeability are less affected by water intrusion. During high velocity injection and production, the gas preferentially percolates into the high permeability formation, resulting in gas saturation recovery in the high permeability reservoir.

The gas permeability of the reservoir may also decrease due to clay minerals in the reservoir, which may expand when they encounter water, thereby occupying the pore space. For this reason, gas production and gas content are relatively low in the first injection and production cycle. Meanwhile, the gas recovery level and gas production level are low due to lower vertical permeability [6]. Furthermore, gas production and injection increase with increasing permeability.

The transition of the reservoir characteristic of the gas-water transition may indicate that water intrusion is caused not only by exploitation of the gas field but also by the gas storage cycle in the operational phase. In general, the gas saturation decreases both in operational zones of the reservoir and in the gas-water zone. However, the effects of physical properties and heterogeneity, wettability, and capillary pressure of the reservoir make the production mechanism of multicycle injection complex. The relationship between gas and water in the gas-water transition zone tends to be rational, and the gas saturation tends to plateau during multicycle injection production [6]. Alternatively, residual gas and trapped gas are effectively reduced due to the back-and-forth movement of gas and water through the gas-containing spaces, but the changes in gas saturation are small throughout the reservoirs. Conversely, in high-speed injection and production, water is still continuously transported into the pore throat for production, but in gas storage, the main discharge concerns water in the large pore throat because it is difficult to displace water in the micropores. For this reason, gas saturation tends to plateau in the gas-water transition zone. Due to the reduction in water intrusion energy, gas saturation increases slightly in the subsequent phases of injection and production operations.

These results show that low gas saturation and poor recovery are observed in the gas-water transition zone of UGS and that effective exploitation of the gas-water transition zone is affected by reservoir heterogeneity [6]. Therefore, the effective utilization of the gas-containing pore space under the high-speed injection and production conditions of gas storage is significantly different from the utilization of the gas reservoir. For that reason, the optimal design of UGS capacity and calculation of gas storage parameters should be founded on the effective utilization of the gas-containing pore space, as well as on the effective pore space of the reservoir structure in different zones [6].

7.2. THM Coupled Modelling

The geomechanical approach presented in this paper allows for characterization, from a geomechanical point of view, of the target reservoir in a very understandable way. The coupled simulation of one-way flow and geomechanics (VISAGE™) focuses on understanding the variation in effective stresses due to pressure changes associated with gas injection into the subsurface reservoir. In other words, the focus is on the pore pressure changes during operation. The VISAGE™ model is created directly from the geological model and can therefore be considered more accurate in terms of geometry compared to another finite element model, such as an ANSYS™ model. The material properties in the VISAGE™ model are scaled up from borehole logs, and their population contains information about the original vertical and horizontal variations. The VISAGE™ simulator treats faults as 3D cells with different material parameters in relation to the surrounding rock. The output of the VISAGE™ model is continuous through space, and the presence of faults can be detected by an abrupt change in depth at the point of displacement, even if the horizon is technically not present at that point. Therefore, VISAGE™ is used for ECLIPSE™-VISAGE™ coupling to build a coupled THM model to achieve the desired geomechanical results.

THM coupled modelling involves coupling of static geomechanical model with dynamic fluid (hydraulic) modelling. The geomechanical stresses of the reservoir change in space and time when coupled with pore pressure, either from reservoir history or from future testing scenarios. In general, these dynamic models are based on the concept of effective stress to characterize the effects of pore pressure changes on reservoir and caprock integrity, as well as fault stability. Simplified geomechanical models rely mainly on the concept of effective stress to distinguish the effects of pore fluid changes on reservoir integrity and fault stability. Terzaghi [36] proposed this effective stress concept for soil systems. The result of the increase in pore fluid pressure is equal to the decreased effective stress on the rock mass, and conversely, this concept implies that the effective stress experienced by the rock framework is due to the changes in fluid pressure, with the absolute magnitudes of the principal in situ stresses themselves remaining unaltered.

From a classical point of view from Terzaghi [36], changes in vertical stress can be predicted as the effective vertical stress increases and decreases during withdrawal and injection, while the reservoir stress path effect can clearly distinguish the changes in the effective minimum horizontal stresses. As a result of the increased fluid pressure, the Mohr circle drifts to the left, in turn reducing the diameter of the circle, implying that the fault is under a lower differential stress. The theoretical viewpoint consolidates only the concept of effective stress, and the assumption can be made that the stress path during compression is parallel to the failure envelope rather than towards it [2]. A steeper stress path during the injection process is the result of a small Poisson's ratio, a large Biot coefficient, a large shear modulus, and a shallower reservoir, further causing drift away from the failure envelope.

At the case study reservoir, reactivation of the fault with increasing fluid pressure would not be possible if the stress path could be assumed to be exactly linear to higher fluid pressures. The magnitude of the stress path in the reservoir is an absolute criterion to ensure that the faults are not tapered by pressurization or depressurization. Furthermore, this linear extrapolation of the stress path is obviously only suitable for moderate fluid pressure levels. During pressurization of the system, there are some points at which the horizontal stress becomes the maximum principal stress instead of the vertical stress. This outcome shows that, in the absence of stress path effects, the stress path changes and moves towards the failure criterion.

One of the main aspects of the THM modelling results is the deformation within and outside of the reservoir formation because of injection and withdrawal of fluid/gas in reservoirs. Pressurization and depressurization cause deformations in both the reservoir and the overburden, which are part of the effective results of geomechanical modelling. During overpressure or underpressure, the elastic or non-elastic expansion or contraction of the rock mass causes such deformations. The reactivation of faults and the integrity of the rock may be affected by deformations within the reservoir. In addition, deformations associated with reservoir depletion can also pose a serious problem in terms of casing collapse and can become a major obstacle to subsequent drilling [37].

The main risks related to pressurization and depressurization of a field during injection and withdrawal periods, respectively, are fault reactivation and associated permeable leakages. During gas injection and production phases, reactivation of the fault and associated permeable pathways is one of the main risks associated with pressurization and depressurization. The classical fault model would predict that, as the fault approaches shear reactivation, there will be an increase in fluid pressure that would reduce the effective normal stress on the fault. However, the stability of the fault is significantly affected by the progression of reservoir stresses through the effects on horizontal stress magnitudes. As discussed in the above section, the simulation results of the THM model predict decreases in effective stresses but increases in pore pressure in various short-term storage cases. If this difference in effective stresses and consequently in pore pressures is very small, no significant stress perturbations would occur. However, if this difference is large, i.e., if the pore pressure in the reservoir is large, then there is a high probability of fault reactivation with increasing pore pressure, as described in the above regarding the various fault reactivation cases. Fault reactivation occurs in the reservoir already at 1.25 times the initial pore pressure of the reservoir, indicating that the Mohr circle reaches the failure line at this pressure.

Distinguishing deformations above and at the surface of the reservoir is also important for surface monitoring and for potential reactivation of faults in the reservoir section. These deformations (either elastic or sometimes plastic) occur due to contraction and expansion of the rock mass during over-pressuring and under-pressuring phases. Over- and under-pressuring of reservoirs are directly related to large amounts of gas injection and production rates, respectively. For this reason, the upper (18.8 MPa) and lower (13.8 MPa) limits of bottom hole pressure (BHP) are considered, which do not allow the pressure to rise or drop below the designated pressure limits. The deformation caused by this over-pressuring and under-pressuring of reservoirs could initiate fault reactivation and may affect caprock integrity. The significant drop in reservoir pressure due to rapid depletion may damage the casing and induce drilling complications for newly planned wells in the reservoir [37].

This study does not include a sensitivity analysis to determine the stress field or displacement of the ground surface when elastic properties vary. Due to the poor correlation between the mechanical properties of the rock and the porosity in the individual wells, there can be considerable variation in the construction of the static geomechanical model, which can affect the distribution of the mechanical properties, rather than the magnitudes. In contrast, the successive calibration of the sonic logs and the mechanical tests themselves appear to have close correlations, demonstrating their accuracy. However, it is possible that the actual values for Poisson's ratio and Young's modulus may differ slightly from those used in the modelling. There are several studies have already been performed on these parameter studies and the effects of parameter variation on geomechanical stresses and ground surface displacement [38]. Chen (2012) modelled the pressurization of a reservoir at 600 m of depth by 1 MPa using Young's modulus and Poisson's ratio values close to our case study reservoir. This modelling study suggests reducing the stiffness of the reservoir; i.e., reducing the value of Young's modulus from 5 GPa to 3 GPa, the vertical displacement increases from 0.08 to 0.12 mm [38]. Similarly, a reduction in the Poisson ratio from 0.3 to 0.2 results in an increase in the ground surface displacement from 0.11 to 0.13 mm [38]. It should be borne in mind that the sensitivity of these parameters is highly dependent on several different factors, including the depth and size of the reservoir, the stress field, and the properties attributed to the surrounding formations [38]. However, the work presented by Chen gives a rough idea of the types of variations that might be expected from such modelling.

8. Conclusions

THM modelling provide important results for short term gas storage, e.g., conversion of the former gas field to a storage site with short-term (weekly) injection-production schedules. There have been three cases tested for future scenarios with short-term operations. German surplus energy data from renewable energy sources (such as wind and solar energy) from 2017 have also been incorporated into these future testing cases to implement PtG technology on former gas reservoirs to meet Germany's future energy demand. The operating volumes in these cases are controlled by the upper (18.8 MPa) and lower limits (13.8 MPa) of well bottom hole pressure (WBHP). The results of the coupled THM modelling of these cases show that the maximum difference in pore pressure and subsequently effective stress is observed to be 0.6 MPa among all tested cases.

Fault reactivation analyses are performed on the THM models, which evaluate various gas injection scenarios without any upper limit of WBHP. The quantified stresses are then examined against Mohr–Coulomb failure criterion to estimate fault stability. This goal is achieved by including faults with specific properties in the 3D geomechanical model. The stress state of the 3D cells with fault properties exceeding the shear failure line is observed against different gas injection volumes. Fault reactivation occurred at the gas injection rate of 240,000 m^3/day, yielding a value of critical pore pressure. This critical pore pressure was observed to be 1.25 times the original pore pressure, which is equivalent to the WBHP value of 20.25 MPa. With the determination of pore pressure, it is also possible to estimate an upper limit of the gas injection rate for safe storage operation, which in this case study is observed to be in the range between 100,000 and 150,000 m^3/day.

In this study, the critical pore pressure is calculated at the well location, whereas the reference location for further studies can be anywhere in the reservoir, and the high permeability of the reservoir can reduce possible differences in the reference sites. Analysis of the case study numerical modelling results shows that a 4-MPa increase in pore pressure would be sufficient to trigger fault failure.

Thermal stresses depend merely upon the thermal-dynamic processes in the system. Therefore, one case (seasonal) has been tested for thermal stress analysis using the same THM model. The original reservoir temperature is 45 °C, and the injected gas temperature is set at 25 °C. The thermal changes are small (0.5 °C) and occur only near and around the wellbore. The thermal effects on stress are not significant in the reservoir, even after injecting 100,000 m^3/day of colder gas for about half a year. This outcome suggests that the thermal changes in the short-term cases are negligible to be considered in the analysis of geomechanical stresses in the reservoir during storage operations. This finding is due to the low operating volumes and short time cycles for injection production (weekly schedule) in short-term cases.

In terms of storage capacity of PtG technologies, the case study storage can store up to maximum of 1,322,400 kWh/d of electricity from renewable or other resources, converting the maximum gas volume of 150,000 m^3/day. This amount of energy storage can at least contribute to Germany becoming a greenhouse gas neutral country (GHGND) in 2050, corresponding to an energy demand of 1600 TWh of electricity.

The entire workflow outlined and tested in this study is not site specific but generally applicable to any gas storage in a porous medium including methane, CO_2, and hydrogen.

Perspectives

The modelling schemes described in this study provide a thermal-hydraulic-geomechanical assessment consistent with the observational data presently available. This THM modelling study can be used to assess borehole stability, which can be achieved by analysing the hydraulic and mechanical limits of the borehole walls. Another application of the model can be a fully coupled model, in which geomechanical and hydraulic models are coupled in such a way that both can receive and process the required parameters simultaneously (explained in the last paragraph). This modelling study can also serve as a benchmark for the use of UGS as a seasonal battery for PtG technologies. In this way, the excess energy can be stored as gas in the UGS and later converted back into electricity when needed.

The rock properties of the reservoir have an important influence on the estimated amount of subsidence. For future modelling, it would be useful to conduct detailed rock mechanics experiments with representative samples from the deposits in the region. It is important to have as much reliable data as possible on vertical elevation changes in the region to test different hypotheses related to subsidence. Such data could come from repeated surveys of existing first-order contour lines, GPS, or InSAR observations. In addition, sensitivity analyses or parameter studies to determine the stress field or displacement of the ground surface as elastic properties vary would be useful to understand the reservoir behaviour in response to changes in, for example, Young's modulus or Poisson's ratio.

THM modelling in this study uses a one-way coupling simulation approach. The coupling between dynamic fluid flow and geomechanical simulation can improved by adopting two-way or even fully coupled simulations. Such a simulation means that not only the pore pressure is passed from the fluid flow model to the geomechanical model, but also properties from the geomechanical model, such as porosity, permeability, and rock compressibility, are passed back to the fluid flow model to obtain continuously updated accurate modelling results. Two-way and fully coupled simulation approaches would also be useful to analyse ageing effects in reservoirs due to long- and short-term storage operations.

Author Contributions: Conceptualization, M.Z.-U.-A. and A.H.; methodology, M.Z.-U.-A.; software, M.Z.-U.-A.; validation, M.Z.-U.-A. and A.H.; formal analysis, M.Z.-U.-A.; investigation, M.Z.-U.-A.; resources, M.Z.-U.-A.; data curation, M.Z.-U.-A.; writing—original draft preparation, M.Z.-U.-A.; writing—review and editing, A.H.; visualization, M.Z.-U.-A.; supervision, A.H.; project administration, A.H.; funding acquisition, A.H. All authors have read and agreed to the published version of the manuscript.

Funding: This research was funded by BMBF (Federal Ministry of Education and Research), grant number 03G0869.

Data Availability Statement: Data available in a publicly accessible repository. The data presented in this study are openly available in [TUprints, Darmstadt] at [DOI: 10.26083/tuprints-00022537, CC-BY-SA 4.0 International.

Acknowledgments: The authors thank Uniper SE for providing data for this research. We acknowledge the German Federal Ministry of Education and Research (BMBF) for providing financial support for the SUBI research project. In addition, we acknowledge support from the Deutsche Forschungsgemeinschaft (DFG, German Research Foundation) and the Open Access Publishing Fund of the Technical University of Darmstadt.

Conflicts of Interest: The authors declare no conflict of interest.

References

1. Teatin, P.; Castelletto, N.; Ferronato, M.; Gambolati, G.; Janna, C.; Cairo, E.; Marzorati, D.; Colombo, D.; Ferretti, A.; Bagliani, A.; et al. Geomechanical response to seasonal gas storage in depleted reservoirs: A case study in the Po River basin, Italy. *J. Geophys. Res. AGU J.* **2011**, *116*, 21. [CrossRef]
2. Tenthorey, E.; Vidal-Gilbert, S.; Backe, G.; Puspitasari, R.; Pallikathekathil, Z.; Maney, B.; Dewhurst, D. Modelling the geomechanics of gas reservoir: A case study from the Iona gas field, Australia. *Int. J. Greenh. Gas Control* **2013**, *13*, 138–148. [CrossRef]
3. Aminian, K.; Brannon, A.; Ameri, S. Gas storage in a depleted gas/condensate reservoir in the appalachian basin. In Proceedings of the SPE Eastern Regional Meeting, Canton, OH, USA, 11–13 October 2006.
4. Juez-Larre, J.; Remmelts, G.; Breunese, J.; Van Gessel, S.Q.; Leeuwenburgh, O. Using underground gas storage to replace the swing capacity of the giant natural gas field of Groningen in the Netherlands. A reservoir performance feasibility study. *J. Pet. Sci. Eng.* **2016**, *145*, 34–53. [CrossRef]
5. Industry, D.O.T.A. *Meeting the Energy Challenge: A White Paper on Energy*; Her Majesty's Stationary Off.: Norwich, UK, 2007.
6. Zhang, J.; Fang, F.; Lin, W.; Gao, S. Research on Injection-Production Capability and Seepage Characteristics of Multi-Cycle Operation of Underground Gas Storage in Gas Field—Case Study of the Wen 23 Gas Storage. *Energies* **2020**, *13*, 17. [CrossRef]
7. Kuncir, M.; Chang, J.; Mansdorfer, J.; Dougherty, E. Analysis and optimal design of gas storage reservoirs. In Proceedings of the SPE Eastern Regional Meeting, Pittsburgh, PA, USA, 6–10 September 2003; pp. 1066–1076.
8. Mazarei, M.; Davarpanah, A.; Ebadati, A.; Mirshekari, B. The feasibility analysis of underground gas storage during an integration of improved condensate recovery processes. *J. Pet. Explor. Prod. Technol.* **2019**, *9*, 397–408. [CrossRef]
9. Dharmananda, K.; Kingsbury, N.; Singh, H. Underground gas storage: Issues beneath the surface. In Proceedings of the SPE Asia Pacific Oil and Gas Conference and Exhibition, Perth, Australia, 18–20 October 2004; pp. 10–17.
10. Abedin, M.Z.; Henk, A. Building 1D and 3D Mechanical Earth Models for Underground Gas Storage—A Case Study from the Molasse Basin, Southern Germany. *Energies* **2020**, *13*, 21.
11. Khaksar, A.; White, A.; Rahman, K.; Burgdorff KOllarves, R.; Dunmore, S. Geomechanical Evaluation for Short Term Gas Storage in Depleted Reservoirs. In Proceedings of the 46th US Rock Mechanics/Geomechanics Symposium, Chicago, IL, USA, 24–27 June 2012.
12. Bachmann, G.H.; Müller, M.; Weggen, K. Evolution of the Molasse Basin (Germany, Switzerland). *Tectonophysics* **1987**, *137*, 77–92. [CrossRef]
13. Dake, L.P. *Fundamentals of Reservoir Engineering*; Elsevier Scientific Pub. Co.: Amsterdam, The Netherlands, 1978.
14. Dai, W.J.; Gan, Y.X.; Hanaor, D. Effective Thermal Conductivity of Submicron Powders: A Numerical Study. *Appl. Mech. Mater.* **2015**, *846*, 500–505. [CrossRef]
15. Settari, A.; Mourits, F. A Coupled Reservoir and Geomechanical Modeling System. *SPE J.* **1998**, *SPE 50939*, 219–226. [CrossRef]
16. Tortike, W.; Farouq, A.S. Reservoir Simulation Integrated with Geomechanics. *J. Can. Pet. Technol.* **1993**, *5*, 28–37. [CrossRef]
17. Jalali, M.R.; Dusseault, M.B. Coupling Geomechanics and Transport in Naturally Fractured Reservoirs. *Int. J. Min. Geol. Eng. (IJMGE)* **2012**, *46*, 105–131.
18. Mainguy, M.; Longuemare, P. Coupling Fluid Flow and Rock Mechanics: Formulation of the Partial Coupling between Reservoir and Geomechanical Simulators. *Oil Gas Sci. Technol.* **2002**, *57*, 355–367. [CrossRef]
19. Perkins, T.K.; Gonzalez, J.A. The Effect of Thermoelastic Stresses on Injection Well Fracturing. *Soc. Pet. Eng.* **1985**, *25*, 78–88. [CrossRef]
20. Geertsma, J. Problems of Rock Mechanics in Petroleum Production Engineering. *First Congr. Int. Soc. Rock Mech.* **1966**, *1*, 585–594.
21. Skempton, A. The Pore Pressure Coefficients A and B. *Geotechnique* **1954**, *4*, 143–147. [CrossRef]
22. Geertsma, J. The Effect of Pressure Decline on Volumetric Changes of Porous Rocks. *Trans. AIME* **1957**, *201*, 331–340. [CrossRef]
23. Van der Knaap, W. Nonlinear Behavior of Elastic Porous Media. *Trans. AIME* **1959**, *216*, 179–187. [CrossRef]
24. Nur, A.; Byerlee, J. An Exact Effective Stress Law for Elastic Deformation of Rock with Fluid. *J. Geophys. Res.* **1971**, *76*, 6414–6418. [CrossRef]
25. Ghaboussi, J.; Wilson, E. Flow of Compressible Fluid in Porous Elastic Media. *Int. J. Numer. Methods Eng.* **1973**, *5*, 419–442. [CrossRef]
26. Rice, J.; Cleary, M. Some Basic Stress-Diffusion Solutions for Fluid Saturated Elastic Porous Media with Compressible Constituents. *Rev. Geophys. Space Phys.* **1976**, *14*, 227–241. [CrossRef]
27. Huan, X.; Xu, G.; Zhang, Y.; Sun, F.; Xue, S. Study on Thermo-Hydro-Mechanical Coupling and the Stability of a Geothermal Wellbore Structure. *Energies* **2021**, *14*, 649. [CrossRef]

28. Rutqvist, J.; Wu, Y.; Tsang, C.F.; Bodvarsson, G. A modeling approach for analysis of coupled multiphase fluid flow, heat transfer, and deformation in fractured porous rock. *Int. J. Rock Mech.* **2002**, *39*, 429–442. [CrossRef]
29. Pang, M.; Xu, G.; Sun, F.; Xue, S.; Wang, Y. Formation Damage and Wellbore Stability of Soft Mudstone Subjected to Thermal–Hydraulic–Mechanical Loading. *J. Eng. Sci. Technol.* **2019**, *12*, 95–102. [CrossRef]
30. Hu, L.; Winterfeld, P.H.; Fakcharoenphol, P.; Wu, Y.S. A novel fully-coupled flow and geomechanics model in enhanced geothermal reservoirs. *J. Pet. Sci. Eng.* **2013**, *107*, 1–11. [CrossRef]
31. Tran, D.; Nghiem, L.; Buchanan, L. *Improved Iterative Coupling of Geomechanics with Reservoir Simulation*; Society of Petroleum Engineers: Houston, TX, USA, 2005.
32. Frey, C. Ministerin Hendricks: Wir Haben im Gegenteil Sogar Gigantische Stromüber-Schüsse. 2018. Available online: https://eike-klima-energie.eu/2018/01/07/ministerin-hendricks-wir-haben-im-gegenteil-sogar-gigantische-stromueberschuesse/?print=pdf (accessed on 1 June 2020).
33. Burger, B. Fraunhofer ISE Fraunhofer ISE, Freiburg. 2018. Available online: https://www.ise.fraunhofer.de/content/dam/ise/de/documents/publications/studies/daten-zu-erneuerbaren-energien/Stromerzeugung_2017.pdf (accessed on 1 June 2020).
34. Carter, G.F.; Paul, D.E. *Materials Science & Engineering, Materials Park Ohio*; ASM International: Novelty, OH, USA, 1991.
35. Fjaer, E.; Holt, R.M.; Horsrud, P.; Raaen, A.M.; Risnes, R. *Petroleum Related Rock Mechanics*; Elsevier Science: Oxford, UK, 2008; p. 514.
36. Terzaghi, K. *Theoretical Soil Mechanics*; Chapman and Hall Limited: London, UK, 1948.
37. Zoback, M. *Reservoir Geomechanics*; Cambridge University Press: Cambridge, UK, 2007; p. 449.
38. Chen, Z.R. Poroelastic model for induced stresses and deformations in hydrocarbon and geothermal reservoirs. *J. Pet. Sci. Eng.* **2012**, *80*, 41–52. [CrossRef]

energies

MDPI

Article

Exploring the Impact of Regional Integrated Energy Systems Performance by Energy Storage Devices Based on a Bi-Level Dynamic Optimization Model

Baohong Jin *, Zhichao Liu and Yichuan Liao

College of Civil Engineering, Hunan University, Changsha 410082, China
* Correspondence: jinbh@hnu.edu.cn; Tel.: +86-18829344979

Abstract: In the context of energy transformation, the importance of energy storage devices in regional integrated energy systems (RIESs) is becoming increasingly prominent. To explore the impact of energy storage devices on the design and operation of RIESs, this paper first establishes a bi-level dynamic optimization model with the total system cost as the optimization objective. The optimization model is used to optimize the design of three RIESs with different energy storage devices, including System 1 without an energy storage device, System 2 with a thermal energy storage (TES) device, and System 3 with TES and electrical energy storage (EES) devices. According to the design and operation results, the impact of energy storage devices on the operational performance of RIESs is analyzed. The results show that under the design conditions, energy storage devices can significantly increase the capacity of the combined heating and power units and absorption chillers in System 2 and System 3 and reduce the capacity of the ground source heat pumps and gas boilers; the impact of the TES device on System 3 is more significant. Affected by systems' configuration, the operating cost, carbon tax, and total cost of System 2 are reduced by 2.9%, 5.5%, and 1.5% compared with System 1, respectively. The EES device can more significantly reduce the operating cost of System 3, with a reduced rate of 5.7% compared with that in System 1. However, the higher equipment cost makes the total cost reduction rate of System 3 less than that of System 1, which is 1.75%. Similar to the design conditions, under the operation conditions, the TES device can effectively reduce the carbon tax, operating cost, and total cost of System 2, while System 3 with an EES device can significantly reduce its operating cost regardless of whether the energy price changes or not. To some extent, this study systematically elucidated the impact of TES and EES devices on the optimal design and operation performance of RIESs and provided a certain reference for the configuration of energy storage devices.

Keywords: regional integrated energy system; energy storage device; Bi-level dynamic optimization model; optimal design; operational analysis

Citation: Jin, B.; Liu, Z.; Liao, Y. Exploring the Impact of Regional Integrated Energy Systems Performance by Energy Storage Devices Based on a Bi-Level Dynamic Optimization Model. *Energies* **2023**, *16*, 2629. https://doi.org/10.3390/en16062629

Academic Editors: Luis Hernández-Callejo, Jesús Armando Aguilar Jiménez and Carlos Meza Benavides

Received: 10 February 2023
Revised: 2 March 2023
Accepted: 8 March 2023
Published: 10 March 2023

1. Introduction

In recent years, the sustainable development of energy has received extensive attention. Regional integrated energy systems (RIESs) are expected to become an important way to improve the energy structure and achieve sustainable energy development [1]. However, the intermittency and volatility of renewable energy have brought certain challenges to the stable operation of RIESs [2,3]. With the rapid development of energy storage technology, the development of RIESs with hybrid energy storage has become the main way to solve the volatility of renewable energy and alleviate the contradiction between supply and demand [4,5].

To explore the performance of the integrated energy system with hybrid energy storage, the studies shown in Table 1 have conducted in-depth research on RIESs from aspects of system structure, operation strategy, and optimization model. Different from the traditional energy system, the RIES involves the deep coupling of multiple heterogeneous energy sources. The modeling idea of the Energy Hub (EH) proposed by Geidl et al. [6] was used to describe the relationship between energy conversion and conservation. However, this modeling idea is not conducive to the RIES's extension and matrix representation in the model. To improve the model portability, an EH modeling method based on graph theory was proposed [7]. Based on this modeling idea, Ma et al. [8] adopted the static equipment model to establish the coupling system optimization model with the total system cost as the optimization objective. User-side energy saving and load management are also one of the main ways to reduce the total cost of RIESs [9–11]. To make full use of the flexibility of loads, Liu et al. [12] established a coupling system optimization model considering the comprehensive demand response. At the same time, this model is also used to explore the impact of energy storage devices on the design and operation of a RIES [13]. The static system optimization model cannot reflect the off-design characteristics of the equipment. To address the issue, a dynamic system optimization model, considering the off-design characteristics of the equipment, was established [14]. On this basis, Mansouri et al. [15] established a dynamic multi-objective optimization model and used it to optimize the design of a RIES with power-gas (P-G) technology. The results showed that the gas storage device could effectively improve the utilization of renewable energy. Unfortunately, none of the above optimization models realizes the decoupling of design and operation.

To achieve the decoupling of design and operation, Mago et al. proposed the following electric load (FEL), following thermal load (FTL), and following hybrid electric-heating load (FHL) strategies according to the role of the combined heating and power (CHP) unit in RIESs [16,17]. Based on the above strategies, Kang et al. [18] explored the operational performance of a RIES under different loads. Wang et al. [19] investigated the impact of energy storage characteristics on the system optimization results based on the system optimization model with the total system cost as the optimization objective. To take the economic, energy-saving, and environmental performance of systems into account, a weighted multi-objective optimization model was established for the optimal design of RIESs [20]. Based on the weighted multi-objective optimization model, Zeng et al. [21,22] used the static and dynamic equipment models to optimize the coupling systems of CCHP and ground source heat pump(GSHP), respectively. However, the value of weight is often subjective. Thus, a multi-objective optimization model with the optimization objectives of cost-saving rate, primary energy saving rate, and CO_2 emission reduction rate was proposed for the optimal design of a RIES [23]. Zhai et al. [24,25] used this model to explore the impact of building types on the operational performance of RIESs. Different types of building loads have certain complementary characteristics. For this reason, Li et al. [26] explored the impact of loads' complementary characteristics on optimization results and operational performance of a RIES. The research showed that the complementary characteristics of loads could reduce the capacity of energy storage devices to a certain extent. In addition, the equipment model and operation strategy also have a certain impact on the optimization results of RIESs. Therefore, Deng et al. [27] established a dynamic multi-objective optimization model based on the dynamic equipment model. Han et al. [28] used a dynamic multi-objective optimization model to optimize the design of a RIES with hybrid, compressed air energy storage. To improve the operational performance of RIESs, an improved FEL strategy was used in the optimal design of a RIES [29]. Compared with the traditional FEL strategy, the improved operation strategy could effectively reduce the energy consumption, operating cost, and CO_2 emission of the RIES. At the same time, the adaptive operation strategy, based on user load, was proposed successively to improve the operational performance of RIESs [30]. However, the relatively fixed operation strategy could not realize the flexible scheduling of RIESs. Hence, Luo et al. [31] adopted the decision tree method to formulate the operation strategy of RIESs. Nonetheless, machine learning greatly relies on building historical load data. To avoid this problem and achieve the flexible scheduling of RIESs, a bi-level optimization model was proposed where the upper-level optimization model is used to determine the optimal configuration of systems, and the lower-level optimization model is used to realize the flexible scheduling of systems [32]. Based on the bi-level optimization model, Luo et al. [33] optimized the standalone renewable energy system with the total system cost as the optimization objective. Energy storage devices can improve the penetration rate of renewable energy. Li et al. [34] used the bi-level optimization model to optimize the design of an electricity-hydrogen RIES. Ma et al. [35] used this model to explore the impact of shared energy storage on the renewable energy utilization rate and operating cost of RIESs. Although the bi-level optimization model has been widely used in the optimization design and operation analysis of the integrated energy system with energy storage devices, few studies have systematically explored the effects of different energy storage devices on the optimal design and operational performance of the system by the bi-level dynamic optimization model.

Table 1. Literature review of the integrated energy system with energy storage.

Ref.	Renewable Energy	Energy Storage Device	Operation Strategy	Equipment Model	System Optimization Model
[8]	Solar and wind energy	EES, TES	/	Static model	Coupled single objective optimization
[13]	Wind energy	EES, TES	/	Static model	Coupled single objective optimization
[14]	Solar and wind energy	EES, TES, Fuel cell	/	Dynamic model	Coupled single objective optimization
[15]	Wind energy	EES, HES	/	Dynamic model	Coupled multi-objective optimization
[18]	Geothermal energy	/	FEL FTL	Static model	Weak decoupling single objective optimization
[21]	Geothermal energy	TES	FEL FTL	Static model	Weak decoupling weighted multi-objective optimization
[22]	Geothermal energy	TES	FEL FTL	Dynamic model	Weak decoupling weighted multi-objective optimization
[24]	Solar energy	TES	FEL FTL	Static model	Weakly decoupled multi-objective optimization
[25]	Solar energy Geothermal energy	EES TES	FEL FTL FHL	Static model	Weakly decoupled multi-objective optimization
[27]	Geothermal energy	TES	FEL FTL	Dynamic model	Weakly decoupled multi-objective optimization
[30]	Solar, wind and geothermal energy	EES TES	FSF	Dynamic model	Weakly decoupled multi-objective optimization
[31]	Solar energy	EES TES	Dynamic strategy	Dynamic model	Weakly decoupled multi-objective optimization
[34]	Solar, wind and geothermal energy	EES, hydrogen storage	FOF	Static model	Bi-level optimization model
[35]	Solar and wind energy	SES	FOF	Static model	Bi-level optimization model

Different energy storage devices can realize the time-series transfer of different energies. To explore the impact of energy storage devices on the design and operation of RIESs, this paper optimizes three RIESs with different energy storage devices and compares their operational performance according to a public building load in Changsha. The main contributions of this paper are as follows: (1) A bi-level dynamic optimization model is established based on the dynamic equipment model; (2) Three RIESs with different energy storage devices are optimally designed; (3) According to the optimization results, the operational performance of three RIESs with different energy storage devices is compared. The remainder of this study is organized as follows: Part II is the introduction and equipment modeling of RIESs with different energy storage devices; Part III is the establishment of the bi-level system optimization model; Part IV presents the impact of different energy storage devices on the optimal design and operational performance of the integrated energy system based on the case results; and the conclusions of this work are drawn in Part V.

2. Modeling of RIESs

2.1. Basic structure of RIESs

RIESs can not only realize efficient energy conversion but also gratify the cooling, heating, and electricity needs of users simultaneously. In the RIES shown in Figure 1, the input energy mainly comes from grid electricity, municipal gas, and regional renewable energy (such as solar and geothermal energy). Energy conversion equipment is used to realize the conversion of input energy to output energy, mainly including the transformer, photovoltaic (PV) arrays, CHP unit, gas boiler, GSHP, and absorption chiller (ABC). Therefore, the power, heating, and cooling hubs are introduced in the modeling idea of EH to realize the collection and distribution of different energies and ensure the balance of the supply and demand for energy. In addition, to explore the influence of energy storage devices on the optimal design and operational performance of RIESs, three RIESs with different energy storage devices are considered, and their energy storage configurations are shown in Table 2.

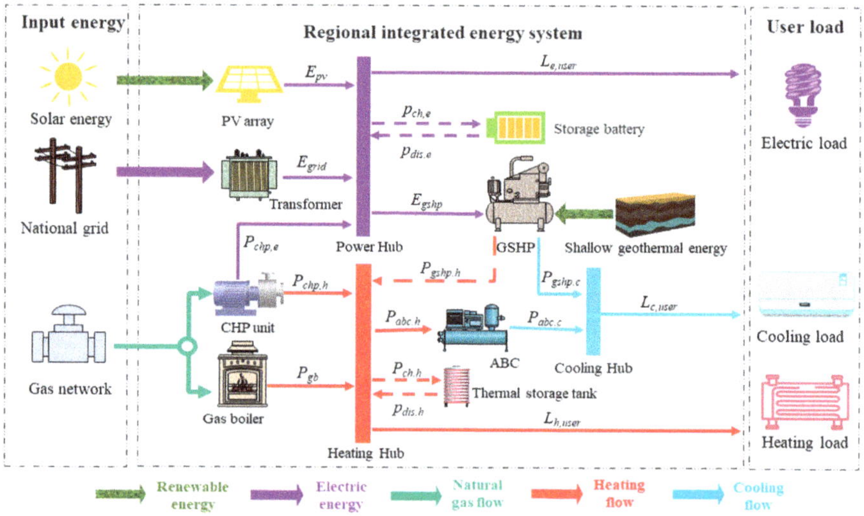

Figure 1. The basic structure of RIESs.

Table 2. Three RIESs with different energy storage devices.

System Name	TES Device	EES Device
System 1	×	×
System 2	✓	×
System 3	✓	✓

2.2. Equipment Mathematical Model

2.2.1. Energy Conversion Device

The equipment model is the basis of system design and operation optimization. In the RIES shown in Figure 1, the PV arrays are one of the effective ways to utilize solar energy, and its power generation is usually affected by the ambient temperature, solar irradiation intensity, and power generation efficiency. Compared with PV arrays, the CHP unit can not only generate electricity, but also the waste heat can be used to meet the cooling and heating needs of buildings. Other than that, the rest of the building's cooling and heating loads are met by the GSHP and gas boiler. When constructing mathematical models for the CHP unit and cooling and heating equipment, relevant studies generally adopt the black-box model based on energy efficiency. The model is usually divided into two types: the static equipment model and the dynamic equipment model. The static model assumes that the operating efficiency of the equipment is constant. The dynamic equipment model considers the influence of equipment partial load rate on its efficiency. To accurately describe the operation performance of the equipment, the dynamic equipment model will be established in this paper, and the specific expressions for different equipment dynamic models are shown in Table 3.

Table 3. Equipment dynamic model.

Items	Mathematical Models	Ref.
PV	$P_{pv}(t) = A_{pv} \cdot \eta_{pv} \cdot I(t) \cdot (1 - 0.005 \times (t_a(t) - 25))$, $\eta_{pv} = 0.165$.	[36]
CHP unit	$P_{chp,e} = F_{chp} \cdot \eta_{chp,e}$, $P_{chp,h} = F_{chp} \cdot \left(1 - \eta_{chp,e} - \eta_{loss}\right)$ $\eta_{chp,e} = \begin{cases} 0 & PLR_{chp} < 0.2 \\ a_0 + a_1 \cdot PLR_{chp} + a_2 \cdot PLR_{chp}^2 & PLR_{chp} \geq 0.2 \end{cases}$ $PLR_{chp} = E_{chp}/E_{chp,r}$, $a_0 = 0.1$, $a_1 = 0.4$, $a_2 = -0.2$.	[37]
GB	$P_{gb} = G_{gb} * \eta_{gb}$, $PLR = P_{gb}^{out}/P_{gb,r}^{out}$, $PLF = \eta_{gb}/\eta_{gb,r}$, $\eta_{gb,r} = 0.9$, $PLF_{gb} = -0.046PLR_{gb} + 1.046$, $0.1 \leq PLR_{gb} \leq 1$.	[38]
GSHP	$P_{gshp} = E_{gshp} * COP_{gshp}$, $PLR_{gshp} = P_{gshp}/P_{gshp,r}$, $PLF_{gshp} = COP_{gshp}/COP_{gshp,r}$, $COP_{gshp,r} = 4.5$, $PLF_{gshp} = PLR_{gshp}/\left(-0.2137PLR_{gshp}^2 + 1.119PLR_{gshp} + 0.1007\right)$, $0.1 \leq PLR_{gshp} \leq 1$.	[38]
ABC	$P_{abc,c} = P_{abc,h} * \eta_{abc}$, $PLR_{abc} = P_{abc}/P_{abc,r}$, $PLF = \eta_{abc}/\eta_{abc,r}$, $\eta_{abc,r} = 0.9$, $PLF_{abc} = PLR_{abc}/(0.75PLR_{abc}^2 + 0.0195PLR_{abc} + 0.213)$, $0.2 \leq PLR_{abc} \leq 1$.	[39]

2.2.2. Energy Storage Devices

On the basis of System 1, this paper investigates the impact of energy storage devices on the optimization and operation of RIESs by sequentially configuring TES and EES devices in System 2 and System 3. Different from the energy conversion equipment, the source-load duality of energy storage devices allows it to achieve the time-series transfer of energy to meet the supply-demand balance of RIESs. Therefore, the mathematical model of energy storage devices can be expressed by the charging and discharging state and power, and its specific expression is shown as follows [40]:

$$S_k(t+1) = S_k(t) + \left(P_{ch,k}(t)\eta_{ch,k} - \frac{P_{dis,k}(t)}{\eta_{dis,k}}\right)\Delta t \quad (1)$$

where $S_k(t+1)$ and $S_k(t)$ are the energies stored in energy storage device k at time $t + 1$ and t, respectively; $\eta_{ch,k}$ and $\eta_{dis,k}$ are the charging and discharging efficiency of energy storage device k; and $P_{ch,k}(t)$ and $P_{dis,k}(t)$ are the charging and discharging powers of energy storage device k at time t.

3. Bi-Level Optimization Model

Energy storage devices not only affect the optimal design of RIESs but also affect their operational performance. To explore the impact of energy storage devices on the optimal design and operation of RIESs, a bi-level dynamic optimization model is established in this paper. In this model, the upper-level optimized configuration model takes the system's total cost as the optimization objective to determine the equipment capacity of RIESs. The lower-level optimal scheduling model takes the operating cost as the optimization objective to determine the reasonable scheduling scheme. To facilitate the understanding of the solution process of this model, the optimization process is drawn in Figure 2. Firstly, based on the outdoor design parameters, the design and operating loads of the building are calculated by Energy Plus. Secondly, the constraints of the upper-level and lower-level optimization models are established according to the design loads and the EH model. Among them, the main constraints of the upper-level optimized configuration model include the maximum equipment capacity and the design load in winter. And the constraints of the lower-level scheduling model include energy conservation and equipment operating power. Finally, the design load, outdoor parameters, and economic parameters are imported into the bi-level optimization model for solving to obtain the equipment capacity of three RIESs. With the equipment capacity known, this paper uses the lower-level scheduling model to optimize the operation of three RIESs and analyzes the impact of energy storage devices on their operational performance.

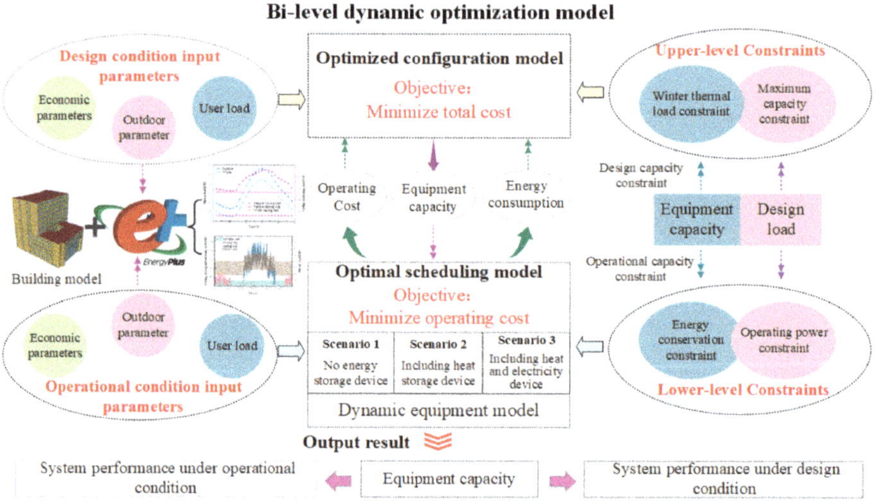

Figure 2. Flow chart of system optimization configuration and operation analysis.

3.1. Upper-Level Optimal Configuration Model

3.1.1. Optimization Objective

To determine the equipment capacity of the three RIESs, this paper optimizes them with the minimum total cost as the optimization objective. In this study, the total cost mainly includes the system equipment cost, operating cost, and carbon tax. The specific calculation expressions are shown below:

$$\min C_{total} = C_{equ} + C_{tax} + C_{op} \quad (2)$$

where C_{total} is the total cost; C_{equ} is the equipment cost; C_{tax} is the carbon tax; and C_{op} is the operating cost, the values can be obtained from the lower-level scheduling model.

The system equipment cost mainly includes the initial investment and equipment maintenance cost. The initial investment in equipment depends on its capacity and the initial unit investment. The initial unit investment in the RIESs, showed Figure 1, is listed in Table 4. Under the condition that the initial unit investment of equipment is known, the initial investment of RIESs on the design day can be determined by the following formula [41].

$$C_{inv} = \sum_{k=1}^{k} \left(\frac{i(1+i)^n}{(1+i)^n - 1} \cdot Cap_k \cdot C_k \right) / 365 \quad (3)$$

where i is the annual interest rate, which is 0.08 in this paper; n is the planning period, which is 20 years; Cap_k is the design capacity of the equipment k; and C_k is the initial unit investment in the equipment k. As the initial investment in the system's equipment is known, the maintenance cost of the system's equipment can be estimated at 2% of its initial investment [42].

Table 4. The initial unit investment in equipment Reproduced from [43,44].

Equipment Name	Unit Price	Equipment Name	Unit Price
PV	2315 (CNY/m^2)	CHP unit	6812 (CNY/kW)
Boiler	790 (CNY/kW)	GSHP	2782 (CNY/kW)
ABC	1436 (CNY/kW)	TES device	358 (CNY/kW)
EES device	1794 (CNY/kW)		

In the context of peak carbon dioxide emissions and carbon neutrality, carbon tax compensation has become an effective means to limit greenhouse gas emissions. In the RIESs shown in Figure 1, CO_2 emissions mainly come from grid power and gas, so the carbon tax cost can be calculated by the following formula.

$$C_{tax} = \vartheta_{tax} \sum_{t=1}^{24} \left(P_{grid}(t) \cdot \lambda_{CO_2,grid} + P_{gas}(t) \cdot \lambda_{CO_2,gas} \right) \quad (4)$$

where ϑ_{tax} is the carbon tax price; $\lambda_{CO_2,grid}$ and $\lambda_{CO_2,gas}$ are the equivalent CO_2 emissions of coal power and gas, which are 0.968 kg/kWh and 0.220 kg/kWh, respectively [45]; and $P_{grid}(t)$ and $P_{gas}(t)$ are the consumption of grid power and gas at time t, which can be obtained from the lower-level scheduling model.

3.1.2. Optimization Variables and Constraints

In the upper-level optimization model, the optimization variable is the capacity of the candidate equipment. Considering the equipment installation conditions and building loads, its optimization variables must satisfy the following constraints.

$$0 \le Cap_k \le Cap_k^{max} \quad (5)$$

where Cap_k^{max} is the maximum design capacity of equipment k, whose value is usually the maximum value of the corresponding load.

In addition, this study uses the design-daily load in summer as the design parameter. To gratify the load demand in winter, the maximum heat production capacity of the system must be greater than the maximal heating load.

$$P_{gshp,h}^{max} + P_{chp,h}^{max} + P_{gb,h}^{max} \ge L_{user,h}^{max} \quad (6)$$

where $P_{gshp,h}^{max}$ is the maximum heat production of the GSHP; $P_{chp,h}^{max}$ is the maximum heat production of the CHP unit; $P_{gb,h}^{max}$ is the maximum heat production of the gas boiler; and $L_{user,h}^{max}$ is the maximum heating load in winter.

3.2. Lower-Level Optimal Scheduling Model

3.2.1. Optimization Objective

To achieve flexible scheduling, the minimum operating cost is used as the optimization objective to optimize the operation of three RIESs. The operating cost mainly comes from the electricity and gas costs, whose values can be determined by the following formula:

$$\min C_{op} = \sum_{t=1}^{24} \left(P_{grid}(t) \cdot \vartheta_{grid}(t) + P_{gas}(t) \cdot \vartheta_{gas} \right) \tag{7}$$

where ϑ_{grid} is the time-of-use (TOU) electricity price; and ϑ_{gas} is the gas price.

3.2.2. Optimization Variables and Constraints

In the lower-level optimization model, the main optimization variable is the operating power of the equipment. The operating power should not only gratify the capacity constraints of the upper-level equipment but also meet the supply-demand balance constraints of the EH.

1. Equipment operating power constraints

(1) Energy conversion equipment

The operating power of the energy conversion equipment is both restricted by the capacity of the upper-level equipment and affected by the start-up and shutdown of the equipment. Therefore, the operating power is a semi-continuous variable whose range is shown in the following formula:

$$\begin{cases} 0 & PLR_k < PLR_k^{min} \\ P_k^{min} \le P_k \le P_k^{max} & PLR_k \ge PLR_k^{min} \end{cases} \tag{8}$$

where PLR_k is the part-load ratio of equipment k; PLR_k^{min} is the minimum part-load ratio for the start-up of equipment k; and P_k^{min} and P_k^{max} are the minimum and maximum output power of equipment k in the operating state.

(2) Energy storage equipment

Different from energy conversion equipment, energy storage devices should not only gratify the charging and discharging power constraints but also the charging and discharging state constraints, which are specifically expressed as follows:

$$0 \le P_{ch,k}(t) \le u_k \cdot \gamma_{ch,k}^{max} \cdot S_k \tag{9}$$

$$0 \le P_{dis,k}(t) \le (1 - u_k) \cdot \gamma_{dis,k}^{max} \cdot S_k \tag{10}$$

$$\alpha_k^{min} \cdot S_k \le S_k(t) \le \alpha_k^{max} \cdot S_k \tag{11}$$

where u_k is a variable of 0 or 1, which is introduced to ensure that the charging process and discharging process will not happen simultaneously; $\gamma_{ch,k}^{max}$ and $\gamma_{dis,k}^{max}$ are the maximum charging and discharging ratios of energy storage device k; α_k^{min} and α_k^{max} are the minimum and the maximum energy storage ratios of energy storage device k, respectively; and S_k is the capacity of energy storage device k.

2. Energy balance constraint

During the operation of RIESs, the EH is only used for energy collection and distribution. Therefore, the three energy hubs must maintain a balance between supply and demand, with the balance constraint shown below.

$$P_{grid}(t) + P_{chp,e}(t) + P_{pv}(t) + P_{dis,ees}(t) = L_{user,e}(t) + P_{gshp,e}(t) + P_{ch,ees}(t) \tag{12}$$

$$P_{gshp,c}(t) + P_{abc,c}(t) = L_{user,c}(t) \tag{13}$$

$$P_{chp,h}(t) + P_{gb}(t) + P_{dis.tes}(t) = L_{user,h}(t) + P_{abc,h}(t) + P_{ch.tes}(t) \quad (14)$$

where $P_{gshp,e}$ is the electricity consumption of GSHP; $P_{gshp,c}$ is the cooling power of GSHP; $P_{abc,c}$ is the cooling power of ABC; $P_{abc,h}$ is the heat consumption of ABC; and P_{gb} is the heat production of the gas boiler.

3.3. Model Solving

The solution methods of the bi-level optimization model usually include classical mathematical programming theory and the combination of intelligent optimization algorithms and classical mathematical programming theory [46,47]. In this study, the lower-level scheduling model takes into account the off-design characteristics of the equipment, which makes the lower-level scheduling model non-convex and nonlinear, and thus makes it difficult for classical mathematical programming theory to solve the bi-level dynamic optimization model. Therefore, this study will adopt the method of combining an intelligent optimization algorithm and classical mathematical programming theory to solve it, in which the upper-level optimization model is solved by a genetic algorithm. However, the calculation of the upper-level optimization objective often depends on the solution of the lower-level model. To realize the fast and accurate solution of the lower-level optimization model, this paper performs piecewise linearization on the performance curve of the equipment and calls Gurobi's non-convex solver to solve it to obtain the minimum operating cost and operating energy consumption. The lower-level optimization model transfers the optimization results to the upper-level optimization model to calculate the total cost of the system, while the upper-level optimization model transfers the optimized equipment capacity to the lower optimization model to constrain its scheduling. After repeated iterations, the optimal configuration and scheduling schemes of three RIESs can be obtained. Figure 3 shows the flow chart of the bi-level dynamic optimization model.

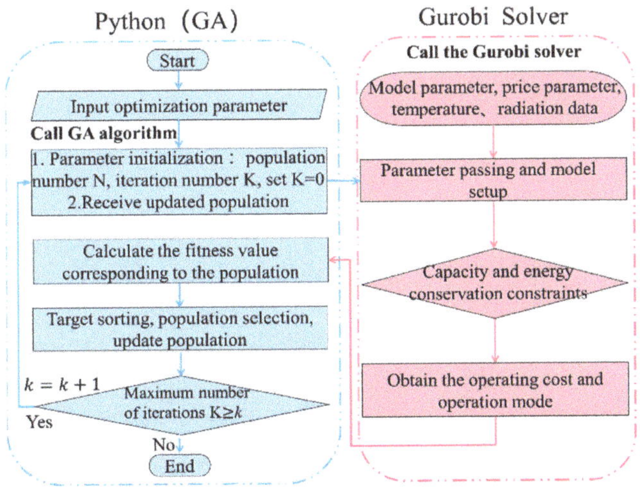

Figure 3. Solution flow chart of the bi-level optimization model.

4. Case Study

This paper takes a public building in Changsha as an example to explore the impact of energy storage equipment on the optimal design and operation results of RIESs. The building consists of two parts, the main building and the podium building, of which the main building has twelve floors, and the podium building has five floors, covering a total area of 2500 m^2. Considering the energy-saving requirements of the building, its envelope adopts the standard building envelope structure in hot-summer and cold-winter climates.

4.1. System Design Parameters

In the process of the optimization of RIESs, outdoor meteorological parameters and design load are the basis of system optimization design. Therefore, through relevant literature, this paper determines the outdoor design temperature of air conditioning in Changsha and the average water temperature of the Xiangjiang River, whose values are shown in Table 5. Unlike the heat load in winter, the cooling load calculation in summer is usually transient. For this reason, this paper corrects the outdoor design temperature of air conditioning in summer, and the hourly outdoor design temperatures and solar radiation intensities obtained from the correction are shown in Figure 4a. Based on the above design parameters, this paper uses Energy Plus to calculate the design load of the building, and the result is shown in Figure 4b. On summer design days, the heating load is mainly domestic hot water load, while the winter heating load includes air conditioning heating load and domestic hot water load. When the design load is known, this paper determines the equipment capacity optimization range, shown in Table 6, according to the design load and equipment installation requirements.

Table 5. Air conditioning outdoor design temperature and groundwater temperature Reproduced from [48,49].

Design Conditions	Design Dry-Bulb Temperature	Groundwater Temperature
Summer	36 °C	17 °C
Winter	−1 °C	11 °C

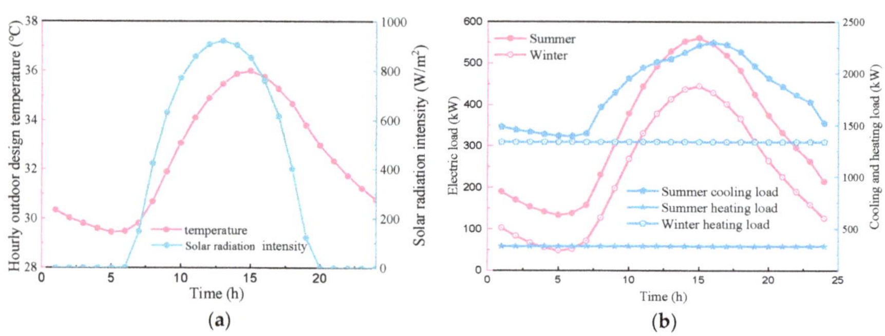

Figure 4. Outdoor parameters and building load under design conditions: (**a**) Hourly outdoor design temperature and solar radiation intensity in summer; (**b**) Cooling, heating, and electric load on the design day.

Table 6. Optimization range of equipment capacity.

Equipment Name	Symbol	Unit	Limitations
CHP unit	Cap_{chp}	kW	[0,2000]
Boiler	Cap_{gb}	kW	[0,1000]
GSHP	Cap_{gshp}	kW	[0,2300]
TES device	Cap_{tes}	kW·h	[0,1000]
EES device	Cap_{ees}	kW·h	[0,1000]
PV	A_{pv}	m^2	[0,1500]

In addition, the energy price and carbon tax price are also indispensable input parameters for calculating the optimization objectives. For this reason, this paper determines the energy and carbon tax prices shown in Table 7 according to relevant literature.

Table 7. Energy price and carbon tax Reproduced from [50–52].

Item	Unit Price (CNY/kWh)	Time Period	Description
Gas	0.3275	–	–
Electricity	1.224	20:00–23:00	Peak hours
	0.911	9:00–12:00, 16:00–20:00	High hours
	0.68	8:00–9:00, 12:00–16:00	Flat hours
	0.306	0:00–8:00, 23:00–24:00	Valley hours
Carbon tax	0.3 (CNY/kgCO_2)	–	–

4.2. *System Optimization Results and Analysis*

4.2.1. Optimization Results

Based on the above inputs, this paper takes the total cost as the optimization objective and adopts the bi-level dynamic optimization model to optimize the design of three RIESs. The equipment capacity obtained by optimization is shown in Table 8. Compared with System 1, the TES device can significantly increase the capacity of the CHP unit and ABC in System 2 and reduce the capacity of the GSHP and boiler. Identically to the TES device, the EES device can also increase the capacity of the CHP unit and ABC in System 3. However, the EES device has less impact on the capacity of energy conversion equipment than the TES device.

Table 8. System equipment capacity.

System Name	Cap_{chp} (kW)	Cap_{gb} (kW)	Cap_{gshp} (kW)	Cap_{abc} (kW)	Cap_{tes} (kW · h)	Cap_{ees} (kW · h)	A_{pv} (m^2)
System 1	1122	205	1780	520	0	0	1500
System 2	1265	0	1682	618	961	0	1500
System 3	1272	0	1678	622	947	926	1500

The difference in equipment capacity affects the equipment cost and operational performance of RIESs. Figure 5 shows the equipment costs, carbon taxes, operating costs, and total costs of the three systems under the design conditions. The difference in equipment capacity makes the equipment cost of System 2 and System 3 larger than that of System 1, with increasing rates of 5.7% and 17.8%, respectively. In comparison to the TES device, the EES device will significantly increase the equipment cost of System 3. Different from the equipment cost, the carbon tax, operating cost, and total cost of System 2 and System 3 are all less than that of System 1. Compared with System 1, the carbon tax, operating cost, and total cost of System 2 decreased by 5.5%, 2.9%, and 1.5%, respectively.The EES device can significantly reduce the operating cost of System 3, with a reduction of 5.7% compared with System 1. However, affected by the equipment cost, the total cost reduction rate of System 3 compared with System 1 is only 1.75%. It can be seen that under the design conditions, the TES and EES devices can reduce the operating cost, carbon tax, and total cost of the RIES to different degrees. However, the EES device has less impact on the carbon tax and total cost of RIESs.

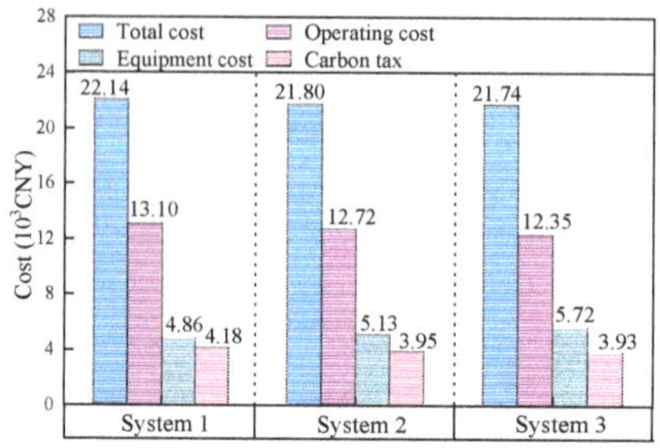

Figure 5. Cost differences between the three systems.

4.2.2. Analysis of Results

The energy utilization cost directly affects the operational performance of RIESs. To analyze the influence of energy storage devices on the operating cost, this paper first calculates the electric load supply-demand relationship and the part-load ratio of the CHP unit in System 1 under design operating conditions. As seen from Figure 6, during the low tariff hours, the grid power consumption is larger than gas-fired generation, and the CHP unit is at the minimum part-load ratio. However, the CHP unit is operating at full load during other hours. This shows that the electricity cost from the grid is less than the utilization cost of gas during the low tariff hours, while during other hours, the electricity cost from the grid is greater than the utilization cost of gas.

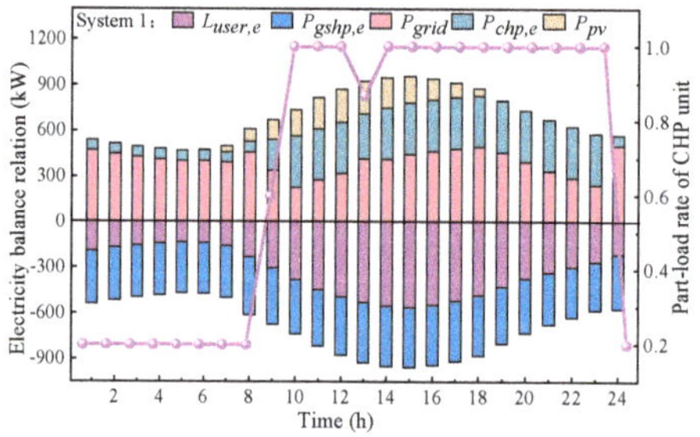

Figure 6. Electric load supply−demand relationship and part-load rate of CHP unit in System 1.

On the premise that the difference in energy utilization cost is known, this paper compares the scheduling process of System 1 and System 2 under the design conditions, as displayed in Figure 7. Figure 7a depicts the cooling load supply-demand relationship of the two systems. First, during the low tariff hours, the cooling load of the two systems is gratified by the GSHP. During other hours, the cooling load is met by the GSHP and ABC. However, the difference in equipment capacity makes the cooling power of the GSHP in System 2 less than that in System 1. Secondly, as seen in Figure 8, the operating efficiency of the GSHP in System 2 is greater than in System 1. The electric load of System 2 is less than System 1 due to the higher operating efficiency and lower cooling power. As shown in Figure 7b, the heating load of System 1 is gratified by the CHP unit and the gas boiler, while the heating load of System 2 is gratified by the CHP unit under the action of the TES device. This ensures the efficient use of gas while avoiding the use of the gas boiler in System 2. When combined with the relationship between supply-demand of electric load shown in Figure 7c, the larger capacity of the CHP unit can effectively reduce the power purchased from the grid during other hours for System 2. Based on the above analysis, the TES device can improve the overall energy efficiency of System 2 and reduce the electric load. Moreover, it increases the power generation of the CHP unit and reduces the grid power consumption, especially during other hours. Therefore, the carbon tax and operating costs of System 2 are less than System 1.

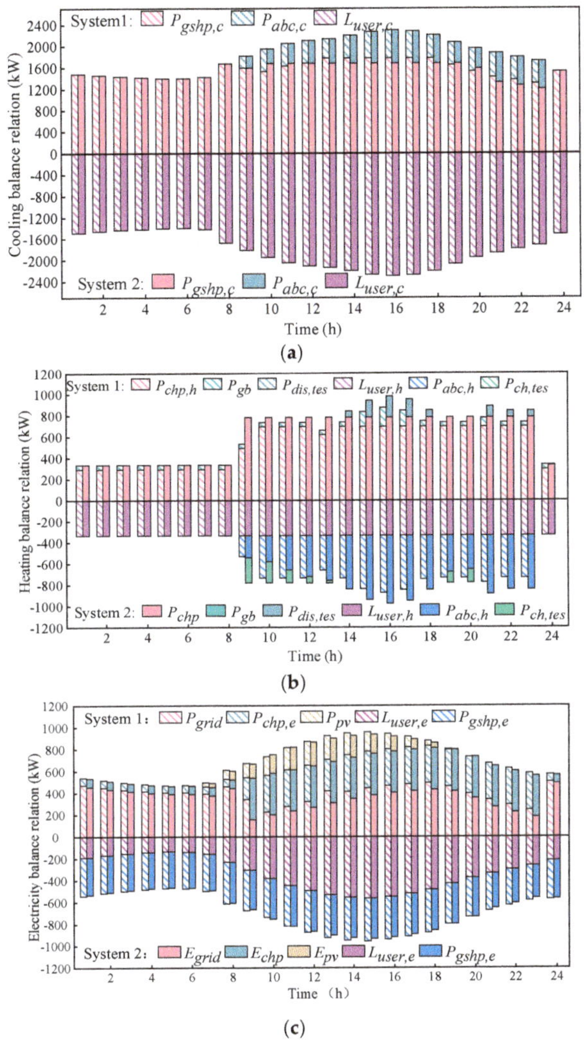

Figure 7. Scheduling process of System 1 and System 2 in design conditions: (**a**) Cooling load; (**b**) Heating load; (**c**) Electric load.

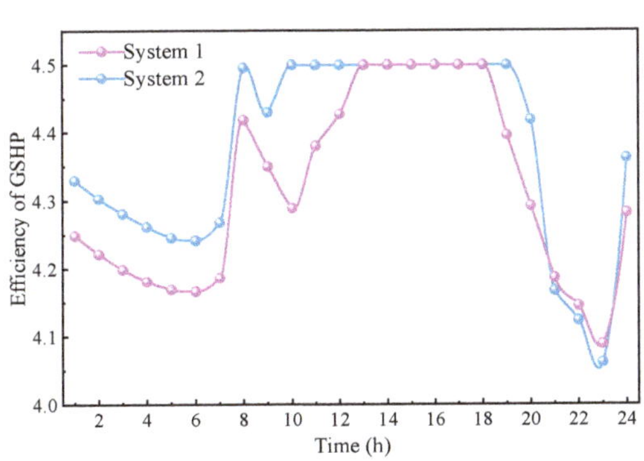

Figure 8. The efficiency of GSHP in System 1 and System 2.

To analyze the difference between the impact of the TES and EES devices on RIESs, this paper compares the electric load supply-demand relationship in System 2 and System 3 under the design conditions, as illustrated in Figure 9. First, the smaller capacity of the GSHP makes the electric load of System 3 less than that of System 2, while the larger CHP unit increases the power generation of the CHP unit in System 3. However, the carbon tax of System 3 is slightly less than that of System 2 due to the slight capacity difference between the GSHP and CHP units between System 2 and System 3. Secondly, the EES device can increase the grid power consumption during flat and valley hours and reduce the grid power consumption during high and peak hours. The difference in grid power consumption during different hours makes the operating cost of System 3 less than that of System 2.

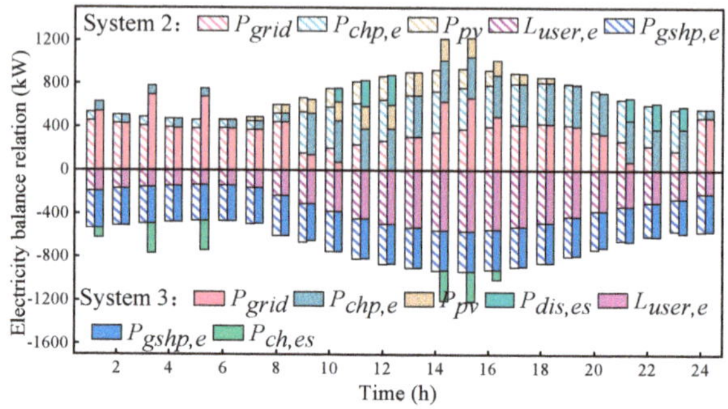

Figure 9. Electricity scheduling process of system 2 and system 3.

4.3. System Operation Result and Analysis

4.3.1. Calculation of Operating Load

The purpose of optimization design is to improve the operational performance of RIESs. To compare the operational performance of the three systems, the annual operating load was calculated by Energy Plus based on the typical annual outdoor meteorological parameters shown in Figure 10a, as shown in Figure 10b. In Changsha, which is hot in summer and cold in winter, the cooling load of the building is much greater than the heating load. Under summer operating conditions, the system cooling load is the building cooling load, and the heating load is the domestic hot water load. However, under winter operating conditions, the system heating load includes the building heating load and domestic hot water load. In addition, since the design parameters are determined by the method of non-guarantee days, the operating load may be greater than the design load most of the time. However, to ensure the feasibility of operation optimization, this study only considers days when the operating load is less than or equal to the design load.

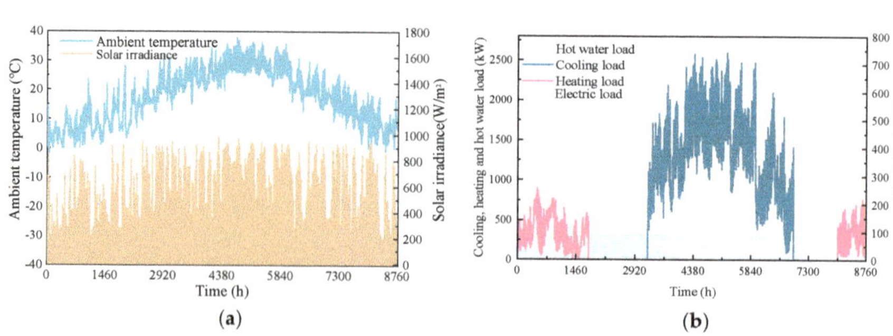

Figure 10. Outdoor parameters and building load under the operating conditions: (**a**) Annual outdoor meteorological parameters; (**b**) Building load under operating conditions.

4.3.2. Operation Results

Based on the above operating loads, the carbon taxes, operating costs, and total costs of the three systems are calculated under winter and summer operating conditions, and the results are shown in Table 9. Under the summer operating conditions, the carbon tax, operating cost, and total costs of System 2 and System 3 are less than those of System 1. Compared with the carbon tax, operating cost, and total cost of System 1, the corresponding costs in System 2 are reduced by 3.2%, 1.5%, and 0.3%, respectively. Different from System 2, since the carbon tax of System 3 is larger than that of System 2, the carbon tax reduction rate of System 3 compared with System 1 is smaller, only 2.7%. However, the EES device can significantly reduce the operating cost of System 3, which results in a 6.2% reduction in the operating cost of System 3 compared to System 1. Affected by the difference in building loads between winter and summer, the carbon taxes, operating costs, and total costs of the three systems under winter operating conditions are less than those under summer operating conditions. Under winter operating conditions, the carbon tax, operating cost, and total cost reduction rates of System 2 compared with System 1 are greater, whose values are 5.5%, 7.3%, and 2.8%, respectively. Similar to the summer operating conditions, System 3 can significantly reduce operating costs compared to System 2 but can also increase its carbon tax. Therefore, under operating conditions, compared with System 1, the TES device can significantly reduce the carbon tax, operating cost, and total cost of System 2, while the EES device can reduce the operating cost of System 3 even more significantly.

Table 9. Costs of three systems under winter and summer operating conditions.

Operating Condition	Summer			Winter		
	C_{tax} (CNY)	C_{op} (CNY)	C_{total} (CNY)	C_{tax} (CNY)	C_{op} (CNY)	C_{total} (CNY)
System 1	526,636	1,582,405	2,672,662	301,465	774,952	1,581,733
System 2	509,527	1,558,375	2,663,411	285,012	718,724	1,537,640
System 3	512,367	1,484,274	2,660,567	294,405	688,785	1,578,434

4.3.3. Operation Result Analysis

According to the analysis of the design condition results, the performance difference of RIESs is mainly affected by energy consumption and the consumption of different types of energy. To analyze the energy consumption of the three systems, this paper compares the average operation efficiency of the GSHPs and CHP units under the operating conditions, as shown in Figure 11. Firstly, under winter and summer operating conditions, the average operation efficiency of the GSHP in System 1, System 2, and System 3 increases sequentially due to the influence of energy storage devices. However, there is little difference between the average operation efficiency of the GSHP in System 2 and System 3. Secondly, the average operation efficiency of the CHP unit in System 2 is greater than in System 1 and System 3, especially in winter operating conditions. This shows that the TES device can significantly improve the operation efficiency of the GSHP and the CHP unit. In addition, the average operation efficiency of the CHP unit in System 3 is greater than that in System 1 under summer operating conditions, while under winter operating conditions, the average operating efficiency of the CHP unit in System 3 is approximately the same as that in System 1. The higher the average operation efficiency of the GSHP and the CHP units, the lower the energy consumption. Therefore, under the operating conditions, the energy consumption of System 2 and System 3 is less than that of System 1.

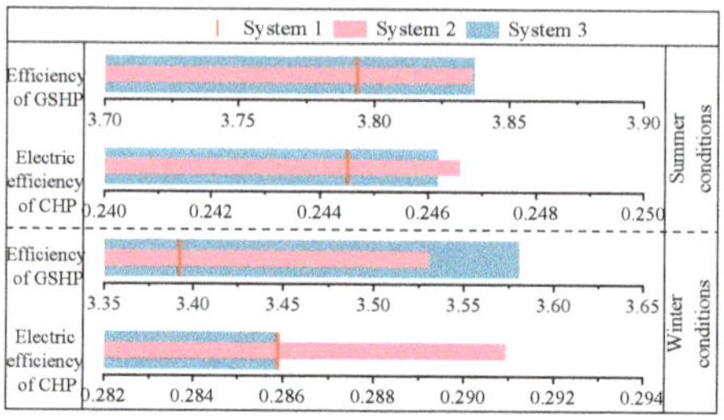

Figure 11. The average efficiency of GSHPs and CHP units under two operating conditions.

In the RIESs, carbon tax not only depends on the energy consumption of systems but is also affected by the ratio of different energy consumptions. To this end, different power consumption ratios of the three systems under winter and summer operating conditions are counted in this paper, and the results are shown in Figure 12. Under summer and winter operating conditions, the ratio of grid power consumption in System 1 is greater than that of System 2 and System 3; hence the carbon tax of System 2 and System 3 is less than that of System 1. Although there is a slight difference in the ratio of grid power consumption between System 2 and System 3 under summer operating conditions, the average operation efficiency of the CHP unit in System 3 is lower than that in System 2. Therefore, under summer operating conditions, the carbon tax of System 3 is slightly larger than System 2. Different from the summer operating conditions, the grid power consumption ratio of System 2 is significantly smaller than for System 3 under the winter operating conditions. As a result, the carbon taxes of System 2 and System 3 are greater than that of System 1, especially System 2.

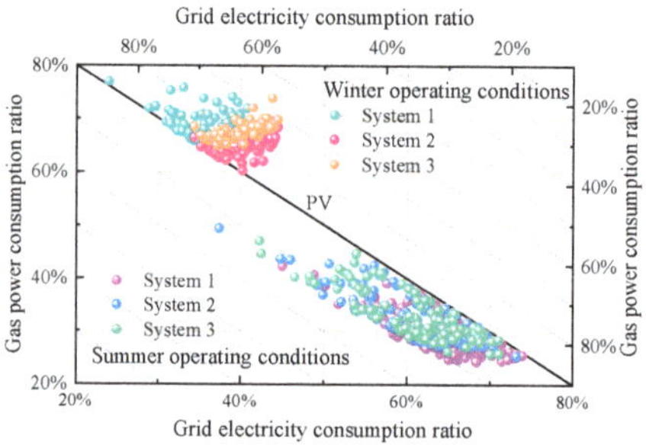

Figure 12. Different power consumption ratios under two operating conditions.

The system's operating cost is closely related to the TOU power price. To analyze the operating cost difference between the three systems, this paper draws the power consumption ratio of the three systems under TOU price hours, as shown in Figure 13. Under the summer operating conditions, the grid power consumption ratios of System 2 and System 3 during high and peak hours are less than that of System 1, especially for System 3, which accounts for only 22.12% during high and peak hours. Similar to the summer operating conditions, the power consumption ratios of System 2 and System 3 during high and peak hours are less than that of System 1 under winter operating conditions. However, under winter operating conditions, the difference in the grid power consumption ratio between System 1 and System 2 is greater than that between System 2 and System 3 during high and peak hours. Therefore, under winter and summer operating conditions, the operating costs of System 2 and System 3 are greater than that of System 1, and the operating cost of System 3 is the lowest, especially under winter operating conditions.

System	Valley	Flat	High	Peak	
System 1	38.64%	25.7%	27.53%	8.13%	Summer conditions
System 2	39.1%	26.58%	27.09%	7.23%	
System 3	45.29%	32.58%	20.16%	1.96%	
System 1	51.33%	36.47%	11.11%	1.1%	Winter conditions
System 2	58.07%	38.54%	3.27%	0.13%	
System 3	63.16%	36.71%	0.11%	0.02%	

Figure 13. The ratio of grid power consumption under the TOU power price.

The above analysis shows that the carbon taxes and operating costs of System 2 and System 3 are less than those of System 1 under winter and summer operating conditions due to the impact of the average efficiency of the equipment, the ratio of different energy consumptions, and the grid power consumption ratio under the TOU power price. The smaller carbon tax and operating cost make the total cost of Systems 2 and System 3 less than that of System 1. However, affected by the equipment cost, the total cost reduction rate difference between System 2 and System 3 compared with System 1 is not so obvious. In addition, under winter operating conditions, the total cost of System 3 is greater than that of System 2 due to the lower operating cost difference between System 2 and System 3.

4.4. Uncertainty Analysis of Energy Price

In the actual operation process, the price of grid power and gas directly affected the operation result of the system. To explore the impact of energy price uncertainty on system operation results, this study increases or decreases the energy prices to ±30% in a 10% step, based on the energy prices in the design conditions. The three systems are optimized for operation according to different energy prices. Based on optimization results, the cost reduction rates of System 2 and System 3 compared with System 1 are calculated, as shown in Figure 14. In the case of changes in energy prices, the reduction rates of the carbon tax, operating cost, and total cost of System 2 compared with System 1 are greater than zero. Different from System 2, the operating cost reduction rate of System 3 compared with System 1 is larger, while the carbon tax and total cost reduction rates are lower. Even under partial energy prices, the carbon tax and total cost of System 3 are greater than they are for System 1. Under different energy prices, the TES device can effectively reduce the operating cost, carbon tax, and total cost of the RIES, while the EES device can significantly reduce the system operating cost.

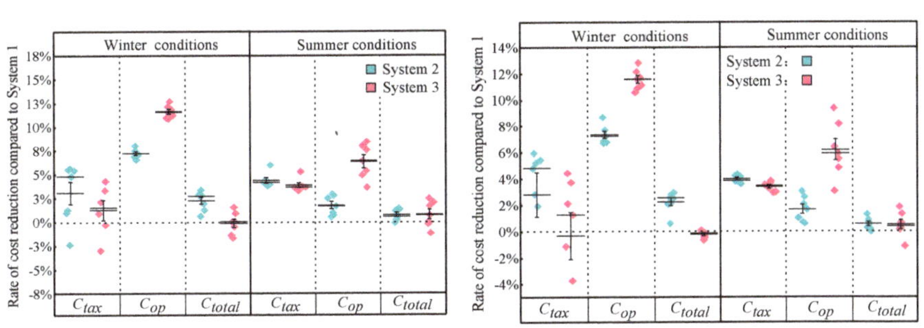

(a) (b)

Figure 14. Cost reduction rates of system 2 and system 3 compared with system 1 under changes in electricity and gas prices: (**a**) Affected by changes in electricity prices; (**b**) Affected by changes in gas prices.

5. Conclusions

The purpose of this study is to explore the impact of energy storage devices on the optimal design and operation of RIESs. For this purpose, a bi-level dynamic optimization model was first established based on the dynamic equipment model. Then, the bi-level dynamic optimization model was used to optimize the design of RIESs with different energy storage devices, and the optimization results are System 1, System 2, and System 3, respectively. According to the optimization results, the impact of energy storage devices on system performance is compared and analyzed. The main conclusions of this study include the following three points:

1. Compared with System 1 without energy storage devices, energy storage devices can increase the capacity of CHP units and ABCs in System 2 and System 3 and reduce the capacity of GSHPs and gas boilers, especially the TES device. Affected by the equipment capacity, the equipment cost increase rates of System 2 and System 3 compared with System 1 are 5.7% and 17.8%, respectively. This shows that the EES device will significantly increase the equipment cost of System 3.
2. The difference in equipment capacity affects not only the equipment cost but also the operation performance of the system. Under the design conditions, higher equipment operation efficiency and lower grid power consumption make the operation cost, carbon tax, and total cost of System 2 lower than that of System 1, with reductions of 2.9%, 5.5%, and 1.5%, respectively. Under the influence of TOU electricity price, the EES device can significantly reduce the operating cost of System 3, which is 5.7% lower than that of System 1.
3. Under the operating conditions, the operating cost, carbon tax, and total cost of System 2 and System 3 remain lower than that of System 1, even if the energy price changes. Therefore, in the design of future RIESs, energy storage devices, especially TES devices, can be used to improve the energy efficiency of RIESs and reduce the operation cost and total cost.

Author Contributions: Conceptualization, B.J.; methodology, B.J.; software, B.J. and Z.L.; validation, B.J. and Z.L.; formal analysis, B.J. and Z.L.; investigation, B.J.; data curation, B.J. and Z.L.; writing—original draft preparation, B.J. and Z.L.; writing—review and editing, B.J. and Y.L.; visualization, B.J.; supervision, B.J.; project administration, B.J. All authors have read and agreed to the published version of the manuscript.

Funding: This research received no external funding.

Institutional Review Board Statement: Not applicable.

Informed Consent Statement: Not applicable.

Data Availability Statement: Not applicable.

Conflicts of Interest: The authors declare no conflict of interest.

Nomenclature

Nomenclature		**Greek symbols**	
Abbreviation		η	Charging and discharging efficiency
A	Area	ϑ	Energy price/ carbon tax price
ABC	Absorption chiller	α	Energy storage ratios
Cap	Capacity	λ	Carbon dioxide emissions factor
CCHP	Combined cooling heating and power	γ	Charging and discharging ratios
CHP	Combined heating and power	**Subscript**	
COP	Coefficient of performance	a	Ambient

EES	Electrical energy storage	abc	Absorption chiller
EH	Energy hub	c	Cooling
FEL	Following electric load	ch	Charge
FHL	Following hybrid electric-heating load	chp	Combined heating and power
FOF	Following objective function	CO_2	carbon dioxide
FSF	Following system flexibility	dis	Discharge
FTL	Following thermal load	e	Electricity
GA	Genetic algorithm	equ	Equipment
GB	Gas boiler	ees	Electrical energy storage
GSHP	Ground source heat pump	gas	Natural gas
HES	Hydrogen energy storage	gb	Gas boiler
L	Load	gshp	Ground source heat pump
P	Power	grid	Grid power
P-G	Power-gas	h	Heating
PLF	Part-load ratio	inv	Initial investment
PLR	Part-load factor	k	Device type
PV	Photovoltaic	op	Operating
RIES	Regional integrated energy system	pv	Photovoltaic
S	Energy storage device status	r	Rated
SES	Share energy storage	tes	Thermal energy storage
t	Temperature/Time	**Superscript**	
TES	Thermal energy storage	max	Maximum
TOU	Time-of-use	min	Minimum

References

1. Zhang, Q.; Ren, Z.; Ma, R.; Tang, M.; He, Z. Research on Double-Layer Optimized Configuration of Multi-Energy Storage in Regional Integrated Energy System with Connected Distributed Wind Power. *Energies* **2019**, *12*, 3964. [CrossRef]
2. Al-Najjar, H.M.T.; Mahdi, J.M.; Bokov, D.O.; Khedher, N.B.; Alshammari, N.K.; Opulencia, M.J.C.; Fagiry, M.A.; Yaïci, W.; Talebizadehsardari, P. Improving the Melting Duration of a PV/PCM System Integrated with Different Metal Foam Configurations for Thermal Energy Management. *Nanomaterials* **2022**, *12*, 423. [CrossRef] [PubMed]
3. Guo, J.; Liu, Z.; Wu, X.; Wu, D.; Zhang, S.; Yang, X.; Ge, H.; Zhang, P. Two-Layer Co-Optimization Method for a Distributed Energy System Combining Multiple Energy Storages. *Appl. Energy* **2022**, *322*, 119486. [CrossRef]
4. Wei, W.; Guo, Y.; Hou, K.; Yuan, K.; Song, Y.; Jia, H.; Sun, C. Distributed Thermal Energy Storage Configuration of an Urban Electric and Heat Integrated Energy System Considering Medium Temperature Characteristics. *Energies* **2021**, *14*, 2924. [CrossRef]
5. Ben Khedher, N.; Bantan, R.A.; Kolsi, L.; Omri, M. Performance Investigation of a Vertically Configured LHTES via the Combination of Nano-Enhanced PCM and Fins: Experimental and Numerical Approaches. *Int. Commun. Heat Mass Transf.* **2022**, *137*, 106246. [CrossRef]
6. Geidl, M.; Koeppel, G.; Favre-Perrod, P.; Klockl, B. Energy Hubs for the Future. *IEEE Power Energy Mag.* **2007**, *5*, 24–30. [CrossRef]
7. Ma, T.; Wu, J.; Hao, L.; Li, D. Energy Flow Matrix Modeling and Optimal Operation Analysis of Multi Energy Systems Based on Graph Theory. *Appl. Therm. Eng.* **2019**, *146*, 648–663. [CrossRef]
8. Ma, T.; Wu, J.; Hao, L.; Lee, W.J.; Yan, H.; Li, D. The Optimal Structure Planning and Energy Management Strategies of Smart Multi Energy Systems. *Energy* **2018**, *160*, 122–141. [CrossRef]
9. Ben Salah, S.; Ben Hamida, M.B. Alternate PCM with Air Cavities in LED Heat Sink for Transient Thermal Management. *Int. J. Numer. Methods Heat Fluid Flow* **2019**, *29*, 4377–4393. [CrossRef]
10. Ben Hamida, M.B.; Hajlaoui, K.; Almeshaal, M.A. A 3D Numerical Analysis Using Phase Change Material for Cooling Circular Light Emitting Diode. *Case Stud. Therm. Eng.* **2023**, *43*, 102792. [CrossRef]
11. Waseem, M.; Lin, Z.; Liu, S.; Sajjad, I.A.; Aziz, T. Optimal GWCSO-Based Home Appliances Scheduling for Demand Response Considering End-Users Comfort. *Electr. Power Syst. Res.* **2020**, *187*, 106477. [CrossRef]
12. Liu, C.; Li, Y.; Wang, Q.; Wang, X.; Chen, C.; Lin, Z.; Yang, L. Optimal Configuration of Park-Level Integrated Energy System Considering Integrated Demand Response and Construction Time Sequence. *Energy Rep.* **2022**, *8*, 1174–1180. [CrossRef]
13. Ding, Y.; Xu, Q.; Yang, B. Optimal Configuration of Hybrid Energy Storage in Integrated Energy System. *Energy Rep.* **2020**, *6*, 739–744. [CrossRef]
14. Deng, Z.G.; Yang, J.H.; Dong, C.L.; Xiang, M.Q.; Qin, Y.; Sun, Y.S. Research on Economic Dispatch of Integrated Energy System Based on Improved Krill Swarm Algorithm. *Energy Rep.* **2022**, *8*, 77–86. [CrossRef]
15. Mansouri, S.A.; Nematbakhsh, E.; Ahmarinejad, A.; Jordehi, A.R.; Javadi, M.S.; Matin, S.A.A. A Multi-Objective Dynamic Framework for Design of Energy Hub by Considering Energy Storage System, Power-to-Gas Technology and Integrated Demand Response Program. *J. Energy Storage* **2022**, *50*, 104206. [CrossRef]
16. Mago, P.; Chamra, L. Analysis and optimization of CCHP systems based on energy, economical, and environmental considerations. *Energy Build.* **2009**, *41*, 1099–1106. [CrossRef]

17. Mago, P.J.; Chamra, L.M.; Ramsay, J. Micro-combined cooling, heating and power systems hybrid electric-thermal load following operation. *Appl. Therm. Eng.* **2010**, *30*, 800–806. [CrossRef]
18. Kang, S.; Li, H.; Liu, L.; Zeng, R.; Zhang, G. Evaluation of a Novel Coupling System for Various Load Conditions under Different Operating Strategies. *Energy Convers Manag* **2016**, *109*, 40–50. [CrossRef]
19. Wang, Y.; Zhang, Y.; Xue, L.; Liu, C.; Song, F.; Sun, Y.; Liu, Y.; Che, B. Research on Planning Optimization of Integrated Energy System Based on the Differential Features of Hybrid Energy Storage System. *J. Energy Storage* **2022**, *55*, 105368. [CrossRef]
20. Duan, Z.; Yan, Y.; Yan, X.; Liao, Q.; Zhang, W.; Liang, Y.; Xia, T. An MILP Method for Design of Distributed Energy Resource System Considering Stochastic Energy Supply and Demand. *Energies* **2018**, *11*, 22. [CrossRef]
21. Zeng, R.; Li, H.; Jiang, R.; Liu, L.; Zhang, G. A Novel Multi-Objective Optimization Method for CCHP-GSHP Coupling Systems. *Energy Build.* **2016**, *112*, 149–158. [CrossRef]
22. Zeng, R.; Zhang, X.; Deng, Y.; Li, H.; Zhang, G. An Off-Design Model to Optimize CCHP-GSHP System Considering Carbon Tax. *Energy Convers. Manag.* **2019**, *189*, 105–117. [CrossRef]
23. Yang, G.; Zhai, X. Optimization and Performance Analysis of Solar Hybrid CCHP Systems under Different Operation Strategies. *Appl. Therm. Eng.* **2018**, *133*, 327–340. [CrossRef]
24. Yang, G.; Zhai, X.Q. Optimal Design and Performance Analysis of Solar Hybrid CCHP System Considering Influence of Building Type and Climate Condition. *Energy* **2019**, *174*, 647–663. [CrossRef]
25. Ren, F.; Wei, Z.; Zhai, X. Multi-Objective Optimization and Evaluation of Hybrid CCHP Systems for Different Building Types. *Energy* **2021**, *215*, 119096. [CrossRef]
26. Li, Y.; Tian, R.; Wei, M. Operation Strategy for Interactive CCHP System Based on Energy Complementary Characteristics of Diverse Operation Strategies. *Appl. Energy* **2022**, *310*, 118415. [CrossRef]
27. Deng, Y.; Zeng, R.; Liu, Y. A Novel Off-Design Model to Optimize Combined Cooling, Heating and Power System with Hybrid Chillers for Different Operation Strategies. *Energy* **2022**, *239*, 122085. [CrossRef]
28. Han, Z.; Ma, F.; Wu, D.; Zhang, H.; Dong, F.; Li, P.; Xiao, L. Collaborative Optimization Method and Operation Performances for a Novel Integrated Energy System Containing Adiabatic Compressed Air Energy Storage and Organic Rankine Cycle. *J. Energy Storage* **2021**, *41*, 102942. [CrossRef]
29. Ghersi, D.E.; Amoura, M.; Loubar, K.; Desideri, U.; Tazerout, M. Multi-Objective Optimization of CCHP System with Hybrid Chiller under New Electric Load Following Operation Strategy. *Energy* **2021**, *219*, 119574. [CrossRef]
30. Zhou, Y.; Wang, J.; Dong, F.; Qin, Y.; Ma, Z.; Ma, Y.; Li, J. Novel Flexibility Evaluation of Hybrid Combined Cooling, Heating and Power System with an Improved Operation Strategy. *Appl. Energy* **2021**, *300*, 117358. [CrossRef]
31. Luo, X.; Xia, J.; Liu, Y. Extraction of Dynamic Operation Strategy for Standalone Solar-Based Multi-Energy Systems: A Method Based on Decision Tree Algorithm. *Sustain. Cities Soc.* **2021**, *70*, 102917. [CrossRef]
32. Lu, S.; Li, Y.; Xia, H. Study on the Configuration and Operation Optimization of CCHP Coupling Multiple Energy System. *Energy Convers. Manag.* **2018**, *177*, 773–791. [CrossRef]
33. Luo, X.; Zhu, Y.; Liu, J.; Liu, Y. Design and Analysis of a Combined Desalination and Standalone CCHP (Combined Cooling Heating and Power) System Integrating Solar Energy Based on a Bi-Level Optimization Model. *Sustain. Cities Soc.* **2018**, *43*, 166–175. [CrossRef]
34. Li, K.; Wei, X.; Yan, Y.; Zhang, C. Bi-Level Optimization Design Strategy for Compressed Air Energy Storage of a Combined Cooling, Heating, and Power System. *J. Energy Storage* **2020**, *31*, 101642. [CrossRef]
35. Ma, M.; Huang, H.; Song, X.; Peña-Mora, F.; Zhang, Z.; Chen, J. Optimal Sizing and Operations of Shared Energy Storage Systems in Distribution Networks: A Bi-Level Programming Approach. *Appl. Energy* **2022**, *307*, 118170. [CrossRef]
36. Song, Z.; Liu, T.; Lin, Q. Multi-Objective Optimization of a Solar Hybrid CCHP System Based on Different Operation Modes. *Energy* **2020**, *206*, 118125. [CrossRef]
37. Guan, T.; Lin, H.; Sun, Q.; Wennersten, R. Optimal Configuration and Operation of Multi-Energy Complementary Distributed Energy Systems. *Energy Procedia* **2018**, *152*, 77–82. [CrossRef]
38. Biglia, A.; Caredda, F.V.; Fabrizio, E.; Filippi, M.; Mandas, N. Technical-Economic Feasibility of CHP Systems in Large Hospitals through the Energy Hub Method: The Case of Cagliari AOB. *Energy Build.* **2017**, *147*, 101–112. [CrossRef]
39. Jian, W.; Yuan, R. *Design and Optimization on Building Distributed Energy System*; Tong Ji University Press: Shanghai, China, 2016; pp. 37–56.
40. Waseem, M.; Lin, Z.; Liu, S.; Zhang, Z.; Aziz, T.; Khan, D. Fuzzy Compromised Solution-Based Novel Home Appliances Scheduling and Demand Response with Optimal Dispatch of Distributed Energy Resources. *Appl. Energy* **2021**, *290*, 116761. [CrossRef]
41. Liu, Y.; Wang, Y.; Luo, X. Design and Operation Optimization of Distributed Solar Energy System Based on Dynamic Operation Strategy. *Energies* **2021**, *14*, 69. [CrossRef]
42. Ma, W.; Fang, S.; Liu, G. Hybrid Optimization Method and Seasonal Operation Strategy for Distributed Energy System Integrating CCHP, Photovoltaic and Ground Source Heat Pump. *Energy* **2017**, *141*, 1439–1455. [CrossRef]
43. Luo, Z.; Yang, S.; Xie, N.; Xie, W.; Liu, J.; Souley Agbodjan, Y.; Liu, Z. Multi-Objective Capacity Optimization of a Distributed Energy System Considering Economy, Environment and Energy. *Energy Convers. Manag.* **2019**, *200*, 112081. [CrossRef]
44. Ren, F.; Wang, J.; Zhu, S.; Chen, Y. Multi-Objective Optimization of Combined Cooling, Heating and Power System Integrated with Solar and Geothermal Energies. *Energy Convers. Manag.* **2019**, *197*, 111866. [CrossRef]

45. Zhang, T.; Wang, M.; Wang, P.; Liang, J. Optimal Design of a Combined Cooling, Heating, and Power System and Its Ability to Adapt to Uncertainty. *Energies* **2020**, *13*, 3588. [CrossRef]
46. Guo, S.; Song, G.; Li, M.; Zhao, X.; He, Y.; Kurban, A.; Ji, W.; Wang, J. Multi-Objective Bi-Level Quantity Regulation Scheduling Method for Electric-Thermal Integrated Energy System Considering Thermal and Hydraulic Transient Characteristics. *Energy Convers. Manag.* **2022**, *253*, 115147. [CrossRef]
47. Ju, L.; Yin, Z.; Yang, S.; Zhou, Q.; Lu, X.; Tan, Z. Bi-Level Electricity–Carbon Collaborative Transaction Optimal Model for the Rural Electricity Retailers Integrating Distributed Energy Resources by Virtual Power Plant. *Energy Rep.* **2022**, *8*, 9871–9888. [CrossRef]
48. Air Conditioning Outdoor Design Parameters, Energy Plus Is Available Online. 2022. Available online: https://energyplus.net/weather-location/asia_wmo_region_2/CHN/CHN_Hunan.Changde.576620_CSWD (accessed on 15 June 2022).
49. Zou, X. Study on Optimization and Control Strategies of Solar-Assisted Ground-Coupled Heat Pump Domestic Hot Water System. Master's Thesis, Hunan University, Changsha, China, 2016.
50. Changsha ENN Gas Co., Ltd. Natural Gas Price. 2022. Available online: http://www.cs95158.cn/contents/68/1594.html (accessed on 15 June 2022).
51. Zou, B.; Peng, J.; Li, S.; Li, Y.; Yan, J.; Yang, H. Comparative Study of the Dynamic Programming-Based and Rule-Based Operation Strategies for Grid-Connected PV-Battery Systems of Office Buildings. *Appl. Energy* **2022**, *305*, 117875. [CrossRef]
52. Chen, Y.; Xu, J.; Wang, J.; Lund, P.D.; Wang, D. Configuration Optimization and Selection of a Photovoltaic-Gas Integrated Energy System Considering Renewable Energy Penetration in Power Grid. *Energy Convers. Manag.* **2022**, *254*, 115260. [CrossRef]

energies

MDPI

Review

Maximizing the Integration of a Battery Energy Storage System–Photovoltaic Distributed Generation for Power System Harmonic Reduction: An Overview

Adedayo Owosuhi [1,*], Yskandar Hamam [1,2] and Josiah Munda [1]

[1] Department of Electrical Engineering, French South African Institute of Technology (F'SATI), Tshwane University of Technology, Pretoria 0001, South Africa
[2] École Supérieure d'Ingénieurs en Électrotechnique et Électronique, Cité Descartes, 2 Boulevard Blaise Pascal, Noisy-le-Grand, 93160 Paris, France
* Correspondence: owosuhiadedayo@gmail.com; Tel.: +2348032381702

Abstract: The highly variable power generated from a battery energy storage system (BESS)–photovoltaic distributed generation (PVDG) causes harmonic distortions in distribution systems (DSs) due to the intermittent nature of solar energy and high voltage rises or falls in the BESS. Harmonic distortions are major concerns in the DS, especially when the sizes and locations of these resources are sub-optimal. As a result, many studies are being performed on the optimal allocation of BESS/PVDG systems in distribution network systems. In this regard, this paper seeks to review the existing planning models, optimization methods and renewable energy resources that uncertainty models have employed in solving BESS/PVDGs allocation problems in terms of obtaining optimal solutions/allocations and curtailing the harmonic contents of the DSs. However, studies on optimal allocation planning of BESS/PVDGs have achieved minimum cost but were not able to meet the standard harmonic level of the DSs. The results identified GA, PSO and AIS as some of the most used methodologies while LP, MILP and different configurations of NLP were used in the model formulations of BESS/PVDGs problems. The results also revealed DC-link voltage and switching and grid voltage harmonics as the notable causes and sources of harmonic distortions in BESS/PVDG systems. The current allocation models presented in the recent literature for the planning of BESS/PVDGs do not include the variables necessary for curtailing the harmonic contents in their planning formulations. This paper, therefore, recommends an improved and all-encompassing planning model with an efficient intelligent search algorithm capable of obtaining a global optimum solution and curtailing harmonic distortions from the BESS/PVDG-connected DSs.

Keywords: photovoltaic distributed generation; battery energy storage system; distribution network system; optimization methodologies; harmonic distortions

Citation: Owosuhi, A.; Hamam, Y.; Munda, J. Maximizing the Integration of a Battery Energy Storage System–Photovoltaic Distributed Generation for Power System Harmonic Reduction: An Overview. *Energies* **2023**, *16*, 2549. https://doi.org/10.3390/en16062549

Academic Editors: Branislav Hredzak, Luis Hernández-Callejo, Jesús Armando Aguilar Jiménez and Carlos Meza Benavides

Received: 20 December 2022
Revised: 17 February 2023
Accepted: 24 February 2023
Published: 8 March 2023

1. Introduction

The rapid expansion in socioeconomics has led to increasing global energy demand and usage. To balance the resulting widening energy deficiency, renewable energy distributed generation (REDG) is considered as an effective approach to solve the rising energy demand and other power system issues that are technical, economic and environmental in nature [1–3]. REDGs are generation technologies integrated at distribution networks near load users to satisfy immediate power demand, defer network upgrade, enhance power quality and reliability, diversify energy resources, and to reduce power losses, distribution and transmission loading, distribution and transmission costs and on-peak operating costs [4].

The battery energy storage system–photovoltaic DG (BESS/PVDG) is a viable renewable option because the resources are inexhaustible, complementary, economically profitable, environmentally friendly and bi-directional [5–8]. However, the power generated from BESS/PVDG depends on charge and discharge schedules of BESS, which is associated with high voltage rise or fall, and temperature and irradiation of solar energy that is intermittent in nature [6,9–11]. Hence, a substantial number of research studies have unanimously agreed/concurred that harmonics occur in the distribution system when BESS/PVDG units are absorbed due to the intermittent and variable nature of PVDG output power and the high variability of voltage and frequency of BESS schedules. In essence, current harmonics occur as a result of sudden disparity between the aggregate output power of BESS/PVDGs and other generations and the total power demand at an instant in a distribution system. The high rises and falls of the voltage and frequency from battery charge/discharge schedules may result in voltage harmonics [11].

The harmonic distortions are a troubling power quality issue for BESS/PVDG power generation, and they have significant consequences on the DNs. The extent of current harmonics is determined by the active output power from BESS/PVDGs. Thus, the magnitudes of current harmonics are enormous at utility-scale BESS/PVDGs penetration levels. The intermittency of PVDG units and the high voltage rise or fall from BESS/PVDG raise concerns on distribution system harmonic distortions, which have negative effects on power quality, stability and reliability of distribution systems [6,12,13]. The high harmonic contents in the power system lead to increased losses in system elements such as transformers and generating plants; economic costs such as productivity, energy and device/equipment losses; and fire hazards due to overheating of system elements [7,14,15]. The issues mentioned make the integration of a large-scale BESS/PVDG into the distribution systems difficult [6,15,16]. Meanwhile, the locations and sizes of BESS/PVDG units could either improve or impair the magnitudes current and voltage harmonic levels of the networks [17–19]. The mentioned issues make the solution of BESS/PVDG allocation problems formulated using simple mathematical models unrealistic. A realistic model, therefore, requires a dynamic model representation of the network, the use of multi-period planning horizon as well as all the necessary constraints. The problem then becomes a multi-objective one with a maximisation of renewable active and reactive powers into the DNs and a minimisation of the total cost subject to the capacity, investment, technical, stability and harmonic constraints throughout the planning horizon.

Several studies have been performed to proffer optimal solutions for the planning allocation of BESS/PVDG in distribution systems [11,13,16,20–23]. The studies on optimal planning of REDG allocation warrant detailed investigations on the prospects of BESS/PVDGs for generating power, the impact on the DNs, and the effects on the inadequate availability and rising cost of energy, the global economy and environment. Various researchers have reviewed some aspects of the BESS/PVDG allocation planning (BESS/PVDGs-AP) problem. Many solution algorithms, planning models, and emerging technologies deployed in BESS/PVDG-AP have been presented [24–28]. Zahraee et al. [24] presented an analysis of some artificial intelligence optimum plans used in the optimization and sizing of hybrid renewable energy systems. The main contribution of this work is the extensive penetration of renewable energy features for economic performance of the systems. The authors in [25] dealt with the review of some solutions that were used to improve the ability of the distribution system to cope with variable renewable energy source unpredictability such as energy storage technologies, PV and wind energy systems. This study concluded that battery energy storage and pump hydro energy storage are the most used technologies to improve the impact of the variable renewable power on distribution systems. A review presented by Hannan et al. [26] on the planning of BESS and renewable energy hybrid DGs discussed the optimal sizing objectives, various optimization models, the BESS system constraints together with their advantages and weaknesses. A detailed discussion of the BESS applications and shortage of optimal BESS sizing models could be identified as the strong point of this study. In [27], a review of the latest research developments and challenges on optimal planning of a BESS-PVDG connected distribution system was presented. The authors suggested key parameters in the process of optimal planning for a PV–battery system such as economic and technical data, objective functions, energy management schemes, design constraints, optimization algorithms, and electricity pricing regimes.

In view of the contribution of the existing review works on the BESS/PVDGs-AP problem, this study is distinct in these ways:

- Based on the authors' awareness, no literature has presented the evaluation of harmonic components of BESS/PVDGs during integration into distribution networks/systems.
- Unlike the existing reviews, this review presents an overview of harmonic distortions in battery energy storage–photovoltaic hybrid distributed generation systems.
- This study provides a methodology for curtailing harmonic distortions from the BESS/PVDGs-connected distribution systems.
- Moreover, a substantial and diverse number of optimization/solution algorithms deployed in solving the BESS/PVDGs allocation problem is surveyed, comparing all their characteristics to assist the researchers to utilize them successfully and in a cost-effective way.

Despite numerous reviews and studies on BESS/PVDGs, some aspects have not been adequately captured for investigation, review and research. These themes, bulleted above, are comprehensively treated in this paper.

The remaining parts of this paper are organized as follows: Section 2 presents the overview of harmonic components in the BESS/PVDG connected distribution networks. Section 3 details a review of various optimization models and techniques published in the existing research works and some promising algorithms that are recently introduced and used for solving BESS/PVDG allocation optimization problems. The methodological approach for curtailing the harmonic distortions in a BESS/PVDG connected distribution system is presented in Section 4. The characteristics of all the models and techniques are compared, and their shortcomings are discussed under Results and Discussion in Section 5, to assist the researchers in choosing and applying them successfully and in a cost-effective way. Section 6 is the concluding part of the paper, and the recommendations for future research directions are also presented here.

2. Overview of Harmonic Components in BESS/PVDG Systems

Power system harmonic distortion is a major issue for power utilities throughout the world. In recent times, statistical analysis reports have revealed that power system harmonics has become a very troubling power quality issue in BESS/PVDG systems. These harmonics have resonating impacts in generating other power quality problems in large-scale BESS/PVDG [7,15,29–31]. The sources of harmonics produced in BESS/PVDGs are broadly classified into DC-link voltage harmonics, switching harmonics and grid voltage harmonics [7,32].

2.1. DC-Link Voltage Harmonics

The DC-link voltage ripples have become a major source of harmonics produced by BESS/PVDGs [32]. The DC-link voltage harmonics are generated by PVDGs due to solar irradiation intermittency and the high rises or falls of BESS voltage. Du et al. [32] illustrated this phenomenon with the experimental setup simulated in MATLAB Simulink. The experimental results in Figure 1 show that the harmonic distortion increases as DC-link voltage increases. However, these harmonics are usually taken as constant in the analyses and designs of BESS/PVDG inverters. They are not always so in the practical sense. This accounts for the odd harmonic frequencies discovered in the spectrum of BESS/PVDG inverter's output current [33]. In addition, Mansor et al. [34] investigated harmonic generation in three-phase BESS/PVDG inverters and found that the second-order harmonics in the DC link produced the third-order harmonic discovered on the AC side of the inverter. [34]. Many methods have been proposed by the researchers to eliminate the current harmonics generated by the DC-link voltage ripple [35–39]. Some of the proposed methods reduced the dynamic performance of the system, and many lack quality information on the connection between the output current harmonics and DC-link voltage ripples [32].

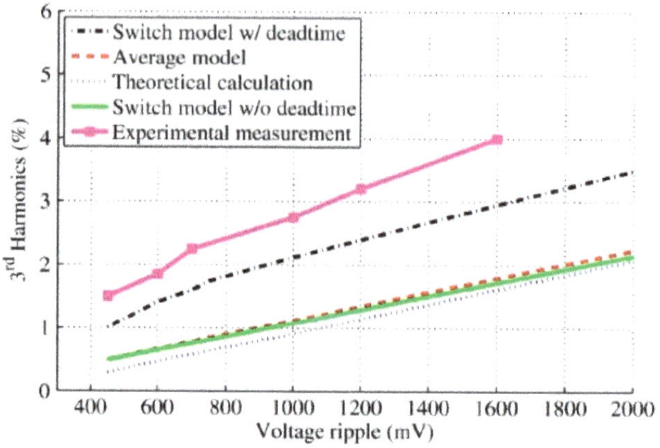

Figure 1. Impact of BESS/PVDG DC-link voltage ripples on harmonics [32].

2.2. *Switching Harmonics*

Switching harmonics is one other cause of current harmonics in BESS/PVDG inverter output. It occurs due to a mismatch in thee generation of switching pulses. The switching harmonics in PWM inverters always double their switching frequency [40,41]. Switching harmonics are very difficult to control and require an appropriate control strategy and optimized BESS/PVDG units; otherwise, system instabilities, harmonic generation and power losses ensue [32,42,43]. Various researchers have presented different methods to control or eliminate the switching harmonics of BESS/PVDG inverters [40,44,45].

Other research works maintained that the effects of quantization and resolution on control systems' measuring instruments are another potent source of harmonics in BESS/PVDG systems [44,46]. Also listed are the inadequacies of the current controllers of inverters in reducing harmonic contents and the positioning of sensors and locations of BESS and PVDG units in the distribution networks [47–49]. The outer voltage control loop of a two-series control algorithm and the PLL system could be another cause of reference current harmonics. In addition, output current harmonics could emerge from the dead time for switching pulse of the BESS/PVDG inverters [32,43].

2.3. *Grid Voltage Harmonics*

The BESS/PVDG inverter output current is produced due to the variation between the inverter's AC output voltage and the distribution network voltage. The output current harmonics are generated from the grid voltage when the grid voltage waveform includes harmonic components. The field measurements and research literature revealed that the grid voltages consistently have harmonics in varying degrees at different locations of the network [7,31,42,43]. For example, Figure 2a,b show the measured individual voltage harmonics up to the 31st order for one PVDG inverter at phase B of the grid and the combination of one PVDG and one BESS inverter for harmonics up to order 25 at phase B [31]. Grid voltage harmonics are usually low orders and are very difficult to annihilate by the filters. Numerous methods have been presented to control current harmonics generated from the grid voltage harmonics [50–53]. Du et al. [43] stated that the current harmonics sourced from grid background voltage do not depend on the magnitude of inverter output power. The grid voltage harmonics only reflect the magnitude of output current harmonics [43].

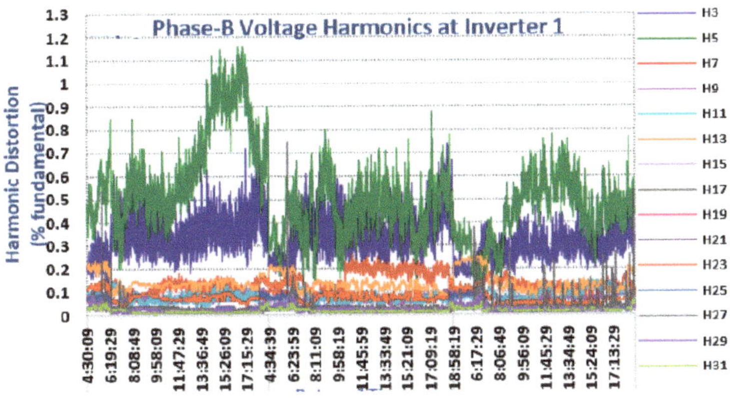

(a)

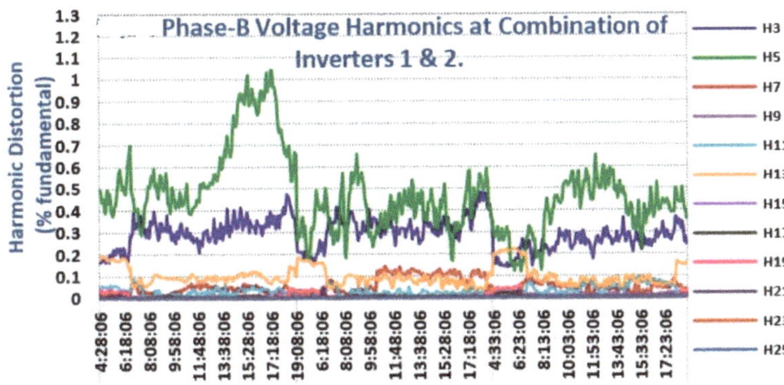

(b)

Figure 2. Individual voltage harmonics of phase B at (**a**) PVDG, (**b**) BESS outputs [31].

2.4. Harmonic Standards for Large-Scale BESS/PVDGs

Power quality is a power system requirement stipulated in all the international standards governing the grid connection of BESS/PVDG systems. Table 1 shows the IEEE 1547 and IEC 61727 standards as related to the requirements for current harmonics of the grid-connected BESS/PVDG systems [42,54,55]. The total harmonic distortion (THD) of generated current should not exceed 5% limit.

Table 1. Current harmonics limits by IEEE 1547 and IEC 61727 standards [54].

Harmonics Orders (I_H)	Corresponding to Fundamental (%)
A. Odd Harmonics	
3, 5, 7, 9	Less than 4%
11, 13, 15	Less than 2%
17, 19, 21	Less than 1.5%
23, 25, 27, 29, 31, 33	Less than 0.6%
>33	Less than 0.3%
B. Even Harmonics (All)	Less than 25% of various Odd harmonics
Total Harmonic Distortion (THD)	Less than 5%

During the conversion processes, the total harmonics produced by a BESS/PVDG system are in high quantity, despite that the inverters are parallel connected and multileveled [48,54]. This is a big issue when such inverter outputs are delivered into the distribution network. The current magnitude of many high-power inverters together with their harmonic contents can release large quantities of harmonics into a distribution system. This is because the magnitudes of current harmonics is proportional to the active output power of the BESS/PVDG system [7,31]. The loss of power in BESS/PVDG is mostly due to harmonics produced during the BESS/PVDG power conversions. In this sense, the proper location of BESS/PVDG units in the DN will result in network harmonic reduction due to harmonic cancellation effects. Power losses as a result of harmonics is seen as a very challenging issue worldwide due to technical damage and economic losses it causes. The economic losses related to harmonics have been geometrically growing at a high rate in recent years because of the high penetration of large-scale BESS/PVDGs into the distribution system. Consequently, re-evaluating the existing optimization models and algorithms used in the planning allocation of BESS/PVDGs to determine their effectiveness in curtailing the harmonics produced by the BESS/PVDGs is important, while taking cognizance of the huge amount of technical damage and economic losses occasioned by the harmonics.

3. Framework for Optimizing BESS/PVDGs into Distribution Networks

BESS/PVDG optimization is the methodological approach for obtaining optimal locations, sizes and times of BESS and PVDG units and installing them in a distribution network under network operating, investment and BESS/PVDG capacity constraints. The sizing and placement of BESS/PVDG units is a highly constrained, complex, nonlinear, mixed-integer and multi-objective optimization problem whose global optimum solution is very hard to find. The optimization of hybrid BESS/PVDGs involves considering contradicting objective functions such as maximising BESS/PVDG capacity and minimising power quality index; complex decision variables such as DG type, size, location and time; constraints such as network harmonic limits, DG voltage limit and power flow constraint; and the required conditions for modelling the uncertainties, especially the intermittency of the constituent distributed units (inaccurate mathematical model) [4,6,56]. Figure 1 provides the framework for optimizing BESS/PVDG into the distribution networks.

3.1. Optimization Objectives

The BESS/PVDG optimization objective functions can be either a single objective or multi-objective. The common single-objective functions used in the recent research works are minimisation of costs, energy losses, power losses, copper losses, emissions, voltage deviations, total harmonic distortions level (voltage and current); maximisation of benefits, profits, revenue of distribution system, DG capacity, reliability metric; enhancement of voltage profile, voltage stability; etc. The formulation of single-objective optimization problem can be from the perspectives of distribution system operator (DSO), the distribution energy resources developer, etc. [2,4,6,57]. A multi-objective function optimization problem requires the addition or combination of many single objectives that are conflicting and from which a single solution obtained may not be able to solve all the different objectives. The multi-objective function optimization involves simultaneous minimisation or maximisation of decision variables to obtain a single-objective formulation.

3.2. Decision Variables for BESS/PVDG Optimization

The decision variables are the unknown design variables that are determined during BESS/PVDG optimization procedures. The BESS/PVDG decision variables are formed from one or an amalgamation of size, location, number of DG, DG type, generated power of DG, installation year, real power and reactive power of DG or storage device, bus voltage angle and bus voltage magnitude [2,4,6]. The bus voltage angle and magnitude are the variables used for the decisions on the stability and power quality of the network.

3.3. Constraints for BESS/PVDG Optimization

Constraints are used in DG optimization problems to impose restrictions on some decision variables during the optimization of the objective function. Some of the commonly applied constraints in the formulation of DG allocation problems are as grouped [2,4,57].

3.3.1. Investment Constraints

They are constraints enforced on investment variables. Investment constraints can take on continuous, discrete or binary values. For example, the inequality constraints imposed on budget limit, divestment and investment options.

3.3.2. Safety Constraints

These are constraints to guarantee network and people's safety. Examples are the inequality constraints imposed for right of way in the installation of DG units, etc.

3.3.3. Technical Constraints

These are the power generation, network power flow and reliability constraints. These guarantee constant and continuous generation, transmission and distribution of power to the consumers. Some of the technical constraints are:

- The equality constraints for power balance that are imposed on active and reactive power of each network bus.
- The inequality constraints imposed on generations from DG units. e.g., DG penetration limits, discrete sizes of DG units, DG capacity limits, DG unit's constant power factor, maximum number of DGs, etc.
- The inequality constraints imposed on transmission lines and other network equipment/elements, e.g., transmission supply limits, transformer or line-overloading limits, dedicated buses for DG installations, transformer or line capacity limit, etc.
- The inequality constraints imposed on the transmission of power to the consumers, e.g., short-circuit constraints, maximum SAIDI, and radiality constraints.

3.3.4. Network Stability Constraints

Network stability constraints are imposed on the system to ensure power system stability. They are the constraints imposed on voltage drop, bus voltage magnitude, voltage angle, etc. The network stability constraints are formulated based on two network variables—voltage magnitude and voltage angle.

- The voltage magnitude constraints are imposed in the networks to ensure voltage stability. Inappropriate voltage magnitude could lead to voltage instabilities in power systems and cause damage to customers' devices, equipment and apparatuses.

$$V_{i(min)} \leq V_i \leq V_{i(max)} \text{ OR } \Delta V_{i(min)} \leq \Delta V_i \leq \Delta V_{i(max)};\ i = 1, 2, \ldots n. \quad (1)$$

The inequality constraint presented in (1) is imposed on all the network buses to enforce voltage stability of the network.

- The phase angle constraints are imposed on the network based on some stability conditions to ensure dynamic stability such as small signal stability of the network. Voltage angle limits are crucial to dynamic stability, as the voltage magnitude is related to voltage stability of the network. Failure to maintain appropriate voltage angle limits can cause enormous dynamic instabilities that can result in total power outage and other serious economic losses. However, almost all the works on distributed generation allocation expansion planning do not utilize voltage angle constraints in the formulation models.

$$\theta_{min} \leq \mid \angle V_i - \angle V_j \mid \leq \theta_{max};\ \text{OR } \theta_{min} \leq \theta_{ij} \leq \theta_{max} \quad (2)$$

This constraint (2) is imposed on all the network buses to enforce some stability criteria.

3.3.5. Power Quality Constraints

These power quality constraints are imposed to ensure the quality of power integrated into the distribution system. Different power quality indices such as total harmonic distortion (THD), total demand distortion (TDD), displacement power factor (DPF), oscillation power factor (OsPF) and transmission efficiency power factor (TEPF) could be used for power quality evaluation. A single power quality index that represents these indices could be formulated to evaluate the power quality of the distribution systems.

- The inequality constraints include voltage rise limits, voltage and current total harmonic distortion (THD) bounds, voltage sag bounds, etc.
- The harmonic constraints can be formulated based on the most important distribution network' constraints such as the voltage magnitude limits and voltage angle constraints.

The voltage magnitude constraints of the system can be reformulated and extended to impose constraints on the voltage harmonics of the distribution system during the integration of BESS/PVDG systems.

$$THD_v = \frac{\sqrt{\sum_{h=2}^{\infty} V_h^2}}{V_1} \tag{3}$$

$$V_{h(\min)} \leq V_h \leq V_{h(\max)}\,;\ \ h = 1,\ \ldots N \tag{4}$$

Similarly, the phase angle constraints could be formulated considering some parameters and assumptions that relate phase angle to active power (current) and can be extended to distribution networks if current harmonics are expected to be curtailed.

$$THD_I = \frac{\sqrt{\sum_{h=2}^{\infty} I_h^2}}{I_1} \tag{5}$$

$$\theta_{h(\min)} \leq \theta_h \leq \theta_{h(\max)}\,;\ \ h = 1,\ \ldots N \tag{6}$$

3.4. Modelling the Uncertainty of BESS/PVDGs

Modelling the uncertainties of BESS/PVDG units, including BESS and solar PV units, and the uncertainties of loads are very important to obtaining accurate solutions for a BESS/PVDG optimal allocation problem. The uncertain parameters that can be modelled in the planning of an electric power system for accounting the uncertainties in the distribution system are also presented in Figure 3. However, several previous research works place the uncertainties of these resources into consideration in their formulation models. Some of the uncertainties that are being considered and modelled in BESS/PVDG optimization studies include uncertainties of solar irradiance, wind speed, PV modules, wind and solar DG units, uncertainties of fuel, generated power, electricity market price, uncertainty of BESS and uncertainty of loads [1,4,14].

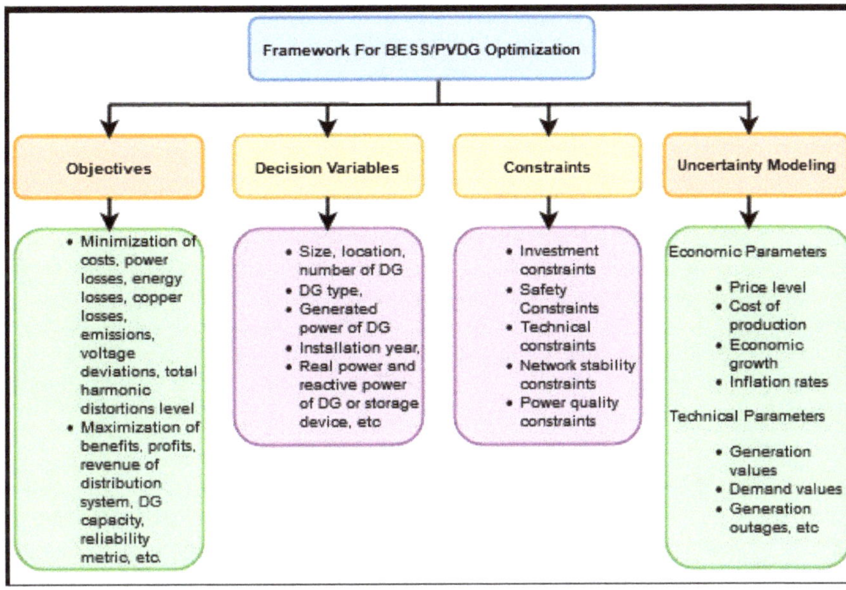

Figure 3. Framework for optimizing BESS/PVDG into distribution networks.

4. Optimization Models and Methods for BESS/PVDGs Allocation

The design and planning of battery energy storage system–photovoltaic distributed generation system is a research area that has continued to generate a lot of interest from many researchers, hence the large number of literature studies on the topic. The planning problem mentioned above concerns the hybrid energy systems that have optimal patterns and whose optimal sizes, placement/location and type of generation components/units can be assigned with minimum costs over the lifetime of the technologies. Therefore, the planning by the minimum net present value (NPV) of cost is called the optimal planning or optimal allocation of all probable hybrid technologies that are in optimal transition [11,24,56,58].

There are several methods for obtaining an optimal planning solution and many real-time, commercially available software applications for energy systems integration. In addition, various researchers have applied different optimal techniques to solve BESS/PVDG allocation problems. Different optimization methods, such as conventional methods, population-based intelligence search methods, some promising heuristic intelligence search approaches and commercial software applications, have been applied by the researchers to optimize hybrid BESS/PV distributed generation systems.

4.1. Conventional Optimization Methods

Conventional optimization methods are analytical and numerical techniques that usually present numerical equations to resolve optimal allocation problems. The methods involve computations, mathematical and theoretical analysis. The accuracy of these methods greatly depends on the efficacy of the model formulated. The advantages of these methods are the ease of implementation and short computation time to obtain convergence for the problem. However, under a complex problem, the accuracy of the solution may not be satisfactory because of the hypotheses used in simplifying the problem. Some of the conventional methods are discussed as [2,57–59].

4.1.1. Sensitivity Analysis Methods

Sensitivity-analysis-based methods use sensitivity indices used to optimally allocate DG units. In these methods, the original nonlinear equations are linearized about their starting operating points to lower the numbers of feasible solutions in the search space. The advantages of sensitivity analysis methods are reduced computation time, which is critical for large practical systems, and good ability to assess the uncertainties of renewable energy resources. Anuradha et al. [60] present a loss-voltage sensitivity index for optimizing the renewable DG size, BESS capacities and power dispatch in distribution networks. The objective is to simultaneously evaluate both minimum effects of network losses and voltage variations for optimizing the DG size [60]. A hybrid of loss sensitivity analysis methods and novel voltage stability index is applied by Murty and Kumar [61] to find optimal sizes and locations of active and reactive power DGs. The objective is to minimise copper losses and enhance network voltage profile. In Saini and Gidwani [62], a comprehensive assessment of battery energy storage system installation and the placement of photovoltaic (PV) units in a radial distribution network is performed utilizing different load models. The objective is to minimise annual energy losses, control overvoltage and reverse power flow problems in a distribution network. Nevertheless, the solutions obtained from the sensitivity analysis methods solely found optimal placements of distributed generators, but the levels of optimality of such solutions are not known [4,58].

4.1.2. Linear Programming

Linear programming (LP) is a method that uses a mathematical model with linear mathematical relationships for optimizing the objective function(s). LP is used in power system optimization problems to obtain optimal sizes of DG units, because it provides precise solutions [2,56,57]. In Altintas et al. [63], the authors proposed a two-objective LP algorithm to incorporate solar and wind renewable DGs as well as BESS into distribution system expansion planning. The objective minimises the total cost of investment and carbon emissions. This algorithm performed a sensitivity analysis test on the effect of investment costs with respect to wind and solar DGs and BESS. Alturki et al. [64] presented an LP method to obtain optimal hosting capacity of a distribution grid with the objective to maximise the PVDG power using some fundamental variables and to minimise total cost using some uncertain criteria. The results revealed that the computation time for the proposed LP algorithm was very small, especially for large-scale problems. However, the network harmonic level and stability were not considered for evaluation in these works.

4.1.3. Mixed-Integer Linear Programming

The mixed-integer linear programming (MILP) method uses a mathematical model with linear objective function and linear constraints in which, at the minimum, one design variable must be an integer. The implementation of MILP is difficult in large-scale problems because it uses too much computation time. In Santos et al. [1], MILP is applied to determine the optimal locations, sizes and timing of smart-grid technologies for minimising the net present value of the total cost and for maximising the renewable DG integration. In Mishra et al. [65], a chance-constrained stochastic MILP algorithm is modelled to determine optimal investment decisions of DGs considering operational uncertainties, while an evolutionary vertical sequencing protocol algorithm is used to further optimize the objective function that minimises the total cost of investment and operation. Santos et al. [66] proposed an improved model aimed at optimizing the system operation in a coordinated way, where distributed renewable energy sources (DRES), energy storage systems (ESS) and distribution network system reconfiguration (DNSR) are considered along with the uncertainty of the resources. The objective function was modelled to incentivize the uptake of DRES by considering the cost of emissions to decarbonize the power system. In Ajeigbe et al. [67,68], the authors applied the MILP algorithm to maximise the optimal allocation of solar, wind and biomass DGs into the distribution system by minimising the NPV of total cost and by confining the small signal stability of the networks to a required level. All the works reviewed here modelled uncertainties of renewable energy resources and evaluated voltage stability of the network but were not able to evaluate the impact of BESS/RERDG powers on the harmonic contents of the networks. Likewise, their results did not report global optimal solutions to BESS/PVDG optimal allocation problems.

LP and MILP suffer from a lack of flexibility. They normally require pre-conditions such as convexity, linearity and continuity of objective functions, which are difficult to meet in practice [2,57].

4.1.4. Nonlinear Programming

Nonlinear programming (NLP) is a mathematical programming method that uses nonlinear objective function and solely continuous variables and constraints. The NLP computation involves the differentials of objective functions and constraints. In solving nonlinear problems, a search path is selected iteratively by defining the starting partial differentials of the problem equation. This approach could be based on first-order or higher-order methods such as the reduced gradient method [69,70] and other search methods [71,72], Newton Raphson method [73] and successive quadratic programming [74,75] which are used for solving DG allocation planning problems.

4.1.5. Mixed-Integer Nonlinear Programming

Mixed-integer nonlinear programming (MINLP) utilizes a mathematical model with nonlinear objective functions and constraints and both continuous and discrete variables. MINLP algorithms have been applied in power systems to determine the optimal sizes and locations of DGs and BESSs. Some of the disadvantages of MINLP are long computation time and a very large number of decision variables [2,56,57]. Salyani et al. [76] applied MINLP in the mathematical modelling for the simultaneous optimal allocation planning of high- and medium-voltage substations, robust medium-voltage feeder routing and renewable DG units. The authors used adaptive GA to find optimal locations and sizes while the uncertainties of renewable DGs, fuel prices, electricity and demand were evaluated. A mixed-integer nonlinear programming-model-based methodology is presented in Valencia et al. [11] for the optimal location, selection, and operation of BESSs and renewable distributed generators (DGs) in medium–low-voltage distribution systems.

4.1.6. Fuzzy Logic

The fuzzy logic (FL) method was developed in 1979 to solve power system problems. The FL method is based on the concept of a classical set, such as the identification of a membership function that is associated with each member as indicated by a binary number 0 and 1 [77]. The membership function dictates the resemblance level of a member in a fuzzy subset. Some of the common membership functions are the triangular, trapezoidal, piecewise-linear and Gaussian functions [2,57,59]. In Injeti and Kumar [78], FL is applied to DG allocation problems, with minimisation of power losses and improvement in voltage profiles as the objective function. Sharma et al. [79] proposed a FL controller in determining the optimal sizes and locations of DGs in order to minimise power losses and to enhance loadability and voltage profiles of distribution networks. However, the results from these works did not report the optimality of their solutions, the evaluation of network stability or harmonic contents.

The works discussed thus far on FL have not considered the impact of DGs and BESS on the oscillatory modes and harmonic contents of the distribution networks. To achieve practical solutions, dynamic networks must be simulated for the evaluation of distribution system stability and harmonic contents.

4.2. Intelligence Search Methods for BESS/PV Distributed Generations

Artificial intelligence (AI) is the application of human intelligence to perform tasks in machines [59]. AI is applied in the intelligence search methods (ISM) used in power systems for optimal sizing and placement of DGs. Intelligence search methods are heuristics algorithms that fasten up the processes of obtaining near-optimal solutions for complex and large DG problems. The advantages of intelligence search methods over other conventional methods is the simplicity of implementation and robustness. However, the accuracy and precision of ISMs are not reliable. They usually take much computation effort [2,56,57,80]. Some of intelligence search methods are presented below.

4.2.1. Genetic Algorithm

Genetic algorithm (GA) is an intelligence search algorithm that was introduced earlier to solve optimization problems. GA is developed from natural selection and genetics principles such as selection, mutation, inheritance and crossover [56,57]. In GA, a set of selection rules is specified to allow a population to achieve a maximum state of fitness. Then, the elements in a population are integrated into chromosomes to enable the potential elements to achieve a better state. The first population of elements evolved through the evolution of generations. The principle of mutation is applied to modify the chosen element to evolve into a new population. The algorithm repeats this procedures until an acceptable solution or the highest number of iterations is attained. [4,6,56]. Genetic algorithms utilize continuous and discrete variables for implementation and work better at obtaining global optimums of various functions. GAs can effectively solve poorly defined and complex problems. GA is the most used optimization method to find optimal locations and sizes of DGs in the literature [22,81,82]. In Liu et al. [22], the authors presented a mixed-integer GA to obtain optimal sizes and locations of hybrid battery energy storage and renewable energy DGs units with objective aiming to minimise system total cost, end-user satisfaction loss caused by demand side management, and tie-line power fluctuation. The methodology in Liu et al. effectively determined the solution of the multi-objective optimization problem compared to others validated with it. However, neither uncertainties of the renewable energy sources nor the voltage variability of the BESS were modelled. In addition, the requirements for the evaluation of network stability and harmonic contents were not included in the proposed methodology. Moreover, genetic algorithms have the disadvantage of evaluating the repeated fitness functions that are time intensive for large and complex problems. The various configurations of GA that are proposed to improve the performance of the GA method in the DG allocation problems are quantum GA (QGA) [83], adaptive genetic algorithm (AGA) [84], etc.

4.2.2. Simulated Annealing

Simulated annealing (SA) uses an iterative procedure for solving combinatory optimization problems. SA employs the process of crystallization at a discrete search space of a physical system [57]. The SA algorithm depends on the cooling criterion and uses initial temperature (T), final temperature (T_{min}) and cooling rate (β) variables. SA algorithms are extensively proposed in the literature to allocate DG units at lower computational time. Simulated annealing algorithms perform effectively in solving reliability-criteria-based optimization problems [2,57]. The advantages of SA algorithms are robustness, simplicity of implementation, and capability to provide feasible solutions to combinatorial problems. Nevertheless, SA algorithms have large computation times without upper limits, terminate at local minimums and lack details on the level of variation between a local minimum and global minimum [56,85]. In Koziel et al. [86], the authors presented a feasibility-preserving SA algorithm to obtain DN reconfiguration with the objective to minimise power loss and improve voltage profile. This study concluded that the proposed algorithm was more efficient than some published population-based intelligence search methods with respect to computational cost and solution repeatability. However, the optimality of the solution was not reported, and the harmonic contents and dynamic stability of the networks were not evaluated in the proposed work.

4.2.3. Particle Swarm Optimization

Particle swarm optimization (PSO) methods are developed based on the social adaptation of flocking bird and schooling fish. In PSO, single intersection of all dimensions produces a particle, and these particles move randomly in a complex search space. The system is then adjusted using a number of solutions that are randomly selected. During each iteration, the particles use their fitness level to assess their positions. Then, the contiguous particles update their previous "best" position to upgrade the final solution [2,57,87]. The advantages of PSO are robustness, simple implementation and running simultaneous computations in less computation time. PSO algorithms use a couple of parameters to modify and converge faster. PSO can also be effectively used to solve DG allocation problems with inaccurate mathematical models. However, the initial design parameter are difficult to define with PSO. During complex DG allocation problems, PSO may converge prematurely and terminate at the local minimum [6,56]. In Prabpal et al. [88], the PSO technique was applied to obtain optimal sizes and locations of multiple BESS and PVDG units with the objective to minimise total cost, minimise the impact of large-scale penetration of BESS, improve voltage profile and increase the stability of the power system. The results showed that PSO and GA methods equally performed better in achieving fewer numbers of iterations and quality of solutions. Shahzad et al. [23], Jamian et al. [89], Rathore et al. [90] and Zeinalzadeh et al. [91] proposed multi-objective PSO methods for determining optimal locations and sizes of BESSs/PVDGs to minimise power losses and improve voltage profiles. However, the uncertainties of the intermittent DGs and BESSs were not modelled, and the impact of their variable output power on the dynamic stabilities and harmonic contents of the distribution networks was not considered. Only the uncertainties related to BESS/PVDG market scenarios were evaluated in Rathore et al. [90].

4.3. *Promising Intelligence Search Methods*

Promising intelligence search methods are the additional optimization algorithms developed to effectively solve distributed generation optimization problems. Some of these methods are as stated [2,57,59].

4.3.1. Artificial Bee Colony Algorithm

The artificial bee colony (ABC) algorithm was developed from the searching behaviour of a swarm of honeybees. Khasanov et al. [16] proposed an application of hybrid teaching-learning and artificial bee colony (TLABC) technique for determining the optimal allocation of PV-based distributed generation and battery energy storage units in a distribution system with the aim of minimising the total power losses. ABC algorithms are applied in Mohandas et al. [92] and Dixit et al. [93] to find optimal DGs locations and sizes with the objective of minimising power losses and of improving voltage stability of the network. In Abu-Mouti and El-Hawary [94], the authors proposed an algorithm of ABC to adjust the control inputs, iteration number and colony size in the DG allocation optimization. El-Zonkoly and Kefayat et al. [95,96] utilized ABC algorithms to solve distribution expansion planning problems and to obtain optimal reinforcement and commitment scheduling for PVDG allocation. Padma Lalitha et al. [97] presented and compared the ABC and PSO algorithms. The authors observed that the ABC algorithm outperformed PSO, having better solutions and convergence. Notwithstanding, the works discussed here do not provide indices to evaluate harmonic contents and dynamic stabilities of the systems.

4.3.2. Ant Colony Algorithm

The ant colony (AC) algorithm is adapted from ants' social behaviours in searching for the shortest route to obtain food. The AC algorithm process begins with random solutions obtained from the ants' random searches in their movements. Ants share information about their movements by leaving chromosome trails behind during their movements. Consequently, a path with trail density becomes the shorter path. This knowledge is utilized in the optimization search to obtain feasible solutions [57]. The advantages of AC algorithms are the ability to discover good solutions and guarantee convergence and the ability to search among a population simultaneously and adapt to changes such as new distances. However, AC optimization algorithms are weak in changing probability distribution, uncertainty of convergence time, sequences of random decisions and theoretical analysis, since they are highly experimental researches. These algorithms are variously used in the literature for optimal allocation of DGs [6,56]. In Gomez et al. [98], Vlachogiannis et al. [99], Wang and Singh [100] and Amohadi and Fotuhi-Firuzabad [101], the variant of AC and ant colony system (ACS) algorithms were presented. They found optimal sizes of DGs, locations of DGs and re-closers in the radial DNs with an objective to use the composite reliability index. Transient stability and reliability of the distribution systems were evaluated to validate the proposed methods. ACS algorithms were observed to be more satisfactory in many engineering applications. However, these works did not include the installation of renewable DGs and could not access the impacts of integrating BESS/PV-distributed generations on the harmonic distortion and oscillation of the networks.

4.3.3. Artificial Immune System Algorithm

The artificial immune system (AIS) algorithm is adapted from immunology, the importance of the immune system and their values in the natural world [102]. The immune system is an indispensable defence against self-approach to protect human health from pathogens such as viruses and microbes. The procedure differentiates between self-cells and non-self-cells. Thereafter, the immune system effects immune actions to destroy the non-self-cells [103–105]. To apply the AIS optimization process in solving DG allocation problems, the instructions in the search area (objective functions, design variables, constraints, etc.) are encrypted in an antigen population of an AIS algorithm. AIS algorithms are proposed in Aghaebrahimi et al. [106] and Hatata et al. [107] to find the optimal locations and sizes of the DGs, with the objective to minimise the power losses of the DN considering bus voltage limits and line current. Souza et al. [108] proposed an AIS algorithm in expansion planning to allocate DG units into distribution network considering the uncertainty of load demands.

4.4. *Probable Hybrid Intelligence Search Methods*

Hybrid optimization methods are a useful combination or collaboration of more than one different intelligence search method. These approaches extract the benefits of the component methods to obtain an optimum solution for a specific planning problem. The allocation expansion planning of BESS/PVDGs problems is multi-objective in nature. Hence, applying a hybrid method in their investigation begets an excellent planning objective and a suitable alternative algorithm to solve the problems that involve better understanding of the methods.

A summary of the various optimization techniques that are developed and applied by the researchers for BESS/PVDGs allocation is presented in Table 2.

Table 2. Summary of optimization methods.

Optimization Method	Optimized Factor	Comment
Conventional Method • Sensitivity Analysis [60–62] • Linear Programming (LP) [63,64] • Mixed Integer Linear Programming [65–68] • Nonlinear Programming (NLP) [69–75] • Mixed-Integer Nonlinear Programming (MINLP) [76] • Fuzzy Logic [77–79]	Hybrid renewable energy sources (solar, wind) and battery energy storage, and cost	Using numerical equations that can be applied to optimization problems due to their capability to provide accurate mathematical model formulation
Intelligence Search • Genetic Algorithm [81–84] • Simulated Annealing [85,86] • Particle Swamp [87–91] • Artificial Bee Colony [92–97] • Artificial Immune System [102–108] • Ant Colony [98–101]	Hybrid renewable energy sources (solar, wind) and battery energy storage, and cost	Using the exhibition of intelligence in machines to determine optimal locations and sizes of hybrid DGs in power system
Deterministic Approaches [59–63]	Standalone renewable energy sources (solar, wind) with battery energy storage, and cost	Using mathematical equations for determining particular values when fixed factors are set

Table 2. *Cont.*

Optimization Method	Optimized Factor	Comment
Probabilistic Approaches [63–67,74,75]	• Efficiency of hybrid renewable energy systems, and cost • Uncertain parameters in power system	Using statistical data gathering methods for finding optimized factors
Software Based Methods [109–116] • HOMER • HYBRIDS, etc.	Hybrid solar/wind and or diesel generators with battery energy storage	Using software applications that uses input file with all necessary data

4.5. Commercial Software Applications for Allocation of (BESS/PV) Hybrid DG Systems

Several software applications have been developed and applied for the sizing of hybrid renewable energy systems (HRESs) such as HOMER [109–111], HYBRIDS [112], HYBRID 2 [113], RET Screen [114], TRNSYS [115] and IHOA [116].

Comparatively, HOMER has a significant application in optimal sizing of HRESs because of its capacity to quickly obtain optimal sizes of energy systems. In addition, it is useful in investigating sensitivity analyses of some uncertainty parameters and changing factors related to the HRESs. However, the mentioned software tools are incapacitated to investigate major network system issues related to the integration of distributed HRESs (DHRESs) such as harmonics and small signal and transient stabilities. A list of commercially available software for the planning of HRES is presented in Table 3.

Table 3. Software applications for optimizing BESS/PVDGs.

Name of Software	Optimization Input	Optimized Output
HOMER	• Load command • Resource input • Cost details (capital, O&M, replacement costs	• Optimize unit size(s) • NPV and energy cost
HYBRIDS	• System control • Wind turbine size(s) and type • Solar size(s) • Type and number of battery storage	• NPV and energy cost • Amount of green-house gases
HYBRID 2	• Resources input • Load demand • Cost details (O&M, investment, components costs)	• Optimize unit size(s) • NPV and energy cost • Proportion of green-house gases released. • System payback time
RET SCREEN	• Load command • Solar size(s) • Climate data input • Invention and hydrology data input	• NPV and energy costs • Economic capability • Production rate • Risk analysis • Energy used and saved
IHOGA	• Load command • Resources data input • Components and economic factors	• Improve multi-objective optimization • Cost of energy • Life cycle release
TRYSYS	• Climate data • Ingrained models	• Dynamic simulation of renewable energy resources

5. Results and Discussion

The increasing needs for energy and the resultant environmental issues arising from fossil energy utilization have encouraged the extensive study of renewable energy technologies in place of traditional fossil fuels. Precisely, hybrid distributed generations, which have been described as a collaboration of renewable energies and support systems, are a significant alternative to confront the concerns over sustainability of energy demands and environmental safety. The planning and optimization of hybrid distributed power systems can meet the essential requirements of a geographical location in terms of availability of potential energy resources, area topography and various kinds of energy demands. Consequently, the optimal allocation of renewable energy sources and storage systems relating to environmentally friendly hybrid distributed systems considerably improves the technical and economic aspects of the power supply system. The addition of storage technologies in the allocation of distributed generations can smoothen output power and reduce REHDG intermittent effects in the network. Including storage devices in the DGs allocation problems provides supporting services to the optimal solutions by eliminating the effects of intermittency in the renewable sources power output. Several allocation methodologies have been proposed to determine the best hybrid renewable energy system with respect to the economy and technology. Determining the optimal allocation of hybrid battery storage and PV-distributed generation systems and other hybrid renewable energy systems is important to increase the technical and economic efficiency of the power distribution system and to encourage the extensive use of environmentally friendly resources.

Various allocation methodologies presented in the recent literature with different optimization algorithms are reviewed here. The GA, PSO, SA and AIS are some of the feasible artificial intelligence algorithms used to investigate the planning and optimization of DG allocation problems. The most important benefit of GAs are the ordered capability to find the global optimal and the ease of achieving a local minimum when used in hybrid system allocation. Another advantage that makes GA suitable for allocation planning studies is code-ability because it is not accessible in other methods such as PSO. For instance, when at most three parameters are to be coded such as in a wind/PV/BESS system, both GA and PSO can perform effectively. However, when more than three elements are involved, only the GA method would be more capable of obtaining optimal solutions. Some other times, PSO has some advantages over GA, although both are very effective in utilizing the same repeatable search approach. Moreover, employing SA in hybrid distributed systems is not as common as GA and PSO methods, but presently, SA is generating more research interest in some approved areas of application. The ACS algorithms have been presented to reduce power losses and to improve power system factors of a radial distributed system. Similar to GA, the AIS optimization algorithm has "collection" and "transformation" operatives which improve the probability of the algorithm to find the global optimum point.

AIS is bound to have a high application in sizing studies because it is similar to GA and can be effective in finding the global optimum in difficult problems. However, GA has greater application than AIS, especially in addressing a large number of parameters. In addition, conventional methods such as LP, MILP and NLP are still being applied in existing studies to detail the features of any physical system into a mathematical model formulation. Often, hybrid optimization methods are applied by combining two or more methods to take beneficial advantage of them in terms of their convergence time during the optimization process. Hybrid methods are characterized due to their dynamic flexibility during the allocation process. Hence, they are the most applied allocation methods.

The intermittent nature of photovoltaic and wind output power and the high voltage rise and fall from BESS cause harmonic distortions which have a negative impact on the power quality, reliability and stability of the distribution networks. The majority of the current works do not include the uncertainties of the renewable and battery storage power sources in their formulation models. They did not combine all the associated investment, technical, safety, DG capacity, network stability, power quality and reliability constraints in the formulation models for the DG allocation problems. In most of these works, the minimum harmonic level and dynamic stability of the network are not constrained but are only assumed, while the constraints for the right of way are neglected for the required buses. All these necessary and associated constraints need to be incorporated to obtain a practical solution from the REHDG allocation models. In essence, future research studies should give adequate consideration to modelling of the impacts of renewable energy intermittencies and the resulting variable output power to culminate in more feasible solutions to BESS/PVDG optimization problems.

In addition, the operations of hybrid DG systems are dynamic. Hence, the planning and design of optimal sizes and placement of RERDGs should be optimized on dynamic networks but not on static ones, as they are mostly performed in the existing planning models. The dynamical issues such as harmonic and system instabilities are very visible while using dynamic networks, since the real power networks are dynamic networks whose load profile periods are estimated hourly during a dynamic planning horizon. Future research needs to focus on the use of dynamic networks to entirely incorporate the intrinsic characteristics of the distribution network such as the harmonic components and dynamic stability of the network.

Moreover, the sizes and locations of battery energy storage, photovoltaic and wind DG units in the distribution network (DN) affect the network harmonic contents by having either positive or negative impacts on the magnitude of the current and voltage harmonics of the networks.

6. Conclusions

This study presents a review of prior research on the optimization methodologies for designing and planning hybrid renewable energy resource distributed generation such as hybrid battery energy storage–photovoltaic DG and other hybrid distributed systems. This paper reviewed more than one hundred papers published by renowned referenced journals on battery energy storage systems and renewable energy resources as well as on robust and efficient optimization methods for solving hybrid DG allocation planning problems. Optimization studies, in the last decade, on DG allocation planning using conventional and intelligence search methods have been analysed, and hybrid optimization algorithms have been presented.

Intelligence search methods have been mostly used in the last decade due to their capacity for shorter computation times, and because they provide better accuracy and have better convergence than the conventional methods. In conclusion, at the beginning, this study investigated a number of research works that have applied optimization methods to solve renewable energy DG allocation problems, including solar, wind and battery energy systems. Many research works use intelligence search methods, most especially GA, PSO and AIS, to solve these allocation problems. Notwithstanding, conventional methods, especially LP and MILP and different configurations of NLP methods are still being used in current studies. In the case of curtailing harmonic distortions of the DNs, which indicate the strength of this study, an optimal planning model is yet to be developed for optimal sizing, placement and timing of renewable DGs and battery energy storage systems. Although, in most cases, the optimal sizing and placement of BESS/REDGs may have attained a minimum cost, the requirements for minimum harmonic levels are yet to be achieved. These requirements are merely presumed in the existing works. Further research is required in this regard to improve the current expansion planning model to obtain optimal allocation of BESS and renewable energy DGs and to constrain the decision variables related to harmonic distortions to a required level. A more comprehensive expansion planning model together with an efficient intelligence search algorithm that has that capability to obtain a global optimum solution is an important approach towards solving optimal BESS/PVDG allocation problems and towards reducing harmonic components of distribution systems during the integration of hybrid battery energy storage systems and photovoltaic DGs.

Author Contributions: The mathematical formulations and programming in this work were developed within the framework of the doctorate of A.O. He is supervised by Y.H. and is co-supervised by J.M. The written manuscript was extensively discussed with the supervisors. All authors have read and agreed to the published version of the manuscript.

Funding: This research received no external funding.

Institutional Review Board Statement: Not applicable.

Informed Consent Statement: Not applicable.

Data Availability Statement: Data sharing is not applicable to this article.

Acknowledgments: The authors would like to acknowledge the research support received from Tshwane University of Technology (TUT), Pretoria, South Africa.

Conflicts of Interest: The authors declare no conflict of interest.

References

1. Santos, S.F.; Fitiwi, D.Z.; Shafie-Khah, M.; Bizuayehu, A.; Catalão, J. Optimal sizing and placement of smart-grid-enabling technologies for maximizing renewable integration. In *Smart Energy Grid Engineering*; Elsevier: Amsterdam, The Netherlands, 2017; pp. 47–81.
2. Ajeigbe, O.A.; Munda, J.L.; Hamam, Y. Towards maximising the integration of renewable energy hybrid distributed generations for small signal stability enhancement: A review. *Int. J. Energy Res.* **2020**, *44*, 2379–2425. [CrossRef]
3. Babatunde, O.M.; Munda, J.L.; Hamam, Y. A comprehensive state-of-the-art survey on hybrid renewable energy system operations and planning. *IEEE Access* **2020**, *8*, 75313–75346. [CrossRef]
4. Jordehi, A.R. Allocation of distributed generation units in electric power systems: A review. *Renew. Sustain. Energy Rev.* **2016**, *56*, 893–905. [CrossRef]
5. Hosenuzzaman, M.; Rahim, N.A.; Selvaraj, J.; Hasanuzzaman, M.; Malek, A.A.; Nahar, A. Global prospects, progress, policies, and environmental impact of solar photovoltaic power generation. *Renew. Sustain. Energy Rev.* **2015**, *41*, 284–297. [CrossRef]
6. Bullich-Massagué, E.; Cifuentes-García, F.-J.; Glenny-Crende, I.; Cheah-Mañé, M.; Aragüés-Peñalba, M.; Díaz-González, F.; Gomis-Bellmunt, O. A review of energy storage technologies for large scale photovoltaic power plants. *Appl. Energy* **2020**, *274*, 115213. [CrossRef]
7. Ajeigbe, O.A.; Chowdhury, S.P.; Olwal, T.O.; Abu-Mahfouz, A.M. Harmonic Control Strategies of Utility-Scale Photovoltaic Inverters. *Int. J. Renew. Energy Res. (IJRER)* **2018**, *8*, 1354–1368.
8. Babatunde, D.E.; Babatunde, O.M.; Emezirinwune, M.U.; Denwigwe, I.H.; Okharedia, T.E.; Omodara, O.J. Feasibility analysis of an off-grid photovoltaic-battery energy system for a farm facility. *Int. J. Electr. Comput. Eng.* **2020**, *10*, 2874–2883. [CrossRef]
9. do Nascimento, Á.D.J.; Rüther, R. Evaluating distributed photovoltaic (PV) generation to foster the adoption of energy storage systems (ESS) in time-of-use frameworks. *Sol. Energy* **2020**, *208*, 917–929. [CrossRef]
10. Lee, M.; Park, J.; Na, S.-I.; Choi, H.S.; Bu, B.-S.; Kim, J. An analysis of battery degradation in the integrated energy storage system with solar photovoltaic generation. *Electronics* **2020**, *9*, 701. [CrossRef]
11. Valencia, A.; Hincapie, R.A.; Gallego, R.A. Optimal location, selection, and operation of battery energy storage systems and renewable distributed generation in medium-low voltage distribution networks. *J. Energy Storage* **2021**, *34*, 102158. [CrossRef]
12. Mirhassani, S.; Ong, H.C.; Chong, W.; Leong, K. Advances and challenges in grid tied photovoltaic systems. *Renew. Sustain. Energy Rev.* **2015**, *49*, 121–131. [CrossRef]
13. Kumar, V.V.; Babulal, C. Application of Fuzzy Logic in Power Quality Assessment of Modern Power Systems. *Microgrid Technol.* **2021**, *1*, 377–403.
14. Manoj Kumar, N.; Ghosh, A.; Chopra, S.S. Power resilience enhancement of a residential electricity user using photovoltaics and a battery energy storage system under uncertainty conditions. *Energies* **2020**, *13*, 4193. [CrossRef]
15. Karimi, M.; Mokhlis, H.; Naidu, K.; Uddin, S.; Bakar, A.A. Photovoltaic penetration issues and impacts in distribution network—A review. *Renew. Sustain. Energy Rev.* **2016**, *53*, 594–605. [CrossRef]
16. Khasanov, M.; Kamel, S.; Ayman, A.; Jurado, F. Optimal planning DG and BES units in distribution system considering uncertainty of power generation and time-varying load. *Turk. J. Electr. Eng. Comput. Sci.* **2021**, *29*, 773–795. [CrossRef]
17. Akagi, S.; Yoshizawa, S.; Ito, M.; Fujimoto, Y.; Miyazaki, T.; Hayashi, Y.; Tawa, K.; Hisada, T.; Yano, T. Multipurpose control and planning method for battery energy storage systems in distribution network with photovoltaic plant. *Int. J. Electr. Power Energy Syst.* **2020**, *116*, 105485. [CrossRef]
18. Duong, M.Q.; Pham, T.D.; Nguyen, T.T.; Doan, A.T.; Tran, H.V. Determination of optimal location and sizing of solar photovoltaic distribution generation units in radial distribution systems. *Energies* **2019**, *12*, 174. [CrossRef]
19. Caballero-Peña, J.; Cadena-Zarate, C.; Parrado-Duque, A.; Osma-Pinto, G. Distributed energy resources on distribution networks: A systematic review of modelling, simulation, metrics, and impacts. *Int. J. Electr. Power Energy Syst.* **2022**, *138*, 107900. [CrossRef]
20. Guo, C.; Luo, F.; Cai, Z.; Dong, Z.Y.; Zhang, R. Integrated planning of internet data centers and battery energy storage systems in smart grids. *Appl. Energy* **2021**, *281*, 116093. [CrossRef]
21. Emad, D.; El-Hameed, M.; El-Fergany, A. Optimal techno-economic design of hybrid PV/wind system comprising battery energy storage: Case study for a remote area. *Energy Convers. Manag.* **2021**, *249*, 114847. [CrossRef]
22. Liu, B.; Zhou, B.; Yang, D.; Li, G.; Cao, J.; Bu, S.; Littler, T. Optimal planning of hybrid renewable energy system considering virtual energy storage of desalination plant based on mixed-integer NSGA-III. *Desalination* **2022**, *521*, 115382. [CrossRef]
23. Shahzad, K.; Amin, A.A. Optimal Planning of Distributed Energy Storage Systems in Active Distribution Networks using Advanced Heuristic Optimization Techniques. *J. Electr. Eng. Technol.* **2021**, *16*, 2447–2462. [CrossRef]
24. Zahraee, S.; Assadi, M.K.; Saidur, R. Application of artificial intelligence methods for hybrid energy system optimization. *Renew. Sustain. Energy Rev.* **2016**, *66*, 617–630. [CrossRef]

25. Groppi, D.; Pfeifer, A.; Garcia, D.A.; Krajačić, G.; Duić, N. A review on energy storage and demand side management solutions in smart energy islands. *Renew. Sustain. Energy Rev.* **2021**, *135*, 110183. [CrossRef]
26. Hannan, M.; Wali, S.; Ker, P.; Abd Rahman, M.; Mansor, M.; Ramachandaramurthy, V.; Muttaqi, K.; Mahlia, T.; Dong, Z. Battery energy-storage system: A review of technologies, optimization objectives, constraints, approaches, and outstanding issues. *J. Energy Storage* **2021**, *42*, 103023. [CrossRef]
27. Khezri, R.; Mahmoudi, A.; Aki, H. Optimal planning of solar photovoltaic and battery storage systems for grid-connected residential sector: Review, challenges and new perspectives. *Renew. Sustain. Energy Rev.* **2022**, *153*, 111763. [CrossRef]
28. Ajeigbe, O.A.; Munda, J.L.; Hamam, Y. Renewable Distributed Generations' Uncertainty Modelling: A Survey. In Proceedings of the 2020 IEEE PES/IAS PowerAfrica, Nairobi, Kenya, 25–28 August 2020; pp. 1–5.
29. Ortega, M.; Hernández, J.; García, O. Measurement and assessment of power quality characteristics for photovoltaic systems: Harmonics, flicker, unbalance, and slow voltage variations. *Electr. Power Syst. Res.* **2013**, *96*, 23–35. [CrossRef]
30. Umamaheswararao, C. Modeling and Simulation of Statcom for Power Quality Improvement. *Inf. Technol. Ind.* **2021**, *9*, 217–229.
31. Chidurala, A.; Saha, T.; Mithulananthan, N. Harmonic characterization of grid connected PV systems & validation with field measurements. In Proceedings of the 2015 IEEE Power & Energy Society General Meeting, Denver, CO, USA, 26–30 July 2015; pp. 1–5.
32. Du, Y.; Lu, D.D.-C.; Chu, G.M.; Xiao, W. Closed-form solution of time-varying model and its applications for output current harmonics in two-stage PV inverter. *IEEE Trans. Sustain. Energy* **2014**, *6*, 142–150. [CrossRef]
33. Reddy, G.R.; Rayaguru, N.; Karthikumar, K.; Chandrasekar, P.; Murthy, P. Enhancement of Power Quality With Fuzzy Based UPQC in Grid Integrated and Battery Assisted PV System. In Proceedings of the 2021 2nd Global Conference for Advancement in Technology (GCAT), Bangalore, India, 1–3 October 2021; pp. 1–8.
34. Mansor, M.A.; Hasan, K.; Othman, M.M.; Noor, S.Z.B.M.; Musirin, I. Construction and performance investigation of three-phase solar PV and battery energy storage system integrated UPQC. *IEEE Access* **2020**, *8*, 103511–103538. [CrossRef]
35. Sanchez-Ruiz, A.; Abad, G.; Echeverria, I.; Torre, I.; Atutxa, I. Continuous phase-shifted selective harmonic elimination and dc-link voltage balance solution for H-bridge multilevel configurations, applied to 5L HNPC. *IEEE Trans. Power Electron.* **2016**, *32*, 2533–2545. [CrossRef]
36. Freddy, T.K.S.; Rahim, N.A.; Hew, W.-P.; Che, H.S. Modulation techniques to reduce leakage current in three-phase transformerless H7 photovoltaic inverter. *IEEE Trans. Ind. Electron.* **2014**, *62*, 322–331. [CrossRef]
37. Vahedi, H.; Shojaei, A.A.; Dessaint, L.-A.; Al-Haddad, K. Reduced DC-link voltage active power filter using modified PUC5 converter. *IEEE Trans. Power Electron.* **2017**, *33*, 943–947. [CrossRef]
38. Deshmukh, S.; Thorat, A.; Korachagaon, I. Modelling and Analysis of PV Standalone System With Energy Management Scheme. In Proceedings of the 2020 IEEE International Conference on Electronics, Computing and Communication Technologies (CONECCT), Bangalore, India, 2–4 July 2020; pp. 1–5.
39. Kumar, M. Modelling and Simulation of Compound PV-BESS Systems. 2018. Available online: https://trepo.tuni.fi/bitstream/handle/123456789/25480/kumar.pdf?sequence=4&isAllowed=y (accessed on 18 December 2022).
40. Rashmi, V.; Khare, M.; Lnct, B. Study of Cascaded H-Bridge Converter Control Strategies and their Impact on Switching Harmonics. *Int. J. Online Sci.* **2018**, *4*, 8. [CrossRef]
41. Ajeigbe, O.; Munda, J.; Hamam, Y. Characterisation of harmonic distortions produced by small domestic back-up generators. In Proceedings of the 2018 IEEE PES/IAS PowerAfrica, Cape Town, South Africa, 26–30 June 2018; pp. 432–437.
42. Shabbir, N.; Kütt, L.; Jarkovoi, M.; Iqbal, M.N.; Rassõlkin, A.; Daniel, K. An overview of measurement standards for power quality. *Agron. Research.* **2021**, *19* (Suppl. S1), 944–960.
43. Du, Y.; Lu, D.D.-C.; James, G.; Cornforth, D.J. Modeling and analysis of current harmonic distortion from grid connected PV inverters under different operating conditions. *Sol. Energy* **2013**, *94*, 182–194. [CrossRef]
44. Sengamalai, U.; Ramasamy, P.; Thentral, T.; Balasubramani, K.; Alagarsamy, M.; Muthusamy, S.; Panchal, H.; Sachithanandam, M.P.; Sadasivuni, K.K. A simplified methodology for mitigating the harmonics and common-mode voltage using multi-level inverters for renewable energy applications. *Energy Sources Part A Recovery Util. Environ. Eff.* **2021**, 1–23. [CrossRef]
45. Rajput, S.; Averbukh, M.; Yahalom, A.; Minav, T. An approval of MPPT based on PV Cell's simplified equivalent circuit during fast-shading conditions. *Electronics* **2019**, *8*, 1060. [CrossRef]
46. Kerekes, T.; Séra, D.; Mathe, L. Three-phase photovoltaic systems: Structures, topologies, and control. In *Renewable Energy Devices and Systems with Simulations in MATLAB® and ANSYS®*; CRC Press: Boca Raton, FL, USA, 2017; pp. 67–90.
47. de Jesus, V.M.R.; Cupertino, A.F.; Xavier, L.S.; Pereira, H.A.; Mendes, V.F. Operation Limits of Grid-Tied Photovoltaic Inverters With Harmonic Current Compensation Based on Capability Curves. *IEEE Trans. Energy Convers.* **2021**, *36*, 2088–2098. [CrossRef]
48. Eroğlu, H.; Cuce, E.; Cuce, P.M.; Gul, F.; Iskenderoğlu, A. Harmonic problems in renewable and sustainable energy systems: A comprehensive review. *Sustain. Energy Technol. Assess.* **2021**, *48*, 101566. [CrossRef]
49. Todeschini, G. Control and derating of a PV inverter for harmonic compensation in a smart distribution system. In Proceedings of the 2017 IEEE Power & Energy Society General Meeting, Chicago, IL, USA, 16–20 July 2017; pp. 1–5.
50. Hrishikesan, V.; Kumar, D.; Kumar, C. Dynamic Voltage Control Using Unified Power Quality Conditioner with Storage. In Proceedings of the 2021 IEEE 15th International Conference on Compatibility, Power Electronics and Power Engineering (CPE-POWERENG), Florence, Italy, 14–16 July 2021; pp. 1–6.

51. Lakshmi, S.; Ganguly, S. Multi-objective planning for the allocation of PV-BESS integrated open UPQC for peak load shaving of radial distribution networks. *J. Energy Storage* **2019**, *22*, 208–218. [CrossRef]
52. Xu, J.; Xie, S.; Qian, Q.; Zhang, B. Adaptive feedforward algorithm without grid impedance estimation for inverters to suppress grid current instabilities and harmonics due to grid impedance and grid voltage distortion. *IEEE Trans. Ind. Electron.* **2017**, *64*, 7574–7586. [CrossRef]
53. Yan, Q.; Wu, X.; Yuan, X.; Geng, Y. An improved grid-voltage feedforward strategy for high-power three-phase grid-connected inverters based on the simplified repetitive predictor. *IEEE Trans. Power Electron.* **2015**, *31*, 3880–3897. [CrossRef]
54. Parvez, M.; Elias, M.; Rahim, N.; Osman, N. Current control techniques for three-phase grid interconnection of renewable power generation systems: A review. *Sol. Energy* **2016**, *135*, 29–42. [CrossRef]
55. Nel, C.; Ajeigbe, O.; Chowdhury, S.D. Taming the total harmonic distortion on the 132kV Arlington Traction Station. In Proceedings of the 2017 IEEE PES PowerAfrica, Accra, Ghana, 27–30 June 2017; pp. 550–554.
56. Zubo, R.H.; Mokryani, G.; Rajamani, H.-S.; Aghaei, J.; Niknam, T.; Pillai, P. Operation and planning of distribution networks with integration of renewable distributed generators considering uncertainties: A review. *Renew. Sustain. Energy Rev.* **2017**, *72*, 1177–1198. [CrossRef]
57. Abdmouleh, Z.; Gastli, A.; Ben-Brahim, L.; Haouari, M.; Al-Emadi, N.A. Review of optimization techniques applied for the integration of distributed generation from renewable energy sources. *Renew. Energy* **2017**, *113*, 266–280. [CrossRef]
58. Shuaibu Hassan, A.; Sun, Y.; Wang, Z. Optimization techniques applied for optimal planning and integration of renewable energy sources based on distributed generation: Recent trends. *Cogent Eng.* **2020**, *7*, 1766394. [CrossRef]
59. Theo, W.L.; Lim, J.S.; Ho, W.S.; Hashim, H.; Lee, C.T. Review of distributed generation (DG) system planning and optimisation techniques: Comparison of numerical and mathematical modelling methods. *Renew. Sustain. Energy Rev.* **2017**, *67*, 531–573. [CrossRef]
60. Anuradha, K.; Jayatunga, U.; Perera, H.R. Loss-Voltage Sensitivity Analysis Based Battery Energy Storage Systems Allocation and Distributed Generation Capacity Upgrade. *J. Energy Storage* **2021**, *36*, 102357. [CrossRef]
61. Murty, V.V.; Kumar, A. Optimal placement of DG in radial distribution systems based on new voltage stability index under load growth. *Int. J. Electr. Power Energy Syst.* **2015**, *69*, 246–256. [CrossRef]
62. Saini, P.; Gidwani, L. An investigation for battery energy storage system installation with renewable energy resources in distribution system by considering residential, commercial and industrial load models. *J. Energy Storage* **2022**, *45*, 103493. [CrossRef]
63. Altıntaş, O.; Okten, B.; Karsu, Ö.; Kocaman, A.S. Bi-objective optimization of a grid-connected decentralized energy system. *Int. J. Energy Res.* **2018**, *42*, 447–465. [CrossRef]
64. Alturki, M.; Khodaei, A.; Paaso, A.; Bahramirad, S. Optimization-based distribution grid hosting capacity calculations. *Appl. Energy* **2018**, *219*, 350–360. [CrossRef]
65. Mishra, S.; Bordin, C.; Tomasgard, A.; Palu, I. A multi-agent system approach for optimal microgrid expansion planning under uncertainty. *Int. J. Electr. Power Energy Syst.* **2019**, *109*, 696–709. [CrossRef]
66. Santos, S.F.; Gough, M.; Fitiwi, D.Z.; Pogeira, J.; Shafie-khah, M.; Catalão, J.P. Dynamic Distribution System Reconfiguration Considering Distributed Renewable Energy Sources and Energy Storage Systems. *IEEE Syst. J.* **2022**, *16*, 3723–3733. [CrossRef]
67. Ajeigbe, O.A.; Munda, J.L.; Hamam, Y. Optimal allocation of renewable energy hybrid distributed generations for small-signal stability enhancement. *Energies* **2019**, *12*, 4777. [CrossRef]
68. Ajeigbe, O.A.; Munda, J.L.; Hamam, Y. Enhancing Small-Signal Stability of Intermittent Hybrid Distributed Generations. In Proceedings of the 2020 5th International Conference on Renewable Energies for Developing Countries (REDEC), Marrakech, Morocco, 29–30 June 2020; pp. 1–6.
69. Wu, F.F.; Gross, G.; Luini, J.F.; Look, P.M. A two-stage approach to solving large-scale optimal power flows. In Proceedings of the IEEE Conference Proceedings Power Industry Computer Applications Conference (PICA-79), Cleveland, OH, USA, 15–19 May 1979; pp. 126–136.
70. Vovos, P.N.; Bialek, J.W. Direct incorporation of fault level constraints in optimal power flow as a tool for network capacity analysis. *IEEE Trans. Power Syst.* **2005**, *20*, 2125–2134. [CrossRef]
71. Pesaran, M.; Mohd Zin, A.A.; Khairuddin, A.; Shariati, O. Optimal sizing and siting of distributed generators by a weighted exhaustive search. *Electr. Power Compon. Syst.* **2014**, *42*, 1131–1142. [CrossRef]
72. Khan, H.; Choudhry, M.A. Implementation of Distributed Generation (IDG) algorithm for performance enhancement of distribution feeder under extreme load growth. *Int. J. Electr. Power Energy Syst.* **2010**, *32*, 985–997. [CrossRef]
73. Van Cutsem, T. A method to compute reactive power margins with respect to voltage collapse. *IEEE Trans. Power Syst.* **1991**, *6*, 145–156. [CrossRef]
74. Sfikas, E.; Katsigiannis, Y.; Georgilakis, P. Simultaneous capacity optimization of distributed generation and storage in medium voltage microgrids. *Int. J. Electr. Power Energy Syst.* **2015**, *67*, 101–113. [CrossRef]
75. Lazzeroni, P.; Repetto, M. Optimal planning of battery systems for power losses reduction in distribution grids. *Electr. Power Syst. Res.* **2019**, *167*, 94–112. [CrossRef]
76. Salyani, P.; Salehi, J.; Gazijahani, F.S. Chance constrained simultaneous optimization of substations, feeders, renewable and non-renewable distributed generations in distribution network. *Electr. Power Syst. Res.* **2018**, *158*, 56–69. [CrossRef]
77. Zadeh, L.A. Information and control. *Fuzzy Sets* **1965**, *8*, 338–353.

78. Injeti, S.K.; Kumar, N.P. Optimal planning of distributed generation for improved voltage stability and loss reduction. *Int. J. Comput. Appl.* **2011**, *15*, 40–46.
79. Sharma, S.K.; Palwalia, D.; Shrivastava, V. Distributed generation integration optimization using fuzzy logic controller. *AIMS Energy* **2019**, *7*, 337–348. [CrossRef]
80. Sharma, A.; Sharma, A.; Chowdary, V.; Srivastava, A.; Joshi, P. Cuckoo search algorithm: A review of recent variants and engineering applications. *Metaheuristic Evol. Comput. Algorithms Appl.* **2021**, *1*, 177–194.
81. Mohamad, F.; Teh, J.; Lai, C.-M. Optimum allocation of battery energy storage systems for power grid enhanced with solar energy. *Energy* **2021**, *223*, 120105. [CrossRef]
82. He, Y.; Guo, S.; Zhou, J.; Ye, J.; Huang, J.; Zheng, K.; Du, X. Multi-objective planning-operation co-optimization of renewable energy system with hybrid energy storages. *Renew. Energy* **2022**, *184*, 776–790. [CrossRef]
83. Liao, G.-C. Solve environmental economic dispatch of Smart MicroGrid containing distributed generation system–Using chaotic quantum genetic algorithm. *Int. J. Electr. Power Energy Syst.* **2012**, *43*, 779–787. [CrossRef]
84. Ma, Y.; Yang, P.; Guo, H.; Wu, J. Power source planning of wind-PV-biogas renewable energy distributed generation system. *Power Syst. Technol.* **2012**, *36*, 9–14.
85. HA, M.P.; Huy, P.D.; Ramachandaramurthy, V.K. A review of the optimal allocation of distributed generation: Objectives, constraints, methods, and algorithms. *Renew. Sustain. Energy Rev.* **2017**, *75*, 293–312.
86. Koziel, S.; Rojas, A.L.; Moskwa, S. Power loss reduction through distribution network reconfiguration using feasibility-preserving simulated annealing. In Proceedings of the 2018 19th International Scientific Conference on Electric Power Engineering (EPE), Brno, Czech Republic, 16–18 May 2018; pp. 1–5.
87. Ashoornezhad, A.; Asadi, Q.; Falaghi, H.; Hajizadeh, A. Optimal Battery Energy Storage Placement in PV-connected Network Considering Uncertainty. In Proceedings of the 2021 11th Smart Grid Conference (SGC), Tabriz, Iran, 7–9 December 2021; pp. 1–5.
88. Prabpal, P.; Kongjeen, Y.; Bhumkittipich, K. Optimal Battery Energy Storage System Based on VAR Control Strategies Using Particle Swarm Optimization for Power Distribution System. *Symmetry* **2021**, *13*, 1692. [CrossRef]
89. Jamian, J.J.; Mustafa, M.W.; Mokhlis, H. Optimal multiple distributed generation output through rank evolutionary particle swarm optimization. *Neurocomputing* **2015**, *152*, 190–198. [CrossRef]
90. Rathore, A.; Patidar, N. Optimal sizing and allocation of renewable based distribution generation with gravity energy storage considering stochastic nature using particle swarm optimization in radial distribution network. *J. Energy Storage* **2021**, *35*, 102282. [CrossRef]
91. Zeinalzadeh, A.; Mohammadi, Y.; Moradi, M.H. Optimal multi objective placement and sizing of multiple DGs and shunt capacitor banks simultaneously considering load uncertainty via MOPSO approach. *Int. J. Electr. Power Energy Syst.* **2015**, *67*, 336–349. [CrossRef]
92. Mohandas, N.; Balamurugan, R.; Lakshminarasimman, L. Optimal location and sizing of real power DG units to improve the voltage stability in the distribution system using ABC algorithm united with chaos. *Int. J. Electr. Power Energy Syst.* **2015**, *66*, 41–52. [CrossRef]
93. Dixit, M.; Kundu, P.; Jariwala, H.R. Integration of distributed generation for assessment of distribution system reliability considering power loss, voltage stability and voltage deviation. *Energy Syst.* **2019**, *10*, 489–515. [CrossRef]
94. Abu-Mouti, F.S.; El-Hawary, M. Optimal distributed generation allocation and sizing in distribution systems via artificial bee colony algorithm. *IEEE Trans. Power Deliv.* **2011**, *26*, 2090–2101. [CrossRef]
95. El-Zonkoly, A.M. Multistage expansion planning for distribution networks including unit commitment. *IET Gener. Transm. Distrib.* **2013**, *7*, 766–778. [CrossRef]
96. Kefayat, M.; Ara, A.L.; Niaki, S.N. A hybrid of ant colony optimization and artificial bee colony algorithm for probabilistic optimal placement and sizing of distributed energy resources. *Energy Convers. Manag.* **2015**, *92*, 149–161. [CrossRef]
97. Padma Lalitha, M.; Veera Reddy, V.; Sivarami Reddy, N. Application of fuzzy and ABC algorithm for DG placement for minimum loss in radial distribution system. *Iran. J. Electr. Electron. Eng.* **2010**, *6*, 248–257.
98. Gomez, J.; Khodr, H.; De Oliveira, P.; Ocque, L.; Yusta, J.; Villasana, R.; Urdaneta, A. Ant colony system algorithm for the planning of primary distribution circuits. *IEEE Trans. Power Syst.* **2004**, *19*, 996–1004. [CrossRef]
99. Vlachogiannis, J.G.; Hatziargyriou, N.D.; Lee, K.Y. Ant colony system-based algorithm for constrained load flow problem. *IEEE Trans. Power Syst.* **2005**, *20*, 1241–1249. [CrossRef]
100. Wang, L.; Singh, C. Reliability-constrained optimum placement of reclosers and distributed generators in distribution networks using an ant colony system algorithm. *IEEE Trans. Syst. Man Cybern. Part C (Appl. Rev.)* **2008**, *38*, 757–764. [CrossRef]
101. Amohadi, M.; Fotuhi-Firuzabad, M. Optimal placement of automatic switching equipment in radial distribution networks based on protective coordination. *J. Electr. Eng. Technol.* **2019**, *14*, 1127–1137. [CrossRef]
102. Prakash, A.; Deshmukh, S. A multi-criteria customer allocation problem in supply chain environment: An artificial immune system with fuzzy logic controller based approach. *Expert Syst. Appl.* **2011**, *38*, 3199–3208. [CrossRef]
103. Aydin, I.; Karakose, M.; Akin, E. A multi-objective artificial immune algorithm for parameter optimization in support vector machine. *Appl. Soft Comput.* **2011**, *11*, 120–129. [CrossRef]
104. Gong, M.; Jiao, L.; Zhang, X. A population-based artificial immune system for numerical optimization. *Neurocomputing* **2008**, *72*, 149–161. [CrossRef]
105. Basu, M. Artificial immune system for dynamic economic dispatch. *Int. J. Electr. Power Energy Syst.* **2011**, *33*, 131–136. [CrossRef]

106. Aghaebrahimi, M.; Amiri, M.; Zahiri, S. An immune-based optimization method for distributed generation placement in order to optimize voltage profile. In Proceedings of the 2009 International Conference on Sustainable Power Generation and Supply, Nanjing, China, 6–7 April 2009; pp. 1–7.
107. Hatata, A.; Osman, G.; Aladl, M. An optimization method for sizing a solar/wind/battery hybrid power system based on the artificial immune system. *Sustain. Energy Technol. Assess.* **2018**, *27*, 83–93. [CrossRef]
108. Souza, B.B.; Carrano, E.G.; Neto, O.M.; Takahashi, R.H. Immune system memetic algorithm for power distribution network design with load evolution uncertainty. *Electr. Power Syst. Res.* **2011**, *81*, 527–537. [CrossRef]
109. Ma, J.; Yuan, X. Techno-economic optimization of hybrid solar system with energy storage for increasing the energy independence in green buildings. *J. Energy Storage* **2023**, *61*, 106642. [CrossRef]
110. Chisale, S.W.; Eliya, S.; Taulo, J. Optimization and design of hybrid power system using HOMER pro and integrated CRITIC-PROMETHEE II approaches. *Green Technol. Sustain.* **2023**, *1*, 100005. [CrossRef]
111. Masih, A.; Verma, H. Optimum sizing and simulation of hybrid renewable energy system for remote area. *Energy Environ.* **2022**, *33*, 933–951. [CrossRef]
112. Masrur, H.; Howlader, H.O.R.; Elsayed Lotfy, M.; Khan, K.R.; Guerrero, J.M.; Senjyu, T. Analysis of techno-economic-environmental suitability of an isolated microgrid system located in a remote island of Bangladesh. *Sustainability* **2020**, *12*, 2880. [CrossRef]
113. Akorede, M.F. Design and performance analysis of off-grid hybrid renewable energy systems. In *Hybrid Technologies for Power Generation*; Elsevier: Amsterdam, The Netherlands, 2022; pp. 35–68.
114. Yakub, A.O.; Same, N.N.; Owolabi, A.B.; Nsafon, B.E.K.; Suh, D.; Huh, J.-S. Optimizing the performance of hybrid renewable energy systems to accelerate a sustainable energy transition in Nigeria: A case study of a rural healthcare centre in Kano. *Energy Strategy Rev.* **2022**, *43*, 100906. [CrossRef]
115. Cao, Y.; Taslimi, M.S.; Dastjerdi, S.M.; Ahmadi, P.; Ashjaee, M. Design, dynamic simulation, and optimal size selection of a hybrid solar/wind and battery-based system for off-grid energy supply. *Renew. Energy* **2022**, *187*, 1082–1099. [CrossRef]
116. Shaheen, A.M.; El-Sehiemy, R.A.; Hasanien, H.M.; Ginidi, A.R. An improved heap optimization algorithm for efficient energy management based optimal power flow model. *Energy* **2022**, *250*, 123795. [CrossRef]

applied sciences

MDPI

Article

Parallel Water Column Technique for Obtaining a Smooth Output Power of the Pump as a Turbine at a Variable Water Flow Rate

Sajjad Hussain [1], Muhammad Humza [2,*], Tanveer Yazdan [3], Ghulam Abbas [3] and Han-Wook Cho [2,*]

[1] Department of Electrical Engineering, Institute of Southern Punjab, Multan 59300, Pakistan
[2] Department of Electrical, Electronics, and Communication Engineering Education, Chungnam National University, Daejeon 34134, Republic of Korea
[3] Department of Electrical Engineering, The University of Lahore, Lahore 54000, Pakistan
* Correspondence: engr.humza7@gmail.com (M.H.); hwcho@cnu.ac.kr (H.-W.C.)

Abstract: Hydro generation is the simplest and oldest method of electricity generation, with a century of successful operation. Using a pump as a turbine (PAT) is an optimal solution for minimizing the cost, particularly in low-head and small-scale hydro plants. Commercially available centrifugal pumps have become a popular solution for small-scale hydro and pumped-hydro facilities owing to their simple geometry, ease of operation, maintenance, and abundant availability in local markets. Variations in the water flow in hydro facilities, such as pumped-hydro and small-scale hydro facilities, are common; however, a PAT is unable to respond to variable flows because it is a fixed-speed device. To overcome this problem, different techniques have been suggested by researchers: (a) a system of parallel PATs; (b) geometrical modifications in the impeller of the PAT; and (c) power electronics-based variable frequency drives. All the aforementioned techniques have limitations, such as low output, high cost, complexity, transportation, and operation and maintenance. In this study, a simple and economical technique is proposed to smooth the output of a PAT on variable/decreasing water flow profiles. In the proposed technique, water columns connected in parallel (PWCs) are used to produce a pressurized water flow, as they have a convergent nozzle at the outlet. The PWC creates more space for water, and this additional mass of water boosts the water flow at the outlet. In this manner, the PWC technique maintains the flow at the inlet of the turbine. A serial integration of five PWCs with the same dimensions was conducted to inject the additional flow into the existing PAT penstock, governing a 37-kW generator. The design flow was maintained at 192.1 L/s without any additional power usage at the inlet of the PAT, and the output was smoothed even at the minimum water flow/head. Pump design and computational fluid dynamics simulations were performed using ANSYS software, whereas generator simulations were performed using MATLAB/Simulink software.

Keywords: energy resources; hydropower generation; small hydro; pump as turbine; PWC technique; smooth output power

Citation: Hussain, S.; Humza, M.; Yazdan, T.; Abbas, G.; Cho, H.-W. Parallel Water Column Technique for Obtaining a Smooth Output Power of the Pump as a Turbine at a Variable Water Flow Rate. *Appl. Sci.* **2023**, *13*, 3232. https://doi.org/10.3390/app13053232

Academic Editor: Luis Hernández-Callejo

Received: 31 January 2023
Revised: 27 February 2023
Accepted: 1 March 2023
Published: 2 March 2023

1. Introduction

Environmental effects, fear of the extinction of conventional energy sources (such as oil, gas, nuclear, and coal), and the high running costs of these costs have compelled planners toward renewable energy sources [1–4]. Among these resources, hydro energy contributes to nearly 16% of the global energy mix [5,6]. However, extensive capital costs, danger to aquatic life, significantly long construction times, and limited locations are the drawbacks of hydro-generation technology. To overcome these drawbacks, new arrangements for hydro facility construction, such as run-of-river, small hydro, and mini, micro, and pico-hydro plants, have been studied [7–9]. These new arrangements of hydro technology have reduced the initial cost, construction time, and environmental effects while increasing the location spectrum for the construction of hydro plants.

In pumped hydro storage (PHS) and small-scale hydro plants, using pumps in reverse (PAT) instead of conventional turbines is preferable to minimize the capital cost [10–12]. However, commercially available pumps are not designed to function in reverse mode, and if this mode is exercised, the pumps prove to be low-efficiency devices. Despite their low efficiency, PATs are used in small-scale hydro-generation and PHS facilities [13]. The use of PATs is abundant for numerous reasons, such as their low cost, bulk availability in local markets, and ease of installation. Furthermore, no specialty is required for operation and maintenance [14]. However, small-scale PHS and hydro-generation plants encounter flow fluctuations for several reasons, including intermittent upstream inflows and the decrement of the head in the upper reservoir of the PHS facility during generation. A PAT is unable to respond to these flow fluctuations owing to its simple geometry, and the output of the plant decreases. The design flow, which is obtained by the design head, determines the maximum efficiency of a PAT. Variations in the flow from its original design flow, for any reason, have an adverse effect on the output [15].

Researchers have adopted various methods to mitigate the adverse effects of flow fluctuations on the output, such as parallel PATs of different capacities, geometrical modifications of a PAT, and power electronics-based variable frequency drives (VFDs). In parallel PATs, when the flow is reduced and a PAT with a large capacity cannot deliver its rated output, it is either manually or automatically turned off. A PAT with a lower capacity is operated to minimize the adverse effects of flow fluctuations. However, the output is reduced by the newly available water flow, and the cost increases owing to paralleling, thereby making the design more complex [13,16]. Regarding the geometrical modifications of PATs, different researchers have suggested certain design modifications of the impeller [13,17–22], which better manage the variable flows. However, these modifications are unable to manage all the flow variations, and the cost is increased. Power electronics-based VFDs can provide any speed profile of variable flow inputs [23,24]. When the flow decreases, which affects the output of the plant, the VFD changes the speed of the generator, and the output of the plant becomes smooth. However, this method is significantly costly because devices with high ratings are required, causing installation, operational, maintenance, and transportation problems.

In this study, a simple technique is proposed to mitigate the adverse effects of flow variations on the output of a plant. This study is novel because a parallel water column (PWC) technique is used to smooth the output of a PAT system while maintaining the water flow at the inlet of the PAT by acting as an auxiliary penstock. With this technique, water columns connected in parallel are used with the existing penstock of a plant to inject additional flow at the inlet of the PAT in the event of a decreasing flow. This additional flow will maintain the smooth rated output of the plant, even at the minimum head. The effectiveness of the PWC technique was confirmed by performing a simulation using ANSYS software for the case study of a variable flow profile. Subsequently, based on the output obtained from the PAT, the electrical output of the generator was analyzed using MATLAB/Simulink software.

2. Materials and Methods

The details of the adverse effects of the flow fluctuations on the output of the PAT are discussed herein, along with the basic concepts of the PWC theory for a better understanding of the proposed PWC technique.

2.1. Effect of Flow Variations on the Output of PAT

The output of any hydro system is solely determined by the water flow, as given below [14,25]:

$$P_{out} = \eta \rho g H Q \quad (1)$$

where P_{out} is the output power, η is the efficiency, ρ is the density of water, g is the gravitational acceleration due to gravity (9.81 m/s^2), H is the head, and Q is the water flow.

The fluctuation of the flow "Q" varies the output of the hydro system. A smooth output can be achieved while maintaining a constant flow, which can be accomplished using the aforementioned techniques. A smooth output makes the system more efficient. The effects of flow fluctuations on the efficiency of PAT are shown in Figure 1, in which five scenarios were considered for the calculations based on a previous study [15]. Notably, the total reduction in flow is 38.4 L/s, which causes a significant decrease of 6.62% in the efficiency of the system. This decrease in plant efficiency is dangerous for the stability of the system.

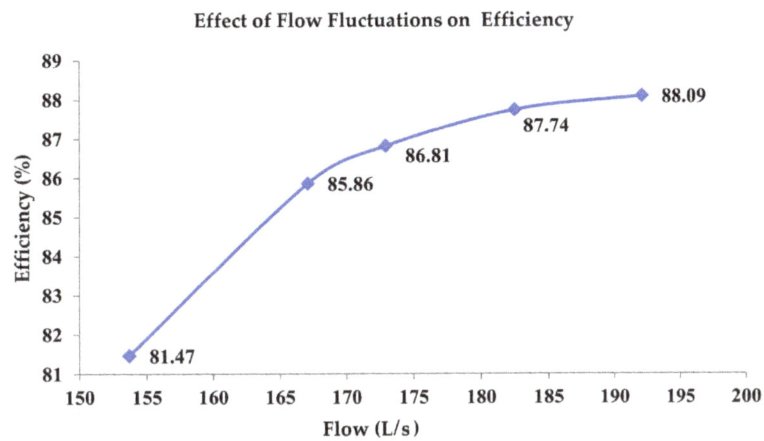

Figure 1. Effect of flow variations on efficiency.

The reduction in flow can be eliminated using the PWC technique. The basics of this newly proposed technique are discussed herein.

2.2. Water Column Theory

The atmospheric pressure (0.1 MPa) is created by a column of air over the surface of the earth. Similarly, water exerts pressure at its bottom, owing to the weight of water acting vertically downward. A water column having a height of 10.3 m creates pressure that is equal to the atmospheric pressure, that is, 101 kPa [26]; pressure is the force acting on an area. Mathematically,

$$P = F/A \tag{2}$$

where P is the pressure, F is the force, and A is the area. In this study, "A" indicates the area of the water discharge. If the area of discharge decreases, then the pressure will increase while the force remains constant, which is the weight of water working vertically downward and is indicated by the following:

$$w = mg \tag{3}$$

where w is the weight of water, m is the mass of water, and g is the gravitational acceleration. Here, "g" is a constant, and the mass of water should be increased to increase the weight. Traditionally, no arrangement can be used for the increment of the mass of water because the penstock is a fixed entity and functions as a singular body. Therefore, flow fluctuations threaten the stability of the system. Therefore, the PWC technique is presented in this study. Increasing the number of PWCs allows more space for a larger mass of water, which increases the force at the same height. In particular, the PWC technique makes the penstock more flexible and can manage increased flows per the requirements.

Alternatively, the pressure is inversely proportional to the area of the discharge area. Therefore, to increase the pressure, the discharge area must be reduced. This reduction was achieved using a nozzle to increase the pressure, which boosted the velocity of the water.

As the PAT is a low-head device, and this study is regarding low heads, only the Q parameter, which indicates the discharge/flow of water, in Equation (1) can increase the output of the PAT system. Mathematically,

$$Q = Av \tag{4}$$

where Q is the water discharge, A is the area of the water discharge, and v is the velocity of the water. According to the continuity equation, the flow in a closed system is always constant. Mathematically,

$$Q_1 = Q_2 \tag{5}$$

where Q_1 is the flow through point 1, and Q_2 is the flow through point 2. Furthermore,

$$A_1 v_1 = A_2 v_2 \Rightarrow v_2 = A_1 v_1 / A_2 \tag{6}$$

The inlet of the PAT is constant since it is a fixed body, and the only option for increasing the value of Q through the PAT while maintaining the same height is by increasing the velocity. The nozzle at the bottom, which has a velocity of "v_2", as indicated in Equation (6), will increase the velocity of water, which will in turn increase the value of 'Q'.

In this manner, as the number of water columns increases, A_1 increases, which increases v_2. The PWC technique injects the additional flow at the inlet of the PAT in the event of a flow decrease with the existing penstock of a plant; the adverse effects of the flow fluctuations are filtered, which smooths the output of the PAT.

2.3. PWC Technique with the Design of Experiment

The PWC follows certain rules to produce an output. To understand the behavior of the PWC, an experiment in which five water columns were connected in parallel was conducted. This experimental design is called a "double-nozzle setup", in which one nozzle makes the main outlet, which injects the additional water into the existing penstock of the PAT, while each water column has a separate nozzle. Computational fluid dynamics (CFD) simulations were performed using ANSYS software R1 2021 [27]. The dimensions of the PWC and specifications of the FLUENT solver are listed in Table 1.

Table 1. Dimension of the water columns and solver settings.

Dimensions/Settings	Values
Head	10 m
Diameter of the outlet nozzle	0.0635 m
Upper diameter of each water column	0.4572 m
Diameter of the nozzle used at the bottom of each water column	0.2286 m
Working fluid	Water (liquid)
Outlet pressure	0 Pascals
Turbulence model	k-ω (SST)
Density of water	1000 kg/m^3
Boundaries	
Inlet	Velocity inlet
Outlet	Pressure outlet
Wall function	Standard

The geometrical dimensions of the PWC were composed in a design modeler, and the settings were changed from solid to fluid, while meshing of the PWC was performed using ANSYS meshing. A tetrahedral mesh was created using a patch-confirming algorithm, and the span angle center was set to fine. Meshing of the PWC is shown in Figure 2, the details of which are provided in Table 2, including the meshing method, inflation layers, element size, and number of elements per PWC.

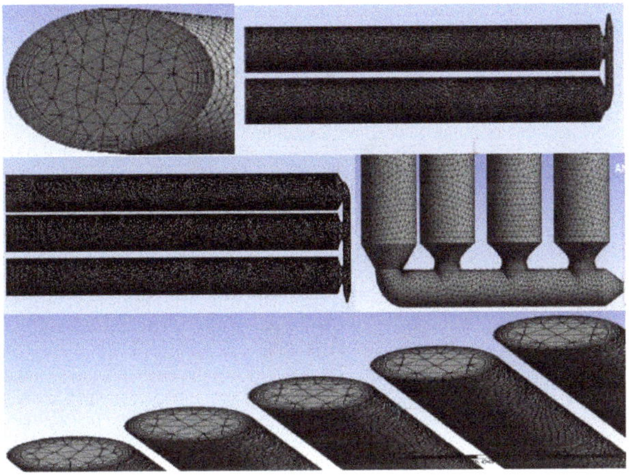

Figure 2. Meshing of PWCs.

Table 2. Mesh report of PWC.

No. of PWC	Element Size	Method of Meshing	Inflation Layers	No. of Elements
1	40	Tetrahedrons	7	274,274
2	40	Tetrahedrons	7	358,456
3	40	Tetrahedrons	7	543,020
4	40	Tetrahedrons	7	736,088
5	40	Tetrahedrons	7	926,713

A mesh independence test was performed for two PWCs to obtain the optimal element size and number of elements. Element sizes of 20, 30, 40, 50, 60, 70, 80, and 90 mm were applied to create 1,477,215, 655,237, 358,456, 237,308, 174,459, 138,933, 100,099, and 77,206 mesh elements, respectively. The optimal element size was 40 mm, which was selected for all the PWC simulation cases. A size of 40 mm yielded 358,456 elements for the two PWCs, as proven by the grid independence test shown in Figure 3.

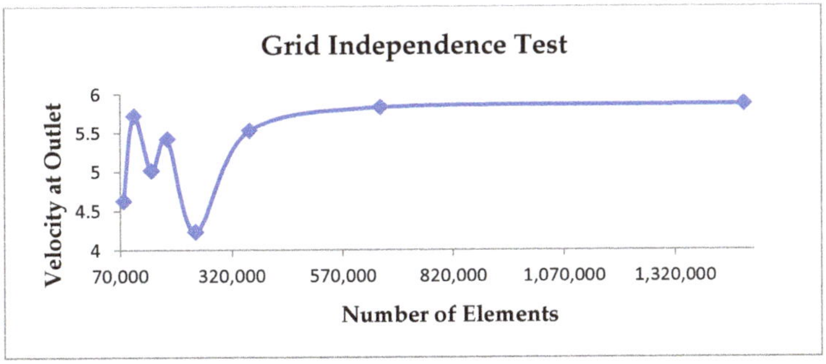

Figure 3. Grid independence test of the two PWCs.

In the experimental design, the dimensions of the PWCs were kept constant, while the number of PWCs was increased individually. The velocity inlet was assigned as the boundary condition at the inlet of the PWCs, whereas the pressure outlet was assigned at the outlet. The velocity and pressure produced in each case, which were processed in CFD-Post, are illustrated in (a–e) of Figure 4.

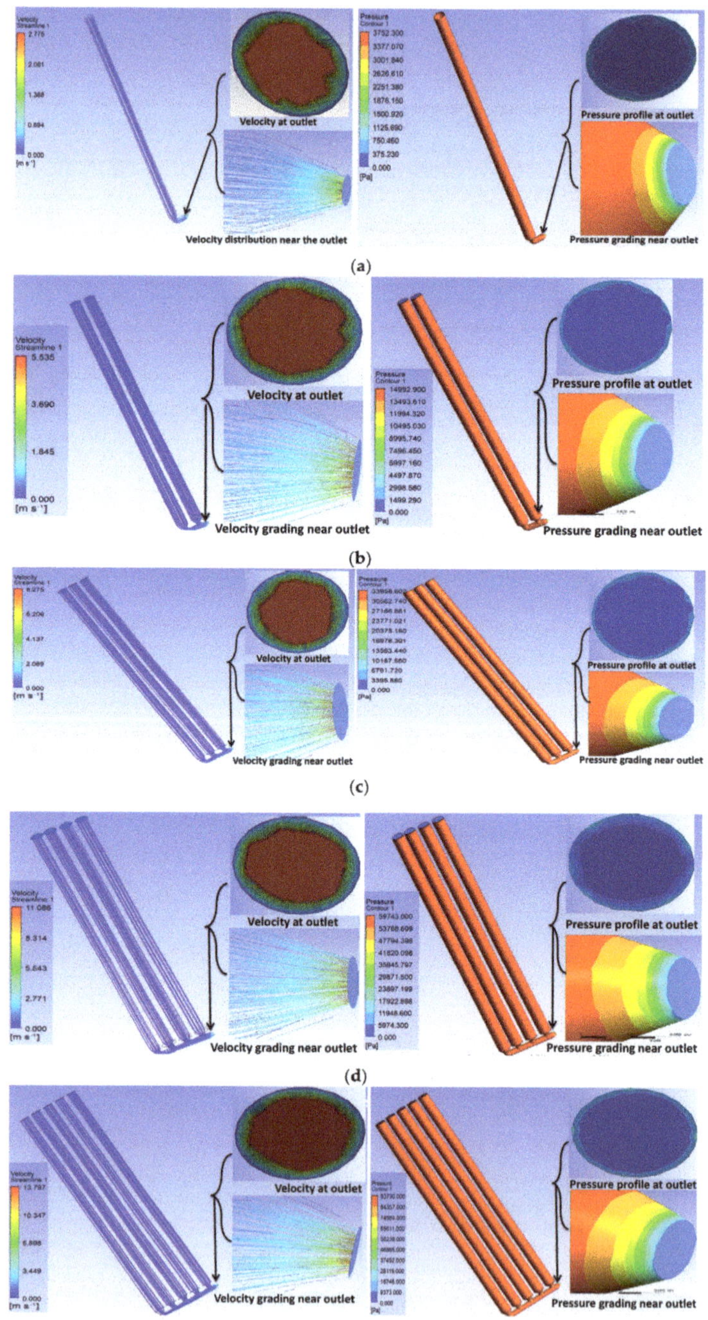

Figure 4. Velocity and pressure produced by (**a**) one water column, (**b**) two water columns, (**c**) three water columns, (**b**) four water columns, and (**e**) five water columns.

Two types of results were obtained from the experimental design study. First, the output behavior of the PWC was determined, and a mathematical model was developed. If the dimensions and output of the water column are known, the optimal number of PWCs can be derived for any flow, velocity, or pressure profile. Second, the optimal number of PWCs was derived for the aforementioned PAT, the water flow of which was variable at its inlet owing to the variations in the head of the upper reservoir during generation.

The results of the flow, outlet velocity, and pressure were calculated in CFD-Post with a function calculator, while the dimensions of the PWC were maintained to be the same; only the number of PWCs was increased individually according to the requirements of the design flow, as shown in Figure 5.

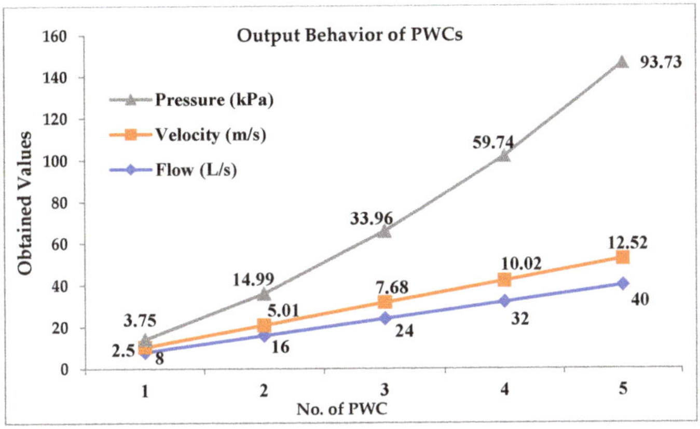

Figure 5. Variations in the pressure, flow, and velocity as the number of PWCs increases.

2.4. *Findings from the Design Experiment*

The findings obtained from the design experiment are presented (Figure 5) while developing a mathematical model to determine the output of the PWC.

Finding 1: Flow

The flow graph shown in Figure 5 demonstrates that the flow increases linearly as the number of water columns increases. Mathematically,

$$F = F_i \times n \tag{7}$$

where F is the flow of water, F_i is the flow obtained from one water column, and n is the number of PWCs used.

Finding 2: Velocity

The velocity of water also increases linearly as the number of water columns increases, as shown in the velocity graph (Figure 5). The mathematical relationship between the velocity and PWCs is as follows:

$$v = v_i \times n \tag{8}$$

where v is the velocity at the outlet, v_i is the velocity obtained from one water column, and n is the number of PWCs used.

Finding 3: Pressure

The water pressure at the nozzle inlet increases with the square of the number of water columns, which increases the water flow at the nozzle outlet. The pressure graph (Figure 5) describes the mathematical relationship between the PWC and the generated pressure. Mathematically,

$$P = P_i \times n^2 \tag{9}$$

where P is the total generated pressure, P_i is the pressure obtained from one water column, and n is the number of PWCs used in any setup.

3. Optimal Number of PWCs for a Smooth Output of the PAT

Based on Section 2.1, the decrease in the flow of the PAT is 38.4 L/s, for which, according to Equation (7), a total of five PWCs with a flow of 8 L/s each are required to smooth the flow at the inlet of the PAT. The PWC dimensions used for this purpose were the same as those shown in Table 1. Thus, the outlet nozzle of the PWC design becomes an additional inlet of the PAT, injecting the flow into the existing penstock, which is required for a particular stage of operation. Initially, the flow through the turbine was a design flow, and no change was required. As the flow decreases, the output also decreases. One water column is added by the system to acquire the design flow; if the new flow through the turbine meets the design flow, no further water column will be added by the system. However, considering a further deviation from the smooth output, another water column is added to the system to check whether the new scenario provides a smooth output. No changes are required if the new output is smooth. These five PWC were added to the system according to the flowchart shown in Figure 6.

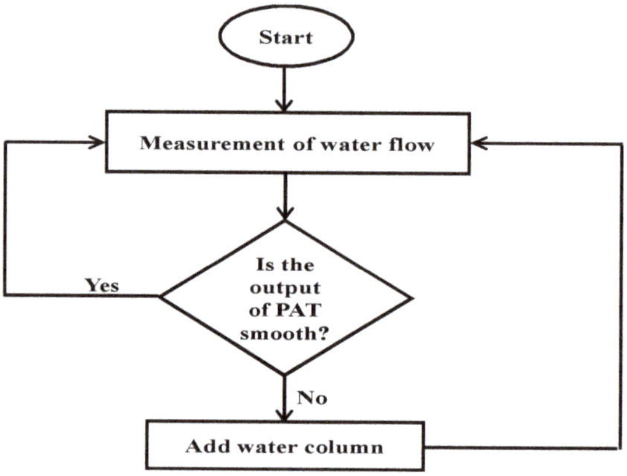

Figure 6. Integration of the PWC to maintain the flow.

The integration of the five PWCs with the existing PAT penstock and the injection of an additional flow at the inlet were conducted in ANSYS, as shown in Figure 7.

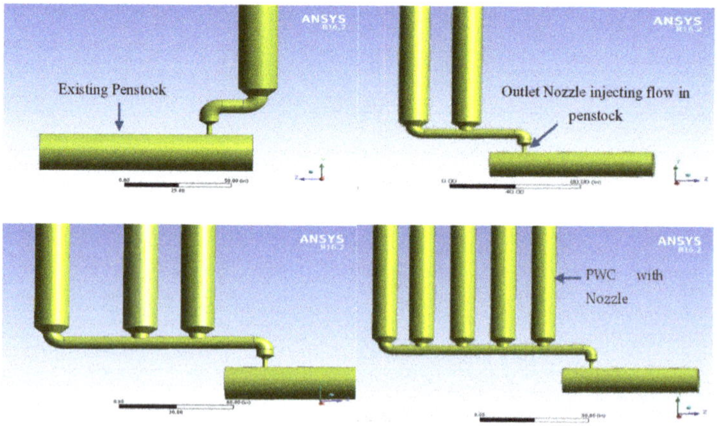

Figure 7. Integration of the PWC with the existing penstock.

As the number of PWCs is increased according to the demand, the flow is maintained at approximately the designed value, as listed in Table 3.

Table 3. Flow profile of PAT before and after the addition of the PWCs.

Flow (L/s)	Reduced Flow (L/s)	PWC Added (8 L/s Each)	Additional Flow (L/s)	New Flow (L/s)
192.1	Nil	No	Nil	192.1
182.2	9.9	1	8	190.2
172.9	19.2	2	16	188.9
167.1	25	3	24	191.7
153.7	38.4	5	40	193.7

4. Torque Calculation and Generator Behavior

The aforementioned flow profile produces torque in the PAT, which is applied as an input to the generator to monitor its output. Simulations were performed using ANSYS and MATLAB software.

4.1. Torque Produced by PAT on Different Flow Values

A pump with a flow of 192.1 L/s was designed using the ANSYS (Vista CPD) tool, consisting of a volute and impeller with six blades, as shown in Figure 8. A flow analysis through PAT was conducted in ANSYS CFX [13,28], in which water was used as the material, and the standard atmospheric pressure was considered to be one atm.

Figure 8. Side views of the pump designed in Vista CPD.

The pump was meshed using TURBO GRID, while the impeller and volute were separately meshed, achieving a good quality mesh, as shown in Figure 9.

Figure 9. Meshing of the impeller and volute.

A detailed mesh report is provided in Table 4, which demonstrates the domains of the analysis, location of the domains, number of nodes, and number of elements.

Table 4. Mesh report of the pump.

Domain	Nodes	Elements	Tetrahedra	Wedges	Hexahedra
R1(Passage)	663,175	627,640	0	0	627,640
S1(B75)	82,927	258,746	165,196	93,550	0
All Domains	746,102	886,386	165,196	93,550	627,640

The mesh independence is checked for the pump and eight different cases with regard to. element numbers, which are obtained by changing the element size and global size factor for the volute and impeller, respectively. Element sizes of 15, 20, 25, 30, 35, 40, 45, and 50 mm were applied for the volute in meshing, yielding 330,438, 281,176, 258,746, 248,172, 240,762, 237,168, 234,902, and 232,452 elements, respectively. In contrast, the global size factor was the main parameter that was manipulated for impeller meshing, achieving a range of eight element numbers. Different global size factors for the impeller were assigned in TURBO GRID; the 0.8, 0.9, 1.0, 1.1, 1.2, 1.3, 1.4, and 1.5 sizes produced 141,792, 203,553, 280,080, 385,118, 501,128, 781,482, and 1,019,900 elements, respectively. The output parameter was the velocity at the outlet, which is demonstrated by the mesh independence test shown in Figure 10; 886,386 elements were selected for the simulations, which were produced by a 25-mm element size and a 1.3 global size factor.

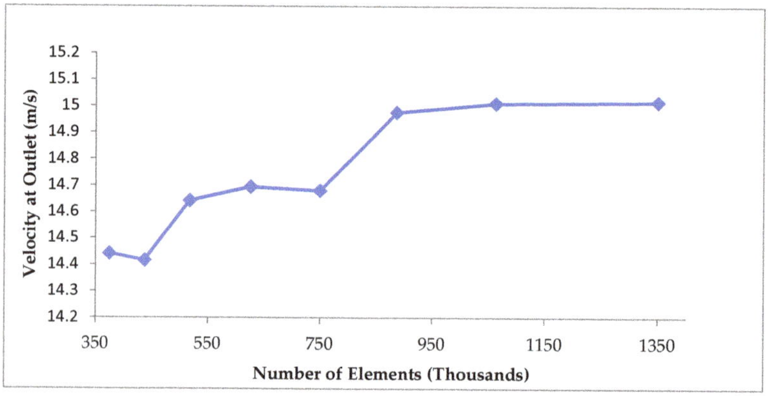

Figure 10. Meshing independency test of the pump.

The steady-state boundary conditions for this simulation were set for the PAT. The stationary (S1) domain is the volute, while the rotating (R1) domain is the impeller of the pump. The inlet was assigned a boundary condition for the mass flow, while the outlet was the pressure, and the standard atmospheric pressure was assigned. The blade and all other boundary walls were assigned a boundary condition (no slip). A frozen-rotor mixing model was adopted for the interface between S1 and R1. The SST turbulence model was applied for the flow analysis, while in the solver control panel, a high resolution was applied as an advection scheme, 1.0×10^{-5} for convergence, first order for numeric turbulence, and SIMPLEC [27,29,30]. Transient or unsteady simulations were also performed after steady simulations. The total time was set to 1 s, the time step to 0.0001 s, and the initial time value to 0 s for the transient case. The transient rotor-stator mixing model was assigned to the interface between S1 and R1. Cartesian velocity components were selected as u = 0 m/s, v = 0 m/s, and z = 0 m/s for the global initialization [31].

Here, the pump functions as a turbine; therefore, the outlet of the pump is the inlet of the PAT, and the direction of the flow of water through the PAT is shown in Figure 11.

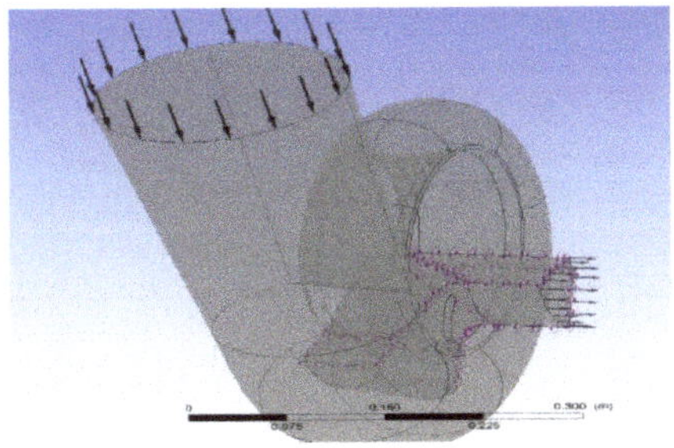

Figure 11. Direction of the flow of water for the PAT.

As indicated in Section 2.1, the torque produced by the variable water flow was calculated using ANSYS CFD-Post, which is a function calculator. A graphical representation of the flow and the produced torque is shown in Figure 12a. When the PWC is integrated, and a new range of flows at the inlet of the PAT is obtained, the model is simulated once again to calculate the new torque produced by the new values of the water flow. A graphical representation of the new flow and torque produced is shown in Figure 12b.

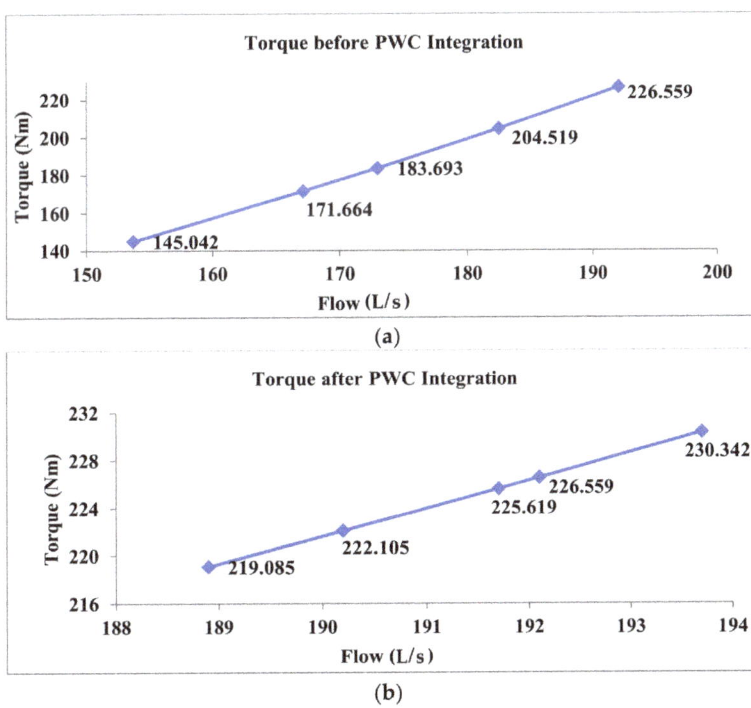

Figure 12. Torque produced by PAT (**a**) before PWC integration and (**b**) after PWC integration.

The torque produced by the PAT at variable and smooth flows is compared in Table 5, along with the deviations given before and after the integration of the PWC. These results clearly demonstrate that the values of the produced torque decrease with decreasing flow. Hence, fluctuations in the flow rate significantly affect the torque of the PAT. When the flow rate is kept constant with the integration of the PWC, the torque of the PAT is smoothed and maximized according to the design flow.

Table 5. Torque of PAT regarding different flow values.

Before PWC Integration			After PWC Integration		
Flow (L/s)	Torque (Nm)	Deviation (%)	Flow (L/s)	Torque (Nm)	Deviation (%)
192.1	226.559	0	192.1	226.559	0
182.5	204.519	9.73	190.2	222.105	1.97
172.9	183.693	18.92	188.9	219.085	3.3
167.1	171.664	24.23	191.7	225.619	0.41
153.7	145.042	35.98	193.7	230.342	1.64

4.2. Generator Behavior on the Variable and Smooth Torque

Although a smooth torque is obtained, which ultimately smooths the output of the PAT, in most cases, the PAT and generator are directly coupled. However, a Simulink model was developed to observe the behavior of the generator with variable and smooth torques, as shown in Figure 13.

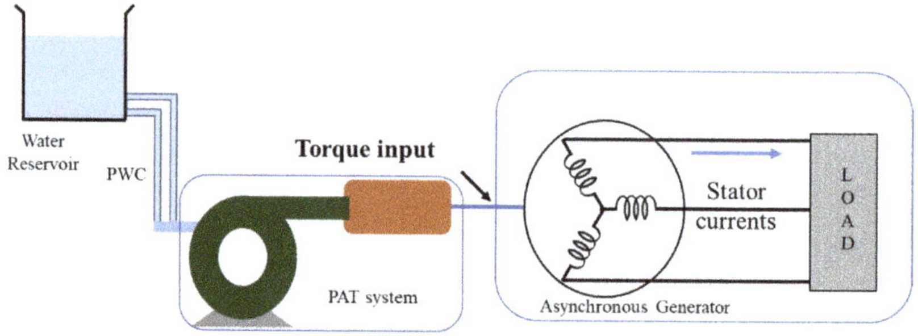

Figure 13. Overall system to observe the behavior of the generator.

A built-in asynchronous machine block rated at 37 kW was selected from the Simulink library to simulate the behavior of the generator. This block can simulate both motors and generators, using (+) and (−) signs with the torque, respectively. The variable and smoothed torques obtained from ANSYS CFX, as shown in Figure 14, were used as the input signals in the Simulink model [32,33]. The model was simulated for 25 s, and each duration of five seconds corresponds to a different torque value.

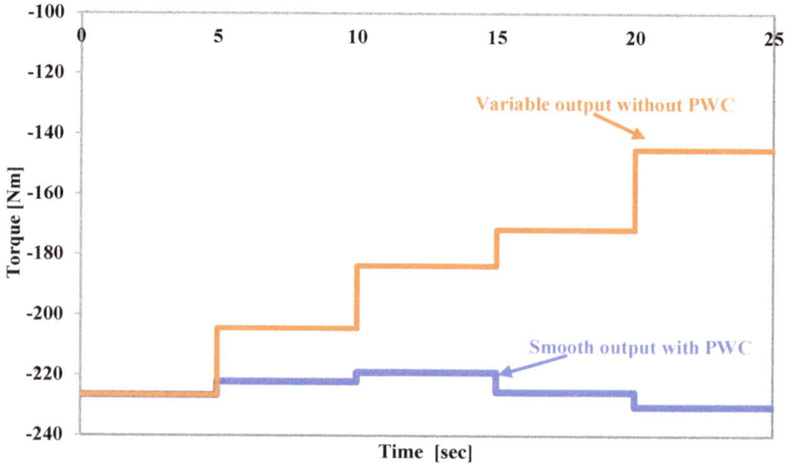

Figure 14. Torque input signal to the generator.

The simulation results of the peak value of the generator current are shown in Figure 15, which demonstrates the variations in the current with respect to the torque obtained from the PAT. In a hydro system, the torque varies with the flow variation. The current depends on the torque value. Therefore, any fluctuation in the flow rate affects the output current of the generator. However, the PWC technique provides a smooth flow, which results in a smooth output.

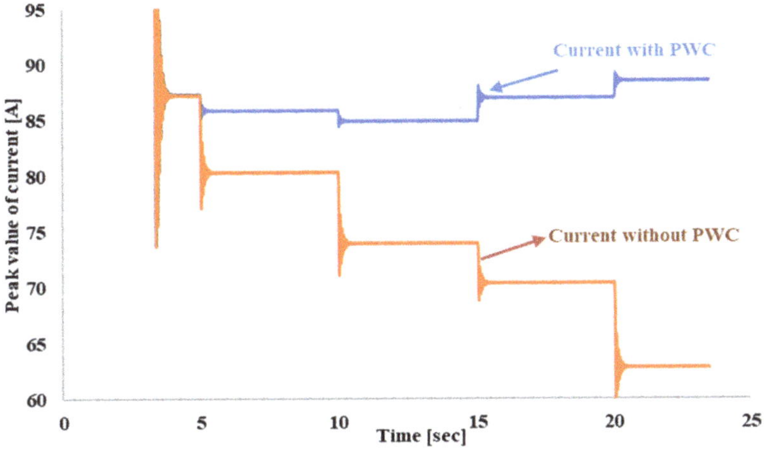

Figure 15. Output current of the generator for both variable and smooth torques.

5. Conclusions

A simple and economical PWC technique was proposed in this study to maintain the design flow at the inlet of a PAT under variable flow conditions. The proposed technique employs the series integration of parallel water columns with the same dimensions using a double-nozzle design. Paralleling the water columns creates more space for water to provide the design flow at the outlet. When water flow is decreased owing to a decrease in the upstream flow or in a PHS facility, the PWC maintains a flow at the inlet of the PAT by acting as an axillary penstock.

The design of the experiment and its findings indicate that, for a given case, the designed flow of 192.1 L/s can create a maximum torque of 226.559 Nm. The reduction in the flow to 153.7 L/s produced a reduced torque of 145.042 Nm. A total reduction of 38.4 L/s in the water flow was compensated for by integrating five PWCs, each having a flow capacity of 8 L/s. After the integration, the new flow and torque values were 193.7 L/s and 230.342 Nm, respectively, which were nearly identical to the designed values with a deviation of only 1.64%. Furthermore, based on the smooth output of the PAT, the generator output was confirmed. Thus, integration of the PWC using a double-nozzle design is an effective technique for maintaining smooth output of the PAT and generator.

Author Contributions: Conceptualization, S.H. and M.H.; methodology, M.H. and T.Y.; software, S.H. and G.A.; formal analysis, M.H., T.Y. and G.A.; writing and draft preparation, S.H., M.H. and T.Y.; review and editing, T.Y. and G.A.; supervision, H.-W.C. and M.H.; funding acquisition, H.-W.C. All authors have read and agreed to the published version of the manuscript.

Funding: This research received no external funding.

Institutional Review Board Statement: Not applicable.

Informed Consent Statement: Not applicable.

Data Availability Statement: Not applicable.

Acknowledgments: This work was supported by the National Research Foundation of Korea (NRF) grant funded by the South Korean government (No. 2022R1I1A3072104).

Conflicts of Interest: The authors declare no conflict of interest.

Nomenclature

A	area of water discharge [m^2]
F	force [N]
g	gravitational acceleration [m/s^2]
H	head height [m]
m	mass of water [kg]
n	number of water columns
P	pressure [Pa]
P_{out}	output power [W]
PAT	pump as turbine
PWC	parallel water columns
Q	flow of water [L/s]
v	velocity of water [m/s]
w	weight of water
η	efficiency [%]
ρ	density of water [kg/m^3]

References

1. Olabi, A.; Abdelkareem, M.A. Renewable energy and climate change. *Renew. Sustain. Energy Rev.* **2022**, *158*, 112111. [CrossRef]
2. Bilgili, F.; Lorente, D.B.; Kuşkaya, S.; Ünlü, F.; Gençoğlu, P.; Rosha, P. The role of hydropower energy in the level of CO2 emissions: An application of continuous wavelet transform. *Renew. Energy* **2021**, *178*, 283–294. [CrossRef]
3. Jurasz, J.; Ciapala, B. Solar–hydro hybrid power station as a way to smooth power output and increase water retention. *Sol. Energy* **2018**, *173*, 675–690. [CrossRef]
4. Wang, Z.; Yen-Ku, K.; Li, Z.; An, N.B.; Abdul-Samad, Z. The transition of renewable energy and ecological sustainability through environmental policy stringency: Estimations from advance panel estimators. *Renew. Energy* **2022**, *188*, 70–80. [CrossRef]
5. International Energy Agency. Available online: https://www.iea.org/reports/renewables-information-overview/supply (accessed on 6 January 2022).
6. International Hydro Power Association (IHA). Available online: https://www.hydropower.org/iha/discover-facts-about-hydropower (accessed on 6 January 2022).
7. Tkac, S. Hydro power plants, an overview of current types and technology. *SSP-J. Civ. Eng.* **2018**, *13*, 115–126.
8. Paish, O. Small hydro power: Technology and current status. *Renew. Sustain. Energy Rev.* **2002**, *6*, 537–556. [CrossRef]
9. Kaunda, C.S.; Kimambo, C.Z.; Nielsen, T.K. Hydropower in the Context of Sustainable Energy Supply: A Review of Technologies and Challenges. *Int. Sch. Res. Not.* **2012**, *2012*, 730631. [CrossRef]
10. Alatorre-Frenk, C. Cost Minimization in Micro-Hydro System Using. PAT. Dissertation, University of Warwickshire, Coventry, UK, 1994.
11. Aggidis, G.A.; Luchinskaya, E.; Rothschild, R.; Howard, D.C. The costs of small-scale hydro power production: Impact on the development of existing potential. *Renew. Energy* **2012**, *35*, 2632–2638. [CrossRef]
12. Hossain, I.M.; Ferdous, S.M.; Salehin, S.; Saleque, A.M.; Jamal, T. Pump-as-turbine (PAT) for small scale power generation: A comparative analysis. In Proceedings of the 2014 3rd International Conference on the Developments in Renewable Energy Technology (ICDRET), Dhaka, Bangladesh, 29–31 May 2014.
13. Binamaa, M.; Sua, W.-T.; Li, X.-B.; Weia, X.-Z.; Anc, S. Investigation on pump as turbine (PAT) technical aspects for micro hydropower schemes: A state-of-the-art review. *Renew. Sustain. Energy Rev.* **2017**, *79*, 148–179. [CrossRef]
14. Morabito, A.; Furtado, G.A. Variable speed regulation for pump as turbine in micro pumped hydro energy storage application. In Proceedings of the 38th IAHR Congress, Panama City, Panama, 1–6 September 2019.
15. Deyou, L.; Hongji, W.; Goaming, X.; Ruxhi, G.; Zhansheng, L. Unsteady simulation and analysis for hump characteristics of a pump turbine model. *Renew. Energy* **2015**, *77*, 32–42. [CrossRef]
16. Nobile, E.; Sari, G.; Schwery, A. Variable Speed Hydro Pumped Storage as Flexible Enabler of Intermittent Renewable Penetration. In Proceedings of the 2018 IEEE Power & Energy Society General Meeting (PESGM), Portland, OR, USA, 5–10 August 2018; pp. 1–5.
17. Elbatran, A.H.; Yakoob, O.B.; Ahmed, Y.M. Operation, performance and economic analysis of low head micro-hydropower turbines for rural and remote areas: A review. *Renew. Sustain. Energy Rev.* **2015**, *43*, 40–50. [CrossRef]
18. Derakhshan, S.; Kasaeian, N. Optimization, Numerical, and Experimental Study of a Propeller Pump as Turbine. *J. Energy Resour. Technol.* **2014**, *136*, 012005. [CrossRef]

19. Yang, S.-S.; Kong, F.-Y.; Jiang, W.-M.; Qu, X.-Y. Effects of impeller trimming influencing pump as turbine. *Comput. Fluids* **2012**, *67*, 72–78. [CrossRef]
20. Yang, S.-S.; Liu, H.-L.; Kong, F.-Y.; Dai, C. Experimental, Numerical, and Theoretical Research on Impeller Diameter Influencing Centrifugal Pump-as-Turbine. *J. Energy Eng.* **2013**, *139*, 299–307. [CrossRef]
21. Yang, S.-S.; Wang, C.; Chen, K.; Yuan, X. Research on Blade Thickness Influencing Pump as Turbine. *Adv. Mechan. Eng.* **2014**, *6*, 190530. [CrossRef]
22. Yang, S.-S.; Kong, F.-Y.; Qu, X.-Y.; Jiang, W.-M. Influence of blade number on the performance and pressure pulsations in a pump used as a turbine. *J. Fluids Eng.* **2012**, *134*, 124503. [CrossRef]
23. Lugauer, F.J.; Kainz, J.; Gehlich, E.; Gaderer, M. Roadmap to Profitability for a Speed-Controlled Micro-Hydro Storage System Using Pumps as Turbines. *Sustainability* **2022**, *14*, 653. [CrossRef]
24. Binama, M.; Kan, K.; Chen, H.-X.; Zheng, Y.; Zhou, D.-Q.; Su, W.-T.; Ge, X.-F.; Ndayizigiye, J. A Numerical Investigation into the PAT Hydrodynamic Response to Impeller Rotational Speed Variation. *Sustainability* **2021**, *13*, 7998. [CrossRef]
25. Hunt, J.D.; Zakeri, B.; Lopes, R.; Barbosa, P.S.F.; Nascimento, A.; de Castro, N.J.; Brandão, R.; Schneider, P.S.; Wada, Y. Existing and new arrangements of pumped-hydro storage plants. *Renew. Sustain. Energy Rev.* **2020**, *129*, 109914. [CrossRef]
26. Bansal, R.K. *A Textbook of Fluid Mechanics and Hydraulic Machines*; Laxmi Publication (P) Ltd.: New Delhi, India, 2010.
27. ANSYS Fluent Tutorial Guide. Release 18.0, January 2017. Available online: www.ansys.com (accessed on 5 December 2021).
28. Zhou, L.; Hang, J.; Bai, L.; Krzemianowski, Z.; El-Emam, M.A.; Yasser, E.; Agarwal, R. Application of entropy production theory for energy losses and other investigation in pumps and turbines: A review. *Appl. Energy* **2022**, *318*, 119211. [CrossRef]
29. Weidong, C.; Lingiun, Y.; Bing, L.; Yining, Z. The influence of impeller eccentricity on centrifugal pump. *Adv. Mech. Eng.* **2017**, *9*, 1–17. [CrossRef]
30. El-Emam, M.; Zhou, L.; Yasser, E.; Bai, L.; Shi, W. Computational Methods of Erosion Wear in Centrifugal Pump: A State-of-the-Art Review. *Arch. Comput. Methods Eng.* **2022**, *29*, 3789–3814. [CrossRef]
31. Wei, Y.; Shi, Y.; Shi, W.; Pan, B. Numerical Analysis and Experimental Study of Unsteady Flow Characteristics in an Ultra-Low Specific Speed Centrifugal Pump. *Sustainability* **2022**, *14*, 16909. [CrossRef]
32. Simulink Getting Started Guide, Release 2015. Available online: www.mathworks.com (accessed on 15 December 2021).
33. Alberizzi, J.C.; Renzi, M.; Nigro, A.; Rossi, M. Study of a Pump-as-Turbine (PAT) speed control for a Water Distribution Network (WDN) in South-Tyrol subjected to high variable water flow rated. *Energy Procedia* **2018**, *148*, 226–233. [CrossRef]

energies

MDPI

Article

Optimal Configuration of Energy Storage Systems in High PV Penetrating Distribution Network

Jinhua Zhang [1], Liding Zhu [1,*], Shengchao Zhao [1], Jie Yan [2] and Lingling Lv [1]

1 School of Electric Power, North China University of Water Resources and Electric Power, Zhengzhou 450045, China
2 State Key Laboratory of New Energy Power System, School of New Energy, North China Electric Power University, Beijing 100096, China
* Correspondence: z20211050617@163.com

Abstract: In this paper, a method for rationally allocating energy storage capacity in a high-permeability distribution network is proposed. By constructing a bi-level programming model, the optimal capacity of energy storage connected to the distribution network is allocated by considering the operating cost, load fluctuation, and battery charging and discharging strategy. By constructing four scenarios with energy storage in the distribution network with a photovoltaic permeability of 29%, it was found that the bi-level decision-making model proposed in this paper saves 2346.66 yuan and 2055.05 yuan, respectively, in daily operation cost compared to the scenario without energy storage and the scenario with single-layer energy storage. After accessing IEEE-33 nodes for simulation verification, it was found that the bi-level decision-making model proposed in this paper has a good inhibition effect on voltage fluctuation and load fluctuation after energy storage configuration. In addition, this paper analyzes the energy storage that can be accessed by photovoltaic distribution networks with different permeability and finds that when photovoltaic permeability reaches 45% and corresponding energy storage is configured, the economic and energy storage benefits of the system are the best.

Keywords: high PV penetration; energy storage; optimal configuration; bi-level decision-making models

Citation: Zhang, J.; Zhu, L.; Zhao, S.; Yan, J.; Lv, L. Optimal Configuration of Energy Storage Systems in High PV Penetrating Distribution Network. *Energies* **2023**, *16*, 2168. https://doi.org/10.3390/en16052168

Academic Editors: Luis Hernández-Callejo, Jesús Armando Aguilar Jiménez and Carlos Meza Benavides

Received: 31 January 2023
Revised: 16 February 2023
Accepted: 20 February 2023
Published: 23 February 2023

1. Introduction

In recent years, with global carbon dioxide emissions hitting record highs, China has proposed a "two-carbon" target to tackle environmental problems. Promoting the development of new energy and the transformation of energy structures has become an important part of global development. Due to abundant reserves and easy access, solar energy has been developing rapidly in recent years, and its proportion in the power grid has been increasing year by year [1]. While improving energy utilization, this has brought a lot of trouble to the power distribution network. With the continuous increase in the penetration rate of photovoltaics integrated into the power distribution network, problems such as voltage collapse may occur, which has a serious impact on the safe and stable operation of the system [2].

Studies have shown that a large number of photovoltaics connected to the distribution network will also increase the number of system equipment, which will bring a burden to the system and easily generate harmonic interference. In addition, the retrograde power generated by the grid connection is prone to exceed the limit of the system node voltage, which not only reduces the power quality but also deteriorates the user experience. When high-penetration photovoltaics are connected to the grid, the uncertainty of output cannot be matched with the load of the distribution network in real-time, which will affect the power balance of the system. When the photovoltaic output fluctuates greatly due to the change in environment and climate, the stability of the system will be affected [3]. In addition, the high-penetration photovoltaic grid connection requires a large number of power electronic equipment to join the distribution network, which leads to the existence of harmonics and affects the power quality [4], and the dispatching flexibility of the distribution network is greatly reduced, which makes it more difficult for the power system to develop a power generation plan. If such problems cannot be properly solved, it will not only seriously threaten the safe and stable operation of the system but also cause a waste of energy and limit the future development of photovoltaic power generation [5].

The authors in [6–8] analyze the influence of photovoltaic systems from the aspects of voltage fluctuation, voltage amplitude, and frequency. From the perspective of stability, Rasoul proposed a new framework to analyze the influence of different photovoltaic permeability on voltage stability [9]. During the study, Zetty found that in a high permeability renewable energy distribution network, load fluctuation is the main factor leading to the voltage fluctuation of the system, and the realization of various fluctuations in the high-light voltage permeability distribution network is important content to achieve the increase of photovoltaic permeability in the distribution network [10]. The introduction of energy storage devices improves the power quality while improving the photovoltaic stable output [11–13]. Through reasonable regulation and control of a BESS, the absorption of new energy on the power generation side can be completed, the permeability of distributed power supply on the transmission side and distribution side can be improved, and the safe, stable, and economic operation of the system can be ensured [14,15]. The authors in [16–18] studied the working principle and characteristic analysis of different types of energy storage devices and different types of BESSs and discussed the practicability of combining BESS energy storage and generation measurement. From the perspective of photovoltaic and load output prediction, Rahman and Zhao verified the feasibility of combining energy storage optimization configuration with the prediction by comparing scenarios with or without prediction [19,20]. In order to meet the photovoltaic energy storage demand in the distribution network, Wang's multiple operation scenarios of energy storage were divided into grid scenarios to obtain the demand relationship of energy storage capacity under different operating conditions and to complete the calculation of energy storage capacity [21].

Access to energy storage equipment requires considerable capital investment in actual project construction and operation and maintenance. Therefore, the demand response for energy storage capacity is important content in optimizing energy storage configuration. In [22], Balouch proposed an optimization goal of matching demand and supply. Based on the analysis of line planning, low-cost scheduling, and demand response, the energy utilization efficiency and comprehensive operating cost of a smart grid were optimized. The authors in [23,24] introduced the improved optimization algorithm to improve the optimization ability so as to determine the optimal scheme of energy storage optimization configuration and realize a higher degree of response between demand and supply by analyzing various indicators of access to the power grid. In [23], Balouch optimized a response scheduling scheme by introducing the GWCSO algorithm. Higher robustness and computational efficiency of the algorithm make the optimization results more advantageous in power cost and peaking ratio. In [24], Mostafa improves the PSO algorithm, improves the accuracy and effectiveness of the algorithm, and optimizes the location and capacity allocation of energy storage in distributed networks. While the optimization objects are complex and diverse when connecting to the power system, the choice and update of the Pareto optimal solution will determine the quality of the final optimization result [25].

In the existing studies, it seems obvious for everyone to apply energy storage in high-permeability photovoltaic distribution networks [26–32]. In the case of low photovoltaic permeability, access to energy storage can indeed improve photovoltaic output and power quality. However, few researchers have analyzed whether energy storage can still meet expectations in the scenario of high photovoltaic permeability, and how to rationally allocate energy storage in a distribution network with high photovoltaic permeability. In this paper, the application of energy storage in a high permeability photovoltaic scenario is analyzed, and the energy storage in a high-light volt distribution network is configured by establishing a two-layer planning model of the distribution network. The optimal size of energy storage was configured considering the fluctuation of power grid voltage and load, economic benefits and energy storage benefits, and the working condition of energy storage in the scenario of high-light voltage permeability, and the improvement of benefits in all aspects of the distribution network were studied. Finally, the feasibility of the proposed method was verified in the IEEE-33 node system.

The main contributions of this study are summarized below:

- Proposed a method for optimal allocation of energy storage capacity of a distribution network based on a two-layer programming model and verified its feasibility.
- Used the K-means method to complete an analysis of the uncertain photovoltaic output into the deterministic scenario.
- The multi-objective particle swarm optimization algorithm was improved to solve the optimal configuration, and the advantages of the improved algorithm were compared.
- By constructing different scenarios, it was verified that energy storage can still improve the power quality in the distribution network with high-light voltage and permeability.
- Through analysis of the optimal configuration of energy storage in the distribution network with different photovoltaic permeabilities, the optimal economic photovoltaic permeability was concluded.

The rest of this paper is organized as follows: The treatment method for the PV uncertainty and the selection of the PV working curve is introduced in Section 2. In Sections 3 and 4, the bi-level decision-making programming model is constructed and solved to realize the addressing and capacity selection of the energy storage device. At the same time, in the fourth section, the specific content of particle swarm optimization is described. In Section 5, four scenarios are constructed to discuss the benefits generated by energy storage configuration and optimization benefits brought by algorithm improvement. Finally, in Section 6, we summarize the content of the thesis.

2. Analysis of Photovoltaic Output Characteristic

Due to the great influence of light and the environment, photovoltaic power generation is full of uncertainties. For further analysis, we collected the annual daily output data of photovoltaic power stations (annual output of an operational photovoltaic power station in Henan Province from June 2019 to July 2020), as shown in Figure 1. When considering energy storage benefits, excessive uncertainty in output will lead to uncertainty in energy storage benefits. In order to avoid this influence, this paper will process various output curves by clustering the division method, summarizing the photovoltaic output with high uncertainty into six typical output scenarios and analyzing them, and transforming the uncertainty into a deterministic analysis.

2.1. K-Means Cluster Analysis Method

The K-means clustering method is a classical clustering analysis method based on the iterative method which has the advantages of high efficiency and convenience in processing large-scale uncertain data [33]. Through the K-means clustering method, a large number of output data can be refined and extracted, and fewer typical output scenarios can be obtained that can represent the output of photovoltaic power stations.

The K-means algorithm sets an initial cluster center in all scenes and iterates clustering for a large number of scenes based on the optimal distance. The iteration is not finished until the clustering presents a steady-state equilibrium. The iteration results are shown in Figure 2. After the whole process is complete, the center of each cluster scene is set as the partition scene, and the probability of each cluster scene is set as the required partition probability $P_r(s)$.

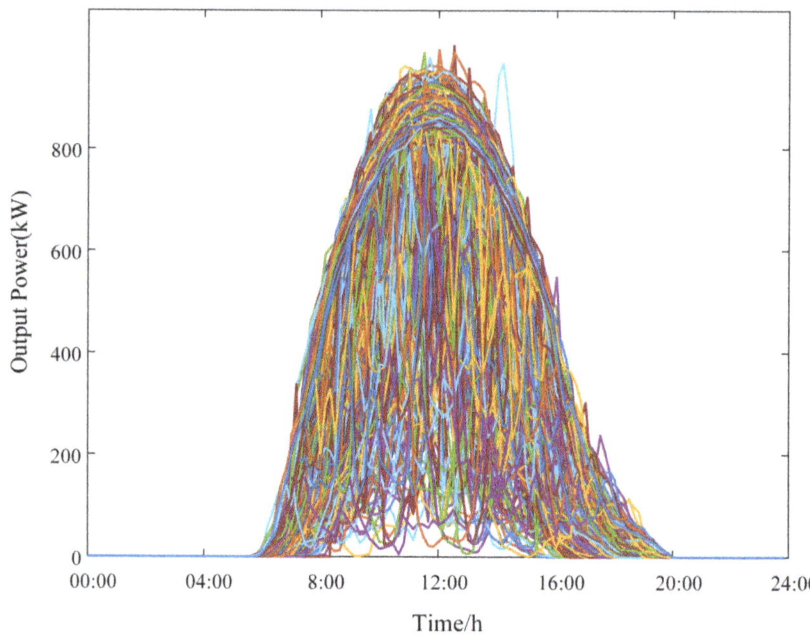

Figure 1. Annual daily output curve of photovoltaic power plants.

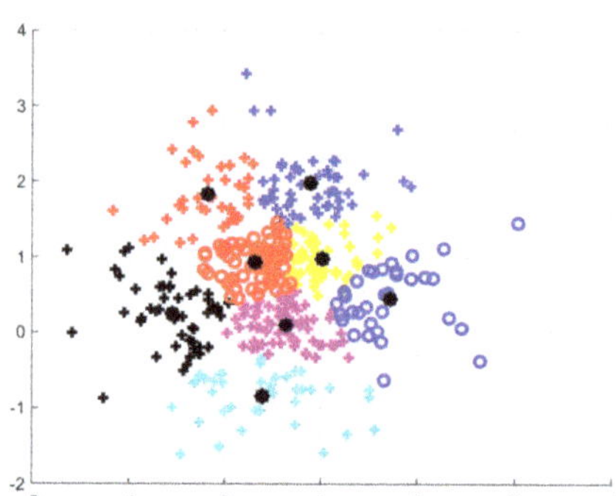

Figure 2. K-means clustering method results.

Set the number of initial scenarios $\xi_s(s = 1,2\ldots N)$ to N_s. The number of target scenarios is M_s, and the entire calculation procedure is as follows:

① M_s target scenes with random data are set as the cluster center, and the set of these scenes is $Center = \{\xi_s^{Center}\}$ $(s = 1,2\ldots M_s)$.

② Excluding the cluster center set, the other scene set is set as $Member = \{\xi_s^{Center}\}$ $(s = 1, 2\ldots M_s)$, and the distance from the other scene set to the cluster center scene set is calculated: $\begin{cases} DT_{s,s'} = DT\left(\xi_s^{Center}, \xi_s^{Member}\right) = ||\xi_s^{Center} - \xi_s^{Member}||_2 \\ s = 1,2\ldots M_s, s' = 1,2\ldots N_s - M_s \end{cases}$

③ The other scene sets excluding the cluster center set are divided into the nearest cluster center according to the distance calculated in ②. We obtain the cluster set $Cluster = \{C_i\}, i = 1,2\ldots M_s$, where C_i is a set of similar scenarios.

④ Set the same cluster C_i including L_s scenarios. Add the distances from each scenarios to the others: $CT_s = \sum_{\substack{s'=1 \\ s' \neq s}}^{L_s} ||\xi_s - \xi_{s'}||_2$, and scene ξ_k in $CT_s = Min(CT_s)$ is selected as the clustering center of the next iteration. This is used to calculate the next iteration cluster center set.

⑤ At this point, stable cluster centers and clustering results can be obtained by repeating steps ②–④. The probability number of each type of scenario is the probability number of a single scenario in that type of scenario.

The process of the clustering algorithm to reduce the scene is shown in Figure 3.

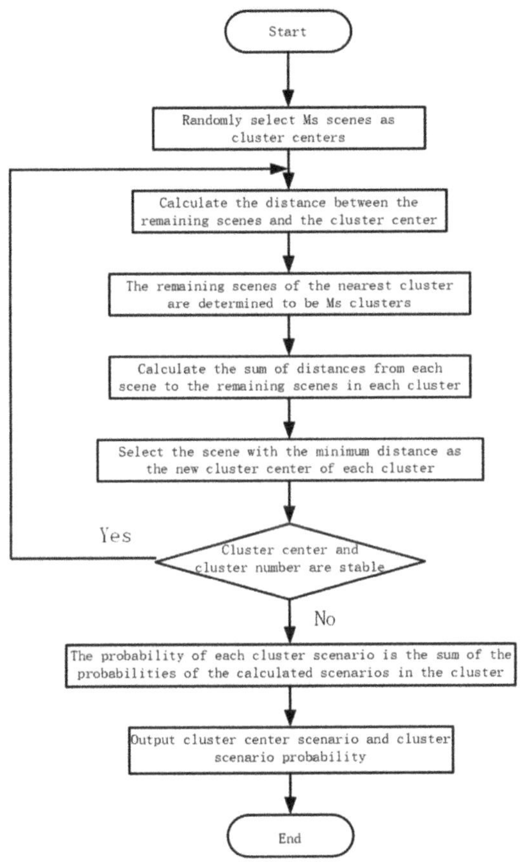

Figure 3. Cluster reduction flowchart.

2.2. Selection of Typical Output Scenarios

After data processing and division by the K-means clustering method, six output scenarios as shown in Figure 4 can be obtained. The occurrence probability and the number of curves of each output scenario are shown in Table 1.

Table 1. Typical scenario probability.

Scenario	Number of Curves	Probability	Scenario	Number of Curves	Probability
1	35	0.0959	4	24	0.0658
2	7	0.0192	5	127	0.3479
3	75	0.2055	6	97	0.2658

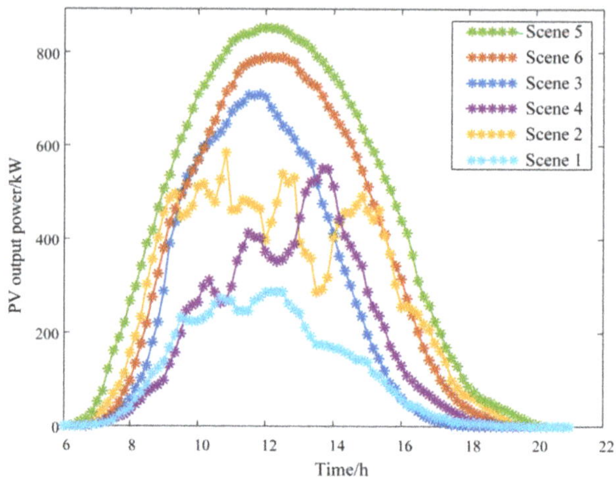

Figure 4. Clustering of typical scenarios.

As can be seen from Figure 4 and Table 1, the photovoltaic output power in Scenario 1 is low, while the output in Scenario 2 has great fluctuation and uncertainty. The intermediate level of output in Scenarios 3 and 4 cannot represent photovoltaic output, and Scenarios 5 and 6 have a high probability and good output curve. In contrast, Scenario 5 with the maximum annual output is selected as the typical photovoltaic output curve, which can better reflect the output characteristics of photovoltaic power generation. In order to facilitate the analysis of the combined effect of photovoltaic and energy storage under different permeability, in this paper, we will only select Scenario 5, which is the most representative and has the highest probability of occurrence at the same time, as the analysis object to study the influence of energy storage access to the power grid during daily operation on voltage fluctuation, operation cost, and other benefits of the distribution network.

3. BESS Bi-Level Decision-Making Model Configuration

Due to the mutual influence between the optimal configuration of the energy storage system and the stable operation of the distribution network, this will bring difficulties to the dispatching of the energy storage devices and will cause the operation stability of the distribution network to decline. Therefore, it is necessary to consider a reasonable location and capacity while taking into account the operation economy of the distribution network. The bi-level decision-making model relies on its own two-level hierarchical structure to optimize the system objectives hierarchically. The upper and lower levels influence each other and seek the overall optimal solution according to the independent objective function and the corresponding constraints [34,35].

3.1. Upper-Level Model Objective Function

In the upper-level optimization, energy storage configuration location, rated power, and installed capacity are considered to reduce the total cost of the energy storage system and distribution network investment and maintenance. The installation location and capacity of the BESS are optimized. After the optimal configuration of energy storage is obtained, the information is transmitted to the lower level to adjust the charge and discharge power of energy storage.

$$\begin{cases} F_{min} = f_{sto} + f_{ope} \\ s.t. g(X) \leq 0 \end{cases} \tag{1}$$

where F_{min} is the minimum daily total cost after the energy storage is connected; f_{ope} is the total cost of distribution network operation investment. f_{sto} is the input costs for energy storage construction. $X = [x_1, x_2, x_3]$, and x_1, x_2, x_3, respectively, represent the BESS input node, power, and capacity.

3.2. Lower Objective Function

In the lower-level optimization, due to the influence of the energy storage installation location and capacity selection on the energy storage life, the lower-level decision-making model fully considers the change of the energy storage charging and discharging power to realize the economical operation mode of the distribution network and achieve the smallest fluctuation range of node voltage and load. The lower optimization objective function is as follows:

$$\begin{cases} min(f_{ope}, f_2, f_3) \\ s.t.h(y) \leq 0 \end{cases} \tag{2}$$

where f_2 is the amplitude of the voltage fluctuation of the distribution network node caused by the access to energy storage, f_3 is the amplitude of the load fluctuation of the distribution network, and $y = [y_1, y_1, y_2 \ldots y_{24}]$ is the average hourly charging and discharging power of the energy storage system throughout the day.

(1) The voltage fluctuation of distribution network nodes caused by energy storage access can be expressed as:

$$min\ f_2 = \sum_{t=1}^{24} \sum_{k=1}^{N} \left[\frac{u_k(t) - u_{kn}}{\Delta u_{kmax}} \right]^2 \tag{3}$$

where N indicates the number of system nodes, and $u_k(t)$ indicates the voltage value of node k at time t.

(2) The load fluctuation of the distribution network caused by access to energy storage can be expressed as:

$$min\ f_3 = \frac{1}{T} \sum_{t=1}^{T} \left[P'_{load}(t) - P'_{ave} \right]^2 \tag{4}$$

where P'_{ave} represents the average load in a period of time when energy storage is connected.

3.3. Constraints

(1) Constraints at the BESS access node

$$1 \leq x_1 \leq N_{max} \tag{5}$$

where x_1 are nodes invested in energy storage, and N_{max} is the maximum number of nodes expressed as energy storage input.

(2) The constraints of power rating and capacity energy storage devices can be expressed as

$$\begin{cases} P_{ess}^{min} \leq P_{essn} \leq P_{ess}^{max} \\ E_{ess}^{min} \leq E_{essn} \leq E_{ess}^{max} \end{cases} \tag{6}$$

where P_{ess}^{max} and P_{ess}^{min} is the maximum and minimum value of the rated output of the energy storage, and E_{ess}^{max} and E_{ess}^{min} are the maximum and minimum values of the energy storage input capacity.

(3) Power balance constraints

$$P_{grid} + P_{gv} = P_{load} + P_{loss} + P_{ess} \tag{7}$$

where P_{grid} is the power value received by the grid, P_{gv} is the PV output power, P_{load} is the output power, P_{loss} is the network loss, and P_{ess} is the BESS input power.

(5) BESS charge and discharge power constraints

$$\begin{cases} 0 \leq P_{ess}^{cha}(t) \leq P_{essn} \\ -P_{essn} \leq P_{ess}^{cha}(t) \leq 0 \end{cases} \tag{8}$$

(6) Voltage constraints in distribution network nodes

$$u_k^{min} \leq u_k(t) \leq u_k^{max} \tag{9}$$

where u_k^{min} and u_k^{max} are the minimum and maximum voltages of node k at time t. Energy storage system SOC constraints

$$SOC(t) = \frac{E(t)}{E_{sn}} = SOC_0 + \frac{\sum_{k=1}^{t}\left\{d_1(t)P_{ess}^{cha}(t)n_c\right\}\Delta t + \sum_{k=1}^{t}\left\{d_2(t)P_{ess}^{cha}(t)/n_d\right\}\Delta t}{E_{ssn}} \tag{10}$$

where SOC_0 is the initial state of the energy storage system, including power and capacity. E_{ssn} is the rated capacity of the energy storage battery.

(7) Supplementary constraints

① Due to the limitation of the SOC range of the BESS, there will be a large number of infeasible solutions during the recovery of its all-day charging and discharging power. If its charge and discharge power is processed, this will greatly improve the convergence rate in the solution process and reduce the amount of calculation.

$$P'_{ess} = \begin{cases} P_{ess}(t)\ SOC_{min} \leq SOC(t) \leq SOC_{max} \\ 0\ otherwise \end{cases} \tag{11}$$

where $P'_{ess}(t)$ is the energy storage charge and discharge power value that has been processed at time t. In this way, the infeasible solution is transformed into an effective feasible solution, and the charging and discharging power that is not within the SOC range of the energy storage is changed to 0.

② Using the penalty function method to deal with the constraints that are not within the valid range:

$$F(x,M) = f(x) + M\sum_{i=1}^{r} max(g_i(x),0) - M\sum_{i=1}^{s} min(h_i(x),0) + M\sum_{i=1}^{t}|k_i(x)| \tag{12}$$

where M is the penalty coefficient. $g_i(x)$ indicates the negative inequality constraint. $h_i(x)$ indicates the positive inequality constraints. $k_i(x)$ indicates the equality constraint at zero. r, s, t are the number of constraints.

4. Solution of Model

For the bi-level programming model, this paper selects the genetic algorithm (GA) for the optimization of the upper layer and improved multi-objective particle swarm optimization (IMOPSO) for the optimization of the lower level. The calculation process is as follows:

① Input the demand parameters of the distribution network into the system.

② Initialize the decision variables for the upper level (including BESS installation location, power rating, and capacity). Under the constraint, the population and other parameters of the GA algorithm are initialized.

③ Initialize the decision variables of the lower level, including the BESS charge–discharge method. Under its constraints, the IMOPSO algorithm population and other related parameters are initialized to solve the initial fitness of each optimization objective.

④ After the optimization of the lower layer is completed, the TOPSIS multi-attribute decision-making method is used to select the upper Pareto solution set obtained, and the best scheme is selected and fed back to the upper layer to solve the fitness of the upper layer target.

⑤ The upper-level GA algorithm population is updated, and the third and fourth steps are continuously executed until the upper-level optimization is completed.

⑥ The optimal BESS configuration scheme of the upper layer, the corresponding optimal charge-discharge method of the lower layer, and the optimal Pareto solution set are obtained.

The calculation process is shown in Figure 5. In the following sections, we will give a specific description of the improvement content of the multi-objective particle swarm optimization.

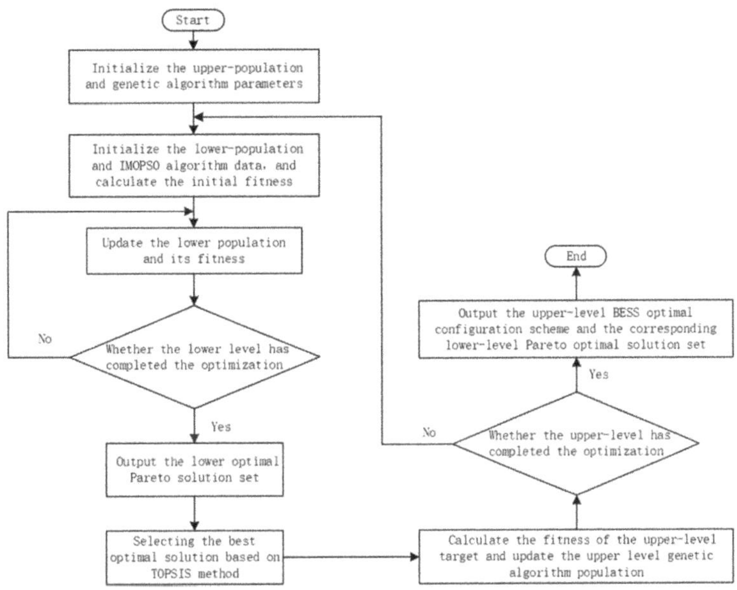

Figure 5. Calculation process.

4.1. Improved PSO Algorithm

The particle swarm optimization (PSO) algorithm is an iterative optimization algorithm. By converting the target into a certain number of particles, the position is updated in each iteration, and the optimal solution is searched through continuous iteration. The updating method is as follows:

$$v_{id}^{k+1} = wv_{id}^{k} + c_1 r_1 \left(p_{id}^{(k)} - x_{id}^{(k)}\right) + c_2 r_2 \left(g_d^{(k)} - x_{id}^{(k)}\right) \tag{13}$$

$$x_{id}^{(k+1)} = x_{id}^{(k)} + v_{id}^{(k+1)} \tag{14}$$

where w represents inertia weight; c_1, c_2 are acceleration coefficients of particle motion respectively. r_1, r_2 are randomly selected in the range of 0 to 1; $P_{id}^{(k)}$ represents the d-dimensional component of the particle numbered i in the optimal position vector at time k; and $g_d^{(k)}$ represents the d-dimensional component of the optimal position of all particles at time k.

In the process of iterative optimization, the traditional muti-objective particle swarm optimization algorithm is prone to fall into local optimal and appear 'Premature convergence'. Therefore, this paper cross-mutates the prescribed bit vectors of particles to prevent them from falling into the local optimum.

$$X_{i,d} = X_{min} + (X_{max} - X_{min})r \tag{15}$$

where X_{min} represents the minimum position variable of the particle; r represents any value between 0 and 1.

Random cross variation is carried out when the probability distribution $P < P_m$ is satisfied. When the fitness of particles tends to be the same or locally optimal, the w will increase. Otherwise, the w will decrease as the particles tend to disperse [36].

$$w = \begin{cases} w_{max} - \frac{(w_{max}-w_{min})(f_i-f_{avg})}{f_{max}-f_{avg}} \; f_i \geq f_{avg} \\ w_{max} \; f_i \leq f_{avg} \end{cases} \tag{16}$$

where f_i is the fitness value of the particle i, w_{max},w_{min} are the maximum and minimum of w, and f_{max},f_{min},f_{avg} are the average and maximum, minimum, and average fitness of all particles at present, respectively.

In order to make the optimal solution distributed evenly in a certain range, the Pareto solution set should be optimized step by step. In this paper, a dynamic image Pareto solution set updating method is adopted, as is shown in Figure 6. In the initial distribution stage of particles, referring to all particles of the original comparison rule, an image V_{iD} is established with the optimal particle as an optimal Pareto solution set. With the progress of iteration, the average value of particles is obtained between every two images, and the optimal solution is used to make the speed and position of particles updated at this time so that the particle population moves towards the target direction. In the whole process, new particles constantly exchange information with image particles and update data in image particles continuously until the end of iteration:

$$V_{iD}^{t+1} = wv_{iD}^{t+1} + c_1 rand_1\left(p_{iD} - X_{iD}^{t+1}\right) + c_2 rand_2(Average(pbest_p) - x_{iD}^{t}) \tag{17}$$

$$x_{iD}^{k+1} = x_{iD}^{k} + v_{iD}^{k+1} \tag{18}$$

where t is the number of iterations; D is the dimension of the decision variable; p_{iD} is the best historical value of single particle i; and $pbest_p$ is the optimal particle of the current Pareto solution set in the image. $Average(pbest_p)$ is the superior particle of the optimal Pareto solution set in the two images.

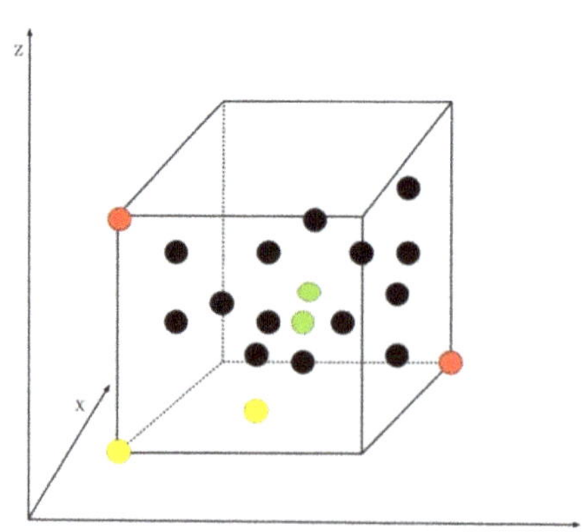

Figure 6. Dynamic image Pareto solution updates schematic.

There are differences in the selection of the optimal solutions between the multi-objective particle swarm optimization and single-objective particle swarm optimization, and the results obtained are complementary dominated Pareto solutions, which cannot be obtained by direct comparison of the particle fitness function. Therefore, in this section, the first 20% Pareto solutions with lower crowding distances and higher priority orders are randomly selected to guide the iterative updating of the particle population.

4.2. Multi-Attribute Decision-Making Based on TOPSIS Method

After solving the lower multi-objective optimization problem, the optimal solution obtained by the IMOPSO algorithm is a set of Pareto solutions, and the selection of the optimal solution is essentially a multi-attribute decision problem. TOPSIS is a method for ordering by similarity to an ideal solution, it selects the optimal solution set and the worst solution set through the established initial decision matrix, then compares the distance between the two solution sets and the evaluation index with the optimal solution set and finally sorts them to evaluate the pros and cons of the scheme. The TOPSIS method has high strict requirements in selecting weights. In this paper, the information entropy method is used to determine the weight of each target value. The information entropy method determines w by the difference of the target value in the Pareto solution, improves the accuracy of the final decision, reduces the difference, and ensures the objectivity of the decision. By using the TOPSIS method, we can determine a set of optimal Pareto solutions to guide us to choose an energy storage configuration scheme.

The optimal solution of the Pareto solution set obtained is selected from $X_1 \sim X_N$ and combined into N alternative schemes. The scheme X_i is selected from N records. It is the composition of some optimal solutions in the Pareto solution set. $g_m(X_i)$ represents the value of the mth attribute of the scheme X_i. Since each attribute is different, it should be unified and changed into the same type. The new attribute value is $G_m(X_i)$, which can be expressed as:

$$G_m(X_i) = \frac{g_m(X_i)}{\sqrt{\frac{1}{N}\sum_{i=1}^{N} g_m^2(X_i)}} \tag{19}$$

$$d(x_i) = \frac{d_+(x_i)}{d_+(x_i) + d_-(x_i)} \tag{20}$$

$$d_+(x_i) = \sqrt{\sum_{m=1}^{n} \left[\lambda_m g'_m(x_i) - \lambda_m g'_{m+}\right]^2} \tag{21}$$

$$d_-(x_i) = \sqrt{\sum_{m=1}^{n} \left[\lambda_m g'_m(x_i) - \lambda_m g'_{m-}\right]^2} \tag{22}$$

where $d(x_i)$ is the relative distance of scheme x_i; $d_+(x_i)$ represents the distance between scheme x_i and the optimal solution. $d_-(x_i)$ represents the distance from solution x_i, the negative worst solution. λ_m indicates the weight value of $g_m(X_i)$, which is randomly set between 0 and 1. g'_{m+} and g'_{m-} indicate the optimal and worst values of all schemes g_m.

5. Analysis and Discussion

5.1. Case Description

In this paper, the proposed scheme is tested on the IEEE-33 node distribution network [37]. In addition, the structure of the system is shown in Figure 7.

In this paper, the rated voltage of the selected distribution network is 12.66 kV, and the total load is 3715 kW + j2300 kvar. The upper and lower limits of the node voltage are specified as not exceeding ±5% of the rated voltage. Node 1 is a balance node, which is connected to the upper-level distribution network for power transmission. Taking into account the actual work and construction of photovoltaic power generation connected to the distribution network, photovoltaic power generation is connected to Node 9, and its installed power generation capacity is 1.077 MW (29% penetration rate).

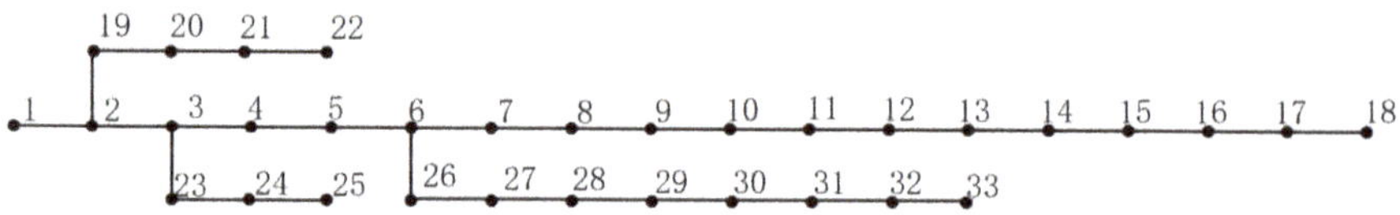

Figure 7. The topology diagram of IEEE-33 bus system.

The typical PV output curve selected by the method in Section 2 is shown in Figure 8. The typical daily load curve in this area is shown in Figure 9.

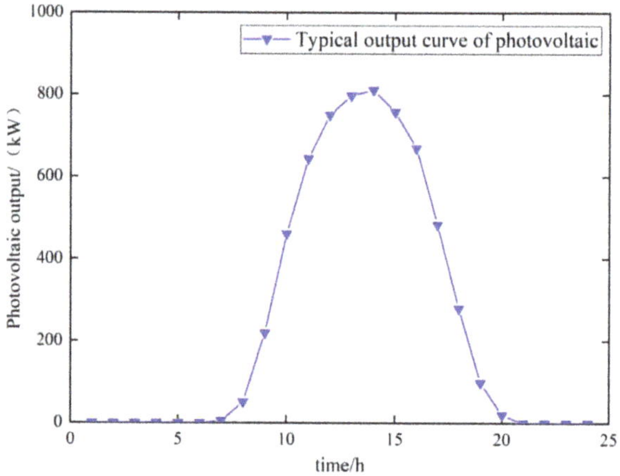

Figure 8. Photovoltaic typical sunrise force curve.

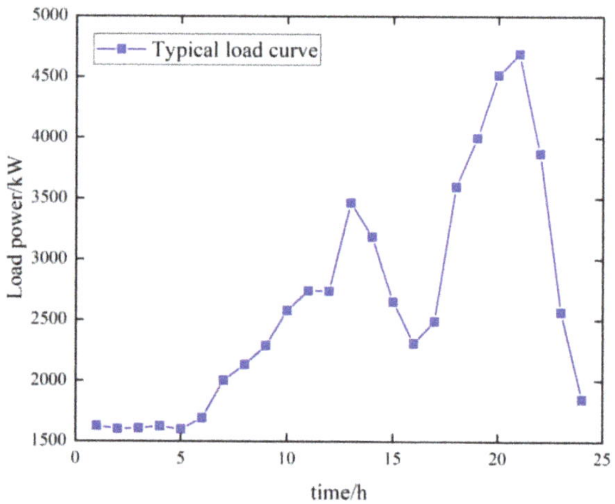

Figure 9. Typical daily characteristic curve of load.

In this paper, the battery is used as the energy storage system for research and introduces the time-of-use pricing strategy proposed in [38]. The specific time-of-use price is shown in Table 2. The energy storage control parameters are shown in Table 3. The specific setting parameters of the energy storage configuration optimization simulation are shown in Table 4.

Table 2. Time-of-use electricity price table.

Type	Period	Electricity Price (yuan/kWh)
Peak time	17:00–22:00	0.9796
Normal time	8:00–17:00 22:00–00:00	0.6570
Trough time	00:00–8:00	0.35

Table 3. Energy storage control parameter table.

Energy Storage Control Parameters	Data
Service life (year)	11
Discount Rate	0.02
Rated power cost (yuan/kW)	1000
Installation cost (yuan/kW)	2500
Operation and maintenance cost (yuan/kW)	0.05
State of charge SOC range	20–90%
Rated power upper limit (MW)	1
Maximum installed capacity (MWh)	5

Table 4. Simulation parameter settings.

Parameter	Data
Power purchase cost of grid	0.35
Expansion cost	1000
Expansion annual profit margin	8%
Load annual growth rate	1.5%
Genetic algorithm population size/number of iterations	60/200
IMOPSO algorithm population size/number of iterations	100/200
Crossover/variation rate	0.1/0.05
Inertia weight range	0.4–0.9
Threshold for difference X	0.1
The size of the Pareto solution set	100

In order to study the actual effect of energy storage configuration, we first analyzed the specific benefits of a photovoltaic distribution network connecting to energy storage configuration and demonstrated that energy storage still has good benefits in the high-light volt distribution network. Then, we compared the photovoltaic distribution network scenarios under different permeability and analyzed and compared the change of photovoltaic permeability with the corresponding change of optimal energy storage configuration scheme. The specific analysis content is introduced in the following section.

5.2. Energy Storage Optimization Scenario Division

Analyze the effectiveness of the method proposed in this paper, set different conditions, divide it into four scenarios, and compare them one by one to verify the feasibility of the method:

Scenario 1: No energy storage.

Scenario 2: With access to energy storage, use the IMOPSO algorithm in this paper to solve the optimization objective of lower-level model in the bi-level decision-making model; introduce the charging and discharging strategy of the energy storage system to simulate and analyze it.

Scenario 3: When solving its single-level model, ignore the charging and discharging management strategy of energy storage in the lower model, and only the energy storage system and distribution network are considered to have the lowest total cost. At the same time, in the time-of-use electricity price model, the energy storage system is charged and discharged at a constant power regardless of the high or low electricity price.

Scenario 4: The optimal configuration result of energy storage in Scenario 2 is used as the constraint condition of this scenario, and the traditional multi-objective PSO algorithm is used to simulate and analyze the lower model in the optimal configuration model of the energy storage double-level. Node voltage curves and load curves in different scenarios are shown in Figures 10 and 11 below, and Table 5 shows the optimization results of different scenarios.

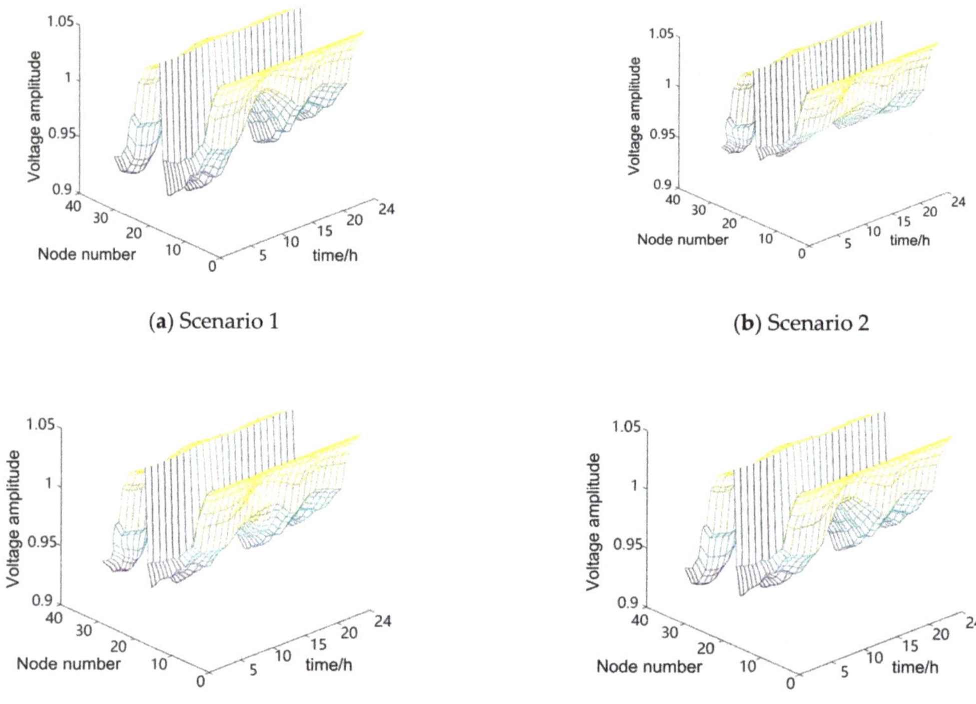

(**a**) Scenario 1 (**b**) Scenario 2

(**c**) Scenario 3 (**d**) Scenario 4

Figure 10. System node voltage curve in different scenarios.

Table 5. Optimization results in different scenarios.

Scenario	Node/Power (kw)/Capacity (kwh)	Cost of Investment	Distribution Network Operating Costs (Yuan)	Voltage Fluctuation Value	Load Variance	Total Cost (Yuan)	Degrees of Savings
1	-	0	16,703.6	70.21	36,5721.76	16,703.6	1.77%
2	14,650,3392	1879.32	14,532.67	63.52	15,3971.12	16,411.99	0%
3	20,273,2733	1497.62	15,717.64	64.98	24,9754.21	17,215.26	4.89%
4	14,650,3392	1879.32	14,767.71	64.85	15,9894.09	16,647.03	1.43%

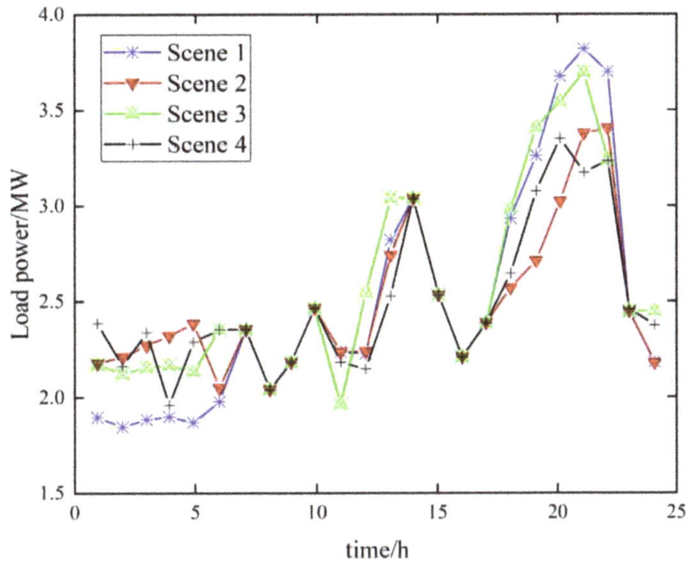

Figure 11. Load curve in different scenarios.

By comparing Scenario 1 and Scenario 2, it can be found that the voltage amplitude curve of the photovoltaic distribution network is smoother after the energy storage is connected, and the voltage fluctuation and load fluctuation are reduced to a large extent, which indicates that the BESS plays a good role in suppressing the node voltage fluctuation and load fluctuation when it is connected to the distribution network. Compared with Scenario 3, the load fluctuation range of Scenario 2 is smaller, and the load smoothing capacity is better. At the same time, the total cost of Scenario 2 is 803.27 yuan lower than that of Scenario 3 economically, which verifies the good characteristics of the model proposed in this paper.

The optimal Pareto solution set distribution of Scenario 2 and Scenario 4 is shown in Figure 12. Scenario 2 adopts the improved IMOPSO algorithm in this paper to solve the inner model, and Scenario 4 adopts the unimproved MOPSO algorithm to solve it. The Pareto solution set in Scene 2 is more evenly distributed than that in Scene 4 due to the introduction of particle cross mutation, adaptive inertia weight, and the Pareto solution set update method of the dynamic image. Moreover, Scene 2 adopts multi-attribute decisions based on the TOPSIS method, resulting in a more diverse solution set.

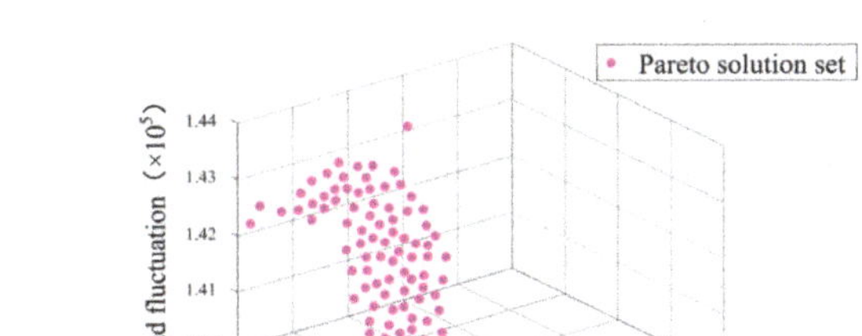

(a) Scenario 2

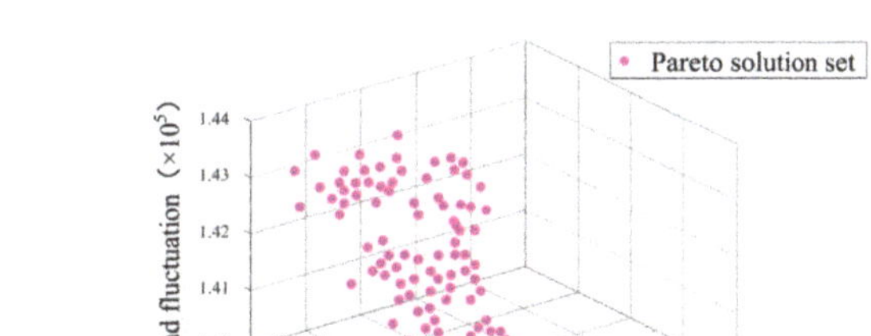

(b) Scenario 4

Figure 12. Pareto solution set distribution.

Through the comparison of scenes, it is obvious that the optimization results of the IMOPSO algorithm are obviously better than the MOPSO algorithm, and the search accuracy is higher. In order to compare the performance of the two algorithms, the external solution set and the spacing S are used in this paper to measure the optimization performance of the two algorithms. The S index refers to whether the particles in the Pareto solution set are evenly distributed in space. The mean variance of the particle density distance is used in this paper to characterize the uniformity and global nature of the population particles, as shown in Equation (23).

$$S = \sqrt{\frac{1}{N}\sum_{i=1}^{N}[I(x_i) - \overline{I}]^2} \tag{23}$$

where $\overline{I}$ represents the average of all particles $I(x_i)$ in the Pareto solution set.

According to the different internal environments of the two algorithms, after 20 cycles, the node voltage and load fluctuations in the optimization target are taken as the research object, as shown in Figure 13 and Table 6.

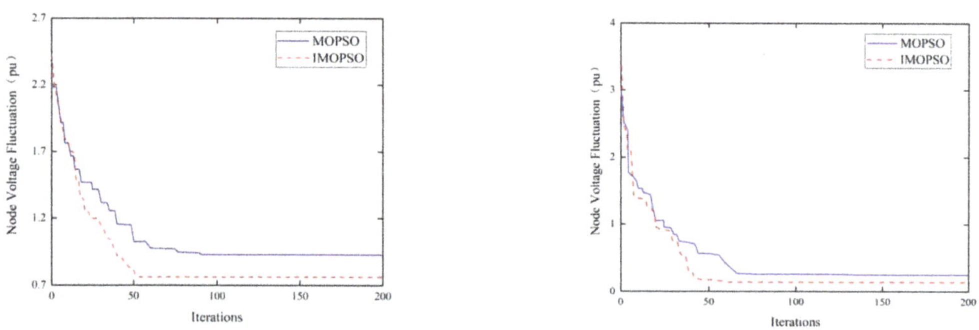

(**a**) Node Voltage Fluctuation (**b**) Load fluctuation

Figure 13. Convergence curves of external solutions for different objectives.

Table 6. Algorithm performance comparison.

Algorithm	External Solution		Distance 'S'
	Node Voltage Fluctuation	Load Fluctuation	
MOPSO	0.8869	0.2103	0.0485
IMOPSO	0.7154	0.1226	0.0317

By combining and comparing the charts, it was concluded that the IMOPSO algorithm proposed in this paper reduces the number of iterations in the node voltage fluctuation and load fluctuation, and the convergence performance is obviously better than the MOPSO algorithm. In addition, the improved algorithm and Pareto solution set update strategy make the solution set distribution more uniform and the type of solution set more diverse, and the improved MOPSO has better robustness and convergence.

5.3. Energy Storage Benefit Analysis under Different Photovoltaic Permeability

In order to verify the effectiveness of the dual-layer multi-objective optimal configuration model of the energy storage system proposed in this paper in the high-light volt permeability distribution network, the upper limit of photovoltaic power generation permeability was set at 60%, and the verification started from 30% permeability. Using the optimal configuration strategy of the BESS, the curve as shown in Figure 14 was obtained.

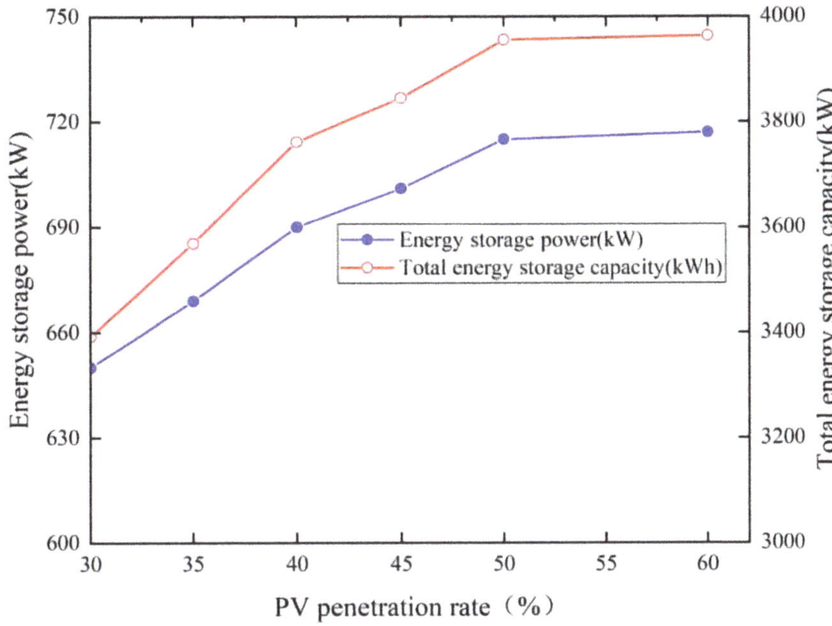

Figure 14. Changes in energy storage capacity and power under different photovoltaic penetration rates.

When the photovoltaic permeability increases from 30% to 60%, the capacity and power of the energy storage system have an obvious rising trend. When the photovoltaic permeability reaches 50%, the growth slows down and tends to remain unchanged. In other words, it is of little significance to increase the capacity of the energy storage system when the permeability reaches a certain level.

Figure 15 below shows the variation trend of energy storage investment and the total cost of distribution network operation under different photovoltaic permeability. It can be clearly seen that the total cost of the system decreases first and then increases when the photovoltaic permeability increases, and the total cost is the minimum when the permeability is 45%. As the cost of photovoltaic power generation decreases with the continuous increase of the permeability but is limited by the load level, the cost of the energy storage system increases with the increase of the capacity. The interaction between the two makes the total cost of the system decrease to the minimum when the photovoltaic permeability is 45%. When the permeability increases again, the system's total cost will keep rising, and the system operation economy will be seriously affected.

Based on the discussion of the above two legends, it is found that the total capacity of the BESS should be controlled in the optimal range according to the actual situation, and the photovoltaic permeability should also be controlled at a certain value so as to ensure the system operation economy while ensuring the safe and stable operation of the system. In order to improve the overall economy of the system, this paper selected 45% photovoltaic permeability to verify and analyze the two-layer programming model of the energy storage system proposed in this paper.

As shown in Figure 16 below, after optimizing the configuration of the energy storage system with 45% photovoltaic permeability, the load curve of the distribution network presents an obvious smoothing trend, and the peak–valley difference decreases. The sufficiency proves that the two-layer optimal configuration model of energy storage can still effectively improve the off-peak load, reduce the peak load of the distribution network, and increase the scheduling flexibility of the distribution network under the condition of high photovoltaic permeability.

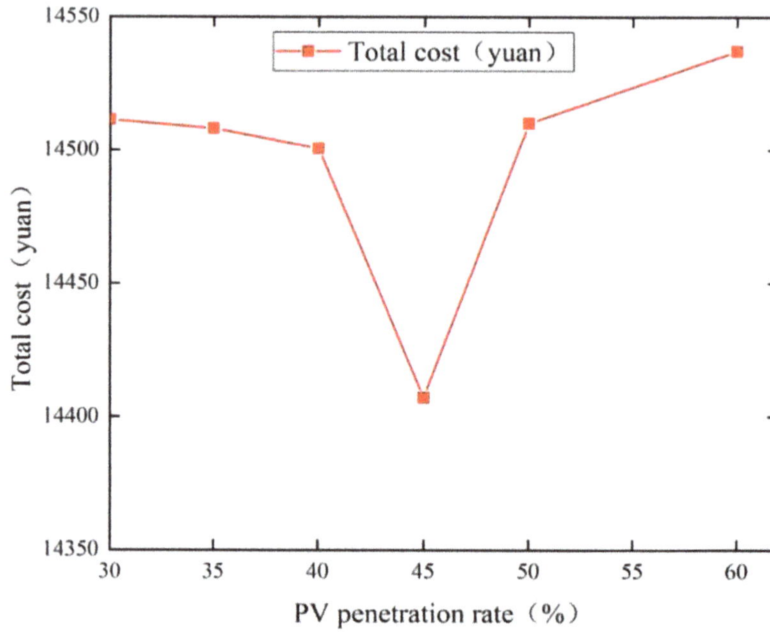

Figure 15. Total cost curve under different PV penetration rates.

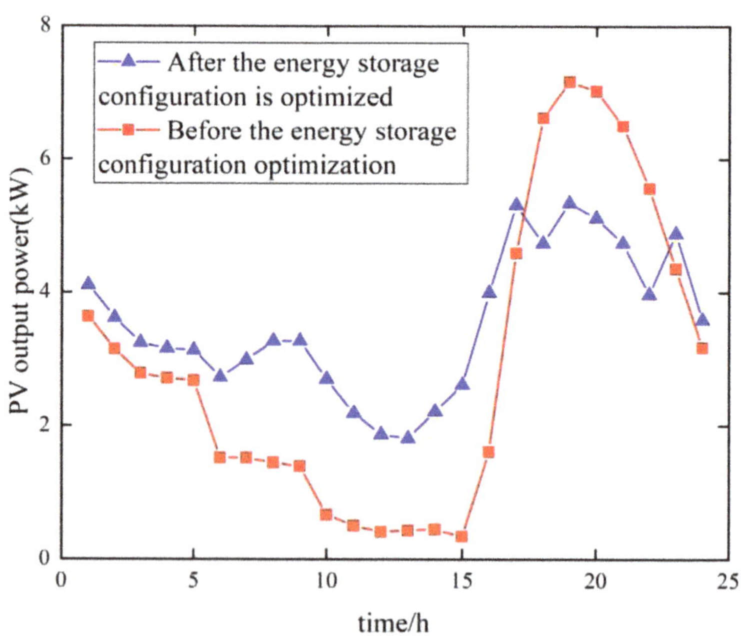

Figure 16. Distribution network load curve before and after energy storage configuration optimization under 45% photovoltaic penetration rate.

6. Conclusions

In order to ensure the power quality and scheduling flexibility of the photovoltaic distribution network with increasing permeability, this paper proposes a joint optimization operation mode of optical storage. Firstly, the PV output model was analyzed, and the scenario planning method was applied. The K-means clustering algorithm was used to divide the output scenarios, and the typical output scenarios were selected for analysis. A BESS two-layer decision model was established, and the improved IMOPSO algorithm was used to solve the two-layer model. The IEEE-33 node example was adopted, and the simulation verification was carried out based on the current feed-in price, selected energy storage parameters, and other parameters. The simulation analysis results are as follows:

① Access to energy storage can effectively smooth the load fluctuation and voltage fluctuation of system nodes in a photovoltaic distribution network. In a distribution network with high-light volt permeability, energy storage can effectively improve the off-peak load of the distribution network and reduce the peak load, thus increasing the scheduling flexibility of the distribution network.

② The bi-level programming model proposed in this paper has a good optimization ability for the rational allocation of energy storage.

③ The improved IMOPSO in this paper has good convergence performance and robustness and has good applicability in application optimization

④ When the optimal energy storage capacity under different photovoltaic permeability is configured, the total cost of the system is optimal when the photovoltaic permeability is 45%, and when the permeability increases again, the total cost of the system will keep rising and seriously affect the operation economy of the system. The analysis of this paper provides a theoretical basis for the optimal configuration of the energy storage system and an important reference for the safe, stable, and economic operation of a high permeability photovoltaic distribution network.

7. Future Work

In this paper, only batteries are considered in the selection of batteries in the energy storage system. However, with a wider application of energy storage, a single energy storage system may not be able to meet the actual demand in the future. In subsequent research, we will combine other types of energy storage for optimization analysis of hybrid energy storage.

At the same time, because of the variety of renewable energy, more and more distributed power is connected to the distribution network. This paper only analyzes access to photovoltaic power generation. In a follow-up study, we will conduct a further study on scenarios with access to various energy sources.

In addition to the voltage fluctuation and load fluctuation considered in this paper, the power system with energy storage access has more indicators to measure security. In a follow-up study, we will also analyze the improvement and influence of energy storage access on various indicators.

Author Contributions: Conceptualization, J.Z. and L.Z.; methodology, S.Z.; software, S.Z. and L.Z.; validation, J.Z., S.Z. and L.Z.; formal analysis, J.Y.; investigation, L.L.; resources, J.Y.; data curation, J.Y.; writing—original draft preparation, L.Z.; writing—review and editing, J.Z.; visualization, L.L.; supervision, J.Y. All authors have read and agreed to the published version of the manuscript.

Funding: This work is supported in part by the National Key Research and Development Program Project (Grant number: 2019YFE0104800), Natural Science Foundation of Henan Province (Grant number: 202300410271) and the Key Project of Science and Technology in Colleges and Universities of Henan Province (Grant number: 162102110130).

Data Availability Statement: The original contributions presented in the study are included in the article; further inquiries can be directed to the corresponding author.

Conflicts of Interest: The authors declare no conflict of interest.

References

1. Yao, M.; Cai, X. An Overview of the Photovoltaic Industry Status and Perspective in China. *IEEE Access* **2019**, *7*, 181051–181060. [CrossRef]
2. Jothibasu, S.; Dubey, A.; Santoso, S. Two-Stage Distribution Circuit Design Framework for High Levels of Photovoltaic Generation. *IEEE Trans. Power Syst.* **2019**, *34*, 5217–5226. [CrossRef]
3. Pan, W.; Mao, M.; Zhou, Y.; Quan, X.; Li, Y. The impact of extreme weather condition on the voltage regulation in distribution systems with high penetration of roof-top photovoltaic. *Energy Rep.* **2021**, *7*, 320–331. [CrossRef]
4. Chidurala, A.; Saha, T.K.; Mithulananthan, N. Harmonic impact of high penetration photovoltaic system on unbalanced distribution networks—Learning from an urban photovoltaic network. *IET Renew. Power Gener.* **2016**, *10*, 485–494. [CrossRef]
5. Karimi, M.; Mokhlis, H.; Naidu, K.; Uddin, S.; Bakar, A. Photovoltaic penetration issues and impacts in distribution network – A review. *Renew. Sustain. Energy Rev.* **2016**, *53*, 594–605. [CrossRef]
6. Gandhi, O.; Rodríguez-Gallegos, C.D.; Zhang, W.; Reindl, T.; Srinivasan, D. Levelised cost of PV integration for distribution networks. *Renew. Sustain. Energy Rev.* **2022**, *169*, 112922. [CrossRef]
7. Dhivya, S.K.; Anurag, S.; Dipti, S.; Reindl, T. Stability implications of bulk power networks with large scale PVs. *Energy* **2019**, *187*, 115927.
8. Maharjan, S.; Dhivya, S.K.; Khambadkone, A.M. Enhancing the voltage stability of distribution network during PV ramping conditions with variable speed drive loads. *Appl. Energy* **2020**, *264*, 114733. [CrossRef]
9. Garmabdari, R.; Moghimi, M.; Yang, F.; Lu, J. Multi-objective energy storage capacity optimisation considering Microgrid generation uncertainties. *Int. J. Electr. Power Energy Syst.* **2020**, *199*, 105908. [CrossRef]
10. Kamaruzzaman, Z.A.; Mohamed, A. Dynamic voltage stability of a distribution system with high penetration of grid-connected photovoltaic type solar generators. *J. Electr. Syst.* **2016**, *12*, 239–248.
11. Xu, Z.; Yuan, B.; Liu, J.; Zheng, K.; Xu, S. Energy storage configuration for smoothing the output volatility of PV power generation. *IOP Conf. Ser. Earth Environ. Sci.* **2019**, *295*, 052045. [CrossRef]
12. Rana, M.M.; Uddin, M.; Sarkar, M.R.; Shafiullah, G.M.; Huadong, M.; Atef, M. A review on hybrid photovoltaic—Battery energy storage system: Current status, challenges, and future directions. *J. Energy Storage* **2022**, *51*, 104597. [CrossRef]
13. Haytham, M.A.A.; Ahmed, S.A.A.; Mohamed, H.A.; Salama, M.M.A. Mitigating voltage-sag and voltage-deviation problems in distribution networks using battery energy storage systems. *Electr. Power Syst. Res.* **2020**, *184*, 106294.
14. Amini, M.; Khorsandi, A.; Vahidi, B.; Hosseinian, S.H.; Malakmahmoudi, A. Optimal sizing of battery energy storage in a microgrid considering capacity degradation and replacement year. *Electr. Power Syst. Res.* **2021**, *195*, 107170. [CrossRef]
15. Hossain, S.J.; Biswas, B.D.; Bhattarai, R.; Ahmed, M.; Abdelrazek, S.; Kamalasadan, S. Operational Value Based Energy Storage Management for Photo-Voltaic(PV) Integrated Active Power DistributionSystems. *IEEE Trans. Ind. Appl.* **2019**, *99*, 5320–5330. [CrossRef]
16. Koohi-Fayegh, S.; Rosen, M.A. A review of energy storage types, applications and recent developments. *J. Energy Storage* **2020**, *2*, 101047. [CrossRef]
17. Chang, L.; Zhuo, J.; Zhao, D.; Li, S.; Chen, J.; Wang, J.; Yao, Q. A review on Flexible and safe operation of renewable energy microgrid using energy storage system. *Proc. CSEE* **2020**, *1*, 1–18+369.
18. Rajamand, S.; Miadreza, S.-K.; Catalão, J.P.S. Energy storage systems implementation and photovoltaic output prediction for cost minimization of a Microgrid. *Electr. Power Syst. Res.* **2022**, *202*, 107596. [CrossRef]
19. Rahman, S.; Saha, S.; Haque, M.E.; Islam, S.N.; Arif, M.T.; Mosadeghy, M.; Oo, A.M.T. A framework to assess voltage stability of power grids with high penetration of solar PV systems. *Int. J. Electr. Power Energy Syst.* **2022**, *139*, 107815. [CrossRef]
20. Zhao, L.; Zhang, T.; Peng, X.; Zhang, X. A novel long-term power forecasting based smart grid hybrid energy storage system optimal sizing method considering uncertainties. *Inf. Sci.* **2022**, *610*, 326–344. [CrossRef]
21. Liang, J.; Lan, F.; Li, J. A grid scenario evaluation method for energy storage capacity demand of photovoltaic-based distribution network. *Autom. Electr. Power Syst.* **2018**, *42*, 40–47+85.
22. Shoaib, B.; Muhammad, A.; Muqeet Hafiz, A.; Muhammad, S.; Harun, J.; Monia, H.; Sattar, M.A.; Habib, H. Optimal Scheduling of Demand Side Load Management of Smart Grid Considering Energy Efficiency. *Front. Energy Res.* **2022**, *10*, 861571.
23. Waseem, M.; Lin, Z.; Liu, S.; Sajjad, I.A.; Aziz, T. Optimal GWCSO-based home appliances scheduling for demand response considering end-users comfort. *Electr. Power Syst. Res.* **2020**, *187*, 106477. [CrossRef]
24. Rezaeimozafar, M.; Eskandari, M.; Hadi Amini, M.; Moradi, M.H.; Siano, P. A Bi-Layer Multi-Objective Techno-Economical Optimization Model for Optimal Integration of Distributed Energy Resources into Smart/Micro Grids. *Energies* **2020**, *13*, 1706. [CrossRef]
25. Waseem, M.; Lin, Z.; Liu, S.; Zhang, Z.; Aziz, T.; Khan, D. Fuzzy compromised solution-based novel home appliances scheduling and demand response with optimal dispatch of distributed energy resources. *Appl. Energy* **2021**, *290*, 116761. [CrossRef]
26. Li, D.; Cai, W. Optimal configuration of photovoltaic energy storage capacity for large power users. *Energy Rep.* **2021**, *7* (Suppl. S7), 468–478. [CrossRef]
27. Yang, P.; Nehorai, A. Joint Optimization of Hybrid Energy Storage and Generation Capacity with Renewable Energy. *IEEE Trans. Smart Grid* **2014**, *5*, 1566–1574. [CrossRef]
28. Wu, J.; Wen, C.; Li, S.; Shi, Q. Capacity optimization allocation of photovoltaic energy storage system based on TOU. *Adv. Technol. Electr. Eng. Energy* **2018**, *37*, 23–30.

29. Lima, D.A.; Feijão, V.R. Stochastic approach for economic viability of photovoltaic systems with battery storage for big electricity consumers in the regulated marketin Brazil. *Electr. Power Syst. Res.* **2022**, *205*, 107744. [CrossRef]
30. Guo, Y.; Xiang, Y. Cost–benefit analysis of photovoltaic-storage investment in integrated energy systems. *Energy Rep.* **2022**, *8* (Suppl. S5), 66–71. [CrossRef]
31. Lu, S.; Wang, X.; Wu, J. Location and Size Planning of Distributed Photovoltaic Generation in Distribution network System Based on K-means Clustering Analysis. *IOP Conf. Ser. Earth Environ. Sci.* **2018**, *108*, 052022. [CrossRef]
32. Home-Ortiz, J.M.; Pourakbari-Kasmaei, M.; Lehtonen, M.; Sanches Mantovani, J.R. Optimal location-allocation of storage devices and renewable-based DG in distribution systems. *Electr. Power Syst. Res.* **2019**, *172*, 11–21. [CrossRef]
33. Wang, W.; Dou, F.; Yu, X.; Liu, G. Scenario analysis of wind power output based on improved k-means algorithm. *IOP Conf. Ser. Earth Environ. Sci.* **2021**, *675*, 012092. [CrossRef]
34. Liang, G.; Sun, B.; Zeng, Y.; Ge, L.; Li, Y.; Wang, Y. An Optimal Allocation Method of Distributed PV and Energy Storage Considering Moderate Curtailment Measure. *Energies* **2022**, *15*, 7690. [CrossRef]
35. Feng, J.; Zhou, H. Bi-Level Optimal Capacity Planning of Load-Side Electric Energy Storage Using an Emission-Considered Carbon Incentive Mechanism. *Energies* **2022**, *15*, 4592. [CrossRef]
36. Feng, H.; Ma, W.; Yin, C.; Cao, D. Trajectory control of electro-hydraulic position servo system using improved PSO-PID controller. *Autom. Constr.* **2021**, *127*, 103722. [CrossRef]
37. Wang, G.; Wan, Z.; Wang, X. Review of two (double) level planning. *Adv. Math.* **2007**, *5*, 513–529.
38. Gong, Z. Study on Two-Stage Optimal Scheduling Strategy and Time-of-Use Price Optimization of Optical Storage System. Master's Thesis, Hefei University of Technology, Hefei, China, 2021.

applied sciences

MDPI

Communication

Novel Pr-Doped $BaLaInO_4$ Ceramic Material with Layered Structure for Proton-Conducting Electrochemical Devices

Nataliia Tarasova [1,2,*], Anzhelika Bedarkova [1,2] and Irina Animitsa [1,2]

1 The Institute of High Temperature Electrochemistry of the Ural Branch of the Russian Academy of Sciences, Yekaterinburg 620000, Russia
2 Institute of Hydrogen Energy, Ural Federal University, Yekaterinburg 620000, Russia
* Correspondence: natalia.tarasova@urfu.ru

Abstract: One of the urgent tasks of applied materials science is the creation of novel high-effective materials with target properties. In the area of energy systems, there is a problem in the conversion of chemical energy to electricity without mechanical work. Hydrogen energy provides a way using electrochemical devices such as protonic ceramic fuel cells. Novel advanced proton-conducting materials with the top characteristics of target properties are strictly needed. Layered perovskites are a novel and promising class of protonic conductors. In this work, the layered perovskite $BaLa_{0.9}Pr_{0.1}InO_4$ was obtained and investigated as a protonic conductor for the first time. The possibility for water intercalation and proton transport is proved. It was shown that isovalent doping $Pr^{3+} \rightarrow La^{3+}$ leads to an increase in the crystal lattice size, proton concentration and proton mobility. The proton conductivity value for doped $BaLa_{0.9}Pr_{0.1}InO_4$ composition is 18 times greater than for undoped $BaLaInO_4$ composition. Layered perovskites based on $BaLaInO_4$ are promising materials for application in proton-conducting electrochemical devices.

Keywords: layered perovskite; oxygen-ion conductivity; proton conductivity; hydrogen energy; Ruddlesden–Popper structure

Citation: Tarasova, N.; Bedarkova, A.; Animitsa, I. Novel Pr-Doped $BaLaInO_4$ Ceramic Material with Layered Structure for Proton-Conducting Electrochemical Devices. *Appl. Sci.* **2023**, *13*, 1328. https://doi.org/10.3390/app13031328

Academic Editors: Luis Hernández-Callejo, Jesús Armando Aguilar Jiménez and Carlos Meza Benavides

Received: 13 December 2022
Revised: 17 January 2023
Accepted: 17 January 2023
Published: 19 January 2023

1. Introduction

Applied materials science plays an important role in the development of various areas of human life. Critical areas such as medicine, energy and mechanical engineering cannot develop without the creation of new materials with targeted properties. Ceramic materials are required for very different needs such as medical applications (endoprosthetics) [1–6] and components for various electrochemical devices [7–12], as examples. The high priority of sustainable development necessitates the development of advanced energy technologies, one of them being hydrogen energy [13–16]. The switchover to renewable energy is, actually, not only for increasing energetically efficiency, but, also, due to the pursuit in decreasing climate changes and the lowering of cases of respiratory diseases [17–23]. Hydrogen energy is considered as an actual, sustainable and promising approach for energy generation [24–35]. This industry includes aspects such as the production, transportation and use of hydrogen as an energy source. The conversion of the chemical energy of hydrogen oxidation to electrical energy can be implemented using electrochemical devices such as protonic ceramic fuel cells [36–41]. This requires new high-effective materials such as electrodes, electrolytes and interconnectors with good compatibility between each other. The traditional and most investigated proton-conducting materials are barium cerate zirconates $BaCeO_3$-$BaZrO_3$, which are characterized by perovskite structure [42,43]. However, they are characterized by a relatively low concentration of protons and low chemical resistance to carbon dioxide. Consequently, novel advanced proton-conducting materials with top characteristics of target properties are strictly needed.

Layered perovskite structure is related to classical perovskite structure. The main difference is the separation of the perovskite framework by the layers with different structures.

Layered perovskites with the general formula $AA'_nB_nO_{3n+1}$ belong to members of layered perovskite family such as compositions with the Ruddlesden–Popper structure. Perovskite slabs $A'BO_3$ are divided by the rock salt layers AO in this structural type, and the existence of compositions with a different "n" is possible. Layered structures provide a concentration of protons an order of magnitude higher than that of classic perovskites. The realization of proton transport was proved for monolayer perovskites such as $BaNdInO_4$ [44–48], $BaNdScO_4$ [49], $SrLaInO_4$ [50–54], $BaLaInO_4$ [55–60] and compositions based on them in the last few years [61]. Such compositions can be potentially used as electrolytic materials in the proton-conducting fuel cells. Various types of doping, including cationic [62] and oxyanionic [63], were investigated. However, the substitutions were implemented by the ions with a stable oxidation state. Meanwhile, the introduction of ions capable of changing their valence can provide control over the contributions of the electronic and ionic components of conductivity. In the future, this can ensure the creation of electrode and electrolyte materials with the same crystal structure and similar chemical composition, which should provide excellent compatibility. In this work, a Pr-doped ceramic material based on $BaLaInO_4$ was obtained and investigated for the first time. The possibility of protonic transport was revealed.

2. Experimental Procedure

The phase $BaLa_{0.9}Pr_{0.1}InO_4$ was synthesized via the nitrate–citrate route according to [52]. The starting reagents $Ba(NO_3)_2$, $In(NO_3)_3{\cdot}6H_2O$, $Pr(NO_3)_3{\cdot}6H_2O$ and $La(NO_3)_3{\cdot}9H_2O$ were used. The X-ray diffraction analysis (XRD) was implemented at the Cu K_α diffractometer Bruker Advance D8. The full profile Le Bail regiments were implemented via the FullProf Suite software. The method of scanning electron microscopy (SEM) of powder and ceramic samples was realized using microscope VEGA3 TESCAN coupled with an energy-dispersive X-ray spectroscopy system (EDS).

The thermogravimetry (TG) measurements were implemented by the Netzsch Analyser STA 409 PC. The samples were initially hydrated using the method of cooling from 1000 to 150 °C (0.5 °C/min) at wet Ar flow. During TG-measurements, the samples were heated from 40 to 1000 °C with the speed 10 °C/min at dry Ar flow.

The resistance of ceramic samples was collected via the impedance spectrometer Elins Z-1000P. The ceramic pellets with a 10 mm diameter and 2 mm thickness were pressed for the investigations. Pt-electrodes were applied on the surfaces of the samples. The temperature range 200–1000 °C was covered, and the speed of cooling was 1°/min. The dry atmosphere was obtained by the circulating of air or Ar through phosphorus pentoxide ($pH_2O = 3.5 \times 10^{-5}$ atm). The wet atmosphere was obtained by the passing of air or Ar through a saturated solution of potassium bromide ($pH_2O = 2 \times 10^{-2}$ atm).

3. Results and Discussions

The phase characterization of composition $BaLa_{0.9}Pr_{0.1}InO_4$ was made using XRD analysis. Figure 1a represents the Le Bail refinement of the X-ray obtained data. The composition is single phase and it has orthorhombic symmetry (*Pbca* space group). The image of the crystal structure of the monolayer perovskite is presented in the inset of Figure 1a. The introduction of Pr^{3+} ions into the La^{3+} sublattice (isovalent doping) leads to the expansion of the crystal lattice (Table 1). An increase in the size of the unit cell is observed during doping despite the close ionic radii of trivalent metals Pr^{3+} and La^{3+} ($r_{La^{3+}} = 1.216$ Å, $r_{Pr^{3+}} = 1.179$ Å [64]). The presence of ions with different electronegativity ($\chi_{La} = 1.10$, $\chi_{Pr} = 1.13$ [65]) could be the cause of the changes in the local structure and in the interatomic distances due to the additional repulsion effects. A similar effect was observed for another doped composition based on $BaLaInO_4$ [59,60], and the presence of ions with different electronegativity in the same sublattice can be considered as the reason of the changes in the crystal lattice size.

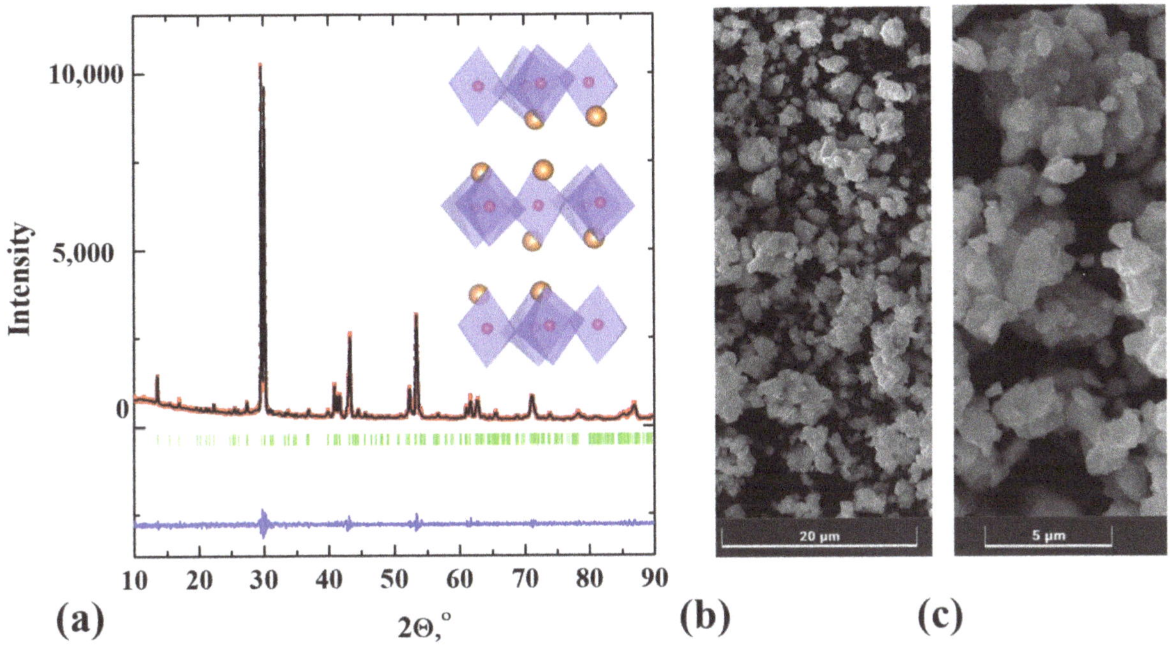

Figure 1. The results of the XRD—(**a**) (R_p = 2.11, R_{wp} = 2.25, χ^2 = 1.32) and SEM-investigations (**b**,**c**) for the composition $BaLa_{0.9}Pr_{0.1}InO_4$.

Table 1. The geometric characteristics of the crystal lattice for the compositions $BaLa_{0.9}Pr_{0.1}InO_4$ and $BaLaInO_4$.

Composition	*a*, Å	*b*, Å	*c*, Å	V, (Å^3)
$BaLa_{0.9}Pr_{0.1}InO_4$	12.968 (1)	5.911 (9)	5.917 (9)	453.17 (7)
$BaLaInO_4$ [56]	12.932 (3)	5.906 (0)	5.894 (2)	450.19 (5)

The morphology of the obtained sample was checked using scanning electron microscopy. Composition $BaLa_{0.9}Pr_{0.1}InO_4$ consists of grains (~3–5 μm) agglomerated in the particles with the size ~10 μm (Figure 1b,c). The elemental composition was proved via EDS analysis. The average element ratios for $BaLa_{0.9}Pr_{0.1}InO_4$ composition are presented in Table 2. The good agreement between theoretical and experimental values was confirmed.

Table 2. The results of the energy-dispersive analysis for the composition $BaLa_{0.9}Pr_{0.1}InO_4$ (theoretical values in atomic % are provided in the brackets).

Metal	Barium	Lanthanum	Praseodymium	Indium
Content	33.4	29.9	3.2	33.5
	(33.3)	(30.0)	(3.3)	(33.4)

The amount of water uptake was measured via thermogravimetry (TG) coupled with the differential scanning calorimetry (DSC) method. The results are presented in Figure 2. As we can see, the initially hydrated composition loses mass due to water release that was confirmed by MS-results. No other volatile components were detected. The main mass loss happens at ~ 200–400 °C, which is confirmed by the signal on the DSC-curve (green line in Figure 2). The dissociative water intercalation into the crystal structure of layered perovskites is possible due to the placement of hydroxyl groups in the rock salt space of the layered framework [62]:

$$H_2O + O_o^x \Leftrightarrow (OH)_o^{\bullet} + (OH)_i^{'} \quad (1)$$

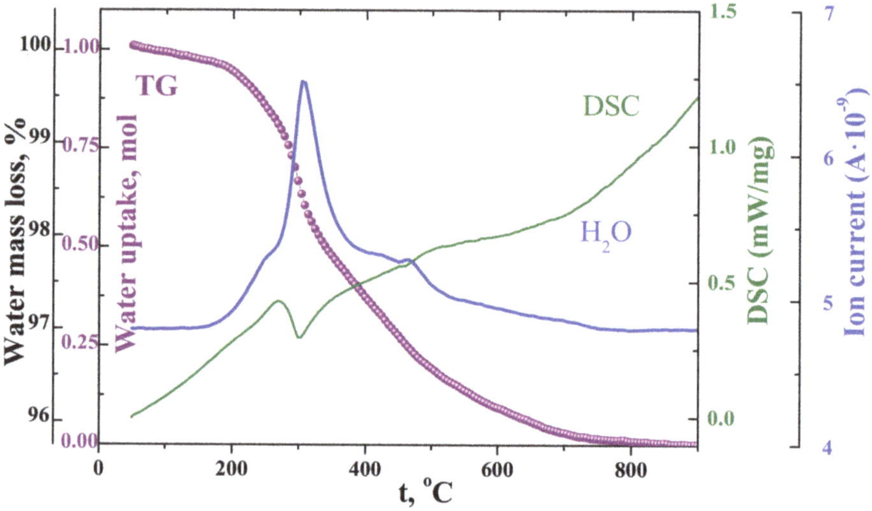

Figure 2. The TG, DSC and MS results for the composition $BaLa_{0.9}Pr_{0.1}InO_4$.

Consequently, the increasing of the crystal lattice size should lead to the increasing of the water uptake [62]. As we can see, the water uptake for the Pr-doped composition is about 1 mol water per formula unit (Figure 2), which is bigger than 0.62 mol registered for undoped $BaLaInO_4$ composition [62]. In other words, a good correlation between the changes of the geometric characteristics of the unit cell and water uptake is observed.

The electrotransport properties of the Pr-doped composition were investigated via the impedance spectroscopy method. Nyquist plots under dry and wet air at 400 °C are presented in Figure 3a as an example of collected data. The calculation of conductivity values was made using the resistance value obtained by extrapolating the high-frequency semicircle to the abscissa axis (capacitance ~10^{-12} F/cm). The effect of variation in oxygen partial pressure to the conductivity values is presented in Figure 3b. As we can see, the electrical conductivity is mixed hole ionic at dry oxidizing conditions. The share of oxygen ionic transport does not change at the temperature variation, and it is around 25%, which is comparable with the value (20%) for undoped $BaLaInO_4$ composition [56]. We can suggest that the dopant concentration of 0.1 mol is not enough to have a meaningful impact on the conductivity nature. At the same time, the nature of dopant is also a possible reason for the absence of significant changes in the nature of conductivity. However, a significant increase in the conductivity values (~1.5 orders of magnitude) during doping is observed (Figure 3c,d). The most probable reason for the increase in the mobility of oxygen ions is due to the increase in the size of the crystal lattice and space for ionic transport. It should be noted that the conductivity values from the electrolytic area (oxygen ionic conductivity, $pO_2 < 10^{-5}$) do correlate well with the conductivity values obtained at the Ar atmosphere. This allows us to consider the values obtained in argon as ionic conductivity values.

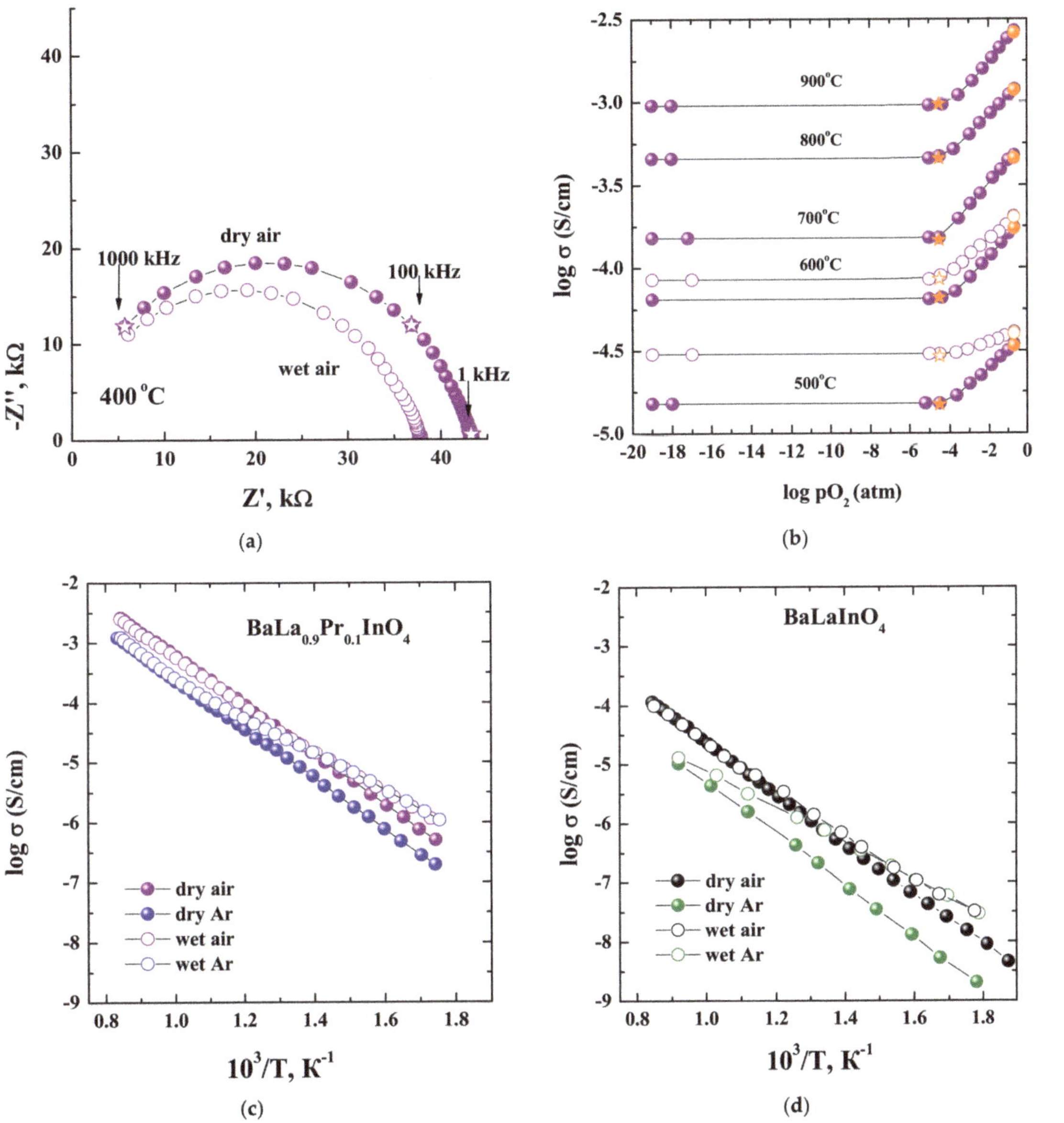

Figure 3. The Nyquist plots for the composition $BaLa_{0.9}Pr_{0.1}InO_4$ obtained under dry and wet air at 400 °C (**a**), the dependencies σ – pO_2 for the composition $BaLa_{0.9}Pr_{0.1}InO_4$ (violet symbols) at dry (closed symbols) and wet (open symbols) atmospheres (**b**), the dependencies σ –1000/T for the compositions $BaLa_{0.9}Pr_{0.1}InO_4$ (**c**) and $BaLaInO_4$ [56] (**d**).

The humidity of the atmosphere affects the conductivity values below 600 °C. The proton concentration increases with temperature decreases. This is because of the increase of the conductivity under wet conditions in comparison with dry conditions (Figure 3b,c). The proton conductivity was calculated as the difference between conductivity values obtained under *wet Ar* and *dry Ar*:

$$\sigma_{H^+} = \sigma_{wet\ Ar} - \sigma_{dry\ Ar} = \sigma_{wet}^{ion} - \sigma_{dry}^{ion} \tag{2}$$

and its temperature dependencies are shown in Figure 4a. As we can see, the protonic conductivity for the Pr-doped composition is higher than the undoped sample by ~1.5 orders of magnitude. This increase is due to the increase in both the proton concentration and proton mobility (Figure 4b). The proton conductivity value for doped $BaLa_{0.9}Pr_{0.1}InO_4$ composition is 5.0×10^{-6} S/cm (T = 400 °C) in comparison with $2.7 \times \cdot 10^{-7}$ S/cm for $BaLaInO_4$ composition that is 18 times greater. It can be suggested that the change in the dopant concentration and dopant nature can lead to significant changes in the nature of electrical conductivity.

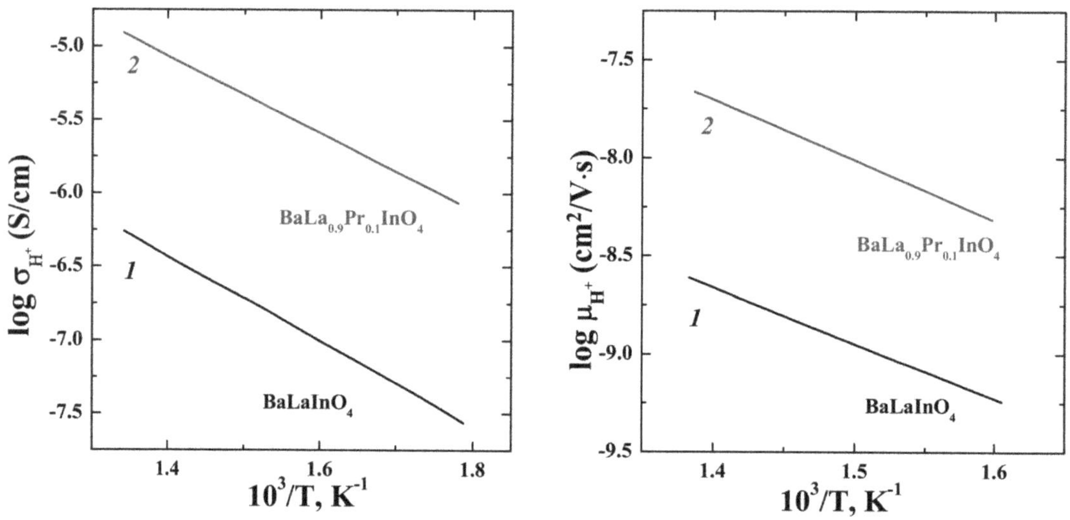

(a) (b)

Figure 4. The dependencies of conductivity (**a**) and mobility (**b**) of protons vs. temperature for the compositions $BaLa_{0.9}Pr_{0.1}InO_4$ and BaLaInO4 [56].

4. Conclusions

The layered perovskite $BaLa_{0.9}Pr_{0.1}InO_4$ was obtained and investigated as a protonic conductor for the first time. The possibility of proton transport is shown. It was proved that isovalent doping $Pr^{3+} \rightarrow La^{3+}$ leads to an increase in the crystal lattice size, proton concentration and proton mobility. The proton conductivity value for the doped $BaLa_{0.9}Pr_{0.1}InO_4$ composition is $5.0 \times \cdot 10^{-6}$ S/cm (T = 400 °C), in comparison with the $2.7 \cdot \times 10^{-7}$ S/cm for $BaLaInO_4$ composition that is 18 times greater. Further research for a dopant capable of a significant change in its electrical conductivity nature is relevant. Layered perovskites based on $BaLaInO_4$ are promising materials for application in proton conducting
electrochemical devices.

Author Contributions: Conceptualization, I.A. and N.T.; methodology, I.A. and N.T.; investigation, A.B.; data curation, N.T. and A.B.; writing—original draft preparation, N.T.; writing—review and editing, N.T. and I.A. All authors have read and agreed to the published version of the manuscript.

Funding: This research was supported by the Russian Science Foundation (grant no 22-79-10003).

Institutional Review Board Statement: Not applicable.

Informed Consent Statement: Not applicable.

Data Availability Statement: Not applicable.

Conflicts of Interest: The authors declare no conflict of interest.

References

1. Punj, P.; Singh, J.; Singh, K. Ceramic biomaterials: Properties, state of the art and future prospectives. *Ceram. Int.* **2021**, *47*, 28059–28074. [CrossRef]
2. Koons, G.L.; Diba, M.; Mikos, A.G. Materials design for bone-tissue engineering. *Nat. Rev. Mater.* **2020**, *5*, 584–603. [CrossRef]
3. Jodati, H.; Yilmaz, B.; Evis, Z. Calcium zirconium silicate (baghdadite) ceramic as a biomaterial. *Ceram. Int.* **2022**, *46*, 21902–21909. [CrossRef]
4. Weng, W.; Wu, W.; Hou, M.; Liu, T.; Wang, T.; Yang, H. Review of zirconia-based biomimetic scaffolds for bone tissue engineering. *J. Mater. Sci.* **2022**, *56*, 8309–8333. [CrossRef]
5. Bunpetch, V.; Zhang, X.; Li, T.; Lin, J.; Maswikiti, E.P.; Wu, Y.; Cai, D.; Li, J.; Zhang, S.; Wu, C.; et al. Silicate-based bioceramic scaffolds for dual-lineage regeneration of osteochondral defect. *Biomaterials* **2019**, *192*, 323–333. [CrossRef]
6. Tarasova, N.; Galisheva, A.; Belova, K.; Mushnikova, A.; Volokitina, E. Ceramic materials based on lanthanum zirconate for the bone augmentation purposes: Materials science approach. *Chim. Techno Acta* **2022**, *9*, 20229209. [CrossRef]
7. Zvonareva, I.; Fu, X.-Z.; Medvedev, D.; Shao, Z. Electrochemistry and energy conversion features of protonic ceramic cells with mixed ionic-electronic electrolytes. *Energy Environ. Sci.* **2022**, *15*, 439–465. [CrossRef]
8. Shim, J.H. Ceramics breakthrough. *Nat. Energy* **2018**, *3*, 168–169. [CrossRef]
9. Bello, I.T.; Zhai, S.; He, Q.; Cheng, C.; Dai, Y.; Chen, B.; Zhang, Y.; Ni, M. Materials development and prospective for protonic ceramic fuel cells. *Int. J. Energy Res.* **2021**, *46*, 2212–2240. [CrossRef]
10. Zhang, W.; Hu, Y.H. Progress in proton-conducting oxides as electrolytes for low-temperature solid oxide fuel cells: From materials to devices. *Energy Sci. Eng.* **2021**, *9*, 984–1011. [CrossRef]
11. Meng, Y.; Gao, J.; Zhao, Z.; Amoroso, J.; Tong, J.; Brinkman, K.S. Review: Recent progress in low-temperature proton-conducting ceramics. *J. Mater. Sci.* **2019**, *54*, 9291–9312. [CrossRef]
12. Medvedev, D. Trends in research and development of protonic ceramic electrolysis cells. *Int. J. Hydrogen Energy* **2019**, *44*, 26711–26740. [CrossRef]
13. Malerba, D. Poverty-energy-emissions pathways: Recent trends and future sustainable development goals. *Int. J. Sustain. Energy Dev.* **2019**, *49*, 109–124. [CrossRef]
14. Buonomano, A.; Barone, G.; Forzano, C. Advanced energy technologies, methods, and policies to support the sustainable development of energy, water and environment systems. *Energy Rep.* **2022**, *8*, 4844–4853. [CrossRef]
15. Chu, S.; Majumdar, A. Opportunities and challenges for a sustainable energy future. *Nature* **2012**, *48*, 294–303. [CrossRef] [PubMed]
16. Østergaard, P.A.; Duic, N.; Noorollahi, Y.; Mikulcic, H.; Kalogirou, S. Sustainable development using renewable energy technology. *Renew. Energy* **2020**, *146*, 2430–2437. [CrossRef]
17. Olabi, A.G.; Abdelkareem, M.A. Renewable energy and climate change. *Renew. Sustain. Energy Rev.* **2022**, *158*, 112111. [CrossRef]
18. Corvalan, C.; Prats, E.V.; Sena, A.; Campbell-Lendrum, D.; Karliner, J.; Risso, A.; Wilburn, S.; Slotterback, S.; Rathi, M.; Stringer, R.; et al. Towards climate resilient and environmentally sustainable health care facilities. *Int. J. Environ. Res. Public Health* **2020**, *17*, 8849. [CrossRef]
19. Watts, N.; Amann, M.; Arnell, N.; Ayeb-Karlsson, S.; Beagley, J.; Belesova, K.; Boykoff, M.; Byass, P.; Cai, W.; Campbell-Lendrum, D.; et al. The 2020 report of the Lancet countdown on health and climate change: Responding to converging crises. *Lancet* **2021**, *397*, 129–170. [CrossRef]
20. Kats, G.H. Slowing global warming and sustaining development: The promise of energy efficiency. *Energy Policy* **1990**, *18*, 25–33. [CrossRef]
21. Afroze, S.; Reza, M.S.; Cheok, Q.; Taweekun, J.; Azad, A.K. Solid oxide fuel cell (SOFC); A new approach of energy generation during the pandemic COVID-19. *Int. J. Integr. Eng.* **2020**, *12*, 245–256. [CrossRef]
22. Stambouli, A.B.; Traversa, E. Solid oxide fuel cells (SOFCs): A review of an environmentally clean and efficient source of energy. *Renew. Sustain. Energy Rev.* **2002**, *6*, 433–455. [CrossRef]
23. Afroze, S.; Reza, M.S.; Cheok, Q.; Islam, S.N.; Abdalla, A.M.; Taweekun, J.; Azad, A.K.; Khalilpoor, N.; Issakhov, A. Advanced applications of fuel cells during the COVID-19 Pandemic. *Int. J. Chem. Eng.* **2021**, *2021*, 5539048. [CrossRef]
24. Dincer, I. Renewable energy and sustainable development: A crucial review. *Renew. Sustain. Energy Rev.* **2000**, *4*, 157–175. [CrossRef]
25. Panwar, N.L.; Kaushik, S.C.; Kothari, S. Role of renewable energy sources in environmental protection: A review. *Renew. Sustain. Energy Rev.* **2011**, *15*, 1513–1524. [CrossRef]
26. Dincer, I.; Rosen, M.A. Sustainability aspects of hydrogen and fuel cell systems. *Int. J. Sustain. Energy Dev.* **2011**, *15*, 137–146. [CrossRef]
27. Branco, H.; Castro, R.; Lopes, A.S. Battery energy storage systems as a way to integrate renewable energy in small isolated power systems. *Int. J. Sustain. Energy Dev.* **2018**, *43*, 90–99. [CrossRef]
28. International Energy Agency. *The Future of Hydrogen: Seizing Today's Opportunities*; OECD: Paris, France, 2019. [CrossRef]
29. Abe, J.O.; Popoola, A.P.I.; Ajenifuja, E.; Popoola, O.M. Hydrogen energy, economy and storage: Review and recommendation. *Int. J. Hydrogen Energy* **2019**, *44*, 15072–15086. [CrossRef]
30. Dawood, F.; Anda, M.; Shafiullah, G.M. Hydrogen production for energy: An overview. *Int. J. Hydrogen Energy* **2019**, *45*, 3847–3869. [CrossRef]

31. Easily, R.R.; Chi, Y.; Ibrahiem, D.M.; Chen, Y. Hydrogen strategy in decarbonization era: Egypt as a case study. *Int. J. Hydrogen Energy* **2022**, *47*, 18629–18647. [CrossRef]
32. Arsad, A.Z.; Hannan, M.A.; Al-Shetwi, A.Q.; Mansur, M.; Muttaqi, K.M.; Dong, Z.Y.; Blaabjerg, F. Hydrogen energy storage integrated hybrid renewable energy systems: A review analysis for future research directions. *Int. J. Hydrogen Energy* **2022**, *47*, 17285–17312. [CrossRef]
33. Scovell, M.D. Explaining hydrogen energy technology acceptance: A critical review. *Int. J. Hydrogen Energy* **2022**, *47*, 10441–10459. [CrossRef]
34. Abdalla, A.M.; Hossain, S.; Nisfindy, O.B.; Azad, A.T.; Dawood, M.; Azad, A.K. Hydrogen production, storage, transportation and key challenges with applications: A review. *Energy Convers. Manag.* **2018**, *165*, 602–627. [CrossRef]
35. Hossain, S.; Abdalla, A.M.; Jamain, S.N.B.; Zaini, J.H.; Azad, A.K. A review on proton conducting electrolytes for clean energy and intermediate temperature-solid oxide fuel cells. *Renew. Sustain. Energy Rev.* **2017**, *79*, 750–764. [CrossRef]
36. Kasyanova, A.; Zvonareva, I.; Bi, L.; Medvedev, D.; Shao, Z. Electrolyte materials for protonic ceramic electrochemical cells: Main limitations and potential solutions. *Mater. Rep. Energy* **2022**, *2*, 100158. [CrossRef]
37. Kim, J.; Sengodan, S.; Kim, S.; Kwon, O.; Bu, Y.; Kim, G. Proton conducting oxides: A review of materials and applications for renewable energy conversion and storage. *Renew. Sustain. Energy Rev.* **2019**, *109*, 606–618. [CrossRef]
38. Chiara, A.; Giannici, F.; Pipitone, C.; Longo, A.; Aliotta, C.; Gambino, M.; Martorana, A. Solid-solid interfaces in protonic ceramic devices: A critical review. *ACS Appl. Mater. Interfaces* **2020**, *12*, 55537–55553. [CrossRef] [PubMed]
39. Cao, J.; Ji, Y.; Shao, Z. New insights into the proton-conducting solid oxide fuel cells. *J. Chin. Ceram. Soc.* **2021**, *49*, 83–92. [CrossRef]
40. Bello, I.T.; Zhai, S.; Zhao, S.; Li, Z.; Yu, N.; Ni, M. Scientometric review of proton-conducting solid oxide fuel cells. *Int. J. Hydrogen Energy* **2021**, *46*, 37406–37428. [CrossRef]
41. Colomban, P. Proton conductors and their applications: A tentative historical overview of the early researches. *Solid State Ion.* **2019**, *334*, 125–144. [CrossRef]
42. Syafkeena, M.A.N.; Zainor, M.L.; Hassan, O.H.; Baharuddin, N.A.; Othman, M.H.D.; Tseng, C.-J.; Osman, N. Review on the preparation of electrolyte thin films based on cerate-zirconate oxides for electrochemical analysis of anode-supported proton ceramic fuel cells. *J. Alloys Compd.* **2022**, *918*, 165434. [CrossRef]
43. Rasaki, S.A.; Liu, C.; Lao, C. A review of current performance of rare earth metal-doped barium zirconate perovskite: The promising electrode and electrolyte material for the protonic ceramic fuel cells. *Prog. Solid State Chem.* **2021**, *63*, 100325. [CrossRef]
44. Fujii, K.; Shiraiwa, M.; Esaki, Y.; Yashima, M.; Kim, S.J.; Lee, S. Improved oxide-ion conductivity of $NdBaInO_4$ by Sr doping. *J. Mater. Chem. A* **2015**, *3*, 11985–11990. [CrossRef]
45. Ishihara, T.; Yan, Y.; Sakai, T.; Ida, S. Oxide ion conductivity in doped $NdBaInO_4$. *Solid State Ion.* **2016**, *288*, 262–265. [CrossRef]
46. Yang, X.; Liu, S.; Lu, F.; Xu, J.; Kuang, X. Acceptor doping and oxygen vacancy migration in layered perovskite $NdBaInO_4$- based mixed conductors. *J. Phys. Chem. C* **2016**, *12*, 6416–6426. [CrossRef]
47. Fujii, K.; Yashima, M. Discovery and development of $BaNdInO_4$–A brief review. *J. Ceram. Soc.* **2018**, *126*, 852–859. [CrossRef]
48. Zhou, Y.; Shiraiwa, M.; Nagao, M.; Fujii, K.; Tanaka, I.; Yashima, M.; Baque, L.; Basbus, J.F.; Mogni, L.V.; Skinner, S.J. Protonic conduction in the $BaNdInO_4$ structure achieved by acceptor doping. *Chem. Mater.* **2021**, *33*, 2139–2146. [CrossRef]
49. Shiraiwa, M.; Kido, T.; Fujii, K.; Yashima, M. High-temperature proton conductors based on the (110) layered perovskite $BaNdScO_4$. *J. Mat. Chem. A* **2021**, *9*, 8607–8619. [CrossRef]
50. Kato, S.; Ogasawara, M.; Sugai, M.; Nakata, S. Synthesis and oxide ion conductivity of new layered perovskite $La_{1-x}Sr_{1+x}InO_{4-d}$. *Solid State Ion.* **2002**, *149*, 53–57. [CrossRef]
51. Troncoso, L.; Alonso, J.A.; Aguadero, A. Low activation energies for interstitial oxygen conduction in the layered perovskites $La_{1+x}Sr_{1-x}InO_{4+d}$. *J. Mater. Chem. A* **2015**, *3*, 17797–17803. [CrossRef]
52. Troncoso, L.; Alonso, J.A.; Fernández-Díaz, M.T.; Aguadero, A. Introduction of interstitial oxygen atoms in the layered perovskite $LaSrIn_{1-x}B_xO_{4+\delta}$ system (B = Zr, Ti). *Solid State Ion.* **2015**, *282*, 82–87. [CrossRef]
53. Troncoso, L.; Mariño, C.; Arce, M.D.; Alonso, J.A. Dual oxygen defects in layered $La_{1.2}Sr_{0.8-x}Ba_xInO_{4+d}$ (x = 0.2, 0.3) oxide-ion conductors: A neutron diffraction study. *Materials* **2019**, *12*, 1624. [CrossRef] [PubMed]
54. Troncoso, L.; Arce, M.D.; Fernández-Díaz, M.T.; Mogni, L.V.; Alonso, J.A. Water insertion and combined interstitial-vacancy oxygen conduction in the layered perovskites $La_{1.2}Sr_{0.8-x}Ba_xInO_{4+d}$. *New J. Chem.* **2019**, *43*, 6087–6094. [CrossRef]
55. Tarasova, N.; Animitsa, I.; Galisheva, A. Effect of acceptor and donor doping on the state of protons in block-layered structures based on $BaLaInO_4$. *Solid State Commun.* **2021**, *323*, 114093. [CrossRef]
56. Tarasova, N.; Galisheva, A.; Animitsa, I. Improvement of oxygen-ionic and protonic conductivity of $BaLaInO_4$ through Ti doping. *Ionics* **2020**, *26*, 5075–5088. [CrossRef]
57. Tarasova, N.; Galisheva, A.; Animitsa, I.; Korona, D.; Davletbaev, K. Novel proton-conducting layered perovskite based on $BaLaInO_4$ with two different cations in B-sublattice: Synthesis, hydration, ionic (O^{2+}, H^+) conductivity. *Int. J. Hydrogen Energy* **2022**, *47*, 18972–18982. [CrossRef]
58. Tarasova, N.; Galisheva, A.; Animitsa, I.; Anokhina, I.; Gilev, A.; Cheremisina, P. Novel mid-temperature $Y^{3+} \rightarrow In^{3+}$ doped proton conductors based on the layered perovskite $BaLaInO_4$. *Ceram. Int.* **2022**, *48*, 15677–15685. [CrossRef]
59. Tarasova, N.; Bedarkova, A.; Animitsa, I. Proton transport in the gadolinium-doped layered perovskite $BaLaInO_4$. *Materials* **2022**, *15*, 7351. [CrossRef]

60. Tarasova, N.; Bedarkova, A. Advanced proton-conducting ceramics based on layered perovskite $BaLaInO_4$ for energy conversion technologies and devices. *Materials* **2022**, *15*, 6841. [CrossRef]
61. Tarasova, N.; Animitsa, I.; Galisheva, A.; Medvedev, D. Layered and hexagonal perovskites as novel classes of proton-conducting solid electrolytes: A focus review. *Electrochem. Mater. Technol.* **2022**, *1*, 20221004. [CrossRef]
62. Tarasova, N.; Animitsa, I. Materials $A^{II}LnInO_4$ with Ruddlesden-Popper structure for electrochemical applications: Relationship between ion (oxygen-ion, proton) conductivity, water uptake and structural changes. *Materials* **2022**, *15*, 114. [CrossRef] [PubMed]
63. Tarasova, N.; Galisheva, A. Phosphorus-doped protonic conductors based on $BaLa_nIn_nO_{3n+1}$ (n = 1, 2): Applying oxyanion doping strategy to the layered perovskite structures. *Chim. Techno Acta* **2022**, *9*, 20229405. [CrossRef]
64. Shannon, R.D. Revised effective ionic radii and systematic studies of interatomic distances in halides and chalcogenides. *Acta Cryst.* **1976**, *A32*, 751–767. [CrossRef]
65. Allred, A.L. Electronegativity values from thermochemical data. *J. Inorg. Nucl. Chem.* **1961**, *17*, 215–221. [CrossRef]

applied sciences

MDPI

Article

Energy Cost Minimization with Hybrid Energy Storage System Using Optimization Algorithm

Krzysztof Rafał *, Weronika Radziszewska, Oskar Grabowski, Hubert Biedka and Jörg Verstraete

Institute of Fluid Flow Machinery, Polish Academy of Sciences, 80-231 Gdańsk, Poland
* Correspondence: krafal@imp.gda.pl

Featured Application: Energy storage systems, microgrids.

Abstract: The purpose of this study is to develop an effective control method for a hybrid energy storage system composed by a flow battery for daily energy balancing and a lithium-ion battery to provide peak power. It is assumed that the system operates behind the meter, the goal is to minimize the energy cost in the presence of a PV installation (as an example of a local renewable source) and energy prices are determined by 3-zone tariffs. The article presents the application of an optimization method to schedule the operation of each battery in the system. The authors have defined an optimization method aimed at minimizing the total cost of the system, taking into account energy costs and batteries depreciation. The techno-economical model of the system, including battery degradation, is constructed and the cost optimization methods are implemented in Python. The results are validated with real energy and price profiles and compared with conventional control strategies. The advantages of optimization in terms of energy cost are discussed. The experiment shows that not only is a hybrid energy system successful in lowering the total operation cost and in increasing self-consumption but also that the implemented methods have slightly different properties, benefits and issues.

Keywords: hybrid energy storage system; optimization algorithm; peak shaving

Citation: Rafał, K.; Radziszewska, W.; Grabowski, O.; Biedka, H.; Verstraete, J. Energy Cost Minimization with Hybrid Energy Storage System Using Optimization Algorithm. *Appl. Sci.* **2023**, *13*, 518. https://doi.org/10.3390/app13010518

Academic Editors: Luis Hernández-Callejo, Jesús Armando Aguilar Jiménez and Carlos Meza Benavides

Received: 29 November 2022
Revised: 19 December 2022
Accepted: 21 December 2022
Published: 30 December 2022

1. Introduction

Large scale deployment of energy storage systems (ESS) is seen as a cost-effective solution for deep decarbonization of electric power systems, which also allows the system stability in the presence of intermittent renewable energy sources (RES) to be maintained [1]. ESS are also seen as part of the solution to reduce the reliance on external fossil fuel imports and high electricity prices [2]. Among available technologies, pumped hydro is still leading; however, grid-scale battery storage is gaining momentum with lithium-ion technology leading and flow batteries emerging [3]. In [4], the overview of technologies, optimization objectives and approaches regarding battery energy storage systems are presented.

Each battery technology is suited for different applications, ranging from short term power system stabilization requiring high power [5] to energy balancing on a daily basis that requires high capacity [6]. The most popular goal of using ESS is the reduction of the operation costs and the maximization of the self-consumption from RES. The profitability aspects are key for the practical popularization of energy storage. An ESS enhances the possibilities of the system by introducing the possibility of shifting part of the energy usage from the moments when cheap or surplus energy is available to the times when it is most needed and/or most costly. The economic outcome is one of the most popular optimization goals realized by a number of methods using classic, e.g., mixed-integer linear programming [7] and heuristic methods, e.g., deep reinforcement learning [8], genetic algorithm and particle swarm optimization [9].

Energy storage is usually just an element of a more complex system that manages the installations and the whole microgrid. Such systems need to be governed by so called energy management systems (EMS), which have been developed for a long time in various configurations. The key to battery functionality and long life depends largely on the battery management system (BMS) and the EMS [9]. Many such models have been developed; some focus on a very detailed model that includes temperatures, voltages, currents and state of charge, e.g., [10], while others use less detailed models but consider the long-term behavior of the system [11].

Basic concepts and different topologies of EMS are presented in [12]; the review of the different approaches can be found in [13,14]. In [15], an EMS, including battery management, which was tested in a real-life environment, is presented.

Hybrid energy storage systems (HESS), consisting of at least two battery types with complementary characteristics, are seen as a comprehensive solution in many applications [16]. Specifically, stationary microgrids seem to benefit from HESS integration [17]; their role may include energy balancing [18], power quality improvement [19] and off-grid operation [20]. Although most of the articles that focus on the energy management of HESS refer to electric vehicles and battery/supercapacitor combination, stationary applications are less explored.

In the review [21], the general classification of energy management systems that are focused on hybrid energy storages is presented. The types of management systems are divided by the objectives and by types, with two general categories: a classical approach and an intelligent approach. There are a few publications regarding EMS that consider hybrid energy storage. In [22], the application of a hybrid storage, consisting of a flow battery and a lithium battery, were simulated for a setup with off-grid renewable power. Ref. [23] shows the advantages of heterogeneous energy storage systems but also explains the possible problems with the implementation of EMS in such a setup, especially when the characteristics of the batteries are substantially different.

The challenge in developing a HESS is to create a more optimal environment for batteries of different technologies in order to maximize their lifetime and the benefits resulting from their operation. We approach this by enclosing the different battery technologies in a single system, a sort of energy storage black box from outside, which automatically creates the best possible environment and usage patterns for the contained technologies.

In order to make the hybrid battery work, a modular EMS is created that consists of two parts: an optimizer that uses predictions and an online controller that copes with deviations from the predictions. The optimizer needs the prediction of both the energy usage and the energy production in order to determine the charging schedule for the next time frame. However, predictions are always prone to uncertainty, so an online controller considers the predictions and adjusts them for the reality of the moment.

In this work, the focus is on a method to optimally determine the operation of the HESS applied to historical data rather than predictions. The reason for this is that it will show us the possibilities and impacts of the batteries, not only for peak-shaving purposes but also for economic purposes. In particular, that latter aspect is of importance: showing the economic viability of the hybrid battery on historical data allows for verification against the true results and can be used as justification for the hybrid battery concept.

The goal of the presented work is to develop a control strategy for an EMS that schedules HESS with the aim of minimizing energy cost. The original contribution of this work is:

- Design and implementation of a techno-economical model of a HESS operating in a microgrid;
- The creation of a model that includes two battery types with their respective round trip efficiencies and costs of depreciation related to battery degradation during cycling;
- The design of an optimization method that calculates a schedule for each battery in a 24 h window;
- The validation and comparative analysis of a proposed method with a benchmark approach based on real life energy usage and production data of a research centre in Poland;

- The novelty of the proposed method is the considering of the multi-battery setup and the inclusion of battery depreciation cost related to its degradation, so that total operating costs are minimized.

2. Materials and Methods

2.1. HESS Model

The setup considered in this work was a scenario of a microgrid that is equipped with RES; in the presented case, this was in the form of a photovoltaic (PV) installation. The system was equipped with a behind-the-meter HESS as shown in Figure 1.

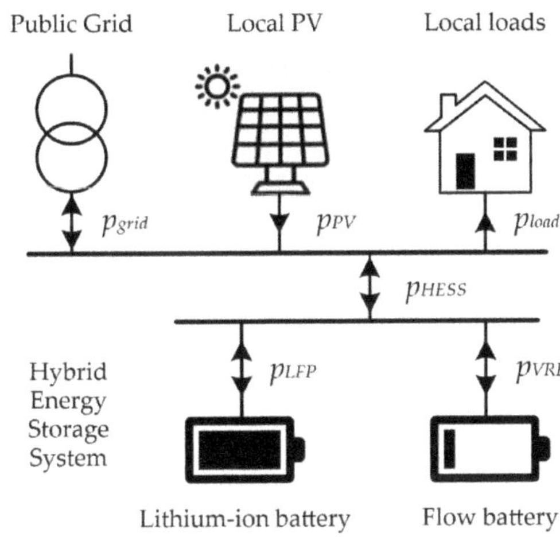

Figure 1. The schematic model of the HESS operating in a microgrid.

Energy profiles were extracted from the historical data recorded by power analyzers at the KEZO Research Centre, Jabłonna, Poland. The usage was gathered from 3 buildings, which had laboratories, conference rooms, kitchens, bathrooms, administration offices, hotel rooms and a server room. The annual consumption was around 221 MWh; it was characterized with irregular patterns of usage—there were long intervals with a very high level of usage, but also non regularly appearing peaks. The usage was variable, ranging from 10 kW to 60 kW, with two dominating values around 20 kW and 38 kW (Figure 2). The centre was equipped with PV installations amassing 180 kWp in total, generating 144 MWh annually. The profiles of power generation and usage were aggregated in 15 min intervals, which is a commonly used interval of energy data gathering and it is consistent with the standard output of the energy meters. The usage and production aggregated per month is presented in Figure 3.

The HESS installed and used in the KEZO power system and modelled in this work consisted of a vanadium redox flow battery (VRFB) and a lithium-iron-phosphate (LFP) battery. The parameters of HESS batteries used for simulations are given in Table 1. Generalized battery cost and performance (number of cycles, efficiency) data were based on [24]. The batteries were connected logically in one network and were managed by the EMS system installed on an industrial computer.

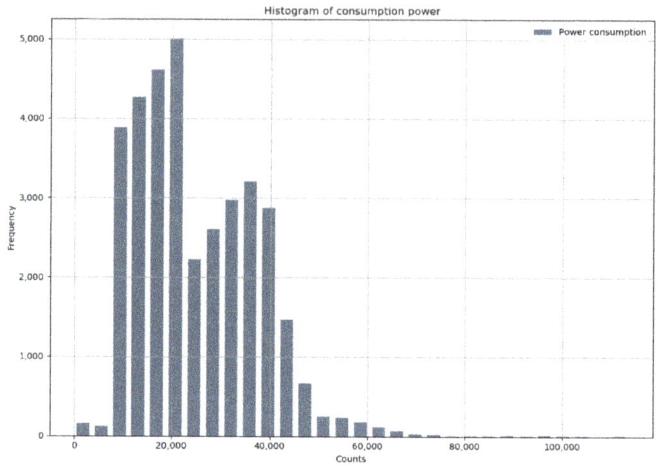

Figure 2. The histogram of power of consumption in the used example in KEZO Research Centre.

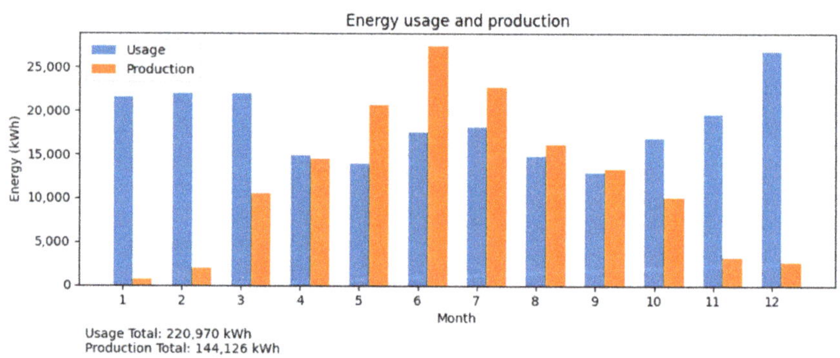

Figure 3. The energy of usage and production aggregated by month (data from the KEZO Research Centre).

Table 1. Hybrid energy storage system parameters assumed for the simulation.

Parameter	Symbol	Unit	VRFB	LFP
Installed capacity	E_{bol}	kWh	100	54
Max. continuous power	P_{max}	kW	15	32
Allowed depth of discharge	DoD	%	100	80
Nominal number of cycles	NoC	-	5200	2000
Round trip efficiency	RTE	%	68	86
Battery block replacement cost	ReC	PLN/system	166,000	60,750

The model included linear battery degradation, which reflected the battery capacity loss during its usage and the cost of replacing the battery block (ReC_i) after exceeding nominal energy throughput (Th_i). To evaluate depreciation cost of operating the battery, the degradation cost was calculated by multiplying discharge energy by a degradation cost factor (DCF). This factor was calculated as follows for each of the batteries (*i*—index of each battery in the HESS):

$$DCF_i = ReC_i / Th_i \tag{1}$$

where:

$$Th_i = NoC_i * (DoD_i * E_{bol,i}) \quad (2)$$

The Th_i is the nominal throughput of the storage calculated as the multiplication of number of cycles (*NoC*), depth of discharge (*DoD*) and the nominal capacity (E_{bol})—the values for the considered setup are presented in Table 1.

The battery model also accounted for energy losses in the batteries. At the battery charging instants, a round trip efficiency (*RTE*) factor was applied to calculate the energy available for discharge.

For the sake of model simplification and reduction of computational burden, efficiency and degradation remained independent of the battery's operational parameters, such as temperature, current and *DoD*.

The model was simulated with up-to-date Polish market prices in tariff B23, which is a real tariff for this type of building and usage. The energy tariff has three zones: the morning peak (from 7:00 until 12:00)—2.53 PLN/kWh, the evening peak (varies between seasons but ranges from 16:00 until 22:00)—3.43 PLN/kWh and the off-peak—1.96 PLN/kWh [25]. Surplus of energy was sold at the flat rate of 0.472 PLN/kWh [26]. The price profiles are depicted in Figure 4.

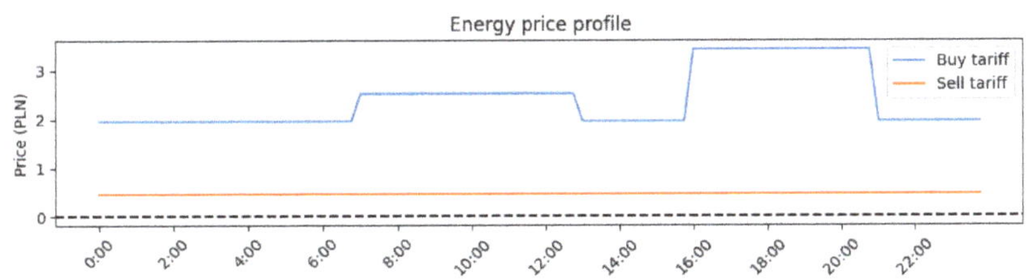

Figure 4. The price tariff for the selling and purchase of energy to/from the grid for months October–March.

The aim was to show that the HESS can be economically justifiable and that it increases the self-consumption of the energy produced from the RES. The aim of the implemented optimization method was solely to minimize the cost of operation, a problem for which a solver was used. A simple control method—energy balancing—was given as a benchmark. Control methods are described in detail in the following sections. We assumed that the usage and production of the energy was not subject to changes. The only thing that changed the balance at the point of coupling with the grid was the operation of the HESS.

2.2. Energy Balancing

The benchmark method was chosen to represent the most simplistic operation of the ESS in the given setup. The idea of the basic energy balancing was very simple: the energy storage charges when production exceeds load and discharges when power was imported from the grid [27]. This control method minimized the exchange of energy with the external grid and did so without considering tariffs or even energy prices, thus foregoing possibilities for price arbitrage. On the other hand, such a method was guaranteed to use as much of the local RES overproduction as possible, in which case it potentially managed to increase self-consumption. In periods with low production from RES, the energy storage may not have a chance to charge. In our model, which assumed PV sources, this occurred in times with less sunshine, and, as a consequence, the HESS spent most of the winter time in a discharged state, thus not helping to decrease the costs of energy. On the upside, not using the battery implied that there is no degradation cost—the model is simplified and did not consider any degradation that might be caused by leaving the battery in an extremely low state of charge.

2.3. Economic Optimization

The operation of the HESS is described by its average power in a time period t and denoted as $p_{HESS}(t) = \sum_{i\in\{LFP,\ VRFB\}} p_i(t)$. The optimization happened in a time window, which had T periods. Any length of time window can be chosen, but the size of this time window correlated with the duration of the computation: the longer the total time, the longer the computation will be. As we assumed that energy comes from PV and, in general, daily cycles are observed in usage patterns, it made sense to choose the total duration of T as 24 h. To speed up the calculation of the solver, an initial solution was calculated and passed to the optimizer; this was a vector of length T of values $p_i(t)$, where $i \in \{LFP,\ VRFB\}$. This initial solution suggested discharging at times of high-price tariff and charging in the low-price tariff. The gain in the time of calculation was small and neglectable. The grid balance at time t was denoted by $p_{grid}(t)$. The general rule was that the energy sent from the installation to the grid and the energy from the installation to the battery was positive; the energy taken from grid and the energy taken from the battery was negative.

The aim was to minimize the total balance that is influenced by the battery operation; therefore, the goal function f_i was:

$$f_i = \sum_{t=0}^{T} \begin{cases} \left|p_{grid}(t) + p_i(t)\right| * c_{buy}(t) + degr_i(t),\ if\ \left(p_{grid}(t) + p_i(t)\right) > 0 \\ \left|p_{grid}(t) + p_i(t)\right| * c_{sell}(t) + degr_i(t),\ if\ \left(p_{grid}(t) + p_i(t)\right) \leq 0 \end{cases} \tag{3}$$

$$degr(t) = \begin{cases} p_i(t) * DCF_i, & if\ x_i(t) < 0 \\ 0, & if\ x_i(t) > 0 \end{cases} \tag{4}$$

The goal function includes the tariffs: $c_{buy}(t)$ is the unit cost for buying energy from the grid for the given time frame t, and $c_{sell}(t)$ is the unit profit for selling energy to the grid. As we assumed that the tariff can change in each t, the main difference with peak shaving was that now it was not just the amount of energy that mattered at a time t, but this amount was linked via the tariff with the time at which it occurred. This combined energy with price, allowing an optimization for the general idea of reducing the amount of energy bought during price peaks with the aim of reducing the cost.

The constraints included the limitation for charging power:

$$\forall t : p_i(t) + p_{chg,i} > 0 \tag{5}$$

where $p_{chg,i}$ is the maximum power allowed for charging the battery.

Symmetrically there is a limitation for discharging power:

$$\forall t : p_{dchg,i} - p_i(t) > 0 \tag{6}$$

where $p_{dchg,i}$ is the maximum power allowed for discharging the battery. In the examples, $p_{chg,i} = p_{dchg,i} = P_{max,i}$.

Constraints of the state-of-charge (SOC) values were defined for each of the consecutive power values for energy storage at each time t. For not exceeding maximum state of charge, constraints were defined as follows:

$$\forall t : (E_{bol,i} - e_i(t)) - \sum_{j=0}^{t} e_i(j) > 0 \tag{7}$$

There are T-1 such constraints for charging the energy storage as power had to be recalculated to the equivalent energy $e(t)$ that was inserted into the battery. The charging energy was multiplied by the RTE value.

For not exceeding minimum state of charge, constraints were defined as follows:

$$\forall t : \left(e_i(t) - E_{bol,i} * \left(\frac{SOC_{min,i}}{100}\right)\right) + \sum_{j=0}^{t} e_i(j) > 0 \tag{8}$$

In this equation $e(t)$ is the energy that is discharged or charged, in case the charging of the RTE value is considered.

2.4. Implementation

The model was implemented in Python [28] as it is one of the popular languages with many well-implemented libraries. The libraries used were: pandas (which contains the very useful DataFrame structure that allows for database-like operation on data), numpy [29] (a package that contains a wide number of data structures and functions for data analysis and scientific methods), scipy [30] (which is equipped with a set of well-known optimization methods and also methods for interpolation and statistical analysis) and datetime (library for advanced operations on date–time formats).

The scheme of the software developed by the authors to run the technical and economic models, to calculate the optimization algorithm and to analyze and plot the results is illustrated in Figure 5. The model was designed to analyze the long-term (yearly) operation of the installation, and for that purpose a 15 min time-step was chosen, as it is a standard resolution of energy meters in Poland. This strictly binds the model to the data gathered in real systems.

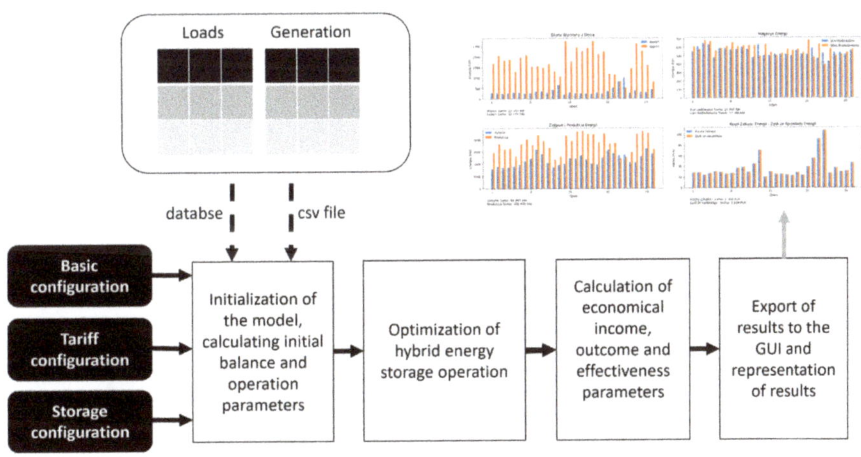

Figure 5. The general scheme of the operation of the optimization method.

The input data for the program were:

- The initial setup parameters, which included the general description of the microgrid parameters and date range for the simulation—the program allowed us to calculate the optimization for any data from a database or csv files.
- The information regarding energy prices—for the calculation of costs and revenues, it was necessary to have the full information regarding the zones, which can change monthly, and the prices of tariffs. The program has the ability to read the prices from a *csv* file in case there are dynamic tariffs; for the purpose of the project, the most typical Polish tariffs were implemented.
- The setup of the HESS—the parameters relevant for cost calculation and optimization of each battery that constitutes HESS had to be defined. The parameters were: the capacity, the maximum power of charging and of discharging, depth-of-discharge (*DoD*), number of cycles limit (*NoC*), round-trip efficiency (*RTE*), the capex cost and the cost *ReC* of replacing the battery unit when it reaches the end of its life.
- Time series of load and generation values for the installation—the required format consisted of separate files with a timestamp and average power in a row of csv files.
- The tariff profile file that consisted of a timestamp, the price for purchasing energy from the grid (or other entity in future, e.g., an aggregator) in PLN per kWh, the price

for selling energy to the grid (or other) in PLN per kWh. It can represent dynamic tariffs [31] related to market or fixed peak hours tariffs. We assumed that changes can occur after a 15 min interval.

The data flow of the implemented model is schematically presented in Figure 5. The first stage is the initialization of a single DataFrame type structure with all the data indexed by a timestamp. Thanks to this, it is possible to make fast operations on columns and rows without the danger of missing the time alignment of the data. The program calculates initial balances of the energy p_{grid} that result from local load p_{load} and generation p_{PV} profiles and initializes structures for energy storage operation. The output data are the input to the optimization algorithm, which sequentially calculates the battery levels.

The program divides the data into 1-day-long chunks (starting from midnight) as daily patterns in usage are very often present and the production from PV modules has, by default, strong daily patterns.

Many different solvers were tested, the COBYLA [32] solver was the fastest in execution, but there were small fluctuations in the result that were not explainable in context of the test case. We chose SLSQP [33] as it is based on verified methods; for the considered problem, it should be effective in finding an optimal solution—if there is one—in reasonable time, without the need for extensive adjustments of the solver's parameters. The SLSQP optimizer is a sequential least squares programming algorithm, it applies the Han–Powell quasi-Newton method with a Broyden–Fletcher–Goldfarb–Shanno algorithm update of the B-matrix and an L1-test function in the step-length algorithm. Its implementation contains a modified version of Lawson and Hanson's nonlinear least-squares solver. The original source code was provided by Dieter Kraft in 1991, based on his work presented in [34]. The SLSQP was used in many publications in the domain of energy, e.g., [35,36]. The convergence and properties of the SLSQP are described in [37] and [38]. The parameters of the solver in this implementation were: maximum 500 iterations and the precision goal for the value of the goal function in the stopping criterion equals 0.1.

The optimization algorithm was called for each 24 h time window independently; only the final state of charge of each battery within HESS was passed to the next iteration. The outcome of the optimization was a time-series of the HESS operation—its discharge and charging power, change in the state of charge and amount of losses. Losses of the battery were defined as impacting the energy while charging—the program subtracts part of the energy that was equivalent to the *1–RTE*.

The order of calculating the energy storage was defined by the durability of the battery, which was the nominal number of cycles *NoC*. The storage was ranked with this parameter, and the one with the highest value was considered first. This approach was chosen as a method to maximize overall durability of the HESS. When the first energy storage was calculated, the program recalculated the balances to include the energy storage impact on the installation. Then, the optimizer started with the next battery in line. Due to the nature of applied batteries it was assumed that priority would go to VRFB because of its high cycle life, low power and large capacity. The LFP battery was calculated in the second step to supplement operation of energy storage when higher power was needed. After each round, the details on the operation of each battery were saved and were also aggregated to the operation of the whole hybrid storage, which consisted of the sum of charging power, discharging power and losses.

The last step was calculating the economic balance of the costs and revenues from the installation. This was calculated for two cases: (a) with HESS and (b) without HESS. On the side of the revenues was the energy that was sold to the grid, which was solely overproduction from the local energy sources. On the cost side was the energy that had to be bought from the grid, as well as the cost of the degradation of the batteries, which was calculated proportionally to the battery throughput.

The analysis of the complexity was performed. We used the O-notation, as defined in [39], as it is a known standard in the defining of algorithm's complexity. In the balancing algorithm, the complexity depended only on the number of considered batteries (*Nb*) and number of days the calculation was performed for (n), the complexity is $O(Nb \times n)$ as the time of calculation of the operation plan for a battery was constant. The optimization method was different—it used a solver and the number of iterations varied depending on the shape of the goal function. The worst case scenario was when there was no convergence and solver reached the maximum number of iterations, which was set to 500. Given that the maximum number of iterations was limited, this can be treated as a constant factor, and according to the O-notation this means the complexity of the operation was also $O(Nb \times n)$.

To estimate the mean time of calculation, we checked the distribution of the number of iterations—the histogram is presented in Figure 6. The number of iterations in time for both batteries is presented in Figure 7. On average, the simulation takes 186 iterations for each 24h window, per battery.

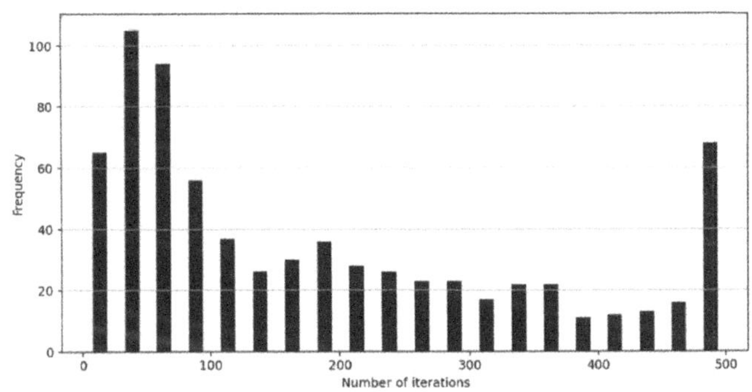

Figure 6. The histogram of the number of iterations for the cost optimization method.

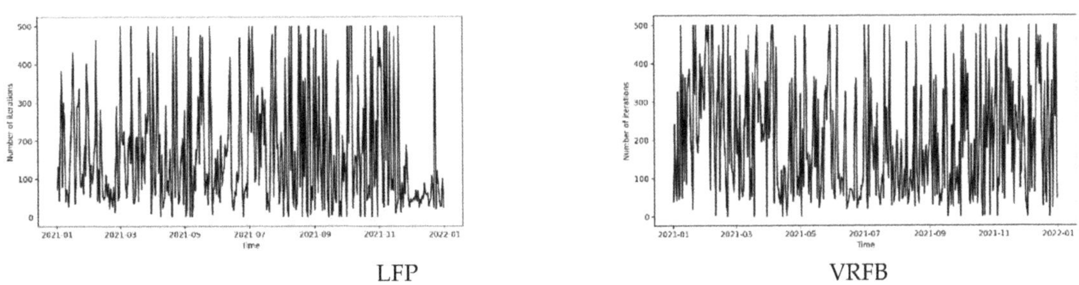

Figure 7. The number of iterations for each day of calculation for the cost optimization method.

3. Results

The annual energy balance without an energy storage system is presented in Figure 3. In the following sections, a series of the graphs to visualize the performance of the behavior of the HESS operating for a full year are presented. The performance of HESS governed by different algorithms is analyzed in daily and seasonal horizons. Annual results for each algorithm are summarized and discussed in Section 4.

To describe the context, the situation of energy flows and costs without any battery system is presented in Figure 8. It is clearly visible that in winter, the purchase of energy in different tariff zones is comparable to the duration of those zones in a day. In the summer months, the photovoltaic installation is producing enough energy to cover the morning peaks almost completely. However, the production cannot cover the evening peak tariffs. The overall cost balance for the year is PLN 295,000, which includes the purchase of energy and the selling of surpluses from PV production.

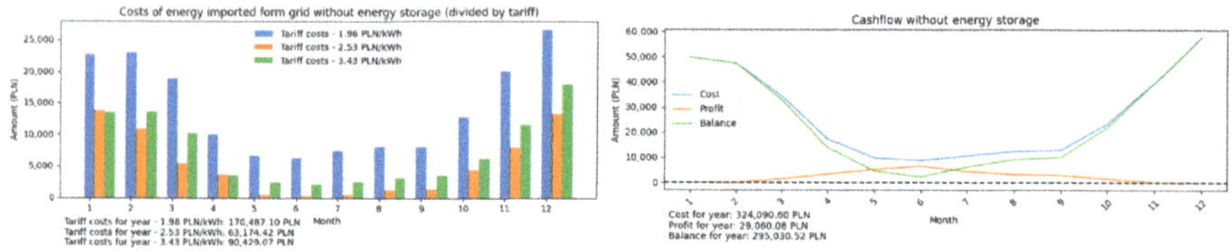

(a) (b)

Figure 8. Monthly economic indicators for the setup without HESS or any other energy storage: (**a**) purchase cost classified by tariff zones (tariff prices), (**b**) costs, profits and financial balance.

3.1. Energy Balancing

The first set presents the operation of the simple energy balancing algorithm. Figure 9 presents the performance indicators aggregated per month to give an overview of the whole year's operation. It is clearly visible that in winter months (November to February) the role of the production from photovoltaic is marginal and almost all of the produced energy is used for the self-consumption. In winter, the energy storage has no chance to charge as there is no surplus of energy. From March to October the activity of the energy storage is substantial, yet still, the storage is not able to completely rule out any purchase energy from the grid for June and July. Figure 9d represents the monthly aggregation of the energy bought and sold; in summer both can occur: sale and purchase can be present in the same month as one month can contain days when there is enough power from RES to cover the usage and sell, but the month can also contain days that might be too cloudy, making it necessary to buy energy.

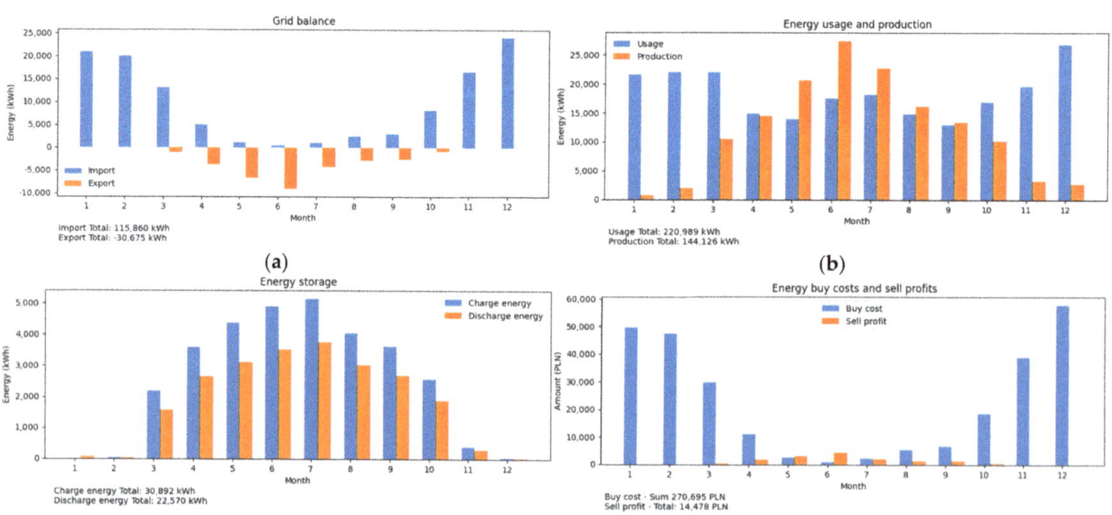

(c) (d)

Figure 9. Results for the energy balancing method, aggregated monthly: (**a**) grid energy balance, (**b**) energy usage and production (input data), (**c**) activity of the energy storage, (**d**) cost of purchased energy and profit for sold energy.

The more detailed view of each month gives a much better picture of the actual HESS performance. In Figures 10 and 11, the two different months are presented—January is the month with minimal energy storage activity, and July is the month when a surplus of produced energy allows the storage to cover the energy use in the evenings. On the 1 January, due to the initial settings of the simulation, the batteries are charged up to 50%, which causes the discharge of the battery immediately. In the other days in January, PV generation does not exceed the load, so there is no ESS activity. In July, however, the situation is much more interesting—there are days where there is no need to buy energy from the grid (Figure 10d), which shows the real usefulness of the energy storage.

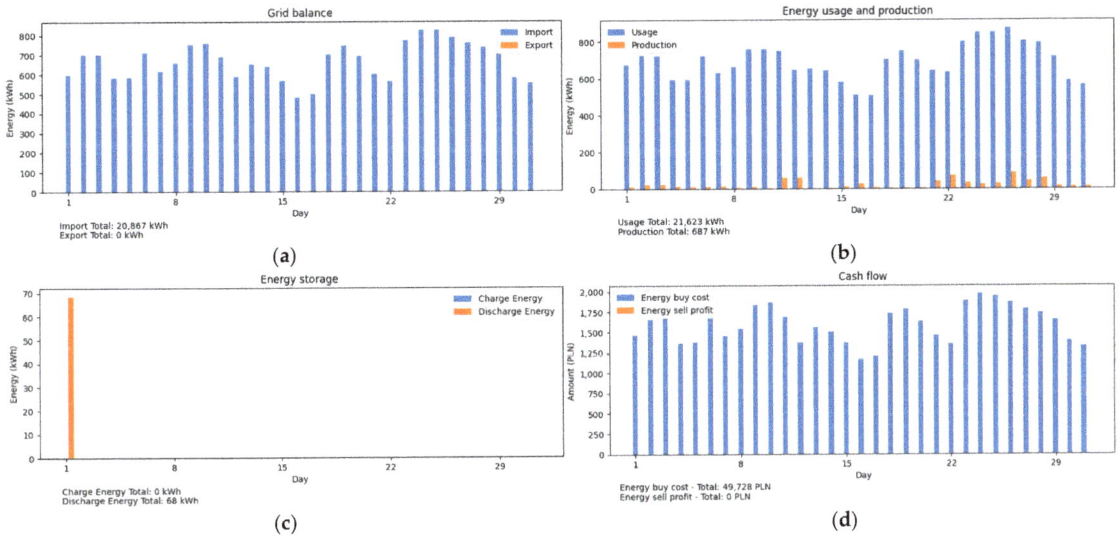

Figure 10. Results for the energy balancing method—data for the month of January, aggregated per day: (**a**) grid energy balance, (**b**) energy usage and production (input data), (**c**) activity of the energy storage, (**d**) cost of purchased energy and profit for sold energy.

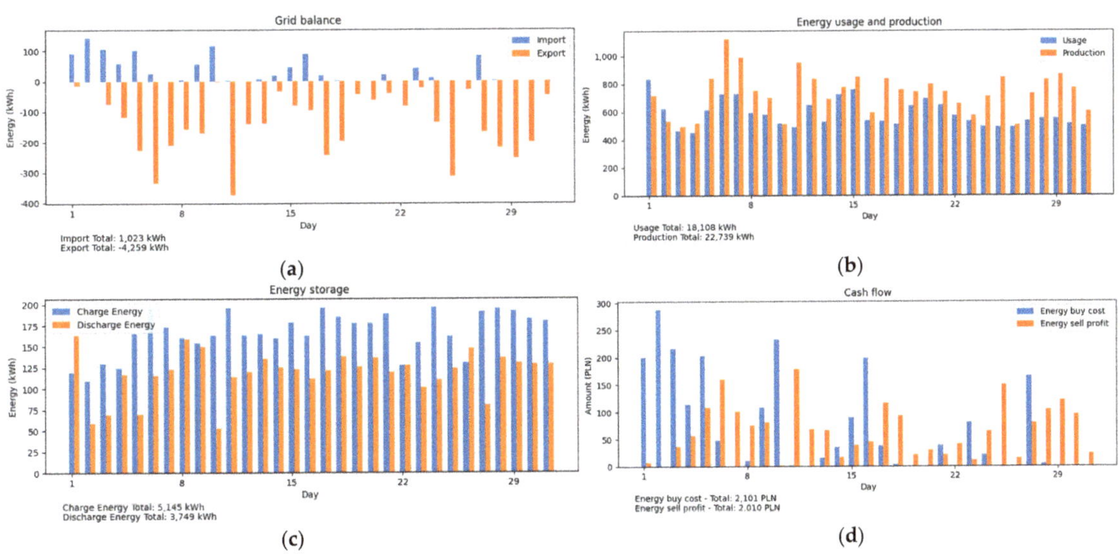

Figure 11. Results for the energy balancing method—data for the month of July, aggregated per day: (**a**) grid energy balance, (**b**) energy usage and production (input data), (**c**) activity of the energy storage, (**d**) cost of purchased energy and profit for sold energy.

The following figures illustrate daily power profiles of the microgrid, including HESS. To fully illustrate behavior of control algorithms under different conditions, three days are chosen: 27 July with high PV generation (Figure 12), 4 October with medium PV generation

(Figure 13) and 6 February with almost no PV generation (Figure 14). The dynamics of the different types of batteries are clearly visible—the VRFB has more capacity, which cannot be fully used due to its lower power. Very visible is the non-optimal behavior of immediately discharging the storage at the beginning of the day in case the batteries have not discharged during previous day. Night time is related to lowest energy prices, so such behavior is not economically justifiable.

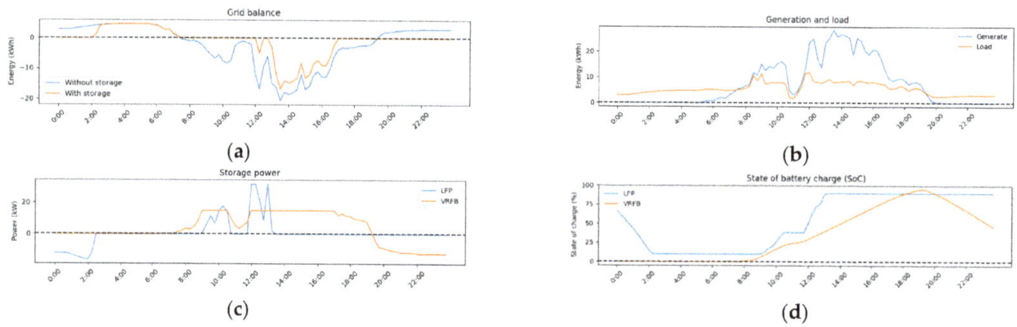

Figure 12. Results for the energy balancing method—data for 27 July: (**a**) grid power balance, (**b**) energy usage and production (input data), (**c**) batteries power, (**d**) batteries state of charge.

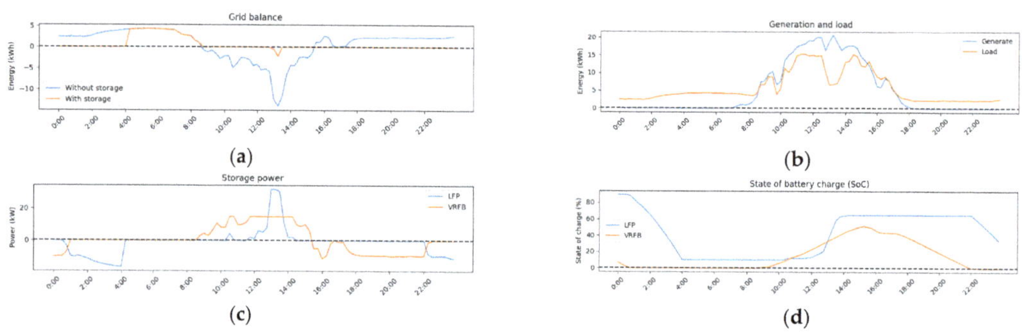

Figure 13. Results for the energy balancing method—data for 4 October: (**a**) grid power balance, (**b**) energy usage and production (input data), (**c**) batteries power, (**d**) batteries state of charge.

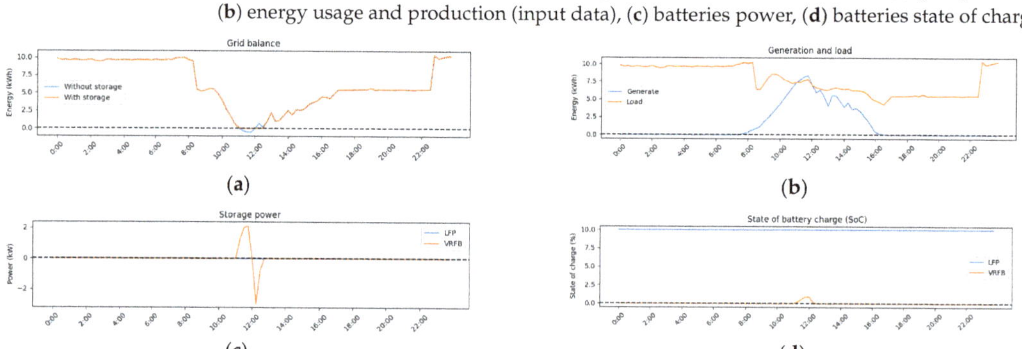

Figure 14. Results for the energy balancing method—data for 6 February: (**a**) grid power balance, (**b**) energy usage and production (input data), (**c**) batteries power, (**d**) batteries state of charge.

In October, the situation is similar to the summer time—there is still enough PV production that allows using both batteries to reduce the power exchange with the grid—first the VRFB is discharging, later the LFP takes over.

In winter (Figure 14), the energy storage is not active as there is little overproduction to be used.

Figure 15 summarizes the financial aspect of the HESS operation. Figure 15a depicts cost of energy purchased in each tariff zone, which clearly demonstrates that the high afternoon peak is not avoided, especially in winter months. In summer months, the PV production combined with the storage can substantially reduce the exchange with the grid. The overall costs and profits are presented in Figure 15b showing that the overall financial outcome is positive in June. Figure 15c gives the values of the surplus RES energy used directly or captured by the HESS. It is calculated as the costs that would have been if there was no energy production. In Figure 15d, the costs saved by the energy storage are presented, this includes the costs that are caused by the degradation of the storage—it is calculated from the difference of the total outcome with and without HESS. The overall cost balance has substantially decreased compared to the situation without the batteries at all, the value is PLN 266,028. This clearly shows that even the simple algorithm of battery management can lower the yearly costs of the operation of the facility.

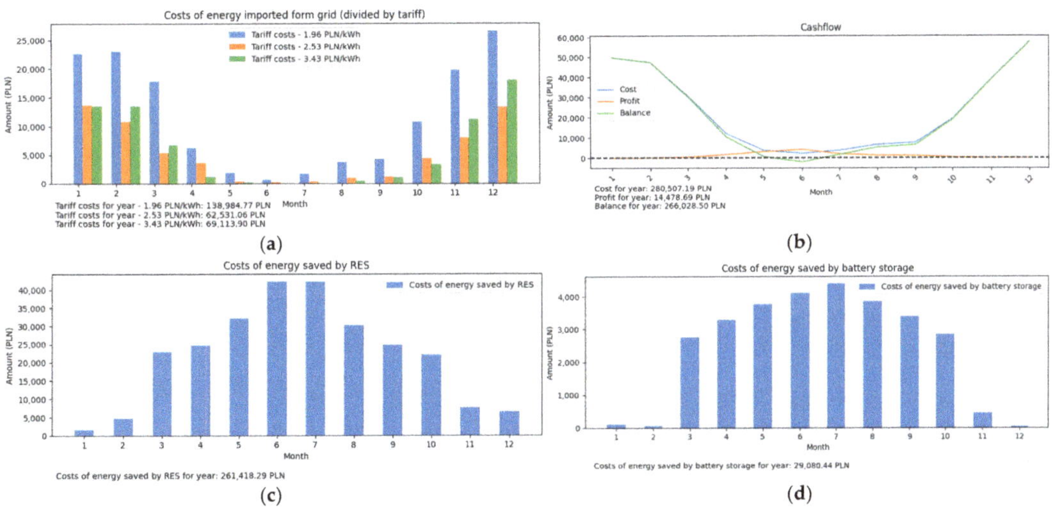

Figure 15. Results for the energy balancing method—monthly economic indicators: (**a**) purchase cost classified by tariff zones (tariff prices), (**b**) costs, profits and financial balance, (**c**) costs of energy saved by PV generation, (**d**) costs of energy saved by HESS operation.

3.2. Economic Optimization

The method of economic optimization is focused on reducing the overall cost of operating of the facility; it should use the energy storage to decrease the usage during peak times and charge in case of surplus or during low-zones of the tariffs. Before the run of the experiments, an analysis of the tariff prices and the degradation costs showed that, according to the calculations, the VRFB should be profitable to use for arbitrage when low tariff and evening peak tariff is considered. It is not profitable to use this storage to move energy from the morning peak to evening peak. The LFP battery has a different degradation cost and RTE, which makes it useful to move energy from the low and morning peak tariff to the evening peak. The using of batteries to increase the surplus of energy produced by PV is always profitable. The execution of the program showed that, very rarely, there is problem with convergence, especially in days with high production from PV modules. On such days there is an excess of energy, which generally creates the multiple solutions that are equivalent for the optimizer.

A comparison of the balancing method and the economic optimization method showed that the latter uses HESS all year round (see Figure 16c). However, the energy purchase costs and sell profits are not much different compared to the balancing algorithm (Figure 16d).

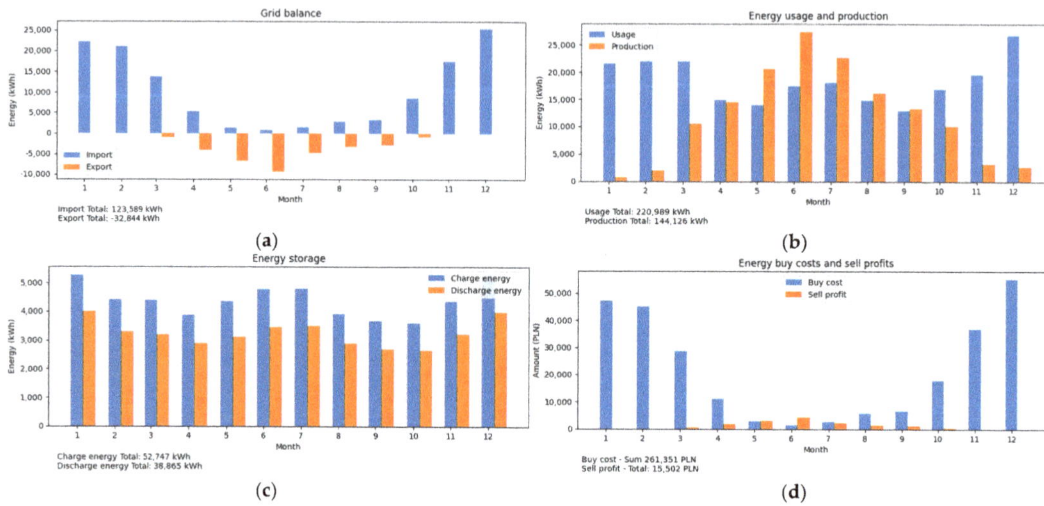

Figure 16. Results for the economic optimization method, aggregated monthly: (**a**) grid energy balance, (**b**) energy usage and production (input data), (**c**) activity of the energy storage, (**d**) cost of purchased energy and profit for sold energy.

The more detailed view of each month provides a much better picture of the actual performance. In January, the HESS uses a lot of the batteries but only to move energy from the expensive time of the day to the cheaper tariff times. There is almost no difference in the cash flow of selling energy, but there is a visible difference when the cost of buying energy from the grid is considered (Figure 17d).

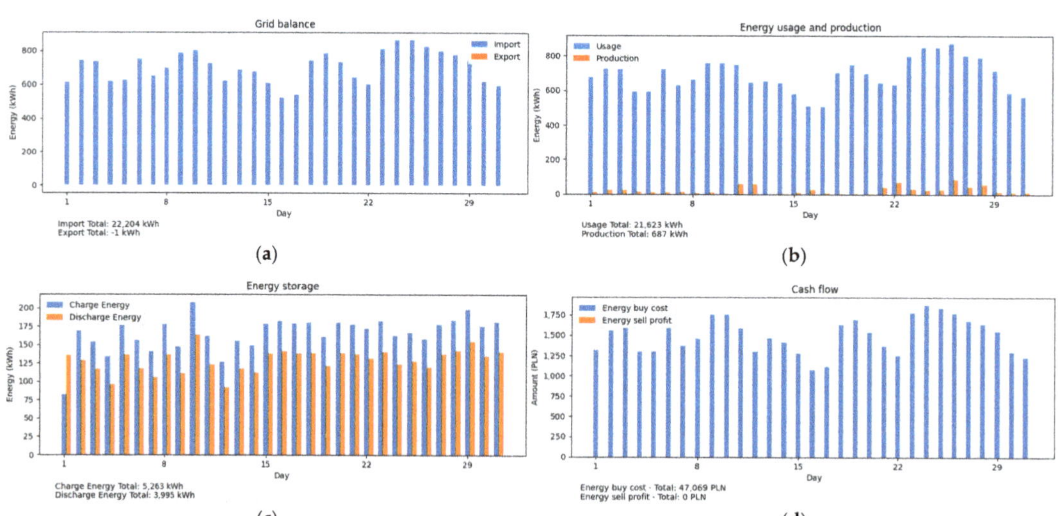

Figure 17. Results for the economic optimization method—data for the month of January, aggregated per day: (**a**) grid energy balance, (**b**) energy usage and production (input data), (**c**) activity of the energy storage, (**d**) cost of purchased energy and profit for sold energy.

In July, the situation is very different—the energy storage reduces bought energy and consumes less from the surplus of the produced energy (Figure 18a). The total sell profits are higher, but also the cost of import from the grid is higher (Figure 18d). The economic optimization method in certain situations is more costly compared to the energy balancing method; it is due to the limitation of the optimization process to 24 h; this is the reason for the modified economic optimization, which is described further in the next section.

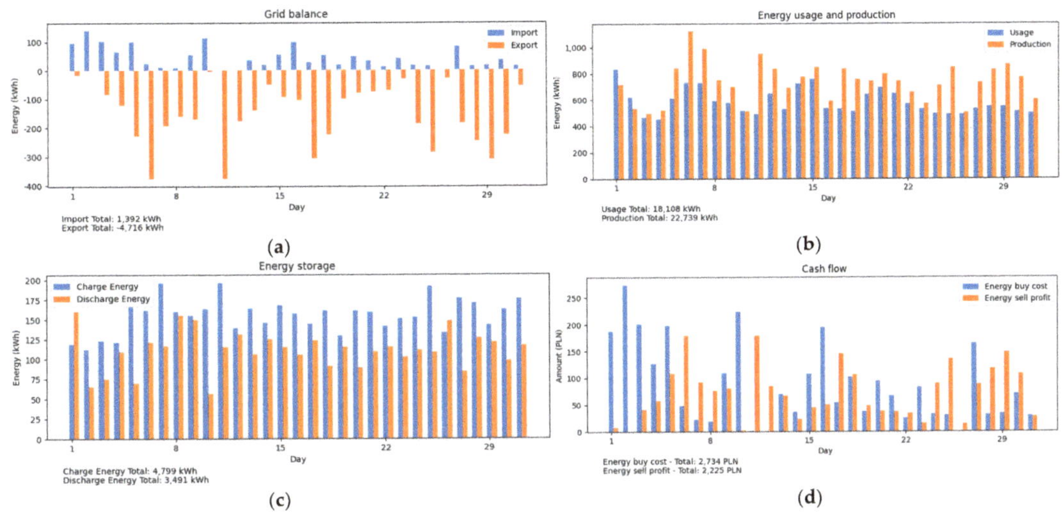

Figure 18. Results for the economic optimization method—data for the month of July, aggregated per day: (**a**) grid energy balance, (**b**) energy usage and production (input data), (**c**) activity of the energy storage, (**d**) cost of purchased energy and profit for sold energy.

Checking the details of the algorithms, a very clear difference can be seen in the example day in July (Figure 19)—the batteries are discharged slowly and their energy is almost uniformly distributed over the whole period where there is a need to decrease energy usage. Similarly, charging shows no extremes, the speed of charging is decreased but maintained for a longer time. Analyzing the behavior day by day, it is clear that the optimizer is considering the degradation costs—the batteries are charged only to the point that is necessary to cover the single day. On 27 July and 4 October, it is especially visible (Figure 20) that the state of charge of the batteries is not reaching the maximum levels, even when it is possible to fully charge from the surplus of produced energy. Generally, the optimizer finds solutions that require smaller power to charge or discharge batteries, which is positive from the durability of the batteries but does not use the full capacity of the batteries when it is possible to charge from the PVs.

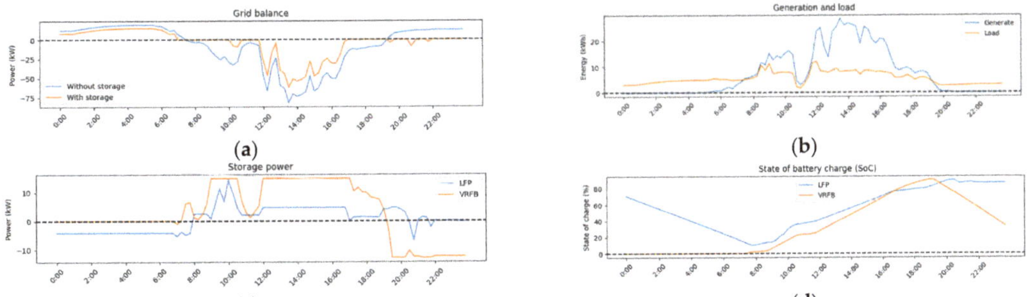

Figure 19. Results for the economic optimization method—data for 27 July: (**a**) grid power balance, (**b**) energy usage and production (input data), (**c**) batteries power, (**d**) batteries state of charge.

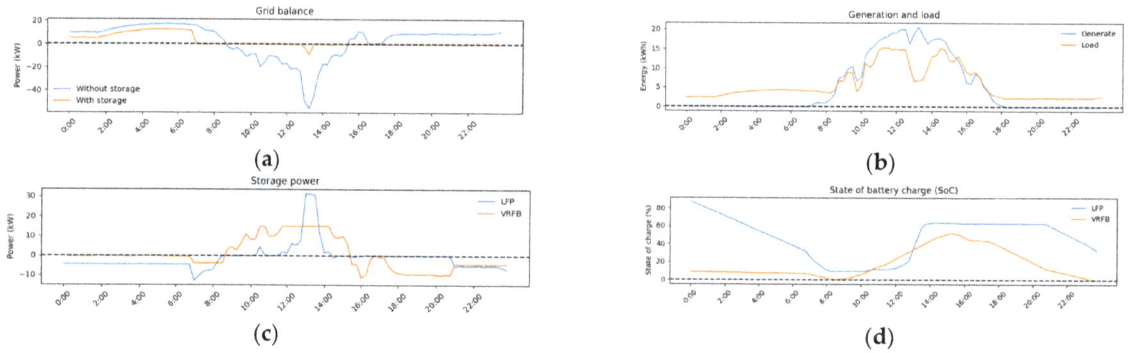

Figure 20. Results for the economic optimization method—data for 4 October: (**a**) grid power balance, (**b**) energy usage and production (input data), (**c**) batteries power, (**d**) batteries state of charge.

During winter (Figure 21), the behavior of the batteries shows the typical schedule of the HESS operation used for price arbitrage. Batteries are being charged during off-peak hours, even if it means importing extra amounts of energy from the grid. During the morning peak, the HESS power remains close to zero, while in the evening peak the highest priced batteries are being discharged. Both the VRFB and the LFP battery follow similar patterns. This behavior is consistent with our assumptions, based on the difference in prices between the price zones in the tariff.

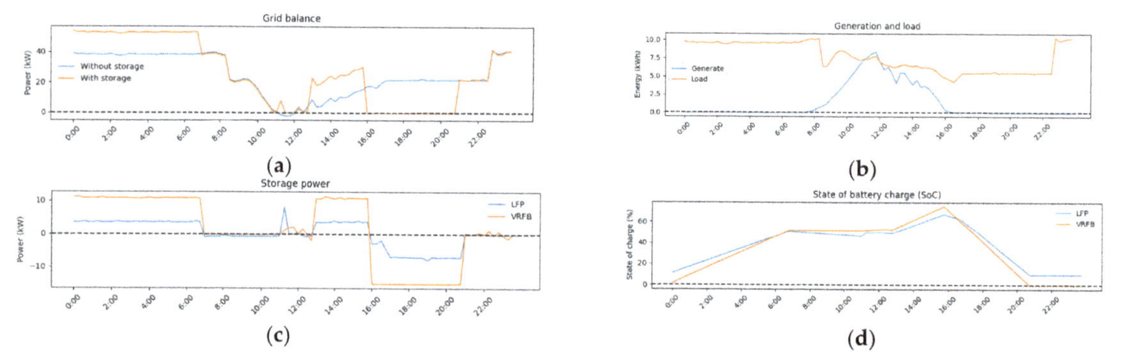

Figure 21. Results for the economic optimization method—data for 6 February: (**a**) grid power balance, (**b**) energy usage and production (input data), (**c**) batteries power, (**d**) batteries state of charge.

The economic factors clearly show that the economic algorithm that reduces the purchase of energy is the highest tariff (Figure 22a), at the same time increasing the use of energy from the lowest tariff, especially in winter months. The general costs of the system are smaller when compared to the situation without any energy storage and also lower than the benchmark. The arbitrage in winter is decreasing the overall cost of operation of the whole facility. The experiments revealed a problem in which the batteries do not charge fully in case of a surplus of energy (this situation is visible on Figure 19b). This is caused by the fact that there is no value for the optimizer to keep a higher state of charge of the batteries at the end of the day. To solve the issue, the obvious action would be to run the optimization for a longer period (e.g., a week, a month) but then the number of changing variables would be significantly increased, which would bring two problems: the problem with convergence and the extended time of computations.

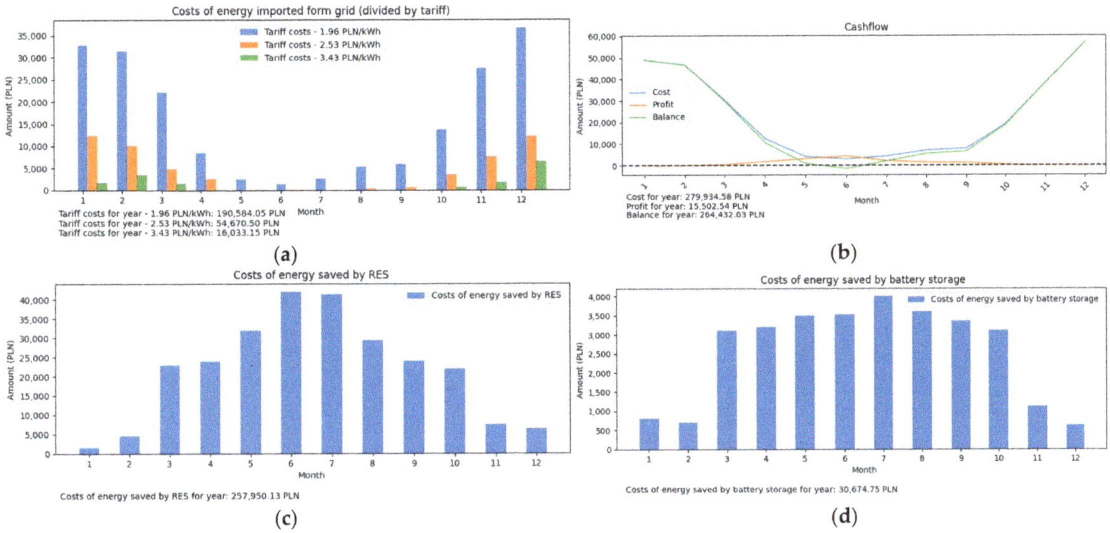

Figure 22. Results for the economic optimization method—monthly economic indicators: (**a**) purchase cost classified by tariff zones (tariff prices), (**b**) costs, profits and financial balance, (**c**) costs of energy saved by PV generation, (**d**) costs of energy saved by HESS operation.

3.3. Modified Economic Optimization

To deal with the problem of single-day optimization, a modification is proposed—the optimizer does not care about the state of charge of the battery at the end of the day, but for the next day it would be generally beneficial to have a higher state of charge, especially when there is a surplus of production that could have been used. To implement that, an extra rule was enforced after the optimization stage: when there is surplus of energy from the renewable sources, the battery will always try to use as much of this surplus as possible to charge. This modifies the solution returned by the solver in a way that the energy storage reaches the full state of charge faster and more often, which, in consequence, causes an increased state of charge of the energy storage at the end of the day. For the following day, the optimizer will be given this raised state of charge to start its calculations. We tested the solution and present the outcome on the following graphs—Figures 23–29. Figure 23 presents the monthly aggregated data—the increase in the operation of the energy storage in the summer months is clearly visible in comparison with the optimization method without modification. What is more, the import of energy has decreased, and, as a consequence, the cost of imported energy is also lower.

The more detailed view of each month gives a much better picture of the actual performance. In January (Figure 24), there are no changes in comparison to the optimization method as there is no surplus of energy.

For the summer months, the differences are much more visible; the graph for July shows the reduction in import and export of energy from and to the grid (Figure 25a). The batteries are charging and discharging more (Figure 25c) and the cost of energy import has dropped (Figure 25d). In this case, the difference between the energy balancing method and the modified economic optimization is very small and is mainly caused by small oscillations of the solutions given by the optimizer when it failed to reach the optimum solution in the defined number of iterations.

In July, the increase in the amount and duration of the charging of both batteries is visible. More of the surplus from the PV production is used. By the end of the day, the state of charge of the VRFB is higher in comparison to economic optimization (Figure 26).

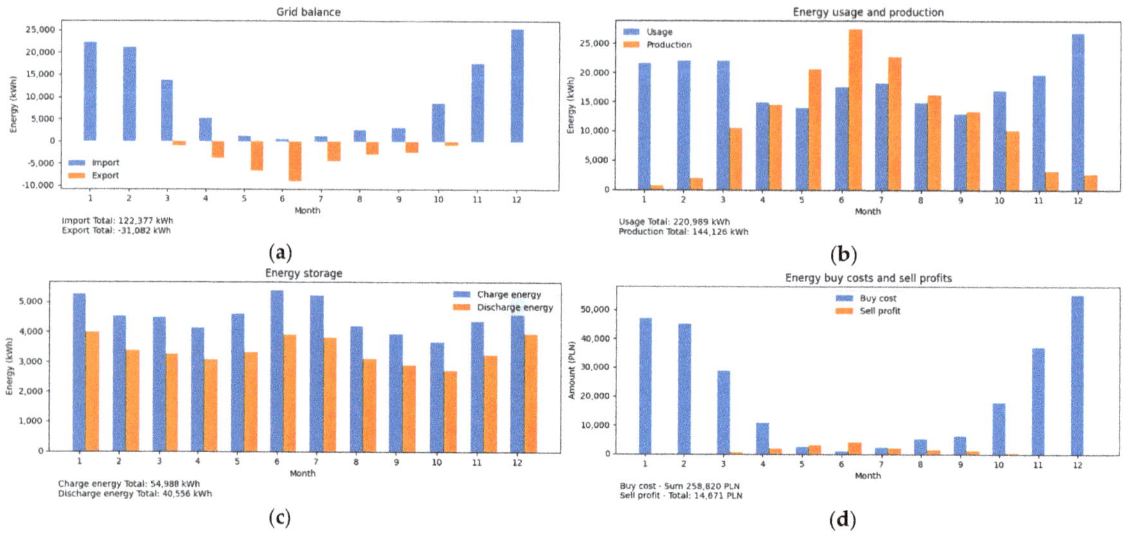

Figure 23. Results for the economic optimization method with modifications, aggregated monthly: (**a**) grid energy balance, (**b**) energy usage and production (input data), (**c**) activity of the energy storage, (**d**) cost of purchased energy and profit for sold energy.

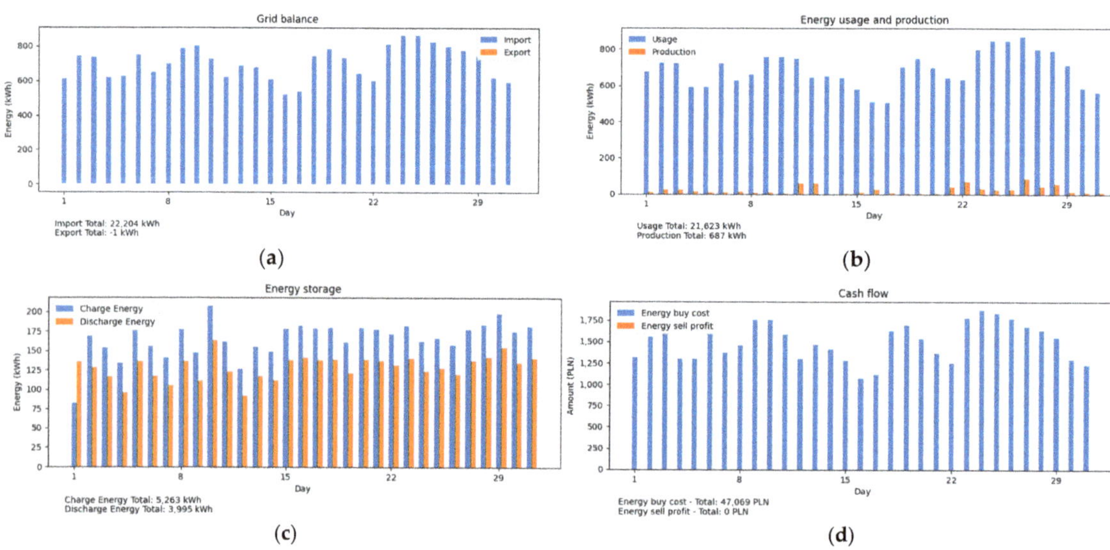

Figure 24. Results for the economic optimization method with modifications—data for the month of January, aggregated per day: (**a**) grid energy balance, (**b**) energy usage and production (input data), (**c**) activity of the energy storage, (**d**) cost of purchased energy and profit for sold energy.

The operation on 4 October (Figure 27) and 6 February (Figure 28) is almost the same as in the economic optimization without modifications.

The economic indicators show very interesting changes—the modification is reducing the use of the energy from the evening peak tariff, decreasing the overall energy costs for the facility and increasing the savings from the PV production and the operation of the HESS. The differences are not very significant, but the improvement is very clear, albeit only in summer months.

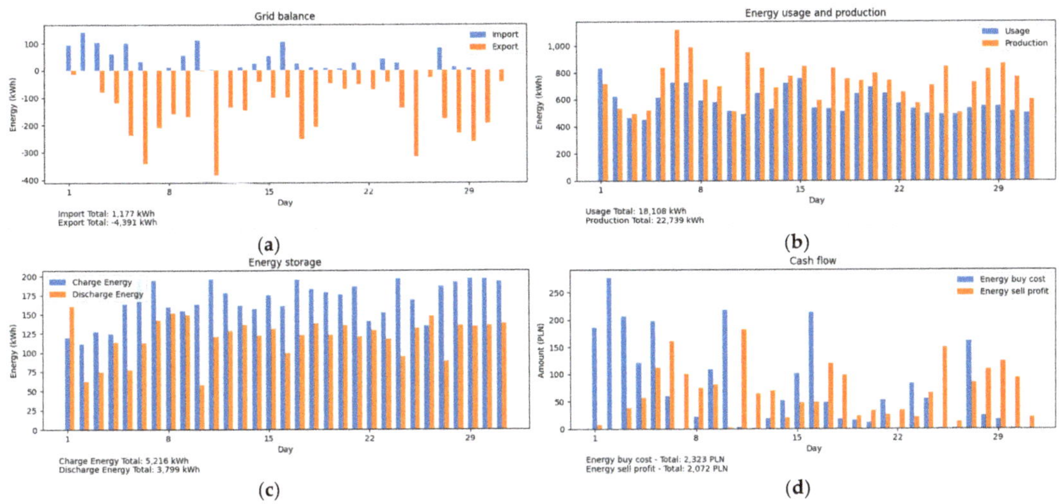

Figure 25. Results for the economic optimization method with modifications—data for the month of July, aggregated per day: (**a**) grid energy balance, (**b**) energy usage and production (input data), (**c**) activity of the energy storage, (**d**) cost of purchased energy and profit for sold energy.

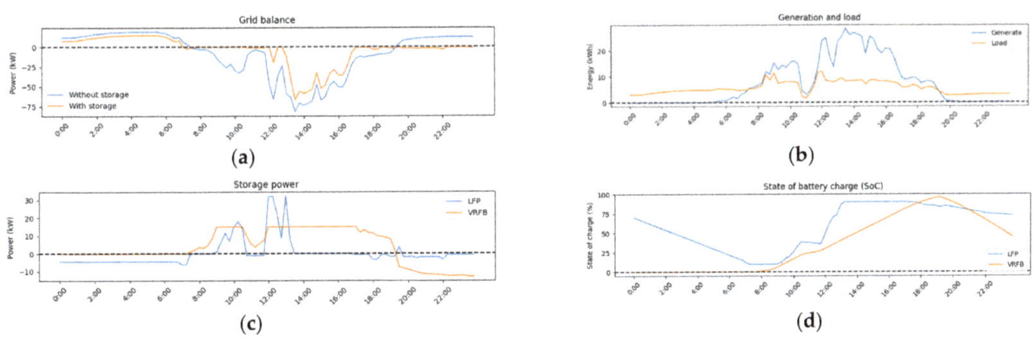

Figure 26. Results for the economic optimization method with modifications—data for 27 July: (**a**) grid power balance, (**b**) energy usage and production (input data), (**c**) batteries power, (**d**) batteries state of charge.

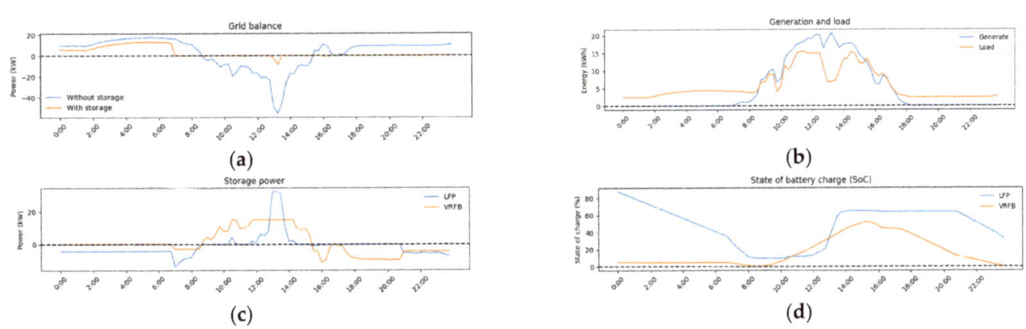

Figure 27. Results for the economic optimization method with modifications—data for 4 October: (**a**) grid power balance, (**b**) energy usage and production (input data), (**c**) batteries power, (**d**) batteries state of charge.

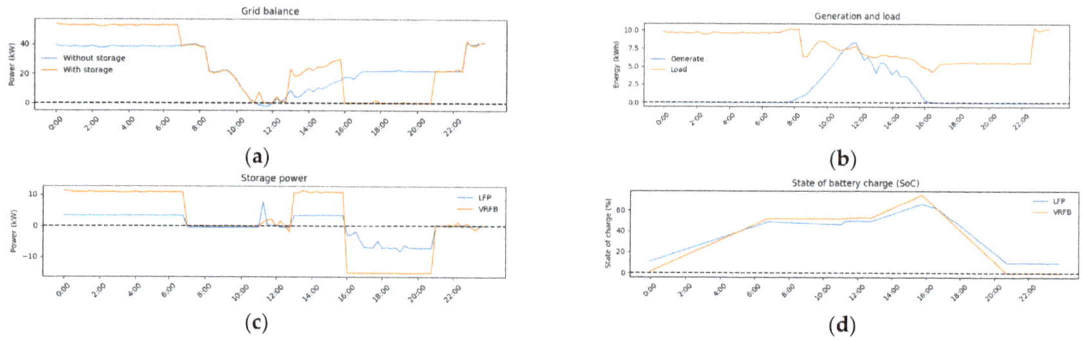

Figure 28. Results for the economic optimization method with modifications—data for 6 February: (**a**) grid power balance, (**b**) energy usage and production (input data), (**c**) batteries power, (**d**) batteries state of charge.

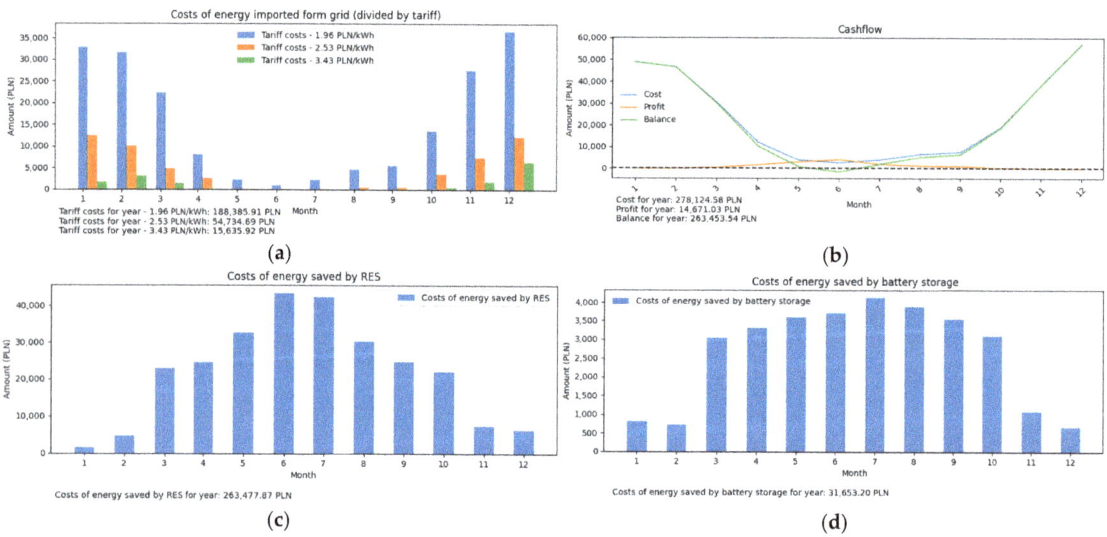

Figure 29. Results for the economic optimization method with modifications—monthly economic indicators: (**a**) purchase cost classified by tariff zones (tariff prices), (**b**) costs, profits and financial balance, (**c**) costs of energy saved by PV generation, (**d**) costs of energy saved by HESS operation.

4. Discussion

The overall aim of this work is to present an economic optimization algorithm for hybrid energy storage that will improve the financial outcome of the setup and show that the hybrid energy storage is a feasible solution to improve the self-consumption of energy from PV installation. The results of the simulations for the benchmark and the proposed HESS control strategies are summarized in Table 2.

The first part of the table focuses on the energy between the facility and the power grid. The batteries generally decrease the import and export of energy; the energy balancing approach is the most limited, due to the fact that it only operates on the surplus of the energy produced by the PV installation. The economic optimizations are realizing arbitrage all year round—buying energy when it is cheap and using it during peak times. The self-consumption rates are interesting—this is the percentage of the energy produced by the PV installation that was used within the installation, which show that, thanks to the batteries, over 77% of the produced energy is either directly consumed or stored for later use. The difference between the storage management methods in this context is very small, which shows that any storage increases the use of energy from PV and that the optimization algorithms are rationally using the surplus from the PV production.

Table 2. Comparison of the control methods and the setup without storage.

	Unit	Without HESS	Energy Balancing	Economic Optimization	Modified Economic Optimization
Import of energy	[MWh]	138.4	115.86	123.59	122.38
Export of energy	[MWh]	−61.5	−30.68	−32.84	−31.08
Self-consumption rate	[%]	57.3	78.7	77.8	79.3
Energy balance	[MWh]	76.8	85.19	90.75	91.3
Cost of import	[PLN]	324,091	270,695	261,351	258,820
Profit from export	[PLN]	29,060	14,478	15,502	14,671
VRFB Charge energy	[MWh]		23.11	35.88	37.71
VRFB Discharge energy	[MWh]	–	15.78	24.45	25.71
VRFB Equivalent cycles	-	–	158	244	257
VRFB Expected Lifetime	[years]	–	33	21	20
VRFB Depreciation cost	[PLN]		5038	7804	8207
LFP Charge energy	[MWh]		7.78	17.78	18.21
LFP Discharge energy	[MWh]	–	6.79	15.33	15.78
LFP Equivalent cycles	-	–	126	284	292
LFP Expected Lifetime	[years]	–	16	7	7
LFP Depreciation cost	[PLN]		4773	10,778	11,096
Energy cost	[PLN]	295,031	256,217	245,850	244,150
Financial outcome (including battery depreciation)	[PLN]	295,031	266,028	264,432	263,453

The second section of Table 2 presents the costs of import and profit from the export of energy. Although profits from exports are clearly correlated with the self-consumption rate, the cost of imports are affected by the cost in tariff zones. Here, the energy balancing method has the highest cost but uses the lowest amounts of import, which clearly demonstrates that economic optimization methods manage to shift energy between the zones.

The subsequent section presents the summary of the batteries' operation. Battery charge and discharge energy is accumulated over the year to estimate the annual throughput and calculate energy losses. Equivalent cycles are calculated using discharge energy, nominal capacity and *DoD*. This, in turn, is used to estimate the lifetime of each battery within HESS, allowing the estimation of the point in the investment horizon that replaces the battery blocks. The economic optimization approaches make much more use of the energy storage, and thus also shorten its lifetime.

The financial outcome accounts for energy trading costs (which include battery losses) and for the depreciation of each battery (to account for the cost of battery block replacement at the end of the expected battery lifetime). For comparison, the values of financial outcome without depreciation costs was presented as this better shows how much the depreciation of the battery costs.

Energy balancing uses every opportunity to charge batteries with surplus generation that would be exported otherwise. As soon as the load is larger than generation, stored energy is used to supply loads. Energy balancing does not cycle batteries at all in winter time, when the PV installation does not generate surplus energy. It can be assumed that a single battery performs on average one cycle every two days. As a result, the balancing algorithm generates almost no cost savings in the winter months when the HESS stays in an idle state.

By contrast, the economic optimization methods leads to heavy balancing of the batteries, resulting in the shortening of the expected lifetime. Additional cycles are caused by the fact that cost optimization implies time-of-use strategy that charges a battery in an off-peak tariff to use it during peaks. This results in increased energy losses and battery depreciation. The advantages of including time-of-use strategy are seen in the financial outcome. Figures 21 and 29 confirm that the majority of energy consumed is drawn from the grid in the off-peak tariff. The HESS leads to cost savings all year round.

The optimization method has another significant advantage over the balancing algorithms that was not reflected in the costs: it results in the operation of the LFP battery with relatively lower power. This leads to operation of the battery at lower temperature, and thus to an increased lifespan. This phenomenon has not been captured by the model applied but is an important point to investigate in future works.

The modification of the optimization method was introduced to partly overcome the problems connected to optimizing in 24 h windows. This 24 h window limits the horizon of the optimizer to the end of the day, and as such the optimizer is unable to increase the state of charge of the HESS, even if this would be beneficial for the next day. The implemented modification improved the solution, but there might still be a slight improvement if the optimization was calculated using longer time windows.

In this work, we limited the calculations to a 24 h window because the optimizer was also intended to perform the on-line optimization for the continuous management of an energy storage using forecasts; reliable forecasts can, however, only be obtained for the next day. Additionally, we considered a single day time rational in the case of a setup with a PV installation. The calculations were performed on standard desktop computers (CPU i5 3.2 GHz, 16 GB RAM) and while the balancing algorithm calculation time was below a minute, the simulation of an entire year using the economic optimization methods took 24 h. There is no possibility to parallelize the computations for this simulation as the state of charge of the storage at the end of the day is an input for the next-day computation.

5. Conclusions

A valuable tool has been implemented to test, simulate and analyze the behavior of the modelled HESS with battery models. The tool integrates a techno-economical model of a microgrid, including loads and RES. The model includes two battery types with their respective round trip efficiencies and costs of depreciation related to battery degradation during cycling. This simulation tool facilitates the sizing of the HESS installation, as well as the development and testing of control algorithms for scheduling the HESS operation. The graphical interface allows easy provisioning of input data while also allowing a visualization and analysis of the output data. The authors have implemented HESS control methods, including a simple energy balancing algorithm and using an energy cost optimization. The model and methods have been tested with real energy profiles recorded at a research centre.

The results explain the difference between the tested methods. The simple balancing algorithm stores surplus RES energy in the HESS and increases self-consumption rate to reduce the cost of energy; it does not do arbitrage as such, as it does not care about prices or tariffs. The advantage of this algorithm is its simplicity and moderate financial outcomes–using energy storages with such an algorithm brings profits in comparison to the setup without storage. The disadvantage is that this approach relies on the surplus of RES energy, otherwise the batteries are not used at all. Such an algorithm can be profitable when the averaged production from RES exceeds the usage of the facility. The economical optimization method on the other hand minimizes the costs of the operation during a single day. It uses the fact that there is a sufficient difference in cost of energy in different zones of the tariffs–using batteries for arbitrage can become profitable.

The total cost of using the VRFB battery (taking into account depreciation cost and losses related to the round trip efficiency) is not compensated by the difference between the prices in morning peak and off peak tariff zones. Adding an LFP battery, which has different properties and therefor cost balance, allows for reducing usage in all price peaks.

The combination of both batteries allow to improve the cost balance of the operation and prolong the lifetime of the batteries.

In the presence of overproduction of PV installation, the simple balancing algorithm at times outperformed the economical optimization method. The reason for this is that the methods employ a 24 h time window, during which the simple balancing algorithm tried to charge the battery as much as possible (it did not count the cost), whereas the economical optimization limited the charging to what was necessary for this day. This meant that in the economical optimization, a subsequent day might start with a lower SOC level, even though there was the potential to charge them the day before. To compensate for this, a modification of the economic optimization was performed, where the surplus of energy in a day was used to charge the HESS as much as possible. This modification improved the economical optimization method compared to the simple balancing algorithm.

The economical optimization method uses the HESS for arbitrage and, as a result, causes a more intense cycling of both batteries within the HESS. The simulation with realistic technical and economic data show that the arbitrage introduced by the economic optimization method has a small effect on the overall financial result. Although the energy consumption in the peak hours, and thus the energy cost, is reduced, there are additional costs of battery depreciation and energy losses in the batteries. The potential for energy price reduction comes at a cost—the final economic result is only slightly better than the simple energy balancing when the battery depreciation cost is included. It should be noted that the impact of leaving a battery in a discharged state, which happens during winter months in the simple energy balancing, is not considered in the model.

Obviously, the financial results depend on a variety of factors, such as battery performance and cost, energy usage and generation patterns and most of all energy price profiles. For this reason, future work includes analyzing different HESS operating scenarios and adjusting the optimization method to take into account additional services that the HESS can provide.

The designed methods will be tested in a real environment with forecasted profiles being the basis for the optimizer. What is more, we plan to test the optimizer on the prices from the day-ahead market, where both the purchase and selling prices are changing every hour. The tool is intended to be further developed into a commercial tool for ESS installation planning and management. The tool will be modified to work with the predictions of load and production rather than with historic data, then it will become a scheduler that can be used to manage the operation of the energy storage, together with a real-time controller.

Author Contributions: Conceptualization, supervision, funding acquisition, validation, K.R.; methodology, software, formal analysis, investigation, writing—original draft preparation W.R.; software, data curation, visualization, O.G.; supervision, resources, review and editing, H.B., writing—review and editing, J.V. All authors have read and agreed to the published version of the manuscript.

Funding: The work was financially supported by The National Centre for Research and Development via grant No. LIDER/30/0166/L-10/18/NCBR/2019.

Institutional Review Board Statement: Not applicable.

Informed Consent Statement: Not applicable.

Data Availability Statement: Not applicable.

Conflicts of Interest: The authors declare no conflict of interest.

References

1. Armstrong, R.; Chiang, Y.-M.; Gruenspecht, H.; Brushett, F.; Deutch, J.; Engelkemier, S.; Gençer, E.; Jaffe, R.; Joskow, P.; Mallapragada, D.; et al. The Future of Energy Storage—An Interdisciplinary MIT Study. Report. 2022. Available online: https://energy.mit.edu/research/future-of-energy-storage/ (accessed on 30 August 2022).
2. European Association of Storage for Energy. Energy Storage Targets 2030 and 2050 Ensuring Europe's Energy Security in a Renewable Energy System. 2022. Available online: https://ease-storage.eu/publication/energy-storage-targets-2030-and-2050/ (accessed on 30 August 2022).

3. IEA. *Grid-Scale Storage*; IEA: Paris, France, 2022; Available online: https://www.iea.org/reports/grid-scale-storage (accessed on 30 August 2022).
4. Hannan, M.A.; Wali, S.B.; Ker, P.J.; Rahman, M.S.A.; Mansor, M.; Ramachandaramurthy, V.K.; Muttaqi, K.M.; Mahlia, T.M.I.; Dong, Z.Y. Battery energy-storage system: A review of technologies, optimization objectives, constraints, approaches, and outstanding issues. *J. Energy Storage* **2021**, *42*, 103023. [CrossRef]
5. Miao, Y.; Chen, T.; Bu, S.; Liang, H.; Han, Z. Co-Optimizing Battery Storage for Energy Arbitrage and Frequency Regulation in Real-Time Markets Using Deep Reinforcement Learning. *Energies* **2021**, *14*, 8365. [CrossRef]
6. Yang, Y.; Lian, C.; Ma, C.; Zhang, Y. Research on Energy Storage Optimization for Large-Scale PV Power Stations under Given Long-Distance Delivery Mode. *Energies* **2020**, *13*, 27. [CrossRef]
7. Chen, Y.-T.; Kuo, C.-C.; Jhan, J.-Z. Research on Energy Storage Optimization Operation Schedule in an Island System. *Appl. Sci.* **2021**, *11*, 3690. [CrossRef]
8. Luo, L.; Abdulkareem, S.S.; Rezvani, A.; Miveh, M.R.; Samad, S.; Aljojo, N.; Pazhoohesh, M. Optimal scheduling of a renewable based microgrid considering photovoltaic system and battery energy storage under uncertainty. *J. Energy Storage* **2020**, *28*, 101306. [CrossRef]
9. Wang, Y.; Tian, J.; Sun, Z.; Wang, L.; Xu, R.; Li, M.; Chen, Z. A comprehensive review of battery modeling and state estimation approaches for advanced battery management systems. *Renew. Sustain. Energy Rev.* **2020**, *131*, 110015. [CrossRef]
10. Wang, L.; Li, M.; Chen, Z. An Energy Management Strategy for Hybrid Energy Storage Systems coordinate with state of thermal and power. *Control Eng. Pract.* **2022**, *122*, 105122. [CrossRef]
11. Lai, C.S.; Locatelli, G.; Pimm, A.; Wu, X.; Lai, L.L. A review on long-term electrical power system modeling with energy storage. *J. Clean. Prod.* **2021**, *280*, 124298. [CrossRef]
12. Katiraei, F.; Iravani, R.; Hatziargyriou, N.; Dimeas, A. Microgrids management. *IEEE Power Energy Mag.* **2008**, *6*, 54–65. [CrossRef]
13. Zia, M.F.; Elbouchikhi, E.; Benbouzid, M. Microgrids energy management systems: A critical review on methods, solutions, and prospects. *Appl. Energy* **2018**, *222*, 1033–1055. [CrossRef]
14. Meng, L.; Sanseverino, E.R.; Luna, A.; Dragicevic, T.; Vasquez, J.C.; Guerrero, J.M. Microgrid supervisory controllers and energy management systems: A literature review. *Renew. Sustain. Energy Rev.* **2016**, *60*, 1263–1273. [CrossRef]
15. Menniti, D.; Pinnarelli, A.; Sorrentino, N.; Vizza, P.; Burgio, A.; Brusco, G.; Motta, M. A Real-Life Application of an Efficient Energy Management Method for a Local Energy System in Presence of Energy Storage Systems. In Proceedings of the IEEE International Conference on Environment and Electrical Engineering and 2018 IEEE Industrial and Commercial Power Systems Europe (EEEIC/I&CPS Europe), Palermo, Italy, 12–15 June 2018. [CrossRef]
16. Wang, Y.; Liu, K.; Tang, X.; Dong, G. Editorial: Hybrid energy storage systems: Materials, devices, modeling, and applications. *Front. Energy Res.* **2022**, *10*, 1–3. [CrossRef]
17. Hajiaghasi, S.; Salemnia, A.; Hamzeh, M. Hybrid energy storage system for microgrids applications: A review. *J. Energy Storage* **2019**, *21*, 543–570. [CrossRef]
18. Tahir, Y.; Nadeem, M.F.; Ahmed, A.; Khan, I.A.; Qamar, F. A Review on Hybrid Energy Storage Systems in Microgrids. In Proceedings of the 3rd International Conference on Computing, Mathematics and Engineering Technologies (iCoMET), Sukkur, Pakistan, 29–30 January 2020. [CrossRef]
19. Tabart, Q.; Vechiu, I.; Etxeberria, A.; Bacha, S. Hybrid Energy Storage System Microgrids Integration for Power Quality Improvement Using Four-Leg Three-Level NPC Inverter and Second-Order Sliding Mode Control. *IEEE Trans. Ind. Electron.* **2018**, *65*, 424–435. [CrossRef]
20. Li, F.; Xie, K.; Yang, J. Optimization and Analysis of a Hybrid Energy Storage System in a Small-Scale Standalone Microgrid for Remote Area Power Supply (RAPS). *Energies* **2015**, *8*, 4802–4826. [CrossRef]
21. Bharatee, A.; Ray, P.K.; Subudhi, B.; Ghosh, A. Power Management Strategies in a Hybrid Energy Storage System Integrated AC/DC Microgrid: A Review. *Energies* **2022**, *15*, 7176. [CrossRef]
22. Gan, L.K.; Reniers, J.; Howey, D. A hybrid vanadium redox/lithium-ion energy storage system for off-grid renewable power. In Proceedings of the IEEE Energy Conversion Congress and Exposition (ECCE), Cincinnati, OH, USA, 1–5 October 2017. [CrossRef]
23. Qiu, X.; Nguyen, T.A.; Crow, M.L. Heterogeneous Energy Storage Optimization for Microgrids. *IEEE Trans. Smart Grid* **2016**, *7*, 1453–1461. [CrossRef]
24. Viswanathan, V.; Mongird, K.; Franks, R.; Li, X.; Sprenkle, V.T. 2022 Grid Energy Storage Technology Cost and Performance Assessment. Available online: https://www.pnnl.gov/sites/default/files/media/file/ESGC%20Cost%20Performance%20Report%202022%20PNNL-33283.pdf (accessed on 30 August 2022).
25. Energa Obrót. Standard Energy Prices for Commercial Customers Valid from 17.08.2022. Available online: https://www.energa.pl/dam/jcr:ae1072bf-f32f-4a83-8a89-dda88aa03ffd/Cennik%20standardowy%20dla%20przedsi%C4%99biorstw%20obowi%C4%85zuj%C4%85cy%20od%2017%20sierpnia%202022.pdf (accessed on 30 August 2022).
26. Urząd Regulacji Energetyki. Energy Market: The Average Energy Price on the Competitive Market in the Second Quarter of 2022 Was almost PLN 472/MWh. Available online: https://www.ure.gov.pl/pl/urzad/informacje-ogolne/aktualnosci/10551,Rynek-energii-srednia-cena-energii-na-rynku-konkurencyjnym-w-drugim-kwartale-202.html (accessed on 30 September 2022).
27. Yang, Y.; Li, H.; Aichhorn, A.; Zheng, J.; Greenleaf, M. Sizing Strategy of Distributed Battery Storage System With High Penetration of Photovoltaic for Voltage Regulation and Peak Load Shaving. *IEEE Trans. Smart Grid* **2014**, *5*, 982–991. [CrossRef]
28. Available online: https://www.python.org/ (accessed on 30 August 2022).

29. Available online: https://numpy.org/ (accessed on 30 August 2022).
30. Available online: https://scipy.org/ (accessed on 30 August 2022).
31. Eid, C.; Koliou, E.; Valles, M.; Reneses, J.; Hakvoort, R. Time-based pricing and electricity demand response: Existing barriers and next steps. *Util. Policy* **2016**, *40*, 15–25. [CrossRef]
32. Powell, M.J. A Direct Search Optimization Method That Models the Objective and Constraint Functions by Linear Interpolation. In *Advances in Optimization and Numerical Analysis: Mathematics and Its Applications*; Springer: Dordrecht, The Netherlands, 1994; pp. 51–67. [CrossRef]
33. SLSQP Documentation of Python Scipy Package. Available online: https://docs.scipy.org/doc/scipy/reference/optimize.minimize-slsqp.html#optimize-minimize-slsqp (accessed on 30 August 2022).
34. *Dieter Kraft: A Software Package for Sequential Quadratic Programming*; DFVLR-FB 88-28; DLR German Aerospace Center: Cologne, Germany, 1988.
35. Fracas, P.; Camarda, K.V.; Zondervan, E. Shaping the future energy markets with hybrid multimicrogrids by sequential least squares programming. *Phys. Sci. Rev.* **2021**, *6*, 20200050. [CrossRef]
36. Fracas, P.; Zondervan, E.; Franke, M.; Camarda, K.; Valtchev, S.; Valtchev, S. Techno-Economic Optimization Study of Interconnected Heat and Power Multi-Microgrids with a Novel Nature-Inspired Evolutionary Method. *Electronics* **2022**, *11*, 3147. [CrossRef]
37. Carrier, G.; Destarac, D.; Dumont, A.; Méheut, M.; El Din, I.S.; Jacques, P.; Saloua, K.; Brezillon, J.; Pestana, M. Gradient-Based Aerodynamic Optimization with the elsA Software. In Proceedings of the 52nd AIAA Aerospace Sciences Meeting—AIAA Science and Technology Forum and Exposition, SciTech 2014, National Harbor, MD, USA, 13–17 January 2014. [CrossRef]
38. Zahery, M.; Maes, H.; Neale, M. CSOLNP: Numerical Optimization Engine for Solving Non-linearly Constrained Problems. *Twin Res. Hum. Genet.* **2017**, *20*, 290–297. [CrossRef] [PubMed]
39. Cormen, T.H.; Leiserson, C.E.; Rivest, R.L.; Stein, C. *Introduction to Algorithms*; The MIT Press: Cambridge, MA, USA, 2001, ISBN 0262032937.

energies

MDPI

Article

Application of Paraffin-Based Phase Change Materials for the Amelioration of Thermal Energy Storage in Hydronic Systems

Dua'a S. Malkawi [1,*], Rabi Ibrahim Rabady [2], Mosa'b S. Malkawi [3] and Said Jereis Al Rabadi [4]

1 Engineering Department, Al-Balqa Applied University, P.O. Box 7041, Al-Salt 19117, Jordan
2 Electrical Engineering Department, Jordan University of Science and Technology, P.O. Box 3030, Irbid 22110, Jordan
3 Mechanical Engineering Department, Jordan University of Science and Technology, P.O. Box 3030, Irbid 22110, Jordan
4 Chemical Engineering Department, Al-Balqa Applied University, P.O. Box 50, Al-Huson 21510, Jordan
* Correspondence: duaa.m@bau.edu.jo

Abstract: This study aims at investigating the improvement in the thermal performance of energy storage for a hydronic system when it is equipped with evacuated tubes integrated within a hot water tank. The PCM shell in the bottom section is thicker than at the top to maintain a uniform, minimal water temperature difference of 5 °C between the top and bottom sections of the hot water tank. The thermal performance of the system was analyzed in diverse months when the ambient temperature fluctuated. The results have revealed that the thermal performance in December, March, and June was 80%, 81%, and 84%, respectively, meaning that the thermal performance is optimal in warm weather. The results confirmed that the system had boosted the presence of hot water throughout the whole day, including the time of the sun's absence, due to the release of stored PCM latent heat. The designed system solves the overheating problem and expands the availability of hot water through the cold weather. The system is characterized by lower heat losses because the average water temperature has decreased.

Keywords: phase change material; solar energy; thermal energy storage; hydronic solar system; paraffin wax

Citation: Malkawi, D.S.; Rabady, R.I.; Malkawi, M.S.; Al Rabadi, S.J. Application of Paraffin-Based Phase Change Materials for the Amelioration of Thermal Energy Storage in Hydronic Systems. *Energies* **2023**, *16*, 126. https://doi.org/10.3390/en16010126

Academic Editor: Gianpiero Colangelo

Received: 4 December 2022
Revised: 16 December 2022
Accepted: 19 December 2022
Published: 22 December 2022

1. Introduction

Intensive combustion processes of conventional petroleum-based fuels pose a significant impact on the environment in the long term and in the vicinity of residential areas due to exposure to harmful concentrations of gaseous emissions, namely COx, SOx, and NOx. As a result, stringent environmental regulations against these gaseous emissions and the operability of thermal combustion facilities to reduce environmental impacts are legitimized [1]. Due to the rising costs of petroleum end-products and increased demands in thermal applications by the residential and industrial sectors, researchers have been encouraged to investigate new resources of renewable energies (REs) and develop eco-environmentally friendly REs such as photovoltaics (PVs) and thermal solar panels (SPs). The installation of solar panels on wide terrain is intended to collect solar energy during the day. Moreover, the current thermal energy demands strongly encourage researchers to explore promising engineering solutions for effective thermal energy depots and dispatching solutions. Thermal energy storage has become increasingly crucial, owing to its interaction with variable production resources, the increase in the demand for conventional fuels for the combustion process, and the adverse environmental impact of other RE sources. Therefore, the ideal way to balance thermal energy is for it to be stored in conservative depots utilizing phase change materials such as paraffin based PCMs, which are ecologically and economically ideal.

Thermal energy storage is a feasible compensation for fluctuations between production and consumption rates during peak demand periods through thermal energy depot facilities that could be integrated within RE producers' and consumers' buildups. The integration of PCMs with an energy storage system has several potential applications, including the intensive and cumulative latent heat of phase changes. Furthermore, the phase change process is compatible and better monitored, since it occurs ideally at isothermal temperatures [2]. Despite these REs' potential, they possess a few deficiencies, such as crisp efficiency and less availability than other RE sources such as wind, traditional solar, and substrates for biofuel production [3]. The availability of sunlight varies across continents and between the earth's upper and lower hemispheres, potentially influencing energy availability.

The PCM products can be classified into three categories: eutectic, organic, and inorganic materials [4–9]. Organic PCMs include paraffin and non-paraffin. The main advantages of organic materials are changing their phase without segregation and latent heat degradation; self-nucleation; non-corrosiveness; chemical stability and safety. Inorganic PCMs include salts, hydrates, and metallic materials. They have a high storage density, high thermal conductivity, are non-flammable, and are readily available, but they need a nucleation agent and have a super-cooling problem in the phase transition. Eutectics are mixtures of two or more components [4–9].

Hydronic systems are usually associated with liquid water as a heat transfer medium for the cooling and heating processes. A hydronic system typically includes both cooled and heated water cycles to allow for separate heat transfer. Typical temperature differences of such systems are within the range of 0 and 15 °C for cooling and between 20 and 100 °C for heating [10–12]. Recently, solar water collectors have been considered a significant alternative to traditional electric heaters in meeting domestic hot water requirements. Although solar water heaters are composed of various types, passive or natural convection types are used widely due to their simplicity and operational efficiency [13,14].

The development of traditional solar heating and cooling systems was reviewed in Ge et al., 2018; storing excess heat for further applications was recommended, and enhancements to the solar energy storage system were highlighted [15]. Moreover, Buker et al., 2015 discussed improvements in solar panel design, such as panel surface, tilt, and shading, that could have a significant influence on the performance of the integrated hydronic systems [16]. Nevertheless, the obstacle that limits the solar water collectors is the scarcity of matching demand and supply throughout the day. The operation of solar water collectors depends on the availability of the sun [17] and heat losses [18].

Several researchers have confirmed that thermal energy storage is an essential issue by using appropriate thermal storage material within the solar energy system, which could be incorporated in a storage tank [19–21] or with collector tubes [22,23]. Recently, the heat that is absorbed or released during a phase change of PCMs has been employed as a thermal storage battery, due to its higher latent heat, wide operating temperatures, and very good thermal properties [24–31]. A PCM absorbs and stores thermal energy during the sunny hours of the day; later, it releases the stored energy after the sun's absence, which improves the solar system's efficiency. Organic PCMs, such as paraffin wax, are best known for storing a large amount of energy due to their high latent heat, thermal and chemical durability, little sub-cooling, and non-toxicity [32,33]. In the recent literature, the thermal behavior of paraffin-based PCMs was studied for the energy depot process. Murali et al., 2015 have examined the effectiveness of flat-plate solar water collectors incorporating paraffin as a PCM in a container placed in the top section of the water tank. Their findings appear to improve the performance of the solar system [34]. Kumar et al., 2020 have investigated the behavior and effect of applying synthesized nano-PCMs on the energy storage of evacuated solar water heating systems. According to their findings, PCMs were filled in evacuated solar tubes, which were connected to cylindrical containers placed inside the water tank [21], and such PCMs flowed as liquid inside and served as an energy storage medium to heat the water inside the main tank. In a previous study, we investigated the thermo-physical properties of PCMs by studying the enhancement of the thermal conductivity of the heat transfer medium of a PCM with the addition of carbon nanotubes (CNT) and graphite nanoparticles (GNP) as nanofillers to PCM composites [35]. So, future outcomes will focus on the enhancement of the performance efficiency of the solar system by adding nanoparticles to the PCM, which are then incorporated into the system. In this context, prior studies [36–38] have addressed the application of a shell and tube thermal storage heat exchanger equipped with finned outer walls for the tubes, and the enthalpy-porosity method was utilized to reveal the transient behavior of the PCMs' melting process. This approach could be subject to various complexities, and several criteria must be met to apply the proposed enthalpy-porosity method. In addition, the wavy annulus tubes could cause apparent vortices inside the heat exchanger that affect the natural convection of heat transfer.

Generally, the reviewed studies imply that the integration of PCMs within a solar system could ameliorate the performance of the thermal mass, maximize operational simplicity, and recover the thermal energy of the hydronic solar system for off-peak periods. It could be understood that few attempts were made to establish an in-field hydronic system that has a potential application of heating water in residential and industrial premises and to replace conventional electrical/fuel-based water heating systems, thereby improving energy storage efficiency during off-peak periods and reducing relevant energy expenses in premises. This study may offer guidance for future research and the thermal design of domestic hydronic solar systems. The performance of the system is assessed with an integrated PCM that is distributed on the shell side of the water storage tank, such that the PCM shell has a different thickness at the top and bottom of the storage tank (the bottom portion is thicker than the top). The effect of the PCM in a natural circulation solar water collector was examined through normal domestic hot water consumption, complete and sudden emptying of the hot water storage tank, and no hot water consumption.

2. Methodology

2.1. Experimental Setup

The manufactured hydronic solar system is located on the Jordan University of Science and Technology campus. Its geographic coordinates are 32.49° N (latitude) and 35.99° E (longitude). The solar collector that was used is an evacuated tube setup with an inclined angle of 45°. The inclined angle was chosen after making calculations to obtain higher gains in energy in the winter and solve the overheating problem in the summer. The main features of the evacuated tube of the solar collector are presented in Table 1.

Table 1. Features of the evacuated tube of the solar collector.

Parameter	Value
Number of tubes	20
Outer diameter	0.058 m
Inner diameter	0.047 m
Length	1.8 m
Tube material	Borosilicate glass
Absorptivity coefficient	95%
Emissivity coefficient	5%

In addition to evacuated tubes, the system contains a water storage tank with a total capacity of 0.200 m^3, a length of 1.6 m, and a diameter of 0.45 m. It is made of galvanized steel with an outer shell with a diameter of 0.53 m and contains paraffin wax as PCM with a thickness of 2 cm at the top and 4 cm in the bottom portion; such an asymmetric design is believed to assist in charging and discharging heat into the system since it provides more effective buoyancy motion for the liquid PCM. The tank was thermally insulated by rock wool to reduce the loss of energy. The insulation shell is covered with galvanized steel sheet. All specifications of the storage tank are shown in Figure 1.

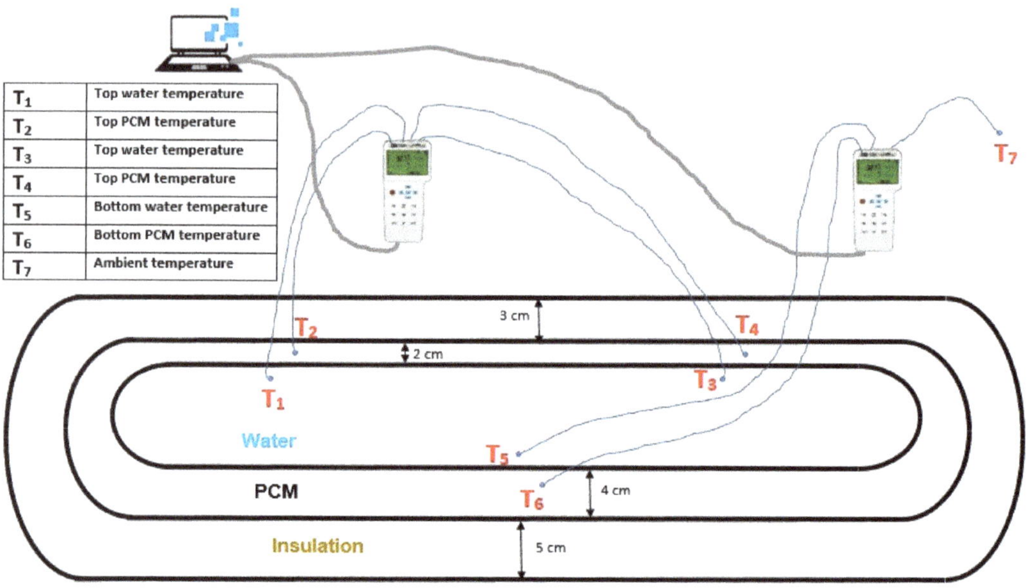

Figure 1. Schematic representation of the storage tank with instrumentation.

The phase change material that was used in this system was paraffin wax with a melting temperature of 48 °C. It was chosen due to its thermal stability, low price, no sub-cooling problem, and suitable latent heat. The thermal specifications of paraffin wax are presented in Table 2.

Table 2. Thermal specifications of paraffin wax [39].

PCM	Melting Temperature [°C]	Latent Heat [kJ/kg]	Specific Heat [kJ/kg·°C]	Thermal Conductivity [W/m·°C]
Paraffin wax	48	210	2.4 (liquid) 2.1 (Solid)	0.24

Additionally, the system consists of a solenoid valve that is programmed to meet the level of family consumption of hot water throughout the day. This valve withdraws hot water at specific times; the following diagram shows the proposed water consumption pattern, which presents the distribution of hot water throughout the day. Figure 2 shows a daily water consumption pattern according to real observed consumption and required estimations.

The system contains a cold water tank to recover hot water discharged from the hot water tank. Thermocouples (Type K) were fixed through the storage tank to notice and record water and PCM temperatures during the heating process. Thermocouples were installed in the system to detect the temperatures of the water, PCM, and ambient. They were placed in the water region in three positions: two at the top and one at the bottom. Additionally, other thermocouples were placed in three positions throughout the PCM region: two at the top and one at the bottom. One thermocouple reads the ambient temperature. All thermocouples were connected to a converter that gives temperature readings in Celsius. The data logger was connected to read and record the temperatures with Windows software easily plugged into a computer. For measuring irradiance (w/m^2), a pyranometer was used. The data was acquired and stored every 4 min. An additional experiment was performed every 5 s and the reading was recorded. The hot water tank was discharged completely in the evening (specifically at sunset) to investigate the water and PCM temperature behavior in this case.

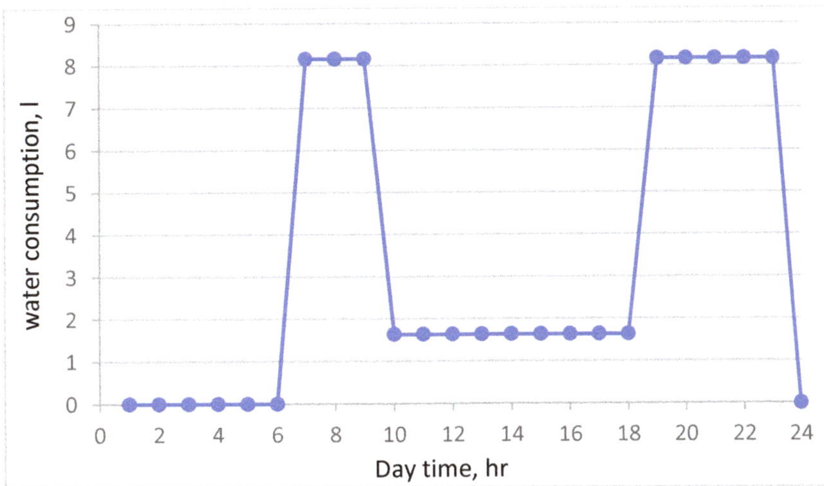

Figure 2. The hot water consumption pattern throughout the day.

As for the hot water region, the PCM region was also equipped with two holes and a lid to fill and discharge the PCM at any time based on necessity. Moreover, the problem of high pressure throughout the system is resolved by setting up vents for both the water and the PCM regions. A photographic view and schematic diagram of the system are presented in Figure 3. The water is replenished from the water supply tank. The hot water storage cylinder receives hot water passively from the evacuated tube, whereas the hot water flows up to the tank naturally due to thermosiphon circulation. The hot water was then used for domestic use, and hot water consumption was recovered by the water supply tank. When water gains heat from solar energy, it conductively exchanges this energy with the PCM. Conversely, as the temperature of water decreases, the latent heat will be released to the water from the PCM during the liquid phase until solidification in the absence of the sun.

2.2. Thermal Model

Energy balance is applied to both parts of the hydronic solar system under steady-state conditions: the evacuated tube and hot water storage tank. The useful energy gained from solar radiation by evacuated tubes can be expressed by [40,41]:

$$Q_{useful} = I\,A_c\,(\tau\alpha)_{eff}\,k_{\theta i} - Q_{loss} \quad (1)$$

and

$$Q_{loss,\ tube} = U_{L,tube}\,A_c\,(T_w - T_a), \quad (2)$$

where I represents a global solar irradiance, A_c represents a solar collector area, $(\tau\alpha)_{eff}$ represents an effective transmissivity-absorptivity product coefficient, $k_{\theta i}$ represents an incident angle modifier, $U_{L,tube}$. represents an over-all heat transfer coefficient of heat loss from the evacuated tubes, and T_w and T_a. represent water and ambient temperatures, respectively.

The solar collector's efficiency η. is determined by the value of the ratio between useful energy and solar radiation that falls on the collector. This can be expressed by:

$$\eta = \frac{Q_{useful}}{I\ A_c} \tag{3}$$

Solar collector efficiency (evacuated tubes) can be explained by:

$$\eta_{collector} = (\tau\alpha)_{eff}\ k_{\theta i} - \frac{U_{L,tube}\ (T_w - T_a)}{I} \tag{4}$$

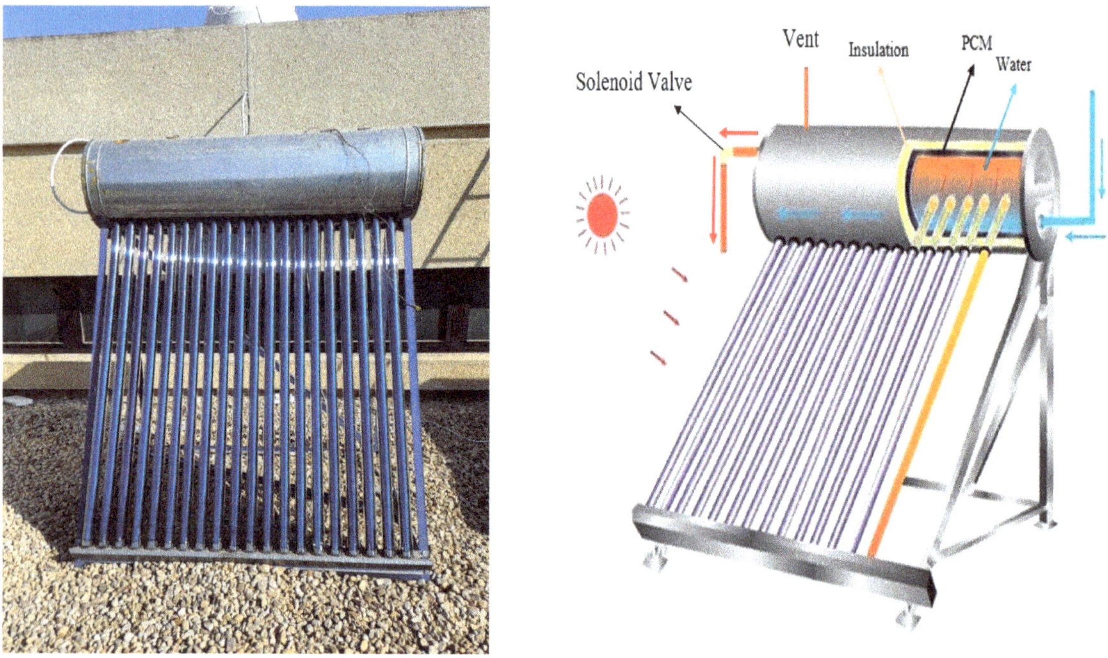

(a) (b)

Figure 3. Hydronic evacuated tube solar system with a PCM: (**a**) photographic view; (**b**) schematic diagram.

The following equation clarifies how the useful energy leaving the evacuated tubes moves to the water tank, which transfers to paraffin, giving rise to temperature changes:

$$Q_{PCM} = \left(m_{PCM}\ c_{p,\ PCM}\ \Delta T\right)_{solid} + m_{PCM}\ \lambda_{PCM} + \left(m_{PCM}\ c_{p,\ PCM}\ \Delta T\right)_{liquid} \tag{5}$$

Energy balance in the water tank can be expressed by:

$$E_{accumulation} = Q_{useful} \pm Q_{PCM} - Q_{load} - Q_{loss,tank} \tag{6}$$

Useful energy, load energy, and the heat loss of the water tank can be calculated by:

$$Q_{useful} = m_{w,\ tank} \cdot c_{p,\ w} \cdot (T_{out,\ w} - T_{in,\ w}) \tag{7}$$

$$Q_{load} = m_{w,load} \cdot c_{p,\ w} \cdot (T_w - T_a) \tag{8}$$

$$Q_{loss,\ tank} = U_{L,\ tank}\ A_{tank}\ (T_w - T_a) \tag{9}$$

The overall heat transfer coefficient of energy losses in the system (U_L,sys) is equivalent to the losses of both the evacuated tube and the water tank. This can be expressed by:

$$U_{L,sys} = U_{L,tube} + U_{L,\ tank} \tag{10}$$

The efficiency of the system with paraffin as the PCM is:

$$\eta_{system} = \eta_{collector} \cdot \eta_{PCM} \tag{11}$$

$$\eta_{system} = \left[(\tau\alpha)_{eff}\, k_{\theta i} - \frac{U_{L,sys}\,(T_{H_2O} - T_a)}{I} \right] \cdot \left[\frac{Q_{PCM}}{I\,A_c} \right] \tag{12}$$

The domestic hydronic solar system was evaluated according to EN 12976 standards, where the solar radiation, water temperature, ambient, and PCM temperatures were recorded for more than 9 months consecutively under two test types: with PCMs and without PCMs. According to ISO 9459-5 DST, the withdrawals of hot water from the storage tank depended on family consumption patterns throughout the testing period. Thermal output characterization tests were conducted according to the results of calculating instantaneous system performance experimentally and theoretically and calculating water storage tank heat losses. The hydronic solar system's thermal performance was measured on days with daily solar radiation and temperatures recorded over consecutive months at different water storage inlet temperatures. Protection against overheating and pressure resistance standards were considered necessary to save the system from deformation.

3. Results and Discussion

The average values of the solar radiation at the JUST campus throughout the year are shown in Figure 4. The temperature distributions of the ambient, water, and PCM at the storage tank were recorded during system testing. All parameters of the system were studied for many months over a year to investigate the effect under different weather conditions.

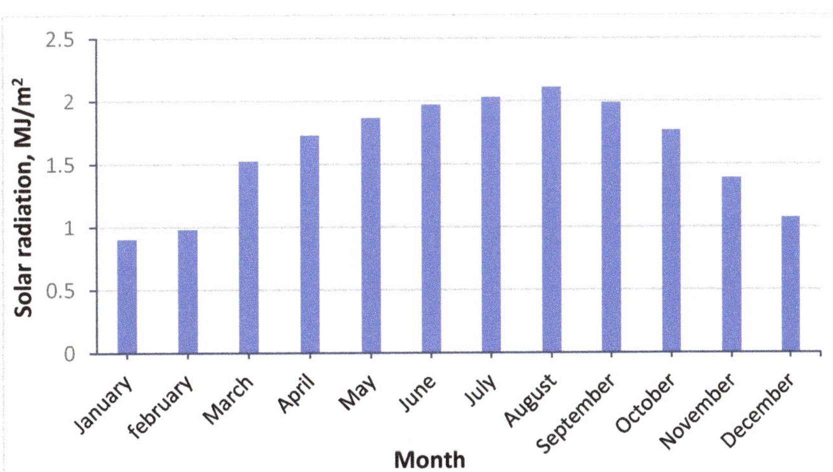

Figure 4. Measured monthly radiation data.

3.1. Temperature Distributions

The temperature distributions of the ambient, PCM, and water at the storage tank (average) with and without PCMs are shown in Figures 5–7. In these figures, it is noticeable that the water temperature increases from sunrise until it reaches the melting temperature of paraffin. The water temperature remains at a fixed value until the paraffin melts completely, at which point it increases to a specific value. The temperatures decrease with the decrease in energy gained from the sun at the end of the day. While the water temperature rises through circulation in the evacuated tubes, water flows into a storage tank where thermal energy exchange starts between hot water and paraffin, which further raises the paraffin's temperature. So, the temperature of the paraffin at the beginning of the day increases gradually with the increasing water temperature that comes from the evacuated tubes. When paraffin reaches its melting point, the temperature stays constant, increases to the maximum specified value, and then gradually decreases at night as a result of the absence of energy from solar radiation. At the melting and solidification temperature of the PCM, the water temperature stays at a fixed value, which can be observed in Figures 5–7.

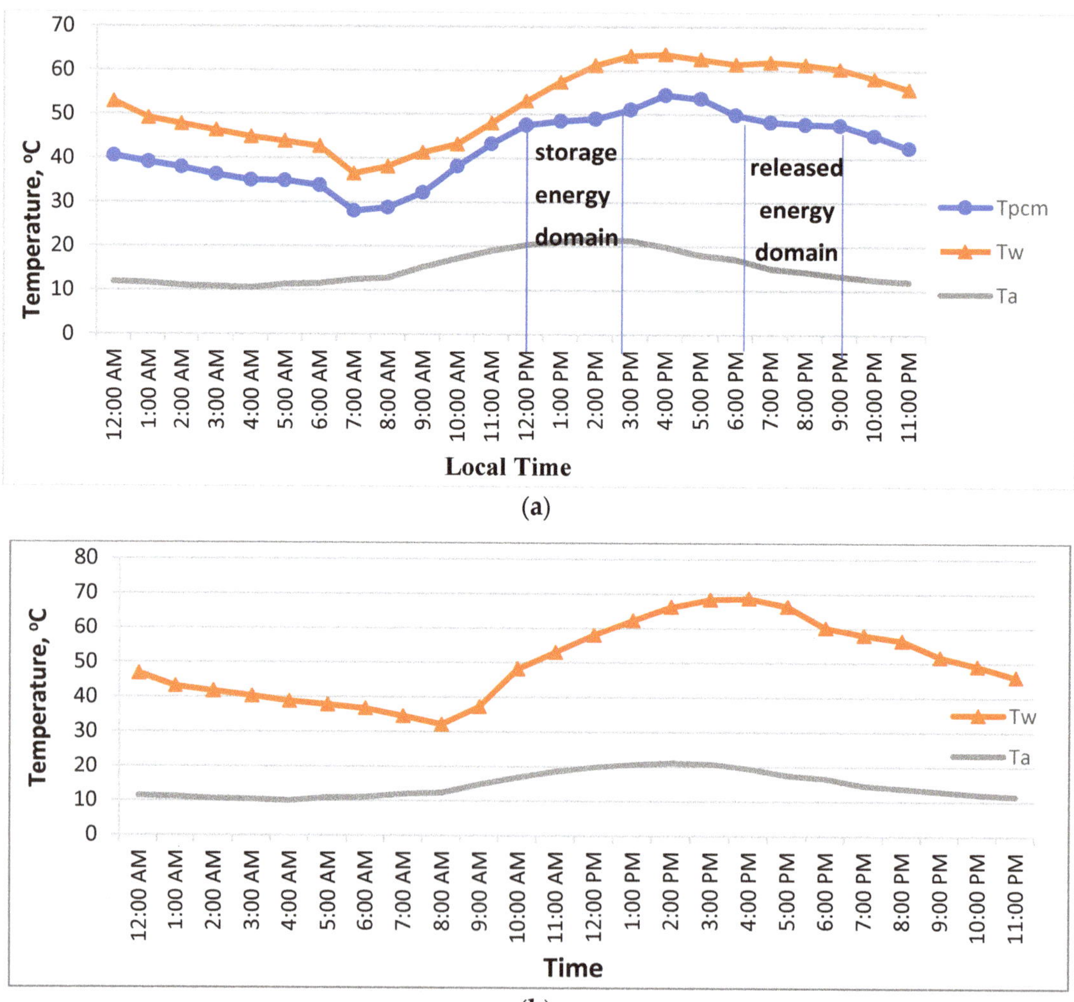

Figure 5. Temperature distribution of the system through December 2021: (**a**) using paraffin as PCM; (**b**) without PCM.

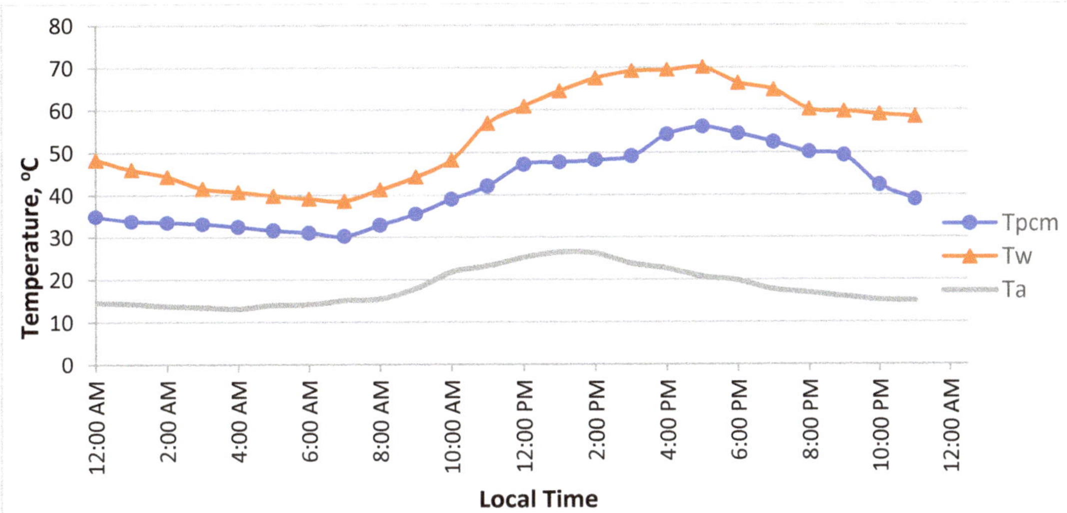

(a)

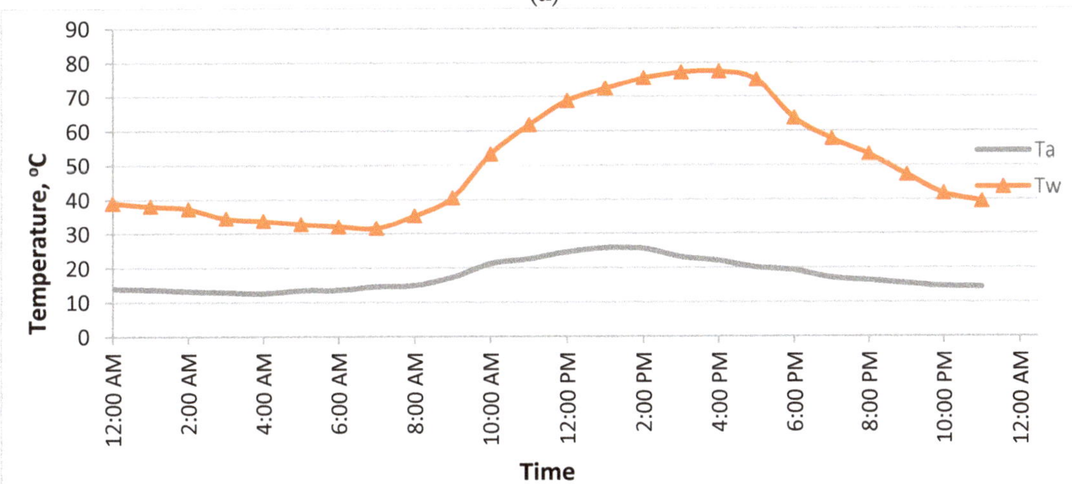

(b)

Figure 6. Temperature distribution of the system through March 2022: (**a**) using paraffin as PCM; (**b**) without PCM.

The following can be observed by using a PCM case: when the PCM temperature reaches its melting point throughout the day, the stored energy begins the PCM phase change from a solid to a liquid. This stored energy is used as released energy in water to maintain its temperature within the domestic usage range.

The temperatures of paraffin decrease constantly in the afternoon to reach a solidification point and stay at the same temperature for a short period, with an exchange of latent heat that is released in water. This process is reflected in the values of water temperatures, where the decrease is very small. Energy loss during the late afternoon and night hours is higher than at any other time during the day. The water temperatures are in the range for domestic use, which is the main goal of the system. The thermal energy that transfers between water and paraffin depends on the temperature difference between them and on the phase of paraffin (liquid or solid). Throughout the day, with the presence of solar energy, the glazing temperature, energy collected, and water temperature increase. Approximately at solar noon, water temperatures reach their maximum.

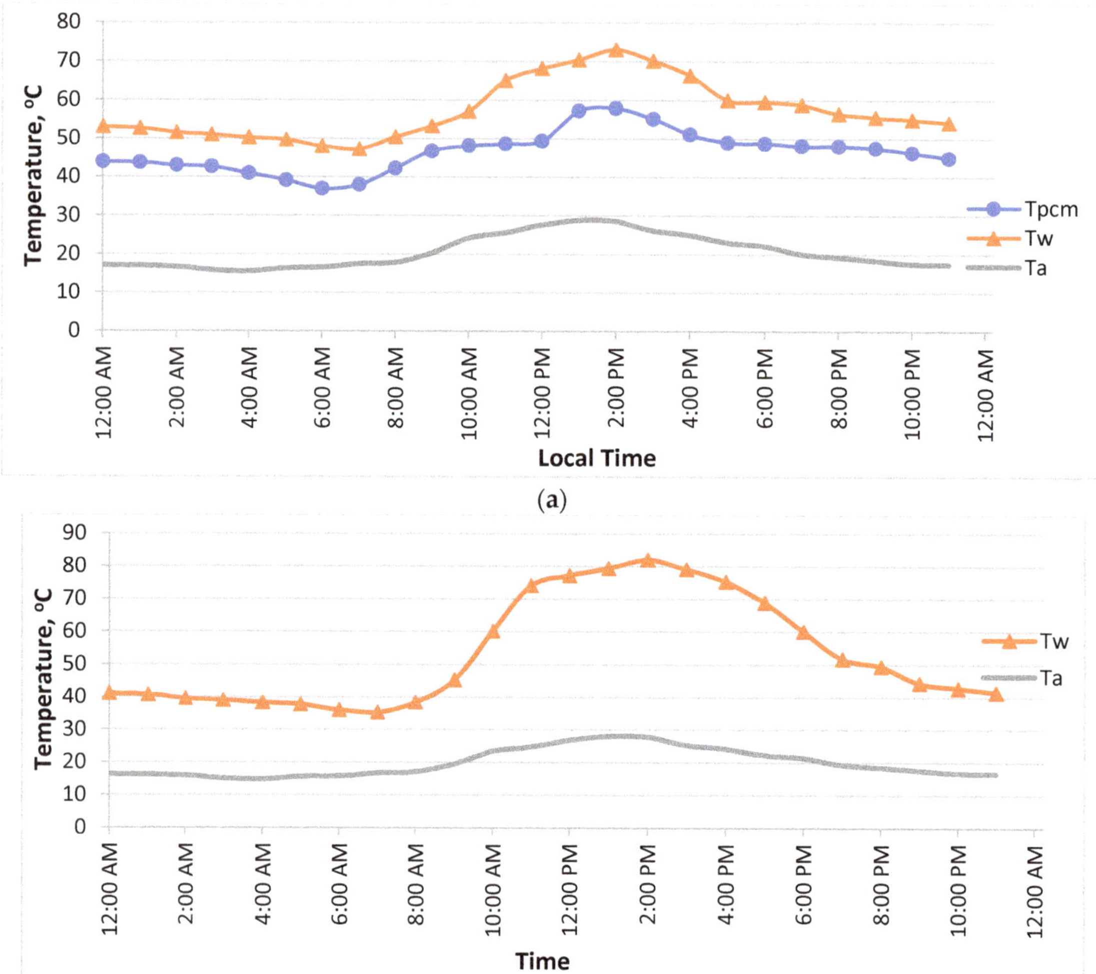

(**b**)

Figure 7. Temperature distribution of the system through June 2022: (**a**) using paraffin as PCM; (**b**) without PCM.

It can be observed in Figures 5–7 that water temperatures at solar noon without using a PCM case are higher than those with PCMs. Higher values due to the transfer of energy from water to PCMs mean a reduction in overheating problems in the water tank. Conversely, through early morning and late afternoon, the water temperatures are lower than those reached when using paraffin as a PCM.

Figure 8 shows different temperature distributions, and the experiment of discharging the storage tank of hot water completely was performed. This experiment was conducted to study the behavior of the PCM and heat exchange with water by discharging all amounts of hot water in the water tank at 4:00 PM. The withdrawal of hot water is replaced by cold water. It is evident from Figure 8 that the water temperature decreases sharply through the discharge process, along with the PCM temperature. After that, the water temperature begins to rise as a result of the heat exchange from the PCM.

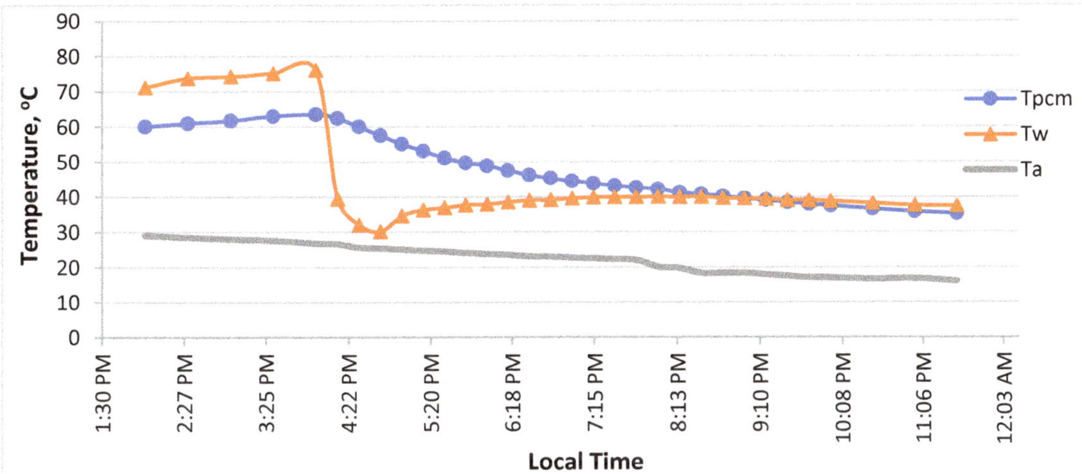

Figure 8. Temperature history of the system through complete hot water consumption.

The water and PCM reach the same temperature at a specific point in time. Additionally, the temperature of the water in the domestic use range can be considered optimal. This experiment shows the exchange of stored thermal energy in PCMs with water and its effect on water temperature. This experiment explains the family's sudden and complete drain of the hot water from the water storage tank and how the PCM raises the water temperature by 10 °C over a short period of time.

Furthermore, Figure 9 shows the temperature distributions of hot water, PCM, and ambient temperature in the absence of hot water consumption throughout the day. This experiment was performed in approximately similar weather conditions to the previous one. It can be noticed that higher values of hot water and PCM temperatures are due to the absence of load energy. Moreover, it is clear that the temperature difference between water and paraffin is small; this difference is less than 1 °C in the morning hours with increasing gains in energy.

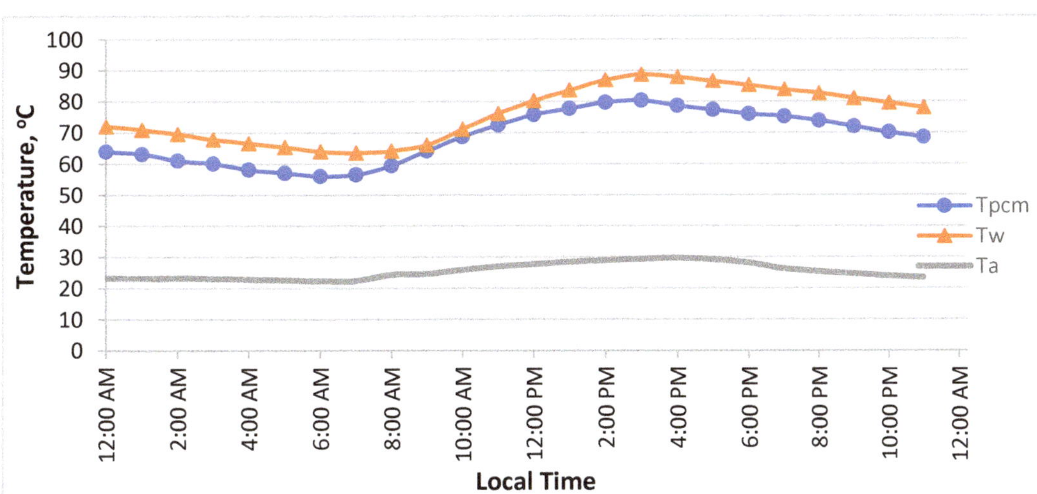

Figure 9. Temperature history of the system through no hot water consumption.

Due to the design of the storage tank, the thickness of the PCM layer on the top and bottom of the tank was different. The thickness at the bottom is higher than the top, which means more mass of PCM and more stored energy through the sun's presence. This stored energy is released in the water at the bottom of the tank, which has a lower temperature than that in the top region. Releasing energy from PCMs means heating water, which makes the water through the whole tank have similar or small differences in temperature, especially in the late afternoon hours. Figure 10 presents the temperature distribution of water on the top and bottom regions; as can be seen, the maximum difference is approximately 5 °C during the daybreak hours. It can be observed that the water temperatures at the top and bottom of the tank are the same at solar noon. This study shows a decrease in water temperature compared with a hydronic solar system without a PCM, which is an advantage to reducing heat losses from the system and avoiding superheating through the tank. Furthermore, Azimi et al., 2015 found that the water temperature at the bottom of the tank is close to ambient temperature without PCMs [42].

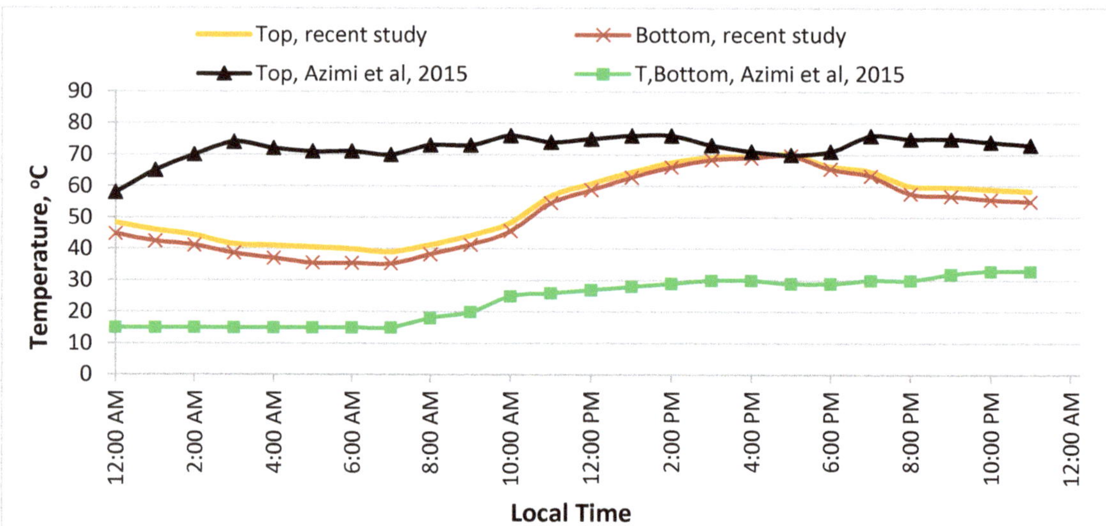

Figure 10. Temperature distribution of water through the top and bottom of the storage tank [42].

Figure 10 presents a comparison and disparity between our study and that of Azimi et al. Not only do our results demonstrate a decrease in the difference between hot water and temperature at the top and bottom of the storage tank greater than that of Azimi et al., but they also show a decrease in the hot water in the system (within domestic use), which means less thermal energy losses and covers the hours of solar absence.

3.2. System Efficiency

The results of our experiments and theoretical calculations are summarized in Tables 3–5. The results show useful gains and losses in energy with the water temperature in the water storage tank for the clear days of December, March, and June. The lowest rate of useful energy is in the evening and morning. However, the highest value of useful energy at solar noon is due to the increasing gain in solar energy. Due to the greater temperature difference between the water and the ambient, heat loss is greater at noon than in the morning and evening. The experimental efficiency of the hydronic solar system was estimated by ($\frac{Q_{useful}}{I\,A_c}$). It is clear from the tables that the system efficiency rises progressively from sunrise to solar noon, the maximum value, then falls off. The maximum experimental efficiency is around 80% in December, 81% in March, and 84% in June. The instantaneous efficiency of the system without using PCMs is around 60%. So, the positive effect of using PCMs on the performance of the hydronic solar system is clear.

Figure 11 illustrates the relationship between the system's thermal efficiency (η_{system}) and the temperature difference between hot water and ambient ($\Delta T/I$), based on the data for December, March, and June. The relationship shows a straight line with the overall heat loss coefficient as a slope. The optical efficiency of the hydronic system is the intercept of the straight line. The values of the determining factor of the relation ($R^2 > 0.9$) point out an intense correlation between both parameters. Each figure shows the comparison of the relationship between systems with and without PCM. It can be concluded from Figure 11 that the case of PCM has higher efficiency than that without PCM.

Table 3. Experimental and calculated results for December 2021 with paraffin.

Local Time	Tw (K)	Ta (K)	I (w/m$^{2.5}$)	Q_{useful} (w)	$Q_{loss, sys}$ (w)	$\Delta T/I$ (K·m^2/w)	η_{system}	
							Experimental	Theoretical
7:00 AM	313.65	285.55	42.09	201.25	92.41	0.37	0.43	0.14
8:00 AM	311.25	285.95	206.47	527.41	82.98	0.12	0.62	0.39
9:00 AM	314.45	288.35	449.72	1170.86	86.23	0.06	0.75	0.46
10:00 AM	316.35	290.35	657.35	1730.45	86.37	0.04	0.77	0.58
11:00 AM	321.15	292.15	660.99	2017.54	97.07	0.04	0.79	0.61
12:00 PM	326.15	293.45	770.52	1540.62	110.55	0.04	0.8	0.6
1:00 PM	330.45	294.25	592.19	924.34	123.29	0.06	0.73	0.55
2:00 PM	334.35	294.75	361.18	311.50	135.52	0.11	0.64	0.47
3:00 PM	336.55	294.45	131.38	194.21	104.64	0.32	0.51	0.1
4:00 PM	336.15	293.05	88.21	141.14	92.41	0.5	0.38	0.05

Table 4. Experimental and calculated results for March 2022 with paraffin.

Local Time	Tw (K)	Ta (K)	I (w/m^2)	Q_{useful} (w)	$Q_{loss, sys}$ (w)	$\Delta T/I$ (K·m^2/w)	η_{system}	
							Experimental	Theoretical
7:00 AM	312.15	290.65	67.82	162.47	63.84	0.32	0.48	0.17
8:00 AM	314.35	291.05	121.32	302.60	69.39	0.19	0.55	0.32
9:00 AM	317.35	293.45	239.80	615.55	71.69	0.1	0.64	0.47
10:00 AM	321.35	297.35	465.14	1211.92	72.73	0.05	0.7	0.55
11:00 AM	329.95	298.75	671.44	1751.12	95.73	0.05	0.8	0.55
12:00 PM	333.95	300.75	795.03	2075.75	102.87	0.04	0.81	0.6
1:00 PM	337.55	301.95	798.70	2082.36	111.08	0.04	0.79	0.62
2:00 PM	340.55	301.75	681.87	1769.23	121.40	0.06	0.77	0.61
3:00 PM	342.25	299.35	480.51	1231.29	134.41	0.09	0.66	0.56
4:00 PM	342.55	298.15	256.79	636.88	119.08	0.17	0.58	0.34
5:00 PM	343.15	296.25	111.33	249.13	96.66	0.42	0.36	0.08
6:00 PM	339.45	295.35	89.32	190.98	67.18	0.58	0.3	0.03

Table 5. Experimental and calculated results for June 2022 with paraffin.

Local Time	*Tw* (K)	*Ta* (K)	*I* (w/m^2)	Q_{useful} (w)	$Q_{loss,\ sys}$ (w)	$\Delta T/I$ (K·m^2/w)	η_{system}	
							Experimental	**Theoretical**
7:00 AM	320.45	290.65	124.60	305.87	101.62	0.41	0.21	0.04
8:00 AM	323.55	291.05	138.30	339.54	104.03	0.24	0.46	0.13
9:00 AM	326.45	293.45	145.10	356.24	104.86	0.23	0.55	0.15
10:00 AM	330.25	297.35	401.15	1733.79	127.29	0.1	0.76	0.5
11:00 AM	338.25	298.75	665.25	2138.88	132.21	0.08	0.82	0.54
12:00 PM	341.35	300.75	820.88	2132.10	136.54	0.06	0.84	0.59
1:00 PM	343.65	301.95	819.26	1719.74	145.85	0.05	0.81	0.57
2:00 PM	346.25	301.75	664.39	1045.19	143.57	0.05	0.76	0.5
3:00 PM	343.35	299.35	410.99	565.76	134.29	0.07	0.67	0.46
4:00 PM	339.55	298.15	229.66	272.33	118.24	0.11	0.62	0.42
5:00 PM	333.15	296.25	117.68	248.26	119.90	0.18	0.51	0.3
6:00 PM	332.75	295.35	106.39	117.31	83.18	0.31	0.42	0.04

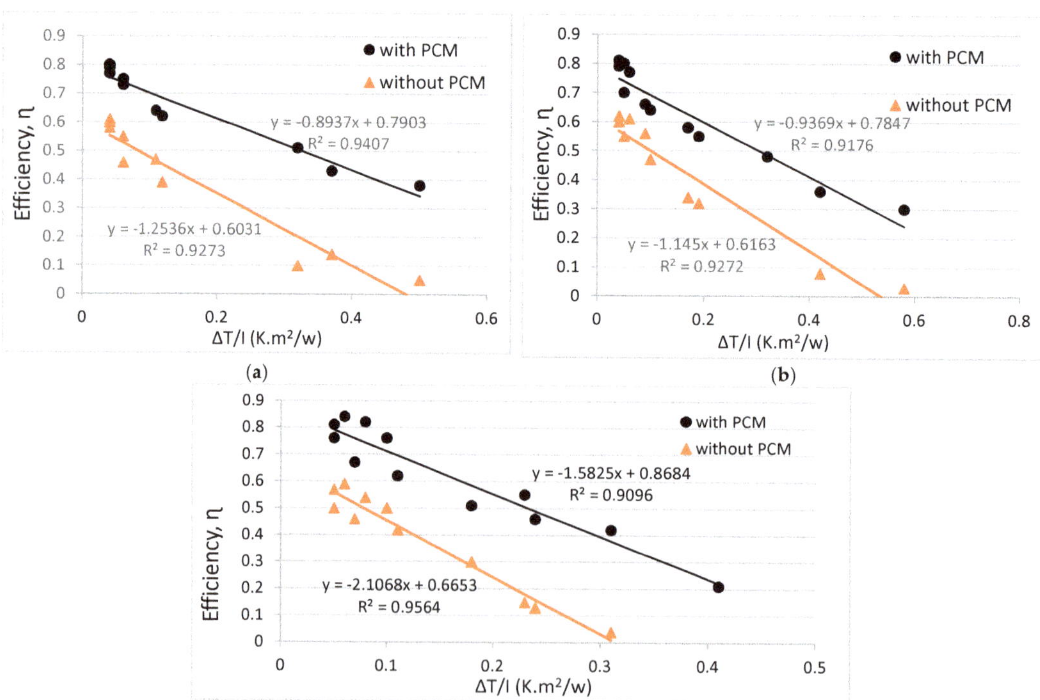

Figure 11. Efficiency of the system versus ΔT/I with and without PCM; (**a**) December, (**b**) March and (**c**) June.

Situta-Olcha et al., 2021 [43] examined the thermal efficiency of the solar system with a heat pipe without a PCM under weather conditions similar to December. Their results show less than 40% thermal efficiency. Moreover, the results of Azimi et al., 2015 and Tong et al., 2016 [42,44] present lower thermal efficiency than that achieved in this study, which was less than 70% and 80%, respectively. Kumar et al., 2020 [21] investigated adding paraffin wax as a PCM inside a cylindrical container through a hot water storage tank. One case of their study was carried out under conditions similar to our study through June; the daily efficiency was less than 70%.

Some of the virtues of the current work are attracting promising investment opportunities, improving the infrastructure of local residential areas and rural regimes, and providing customizability in the energy sector for the benefit of the individual market with innovative products and dependable services.

4. Conclusions

A hydronic evacuated tube solar heating system is fabricated and installed to match the domestic requirements and uses throughout the day by using PCM latent heat. In the analysis of the hydronic solar system, the influence of the weather and operating conditions was considered. The thickness of the PCM shell in the bottom portion is greater than that of the top, minimizing the water temperature difference at the top and bottom to 5 °C.

The water and ambient temperatures through the system testing are presented and discussed in three cases: the first one was at normal conditions on clear days with water consumption, the second was without water consumption, and the final case was the effect of sudden and complete consumption of hot water while observing the behavior of PCMs in heated water. The results show increasing water temperature after a short period for domestic water temperature values.

The thermal efficiency of the hydronic solar system in December, March, and June was 80%, 81%, and 84%, respectively. This conclusion is higher than that of the experiments conducted by Kumar et al., 2020 [21]. Thermal efficiency depends linearly on $(\frac{\Delta T}{I})$. The hot water for domestic use is available throughout the day, which is achieved by using PCMs.

Author Contributions: Methodology, D.S.M. and R.I.R.; Software, M.S.M.; Formal analysis, D.S.M. and M.S.M.; Investigation, D.S.M. and R.I.R.; Data curation, D.S.M., M.S.M. and S.J.A.R.; Writing—original draft, D.S.M., R.I.R., M.S.M. and S.J.A.R.; Writing—review & editing, R.I.R. and S.J.A.R.; Supervision, R.I.R.; Funding acquisition, R.I.R. All authors have read and agreed to the published version of the manuscript.

Funding: This research was funded by the Jordan University of Science and Technology/Deanship of Research grant number [50/2018].

Institutional Review Board Statement: Not applicable.

Acknowledgments: The authors are immensely grateful to the Jordan University of Science and Technology/Deanship of Research for their financial support of this project.

Conflicts of Interest: The authors declare no conflict of interest.

Nomenclature

T	Temperature: K
Ac	Solar collector area, m^2
Cp	Specific heat, kJ/kg·K
I	Global solar irradiance, w/m^2
$k_{\theta i}$	Incident angle modifier
m	Mass flow rate, kg/s
Q_{PCM}	Phase change material energy, w

Q_{Load}	Load energy, w
Q_{loss}	energy losses, w
Q_{useful}	Useful energy, w
U_L	Over-all heat transfer coefficient of heat loss, $W/m^2 \cdot K$
r_{in}	Inner radius of the hot water tank, m
r_o	Outer radius of the hot water tank, m
$(\tau\alpha)_{eff}$	Effective transmissivity-absorptivity product coefficient
λ	Latent heat, kJ/kg
η	Thermal efficiency, %
Subscripts	
a	Ambient
w	water
tube	Evacuated tube solar collector
tank	Hot water tank
sys	system

References

1. Al Rabadi, S. Theoretical Investigations on The Optimal Topology of Crude Oil Desulfurization Process. In Proceedings of the 9th Jordan International Chemical Engineering Conference (JICHEC9), Online, 12–14 October 2021; pp. 225–238. Available online: https://jeaconf.org/UploadedFiles/AssetsManagement/JIChECIX/JIChECIX.pdf (accessed on 1 October 2022).
2. Koželj, R.; Osterman, E.; Leonforte, F.; Del Pero, C.; Miglioli, A.; Zavrl, E.; Stropnik, R.; Aste, N.; Stritih, U. Phase-change materials in hydronic heating and cooling systems: A literature review. *Materials* **2020**, *13*, 2971. [CrossRef] [PubMed]
3. Al Rabadi, S.; Al-Zboon, K.; Shawaqfah, M.; Damseh, R.; Al Zoubi, O. Biogas Production through Co-Digestion of Olive Mill with Municipal Sewage Sludge and Cow Manure. *Environ. Nat. Resour. J.* **2022**, *20*, 137–147. [CrossRef]
4. Chiradeja, P.; Pothisarn, C.; Jettanasen, C.; Yoomak, S.; Songsukthawan, P.; Bunjongjit, S.; Leelajindakrairerk, M.; Ngaopitakkul, A. Solar water heating in residential building. *Int. J. Smart Grid Clean Energy* **2019**, *8*, 422–429. [CrossRef]
5. Stropnik, R.; Stritih, U. Increasing the efficiency of PV panel with the use of PCM. *Renew. Energy* **2016**, *97*, 671–679. [CrossRef]
6. Browne, M.C.; Norton, B.; McCormack, S.J. Phase change materials for photovoltaic thermal management. *Renew. Sustain. Energy Rev.* **2015**, *47*, 762–782. [CrossRef]
7. Liu, C.; Xu, D.; Weng, J.; Zhou, S.; Li, W.; Wan, Y.; Jiang, S.; Zhou, D.; Wang, J.; Huang, Q. Phase change materials application in battery thermal management system: A review. *Materials* **2020**, *13*, 4622. [CrossRef]
8. Yang, L.; nan Huang, J.; Zhou, F. Thermophysical properties and applications of nano-enhanced PCMs: An update review. *Energy Convers. Manag.* **2020**, *214*, 112876. [CrossRef]
9. Faraj, K.; Khaled, M.; Faraj, J.; Hachem, F.; Castelain, C. A review on phase change materials for thermal energy storage in buildings: Heating and hybrid applications. *J. Energy Storage* **2021**, *33*, 101913. [CrossRef]
10. Rezaie, B.; Rosen, M.A. District heating and cooling: Review of technology and potential enhancements. *Appl. Energy* **2012**, *93*, 2–10. [CrossRef]
11. Li, Y.; Huang, G.; Wu, H.; Xu, T. Feasibility study of a PCM storage tank integrated heating system for outdoor swimming pools during the winter season. *Appl. Therm. Eng.* **2018**, *134*, 490–500. [CrossRef]
12. Venegas-Troncoso, T.; Ugarte-Larraguibel, G.; Vasco, D.A.; Rouault, F.; Pérez, R.A. Feasibility study of the application of a cooling energy storage system in a chiller plant of an office building located in Santiago, Chile. *Int. J. Refrig.* **2019**, *102*, 142–150. [CrossRef]
13. Mazarrón, F.R.; Porras-Prieto, C.J.; García, J.L.; Benavente, R.M. Feasibility of active solar water heating systems with evacuated tube collector at different operational water temperatures. *Energy Convers. Manag.* **2016**, *113*, 16–26. [CrossRef]
14. Liu, Z.; Li, H.; Liu, K.; Yu, H.; Cheng, K. Design of high-performance water-in-glass evacuated tube solar water heaters by a high-throughput screening based on machine learning: A combined modeling and experimental study. *Sol. Energy* **2017**, *142*, 61–67. [CrossRef]
15. Ge, R.Z.T.S.; Wang, Z.Y.; Xu, Q.W.; Pan, S.; Du, X.M.; Chen, T.; Ma, X.N.; Wu, X.L.; Sun, J.F.C. Solar heating and cooling: Present and future development. *Renew. Energy* **2018**, *126*, 1126–1140. [CrossRef]
16. Buker, M.S.; Riffat, S.B. Building integrated solar thermal collectors-A review. *Renew. Sustain. Energy Rev.* **2015**, *51*, 327–346. [CrossRef]
17. Sobhansarbandi, S.; Martinez, P.M.; Papadimitratos, A.; Zakhidov, A.; Hassanipour, F. Evacuated tube solar collector with multifunctional absorber layers. *Sol. Energy* **2017**, *146*, 342–350. [CrossRef]
18. Krishnavel, V.; Karthick, A.; Murugavel, K.K. Experimental analysis of concrete absorber solar water heating systems. *Energy Build.* **2014**, *84*, 501–505. [CrossRef]
19. Malkawi, D.S.; Tamimi, A.I. Comparison of phase change material for thermal analysis of a passive hydronic solar system. *J. Energy Storage* **2021**, *33*, 102069. [CrossRef]
20. Gürtürk, M.; Koca, A.; Öztop, H.F. Energy and exergy analysis of a heat storage tank with a novel eutectic phase change material layer of a solar heater system. *Int. J. Green Energy* **2017**, *14*, 1073–1080. [CrossRef]

21. Manoj Kumar, P.; Mylsamy, K.; Alagar, K.; Sudhakar, K. Investigations on an evacuated tube solar water heater using hybrid-nano based organic phase change material. *Int. J. Green Energy* **2020**, *17*, 872–883. [CrossRef]
22. Feliński, P.; Sekret, R. Effect of PCM application inside an evacuated tube collector on the thermal performance of a domestic hot water system. *Energy Build.* **2017**, *152*, 558–567. [CrossRef]
23. Papadimitratos, A.; Sobhansarbandi, S.; Pozdin, V.; Zakhidov, A.; Hassanipour, F. Evacuated tube solar collectors integrated with phase change materials. *Sol. Energy* **2016**, *129*, 10–19. [CrossRef]
24. Farid, M.M.; Khudhair, A.M.; Razack, S.A.K.; Al-Hallaj, S. A review on phase change energy storage: Materials and applications. *Energy Convers. Manag.* **2004**, *45*, 1597–1615. [CrossRef]
25. Cárdenas, B.; León, N. High temperature latent heat thermal energy storage: Phase change materials, design considerations and performance enhancement techniques. *Renew. Sustain. Energy Rev.* **2013**, *27*, 724–737. [CrossRef]
26. Pereira da Cunha, J.; Eames, P. Thermal energy storage for low and medium temperature applications using phase change materials—A review. *Appl. Energy* **2016**, *177*, 227–238. [CrossRef]
27. Rousse, D.R.; Ben Salah, N.; Lassue, S. An overview of phase change materials and their implication on power demand. In Proceedings of the 2009 IEEE Electrical Power & Energy Conference (EPEC), Montreal, QC, Canada, 22–23 October 2009; pp. 1–6. [CrossRef]
28. Sharma, A.; Tyagi, V.V.; Chen, C.R.; Buddhi, D. Review on thermal energy storage with phase change materials and applications. *Renew. Sustain. Energy Rev.* **2009**, *13*, 318–345. [CrossRef]
29. Manoj Kumar, P.; Anandkumar, R.; Mylsamy, K.; Prakash, K.B. Experimental investigation on thermal conductivity of nanoparticle dispersed paraffin (NDP). *Mater. Today Proc.* **2021**, *45*, 735–739. [CrossRef]
30. Abhat, A. Low temperature latent heat thermal energy storage: Heat storage materials. *Sol. Energy* **1983**, *30*, 313–332. [CrossRef]
31. Khare, S.; Dell'Amico, M.; Knight, C.; McGarry, S. Selection of materials for high temperature latent heat energy storage. *Sol. Energy Mater. Sol. Cells* **2012**, *107*, 20–27. [CrossRef]
32. Sharma, R.K.; Ganesan, P.; Tyagi, V.V. Long-term thermal and chemical reliability study of different organic phase change materials for thermal energy storage applications. *J. Therm. Anal. Calorim.* **2016**, *124*, 1357–1366. [CrossRef]
33. Tyagi, V.V.; Chopra, K.; Sharma, R.K.; Pandey, A.K.; Tyagi, S.K. Muhammad Shakeel Ahmad, Ahmet Sarı, R.K. A comprehensive review on phase change materials for heat storage applications: Development, characterization, thermal and chemical stability. *Sol. Energy Mater. Sol. Cells* **2022**, *234*, 111392. [CrossRef]
34. Murali, G.; Mayilsamy, K.; Arjunan, T.V. An Experimental Study of PCM-Incorporated Thermosyphon Solar Water Heating System. *Int. J. Green Energy* **2015**, *12*, 978–986. [CrossRef]
35. Rabady, R.I.; Malkawi, D.S. Thermal conductivity enhancement of sodium thiosulfate pentahydrate by adding carbon nano-tubes/graphite nano-particles. *J. Energy Storage* **2020**, *27*, 101166. [CrossRef]
36. Mourad, A.; Abderrahmane, A.; Said, Z.; Younis, O.; Iqbal, M.; Alazzam, A. Recent advances on the applications of phase change materials for solar collectors, practical limitations, and challenges: A critical review. *J. Energy Storage* **2022**, *49*, 104186. [CrossRef]
37. Jirawattanapanit, A.; Abderrahmane, A.; Mourad, A.; Guedri, K.; Younis, O.; Bouallegue, B.; Subkrajang, K.; Rajchakit, G.; Shah, N.A. A Numerical Investigation of a Melting Rate Enhancement inside a Thermal Energy Storage System of Finned Heat Pipe with Nano-Enhanced Phase Change Material. *Nanomaterials* **2022**, *12*, 2519. [CrossRef] [PubMed]
38. Ahmed, S.; Abderrahmane, A.; Saeed, A.; Guedri, K.; Mourad, A.; Younis, O.; Botmart, T.; Shah, N. Melting enhancement of PCM in a finned tube latent heat thermal energy storage. *Sci. Rep.* **2022**, *12*, 1–14. [CrossRef] [PubMed]
39. Lachheb, M.; Karkri, M.; Albouchi, F.; Ben Nasrallah, S.; Fois, M.; Sobolciak, P. Thermal properties measurement and heat storage analysis of paraffin/graphite composite phase change material. *Compos. Part B Eng.* **2014**, *66*, 518–525. [CrossRef]
40. Duffie, J.A.; Beckman, W.A. *Solar Engineering of Thermal Processes*, 4th ed.; John Wiley and Sons: Hoboken, NJ, USA, 2013; ISBN 9780470873663.
41. Holman, V. *Heat Transfer*, 10th ed.; McGraw-Hill: New York, NY, USA, 2010.
42. Azimi, M.; Mirjavadi, S.S.; Mohammadkarim, A. Simulation and Optimization of Vacuum Tube Solar Collector Water Heating System in Iran. *J. Sci. Eng.* **2015**, *7*, 001–019.
43. Siuta-Olcha, A.; Cholewa, T.; Dopieralska-Howoruszko, K. Experimental studies of thermal performance of an evacuated tube heat pipe solar collector in Polish climatic conditions. *Environ. Sci. Pollut. Res.* **2021**, *28*, 14319–14328. [CrossRef]
44. Tong, Y.; Kim, H.M.; Cho, H.H. Theoretical investigation of the thermal performance of evacuated heat pipe solar collector with optimum tilt angle under various operating conditions. *J. Mech. Sci. Technol.* **2016**, *30*, 903–913. [CrossRef]

batteries

MDPI

Review

Optimal Planning of Battery Energy Storage Systems by Considering Battery Degradation due to Ambient Temperature: A Review, Challenges, and New Perspective

Chico Hermanu Brillianto Apribowo [1,2], Sarjiya Sarjiya [1,3,*], Sasongko Pramono Hadi [1] and Fransisco Danang Wijaya [1]

1 Department of Electrical and Information Engineering, Universitas Gadjah Mada, Yogyakarta 55281, Indonesia
2 Department of Electrical Engineering, Universitas Sebelas Maret, Surakarta 57126, Indonesia
3 Center for Energy Studies, Universitas Gadjah Mada, Yogyakarta 55281, Indonesia
* Correspondence: sarjiya@ugm.ac.id

Abstract: In recent years, the goal of lowering emissions to minimize the harmful impacts of climate change has emerged as a consensus objective among members of the international community through the increase in renewable energy sources (RES), as a step toward net-zero emissions. The drawbacks of these energy sources are unpredictability and dependence on nature, leading to unstable load power supply risk. One way to overcome instability in the power supply is by using a battery energy storage system (BESS). Therefore, this study provides a detailed and critical review of sizing and siting optimization of BESS, their application challenges, and a new perspective on the consequence of degradation from the ambient temperature. It also reviews advanced battery optimization planning that considers battery degradation, technologies, degradation, objective function, and design constraints. Furthermore, it examines the challenges encountered in developing the BESS optimization model and evaluates the scope of the proposed future direction to improve the optimized BESS, especially its battery.

Keywords: battery energy storage system; sizing; optimal planning; battery degradation; ambient temperature; renewable energy sources

Citation: Apribowo, C.H.B.; Sarjiya, S.; Hadi, S.P.; Wijaya, F.D. Optimal Planning of Battery Energy Storage Systems by Considering Battery Degradation due to Ambient Temperature: A Review, Challenges, and New Perspective. *Batteries* **2022**, *8*, 290. https://doi.org/10.3390/batteries8120290

Academic Editors: Luis Hernández-Callejo, Jesús Armando Aguilar Jiménez and Carlos Meza Benavides

Received: 26 October 2022
Revised: 22 November 2022
Accepted: 10 December 2022
Published: 16 December 2022

1. Introduction

Lately, there has been a growing consensus among people worldwide regarding the importance of reducing emissions to mitigate the adverse effects of climate change. Several nations and companies globally are beginning to commit to net-zero emissions. Despite its vulnerability to climate change, it is also realized by Indonesia, which is Although vulnerable to climate change, this is also realized by Indonesia, which is an archipelago country country [1]. The utilization of alternative or renewable energy sources (RES) is one of the most effective ways to reduce emissions generated from fossil fuels. Solar photovoltaic (PV) is the most extensively utilized RES owing to its installation simplicity, low cost, and scalability [2]. However, problems arise because the RES generation is unpredictable and highly dependent on nature, resulting in an unstable power supply to the load [3]. Due to its high penetration, the uncertainty of PV plants expose the power grid to many challenges, such as voltage, frequency fluctuations, reverse power flow, and harmonics [4]. The successful integration of RES into the planning and operating model of an electric power system on a grid-scale increases the flexibility of the battery [5].

The battery energy storage system (BESS) helps ease the unpredictability of electrical power output in RES facilities which is mainly dependent on climatic conditions. The integration of BESS in RES power plants boost PV penetration rates [6], thereby improving the efficiency and reliability of the generating system [7]. Furthermore, BESS plays an essential role in distribution networks, where it is used to assist auxiliary services, load shifting and leveling, backup power, peak shaving, demand response, renewable energy integration, frequency control, voltage management, long-term, and seasonal storages [8–10]. Therefore, its optimization is essential.

BESS capacity and its ideal location are both determined by its optimization indicator. The performance of the electric power system is also significantly improved by its optimization in terms of establishing the appropriate capacity and rating. Meanwhile, inadequate capacities and ratings tend to result in greater power losses and increased costs for both the investment and operation of the power system [11]. BESS capacity needs to be optimized to ensure continuous electric power alongside robust and economical operation [12]. Its optimal placement is also extremely relevant on grid-scale networks. This is because it affects BESS costs and services by delaying investment from peak loads, improving the response to changes in electrical energy generation and demand, reducing transmission and distribution losses, as well as restrictions on RES generation [13]. One of the most significant decisions to make is planning to optimize the performance of the RES system to achieve profitable investments. The optimization of BESS capacity and placement is a significant problem due to the need for ideal energy exchange equilibrium [14] and the total cost of installation [15].

BESS technology includes the use of lithium-ion (Li-Ion), lead-acid (LA), sodium-sulfur (NaS), zinc-bromine (ZBB), nickel-cadmium (Ni-Cd), vanadium-redox (VRB), and polysulfide bromine batteries (PSB) [16,17]. These are typically used for load leveling, power quality, grid extension and support, demand management, and voltage regulation. One of the major advantages of LA is that it has relatively low investment opportunities, and expensive to operate with limited energy density. Although the Li-Ion batteries have high energy and power densities with long-lasting life cycle and excellent efficiency, it is an expensive investment [18]. This battery type is also manufactured as packs, organized in series or parallel to realize the necessary current, voltage, and power. Throughout the development of this battery, large-scale battery packs were built as power walls [19].

Li-Ion batteries' performance deteriorated over time and is referred to as calendar and cycle life [20]. This is due to two causes, first is the loss of Li-Ion triggered by the formation of a solid electrolyte contact (SEI). Second is the loss of electrode sites [21], which increases internal resistance, lowers capacitance and efficiency, and diminishes battery life [22,23]. Consequently, battery deterioration always impacts the optimal operation and longevity of Li-Ion battery energy storage, particularly the percentage of power systems [24]. It also predicts battery life, maximum charge or discharge cycles, or Ah-overall. The data is then used for cost or benefit analysis [25].

The degradation costs for a charge or discharge cycles need to be considered when analyzing real-time energy management challenges. In this case, the energy management running expenditures tend to grow because of battery life and actual unrepresented electricity prices [26]. According to Cardoso et al. [27] the overall annual power cost reductions from PV and storage systems can be reduced by 5–12% if the battery deterioration limits are considered. Ren et al. [28] stated that it significantly reduces the system's electrical performance and increases unanticipated maintenance expenditures. Battery failure is usually due to deterioration caused by increased rate of usage, and this can limit its lifespan and potentially lead to significant accidents. Likewise, battery degradation significantly reduces the system's electrical performance and increases unanticipated maintenance expenditures. Severson et al. [29] stated that the prediction of battery life facilitates new production, use, and optimization opportunities. If one can accurately anticipate the lifespan of a battery, then they can create new uses as well as optimize its performance. This leads to innovative opportunities for the manufacturing process and optimization.

The present study examines the optimization plan for the BESS system problem by considering battery degradation due to ambient temperature. It serves as a reference for investigating areas of electrification using renewable energy sources. This engineering topic covers BESS planning in relation to deterioration from a practical standpoint. However, this static problem involves battery capacity and location to attain the desired goals. These tend to be influenced by technological and economic concerns, as well as other factors such as reliability. As a result, BESS planners encounter certain challenges in gathering and inputting data, dealing with design constraints, and implementing effective energy management.

The following are the key contributions of this research:

- Explain the state-of-the-art expansion planning with BESS optimization.
- Explain how battery degradation due to ambient temperature can affect BESS.
- To study different technologies, objectives, and constraints of BESS.
- Review the challenges and future scopes encountered in developing BESS optimization.

The present research is arranged as follows. Section 2 outlines the methods used to review the literature. Section 3 investigates BESS with respect to expansion planning. Sections 4 and 5 reviewed its application and battery technology, respectively. Section 6 focuses on the study of battery degradation. Meanwhile, Section 7 reviews the objective function, design constraint, and algorithm of BESS optimization. Section 8 discusses the issues and challenges of BESS, while Section 9 concludes the research and provides areas for future works.

2. Methodology

The systematic literature review (SLR) was summarized using the preferred reporting items for systematic reviews and meta-analyses (PRISMA) approach. Data were selected from the Scopus, Science Direct, IEEE Xplore and Web of Science databases in three stages, namely identification, screening, and reporting. Figure 1, shows the identification stage, which is carried out by searching for related articles in each database, as illustrated in Table 1. The strategy adopted at the time of initial screening on the database is in accordance with the provision of the title, abstract, and keyword. This led to the realization of 1584 articles, of which 824, 352, 187, and 221 were from Scopus, Science Direct, IEEE Xplore, and Web of Science concerning the optimization of BESS and battery degradation, respectively.

Table 1. Search term selection.

Search Term	Descriptor
Database	Scopus, Science Direct, IEEE Xplore, and Web of Science
Keyword Fields	Battery Energy Storage System; Sizing; Battery Degradation; Battery Aging
Year Publication	2018–2022
Document Type	Article

After checking and removing duplicate reports and records marked as ineligible by automation tools, 139 papers were obtained for screening. The papers were selected in accordance with exclusion and inclusion criteria based on Table 2. Incidentally, 42 records were excluded, 12 were not retrieved, and 15 reports were omitted due to inclusion and exclusion criteria at the screening stage. Finally, the total number of comprehensive SLR articles to be reviewed are 69.

Table 2. Criteria for the systematic literature review.

Criteria	Description
Inclusion	A journal that has the highest relevance with BESS and battery degradation due to ambient temperature Has an impact factor Q1 Paper publication 2018 to 2022
Exclusion	Studies that have information relatable to support state-of-the-art BESS or battery degradation Paper publication 2018 to 2022

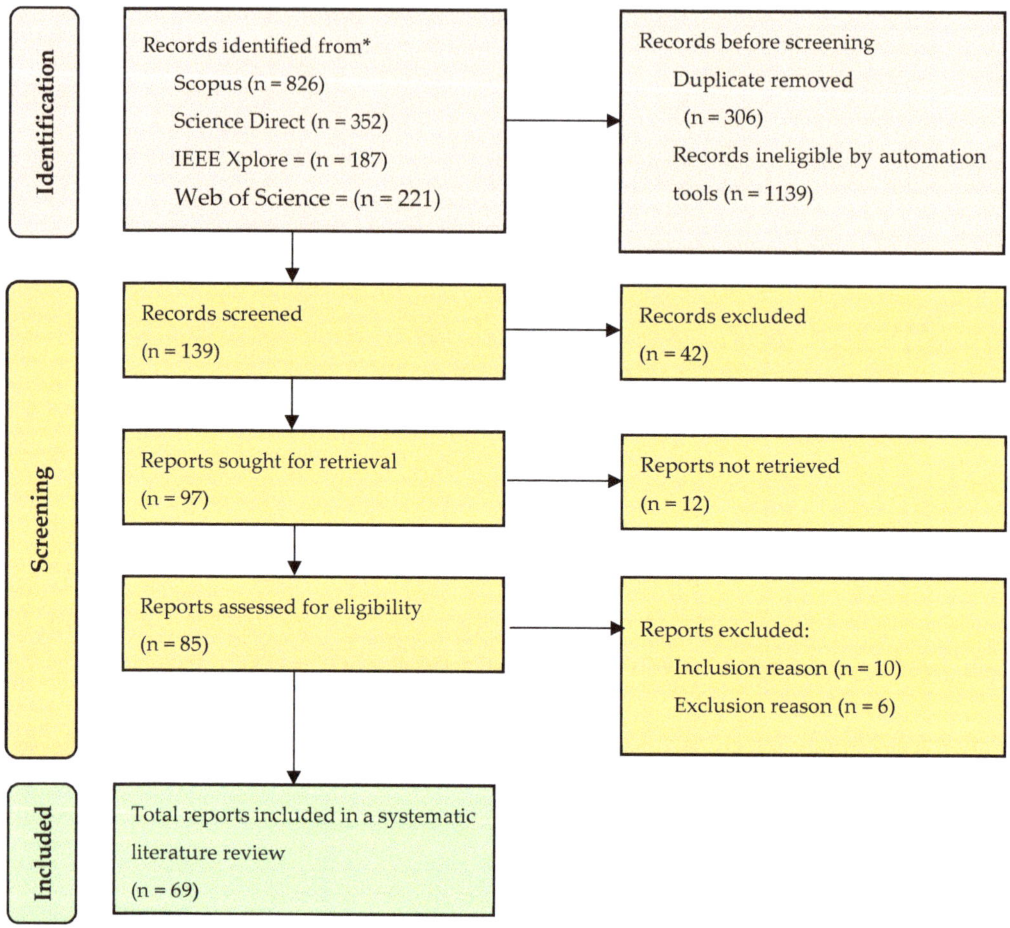

Figure 1. Block diagram selection based on PRISMA flow diagram approach [30].

As a result, this SLR was carried out to respond to the following research objectives and questions.

1. How does the development of BESS optimization affect expansion planning and the impact of the BSS applications on the grid or microgrid?
2. How does the battery technologies use affect BESS? And what can affect battery degradation?
3. How does battery degradation due to ambient temperature affect BESS optimization?
4. What are the main parameters and variables in BESS optimization planning?

The number of publications on this topic has increased over the past five years, as shown in Figure 2. For example, from 2018 to 2021 there were 53 articles, with 16 new publications in October 2022.

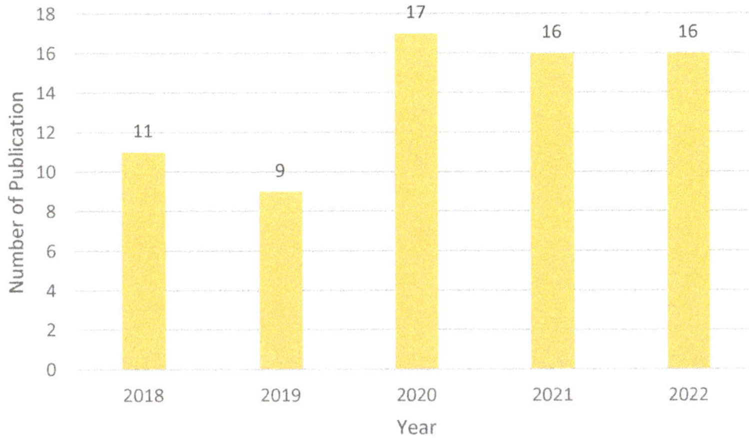

Figure 2. Number of BESS-related publications eligible for review in the last five years.

Meanwhile, 69 comprehensive articles have been selected for review. The acquired data has a Q1 journaling tool from the Scimago Journal Rank (SJR). Table 3 shows the list of publications or journals selected for review.

Table 3. Distribution of articles in each journal.

Journal Name	Scimago Journal Rank	Impact Score	Number of Articles
IEEE Transactions on Smart Grid	5.25	11.95	1
IEEE Transactions on Power Systems	4.64	8.42	2
IEEE Transactions on Industrial Informatics	4.33	12.03	2
IEEE Transactions on Sustainable Energy	4.16	9	3
Applied Energy	3.06	11.3	9
IEEE Transactions on Transportation Electrification	2.17	7.64	1
IEEE Transactions on Energy Conversion	2.09	5.79	1
Energy	2.04	8.51	4
IEEE Transactions on Industry Applications	1.98	5.21	2
Journal of Power Sources	1.98	9.07	2
Journal of Cleaner Production	1.92	10.96	3
Renewable Energy	1.88	8.65	3
IEEE Transactions on Green Communications and Networking	1.87	3.88	1
Energy and Buildings	1.68	7.13	1
International Journal of Electrical Power and Energy Systems	1.54	6.06	1
Journal of Energy Storage	1.35	8.78	5
Electric Power Systems Research	1.11	4.39	3
MRS Energy and Sustainability	1.03	2.2	1
IEEE Access	0.93	4.3	4
Batteries	0.87	5.77	4
PLoS ONE	0.85	3.58	1
International Journal of Energy Research	0.81	5.81	1
Sustainability (Switzerland)	0.66	4.17	1
Energies	0.65	3.54	11
Automotive Innovation	0.4	1.99	1
International Journal of Renewable Energy Research	0.3	1.61	1

Brief Review

Until now, the trends of BESS have been widely studied in several aspects. As explained in Table 4, a BESS is often applied to solve microgrid, grid-scale, and hybrid renewable energy system (HRES) problems. However, to obtain economical results, its sizing and siting was optimally analyzed with a significant dependence on the problem to be solved. BESS is usually used to solve problems related to system flexibility, such as demand load shifting, loss of load, avoidance of RES curtailment, and RES peak shaving. As its research progresses, it becomes increasingly important to consider the impact on battery health, as well as the choice of battery technology used, which can affect the system and its economic value. Battery health needs to be considered to ensure it does not experience degradation, when the BESS needs to be replaced. In general, the battery degradation factors considered during the optimization process are SOC, DOD, cycle number, and battery lifetime. Furthermore, studies have also been developed on the use of recycled batteries from electric vehicles with BESS integrated into the microgrid system. Research on the effect of temperature on the optimization of BESS was also considered recently. The temperature factor that affects BESS consists of operating temperature and ambient temperature. However, little research has been carried out on the effect of BESS environmental temperature optimization. Yuhan Wu et al. [31] conducted research on optimizing BESS considering the ambient temperature. However, in this research the temperature variable was not explained in sufficient detail.

Table 4. Review of a recently published article on BESS optimization.

Ref	Research Topics	Research Gaps
Cardoso et al. (2018) [27]	BESS optimization was discussed while taking battery degradation and micro sizing problems into account	Investigate the operating temperature of the BESS because it has a significant impact on battery health
Alsaidan et al. (2018) [32]	Using BESS to find a solution to the specific problem of microgrid expansion. Considering the characteristics of various technologies, a distributed deployment, considering the impact of in-depth discharge, and the number of charging and discharging cycles	The challenges in BESS optimal sizing are brought on by the need to use the it for multiple applications and the use of linear power flow model to calculate the angle and voltage magnitude at each bus as well as the active and reactive power flow
Talal Alharbi, et al. (2019) [33]	Framework for the planning and operation of the BESS is based on recycled batteries from electric vehicles	The problem of optimizing BESS requires reducing the computation complexity and incorporating more dynamic decision variables, both of which can benefit from the application of decomposition methods
A. Pena-Bello et al. (2019) [34]	Develop an optimization framework to determine the most suitable battery PV self-consumption. The avoidance of PV curtailment, demand peak shaving, demand load-shifting, and technology depending on the size	The proposed challenge is to extend the optimization framework to more regions, while considering transport demand and trade-offs as well as incorporating heat and electric vehicles
Timur Sayfutdinov et al. (2020) [35]	The most optimal placement, sizing, and technology choice for BESS was discussed, by considering the degradation obtained from the state of charge and the depth of discharge	Although the constraint of BESS degradation taking SOC and DOD into consideration has been provided, the temperature value was still fixed when the model was developed
G. Mohy-Ud-Din et al. (2020) [36]	The energy management strategy that has been described is used to optimize the functioning industrial microgrids, with the BESS scalability serving as a limiting factor due to the presence of uncertainties	The challenges of integrating many decentralized energy sources into a microgrid controller in a way that allows it to be used in an economic dispatch
Yunfang Zhang et al. (2021) [37]	An optimal sizing model was presented for grid-scale BESS, taking into consideration its operation under uncertainties induced by volatile wind generation. The cycle life model of batteries was evaluated, and marginal economic utility analysis performed	Studies on BESS allocation planning needs to consider the decision regarding installation location
Mohammad Amini et al. (2021) [38]	A description of the optimal BESS size, technology, depth of discharge, and replacement year was provided, reckoning the system's technical characteristics, service life, and capacity degradation. This was conducted to reduce the total cost of MG scheduling while simultaneously improving the BESS's precision and economic feasibility	The temperature factor has not been taken into consideration in the BESS degradation model

Table 4. *Cont.*

Ref	Research Topics	Research Gaps
Rehman et al. (2022) [39]	Presented optimal sizing for a BESS and PV system in an extremely fast charging station (XFCS) to reduce the annualized total cost. This was carried out with consideration given to evaluating optimal energy management for the station as well as energy arbitrage	This research proposed a model of battery degradation; however, the lifetime project only used one year and did not consider replacement batteries
Yuhan Wu et al. (2022) [31]	Examined the algorithm for optimal capacity allocation of BESS in contemporary distribution networks, while considering the ambient temperature	A model of battery degradation, which concerns the ambient temperature has been developed. However, the variable of temperature has not been described in sufficient detail

This review provides a discussion about the expansion planning with BESS optimization by considering battery degradation due to ambient temperature to fill in the research gaps. Figure 3 shows the mind map of BESS relating to the application, batteries energy storage technologies, battery degradation, objective function, design constraints, optimization algorithms, and challenges used in this review.

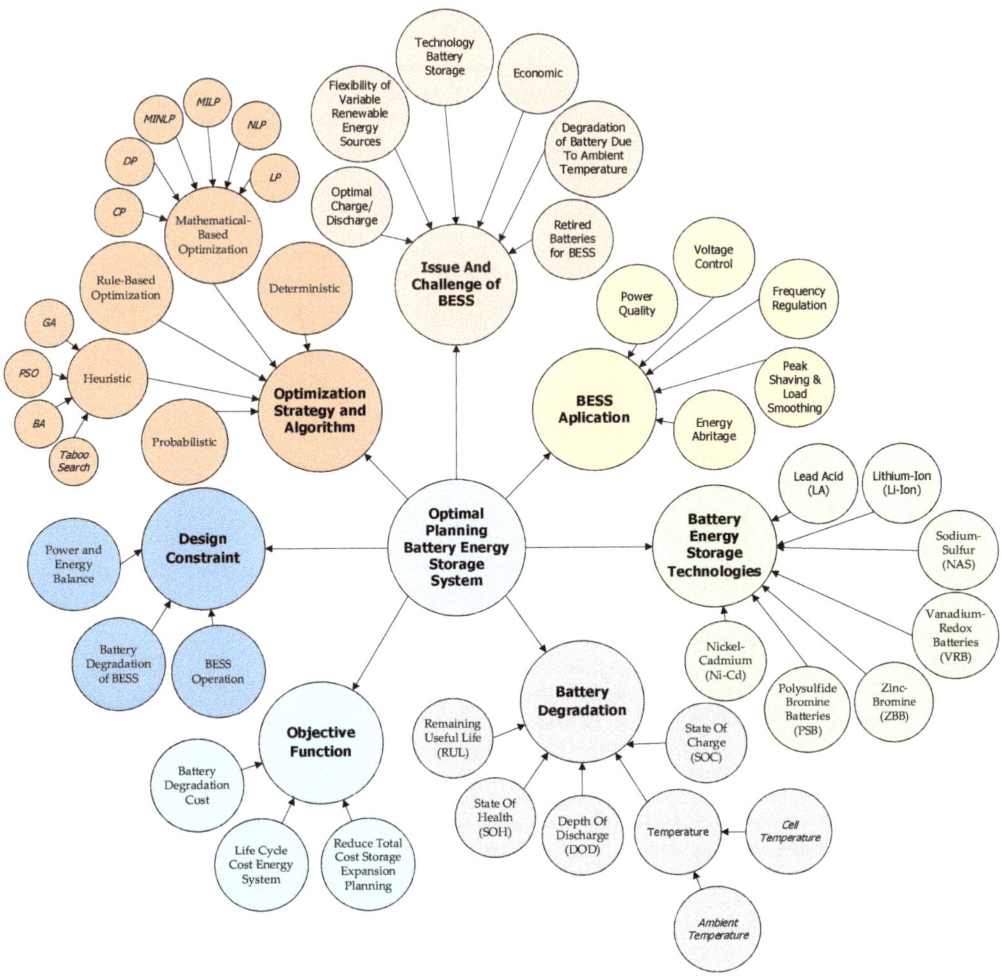

Figure 3. Mind map of BESS optimization.

3. Expansion Planning Overview

A combination of BESS technology and expansion planning is frequently adopted to overcome the issues of VRE integration. For example, generation expansion planning (GEP) tries to meet energy demands alongside several economic and technological restrictions. It determines the generating capacity of an ideal investment plan during a specific study period. Governments and decision-makers routinely utilize GEPs to select when and where to invest in generating technologies. Based on the decision factors, energy expansion approaches are broadly classified as GEP and transmission expansion planning (TEP). However, storage expansion planning (SEP) is widely used when dealing with BESS investment choices. In reality, creating, transmitting, and storing processes tend to be synchronized [5].

The main challenge of GEP is determining the appropriate capacity size, generating unit, and timing of a new facility's building to fulfill the electric power requirement, at least during the planning period. GEP models are made more versatile by considering numerous goal functions and constraints as shown in Table 5. Its goal function typically consists of two major components, namely, investment and operation. To establish an optimal GEP strategy, different restrictions that impact the execution of the plan must be considered. There are two types of constraints, namely required and discretionary. One of the relevant limitations is ensuring the balance of electricity demand. Therefore, there is a possibility that minimizing total expenditures for a GEP project is not an effective target function, especially if there are other fascinating aspects that compete for attention. Consequently, issues related to GEP are frequently posed as a multi-objective optimization process. This approach can handle the simultaneous compromising of multiple goal-planning functions to determine which alternative capacity is the most effective. Several of these goals are intertwined, such as incorporating DSM and RES in the generating mix, reducing pollution, reliability, fuel consumption, costs associated with the intermittent nature of RES, and the risk of fluctuations in energy expenditure. All these are carried out to improve the flexibility of the GEP model [40–42].

Table 5. Generic objective function, constraint, and uncertatnties in GEP [40–42].

Categories in GEP Problem	Objectives	Constraint	Uncertainties
Social-Economic	Emission Cost Energy Cost Emission Cost Fuel cost Electric Vehicle Cost Storage Cost Electricity Price Renewable Cost Social acceptance	Peak Demand Spin Reserve Emission Level Generator Capacity Renewable Penetration Level	Electrical price variability Public Health Social Acceptance Behavior Shift Demand Growth Rates Interest Rates Fuel Cost fluctuation Carbon Prices
Policy	Target Energy Target Renewable Penetration Target Environmental Regulation Target Access to Energy Resources	Governmental Policy Industrial Policy Carbon Market Environmental Regulation Renewable Supporting Schemes	National Energy Policies International climate agreements Taxation regime Energy Security International Climate Agreements
Technical	Increasing Energy Penetration With Other Energy Sectors Ancillary Services Target Ageing Infrastructure	Renewable Curtailment Flexibility and Reliability Grid curtailment Forced outages Reliability Margin Energy Balance Network Constraint	Ramping Capability Learning Rate Evolution For Energy Supply Technologies Flexibility and Reliability Needs

Table 5. *Cont.*

Categories in GEP Problem	Objectives	Constraint	Uncertainties
Climate/Environmental			
	Target Renewable Energy Generation	Renewables Availability	Renewables Variability
	Target Life-Cycle Infrastructure	Climate Change	Natural Disaster
	Use of Fossil Fuels	Life-Cycle Assessment	Extreme Climatic Events
		Resource Allocation	
		Retirement or Lifespan	
		Peak Energy Generation	

SEP can be categorized by its storage capacity, geographical distribution, and mobility, in addition to the kind and quantity of BESS. Furthermore, energy storage systems are classified as either short or long-term, depending on their capacity. Short-term appliances, such as capacitors, flywheels, compressed air energy techniques, and BESS, stores energy from seconds to days. Certain long-term appliances, such as hydrogen storage and water reservoirs, can supply energy from one week to an entire season. BESS can also be classified as centralized or dispersed. When categorized by centralized, it refers to a single place. Even though most BESS are either centralized or dispersed, BESS can categorized by mobility such as on electric vehicles (EVs) [5].

The primary goal of decoupling is to ensure that cost-cutting initiatives are carried out by central planners (vertically integrated electrical firms) or politicians, as opposed to private investors. In the SEP model, reliability indices account for expected energy not served (EENS) or loss of load probability or expectation (LOLP/LOLE). There is also a possibility of adding any necessary technical constraints for unit commitment (UC) that are essential for scheduling the operation of the producing sector. These include minimal timeframes between turning on and off, beginning and shutting down, ramping up and down, as well as the least power outputs. There is a possibility that further operational reserve limits, such as the spinning types, alongside frequency and voltage support replacements, are imposed on the way the system operates [5].

4. BESS Application Overview

BESS delivers various services to network operators, DG plants, energy retailers, and consumers. Figure 4 categorizes its applications in in the grid based time scale. Additionally, BESS consumption is classified in accordance with the time scale of its deployment, which ranges from milliseconds to hours. Its applications in grids or microgrids tend to improve power quality, voltage management, peak shaving, load smoothing, frequency control, and energy arbitrage [43].

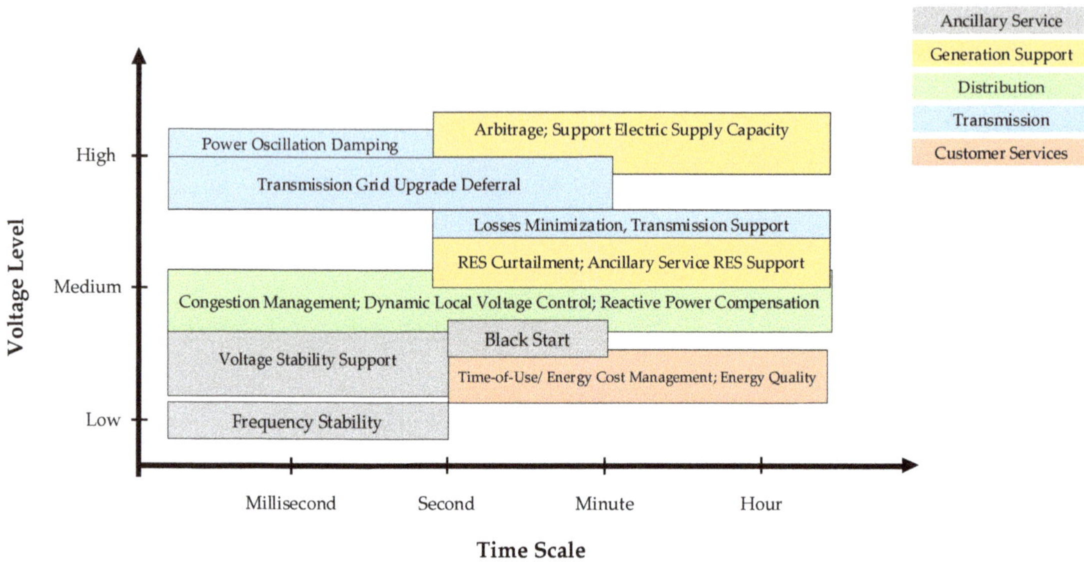

Figure 4. Application of BESS based on time scale [43,44].

4.1. *Power Quality*

The power quality index is used to measure voltage and current waveform distortions in pure sinusoidal ideals [43]. Variations in solar irradiance and wind speed trigger the negative effect of high-variance DG plants. Consequently, the BESS added to the DG plant has the potential to smoothen temporary power fluctuations. In this situation, it is viewed as an extra cost component with respect to the RES plant that serves as a revenue system. The provision of economic incentives to plant owners to reduce power fluctuations is a technique used to compensate for revenue losses [45,46].

4.2. *Voltage Control*

Capacitor banks, tap changers, voltage regulators, and static VAR compensators are equipment used to manage voltage during grid distribution. This is because DG injection makes the regulation of equipment at the substations useless, such as transformer tap changers, with many units scattered around the network selectively creating reactive power to allow for simpler voltage management. For example, a PV generator produces overvoltage at the network's end [43]. Therefore, implementing BESS in such cases has been proven to be effective and potentially reduce overvoltage [47,48].

4.3. *Peak Shaving and Load Smoothing*

Both peak shaving and load smoothing aim to reduce the maximum amount of power visible to the system by striking a balance between the generation profile and demand. This approach produces real-time network congestion solutions by minimizing conductor overloads caused by the generation of peak power loads. Furthermore, peak shaving and load smoothing help to reduce network losses. BESS operations also reduce system losses by increasing load-to-local-generation profile matching [43,49].

4.4. *Frequency Regulation*

In an auxiliary service market, frequency regulation is typically provided by generators connected to a transmission network. Interestingly, it is described as a commercial offering. However, in recent years, generators and energy storage devices connected to the distribution network also provided this service. This is possible because the distribution network has become more decentralized. Additionally, the increasing demand for renewable energy brought about the modification of this policy. Both the generator and BESS use drop control to monitor the frequency and adjust the power output appropriately. In this scenario, BESS allows restrictions to be specified by the state charges (SOC) [50,51].

4.5. Energy Arbitrage

Energy arbitrage is the process of simultaneously purchasing and offering energy supplies in the marketplace. It was only initiated by commercial users because the power sectors of most countries do not have any form of regulation. The application of BESS pairs with DG or load, in which storage units are utilized to redirect energy production or generation, is aimed at maximizing profit irrespective of the fluctuations in market prices [43,52]

5. Battery Energy Storage Technologies

LA, Li-Ion, NaS, and RF are grid applications' most common battery technologies. These are classified according to their energy density, efficiency, lifespan, and cost when coupled to a storage network, as shown in Tables 6–8. The LA battery has high efficiency between 80 and 90% and low costs within the range of 50 to 600 $/kWh [52,53] However, when compared to other technologies, it has a significant disadvantage in terms of lifespan (approximately 2500 cycles) [54] and low energy density (within the range of 20 and 30 Wh/kg). A high discharge depth shortens an LA battery's life [52,55].

The characteristics of Li-Ion batteries are based on the chemical composition of both the cathode and anode, which typically consists of graphite and lithium metal oxide. Interestingly, the cathode and anode give the battery its name and power, respectively. This technique is highly efficient, with a maximum efficiency of approximately 90%. On the other hand, some commercial devices boast reported round trip efficiencies of more than 95% with energy density within the range of 90 to 190 Wh/kg [56] and extended service life of relatively 10,000 cycles [54]. Cell temperature, an essential element in the deterioration process, significantly affects the battery life [30]. Li-Ion batteries are commonly found in electronic devices and recently emerged as the industry standard for EV. This technology is suitable for grid-connected network applications, even though it is still somewhat expensive. Presently, there are several Li-Ion technologies, for example, lithium manganese oxide ($LiMn_2O_4$), lithium cobalt oxide ($LiCoO_2$), lithium nickel cobalt aluminum oxide ($LiNiCoAlO_2$), lithium iron phosphate ($LiFePO_4$), and cobalt-based Lithium nickel manganese oxide ($LiNiMnCoO_2$) [57]. Tables 7 and 8 show details of the Li-Ion and nickel-based battery specifications, respectively.

NaS batteries have a high working temperature (approximately 300 °C), efficiency (>80%), energy density within the range of 150 to 240 Wh/kg, and a long lifespan of relatively 4500 cycles [58,59] As a result, this technique has been utilized to lessen the effect of renewable energy-based generators as an in-grid [58,60]. Vanadium redox flow batteries (VRB) batteries comprise two containers, one containing two chemical reagents and the other two electrodes partitioned by a membrane. Incidentally, when the two components combine, it results in an oxidation reaction. One of the containers holds the chemical reagents, while the other contains the electrodes. The amount of stored chemicals contributes to the flow cell's total energy capacity. Meanwhile, the electrodes and membrane filtering system are responsible for individual energy capacity flow cell. The power and energy ratings are separated, resulting in the increased design and operational flexibility. The energy density of VRB is relatively low, ranging from 15 to 30 Wh/kg, and its efficiency is approximately 75% in some cases [61]. On the other hand, they are not constrained by reactant life cycles or discharge depth [62]. Due to the low costs involved in their maintenance and operation, VRB have been suggested as viable options for large-scale grid-based energy storage [63]. The reactants have been investigated, and several chemical compositions have been proposed. The most utilized ones are vanadium and Zn-Br [64].

Table 6. Review of technology BESS [65–69].

Technology	Efficiency (%)	Life Cycle (DOD 80%)	Battery Energy Density (Wh/L)	Battery Power Density (W/L)	Application Battery	Benefits	Disadvantage
Lead Acid (LA)	75–85	300–3000	50–90	10–400	Diesel electric-powered submarines, electric motors	Cheap	Low energy density, limited cycling ability
Lithium Ion (Li-Ion)	90–99	3000–10,000	200–500	1500–10,000	Laptops, mobile phones, EV	Fast response time, high efficiency, and energy density	Some security issues depend on the type
Sodium Sulfur (NaS)	75–90	4500	150–300	140–180	Load residential, support ups	High efficiency and life cycle	High maintenance and operating temperatures
Nickel Batteries	15–400	500–3000	10–150	50–1200	Mobile phones, emergency lighting	High reliability and energy density, long cycle life,	Environmental hazards, influenced by the memory effect
Zinc Bromine (ZnBr)	2000	30–65	<25	65–80	Diesel electric-powered	Long lifetime, high energy density, and deep discharge capacity,	Dendrite formation, corrosivity, require working temperature, and low cycle efficiency
Polysulfide Bromine (PSB)	-	20–30	<2	60–75	Electrical vehicle, support ups	Fast reaction speed	No large-scale application experience, and environmental issues,
Vanadium Redox Flow (VRB)	65–85	2000–20,000	40	-	Electrical vehicle, support ups	Stability for large scale	Difficult maintenance, complex battery

Table 7. Specification of technology lithium-ion batteries [70,71].

Technology	Efficiency (%)	Life Cycle (DOD 80%)	Battery Energy Density (Wh/L)	Battery Power Density (W/L)
Lithium Iron Phosphate ($LiFePO_4$)	92	>2000	90–120	1932
Lithium Cobalt Oxide ($LiCoO_2$)	95.7–98.4	500–1000	150–200	2710
Lithium Nickel Manganese Cobalt Oxide ($Li(Ni_xMn_yCo_{1-x-y})O_2$)	90	1000–2000	150–220	-
Lithium Nickel Cobalt Aluminum Oxide ($Li(NixCoyAl_{1-x-y})O_2$)	-	500	200–260	-
Lithium Manganese Oxide ($LiMn_2O_4$)	-	300–700	100–150	-
Lithium Titanate ($Li_4Ti_5O_{12}$)	98	3000–7000	50–80	-

Table 8. Specification of technology nickel batteries [69].

Technology	Efficiency (%)	Life Cycle (DOD 80%)	Battery Energy Density (Wh/L)	Battery Power Density (W/L)
Ni-Cd	70–90	2000–2500	15–150	75–700
Ni-MH	90	700–1000	38.9–350	7.8–588
Ni-Zn	<87	>5000	80–400	121.38
Ni-Fe	<65	-	25–80	12.68–35.18

6. Battery Degradation

Battery degradation leads to a reduction in its capacity and efficiency and even safety problems. The term cycle life refers to the total number of times a battery can be discharged or charged before it is replaced [72]. Nonlinearity in battery degradation can be traced to a variety of causes, such as SOC, high temperature, depth of discharge (DOD), and charge or discharge current rate [73], as shown in Figure 5. One of the issues contributing to the short lifespan of Li-Ion batteries, for example, is the highly utilized DOD, which tends to significantly reduce the total number of cycles [74,75].

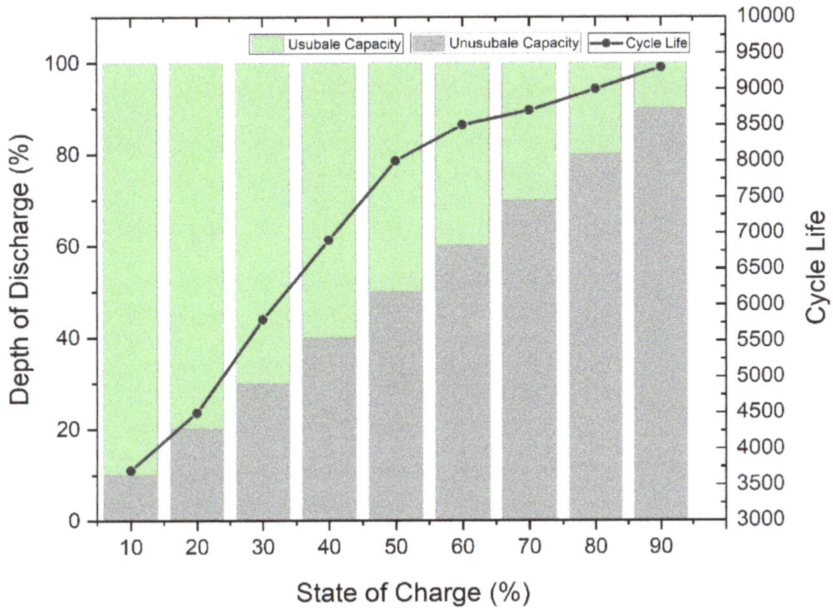

Figure 5. Relationship between battery capacity and SOC, DOD, and cycle life Li-Ion battery [38].

The remaining useful life (RUL) and state of health (SOH) are the most critical factors in predicting Li-Ion battery degeneration. Generally, usage capacity, energy, and accessible power, which diminish with battery age, influence SOH and RUL [76]. Although SOH tests detect a decrease in performance, they also prevent potential accidents [77]. The accuracy with which one may anticipate the RUL of a given battery capacity relies on several factors, and the most important is the ability to calculate the SOH. Managing discharge problems, improved performance, and optimized operation requires precise and reliable prediction algorithms to determine a battery SOH and RUL.

SOH refers to the percentage of a battery cell's capacity that is still usable and used to quantify the entire aging degree. This value is expressed as a percentage [78] and ideally, the SOH of the new battery should be 100%. The decreasing trend of SOH is due to the accelerated aging of the battery, which is one of the reasons of the increased cycle times. When the state of health reaches the failure threshold, the battery becomes ineffective [79]. The formula for SOH is written in Equation (1).

$$\mathrm{SOH}(t) = \frac{C_t}{C_0} \quad (1)$$

where C_t and C_0 denote the t-th cycle and initial battery capacity. The maximum capacity of the battery tends to drop in accordance with the number of times it is cycled, with continuous increase in the battery's internal resistance. Generally, a battery fails when its internal impedance increases to a level that is twice as high as its initial impedance.

Several performance parameters, such as power and the number of charge and discharge cycles, can also be used to define SOH. Further studies must utilize a wide variety of methods or models to estimate SOH, such as the use of direct measurement and indirect analysis. By measuring the standard aging characteristic parameters of the battery, the direct measurement technique determines the value of its current capacity, internal resistance, cycle times, etc. This is the technique through which the values of the current state's identifying parameters are determined. Examples of direct measurements are counting ampere hours, cycle numbers, measuring internal resistance and impedance. The indirect analysis consists of obtaining the SOH value by estimation based on online observable data from health indicators that have a high link with the performance and characteristic parameter degradation that occurs with the SOH condition. Model-based analysis, data-driven analysis, and hybrid analysis are examples of indirect analysis [80].

Wei J et al. [81] monitored the estimated diagnosis of battery SOH with three stages. In the initial stage, a particle filter (PF) technique was initiated, followed by the execution of a procedure to update the particle's time. The support vector regression (SVR) model was also used to estimate the capacity in each battery cycle number in the second stage. This SVR model is trained with characteristics collected from sensor data during constant-voltage (CV) charging mode at cycle number, to determine the charged capacity. The third stage updated the particle constitutes, which can be resampled based on their normalized importance weights. In accordance with the PF-based estimator, the anticipated capacity at the cycle number is considered as a Gaussian distribution, whose variance and mean are obtained. SOH is further defined as the ratio between the capacity of a new battery and the expected capacity. In general, the SOH estimation flowchart can be seen in the flowchart in Figure 6.

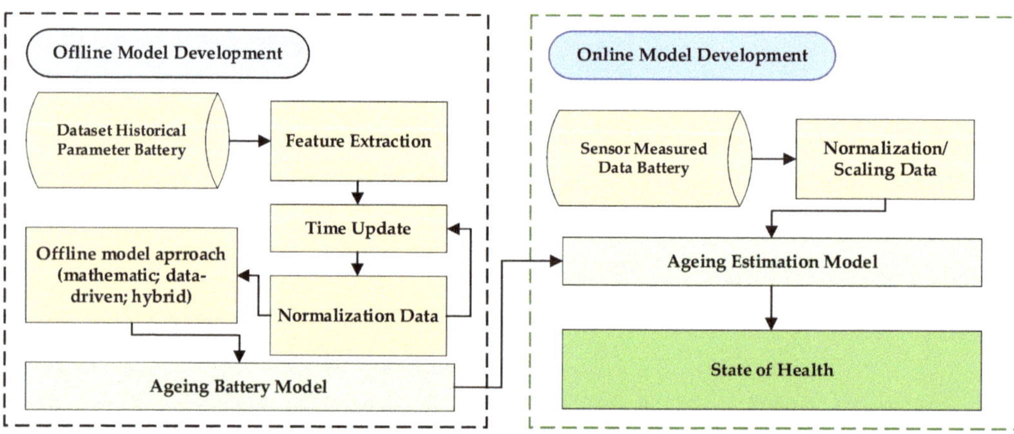

Figure 6. Block diagram of SOH estimation in general.

RUL refers to the information on the remaining life of a battery. It is imperative to change old and damaged batteries whose SOH has reached 0%, to guarantee the safety of the system and hence prevent problems [80,82]. The formula for RUL is written in Equation (2):

$$\mathrm{RUL}(t) = t - t_{eol} \tag{2}$$

where t and t_{eol} denote the t-th and number of cycles remaining at the completion of a battery's life. It is difficult to compute the RUL of a battery due to several variables, such as its present health condition, historical data, and failure. Therefore, further study needs to be conducted on the prediction of batteries' RUL. Presently, there is no standard framework that is considered the optimal model for estimating RUL due to a lack of available data, model complexity, and system limitations. In general, RUL prediction methods can be categorized as physics-based, mathematical, data-driven, or hybrids [80].

Wei J et al. [81] also predicted the RUL of a battery using the SVR-based model using a flowchart as shown in Figure 7. Monitoring the prediction of RUL starts with developing a model that has been trained using extracted sensor data features and predicted capacity for SVR-based input models. Wei J. et al. applied the average degradation parameter to characterize the expected capacity distribution in this section. The result showed that RUL is considered the n + 1 after the predicted capacity has reached the EOL threshold.

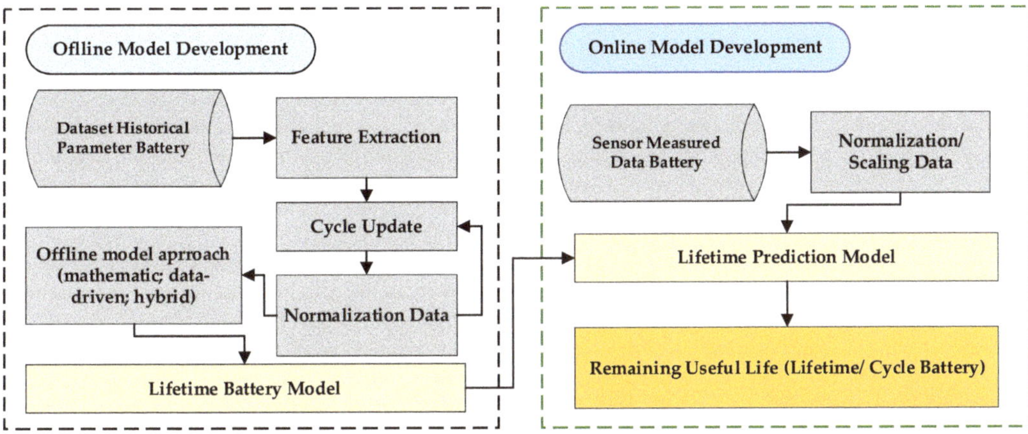

Figure 7. Block diagram of RUL prediction in general.

The diagram in Figure 8 illustrates the connection between SOH, RUL, and the modeling of battery degradation. Some preliminary research developed a battery deterioration mechanism model using a framework that incorporated SOH and RUL [76]. The elements that influence general battery deterioration and failure were further explained in the SOH estimation model. Furthermore, its diagnostics and estimation help boost RUL battery modeling by determining how much time or cycles are left to attain 80% SOH. As a result, the reliable prediction of SOH and RUL is required for modeling battery deterioration behavior.

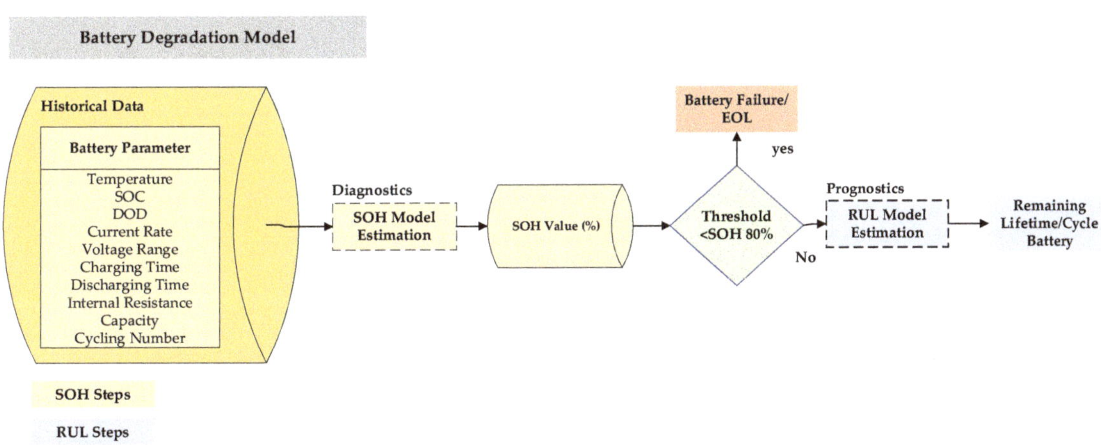

Figure 8. Relationship battery degradation models.

The SOH of a battery is measured in terms of its present ability to supply a certain quantity of energy in comparison to the initial capacity. At the same time, the RUL is helpful for monitoring the state of the battery and is also essential for executing operations that evaluate its degeneration. Due to the nonlinear nature of battery deterioration, it is necessary to have appropriate RUL estimations that are based on aging processes and suitable life models at various fading stages [76]. This entails calculating the time until a battery reaches its EOL. It tends to occur when the battery has reached the failure threshold. Moreover, the time left and the total number of charge-discharge cycles are considered [83]. The RUL estimation and degradation process are intimately linked to the working circumstances and dependability of Li-Ion batteries. Previous studies have reported that the successful prediction of the RUL prevents failure and timely functional maintenance without irreversibly harming the battery [84].

Scholars estimated the RUL using several different methodologies, as shown in Figure 9. These tend to be broken down into one of the four categories, namely based on physics, mathematics, data, or hybrid models. The amount of time a battery is going to be valuable is evaluated using a model-based technique. Therefore, a model that is representative of a battery application found in the real world, as well as an estimated algorithm used to predict voltage or other characteristics, needs to be developed. Empirical, analogous circuit and electrochemical models, including Kalman filters, are a few examples of the various methods that fall under this category. Data-driven RUL estimation is a prediction method that collects excess information and continues recording until battery health reaches its limit. Meanwhile, applying a hybrid model implies combining a model-based method with a data-driven model [76].

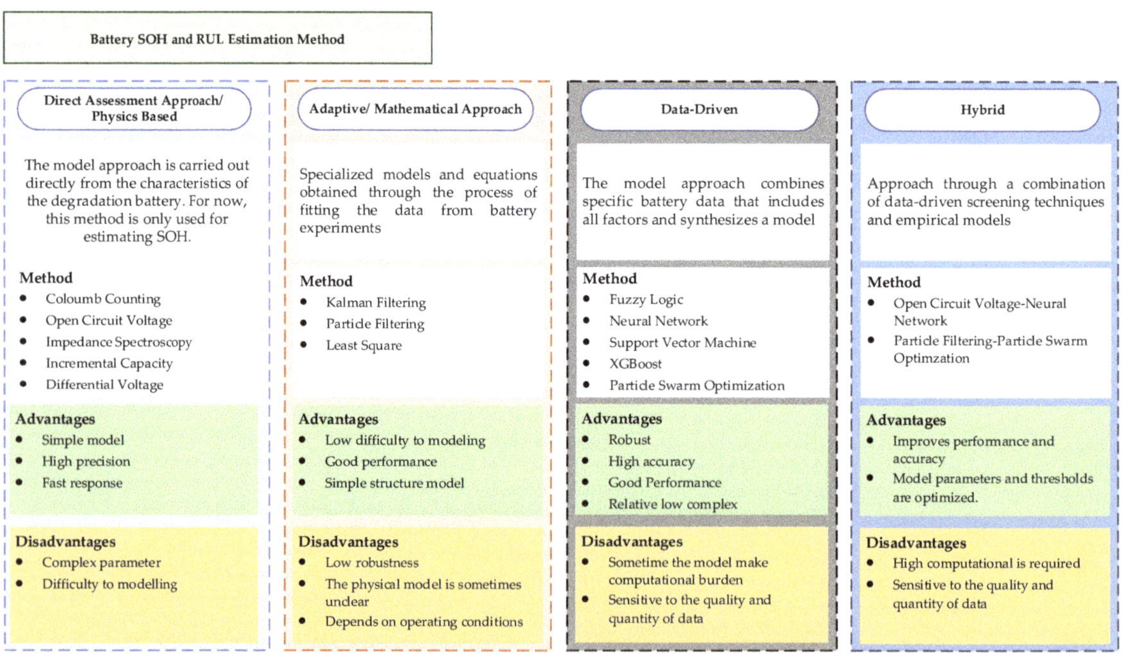

Figure 9. Classification method estimation RUL battery [76,82].

Table 9 reviews variables used to optimize BESS capacity size and placement with battery degradation models, which vary in different studies. Aside from the SOH and RUL models, preliminary research also used fading capacity and residual battery life for BESS optimization. Table 10 reviews the algorithm used for battery degradation models for BESS optimization.

Table 9. Review of variable features used in battery degradation models for optimization of BESS sizing and siting.

Author	Feature Variables for Battery Degradation Model					Model Battery Degradation
	SOC	DOD	Temperature	Cycle Life	Charging/ Discharging	
Alsaidan I et al. [32]	✓			✓		Lifetime
Khezri R, et al. [85]	✓				✓	Capacity fade
Sayfutdinov T, et al. [86]	✓	✓	✓			Capacity fade
Cardoso G, et al. [27]			✓		✓	Capacity fade, lifetime
Hernandez J, C, et al. [87]	✓	✓	✓	✓	✓	Lifetime
Garrido A.G, et al. [88]	✓		✓			Capacity fade, SOH, lifetime
Shin H et al. [46]	✓	✓	✓			Capacity fade, SOH
Arias N.B et al. [89]		✓				RUL
Amini M, et al. [38]		✓		✓		Capacity fade, lifetime
Mulleriyawage U.G.K, et al. [90]	✓			✓		SOH
Wu Y, et al. [91]	✓			✓	✓	SOH, capacity fade

Table 10. Advantages and disadvantages of battery degradation algorithm for BESS optimization.

Author	Battery Degradation Factors	Algorithm Battery Degradation	Advantages	Disadvantages
Alsaidan, et al. (2018) [32]	Energy capacity fading, cycle battery	Piecewise linear approximation	Easy to apply in small data on time to the events provided	When much data requires many limits
Timur Sayfutdinov, et al. (2020) [86]	Energy capacity fading, calendar aging, cycling aging	Least-squares fitting	Simple, easy to apply	Very sensitive to outliers, tendency to overfit, unreliable when the data distribution is not normal
Mohammad Amini et al. (2021) [38]	Energy capacity fading, calendar aging, cycling aging, lifetime battery	Mathematical model	Simple structure, low model difficulty, and fast performance	Less robust and significantly affected by operating conditions
Hunyong Shin, et al. (2022) [92]	Energy capacity fading, SOH, operating temperature, cycle battery	Rainflow-counting algorithm	Estimation of model parameters is based on linear regression analysis, which can be carried out with simple hand calculations.	Requires a lot of experimental data application of parameters based on estimates

Battery lifespan is influenced by calendar and cycling aging. However, this is also determined by cycle or float lives [93]. Even though the computation of the BESS life value tends to be inaccurate, its datasheet is dependent on two limits, cycle and float lives. Both restrictions are measured in years, and when the BESS maximum life is equal to or exceeded by its float life, it is said to have a floating life equal to or exceeds its maximum life. The cycle life is represented as the maximum number of charge and discharge cycles that can occur prior to the BESS failing, and it varies depending on the technology of both the BES and the DOD [38].

6.1. Battery Degradation Due to Changes in Ambient Temperature

The performance of lithium-ion batteries and their lifespan is significantly influenced by temperature. When exposed to high temperatures, its rate of degradation is significantly accelerated. Li-Ion batteries are temperature-sensitive [9], and their performance is affected not only by the temperature of the cell itself but also by the environment in which it is located. Battery degradation is caused by a combination of the SEI and the loss of active material. The one brought about by SEI is the most common and fundamental cause of capacity fade rate in batteries. As a result of the high temperature, the surface particles of the electron undergo a rapid development of SEI, thereby causing the battery's capacity to reduce [94]. According to some literature [95] on the systematic establishment of the theory on SEI growth and reduction in battery capacity, it was asserted that temperature changes trigger capacity fade due to alterations in the SEI layer. Incidentally, SEI growth can occur in idle situations, during the cycle, and during temperature changes. Some literature [96] clearly stated that temperature changes severely affect battery degradation. This process is of two types, namely actual and temporary capacity fading and loss. The actual capacity fading suggests that there has been irreversible cell loss due to the ingestion of lithium-ion. The high temperature of the battery accelerates the rapid rate of cell deterioration. On the other hand, a temporary capacity loss is caused by a drop in temperature during a specific cycle. It can be restored if the battery temperature returns to a certain level.

The literature [97] focuses on the ambient temperature impact on a battery's lifespan. The formation of the film on the electrodes of Li-Ion batteries explains the effect that the surrounding temperature has on its lifespan. This is because of the oxidation of the cell, proven by the film produced on the electrodes. It causes an irreversible increase in the Li-Ion battery's internal resistance, ultimately leading to damage. The findings on the simulation process show that higher temperatures during idle battery scenarios resulted in extreme capacity loss and self-discharge.

Some studies on calendar aging reported that it is related to temperature. Battery aging testing is performed at different temperatures, SOC, and end-of-life. The tests were conducted in a laboratory with temperature control facilities and charging or discharge operations. In reality, the battery is in extremely harsh operational conditions. The results of Li-Ion testing for EVs are reported to last 2000 and 800 cycles at temperatures of 25 °C and 55 °C, respectively [98]. Additionally, testing the influence of battery temperature due to discharge rate differences such as 1C, 2C, 3C, and 4C was also conducted [99]. It is possible to determine the varying contours due to the changing temperatures of the battery cells and their discharge at a consistent rate.

The pace at which capacity is lost is significantly affected by the temperature of the surrounding environment. Meanwhile, when it is greater than 35 degrees Celsius, it triggers more changes in the composition of the electrolyte due to the substantial temperature rise. This causes the process at which active lithium is utilized to quickly move forward [100]. As a result, the battery's capacity starts to decrease at various room temperatures, as shown in Figure 10. It is evident that when the perimeter temperature is greater than 35 °C, the capacity fades level drops significantly during the first 50 cycles. This phenomenon occurs while the battery is being used. When the temperature is 55 °C, the maximum capacity fades, while the temperatures of 25 °C and 35 °C are projected to be the same [100].

Characteristics of the capacity fade rate of the battery which is affected by the ambient temperature as shown in Figure 11. Yuhan Wu et al. [31] stated that $LiFePO_4$ battery degradation caused by the average temperature in BESS is modeled by combining calendar and cycle aging. This model is depicted by a single operating cycle, as shown in Equations (3)–(8). By knowing the characteristics of the battery aging cycle to set the optimal operating temperature of BESS, it can reduce the battery degradation rate so that the battery life is longer.

$$\xi = \xi_{cal} + \xi_{cyc} \tag{3}$$

$$\xi_{cyc} = f_{d,soc}(SOC_{avg}) \tag{4}$$

$$\xi_{cal} = \sum_{i=1}^{n} f_{d,dod}(DOD_i) f_{d,T}(T_{i,avg}) \tag{5}$$

$$f_{d,soc}(SOC_{avg}) = k_1 SOC_{avg}^2 + k_2 SOC_{avg} \tag{6}$$

$$f_{d,dod}(DOD_i) = k_3 DOD^2 + k_4 DOD \tag{7}$$

$$f_{d,T}(T_{i,avg}) = \begin{cases} e^{k_5/T}/k_6, & 298K \geq T \geq 273K \\ e^{k_7/T}/k_8, & 333K \geq T \geq 298K \end{cases} \quad (8)$$

where ξ represent of battery degradation from calendar aging (ξ_{cal}) and cyclic aging (ξ_{cyc}). n is the number of cycles charged or discharged in one day. SOC_{avg} represents the average SOC, DOD_i depicts the difference between the i-th charge and discharge cycles DOD, and $T_{i,avg}$ is the average temperature in BESS. In most cases, the value of the k parameter is determined by the experimental observation [31,35].

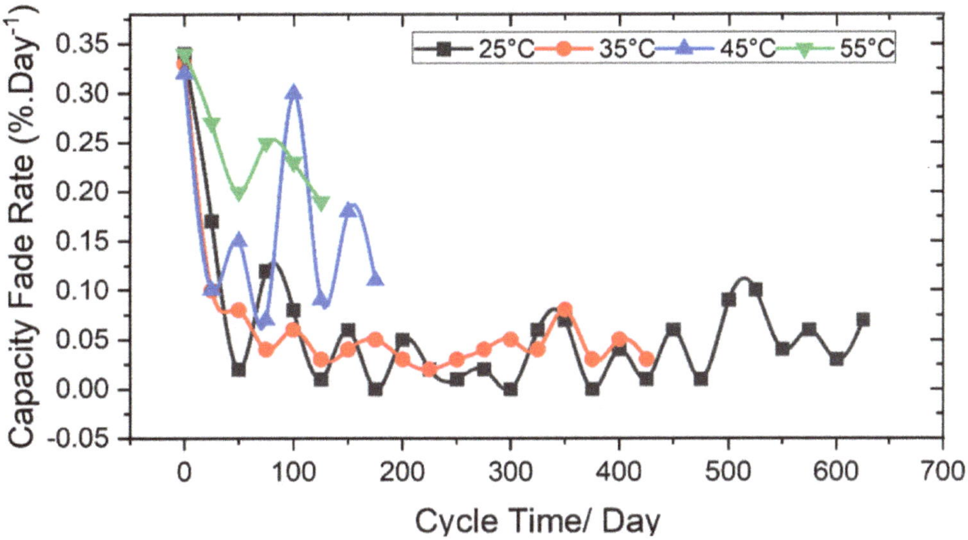

Figure 10. Capacity fade rate of $LiFePO_4$ battery at each temperature during cycling [100].

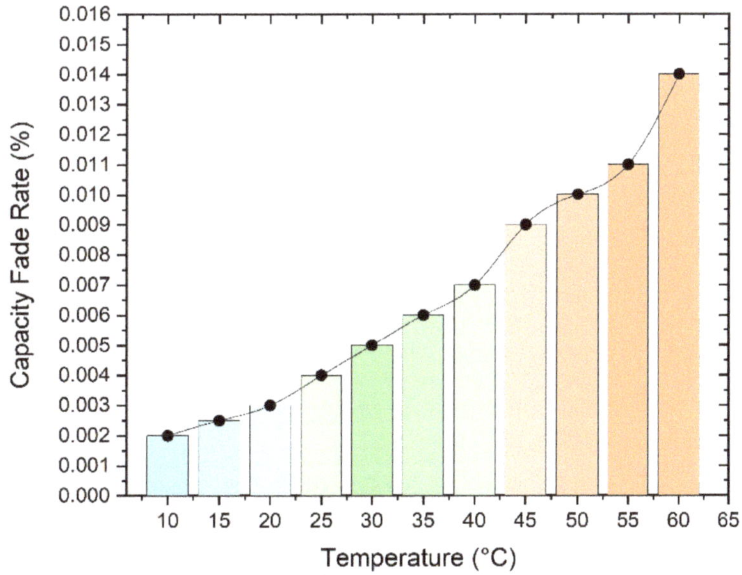

Figure 11. Characteristic cycle aging battery [31].

6.2. Battery Thermal Management

Complex electrochemical reactions and electric-to-thermal conversion determine the thermal characteristics of a battery [101]. The production of heat by Li-Ion batteries is a complex process that involves a knowledge of how the rate of electrochemical reaction varies with time and temperature, in addition to how current flows within the battery [102]. Simply, heat generation of the battery is written as Equation (9):

$$Q = I(U - V) - I\left(T\frac{dU}{dT}\right) \tag{9}$$

where Q denotes the rate of heat generation, I denotes the electric current flowing through the cell, U denotes the open-circuit voltage, and V represents the voltage of each individual cell in the Li-Ion batteries. In general, the thermal model of a battery has been examined according to the dimensions of the battery as well as the physical mechanism (electro-thermal model, electrochemical thermal model, and thermal runaway propagation model) (lumped model, 1D, 2D, and 3D). In most cases, the charging and discharging procedures for Li-Ion batteries result in the production of three distinct types of heat. These forms of heat include activation of irreversible heat as a result of the polarization of an electrochemical reaction, joule heating as a result of ohmic losses, and reversible reaction heat as a result of the change in entropy that takes place during the charging and discharging processes. Consequently, if the heat created by the battery while charging or discharging is not correctly dissipated, the temperature of the battery may grow because of heat accumulation, which may have a severe influence on the battery's performance, life, and safety [102].

The thermal management process, which is a critical component of the battery management system, is most concerned with estimating the precise state of temperature (SOT). Using more traditional measurement methods, such as thermocouples, it is simple to obtain an accurate reading of the temperature at the surface of the battery. Nevertheless, the temperature on the inside of the cell during transients is significantly different [103]. In general, the SOT estimation methods can be broken down into four categories: the direct measurement method, the electrochemical impedance-based method, the model-based estimation method, and the data-driven method.

Using a direct measurement methodology, researchers proposed ways for monitoring the temperature of a battery's internal layers. Temperature micro-sensors are integrated into the interior layers of the battery cells in these technologies. Thermocouples and resistance thermometers are the two most common types of sensors used to indicate the temperature of a battery's interior. The model-based estimation approach typically makes extensive use of numerical thermoelectric and thermal models when attempting to determine an object's internal temperature. To construct thermoelectric and thermal models such as the lumped-parameter battery model and the distributed battery thermal model, it is very required to understand heat generation, conduction, dissipation, balancing, and thermal boundary conditions. A few different approaches for calculating the temperature of a battery based on electrochemical impedance spectroscopy EIS measurements have been proposed in the electrochemical impedance-based approach without first constructing a thermal model. Temperature can be linked to impedance indicators acquired via EIS. These indicators include phase shift, real part amplitude, and imaginary part amplitude, per the most recent data-driven strategies. Data-driven approaches were used to estimate the temperature of the batteries inside [103].

7. Objective, Design Constraint, and Algorithm BESS Optimization

This section explains the objective functions frequently reported by previous studies, design constraints, algorithms used for BESS optimization, and a review of its state-the-art development. The steps involved in BESS optimization are depicted in the flowchart shown in Figure 12. This starts with collecting input system data, then determining the direction of the model development, selecting an objective function and design constraints, optimizing strategy and algorithm, and finally evaluating the optimization results.

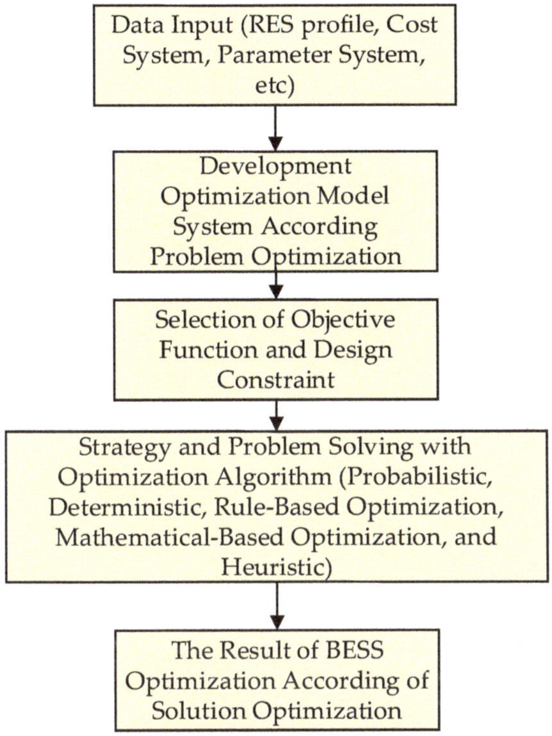

Figure 12. Flowchart of optimization of BESS.

7.1. Objective Function BESS

Since BESS plays an important role, its sizing is essential to ensure the normal functioning of distribution networks. An accurate and realistic model improves the operating systems from an economic and safety standpoint [104]. BESS optimum sizing is centered on finding its optimal capacity and the ability to minimize distribution network operating costs while meeting performance goals. Its investment cost is an essential component in calculating the distribution network operating expense. Moreover, this is affected by the investment payback period. As a result, BESS life is significant, and the number of cycles it can complete as well as the SOC at which it runs, are the two most important parameters used to determine the longevity of the battery. To assess the expenses linked to BESS, the anticipated lifespan was used [105]. In [106], the lifetime was determined by predictive models. The main objective of the study is to reduce costs, integrate RES, analyze its effects, and obtain benefits for the network.

7.1.1. Objective Function BESS to Reduce Total Cost Storage Expansion Planning

In the literature [32] the objective function was considered to reduce the total cost of storage expansion planning on the microgrid. It is defined as follows

$$\begin{aligned} Min \sum_{i\in G}\sum_{d}\sum_{h} F_i\left(P_{idh()}, I_{idh()}\right) + \sum_{d}\sum_{h}\rho_{dh}P^{M}_{dh()} \\ + \sum_{s} pr_s \sum_{b\in K}\sum_{d}\sum_{h} LS_{bdhs}v \\ + \textstyle\sum_{i\in B}\sum_{b\in K}\left(P^{R}_{ib}\left(CP^{a}_{i} + CM_i\right) + C^{R}_{ib}\left(CE^{a}_{i} + CI^{a}_{i}\right)\right) \end{aligned} \tag{10}$$

The first two-term Equation (10) indicates the operating cost of the microgrid when connected to the grid. Where b, d, h, i, l, s and B are the bus, day, hour, distributed energy resources, lines, scenarios, and battery technologies indices, respectively. F_i represents the microgrid local DG units cost function, $P_{idh()}$ is DG output power, $I_{idh()}$ depicts the commitment state of dispatchable units, ρ_{dh} is electricity market price ($/kWh), and $P^M_{dh()}$ illustrates the power transferred to and from the utility grid. The third term accounts for the costs of dissatisfying the requirements of the MG demand. Due to insignificant changes in the demand for microgrids, the output of generators distributed at the price of electricity during the planning period need to consider the historical data of one year. Where pr_s is the probability of islanding scenarios, LS_{bdhs} depicts load curtailment, and v represents the value of lost load ($/kWh). Incidentally, the value of lost load (VOLL) measures the economic losses associated with underserved energy. It depicts the willingness of customers to pay for reliable electrical services. This number is not dependent on the time or length of the outage rather, it is determined by the kind of consumer and location. The last term reflects the costs of BESS. Where C^R_{ib}, P^R_{ib} is BESS rated energy and power, CE^a_i, CP^a_i depicts annualized energy or power investment cost of BESS, CI^a_i is the cost of BESS installation on an annualized basis and CM_i represents the annual operating and maintenance cost of BESS.

In addition, there is also a BESS objective function to be applied in storage expansion planning on the grid. Based on the literature [35], it is stated as follows

$$\min \sum_{s\in S} \pi_s \sum_{t\in T} \left[\sum_{i\in I} A^G_i P^G_{s,i,t}{}^2 - B^G_i P^G_{s,i,t} + \sum_{km\in Br} \left(F^2_{s,km,t} \frac{R_{km}}{V^2_{km}} C_{APL} \right) \right] \Delta t + \sum_{k\in K} \sum_{j\in J} \frac{\overline{E}^{ES}_{j,k} C^E_j + \overline{P}^{ES}_{j,k} C^P_j}{365\ T^{Lt}_j} \quad (11)$$

Equation (11) shows the objective function that considers the exchange between investment costs and BESS operations. Due to this, BESS can demonstrate energy time-shift applications, which, in turn, contributes to the reduction in the day-to-day running expenses of the network. This is accomplished through a series of hypothetical situations that reflects the whole life span of BESS. The first group indicates the total operating cost of DG, where S represent the set of future network operation scenarios, T is the time intervals, π_s depicts the probability value of the scenario s, I represent the generation units, A^G_i, B^G_i illustrates a generation cost function, and $P_{s,i,}$ it is the scheduled power output of a thermal unit. The second term shows active power losses on the network, F_{km}, R_{km}, V_{km} depicting thermal limit, resistance, and the voltage level of the line. Br is an index of branches connecting pairs of nodes km, while C_{APL} represent energy price for active power losses. The last term illustrates the investment cost of BESS, where K represent of index of transmission grid nodes, J is the set of energy storage technologies, $\overline{P}^{ES}_{j,k}, \overline{E}^{ES}_{j,k}$ represents the rated power and energy capacity of BESS, C^E_j, C^P_j depicts the investment costs of battery technology, and T^{Lt}_j is the service lifetime battery.

7.1.2. Objective Function BESS of Life Cycle Cost Energy System

This energy system objective Life Cycle Cost (LCC) is used to minimize the total planning costs calculated only from BESS [91]. It is defined by some literature as follows:

$$\text{Min LCC} = C_{batt} + C_{O-M} \quad (12)$$

$$C_{O-M} = \frac{\sum_{y=1}^{Y} (1+r)^{Y-y} [\sum_{t=1}^{8760} (C_{out.y}(t) + C_{fit.y}(t) + \xi C_{batt}]}{(1+r)^Y} \quad (13)$$

$$C_{batt} = Cap_{bat}\, \mu_{batt} \quad (14)$$

$$C_{out.y}(t) = \left(P^{g-b}_y(t) + P^{g-l}_y(t) \right) \Delta t\, \varnothing_{buy} \quad (15)$$

$$C_{fit.y}(t) = \left(P^{b-g}_y(t) + P^{pv-g}_y(t) \right) \Delta t\, \varnothing_{sell} \quad (16)$$

Equation (12) is an LCC consisting of the initial investment cost of BESS (C_{batt}), including the cost of operation and maintenance BESS (C_{O-M}). Furthermore, Equation (13) is used to obtain the operation and maintenance costs where y and t is the index year, and time interval respectively, $C_{out.y}(t)$ depicts electricity bills, and $C_{fit.y}(t)$ is the benefit from selling electricity to the grid. Equation (14) represents the initial investment cost of BESS, where Cap_{bat} depicts the capacity of the battery, and μ_{batt} is the unit capacity price. Additionally, Equation (15) is used to calculate the electricity bills where $P_y^{g-b}(t)$ represents the power flow from grid to BESS (kW), $P_y^{g-l}(t)$ is the power flow grid to the line, and $\varnothing_{buy}$ depicts electricity price. Equation (16) is the profit realized from selling electricity to the grid, where $P_y^{b-g}(t)$ represents power flow battery to the grid, $P_y^{pv-g}(t)$ illustrates the power flow PV to the grid, and $\varnothing_{sell}$ is feed-in tariff.

7.1.3. Objective Function BESS for Battery Degradation Cost

According to the literature [107], the optimal scheduling of BESS is supposed to minimize the degradation costs, which are the proposed objective function. The intended degradation charge model accounts for the nonlinearities of battery life. As a result, the ideal SOC profile is the same regardless of the degradation cost model if the pricing pattern is either too flat or there are excessive disparities between the maximum and minimum prices. The objective function is stated in the following equation:

$$Min \sum_t \left(\left(\lambda_t P_{grid,t} \right) + C_E\left(SoC_t^{aux}\right) - C_E\left(SoC_{t-1}\right) \right) \tag{17}$$

Equation (17), is the optimal cost scheduling of BESS. It consists of power grid expense and degradation cost function for optimal scheduling, where t represents index of time interval, λ_t is electricity price, $P_{grid,t}$ represents the power from the grid, C_E denotes degradation cost for scheduling, SoC_t^{aux}, SoC is auxiliary and actual SOC BESS.

7.2. Design Constraint

In an arbitrary situation, the requirements or needs that must be considered are referred to as constraints. The power balance between the consumption and generation aspect is the most important constraint [108]. In distribution networks, electricity is imported or exported to the major grid, although this is often limited [109], to BESS-based operations [31]. The following are the most important limitations in maximizing the BESS size.

7.2.1. BESS Operation Constraint

The most common operational constraints when sizing BESS optimization techniques are charge or discharge or SOC constraints. In addition, battery degradation rate and life span needs to be regarded. The literature published by [110–113] reported otherwise, that the optimization of the BESS must consider the SOC. This constraint was taken into [114–118] consideration by maximizing BESS power loss, capacity, method, power balance, and battery lifecycle. In [32], the impact of BESS operation constraints is analyzed based on microgrid application and stated as follows

$$P_i^{min} x_{ib} \leq P_{ib}^R \leq P_i^{max} x_{ib} \tag{18}$$

$$\alpha_i^{min} P_{ib}^R \leq C_{ib}^R \leq \alpha_i^{max} P_{ib}^R \tag{19}$$

P_{ib}^R, C_{ib}^R denote power and energy rating BESS. The maximum and lowest BESS power ratings of P_i^{min}, P_i^{max} are represented by Equation (18). To determine the current investment status of BES technology, the binary variable x is used. Equation (19) utilized the power capacity to compute the maximum discharge time and measure the BESS capacity, where α_i^{max}, α_i^{min} indicates the highest and lowest possible energy to power rating ratios for the BES.

$$0 \leq P_{ibdhs}^{dch} \leq P_{ib}^R u_{ibdhs} \tag{20}$$

$$- P_{ib}^{R}(1 - u_{ibdhs}) \leq P_{ibdhs}^{ch} \leq 0 \quad (21)$$

The charging or discharge power of BESS P_{ibdhs}^{ch}, P_{ibdhs}^{dch} is limited depicted in Equations (20) and (21), where i, b, d, h, and s denote the distributed energy resources, bus, day, hour, and scenarios indices, respectively. u_{ibdhs} is BES operating state. BESS power turns negative and positive while charging and discharging, respectively. The current state of the BESS operation is determined by the value of the binary variable u. BESS can only flow when it is equal to one, and charges when it is equivalent to zero. The magnitude of the discharge has a direct bearing on the BESS life cycle, which varies from the diverse technologies. The BESS cycle refers to a complete one that includes both charging and discharging of the battery.

$$\xi_{ibdhs} = \left(u_{ibdhs} - u_{ibd(h-1)s}\right)u_{ibdhs} \quad (22)$$

$$\sum_{d}\sum_{h} pr_s \xi_{ibdhs} \leq \frac{1}{T}\sum_{m \in N} K_{im} W_{ibm} \quad (23)$$

Equation (22) is used to determine the BESS cycle, where ξ_{ibdhs} is BESS cycle indicator. Every time the charging process begins, the value is bound to be one, otherwise, it is zero. During the planned time horizon, the total BES cycle need not exceed the specified lifespan regarding the determined maximum DOD and the life project stated in Equation (23), where K_{im} is BESS lifecycle, and W_{ibm} represents a binary variable that reflects the value of the BESS maximum DOD.

$$\sum_{m \in N} W_{ibm} \leq x_{ib} \quad (24)$$

$$C_{ibdhs} = C_{ibd(h-1)s} - \frac{P_{ibdhs}^{dch} T}{\eta_i} - P_{ibdhs}^{ch} T \quad (25)$$

$$(1 - \sum_{m \in N} Y_{ibm} W_{ibm}) C_{ib}^{R} \leq C_{ibdhs} \leq C_{ib}^{R} \quad (26)$$

Equation (24), assures that for each BESS deployed, only one maximum depth of discharge value is evaluated. According to Equation (25), the energy stored at each time interval is equal to the preceding period minus the discarded or charged energy, where C_{ibdhs} is stored energy BESS during each interval. Meanwhile in Equation (26), BESS cannot be discharged with less energy than the minimum value specified by the maximum depth. This is not indicated by the discharge, nor can it be charged with more energy than its rated capacity allows during the process. Where Y_{ibm} is maximum DOD BESS.

7.2.2. Battery Degradation of BESS Constraint

Battery degradation in BESS is important to consider. Cardoso et al. [27], stated that the total annual electricity cost savings from PV and BESS can be reduced by 5–12% by solely considering the battery degradation constraint limitations. Furthermore, some literature [35] stated that a battery degradation model is based on cycling and aging conditions. Afterwards, it is used in the BESS operation constraint to support its optimization by lowering the planning cost of energy storage.

$$\gamma^{Idl}\left(SoC_{j,k}\right) = A_j^{Idl} SoC_{j,k}^2 + B_j^{Idl} SoC_{j,k} + C_j^{idl} \quad (27)$$

$$\gamma^{Cyc}\left(DoD_{j,k,n}\right) = A_j^{Cyc} DoD_{j,k,n}^2 + B_j^{Cyc} DoD_{j,k,n} \quad (28)$$

Equations (27) and (28) are capacity fade rates during idling and cycling conditions resulting from historical data on battery characteristics and adjusted to the least squares fitting method [35]. Where j, k, n are the battery technology, transmission grid nodes, and charge/discharge cycles indices, respectively. $\gamma^{Idl}, \gamma^{Cyc}$ is the capacity fade rate during the idling condition, and A_b^{Idl}, B_b^{Idl}, C_b^{Idl}, A_b^{Cyc}, B_b^{Cyc} is a quadratic, linear, and constant of the degradation functions during idling and cycling.

$$0 \leq E_{s,j,k,t}^{BESS} \leq \overline{E}_{j,k}^{BESS}\left[1 - \left(\gamma^{Idl}\left(SoC_{j,k}\right) + \sum_n y_n \gamma^{Cyc}\left(DoD_{j,k,n}\right)\right)Y(s)\right] \quad (29)$$

BESS charging is limited to the energy rating of those batteries which continues to fade due to the life horizon, depicted in the Equation (20), where $E_{b,i,y,d,t}^{BESS}$ is the BESS continuity energy, and $\overline{E}_{b,i}^{BESS}$ represents the installed BESS Energy. The value can be 0.5 for half cycles and 1.0 for full ones y_n. Y represents years for the number of the scenario s.

$$rem_{j,k} = 1 - \left(\gamma^{Idl}\left(SoC_{j.k}\right) + \sum_n y_n \gamma^{Cyc}\left(DoD_{j.k,n}\right)\right)T_j^{Lt} \quad (30)$$

$$EoL_j \leq rem_{j.k} \leq 1 \quad (31)$$

Equation (30), $rem_{j,k}$ is a formulation of the remaining BESS capacity at the end of battery service life due to idling degradation and cycling. T_j^{Lt} represents service lifetime period BESS of a manufacturer. The selected operating strategy is dependent on the remaining BESS capacity. $rem_{j.k}$ ensures that the remaining capacity is not less than the EOL threshold, moreover a constraint is applied in Equation (31).

7.2.3. Power and Energy Balance Constraint

When it comes to BESS size, the power, and energy balance between demand and generation is crucial. In the following literatures [112,116,118–122], the energy and power balance are constraints in the process of optimizing the size of the BESS. Based on [32], the power and energy balance constraints are expressed as follows

$$\sum_{g\in[G,W]} \mu_{ib} P_{idhs} + \sum_{b\in B}\left(P_{ibdhs}^{ch} + P_{ibdhs}^{dch}\right) + \sum_{i\in I} \psi_{ib} f_{idhs} + P_{dhs}^{M} + LS_{bdhs} = D_{bdh} \quad (32)$$

The balance of power and energy constraints are stated in Equation (32). This guarantees the amount of power provided by the distributed energy resources (DER) installed on that bus, plus or minus the amount of electricity going into or emanating from it, is equal to the quantity of power locally needed on that bus. If there is not enough generation to maintain BESS balance, the load is reduced, and the strength tends to be positive while the system is discharging and negative while it is charging. However, if the power is flowing from the utility grid into the microgrid, then it has a positive value, otherwise, it is negative. Where i, b, d, h, and s are the distributed energy resources, bus, day, hour, and scenarios indices, respectively. μ_{ib} is a generation-bus incidence matrix element, P_{idhs} is DER output power, P_{ibdhs}^{ch}, P_{ibdhs}^{dch} depicts BESS charging and discharging power, ψ_{ib} represents a line-bus matrix element (one if line l is connected to bus b, 0 if otherwise), f_{idhs} denotes distribution line power flow, P_{dhs}^{M} is electricity moved to and from the utility grid, LS_{bdh} is the load shedding cost, and D_{bdh} is total load demand.

$$-P^{M,max} z_{dhs} \leq P_{ds}^{M} \leq P^{M,max} z_{dhs} \quad (33)$$

$$0 \leq LS_{bdh} \leq (D_{bdh} - CD_{bdh}) \quad (34)$$

$$-f_l^{max} \leq f_{idhs} \leq f_l^{max} \quad (35)$$

Equation (33) is the limitation of a microgrid network of power transfer to the grid. Furthermore, Equation (34) is the limit for load reduction, where $P^{M,max}$ denotes the maximum power capacity of the microgrid to the utility grid, z_{dhs} is microgrid/utility grid status, D_{bdh}, CD_{bdh} represents the sum of all load demands as well as the critical load demand. Equation (35) is the amount of power that flows through a distribution network microgrid due to channel capacity constraints, where f_l^{max} is the maximum power capacity of distribution line.

7.3. Optimization Strategy and Algorithm

Size, capacity, cost, and lifetime are all aspects of the BESS that need to be improved. Existing research on BESS sizing-related problems is categorized according to grid scenario, goals that need to be achieved, the strategy applied, test bus, and various advantages and limitations to optimize the different algorithms. These include genetic algorithms (GA), particle swarm optimization (PSO), dynamic programming (DP), taboo search, fuzzy PSO, and bat algorithm. Simulation and modeling technologies such as PSLF, MATLAB, CPLEX, OpenDSS, GAMS, Gurobi, PowerFactory, and DIgSILENT are extensively used to improve BESS sizes. MATLAB is also a viable choice. Moreover, several research use a test bus from the IEEE study case to evaluate the system's performance instead of the current test systems [44]. The following are some of the most often used algorithms for predicting BESS size.

7.3.1. Probabilistic

Since several parameters tend to be improved, the probabilistic technique is regarded as one of the simplest ways of measuring BESS. The fundamental constraint of such a method is the number of parameters that need to be examined. Based on preliminary research, the probabilistic method was discovered to be the most useful approach for calculating the uncertainty parameter of the optimization process to obtain the best BESS measure [123–129]. Its key benefit is the need for a small amount of data to conclude. As a result, probabilistic approaches are excellent in circumstances where information is scarce.

7.3.2. Deterministic

The deterministic techniques examine various electrical configurations, system components being altered, and how they need to be optimized based on preset principles. A deterministic technique is a direct approach to cost [130] and capacity [131,132] alongside the optimization process investigated by some other analysis.

7.3.3. Rule-Based Optimization

The rule-based optimization (RBO) method defines an expected solution, such as fuzzy logic. In accordance with the following literature [131,133–136], optimization of BESS sizing is realized using fuzzy logic. Based on the research, a fuzzy-based method was adopted to reduce both the RES and the cost of BESS [137]. According to the data, an ideal BESS reduces microgrid costs by 3.2 percent, and battery longevity significantly affects MG costs. The primary advantage of utilizing a fuzzy optimizer is that either the total number of parameters is unknown or the scale of the optimization issue is unaffected by any change [138].

7.3.4. Mathematical-Based Optimization

The most comprehensive method is mathematical modeling when it comes to finding the solution to the BESS sizing-related problem. This approach for determining the optimal size of the BESS is categorized as linear programming (LP), nonlinear, or mixed-integer programming (MILP). Mathematical optimization is approached in three different ways, namely DP, convex programming (CP), and second-order cone programming (SOCP). Since the DP model separates this process into several different time slots, and the solutions are recognized at each level, it is both possible and advantageous to combine time-varying elements. In some literature, this model was used to maximize BESS size [111,139–141]. The CP technique also has the advantage of discretionary independence. Furthermore, its optimization strategy is employed in [142,143], to achieve the best possible results in minimizing the linear objective function. It is necessary to intersect the affine linear manifold with the product of second-order cones. Based on the literature published by [144,145], SOCP is used to size BESS.

7.3.5. Heuristics

Heuristic strategies allow suitable, non-ideal arrangements to be applied in real time. There is no mathematical foundation that is effective in obtaining optimal solutions, instead, approaches such as nature-inspired algorithms are used. These include GA [146], PSO [147], bat algorithm [148], and taboo search [149]. The key benefits of using heuristic approaches are flexibility, high accuracy, and computation timelessness.

7.4. Review of Existing Studies BESS

A state-of-the-art review of BESS optimization considering battery degradation was conducted to discover new perspectives in terms of developing its models. Table 11 summarizes several selected studies that can be distinguished based on main objectives, design constraints, algorithms, and battery degradation factors. It is evident that the perspective of battery degradation in BESS optimization is getting deeper. Its factors vary, such as energy capacity fading, calendar, and cycling aging, battery lifetime, cycle battery, and temperature. The development of the BESS optimization model considering battery degradation due to temperature is an interesting and rare study. There are certain related studies [27,35] in terms of developing a battery degradation model for optimal BESS using a fixed value of battery temperature. Meanwhile, literature [31] tends to develop a degradation battery model due to ambient temperature with dynamic values during the winter. Based on the study of the optimal BESS, ambient temperature affects battery degradation, according to the literature [100] The capacity fade level drops significantly when the perimeter temperature exceeds 35 °C. Therefore, the development of a battery degradation model due to ambient temperature is a new perspective in optimizing BESS.

Table 11. Literature review of studies of the BESS optimization effect considering battery degradation.

Author	Main Objective	Constraint	Battery Technology	Case Study	Algorithm/Method Optimization BESS	Battery Degradation Factors	Algorithm Battery Degradation
Ting Qiu, et al. (2017) [150]	Sizing BESS for co-planning the transmission model of expansion	CDC, CC, PEBC, PELC, RCC	Li-Ion	Modified IEEE-RTS 24-bus system	MILP	Energy capacity fading, calendar, and cycling aging	Flat rate degradation
Cardoso, et al. (2018) [27]	Sizing BESS by considering the linear battery degradation model for microgrid problem	CDC, PEBC, FC	Li-Ion	San Francisco	MILP	Capacity loss, battery lifetime, and cycle, operating temperature	Mathematical model
Alsaidan, et al. (2018) [32]	Optimal sizing BESS for microgrid expansion problem by considering technology, cycle life, and maximum depth of discharge	CDC, CC, PELC, PEBC, RCC	Li-Ion	Modified IEEE-5 bus	MINLP	Energy capacity fading, cycle battery	Piecewise linear approximation
A. Pena-Bello, et al. (2019) [34]	Sizing BESS by considering self-consumption, demand load-shifting, demand peak shaving and avoidance of PV curtailment.	CDC, PELC, PEBC, EFC	NCA, NMC, LFP, LTO, VRLA, & ALA	Austin (US), Geneva (Switzerland)	MILP	N/A	N/A
G. Mohy-Ud-Din, et al. (2020) [36]	Energy management system for industrial microgrids with optimal size BESS	CDC, PELC, EFC	Li-Ion	Australia	two–stage energy management strategy (single-stage linear program)	Energy capacity fading, cycle battery	Mathematical model
V.V. S. N. Murty, et al. (2020) [151]	Microgrid energy management by considering multi-objective solution and optimal sizing BESS	CDC, PELC, PEBC	Li-Ion	N/A	Multi-Objective (MILP, Fuzzy)	cycling aging	Mathematical model

Table 11. *Cont.*

Author	Main Objective	Constraint	Battery Technology	Case Study	Algorithm/Method Optimization BESS	Battery Degradation Factors	Algorithm Battery Degradation
Timur Sayfutdinov, et al. (2020) [35]	Optimal siting, sizing, and technology selection of BESS	CDC, CC, PEBC, PELC	LFP, LMO, NMC, LTO	Modified IEEE-9 bus, 14 bus, 24 bus, 39 bus	Mixed Integer Convex Programming (MICP)	Energy capacity fading, calendar, and cycling aging, cycle battery	Least-squares fitting
Yang Li, et al. (2020) [25]	Application of Li-Ion for optimal sizing of BESS in renewable power plant	CDC, CC	Li-Ion	A hypothetical 100-MW wind farm	Particle Swarm Optimization	Energy capacity fading, SOH, state of energy (SOE)	Physics-based
Hunyong Shin, et al. (2020) [92]	The process of sizing BESS for renewable power plant is becoming economical	CDC, CC, FC	Li-Ion	RES with storage power plants in South Korea	battery augmentation scheme (BAS)	Energy capacity fading, SOH, operating temperature, cycle battery	Rainflow-counting algorithm
Mattia Secchi, et al. (2021) [152]	Multi-objective sizing BESS for renewable energy with communities	CDC, PEBC	Li-Ion	Modified IEEE 906-bus European Low Voltage	NSGA-II	N/A	Mathematical model
Farihan Mohamad, et al. (2021) [153]	Sizing and Siting BESS to minimize solar energy curtailment	CDC, CC, PEBC, PELC	Li-Ion	IEEE 24-bus reliability test network (RTN)	GA dan Sequential Monte Carlo (SMC)	N/A	Mathematical model
Nataly Bañol Arias, et al. (2021) [89]	Sizing BESS by considering frequency regulation and peak shaving	CC, PEBC, PELC	Li-Ion	240-node three-phase distribution system	Pareto optimal	Energy capacity fading, cycle battery	Mathematical model
Yunfang Zhang, et al. (2021) [37]	Optimal Sizing BESS for grid scale by considering uncertainties and wind generation	CDC, CC, PEBC, PELC	Li-Ion	Modified IEEE RTS-24	Two-level model (MILP)	N/A	N/A
Mohammad Amini, et al. (2021) [38]	Sizing BESS for flexible, effective, efficient and better microgrid performance	CDC, CC, PEBC, RCC, PELC	NaS, Li-Ion, Lead-Acid, Nicd	Connected/Islanded Microgrid	MILP	Energy capacity fading, calendar, and cycling aging, battery lifetime	Mathematical model
U.G.K. Mulleriyawage, et al. (2021) [154]	Optimal sizing BESS by considering the demand and management attributes	CDC, CC, PELC, PEBC	Li-Ion	A grid-connected residential DC microgrid	MILP	Energy capacity fading, calendar, and cycling aging, SOH, EOL	Physics-based
Yuhan Wu, et al. (2021) [31]	Optimal capacity location BESS by considering the ambient temperature	CDC, CC, PEBC, PELC	$LiFePO_4$	modified IEEE 33 distribution network	Bi-level (GA, simulated annealing algorithm (SA))	Energy capacity fading, calendar and cycling aging, ambient temperature	Rainflow-counting algorithm
Yaling Wu, et al. (2022) [91]	Sizing BESS by considering the long-term battery degradation	CDC, CC, PEBC, PELC	Li-Ion	Connected/Islanded Microgrid	two-layer optimization method (MINLP)	Energy capacity fading, calendar, and cycling aging, SOH	Mathematical model
Davide Fioriti, et al. (2022) [155]	Multi-year sizing BESS for residential applications	CDC, CC, PEBC, PELC	Li-Ion	Residential grid-connected (399 Italian households in different regions (North, Center, South, and islands))	Heuristic optimization	Energy capacity fading, calendar and cycling aging, operating temperature	Rainflow-counting algorithm
Waqas ur Rehman et al. (2022) [39]	Optimal sizing BESS and solar generation system in an extreme fast charging station to reduce the annualized cost	CDC, CC, PEBC, PELC	Li-Ion	Extreme fast charging station (XFCS) demand modeling	MILP	Energy capacity fading, cycle battery	Mathematical model

Table 11. *Cont.*

Author	Main Objective	Constraint	Battery Technology	Case Study	Algorithm/Method Optimization BESS	Battery Degradation Factors	Algorithm Battery Degradation
Mohammad-Ali Hamidan, et al. (2022) [156]	Optimal sizing BESS for loss reduction and reliability improvement	CDC, CC, PEBC, PELC	Li-Ion	30-bus radial distribution network, 69-bus radial distribution network	Evolutionary algorithm based on decomposition (MOEA/D)	N/A	N/A
Noman Shabbir, et al. (2022) [157]	Optimal sizing BESS for solar PV systems to be self-sufficient and sustainable	CDC, CC, PEBC, PELC, FC	Li-Ion	Estonian low-distribution network	Heuristic optimization	Energy capacity fading, cycle battery	Mathematical model

CC, capacity constraint, CDC, charging and discharging constraint, PEBC, power and energy balance constraint, PELC, power and energy limit constraint, EC, environmental constraint, RCC, ramping capability, EFC; efficiency losses constraint, FC, financial constraint.

In addition, the battery degradation algorithm needs to be considered. Similar models are generally mathematical, physics-based, data-driven, and hybrid. Algorithm battery degradation affects the speed and convergence of BESS optimization. Therefore, several studies still utilize mathematical algorithm models because they are simple and exhibit rapid performance. However, data-driven models are flexible in modeling battery degradation due to several factors. Examples are piecewise linear approximation, least-squares fitting, and the rainflow-counting algorithm.

8. Issues and Challenge BESS

In terms of optimizing BESS sizing and location, several factors need to be considered by the expected operating objectives. To reduce the investment cost BESS not only makes it cost-effective. But, can be adjusted to boost reliability, power and voltage quality, peak shaving, load smoothing, frequency control, and energy arbitrage. One of the challenges of BESS optimization is battery degradation. The selection of battery technology is essential and BESS optimization solutions need to be assessed.

8.1. Economic Analysis

The economic aspect of building a BESS system is perhaps the most challenging. Preliminary studies created a BESS sizing and siting system to reduce investment costs or optimize profits received once it was implemented. Its cost is determined by numerous aspects, including the type of BESS technology selected, the number of energy source integrations, geographical conditions, features of the deployed region, installation expenses, and maintenance expenses. Technology types differ depending on energy density, efficiency, battery longevity, and cost. Installation and maintenance expenses include the capital for converter interface power, such as energy costs for storage capacity investment, replacement, annual operating and maintenance expenditures. Furthermore, various factors influence the cost of the BESS system, including service life, battery capacity, degradation rate, power loss, and SOC. As a result, its capacity and placement must be properly specified to minimize the installation cost. A BESS capacity that is extremely large is bound to raise the total cost of the system, thereby resulting in power loss. Assuming it is extremely tiny, it reduces efficiency and creates an imbalance in supply and demand.

The uncertainty of the RES system influences BESS cost optimization, such as peak shaving and load shifting. Peak shaving is an efficient method of lowering demand costs by leveling the highest electricity consumption. Meanwhile, load shifting is a temporary reduction in power used followed by subsequent production increases when prices are low. As a result, advanced optimization of the BESS model is required in conjunction with the uncertainty of RES to achieve optimal system planning and operational costs.

8.2. Technology Battery Storage Selection

Some of the battery technologies for BESS include LA, Li-Ion, Nickel Batteries, ZnBr, NaS, PSB and VRB. The appropriate one can be employed to optimize the system planning or operational costs. Energy density, extended discharge time, battery efficiency, longevity, and life cycle are all factors that determine technology selection. This battery is great for power quality and frequency management applications. It is due to the high-power density possessed as well as the lightning-fast response time. Although this type of battery, with its high energy density and longer discharge time, is ideally suited for long-term applications, it can also be used in certain circumstances to enable peak shaving and load shifting. This is because of the battery's extended discharge period. Therefore, the selection of battery technology is critical to supporting its applications and indirectly impacts the cost of installing BESS.

8.3. Optimal Charge or Discharge

Selecting the optimal BESS charge or discharge strategy is an important aspect of optimal sizing and tends to influence the life cycle of the battery. When determining the ideal size of a BESS, the most important parameters to take into consideration are speed of charging, rate of discharging, efficiency, and length of service life. Additionally, the effective control of the BESS charge and discharge can contribute to developing more advanced models.

8.4. Degradation of Battery Due to Ambient Temperature

Due to calendar and cycle aging, the amount of time a battery has been in use impacts how old it appears. Even though its life is determined by calendar aging, the BESS datasheet includes two limits cycle and float life. The likely computation of the BESS life value being accurate is low since battery life is dependent on cycle or float life. This is unlikely to affect the computation process. The term float life refers to the length of time that a BES is guaranteed to operate at its maximum capacity. When designing a BES system, the impacts of battery aging need to be considered with respect to the overall cost. High operating temperature, SOC, DOD, and charge or discharge current rate are all nonlinear factors that influence battery degeneration. The aging of the battery has an impact on the BESS performance and the cost of the electric power system. The major parameters of its deterioration capacity are voltage, current, charge or discharge cycle, and battery life. Generally, two things contribute to battery degeneration. First, there is loss of lithium ions as a result of SEI production. Second, it is caused by the loss of electrode particles. This is because the battery experiences an increase in its internal resistance. It causes a decrease in the battery's capacity as well as its efficiency, which eventually results in a shorter lifespan.

The battery performance and life cycle of Li-Ion batteries are susceptible to high temperatures, which tend to accelerate degradation significantly. This triggers the rapid growth of SEI on the surface of electron particles, leading to a loss in battery capacity. It is since the rapid growth of SEI on the surface of electron particles causes a decrease in battery capacity. In addition to this, the temperature of the surrounding environment has a significant bearing on the rate at which capacity is lost. The temperature of the battery cell and the high ambient contribute to the rapid growth of SEI on the surface of electron particles. Its development also contributes to a decrease in the capacity of the battery. According to the literature [100], when the ambient temperature exceeds 35 °C, changes in electrolyte composition increase. This is due to a significant temperature rise, accelerating active lithium consumption rate. Therefore, ambient temperature considerations can be challenging in influencing BESS battery degradation.

8.5. Retired Batteries for BESS

Hazardous chemical waste on BESS construction cells significantly affects the environment. Damaged batteries can be recycled and reused. Approximately 95% of the main material in LA batteries are recyclable and reusable [15,158]. In the past ten years, approximately five million EVs and 400 GWh of lithium-ion batteries have been sold all over the world [159]. The development of the EV market will eventually result in a large flow of retired batteries. Meanwhile, Li-Ion recycling is likely feasible, battery reuse and recycling are complementary processes that only slow down the cycle of excess resources. lion recycling has proven to be uneconomical [160]. The repurposing of retired batteries from EVs as BESS is a new challenge. To reduce battery disposal problems due to EOL [161] in electric power systems, BESS can be built to provide related services from EOL batteries. This is because these batteries tend to qualify for less-demanding grid services [162]. Retired BESS can increase the RES penetration of the electric power system for reverse spinning [163] with relatively cheaper installation costs.

8.6. Flexibility of Variable Renewable Energy Sources

Because of nature intermittency, RES such as solar PV and wind energy are inextricably connected to uncertainty. Higher renewable penetration rates substantially influence microgrid or grid system operation, data transfer, and handling, including remote sensing, decision-making, and system control. Therefore, this RES requires storage facilities such as BESS to store and supply electricity as needed. Most studies generate RES variability data using probabilistic methods such as Monte Carlo simulations, analytical and approximation models. However, these methods are insufficient for expressing random variables. These processes are also computationally challenging and need large amounts of historical data, extended run times, and precise mathematical premises. As a result, precise modeling and analytical treatment of this uncertainty while considering the geographic situation are crucial to making the best operational and financial decisions during microgrid or grid applications.

9. Conclusions

This study reviews the state-of-art BESS optimization methods considering battery degradation in connection to its diverse technologies. A comprehensive analysis of the development of the current BESS modeling approach with the objective function, battery degradation characteristics, and design constraints was employed. BESS is related to expansion planning, often called SEP. Its primary goal is to ensure that central planners, such as vertically integrated power companies and policymakers from governments or groups of countries responsible for minimizing costs rather than maximizing the benefits to private investors. Additionally, the use of BESS on the grid or microgrid is adopted to improve power quality, voltage and frequency control, peak shaving, load smoothing, and energy arbitrage.

LA, Li-Ion, NaS, and VRB are grid applications most common battery technologies. The energy density, efficiency, longevity, and cost of batteries linked to a storage network are all classed. Battery degradation reduces power efficiency in BESS. As a result, its deterioration needs to be considered during BESS optimization. The degradation of batteries owing to ambient temperature is currently understudied. Lithium-ion batteries' performance and life cycle are extremely temperature sensitive. In addition, high temperatures greatly accelerate battery degradation. The ambient temperature has a significant influence on the capacity fading rate, especially when it surpasses 35 °C, the composition of the electrolyte changes because of the large increase in temperature.

Generally, the objective function of optimizing BESS is to reduce the total cost of planning. The objective function and design constraints of BESS are highly dependent on the purpose for which BESS is used. BESS objective function is used to reduce LCC and battery degradation costs to minimize the total cost of system planning. The only components that make up this LCC are the costs of operation and maintenance, as well as the initial investment in the BESS. Based on the study of the optimal BESS, ambient temperature affects battery degradation. The development of its model due to ambient temperature can be a new perspective in optimizing BESS. The battery degradation algorithm affects the speed and convergence of BESS optimization. The determination of the model algorithm and battery degradation factors needs to be appropriately considered.

The challenges that need to be faced and the scope of future research in optimizing BESS by considering battery degradation of ambient temperature are the economic analysis, utilizing proper battery storage technology, and developing optimal charge or discharge model. Others include developing model degradation due to ambient temperature of BESS, considering retired batteries for BESS, and using the RES variable due to the uncertainty of natural conditions.

Author Contributions: Conceptualization, C.H.B.A., S.S., S.P.H. and F.D.W.; methodology, C.H.B.A. and S.S.; software, C.H.B.A.; validation, S.S., S.P.H. and F.D.W.; formal analysis, C.H.B.A.; investigation, C.H.B.A.; resources, C.H.B.A. and S.S.; data curation, C.H.B.A., S.S., S.P.H. and F.D.W.; writing—original draft preparation, C.H.B.A., S.S., S.P.H. and F.D.W.; writing—review and editing, C.H.B.A., S.S., S.P.H. and F.D.W.; visualization, C.H.B.A.; supervision, S.S., F.D.W. and S.P.H.; project administration, C.H.B.A.; funding acquisition, S.S. All authors have read and agreed to the published version of the manuscript.

Funding: This research was funded by Directorate General of Higher Education (DIKTI), Ministry of Education, Culture, Research and Technology, Research Grant: Penelitian Disertasi Doktor (PDD) with contract number 1929/UN1/DITLIT/Dit-Lit/PT.01.03/2022.

Institutional Review Board Statement: Not appliable.

Informed Consent Statement: Not appliable.

Data Availability Statement: Not appliable.

Acknowledgments: The authors are grateful to the Center for Education Financial Services (PUSLAPDIK), Ministry of Education, Culture, Research, and Technology and Indonesia Endowment Fund for Education (LPDP), Ministry of Finance of Republic of Indonesia: Beasiswa Pendidikan Indonesia (BPI) for supporting the funding of doctoral studies scholarship through contract number 1358/J5.2.3./BPI.06/10/2021.

Conflicts of Interest: The authors state that there is no conflict of interest. The research initiatives used as support had no part in the planning, collecting, analyzing, and interpreting data, as well as in compositing the paper and publishing the results.

Abbreviations

The following are some of the abbreviations that can be found in this manuscript:

BESS	Battery Energy Storage System
CV	Constant-Voltage
CP	Convex Programming
DER	Distributed Energy Resources
DG	Diesel Generator
DOD	Depth Of Discharge
DP	Dynamic Programming
EENS	Expected Energy Not Served
EIS	Electrochemical Impedance Spectroscopy
EOL	End-Of-Life
EV	Electric Vehicles
GA	Genetic Algorithms
GEP	Generation Expansion Planning
LA	Lead-Acid
LCC	Life Cycle Cost
$LiCoO_2$	Lithium Cobalt Oxide
$LiFePO_4$	Lithium Iron Phosphate
Li-Ion	Lithium-Ion

$LiMn_2O_4$	Lithium Manganese Oxide
$LiNiCoAlO_2$	Lithium Nickel Cobalt Aluminum Oxide
$LiNiMnCoO_2$	Cobalt-Based Lithium Nickel Manganese Oxide
LOLP/LOLE	Loss Of Load Probability Or Expectation
LP	Linear Programming
MILP	Mixed-Integer Programming
NAS	Sodium-Sulfur
Ni-Cd	Nickel-Cadmium
PRISMA	Preferred Reporting Items For Systematic Reviews And Meta-Analyses
PSB	Polysulfide Bromine Batteries
PSO	Particle Swarm Optimization
PF	Particle Filter
PV	Photovoltaic
RBO	Rule-Based Optimization
RES	Renewable Energy Sources
RF	Redox Flow
RUL	Remaining Useful Life
SEP	Storage Expansion Planning
SJR	Scimago Journal Rank
SLR	Systematic Literature Review
SOC	State Of Charges
SOCP	Second-Order Cone Programming
SOH	State Of Health
SOT	State Of Temperature
SVR	Support Vector Regression
TEP	Transmission Expansion Planning
UC	Unit Commitment
VOLL	Value Of Lost Load
VRB	Vanadium-Redox
ZBB	Zinc-Bromine

References

1. IESR. *Indonesia Energy Transition Outlook 2022. Tracking Progress of Energy Transition in Indonesia: Aiming for Net-Zero Emissions by 2050*; Institute for Essential Services Reform (IESR): Jakarta, Indonesia, 2022.
2. Parmeshwarappa, P.; Gundlapalli, R.; Jayanti, S. Power and Energy Rating Considerations in Integration of Flow Battery with Solar PV and Residential Load. *Batteries* **2021**, *7*, 62. [CrossRef]
3. Tsai, C.-T.; Beza, T.M.; Molla, E.M.; Kuo, C.-C. Analysis and Sizing of Mini-Grid Hybrid Renewable Energy System for Islands. *IEEE Access* **2020**, *8*, 70013–70029. [CrossRef]
4. Hao, H.; Wu, D.; Lian, J.; Yang, T. Optimal Coordination of Building Loads and Energy Storage for Power Grid and End User Services. *IEEE Trans. Smart Grid* **2018**, *9*, 4335–4345. [CrossRef]
5. Haas, J.; Cebulla, F.; Cao, K.; Nowak, W.; Palma-Behnke, R.; Rahmann, C.; Mancarella, P. Challenges and trends of energy storage expansion planning for flexibility provision in low-carbon power systems—A review. *Renew. Sustain. Energy Rev.* **2017**, *80*, 603–619. [CrossRef]
6. Killer, M.; Farrokhseresht, M.; Paterakis, N.G. Implementation of large-scale Li-ion battery energy storage systems within the EMEA region. *Appl. Energy* **2020**, *260*, 114166. [CrossRef]
7. Gupta, P.; Pandit, M.; Kothari, D.P. A review on optimal sizing and siting of distributed generation system: Integrating distributed generation into the grid. In Proceedings of the 2014 6th IEEE Power India International Conference (PIICON), Delhi, India, 5–7 December 2014; pp. 1–6. [CrossRef]
8. Eyer, J.; Corey, G. *Energy Storage for the Electricity Grid: Benefits and Market Potential Assessment Guide A Study for the DOE Energy Storage Systems Program*; Sandia National Laboratories (SNL): Albuquerque, NM, USA, 2010.
9. Vazquez, S.; Lukic, S.M.; Galvan, E.; Franquelo, L.G.; Carrasco, J.M. Energy Storage Systems for Transport and Grid Applications. *IEEE Trans. Ind. Electron.* **2010**, *57*, 3881–3895. [CrossRef]
10. Sheibani, M.R.; Yousefi, G.R.; Latify, M.A.; Dolatabadi, S.H. Energy storage system expansion planning in power systems: A review. *IET Renew. Power Gener.* **2018**, *12*, 1203–1221. [CrossRef]
11. de Quevedo, P.M.; Muñoz-delgado, G.; Contreras, J. Impact of Electric Vehicles on the Expansion Planning of Distribution Systems Considering Charging Stations. *IEEE Trans. Smart Grid* **2019**, *10*, 794–804. [CrossRef]
12. Hannan, M.; Faisal, M.; Ker, P.J.; Begum, R.; Dong, Z.; Zhang, C. Review of optimal methods and algorithms for sizing energy storage systems to achieve decarbonization in microgrid applications. *Renew. Sustain. Energy Rev.* **2020**, *131*, 110022. [CrossRef]

13. Bowen, T.; Chernyakhovskiy, I.; Denholm, P.L. *Grid-Scale Battery Storage: Frequently Asked Questions*; National Renewable Energy Lab.(NREL): Golden, CO, USA, 2018; pp. 1–8. [CrossRef]
14. Yang, Y.; Bremner, S.; Menictas, C.; Kay, M. Battery energy storage system size determination in renewable energy systems: A review. *Renew. Sustain. Energy Rev.* **2018**, *91*, 109–125. [CrossRef]
15. Hannan, M.; Wali, S.; Ker, P.; Rahman, M.A.; Mansor, M.; Ramachandaramurthy, V.; Muttaqi, K.; Mahlia, T.; Dong, Z. Battery energy-storage system: A review of technologies, optimization objectives, constraints, approaches, and outstanding issues. *J. Energy Storage* **2021**, *42*, 103023. [CrossRef]
16. Díaz-González, F.; Sumper, A.; Gomis-Bellmunt, O.; Villafáfila-Robles, R. A review of energy storage technologies for wind power applications. *Renew. Sustain. Energy Rev.* **2012**, *16*, 2154–2171. [CrossRef]
17. Akinyele, D.; Belikov, J.; Levron, Y. Battery Storage Technologies for Electrical Applications: Impact in Stand-Alone Photovoltaic Systems. *Energies* **2017**, *10*, 1760. [CrossRef]
18. Rosewater, D.M.; Copp, D.A.; Nguyen, T.A.; Byrne, R.H.; Santoso, S. Battery Energy Storage Models for Optimal Control. *IEEE Access* **2019**, *7*, 178357–178391. [CrossRef]
19. TESLA, "Powerwall". Available online: https://www.tesla.com/powerwall (accessed on 5 September 2022).
20. Schmalstieg, J.; Käbitz, S.; Ecker, M.; Sauer, D.U. A holistic aging model for Li(NiMnCo)O_2 based 18650 lithium-ion batteries. *J. Power Sources* **2014**, *257*, 325–334. [CrossRef]
21. Wang, J.; Liu, P.; Hicks-Garner, J.; Sherman, E.; Soukiazian, S.; Verbrugge, M.; Tataria, H.; Musser, J.; Finamore, P. Cycle-life model for graphite-Li$FePO_4$ cells. *J. Power Sources* **2011**, *196*, 3942–3948. [CrossRef]
22. Smith, K.; Saxon, A.; Keyser, M.; Lundstrom, B.; Cao, Z.; Roc, A. Life prediction model for grid-connected Li-ion battery energy storage system. In Proceedings of the 2017 American Control Conference (ACC), Seattle, WA, USA, 24–26 May 2017; pp. 4062–4068. [CrossRef]
23. Ahmadi, L.; Fowler, M.; Young, S.B.; Fraser, R.A.; Gaffney, B.; Walker, S.B. Energy efficiency of Li-ion battery packs re-used in stationary power applications. *Sustain. Energy Technol. Assess.* **2014**, *8*, 9–17. [CrossRef]
24. Hou, Q.; Yu, Y.; Du, E.; He, H.; Zhang, N.; Kang, C.; Liu, G.; Zhu, H. Embedding scrapping criterion and degradation model in optimal operation of peak-shaving lithium-ion battery energy storage. *Appl. Energy* **2020**, *278*, 115601. [CrossRef]
25. Li, Y.; Vilathgamuwa, M.; Choi, S.S.; Xiong, B.; Tang, J.; Su, Y.; Wang, Y. Design of minimum cost degradation-conscious lithium-ion battery energy storage system to achieve renewable power dispatchability. *Appl. Energy* **2020**, *260*, 114282. [CrossRef]
26. Xu, B.; Zhao, J.; Zheng, T.; Litvinov, E.; Kirschen, D. Factoring the Cycle Aging Cost of Batteries Participating in Electricity Markets. In Proceedings of the 2018 IEEE Power & Energy Society General Meeting (PESGM), Portland, OR, USA, 5–10 August 2018; p. 1. [CrossRef]
27. Cardoso, G.; Brouhard, T.; DeForest, N.; Wang, D.; Heleno, M.; Kotzur, L. Battery aging in multi-energy microgrid design using mixed integer linear programming. *Appl. Energy* **2018**, *231*, 1059–1069. [CrossRef]
28. Ren, L.; Dong, J.; Wang, X.; Meng, Z.; Zhao, L.; Deen, M.J. A Data-Driven Auto-CNN-LSTM Prediction Model for Lithium-Ion Battery Remaining Useful Life. *IEEE Trans. Ind. Inform.* **2021**, *17*, 3478–3487. [CrossRef]
29. Severson, K.A.; Attia, P.M.; Jin, N.; Perkins, N.; Jiang, B.; Yang, Z.; Chen, M.H.; Aykol, M.; Herring, P.K.; Fraggedakis, D.; et al. Data-driven prediction of battery cycle life before capacity degradation. *Nat. Energy* **2019**, *4*, 383–391. [CrossRef]
30. Moher, D.; Liberati, A.; Tetzlaff, J.; Altman, D.G.; Altman, D.; Antes, G.; Atkins, D.; Barbour, V.; Barrowman, N.; Berlin, J.A.; et al. Preferred Reporting Items for Systematic Reviews and Meta-Analyses: The PRISMA Statement. *PLoS Med.* **2009**, *6*, e1000097. [CrossRef] [PubMed]
31. Wu, Y.; Xu, T.; Meng, H.; Wei, W.; Cai, S.; Guo, L. Energy storage capacity allocation for distribution grid applications considering the influence of ambient temperature. *IET Energy Syst. Integr.* **2022**, *4*, 143–156. [CrossRef]
32. Alsaidan, I.; Khodaei, A.; Gao, W. A Comprehensive Battery Energy Storage Optimal Sizing Model for Microgrid Applications. *IEEE Trans. Power Syst.* **2018**, *33*, 3968–3980. [CrossRef]
33. Alharbi, T.; Bhattacharya, K.; Kazerani, M. Planning and Operation of Isolated Microgrids Based on Repurposed Electric Vehicle Batteries. *IEEE Trans. Ind. Inform.* **2019**, *15*, 4319–4331. [CrossRef]
34. Pena-Bello, A.; Barbour, E.; Gonzalez, M.; Patel, M.; Parra, D. Optimized PV-coupled battery systems for combining applications: Impact of battery technology and geography. *Renew. Sustain. Energy Rev.* **2019**, *112*, 978–990. [CrossRef]
35. Sayfutdinov, T.; Patsios, C.; Vorobev, P.; Gryazina, E.; Greenwood, D.M.; Bialek, J.W.; Taylor, P.C. Degradation and Operation-Aware Framework for the Optimal Siting, Sizing, and Technology Selection of Battery Storage. *IEEE Trans. Sustain. Energy* **2020**, *11*, 2130–2140. [CrossRef]
36. Mohy-Ud-Din, G.; Vu, D.H.; Muttaqi, K.M.; Sutanto, D. An Integrated Energy Management Approach for the Economic Operation of Industrial Microgrids under Uncertainty of Renewable Energy. In Proceedings of the 2019 IEEE Industry Applications Society Annual Meeting, Baltimore, MD, USA, 29 September–3 October 2019. [CrossRef]
37. Zhang, Y.; Su, Y.; Wang, Z.; Liu, F.; Li, C. Cycle-Life-Aware Optimal Sizing of Grid-Side Battery Energy Storage. *IEEE Access* **2021**, *9*, 20179–20190. [CrossRef]
38. Amini, M.; Khorsandi, A.; Vahidi, B.; Hosseinian, S.H.; Malakmahmoudi, A. Optimal sizing of battery energy storage in a microgrid considering capacity degradation and replacement year. *Electr. Power Syst. Res.* **2021**, *195*, 107170. [CrossRef]
39. Rehman, W.U.; Bo, R.; Mehdipourpicha, H.; Kimball, J.W. Sizing battery energy storage and PV system in an extreme fast charging station considering uncertainties and battery degradation. *Appl. Energy* **2022**, *313*, 118745. [CrossRef]

40. Babatunde, O.M.; Munda, J.L.; Hamam, Y. A comprehensive state-of-the-art survey on power generation expansion planning with intermittent renewable energy source and energy storage. *Int. J. Energy Res.* **2019**, *43*, 6078–6107. [CrossRef]
41. Koltsaklis, N.E.; Dagoumas, A.S. State-of-the-Art Generation Expansion Planning: A Review. *Appl. Energy* **2018**, *230*, 563–589. [CrossRef]
42. Dagoumas, A.S.; Koltsaklis, N.E. Review of Models for Integrating Renewable Energy in the Generation Expansion Planning. *Appl. Energy* **2019**, *242*, 1573–1587. [CrossRef]
43. Stecca, M.; Elizondo, L.R.; Soeiro, T.B.; Bauer, P.; Palensky, P. A Comprehensive Review of the Integration of Battery Energy Storage Systems into Distribution Networks. *IEEE Open J. Ind. Electron. Soc.* **2020**, *1*, 46–65. [CrossRef]
44. Wüllner, J.; Reiners, N.; Millet, L.; Salibi, M.; Stortz, F.; Vetter, M. Review of Stationary Energy Storage Systems Applications, Their Placement, and Techno-Economic Potential. *Curr. Sustain. Renew. Energy Rep.* **2021**, *8*, 263–273. [CrossRef]
45. Sandelic, M.; Stroe, D.-I.; Iov, F. Battery Storage-Based Frequency Containment Reserves in Large Wind Penetrated Scenarios: A Practical Approach to Sizing. *Energies* **2018**, *11*, 3065. [CrossRef]
46. Shin, H.; Roh, J.H. Framework for Sizing of Energy Storage System Supplementing Photovoltaic Generation in Consideration of Battery Degradation. *IEEE Access* **2020**, *8*, 60246–60258. [CrossRef]
47. Wang, S.; Lu, L.; Han, X.; Ouyang, M.; Feng, X. Virtual-battery based droop control and energy storage system size optimization of a DC microgrid for electric vehicle fast charging station. *Appl. Energy* **2020**, *259*, 114146. [CrossRef]
48. Johnson, R.; Mayfield, M.; Beck, S. Optimal placement, sizing, and dispatch of multiple BES systems on UK low voltage residential networks. *J. Energy Storage* **2018**, *17*, 272–286. [CrossRef]
49. Martins, R.; Hesse, H.C.; Jungbauer, J.; Vorbuchner, T.; Musilek, P. Optimal Component Sizing for Peak Shaving in Battery Energy Storage System for Industrial Applications. *Energies* **2018**, *11*, 2048. [CrossRef]
50. Vermeer, W.; Mouli, G.R.C.; Bauer, P. Optimal Sizing and Control of a PV-EV-BES Charging System Including Primary Frequency Control and Component Degradation. *IEEE Open J. Ind. Electron. Soc.* **2022**, *3*, 236–251. [CrossRef]
51. Engels, J.; Claessens, B.; Deconinck, G. Techno-economic analysis and optimal control of battery storage for frequency control services, applied to the German market. *Appl. Energy* **2019**, *242*, 1036–1049. [CrossRef]
52. Luo, X.; Wang, J.; Dooner, M.; Clarke, J. Overview of current development in electrical energy storage technologies and the application potential in power system operation. *Appl. Energy* **2015**, *137*, 511–536. [CrossRef]
53. Chen, H.; Cong, T.N.; Yang, W.; Tan, C.; Li, Y.; Ding, Y. Progress in electrical energy storage system: A critical review. *Prog. Nat. Sci.* **2009**, *19*, 291–312. [CrossRef]
54. IRENA. *Electricity Storage and Renewables: Costs and Markets to 2030*, no. October. 2017. Available online: http://irena.org/publications/2017/Oct/Electricity-storage-and-renewables-costs-and-markets%0Ahttps://www.irena.org/-/media/Files/IRENA/Agency/Publication/2017/Oct/IRENA_Electricity_Storage_Costs_2017.pdf (accessed on 5 September 2022).
55. Ibrahim, H.; Ilinca, A.; Perron, J. Energy storage systems—Characteristics and comparisons. *Renew. Sustain. Energy Rev.* **2008**, *12*, 1221–1250. [CrossRef]
56. Diouf, B.; Pode, R. Potential of lithium-ion batteries in renewable energy. *Renew. Energy* **2015**, *76*, 375–380. [CrossRef]
57. Stan, A.-I.; Swierczynski, M.; Stroe, D.-I.; Teodorescu, R.; Andreasen, S.J. Lithium ion battery chemistries from renewable energy storage to automotive and back-up power applications—An overview. In Proceedings of the 2014 International Conference on Optimization of Electrical and Electronic Equipment (OPTIM), Bran, Romania, 22–24 May 2014; pp. 713–720. [CrossRef]
58. Kawakami, N.; Iijima, Y.; Fukuhara, M.; Bando, M.; Sakanaka, Y.; Ogawa, K.; Matsuda, T. Development and field experiences of stabilization system using 34 MW NAS batteries for a 51 MW wind farm. In Proceedings of the 2010 IEEE International Symposium on Industrial Electronics, Bari, Italy, 4–7 July 2010; pp. 2371–2376. [CrossRef]
59. Liao, Q.; Sun, B.; Liu, Y.; Sun, J.; Zhou, G. A techno-economic analysis on NaS battery energy storage system supporting peak shaving. *Int. J. Energy Res.* **2016**, *40*, 241–247. [CrossRef]
60. Tewari, S.; Mohan, N. Value of NAS Energy Storage Toward Integrating Wind: Results From the Wind to Battery Project. *IEEE Trans. Power Syst.* **2013**, *28*, 532–541. [CrossRef]
61. Leung, P.; Shah, A.; Sanz, L.; Flox, C.; Morante, J.; Xu, Q.; Mohamed, M.; de León, C.P.; Walsh, F. Recent developments in organic redox flow batteries: A critical review. *J. Power Sources* **2017**, *360*, 243–283. [CrossRef]
62. de León, C.P.; Frías-Ferrer, A.; González-García, J.; Szánto, D.A.; Walsh, F.C. Redox flow cells for energy conversion. *J. Power Sources* **2006**, *160*, 716–732. [CrossRef]
63. Leung, P.; Li, X.; de León, C.P.; Berlouis, L.; Low, C.T.J.; Walsh, F.C. Progress in redox flow batteries, remaining challenges and their applications in energy storage. *RSC Adv.* **2012**, *2*, 10125–10156. [CrossRef]
64. Dunn, B.; Kamath, H.; Tarascon, J.-M. Electrical Energy Storage for the Grid: A Battery of Choices. *Science* **2011**, *334*, 928–935. [CrossRef]
65. May, G.J.; Davidson, A.; Monahov, B. Lead batteries for utility energy storage: A review. *J. Energy Storage* **2018**, *15*, 145–157. [CrossRef]
66. Das, C.K.; Bass, O.; Kothapalli, G.; Mahmoud, T.S.; Habibi, D. Overview of energy storage systems in distribution networks: Placement, sizing, operation, and power quality. *Renew. Sustain. Energy Rev.* **2018**, *91*, 1205–1230. [CrossRef]
67. Baumann, M.; Weil, M.; Peters, J.F.; Chibeles-Martins, N.; Moniz, A.B. A review of multi-criteria decision making approaches for evaluating energy storage systems for grid applications. *Renew. Sustain. Energy Rev.* **2019**, *107*, 516–534. [CrossRef]

68. Zhang, Z.; Ding, T.; Zhou, Q.; Sun, Y.; Qu, M.; Zeng, Z.; Ju, Y.; Li, L.; Wang, K.; Chi, F. A review of technologies and applications on versatile energy storage systems. *Renew. Sustain. Energy Rev.* **2021**, *148*, 111263. [CrossRef]
69. Georgious, R.; Refaat, R.; Garcia, J.; Daoud, A.A. Review on Energy Storage Systems in Microgrids. *Electronics* **2021**, *10*, 2134. [CrossRef]
70. Miao, Y.; Hynan, P.; von Jouanne, A.; Yokochi, A. Current Li-Ion Battery Technologies in Electric Vehicles and Opportunities for Advancements. *Energies* **2019**, *12*, 1074. [CrossRef]
71. Du, W.; Xue, N.; Sastry, A.M.; Martins, J.R.R.A.; Shyy, W. Energy Density Comparison of Li-ion Cathode Materials Using Dimensional Analysis. *J. Electrochem. Soc.* **2013**, *160*, A1187–A1193. [CrossRef]
72. Wang, Y.; Zhou, Z.; Botterud, A.; Zhang, K.; Ding, Q. Stochastic coordinated operation of wind and battery energy storage system considering battery degradation. *J. Mod. Power Syst. Clean Energy* **2016**, *4*, 581–592. [CrossRef]
73. Koller, M.; Borsche, T.; Ulbig, A.; Andersson, G. Defining a degradation cost function for optimal control of a battery energy storage system. In Proceedings of the 2013 IEEE Grenoble Conference, Grenoble, France, 16–20 June 2013.
74. Kintner-Meyer, M.C.; Balducci, P.J.; Jin, C.; Nguyen, T.B.; Elizondo, M.A.; Viswanathan, V.V.; Guo, X.; Tuffner, F.K. *Energy Storage for Power Systems Applications: A Regional Assessment for the Northwest Power Pool (NWPP)*; Pacific Northwest National Lab.(PNNL): Richland, WA, USA, 2010. [CrossRef]
75. Aurbach, D.; Zinigrad, E.; Teller, H.; Dan, P. Factors Which Limit the Cycle Life of Rechargeable Lithium (Metal) Batteries. *J. Electrochem. Soc.* **2000**, *147*, 1274–1279. [CrossRef]
76. Rauf, H.; Khalid, M.; Arshad, N. Machine learning in state of health and remaining useful life estimation: Theoretical and technological development in battery degradation modelling. *Renew. Sustain. Energy Rev.* **2022**, *156*, 111903. [CrossRef]
77. Liu, D.; Pang, J.; Zhou, J.; Peng, Y.; Pecht, M. Prognostics for state of health estimation of lithium-ion batteries based on combination Gaussian process functional regression. *Microelectron. Reliab.* **2013**, *53*, 832–839. [CrossRef]
78. Meng, J.; Boukhnifer, M.; Diallo, D. Lithium-Ion Battery Monitoring and Observability Analysis with Extended Equivalent Circuit Model. In Proceedings of the 2020 28th Mediterranean Conference on Control and Automation (MED), Saint-Raphaël, France, 15–18 September 2020; pp. 764–769. [CrossRef]
79. Zhou, D.; Li, Z.; Zhu, J.; Zhang, H.; Hou, L. State of Health Monitoring and Remaining Useful Life Prediction of Lithium-Ion Batteries Based on Temporal Convolutional Network. *IEEE Access* **2020**, *8*, 53307–53320. [CrossRef]
80. Ge, M.-F.; Liu, Y.; Jiang, X.; Liu, J. A review on state of health estimations and remaining useful life prognostics of lithium-ion batteries. *Measurement* **2021**, *174*, 109057. [CrossRef]
81. Wei, J.; Dong, G.; Chen, Z. Remaining Useful Life Prediction and State of Health Diagnosis for Lithium-Ion Batteries Using Particle Filter and Support Vector Regression. *IEEE Trans. Ind. Electron.* **2018**, *65*, 5634–5643. [CrossRef]
82. Lipu, M.H.; Hannan, M.; Hussain, A.; Hoque, M.; Ker, P.J.; Saad, M.; Ayob, A. A review of state of health and remaining useful life estimation methods for lithium-ion battery in electric vehicles: Challenges and recommendations. *J. Clean. Prod.* **2018**, *205*, 115–133. [CrossRef]
83. Wu, L.; Fu, X.; Guan, Y. Review of the Remaining Useful Life Prognostics of Vehicle Lithium-Ion Batteries Using Data-Driven Methodologies. *Appl. Sci.* **2016**, *6*, 166. [CrossRef]
84. Zhang, J.; Lee, J. A review on prognostics and health monitoring of Li-ion battery. *J. Power Sources* **2011**, *196*, 6007–6014. [CrossRef]
85. Khezri, R.; Member, S.; Mahmoudi, A.; Member, S.; Haque, M.H.; Member, S. A Demand Side Management Approach For Optimal Sizing of Standalone Renewable-Battery Systems. *IEEE Trans. Sustain. Energy* **2021**, *12*, 2184–2194. [CrossRef]
86. Sayfutdinov, T.; Ali, M.; Khamisov, O. Alternating direction method of multipliers for the optimal siting, sizing, and technology selection of Li-ion battery storage. *Electr. Power Syst. Res.* **2020**, *185*, 106388. [CrossRef]
87. Hernández, J.; Sanchez-Sutil, F.; Muñoz-Rodríguez, F. Design criteria for the optimal sizing of a hybrid energy storage system in PV household-prosumers to maximize self-consumption and self-sufficiency. *Energy* **2019**, *186*, 115827. [CrossRef]
88. González-Garrido, A.; Gaztañaga, H.; Saez-De-Ibarra, A.; Milo, A.; Eguia, P. Electricity and reserve market bidding strategy including sizing evaluation and a novel renewable complementarity-based centralized control for storage lifetime enhancement. *Appl. Energy* **2020**, *262*, 114591. [CrossRef]
89. Arias, N.B.; Lopez, J.C.; Hashemi, S.; Franco, J.F.; Rider, M.J. Multi-Objective Sizing of Battery Energy Storage Systems for Stackable Grid Applications. *IEEE Trans. Smart Grid* **2021**, *12*, 2708–2721. [CrossRef]
90. Mulleriyawage, U.G.K.; Shen, W.X. Optimally sizing of battery energy storage capacity by operational optimization of residential PV-Battery systems: An Australian household case study. *Renew. Energy* **2020**, *160*, 852–864. [CrossRef]
91. Wu, Y.; Liu, Z.; Liu, J.; Xiao, H.; Liu, R.; Zhang, L. Optimal battery capacity of grid-connected PV-battery systems considering battery degradation. *Renew. Energy* **2022**, *181*, 10–23. [CrossRef]
92. Shin, H.; Hur, J. Optimal Energy Storage Sizing With Battery Augmentation for Renewable-Plus-Storage Power Plants. *IEEE Access* **2020**, *8*, 187730–187743. [CrossRef]
93. He, G.; Chen, Q.; Kang, C.; Pinson, P.; Xia, Q. Optimal Bidding Strategy of Battery Storage in Power Markets Considering Performance-Based Regulation and Battery Cycle Life. *IEEE Trans. Smart Grid* **2016**, *7*, 2359–2367. [CrossRef]
94. Jiang, Z.; Li, H.; Qu, Z.; Zhang, J. Recent progress in lithium-ion battery thermal management for a wide range of temperature and abuse conditions. *Int. J. Hydrogen Energy* **2022**, *47*, 9428–9459. [CrossRef]
95. Pinson, M.B.; Bazant, M.Z. Theory of SEI Formation in Rechargeable Batteries: Capacity Fade, Accelerated Aging and Lifetime Prediction. *J. Electrochem. Soc.* **2013**, *160*, A243–A250. [CrossRef]

96. Fellner, J.; Loeber, G.; Sandhu, S. Testing of lithium-ion 18650 cells and characterizing/predicting cell performance. *J. Power Sources* **1999**, *81–82*, 867–871. [CrossRef]
97. Qian, K.; Zhou, C.; Yuan, Y.; Allan, M. Temperature effect on electric vehicle battery cycle life in vehicle-to-grid applications. In Proceedings of the CICED 2010 Proceedings, Nanjing, China, 13–16 September 2010. Available online: https://ieeexplore.ieee.org/abstract/document/5736181/ (accessed on 18 August 2022).
98. Ramasamy, R.P.; White, R.E.; Popov, B.N. Calendar life performance of pouch lithium-ion cells. *J. Power Sources* **2005**, *141*, 298–306. [CrossRef]
99. Tarascon, J.-M.; Gozdz, A.; Schmutz, C.; Shokoohi, F.; Warren, P. Performance of Bellcore's plastic rechargeable Li-ion batteries. *Solid State Ion.* **1996**, *86–88*, 49–54. [CrossRef]
100. Sun, S.; Guan, T.; Shen, B.; Leng, K.; Gao, Y.; Cheng, X.; Yin, G. Changes of Degradation Mechanisms of LiFePO4/Graphite Batteries Cycled at Different Ambient Temperatures. *Electrochimica Acta* **2017**, *237*, 248–258. [CrossRef]
101. Zichen, W.; Changqing, D. A comprehensive review on thermal management systems for power lithium-ion batteries. *Renew. Sustain. Energy Rev.* **2021**, *139*, 110685. [CrossRef]
102. Kim, J.; Oh, J.; Lee, H. Review on battery thermal management system for electric vehicles. *J. Appl. Therm. Eng.* **2019**, *149*, 192–212. [CrossRef]
103. Wang, Y.; Tian, J.; Sun, Z.; Wang, L.; Xu, R.; Li, M.; Chen, Z. A comprehensive review of battery modeling and state estimation approaches for advanced battery management systems. *Renew. Sustain. Energy Rev.* **2020**, *131*, 110015. [CrossRef]
104. Marwali, M.; Haili, M.; Shahidehpour, S.; Abdul-Rahman, K. Short term generation scheduling in photovoltaic-utility grid with battery storage. *IEEE Trans. Power Syst.* **1998**, *13*, 1057–1062. [CrossRef]
105. Jayasekara, N.; Masoum, M.A.S.; Wolfs, P.J. Optimal Operation of Distributed Energy Storage Systems to Improve Distribution Network Load and Generation Hosting Capability. *IEEE Trans. Sustain. Energy* **2016**, *7*, 250–261. [CrossRef]
106. Jenkins, D.; Fletcher, J.; Kane, D. Lifetime prediction and sizing of lead–acid batteries for microgeneration storage applications. *IET Renew. Power Gener.* **2008**, *2*, 191–200. [CrossRef]
107. Lee, J.-O.; Kim, Y.-S. Novel battery degradation cost formulation for optimal scheduling of battery energy storage systems. *Int. J. Electr. Power Energy Syst.* **2022**, *137*, 107795. [CrossRef]
108. Mohamed, S.; Shaaban, M.F.; Ismail, M.; Serpedin, E.; Qaraqe, K.A. An Efficient Planning Algorithm for Hybrid Remote Microgrids. *IEEE Trans. Sustain. Energy* **2019**, *10*, 257–267. [CrossRef]
109. Atia, R.; Yamada, N. Sizing and Analysis of Renewable Energy and Battery Systems in Residential Microgrids. *IEEE Trans. Smart Grid* **2016**, *7*, 1204–1213. [CrossRef]
110. Jannesar, M.R.; Sedighi, A.; Savaghebi, M.; Guerrero, J.M. Optimal placement, sizing, and daily charge/discharge of battery energy storage in low voltage distribution network with high photovoltaic penetration. *Appl. Energy* **2018**, *226*, 957–966. [CrossRef]
111. Kim, M.; Kim, K.; Choi, H.; Lee, S.; Kim, H. Practical Operation Strategies for Energy Storage System under Uncertainty. *Energies* **2019**, *12*, 1098. [CrossRef]
112. Zheng, Y.; Zhao, J.; Song, Y.; Luo, F.; Meng, K.; Qiu, J.; Hill, D.J. Optimal Operation of Battery Energy Storage System Considering Distribution System Uncertainty. *IEEE Trans. Sustain. Energy* **2018**, *9*, 1051–1060. [CrossRef]
113. Jacob, R.A.; Bhattacharya, A.; Sharma, S. Planning of battery energy storage system in distribution network considering uncertainty. In Proceedings of the 2017 International Conference on Technological Advancements in Power and Energy (TAP Energy), Kollam, India, 21–23 December 2017; pp. 1–6. [CrossRef]
114. Cao, M.; Xu, Q.; Qin, X.; Cai, J. Battery energy storage sizing based on a model predictive control strategy with operational constraints to smooth the wind power. *Int. J. Electr. Power Energy Syst.* **2020**, *115*, 105471. [CrossRef]
115. Moghaddam, I.N.; Chowdhury, B.H.; Mohajeryami, S. Predictive Operation and Optimal Sizing of Battery Energy Storage With High Wind Energy Penetration. *IEEE Trans. Ind. Electron.* **2018**, *65*, 6686–6695. [CrossRef]
116. Caro-Ruiz, C.; Lombardi, P.; Richter, M.; Pelzer, A.; Komarnicki, P.; Pavas, A.; Mojica-Nava, E. Coordination of optimal sizing of energy storage systems and production buffer stocks in a net zero energy factory. *Appl. Energy* **2019**, *238*, 851–862. [CrossRef]
117. Babacan, O.; Torre, W.; Kleissl, J. Siting and sizing of distributed energy storage to mitigate voltage impact by solar PV in distribution systems. *Sol. Energy* **2017**, *146*, 199–208. [CrossRef]
118. Tang, W.-J.; Yang, H.-T. Optimal Operation and Bidding Strategy of a Virtual Power Plant Integrated With Energy Storage Systems and Elasticity Demand Response. *IEEE Access* **2019**, *7*, 79798–79809. [CrossRef]
119. Keskamol, K.; Hoonchareon, N. Sizing of battery energy storage system for sustainable energy in a remote area. In Proceedings of the 2015 IEEE Innovative Smart Grid Technologies—Asia (ISGT ASIA), Bangkok, Thailand, 3–6 November 2015. [CrossRef]
120. He, G.; Kar, S.; Mohammadi, J.; Moutis, P.; Whitacre, J.F. Power System Dispatch With Marginal Degradation Cost of Battery Storage. *IEEE Trans. Power Syst.* **2021**, *36*, 3552–3562. [CrossRef]
121. Choi, Y.; Kim, H. Optimal Scheduling of Energy Storage System for Self-Sustainable Base Station Operation Considering Battery Wear-Out Cost. *Energies* **2016**, *9*, 462. [CrossRef]
122. Li, J.; Niu, D.; Wu, M.; Wang, Y.; Li, F.; Dong, H. Research on Battery Energy Storage as Backup Power in the Operation Optimization of a Regional Integrated Energy System. *Energies* **2018**, *11*, 2990. [CrossRef]
123. Kabir, M.N.; Mishra, Y.; Ledwich, G.; Dong, Z.Y.; Wong, K.P. Coordinated Control of Grid-Connected Photovoltaic Reactive Power and Battery Energy Storage Systems to Improve the Voltage Profile of a Residential Distribution Feeder. *IEEE Trans. Ind. Inform.* **2014**, *10*, 967–977. [CrossRef]

124. Baker, K.; Hug, G.; Li, X. Energy Storage Sizing Taking Into Account Forecast Uncertainties and Receding Horizon Operation. *IEEE Trans. Sustain. Energy* **2017**, *8*, 331–340. [CrossRef]
125. Mashlakov, A.; Lensu, L.; Kaarna, A.; Tikka, V.; Honkapuro, S. Probabilistic Forecasting of Battery Energy Storage State-of-Charge under Primary Frequency Control. *IEEE J. Sel. Areas Commun.* **2020**, *38*, 96–109. [CrossRef]
126. Kabir, M.N.; Mishra, Y.; Ledwich, G.; Xu, Z.; Bansal, R.C. Improving voltage profile of residential distribution systems using rooftop PVs and Battery Energy Storage systems. *Appl. Energy* **2014**, *134*, 290–300. [CrossRef]
127. Baziar, A.; Kavousi-Fard, A. Considering uncertainty in the optimal energy management of renewable micro-grids including storage devices. *Renew. Energy* **2013**, *59*, 158–166. [CrossRef]
128. Awad, A.S.A.; El-Fouly, T.H.M.; Salama, M.M.A. Optimal ESS Allocation for Load Management Application. *IEEE Trans. Power Syst.* **2015**, *30*, 327–336. [CrossRef]
129. Zamee, M.A.; Won, D. Novel Mode Adaptive Artificial Neural Network for Dynamic Learning: Application in Renewable Energy Sources Power Generation Prediction. *Energies* **2020**, *13*, 6405. [CrossRef]
130. Grantham, A.; Pudney, P.; Ward, L.A.; Whaley, D.; Boland, J. The viability of electrical energy storage for low-energy households. *Sol. Energy* **2017**, *155*, 1216–1224. [CrossRef]
131. Bel, A.M.; Aldik, A.; Al-Awami, A.T.; Alismail, F. Fuzzy Optimization-based Sizing of a Battery Energy Storage System for Participating in Ancillary Services Markets. In Proceedings of the 2018 IEEE Industry Applications Society Annual Meeting (IAS), Portland, OR, USA, 23–27 September 2018. [CrossRef]
132. Atwa, Y.M.; El-Saadany, E.F. Optimal Allocation of ESS in Distribution Systems With a High Penetration of Wind Energy. *IEEE Trans. Power Syst.* **2010**, *25*, 1815–1822. [CrossRef]
133. Pashaei-Didani, H.; Nojavan, S.; Nourollahi, R.; Zare, K. Optimal economic-emission performance of fuel cell/CHP/storage based microgrid. *Int. J. Hydrogen Energy* **2019**, *44*, 6896–6908. [CrossRef]
134. Nojavan, S.; Majidi, M.; Esfetanaj, N.N. An efficient cost-reliability optimization model for optimal siting and sizing of energy storage system in a microgrid in the presence of responsible load management. *Energy* **2017**, *139*, 89–97. [CrossRef]
135. Brivio, C.; Mandelli, S.; Merlo, M. Battery energy storage system for primary control reserve and energy arbitrage. *Sustain. Energy Grids Netw.* **2016**, *6*, 152–165. [CrossRef]
136. Fossati, J.P.; Galarza, A.; Martín-Villate, A.; Fontán, L. A method for optimal sizing energy storage systems for microgrids. *Renew. Energy* **2015**, *77*, 539–549. [CrossRef]
137. Hannan, M.A.; Young, Y.S.; Hoque, M.M.; Ker, P.J.; Uddin, M.N. Lithium Ion Battery Thermal Management System Using Optimized Fuzzy Controller. In Proceedings of the 2019 IEEE Industry Applications Society Annual Meeting, Baltimore, MD, USA, 29 September–3 October 2019. [CrossRef]
138. Hannan, M.A.; Ali, J.A.; Lipu, M.S.H.; Mohamed, A.; Ker, P.J.; Mahlia, T.M.I.; Mansor, M.; Hussain, A.; Muttaqi, K.M.; Dong, Z.Y. Role of optimization algorithms based fuzzy controller in achieving induction motor performance enhancement. *Nat. Commun.* **2020**, *11*, 3792. [CrossRef]
139. Kalavani, F.; Mohammadi-Ivatloo, B.; Karimi, A. Stochastic optimal sizing of integrated cryogenic energy storage and air liquefaction unit in microgrid. *Renew. Energy* **2019**, *136*, 15–22. [CrossRef]
140. Sui, X.; Tang, Y.; He, H.; Wen, J. Energy-Storage-Based Low-Frequency Oscillation Damping Control Using Particle Swarm Optimization and Heuristic Dynamic Programming. *IEEE Trans. Power Syst.* **2014**, *29*, 2539–2548. [CrossRef]
141. Nguyen, T.A.; Crow, M.L.; Elmore, A.C. Optimal Sizing of a Vanadium Redox Battery System for Microgrid Systems. *IEEE Trans. Sustain. Energy* **2015**, *6*, 729–737. [CrossRef]
142. Hu, X.; Murgovski, N.; Johannesson, L.M.; Egardt, B. Optimal Dimensioning and Power Management of a Fuel Cell/Battery Hybrid Bus via Convex Programming. *IEEE/ASME Trans. Mechatron.* **2015**, *20*, 457–468. [CrossRef]
143. Pham, C.M.; Tran, Q.T.; Bacha, S.; Hably, A.; Nugoc, A.L. Optimal sizing of battery energy storage system for an island microgrid. In Proceedings of the IECON 2018—44th Annual Conference of the IEEE Industrial Electronics Society, Washington, DC, USA, 21–23 October 2018; pp. 1899–1903. [CrossRef]
144. Grover-Silva, E.; Girard, R.; Kariniotakis, G. Optimal sizing and placement of distribution grid connected battery systems through an SOCP optimal power flow algorithm. *Appl. Energy* **2018**, *219*, 385–393. [CrossRef]
145. Nick, M.; Cherkaoui, R.; Paolone, M. Optimal Allocation of Dispersed Energy Storage Systems in Active Distribution Networks for Energy Balance and Grid Support. *IEEE Trans. Power Syst.* **2014**, *29*, 2300–2310. [CrossRef]
146. Herrera, V.I.; Gaztanaga, H.; Milo, A.; Saez-De-Ibarra, A.; Etxeberria-Otadui, I.; Nieva, T. Optimal Energy Management and Sizing of a Battery–Supercapacitor-Based Light Rail Vehicle With a Multiobjective Approach. *IEEE Trans. Ind. Appl.* **2016**, *52*, 3367–3377. [CrossRef]
147. Kerdphol, T.; Qudaih, Y.; Mitani, Y. Battery energy storage system size optimization in microgrid using particle swarm optimization. In Proceedings of the IEEE PES Innovative Smart Grid Technologies, Europe, Istanbul, Turkey, 12–15 October 2014. [CrossRef]
148. Bahmani-Firouzi, B.; Azizipanah-Abarghooee, R. Optimal sizing of battery energy storage for micro-grid operation management using a new improved bat algorithm. *Int. J. Electr. Power Energy Syst.* **2014**, *56*, 42–54. [CrossRef]
149. Chakraborty, S.; Funabashi, T.; Saber, A.; Toyama, H.; Senjyu, T. Determination methodology for optimising the energy storage size for power system. *IET Gener. Transm. Distrib.* **2009**, *3*, 987–999. [CrossRef]

150. Qiu, T.; Xu, B.; Wang, Y.; Dvorkin, Y.; Kirschen, D.S. Stochastic Multistage Coplanning of Transmission Expansion and Energy Storage. *IEEE Trans. Power Syst.* **2017**, *32*, 643–651. [CrossRef]
151. Murty, V.V.S.N.; Kumar, A. Multi-objective energy management in microgrids with hybrid energy sources and battery energy storage systems. *Prot. Control Mod. Power Syst.* **2020**, *5*, 1–20. [CrossRef]
152. Secchi, M.; Barchi, G.; Macii, D.; Moser, D.; Petri, D. Multi-objective battery sizing optimisation for renewable energy communities with distribution-level constraints: A prosumer-driven perspective. *Appl. Energy* **2021**, *297*, 117171. [CrossRef]
153. Mohamad, F.; Teh, J.; Lai, C.-M. Optimum allocation of battery energy storage systems for power grid enhanced with solar energy. *Energy* **2021**, *223*, 120105. [CrossRef]
154. Mulleriyawage, U.; Shen, W. Impact of demand side management on optimal sizing of residential battery energy storage system. *Renew. Energy* **2021**, *172*, 1250–1266. [CrossRef]
155. Fioriti, D.; Pellegrino, L.; Lutzemberger, G.; Micolano, E.; Poli, D. Optimal sizing of residential battery systems with multi-year dynamics and a novel rainflow-based model of storage degradation: An extensive Italian case study. *Electr. Power Syst. Res.* **2022**, *203*, 107675. [CrossRef]
156. Hamidan, M.-A.; Borousan, F. Optimal planning of distributed generation and battery energy storage systems simultaneously in distribution networks for loss reduction and reliability improvement. *J. Energy Storage* **2022**, *46*, 103844. [CrossRef]
157. Shabbir, N.; Kutt, L.; Astapov, V.; Jawad, M.; Allik, A.; Husev, O. Battery Size Optimization With Customer PV Installations and Domestic Load Profile. *IEEE Access* **2022**, *10*, 13012–13025. [CrossRef]
158. Khezri, R.; Mahmoudi, A.; Aki, H. Optimal planning of solar photovoltaic and battery storage systems for grid-connected residential sector: Review, challenges and new perspectives. *Renew. Sustain. Energy Rev.* **2022**, *153*, 111763. [CrossRef]
159. Steckel, T.; Kendall, A.; Ambrose, H. Applying levelized cost of storage methodology to utility-scale second-life lithium-ion battery energy storage systems. *Appl. Energy* **2021**, *300*, 117309. [CrossRef]
160. Horesh, N.; Quinn, C.; Wang, H.; Zane, R.; Ferry, M.; Tong, S.; Quinn, J.C. Driving to the future of energy storage: Techno-economic analysis of a novel method to recondition second life electric vehicle batteries. *Appl. Energy* **2021**, *295*, 117007. [CrossRef]
161. Chai, S.; Xu, N.Z.; Niu, M.; Chan, K.W.; Chung, C.Y.; Jiang, H.; Sun, Y. An Evaluation Framework for Second-Life EV/PHEV Battery Application in Power Systems. *IEEE Access* **2021**, *9*, 152430–152441. [CrossRef]
162. Martinez-Laserna, E.; Gandiaga, I.; Sarasketa-Zabala, E.; Badeda, J.; Stroe, D.I.; Swierczynski, M.; Goikoetxea, A. Battery second life: Hype, hope or reality? A critical review of the state of the art. *Renew. Sustain. Energy Rev.* **2018**, *93*, 701–718. [CrossRef]
163. Ortega-Vazquez, M.A.; Kirschen, D.S. Estimating the Spinning Reserve Requirements in Systems With Significant Wind Power Generation Penetration. *IEEE Trans. Power Syst.* **2009**, *24*, 114–124. [CrossRef]

applied sciences

Review

Voltage Optimization in PV-Rich Distribution Networks—A Review

Marina Dubravac, Krešimir Fekete *, Danijel Topić and Marinko Barukčić

Faculty of Electrical Engineering, Computer Science and Information Technology Osijek, 31000 Osijek, Croatia
* Correspondence: kresimir.fekete@ferit.hr

Abstract: There is a rising trend to integrate different types of distributed generation (DG), especially photovoltaic (PV) systems, on the roofs of existing consumers, who then become prosumers. One of the prosumer impacts is voltage violations, which conventional strategies find hard to solve. However, some prosumers, such as those with PV with inverters in their configurations, can actively participate in voltage optimization. To help find the optimal PV inverter setting with the objective of voltage optimization, an optimal power flow (OPF) can be a promising and reliable tool. This paper tries to shed light on the complex problem of voltage optimization in distribution networks (DNs) with PV prosumers. Relevant scientific papers are analyzed and optimization characteristics such as objective functions, variables, and constraints are summarized. Special attention is given to the systematization and classification of papers according to the mathematical formulation of the optimization problem (linear, nonlinear, integer, etc.) and the applied solving methods. Both analytical and computational intelligence optimization methods as well as their advantages and limitations are considered. Papers are also categorized according to the distribution network model used for testing the developed solutions.

Keywords: active distribution networks; optimal power flow; prosumers; PV inverter volt/VAR optimization; review

Citation: Dubravac, M.; Fekete, K.; Topić, D.; Barukčić, M. Voltage Optimization in PV-Rich Distribution Networks—A Review. *Appl. Sci.* **2022**, *12*, 12426. https://doi.org/10.3390/app122312426

Academic Editors: Luis Hernández-Callejo, Jesús Armando Aguilar Jiménez and Carlos Meza Benavides

Received: 4 November 2022
Accepted: 1 December 2022
Published: 5 December 2022

1. Introduction

The participation of renewable energy sources (RESs), battery storage systems, and other flexible loads, commonly referred to as prosumers, changes the character of distribution networks (DNs) from passive to active. For this reason, both generating units and loads are included in the determination of power flow and voltage profile. The nature of prosumers is unpredictable and intermittent, so existing DNs are not adapted to their influence. This is particularly evident in frequent voltage violations [1]. A voltage rise is addressed as a major issue caused by prosumers and impacts DNs [2]. The distribution system operator (DSO) is responsible for maintaining voltage within the allowable limits for the secure operation of DNs [3]. However, voltage control mechanisms that were once applicable in passive DNs become less valid and new mechanisms are required [4–6].

While DG can cause voltage violation in the DN, the same DG can help solve the problem of voltage violation, for example by managing active or reactive power. This problem/solution principle is especially interesting in the case of reactive power management [7]. When it comes to RESs in DNs, the main representative is a photovoltaic (PV) system [8,9]. Traditionally, most DSOs require PVs to operate with the unit or fixed power factor [1]. PV inverters have several modes of operation, but volt–VAR control has become certainly significant for voltage optimization. A major advantage of using a PV inverter and volt–VAR control is that reactive power can be injected/absorbed even where there is no production. Since prosumers contribute to the complexity and unpredictability of such DNs, it is crucial to use optimization methods and analyzed software tools that allow DN monitoring and finding suitable and optimal set points for PV inverters. In addition, optimal power flow (OPF) has proven to be an efficient tool for the operation of active DNs. In order to illustrate the problem/solution principle of PV inverters, a case study is developed in this paper presenting the low voltage DN with the connected PVs. Different power flow simulations are presented and they include:

- the situation when PVs production is maximal and voltage rise along the feeder is present;
- the situation when PV inverters inject reactive power.

Nowadays, optimal management of active DNs represents an attractive topic, and a large and growing body of literature focuses on this topic. Some related reviews on the optimization DNs are listed in Table 1. Refs. [10,11] show the impact of PV on DNs, the related problems, and possible solutions. Refs. [12–14] study voltage violation mitigation and voltage control strategies. Ref. [12] deals with voltage control methods in DNs with PVs and their advantages and disadvantages. In addition to [12], a comprehensive review of voltage control methods is given in [13]. In [14], voltage control models and methods are divided according to the communication strategy. The application of OPF in DNs is discussed in [15,16]. The researchers in [15] address a probabilistic approach to solving OPF in active DN. The authors in [16] study OPF in smart DNs and microgrids according to objectives, methods, and future challenges.

Table 1. Related review papers on the optimization of DNs.

Review Paper	Year	Focused Topics	Prosumer	Notes
[10]	2016	Impact of distributed generation (DG) on voltage control on DNs	Yes	Reviewed voltage control with DG with a focus on smart network technologies—demand side management (DSM) and energy storage systems (ESS)
[11]	2016	PV impact on DNs including voltage regulation issues, harmonic, and islanding operation	Yes	Reviewed issues caused by PV penetration in DN insight voltage regulation, harmonic, and islanding operation, and proposed technical solution
[14]	2017	Distributed and decentralized voltage control in smart DNs	Yes	Reviewed smart DNs according to communication systems, control models, and methods
[16]	2017	Application of OPF in smart DNs and microgrids	Yes	Reviewed OPF according to objectives, constraints, methods, and challenges
[12]	2018	Mitigation methods for voltage regulation in DNs with PV	Yes	Discussed ESS strategies, active power curtailment-based strategies, and reactive power control strategies
[13]	2020	Mitigation methods for voltage violation in DNs with PV systems	Yes	Presented different mitigation methods for voltage regulation in DNs and their merits and shortcomings
[15]	2022	Probabilistic OPF in active DNs	Yes	Scientometric review of OPF in active DNs—characteristics and challenges

While refs. [15,16] present interesting and useful reviews of OPF application to modern DNs, they are mainly focused on active power objectives [16] and probabilistic OPF [15]. So far, the research gap is present in the area of voltage optimization objectives in the case of using PV inverters for reactive power management. The problem/solution principle of using PV inverter reactive power for voltage mitigation motivated the authors of this paper to focus their review on scientific papers that applies OPF for voltage optimization in the DN using PV inverter reactive powers. The contributions of this review paper are:

1. Summary and classification of OPF objectives and variables in the case of voltage optimization in the DNs using PVs reactive power.

2. Comparison of the used mathematical formulations of the OPFs and their connections to analytical or computational intelligence solution methods.
3. Review of the different DN examples that are used for testing the developed optimization solutions.

This paper aims to provide the readers with starting points for OPF applications in PV-rich DNs and, in some way, to compare with conventional OPF. Therefore, the authors searched several bibliographical databases—IEEE Xplore, ScienceDirect, and MDPI.

The paper is structured as follows: Section 2 gives a prosumer definition, describes the voltage control capabilities of PV inverter, and presents the case study that illustrates the impact of PVs on the DN voltages. Section 3 addresses OPF in PV-rich DN—objectives and variables. OPF formulation and solution methods are discussed in Section 4. Test network models used in the literature are presented in Section 4. Conclusions with a note on future research are given in Section 5.

2. PV Prosumers in Distribution Network

In this section, various definitions and configurations of prosumers are presented. Then, the focus is on PVs and their voltage control capabilities. Finally, theoretical foundations are supported by the case study in which the part of the low voltage DN is presented and various cases are simulated.

2.1. Prosumer Definition and Configuration

Historically, the term "prosumer" was first mentioned in 1980 in Alvin Toffler's book *The Third Wave* [17]. In this book, the author discusses the transition of society from an Industrial Era in which production and consumption were separated to the Information Era. In this new concept of society, the term prosumer is created by combining a producer and a consumer. Nowadays, with the increase of integration of PVs in DNs, the notion of prosumer and prosumerism has become increasingly significant in electrical engineering [18]. Therefore, many authors propose the definition of a prosumer. The authors in [19–21] define a prosumer as an end-user that consumes electrical energy, acts as an energy producer, and shares surplus energy with utility networks and other consumers. In [22], prosumers are described as energy consumers or energy producers in different periods depending on their electricity demand and price. The focus is on the prosumer that only generates electricity. In [23,24], the authors include a facility for electricity generation and energy storage systems in the prosumer definition. The definition of prosumer was expanded in [25,26] and includes consumers that generate electricity and/or contain in their configuration home energy management systems (EMS), ESS, electric vehicles (EVs), and electric vehicle-to-grid (V2G) systems. In [27], a prosumer has been deemed as a consumer or an electricity producer and can be actively managed.

The European Union defined an active consumer uniform in its 2016 directive [28]. According to [28,29], an active consumer is "a customer or a group of jointly acting customers who consume, store or sell electricity generated on their premises, including through aggregators, or participate in demand response or energy efficiency schemes provided that these activities do not constitute their primary commercial or professional activity".

While a variety of definitions of the term prosumer have been suggested, this paper uses the following definition: a prosumer is an entity that not only withdraws/retracts energy from a network but also produces energy that can be consumed, stored or sold to the network and other consumers and actively participates in providing more flexibility such as voltage and reactive power control.

Prosumer configuration and interconnection with DN is shown in Figure 1. The main representative of distributed generation is the PV system. It often includes an ESS such as a battery in its configuration and together forms a hybrid system. EVs, electric V2G vehicles, smart home EMS, and other flexible loads represent demand-side management (DSM).

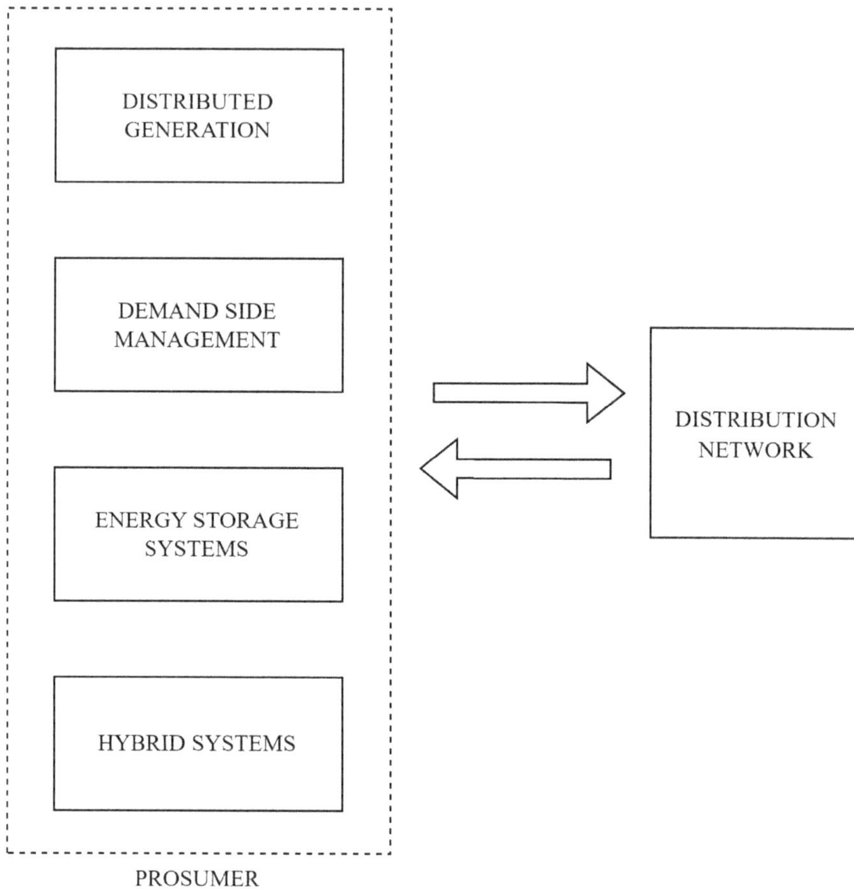

Figure 1. Prosumer configuration and interconnection with DNs.

2.2. PV System Capabilities for Voltage Optimization

This paper deals with PV prosumers with reactive power capability, i.e., other prosumer types are not considered. DSOs have the main responsibility for voltage optimization in DNs. Traditionally, the available variables are limited to the capacitor bank placement, tap changing transformer, network reconfiguration, cross-section enhancement, etc. However, some researchers [30,31] suggest that the aforementioned strategies may not be effective for prosumer-based DNs due to their slow response. The application of PV inverters represents a promising solution and in combination with already present control mechanisms can give results, so several studies have proposed their use for voltage optimization [32–36]. Different modes of operation are possible for PV inverters and the authors of [30] distinguish the following:

- fixed power factor mode;
- volt–VAR control;
- volt–watt control;
- mode for power rate limit;
- voltage balance mode.

In the fixed power factor mode, the power factor is maintained at a constant value and thus voltages are directly affected. Voltage control in the volt–watt control mode is achieved by active power from PVs. In the power rate limit mode, the rate of active power output from the PV inverter is limited. In volt–VAR control, reactive power from the PV inverter is used for voltage optimization. The general operating principle of volt–VAR control is described using a volt–VAR curve shown in Figure 2 [37]. The volt–VAR curve represents a relation between a voltage value at the point of common coupling (PCC) and reactive power from the PV inverter. If the voltage value on PCC is lower than the specific threshold, the PV inverter injects reactive power. On the other hand, in the case of the higher voltage value on PCC, reactive power is absorbed.

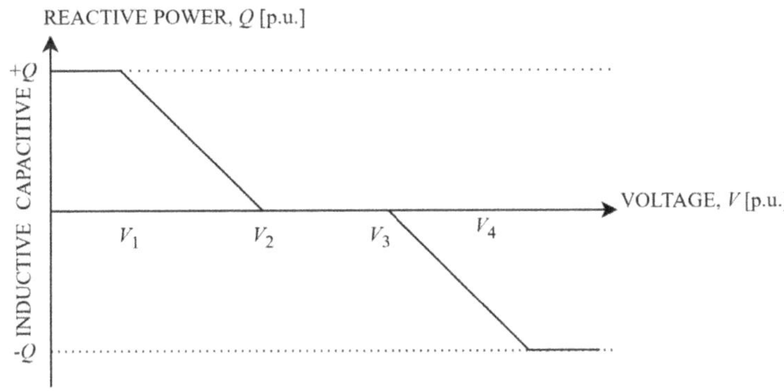

Figure 2. Volt–VAR curve of the PV inverter. Inductive reactive power reduces voltage value while capacitive reactive power increases the voltage value.

Reactive power capability determines the amount of reactive power available from the inverter [38]. Figure 3 represents reactive power capability determined with vectors of apparent power S and active power P [8,38,39]. The reactive power of the PV inverter depends on the active power and can be determined as:

$$Q_1^2 \leq \sqrt{S_1^2 - P_1^2} \tag{1}$$

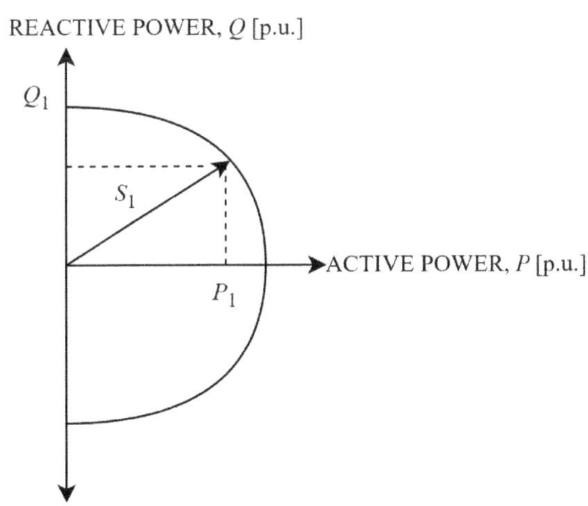

Figure 3. Power capability curve of PV inverter.

2.3. *PV Inverter Impact on Distribution Feeder Voltage Profile*

To provide a better insight into the voltage problem in DN caused by PVs, a case study is carried out. Two PV inverter control modes are chosen to clarify its capabilities for voltage optimization. The DN model, presented in Figure 4 [40], consists of three radial feeders supplied by a 10/0.4 kV substation. Each feeder supplies 20 residential consumers. More information about the network model can be found in [40]. It is assumed that half of the residential consumers have PV systems on their rooftops. The nominal power of each PV plant is 5 kW. The case studied in the simulations corresponds to maximum production and consumption of 0.2 kW with an inductive power factor of 0.9.

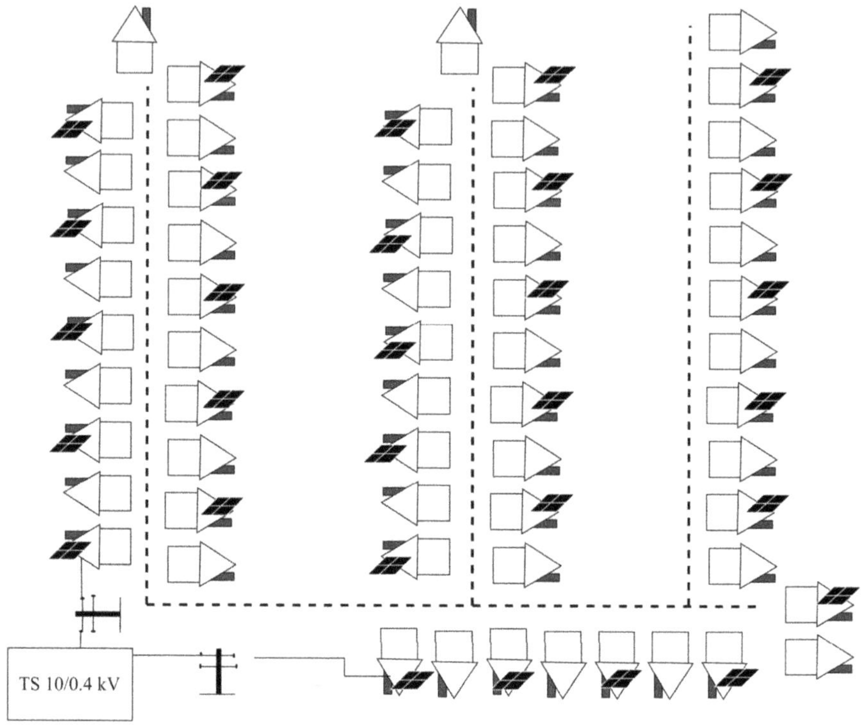

Figure 4. Model of DN used in the case study.

DIgSILENT PowerFactory [41] software is used for case study implementation and the conventional power flow is analyzed. Two modes of PV inverter operation are used in simulations: fixed power factor mode and volt–VAR control mode. The results for the fixed power factor mode are presented in Figure 5. The power factor range is taken from a real-life example of an inverter [42]. The voltage profiles at different power factor values are compared. There is an increase in the voltage profile at the unit power factor. The voltage profile is corrected by changing the power factor.

The comparison of voltage profiles at the unit power factor and the applied volt–VAR control mode is shown in Figure 6. In the case without voltage control (unit power factor), there is a voltage rise in the distribution feeder caused by PVs. The voltage values are in the range of 1.005 p.u. to 1.05 p.u. In the volt–VAR control mode, the voltage values are lower than the unit power factor and are in the narrower range of 0.992 p.u. to 1.005 p.u. These values are more acceptable for the operation of DN, i.e., voltage deviation is smaller.

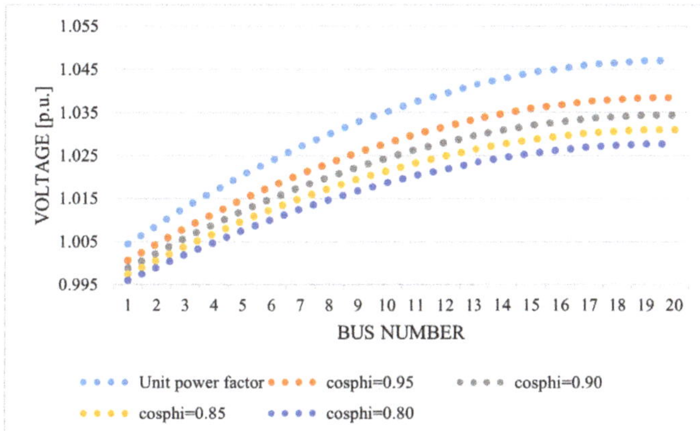

Figure 5. Voltage profile in a distribution feeder obtained using the fixed power factor control mode.

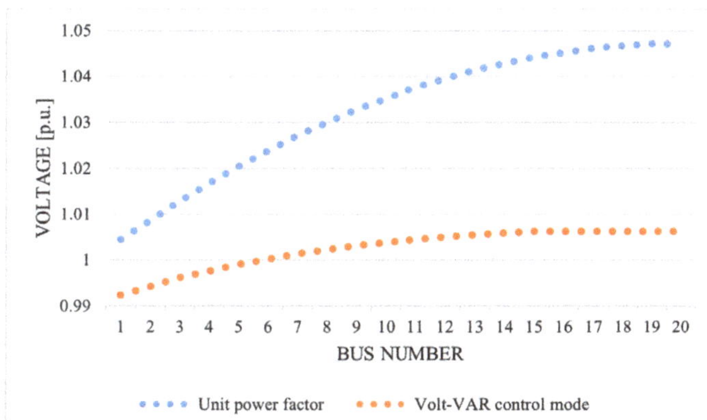

Figure 6. Voltage profile in the distribution feeder obtained using the volt–VAR control mode.

The obtained results show an improvement in the voltage profile of DN compared to the case without voltage control (unit power factor of the PV inverter). Voltage control in PV prosumer-rich DNs has a positive impact. The case study shows the possibilities of PV inverters regarding voltage control and the situation when there are lots of inverters placed at different positions in the DN. Determining the optimal operating point of the PV inverters imposes using optimization algorithms from which the OPF are imposed as a logical solution.

3. Voltage Optimization in PV-Rich Distribution Networks—Objectives and Variables

The OPF concept was proposed in the early 1960s [43] as an enhancement of economic dispatch to find the optimal solution for controlling variable settings under different constraints. The OPF is used as a universal term for problems associated with network optimization [44–47]. The OPF is ordinarily modeled to the appliance on transmission level considering large generating units. Besides the fundamental variables, the OPF model may contain ancillary generation units and variables representing the other segments of the power system used for optimal operation.

The transmission network (TN) diverges from DN in topology, nature, electrical parameters, power flow values, and a number of control devices. Unlike TNs, DNs are inherently unbalanced and more complex [48]. The reason for the imbalance is that the DN supplies unequal single-phase loads and contains unequal conductor interspace of three-phase segments [49,50]. The *R/X* ratio is high in DNs and contributes to the complexity of control and optimization. In contrast, the *R/X* ratio is low for TNs. Compared to DNs, TNs have a few direct consumers. The simple control and a well-built communication system of TNs are the main reasons why OPF has applied only at the transmission level. The integration of DGs and flexible loads such as EVs makes OPF feasible in DN optimization. To incorporate unpredictable DG and to exploit the potential of flexible loads, OPF became imminent for DNs [51]. Although there is no official record in the literature of the beginning of the application of OPF in DNs, it can be said it started with the integration of different types of prosumers in DNs [48].

3.1. General Formulation—Objectives and Variables

The OPF problem can be described as minimizing the objective function while taking equality and inequality constraints into account [48]:

$$minF(\mathbf{x},\mathbf{u}) = 0 \tag{2}$$

$$g(\mathbf{x},\mathbf{u}) = 0 \tag{3}$$

$$h(\mathbf{x},\mathbf{u}) \leq 0 \tag{4}$$

where $F(\mathbf{x},\mathbf{u})$ represents the objective function and $g(\mathbf{x},\mathbf{u})$ represents nonlinear equality constraints i.e., power flow equations, $h(\mathbf{x},\mathbf{u})$ represents nonlinear inequality constraints. The vectors **x** and **u** present state variables, and control variables, respectively.

In [48], the generally used objectives for OPF formulation are given. It should be noted that the objectives and constraints must be modeled accurately to obtain a satisfactory solution.

Scientific papers are included in this review if at least one of the objectives is voltage optimization and if one of the optimization variables is PV inverter reactive power injection. Furthermore, the voltage optimization problem is mostly described as the objective of voltage deviation (VD) minimization, i.e., maintaining voltages within boundaries determined by grid codes. The general mathematical expression for VD is:

$$V_{dev} = \sum_{i \in N} (V_i - V^{nom})^2 \tag{5}$$

where:

V_{dev}—voltage deviation;

V_i—voltage at bus i;

V^{nom}—nominal voltage.

Another objective that appears is related to the voltage unbalance, commonly presented as the voltage unbalance factor (VUF). The definition of VUF is given in [52] as the ratio of negative $V^{-}_{sequence}$ and positive $V^{+}_{sequence}$ voltage sequences and is most often expressed in percentages:

$$VUF = \frac{V^{-}_{sequence}}{V^{+}_{sequence}}. \tag{6}$$

In addition to voltage optimization, the following objectives also appear: (i) power loss minimization [53–55], (ii) on load tap changer (OLTC) switching operation minimization [56], (iii) PV cost minimization [38], (iv) reactive power injection/absorption minimization [57], (v) active power curtailment (APC) minimization [58], (vi) cost of purchased energy minimization [59], (vii) peak shaving minimization [59], and (viii) security margin index (SMI) minimization [59]. The mathematical expressions of the commonly used objectives are given in Table 2.

Table 2. Mathematical expressions of the commonly used objectives in voltage optimization problems.

Objective	Formulation	Explanation
Power loss minimization [53–55]	$\sum_{k=i}^{N} g_{ik}(V_i^2 + V_k^2 - 2V_iV_k \cos\theta_{ik})$ $\sum_{(i,k)\in B} r_{ik} l_{ik,t}$	V_i, V_k—voltage magnitude at i th and k th buses θ_{ik}—phase angle ith and kth elements of conductance g_{ik} $l_{ik,t}$ and r_{ik}—square value of current and resistance of branch line from bus i to bus k
OLTC tap operation minimization [56]	$\sum_{m=1}^{M} s_{m,t}$	M—number of discrete devices $s_{m,t}$—status of discrete device m at time t
APC minimization [58]	$\sum_{i=1}^{N} \lvert P_{t,i}^{PV,curt}\rvert$	$P_{t,i}^{PV,curt}$—curtailed active power of PV at time t
PV inverter loss minimization [38]	$\sum_{p=1}^{3}\sum_{i\in\gamma}(k_{i1}^{p} {S_{PVi}^{p}}^{2} + k_{i2}^{p} S_{PVi}^{p} + k_{i3}^{p})$	p—phases γ—set of buses with PVs S_{PVi}^{p}—apparent power k_{i1}^{p}, k_{i2}^{p}, k_{i3}^{p}—coefficients of each inverter's efficiency data
Reactive power inj./abs. minimization [57]	$\lvert Q_{inj.}\rvert/\lvert Q_{abs.}\rvert$	$Q_{inj.}$ and $Q_{abs.}$—injected/absorbed reactive power
Cost of energy minimization [59]	$\sum_{t\in T}\alpha_t P_{1,t}\Delta t$	α_t—price of energy at tth time $P_{1,t}$—active power imported from the external network at time t Δt—duration of time intervals
Security margin maximization [59]	$\sum_{t\in T}(1 - min\lvert\frac{I_{l,t}-I_l^r}{I_l^r}\rvert)$	$I_{l,t}$—line current in lth line at time t I_l^r—ampacity of line current in lth line

In power systems, the conventional power flow is both nonlinear and nonconvex and commonly solved by the Newton–Raphson iterative method. In constrained OPF applications, equality constraints incorporate conventional power flow equations and other constraints to ensure balance. A detailed version of the power flow is named AC power flow [60]. AC power flow as a constraint in OPF is most often formulated in the polar form [60]:

$$P_i = \sum_{k=i}^{N} |V_i||V_k||Y_{ik}|\cos(\delta_i - \delta_k - \theta_{ik}) \tag{7}$$

$$Q_i = \sum_{k=i}^{N} |V_i||V_k||Y_{ik}|\sin(\delta_i - \delta_k - \theta_{ik}) \tag{8}$$

where:

P_i—active power at bus i;
Q_i—reactive power at bus i;
V_i—voltage magnitude at bus i;
V_k—voltage magnitude at bus k;
Y_{ik}—ikth element of bus admittance matrix $\mathbf{Y}_{bus}$;
δ_i—voltage phase angle at ith bus;
δ_k—voltage phase angle at kth bus;
θ_{ik}—phase angle of ikth element of bus admittance matrix $\mathbf{Y}_{bus}$.

Besides AC power flow, the authors use two other formulation approaches: decoupled AC power flow [49] and DC power flow [50]. In decoupled AC power flow, active and reactive powers are decoupled as a function of voltage angle and voltage magnitude, respectively. Assumptions made for the DC power flow formulation include purely imaginary elements of Y and a small difference between two voltage angles of two adjacent busses.

Various inequality constraints are given in [48,61]:

- control variables limit;
- limits for power generation (active and reactive power upper and lower limits);
- network operational limit determined in the network analysis (e.g., MVA limit).

Voltage optimization needs to meet the following constraint requirements:

1. Power flow equations given as Equations (7) and (8);
2. Voltage constraint
$$V_i^{min} \leq V_i \leq V_i^{max} \quad (9)$$
where V_i^{min} and V_i^{max} are the lower and upper voltage limits.
3. PV active and reactive power constraint
$$0 \leq P_{PV,i} \leq P_{PV,av,i} \quad (10)$$
$$-\sqrt{S_{PV,i}^2 - P_{PV,i}^2} \leq Q_{PV,i} \leq \sqrt{S_{PV,i}^2 - P_{PV,i}^2} \quad (11)$$
where $P_{PV,i}$, , Q_{PV_i}, and $S_{PV,i}$ are active, reactive, and apparent powers at bus i. $P_{PV,av,i}$ is available active power at bus i.
4. Line current (thermal) constraint
$$I_{ik}^{min} \leq I_{ik} \leq I_{ik}^{max} \quad (12)$$
where I_{ik}^{min} and I_{ik}^{max} are the lower and upper limits of the line current between buses i and k.
5. OLTC tap position constraint (if it is included)
$$T_i^{min} \leq T_i \leq T_i^{max} \quad (13)$$
where T_i^{min} and T_i^{max} are the lower and upper positions of OLTC tap at bus i.
6. Capacitor constraint (if it is included)
$$Q_{Ci}^{min} \leq Q_{Ci} \leq Q_{Ci}^{max} \quad (14)$$
where Q_{Ci}^{min} and Q_{Ci}^{max} are the lower and upper limits of capacitor reactive power at bus i.
7. Energy storage constraint (if it is included)
$$SoC_{i,t}^{min} \leq SoC_{i,t} \leq SoCi,t^{max} \quad (15)$$
where $SoC_{i,t}^{min}$ and $SoC_{i,t}^{max}$ are the lower and upper limits of the charge state of the storage system at time t.

The voltage optimization problem can be single-objective or multi-objective. OPF objectives and variables used in the review papers are categorized and summarized in Table 3.

Table 3. Overview of objectives and variables.

Reference	Single/ Multi-Objective	Objectives	Variables
[62]	multi-objective	min VD from 0.95 pu threshold, min losses, min reactive power from capacitors	PV inverter reactive power, OLTC, SC, and
[63]	multi-objective	min losses, min VD—DN, min active power curtailed from available power—prosumer	PV active and reactive power
[64]	multi-objective	min VD from expected CVR voltage, min losses	PV inverter reactive power, OLTC/AVR, and CBs

Table 3. *Cont.*

Reference	Single/ Multi-Objective	Objectives	Variables
[65]	single-objective	min VD	PV inverter reactive power, OLTC
[38]	multi-objective	min losses, min VD, min VUF, min PV generation cost, min PV APC cost	PV inverter reactive power
[66]	multi-objective	min losses, min VD, improvement VSI	PV inverter reactive power and static compensator
[53,54,67–69]	multi-objective	min losses, min VD	PV inverter reactive power
[70]	multi-objective	min VD, min losses	PV inverter reactive power, CBs, and OLTC
[71]	single-objective	min VUF	PV inverter active and reactive power, power injected by TS
[72]	single-objective	min VD	PV inverter reactive power
[73]	multi-objective	min losses, min cost of APC and generated/consumed reactive power, min VD	PV inverter reactive power
[74]	multi-objective	min VD, min voltage unbalance	PV inverter reactive power, OLTC, VR, and CB
[75]	multi-objective	min losses, min VD, min VUF	PV inverter reactive power
[32]	single-objective	min VUF	PV inverter reactive power
[76]	single-objective	min VD	PV inverter reactive power, OLTC
[77]	multi-objective	min losses, min VD, min control action of OLTC and SC	PV inverter reactive power, OLTC, SC
[78]	single-objective	min VD	PV inverter reactive power, OLTC
[79]	multi-objective	min VD, min losses	PV inverter reactive power, OLTC, and SC
[57]	multi-objective	min VD, min losses, min reactive power injection, and absorption	PV inverter reactive power
[80]	multi-objective	min VD, min losses	PV inverter reactive power, OLTC
[81]	single-objective	min VD	PV inverter reactive power, OLTC, and VR
[58]	multi-objective	min VD, min losses, min APC	PV inverter reactive power, OLTC and CB
[55]	multi-objective	min VD, min losses	PV and EV inverter reactive power, the compensation device
[56]	multi-objective	min VD, min OLTC tap operation	PV inverter reactive power, OLTC
[82]	single-objective	min VD	PV inverter reactive power, charge/discharge rate of ESS
[83]	multi-objective	min losses, min VUF	PV inverter reactive power
[34]	multi-objective	min cost, min losses, min cost associated with active power setpoints, min VD	PV inverter active and reactive power
[84]	multi-objective	min active and reactive power output, min VD	PV inverter active and reactive power
[37]	multi-objective	min VD, min losses, min peak of reactive power	PV inverter reactive power
[85]	multi-objective	min VD, min losses	PV inverter reactive power, OLTC, CB
[59]	multi-objective	min VUF, min cost of purchased energy, min peak shaving, min losses, min SMI, min VD	PV inverter reactive power, EV active and reactive power, bus voltages at all time intervals of the day
[86]	multi-objective	min VUF, min losses	PV inverter reactive power, OLTC, CB
[87]	single-objective	min VD	PV inverter reactive power
[88]	multi-objective	min VD, min losses	PV inverter reactive power, SC, OLTC, ESS
[89]	multi-objective	min VD, min losses, min number of switching operations of OLTC and CB, min APC	PV inverter active and reactive power, CB, OLTC
[90]	multi-objective	min VD, min operational cost	PV inverter active and reactive power, CB, OLTC, ESS
[91]	multi-objective	min VD, min losses, min peak of reactive power	PV reactive power
[39]	single-objective	min VD	PV inverter reactive power

The abbreviations are as follows: CVR: conservation voltage reduction; VSI: voltage stability index; SC: shunt capacitor; AVR/VR: automatic voltage regulator; CB: capacitor bank.

According to the literature review, the multi-objective problem prevails.

Besides PV inverter reactive power, other variables include: (i) PV active power, (ii) OLTC, (iii) CB, (iv) static compensator, (v) reactive power from the substation, (vi) VRs, (vii) charge/discharge rate of ESS, (viii) EV active power, and (ix) SC.

3.2. *Objectives and Variables—Discussion*

Figure 7 represents objectives quantitatively. Almost all objectives include VD. In multi-objective problems, VD is most combined with losses, however, many other objectives also appear.

If OPF is regarded as a part of the distribution energy management system (DEMS), the dominance of the multi-objective formulation of OPFs is logical. DSO tries to reach the optimal operation point regarding several objectives and the most commonly used ones are loss minimization together with voltage deviation minimization. Additionally, active power curtailment (APC) minimization is frequently a combined objective with the minimization of voltage deviations. Other objectives are rare and they are used only in a few papers.

The variables are presented quantitatively in Figure 8.

A similar conclusion can be made regarding optimization variables. DSO tries to utilize all the available controls such as OLTC tap settings, CB reactive power, and ESS variables. Some of the variables are continuous but some are discrete (such as OLTC tap settings), which will affect the formulation of the OPF problem (the appearance of integer variables) and largely the choice of the solution method.

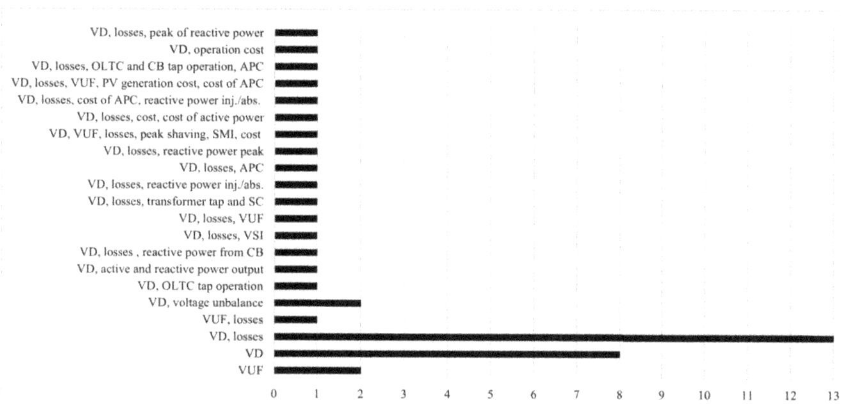

Figure 7. Quantitative presentation of the objectives.

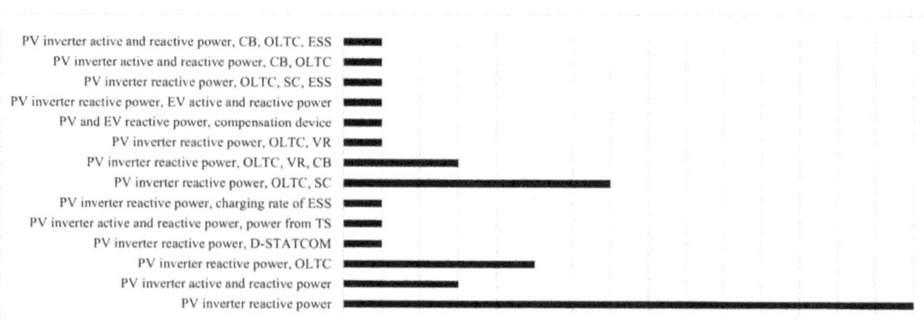

Figure 8. Quantitative presentation of the variables.

4. Voltage Optimization in PV-Rich Distribution Networks—Formulation and Solution Methods

The complexity of the OPF depends on the power flow formulation approach. If the original AC power flow equations are used for OPF formulation, an optimization problem is nonlinear and hard to solve. Thus, many researchers try to simplify OPF formulation in order to obtain a linear or quadratic optimization problem that is easier to solve. According to [60], the OPF formulation can be classified into:

- nonlinear programming (NLP);
- linear programming (LP);
- quadratic programming (QP);
- mixed-integer linear programming (MILP);
- mixed-integer nonlinear programming (MINLP).

In early papers, continuous NLP formulation is used. All discrete variables are approximated as continuous for simplicity. This formulation includes nonlinear objectives and constraints. The LP formulation uses the DC power flow approach, i.e., both the objectives and the constraints are linear. Due to its simplicity, robustness, speed, and well-developed solution methods, it is an attractive OPF formulation, especially for industry [51]. However, due to the modeling assumptions, LP is not adequate for problems such as minimizing power losses, and a global optimum cannot be guaranteed. QP is a special case of NLP with quadratic objective and linear constraints and represents an alternative to LP. The inclusion of discrete variables (transformer tap settings, shunt capacitor settings, etc.) in NLP results in MINLP being the most realistic and accurate formulation of OPF, but also the most complex and difficult to solve. However, there is a trade-off between the system description and the tractability of the problem. One way is to linearize and apply MILP. For more details, see [60,92].

The OPF formulation determines which solution method is used. Figure 9 presents the mathematical formulation and solution methods. For each solution method in Figure 9, a scientific paper in which it is applied is listed. The basic categorization of OPF solution methods is divided into analytical methods and computational intelligence methods. The most commonly used analytical methods for linear OPF are the well-developed simplex methods [93], sequential linear programming (SLP) [94], and interior point methods (IPMs) [95]. In the first period, analytical iterative methods were applied to the NLP OPF. They were Newton-based methods [96]; gradient methods—reduced gradient method (RG) [97], conjugate gradient method (CG) [98], and generalized reduced gradient method (GRG) [99]; IPMs [100]; sequential quadratic programming (SQP) [101]. Recently, computational intelligence methods have been applied to solving OPF problems. Computational intelligence methods have been developed to overcome the weak capabilities of analytical methods for solving global optimization [102]. Although computational intelligence methods do not require a precise mathematical formulation of the OPF problem, the authors include them in the group of solution methods that can solve nonlinear and integer formulations of the OPF since they can take into account nonlinearities in the original problem. Computational intelligence methods include artificial neural networks (ANNs) [103], genetic algorithms (GAs) [104], particle swarm optimization (PSO) [105,106], ant colony optimization (ACO) [107,108], bacterial foraging algorithm (BFA) [109], simulated annealing (SA) [110], tabu search (TS) [111], and fuzzy logic (FL) [112].

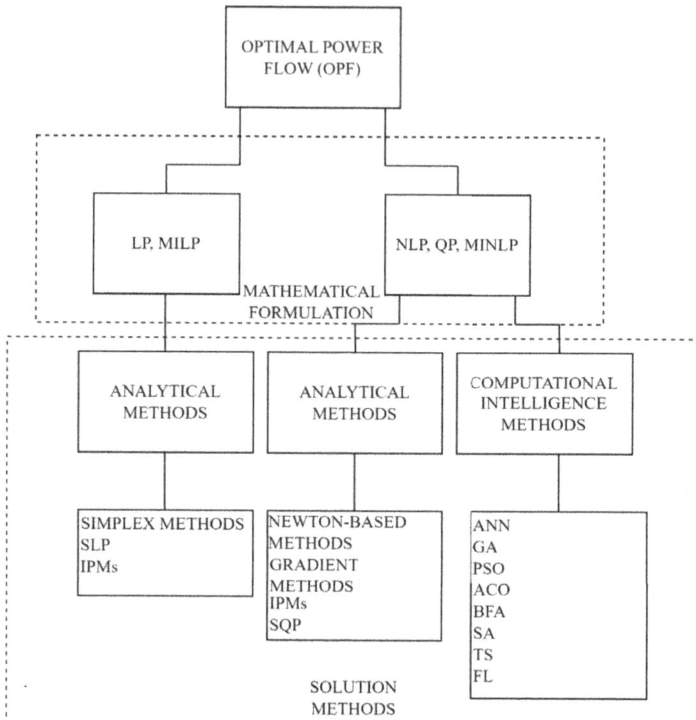

Figure 9. Mathematical formulation and solution methods used in OPF solving.

4.1. Analytical Methods

Classical analytical methods are used in multi-field optimization problems. In addition to the basic methods such as LP, QP [69], NLP, MINLP [76], and MILP, some papers deal with problems that reduce to the basic ones (see Table 4). In [65], SLP is developed to solve the optimization problem in real time. To obtain global optima, SQP has been developed in [38]. The iterative gradient projection method is implemented to specify VAR outputs for voltage optimization in [87] and additionally, active power outputs in [84]. A method for solving linear and nonlinear optimization problems was developed in [34,78]. The alternating direction method of multipliers (ADMM), which is one of the augmented Lagrangian-based methods, was developed as one of the most used methods for a network optimization problem. In ADMM, the optimization problem is decomposed into subproblems to deal with it. The subproblems are coordinated to seek the global optimal solution. The authors in [70,72,73] solve the voltage optimization problem using ADMM.

Table 4. Overview of formulation and analytical methods.

Formulation	Analytical Methods	Reference
NLP	ADMM-based method	[69]
	IPM	[36]
	SQP	[54]
LP	SLP	[63]
QP	QP	[52,53,74]
	Gradient projection methods	[59,82]
	ADMM-based method	[70,71]
MILP	MILP	[58]
MINLP	QP	[74]
	IPM	[76]

4.2. Computational Intelligence Methods

In recent years, computational intelligence methods have been increasingly used to solve voltage optimization problems. One of the main advantages of computational intelligence methods is that they do not require a mathematical formulation of the optimization problem. For the sake of unification and formality, the authors keep the formulation of the OPF problem in Table 5 even though it is not required for computational intelligence methods. The most commonly used methods are GA [37,57,62,83] and PSO [64,82,85,88]. Other computational intelligence methods that appear in the literature are the sine-cosine algorithm (SCA) [66], feasibility pump (FP) method [67], sparrow search algorithm (SSA) [74], pattern search algorithm [80], and grey wolf optimization (GWO) [81]. In some papers, hybrid methods are applied. In [79], a GA is applied to solve the day-ahead scheduling optimization problem in the first stage, while the pattern search algorithm (PSA) is used to solve the real-time optimization problem in the second stage. The branch-bound algorithm is combined with the IPM in [86] to solve the discrete problem. The combination of modified PSO and direct load flow (DLF) is used in [75]. DLF is used for power flow analysis and according to obtained data and PSO is used to evaluate network performance. In [77], the authors used both computational intelligence methods and analytical methods. In [55], five multi-objective evolution algorithms (MOEAs), named promising-region-based evolutionary many-objective algorithm (PREA), strength Pareto evolutionary algorithm 2 (SPEA 2), nondominated sorting genetic algorithm II (NSGA-II), nondominated sorting genetic algorithm III (NSGA-III), and two-phase framework (ToP), are used to determine the reactive power capacity of PVs and EVs. The results obtained by MOEAs are used to train a deep deconvolution neural network (DDNN) to solve the problem of voltage deviation and loss minimization. For inverter coordination, the authors in [113] use deep deterministic policy gradient (DDPG).

Table 5. Overview of formulation and computational intelligence methods.

Formulation	Computational Intelligence Methods	Reference
NLP	GA	[37,57]
MINLP	GA	[62]
NLP	NDSGA II	[83]
NLP	PSO	[68,82]
MINLP	PSO	[64,85,88,114]
NLP	SCA	[66]
MINLP	C&CG algorithm	[53]
NLP	SSA	[74]
MINLP	GWO	[81]
MINLP	Modified PSO, DLF algorithm	[75]
MINLP	MOPSO, IPM	[77]
MINLP	GA,PSA	[79]
NLP	PREA, SPEA2, NSGA-II, NSGA-III, ToP, DDNN	[55]
MINLP	FP	[67]
MINLP	PSA	[80]
MINLP	NSGA-III	[89]
MINLP	ϵ-constrained method and FL	[90]
MINLP	DDPG	[113]
MINLP	ANN	[114]

4.3. Formulation and Solution Methods—Discussion

Analytical solution methods require a strictly mathematical formulation of the OPF problem, which can then be solved by an appropriate analytical method. Since there are a few effective analytical algorithms for solving nonlinear problems (especially with integer variables), most of the papers in which analytical methods are used transform the original NLP (or MINLP) into some of the more convenient forms—usually QP or MIQP. Transformations into a linear form (LP or MILP) are very rare due to the nature of the problem, i.e., the quadratic function of voltage deviation and poor performance of DC power flows in the environment of the DN (ratio *R/X* is not as small as in the TN).

According to Table 5, the popularity of computational intelligence methods can be observed. According to Figure 10, a decision about which solution method would be used depends on the objective and mathematical formulation. To briefly address Table 3, multi-objective optimization problems dominate. In single-objective problems, analytical solution methods prevail. Analytical methods require that a multi-objective (usually known as Pareto optimization) problem transforms (scalarizes) into a single-objective using weighting coefficients, which is not a straightforward procedure. Computational intelligence methods are most used for multi-objective problems. A comparison of analytical and computational intelligence methods is presented in Table 6.

Analytical methods are well-developed and applicable in systems where the requirements of modeling accuracy are low. Analytical methods are able to straightforwardly find an optimal solution but there is no guarantee that the optimum is global. If multiple local optima exist, global optima cannot be guaranteed and the analytical method can stuck in local optima. To apply the analytical method, it is necessary to perform a transformation of the original problem to a level that it can solve. This is where the problem of trade-off comes in. On the one hand, there is an accurate real-life system description and, on the other hand, there is an applicable solution method. Some shortcomings of analytical methods are solved by computational intelligence methods. These methods do not depend on mathematical formulation because they required only parameters that can be calculated separately (for example solution of the power flows). Compared to analytical methods, a hard computational effort is required and there is no guarantee of finding an optimal solution thus some expert knowledge of the system is needed. In recent years, computational intelligence methods are used in co-simulation with proven power flow tools DIgSILENT PowerFactory [41], DLF [75], OpenDSS [115], etc. This approach simplified the application of computational intelligence methods for large-scale DNs.

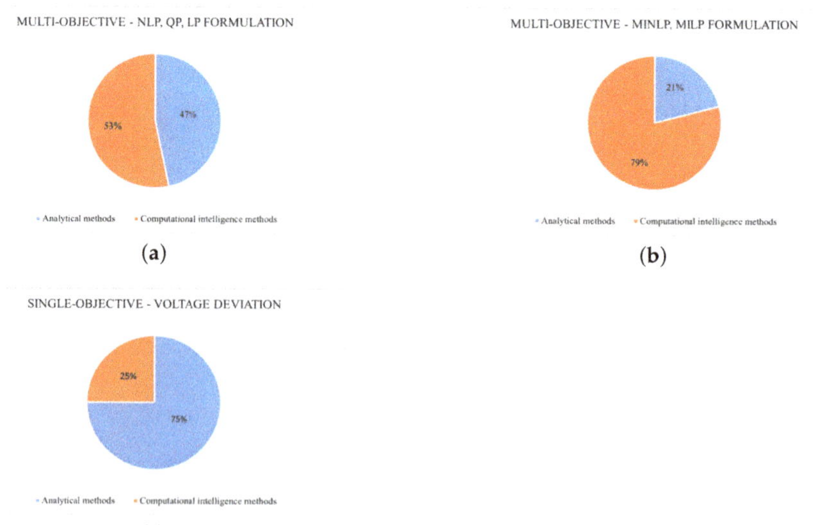

Figure 10. Solution methods for different objectives and formulations. (**a**) Multi-objective problem and continuous formulation. (**b**) Multi-objective problem including integer variables. (**c**) Single-objective problem.

Table 6. Comparison of analytical and computational intelligence methods.

	Advantages	Shortcomings
Analytical methods (simplex method, SLP, SQP, ADMM, gradient projection method, IPM)	- well-developed methods - fast computational performance of linear methods and IPMs	- stuck in local optima - modeling accuracy problem - the sensibility of initial conditions - cannot handle the multi-objective problem properly
Computational intelligence methods (evolutionary and biologically inspired methods, artificial intelligence methods, FL)	- do not depend on mathematical formulation - convergence is easier to set up compared to analytical methods	- hard computational effort - do not guarantee optima

4.4. Test Network Models

To validate the efficiency of different solution methods for the voltage optimization problem, the authors use test network models that can be divided into a standard test model and a test model based on real-life examples. The most common test network models used in literature represent IEEE test network models which, depending on the number of buses, can be IEEE–13 bus, IEEE–15 bus, IEEE–33 bus, IEEE–34 bus, IEEE–37 bus, IEEE–69 bus, IEEE–123 bus, and IEEE–8500 bus. Almost all have radial topology. According to the processed problem, some authors modify standard test models. For instance, the standard test model [70] is modified according to balance. Real-life-based models represent urban residential feeders located in the US, China, Italy, Egypt, Australia, Ireland, and the UK. Unbalanced networks are mostly low voltage and belong to real-life models. In Table 7, test network models, their voltage level, and balance are summarized. According to the reviewed literature, more authors utilize standard test network models IEEE–33 bus, IEEE–69 bus, and IEEE–123 bus node due to their flexibility and robustness. One possible problem that can appear is the OPF application for unbalanced DNs. For instance, the authors in [72] reduce an unbalanced system to a balanced assuming that voltage magnitudes between phases are analogous and phase angles on nodes are not drastic. Therefore, an unbalance between phases is low, and almost balanced. For more, see [72].

Table 7. Overview of test network models.

Reference	Test Network Model	Voltage Level	Balanced/Unbalanced
[32,79,80]	IEEE–34 bus	MV–24.9 kV and 4.16 kV	Balanced
[70]	IEEE–34 bus modified according to balance	MV–24.9 kV and 4.16 kV	Unbalanced
[55,63,67,81,82,85,89,90]	IEEE–33 bus	MV–12.66 kV	Balanced
[63]	Real–266 bus, Shenzen, China	MV–20 kV	Balanced
[53,64,83,88,113]	IEEE-123 bus	MV–4.16 kV	Unbalanced
[70,72]	IEEE–123 bus modified according to balance	MV–4.16 kV	Balanced
[65]	Real distribution feeder–187 bus	MV–12.47 kV and LV–120/240 V	Balanced
[38,75]	Perth Solar City–101 bus	LV–415/240 V	Unbalanced
[66]	Tala City, Egypt–37 bus	MV–11 kV	Balanced
[67,76,78,81]	IEEE–69 bus	MV–12.66 kV	Balanced
[32,57]	IEEE–13 bus	MV–4.16 kV	Balanced
[70]	IEEE–13 bus	MV–4.16 kV	Unbalanced

Table 7. *Cont.*

Reference	Test Network Model	Voltage Level	Balanced/Unbalanced
[71]	Real UK 1	LV–0.4 kV	Unbalanced
[72]	IEEE–15 bus	MV–11 kV	Unbalanced
[74]	IEEE–8500 bus	MV and LV	Both
[75]	Real Australian–565 bus	MV–22 kV and LV–415 V	Both
[68]	22 bus	MV–11.4 kV	Balanced
[78]	17 bus	MV–25 kV	Balanced
[56,113]	IEEE–37 bus	MV–4.16 kV	Unbalanced
[56]	Real Californian utility feeder–2884 bus	N/A	Unbalanced
[34]	Illustrative model	LV	N/A
[84]	K1 feeder–1747 bus in the southeastern US	MV and LV	N/A
[69,116]	33 bus	MV–12.66 kV	N/A
[69]	830 bus	N/A	N/A
[37,91]	Real South Korean–20 bus	MV–22.9 kV	Balanced
[59]	Real South Italian–16 bus	LV	Unbalanced
[39]	Real Irish suburban–85 bus	LV	N/A
[58]	Modified PG&E–69 bus	MV	Balanced
[114]	CIGRE–12 bus	MV	N/A
[116]	118 bus	MV	N/A

5. Conclusions

This paper aims to systematize and categorize scientific papers that are dealing with the optimization of voltage in the DN using the reactive power management of PV inverters. Additionally, the papers are categorized according to the optimization problem formulation and applied solution methods. It can be observed that the original voltage optimization problem is nonlinear due to a quadratic objective function and nonlinear power flow equations. Additionally, some authors propose a mixed-integer nonlinear formulation due to integer variables such as the OLTC tap setting. To solve such complex optimization problems, some authors use analytical methods and some use computational ones. In this review paper, the authors tried to point out the advantages and shortcomings of both approaches without favoring one. When analytical methods are used, the compromise regarding the transformation of the original problem into a standard one is present but the analytical approach enables the straightforward method to find the optimum of a well-defined optimization problem (although special attention is required in order to determine whether a calculated optimum is local or global). On the other hand, computational intelligence methods can solve complex optimization problems without the transformation of the original formulation but they required higher computational performance as well as more computational time. One of the trends in applying computational intelligence methods is using well-known power flow calculation tools in order to feed the computational intelligence method with multiple power flow solutions. This principle is recognized as co-simulation.

The research potential of the reviewed field lies in the fact that more and more inverter-based sources are installed in distribution networks worldwide. The importance of voltage optimization is specifically stressed in microgrids where voltage supports depend mainly on the inverter-based source. Since the PV active power production depends on variable and stochastic sun irradiation, further research direction in the field of voltage optimization will strive to create an adequate probabilistic formulation of the OPF problem which is computationally more demanding since large numbers of possible scenarios need to be analyzed. Some probabilistic OPF solutions are already created for the transmission system environment but their replication in the distribution network (or microgrid) is not straightforward. It is hard to foresee which solution methodology (analytical or computational intelligence) will show better performance in a probabilistic environment and there is still plenty of research challenges and gaps present for further research.

Author Contributions: Conceptualization, M.D. and K.F.; methodology, M.D. and K.F.; formal analysis, M.D., K.F. and D.T.; investigation, M.D.; writing—original draft preparation, M.D., K.F. and D.T.; writing—review and editing, D.T. and M.B.; funding acquisition, D.T. All authors have read and agreed to the published version of the manuscript.

Funding: The APC was funded by Croatian Science Foundation under the project "Prosumer-rich distribution power network" (project number: UIP-2020-02-5796).

Institutional Review Board Statement: Not applicable.

Informed Consent Statement: Not applicable.

Data Availability Statement: Not applicable.

Acknowledgments: The work of doctoral student Marina Dubravac has been fully supported/supported in part by the "Young researchers' career development project—training of doctoral students" (DOK-2021-02-1257) of the Croatian Science Foundation. This work was founded by the Croatian Science Foundation under the project "Prosumer-rich distribution power network" (project number: UIP-2020-02-5796).

Conflicts of Interest: The authors declare no conflict of interest.

References

1. Carvalho, P.M.S.; Correia, P.F.; Ferreira, L.A.F.M. Distributed reactive power generation control for voltage rise mitigation in distribution networks. *IEEE Trans. Power Syst.* **2008**, *23*, 766–772. [CrossRef]
2. Kjaer, S.B.; Lazar, R.D.; Constantin, A.; Ballegaard, H.P.; Yang, G.; Østergaard, J.; Ipsen, H.H.; Frederiksen, K.H.B. Voltage control in low voltage networks by photovoltaic inverters—PVNET.DK. In Proceedings of the 28th European Photovoltaic Solar Energy Conference and Exhibition (28th EU PVSEC), Villepinte, France, 30 September–4 October 2013.
3. Mokhtari, G.; Nourbakhsh, G.; Zare, F.; Ghosh, A. Overvoltage prevention in LV smart grid using customer resources coordination. *Energy Build.* **2013**, *61*, 387–395. [CrossRef]
4. Xu, T.; Taylor, P.C. Voltage control techniques for electrical distribution networks including distributed generation. *IFAC Proc. Vol.* **2008**, *41*, 11967–11971. [CrossRef]
5. Juamperez, M.; Yang, G.; Kjær, S.B. Voltage regulation in LV grids by coordinated volt-var control strategies. *J. Mod. Power Syst. Clean Energy* **2014**, *2*, 319–328. [CrossRef]
6. Jamsek, S.; Tot, S.; Lasic, M.; Perisa, I.; Miklavcic, M. Regionaly important sincro.grid smart grid project. *J. Energy Energ.* **2018**, *67*, 86–93. [CrossRef]
7. Turitsyn, K.; Šulc, P.; Backhaus, S.; Chertkov, M. Options for control of reactive power by distributed photovoltaic generators. *Proc. IEEE* **2011**, *99*, 1063–1073. [CrossRef]
8. Takasawa, Y.; Akagi, S.; Yoshizawa, S.; Ishii, H.; Hayashi, Y. Effectiveness of updating the parameters of the Volt-VAR control depending on the PV penetration rate and weather conditions. In Proceedings of the 2017 IEEE Innovative Smart Grid Technologies–Asia (ISGT-Asia), Asia Smart Grid Smart Community, Auckland, New Zealand, 4–7 December 2017; pp. 1–5.
9. Culjak, M.; Beus, M.; Pandzic, H. Development of a LabVIEW - Based Data Logging and Monitoring Application for a Photovoltaic Power Plant at FER. *J. Energy Energ.* **2022**, *211*, 14–18. [CrossRef]
10. Petinrin, J.O.; Shaabanb, M. Impact of renewable generation on voltage control in distribution systems. *Renew. Sustain. Energy Rev.* **2016**, *65*, 770–783. [CrossRef]
11. Karimi, M.; Mokhlis, H.; Naidu, K.; Uddin, S.; Bakar, A.H.A. Photovoltaic penetration issues and impacts in distribution network—A review. *Renew. Sustain. Energy Rev.* **2016**, *53*, 594–605. [CrossRef]
12. Chaudhary, P.; Rizwan, M. Voltage regulation mitigation techniques in distribution system with high PV penetration: A review. *Renew. Sustain. Energy Rev.* **2018**, *82*, 3279–3287. [CrossRef]
13. Nour, A.M.M.; Hatata, A.Y.; Helal, A.A.; El-Saadawi, M.M. Review on voltage-violation mitigation techniques of distribution networks with distributed rooftop PV systems. *IET Gener. Transm. Distrib.* **2020**, *14*, 349–361. [CrossRef]
14. Antoniadou-Plytaria, K.E.A.; Kouveliotis-Lysikatos, I.N.; Georgilakis, P.S.; Hatziargyriou, N.D. Distributed and decentralized voltage control of smart distribution networks: Models, methods, and future research. *IEEE Trans. Smart Grid* **2017**, *8*, 2999–3008. [CrossRef]
15. Ullah, Z.; Wang, S.; Wu, G.; Hasanien, H.M.; Jabbar, M.W.; Qazi, H.S.; Tostado-Véliz, M.; Turky, R.A.; Elkadeem, M.R. Advanced studies for probabilistic optimal power flow in active distribution networks: A scientometric review. *IET Gener. Transm. Distrib.* **2022**, *16*, 3579–3604. [CrossRef]
16. Abdi, H.; Beigvand, S.D.; La Scala, M. N.D. A review of optimal power flow studies applied to smart grids and microgrids. *Renew. Sustain. Energy Rev.* **2017**, *71*, 742–766. [CrossRef]

17. Toffler, A. *The Third Wave: Learning for Tomorrow, the Futurists, the Schoolhouse and the City*, 1st ed.; William Morrow and Company, Inc.: New York, NY, USA, 1980.
18. Korotko, T.; Rosin, A.; Ahmadiahangar, R. Development of prosumer logical structure and object modeling. In Proceedings of the 2019 IEEE 13th International Conference on Compatibility, Power Electronics and Power Engineering (CPE-POWERENG), Sonderborg, Denmark, 23–25 April 2019.
19. Cao, J.; Bu, Z.; Wang, Y.; Jiang, J.; Li, H.J. Detecting prosumer-community groups in smart grids from the multiagent perspective. *IEEE Trans. Syst. Man. Cybern. Syst.* **2019**, *49*, 1652–1664. [CrossRef]
20. Rathnayaka, A.J.D.; Potdar, V.M.; Dillon, T.S.; Hussain, O.K.; Chang, E. A methodology to find influential prosumers in prosumer community groups. *IEEE Trans. Ind. Inform.* **2014**, *10*, 706–713. [CrossRef]
21. Yang, H.; Xiong, T.; Qiu, J.; Qiu, D.; Dong, Z.Y. Optimal operation of DES/CCHP based regional multi-energy prosumer with demand response. *Appl. Energy* **2016**, *167*, 353–365. [CrossRef]
22. Liu, N.; Cheng, M.; Yu, X.; Zhong, J.; Lei, J. Energy-sharing provider for PV prosumer clusters: A hybrid approach using stochastic programming and Stackelberg game. *IEEE Trans. Ind. Electron.* **2018**, *65*, 6740–6750. [CrossRef]
23. Ghosh, A.; Aggarwal, V.; Wan, H. Exchange of renewable energy among prosumers using blockchain with dynamic pricing. *arXiv* **2018**, arXiv:1804.08184v1.
24. El Rahi, G.; Etesami, S.R.; Saad, W.; Mandayam, N.B.; Poor, H.V. Managing price uncertainty in prosumer-centric energy trading: A prospect-theoretic Stackelberg game approach. *IEEE Trans. Smart Grid.* **2019**, *10*, 702–713. [CrossRef]
25. Garijalva, S.; Tariq, M.U. Prosumer-based smart grid architecture enables a flat, sustainable electricity industry. In Proceedings of the Innovative Smart Grid Technologies (ISGT), Sonderborg, Denmark, 23–25 April 2011.
26. Lavrijssen, S.; Parra, A.C. Radical prosumer innovations in the electricity sector and the impact on prosumer regulation. *Sustainability* **2017**, *9*, 1207. [CrossRef]
27. Merino, J.; Gómez, I.; Fraile-Ardanuy, J.; Santos, M.; Cortés, A.F.; Jimeno, B.J.; Madina, C. Chapter 6—Fostering DER integration in the electricity markets. In *Distributed Energy Resources in Local Integrated Energy Systems*; Graditi, G., Di Somma, M., Eds.; Elsevier: Amsterdam, The Netherlands, 2021.
28. European Parliament DIRECTIVE (EU) 2019/944 of the European Parliament and of the Council on Common Rules for the Internal Market for Electricity and Amending Directive 2012/27/EU. *Off. J. Eur. Union* **2019**, 125–199. Available online: https://eur-lex.europa.eu/legal-content/EN/TXT/?uri=celex%3A32019L0944 (accessed on 15 June 2022).
29. Hrvatski Sabor Zakon o Tržištu Električne Energije. Narodne Novine 111/2021. 2021. Available online: https://narodne-novine.nn.hr/clanci/sluzbeni/2021_10_111_1940.html (accessed on 15 June 2022).
30. Gebbran, D.; Mhanna, S.; Ma, Y.; Chapman, A.C.; Verbič, G. Fair coordination of distributed energy resources with Volt-Var control and PV curtailment. *Appl. Energy* **2021**, *286*, 116546. [CrossRef]
31. Safayet, A.; Fajri, P.; Husain, I. Reactive power management for overvoltage prevention at high PV penetration in a low-voltage distribution system. *IEEE Trans. Ind. Appl.* **2017**, *53*, 5786–5794. [CrossRef]
32. Yung, Y.; Han, C.; Lee, D.; Song, S.; Jang, G. Adaptive volt–var control in smart pv inverter for mitigating voltage unbalance at pcc using multiagent deep reinforcement learning. *Appl. Sci.* **2021**, *11*, 8979.
33. Demirok, E.; Gonzalez, P.C.; Frederiksen, K.H.; Sera, D.; Rodriguez, P.; Teodorescu, R. Local reactive power control methods for overvoltage prevention of distributed solar inverters in low-voltage grids. *IEEE J. Photovolt.* **2011**, *1*, 174–182. [CrossRef]
34. Dall'Anese, E.; Dhople, S.V.; Johnson, B.B.; Giannakis, G.B. Decentralized optimal dispatch of photovoltaic inverters in residential distribution systems. *IEEE Trans. Energy Convers.* **2014**, *29*, 957–967. [CrossRef]
35. Jahangiri, P.; Aliprantis, D.C. Distributed Volt/VAr control by PV inverters. *IEEE Trans. Power Syst.* **2013**, *28*, 3429–3439. [CrossRef]
36. Jabr, R.A. Robust volt/var control with photovoltaics. *IEEE Trans. Power Syst.* **2019**, *34*, 2401–2408. [CrossRef]
37. Lee, H.; Kim, J.C.; Cho, S.M. Optimal volt-var curve setting of a smart inverter for improving its performance in a distribution system. *IEEE Access* **2020**, *8*, 157391–157945. [CrossRef]
38. Su, X.; Masoum, M.A.S.; Wolfs, P.J. Optimal PV inverter reactive power control and real power curtailment to improve performance of unbalanced four-wire LV distribution networks. *IEEE Trans. Sustain. Energy* **2014**, *5*, 967–977. [CrossRef]
39. Connell, A.O.; Keane, A. Volt-var curves for photovoltaic inverters in distribution systems. *IET Gener. Transm. Distrib.* **2017**, *11*, 730–739. [CrossRef]
40. Fekete, K.; Klaic, Z.; Majdančić, L. Expansion of the residential photovoltaic systems and its harmonic impact on the distribution grid. *Renew. Energy* **2012**, *43*, 140–148. [CrossRef]
41. DIgSILENT PowerFactory. Available online: https//www.digsilent.de/en/ (accessed on 4 June 2022).
42. Growatt. Available online: https://www.ginverter.com/ (accessed on 10 June 2022).
43. Carpentier, J. Contribuition to the economic dispatch problem. *Bull. Soc. Fr. Des Electr.* **1962**, *8*, 431–447. (In French)
44. Dommel, H.W.; Tinney, W.F. Optimal power flow solutions. *IEEE Trans. Power Appar. Syst.* **1968**, *PAS-87*, 1866–1876. [CrossRef]
45. Momoh, J.A.; El-Hawary, M.E.; Adapa, R. A review of selected optimal power flow literature to 1993 part i: Nonlinear and quadratic programming approaches. *IEEE Trans. Power Syst.* **1999**, *14*, 96–103. [CrossRef]
46. Momoh, J.A.; El-Hawary, M.E.; Adapa, R. A review of selected optimal power flow literature to 1993 part ii: Newton, linear programming and interior point method. *IEEE Trans. Power Syst.* **1999**, *14*, 96–104. [CrossRef]
47. Huneault, M.; Galiana, F.D. A survey of the optimal power flow literature. *IEEE Trans. Power Syst.* **1991**, *6*, 762–770. [CrossRef]

48. Momoh, J.A. Optimal Power Flow. In *Electric Power System Applications of Optimization*, 2nd ed.; CRC Press: Boca Raton, FL, USA, 2008; Chapter 12.
49. Contaxis, G.C.; Delkis, C.; Korres, G. Decoupled optimal load flow using linear or quadratic programming. *IEEE Power Eng. Rev.* **1986**, *1*, 1–7.
50. Biskas, P.N.; Bakirtzis, A.G. A decentralized solution to the Security Constrained DC-OPF problem of multi-area power systems. In Proceedings of the 2005 IEEE Russia Power Tech, St. Petersburg, Russia, 27–30 June 2005; Volume 18, pp. 1007–1013.
51. Rau, N.S. Issues in the path toward an RTO and standard markets. *IEEE Trans. Power Syst.* **2003**, *18*, 435–443. [CrossRef]
52. El-Hawary, M.E.; Bose, A.; de Barros, M.C.; El-Kieb, A.R.; Hammons, T.J.; Iwamoto, S.I. Power Engineering Letters. *IEEE Power Eng. Rev.* **2001**, *21*, 49–51. [CrossRef]
53. Zhang, C.; Xu, Y.; Dong, Z.Y.; Zhang, R. Multi-objective adaptive robust voltage/VAR control for high-PV penetrated distribution networks. *IEEE Trans. Smart Grid* **2020**, *11*, 5288–5300. [CrossRef]
54. Kundu, S.; Backhaus, S.; Hiskens, I.A. Distributed control of reactive power from photovoltaic inverters. In Proceedings of the 2013 IEEE International Symposium on Circuits and Systems (ISCAS), Beijing, China, 19–23 May 2013.
55. Wu, R.; Liu, S. Deep learning based muti-objective reactive power optimization of distribution network with PV and EVs. *Sensors* **2022**, *22*, 4321. [CrossRef] [PubMed]
56. Li, C.; Disfani, V.R.; Haghi, H.V.; Kleissl, J. Coordination of OLTC and smart inverters for optimal voltage regulation of unbalanced distribution networks. *Electr. Power Syst. Res.* **2020**, *187*, 106498. [CrossRef]
57. Jafari, M.; Olowu, T.O.; Sarwat, A.I. Optimal smart inverters volt-VAR curve selection with a multi-objective volt-VAR optimization using evolutionary algorithm approach. In Proceedings of the 2018 North American Power Symposium (NAPS), Fargo, ND, USA, 9–11 September 2017.
58. Ji, H.; Wang, C.; Li, P.; Zhao, J.; Song, G.; Ding, F.; Wu, J. A centralized-based method to determine the local voltage control strategies of distributed generator operation in active distribution networks. *Appl. Energy* **2018**, *228*, 2024–2036. [CrossRef]
59. Carpinelli, G.; Mottola, F.; Proto, D.; Varilone, P. Minimizing unbalances in low-voltage microgrids: Optimal scheduling of distributed resources. *Appl. Energy* **2017**, *191*, 170–182. [CrossRef]
60. Frank, S.; Steponavice, I.; Rebennack, S. Optimal power flow: A bibliographic survey I: Formulations and deterministic methods. *Energy Syst.* **2012**, *3*, 221–258. [CrossRef]
61. Bacher, R. *Netzleittechnik und Optimierung Elektrischer Netze*; ETH Zürich: Zürich, Switzerland, 2000.
62. Olowu, T.O.; Jafari, M.; Sarwat, A.I. A multi-objective optimization technique for volt-Var control with high PV penetration using genetic algorithm. In Proceedings of the 2018 North American Power Symposium (NAPS), Fargo, ND, USA, 9–11 September 2018.
63. Lu, W.; Liu, M.; Liu, Q. Increment-exchange-based decentralized multiobjective optimal power flow for active distribution grids. *IEEE Syst. J.* **2020**, *14*, 3695–3704. [CrossRef]
64. Shailendra, S.; Pamshetti, V.B.; Thakur, A.K.; Singh, S.P. Multistage multiobjective Volt/VAR control for smart grid-enabled CVR with solar PV penetration. *IEEE Syst. J.* **2020**, *15*, 2767–2778.
65. Zhong, C.; Sakis Meliopoulos, A.P.; Xie, B.; Xie, J.; Liu, K. Multi-stage quadratic flexible optimal power flow with a rolling horizon. *IEEE Trans. Smart Grid* **2021**, *12*, 3128–3137. [CrossRef]
66. Ramadan, A.; Ebeed, M.; Kamel, S. Performance assessment of a realistic Egyptian distribution network including PV penetration with DSTATCOM. In Proceedings of the 2019 International Conference on Innovative Trends in Computer Engineering (ITCE), Aswan, Egypt, 2–4 February 2019.
67. Xu, R.; Zhang, C.; Xu, Y.; Dong, Z.; Zhang, R. Multi-objective hierarchically-coordinated volt/var control for active distribution networks with droop-controlled PV inverters. *IEEE Trans. Smart Grid* **2022**, *13*, 998–1011. [CrossRef]
68. Yang, H.T.; Liao, J.T. MF-APSO-based multiobjective optimization for PV system reactive power regulation. *IEEE Trans. Sustain. Energy* **2015**, *6*, 1346–1355. [CrossRef]
69. Jabr, R.A. Linear decision rules for control of reactive power by distributed photovoltaic generators. *IEEE Trans. Power Syst.* **2018**, *33*, 2165–2174. [CrossRef]
70. Zhang, Q.; Dehghanpour, K.; Wang, Z. Distributed CVR in unbalanced distribution systems With PV penetration. *IEEE Trans. Smart Grid* **2018**, *10*, 5308–5319. [CrossRef]
71. Al-Ja'Afreh, M.A.A.; Mokryani, G. Voltage unbalance mitigation in low voltage distribution networks using time series three-phase optimal power flow. In Proceedings of the 2021 56th International Universities Power Engineering Conference (UPEC), Middlesbrough, UK, 7–10 September 2020.
72. Robbins, B.A.; Domínguez-García, A.D. Optimal reactive power dispatch for voltage regulation in unbalanced distribution systems. *IEEE Trans. Power Syst.* **2016**, *31*, 2903–2913. [CrossRef]
73. Guggilam, S.S.; Dall'Anese, E.; Chen, Y.C.; Dhople, S.V.; Giannakis, G.B. Scalable optimization methods for distribution networks with high PV integration. *IEEE Trans. Smart Grid* **2016**, *7*, 2061–2070. [CrossRef]
74. Der Lee, Y.; Lin, W.C.; Jiang, J.L.; Cai, J.H.; Huang, W.T.; Yao, K.C. Optimal individual phase voltage regulation strategies in active distribution networks with high PV penetration using the sparrow search algorithm. *Energies* **2021**, *14*, 2061.
75. Su, X.; Liu, Y.; Tian, S.; Ling, P.; Fu, Y.; Wei, S.; SiMa, C. A multi-stage coordinated volt-Var optimization for integrated and unbalanced radial distribution networks. *Energies* **2020**, *13*, 4877. [CrossRef]
76. Il Go, S.; Yun, S.Y.; Ahn, S.J.; Choi, J.H. Voltage and reactive power optimization using a simplified linear equations at distribution networks with DG. *Energies* **2020**, *13*, 3334.

77. Jin, D.; Chiang, H.D.; Li, P. Two-timescale multi-objective coordinated volt/var optimization for active distribution networks. *IEEE Trans. Power Syst.* **2019**, *34*, 4418–4428. [CrossRef]
78. Ammar, M.; Sharaf, A.M. Optimized Use of PV distributed generation in voltage regulation: A probabilistic formulation. *IEEE Trans. Ind. Inform.* **2019**, *15*, 247–256. [CrossRef]
79. Chen, Y.; Luckey, B.; Wigmore, J.; Davidson, M.; Benigni, A. Real-time volt/var optimization for distribution systems with photovoltaic integration. In Proceedings of the IECON 2017-43rd Annual Conference of the IEEE Industrial Electronics Society, Beijing, China, 29 October–1 November 2017.
80. Chen, Y.; Strothers, M.; Benigni, A. Day-ahead optimal scheduling of PV inverters and OLTC in distribution feeders. In Proceedings of the 2016 IEEE Power and Energy Society General Meeting (PESGM), Boston, MA, USA, 17–21 July 2016.
81. Ceylan, O.; Liu, G.; Tomasovic, K. Coordinated distribution network control of tap changer transformers, capacitors and PV inverters. *Electr. Eng.* **2018**, *100*, 1133–1146. [CrossRef]
82. Li, C.; Disfani, V.R.; Haghi, H.V.; Kleissl, J. Coordinated voltage control for active distribution network considering the impact of energy storage. *Energy Procedia* **2019**, *158*, 1122–1127.
83. Xu, J.; Wang, J.; Liao, S.; Sun, Y.; Ke, D.; Li, X.; Liu, J.; Jiang, Y.; Wei, C.; Tang, B. Stochastic multi-objective optimization of photovoltaics integrated three-phase distribution network based on dynamic scenarios. *Appl. Energy* **2018**, *231*, 985–996. [CrossRef]
84. Ding, F.; Zhang, Y.; Simpson, J.; Bernstein, A.; Vadari, S. Optimal energy dispatch of distributed PVS for the next generation of distribution management systems. *IEEE Open Access J. Power Energy* **2020**, *7*, 287–295. [CrossRef]
85. Sidea, D.O.; Picioroaga, I.I.; Tudose, A.M.; Bulac, C.; Tristiu, I. Multi-objective particle swarm optimization applied on the optimal reactive power dispatch in electrical distribution systems. In Proceedings of the 2020 International Conference and Exposition on Electrical and Power Engineering (EPE), Iasi, Romania, 22–23 October 2020.
86. Zhao, J.; Liu, X.; Lin, C.; Wei, W. Three-phase unbalanced voltage/VAR optimization for active distribution networks. In Proceedings of the 2016 IEEE Power and Energy Society General Meeting (PESGM), Boston, MA, USA, 17–21 July 2016.
87. Zhu, H.; Liu, H.J. Fast local voltage control under limited reactive power: Optimality and stability analysis. *IEEE Trans. Power Syst.* **2015**, *31*, 3794–3803. [CrossRef]
88. Jin, D.; Chiang, H.D. Multi-objective look-ahead reactive power control for active distribution networks with composite loads. In Proceedings of the 2018 IEEE Power & Energy Society General Meeting (PESGM), Portland, OR, USA, 5–10 August 2018.
89. Ma, W.; Wang, W.Z.; Chen, Z.; Hu, R. A centralized voltage regulation method for distribution networks containing high penetrations of photovoltaic power. *Int. J. Electr. Power Energy Syst.* **2021**, *129*, 106852. [CrossRef]
90. Pamshetti, V.B.; Singh, S.P. Optimal coordination of PV smart inverter and traditional volt-VAR control devices for energy cost savings and voltage regulation. *Int. Trans. Electr. Energy Syst.* **2019**, *29*, 1–24. [CrossRef]
91. Lee, H.J.; Yoon, K.H.; Shin, J.W.; Kim, J.C.; Cho S.M. Optimal parameters of volt-var function in smart inverters for improving system performance. *Energies* **2020**, *13*, 2294. [CrossRef]
92. Frank, S.; Steponavice, I.; Rebennack, S. Optimal power flow: A bibliographic survey II: Non-deterministic and hybrid methods. *Energy Syst.* **2012**, *3*, 259–289. [CrossRef]
93. Stott, B.; Alsac, O.; Monticelli, A.J. Security Analysis and Optimization. *Proc. IEEE* **1987**, *75*, 1623–1644. [CrossRef]
94. Alsac, B.; Bright, O.J.; Praise, M.; Stott, B. Further developments in LP-based optimal power flow. *IEEE Trans. Power Syst.* **2003**, *5*, 435–443. [CrossRef]
95. Unum, M.R.; Box, P. Network constrained security control using an interior point algorithm. *IEEE Trans. Power Syst.* **1993**, *8*, 1068–1076.
96. Tinney, W.F.; Ashley, B.; Brewer, B.; Hughes, A. Optimal power flow by Newton approach. *IEEE Power Eng. Rev.* **1984**, *PER-4*, 2864–2880.
97. Alsac, O.; Stott, B. Optimal load flow with steady-state security. *IEEE Trans. Power Appar. Syst.* **1974**, *PAS-93*, 745–751. [CrossRef]
98. Burchett, R.C.; Happ, H.H.; Vierath, D.R.; Wirgau, K.A. Developments in optimal power flow. *IEEE Trans. Power Appar. Syst.* **1982**, *PAS-101*, 406–414. [CrossRef]
99. Peschon, J.; Bree, D.W.; Hajdu, L.P. Optimal power-flow solutions for power system planning. *Proc. IEEE* **1972**, *6*, 64–70. [CrossRef]
100. Torres, G.L.; Quintana, V.H. On a nonlinear multiple-centrality-corrections interior-point method for optimal power flow. *IEEE Trans. Power Syst.* **2001**, *16*, 222–228. [CrossRef]
101. Chang, S.-K.; Albuyeh, F.; Gilles, M.L.; Marks, G.E.; Kato, K. Optimal real-time voltage control. *IEEE Trans. Power Syst.* **1990**, *5*, 750–758. [CrossRef]
102. AlRashidi, M.R.; El-Hawary, M.E. Applications of computational intelligence techniques for solving the revived optimal power flow problem. *Electr. Power Syst. Res.* **2009**, *79*, 694–702. [CrossRef]
103. Hartati, R.S.; El-Hawary, M.E. Optimal active power flow solutions using a modified Hopfield neural network. In Proceedings of the Canadian Conference on Electrical and Computer Engineering 2001, Toronto, ON, Canada, 13–16 May 2001.
104. Bakirtzis, A.G.; Biskas, P.N.; Zoumas, C.E.; Petridis, V. Optimal power flow by enhanced genetic algorithm. *IEEE Trans. Power Syst.* **2002**, *17*, 229–236. [CrossRef]
105. Gaing, Z.L. Constrained optimal power flow by mixed-integer particle swarm optimization. In Proceedings of the IEEE Power Engineering Society General Meeting, San Francisco, CA, USA, 16 June 2005.
106. Swarup, K.S. Swarm intelligence approach to the solution of optimal power flow. *J. Indian Inst. Sci.* **2006**, *86*, 439–455.

107. Teng, J.; Liu, Y. A novel ACS-based optimum switch relocation method. *IEEE Trans. Power Syst.* **2003**, *18*, 113–120. [CrossRef]
108. Vlachogiannis, J.G.; Hatziargyriou, N.D.; Lee, K.Y. Ant colony system-based algorithm for constrained load flow problem. *IEEE Trans. Power Syst.* **2005**, *20*, 1241–1249. [CrossRef]
109. Mishra, S. Bacteria foraging based solution to optimize both real power loss and voltage stability limit. In Proceedings of the 2007 IEEE Power Engineering Society General Meeting, Tampa, FL, USA, 24–28 June 2007.
110. Roa-Sepulveda, C.A.; Pavez-Lazo, B.J. A solution to the optimal power flow using simulated annealing. In Proceedings of the 2001 IEEE Porto Power Tech Proceedings, Porto, Portugal, 10–13 September 2001.
111. Abido, M.A. Optimal power flow using tabu search algorithm. *Electr. Power Compon. Syst.* **2002**, *30*, 469–483. [CrossRef]
112. Venkatesh, B.; Sadasivam, G.; Khan, M. Optimal reactive power planning against voltage collapse using the successive multiobjective fuzzy LP technique. *IEEE Proc. Gener. Transm. Distrib.* **1999**, *146*, 343–348. [CrossRef]
113. Li, C.; Chen, Y.-A.; Jin, C.; Sharma, R.; Kleissl J. Online PV smart inverter coordination using deep deterministic policy gradient. *Electr. Power Syst. Res.* **2021**, *209*, 107988. [CrossRef]
114. Utama, C.; Meske, C.; Schneider, J.; Ulbrich, C. Reactive power control in photovoltaic systems through (explainable) artificial intelligence. *Appl. Energy* **2021**, *328*, 120004. [CrossRef]
115. OpenDSS. Available online: https://www.epri.com/pages/sa/opendss (accessed on 8 September 2022).
116. Rossoni, P.; Belati, E.A.; da S. Benedito, R. A hybrid approach for optimization of electric power distributed networks with photovoltaic sources. *Electr. Power Syst. Res.* **2022**, *211*, 108183. [CrossRef]

applied sciences

MDPI

Article

Thermal Performance Analysis of a Double-Pass Solar Air Collector: A CFD Approach

Irving A. Chávez-Bermúdez [1], Norma A. Rodríguez-Muñoz [2], Eduardo Venegas-Reyes [3], Loreto Valenzuela [4] and Naghelli Ortega-Avila [2,*]

1 Centro de Investigación en Materiales Avanzados, Calle CIMAV 110, Ejido Arroyo Seco, Durango 34147, Mexico
2 CONACYT-Centro de Investigación en Materiales Avanzados, Calle CIMAV 110, Ejido Arroyo Seco, Durango 34147, Mexico
3 Instituto Mexicano de Tecnología del Agua, Paseo Cuauhnáhuac 8532, Progreso, Jiutepec 62550, Mexico
4 CIEMAT-Plataforma Solar de Almería, Ctra. Senés Km 4.5, Tabernas 04200, Almería, Spain
* Correspondence: naghelli.ortega@cimav.edu.mx

Featured Application: The developed solar air heater can be applied in drying and space heating applications from 50 °C to 90 °C. The solar collector could be modular, so it can be coupled to a variety of processes.

Abstract: Solar air heaters can reduce climate change by replacing conventional fossil fuel-burning technologies in drying and space heating applications. Concentrating solar technologies, such as compound parabolic concentrators, allow air temperatures up to 120 °C; however, it is desirable to improve their heat transfer to reduce the space requirements for their installation. In this work, a parabolic concentrator composed of a flat receiver designed to recover heat from the cover–receiver–reflectors cavity is analyzed, operating it as a U-shape double pass solar heater. With this operation, first, the air flows through the cavity, and then it is incorporated into the duct, where the dominant heat gain occurs due to the capture of solar radiation. Thus, four input–output configurations in the cavity were modeled through dynamic simulations to determine the influence of the inlet and outlet air flow positions on the solar concentrator outlet temperature. Therefore, the incorporation of the first pass has a contribution of between 36% and 45% in useful energy gain, showing that this appropriate and relatively simple strategy can be implemented to improve the thermal performance of solar air collectors, resulting in instantaneous efficiencies higher than 75%. However, the simulation results demonstrate that the position of the inlets and outlets does not significantly impact the efficiency and outlet temperature.

Keywords: solar energy; CPC; solar heating; solar drying; industrial process; solar air heater; space heating

Citation: Chávez-Bermúdez, I.A.; Rodríguez-Muñoz, N.A.; Venegas-Reyes, E.; Valenzuela, L.; Ortega-Avila, N. Thermal Performance Analysis of a Double-Pass Solar Air Collector: A CFD Approach. *Appl. Sci.* **2022**, *12*, 12199. https://doi.org/10.3390/app122312199

Academic Editors: Luis Hernández-Callejo, Jesús Armando Aguilar Jiménez and Carlos Meza Benavides

Received: 22 September 2022
Accepted: 27 November 2022
Published: 29 November 2022

1. Introduction

Air heating is used for various applications, such as heating and air conditioning of buildings or drying of food and industrial products, among others. Air can be heated with electric heaters or by directly burning fuels such as gas; however, their use implies the emission of greenhouse gases and their consequent contribution to climate change. One way to minimize fossil fuel burning is to use solar collectors to directly heat the air, ranging from flat-plate collectors to solar concentrators.

According to the International Energy Agency, 985 MW_{th} of solar air collectors were installed by the end of 2020, and the global market was around 12 MW_{th} [1]. As of March 2022, 41 solar air collector systems producing solar process heat are registered, with a cumulative capacity of 6 MW_{th} [1]. Thus, the direct application of solar collectors for air heating is low due to the boost that low fossil fuel prices give to using conventional technologies [2], so it is essential to develop reliable and economically efficient solar air heating technologies. Flat-plate solar collectors are recommended for temperatures below 70 °C because of their ease of manufacture and operation. For higher temperatures, it is necessary to use some solar concentrating technology, such as compound parabolic concentrators (CPCs), which allow fluid heating temperatures up to 120 °C, depending on their design, and are easy to operate and maintain.

However, to ensure good efficiency of CPCs, it is necessary to perform an optimal optical design and minimize thermal losses or improve heat transfer. Strategies to reduce convection losses in the receiver of a solar collector include using evacuated tubes or filling the CPC cavity with gases such as Argon and Krypton [3,4], which are denser gases and have lower thermal conductivity than air, or even applying a vacuum throughout the cavity [5]. In contrast, double absorbers have been proposed to reduce conduction losses [6].

In other technologies, such as flat-plate solar air heaters, it has been proposed to increase the heat transfer rate by incorporating multiple passages, including extended surfaces, artificial roughness, and packed mesh [7]. This multi-pass strategy is used in hybrid CPCs (PV/T) to cool the photovoltaic cells on the flat-plate receiver with fins on the back side [8].

In the general design of solar collectors, computational fluid dynamics (CFD) tools can be used to reliably estimate their thermo-hydraulic performance before building them, saving time and resources. Several analyses of solar collectors by computational fluid dynamics (CFD) can be found in the literature, both for liquid and air heating. Table 1 provides an overview of the different solar collector models and assumptions found in the literature review.

Table 1. Solar collectors CFD and radiation models in the literature review.

Reference	Solar Technology	TES Fluid	CFD Model	Radiation Model
[9]	Flat-plate	Air in gap of upper part	2D model Newtonian and incompressible gas Boussinesq approximation	ER: Constant radiative flux
[10]	Heat-pipe evacuated tube (HPETC)	Water in heat pipe Air or OCMs in inner evacuated tube	3D model Boussinesq approximation	ER: Time-dependent polynomial function for the solar irradiance data as a boundary condition
[11]	Evacuated tubes with and without circular ribs	Air	3D steady-state model RNG k-ε turbulence model with enhanced wall treatment model under periodic flow conditions	ER: Uniform and constant heat input as a boundary condition
[12]	Flat plates with selective surfaces and rectangular fins	Air	3D steady-state model Incompressible air Boussinesq approximation Standard k–ε turbulence model	IR: Discrete ordinates model ER: Constant solar heat flux
[13]	CPC with a single-pass evacuated tube receiver	Air	3D steady-state model k–ε turbulence model	ER: Ray-tracing and finite volume solver to determine the non-uniform solar flux distribution on the absorber surface
[14]	CPCs with three types of tubular receiver	Water	3D transient-state model Incompressible fluid Spalart-Allmaras turbulence model	ER: Ray-tracing and finite element solver to determine the incident solar flux on the absorber surface
[15]	CPCs with tubular receivers (with and without ETFE foil)	Water in absorber Air in cavity	3D steady-state model Boussinesq approximation Standard k-ε turbulence model	IR: S2S radiation model ER: Ray-tracing used to determine a correlation of absorbed solar energy as a function of angle along perimeter

Table 1. *Cont.*

Reference	Solar Technology	TES Fluid	CFD Model	Radiation Model
[16]	CPCs with tubular receivers and flat-plate receiver	Air in cavity	2D pseudo-transient model Incompressible ideal gas Buoyancy forces k–ε turbulent model	IR: S2S radiation model ER: Solar radiation is completely absorbed by the receiver
[17]	Panels of CPCs with tubular receiver	Air in cavity	2D pseudo-transient model Incompressible ideal gas k-ε turbulent model with the enhanced wall treatment	IR: S2S radiation model ER: Receiver temperature as a boundary condition (solar radiation is not simulated)
[18]	CPC with flat-plate receiver	Air in cavity	3D steady-state model Incompressible ideal gas Thermo-physical properties constants Non-Bousinessq approximation Standard k–ε turbulence model with enhanced wall treatment	IR: Discrete-ordinate radiation model ER: Thermal boundary condition with the surface external emissivity specified
[19]	Multi-pass flat-plates	Air	2D and 3D steady-state models k–ε turbulent model Standard k–ε turbulent model with realizability constraints in the vicinity of walls	IR: S2S radiation model ER: External Radiation Source sub-node is applied to contribute to the incident radiative heat flux on the solar spectral bands
[20]	Three-pass and quadruple-pass flat-plates	Air	3D steady-state model k–ε turbulence model	Not specified
[21]	Double-pass flat-plate	Air	3D steady-state model RNG k–ε turbulence model	IR: Discrete-ordinate radiation model ER: Energy equation and Discrete Transfer radiation model
[22]	Double-pass flat-plates	Air	2D steady-state model Incompressible flow k–ε turbulence model	ER: Uniform heat flux
[23]	Double-pass flat-plates with three fin configurations	Air	2D steady-state model RNG k–ε turbulence model	ER: Constant heat flux
[24]	Double-pass curved collectors	Air	2D steady-state model k-ε turbulence model	ER: Constant solar radiation flux

IR: Internal radiation; ER: External radiation.

Thus, Mekahlia et al. determined the influence of the thickness and number of transparent covers to reduce the heat losses of a flat-plate solar collector [9], and Pawar and Sobhansarbandi modeled an evacuated heat-pipe solar collector with and without integrated phase change materials as a thermal storage medium [10]. In the particular case of solar air collectors, Singla et al. analyzed an evacuated tube collector with ribs of different roughness [11]. At the same time, Ammar et al. performed a three-dimensional CFD model to optimize the design of a solar air collector with an extended surface area by a different number of rectangular fins [12]. In addition, they analyzed the effect of adding a selective surface on the absorber.

Regarding the analysis of CPC collectors, Li et al. analyzed by CFD the thermal behavior of an evacuated tube collector as a receiver of a compound parabolic concentrator, and the simulation was validated with experimental data [13]. Barrón-Díaz et al. performed the numerical simulation of CPCs with tubular receivers, with and without fins, for residential water heating [14]. This study focused on the ray-tracing analysis of radiation and heat transfer by coupled finite element and CFD methods. In addition, Yuan et al. developed two simplified computational fluid dynamics models to determine the temperature and velocity distribution in two almost identical parabolic tube-receiver CPCs [15]. One had a transparent ETFE sheet around the receiver to reduce convective heat losses. The models included the reflector, receiver, cover, and back insulating material and allowed the analysis of both air movement in the cavity and water movement in the absorber tube. Ray tracing was applied to analyze the radiation distribution on the receiver tube at normal incidence, with a correlation of the absorbed solar energy as a function of the angle along the perimeter of the tube. Both models were experimentally validated, and relative errors of less than 3.7% in temperature and 1% in efficiency were obtained.

On the other hand, Antonelli et al. analyzed the air heat transfer inside the cavity of a collector with a tubular receiver and with a flat-plate receiver and developed some correlations to express the Nusselt over the receiver [16]. Subsequently, Francesconi and Antonelli performed the numerical analysis of a panel with several tubular receiver CPCs to determine the influence on the thermal efficiency of the number and position of the CPCs along the panel, the use of a second transparent cover, the spacing between collectors, and the truncation of the reflectors [17]. For their part, Reddy et al. performed three-dimensional modeling of a flat-plate receiver CPC to determine the thermal losses in the cavity as a function of its aspect ratio and tilt, the optical properties of the materials, and the absorber and ambient temperatures [18]. To model the internal radiative heat transfer, they used a discrete ordinary radiation model, and for the external one, they established the thermal boundary conditions and emissivity.

As mentioned above, another strategy to improve the efficiency of solar air heating collectors is to increase the number of passes. Thus, Al-Damook et al. analyzed the effect of double-pass configuration in a solar air heater when operating in concurrent parallel flow, parallel in counterflow, and double U-pass [19]; the latter presented the best thermal performance. Tuncer et al. analyzed, through CFD simulation, two flat-plate solar collectors for air heating with three and four passes and determined which one had the best performance to evaluate it experimentally [20]. In both solar collectors analyzed, air enters through the lower pass and exits through the upper pass, which has the radiant heat gain. They found that the four-pass collector has a heat gain 3 °C higher than that obtained with the three-pass collector and that the maximum deviation between the CFD model and the experimental results was 10%. In addition, Mutabilwa and Nwaigwe performed a CFD analysis of a two-covers, double-pass flat-plate solar collector for air heating, which was validated with experimental results [21]. The air enters through the space between the two covers and returns between the second cover and the absorber plate. The temperatures on the absorber plate obtained with the model had a standard deviation from experimental results between 1.05 K and 4.65 K, while for the cover, it was between 0.1 K and 0.45 K.

Likewise, improved surfaces or novel geometries have been incorporated in multi-passes solar collectors, such as the work of Desisa and Shekata [22]; they analyzed the impact of using smooth, rough, and corrugated surfaces in a double-pass flat-plate air solar collector and obtained average thermal efficiencies of 78%, 62%, and 90%, respectively. On the other hand, Singh determined the performance of double-pass flat-plate air solar collectors with different fin configurations [23]. They varied in size, angle, arrangements (in-line, staggered, and hybrid), and hydraulic diameter. Finally, Kumar et al. proposed a curved air heater with asymmetric double-pass counterflow turbulators, whose design was determined from CFD analysis by comparing various flow configurations and geometric parameters [24].

Two or more pass technologies have been applied in flat-plate solar collectors to improve their efficiency; however, this strategy has not been applied in CPCs for air heating. This study proposes the CFD analysis of a CPC-type solar air heater with U-shape double-pass airflow. The air first circulates through the trapezoidal cavity contained in the volume formed by the cover, the reflecting walls of the CPC, and the flat-plate receiver and then circulates in counterflow through the receiver's duct interior. The objective of the numerical analysis presented is to test different inlet and outlet configurations in the CPC array to determine how these configurations influence the velocity distribution, outlet temperature, and instantaneous efficiency of the U-shape double-pass CPC solar heater.

Section 2 of this manuscript describes the main characteristics of the U-shape double-pass CPC and the four air inlet/outlet configurations considered in its design. It also defines the mesh design to perform the CFD simulation, the mathematical model for such simulation, the boundary conditions applied in the study, and the methodology followed to estimate the thermal efficiency of the U-shape double-pass CPC. Section 3 includes a summary of the simulation results obtained and their discussion and concludes with a summary of the efficiencies calculated for each of the four configurations analyzed.

2. Materials and Methods

2.1. U-Shape Double-Pass CPC Description and Physical Model

The proposed solar air heater is a variant of the flat-plate receiver Compound Parabolic Concentrator (CPC) conceptualized as a U-shape double-pass heat exchanger. Figure 1a shows the evaluated geometry dimensions and the inlet and outlet positions, whereas, in Figure 1b, the CPC cross-section is shown. The CPC is tilted 24° since it is the latitude of the City of interest (Durango, Mexico) and consists of a flat-plate receiver, two reflectors, a cover, and a duct. The first pass of the airflow inside the CPC occurs in the cavity formed by the receiver, two reflectors, and the cover, while the second pass is in the duct section. The aperture area where the solar radiation enters the CPC is 0.42 m^2, while the area where it is absorbed is 0.20 m^2.

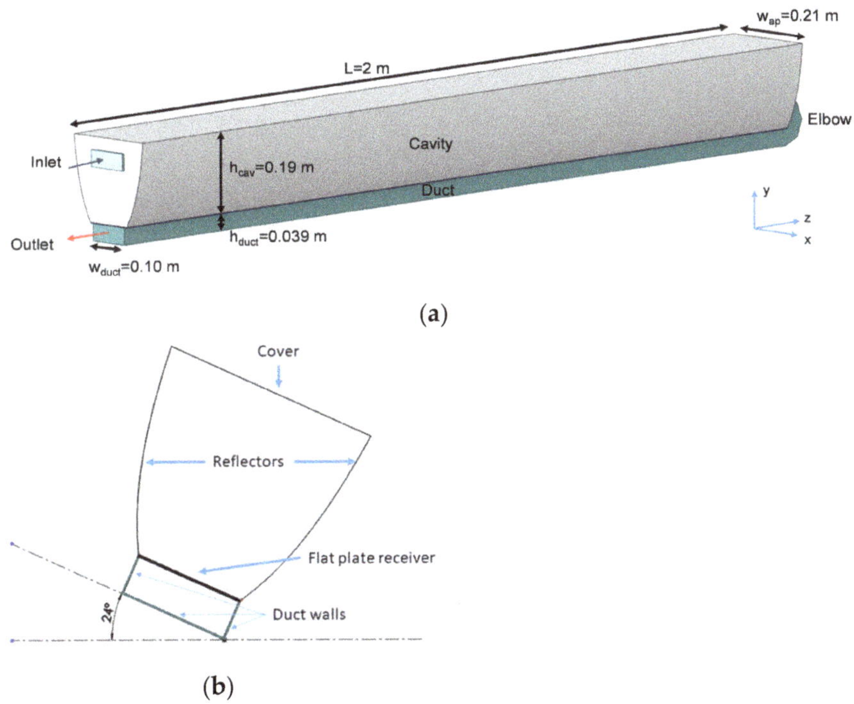

Figure 1. U-shape double-pass CPC solar heater. (**a**) Geometry; (**b**) Cross-section.

The analysis of the position of the air inlet and outlet in the cavity consisted of the study of four configurations that were positioned concerning the height of the cavity (h_{cav}): (a) inlet $\frac{1}{4}h_{cav}$-outlet $\frac{1}{4}h_{cav}$ (Down–Down), (b) inlet $\frac{1}{4}h_{cav}$-outlet $\frac{3}{4}h_{cav}$ (Down–Up), (c) inlet $\frac{3}{4}h_{cav}$-outlet $\frac{3}{4}h_{cav}$ (Up–Up), and (d) inlet $\frac{3}{4}h_{cav}$-outlet $\frac{1}{4}h_{cav}$ (Up–Down). The air inlet and outlet configurations are shown in Figure 2.

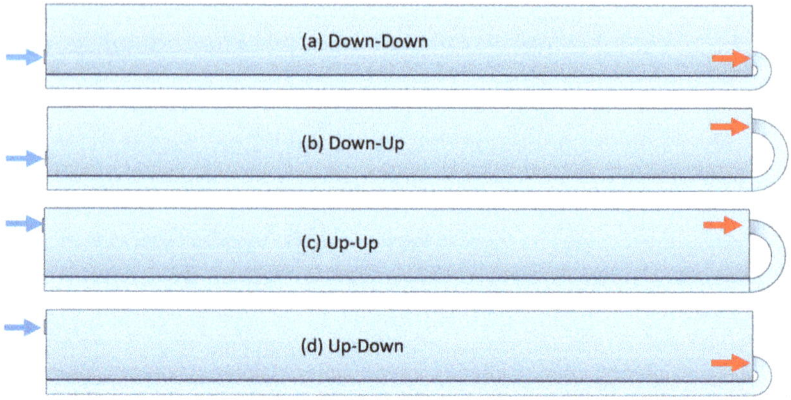

Figure 2. Inlet and outlet studied configurations of U-shape double-pass CPC.

The properties were considered constant in the solid (Table 2). The reflector is made of anodized aluminum, while the other components of the U-shape double-pass CPC, shown in Table 2, were considered in the CFD simulation with a certain thickness to model the conduction. The duct is made of aluminum, and the receiver substrate has a selective surface; this surface has high absorptivity in the solar spectrum and low emissivity in the infrared to avoid losses due to thermal radiation. Finally, the cover is made of solid polycarbonate, and an insulating material (EPS) was considered outside the reflector and the duct to avoid thermal losses from the surface exposed to the environment.

Table 2. Material properties of the U-shape double-pass CPC elements.

Material	Thickness (m)	Thermal Conductivity (W/m·K)	Density (kg/m^3)	Specific Heat Capacity (J/kg·K)
Aluminum (receiver, duct)	0.001	237	2702	903
Expanded polystyrene (EPS) (insulation)	0.050	0.046	14	1210
Polycarbonate (cover)	0.003	0.210	1200	1300

For air, the density, thermal conductivity, and viscosity were considered as polynomial functions of temperature, and the specific heat as a piecewise-linear function (Table 3).

Table 3. Thermal properties of air.

Property	Type	Coefficients/Interval	Temperature Interval of Validity (K)
Density *, ρ, kg/m^3	Polynomial	$2.1781T - 0.0033$	273.15–393.15
Thermal conductivity *, k, W/(m K)	Polynomial	$0.003792T + 7.3e^{-5}$	273.15–393.15
Molecular viscosity *, μ, kg/(m s)	Polynomial	$5.141e^{-6}T + 4.5e^{-8}$	273.15–393.15
Specific heat, C_p, J/(kg K)	Piecewise-linear	1006@273.15 K; 1007@288.15 K; 1008@353.15 K; 1011@398.15 K	273.15–398.15

* The polynomial functions were obtained with data from [25].

2.1.1. Computational Domain

The fluid and solid domains were generated in the SolidWorks 2013 SP2.0 software. The solid domain simulated the absorber plate, while the fluid domain was sectioned into three volumes to facilitate meshing: (a) inlet section, (b) cavity, and (c) elbow-duct.

2.1.2. Mesh

A hexahedral structured mesh was generated according to the proposed computational domain. The near-wall model approach was used to accurately predict the hydrodynamic behavior of the flow and the heat transfer in the system. The method was to implement 15 cells to cover the viscous and buffer sublayer to have accurate results in a reasonable computation time.

The mesh refinement was carried out considering the shear stress for the hydrodynamic phenomenon and the Nusselt number (Nu) for the thermal boundary layer. The Nusselt number represents the dimensionless temperature gradient in the wall of interest.

The mesh size was refined until the variation of the shear stress and the average Nu was less than 1%. Next, the size of the viscous sublayer and the buffer sublayer were calculated for the interval $0.5 < y^+ < 5$ using Equation (1), and for the thermal sublayer, the Nu was monitored. In addition, y^+ values of 35 and 60 were applied to the turbulent sublayer to carry out the mesh independence study; this monitoring was carried out to describe the viscous sublayer. Then, to obtain the final mesh size used in this work, the mesh was refined in the z-axis from 100 divisions to 1600. Through the analysis, Nu varied 0.2% with y^+ values of 0.8 and 0.5, selecting y^+ 0.8.

$$y = \frac{y^+ \mu}{\rho u_T} \tag{1}$$

Additionally, the mesh size was verified in the direction of the entrance flow with a cavity mesh refinement in the longitudinal axis (z-axis), as shown in Figure 3. The analysis found that the Nusselt had a variation of less than 0.1% from 550 divisions onwards. In Figure 3a, a cross-section of the U-shape double-pass CPC solar heater is shown, with the magnified detail of the mesh in the receiver. Figure 3(b1,b2) present the longitudinal section, where the coarse and refined mesh in the cavity are presented.

The k-ω models are y^+ insensitive treatments; therefore, the ω-equation can be integrated without additional terms through the viscous sublayer. Nevertheless, the Transition SST k-ω model requires a more stringent grid resolution to solve the thin laminar boundary layer upstream of the transition location. For this reason, using a near-wall mesh with $y^+ \approx 1$, especially for heat transfer predictions, is recommended [26].

2.2. Mathematical Model

The mathematical model had the subsequent considerations for the governing equations: steady state, Newtonian fluid, incompressible flow, and transition turbulence regime; therefore, the governing equations for the U-shape double-pass solar heater are as follows.

$$\frac{\partial \rho}{\partial t} + \nabla \cdot \left(\rho \vec{v} \right) = 0 \tag{2}$$

$$\frac{\partial}{\partial t}\left(\rho \vec{v} \right) + \nabla \cdot \left(\rho \vec{v} \vec{v} \right) = -\nabla p + \nabla \cdot \left(\overline{\overline{\tau}} \right) + \rho \vec{g} + \vec{F} \tag{3}$$

$$\frac{\partial (\rho E)}{\partial t} + \nabla \cdot \left(\vec{v} (\rho E + P) \right) = \nabla \cdot \left(\sum_j h'_j J_j\right) \tag{4}$$

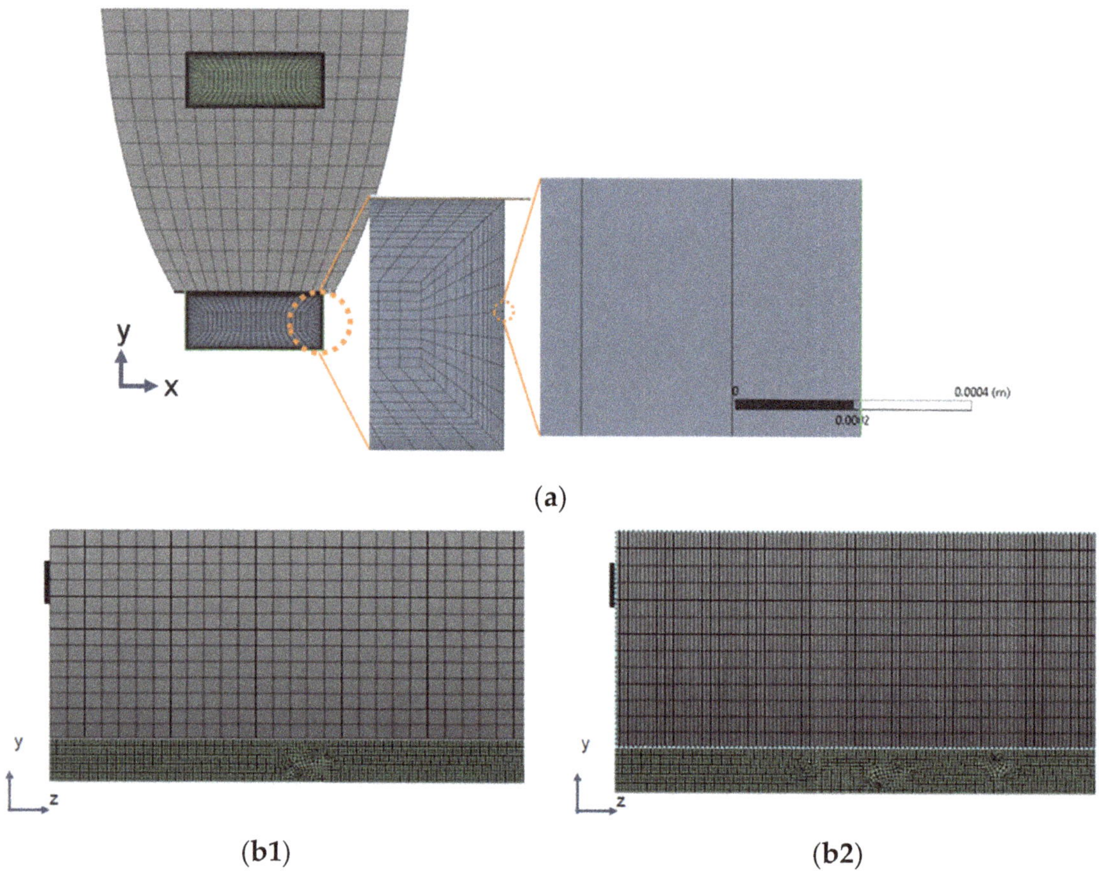

Figure 3. Solar collector mesh. (**a**) Transversal section; (**b1**) coarse mesh in longitudinal section; (**b2**) refined mesh in longitudinal section.

2.2.1. Turbulence Model

A preliminary hydrodynamic analysis performed in the SolidWorks 2013 SP2.0 software determined that the flow separates due to the sudden expansion at the cavity inlet. Furthermore, the flow was found to be under development ($L_{h,turbulent} < L_{collector}$), and the calculated average Reynolds numbers (Re) were very low (for flow 1 (0.01 kg/s) was $Re_{cav\text{-}1} = 2972$, and for flow 2 (0.02 kg/s) was $Re_{cav\text{-}2} = 5961$). The k-ω turbulence models are better at predicting adverse pressure gradient boundary layer flows and separation, and they also have the ability to simulate the laminar–turbulent transition of wall boundary layers [26]. Additionally, the k-ω models have low-Reynolds number terms ($Re < 10^4$) that mimic laminar–turbulent transition processes. However, this function is not widely calibrated in the SST k-ω model; therefore, it is recommended to use the Transition SST k-ω model [27,28].

The Transition SST model was selected based on the described above and considering the required accuracy to predict heat transfer from the absorber plate to the air as the flow. Furthermore, the buoyancy effects were adjusted to full, and the viscous heating was activated. The turbulence modeling consisted of implementing the Transition SST k-omega model described in Equations (5)–(12). Equation (5) corresponds to the transport equation for intermittency (γ), whereas Equations (6) and (7) represent the transition sources $P_{\gamma 1}$ and $E_{\gamma 1}$, respectively; and Equations (8) and (9), the destruction/re-laminarization sources $P_{\gamma 2}$ and $E_{\gamma 2}$.

Flows 1 and 2 were selected based on a preliminary analysis using the thermal model described in [29]; among those flows, the best balance between air outlet temperature and thermal efficiency was found. In the calculation of the Reynolds number, the cavity was approximated as a trapezoidal cross-section duct for calculating the hydraulic diameter (D_h).

$$\frac{\partial(\rho\gamma)}{\partial t} + \frac{\partial(\rho U_j \gamma)}{\partial x_j} = P_{\gamma 1} - E_{\gamma 1} + P_{\gamma 2} - E_{\gamma 2} + \frac{\partial}{\partial x_j}\left[\left(\mu + \frac{\mu_t}{\sigma_\gamma}\right)\frac{\partial}{\partial x_j}\right] \tag{5}$$

$$P_{\gamma 1} = C_{a1} F_{length} \rho S [\gamma F_{onset}]^{c_{\gamma 3}} \tag{6}$$

$$E_{\gamma 1} = C_{e1} P_{\gamma 1} \gamma \tag{7}$$

$$P_{\gamma 2} = C_{a2} \rho \Omega \gamma F_{turb} \tag{8}$$

$$E_{\gamma 2} = C_{e2} P_{\gamma 2} \gamma \tag{9}$$

On the other hand, Equation (10) refers to the interaction of the transition model with the SST turbulence model by modifying equation k, where G_k^* and Y_k^* are the original production and destruction terms of the SST model [28].

$$\frac{\partial(\rho k)}{\partial t} + \frac{\partial(\rho k u_i)}{\partial x_j} = \frac{\partial}{\partial x_j}\left(\Gamma_k \frac{\partial k}{\partial x_j}\right) + G_k^* - Y_k^* \tag{10}$$

$$G_k^* = \gamma_{ef\,f} \widetilde{G}_k \tag{11}$$

$$Y_k^* = \min\left(\max\left(\gamma_{ef\,f}, 0.1\right), 1.0\right) Y_k \tag{12}$$

The pressure-based solver with the Coupled scheme was selected, and a second-order spatial discretization scheme was implemented. The gradient evaluation method selected was based on Least Squares Cell-Based, and the high-order term relaxation option was used. The pressure factor was adjusted to 0.1, and the Flow Courant Number to 4.

2.2.2. Boundary Conditions

The momentum boundary conditions in the walls were considered no-slip stationary with a constant rugosity of 0.5. The thermal boundary conditions modeled the incident solar radiation on the cover and the absorber plate as a heat generation source, calculated using Equations (13) and (14).

$$q_c(t) = \left(I(t)\left[\overline{\alpha}_c + \overline{\alpha}_c \overline{\tau}_c \overline{\varrho}_p \varrho_r^{2\langle n\rangle}\right]\frac{A_c}{A_p}\right)/w_c \tag{13}$$

$$q_p(t) = \left(I(t)\,\overline{\tau}_c \varrho_r^n P_g\left[\overline{\alpha}_p + \overline{\alpha}_p \overline{\varrho}_p \overline{\varrho}_c \overline{\varrho}_r^{2\langle n\rangle}\frac{A_p}{A_c}\right]\frac{A_c}{A_p}\right)/w_p \tag{14}$$

where $q(t)$ is the heat generation, $I(t)$ is the solar irradiance, P_g is the gap loss factor (0.96), α is the absorptivity, τ is the transmissivity, ρ is the reflectivity, A_c is the cover area, A_p is the absorber plate area, and w_c and w_g, are the thickness of the cover and plate, respectively. The optical properties of the cover are α_c = 0.05, τ_c = 0.89 y ρ_c = 0.05, the receiver are α_p = 0.95 and ρ_p = 0.05, and from the reflector ρ_r = 0.91.

The heat transfer coefficient from the cover to the environment ($HTC_{c\text{-}a}$) was obtained by applying the flow around finite flat-plates methodology reported in [30], while the convection losses of the external walls were calculated using the heat transfer coefficient (HTC_b) correlation proposed by [31]. Additionally, in laminar flow, the heat transfer coefficient from the flat-plate receiver to the fluid in the cavity ($HTC_{p\text{-}cav}$) was estimated using the discretized Fourier's law, considering the local temperature normal to the wall [26].

Moreover, in turbulent flow, the $HTC_{p\text{-}cav}$ was determined using the law of the wall for estimating the local temperature of the fluid by applying the Reynolds analogy [26].

The heat conduction in the exterior walls of the CPC was modeled as shell conduction, whereas the radiation losses in the cover were calculated with an emissivity value of $\varepsilon_c = 0.81$, and the sky temperature (T_s) was calculated with the correlation proposed by Swinbank (Equation (15)), reported in [32], where T_a refers to the ambient temperature.

$$T_s = 0.0552 T_a^{1.5} \tag{15}$$

Regarding the turbulence parameters, a turbulence intensity of 5% and a turbulent viscosity ratio of 10 were applied. Table 4 summarizes the parameters of the boundary conditions of the CFD modeling.

Table 4. Boundary conditions considered in the CFD modeling of the U-shape double-pass CPC.

Boundary	Type of Boundary	Characteristics
Inlet	Inlet-vent	$T_{ps1,in}$ = 298.15 K
Outlet	Mass flow	$\dot{m}_1$ = 0.01 kg/s $\dot{m}_2$ = 0.02 kg/s
Cover	Wall Mixed (convection, radiation, and heat generation)	q_c = 37.37 kW/m^3 $HTC_{c\text{–}a}$ = 8.27 W/m·K T_a = 298.15 K T_s = 280.05 K
Bottom duct	Wall Convection Conduction	HTC_b = 0.6 W/m·K
Absorber plate	Wall Heat generation	I = 900 W/m^2 ε_p = 0.35 q_p = 13,481 kW/m^3
Fluid interfaces: - cavity-absorber plate - duct-absorber plate	Coupled wall	-

The pressure-based solver with the Coupled scheme was selected. In addition, a Second-Order scheme for spatial discretization was implemented because of numerical simulation stability. The formal truncation errors of individual terms in the governing equations were calculated; the error for the $HTC_{p\text{-}cav}$ was 1.27%, 0.051% for the shear stress, and 0.05% for the Nusselt number (with a security factor of Fs = 3) [32]. In addition, values of 1×10^{-4} for the mass residual and a mass imbalance of 5.3×10^{-8}% and 2×10^{-7} energy residual were accomplished. The verification of the results was carried out by quantifying the uncertainty of the numerical calculations. For Nu, the spatial error of 0.51% was obtained, while for $HTC_{p\text{-}cav}$ and shear stress were 0.66% and 0.37%, respectively. The grid convergence index was also verified, finding out 0.05% for Nu, 0.19% for shear stress, and 0.09% for $HTC_{p\text{-}cav}$.

2.3. Efficiency Calculation

The thermal efficiency of the collector is calculated as the ratio between the useful energy gained by the fluid on the collector's cavity and absorber and the net solar energy on the collector's aperture, using Equation (16):

$$\eta = \frac{Q_u}{A_c I} \tag{16}$$

The heat transfer of the analyzed solar collector resembles a counterflow heat exchanger, where the cold fluid flows through the cavity, and the hot fluid flows through the duct. Therefore, the useful energy gain of the collector is calculated using Equation (17).

$$Q_u = \dot{m} C_{p,avg} \left(T_{ps2,out} - T_{ps1,in} \right) \quad (17)$$

3. Results and Discussion

Once the simulation model of the four CPC configurations with different inlet and outlet positions was implemented (see Figure 2), simulations were carried out considering the boundary conditions for two different values of mass flow rates to analyze: 0.01 kg/s and 0.02 kg/s. The materials, air properties, and boundary conditions are presented in Tables 2–4, respectively. In addition, the results of the hydraulic and thermal behavior of the air in the cavity of the CPC for each of these configurations are presented below.

3.1. Hydraulic Behavior

Figures 4 and 5 show the velocity streamlines with the two analyzed air mass flow rates (flow 1: 0.01 kg/s and flow 2: 0.02 kg/s, respectively). In Figure 4 (flow 1: 0.01 kg/s), the configurations with an inlet from below (a and b), a sudden expansion occurs near the inlet of the cavity, forming an eddy in the upper part of the collector. While in the configurations with the entrance at the top (c and d), the eddy forms at the bottom. Further, in all configurations except (b), it is observed that after the air enters the cavity, several families of eddies form until the end of the collector (z = 2.0 m). On the other hand, in configuration (b), a large eddy is observed in the first half of the cavity, and then the formation of some smaller eddies. Finally, it is essential to note that the highest magnitude velocities are generated at the elbow and at the beginning of the duct (second pass).

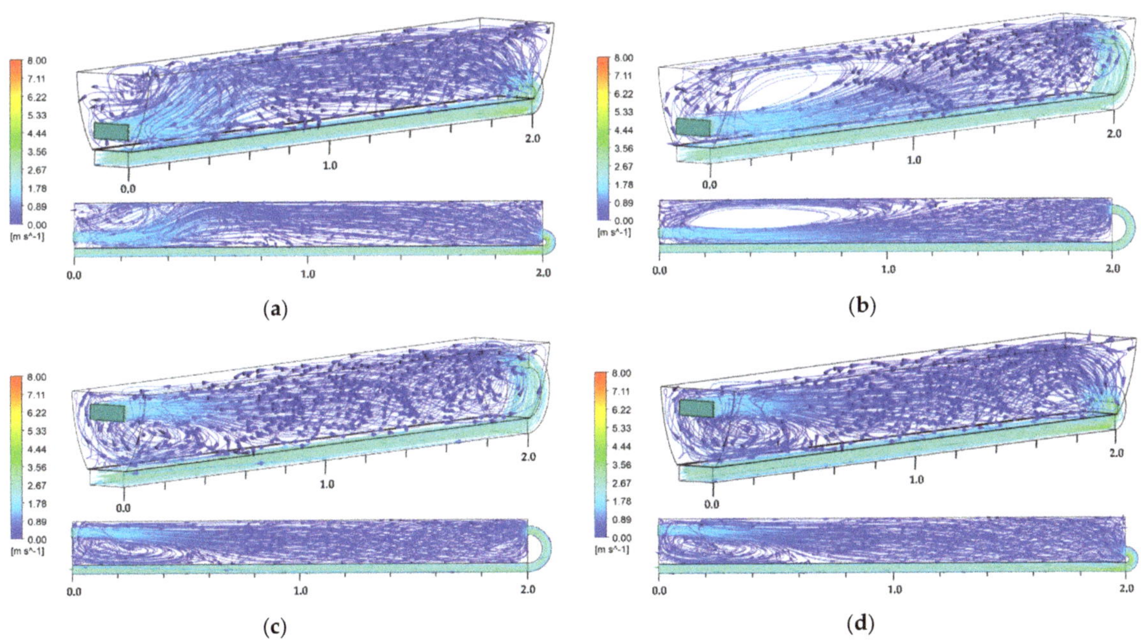

Figure 4. Streamlines at the collector cross-section with air mass flow 1: 0.01 kg/s. (**a**) Down–Down; (**b**) Down–Up; (**c**) Up–Up; (**d**) Up–Down.

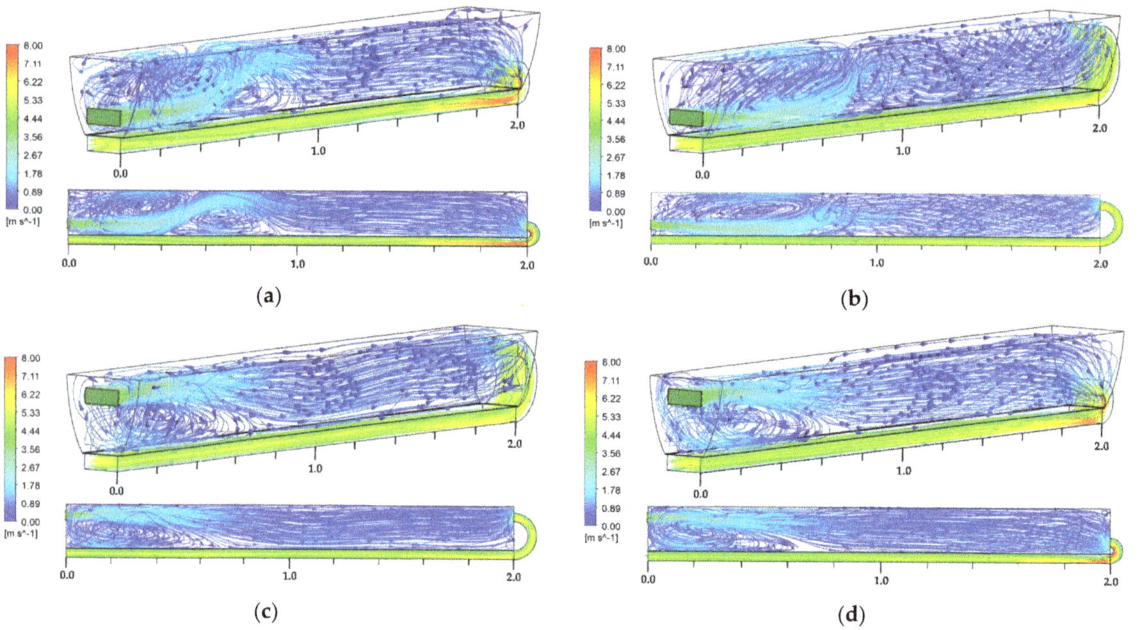

Figure 5. Streamlines at the collector cross-section with air mass flow 2: 0.02 kg/s. (**a**) Down–Down; (**b**) Down–Up; (**c**) Up–Up; (**d**) Up–Down.

In Figure 5, a similar behavior to the one described for flow 1 is shown for flow 2 (0.02 kg/s), where a sudden expansion of air near the inlets occurs. For configurations (a) and (b) (inlet from the bottom), a visible jet can be observed at the bottom of the cavity; here, an eddy at the upper region can also be observed. Contrarily, in the configurations with an inlet from above, the jet is formed at the top, whereas the eddy occurs at the bottom region. Additionally, in configurations (a) and (c) (either both inlet and outlet from above or below), there is a primary air current with high speed. In configuration (a), the current goes up and down the cavity, generating several eddies at the upper and bottom sides opposite to the main flow. Contrarily, in configurations (b) and (d), which have inlet and outlet in opposite positions on the y-axis, a large eddy is formed near the cavity inlet, observing a larger eddy in configuration (b) that moves towards the exit in a disorderly manner. Moreover, the largest speed occurs at the elbow of the collector, having higher speeds in the configurations with a bottom outlet.

One reason that explains a disordered and asymmetric flow is because of the collector tilt (see Figure 1). In addition, the pressure increase justifies the phenomenon of the sudden air contraction caused by the elbow area reduction. Furthermore, a higher mass flow influences the amplitude and turbulence of the eddies found in the first section of the cavity.

3.2. Thermal Behavior

Figures 6 and 7 show the temperature fields of the four analyzed configurations. In Figure 6, for flow 1, it is observed that there is an extended region at low temperatures for all configurations in the cavity inlet, which is related to the air inlet in the form of a jet described in Figure 4. In all configurations, a region with low temperature is generated related to the jet of cold air that enters the cavity. For the configurations with an inlet from the bottom, the zone is located in the upper left corner. This is due to the presence of the primary eddies observed in the hydrodynamic behavior of the fluid (Figures 4 and 5). In configurations with a top inlet, the area is also extended towards the middle of the cavity, which is related to less turbulence in the movement of the fluid.

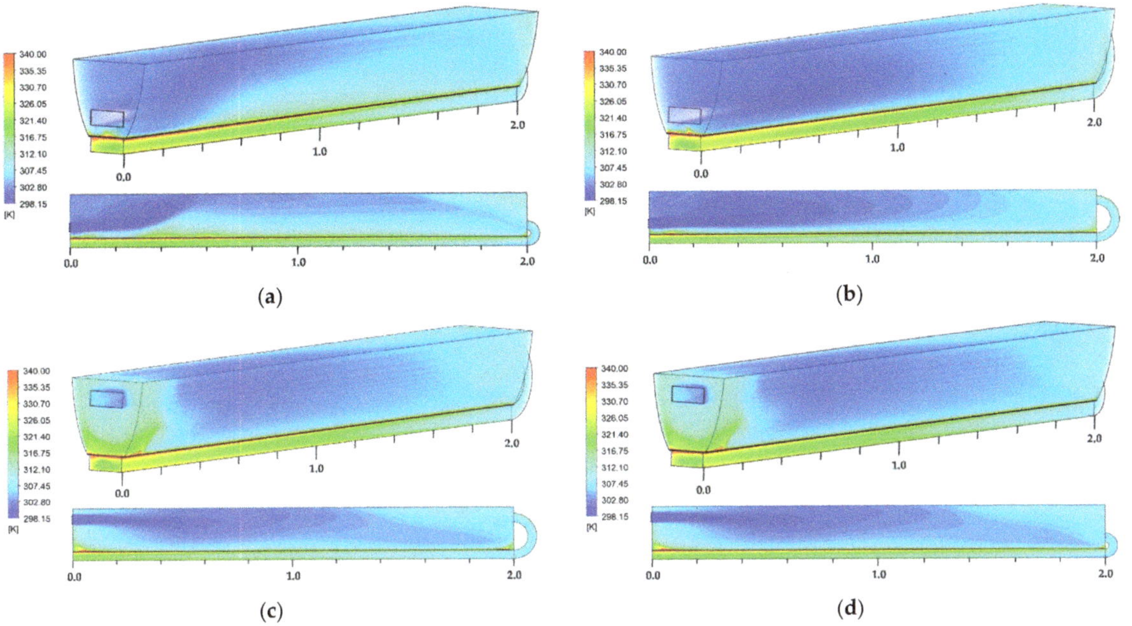

Figure 6. Air temperature fields at the collector cross-section with air mass flow 1: 0.01 kg/s. (**a**) Down–Down; (**b**) Down–Up; (**c**) Up–Up; (**d**) Up–Down.

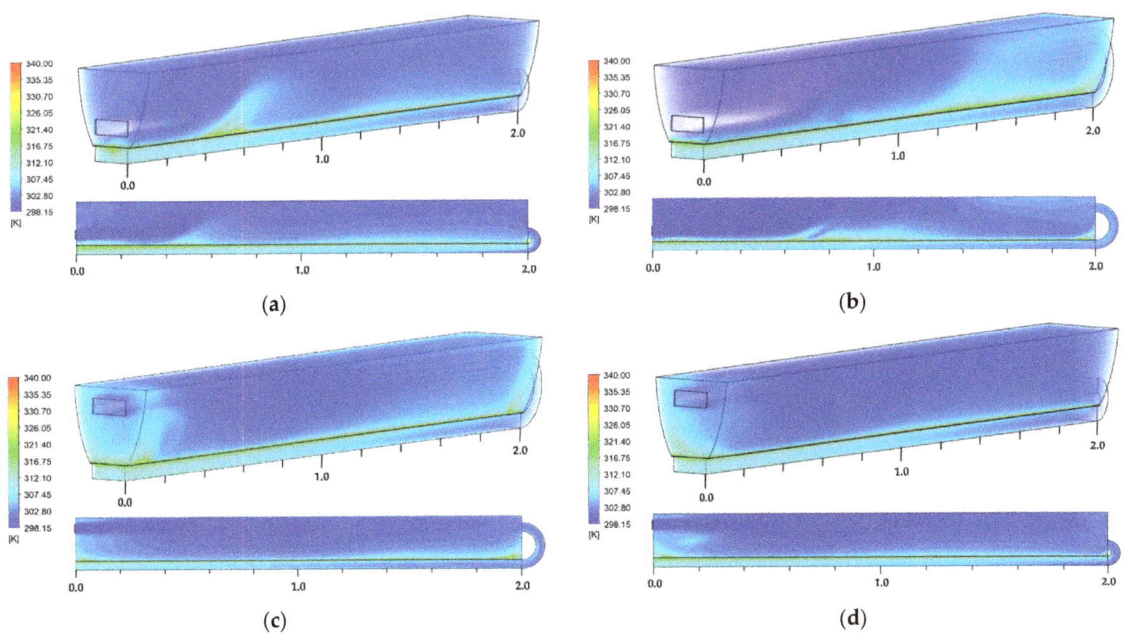

Figure 7. Air temperature fields at the collector cross-section with air mass flow 2: 0.02 kg/s. (**a**) Down–Down; (**b**) Down–Up; (**c**) Up–Up; (**d**) Up–Down.

When the air enters from above, a zone of hot air is generated in the region near the entrance and another near the exit, while configurations with a bottom inlet have a heat recovery since the air enters the cavity.

For all configurations, the region with the highest temperature is located in the proximity of the receiver, and it is related to the heat generation on the receiver plate. Consequently, in the final part of the duct, there is another region with high temperatures. High temperatures are highly desirable since heat is extracted from the plate to the working fluid.

In Figure 7 (flow 2: 0.02 kg/s), the temperatures are lower than the ones observed in flow 1 (0.01 kg/s). In the four configurations, the temperature fields are highly dependent on the movement of the fluid in the cavity. The phenomena that drive the low temperatures in the section near the cavity inlet are the presence of a cold air jet from the inlet air and the consequent formation of eddies throughout the cavity. In the configurations with an inlet from below (a and b), there is another circumstance that causes the low temperatures, and it is due to the formation of the main eddy in the upper part of the cavity that promotes the stagnation of cold air in this section. Similar to the behavior described for temperatures of flow 1, for flow 2, the highest temperatures are always found in the regions neighboring the receiving duct.

Figures 8 and 9 show the four configurations of the flat plate receiver temperature contours at flow 1 (0.01 kg/s) and flow 2 (0.02 kg/s), respectively.

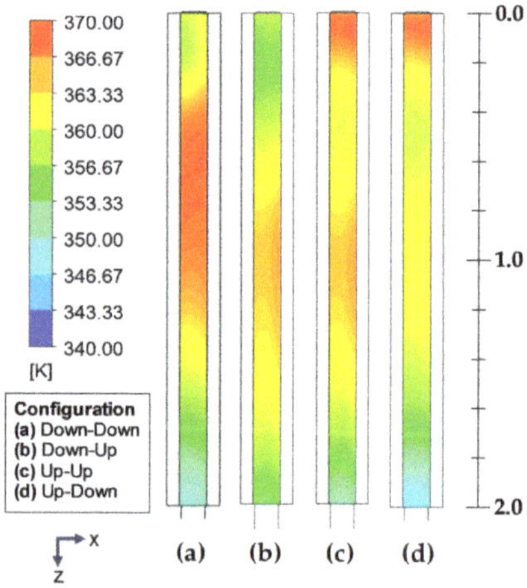

Figure 8. Temperature contours in the flat plate receiver, first pass (Flow 1: 0.01 kg/s).

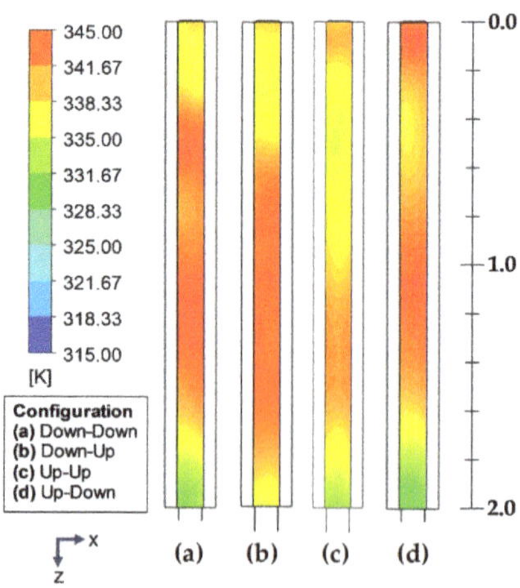

Figure 9. Temperature contours in the flat plate receiver, first pass (Flow 2: 0.02 kg/s).

Figure 8 (flow 1: 0.01 kg/s) shows that when the air enters from below (configurations a and b), it removes heat from the first part of the receiver since it enters the cavity. In contrast, when the air enters from above (configurations c and d), the zones with lower temperatures are displaced from around positions z = 0.30 m to z = 0.65 m. The displacement of the presence of the lower temperature zones is a direct consequence of the formation of the main eddy, which helps to remove heat from the receiver. On the other hand, in configuration (a) Down–Down, a higher temperature of around 370 K is observed, indicating that the heat would not be uniformly removed in the first pass. This high-temperature zone is explained by contrasting with the hydrodynamic behavior observed in Figure 4, since when the air enters from the bottom, it heats up when it comes into contact with the receiver and rises, then continues its movement mainly through the upper part and then descend when looking for the exit that is in the lower part of the cavity. In contrast, configurations (b), (c), and (d) have a medium–high temperature zone in the center of the receiver. When the air leaves the collector cavity (z = 2.0 m), low-temperature zones are generated in the configurations with an outlet from below (configurations (a) and (d)).

In contrast, in configurations (c) and (d), the existence of two zones of low and medium–high temperature indicates that although heat removal is heterogeneous, energy is recovered in the zone between z = 0.30 m and z = 0.8 m, which is related to the presence of a large eddy in the inlet zone surroundings. In configuration (b), two low-temperature zones are observed at the inlet and outlet of the collector and a medium–high temperature zone in the center.

In Figure 9, similar temperatures but lesser magnitude can be observed, corresponding to the prevalence of high velocities due to the application of a large mass flow rate (0.02 kg/s). Low-temperature regions are observed near the inlets in the receiver plates of configurations (a) and (b), where the inlet is from below. Configurations (a) and (d), with outlets from below, have low-temperature regions near the outlets meaning that heat removal mainly occurs in those sections. On the other hand, configurations (a) and (b) have a sizeable high-temperature area in their central zone, noting that this area comprises most of the receiver plate extension. For configuration (c), a zone from z = 0.2 m to z = 1.0 m with lower temperatures is observed, and even though its outlet is from above; it also has a small region with low temperatures near the exit. Contrastingly, configuration (d) has most of its receiver flat-plate with high temperatures, except for its final part (around z = 1.7 m to z = 2.0 m). Still, it is essential to note that the high temperatures observed for the receiver plates of Figure 9 (around 345 K) are of lesser magnitude than those observed in Figure 8 (around 370 K), and thus higher heat removal was accomplished with the application of flow 2 (0.02 kg/s).

Table 5 shows the average temperatures in the flat plate receiver ($T_{p,avg}$) for both airflow rates. The highest temperature, as expected, is obtained with flow 1, 362.2 K with configuration (a); the lowest is obtained with flow 2, with configuration (c), of 338.5 K. The flat-plate temperature differences between flows 1 and 2 are greater than 19.7 K.

Table 5. Mean temperatures in the flat plate receiver of the U-shape double-pass CPC.

		Mean Temperature, $T_{p,avg}$ (K)	
	Configuration	**Flow 1: 0.01 kg/s**	**Flow 2: 0.02 kg/s**
(a)	Down–Down	362.17	340.77
(b)	Down–Up	360.83	341.08
(c)	Up–Up	362.03	338.51
(d)	Up–Down	359.92	340.00

Figure 10 displays the first and second pass air temperature profiles for the four analyzed configurations. The blue line represents the air temperature profile in the first pass, and the red line is the profile of the second pass. In addition, the scale of the horizontal axis of the graphs (z-position) is inverted to facilitate the interpretation of the results since it allows visualizing the air outlet of the collector on the far right.

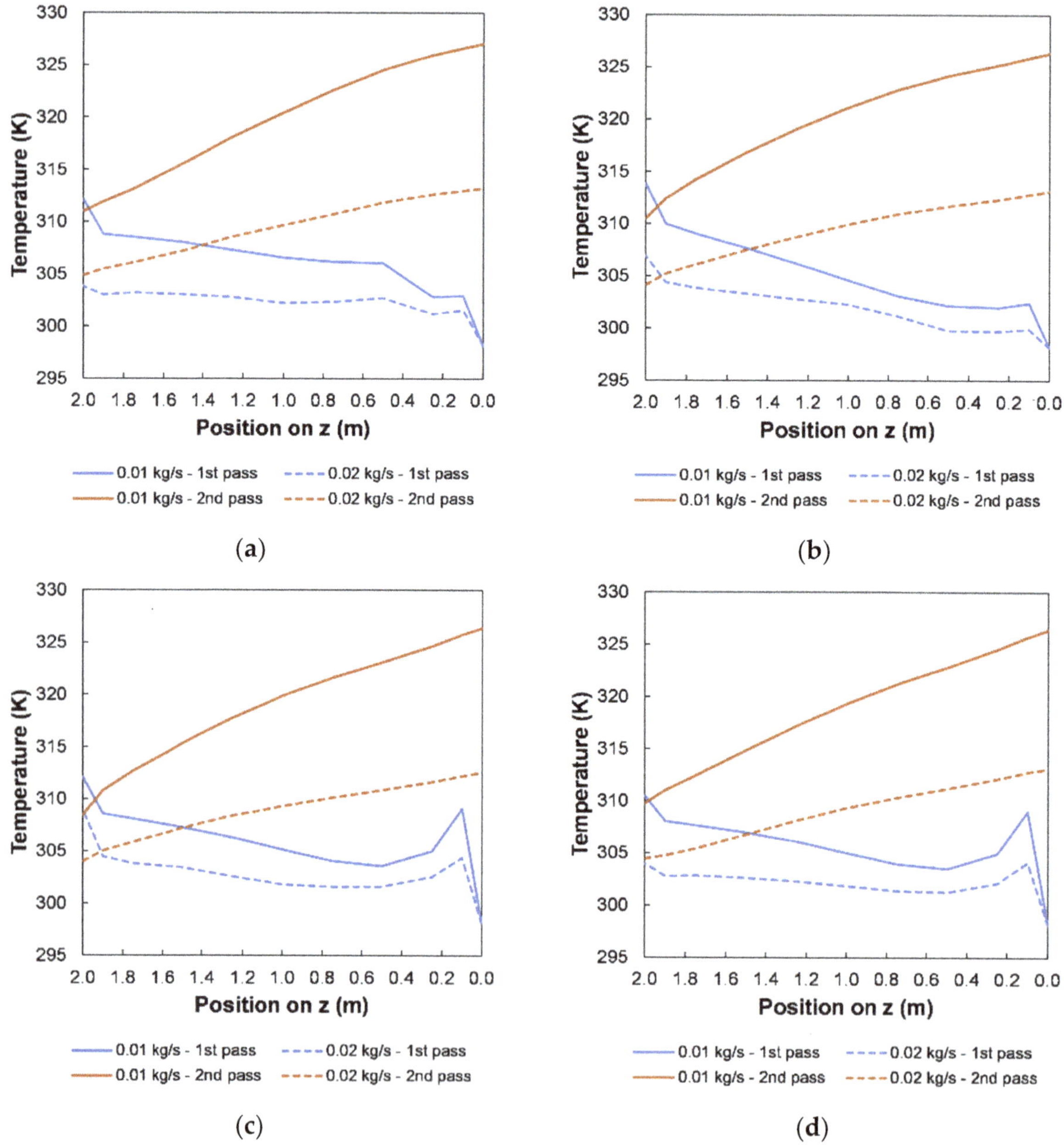

Figure 10. Air temperature profiles. (**a**) Down–Down; (**b**) Down–Up; (**c**) Up–Up; (**d**) Up–Down.

In the first pass for the configurations with the inlet from the bottom, (a) Down–Down and (b) Down–Up, the temperature profile has a slight increase near the entrance (z = 0.1 m), then it continues to increase until near the end of the cavity (z = 1.9 m) where it has another sharp increase. In the configurations that have an entrance to the cavity from the top, (c) Up–Up and (d) Up–Down, it is observed that in the first pass, there is a sudden increase in temperature in the section close to the entrance of the cavity (z = 0.1 m). In addition, all the temperature profiles are smooth for both flow rates; nevertheless, the temperature profiles have a less pronounced slope for configurations (a), (c), and (d).

Furthermore, in the second pass of all configurations, the temperature profile is smooth, with a sustained increase in temperature, reaching similar outlet temperatures of 327 K for flow 1 and 313 K for flow 2.

In the second pass for all the studied configurations, the temperature profile steadily increases from z = 2.0 m to z = 0.0 m, where it is observed that the temperature profiles are smooth for both flow rates. The differences in the behavior among the configurations in the second pass are only slight differences in the temperatures at the inlet and outlet, where the lowest inlet temperature is found in configuration (c) Up–Up, and the highest in (a) Down–Down.

On the other hand, the configurations that have an outlet from above, (b) Down–Up and (c) Up–Up, present a higher temperature at the outlet end of the cavity (end of the first pass) than at the entrance to the duct (beginning of the second pass). The phenomenon is caused by stratification in the cavity, so the air enters the connecting elbow at a temperature lower than that shown in Figure 10 (See Table 6). This occurs with both flows but is more significant with 0.01 kg/s due to the greater air stagnation in the cavity.

Table 6. Air temperatures in the elbow inlet and outlet of the U-shape double-pass CPC.

		Flow 1: 0.01 kg/s		Flow 2: 0.02 kg/s	
	Configuration	$T_{elb,in}$ **(K)**	$T_{elb,out}$ **(K)**	$T_{elb,in}$ **(K)**	$T_{elb,out}$ **(K)**
(a)	Down–Down	308.7	311.0	303.2	304.9
(b)	Down–UP	309.7	310.5	303.8	304.1
(c)	Up–Up	308.2	308.5	303.5	304.0
(d)	Up–Down	307.8	309.8	302.9	304.4

Table 6 shows the elbow inlet and outlet temperatures for both air flow rates. The presented elbow inlet temperature refers to the average temperature at the outlet of the cavity (z = 2.0 m) in positions y = 0.1255 m to y = 0.1645 m, where the height of the elbow is 0.036 m. As expected, the temperatures corresponding to flow 1 (0.01 kg/s) are higher than flow 2 (0.02 kg/s). Therefore, all the air temperatures at the elbow inlet are lower than the air temperatures at the outlet of the cavity z = 2.0 m. On the other hand, the elbow outlet air temperatures and the duct inlet air temperatures are the same.

Figure 11 shows the heat transfer coefficients ($HTC_{p\text{-}cav}$) in the cavity for both air flow rates. As expected, the coefficients are lower when the airflow is lower and higher when the air flow rate is higher. Configurations with an inlet from below ((a) and (b)) have a high coefficient at the entry, which decreases to subsequently increase until it reaches a maximum near the outlet for configuration (a) and at the outlet (z = 2.0 m) for configuration (b). It is also noted that configuration (a) has a sharp increase and then a decrease from z = 0.2 m to z = 0.8 m, which is caused by the rise of the jet until it reaches the cover (see Figures 4a and 5a). In contrast, configurations with an inlet from the top (c and d) have an increase in the $HTC_{p\text{-}cav}$ in the region near the entry; then, it sharply declines to later gradually increase towards the region near the outlet where it reaches its maximum value. For example, the maximum $HTC_{p\text{-}cav}$ values with flow 1 are 15.0 W/(m^2 K), 15.8 W/(m^2 K), 16.0 W/(m^2 K), and 15.9 W/(m^2 K), while for flow 2 of 24.4 W/(m^2 K), 22.3 W/(m^2 K), 21.9 W/(m^2 K), and 25.0 W/(m^2 K), for configurations (a), (b), (c) and (d).

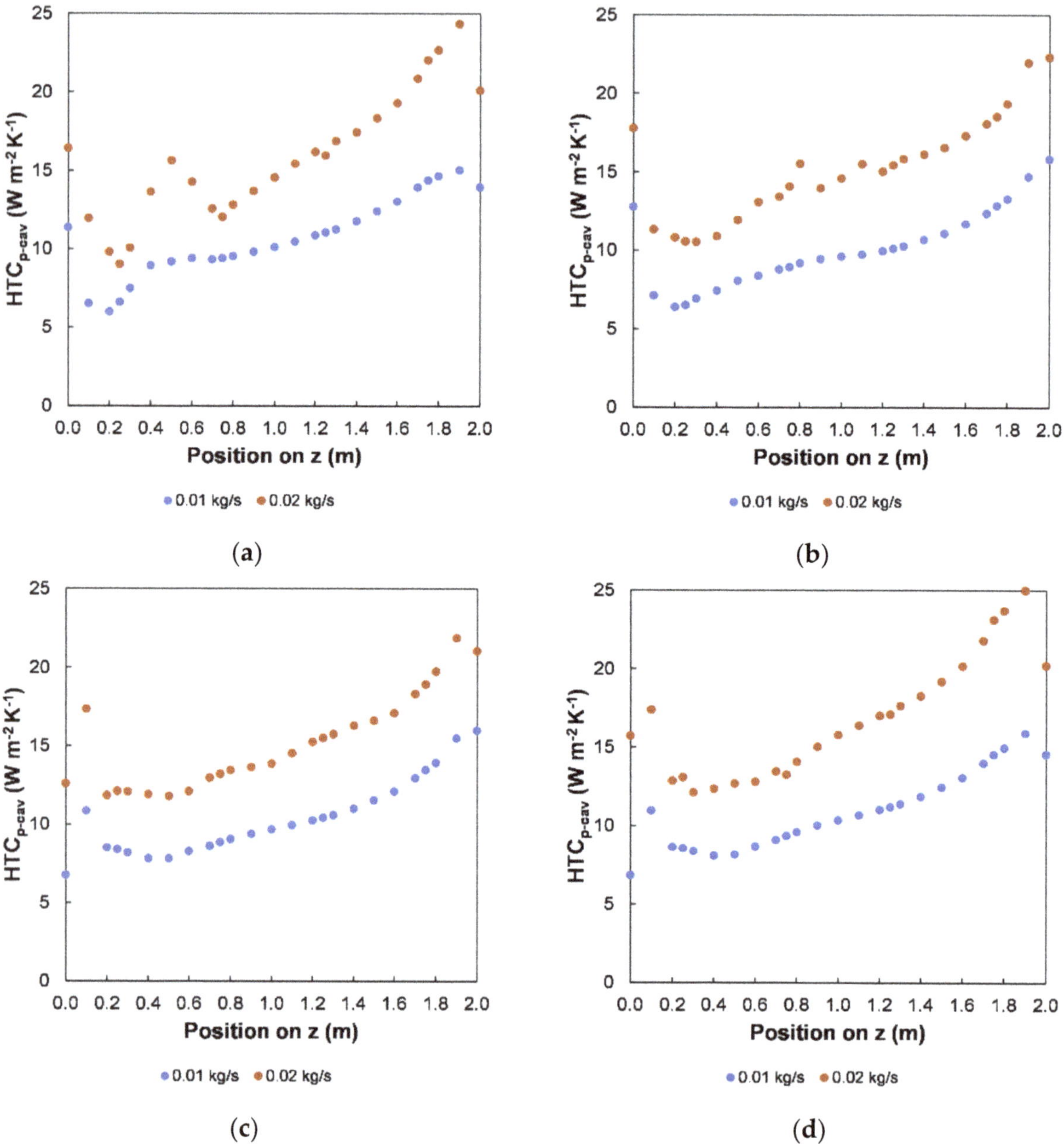

Figure 11. Flat-plate receiver heat transfer coefficients (HTC$_{p\text{-}cav}$). (**a**) Down–Down; (**b**) Down–Up; (**c**) Up–Up; (**d**) Up–Down.

In configurations where the air enters from the top ((c) and (d)), there is a high HTC$_{p\text{-}cav}$ at z = 0.1 m; nevertheless, all configurations have the maximum HTC$_{p\text{-}cav}$ in the region near the outlet. The above observations indicate that the HTC $_{p\text{-}cav}$ maximums correspond to the presence of the eddies produced by the sudden expansion at the inlet. While in the region near the outlet, there are also high HTC$_{p\text{-}cav}$ at z = 2.0 m in configuration (b), but in configurations (a), (c), and (d), it occurs around z = 1.8 m to z = 1.9 m. For configurations

(c) and (d), this occurs because the fluid becomes turbulent in the final region of the cavity as air is forced out of the manifold elbow.

Table 7 shows the pressure drop in each configuration for the two mass flow rates. First, the highest pressure drop is seen in the bottom outlet configurations (a and d). Moreover, the pressure drop increases three to four times with the highest air flow rate.

Table 7. Pressure drop in the system.

		Pressure Drop (Pa)	
	Configuration	**Flow 1: 0.01 kg/s**	**Flow 2: 0.02 kg/s**
(a)	Down–Down	10.97	36.85
(b)	Down–Up	9.01	31.31
(c)	Up–Up	8.84	31.33
(d)	Up–Down	10.70	37.58

3.3. *Efficiency*

Table 8 shows the temperature increments in the first and second passes and the outlet temperature for each configuration. The temperature increase resulting from the first pass (ΔT_{ps1}) and the second pass (ΔT_{ps2}) are more significant with flow 1. Moreover, for each flow, the U-shape double-pass CPC collector outlet temperature ($T_{ps2,out}$) of all the configurations is very similar, with differences between 0.7 K and less.

Table 8. U-shape double-pass CPC temperature increments and outlet temperature ($T_{f1,in}$ = 298.15 K).

		Flow 1: 0.01 kg/s			Flow 2: 0.02 kg/s		
	Configuration	ΔT_{ps1} **(K)**	ΔT_{ps2} **(K)**	$T_{ps2,out}$ **(K)**	ΔT_{ps1} **(K)**	ΔT_{ps2} **(K)**	$T_{ps2,out}$ **(K)**
(a)	Down–Down	12.84	16.10	327.09	6.74	8.33	313.22
(b)	Down–Up	12.31	15.90	326.36	5.98	8.99	313.12
(c)	Up–Up	10.30	17.94	326.39	5.87	8.54	312.57
(d)	Up–Down	11.62	16.60	326.37	6.29	8.66	313.10

Table 9 shows the contribution of the first pass (cavity) and second pass (receiver duct) to the useful energy gain inside the U-shape double-pass CPC collector and the efficiency for each configuration applying Equations (16) and (17). For instance, in configurations (a) and (b), between 40% and 45% of the total heat is extracted in the first pass with both flows. While in configurations (c) and (d), the heat recovery depends on the operating flow and is between 36% and 42% in the first pass.

Moreover, the efficiencies are higher with flow 2 due to the better heat transfer and lower heat losses to the ambient. In addition, it is observed that configuration (a) Down–Down is the most efficient, while configuration (c) Up–Up provides the lower efficiency of the cases analyzed at flow 2.

Configuration (a) has the highest efficiency because the air flows predominantly through the bottom section of the cavity, which is closest to the receiver plate. This surface has the highest temperature, thus is where heat recovery is desired. In addition, as the air descends, the final part of the cavity recovers heat since the fluid is forced to exit from the bottom.

Table 9. U-shape double-pass CPC useful energy gain percentage and thermal efficiency ($T_{f1,in}$ = 298.15 K).

		Flow 1: 0.01 kg/s			Flow 2: 0.02 kg/s		
	Configuration	$Q_{u,ps1}$ (%)	$Q_{u,ps2}$ (%)	η (%)	$Q_{u,ps1}$ (%)	$Q_{u,ps2}$ (%)	η (%)
(a)	Down–Down	44.37	55.63	77.31	44.75	55.25	80.52
(b)	Down–Up	43.64	56.36	75.37	39.93	60.07	80.01
(c)	Up–Up	36.47	63.53	75.44	40.75	59.25	77.02
(d)	Up–Down	41.18	58.82	75.40	42.09	57.91	79.88

4. Conclusions

This work investigates a compound parabolic concentrator (CPC) design for air heating with a double U-pass configuration. The double pass is incorporated to recover part of the heat lost by the flat plate receiver inside the CPC cavity. Overall, four configurations have been studied, and they are differentiated by the position of the air inlet into the cavity and the position of the air outlet of the cavity towards the receiving duct that constitutes the second pass of the collector.

In general, adding the first pass through the CPC cavity significantly increases the air temperature. Hence, by making the air circulate first through the CPC cavity, instead of a conventional manner where it only circulates through the duct, an increase in temperature is accomplished before entering the receiving duct. As a result, an average increase in air temperature of 11.8 K at a mass flow rate of 0.01 kg/s and 6.2 K at a mass flow rate of 0.02 kg/s was achieved. This represents a minimum temperature rise of 36% (0.01 kg/s) and 40% (0.02 kg/s) when only the first pass is used.

In addition, the analysis showed that the positions of the air inlet and outlet in the cavity do not influence the outlet temperature of the U-shape double pass CPC solar heater due to an efficiency difference of up to 3.5% being achieved. The Down–Down configuration is the one that provides slight outlet temperature and thermal efficiency increases. With this configuration, an air temperature increase from the U-shape double-pass CPC inlet to the outlet of 28.9 K and an efficiency of 77.3% are obtained when the airflow rate is 0.01 kg/s, and 15.1 K and 80.5% when the flow rate is 0.02 kg/s. This is because the air heated in the cavity is transported more effectively, as it is a more homogeneous flow and the heat transfer coefficient in the flat-plate receiver is high.

The first pass allows an extraction between 36% and 45% of the total heat, justifying its inclusion into the solar collector. Although configuration (a) Down–Down presents a slightly higher efficiency with both studied flows, the difference between the values of configurations (b), (c), and (d) seems irrelevant. Therefore, the decision of the configuration should be based on other aspects such as the manufacture of the collector, air pumping requirements, and the design of the collectors' array, which are highly dependent on the application conditions and needs.

In the case of the (a) Down–Down configuration, including elements that restrict air circulation from the cavity to the area near the flat-plate receiver could be explored to encourage contact with the hot surface. Strategies to accomplish this are to add a second cover close to the receiver inside the cavity or to increase the truncation of the collector reflectors. Finally, the structural design and financial analysis remain as future work to determine the potential application of this technology.

Author Contributions: Conceptualization, I.A.C.-B., E.V.-R. and N.O.-A.; Data curation, I.A.C.-B.; Formal analysis, I.A.C.-B., N.A.R.-M., L.V. and N.O.-A.; Funding acquisition, N.O.-A.; Investigation, I.A.C.-B.; Methodology, I.A.C.-B. and N.A.R.-M.; Project administration, N.O.-A.; Resources, N.A.R.-M. and L.V.; Software, I.A.C.-B.; Supervision, N.A.R.-M., E.V.-R. and N.O.-A.; Validation, I.A.C.-B., L.V. and N.A.R.-M.; Visualization, I.A.C.-B., N.A.R.-M. and N.O.-A.; Writing—original draft, I.A.C.-B., N.A.R.-M. and N.O.-A.; Writing—review & editing, I.A.C.-B., N.A.R.-M., E.V.-R., L.V. and N.O.-A. All authors have read and agreed to the published version of the manuscript.

Funding: The APC was funded by the Centro de Investigacion en Materiales Avanzados (CIMAV) through project number PI-22-24/2022.

Institutional Review Board Statement: Not applicable.

Informed Consent Statement: Not applicable.

Data Availability Statement: Not applicable.

Acknowledgments: The authors wish to thank the Mexican National Science Council (CONACYT) for the financial support through scholarship 698432. To Durango Science and Technology Council (COCYTED) for the financing of the annual ANSYS software license. Also, to the Laboratorio de Computo Cientifico of FCQ-UJED and the Academic Group UJEDCA129 for the use of software and computational facilities.

Conflicts of Interest: The authors declare no conflict of interest. The funders had no role in the design of the study; in the collection, analyses, or interpretation of data; in the writing of the manuscript; or in the decision to publish the results.

Nomenclature

$<n>$	Average reflections number, dimensionless
A	Area, m^2
C	Constant
C_p	Specific heat, J/(kg K)
D_h	Hydraulic diameter, m
E	Relaminarization source
F	Empirical correlation of length of the transition region, m
$\vec{F}$	Force vector, N
$\vec{g}$	Gravitational acceleration, m/s^2
G_k	Generation of turbulence kinetic energy due to mean velocity gradients, J/kg
G_w	Generation of specific dissipation rate, 1/s
h	Height, m
h'	Species enthalpy, J/kg
I	Solar irradiance, W/m^2
J	Mass flux, kg/m^2
k	Turbulent kinetic energy, J
L	Length, m
$\dot{m}$	Mass flow rate, kg/s
Nu	Nusselt number, dimensionless
P	Pressure, Pa
P_g	Gap loss factor, dimensionless
q	Heat generation, W/m^3
Q_u	Useful heat gain, W
Re	Reynolds number, dimensionless
S	Strain rate magnitude
t	Time, s
T	Temperature, K
u	Velocity magnitude, m/s
U	Local velocity, m/s

u_T	Friction velocity, m/s
V	Free-stream speed, m/s
v	Wind velocity above cover, m/s
w	Width, m
x	Velocity field coordinate
y	Wall-normal distance, m
Y	Destruction term of SST turbulence model, m^2/s^3
y^+	Dimensionless distance in wall coordinates
Greek letters	
α	Absorptance, dimensionless
γ	Intermittency
Γ	Effective diffusivity, dimensionless
ε	Emittance, dimensionless
η	Thermal efficiency, dimensionless
μ	Molecular viscosity of air, kg/(m s)
ρ	Air density, kg/m^3
ρ	Reflectance, dimensionless
τ	Transmittance, dimensionless
$\overline{\overline{\tau}}$	Stress tensor, Pa
υ	Eddy viscosity, m^2/s
Ω	Vorticity magnitude, 1/s
Subscripts	
a	Ambient
$a1$, $a2$	Turbulence damping constants
ap	Aperture
avg	Average
b	Bottom duct
c	Cover
cav	Cavity
$collector$	Collector
$duct$	Duct
elb	Elbow
$e1$, $e2$	Dissipation
h	Hydrodynamic
i, j	Orthogonal coordinate
in	Inlet
k	Turbulence kinetic energy
$length$	Length
out	Outlet
p	Flat plate receiver
$ps1$	Pass 1 (Through the cavity)
$ps2$	Pass 2 (Through the duct)
r	Reflectors
s	Sky
$turb$	Turbulent
$\gamma1$, $\gamma2$	Intermittence
Acronyms	
CFD	Computational fluid dynamics
CPC	Compound parabolic concentrator
EPS	Expanded polystyrene
ETFE	Ethylene tetrafluoroethylene (fluorine-based plastic)
HTC	Heat transfer coefficient, $W/(m^2\ K)$
PV/T	Photovoltaic/Thermal

References

1. Weiss, W.; Spörk-Dür, M. Solar Heat World 2022. Global Market Development and Trends 2021. Detailed Market Figures 2020. 2022. Available online: https://www.iea-shc.org/Data/Sites/1/publications/Solar-Heat-Worldwide-2022.pdf (accessed on 21 September 2022).
2. Bhusal, Y.; Hassanzadeh, A.; Jiang, L.; Winston, R. Technical and Economic Analysis of a Novel Low-Cost Concentrated Medium-Temperature Solar Collector. *Renew. Energy* **2020**, *146*, 968–985. [CrossRef]
3. Buttinger, F.; Beikircher, T.; Pröll, M.; Schölkopf, W. Development of a New Flat Stationary Evacuated CPC-Collector for Process Heat Applications. *Sol. Energy* **2010**, *84*, 1166–1174. [CrossRef]
4. Horta, P.; Henriques, J.C.C.; Collares-Pereira, M. Impact of Different Internal Convection Control Strategies in a Non-Evacuated CPC Collector Performance. *Sol. Energy* **2012**, *86*, 1232–1244. [CrossRef]
5. Zheng, W.; Yang, L.; Zhang, H.; You, S.; Zhu, C. Numerical and Experimental Investigation on a New Type of Compound Parabolic Concentrator Solar Collector. *Energy Convers. Manag.* **2016**, *129*, 11–22. [CrossRef]
6. Nkwetta, D.N.; Smyth, M. Performance Analysis and Comparison of Concentrated Evacuated Tube Heat Pipe Solar Collectors. *Appl. Energy* **2012**, *98*, 22–32. [CrossRef]
7. Alam, T.; Kim, M.-H. A Critical Review on Artificial Roughness Provided in Rectangular Solar Air Heater Duct. *Renew. Sustain. Energy Rev.* **2017**, *69*, 387–400. [CrossRef]
8. Othman, M.Y.; Yatim, B.; Sopian, K.; Abu Bakar, M.N. Performance Analysis of a Double-Pass Photovoltaic/Thermal (PV/T) Solar Collector with CPC and Fins. *Renew. Energy* **2005**, *30*, 2005–2017. [CrossRef]
9. Mekahlia, A.; Boumaraf, L.; Abid, C. CFD Analysis of the Thermal Losses on the Upper Part of a Flat Solar Collector. *Heat Transf.* **2020**, *49*, 2921–2942. [CrossRef]
10. Pawar, V.R.; Sobhansarbandi, S. CFD Modeling of a Thermal Energy Storage Based Heat Pipe Evacuated Tube Solar Collector. *J. Energy Storage* **2020**, *30*, 101528. [CrossRef]
11. Singla, M.; Hans, V.S.; Singh, S. CFD Analysis of Rib Roughened Solar Evacuated Tube Collector for Air Heating. *Renew. Energy* **2022**, *183*, 78–89. [CrossRef]
12. Ammar, M.; Mokni, A.; Mhiri, H.; Bournot, P. Numerical Analysis of Solar Air Collector Provided with Rows of Rectangular Fins. *Energy Rep.* **2020**, *6*, 3412–3424. [CrossRef]
13. Li, X.; Dai, Y.J.; Li, Y.; Wang, R.Z. Performance Investigation on a Novel Single-Pass Evacuated Tube with a Symmetrical Compound Parabolic Concentrator. *Sol. Energy* **2013**, *98*, 275–289. [CrossRef]
14. Barrón-Díaz, J.E.; Flores-Johnson, E.A.; Chan-Colli, D.G.; Koh-Dzul, J.F.; Bassam, A.; Patiño-Lopez, L.D.; Carrillo, J.G. FEM-CFD Simulation and Experimental Study of Compound Parabolic Concentrator (CPC) Solar Collectors with and without Fins for Residential Applications. *Appl. Sci.* **2021**, *11*, 3704. [CrossRef]
15. Yuan, G.; Fan, J.; Kong, W.; Furbo, S.; Perers, B.; Sallaberry, F. Experimental and Computational Fluid Dynamics Investigations of Tracking CPC Solar Collectors. *Sol. Energy* **2020**, *199*, 26–38. [CrossRef]
16. Antonelli, M.; Francesconi, M.; Di Marco, P.; Desideri, U. Analysis of Heat Transfer in Different CPC Solar Collectors: A CFD Approach. *Appl. Therm. Eng.* **2016**, *101*, 479–489. [CrossRef]
17. Francesconi, M.; Antonelli, M. A CFD Analysis to Investigate Thermal Losses in a Panel Composed of Several CPC Concentrators. *Therm. Sci. Eng. Prog.* **2018**, *5*, 278–288. [CrossRef]
18. Reddy, K.S.; Parthiban, A.; Mallick, T.K. Numerical Modeling of Heat Losses in a Line Focusing Solar Compound Parabolic Concentrator with Planar Absorber. *Appl. Therm. Eng.* **2020**, *181*, 115938. [CrossRef]
19. Al-Damook, M.; Obaid, Z.A.H.; Al Qubeissi, M.; Dixon-Hardy, D.; Cottom, J.; Heggs, P.J. CFD Modeling and Performance Evaluation of Multipass Solar Air Heaters. *Numer. Heat Transf. Part A Appl.* **2019**, *76*, 438–464. [CrossRef]
20. Tuncer, A.D.; Sözen, A.; Khanlari, A.; Amini, A.; Şirin, C. Thermal Performance Analysis of a Quadruple-Pass Solar Air Collector Assisted Pilot-Scale Greenhouse Dryer. *Sol. Energy* **2020**, *203*, 304–316. [CrossRef]
21. Mutabilwa, P.X.; Nwaigwe, K.N. Experimental Evaluation of Drying of Banana Using a Double-Pass Solar Collector (DPSC) and Theoretical Analysis Using a CFD Model. *Cogent Eng.* **2020**, *7*, 1789363. [CrossRef]
22. Desisa, D.G.; Shekata, G.D. Performance Analysis of Flat-Plate and V-Groove Solar Air Collector through CFD Simulation. *Int. J. Renew. Energy Dev.* **2020**, *9*, 369–381. [CrossRef]
23. Singh, S. Thermohydraulic Performance of Double Pass Solar Thermal Collector with Inline, Staggered and Hybrid Fin Configurations. *J. Energy Storage* **2020**, *27*, 101080. [CrossRef]
24. Kumar, A.; Singh, A.P.; Akshayveer; Singh, O.P. Performance Characteristics of a New Curved Double-Pass Counter Flow Solar Air Heater. *Energy* **2022**, *239*, 121886. [CrossRef]
25. Cengel, Y.A.; Ghajar, A.J. *Heat and Mass Transfer: Fundamentals and Applications*, 5th ed.; McGraw Hill: New York, NY, USA, 2014; ISBN 978-0073398181.
26. *ANSYS Fluent User's Guide*; 2020R1; ANSYS: Canonsburg, PA, USA, 2020.
27. Langtry, R.B.; Menter, F.R. Correlation-Based Transition Modeling for Unstructured Parallelized Computational Fluid Dynamics Codes. *AIAA J.* **2009**, *47*, 2894–2906. [CrossRef]
28. Menter, F.R.; Langtry, R.B.; Likki, S.R.; Suzen, Y.B.; Huang, P.G.; Völker, S. A Correlation-Based Transition Model Using Local Variables—Part I: Model Formulation. *J. Turbomach.* **2006**, *128*, 413. [CrossRef]

29. Chávez-Bermúdez, I.A.; Ortega-Avila, N.; Venegas-Reyes, E. A Thermal Model for a CPC-Type Double-Pass Solar Air Heater. *AIP Conf. Proc.* **2022**, *2445*, 120007.
30. Santos-González, I.; García-Valladares, O.; Ortega, N.; Gómez, V.H. Numerical Modeling and Experimental Analysis of the Thermal Performance of a Compound Parabolic Concentrator. *Appl. Therm. Eng.* **2017**, *114*, 1152–1160. [CrossRef]
31. Rabl, A. Optical and Thermal Properties of Compound Parabolic Concentrators. *Sol. Energy* **1976**, *18*, 497–511. [CrossRef]
32. Duffie, J.A.; Beckman, W.A. *Solar Engineering of Thermal Processes*; John Wiley & Sons, Inc.: Hoboken, NJ, USA, 2013. ISBN 9781118671603.

applied sciences

MDPI

Article

Research on Dynamic Reserve and Energy Arbitrage of Energy Storage System

Jia-Zhang Jhan [1], Tzu-Ching Tai [2], Pei-Ying Chen [1] and Cheng-Chien Kuo [1,*]

[1] Department of Electrical Engineering, National Taiwan University of Science and Technology, Taipei 10607, Taiwan
[2] Graduate Institute of Energy and Sustainability Technology, National Taiwan University of Science and Technology, Taipei 10607, Taiwan
* Correspondence: cckuo@ntust.edu.tw

Abstract: Replacing the traditional rotating generators with renewable energy will reduce the grid's inertia and with it the minimum frequency when N-1 contingency occurs triggering an Under-Frequency Load Shedding (UFLS). This study proposes a method for the energy storage system (ESS) to simultaneously provide energy arbitrage, reserve capacity, and assist N-1 contingency, by modifying the restriction formula of economic dispatch (ED) and limiting the SOC range of the ESS. Let the ESS join the Spinning Reserve. Through the PSS®E iterative ESS charging power required at moments when the frequency of contingency is too low in the ED. Let the ESS act as a N-1 contingency extra frequency reserve. This would prevent UFLS and still maintain the demand. The proposed method is applicable to different types of ESS. The method allows energy storages, originally designed for energy arbitrage, to participate in frequency support and spinning reserve.

Keywords: energy storage system; renewable energy; economic dispatch; security constraint; PSS®E

Citation: Jhan, J.-Z.; Tai, T.-C.; Chen, P.-Y.; Kuo, C.-C. Research on Dynamic Reserve and Energy Arbitrage of Energy Storage System. *Appl. Sci.* **2022**, *12*, 11953. https://doi.org/10.3390/app122311953

Academic Editors: Luis Hernández-Callejo, Jesús Armando Aguilar Jiménez and Carlos Meza Benavides

Received: 29 August 2022
Accepted: 21 November 2022
Published: 23 November 2022

1. Introduction

When the power generation and the power consumption are not equal, the frequency will deviate from the nominal value [1]. In the power system, if one of the generators fails, it will cause an energy mismatch and the frequency will begin to drop. The system needs to provide for the missing power generation immediately. There are frequency control responses that can be applied during generator contingency. First is the inertia response which comes from the rotating synchronous generator connected to the grid. When the frequency drops, the rotational inertia will be converted from kinetic energy into electrical energy and input into the grid instantly, reducing RoCoF (Rate of Change of Frequency). The second response is the droop control of some generators wherein the governor is controlled automatically to respond when the frequency starts to deviate [2]. This control allows more steam to enter the turbine to generate electricity, matching the grid energy for a few seconds, preventing frequency reduction, and using renewable energy instead of traditional rotation.

In the case of system energy imbalance, the rotational kinetic energy stored in the rotor of the traditional synchronous generator is used to provide inertial support for the power grid, keeping the minimum frequency of accidents at a certain level. However, most inverter-based sources (IBRs) such as wind or solar cannot provide inertia [3]. Replacing traditional generators with IBRs will reduce the inertia of the system, causing a larger RoCoF and a lower minimum frequency when the generator trips [4].

In the past, the power system usually used under-frequency load shedding (UFLS) to balance the insufficient power generation when the generator contingency caused the reduction of supply [4–7]. Some literatures [8–13] used RoCoF and the lowest frequency point to estimate the energy storage system (ESS) capacity and location for the frequency regulation required by the system.

In [10,14], the virtual inertia and the primary frequency response (PFR) provided by the ESS are taken into account to estimate the required ESS power and capacity. These two studies consider ESSs that can quickly respond to frequencies, but according to [15] not all ESSs have the ability to adjust the frequency.

In [16–18], ESS was added to the economic dispatch (ED) to deal with a high penetration of renewable energy. These two papers used ESS for peak shaving but did not consider frequency stability.

In [19], an ED considering automatic generation control (AGC) is proposed when the proportion of renewable energy is high, but peak shaving ESS is not considered. In [20], it is proposed to consider both PFR and secondary frequency response (SFR) in the unit commitment (UC) and consider the cost rate of ancillary services in different regions to propose a new market settlement strategy to compensate for the regional marginal cost of providing frequency reserves.

At present, most of the ED studies that consider security constraints do not include ESS. Further, most of the EDs that include ESS do not consider frequency security constraints. Therefore, this study proposes a method to incorporate both security constraints and ESS into the ED to find a safe and economical schedule.

In this study, the ESS was first added to the spinning reserve in the ED, which could reduce the generator's online time and thus reduce the cost. The base models are described in [21] with the addition of the ESS. With the inclusion of the ESS, the ESS will reduce the cost in the ED through its charging and discharging, and will automatically perform energy arbitrage, thereby providing two functions at the same time. However, due to the high penetration of renewable energy, the occurrence of N-1 contingency during certain periods will result in very low frequency. It is, therefore, important to calculate how much the ESS needs to be charged during these periods.

When an N-1 contingency occurs, the charging of the ESS can be cut off immediately to compensate for insufficient power generation and not trigger the UFLS during the low frequency. The addition of security constraints in the ED would find a safe ED to schedule.

PSS®E is a software from Siemens widely used in the analysis of power systems [22–24]. The N-1 contingency minimum frequency was also calculated using PSS®E in papers [22,23] and was used in this study. PSS®E mainly uses the dynamic reduction method to calculate contingency minimum frequency [25,26].

The content of the paper is as follows: Section 2 describes the proposed strategy; Section 3 describes the constraints of ED; Section 4 describes the simulation scenarios; Section 5 shows the simulation results; Section 6 is the discussions; and Section 7 is the conclusion.

2. Proposed Preventive Control Strategy

The flowchart shown in Figure 1, shows the flow of the simulations for the proposed preventive control strategy. The hourly power generation and hourly forecasted data of renewable energy will be first read to calculate the net load. Then, using the MILP, an ED for the new generator and ESS will be determined for the next 24 h. The scheduled ED will be used to determine the minimum frequency (F_{nadir}) that will be calculated at each hour using the PSS®E software. If the minimum frequency is lower than the set value (F_{min}), then the charging power of the ESS will be calculated by adding 0.1 MW in that hour to meet the F_{min} requirement. This charge will be added to the ED constraint and rescheduled until the minimum frequency per hour is higher than the set value. If the maximum charge of the ESS is reached, one generator will be added to the schedule and the ED will again be computed. An additional charge will again be included in the ESS schedule to make sure that the new minimum frequency is greater than the set value. In other words, the minimum frequency of N-1 contingency should always be higher than the set value every hour to make sure that the charge from the ESS can support a sudden drop in frequency.

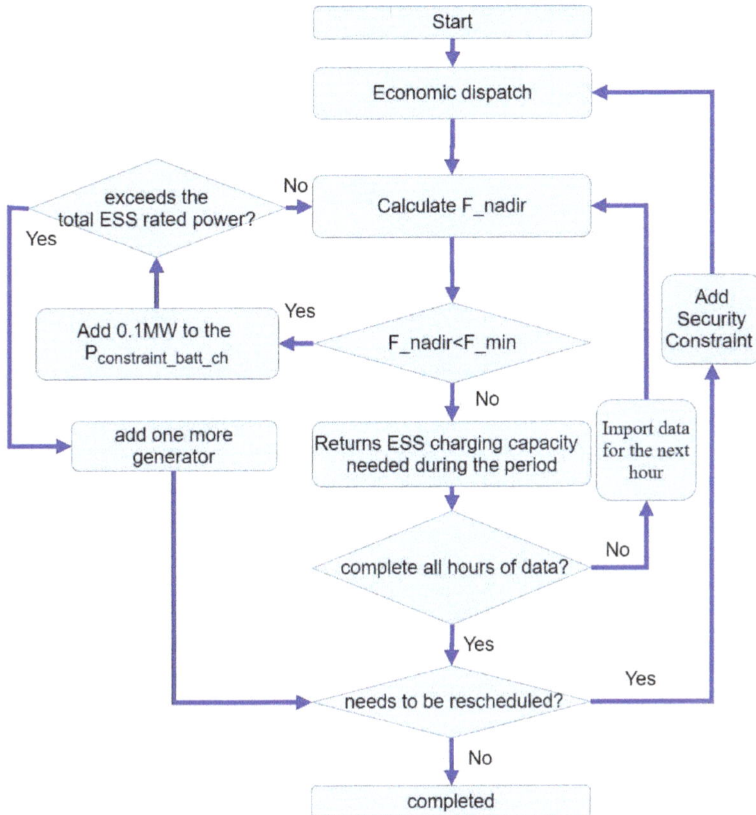

Figure 1. Flow chart.

3. Constraints of Economic Dispatch

3.1. Objective Function

The objective function is to minimize the operating cost, as shown in Equation (1). $C_L(t)$ represents the total fuel cost of all diesel generator sets at time t, $C_{st}(t)$ is the total startup cost of all diesel generator sets at time t, and $C_{batt}(t)$ represents the ESS cost. The power generation cost of PV in this study is set to 0. F represents the total economic cost.

$$Min\ F = \sum_{t=1}^{24}[C_L(t) + C_{st}(t) + C_{batt}(t)] \quad (1)$$

3.2. Diesel Generator

Equation (2) indicates the total fuel cost of a diesel generator in quadratic form, where FC_n represents the fuel cost of the nth diesel generator. The a_n, b_n and c_n represents the quadratic fuel cost constants of the nth diesel generator. $P_n(t)$ is the power generated by the nth diesel generator at time t. Figure 2 shows a typical fuel cost in quadratic form. However, because MILP is used, the quadratic curve needs to be linearized. In order to have a linear equation that is near the quadratic form, the curve is divided into segments and a line is drawn in each segment as the linear representation of the fuel cost curve for the ith segment.

$$FC_n(P_n(t)) = a_n + b_nP_n(t) + c_nP_n(t)^2 \quad (2)$$

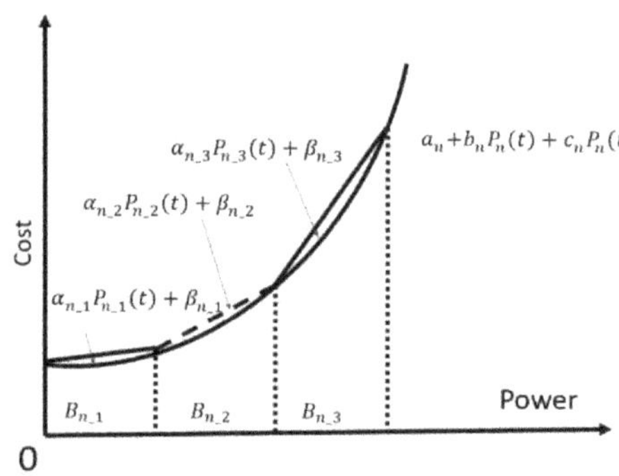

Figure 2. Typical generator cost curve with piecewise linearization.

Equations (3)–(5) are limits on the amount of electricity generated by the generator, where $P_{n\_i\_max}$ and $P_{n\_i\_min}$ are the maximum and minimum power generation in the ith line segment of the n generator set, respectively. $B_{n\_i}(t)$ is a binary integer representing whether diesel generator n is running in the ith linear interval at time t. $P_{n\_i}(t)$ is the amount of electricity generated in line segment i. The i represents the number of the line segment in the quadratic curve. Inequality (5) ensures that only one line segment is selected for the nth generator at any given time t.

$$B_{n_i}(t)P_{n_{im}} \leq P_{n_i}(t) < B_{n\_i}(t)\ P_{n\_i\_max} \tag{3}$$

$$P_n(t) = \sum_{i=1}^{I} P_{n\_i}(t) \tag{4}$$

$$\sum_{i=1}^{I} B_{n_i}(t) \leq 1 \tag{5}$$

The total fuel cost of all generators set at time t is expressed in Equation (6). $U_n(t)$ represents a binary integer variable of whether the nth diesel generator is turned on at time t. N represents the total number of diesel generators. α_{n_i} and β_{n_i} represent the slope and intercept, respectively, of the linear fuel cost when the nth diesel generator operates on line segment i at time t.

$$C_L(t) = \sum_{n=1}^{N}\sum_{i=1}^{I} U_n(t)B_{n_i}(t)\ [\alpha_{n_i}P_{n_i}(t) + \beta_{n\_i}] \tag{6}$$

Equation (7) represents the total startup cost of N diesel generators at time t. ST_{price} represents the startup cost of generators.

$$C_{st}(t) = \sum_{n=1}^{N} ST_{price}(U_n(t) - U_n(t-1)) \tag{7}$$

3.3. ESS

Inequalities (8) and (9) limit the charge and discharge power of the ESS so that it does not exceed the limit of the maximum charge and discharge power. The $B_{batt\_dis}(t)$ and $B_{batt\_ch}(t)$ are binary integers representing the discharge or charge states of the ESS at time t while. The $P_{batt\_dis\_max}$ and $P_{batt\_ch\_max}$ on the other hand represent the maximum discharge power and the maximum charging power of the battery ESS, respectively.

$$0 \leq P_{batt_{dis}}(t) \leq P_{batt\_dis\_max}B_{batt\_dis}(t) \tag{8}$$

$$-P_{batt\_ch\_max}B_{batt\_ch}(t) \leq P_{batt_{ch}}(t) \leq 0 \quad (9)$$

Inequality (10) is used to ensure that a single ESS cannot be charged and discharged at the same time.

$$B_{batt\_dis}(t) + B_{batt\_ch}(t) \leq 1 \quad (10)$$

Equation (11) describes the charge and discharge cost of the ESS. The purpose is to prevent the ESS from charging and discharging at an unnecessary time. Because if the battery ESS does not add the cost of charging and discharging. It may be charged and discharged in two time periods with the same electricity price. For example, if the average power generation cost of the first hour and the fifth hour is both 4. The battery ESS may be fully charged in the first hour and fully discharged in the fifth hour. This will not affect the final cost. However, this phenomenon is unreasonable in scheduling. Therefore, it is appropriate to add some small costs to the charging and discharging of the ESS to resolve this. Where $COST_{batt}$ represents the cost of the battery ESS per unit of charge and discharge. In this study, it is set to 0.1 NTD/kWh

$$C_{batt}(t) = \left(P_{batt_{dis}}(t) + P_{batt_{ch}}(t)\right) \times COST_{batt} \quad (11)$$

Equation (12) is mainly to calculate the power of the battery ESS at time t. $SOC(t)$ represents the state of charge(SOC) of ESS at time t, δ_t is the time interval, η_{Ind} and η_{Inc} represent discharge and charge efficiency of the ESS, respectively, and $P_{batt\_capacity}$ represents the capacity of the ESS.

In inequality (13), SOC_{min} and SOC_{max} represent the minimum and maximum value of SOC. Table 1 shows the specifications of the different ESSs used for this study. As seen in the table, since different energy storages have different capacities and power characteristics, the two energy storages will not be able to charge or discharge full power when they are close to their full energy or almost no energy [27,28].

Table 1. ESS specification.

	Type	Specification	Original Function	SOC_{min}	SOC_{max}
ESS1	lithium ion	2 MW/1 MWh	frequency regulation	x	x
ESS2	sodium-sulfur	1.8 MW/10.8 MWh	energy arbitrage	19%	81%
ESS3	not yet announced	4 MW/24 MWh	energy arbitrage	19%	81%

Originally, the SOC_{min} and SOC_{max} for both ESS2 and ESS3 are 10 and 90%, respectively. However, for this study, it would be set to 19 and 81%, respectively. These two values were adjusted because when an N-1 contingency occurs, the energy storage takes about half an hour (for the case of Kinmen Island) of continuous charging or discharging before a new generator is turned on. This charge or discharge decreases the SOC by 8.33%. Therefore, in order for the ESS2 and ESS3 to join the spinning reserve, their SOC_{min} should be set to 19% and SOC_{max} is set to 81% to make sure that the 8.33% of SOC for the N-1 contingency is always on standby and available at any time period.

In Equation (14), the final SOC must return to its initial value. SOC_{ini} and SOC_{end} represent the initial and end SOC, respectively.

$$SOC(t) = SOC(t-1) - \delta_t \left(\frac{P_{batt_{dis}}(t)}{\eta_{Ind} \times P_{batt_{capacity}}} + \frac{\eta_{Inc} \times P_{batt\_ch}(t)}{P_{batt\_capacity}} \right) \quad (12)$$

$$SOC_{min} \leq SOC(t) \leq SOC_{max} \quad (13)$$

$$SOC_{ini} = SOC_{end} \quad (14)$$

3.4. Power Balance Constraint

Shown in Equation (15) is the power balance constraint of the power system. Where $P_{PV}(t)$ represents the total power generation of renewable energy at time t. $P_{tolerance}(t)$ represents the allowable error value for the solution. $P_L(t)$ is the total load at time t.

$$P_{PV}(t) + P_{Inv}(t) + \sum_{n=1}^{N} P_n(t) + P_{tolerance}(t) = P_L(t) \tag{15}$$

3.5. Spinning Reserve Constraint

In inequality (16), the ESS available power is added to the spinning reserve to improve the reliability of the power grid. $P_{n\_max}(t)$ in (12) represents the maximum power generation of the nth generator while $P_{spin\_reserve}$ is the required standby capacity of the overall system. The current maximum generating capacity of the system at time t is used as the reserve capacity limit, representing the left side of the Equation (16).

$$\sum_{n=1}^{N} U_n(t)(P_{n\_max}(t) - P_n(t)) + P_{batt\_dis\_max}(t) - P_{batt_{dis}}(t) + P_{batt_{ch}}(t) \geq P_{spin\_reserve} \tag{16}$$

3.6. Ramp Rate Constraint

In inequality (17), $R_{rate\_n}(t)$ represents the ramp rate of the nth diesel generator per second at the tth time. R_{rate_min} represents the minimum required ramp-up and ramp-down per second of the grid.

$$\sum_{n=1}^{N} U_n(t) R_{rate\_n}(t) \geq R_{rate\_min} \tag{17}$$

3.7. PV Curtailment Constraint

Curtailment of generated power is required when the penetration rate is too high. The scheduled PV generation needs to be less than the predicted generation, as shown in (18).

$$0 \leq P_{PV}(t) \leq P_{PV\_predict}(t) \tag{18}$$

3.8. Security Constraints

ESS charging power is used to increase the minimum frequency when the grid is vulnerable. The inequality is shown in Equation (19). $P_{batt\_ch\_ESS2}(t)$ and $P_{batt\_ch\_ESS3}(t)$ represent the respective charge amounts of the two ESSs specifically used for energy arbitrage. The $P_{constraint\_batt\_ch}(t)$ represents the minimum total charge required to support the N-1 contingency for these two ESSs at time t.

$$|P_{batt\_ch\_ESS2}(t)| + |P_{batt\_ch\_ESS3}(t)| \geq |P_{constraint\_batt\_ch}(t)| \tag{19}$$

4. Description and Introduction of Simulation Environment

4.1. System Model and Settings

This study takes the Kinmen grid as the system under study. Kinmen Island is a small island west of Taichung City (R.O.C.), Taiwan, very close to mainland China. The winter load is about 21.9 to 42.95 MW while the summer load is about 43.26 to 73.81 MW.

Kinmen Island uses diesel to generate electricity making its cost usually higher because of the cost of transporting fuel. On the other hand, solar energy is a cheaper replacement for fossil fuel.

Currently Kinmen Island has two power plants and a 12.3 MW PV plant. Power plant 1 has 10 heavy oil diesel generators. Power plant 2 has 6 light oil generators and 2 ESSs. ESS1 is 2 MW/1 MWh lithium-ion batteries used for frequency regulation, while ESS2 is 1.8 MW/10.8 MWh sodium-sulfur batteries used for energy arbitrage. It is expected that an additional 4 MW/24 MWh ESS will be built in 2023 for energy arbitrage.

This study considers the future winter conditions of Kinmen Island. Only heavy oil diesel generators will be used because of the low operational cost of heavy oil diesel generator. There is a 27 MW PV plant with the three ESSs mentioned in the previous paragraph. The system has 4–22.8 kV busbars, 4–11.4 kV busbars, 4 main transformer loads, and 2 substations. A simplified schematic diagram of connections between all facilities is shown in Figure 3. The trip settings for the underfrequency relays has four levels, 57.3, 57.0, 56.5, and 56.0 Hz. After triggering the underfrequency relay, it takes about 5–6 cycles to open the circuit breaker [5,29]. In this study, F_{nadir} is set to 57.3 Hz, in order not to trigger UFLS.

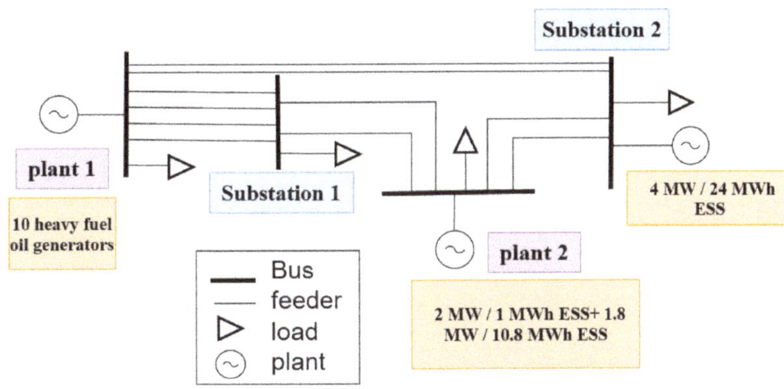

Figure 3. System diagram.

The ED is solved using mixed integer linear programming (MILP) using the IBM CPLEX 12.10.0 solver. The computer used is for these simulations is an Intel Core (TM) i5-7500 CPU @ 3.4 GHz. 16G RAM. PSS®E version is 33.4.0. It is coded in a Python program for automated simulation.

4.2. Generator ED Model

Various parameters of diesel generators and upper and lower limits of power generation in ED is shown in Table 2.

Table 2. Parameters of diesel generators.

Units	Capacity (MVA)	Minimum Power Generation (MW)	Maximum Power Generation (MW)
Plant1 #1–4	10.2	4	7.7
Plant1 #5–8	9.7	4.1	7.8
Plant1 #9–10	13.8	5.5	10.5
Plant2 #1–6	4.36	1.5	3.488

The fuel cost coefficients a_n, b_n, and c_n of the hypothetical heavy oil diesel generator are shown in Table 3. The generator fuel cost after piecewise linearity is shown in Table 4.

Table 3. Diesel generator cost factor and startup cost.

Units	a_n	b_n	c_n	Start-Up Cost
Plant1 #1–4	15	1.9161	0.0661	7
Plant1 #5–8	13	1.8518	0.0657	7
Plant1 #9–10	12	1.7966	0.0615	10

Table 4. Diesel Generator Linear Fuel Cost.

Units	Segment 1	Segment 2	Segment 3
Plant1 #1–4	2.526x + 13.617	2.690x + 12.763	2.853x + 11.709
Plant1 #5–8	2.471x + 11.564	2.634x + 10.699	2.796x + 9.635
Plant1 #9–10	2.576x + 9.576	2.781x + 8.107	2.986x + 6.296

According to the government website [30], it is assumed that the ramp-up and ramp-down rate per sec of each generator set is shown in Table 5. In this study, this value was set to 430 kW/s. Since the highest ramp-up and ramp-down value of a single generator to 420 kW/s, the value of 430 kW/sec is chosen here to ensure that at least two generator sets will be running at any point in time. Two generators are needed because if the system trips contingency, at least one generator can provide the reactive power required for the grid to maintain voltage stability.

Table 5. The ramping rate of the generator set.

Units	Ramping Rate (kW/s)
Plant 1 #1–4	15
Plant 1 #5–8	15
Plant 1 #9–10	420

4.3. Selection of Diesel Generator Model in PSS®E

The generator model selected in PSS®E is as shown in Table 6. All ESS models use second generation ESS general model of Western Electricity Coordinating Council (WECC). It consists of REPC_A, REEC_A and REGC_A.

Table 6. Diesel Generator model in PSS®E.

Units	Dynamic Models	Excitation System Model	Governor Model
Plant1 #1–4	GENSAL	ESAC8B	DEGOV
Plant1 #5–8	GENSAL	IEEEX1	DEGOV1
Plant1 #9–10	GNSAE	AC7B	DEGOV1

4.4. ESS's Response in Trip Contingency

Different ESSs will respond differently when a N-1 contingency fault occurs. The following will show the grid frequency and the actual power output of the two ESSs when a N-1 contingency fault occurs.

4.4.1. ESS1(Frequency Regulation)

ESS1 is used for frequency regulation. When the frequency exceeds the deadband (59.85–60.12 Hz), the ESS will start to discharge power. The rising time from 0 to full output is 167 ms. The response is shown in Figure 4.

Using the calibrated generator parameters set in Section 4.3, as also use in the paper [29], the ESS1 response can be replicated using the PSS®E, as shown in Figure 5. Figure 5 shows the data measured during an N-1 contingency when the load is 36.3 MW on 13 December 2019. At this time, the ESS1 and ESS2 has not been completed. Figure 6 shows the frequency measurement and simulation results for this N-1 contingency. A good match exists of F_{nadir} between calibrated simulation and measurement.

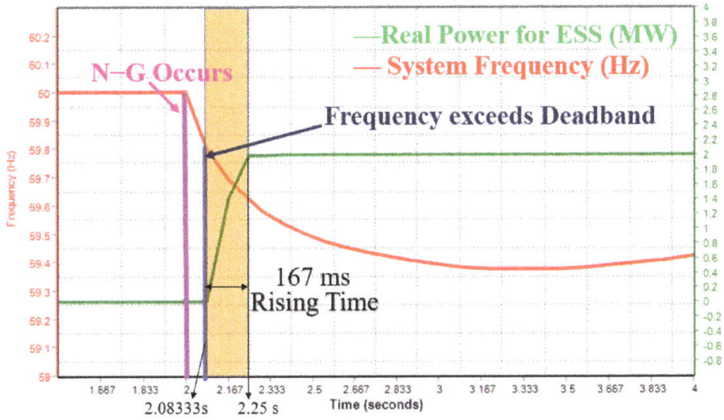

Figure 4. ESS1 real power output and grid frequency when N-1 contingency occurs.

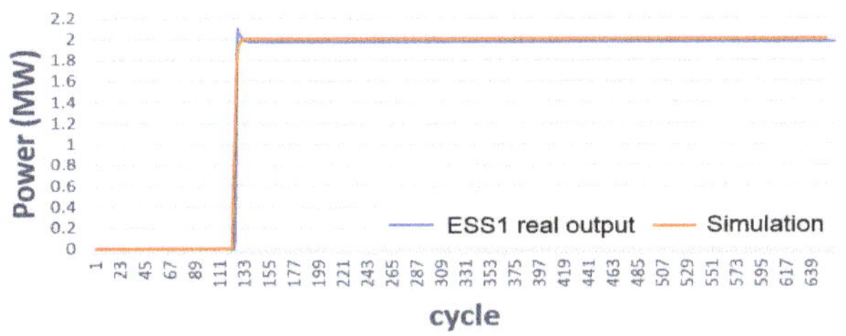

Figure 5. Output response of ESS1 to an N-1 contingency from PSSE and as measured from phasor measurement Unit (PMU).

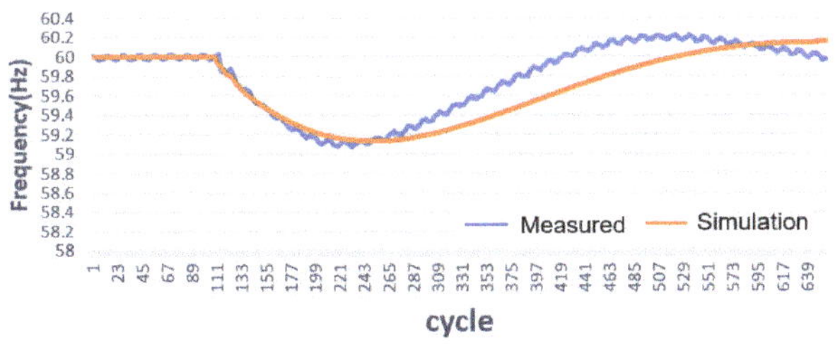

Figure 6. Frequency response simulation (PSSE) of ESS2 and ESS3 to a N-1 contingency and as measured from PMU.

4.4.2. ESS 2 and ESS 3 (Energy Arbitrage)

The responses of ESS2 and ESS3 during charging and discharging are as follows:

When the ESS is charging and a fault occurs, the ESS will quickly stop charging using low frequency relay tripping. As shown in Figure 7.

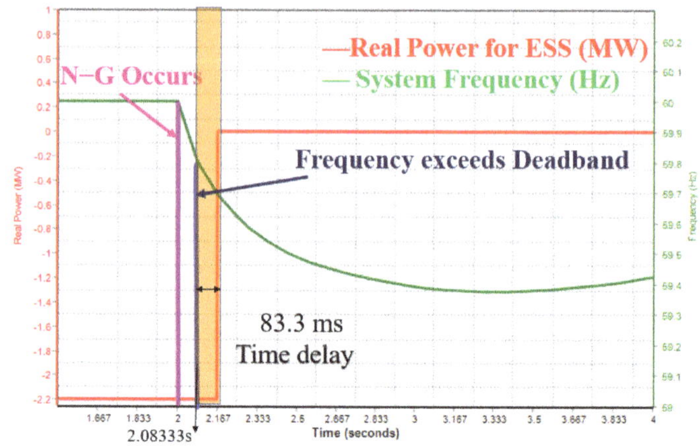

Figure 7. ESS 2 and ESS 3 grid frequency and real power output when N-1 contingency occurs while charging.

When the ESS is discharging and a fault occurs, the ESS will keep discharging. As shown in Figure 8.There would be a time delay of 83.33 ms from exceeding the deadband (59.85–60.12 Hz) to cutting off the ESS.

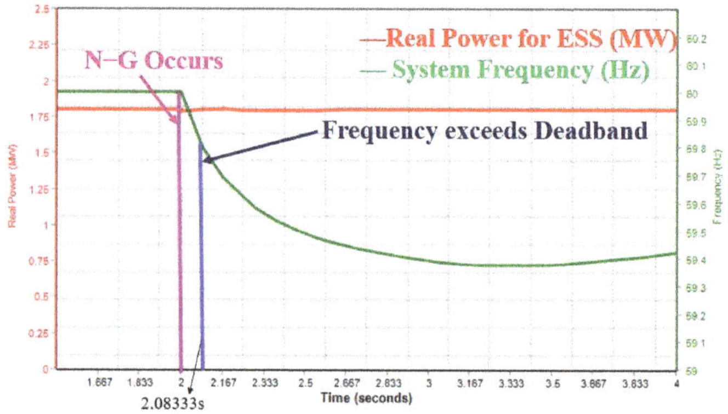

Figure 8. Grid frequency and real power output when contingency occurs during ESS2 and ESS3 discharge.

When ESS2 and ESS3 are discharging and the contingency occurs. As long as the voltage does not exceed the allowable range of high voltage ride through (HVRT) and low voltage ride through (LVRT), the ESS will continue to discharge and not cut off from the grid.

The transient performance of ESS2 and ESS3 depends on the circuit breaker, it is assumed that the local circuit breaker can cut off the electricity after about 5 cycles [27]. It can be seen from the picture that the simulated energy storage transient output and the lowest frequency of the contingency have a good match with the actual measurement.

5. Simulation Results

5.1. Case 1: Multi-Function ESS (Proposed Method)

Figure 9 show the results of the first ED or the initial ED. The N-1 contingency minimum frequency per hour is shown in Figure 10 with a blue line. Two generators are operating on at the 9th hour, with values of 6.4 and 7.1 MW. In the 15th hour, two generators are operating with output of 6.6 and 7.1 MW. The N-1 contingency minimum frequency for the 9th and 15th hours are 56.079 and 56.074 Hz, respectively.

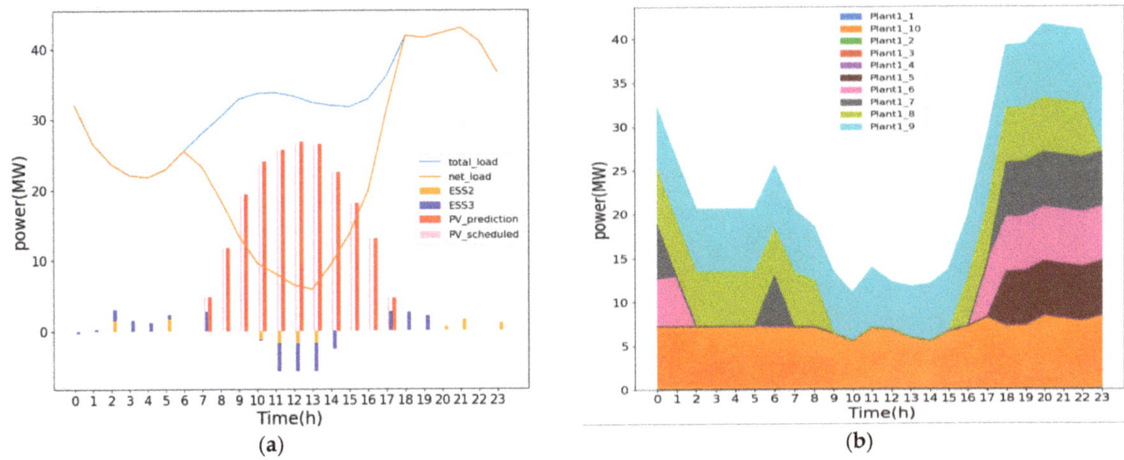

Figure 9. Case 1: result of first ED of (**a**) PV generation and the charge and discharge of ESS and (**b**) the power output of the diesel generators.

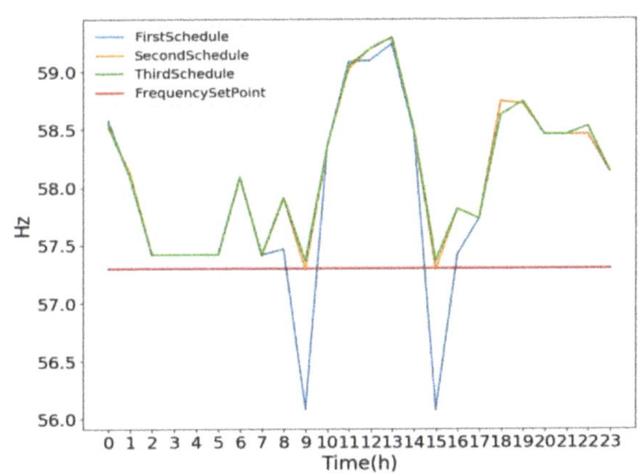

Figure 10. Minimum frequency of N-1 contingency per hour for three schedules of case 1.

The 9th hour frequency simulation results with generator active power outputs are shown in Appendix A. The ESS2 and ESS3 were not charged in both hours. The total cost of this third ED is 2422 kNTD. Using the process flow as discussed in Figure 1, the ESS requires 1.7 MW to charge at 9th hour and 1.7 MW to charge at 15th hour to get the frequency above the set value, equivalent to the 8.33% of SOC. Adding these two limits for the said hours, the second ED is rescheduled.

The results of the second ED are shown in the Figure 11. Case 1: result of second ED of (a) PV generation and the charge and discharge of ESS and (b) the power output of the diesel generators.

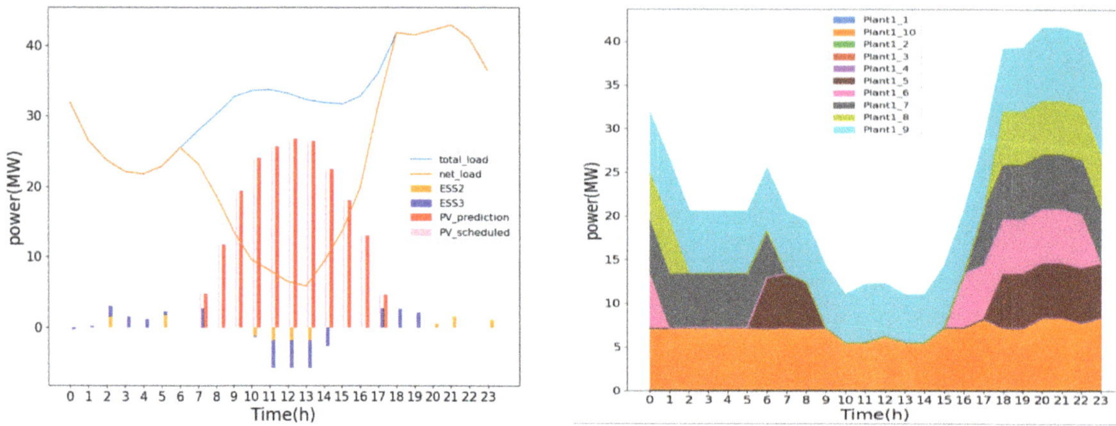

(**a**) (**b**)

Figure 11. Case 1: result of second ED of (**a**) PV generation and the charge and discharge of ESS and (**b**) the power output of the diesel generators.

The minimum frequency of the N-1 contingency per hour is shown in Figure 10 in the orange line. Notice that the minimum frequency touches the set value of the frequency. The minimum frequency of N-1 contingency for the 9th hour and the 15th hour are 57.287 and 57.287 Hz, respectively. Using the proposed procedure to determine the required ESS charging for the 9th and 15th hours, the charging power required are 1.8 MW for the 9th hour and 1.8 MW for the 15th. Again, it is required to add this in the charging limits for the next ED. The total cost of this third ED is 2422 kNTD.

The result of the third ED is shown in Figure 12. The minimum frequency of N-1 contingency per hour is shown in Figure 10 in the green line. The 9th hour frequency simulation results with generator active power outputs are shown in the Appendix A. The minimum frequency of each hour is higher than the set value. Therefore, this is the final ED to support the N-1 contingency. The total cost of this third ED is 2422 kNTD.

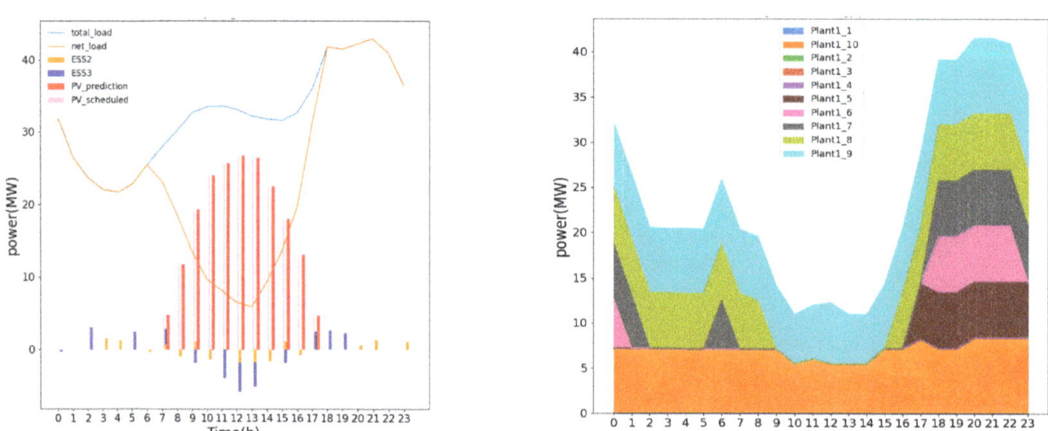

(**a**) (**b**)

Figure 12. Case 1: result of third ED of (**a**) PV generation and the charge and discharge of ESS and (**b**) the power output of the diesel generators.

The Figure 13 shows the changes in ED of the ESS2 and ESS3. In the second and third results, the two energy storages are in the state of charge and discharge at the 9th and 15th hours, so that F_{nadir} can be increased, and keep close to the original total output of that hour.

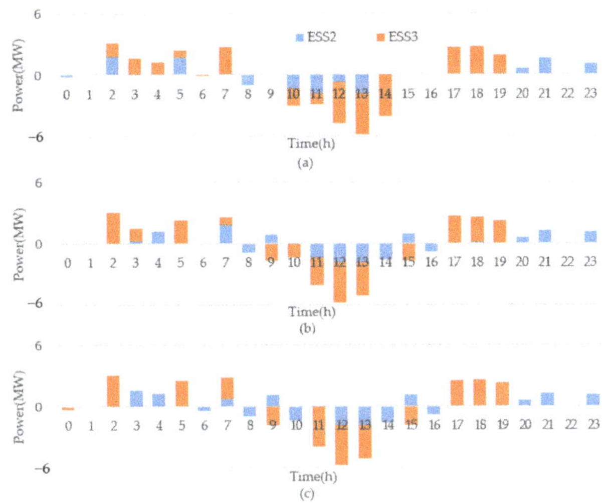

Figure 13. ED results of the ESSs (**a**)Result of 1st ED (**b**) Result of 2nd ED (c) Result of 3rd ED.

5.2. Case 2: ESS Functioning as a Frequency Support

Removing the two ESS, that function as energy arbitrage, verifies energy arbitrage function of our proposed method. The Figure 14 shows the ED result when the ESS2 and ESS3 are removed. According to the system conditions, two generators must be turned on to maintain the system stability.

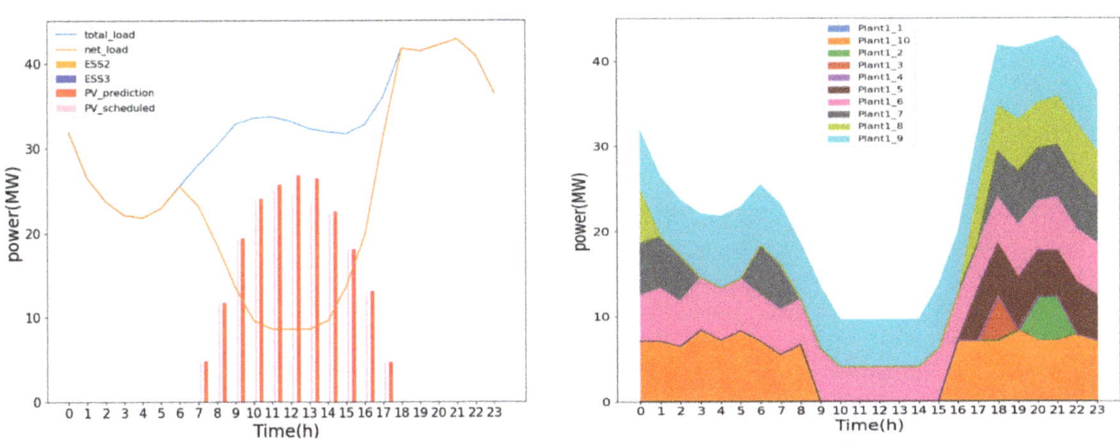

(**a**) (**b**)

Figure 14. Case 2: Result of the scenario without ESS for (**a**) the PV generation and (**b**) the power output of the diesel generators.

Since there are no ESS performing energy arbitrage, low net load causes the PV power generation to be curtailed, as shown in the Figure 14a, as the pink bar is lower than the red bar. Furthermore, without energy arbitrage, the total operating cost raises to 2555 kNTD from 2422 kNTD.

6. Discussion

It can also be found from the simulation results that when N-1 contingency occurs, the most dangerous instance is not necessarily when the penetration of renewable energy is the highest, but tends to occur in the midway from zero penetration to highest penetration.

The reason may be that when the penetration of renewable energy at its peak, the ESS will be charged more. If a contingency occurs, it can quickly trip and reduce a lot of energy use. The lowest point of the frequency is related to the amount of power generation that is tripped. When the penetration rate is the highest, the generators are almost always lightly loaded, so the reduced power generation of the contingency is relatively small. Therefore, even if the system inertia at that time is small, the lowest frequency may not be the lowest during this period.

Furthermore, in theory the outcome of stopping the charging of EVs would be similar to cutting off the ESS from charging in this study, probably also increasing the minimum frequency of N-1 1 contingency. The results of this study can be used to set the time price of electric vehicles to charge, so that more EVs can be charged when the power grid is weak.

In the picture., two ESSs are used to charge and discharge, respectively, to improve the safety of the N-1 contingency. In the real world, energy loss will occur due to the round-trip efficiency of the ESS. This situation can be understood as exchanging energy for N-1 contingency resiliency. May be especially suitable for pumped-storage power plants when rainwater is abundant.

7. Conclusions

This paper proposes a method by which energy arbitrage energy storage can help the N-1 contingency. The frequency regulation ESS and the energy arbitrage ESS are considered in the simulation. PSS®E is used to verify that the energy arbitrage ESS disconnected from charging can increase the minimum frequency when contingency occurs. In this way, the ESS can provide spinning reserve, energy arbitrage, and help N-1 contingency at the same time. This method is also not only suitable for lithium-ion batteries, but for all battery types. It can also be applied to various ESSs, such as flow batteries, pumped storage power, etc. The simulation results show that the proposed method can effectively improve the minimum contingency frequency higher than the set value.

Author Contributions: Conceptualization, T.-C.T.; Methodology, J.-Z.J.; Software, J.-Z.J. and P.-Y.C.; Validation, P.-Y.C.; Visualization, T.-C.T.; Supervision, C.-C.K.; Project administration, C.-C.K.; Funding acquisition, C.-C.K. All authors have read and agreed to the published version of the manuscript.

Funding: The support of this research by the Ministry of Science and Technology of the Republic of China under Grant No. MOST 111-2622-8-011-006-TE1 & MOST 111-3116-F-006-006—are gratefully acknowledged.

Institutional Review Board Statement: Not applicable.

Informed Consent Statement: Not applicable.

Data Availability Statement: All data are provided in this manuscript.

Conflicts of Interest: The authors declare no conflict of interest. The funders had no role in the design of the study; in the collection, analyses, or interpretation of data; in the writing of the manuscript, or in the decision to publish the results.

Appendix A

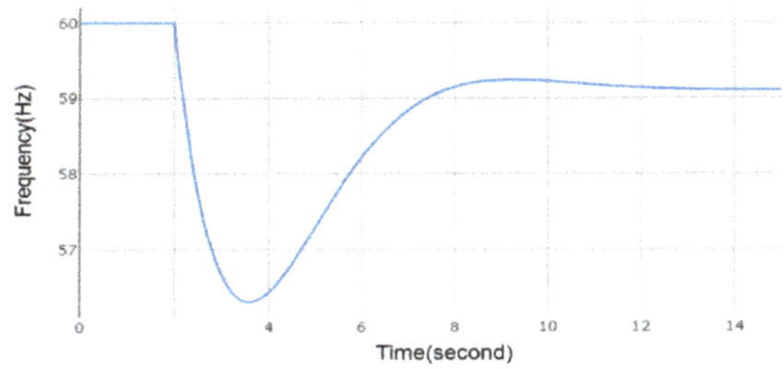

Figure A1. N-1 contingency frequency at 9th hour of first ED.

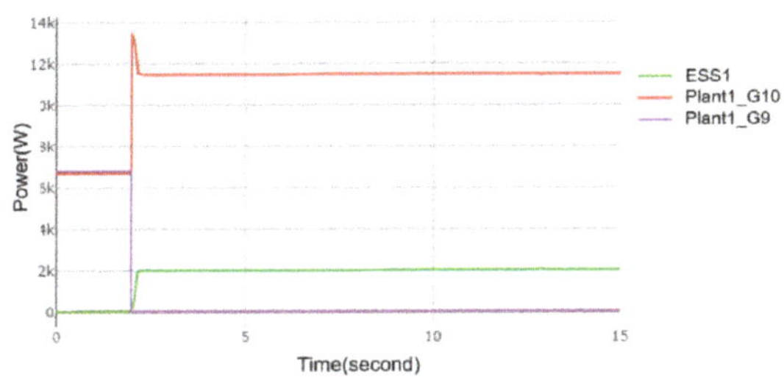

Figure A2. Generators active power at the 9th hour of the first ED.

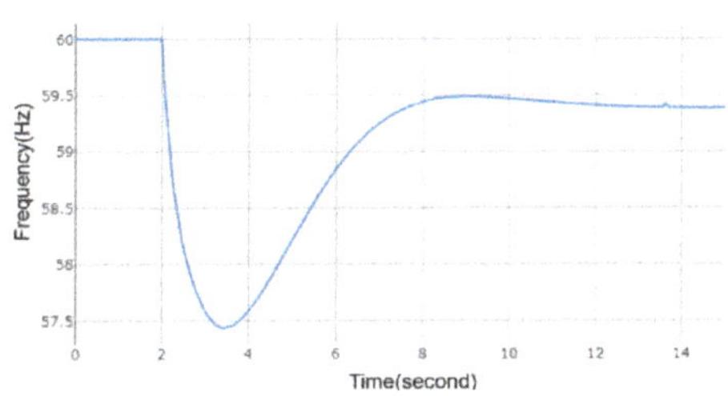

Figure A3. N-1 contingency frequency at 9th hour of 3rd ED.

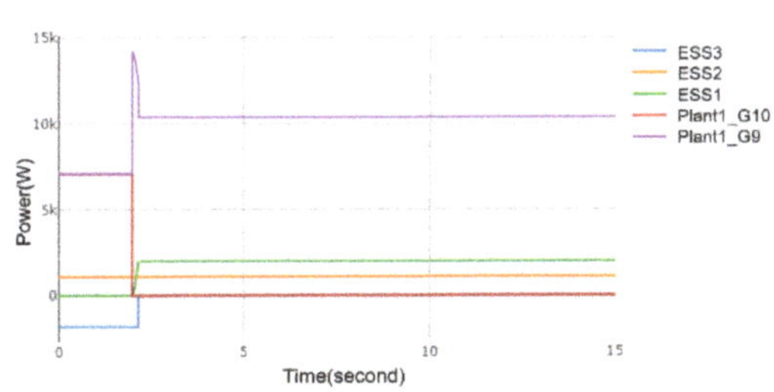

Figure A4. Generator power at the 9th hour of the 3rd ED.

References

1. Kundur, P.; Balu, N.J.; Lauby, M.G. *Power System Stability and Control*; McGraw-Hill Education: New York, NY, USA, 1994.
2. Ela, E.; Milligan, M.; Kirby, B. *Operating Reserves and Variable Generation*; National Renewable Energy Laboratory: Berkeley, CA, USA, 2011.
3. Yan, R.; Saha, T.K.; Modi, N.; Masood, N.-A.; Mosadeghy, M. The combined effects of high penetration of wind and PV on power system frequency response. *Appl. Energy* **2015**, *145*, 320–330. [CrossRef]

4. Fernández-Guillamón, A.; Gómez-Lázaro, E.; Muljadi, E.; Molina-García, Á. Power systems with high renewable energy sources: A review of inertia and frequency control strategies over time. *Renew. Sustain. Energy Rev.* **2019**, *115*, 109369. [CrossRef]
5. Yang, J.-S.; Liao, C.-J.; Wang, Y.-F.; Chu, C.-C.; Lee, S.-H.; Lin, Y.-J. Design and Deployment of Special Protection System for Kinmen Power System in Taiwan. *IEEE Trans. Ind. Appl.* **2017**, *53*, 4176–4185. [CrossRef]
6. Kuo, M.-T.; Lu, S. Experimental Research and Control Strategy of Pumped Storage Units Dispatching in the Taiwan Power System Considering Transmission Line Limits. *Energies* **2013**, *6*, 3224–3244. [CrossRef]
7. Shekari, T.; Aminifar, F.; Sanaye-Pasand, M. An Analytical Adaptive Load Shedding Scheme Against Severe Combinational Disturbances. *IEEE Trans. Power Syst.* **2016**, *31*, 4135–4143. [CrossRef]
8. Akram, U.; Mithulananthan, N.; Shah, R.; Basit, S.A. Energy Storage for Short-Term Frequency Stability Enhancement in Low-Inertia Power Systems. In Proceedings of the 2020 Australasian Universities Power Engineering Conference (AUPEC), Hobart, Australia, 29 November–2 December 2020.
9. Das, C.K.; Mahmoud, T.S.; Bass, O.; Muyeen, S.; Kothapalli, G.; Baniasadi, A.; Mousavi, N. Optimal sizing of a utility-scale energy storage system in transmission networks to improve frequency response. *J. Energy Storage* **2020**, *29*, 101315. [CrossRef]
10. Ramírez, M.; Castellanos, R.; Calderón, G.; Malik, O. Placement and sizing of battery energy storage for primary frequency control in an isolated section of the Mexican power system. *Electr. Power Syst. Res.* **2018**, *160*, 142–150. [CrossRef]
11. Zhao, H.; Wu, Q.; Huang, S.; Guo, Q.; Sun, H.; Xue, Y. Optimal siting and sizing of Energy Storage System for power systems with large-scale wind power integration. In Proceedings of the 2015 IEEE Eindhoven PowerTech, Eindhoven, The Netherlands, 29 June–2 July 2015.
12. Bera, A.; Abdelmalak, M.; Alzahrani, S.; Benidris, M.; Mitra, J. Sizing of Energy Storage Systems for Grid Inertial Response. In Proceedings of the 2020 IEEE Power & Energy Society General Meeting (PESGM), Montreal, QC, Canada, 2–6 August 2020.
13. Knap, V.; Chaudhary, S.K.; Stroe, D.-I.; Swierczynski, M.J.; Craciun, B.-I.; Teodorescu, R. Sizing of an Energy Storage System for Grid Inertial Response and Primary Frequency Reserve. *IEEE Trans. Power Syst.* **2016**, *31*, 3447–3456. [CrossRef]
14. Bera, A.; Chalamala, B.R.; Byrne, R.H.; Mitra, J. Sizing of Energy Storage for Grid Inertial Support in Presence of Renewable Energy. *IEEE Trans. Power Syst.* **2022**, *37*, 3769–3778. [CrossRef]
15. Rahman, M.; Oni, A.O.; Gemechu, E.; Kumar, A. Assessment of energy storage technologies: A review. *Energy Convers. Manag.* **2020**, *223*, 113295. [CrossRef]
16. Ross, M.; Abbey, C.; Bouffard, F.; Joos, G. Microgrid Economic Dispatch With Energy Storage Systems. *IEEE Trans. Smart Grid* **2018**, *9*, 3039–3047. [CrossRef]
17. Liu, J.; Sun, X.-Y.; Bo, R.; Wang, S.; Ou, M. Economic dispatch for electricity merchant with energy storage and wind plant: State of charge based decision making considering market impact and uncertainties. *J. Energy Storage* **2022**, *53*, 104816. [CrossRef]
18. Zeng, Y.; Li, C.; Wang, H. Scenario-Set-Based Economic Dispatch of Power System With Wind Power and Energy Storage System. *IEEE Access* **2020**, *8*, 109105–109119. [CrossRef]
19. Zhang, G.; McCalley, J.D.; Wang, Q. An AGC Dynamics-Constrained Economic Dispatch Model. *IEEE Trans. Power Syst.* **2019**, *34*, 3931–3940. [CrossRef]
20. Zhang, G.; Ela, E.; Wang, Q. Market Scheduling and Pricing for Primary and Secondary Frequency Reserve. *IEEE Trans. Power Syst.* **2019**, *34*, 2914–2924. [CrossRef]
21. Chen, Y.-T.; Kuo, C.-C.; Jhan, J.-Z. Research on Energy Storage Optimization Operation Schedule in an Island System. *Appl. Sci.* **2021**, *11*, 3690. [CrossRef]
22. Xu, X.; Elkhatib, M.; Yousefian, R.; Tang, J.; Choi, B.; Huang, L.; Mao, Y.; Berner, A. Automatic N-1–1 System Stability Study of the PJM System. In Proceedings of the 2020 IEEE Power & Energy Society General Meeting (PESGM), Montreal, QC, Canada, 2–6 August 2020.
23. Xue, N.; Wu, X.; Gumussoy, S.; Muenz, U.; Mesanovic, A.; Heyde, C.; Dong, Z.; Bharati, G.; Chakraborty, S.; Cockcroft, L. Dynamic Security Optimization for N-1 Secure Operation of Hawai'i Island System With 100% Inverter-Based Resources. In Proceedings of the 2021 IEEE Power & Energy Society General Meeting (PESGM), Virtual, 26–29 July 2021.
24. Arava, V.N.; Vanfretti, L. Analyzing the Static Security Functions of a Power System Dynamic Security Assessment Toolbox. *Int. J. Electr. Power Energy Syst.* **2018**, *101*, 323–330. [CrossRef]
25. Singh, R.; Elizondo, M.; Lu, S. A review of dynamic generator reduction methods for transient stability studies. In Proceedings of the 2011 IEEE Power and Energy Society General Meeting, Detroit, MI, USA, 24–28 July 2011.
26. Yang, J.P.; Cheng, G.H.; Xu, Z. Dynamic Reduction of Large Power System in PSS/E. In Proceedings of the 2005 IEEE/PES Transmission & Distribution Conference & Exposition: Asia and Pacific, Dalian, China, 18 August 2005.
27. Gonzalez-Castellanos, A.; Pozo, D.; Bischi, A. Detailed Li-ion battery characterization model for economic operation. *Int. J. Electr. Power Energy Syst.* **2020**, *116*, 105561. [CrossRef]
28. Schaefer, S.; Vudata, S.P.; Bhattacharyya, D.; Turton, R. Transient modeling and simulation of a nonisothermal sodium–sulfur cell. *J. Power Sources* **2020**, *453*, 227849. [CrossRef]
29. Liao, C.-J.; Hsu, Y.-F.; Wang, Y.-F.; Lee, S.-H.; Lin, Y.-J.; Chu, C.-C. Experiences on Remediation of Special Protection System for Kinmen Power System in Taiwan. *IEEE Trans. Ind. Appl.* **2020**, *56*, 2418–2426. [CrossRef]
30. Ministry of Economic Affairs, R.O.C. 2020. Available online: https://www.moea.gov.tw/Mns/Populace/news/News.aspx?kind=1&menu_id=40&news_id=90470 (accessed on 20 November 2022).

sustainability

MDPI

Article

Energy Storage Capacity Planning Method for Improving Offshore Wind Power Consumption

Hao Yu [1], Xiaojuan Yang [2,*], Honglin Chen [1], Suhua Lou [2] and Yong Lin [1]

1 Grid Planning & Research Center of Guangdong Power Grid Co., Ltd., Guangzhou 510060, China
2 School of Electrical and Electronic Engineering, Huazhong University of Science and Technology, Wuhan 430074, China
* Correspondence: xiaojuan_yang@hust.edu.cn

Abstract: This paper proposes a method of energy storage capacity planning for improving offshore wind power consumption. Firstly, an optimization model of offshore wind power storage capacity planning is established, which takes into account the annual load development demand, the uncertainty of offshore wind power, various types of power sources and line structure. The model aims at the lowest cost of investment, operation and maintenance of the system, and takes lower than a certain abandoned wind level as the strict constraint to obtain two parameters of power capacity and energy capacity of energy storage on the source side. Secondly, taking a coastal power grid as a typical case, the energy storage capacity planning method is verified. Finally, the key factors affecting offshore wind power consumption are summarized, and the sensitivity analysis is carried out from the point of view of the transmission protocol of the transmission lines outside the province and the capacity allocation of the tie lines in the province. This study will be helpful for the planning and operation of the high-proportion of offshore wind energy power systems.

Keywords: offshore wind power; energy storage system; wind power consumption; planning optimization model

Citation: Yu, H.; Yang, X.; Chen, H.; Lou, S.; Lin, Y. Energy Storage Capacity Planning Method for Improving Offshore Wind Power Consumption. *Sustainability* **2022**, *14*, 14589. https://doi.org/10.3390/su142114589

Academic Editors: Luis Hernández-Callejo, Jesús Armando Aguilar Jiménez and Carlos Meza Benavides

Received: 7 October 2022
Accepted: 4 November 2022
Published: 6 November 2022

1. Introduction

With the development of the economy, fossil energy is decreasing and environmental pollution is increasing day by day. In order to alleviate the pressure of energy shortages and environmental deterioration, various countries are committed to the development and utilization of clean energy. The proposal of the carbon peaking and carbon neutrality goals demonstrates China's determination to actively respond to climate change and achieve high-quality economic development. To further accelerate the development and utilization of non-fossil energy, especially new energy represented by wind and solar energy, is an important measure to achieve the arduous task of the carbon peaking and carbon neutrality goals.

Compared with other clean energy sources, wind power has greater development advantages and competitive potential. In the last 10 years, global onshore wind power has achieved rapid development, and the development of onshore wind power in some countries has become saturated. At present, there is an urgent demand for offshore wind power development and application all over the world [1]. China's offshore wind power has great development potential and good development prospects. To develop a high-quality offshore wind power industry and accelerate the development of offshore wind power from near-sea to deep-sea to far-sea, promoting the large-scale, intensive and sustainable development of offshore wind power is an important support to promote the adjustment of China's energy structure and achieve the carbon peaking and carbon neutrality goals [2]. Compared with onshore wind power, offshore wind power has three outstanding characteristics: (1) The offshore wind energy resources in the southeast coastal areas of China are abundant and close to the power load center, which is convenient for the consumption and utilization of the power grid. However, transmission cables need to be configured; (2) The overall output level of offshore wind farms is higher, with higher annual power generation utilization hours; (3) The transmission mode of offshore wind power is more flexible and does not need to occupy land resources.

The cost of transmission cable is high, and it is often difficult to send it complete. On the one hand, offshore wind power connected to the grid for operation will cause abandoned wind due to transmission congestion in part of the overall output. On the other hand, it will bring severe challenges to the peak regulation of the power grid. The lack of peak regulation capacity of the power grid leads to abandoned wind. The installation of an energy storage system is flexible, and the configuration of energy storage for an offshore wind power station can promote it to become a high-quality power supply. The source-side energy storage mainly works out a charge and discharge scheme to stabilize the fluctuation of its output power to achieve a higher proportion of offshore wind power consumption. However, there are some problems, such as the high cost of the energy storage configuration, the mismatch between energy storage technology and offshore applications, and so on. The improper use of energy storage equipment limits the appropriate absorption of wind power and increases the cost. The optimal allocation of energy storage capacity has always attracted much attention, and domestic and foreign researchers have carried out a lot of research on this issue.

The current research is mainly focused on energy storage capacity planning [3–6] and wind–storage operation optimization [7–10], and there is little research in [11,12] considering the interaction between the energy storage system planning and operation at the same time; a two-layer collaborative optimization model for energy storage capacity planning and operation is proposed. Starting from the wind power delivery channel, based on multi-stage stochastic programming and sequential Monte Carlo simulation, an optimal energy storage allocation method for wind farm stations considering energy storage life loss and delivery channels is proposed in [13]. Based on the characteristics of the wind farm, references [14–16] determined the energy storage allocation method based on a wind power prediction error, wind farm generation curve deviation correction, and wind farm output smoothing, respectively, but the work done in these references does not take into account the influence of the power grid peak regulation capacity. References [17–19] put forward the optimal allocation method of energy storage from the point of view of the influence of insufficient peak regulation capacity of grid-connected wind power, so that the power grid has a more downward regulation capacity to accept wind power, but only the performance of thermal power peak regulation is considered. The work done in these references does not consider the comprehensive effects of many types of power sources.

Reference [20] studies the feasibility and rationality of wind–storage combined power generation under current market conditions from the point of view of a technical and economic evaluation, which provides a reference for an optimal allocation in a wind–storage combined power generation system. References [21,22] fully consider the operation characteristics of the energy storage system and determine the energy storage allocation method based on the dynamic adjustment of charge state and a variable life model. The joint planning model of energy storage and the transmission network for improving the receptive capacity of wind power is established in [23], but the focus is on the power capacity of the energy storage, and the energy capacity is not mentioned. Based on the characteristics of offshore wind power, an optimal scheduling method for the joint operation of offshore wind power and seawater-pumped storage power stations is proposed in [24], but the work done in the reference only mentions optimization and does not involve the optimal allocation of offshore energy storage units. There is little research on offshore wind power and energy storage. A physical model combining offshore wind power generation with an underwater compressed air energy storage system was established in [25]. In [26], an optimal energy storage allocation model was constructed based on the improved scene clustering algorithm under the application scenario of smoothing the offshore wind power output fluctuation. A new multi-objective programming framework is proposed in [27] to determine the optimal capacity of battery energy storage systems in the cooperative operation of large offshore wind farms and battery energy storage systems. The cited references present the different technologies of energy storage. Their characteristics are shown in Table 1.

Table 1. The characteristics of common energy storage types.

Types	Advantages	Drawbacks	Efficiency	Application
pumped hydro storage	Mature technology, large capacity, low cost and long life.	The installation position has special requirements, and the conveying loss is large.	65–75%	improve the capacity of new energy consumption; smooth the short-term fluctuations of new energy output; provide capacity reserve and power grid peak and frequency modulation services.
lithium battery	High energy density and fast charge and discharge.	High production cost and a special charging circuit is required to avoid overheating.	85–98%	
lead–acid battery	Mature technology, easy availability of raw materials and low cost.	Low energy density and short life during deep charge and discharge.	80–90%	
sodium-sulfur battery	High energy density and fast response time.	Electrode materials are flammable and there are risks in high-temperature operation.	75–90%	
liquid flow battery	Large capacity, good safety and long cycle life.	High maintenance cost and low energy density.	75–85%	

Pumped hydro storage is the most reliable, economical, large capacity and most mature energy storage device in the power system. It has the advantages of flexible start-up, fast climbing speed, peak cutting and valley filling, and it is an important part of the development of new energy. However, it depends on geographical conditions and needs high hydropower stations. It has a great impact on geography and terrain, and the construction cost is high. Large-scale battery energy storage systems can be used for power grid energy management and peak regulation, and the technology is quite mature. Moreover, it has a fast load response and daily regulation ability, so it is suitable for large-scale wind power generation.

This paper presents two innovative points: based on the idea of combining planning and operation through operation simulation, an optimization model of offshore wind energy storage capacity planning is established, which aims to minimize the total national economic expenditure of the system. It considers the offshore wind power transmission channel constraint and uses the abandoned wind rate below a certain level as a strict constraint. Further, based on 24 scenarios for the optimization model and the contour line of annual cost and the contour line of abandoned wind rate, an energy storage capacity planning method for improving offshore wind power consumption is proposed, which can obtain a reasonable economic and optimal energy storage configuration scheme quickly.

The rest of the paper is organized as follows. The optimization model of offshore wind energy storage capacity planning is established and the principal block diagram of the planning and optimization process is shown in Section 2. The case study and data analysis for the optimization model for offshore wind energy storage capacity planning are carried out and an energy storage capacity planning method for improving offshore wind power consumption is proposed in Section 3. Finally, Section 4 concludes the paper.

2. Model and Methods

At present, electrochemical energy storage systems are the most widely used technology on the source side of offshore wind farms. Small-scale battery storage systems are generally used in ships and offshore platforms, while large-scale battery storage systems are mainly used in islands and coastal areas. This paper takes electrochemical energy storage systems as an example to conduct relevant research on the energy storage technology of offshore wind farms [28–33].

The electrochemical energy storage for offshore wind farms is required to meet the applicable conditions of environmental temperature; it is not easy to maintain the working temperature of high-temperature sodium-sulfur batteries and liquid metal batteries in the sea environment. It is required that the mechanical moving parts of the energy storage device should be as few as possible, so as not to be damaged by corrosion or wave impact in the marine environment. The liquid flow battery should not be adopted because of its electrode characteristics with many such parts. It is required that the leakage of battery materials has no obvious harm to prevent pollution of the marine environment, so lead–acid batteries, lithium-ion batteries, seawater batteries and silver–zinc batteries can be used for offshore wind farms [34]. Offshore energy storage needs to be resistant to wind and wave impact, seawater immersion, seawater corrosion, and so on. Therefore, developing offshore energy storage systems tends to be more costly than developing onshore energy storage systems. It is necessary to configure suitable offshore energy storage capacities for offshore wind power to avoid excessive costs.

2.1. Optimization Model of Offshore Wind Energy Storage Capacity Planning

2.1.1. Objective Function

On the premise of satisfying the system demand and all kinds of constraint conditions, the system can minimize the total national economic expenditure in the whole planning period. The objective function of the model can be expressed as:

$$MinF_{\Sigma} = \sum_{t=1}^{N_m} C_t(1+i)^{N_m-t} + \sum_{t=1}^{N_T} \left(F_{gt} + F_{kt} + O_t - B_t\right) \cdot (1+i)^{-N_T} \tag{1}$$

where N_m is the construction cycle of the newly invested energy storage power station, C_t is the investment cost of the newly invested energy storage power station at the beginning of the year t, and F_{gt} and F_{kt} are the fixed operation and maintenance costs and fuel costs of the system in year t, respectively. O_t is the outage loss cost of the system in year t, and B_t is the benefit obtained by the system in year t, except for power generation. N_T is the number of planning years and i is the discount rate.

Taking the first year as the base year, when the construction process of the newly invested energy storage power station is simplified, it can be considered that the power station generates investment costs at the beginning of the first year of the planning period, and the loss of power outage and other benefits are ignored. The total calculated cost of the planning period can be equivalent to the annual cost. It means that the investment cost of the new power station at the beginning of the first year can be evenly allocated to each year of the planning period, and then added to the annual operating cost. Then the objective function can be expressed as:

$$MinF = \frac{i(1+i)^{N_T}}{(1+i)^{N_T}-1} \times C_{ess} + \left(F_g + F_k\right) \tag{2}$$

C_{ess}, the investment cost of the energy storage power station, can be expressed as:

$$C_{ess} = \lambda_p P_{ess} + \lambda_e E_{ess} \tag{3}$$

$$E_{ess} = P_{ess} T_{ess} \tag{4}$$

where P_{ess} and E_{ess} are the rated power capacity and energy capacity of the energy storage, respectively, T_{ess} is the charging and discharging time of energy storage, and λ_p and λ_e are the cost per unit power capacity and the cost per unit energy capacity, respectively.

The annual fixed operation and maintenance cost F_g consists of a conventional thermal power station F_{g1} and an energy storage station F_{g2}, which can be expressed as:

$$F_{g1} = \alpha_{g1} \cdot C_{g1} \tag{5}$$

$$F_{g2} = \alpha_{g2} \cdot P_{ess} \tag{6}$$

$$F_g = F_{g1} + F_{g2} \tag{7}$$

where C_{g1} is the total investment cost of a conventional thermal power station, α_{g1} is the annual fixed operation and maintenance cost rate of the power station, and α_{g2} is the fixed operation and maintenance cost of energy storage per unit power.

The thermal power station's annual operating fuel cost F_k can be expressed as:

$$F_k = \beta_k \cdot E_k \tag{8}$$

where E_k is the annual energy yield of a conventional thermal power station, and β_k is the fuel cost of the unit energy yield of the power station.

2.1.2. Constraint

In comprehensively considering a variety of power supply types, including wind power, photovoltaic, hydropower, thermal power, pumped storage and new energy storage units, the electricity transmitted by the inter-provincial tie lines and the transmission lines outside the province can be classified into the load demand, and the constraint conditions to be met are shown in the following equations.

① Constraints on system power balance:

$$Pess_t + P_{0t} = LD_t(1 + \rho + \sigma) \tag{9}$$

where $Pess_t$ represents the output of the newly invested energy storage system at time t, P_{0t} represents the output of the original power station of the system at time t, LD_t is the load value of the system at time t, and ρ and σ are the power consumption rate and system line loss rate, respectively.

② Maximum and minimum output constraints of power station:

$$P_{\text{kmin}} \leq P_{\text{k}} \leq P_{\text{kmax}} \tag{10}$$

where P_{kmin} and P_{kmax} are the minimum and maximum technical outputs of unit k.

③ Thermal power fuel consumption constraints:

$$\sum_{t=1}^{\tau} E_{it}\beta_i \leq A_{i\tau} \tag{11}$$

where E_{it} is the generating capacity of the thermal power plant i at time t, $A_{i\tau}$ is the fuel consumption limit of power plant i in the period τ, and β_i is the average fuel consumption per unit of power plant i.

④ Climbing constraints of thermal power units:

$$U_i{}^t DR_i \leq P_i(t) - P_i(t-1) \leq U_i{}^t UR_i \tag{12}$$

where UR_i and DR_i are the loading and unloading rate of unit i, respectively, and $U_i{}^t$ represents the start-stop state of thermal power unit i at time t, which is 0–1. The start-up is 1, and others are 0.

⑤ Constraints on the start and stop of thermal power units:

$$\sum_{k=t}^{t+T_S-1}(1-U_i^k)\geq T_S(U_i^{t-1}-U_i^t) \tag{13}$$

$$\sum_{k=t}^{t+T_O-1}U_i^k\geq T_O(U_i^t-U_i^{t-1}) \tag{14}$$

where T_S and T_O are the minimum shutdown and start-up time of the thermal power unit, respectively.

⑥ Constraints on the generating capacity of hydropower units:

$$\sum_{t=1}^{\tau}E_{jt}\beta_j\leq W_{j\tau} \tag{15}$$

where E_{jt} is the generating capacity of the hydropower plant j at time t, $W_{j\tau}$ is the available water limit of power plant j in the period τ, and β_j is the average water consumption per unit of power plant j.

⑦ Constraints on pumped storage units:

$$E_{jG}=\eta_jE_{jP} \tag{16}$$

$$C_{P.t}=mP_{PS.P.N} \tag{17}$$

where η_j is the pumping-power generation conversion efficiency of pumped storage power station j, and E_{jG} and E_{jP} are the generating capacity and pumping load capacity of the pumped storage power station j, respectively, within its dispatching period τ. The pumped power of a pumped storage power station at a certain period of time must be an integer multiple of its single capacity. $C_{P.t}$ is the pumping capacity of the pumped storage power station at time t, and $P_{PS.P.N}$ is the rated pumping capacity of the pumped storage unit.

⑧ Energy storage operation constraints:

$$-P_{cmax\_ESS}\leq P_{out\_ESS}\leq P_{dmax\_ESS} \tag{18}$$

$$E_{min}\leq E_t\leq E_{max} \tag{19}$$

$$E_{ess}(0)=E_{ess}(T) \tag{20}$$

where P_{cmax_ESS} and P_{dmax_ESS} are the maximum charge and discharge power, respectively. P_{out_ESS} is the real-time output power, and E_t is the real-time energy capacity.

⑨ Standby constraints:

$$\sum_{i=1}^{N}U_i^t(P_{i,\max}-P_i(t))\geq ur_N(t) \tag{21}$$

$$\sum_{i=1}^{N}P_{i,\max}\geq\alpha LD_{\max} \tag{22}$$

where N units are providing a certain reserve capacity, $ur_N(t)$ represents the spinning reserve of N units at time t, α is the total reserve rate, and $LD_{\max}$ is the maximum load.

⑩ Offshore wind power transmission channel constraints:

$$P_{pass}\leq\eta P_{WN} \tag{23}$$

where P_{pass} is the maximum transmission capacity of the offshore wind power transmission channel. P_{WN} is the rated installed capacity of the offshore wind farm, and η is the transmission channel ratio.

2.2. Principal Block Diagram of Planning and Optimization Process

A typical case of a coastal power grid is taken to verify the effectiveness of the energy storage capacity planning method. First, the methods of cluster analysis and probabilistic modeling are adopted to consider the uncertainty of offshore wind power, and the annual output characteristic curves are shown in Figure 1.

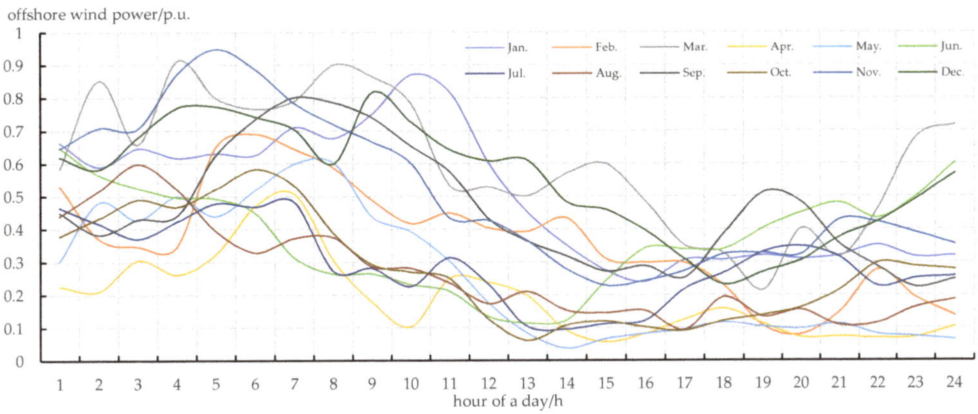

Figure 1. Offshore wind power output curve clustering scenario set.

The principal block diagram of offshore wind power storage capacity planning and optimization is shown in Figure 2. The long-term operation data of the combined wind–storage system can be obtained through operation simulation, and the consumption index of offshore wind power can be calculated. After a comprehensive optimization comparison and sensitivity analysis, the optimal planning results can be outputted.

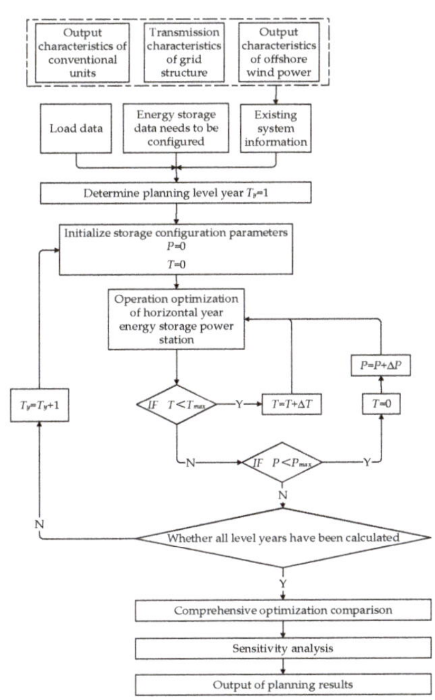

Figure 2. Principal block diagram of offshore wind energy storage capacity planning and optimization.

3. Results and Discussion

3.1. Description of the Basic Conditions of the Example

It is expected that by 2025, the annual maximum load of the power grid in this coastal area will be 0.0111 billion kW, with a total power consumption of 59.2 billion kWh, and the total installed offshore wind power will reach 9176.5 MW. The transmission channel ratio η = 0.8, and this means that the maximum capacity of the transmission channel will be 7341.2 MW.

The multi-type power supply and line structure in this coastal area are shown in Figure 3. The installed capacity of the multi-type power supply corresponding to Figure 3 is shown in Table 2. The installed capacity of offshore energy storage needs to be planned and then configured. Load characteristics are described by an annual maximum load curve, typical weekly maximum load curve and typical daily load curve. Load data are shown in Figure 4.

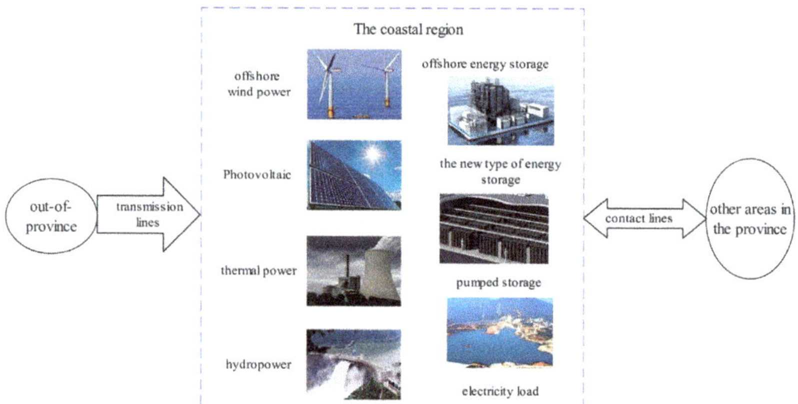

Figure 3. Power supply and line structure in this coastal area.

Table 2. The installed capacity of various power sources of the coastal power grid.

Power Sources	Offshore Wind Power	Photovoltaic	Thermal Power	Hydropower	Pumped Storage
Capacity/MW	9176.5	8220.7	9490	772.5	2400

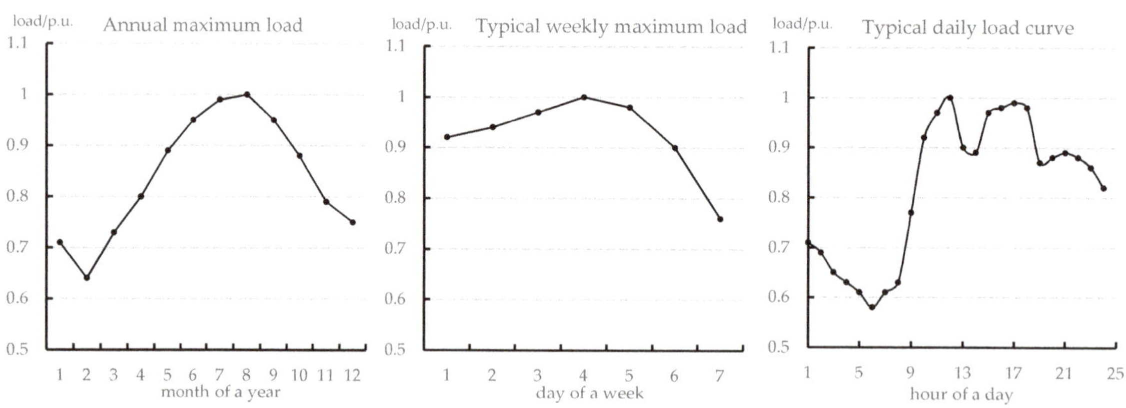

Figure 4. Annual load characteristic curves.

We used the Gurobi solver to solve the model in the MATLAB programming environment. The simulation was carried out with the year as the cycle and the day as the unit. Inputs should be the load curves and offshore wind power output curves of the coastal area based on historical data, combined with the power installation structure and the grid structure inside and outside of the province. The monthly statistics of offshore wind power and abandoned wind power in this coastal area can be obtained without new energy storage, as shown in Figure 5.

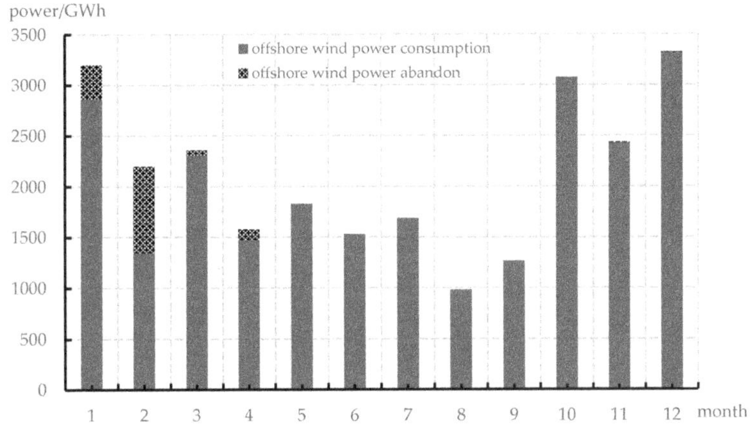

Figure 5. Annual utilization of offshore wind power in this coastal area.

All of the offshore wind farms in this coastal area can generate 25,441.25 GWh of electricity in a year. The practical electricity is 24,085.76 GWh, and the abandoned wind power is 1355.49 GWh. The abandoned wind rate is 5.33%, and the utilization hours of offshore wind power are 2625 h. Further, the utilization hours of the transmission channels are 3281 h. It can be seen from Figure 5 that the abandoned wind power of offshore wind power is mainly concentrated from January to April, with the most serious abandoned wind in February and a little abandoned wind in November and December.

For lead–acid battery and lithium-ion battery energy storage systems, the cost coefficients per unit of energy capacity, per unit power capacity, the operation and maintenance costs and engineering life obtained, are shown in Table 3.

Table 3. Related parameters of energy storage.

Type	λ_p (10^4 Yuan/MWh)	λ_e (10^4 Yuan/MW)	α_{g2} 10^4 Yuan/(MW×Year)	Engineering Life (Years)
rich liquid lead–acid	150	125	15	20
lithium-ion battery	500	175	20	20

According to relevant parameters, the planning period is selected as 20 years, and the comprehensive discount rate for the whole society is 10%. According to the offshore wind energy storage capacity planning optimization model, the next step is to set up the energy storage configuration. The offshore wind farms are configured with an energy storage capacity of 10% to 40% of their rated installed capacity. Therefore, the rated power capacity of the energy storage system is described as 0.1~0.4 in the following. The installed capacity of energy storage under different configuration schemes is shown in Table 4. With daily cycle adjustments of energy storage devices, the charging and discharging time is set from 1 to 6 h, respectively, and the 24 energy storage configuration schemes are combined with different power P and charging and discharging time T.

Table 4. Storage capacity configuration of offshore wind farms.

Configuration Ratio	10%	20%	30%	40%
Capacity (MW)	917.65	1835.3	2752.95	3670.6

3.2. Example Analysis of Simulation Results

Based on the energy storage configuration scheme, the annual electricity balance of operation simulation from the planning level is conducted to obtain the operation simulation results of the coastal area. The relationship between the abandoned wind rate of the offshore wind power and the energy storage configuration scheme is shown in Table 5. Thus, with the further increase in new energy storage power capacity and energy capacity, the abandoned wind rate of offshore wind power gradually decreases.

Table 5. Relationship between the abandoned wind rate of offshore wind power and the energy storage configuration scheme in this region.

(P/T)	Without Storage	0.1	0.2	0.3	0.4
1	5.33%	5.31%	5.08%	4.78%	4.67%
2		5.18%	4.88%	4.67%	4.41%
3		5.10%	4.82%	4.44%	4.04%
4		5.04%	4.62%	4.21%	3.97%
5		5.02%	4.51%	4.15%	4.10%
6		4.90%	4.43%	4.07%	4.11%

Here, when the lithium-ion battery energy storage system with a scale of 917.65 MW/917.65 MWh is configured in the offshore wind farm of this coastal area, the annual cost is analyzed, as shown in Table 6.

Table 6. Composition of annual expenses (10^4 Yuan).

Annual Value of Energy Storage Investment Costs	Annual Operation and Maintenance Cost of Energy Storage	Annual Operating and Maintenance Cost of Thermal Power	Annual Fuel Cost for Thermal Power Operation	Annual Total Cost
72,756.1	13,764.8	1,933,090	5,431,380	7,450,990.9

Based on this, the relationship between different energy storage configuration schemes and the annual costs can be obtained, as shown in Table 7. It can be seen that with the further increase in new energy storage power capacity and energy capacity, the annual system costs gradually increase. Therefore, the decrease in the abandoned wind rate of offshore wind power is accompanied by an increase in the annual system cost. This paper studies the method to achieve the lowest annual cost while meeting the strict constraints below a certain curtailment level.

Table 7. Annual total cost under different schemes.

10^7 Yuan	Without Storage	0.1	0.2	0.3	0.4
1	7343.4	7451	7516.9	7597.8	7684.2
2		7496	7634.6	7778.3	7905.5
3		7550.2	7741.5	7924.8	8118.2
4		7602.2	7847.3	8096.9	8330.4
5		7655.8	7953.1	8255.7	8545.5
6		7708.6	8058.5	8416.0	8761.9

Based on Tables 5 and 7, contour lines of wind curtailment rate and annual cost can be drawn on a two-dimensional plane, as shown in Figures 6 and 7, respectively. The curve of wind curtailment rate indicates that different energy storage configurations can bring the same consumption effect of offshore wind power.

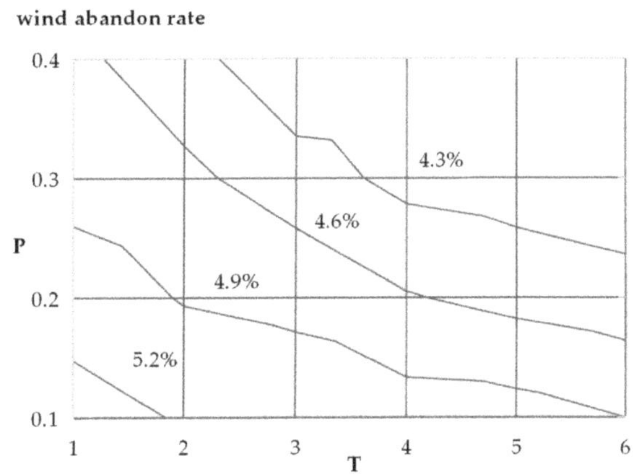

Figure 6. Contour lines of abandoned wind rates of offshore wind power.

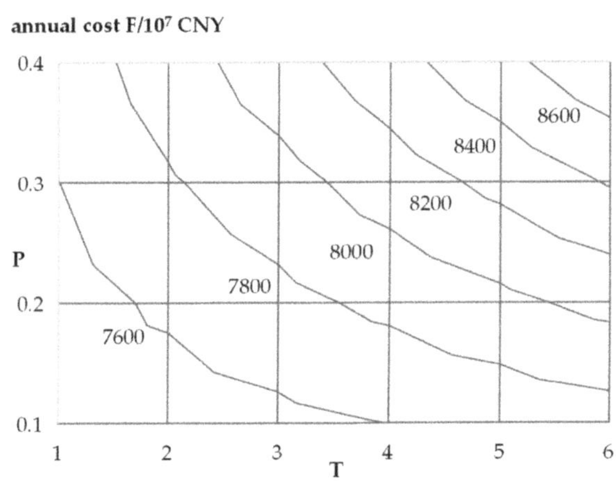

Figure 7. Contour lines of the annual cost of the planning scheme.

In order to find the optimal economic scheme combined with the annual cost contour line, it can be known that when the abandoned wind rate is at a certain standard level, different annual cost contour lines are used to be tangent to the determined abandoned wind rate contour line, and the tangent point (power P, charge and discharge time T) is the best scheme.

In practical application, 5% of new energy is allowed to abandon power, which is scientifically reasonable. Therefore, the alternative energy storage configuration schemes are (0.3, 1), (0.2, 2), (0.1, 6), etc. According to this method, the best energy storage configuration scheme is (0.3, 1). It means that the scale of the lithium-ion battery energy storage system configured for the offshore wind farm with a total installed capacity of 9176.5 MW in the coastal area is 2752.95 MW/2752.95 MWh.

At this time, the practical electrical output of the offshore wind farm is 24,225.85 GWh. The abandoned wind power quantity is 1215.4 GWh, and the abandoned wind rate is 4.78%.

The utilization hours of offshore wind power are 2640 h, and the utilization hours of the transmission channel are 3300 h. Further, the annual cost is 75.978 billion yuan.

For this study, only 24 scenarios, based on the optimization model to present the energy storage capacity allocation method, were used. By using fast computer calculation, the step size of the configuration scheme is further reduced. Based on the energy storage capacity planning method proposed in this paper, the configuration scheme with the best economy and applicability can be obtained more quickly and accurately.

3.3. Sensitivity Analysis

According to the above scheme, the configuration of a 2752.95 MW/2752.95 MWh lithium-ion battery energy storage system is relatively large in terms of the annual cost from 73.434 billion yuan to 75.978 billion yuan. This section studies the factors influencing the abandoned wind rate of offshore wind power from other perspectives, exploring feasible schemes to reduce the abandoned wind rate, and further allocating the source-side energy storage, paving the way to reduce the power capacity and energy capacity of the energy storage system configuration, thus reducing the investment costs and operation and maintenance costs, and improving the economic performance.

As shown in Figure 3, the consumption and utilization of offshore wind power in this coastal area are not only related to the installed scale of the power structure, including offshore wind power and energy storage but it is also affected by the transmission agreement signed with other provinces and the exchange of electricity in contact lines with other regions in the province. Therefore, a sensitivity analysis is carried out from the transmission agreement of the transmission lines outside of the province and the capacity allocation of the link line within the province.

3.3.1. Influence of Transmission Line Agreement

Out-of-province transmission line refers to a power transmission line from another province to the coastal area, with a maximum transmission capacity of 760 MW, which is sent to the coastal area in accordance with the transmission agreement signed with another province and given priority to use. Taking the daily transmission curve as an example, the transmission agreement can be adjusted to 1.1 times the original transmission agreement, the original transmission agreement, 0.9 times the original transmission agreement, and 0.8 times the original transmission agreement, as shown in Figure 8.

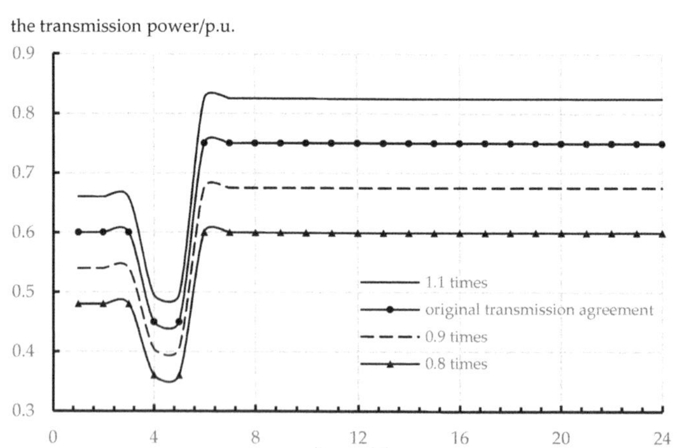

Figure 8. Schematic diagram of different transmission protocols.

The original transmission agreement refers to the existing transmission agreement between the grid in the coastal area and another province. Under the existing transmission agreement, this paper adjusts it to 1.1 times, 0.9 times and 0.8 times, and then obtains the utilization of offshore wind power according to the optimization model, and analyzes the reasons for this situation. After the operation simulation, the changes in the offshore abandoned wind power rate under different transmission agreements can be compared and analyzed, and the results are shown in Table 8.

Table 8. Utilization of offshore wind power under different transmission agreements.

Transmission Scenario	Consumption Power (GWh)	Abandoned Power (GWh)	The Abandoned Wind Rate
1.1 times	24,053.65	1387.61	5.45%
the original agreement	24,085.76	1355.49	5.33%
0.9 times	24,119.29	1321.96	5.20%
0.8 times	24,362.69	1078.56	4.24%

Therefore, it can be seen that the electricity sent by the out-of-province transmission lines in this coastal area is too much, which affects the consumption and utilization of internal offshore wind power. Therefore, the transmission agreement can be optimized in the direction of reduction without additional cost.

3.3.2. Capacity Allocation of Tie Lines

There is a contact exchange between this coastal area and other areas A and B in the province. The maximum exchange capacity of the contact line between this coastal area and area A or area B is 3000 MW. This means that the total capacity of the external contact line in this coastal area is 6000 MW. This is because the total annual load demand power ratio of region A and region B is 1:1.2. Without changing the coastal area foreign link under the premise of a total exchange capacity of 6000 MW, the capacity ratio of the two contact lines is adjusted, and the changes in the abandoned wind rate of the offshore wind power under different capacity ratios of the contact line are compared and analyzed after the operation simulation. The results are shown in Table 9.

Table 9. Utilization of offshore wind power under different capacity ratios of tie lines.

The Total Capacity (MW)	Transmission Ratio		The Total
	The Coastal Area-Area A	The Coastal Area-Area B	Abandoned Wind Rate
6000	1	1	5.33%
	1	1.2	5.27%
	1	1.4	5.23%
	1	1.6	5.20%
	1	2	5.13%
	1	3	5.09%

Therefore, the capacity ratio of the contact line can be optimized according to the load demand of the contact area. This is to reduce the maximum exchange capacity of the contact line between the coastal area and area A with low electrical demand, to limit and reduce the amount of electricity fed back to the coastal area, and promote the absorption and utilization of internal offshore wind power. Increasing the maximum exchange capacity of the link line between the coastal area and region B with a high electricity demand can effectively export the electrical power of the coastal area, and further export and utilize the offshore wind power that is difficult to be absorbed internally when needed.

On one hand, the abandonment of offshore wind power comes from transmission congestion in the transmission channel, and on the other hand, it comes from the lack of peak regulation capacity of the system. When the transmission protocol or tie line capacity ratio is optimized, the source-side energy storage can be further configured according to the method described in this paper, which can reduce the energy storage investment costs and operation and maintenance costs, and improve the economic performance.

4. Conclusions

This paper studies an energy storage capacity planning method for improving offshore wind energy consumption, and the conclusions are as follows:

(1) An optimization model for offshore wind power storage capacity planning is established to seek an economic and reasonable energy storage power construction and configuration scheme within the planning period, on the premise of meeting the system's annual load development needs and other various constraints;

(2) Based on the power supply and line structure of the power grid in a coastal area, an example analysis of offshore wind power storage planning was conducted. According to this method, the best energy storage configuration scheme was (0.3, 1), at an annual cost of 75.978 billion yuan. In order to fully utilize offshore wind power and further improve economic performance, the sensitivity analysis of the abandoned wind rate of offshore wind power in this coastal area was carried out. The result proved that the reasonable optimization of the transmission agreement and the capacity ratio of tie lines can improve the acceptance capacity of the power grid to offshore wind power.

The results of this paper can provide some reference value for further research on capacity planning and the optimal operation of offshore wind energy storage. However, this paper sets a fixed value for the capacity of the offshore transmission channel, without joint planning of the offshore energy storage and the offshore transmission channel.

Author Contributions: Conceptualization, H.Y.; methodology, H.C.; software, X.Y. and S.L.; validation, X.Y. and S.L.; resources, H.C.; data curation, H.Y.; writing—original draft preparation, X.Y.; writing—review and editing, H.Y.; visualization, Y.L.; supervision, Y.L. All authors have read and agreed to the published version of the manuscript.

Funding: This research was funded by the Science and Technology Project of China Southern Power Grid Co., Ltd., grant number 037700KK52190012 (GDKJXM20198282) and supported by the Research and Development Plan in key areas of Guangdong Province, grant number 2021B0101230004.

Institutional Review Board Statement: Not applicable.

Informed Consent Statement: Not applicable.

Data Availability Statement: Not applicable.

Acknowledgments: The authors would like to thank the reviewers for their valuable comments on this research.

Conflicts of Interest: The authors declare no conflict of interest.

References

1. Chi, Y.N.; Liang, W.; Zhang, Z.K.; Li, Y.; Jin, S.L.; Cai, X.; Hu, J.H.; Zhao, S.X.; Tian, W. An Overview on Key Technologies Regarding Power Transmission and Grid Integration of Large Scale Offshore Wind Power. *Proc. CSEE* **2016**, *36*, 3758–3771.
2. Yao, G.; Yang, H.M.; Zhou, L.D.; Li, D.D.; Li, C.B.; Wang, J. Development Status and Key Technologies of Large-capacity Offshore Wind Turbines. *Autom. Electr. Power Syst.* **2021**, *45*, 33–47. [CrossRef]
3. Song, Y.H.; Tan, Z.F.; Li, H.H.; Liu, W.Y. An Optimization Model Combining Generation Side and Energy Storage System With Demand Side to Promote Accommodation of Wind Power. *Power Syst. Technol.* **2014**, *38*, 610–615.
4. Li, J.L.; Guo, B.Q.; Niu, M.; Xiu, X.Q.; Tian, L.T. Optimal Configuration Strategy of Energy Storage Capacity in Wind/PV/Storage Hybrid System. *Trans. China Electrotech. Soc.* **2018**, *33*, 1189–1196. [CrossRef]
5. Dai, L.S.; Ye, C.J.; Fu, X.H.; Sun, K.; Bian, Q.Y.; Xin, H.H. Distributional Robust Joint Chance Constrained Optimal Capacity Installment of Energy Storage in Power System With High Penetration of Wind Power. *Power Syst. Technol.* **2017**, *41*, 769–777.
6. Xu, G.D.; Cheng, H.Z.; Ma, Z.F.; Fan, S.L.; Fang, S.D.; Ma, Z.L. Overview of ESS planning methods for alleviating peak-shaving pressure of grid. *Electr. Power Autom. Equip.* **2017**, *37*, 3–11.
7. Lou, S.H.; Yang, T.M.; Wu, Y.W.; Wang, Y.C. Coordinated Optimal Operation of Hybrid Energy Storage in Power System Accommodated High Penetration of Wind Power. *Autom. Electr. Power Syst.* **2016**, *40*, 30–35.
8. Feng, J.X.; Liang, J.; Zhang, F.; Wang, C.F.; Sun, B.H. An Optimization Calculation Method of Wind Farm Energy Storage Capacity. *Autom. Electr. Power Syst.* **2013**, *37*, 90–95. [CrossRef]
9. Chen, H.H.; Du, H.H.; Zhang, R.F.; Jiang, T.; Li, X. Optimal capacity configuration and operation strategy of hybrid energy storage considering uncertainty of wind power. *Electr. Power Autom. Equip.* **2018**, *38*, 174–182.

10. Shi, L.; Luo, Y.; Tu, G.Y.; Shi, N. Energy Storage Sizing Method Considering Dispatchability of Wind Farm. *Trans. China Electrotech. Soc.* **2013**, *28*, 120–127. [CrossRef]
11. Zheng, L.; Hu, W.; Lu, Q.Y.; Min, Y.; Yuan, F.; Gao, Z.H. Research on Planning and Operation Model for Energy Storage System to Optimize Wind Power Integration. *Proc. CSEE* **2014**, *34*, 2533–2543.
12. Xie, P.; Cai, Z.X.; Liu, P.; Li, X.H.; Zhang, Y.J.; Sun, Y.Y. Cooperative Optimization of Energy Storage Capacity for Renewable and Storage Involved Microgrids Considering Multi Time Scale Uncertainty Coupling Influence. *Proc. CSEE* **2019**, *39*, 7126–7136.
13. Wu, W.P.; Hu, Z.C.; Song, Y.H. Optimal Sizing of Energy Storage System for Wind Farms Combining Stochastic Programming and Sequential Monte Carlo Simulation. *Power Syst. Technol.* **2018**, *42*, 1055–1062.
14. Ye, R.L.; Guo, Z.Z.; Liu, R.Y.; Liu, J.N. Reliable power output based on confidence interval estimation and optimal ESS configuration of wind farm. *Electr. Power Autom. Equip.* **2017**, *37*, 85–91.
15. Cao, M.H.; Yu, J.L. Capacity Planning Method of Electrochemical Energy Storage System for Generation Curve Deviation Correction of Wind Power Farm. *Autom. Electr. Power Syst.* **2022**, *46*, 27–36. [CrossRef]
16. Gan, W.; Guo, J.B.; Ai, X.M.; Yao, W.; Yang, B.; Yao, L.Z.; Wen, J.Y. Multi-scale Multi-index Sizing of Energy Storage Applied to Fluctuation Mitigation of Wind Farm. *Autom. Electr. Power Syst.* **2019**, *43*, 92–98.
17. Xu, G.D.; Cheng, H.Z.; Fang, S.D.; Ma, Z.L.; Zhang, J.P.; Zhu, Z.L. An Optimization Model of Battery Energy Storage System Configuration to Improve Benefits of Wind Farms. *Autom. Electr. Power Syst.* **2016**, *40*, 62–70.
18. Sun, W.Q.; Song, H.; Qin, Y.H.; Li, H. Energy Storage System Optimal Allocation Considering Flexibility Supply and Demand Uncertainty. *Power Syst. Technol.* **2020**, *44*, 4486–4497. [CrossRef]
19. Yan, G.G.; Feng, X.D.; Li, J.H.; Mu, G.; Xie, G.Q.; Dong, X.C.; Wang, Z.M.; Yang, K. Optimization of Energy Storage System Capacity for Relaxing Peak Load Regulation Bottlenecks. *Proc. CSEE* **2012**, *32*, 27–35.
20. Jiang, S.P.; Qiao, Y.; Xu, F.; Nie, H.Z.; Hu, D. Capacity Optimization and Sensitivity Analysis of Cogeneration System of Wind Power and Energy Storage. *Autom. Electr. Power Syst.* **2013**, *37*, 16–21. [CrossRef]
21. Zhang, X.; Zhang, F.; Gong, N.Q.; Liu, H.K.; Liang, J.; Yang, L.B. BESS capacity planning based on dynamic SOC adjustment. *Electr. Power Autom. Equip.* **2015**, *35*, 20–25.
22. Lou, S.H.; Yi, L.; Wu, Y.W.; Hou, T.T.; Yang, Y.F. Optimizing Deployment of Battery Energy Storage Based on Lifetime Predication. *Trans. China Electrotech. Soc.* **2015**, *30*, 265–271. [CrossRef]
23. Huang, Y.; Liu, B.Z.; Wang, K.Y.; Ai, X. Joint Planning of Energy Storage and Transmission Network Considering Wind Power Accommodation Capability. *Power Syst. Technol.* **2018**, *42*, 1480–1489.
24. Wu, H.Y.; Deng, C.H.; Li, D.L.; Yao, W.W.; Zhang, H.; Ying, H.M. A multi-time scale optimal dispatching method for combined operation of offshore wind power and seawater pumped storage. *Eng. J. Wuhan Univ.* **2021**, *54*, 361–368.
25. Liu, Y.B.; Chen, J.S.; Li, Q.J.; Zhang, J.W.; He, T.; Wu, Z.L. Operation and Varying Load Analysis of Offshore Wind-Underwater Compressed Air Energy Storage System. *South. Power Syst. Technol.* **2022**, *16*, 50–59.
26. Yi, J.G.; Zhu, Z.W.; Xie, Q. Research on Optimal Allocation of Energy Storage in Offshore Wind Farm Based on Improved Scene Clustering Algorithm. *China Electr. Power* **2022**, 1–9. Available online: https://kns.cnki.net/kcms/detail/11.3265.TM.20220622.1427.006.html (accessed on 6 October 2022).
27. Paul, S.; Nath, A.P.; Rather, Z.H. A Multi-Objective Planning Framework for Coordinated Generation From Offshore Wind Farm and Battery Energy Storage System. *IEEE Trans. Sustain. Energy* **2020**, *11*, 2087–2097. [CrossRef]
28. Koohi-Fayegh, S.; Rosen, M.A. A review of energy storage types, applications and recent developments. *J. Energy Storage* **2020**, *27*, 101047. [CrossRef]
29. Wang, Z.; Carriveau, R.; Ting, D.S.K.; Xiong, W.; Wang, Z. A review of marine renewable energy storage. *Int. J. Energy Res.* **2018**, *43*, 6108–6150. [CrossRef]
30. Li, B.; DeCarolis, J.F. A techno-economic assessment of offshore wind coupled to offshore compressed air energy storage. *Appl. Energy* **2015**, *155*, 315–322. [CrossRef]
31. Lawder, M.T.; Suthar, B.; Northrop, P.W.C.; De, S.; Hoff, C.M.; Leitermann, O.; Crow, M.L.; Santhanagopalan, S.; Subramanian, V.R. Battery Energy Storage System (BESS) and Battery Management System (BMS) for Grid-Scale Applications. *Proc. IEEE* **2014**, *102*, 1014–1030. [CrossRef]
32. Lu, M.; Chang, C.; Lee, W.; Wang, L. Combining the Wind Power Generation System With Energy Storage Equipment. *IEEE Trans. Ind. Appl.* **2009**, *45*, 2109–2115. [CrossRef]
33. Vazquez, S.; Lukic, S.M.; Galvan, E.; Franquelo, L.G.; Carrasco, J.M. Energy Storage Systems for Transport and Grid Applications. *IEEE Trans. Ind. Electron.* **2010**, *57*, 3881–3895. [CrossRef]
34. Gao, J.; Zhao, B.; Yang, C.; Yang, H.R.; Han, X.G. Development and Prospect of Energy Storage at Sea. *Adv. New Renew. Energy* **2020**, *8*, 136–142. [CrossRef]

energies

MDPI

Article

Influence of Several Phosphate-Containing Additives on the Stability and Electrochemical Behavior of Positive Electrolytes for Vanadium Redox Flow Battery

Xukun Zhang [1,2,3], Fancheng Meng [1,2,3,*], Linquan Sun [1,2,3], Zhaowu Zhu [1,2], Desheng Chen [1,2,4] and Lina Wang [1,2,3,*]

1 Key Laboratory of Green Process and Engineering, Institute of Process Engineering, Chinese Academy of Sciences, Beijing 100190, China
2 National Engineering Research Center of Green Recycling for Strategic Metal Resources, Institute of Process Engineering, Chinese Academy of Sciences, Beijing 100190, China
3 School of Chemical Engineering, University of Chinese Academy of Sciences, Beijing 101408, China
4 Hebei Zhongke Tongchuang Vanadium & Titanium Technology Co., Ltd., Hengshui 053000, China
* Correspondence: fcmeng@ipe.ac.cn (F.M.); linawang@ipe.ac.cn (L.W.)

Abstract: The poor operational stability of electrolytes is a persistent impediment in building redox flow battery technology; choosing suitable stability additives is usually the research direction to solve this problem. The effects of five phosphate containing additives (including 1-hydroxyethylidene-1,1-diphosphonic acid (HEDP), hexamethylene diamine tetramethylene phosphonic acid (HDTMPA), amino trimethylene phosphonic acid (ATMPA), sodium ethylenediamine tetramethylene phosphonate (EDTMPS), and diethyl triamine pentamethylene phosphonic acid (DTPMP)) on the thermal stability and electrochemical performance of the positive electrolyte of vanadium redox flow battery were investigated. With 0.5 wt% addition, most of the selected additives were able to improve the thermal stability of the electrolyte. HEDP and HDTMPA extended the stability time of the pentavalent vanadium electrolyte at 50 °C from 5 days (blank sample) to 30 days and 15 days, respectively. The electrochemical performance of the electrolyte was further investigated by cyclic voltammetry, steady state polarization, and electrochemical impedance spectroscopy tests. It was found that most of the additives enhanced the electrochemical activity of the positive electrolyte, and the diffusion coefficients, exchange current densities, and reaction rate constants of V(IV) species became larger with the addition of these additives. It is verified that the thermal stability and electrochemical stability of the electrolyte are significantly improved by the combination of ATMPA + HEDP or ATMPA + HDTMPA. This study provides a new approach to improve the stability of the positive electrolyte for vanadium redox flow battery.

Keywords: vanadium redox flow battery; positive electrolyte; phosphate containing additives; stability; electrochemical behavior

Citation: Zhang, X.; Meng, F.; Sun, L.; Zhu, Z.; Chen, D.; Wang, L. Influence of Several Phosphate-Containing Additives on the Stability and Electrochemical Behavior of Positive Electrolytes for Vanadium Redox Flow Battery. *Energies* **2022**, *15*, 7829. https://doi.org/10.3390/en15217829

Academic Editors: Luis Hernández-Callejo, Jesús Armando Aguilar Jiménez and Carlos Meza Benavides

Received: 16 September 2022
Accepted: 18 October 2022
Published: 22 October 2022

1. Introduction

The vanadium redox flow battery (VRFB), proposed by Maria Skyllas-Kazacos and dating back to 1970, is considered the most promising renewable energy storage system, with the advantages of high capacity, excellent stability, high operation security, and long cycle, and it has attracted widespread attention and been investigated worldwide [1]. The positive and negative electrolytes of VRFB are stored in two separate tanks, and they flow through a separate half-cell during operation and then return to the tank for recirculation. Each half-cell of VRFB consists of an electrode and bipolar plate, and two half-cells are separated by a membrane that allows selective ion exchange while preventing cross-contamination of the electrolyte [2]. The chemical reactions occurring at the electrodes of positive and negative half-cell, as well as the overall cell reaction, are as follows:

Positive cell reaction:

$$VO_2^+ + 2H^+ + e^- \rightleftharpoons VO^{2+} + H_2O \quad (1)$$

Negative cell reaction:

$$V^{2+} \rightleftharpoons V^{3+} + e^- \quad (2)$$

Overall cell reaction:

$$VO_2^+ + V^{2+} + 2H^+ \rightleftharpoons VO^{2+} + V^{3+} + H_2O \quad (3)$$

Under the fully discharged circumstance, the positive and negative electrolytes contain only V(IV) (VO^{2+}) and V(III), respectively. During charging, the V(III) ions in the negative electrolyte are reduced to V(II), and the VO^{2+} ions in the positive electrolyte are oxidized to V(V) (VO_2^+). The electrons move through the bipolar plate from the positive electrode to the negative electrode, causing hydrogen ions (H^+) to diffuse across the membrane to the negative electrode. The same reaction occurs in reverse when discharging [3].

Despite the rapid development of VRFB in recent years, some issues limiting its commercialization need to be addressed, one of which is the fact that the vanadium compound, as the active substance in the electrolyte, accounts for a substantial part of the capital cost (40%) [4]. In addition, the battery capacity depends on the vanadium concentration, and the pentavalent vanadium species have a low solubility in sulfuric acid (<2 M) and a narrow operating temperature window (10–40 °C), thus limiting the energy density of the battery (typically < 25 Wh L^{-1}) [5]. At higher temperatures (>40 °C), the precipitation of V^{5+} in the positive electrolyte is as follows [6]:

$$2[VO_2(H_2O)_3]^+ \rightarrow 2VO(OH)_3 + 2H_2O + 2H^+ \quad (4)$$

$$2VO(OH)_3 \rightarrow V_2O_5 + 3H_2O \quad (5)$$

The precipitation process of V_2O_5 is irreversible, which is mainly responsible for the loss of battery capacity. In order to improve the solubility of vanadium compounds in the sulfuric acid electrolyte, the introduction of additives is commonly performed. Owing to their being cost-effective and not interfering with electrolyte performance, they have been investigated widely nowadays [7]. Skyllas-Kazacos et al. used phosphoric acid and ammonium phosphate as additives. The results show that phosphate anions enhance the stability of V(V) compounds at high temperatures, whereas for ammonium cations, the opposite is true—ammonium cations stabilize the negative half-cell electrolyte at low temperatures. The effects of sodium triphosphate and sodium hexametaphosphate as additives were also studied; they both retarded the precipitation to a certain extent [8,9]. Roznyatovskaya et al. investigated the mechanism of precipitation retarding by phosphate in the vanadium electrolyte using nuclear magnetic resonance (NMR) spectroscopy and dynamic light scattering (DLS). It was concluded that the electrolyte stabilization mechanism by phosphoric acid at high temperatures could be attributed to the interaction between them and V(V) monomers or dimers forming two phosphate-containing substances, thus retarding the V_2O_5 precipitation [10]. Park et al. used 0.05 M sodium pyrophosphate as an additive in the positive electrolyte with 2.0 M V(V) and 4.0 M H_2SO_4, and the long-term stability of electrolyte was improved compared with the blank solution. In addition, none of the new precipitation was proved to have been generated in the electrolyte. Nonetheless, its electrochemical cycling performance was optimized [11]. Zhang et al. investigated the effect of Na_3PO_4 as an electrolyte additive and found that it indeed delayed the V_2O_5 precipitation, but the $VOPO_4{\cdot}2H_2O$ precipitation was detected on the positive graphite mat after several cycle tests [12]. Li et al. reported some organic additives containing hydroxyl (–OH), such as sorbitol, mannitol, glucose, and fructose, and elaborated their stabilizing mechanism, indicating that these organic additives can clad the hydrated V(V) species and thus inhibit the formation of precipitation [13]. Zhang et al. selected 1 wt% HEDP as an electrolyte additive and confirmed that it can improve the electrolyte thermal stability and battery cycle efficiency of VRFB. Besides, the research confirmed in two ways (the Job plot and the Benesi–Hildebrand plot methods) that HEDP interacts with VO_2^+ in a 1:1 binding stoichiometry, which is the reason for the enhancement in the stability of the electrolyte [14]. Through the above studies, it is found that both phosphate and –OH have a good effect on stabilizing pentavalent vanadium. In summary, some research results on additives of the positive electrolyte are summarized in Table 1.

Table 1. Some research results on additives of the VRFB positive electrolyte.

Additive	Amount	V(V)/M	H_2SO_4/M	Temperature/°C	Effect of Thermal Stability	References
Sodium tripolyphosphate	1 wt%	2	-	44	Improved	[8]
Sodium hexametaphosphate	1 wt%	2	-	44	Improved	[9]
Sodium pyrophosphate	0.05 M	2	4	25	Improved	[11]
glucose	1 wt%	1.8	4.8	20–60	Improved	[13]
K_3PO_4	1 wt%	3	5	30/50	Improved	[15]
Polyacrylic acid	0.5 wt/vol%	4.7	6	50	Slightly improved	[16]
$(NH_4)_2SO_4$	2 wt%	1.8	5	50	Improved	[17]
H_3PO_4	1 wt%	2	5	50	Significantly improved	[17]
CH_3SO_3H	2.1–3 wt%	2	5	40	Improved	[12]
Hexadecyl trimethyl ammonium bromide (CTAB)	0.00106–0.0053 M	1.5	4.5	45	Improved	[18]
Phytic acid	N/A	1.8	3	25–60	Improved	[19]

In the present work, five additives containing both phosphate and more –OH, including 1-hydroxyethylidene-1,1-diphosphonic acid (HEDP), hexamethylene diamine tetramethylene phosphonic acid (HDTMPA), amino trimethylene phosphonic acid (ATMPA), sodium ethylenediamine tetramethylene phosphonate (EDTMPS), and diethyl triamine pentamethylene phosphonic acid (DTPMP), were selected and added into the V(V) electrolyte to investigate their effects on precipitation inhibition and electrochemical behavior, and the obtained results were compared with those of the original blank electrolyte. Among the five selected additives, except HEDP, other additives have not been used and discussed in such studies. The novelty of this paper is that this research has explored five kinds of phosphate containing positive electrolyte additives and their effects on stability and electrochemical performance and found two combinations that can improve the thermal stability and electrochemical performance of the electrolyte at the same time.

2. Materials and Methods

2.1. Materials

$VOSO_4 \cdot 3.5H_2O$ (99%) was provided by Shenyang Haizhongtian Fine Chemical Co., Ltd.(Shenyang, China). The additives are listed in Table 2. HEDP (60% in water), ATMPA (50% in water), EDTMPS (98%), and DTPMP (50% in water) were purchased from Shanghai Aladdin Bio-Chem Technology Co., Ltd., Shanghai, China. HDTPMA (>98%) was obtained from Adamas-beta Co., Ltd.(Shanghai, China). Other chemicals used in the experiment are of analytical grade. The previous experiment shows that, under the experimental conditions selected in this study, the five additives are stable.

2.2. Preparation of the V(V) Electrolyte Solution

$VOSO_4 \cdot 3.5H_2O$ (99%) was dissolved in a 3 M H_2SO_4 solution to prepare 2 M $VOSO_4$ solution, and the prepared V(IV) electrolyte solution was placed in a double-chamber electrolytic cell with 50 mL and 25 mL of positive and negative electrolyte tank capacity, respectively. The electrolyte was then charged (cut-off potential: 1.55 V, stepwise current density: 200−10 mA cm^{-2}) until the V(IV) ions in the positive electrolyte were converted to V(V) and the V(IV) ions in the negative electrolyte were converted to V(II). When the electrolysis was done, the total vanadium concentration in the positive electrolyte was determined by redox titration using a potentiometric titrator (PHS-3C, Shanghai Leici Co., Ltd., Shanghai, China).

Table 2. Molecular structure of the studied organic additives.

Chemical Name (Short Form)	Molecular Structure
Hexamethylenediamine tetramethylene phosphonic acid (HDTMPA)	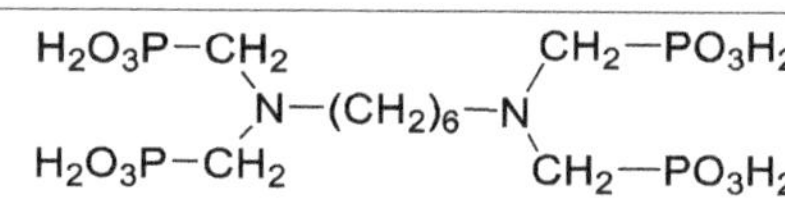
1-hydroxyethylidene-1,1-diphosphonic acid (HEDP)	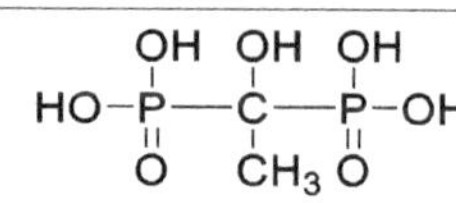
Amino trimethylene phosphonic acid (ATMPA)	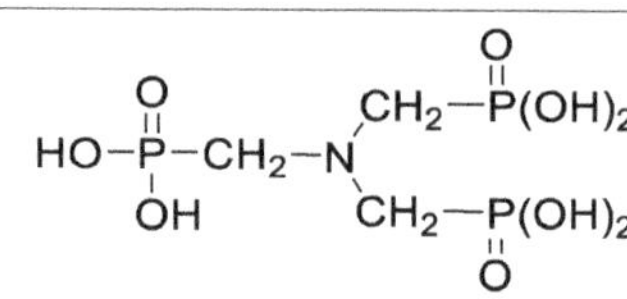
Sodium ethylenediamine tetramethylene phosphonate (EDTMPS)	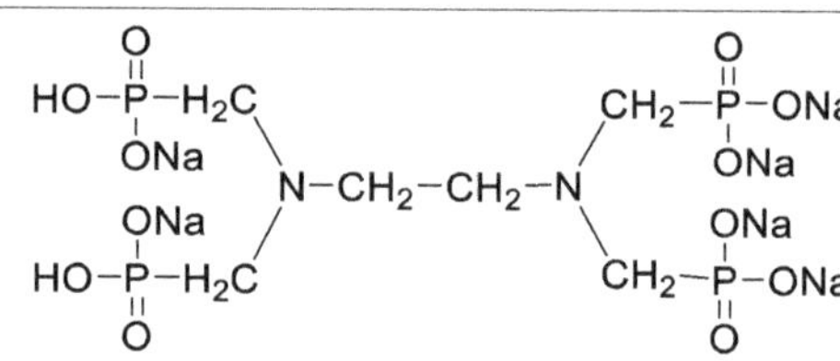
Diethyl triamine pentamethylene phosphonic acid (DTPMP)	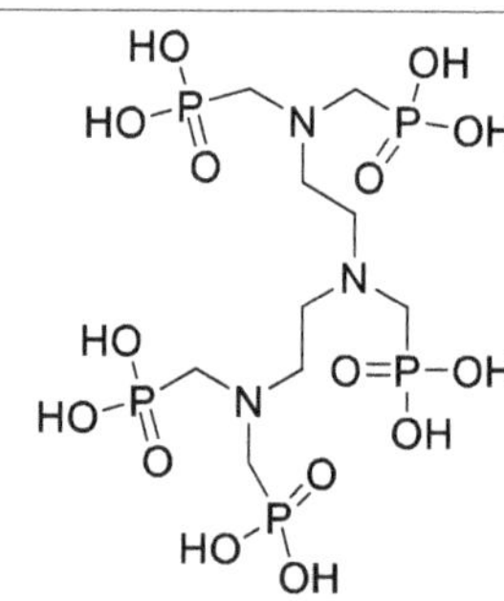

2.3. Thermal Stability Test of V(V)

To investigate the effect of various phosphonates additives on the long-term stability of the V(V) electrolyte, electrolyte samples containing 0.05 wt% additives and blank sample were stored in a sealed oven at 50 °C until measurable orange precipitation was observed. All thermal stability tests were performed without any agitation. Each sample was visually monitored more than twice a day during the test to record the V_2O_5 precipitation and the change in solution color. The samples were filtered and their equilibrium concentrations of vanadium were determined by redox titration again, at the end of the 30-day test.

2.4. Electrochemical Tests

The CV cycle test and steady-state polarization curve test of the electrolyte were performed using a CHI 760B electrochemical workstation (Shanghai Chenhua Instrument, Shanghai, China). We recorded the current versus potential curves using a three-electrode electrochemical cell in the CV cycling test with a scan rate range of 10–200 mV s^{-1} in a potential range of −0.6–1.8 V at 25 °C, in which the graphite electrode (surface area of 3.14 mm^2), saturated calomel electrode, and platinum electrode (surface area of 1 cm^2) are the working electrode, reference electrode, and counter electrode, respectively. The steady-state polarization curve was tested with a potential range of 0.49–0.56 V and a scan rate of 1 mV s^{-1}, because the current and voltage are closer to a straight line under the condition of lower potential. Electrochemical impedance spectroscopy (EIS) was also obtained at room temperature with the sinusoidal excitation voltage of the electrolyte of 5 mV and the frequency range between 0.01 Hz and 100 kHz. Prior to each electrochemical measurement, the working electrode was polished with SiC paper and then washed with distilled water. The reference electrode was washed with distilled water and the solution in salt bridge was replaced before use. Platinum plate electrode is cleaned with distilled water and ultrasonically treated.

3. Results and Discussion

3.1. Effect of Additives on the Stability of the V(V) Electrolyte

The effect of different additives (Table 1) on the thermal stability of the V(V) electrolyte was investigated by adding 0.5 wt% of additives at 50 °C. Table 3 shows the very time when V_2O_5 started to precipitate in the V(V) electrolyte samples with different additives and the V(V) concentration in the positive electrolyte after 30 days.

Table 3. Effect of several additives (dosage 0.5 wt%, 50 °C) on the thermal stability of the 2 M V(V)/3 M H_2SO_4 electrolyte.

	Blank	HDTMPA	HEDP	ATMPA	EDTMPS	DTPMP
Time to precipitation	5 days	15 days	30 days	4 days	10 days	12 days
V(V) concentration after 30 days	1.27 M	1.46 M	1.86 M	1.20 M	1.40 M	1.44 M

It was observed that the selected additives, except for ATMPA, delayed the precipitation of V(V) in the electrolyte under the same experimental conditions. The blank sample started to precipitate after 5 days and the retarding effect for the additives follows the order: HEDP (30d) > HDTMPA (15d) > DTPMP (12d) > EDTMPS (10d) > ATMPA (4d). The remaining V(V) concentration in the electrolyte after 30 days showed the same variation pattern as the initial time of precipitation. The remaining V(V) concentration in the electrolyte with HEDP was 1.86 M, followed by 1.46 M for HDTMPA, and that of the blank electrolyte sample was 1.27 M after 30 days. As for ATMPA, it was 1.20 M, which had a negative effect in this test. The vanadium concentration in the electrolyte directly determines the energy density and capacity of the battery [20], and the experimental results show that the thermal stability of the V(V) electrolyte is improved by the additives (except ATMPA). This means that these additives facilitate the VRFB to improve its energy density and capacity.

The stabilizing mechanism of HEDP for the electrolyte might be attributed to the synergism of –OH and phosphate. –OH could clad the hydrated penta-coordinate V(IV) vanadate, which prevents it from being oxidized at a low concentration and inhibits its precipitation [9,21]. Phosphate could interact with V(V) monomers or dimers, forming a stable phosphate-containing substance, and thus retarding the precipitation [10]. Similarly, the stabilizing capability of HDTMPA and DTPMP is probably due to the presence of more phosphate. The EDTMPS, with good chemical stability and temperature resistance, is soluble in water, non-toxic, and environmentally friendly. It can dissociate into eight anions/cations in aqueous solution, and thus chelate with multiple V(V) ions, forming multiple monomeric structured reticular macromolecular complexes that are loosely dispersed in water, so the normal precipitation process of V(V) was disrupted [22]. Although ATMPA was reported to have low limit inhibition, good chelation, and lattice distortion effects [23], it exhibited the worst effect on the thermal stability, which was likely due to the formation of chelate, which is not conductive to solution stability.

3.2. CV Test

Figure 1 shows the CV curves of V(IV) electrolyte samples with additives and blank one, and it can be observed that all CV curves show the similar peak position and one pair of redox peaks with a similar shape. The additives slightly changed the shape and position of the peak, which means that these additives will affect the reversibility of V(IV)/V(V) redox pairs to some extent [19]. The relevant data derived from Figure 1 are summarized in Table 4. The effect of additives on the V(IV)/V(V) redox coupling is characterized by I_{pO}/I_{pR} (ratio of the oxidation peak current to reduction peak current) and ΔV_p (separation between the oxidation and reduction peak potential). The HDTMPA just incurred a minor decrease in the ΔV_p of the electrolyte and a small increment in the I_{pO}/I_{pR} as if it had little effect on the reversibility of the V(IV)/V(V) redox pair. In addition, it had a small effect on the oxidation peak current and reduction peak current as well as the overall peak shape of the curve, indicating that its effect on the electrode reaction kinetics of the electrolyte was not that significant either [24]. The addition of HEDP increased the ΔV_p of the electrolyte while decreasing I_{pO}/I_{pR} significantly. The addition of HEDP, DTPMP, ATMPA, and EDTMPS had a greater effect on the reversibility of V(V)/V(IV) redox pairs, and the HEDP significantly increased the peak oxidation current and peak reduction current, indicating that it might enhance the electrode reaction kinetics of the electrolyte. The main reason for the improvement in electrode reaction kinetics by HEDP might be attributed to the fact that the –OH could complex with V(IV)/V(V) ions, which provide more available –OH to the stable electrode reaction of V(IV)/V(V) for ion exchange on the electrode surface, thus resulting in a higher oxidation peak current and reduction in peak current [25]. Among all additives, ATMPA best enhanced the electrode reaction kinetics of electrolyte, but it caused a decrease in electrolyte thermal stability. HDTMPA, EDTMPS, and DTPMP probably affect the cyclic reversibility performance and electrode reaction kinetics of the electrolyte by, firstly, phosphate and, secondly, according to calculations, the C atoms adjacent to N atoms have a high positive charge density, counteracting the strong electron affinity of N atoms [26], and the positively charged C atoms activated by the N atoms can work as an active site, affecting electron distribution, thereby improving the electrochemical performance. In addition, EDTMPS benefits from its Na^+ ions, increasing the number of ionizable cations in the solution, which enhances the electrode reaction kinetics [12]. As for HDTMPA, its large groups slightly hinder the ion exchange and charge transfer on the electrode surface owing to the steric hindrance, which is obviously unfavorable [18]. When ATMPA and HDTMPA are used in combination, ΔVp of the electrolyte declined compared with the blank sample and I_{pO}/I_{pR} displayed a small change. The combination of ATMPA and HEDP also showed the same performance. At the same time, compared with the CV curve of the blank sample, the peak current and the peak area of these two complex schemes are larger, which indicates that the electrolytes affected by these two schemes have better electrochemical performance.

A series of CV curves on graphite electrode for the blank electrolyte and the electrolyte with different additives at different scan rates are depicted in Figure 2, which further reveals the effect of additives on the electrode reaction kinetics. The peak potential of the anode and cathode varies gradually with the scan rate, presenting the typical characteristics of a quasi-reversible single-electron process [27]. The diffusion coefficient of the quasi-reversible reaction (D) is between that of the reversible reaction (D_1) and irreversible reaction (D_2) [28]. As for the reversible and irreversible one-step single-electron reactions, their peak current (i_p) can be represented as follows [29]:

$$i_p = 0.4463(F^3/RT)^{1/2} C A n^{3/2} v^{1/2} {D_1}^{1/2} \text{ (Reversible reaction)} \tag{6}$$

$$i_p = 0.4958(F^3/RT)^{1/2} C A \alpha^{1/2} n^{3/2} v^{1/2} {D_2}^{1/2} \text{ (Irreversible reaction)} \tag{7}$$

where R is the universal gas constant; F is the Faraday constant; T is the Kelvin temperature; n is the amount of substance of transferred electrons during electrode reaction; A is the surface area of working electrode; C is the bulk concentration of primary reactant; v is the scanning rate; α is the transfer coefficient for an irreversible reaction; and D_1 and D_2 are the diffusion coefficients of reversible and irreversible reactions, respectively.

For a single-electron reaction at room temperature, Equations (6) and (7) can be simplified as follows:

$$i_p = 2.69 \times 10^5 ACD_1^{1/2}v^{1/2} \quad (8)$$

$$i_p = 2.99 \times 10^5 \alpha^{1/2} ACD_2^{1/2}v^{1/2} \quad (9)$$

According to the present experimental conditions, Equations (10) and (11) can be further derived from Equation (8) and Equation (9), respectively:

$$i_p/A = 538D_1v^{1/2} = v^{1/2}k \quad (10)$$

$$i_p/A = 598D_2^{1/2}v^{1/2} = v^{1/2}k \quad (11)$$

Equations (10) and (11) indicate that the current density (i_p/A) is linearly related to the square root of scan rate ($v^{1/2}$) and k denotes the slope of this line, illustrated in Figure 3, thus D_1 and D_2 were calculated. The values of k for the blank electrolyte and those with 0.5 wt% of different additives are concluded in Table 5.

$$D_1 = 3.45 \times 10^{-6}k^2 \quad (12)$$

$$D_2 = 2.80 \times 10^{-6}k^2 \quad (13)$$

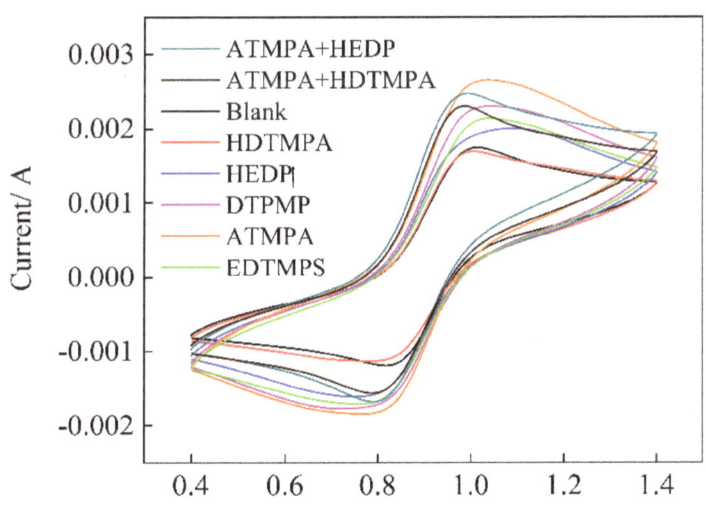

Figure 1. CV curves of the electrolyte (2.0 M V(IV)/3.0 M H_2SO_4) with additives (0.5 wt%) and the blank one at a scan rate of 20 mV s^{-1} at room temperature.

Table 4. ΔV_p and I_{pO}/I_{pR} of the electrolyte (2.0 M V(IV)/3.0 M H_2SO_4) with additives (0.5 wt%) and the blank one on the graphite electrode.

Additives	Blank	HDTMPA	HEDP	DTPMP	ATMPA	EDTMPS	ATMPA + HEDP	ATMPA + HDTMPA
$\Delta V_p/V$	0.241	0.231	0.340	0.320	0.275	0.285	0.208	0.200
I_{pO}/I_{pR}	1.462	1.496	1.242	1.299	1.432	1.246	1.470	1.474

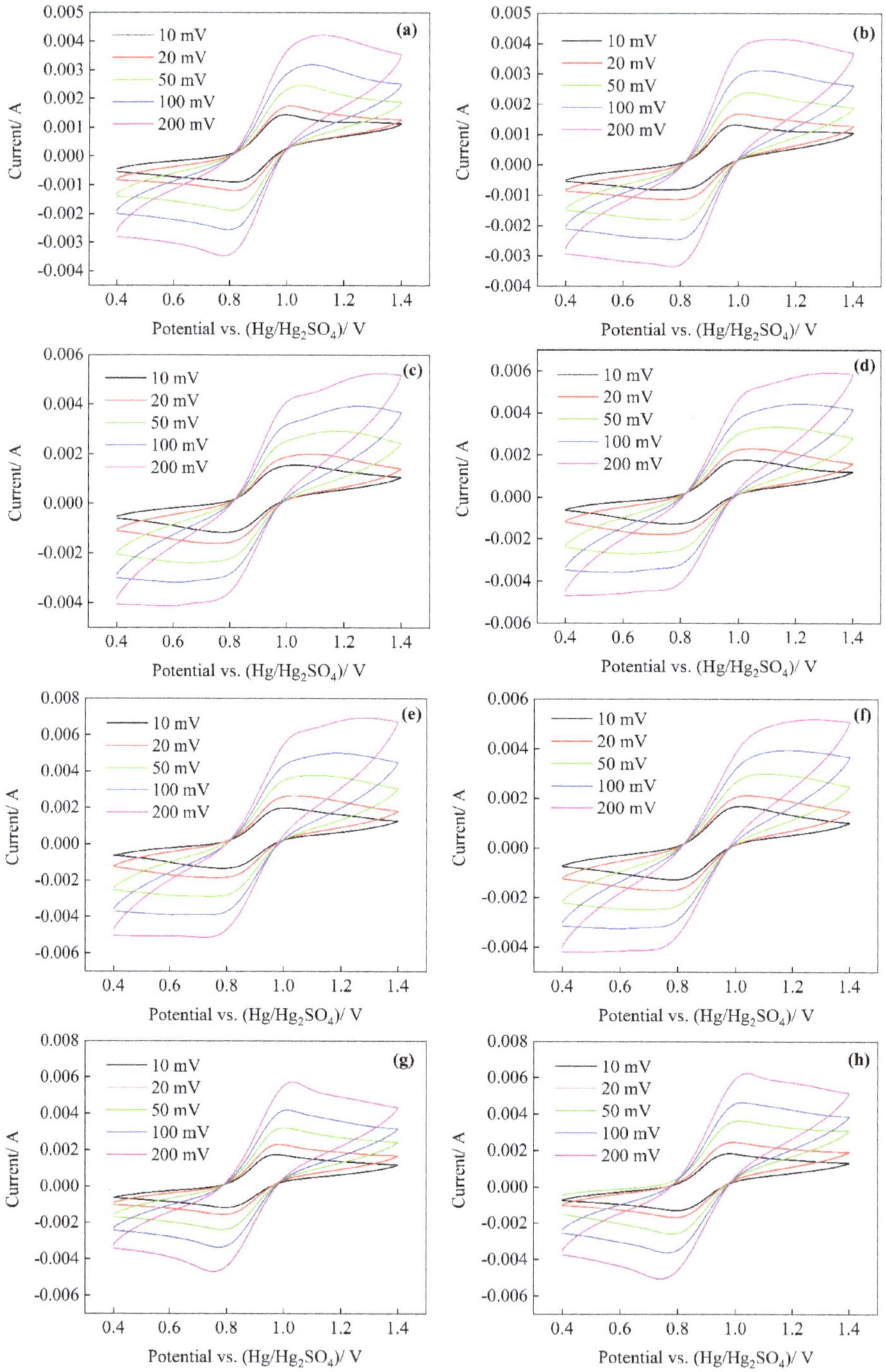

Figure 2. CV curves of the electrolyte (2.0 M V(IV)/3.0 M H_2SO_4) with/without additives ((**a**) blank sample; (**b**) HDTMPA; (**c**) HEDP; (**d**) DTPMP; (**e**) ATMPA; (**f**) EDTMPS; (**g**) ATMPA + HDTMPA; (**h**) ATMPA + HEDP) on the graphite electrode at different scan rates at room temperature.

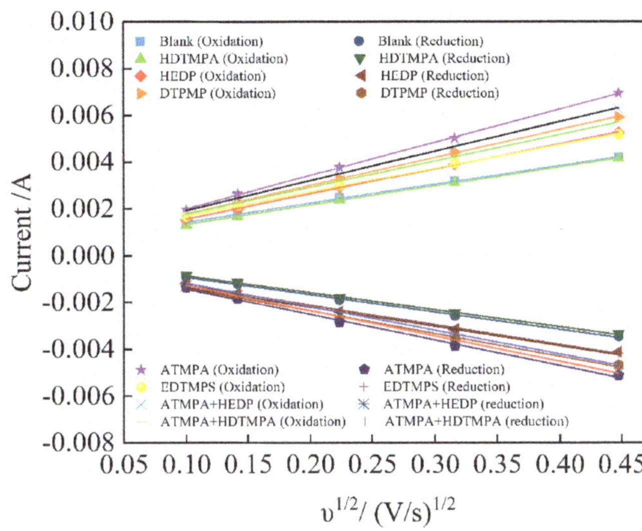

Figure 3. Linear plot of i_p versus $v^{1/2}$.

Table 5. Diffusion coefficients (D_1 and D_2) of V(IV) species at 25 °C.

Additives	Diffusion Coefficient D_1 and D_2 of V(IV) Species ($cm^2\ s^{-1}$)		
	D_1	D_2	Error (%)
Blank	2.20×10^{-7}	1.88×10^{-7}	1.9
HDTMPA	2.26×10^{-7}	1.84×10^{-7}	1.3
HEDP	3.89×10^{-7}	3.15×10^{-7}	1.5
DTPMP	4.86×10^{-7}	3.94×10^{-7}	1.3
ATMPA	6.80×10^{-7}	5.52×10^{-7}	1.3
EDTMPS	3.48×10^{-7}	2.83×10^{-7}	1.6
ATMPA + HDTMPA	4.34×10^{-7}	3.53×10^{-7}	1.8
ATMPA + HEDP	5.43×10^{-7}	4.40×10^{-7}	1.7

In fact, the diffusion coefficient of the electrolyte is between D_1 and D_2 for the quasi-reversible process. Under the present experimental conditions, the diffusion coefficient is 1.88–2.20 $\times$ 10^{-7} for the blank electrolyte and 1.84–2.26 $\times$ 10^{-7} for the sample with HDTMPA, and the result of the latter is close to that of the blank electrolyte. When it comes to the rest, their diffusion coefficients are 3.15–3.89 $\times$ 10^{-7} (HEDP), 3.94–4.86 $\times$ 10^{-7} (DTPMP), 5.52–6.80 $\times$ 10^{-7} (ATMPA), and 2.83–3.48 $\times$ 10^{-7} (EDTMPS), respectively. These increased diffusion coefficients of the electrolyte with additives indicate that the additives (except HDTMPA) can effectively improve the diffusion of vanadium species at the electrodes and enhance the mass transfer and charge transfer of the V(V)/V(IV) redox pair, thus increasing the corresponding reactivity. Compared with the blank sample, the combination of ATMPA + HDTMPA and ATMPA + HEDP also have a larger diffusion coefficient, showing that the compounding scheme has played a positive role in mass transfer and charge transfer in the electrolyte.

3.3. Steady-State Polarization Test

The steady-state polarization curve of the V(IV) electrolyte allows the determination of the polarization resistance, the exchange current density, and the electrochemical reaction rate constant.

In the relatively-low-overvoltage region, the overvoltage and current density are approximately linearly correlated [6]. These parameters can be calculated by Equation (14).

$$R_{ct} = \frac{\eta}{i},\ i_0 = \frac{RT}{nFR_{ct}},\ k_0 = \frac{i_0}{nFC_0} \tag{14}$$

where R_{ct}, i_0, and k_0 refer to the charge-transfer resistance, exchange current density, and rate constant, respectively; R, T, n, F, and C_0 are the universal gas constant, Kelvin temperature, amount of transferred electrons in the electrode reaction, Faraday constant, and solution concentration, respectively [30].

The steady-state polarization curves of the 2.0 M $VOSO_4$/3.0 M H_2SO_4 electrolyte with different additives on graphite electrode are demonstrated in Figure 4, and the corresponding parameters derived from Equation (14) are listed in Table 6. One can see that the charge transfer resistance of electrolyte samples with additives decreased and the electrochemical reaction rate constant and the exchange current density increased compared with the blank sample. The charge transfer resistance of the electrolyte with EDTMPS and DTPMP, for example, decreased from 12.40 Ω cm^2 (blank sample) to 8.15 Ω cm^2 and 8.84 Ω cm^2, respectively, at 25 °C, while the exchange current density of these two samples increased from 2.07 mA cm^2 (blank sample) to 3.15 mA cm^2 and 2.91 mA cm^2, respectively. The corresponding electrochemical reaction rate constant increased from 1.07×10^5 cm s^{-1} (blank sample) to 1.63×10^5 cm s^{-1} and 1.51×10^5 cm s^{-1}, respectively, at 25 °C. The other additives also accelerated the chemical reaction process of V(IV) on the electrode surface to varying degrees. The rest of the selected additives also accelerated the kinetics process of V(IV) species on the electrode surface to a certain level. Compared with the blank sample, the combination of ATMPA + HDTMPA and ATMPA + HEDP also had lower charge transfer resistance and higher exchange current density and electrochemical reaction rate constant, which was consistent with the CV tests.

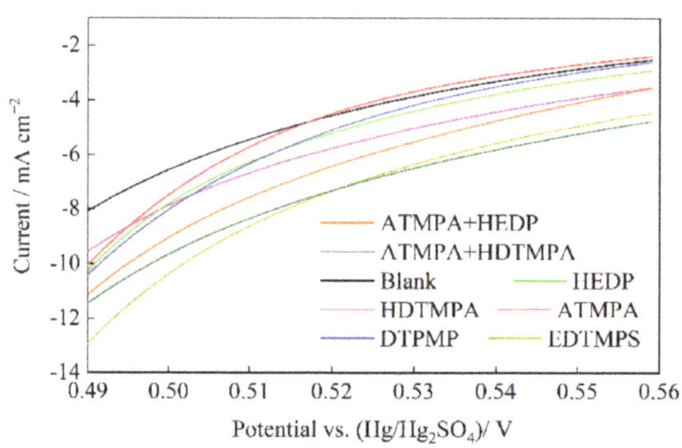

Figure 4. Steady-state polarization curves for the 2.0 M V(IV)/3.0 M H_2SO_4 blank electrolyte and those with 0.5 wt% additives on graphite electrode at a scan rate of 1 mV s^{-1}.

Table 6. Kinetic parameters for the 2.0 M $VOSO_4$/3.0 M H_2SO_4 experimental electrolyte with different additives on the graphite electrode.

Additives	R_{ct} (Ω cm^2)	i_0 (mA cm^{-2})	k_0 (10^{-5} cm s^{-1})
Blank	12.40	2.07	1.07
HDTMPA	11.48	2.24	1.16
HEDP	9.40	2.73	1.42
DTPMP	8.84	2.91	1.51
ATMPA	9.01	2.85	1.48
EDTMPS	8.15	3.15	1.63
ATMPA + HDTMPA	10.35	2.48	1.29
ATMPA + HEDP	9.11	2.82	1.46

3.4. Electrochemical Impedance Spectroscopy Test

For the further analysis of the electrode reaction diffusion kinetics of vanadium and the charge transfer and mass transfer of the V(IV)/V(V) redox pair, Nyquist plots of the eight (including two compound schemes) V(IV) electrolyte samples at room temperature were recorded by electrochemical impedance spectroscopy. Figure 5 shows that each plot consists of a semicircle in the high-frequency region and a diagonal line in the low-frequency region, indicating that the redox reaction of the V(IV)/V(V) pair is controlled by both high-frequency charge transfer and low-frequency diffusion. The radius of the semicircle corresponds to the charge transfer resistance and the linear part relates to the diffusion of vanadium species on the electrode [31]. The equivalent circuits of these Nyquist plots were fitted, and the corresponding parameters were obtained using ZView software, which are listed in Table 7. In the equivalent circuit, R_1 is the resistance consisting of the solution resistance, electrode resistance, and contact resistance, and R_2 and W_0 represent the charge transfer resistance and diffusion impedance, respectively. The constant phase element (CPE) represents the bilayer capacitance at the electrode–electrolyte interface.

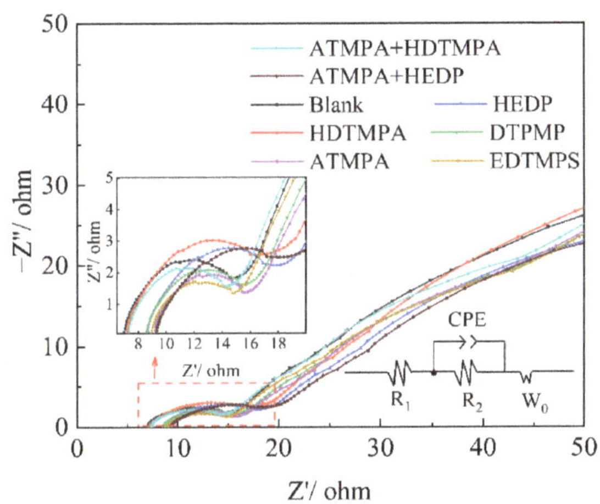

Figure 5. Nyquist plots of the 2.0 M V(IV) electrolyte on the graphite plate and the corresponding equivalent circuit.

Table 7. Model parameters of equivalent circuits corresponding to Nyquist plots.

Additives	$R_1/\Omega\ cm^2$	CPE/S s^n cm^{-2}		$R_2/\Omega\ cm^2$	W_0, $Y_{0,2}$/S s^{-5} cm^{-2}
		$Y_{0,1}$	n		
Blank	0.218	5.33×10^{-3}	0.775	0.168	0.409
HDTMPA	0.223	9.11×10^{-3}	0.727	0.215	0.411
HEDP	0.268	10.85×10^{-3}	0.707	0.205	0.397
DTPMP	0.268	8.18×10^{-3}	0.737	0.152	0.397
ATMPA	0.286	8.70×10^{-3}	0.727	0.154	0.408
EDTMPS	0.279	5.61×10^{-3}	0.769	0.116	0.390
ATMPA + HDTMPA	0.276	16.87×10^{-3}	0.621	0.206	0.283
ATMPA + HEDP	0.282	18.63×10^{-3}	0.552	0.186	0.288

The additives slightly increased the contact resistance of the electrolyte, and all additives except HDTMPA and HEDP decreased the charge transfer resistance of the solution. The decrease in charge transfer resistance implies a faster charge transfer process, which is consistent with the above study. All of the additives except HDTMPA decrease the diffusion resistance of the electrolyte, which facilitates the diffusion process on the electrode surface and enhances the electrochemical reaction kinetics. In addition, all additives lead to an increase in the CPE parameter $Y_{0,1}$, indicating an enhanced bilayer capacitance at the electrode–electrolyte interface. n represents the index of CPE, ranging from 0 to 1. The larger the n, the higher the capacitive property and the lower the resistive property of CPE.

Compared with the samples using HDTMPA and HEDP additives alone, the charge transfer resistance of the two compounding schemes (ATMPA + HDTMPA and ATMPA + HEDP) is reduced, which accelerates the charge transfer in the solution, indicating that the compounding schemes (HDTMPA + ATMPA and HEDP + ATMPA) improve the electrochemical performance of the electrolyte.

After the above experimental investigation, it was found that the different additives selected could have a positive effect on the positive electrolyte of VRFB in terms of thermal stability and electrochemical performance. In future studies, it is expected that these additives may work better if used in combination, rendering these additives potential and promising for development.

4. Conclusions

In this study, five phosphate-containing scale inhibitions, including HDTMPA, HEDP, DTPMP, ATMPA, and EDTMPS, were employed as additives for the VRFB positive electrolyte and their effects on the electrolyte thermal stability and electrochemical performance were investigated. HDTMPA has a great positive effect on the thermal stability of the electrolyte, extending the time at which the electrolyte begins to precipitate to 15 days, while it has a lesser effect on the electrochemical performance; HEDP, DTPMP, and EDTMPS effectively improve the thermal stability of electrolyte and simultaneously accelerate its electrochemical reaction kinetics, but they have a greater effect on the cyclic reversibility of electrolyte, among which HEDP extended the time at which the electrolyte began to precipitate to 30 days, and the charge transfer resistance of the electrolyte decreased to 9.40 Ω cm^2; and the addition of ATMPA greatly improves the mass transfer kinetics of the electrolyte, increasing the electrolyte diffusion coefficient on the electrode surface to 6.80×10^{-7}, simultaneously causing poor thermal stability of the electrolyte. In addition, the compounding effect of additives was studied. It was verified that the compounding combination of ATMPA + HDTMPA and ATMPA + HEDP had a good effect on the electrochemical performance of the electrolyte. In conclusion, most of the additives selected in this study have positive effects on the positive electrolyte of VRFB in terms of thermal stability and electrochemical performance; the diffusion coefficients of the electrolyte were 4.34×10^{-7} and 5.43×10^{-7}, respectively. The necessity of combined utilization of these additives should be recognized, which may work better and has great potential for future development.

Author Contributions: Conceptualization, X.Z. and L.W.; Data curation, X.Z.; Formal analysis, F.M.; Project administration, Z.Z. and D.C.; Resources, L.W.; Writing—original draft, X.Z.; Writing—review and editing, X.Z., F.M., L.S. and Z.Z. All authors have read and agreed to the published version of the manuscript.

Funding: National Key R&D Program of China (2018YFC1900500), the National Natural Science Foundation of China (51804289), General project of Beijing Municipal Natural Science Foundation (2202053), Strategic Priority Research Program of the Chinese Academy of Sciences (XDC04010100), Special Project for Transformation of Major Technological Achievements in Hebei Province (19044012Z), and Province Key R&D Program of Hebei (20374105D).

Data Availability Statement: Not applicable.

Conflicts of Interest: The authors declare no conflict of interest.

References

1. Hsieh, C.; Tsai, P.; Hsu, N.; Chen, Y. Effect of compression ratio of graphite felts on the performance of an all-vanadium redox flow battery. *Energies* **2019**, *12*, 313. [CrossRef]
2. Chen, L.; Liu, T.; Zhang, Y.; Liu, H.; Ding, M.; Pan, D. Mitigating Capacity Decay by Adding Carbohydrate in the Negative Electrolyte of Vanadium Redox Flow Battery. *Energies* **2022**, *15*, 2454. [CrossRef]
3. Aramendia, I.; Fernandez-Gamiz, U.; Martinez-San-Vicente, A.; Zulueta, E.; Lopez-Guede, J.M. Vanadium redox flow batteries: A review oriented to fluid-dynamic optimization. *Energies* **2020**, *14*, 176. [CrossRef]
4. Leung, P.; Mohamed, M.; Shah, A.; Xu, Q.; Conde-Duran, M. A mixed acid based vanadium–cerium redox flow battery with a zero-gap serpentine architecture. *J. Power Sources* **2015**, *274*, 651–658. [CrossRef]
5. Behi, B.; Baniasadi, A.; Arefi, A.; Gorjy, A.; Jennings, P.; Pivrikas, A. Cost–benefit analysis of a virtual power plant including solar PV, flow battery, heat pump, and demand management: A western australian case study. *Energies* **2020**, *13*, 2614. [CrossRef]
6. Rahman, F.; Skyllas-Kazacos, M. Vanadium redox battery: Positive half-cell electrolyte studies. *J. Power Sources* **2009**, *189*, 1212–1219. [CrossRef]
7. Cao, L.; Skyllas-Kazacos, M.; Menictas, C.; Noack, J. A review of electrolyte additives and impurities in vanadium redox flow batteries. *J. Energy Chem.* **2018**, *27*, 1269–1291. [CrossRef]
8. Kazacos, M.; Skyllas-Kazacos, M. High Energy Density Vanadium Electrolyte Solutions, Methods of Preparation Thereof and All-Vanadium Redox Cells and Batteries Containing High Energy Vanadium Electrolyte Solutions. W.O. Patent 9,635,239 A1, 7 November 1996.
9. Skyllas-Kazacos, M.; Kazacos, M. High Energy Density Vanadium Electrolyte Solutions, Methods of Preparation Thereof and All-Vanadium Redox Cells and Batteries Containing High Energy Vanadium Electrolyte Solutions. U.S. Patent 7,078,123 B2, 18 July 2006. Available online: https://patentimages.storage.googleapis.com/eb/d1/4a/83586dc481f159/US7078123.pdf (accessed on 15 September 2022).
10. Roznyatovskaya, N.V.; Roznyatovsky, V.A.; Höhne, C.-C.; Fühl, M.; Gerber, T.; Küttinger, M.; Noack, J.; Fischer, P.; Pinkwart, K.; Tübke, J. The role of phosphate additive in stabilization of sulphuric-acid-based vanadium (V) electrolyte for all-vanadium redox-flow batteries. *J. Power Sources* **2017**, *363*, 234–243. [CrossRef]
11. Park, S.K.; Shim, J.; Yang, J.H.; Jin, C.S.; Lee, B.S.; Lee, Y.S.; Shin, K.H.; Jeon, J.D. Effect of inorganic additive sodium pyrophosphate tetrabasic on positive electrolytes for a vanadium redox flow battery. *Electrochim. Acta* **2014**, *121*, 321–327. [CrossRef]
12. Zhang, J.; Li, L.; Nie, Z.; Chen, B.; Vijayakumar, M.; Kim, S.; Wang, W.; Schwenzer, B.; Liu, J.; Yang, Z. Effects of additives on the stability of electrolytes for all-vanadium redox flow batteries. *J. Appl. Electrochem.* **2011**, *41*, 1215–1221. [CrossRef]
13. Li, S.; Huang, K.; Liu, S.; Fang, D.; Wu, X.; Lu, D.; Wu, T. Effect of organic additives on positive electrolyte for vanadium redox battery. *Electrochim. Acta* **2011**, *56*, 5483–5487. [CrossRef]
14. Zhang, Y.; Xi, J.; Liu, L.; Wu, Z. Boosting the thermal stability of electrolytes in vanadium redox flow batteries via 1-hydroxyethane-1, 1-diphosphonic acid. *J. Appl. Electrochem.* **2020**, *50*, 255–264. [CrossRef]
15. Roe, S.; Menictas, C.; Skyllas-Kazacos, M. A high energy density vanadium redox flow battery with 3 M vanadium electrolyte. *J. Electrochem. Soc.* **2015**, *163*, A5023. [CrossRef]
16. Rahman, F.; Skyllas-Kazacos, M. Evaluation of additive formulations to inhibit precipitation of positive electrolyte in vanadium battery. *J. Power Sources* **2017**, *340*, 139–149. [CrossRef]
17. Kausar, N.; Mousa, A.; Skyllas-Kazacos, M. The effect of additives on the high-temperature stability of the vanadium redox flow battery positive electrolytes. *ChemElectroChem* **2016**, *3*, 276–282. [CrossRef]
18. Xuewen, W.; Suqin, L.; Kelong, H. Characteristics of CTAB as electrolyte additive for vanadium redox flow battery. *J. Inorg. Mater.* **2010**, *25*, 641–646.
19. Wu, X.; Liu, S.; Wang, N.; Peng, S.; He, Z. Influence of organic additives on electrochemical properties of the positive electrolyte for all-vanadium redox flow battery. *Electrochim. Acta* **2012**, *78*, 475–482. [CrossRef]
20. Yang, Z.; Zhang, J.; Kintner-Meyer, M.C.; Lu, X.; Choi, D.; Lemmon, J.P.; Liu, J. Electrochemical energy storage for green grid. *Chem. Rev.* **2011**, *111*, 3577–3613. [CrossRef] [PubMed]
21. Skyllas-Kazacos, M. Stabilized Vanadium Electrolyte Solutions for All-Vanadium Redox Cells and Batteries. U.S. Patent 6,562,514, 13 May 2003.
22. Sun, H.; Yang, B.; Zhu, Z.; Yin, W.; Sheng, Q.; Hou, Y.; Yao, J. New insights into selective-depression mechanism of novel depressant EDTMPS on magnesite and quartz surfaces: Adsorption mechanism, DFT calculations, and adsorption model. *Miner. Eng.* **2021**, *160*, 106660. [CrossRef]
23. Hu, Y.; Xu, Y.; Xie, M.; Huang, M.; Chen, G. Characterization of scalants and strategies for scaling mitigation in membrane distillation of alkaline concentrated circulating cooling water. *Desalination* **2022**, *527*, 115534. [CrossRef]
24. Peng, S.; Wang, N.-F.; Wu, X.-J.; Liu, S.-Q.; Fang, D.; Liu, Y.-N.; Huang, K.-L. Vanadium species in CH_3SO_3H and H_2SO_4 mixed acid as the supporting electrolyte for vanadium redox flow battery. *Int. J. Electrochem. Sci.* **2012**, *7*, 643–649.
25. Wang, G.; Chen, J.; Xu, Y.; Yan, B.; Wang, X.; Zhu, X.; Zhang, Y.; Liu, X.; Wang, R. Several ionic organic compounds as positive electrolyte additives for a vanadium redox flow battery. *RSC Adv.* **2014**, *4*, 63025. [CrossRef]
26. Wu, T.; Huang, K.; Liu, S.; Zhuang, S.; Fang, D.; Li, S.; Lu, D.; Su, A. Hydrothermal ammoniated treatment of PAN-graphite felt for vanadium redox flow battery. *J. Solid State Electrochem.* **2012**, *16*, 579–585. [CrossRef]

27. Skyllas-Kazacos, M.; Menictas, C.; Kazacos, M. Thermal stability of concentrated V (V) electrolytes in the vanadium redox cell. *J. Electrochem. Soc.* **1996**, *143*, L86. [CrossRef]
28. Huang, F.; Zhao, Q.; Luo, C.; Wang, G.; Yan, K.; Luo, D. Influence of Cr^{3+} concentration on the electrochemical behavior of the anolyte for vanadium redox flow batteries. *Chin. Sci. Bull.* **2012**, *57*, 4237–4243. [CrossRef]
29. Bard, A.J.; Faulkner, L.R.; White, H.S. *Electrochemical Methods: Fundamentals and Applications*; John Wiley & Sons: New York, NY, USA, 2022.
30. Xue, F.; Wang, Y.; Wang, W.; Wang, X. Investigation on the electrode process of the Mn (II)/Mn (III) couple in redox flow battery. *Electrochim. Acta* **2008**, *53*, 6636–6642. [CrossRef]
31. Gao, C.; Wang, N.; Peng, S.; Liu, S.; Lei, Y.; Liang, X.; Zeng, S.; Zi, H. Influence of Fenton's reagent treatment on electrochemical properties of graphite felt for all vanadium redox flow battery. *Electrochim. Acta* **2013**, *88*, 193–202. [CrossRef]

applied sciences

MDPI

Review

Applications of Artificial Intelligence to Photovoltaic Systems: A Review

Héctor Felipe Mateo Romero [1,*], Miguel Ángel González Rebollo [1], Valentín Cardeñoso-Payo [2], Victor Alonso Gómez [3], Alberto Redondo Plaza [4], Ranganai Tawanda Moyo [5] and Luis Hernández-Callejo [4,*]

1 Departamento Fisica de la Materia Condensada, Universidad de Valladolid, 47011 Valladolid, Spain
2 Departamento Informatica, Universidad de Valladolid, 47011 Valladolid, Spain
3 Departamento de Física, Universidad de Valladolid, 47002 Valladolid, Spain
4 Departamento Ingeniería Agrícola y Forestal, Universidad de Valladolid, 47002 Valladolid, Spain
5 Department of Mechanical Engineering, Durban University of Technology, P.O. Box 1334, Durban 4000, South Africa
* Correspondence: hectorfelipe.mateo@uva.es (H.F.M.R.); luis.hernandez.callejo@uva.es (L.H.-C.); Tel.: +34-975129418 (L.H.-C.)

Abstract: This article analyzes the relationship between artificial intelligence (AI) and photovoltaic (PV) systems. Solar energy is one of the most important renewable energies, and the investment of businesses and governments is increasing every year. AI is used to solve the most important problems found in PV systems, such as the tracking of the Max Power Point of the PV modules, the forecasting of the energy produced by the PV system, the estimation of the parameters of the equivalent model of PV modules or the detection of faults found in PV modules or cells. AI techniques perform better than classical approaches, even though they have some limitations such as the amount of data and the high computation times needed for performing the training. Research is still being conducted in order to solve these problems and find techniques with better performance. This article analyzes the most relevant scientific works that use artificial intelligence to deal with the key PV problems by searching terms related with artificial intelligence and photovoltaic systems in the most important academic research databases. The number of publications shows that this field is of great interest to researchers. The findings also show that these kinds of algorithms really have helped to solve these issues or to improve the previous solutions in terms of efficiency or accuracy.

Keywords: PV; artificial intelligence; MPPT; forecasting; parameter estimation; faults detection

Citation: Mateo Romero, H.F.; González Rebollo, M.A.; Cardeñoso-Payo, V.; Alonso Gómez V.; Redondo Plaza A.; Moyo, R.T.; Hernandez Callejo, L. Applications of Artificial Intelligence to Photovoltaic Systems: A Review. *Appl. Sci.* **2022**, *12*, 10056. https://doi.org/10.3390/app121910056

Academic Editor: Fabrice Goubard

Received: 7 September 2022
Accepted: 3 October 2022
Published: 6 October 2022

1. Introduction

Energy is essential in our society, being the motor of almost every sector. Fossil-fuels are historically the most important source of energy, representing 80.2% in 2019 [1] These kinds of energies have different problems; one of them is their scarcity, since they are limited resources that have been exploited for a long time. Another critical problem is the pollution caused by the burning and extraction of these fuels, which is hazardous for people [2] and the environment [3]. To solve these problems, other energy sources can be used. These alternative energies, renewable energies, have two main benefits. First of all, they are based on unlimited resources that will not run out, even with extensive exploitation. Their exploitation is also nonpolluting. Investment in these energies has been rising in the last years, even with a crisis such as the COVID-19 pandemic [1,4].

One of the most important green energies is solar energy. This energy is composed of solar, thermal and photovoltaic (PV). The latter has been found to be more useful and profitable for industry production [5,6] and has been growing steadily in recent years. As we can see in Figure 1, the share of PV systems is increasing, and it is expected to be one of the prime energy sources in the next years [7].

PV energy is produced by photovoltaic modules. Each module is composed of different sub-units, called solar cells, which absorb the energy emitted by the sun [8]. PV panels are usually connected in series to each other, this is known as a PV array. Each PV array is connected to a power inverter to control the production and check the performance of the array [9]. PV farms (also known as PV plants) are usually composed of many PV arrays. The maintenance of these factories is extremely complex. The production of the modules depends on different conditions, this makes mechanisms of control to optimize the production necessary. Solar modules are also vulnerable to physical defects, which can reduce or even nullify the production of one cell, or even the whole modules in the worst cases. This is usually dealt with by human labor, checking each module in a certain period of time.

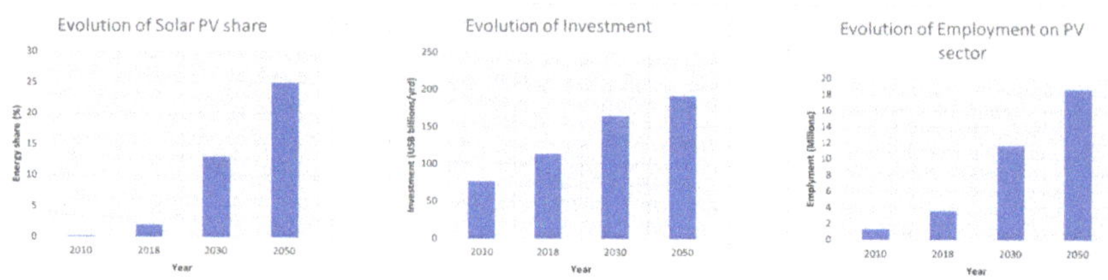

(a) (b) (c)

Figure 1. Evolution of the importance of PV sector (adapted from [7]). (**a**) Evolution of the energy share of PV systems; (**b**) evolution of the investment in PV energy; (**c**) evolution of employment in the PV sector.

Solutions to maintenance problems in PV systems have been traditionally circumscribed either to simplistic automatic supervision approaches [10] or costly direct human supervision.

In the past recent years, however, artificial intelligence (AI)-based approaches have emerged. AI techniques are being applied in almost every research field or industry to improve services or solve problems which are impossible for traditional methods [11].

These techniques can also be applied to solve the problems of PV systems. In this review, we analyze how AI is being applied to the PV sector. PV systems face different problems during installation but also during exploitation, since PV modules are vulnerable to the climate conditions' unpredictable events. An analysis of the problems found in PV plants can be found in [12,13]; also, the maintenance of the modules is key in order to secure maximum production and to improve the security of the installations [14]. In order to reduce the scope of this review, only the four most critical problems related to energy optimization and maintenance are considered:

- **Max Power Point Tracking:** Vital for optimizing the production.
- **Output Power Forecasting:** Critical for predicting possible problems in production related to climate conditions.
- **Parameter Estimation:** Extremely important for optimizing the production of PV modules.
- **Defect Detection:** Important for finding bad-performing modules or faults that can be fatal for overall system performance and security.

The most important problems and the technologies that have been used for dealing with each one of these problems are addressed. The final objective of this review is to analyze the most important techniques used and how they have improved the solutions of the problems in order to have a clear understanding of the state of the art in the area.

The paper is structured as follows: First, an explanation of the problem is described in Section 2, in order to provide more information to readers who are not familiarized with this topic. After that, the artificial intelligence techniques used of each of the problems are explained. The Maximum Power Point Tracking is discussed in Section 3.2; Section 3.3 is about the forecasting; Section 3.4 presents the parameter estimation, and defect detection problems are discussed in Section 3.5. Finally, an analysis of the different problems is performed in Section 4, identifying the tendencies and flaws in the research of each one of them.

2. Relevant PV Problems

As we explained before, PV installations have to face a large amount of problems. The most important ones are related to optimizing energy production, since it is the final objective of an energy installation. They are also related to the maintenance of equipment, mostly the solar modules. In this section, four problems related to this issue are explained in order to give the reader a basic context of the different problems.

2.1. Maximum Power Point Tracking (MPPT)

PV cells have a complex relationship between their environment and the power they can produce. Along the IV curve (Figure 2) of solar cells there is a point where the power will be maximized, this is called the Maximum Power Point (MPP). This point usually changes depending on conditions such as irradiation, temperature or the state of the PV cell. These conditions can change the shape of the curve, making the problem nonlinear and time-varying due to the changes produced by the atmospheric and load conditions.

Another problem is that it is not possible to directly obtain the IV curve of a single PV cell. The IV curves are usually taken from one single module or even from a PV array. The measured curves are more complex than the IV curve of a cell. The more complex a curve is, the harder it is to track the MPP.

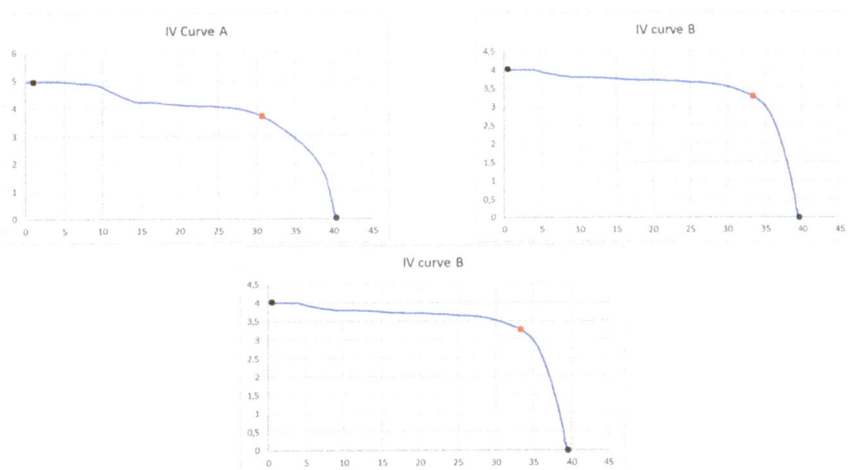

Figure 2. Different IV curves. Green circle: Max Power Point; Red Star: Open-Circuit Voltage (V_{OC}); Yellow Square: Short-Circuit Current (I_{SC}).

The algorithms to solve this problem can be classified according to different criteria; one of the most important ones is according to the number of variables used for measuring the tension. Another interesting approach is to classify the method according to the control strategy used. A brief explanation of the most important traditional methods can be found in Table 1, and more information can be found in works such as [15,16].

Table 1. Traditional Methods for MPPT.

Methods	Features
Perturbation and Observation [15,16]	This algorithm does not depend on previous knowledge, it is the simplest and is widely used due to its balance between efficiency and simplicity. It disturbs the operating point of the system, causing the PV voltage to fluctuate near the MPP voltage.
Incremental Conductance [15,16]	This method is based on the basis that the slope of a PV curve is zero at MPP. The algorithm tracks the MPP by searching for the peak of the PV curves. This algorithm uses the instantaneous conductance I/V and the incremental conductance dI/dV.
Curve Fitting [15,16]	This method implements a mathematical function to describe the output of the generator. The disadvantage of this method is that it requires a high memory capacity and is not optimum for high-speed changes in the irradiance.
Open Circuit [15,16]	This method implements a mathematical function to describe the output of the generator. The disadvantage of this method is that it requires a high memory capacity and is not optimum for high-speed changes in the irradiance. This method is simple, and it uses a single control loop.
Feedback Tension [15,16]	This method can be used with the feedback of the voltage of the panels, which is compared with the tension of constant reference to adjust the word cycle. This system is unable to adapt to changes in irradiance or climate.
Measurement of the Current of the PV Generator [15,16]	This method is based on one variable, the output current of the PV Generator, which is the input current of the generator. The control of the output optimizes the maximum output current.

2.2. *Forecasting Problems*

There are several variants of the forecasting problem which arise in PV: weather forecasting, solar irradiance forecasting and energy production forecasting, which is to estimate the energy production of the system. This is really important to optimize the real-time management of systems that use this kind of energy (smart cities, villages, etc.). This problem has high priority for electric companies because they want a more robust and reliable system to predict the changes in energy loads and demands. Another important aspect is the amount of time that has to be predicted.

- Short-term forecasting is usually from 1 hour to a week ahead and is used for scheduling energy transfer, economic load dispatch and demand response.
- Mid-term forecasting is usually considered between 1 month and 1 year ahead, usually for planning the near-forthcoming power plans and to show the dynamics of the system in that interval.
- Long-term forecasting is considered between 1 year and 10 years. Its function is to plan the generation power plant so as to satisfy future requirements and cost efficiency.

Another important factor for forecasting is the number of parameters, the amount of information and data is key when it comes to obtaining a precise forecasting model, but it is also true that sometimes too much data can provide noise or misleading information that can injure performance.

Each kind of forecasting is usually tacked as a different problem, since the amount of data and precision required are highly different. More information about forecasting can be found in [17].

2.3. Estimation of Parameters of Model Circuits

The simulation of PV systems is important to optimize the production of the real systems. It is know that any PV can be modeled and represented by an equivalent electric circuit, whose parameters control the predicted or estimates operation of the PV cell or module. The single-diode circuit presents five unknown parameters [18,19] ($I_{ph}, I_{sd}, R_l, R_{sh}$ and n), and the output current is evaluated as follows:

$$I = I_{ph} - I_{sd} \times \left[exp\left(\frac{q \times (V_L + R_s \times I_L)}{n \times k \times T} \right) - 1 \right] - \frac{V_L + R_S \times I_L}{R_s h}$$

where I_L, I_{ph}, I_d and I_{sh} are the solar cell output current, total current, diode current and shunt current, respectively. R_s represents the series, and R_{sh} denotes the shunt resistances. In addition, V_L means the cell output voltage; n is the ideal factor of diode. k represents the Boltzmann constant, which is set as 1.3806503×10^{23} J/K; q is set as $1.60217646 \times 10^{19}$ C, which is the electron charge, and T means the cell absolute temperature.

The double-diode model presents seven unknown parameters [18,19] ($I_{ph}, I_{sd1}, I_{sd2}, R_l, R_{sh}$, n_1 and n_2), and the output current is evaluated as follows:

$$\begin{aligned} I_L =& I_{ph} - I_{sd1} \times \left[exp\left(\frac{q \times (V_L + R_s \times I_L)}{n_1 \times k \times T} \right) - 1 \right] \\ &- I_{sd2} \times \left[exp\left(\frac{q \times (V_L + R_s \times I_L)}{n_2 \times k \times T} \right) - 1 \right] - \frac{V_L + R_S \times I_L}{R_s h} \end{aligned}$$

where I_{sd1} and I_{sd2} represent the diffusion and saturation currents, while n_1 and n_2 represent the ideal factors of diffusion and recombination diode. The other parameters have the same meaning as the previous equation.

This problem is presented as a optimization problem, where the output to optimize is I_L, and the variables to be found are the unknown parameters.

2.4. Defect Detection

Solar modules are vulnerable to modifications in their surface; therefore, it is required to have a system to find faults. These kinds of faults and defects affect to the production of the module, making it not work at all in the worst cases. The problem is that the majority of faults are not detected with typical cameras (Figure 3), so it is necessary to apply different techniques such as thermography (Figure 3) or electroluminiscence (Figure 3).

The traditional way of finding faults is by performing a manual visual inspection, but the size of the solar farms has made this method almost unmanageable. In order to solve this problem, different techniques have been proposed, most of them using electroluminiscence.

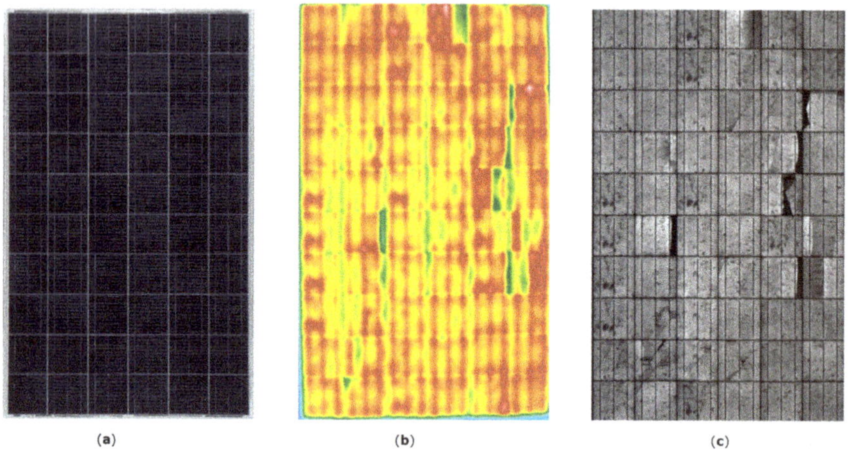

Figure 3. Different techniques for photography modules. (a) Visual spectrum; (b) thermography; (c) electroluminiscence.

3. Artificial Intelligence Applied to PV Systems

3.1. Methodology

With an intention to provide the most relevant and comprehensive review, a proper selection criterion is needed; therefore, different bibliographic databases were searched: Web of Science, Scopus, Google Researcher and Arxiv. With the aim of finding relevant works, a selection was performed searching keywords related with AI and PV systems. The articles with fewer than 8 citations were excluded since they were not considered relevant enough to the state of the art. As a exception to this rule, the articles published in 2021 were selected even if they did not have enough citations. After removing duplicates and nonrelated papers, 250 articles were obtained. The articles tackled different problems found in PV systems, but most of them were focused on four different problems due to their importance:

- Max Power Point Tracking.
- Forecasting of the energy production.
- Estimation of parameters of model circuits.
- Detection of defects and faults in solar modules.

In this section, these different problems are addressed by explaining the contributions of each paper in order to provide a global vision of the state of the art of each problem.

3.2. Maximum Power Point Tracking (MMPT)

The tracking of the Maximum Power Point is vital to optimize the PV systems, and it is probably the most interesting problem for research. Different techniques have been used to solve this problem, as it can be seen in Figure 4. Some classical techniques include Incremental Conductance and Perturb and Observe. Recent trends show that AI techniques are also used to solve this problem. Metaheuristics and Neural Networks were found as the most used techniques after surveying the literature.

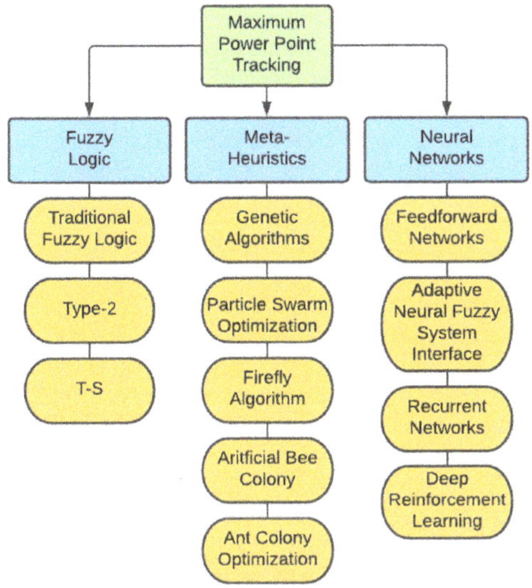

Figure 4. Taxonomy of most used IA method for MPPT.

3.2.1. Fuzzy Logic

Traditional Logic [20] is limited to only two values of truth (True and False), this limits its versatility and makes it difficult to model some systems. Fuzzy Logic (FL) [21] is an extension of Traditional Logic. The main benefit of FL is that it can give a true value between 0 and 1. FL is specialized in addressing uncertainty in inputs and obtaining high performance under rapidly changing conditions, such as atmospheric ones. These kinds of techniques can also be used to aid other systems in order to improve their performance. In this section, the most important techniques related to FL are reviewed.

The first implementation focused on MPPT can be found in [22]. This system uses 7 membership functions for each variable with two input variables: error in the power and the change in error; the output inferred by the fuzzy system is the change in duty cycle that controls the pulse width generation block. The main benefit of FL control is that it does not require changes or variations in its membership functions.

In order to evolve these systems, it was necessary to find a method to modify the parameters of the fuzzy systems. In [23], a Fuzzy Controller that is able to perform online parameter auto-tuning is found. This system used 2 kinds of control. First, a traditional PID control to manage the small deviations and an Adaptive Fuzzy Controller to deal with the larger deviations, since this system is ideal for obtaining rapid responses. A switching function was set to determine the controller to be used. The initial fuzzy controller used triangular-shape functions as membership functions, changing the curve depending on the error. This system reduces the oscillation near the Maximum Power Point, reducing the loss of power.

Another implementation of the adaptive behavior can be found in [24]. This proposal combines Fuzzy Cognitive Networks (FCN) with Fuzzy Logic Control (FLC). FCN is constructed as an extension of another system called Fuzzy Cognitive Maps (FCM) [25]. FCMs are composed of nodes and weighted arcs. Nodes represent the concepts represented, and the arcs represent the causal relationships between them. FCN relies on the knowledge of experts for the description of the nodes and the construction of the graph but does not need an initial estimation of the weights of the arcs. The combination of FLC and FCN makes the system able to track and adapt to any kind of physical variation. A Fuzzy controller needed 12 iterations to reach the same MPP that this algorithm found with 5.

In order to further improve the tracking speed and accuracy of FLC, the classic Open-Circuit Technique [26] is used to find an initial estimation of the MPP voltage. The implemented system showed a good response even under variable atmospheric conditions.

The work in [27] used Type-2 Fuzzy Logic [28], an extension of classical FL [29]. The main feature of Type-2 FL is that membership functions are also fuzzy, and it is used in applications where determining the exact function of a set is difficult [30]. The Type-2 FL functions are three-dimensional, they depend on three different parameters, allowing them to directly model and handle uncertainties. The changes in PV power and voltage are set as the input variables, each one using seven membership functions. The simulation showed that the system tracked MPPT even in irradiance and load variation. Oscillations around MPPT are greatly reduced and are useful for rapidly changing conditions. The overall output energy due to the proposed MPPT method was around 27.7%. The implementation and simulation were performed with MATLAB/Simulink simulation studies [31]. Type-2 Fuzzy Logic is also used for MPPT in [27,32–34].

Another important FL extension used in MPPT is the Takagi–Sugeno Fuzzy Model (T-S FM) [35–37], T-S FM [38] is usually used in approximating complex nonlinear systems [39] and is really important due to the fact that they enable a kind of control called parallel distributed control. The results of these systems present better settling time than classical FL, fewer oscillations and accurate output. The tracking is achieved even for abrupt insulation variations. The implementation and simulation are usually conducted with MATLAB.

One of the most popular trends is to use fuzzy logic as a complement of other techniques. In [40], it is used to tune the PID controller parameter. In [41], the fuzzy behavior of the PSO algorithms improves the system. Fuzzy Logic has also been used along with Neural Networks [42–45]. These works demonstrate that this combination improves the original algorithms, improving the results of the fuzzy systems but reducing the data and time needed to train neural networks.

A summary of the Fuzzy Logic methods applied to MPPT can be seen in Table 2.

Table 2. Fuzzy Methods for MPPT.

Method	Features
FLC [22–24,26,46]	FLC systems provide quick responses to changes and low oscillations near MPPT that reduce the power loss compared with traditional systems. The combination with FCN or the initial estimation of the MPP voltage further improves the results.
Type-2 [27,32–34]	Type-2 FL provides the methods to model and handle uncertainties, boosting the robustness of the system and hence its results.
T-S [35–37]	The parallel distributed control provided by the T-S FL further improves the results of FL systems, having an acceptable settling time, less oscillations and an accurate output.
Combined with other methods [40–45]	Other methods can take advantage of the benefits of FLC systems to improve their results in MPPT.

3.2.2. Metaheuristics

Metaheuristics [47] are algorithmic approaches specialized in solving problems that are not possible to directly find the best solution in a feasible amount of time. They will search the solution space in order to identify the best solution that they can find. In this section, the most important metaheuristics applied to this problem are reviewed. These kinds of algorithms can be used alone or with the aid of other algorithms.

One of the most common classifications for metaheuristics [47] is the differentiation between the algorithms that try to imitate the behavior of animals or things of nature (ants, bees, particles, etc.) and the algorithms which are focused on imitating the basis of genetics.

Genetic Algorithms (GA) [48] are one of the most important genetic metaheuristics because of their capacity to find great solutions to many problems, commonly used as a way to improve the performance of other Artificial Intelligence techniques. In [49], GA are used to optimize the training data for an ANN of 5 hidden nodes. The objective of the GA was to produce a smaller and more effective input dataset. The GA is used to remove the unnecessary data, reducing the error at the end of the network training. This technique can be also used with other techniques, as it is independent of the model that is applied afterward. In [50], a GA is used to determine the number of neurons of an ANN. The number of hidden neurons is one of the key problems of optimizing an ANN since it can improve its performance but can also slow its training. Three different objective functions were tried on the GA. The best neural network was found with 5 hidden nodes. In [51], they are used to optimize the membership functions on an FLC system. The chromosomes decode the shape of these functions; the algorithm will try to minimize the quadratic function based on the error between the desired power and the maximum power delivered by the system. The results show faster convergence and a more stable tracking, which leads to reduced oscillations.

Behavior-oriented metaheuristics can also be used to solve MPPT. Particle Swarm Optimization (PSO) [52] is also used for this. Each particle is initialized with a value in the voltage search interval. Each particle is evaluated by the inverse of the PV power; after that, each particle will have its position and velocity updated. Finally, a new evaluation will be performed until all iterations have been carried out. Ref. [53] showed a better performance than other methods, being able to operate even on rapidly changing atmospheric conditions. The work by [54] proposed the use of an Accelerated PSO. This algorithm combines PSO and Permute and Observe to accelerate MPP searching. It also offers a higher convergence speed and better dynamic response compared with PSO.

Another important metaheuristic is the Firefly Algorithm (FA), developed by [55] for solving multimodal problems, and it has also been used to solve MPPT. The algorithm mimics how fireflies interact with each other using their lights. The attractiveness of the light will depend on its brightness and distance. For solving MPPT, the position of the firefly is related to the PV voltage [56]. This algorithm assures fast convergence, with almost zero steady-state oscillations, providing good tracking speed.

Artificial Bee Colony (ABC) [57] is focused on simulating the behavior of honey bees and was used for MPPT in [58,59]. ABC consists of three different kinds of agents: workers, onlookers and scouts. First of all, worker bees go to the food sources, estimate their utility value and dance back in the hive. Every onlooker observes the dances, chooses one of their sources and goes there. Abandoned food sources are located by scouts and are exchanged with the new food sources found by them. The fitness function is set as the generated power when this algorithm is used for MPPT. The algorithm will continue until the solutions do not change. The main advantage of ABC is that it does not need hyper-parameter tuning as in the case of other metaheuristics such as GA. In MPPT, this algorithm provides quick convergence and accuracy in tracking.

Ant Colony Optimization (ACO) [60] is used in [61] for optimizing neural networks in order to solve the MPPT problem more efficiently. The ACO method was adopted in the learning algorithms for adjusting the weights and biases of the neurons in the process of training. The final network had a single hidden layer with 20 nodes. The results show an improvement over the networks which are not optimized and over other traditional methods.

Other metaheuristics have been used for MPPT as can be seen in other reviews, such as [62,63]. A summary of the commented methods can be found in Table 3.

Table 3. Metaheuristic Algorithms for MPPT.

Algorithm	Features
GA [49–51]	Genetic Algorithms improves the results of other methods such as ANN or FPSO
PSO [53,54]	PSO is used to optimize Neural Network learning
FA [56]	This algorithm is used directly to solve MPPT. It assures fast convergence with almost zero oscillations
ABC [58,59,64]	In MPPT, this algorithm provides quick converge and accuracy in tracking.
ACO [62]	ACO is used in the learning process for adjusting weights and biases or the neural networks in other to improve its results

3.2.3. Neural Networks

Neural Networks have shown excellent adequacy and high capabilities for complex learning problems and, thus, they are ideal for tracking the Max Power Point. They can be used alone or helped by other methods. The hybrid techniques are usually focused on improving the performance of the neural networks by optimizing the hyper-parameters of the networks since it is a really complex task. In [65], the proposed NN was composed of a single hidden layer of twenty nodes, two inputs and one or two outputs, all of them using tangent sigmoid. Two networks are built, one for approximating the voltage and current curves and the second for estimating the optimal voltage factors. The input of both networks is temperature modules and solar irradiation. The outputs for the first networks are the optimal PV voltage and optimal PV and optimal voltage factor for the second. This method improves the deficiency of traditional algorithms and improves its results.

The work in [49] presents an NN composed of 5 nodes on a single hidden layer but with the novelty of prepossessing of the data via genetic algorithms, using the same inputs (irradiance and temperature) but with only one output: the Voltage at V_{MPP}. The model improves the transitional state and reduces the oscillations in the steady state compared with traditional methods.

The approach presented by [66] uses a hyperbolic activation function. The structure is defined by 4 inputs, 1 output and 3 hidden layers with 8,7,7 neurons, respectively. The inputs are composed of three irradiation levels and the temperature. The output is a prediction of the PV voltage corresponding to the MPP; this output goes to a calculation block where it is converted for the traditional P&O algorithm. The training was carried out with a Bayesian regulated back-propagation, which performed better than standard BP. The results provide better efficiency compared with classical methods, even under partial shading. The authors used MATLAB for the implementation and simulations.

The technique presented in [67] combines Fuzzy Logic and Neural Networks, building the system known as the Adaptive Neural Fuzzy System Interface (ANFIS). The ANFIS does not need any prior knowledge of the system like the other NN methods. The structure is composed of 5 layers: inputs (irradiance and temperature), output and three intermediate layers which maintain the fuzzy logic system and provide the output based on the rules. Each input has three membership functions that are generated by the ANFIS method. The results show that the system is efficient to track MPP even under varying weather conditions. The method was designed with MATLAB/Simulink.

The work in [40] provides a different approach using recurrent neural networks along with fuzzy logic. The structure of the networks is composed by a hidden layer, a context layer storing the results of the previous outputs of the hidden layers, the output layer (solar radiation intensity) and the input layer (voltage of PV cell and the current of PV cell at the operational point. With the solar radiation intensity and temperature, the V_{MPP} is computed using the mathematical model. Another improvement found in this work is how a metaheuristic is used to optimize the structure and the weights/bias of the RNN. The results show an improvement over other competitive methods.

Another important hybrid is the method in [68], which combines ANN with Support Vector Machines (SVM) [69]. The Course Gaussian Support Vector Machine (CGSVM) is used to improve the dataset before sending the dataset to the neural network. The CGSVM is a type of nonlinear SVM and is usually used on optimization tasks. The NN was composed of 2 inputs (temperature and irradiance), 1 output (PV current) and a single hidden layer with 13 neurons. The results are slightly worse on power than the ANFIS, but the required time was significantly less than the ANFIS.

The work presented in [41] shows a new way of optimizing the ANN for MPPT. The ANN is composed of a 2-3-3-1 structure in order to make real-time applications and to avoid memorization events. The ANN takes input as irradiance and temperature and gives output as maximum voltage and is optimized with a metaheuristic called FPSOGSA. The method compares different activation functions in order to maximize performance. The results are compared with P&O and traditional NN; it is found that the method provides more stability and efficiency. A similar approach is presented in [70]. A PSO algorithm is used to find the best topology, find the best 20 hidden nodes and to optimize the initial weights of the neural network. Two inputs are used (G—level of irradiance and T—temperature), and a single output (predicting power of PV array at MPP). The model proved to be more effective under various weather conditions than other ANN or FLC techniques.

The work in [61] attempts to optimize an ANN using ACO. Using this algorithm for ANN training results in quicker training. Tangent sigmoid is chosen as the activation function. Six different topologies were evaluated in order to find the best structure, the best being a single hidden layer with 20 neurons. Two inputs are transmitted to the ANN, PV array voltage and current. The output is set as the duty cycle (d). The model tracks MPP efficiently even under irradiation changes.

The approach found in [71] uses the concept of Deep Reinforcement Learning (DRL), which tries to implement Reinforcement Learning (RL) through NN. The advantage of RL for MPPT is that RL techniques are model-free, they do not require knowing the behavior of the PV source or predefining its dynamics. A continuous state space is defined, corresponding to the current (I) values. The action space is also continuous, so it contains all the actions that can be applied to generate a change in the system. Finally, the reward function is computed directly proportional to the power, and no prior knowledge about the system is needed to define it. The system uses four networks, one for computing the policy, one for the critic and two called targets that are used to stabilize the learning procedure. The model can learn highly efficient policies from scratch, and the results show higher performance than other models.

With the aim of improving ANFIS, a new hybrid was proposed in [42]. The ANFIS is trained using the BAT algorithm [72]. The use of metaheuristic improves the training of metaheuristic compared with back-propagation; the BAT algorithm provides better convergence, simplicity and faster tracking speed than other techniques. The results show an improvement over standard ANFIS or ANFIS optimized with other metaheuristics such as PSO. Similar work is found in [73], where an ANFIS-CPHO is presented. The Crowded Plant Height Optimization [74] is in charge of training the ANFIS. The results are compared with standard ANFIS and show an increase in the speed and efficiency of the tracker.

A summary of the analyzed models in this section can be found in Table 4.

Table 4. Neural Network Models for MPPT.

Type	Reference	Features
FeedForward Neural Network	[65]	2 networks. Each one with a single hidden layer of 20 nodes.
	[49]	5 Nodes on a single layer. Data preprocessed by Genetic Algorithm.
	[66]	Three hidden layers with 8,7,7 nodes, respectively. Bayesian-Regulated back-propagation for training.
	[68]	A Single hidden layer with 13 neurons. Data created by a Course Gaussian Support Vector Machine.
	[41]	2-3-3-1 structure. The NN is optimized by FPSOGSA.
	[70]	The topology and best weights are optimized by a PSO algorithm.
	[61]	ACO is used to optimize the neural network.
Adaptive Neural Fuzzy System Interface	[42]	Bat Algorithm is used to train the network.
	[73]	Crowded Plant Height Optimization is in charge of performing the learning of the network.
	[67]	Combines Fuzzy Logic and Neural Networks. Three intermediate layers in which the output is based on fuzzy rules.
Recurrent Neural Network	[40]	A hidden layer and a context layer storing the results of the previous outputs of the hidden layer. A metaheuristic is used to optimize the structure and weights.
Deep Reinforcement Learning	[71]	Four networks, one for computing the policy, one for the critic and two called targets that are used to stabilize the learning procedure

3.3. Forecasting

Energy production forecasting has been an important problem, even in traditional systems, and it has been tackled with different techniques, as it can be seen in Figure 5.

In [75], we found a system that uses Support Vector Machines (SVM) [69]. SVM is mostly used for regression. The model uses two different inputs: solar irradiance and environmental temperature, with energy production as the output. This work included the use of a parameter to tune the number of support vectors during the training. The results show a low error, with a Mean Absolute Percentage Error (MAPE) of 0.1143, but it was really intolerant at errors in the input data. The method was implemented using MATLAB. Another approach related with SVM is found in [76]. The authors propose a multi-input support vector. Three different inputs were tested. Only solar power, solar power and solar irradiance combined and finally solar power, temperature and irradiance. The best predictions were made when the third vector was used to train the network with. The model showed better results than analytical methods with a MAPE of 36%, but it was found that it was weak against changes in the climate. The method was implemented using MATLAB.

In [77], a Neural Network was used for Short-Term Forecasting. The input data were composed of the the deviation of load power and temperature of 30 days before the forecast day and the same data of 60 days before and after the forecast day in the previous year. If the forecast day is changed, the neural network needs to be retrained. The network is composed of 9 inputs nodes, 20 hidden nodes and one output neuron. The results show a Mean Absolute Percentage Error (MAPE) of 1.63% on average.

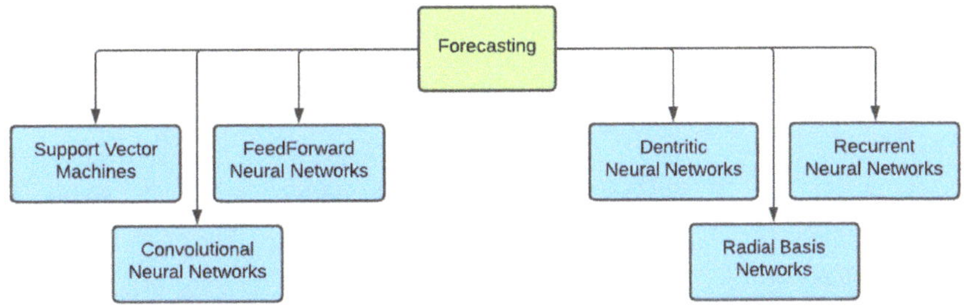

Figure 5. Taxonomy of most used IA methods for forecasting.

The work of [78] tries to go further, presenting a neural network of 2 hidden layers, one of 6 nodes and the second with 4. This model has nine inputs (Day, Time, Cloud Cover Index, Air Temperature, Wind speed, Air Humidity, UV index, precipitation and air pressure) and is trained using a hybrid metaheuristic, which combines PSO and GA [79]. This hybrid is faster and more robust than back-propagation for this problem.

Neural Networks have been found to be sensitive to many factors, including the architecture or the initialization of weights. Combining different NNs in an ensemble has been found to be a strategy to reduce these problems. The work of [80] tested different combinations using temperature and solar irradiance as inputs. Every combination was found to be better than using only a single NN. The data were composed of 7300 data from 365 different days. The findings were that the best architecture for forecasting is the one which uses an iterative methodology to find the outputs, forecasting one at a time with a Mean Absolute Error (MAE) of 51.48% and Mean Relative Error (MRE) of 17.24%.

The work in [81] used a fixed methodology, changing activation functions, learning rules and architecture in order to find the best neural network for their dataset. The data were acquired along a period of 70 days, obtaining 11,200 examples. The best network had 1 hidden layer with a Linear Sigmoid Activation Function. The learning rule as Conjugate Gradient [82], which uses second derivatives to determinate the weight update, inputs temperature and photovoltaic power and outputs next-day forecasting of PV power output. The validation study indicates that this network is simple and versatile and can precisely forecast with a minimum MAPE of 0.8655. The experiments were implemented using the NeuroSolutions [83].

Another problem of NN is that training can be slow since back-propagation is highly demanding. For solving this problem, the work in [84] used the extreme learning machine (ELM) technique to train the network. ELM [85] has a faster learning speed while obtaining better generalization performance, and it also optimizes the number of hidden neurons. The system is composed of three networks, one for each kind of weather. The network is trained with the PV output history and the weather history data. Based on the weather report of the next day, the model is chosen to forecast the day-ahead PV. The results show that ELM networks outperformed BP networks with a MAPE of 2.78% in the best case. The experiments were implemented using MATLAB.

Another improvement can be found in [86]; the neural network is aided by a technique known as Wavelet Transform (WT) [87]. This algorithm is specialized in isolating the spikes produced by continuous fluctuations of the PV data. It has two stages: decomposition of the input signal, which is performed before the neural networks, and reconstruction, which is performed with the output of the NN. The model used is a Radial Basic Neural Networks (RBNN) [88], which needs less computation time and is more effective than Back-propagation Neural Networks and takes as input the PV, solar irradiance and temperature of the current hour, twelve hours before and twenty hours before in order to predict the one-hour-ahead power output. The results show that the proposed model outperformed RBNN without WT for hourly PV for the horizon of 12 hours with a MAPE of 2.38% in the best case.

WT is used along other architectures as in [89]. RNNs are probed to be useful in order to predict from time series and WT deals with the fluctuations on the data provided by the meteorological time series obtained from sampling at intervals of 10 min and stored as time series. This combination proved to be able to forecast 2 days ahead more accurately than other Neural Networks.

A recent use of WT is found in [90]. This work presents a hybrid algorithm composed of WT, PSO and RBFNN used to forecast from 1 to 6 hours ahead. The inputs that are used in the model are set as Actual PV, irradiance and temperature. The WT is used to perform an data filtering on the past 15 days before the forecast day. The RBFNN is optimized by the PSO algorithm. The network performed better than the compared methods, with an MAE of 4.22% on average for a 1-hour-ahead forecast, 7.04% for a 3-hour-ahead one and 9.13% for 6-hour- ahead one.

Recurrent Neural Networks are also used in [91]. Deep Recurrent Neural Networks (DRNN), RNNs with many hidden layers, are used to forecast. These networks are capable of representing complex functions more efficiently than RNNs. The input data are composed of high-resolution time series, which are preprocessed and normalized to obtain a high-resolution time-series dataset of four different days. The architecture used was a DRNN with Long Short-Term Memory (LSMT) [92] units with two hidden layers of 35 neurons. Other models showed lower accuracies and more bias error than the proposed method that obtained an RMSE of 0.086. The experiments were implemented using MATLAB and the Keras library (now on tensorflow) in Python.

Another RNN method is found in [93]. The authors compared 5 different architectures of RNN: A basic LSTM, an LSTM with the window technique, an LSTM with time steps, an LSTM with memory between batches and stacked LSTMs with memory between batches. Two datasets of different cities were used to test the 3 models. The results show the third proposal with an RMSE of 82.15 in the first dataset and an RMSE of 136.87 in the second, which uses prior time steps in the PV series as inputs, is the most accurate and reliable, even compared with other methods such as ANN. The experiments were implemented using Keras.

The authors of [94] present an interesting modification of RNN. This work used the networks know as Echo State Network [95]. ESN presented a dynamical reservoir instead of the traditional hidden layers of RNN. Their main advantage is that only the output weights need to be trained since the reservoir and input ones are random. These networks can obtain better results than typical RNN. A restricted Boltzmann machine (RBM) [96] and principal component analysis (PCA) [97] are used in order to determine the number of reservoirs and inputs. The network parameters are found by a DFP Quasi-Newton algorithm [98]. Compared with other PV forecasting methods, the results show that the proposed model could outperform other forecasting systems with a MAPE of 0.00195%.

A complex hybrid is found in [99]. This system uses NN aided by different algorithms trained on data obtained during a year. Random Forest (RF) [100] is used to rank the different factors that affect PV in order to eliminate the less important ones. This importance degree, computed by RF, is transferred to Improved Gray Ideal Value Approximation (IGIVA) [101] as weights to determine the similar days of different climates type. The objective of this is to improve the quality of datasets. After that, the original sequence is decomposed by Complementary Ensemble Empirical Mode Decomposition (CEEMD) [102] to reduce the fluctuation of the original data. Finally, the neural network is optimized by a modification of PSO known as dynamic inertial factor particle swarm optimization (DIFPSO) [103,104]. The proposed model reduced training time and improved the forecasting accuracy with an MAE of 2.84 on sunny days, 10.12 on cloudy days and 13.01 on rainy or snowy days.

Another interesting approach is the Neuro-Fuzzy hybrid found in [105]. Fuzzy Logic is applied as a filter to the input data obtained in the energy production and weather forecast for 12 months (day, irradiance, temperature, humidity, pressure, wind speed and cloud clover) in order to speed up the system. The neural structure is composed of 7 inputs, 2 hidden layers of 9 and 5 nodes, respectively, and input. The network is trained by BP aided by a combination of PSO and GA, known as Genetic Swarm Optimization [106]. This method improved convergence speed and the predictive performance over other hourly forecast methods. The experiments were implemented using MATLAB Convolutional Neural Networks have also been applied to time-series data since they are able to learn filters that represent repeated patterns in the data without needing any prior knowledge. They also work well with noisy data. In [107], CNNs are applied for forecasting PV power using Solar Data and Electricity Data as inputs. The CNNs used the ReLu activation function, Adam optimizer and dropout to avoid overfitting. The parameters were selected by testing different architectures and choosing the most promising. The models were compared of an FFNN and an RNN of 128 hidden nodes. The results show that CNN performed similarly to LSTM and better than MLP with an MAE of 114.38.

An interesting approach mixing Big Data and Deep Learning is found in [108]. This method was used to next-day-ahead forecast in 30 min intervals. It used a multistep methodology that decomposes the forecasting problem in different subproblems. For the Big Data, Spark Apache was used. The neural Network parameters were searched using the grid search strategy. The best structure was found with 3 hidden layers with between 12 and 32 neurons. The method demonstrated that DL is suitable for big solar data since it has a linear increase in training time and performs better than other methods.

The work of [109] makes use of a new kind of Neural Network, the Dendritic Neuron Network [110], in order to forecast PV power. These kinds of neurons have 4 types of layers: synaptic layer, branch layer, membrane layer and cell-body layer. The input data (temperature and irradiance of the actual moment and the last) are transferred to the synaptic layers where they are converted by the sigmoid function and summarized to the branch layer. The results are transported to the cell-body layer for numerical judgment. This layer will transmit the data thought the axon to other neurons when the data exceed a given threshold. This new kind of network provides higher convergence speed and enhanced fitting ability. The network is also aided by WT. The results show that the model outperformed typical Feed-Forward models with an average MAPE of 10.9, with strong fluctuations and 4.55 on weak fluctuations. The experiments were run using MATLAB.

In Table 5, a summary of the reviewed models is presented.

Table 5. Models for forecasting.

Type	Features
Feed-Forward Neural Network	Nine inputs, 20 hidden nodes on a single layer. [77] Nine inputs, 2 hidden layers with 6 and 4 nodes, respectively. Trained by a hybrid PSO GA algorithm. [78] Two inputs, creates ensembles of neural networks. [80] Two inputs, 1 hidden layer, Conjugate Gradient as learning rule. [81] Three neural networks, one for each kind of weather. Uses Extreme Learning to optimize the parameters and architecture. [84] Fuzzy Logic is applied as a filter to the input data. Seven inputs, 2 hidden layers of 9 and 5 nodes, respectively. Trained by a hybrid of PSO and GA. [105] Uses Big Data. Multistep methodology decomposes the problems into subproblems. [108]

Table 5. *Cont.*

Type	Features
Convolutional Neural Networks	Two inputs. Parameters are selected by testing different combinations. [107]
Dendritic Neural Networks	Aided by WT. Provides better convergence speed and better fitting ability. [109]
Radial Basis Network	Two inputs, aided by Wavelet Transform to preprocess the input data. [86] High-resolution time series as input. Aided by Wavelet Transform to preprocess input data and PSO to optimize the neural network. [90]
Recurrent Neural Network	Aided by Wavelet Transform to deal with fluctuation in time series input data. [89] Preprocessed and normalized high-resolution time series as input. Two hidden layers of 35 neurons. [91] Tested Different RRN architectures. LSTM, which uses previous time steps, found the best one. [93] Uses Echo State Networks aided by Restricted Boltzmann Machine, Principal Component Analysis and DFP Quasi-Newton Algorithm to optimize the network. [94]
Support Vector Machines	Two inputs. A parameter to tune the number of SVM during training. [75] SMV compared with KNN. SMV was found to be better. [111] Multi-input SV. Different combinations of inputs were tested. Three inputs was the best one found. [76]

3.4. Parameter Estimation

Finding the parameters of the PV models is vital to simulate their behavior and to optimize their production. This problem is simplified by finding the unknown parameters in order to optimize the output power. Different techniques, most of them metaheuristics, have been used to solve this problem, as can be seen in Figure 6.

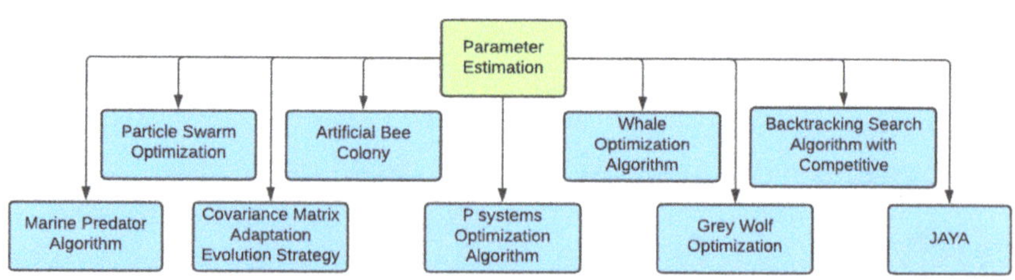

Figure 6. Taxonomy of most used metaheuristics for parameter estimation.

Metaheuristics are the most used techniques to estimate PV parameters. Different kinds of algorithms have been evaluated in recent years. The work in [112] compares different evolutionary algorithms, comparing Genetic Algorithms [48], Particle Swarm Optimization [52] and differential evolution [113]. DE is an evolutionary algorithm similar to Genetic Algorithms but which uses real numbers to codify the problem, this solves the problem of GA when it comes to converging speed. The fitness function was computed as the sum of the absolute errors in current and voltage. The findings showed that the best results were given by DE and the worst ones by GA. The authors also implemented different hybrids: Tabu Search [114] assisted differential evolution to avoid falling in local minimums, PSO assisted DE in which PSO is activated after 5 generations of DE and DE assisted by Tabu Search where DE is used to search for the optimal solution in a subset of the whole search space, while TS is used to move the local search within the global space. These hybrids performed better than the originals and provided more stability. DE assisted TS and provided the best results, and it was the fastest.

In [18], an ABC-based approach is proposed. This method combines Extreme Optimization (EO) with 2 different versions of ABC. EO provides new insights into the optimization of metaheuristics due to the fact that only the worst variables in the suboptimal solutions are selected to be mutated, instead of favoring the good ones; this is provided by the strong local-search capability of EO. The introduction of EO to ABC is applied when the global optimum is not becoming better for several iterations. The results show that the addition of EO to ABC outperformed other metaheuristics such as PSO on the single-diode model (mean RSME of 1.1678×10^{-3}) and on the double-diode model (mean RSME of 1.1479×10^{-3}). The major drawback is that EO has a higher computation cost than other methods. The experiments were run in MATLAB.

The authors of [115] presented a variant of the Covariance Matrix Adaptation Evolution Strategy. CMA-ES is an efficient derivative-free optimization algorithm. It operates using the 3 typical evolutionary operations (recombination, mutation and selection). The proposed variant combines CMA-ES with 2 strategies that can adjust the evolutionary directions and enrich the population diversity (Anisotropic Eigenvalue Adaptation and Local Search). The results show that the algorithm was competitive in terms of convergence efficiency and accuracy, with a mean RSME of 9.8603×10^{-4} and Standard Deviation of 1.6550×10^{-17} on the single-diode model and mean RSME of 9.8402×10^{-4} and Standard Deviation of 1.3398×10^{-12} on the double-diode model. It also had a good balance of exploration and exploitation. The simulation and experiments were implemented with MATLAB.

The Whale Optimization Algorithm is a recent metaheuristic that simulates the hunting behavior of humpback whales. The basic WOA is composed of three consecutive stages: encircling prey, bubble-net attacking and searching for prey. In [116], a variant of WOA is used to estimate the parameters of a PV system. The proposed method, RLWOA, adopts a modified conversion parameter update rule and relies on the Logistic Model to balance between exploration and exploitation. This algorithm mitigates the slow convergence and ease of being trapped in local optima of the original. The results show that RLWOA performed better or at least competitively with standard WOA, other WOA variants and other metaheuristics with a mean RSME of 9.8602×10^{-4} on single-diode. The experiments were run in MATLAB.

The work in [117] presents a new optimization method called backtracking search algorithm with competitive learning (CBSA). The principle basis of BSA is composed of 4 parts: the initialization of the population, selection, genetic operators such as mutation or crossover and second selection in order to select the best candidate. The main idea of CBSA is to increase the chance of the backtracking algorithm to jump out of the local optimum by the designed competitive learning machine. Each population is divided in two subgroups, then each subgroup has three different search operations in order to update its individuals. Unlike other metaheuristics, CSBA does not need any extra control parameters. The results show the superiority of CBSA for complex optimization problems with a mean RSME of 9.8602×10^{-4} on the single-diode model. The experimentation was performed in MATLAB.

Another interesting metaheuristic is found in [118]. The author presents an advanced version of the Gray Wolf Optimizer [119] applied to parameter estimation. GWO is motivated by gray wolf behavior. Wolves are divided into four categories: Alpha wolves, which are dominant, and beta wolves, which are used to assist alpha wolves in decision making or in other activities. The order given by alpha and beta is followed by the delta wolves. Finally, the omega wolves play the role of scapegoat. The presented method is known as the Intelligent Gray Wolf Optimizer, which incorporates sinusoidal truncated functions as a bridging mechanism and opposition-based learning. The results show that the algorithm was competitive with other optimizers, with a mean error of 4.65×10^{-13} on single-diode mono-crystalline, a mean error of 1.07×10^{-12} on double-diode mono-crystalline, a mean error of 8.50×10^{-12} on single-diode poly-crystalline and a mean error of 1.95×10^{-12} on double-diode poly-crystalline. Additionally, execution time was not compromised.

A hybrid between PSO and GWO is found in [120]. The fundamental principle of this hybridization is to ingrate the social thing capability of PSO with the local search ability of GWO. After performing PSO, each particle position with a certain probability is updated using the average of the three best wolves. This method reduces the drawbacks of PSO, increasing the possibility of the running of local optimums and improving the balance between exploration and exploitation. The results confirm the superiority of PSO–GWO compared with other competitive methods, with an RMSE of 3.06×10^{-3} and an MAE of 2.43×10^{-3} on a PV module model.

In [121], a variant of the Chicken Swarm Optimization is used to solve this problem. CSO [122] is inspired by the foraging behavior and hierarchy of chicken flocks. In CSO, each chicken is considered a potential solution. The chicken flock is divided into the rooster subflock, hen subflock and chick subflock according to the fitness of each individual. Each group uses a different update mechanism to update its position. The algorithm is upgraded by using a Spiral Movement Strategy. The spiral movement allows each hen to bypass the rooster and explore a wider space instead of being limited to the search space between them. The experimental results show that the algorithm performs better in robustness and accuracy than other metaheuristics with a mean RSME of 9.8602×10^{-4} and standard deviation of 2.3517×10^{-12} on a single-diode model and a mean RSME of 9.8366×10^{-4} and standard deviation of 1.4171×10^{-6} on single-diode model.The experimentation was performed in MATLAB.

Another interesting proposal is found in [123]. An Enhanced JAYA (EJAYA) is presented. The basis of the original JAYA [124] algorithm is as follows: After initializing the solutions, the algorithm identifies the best and worst solutions and modifies all the solutions based on them. All of the solutions that have better performance than the originals are kept. This process is repeated until the stop criteria are achieved. EJAYA presents three improvements: A modified evolution operator to increase the probability of approaching the victory. A simple deterministic population resizing to control the convergence rate during the search and a generalized opposition-based learning mechanism to avoid being trapped on local optima. The results indicate that the algorithm can estimate the most accurate model parameters with a mean RSME of 9.8602×10^{-4} on a single-diode model and a mean RSME of 9.8248×10^{-4} on a double-diode model. It also provided a high computational efficiency among the compared methods.The method was implemented in MATLAB.

The work in [125] presents a Marine Predator Algorithm [126] applied to parameter estimation. The optimization process of MPA is divided into three main phases: The first is in high-velocity ratio or when prey is moving faster than the predator. The second is when both are moving at the same pace and the third is in a low-velocity ratio when the predator is faster than the prey. The algorithm extracted PV parameters in an accurate manner, fast speed, less time of computation and high reliability and robustness with a mean RSME of 7.73×10^{-4} on a single-diode for a France Solar cell and a mean RSME of 7.65×10^{-4} on double-diode for a France Solar cell. The examination and test occurred via MATLAB.

A novel approach is found in [127] presenting a variant of P system Optimization Algorithms (POAs). POAs are helpful and reliable search techniques that abstract the structure and function of living cells. The proposed Micro-change Field Effect P System is a deeper exploration of the standard POA. The experiments showed that the method can produce solutions of high quality and has great stability with a mean RSME of 9.8606×10^{-4} on the single-diode model and a mean RSME of 9.8256×10^{-4} on the double-diode model. The method was implemented in MATLAB.

In the Table 6, a summary of the reviewed models is presented.

Table 6. Models for parameter estimation. SD: single diode, DD: double diode, MC: monocrystalline, PC: polycrystalline.

Type	Features	Error
ABC and CE	Combines Extreme Optimization with ABC. EO is introduced in ABC when the global optimum is not improving. EO has a high computation cost. [18]	RMSE: SD: 1.1678×10^{-3} DD: 1.1479×10^{-3}
CMA-ES	This evolutionary algorithm brings a good balance between exploration and exploitation and is competitive with other methods. [115]	RMSE: SD: 9.8603×10^{-4} DD: 9.8402×10^{-4}
WOA and LM	A variant RLWOA changes the parameter update rule and relies on the Logistic Model to balance exploration and exploitation [116]	RMSE: SD: 9.8602×10^{-4} DD: —
CBSA	Combines the exploratory capacities of WOA and convergence capacities of Social Group Optimization. [117]	RMSE: SD:9.8602×10^{-4} DD:
GWO	Improves basic GWO with a new bridging mechanism and opposition-based learning. [118]	MAE: SD-MC: 4.65×10^{-13} DD-MC: 1.07×10^{-12} SD-PC: 8.50×10^{-12} DD-PC: 1.95×10^{-12}
GWO and PSO	This method combines the social thing capability of PSO with the local search ability of GWO. [120]	RMSE: SD: 3.06×10^{-3} DD: —
CSO	CSO is improved by a Spiral Movement Strategy in order to improve the results. [121]	RMSE: SD: 1.1678×10^{-3} DD: —
JAYA	This method improves basic JAYA with an improved evolution operator, control of the size of the population and generalized opposition-based learning. [123]	RMSE: SD: 9.8602×10^{-4} DD: 9.8248×10^{-4}
MPA	The algorithm extracted the parameters faster and with high reliability and robustness. [125]	RMSE: SD: 7.73×10^{-4} DD: 7.65×10^{-4}
POA	Proposes an extension of standard POA. The results show that the method produces solutions of high quality. [127]	RMSE: SD: 9.8606×10^{-4} DD: 9.8256×10^{-4}
GA, PSO, DE and others	Propose a comparison of different algorithms and crossover between them; differential evolution assisted by Tabu Search is found to be the best. [112]	RMSE:— SD: — DD: —

3.5. Defects Detection

Finding defects on the surface of the PV cells is a problem completely related to computer vision. As observed in the bibliography, the most used technique for photographing the images is electroluminescence. The datasets are usually private, but there are some exceptions. We can see in Figure 7 the most used techniques for detecting defects.

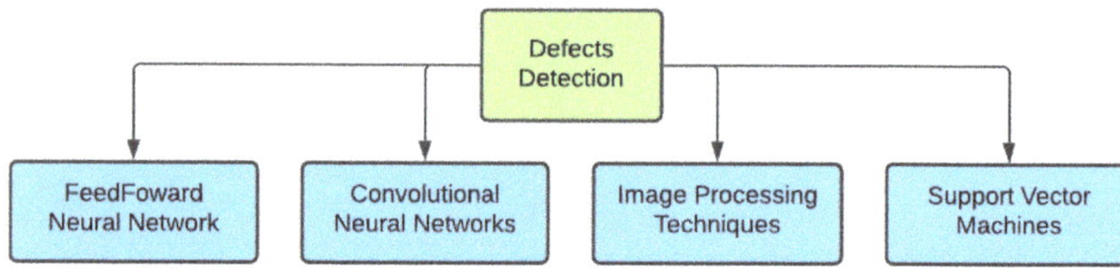

Figure 7. Most used IA method for defect detection.

Classical approaches as found in [128], which tried to detect defects in the solar modules using image processing techniques. In order to segment the different modules, they used the first derivative of the statistic curve in order to find the division line between each chip. After that, they used another technique, the otsu method, to obtain a binary image. Finally, the algorithm tries to identify the state of the module using the geometry of the resulting image. This algorithm produced interesting results, with a recognition rate of 80% on cracked modules, 95% on fragmented and 99% on good state modules. The recognition was also quite fast. The algorithms were implemented and applied via MATLAB.

Another approach is found in [129]. This method combines the image processing techniques with Support Vector Machines. The dataset featured 13,392 samples of EL images of solar cells. The images are preprocessed in order to reduce spatial noises and to accurately highlight crack pixels in images. After that, binary processing is performed, and finally, the features are extracted from the image. These features are used by different SVMs in order to classify the cells. The results present that the SVM with penalty parameter weighting is the best SVM, with a correct detection rate of 91%, with specificity and accuracy of more than 97%. The experiments were run in MATLAB.

In [130], the author compare Convolutional Neural Networks with SVM. The SVM is trained with data from the ELPV dataset, composed of 2624 EL images of solar cells, obtained by finding the features of the images using different feature descriptors. The CNN used was a pretrained VGG19 with the upper layers changed and trained with the examples. The models were tested with both monocrystalline and polycrystalline modules. The results show that both classifiers were useful for visual inspection, both with an average accuracy of 82.4%. The algorithms were implemented in Python, using Keras for the Neural Network.

The work in [131] presented a similar approach using SVM and CNN. The CNN was composed of two convolutional layers with leaky-relu and max-pooling. The convolutional part was aided by two leaky-relu dense layers and the output layer. The SVM was trained with different features extracted from the images. The dataset was built with 90 images of full-sized commercial modules that were segmented afterward, obtaining 540 cells. The results show similar behavior in both methods, with an accuracy of 98%. The article also tackled unsupervised learning, trying to cluster the images by two features. This resulted in a model that was able to assign the correct label in 66% of cases.The algorithms were implemented in Python, using Tensorflow and OpenCV.

The work found in [132] presents a CNN with 13 convolutional layers, an adaptation of the VGG16 architecture. The dataset was obtained by photographing solar modules of 6×12 cells with an EL camera. The network was trained using oversampling and data augmentation in order to reduce the error. The results show that the network performed the best when both oversampling and data augmentation were presented with a Balance Error Rate of 7.73% on binary classification problems of quick convergence. The method was implemented with Keras. The preprocessing was performed with OpenCV.

The authors of [133] present new models that are trained not only with images with cracks but also with corrosion. The images were obtained by photographing modules with the EL technique and performing segmentation afterward, obtaining 5400 images. The models are SVM and CNN. The CNN is composed of two convolutional layers. The SVM parameters are optimized by a grid search. The results show a precision of 99%, an improvement over other methods. The experiments were conducted via Keras and Tensorflow.

A variation of convolutional networks is found in [134]. A multichannel CNN is presented. This network has different convolutional layers for each kind of input. This network also can use inputs of different sizes. After each convolutional layer, a dense layer is applied. Finally, a final dense layer combines all the previous data in order to classify the image. This multichannel CNN improves the feature extraction of single-channel CNNs. The dataset was made by 8301 different EL images of cells. The results show a 96.76% accuracy, much more than the 86% presented by single-channel CNNs. The algorithms were implemented in Python using Keras.

The model presented in [135] is composed of six convolutional layers using different regularization techniques such as batch optimization. The dataset used was the ELPV dataset, with 2624 images. The resulting network is a light architecture that achieved high performance using few parameters with an accuracy of 93%.The experiments were run on Tensorflow.

In order to further improve the results, a new approach is presented in [136]. The authors use Fully Convolutional Neural Networks. An FCNN is a CNN without any dense layer. The model used is the U-net, which has been used previously in biomedical image problems with low data. This dataset was composed of 542 EL images. It is composed of 21 convolutional layers of different sizes. The results show that it was better to accept a slight decrease in the performance in order to improve the speed of the system. The algorithms were implemented in python using Keras and Tensorflow.

Wavelet Transform is used in [137]. This method combines two kinds of WT: Discrete WT and Stationary WT in order to extract textural and edge features from solar cells that have been previously preprocessed. The dataset was composed of 2300 EL images. Finally, two different classifiers are used: An SVM and an FFNN. The best model was the FFNN with 93.6% accuracy, over the 92.6% presented by the SVM.

Another Neural Network used is the Complementary Attention Network in [138]. The CAN is composed of a channel-wise attention subnetwork connected with a spatial attention subnetwork. This CAN can be grouped with any CNN, Fast R CNN [139] being the one chosen by the authors. Two datasets were used, one composed of 2029 images and another of 2129 EL images. The network was used for classification and detection, obtaining an accuracy of 99.17% for classification and a mean average precision of 87.38%. The network was faster and had similar parameter numbers to other commercial methods. The algorithms were implemented using Python.

A very interesting approach is presented in [140]. This method is Deep-Feature-Based, extracting features through convolutional neural networks that are classified afterward for classification algorithms such as SVM, KNN or FNN. The particularity of this system is that it used features from different networks. These features are combined using minimum redundancy and maximum relevance for feature selection. The dataset used was the ELPV dataset, with 2624 images. The selected CNNs for feature extraction are Resnet-50, VGG-16, VGG-19 and DarkNet-19. The best method was found with SVM, selecting 2000 features with an accuracy of 94.52% in two-class classification and 89.63% in four-class classification.

In the Table 7, a summary of the reviewed models is presented.

Table 7. Models for detection of faults.

Type	Features	Accuracy	Dataset Size
Image Processing Techniques	Segmentation + obtention of binary image + classification. [128]	from 80% to 99%	—
SVM + Image Processing Techniques	Images are preprocessed and features are extracted from the image. These features are used in an SVM with penalty parameter weighting. [129]	97%	13,392
SVM and CNN	Pretrained VGG19 using different feature descriptors. Similar results for both methods. [130]	82.4%	2624
	CNN is composed of 2 layers using leaky-relu. SVM trained with different features extracted from the images. Similar behavior in both models. [131]	98%	540
	CNN is composed of 2 convolutional layers. SVM parameters optimized by search grid. [133]	96%.	2840
CNN	Thirteen convolutional layers, an adaptation of VGG16. Uses oversampling and data augmentation. [132]	Uses a different measurement	5400
	Multichannel CNN. Accepts inputs of different sizes. Improves the feature extraction of single-channel CNN. [134]	96.76%	8301
	Six convolutional layers. Regulation techniques such as batch optimization. [135]	93%	2624
	Fully Convolutional Neural Network. Pretrained u-net, composed of 21 convolutional layers. [136]	Uses a different measurement	542
	CNN aided by a Complementary Attention Network, composed of a channel-wise attention subnetwork connected with a spatial attention subnetwork. Usable with different CNNs. [138]	99.17%	2300
WT+ SVM and FFNN	Combines discrete WT and stationary WT to extract features and SVM and FFNN to classify them. [137]	93.6%	2029
CNN + SVM, KNN, etc.	Extracts features from different networks, combining them with minimum redundancy and maximum relevance for feature selection. Uses Resnet-50, VGG-16, VGG-19 and DarkNet-19. [140]	94.52%	2624

4. Discussion

In Section 3, the different IA techniques applied are reviewed for each problem by explaining various important works. In this section, a discussion about the state of the art is performed, summarizing the trends of research and some possible new approaches to consider.

The tracking of the Maximum Power Point has been considered in numerous ways, from traditional and simple methods to methods that use complex technology such as neural networks. The most simple methods still have importance, since a large amount of systems do not need complex MPPT algorithms in order to optimize production. The most complex algorithms are used only in the biggest power plants, where the configuration of the PV arrays, and the large amount of them, makes the process of tracking the Maximum Power Point more complex. Comparing the different techniques presented, it is clear that the most used technologies are the Feed-Forward Neural Networks and the FLC systems (Figure 8). Neural Networks perform better than Logic Systems, but they have some flaws. Neural Networks are highly demanding in terms of computational cost compared with FL systems. The need for large amounts of data is an intrinsic problem of Neural Networks, but it is not as important as it used to be thanks to the high availability of data. Another important problem is the complexity in the optimization of the hyper-parameters, since Neural Networks have a large amount of them. The solution has been found by using optimization algorithms, such as metaheuristics. These algorithms can be used to find the best combinations of parameters or even to find the optimal architecture. As we observed, the problem is still regarded nowadays, since there are ways to improve the efficiency and performance of the most complex systems. To fulfill this objective, more of the newest technologies applied in other sectors should be tried, since as it has been confirmed with the previous works that these kinds of algorithms perform really well on this problem. The FL methods and metaheuristics are usually implemented with MATLAB, but Neural Networks can be implemented with Python as well, using libraries such as Tensorflow [141].

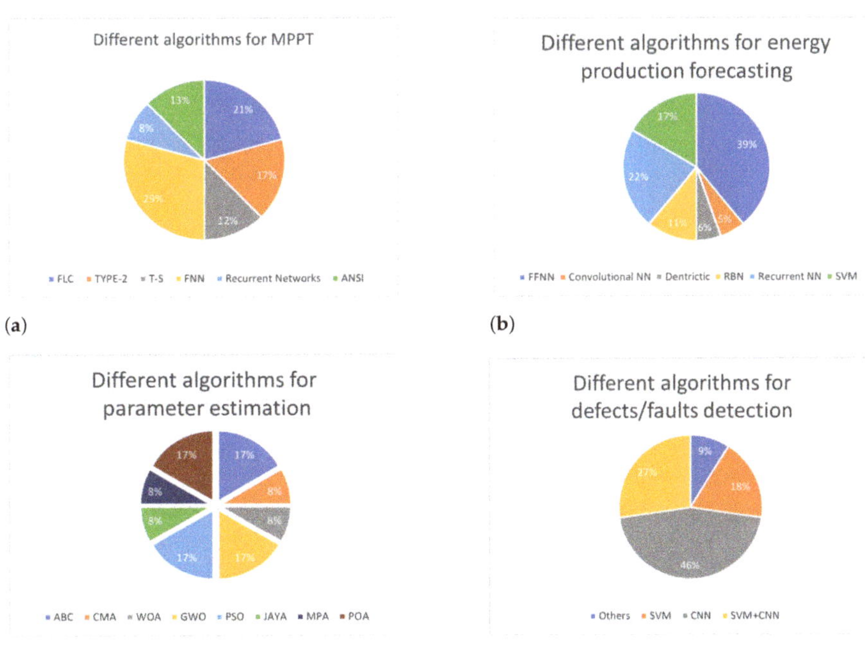

Figure 8. Different algorithms for each of the problems. (**a**) MPPT; (**b**) forecasting; (**c**) parameter estimation; (**d**) defects detection.

Forecasting is a key problem in PV systems. The estimation of the energy produced by solar plants has been approached as a regression problem in the majority of papers. Due to the availability of data, neural models are highly suitable for solving this problem (Figure 8). The trend is to use the newest neural architectures while optimizing their parameters and their architectures using other methods such as metaheuristics. As observed, an increase in the complexity of the networks improves the results, but this is not the only way of increasing performance. The combination with other systems such as Wavelet Transform increases reliability. Alternative network systems such as Recurrent Networks or Dentritic Networks further improve the results of traditional Neural Networks. This area also has some room for improvement since forecasting is a tricky problem due to its dependence on a large amount of variables. For future research, it would be interesting to test a new combination of parameters, improve the datasets or even try the most innovative technologies that have been used in similar problems. Most of the works are run on MATLAB, but there is an increase in the presence of Python in the latest years due to the Deep Learning Libraries.

Estimating the parameters of the PV models has been conducted with a large number of different algorithms, most of them from the family of metaheuristics (Figure 8). The results are quite similar between them in terms of error (Figure 9); this fact shows that trying to further minimize error is not a worthy effort. The most promising works present a mixture of different metaheuristics, solving the problems or flaws of one metaheuristic by using others. The focus of research should be moved towards finding algorithms with lower computational cost while maintaining the same levels of error. The majority of the methods can be found implemented in MATLAB [141].

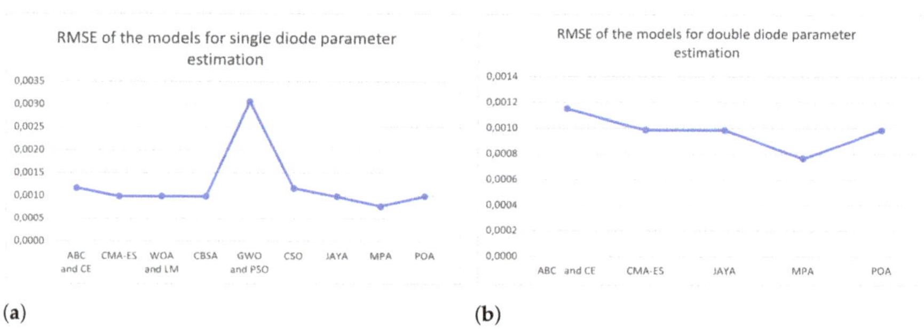

Figure 9. RMSE of the different models for parameter estimation. (**a**) Single-diode; (**b**) double-diode.

Analyzing the state of modules or cells has always been important for optimizing production since damaged modules are not as productive as they should be. As it seen, this problem has been applied mostly to the cell level, segmenting images previously taken of solar modules. Some author have even presented open datasets in order to test the models with a more regular amount of data. The main reason for this problem is related to its nature. The majority of models are trained with unbalanced datasets since the number of damaged modules is usually considerably smaller than good state modules. Another different way of improving the neural models is using pretrained neural networks such as VGG-19 in order to make use of the patterns found in other datasets. Even with their problems, in the bibliography, a considerable amount of models are presented, and they obtain good results (Figure 10); the Convolutional Neural Networks being the most used ones (Figure 8). However, there is a lot to do in this area, mostly, all of the models only use electroluminescence images; the utilization of other techniques such as thermography could bring more information and better results to the models. Another interesting new approach could be fixing the unbalance in the data. Some studies have tried to use simple methods such as flips or rotations, but it is necessary to implement more complex algorithms to generate images that can be used to better train the models. The deep learning methods are mostly implemented with Tensorflow, and OpenCV is usually used for preprocessing the images. MATLAB is used for traditional methods.

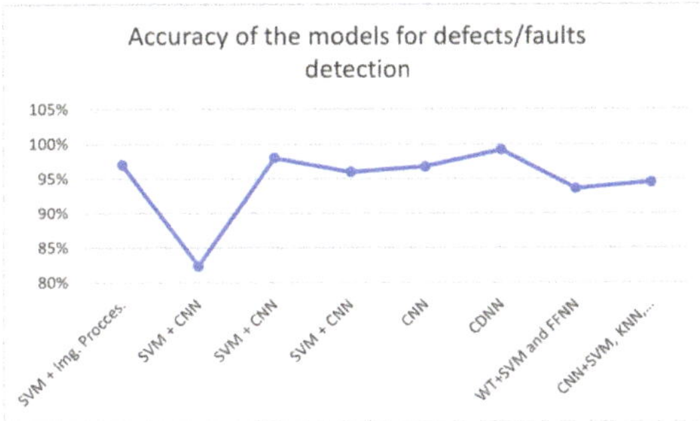

Figure 10. Accuracy of the different models for detection.

5. Conclusions

In this article, the relationship between Artificial Intelligence and Photovoltaic Systems is explained. Numerous problems in this sector can be solved with the use of AI techniques. These techniques present better performance than traditional methods.

Different techniques are applied to the MPPT problem, Neural Networks being the ones which provided better results, even considering their limitations such as high computational requirements or the need for large amounts of data, other approaches involve Fuzzy Logic and Metaheuristics. The forecasting problem is key for PV installations, different models have been created to solve this problem, most of them related to Artificial Neural Networks. These models are usually aided by other algorithms such as metaheuristics to optimize the architecture or the hyperparameters. The estimation of model parameters is also a really important problem, a large variety of metaheuristics have been used to solve this problem with notably good results in terms of error and efficiency. The detection of faults in PV modules has been proved to be vital for the maintenance of PV installations. This is mostly conducted at the cell level and usually only with electroluminescence images. Convolutional Neural Networks are the most used technology for the classification of images, but they need to be empowered with some techniques such as Data Augmentation or Knowledge Transfer.

Research in these areas is not finished, and it is still a hot topic nowadays as it can be seen from the number of publications in recent years; the ways of improving performance and efficiency are still being researched to adapt to every Photovoltaic System. It is observed that one of the most important issues is the quality and quantity of the data. Machine Learning methods need big amounts of data to be able to find patterns for predicting. This issue is even more critical in Deep Learning methods. This research group is addressing this issue for defect detection, creating synthetic EL images of photovoltaic cells to obtain more examples to train models. These images could even be used in other problems of PV systems.

Author Contributions: Conceptualization, H.F.M.R. and L.H.-C.; methodology, H.F.M.R., V.C.-P. and M.Á.G.R.; validation, H.F.M.R., V.A.G. and A.R.P.; writing—original draft preparation, H.F.M.R.; writing—review and editing, H.F.M.R. , L.H.-C., V.C.-P. and R.T.M.; project administration, L.H.-C. All authors have read and agreed to the published version of the manuscript.

Funding: This study was supported by the Universidad the Valladolid with the predoctoral contracts of 2020, cofunded by Santander Bank. It was also supported by Cátedra Cel Caja Rural de Soria.

Institutional Review Board Statement: Not applicable.

Informed Consent Statement: Not applicable.

Data Availability Statement: Not applicable.

Acknowledgments: This study was also supported by the Universidad de Valladolid with ERASMUS+ KA-107. We also appreciate the help of other members of our departments.

Conflicts of Interest: The funders had no role in the design of the study; in the collection, analyses, or interpretation of data; in the writing of the manuscript; or in the decision to publish the results.

References

1. Adib, R.; Zervos, A.; Eckhart, M.; David, M.E.A.; Kirsty, H.; Peter, H.; Governments, R.; Bariloche, F. Renewables 2021 Global Status Report. In *REN21 Renewables Now*; 2021. Available online: https://www.iea.org/reports/renewables-2021 (accessed on 10 September 2022).
2. Kampa, M.; Castanas, E. Human health effects of air pollution. *Environ. Pollut.* **2008**, *151*, 362–367. [CrossRef] [PubMed]
3. Hersbach, H.; Bell, B.; Berrisford, P.; Hirahara, S.; Horányi, A.; Muñoz-Sabater, J.; Nicolas, J.; Peubey, C.; Radu, R.; Schepers, D.; et al. The ERA5 global reanalysis. *Q. J. R. Meteorol. Soc.* **2020**, *146*, 1999–2049. [CrossRef]
4. Eroğlu, H. Effects of Covid-19 outbreak on environment and renewable energy sector. *Environ. Dev. Sustain.* **2021**, *23*, 4782–4790. [CrossRef] [PubMed]
5. Danowitz, A. Solar Thermal vs. Photovoltaic. 2010. Available online: http://large.stanford.edu/courses/2010/ph240/danowitz2/ (accessed on 10 September 2022).
6. GreenMatch. *Differences Between Solar PV and Solar Thermal*; GreenMatch: Copenhagen, Denmark, 2016. Available online: https://www.greenmatch.co.uk/blog/2015/04/solar-panels-vs-solar-thermal (accessed on 10 September 2022).
7. Irena International Renewable Energy Agency. *Future of Solar Photovoltaic Deployment, Investment, Technology, Grid Integration and Socio-Economic Aspects*; A Global Energy Transformation Paper About IRENA; Technical Report; Irena International Renewable Energy Agency: Abu Dhabi, United Arab Emirates, 2019.
8. Markvart, T.; Castañer, L. Principles of Solar Cell Operation. In *Practical Handbook of Photovoltaics: Fundamentals and Applications*; Elsevier: Amsterdam, The Netherlands, 2003; pp. 71–93. [CrossRef]
9. Satpathy, R.; Pamuru, V. *Solar PV Power: Design, Manufacturing and Applications from Sand to Systems*; Elsevier: Amsterdam, The Netherlands, 2020; pp. 1–493. [CrossRef]
10. Yetayew, T.T.; Workineh, T.G. A Comprehensive Review and Evaluation of Classical MPPT Techniques for a Photovoltaic System. In *Advances of Science and Technology*; Springer: Berlin/Heidelberg, Germany, 2021; Volume 384, pp. 259–272. [CrossRef]
11. Pallathadka, H.; Ramirez-Asis, E.H.; Loli-Poma, T.P.; Kaliyaperumal, K.; Ventayen, R.J.M.; Naved, M. Applications of artificial intelligence in business management, e-commerce and finance. *Mater. Today Proc.* 2021, *in press*. [CrossRef]
12. Alshahrani, A.; Omer, S.; Su, Y.; Mohamed, E.; Alotaibi, S. The technical challenges facing the integration of small-scale and large-scale PV systems into the grid: A critical review. *Electronics* **2019**, *8*, 1443. [CrossRef]
13. Valer, L.R.; Manito, A.R.; Ribeiro, T.B.; Zilles, R.; Pinho, J.T. Issues in PV systems applied to rural electrification in Brazil. *Renew. Sustain. Energy Rev.* **2017**, *78*, 1033–1043. [CrossRef]
14. Bosman, L.B.; Leon-Salas, W.D.; Hutzel, W.; Soto, E.A. PV System Predictive Maintenance: Challenges, Current Approaches, and Opportunities. *Energies* **2020**, *13*, 1398. [CrossRef]
15. Baba, A.O.; Liu, G.; Chen, X. Classification and Evaluation Review of Maximum Power Point Tracking Methods. *Sustain. Future* **2020**, *2*, 100020. [CrossRef]
16. Eltawil, M.A.; Zhao, Z. MPPT techniques for photovoltaic applications. *Renew. Sustain. Energy Rev.* **2013**, *25*, 793–813. [CrossRef]
17. Sobri, S.; Koohi-Kamali, S.; Rahim, N.A. Solar photovoltaic generation forecasting methods: A review. *Energy Convers. Manag.* **2018**, *156*, 459–497. [CrossRef]
18. Chen, M.R.; Chen, J.H.; Zeng, G.Q.; Lu, K.D.; Jiang, X.F. An improved artificial bee colony algorithm combined with extremal optimization and Boltzmann Selection probability. *Swarm Evol. Comput.* **2019**, *49*, 158–177. [CrossRef]
19. Villalva, M.G.; Gazoli, J.R.; Filho, E.R. Comprehensive approach to modeling and simulation of photovoltaic arrays. *IEEE Trans. Power Electron.* **2009**, *24*, 1198–1208. [CrossRef]
20. Shapiro, S.; Kouri Kissel, T. Classical Logic. In *The Stanford Encyclopedia of Philosophy*, Spring 2021 ed.; Zalta, E.N., Ed.; Metaphysics Research Lab, Stanford University: Stanford, CA, USA, 2021.
21. Ross, T.J. *Fuzzy Logic with Engineering Applications*, 3rd ed.; John Wiley & Sons: Hoboken, NJ, USA, 2010.
22. Mahmoud, A.M.A.; Mashaly, H.M.; Kandil, S.A.; Khashab, H.E.; Nashed, M.N.F. Fuzzy logic implementation for photovoltaic maximum power tracking. In Proceedings of the 26th Annual Conference of the IEEE-Industrial-Electronics-Society, Nagoya, Japan, 22–28 October 2000; pp. 735–740.
23. Hui, J.; Sun, X. MPPT Strategy of PV System Based on Adaptive Fuzzy PID Algorithm. In *Life System Modeling and Intelligent Computing Pt I, Proceedings of the International Conference on Life System Modeling and Simulation/International Conference on Intelligent Computing for Sustainable Energy and Environment, Wuxi, China, 17–20 September 2010*; Springer: Berlin/Heidelberg, Germany, 2010; Volume 97, pp. 220–228.
24. Kottas, T.L.; Karlis, A.D.; Boutalis, Y.S. Fuzzy Cognitive Networks for Maximum Power Point Tracking in Photovoltaic Arrays. *Fuzzy Cogn. Maps Adv. Theory Methodol. Tools Appl.* **2010**, *247*, 231–257.
25. Stylios, C. Modeling Complex Systems Using Fuzzy Cognitive Maps. *Syst. Man Cybern. Part A Syst. Hum.* **2004**, *34*, 155–162. [CrossRef]

26. Adly, M.; El-Sherif, H.; Ibrahim, M. Maximum Power Point Tracker for a PV Cell using a Fuzzy Agent adapted by the Fractional Open Circuit Voltage Technique. In Proceedings of the IEEE International Conference on Fuzzy Systems (FUZZ 2011), Taipei, Taiwan, 27–30 June 2011; pp. 1918–1922.
27. Altin, N. Interval type-2 fuzzy logic controller based Maximum Power Point tracking in photovoltaic systems. *Adv. Electr. Comput. Eng.* **2013**, *13*, 65–70. [CrossRef]
28. Karnik, N.; Mendel, J.; Liang, Q. Type-2 fuzzy logic systems. *IEEE Trans. Fuzzy Syst.* **1999**, *7*, 643–658. [CrossRef]
29. Zadeh, L.A. The concept of a linguistic variable and its application to approximate reasoning—I. *Inf. Sci.* **1975**, *8*, 199–249. [CrossRef]
30. Mittal, K.; Jain, A.; Vaisla, K.S.; Castillo, O.; Kacprzyk, J. A comprehensive review on type 2 fuzzy logic applications: Past, present and future. *Eng. Appl. Artif. Intell.* **2020**, *95*, 103916. [CrossRef]
31. MATLAB. *Version 7.10.0 (R2010a)*; The MathWorks Inc.: Natick, MA, USA, 2010.
32. Kececioglu, O.F.; Gani, A.; Sekkeli, M. Design and Hardware Implementation Based on Hybrid Structure for MPPT of PV System Using an Interval Type-2 TSK Fuzzy Logic Controller. *Energies* **2020**, *13*, 1842. [CrossRef]
33. Altin, N. Single phase grid interactive PV system with MPPT capability based on type-2 fuzzy logic systems. In Proceedings of the International Conference on Renewable Energy Research and Applications, ICRERA 2012, Istanbul, Turkey, 18–21 September 2012. [CrossRef]
34. Verma, P.; Garg, R.; Mahajan, P. Asymmetrical interval type-2 fuzzy logic control based MPPT tuning for PV system under partial shading condition. *ISA Trans.* **2020**, *100*, 251–263. [CrossRef]
35. Allouche, M.; Dahech, K.; Chaabane, M. Multiobjective maximum power tracking control of photovoltaic systems: T-S fuzzy model-based approach. *Soft Comput.* **2018**, *22*, 2121–2132. [CrossRef]
36. Zayani, H.; Allouche, M.; Kharrat, M.; Chaabane, M. T-S fuzzy Maximum Power Point tracking control of photovoltaic conversion system. In Proceedings of the 16th International Conference on Sciences and Techniques of Automatic Control and Computer Engineering, STA 2015, Monastir, Tunisia, 21–23 December 2016; pp. 534–539. [CrossRef]
37. Dahmane, M.; Bosche, J.; El-Hajjaji, A.; Davarifar, M. T-S implementation of an MPPT algorithm for photovoltaic conversion system using poles placement and H performances. In Proceedings of the 3rd International Conference on Systems and Control, ICSC 2013, Algiers, Algeria, 29–31 October 2013; pp. 1116–1121. [CrossRef]
38. Takagi, T.; Sugeno, M. Fuzzy identification of systems and its applications to modeling and control. *IEEE Trans. Syst. Man Cybern.* **1985**, *SMC-15*, 116–132. [CrossRef]
39. Vaidyanathan, S.; Azar, A.T. Takagi-Sugeno fuzzy logic controller for Liu-Chen four-scroll chaotic system. *Int. J. Intell. Eng. Inform.* **2016**, *4*, 135. [CrossRef]
40. Azali, S.; Sheikhan, M. Intelligent control of photovoltaic system using BPSO-GSA-optimized neural network and fuzzy-based PID for Maximum Power Point tracking. *Appl. Intell.* **2016**, *44*, 88–110. [CrossRef]
41. Duman, S.; Yorukeren, N.; Altas, I.H. A novel MPPT algorithm based on optimized artificial neural network by using FPSOGSA for standalone photovoltaic energy systems. *Neural Comput. Appl.* **2018**, *29*, 257–278. [CrossRef]
42. Sarkar, R.; Kumar, J.R.; Sridhar, R.; Vidyasagar, S. A New Hybrid BAT-ANFIS-Based Power Tracking Technique for Partial Shaded Photovoltaic Systems. *Int. J. Fuzzy Syst.* **2021**, *23*, 1313–1325. [CrossRef]
43. Naveen; Dahiya, A.K. Implementation and Comparison of Perturb Observe, ANN and ANFIS Based MPPT Techniques. In Proceedings of the International Conference on Inventive Research in Computing Applications, ICIRCA 2018, Coimbatore, India, 11–12 July 2018; pp. 1–5. [CrossRef]
44. Arora, A.; Gaur, P. Comparison of ANN and ANFIS based MPPT Controller for grid connected PV systems. In Proceedings of the 12th IEEE International Conference Electronics, Energy, Environment, Communication, Computer, Control: (E3-C3), INDICON 2015, New Delhi, India, 17–20 December 2016. [CrossRef]
45. Padmanaban, S.; Priyadarshi, N.; Bhaskar, M.S.; Holm-Nielsen, J.B.; Ramachandaramurthy, V.K.; Hossain, E. A Hybrid ANFIS-ABC Based MPPT Controller for PV System with Anti-Islanding Grid Protection: Experimental Realization. *IEEE Access* **2019**, *7*, 103377–103389. [CrossRef]
46. Qiao, X.; Wu, B.; Deng, Z.; You, Y. MPPT of photovoltaic generation system using fuzzy/PID control. *Electr. Power Autom. Equip.* **2008**, *28*, 92–95.
47. Gendreau, M.; Potvin, J.Y. (Eds.) *Handbook of Metaheuristics*; Springer: New York, NY, USA, 2010. [CrossRef]
48. Goldberg, D.E. *Genetic Algorithms in Search, Optimization and Machine Learning*, 1st ed.; Addison-Wesley Longman Publishing Co., Inc.: Boston, MA, USA, 1989.
49. Kulaksiz, A.A.; Akkaya, R. Training data optimization for ANNs using genetic algorithms to enhance MPPT efficiency of a stand-alone PV system. *Turk. J. Electr. Eng. Comput. Sci.* **2012**, *20*, 241–254. [CrossRef]
50. Kulaksiz, A.A.; Akkaya, R. A genetic algorithm optimized ANN-based MPPT algorithm for a stand-alone PV system with induction motor drive. *Sol. Energy* **2012**, *86*, 2366–2375. [CrossRef]
51. Larbes, C.; Cheikh, S.M.A.; Obeidi, T.; Zerguerras, A. Genetic algorithms optimized fuzzy logic control for the maximum power point tracking in photovoltaic system. *Renew. Energy* **2009**, *34*, 2093–2100. [CrossRef]
52. Clerc, M. *Particle Swarm Optimization*; Wiley: Hoboken, NJ, USA, 2010. [CrossRef]
53. Azab, M. Optimal power point tracking for stand-alone PV system using particle swarm optimization. In Proceedings of the IEEE International Symposium on Industrial Electronics, Bari, Italy, 4–7 July 2010; pp. 969–973. [CrossRef]

54. Alshareef, M.; Lin, Z.; Ma, M.; Cao, W. Accelerated Particle Swarm Optimization for Photovoltaic Maximum Power Point Tracking under Partial Shading Conditions. *Energies* **2019**, *12*, 623. [CrossRef]
55. Yang, X.S. Firefly Algorithms for Multimodal Optimization. In Proceedings of the International Symposium on Stochastic Algorithms, Sapporo, Japan, 26–28 October 2009; 5792 LNCS, pp. 169–178. [CrossRef]
56. Sundareswaran, K.; Peddapati, S.; Palani, S. MPPT of PV systems under partial shaded conditions through a colony of flashing fireflies. *IEEE Trans. Energy Convers.* **2014**, *29*, 463–472. [CrossRef]
57. Karaboga, D. *An Idea Based on Honey Bee Swarm for Numerical Optimization*; Technical Report-tr06; Erciyes University, Engineering Faculty, Computer Engineering Department: Kayseri, Turkey, 2005.
58. soufyane Benyoucef, A.; Chouder, A.; Kara, K.; Silvestre, S.; Sahed, O.A. Artificial bee colony based algorithm for Maximum Power Point tracking (MPPT) for PV systems operating under partial shaded conditions. *Appl. Soft Comput. J.* **2015**, *32*, 38–48. [CrossRef]
59. Bilal, B. Implementation of Artificial Bee Colony algorithm on Maximum Power Point Tracking for PV modules. In Proceedings of the 8th International Symposium on Advanced Topics in Electrical Engineering, ATEE 2013, Bucharest, Romania, 23–25 May 2013. [CrossRef]
60. Dorigo, M.; Birattari, M.; Stutzle, T. Ant colony optimization. *IEEE Comput. Intell. Mag.* **2006**, *1*, 28–39. [CrossRef]
61. Babes, B.; Boutaghane, A.; Hamouda, N. A novel nature-inspired Maximum Power Point tracking (MPPT) controller based on ACO-ANN algorithm for photovoltaic (PV) system fed arc welding machines. *Neural Comput. Appl.* **2021**, 34, 299–317 [CrossRef]
62. Moreira, H.S.; Silva, J.L.D.S.; Prym, G.C.; Sakô, E.Y.; dos Reis, M.V.G.; Villalva, M.G. Comparison of Swarm Optimization Methods for MPPT in Partially Shaded Photovoltaic Systems. In Proceedings of the 2nd International Conference on Smart Energy Systems and Technologies (SEST), Porto, Portugal, 9–11 September 2019.
63. Telbany, M.E.; Youssef, A.; Zekry, A.A. Intelligent Techniques for MPPT Control in Photovoltaic Systems: A Comprehensive Review. In Proceedings of the 4th International Conference on Artificial Intelligence with Applications in Engineering and Technology, ICAIET 2014, Kota Kinabalu, Malaysia, 3–5 December 2015; pp. 17–22. [CrossRef]
64. Oshaba, A.S.; Ali, E.S.; Elazim, S.M.A. PI controller design using ABC algorithm for MPPT of PV system supplying DC motor pump load. *Neural Comput. Appl.* **2017**, *28*, 353–364. [CrossRef]
65. Habibi, M.; Yazdizadeh, A. New MPPT Controller Design for PV Arrays Using Neural Networks (Zanjan City Case Study). In Proceedings of the 6th International Symposium on Neural Networks, Wuhan, China, 26–29 May 2009; Volume 5552, pp. 1050–1058.
66. Veligorskyi, O.; Chakirov, R.; Vagapov, Y. Artificial neural network-based Maximum Power Point tracker for the photovoltaic application. In Proceedings of the 2015 1st International Conference on Industrial Networks and Intelligent Systems, INISCom 2015, Tokyo, Japan, 2–4 March 2015; pp. 133–138. [CrossRef]
67. Mlakic, D.; Nikolovski, S. ANFIS as a method for determinating MPPT in the photovoltaic system simulated in MATLAB/Simulink. In Proceedings of the 39th International Convention on Information and Communication Technology, Electronics and Microelectronics, MIPRO 2016, Opatija, Croatia, 30 May–3 June 2016; pp. 1082–1086. [CrossRef]
68. Farayola, A.M.; Hasan, A.N.; Ali, A. Efficient photovoltaic mppt system using coarse gaussian support vector machine and artificial neural network techniques. *Int. J. Innov. Comput. Inf. Control.* **2018**, *14*, 323–339. [CrossRef]
69. Kecman, V. *Support Vector Machines—An Introduction*; Springer: Berlin/Heidelberg, Germany, 2005; Volume 177, p. 605. [CrossRef]
70. Al-Majidi, S.D.; Abbod, M.F.; Al-Raweshidy, H.S. A particle swarm optimisation-trained feedforward neural network for predicting the Maximum Power Point of a photovoltaic array. *Eng. Appl. Artif. Intell.* **2020**, *92*, 103688. [CrossRef]
71. Avila, L.; Paula, M.D.; Trimboli, M.; Carlucho, I. Deep reinforcement learning approach for MPPT control of partially shaded PV systems in Smart Grids. *Appl. Soft Comput.* **2020**, *97*, 106711. [CrossRef]
72. Yang, X.S. Bat algorithm: Literature review and applications. *Int. J. Bio-Inspired Comput.* **2013**, *5*, 141–149. [CrossRef]
73. Pachaivannan, N.; Subburam, R.; Padmanaban, M.; Subramanian, A. Certain investigations of ANFIS assisted CPHO algorithm tuned MPPT controller for PV arrays under partial shading conditions. *J. Ambient. Intell. Humaniz. Comput.* **2021**, *12*, 9923–9938. [CrossRef]
74. Pachaivannan, N.; Subburam, R.; Ramkumar, U.; Kasinathan, P. Crowded plant height optimisation algorithm tuned Maximum Power Point tracking for grid integrated solar power conditioning system. *IET Renew. Power Gener.* **2019**, *13*, 2137–2147. [CrossRef]
75. Leone, R.D.; Pietrini, M.; Giovannelli, A. Photovoltaic energy production forecast using support vector regression. *Neural Comput. Appl.* **2015**, *26*, 1955–1962. [CrossRef]
76. Nageem, R.; Jayabarathi, R. Predicting the Power Output of a Grid-Connected Solar Panel Using Multi-Input Support Vector Regression. *Procedia Comput. Sci.* **2017**, *115*, 723–730. [CrossRef]
77. Senjyu, T.; Takara, H.; Uezato, K.; Funabashi, T. One-hour-ahead load forecasting using neural network. *IEEE Trans. Power Syst.* **2002**, *17*, 113–118. [CrossRef]
78. Caputo, D.; Grimaccia, F.; Mussetta, M.; Zich, R.E. Photovoltaic Plants Predictive Model by means of ANN trained by a Hybrid Evolutionary Algorithm. In Proceedings of the World Congress on Computational Intelligence (WCCI 2010), Barcelona, Spain, 18–23 July 2010.

79. Gandelli, A.; Grimaccia, F.; Mussetta, M.; Pirinoli, P.; Zich, R.E. Development and validation of different hybridization strategies between GA and PSO. In Proceedings of the 2007 IEEE Congress on Evolutionary Computation, CEC 2007, Singapore, 25–28 September 2007; pp. 2782–2787. [CrossRef]
80. Rana, M.; Koprinska, I.; Agelidis, V.G. Forecasting solar power generated by grid connected PV systems using ensembles of neural networks. In Proceedings of the International Joint Conference on Neural Networks, Killarney, Ireland, 12–17 July 2015. [CrossRef]
81. Ehsan, R.M.; Simon, S.P.; Venkateswaran, P.R. Day-ahead forecasting of solar photovoltaic output power using multilayer perceptron. *Neural Comput. Appl.* **2016**, *28*, 3981–3992. [CrossRef]
82. Johansson, E.; Dowla, F.; Goodman, D. Backpropagation Learning for Multilayer Feed-Forward Neural Networks Using the Conjugate Gradient Method. *Int. J. Neural Syst.* **2011**, *2*, 291–301. [CrossRef]
83. NeuroDimension, Inc. Neurosolutions. Available online: http://www.neurosolutions.com/neurosolutions/ (accessed on 10 September 2022).
84. Li, Z.; Zang, C.; Zeng, P.; Yu, H.; Li, H. Day-ahead Hourly Photovoltaic Generation Forecasting using Extreme Learning Machine. In Proceedings of the IEEE International Conference on Cyber Technology in Automation, Control, and Intelligent Systems (CYBER), Shenyang, China, 8–12 June 2015; pp. 779–783.
85. Huang, G.B.; Zhu, Q.Y.; Siew, C.K. Extreme learning machine: A new learning scheme of feedforward neural networks. In Proceedings of the IEEE International Conference on Neural Networks—Conference Proceedings, Budapest, Hungary, 25–29 July 2004; Volume 2, pp. 985–990. [CrossRef]
86. Mandal, P.; Madhira, S.T.S.; haque, A.U.; Meng, J.; Pineda, R.L. Forecasting power output of solar photovoltaic system using wavelet transform and artificial intelligence techniques. *Procedia Comput. Sci.* **2012**, *12*, 332–337. [CrossRef]
87. Mallat, S.G. A Theory for Multiresolution Signal Decomposition: The Wavelet Representation. *IEEE Trans. Pattern Anal. Mach. Intell.* **1989**, *11*, 674–693. [CrossRef]
88. Park, J.; Sandberg, I.W. Approximation and Radial-Basis-Function Networks. *Neural Comput.* **1993**, *5*, 305–316. [CrossRef]
89. Capizzi, G.; Napoli, C.; Bonanno, F. Innovative second-generation wavelets construction with recurrent neural networks for solar radiation forecasting. *IEEE Trans. Neural Netw. Learn. Syst.* **2012**, *23*, 1805–1815. [CrossRef]
90. Wen, Y.; AlHakeem, D.; Mandal, P.; Chakraborty, S.; Wu, Y.K.; Senjyu, T.; Paudyal, S.; Tseng, T.L. Performance Evaluation of Probabilistic Methods Based on Bootstrap and Quantile Regression to Quantify PV Power Point Forecast Uncertainty. *IEEE Trans. Neural Netw. Learn. Syst.* **2020**, *31*, 1134–1144. [CrossRef] [PubMed]
91. Alzahrani, A.; Shamsi, P.; Dagli, C.; Ferdowsi, M. Solar Irradiance Forecasting Using Deep Neural Networks. *Procedia Comput. Sci.* **2017**, *114*, 304–313. [CrossRef]
92. Hochreiter, S.; Schmidhuber, J. Long short-term memory. *Neural Comput.* **1997**, *9*, 1735–1780. [CrossRef] [PubMed]
93. Abdel-Nasser, M.; Mahmoud, K. Accurate photovoltaic power forecasting models using deep LSTM-RNN. *Neural Comput. Appl.* **2019**, *31*, 2727–2740. [CrossRef]
94. Yao, X.; Wang, Z.; Zhang, H. A novel photovoltaic power forecasting model based on echo state network. *Neurocomputing* **2019**, *325*, 182–189. [CrossRef]
95. Gallicchio, C.; Micheli, A. Deep Echo State Network (DeepESN): A Brief Survey. *arXiv* **2017**, arXiv:abs/1712.04323.
96. Hinton, G.E. A Practical Guide to Training Restricted Boltzmann Machines. In *Lecture Notes in Computer Science (Including Subseries Lecture Notes in Artificial Intelligence and Lecture Notes in Bioinformatics)*; 7700 Lecture N.; Springer: Berlin/Heidelberg, Germany, 2012; pp. 599–619. [CrossRef]
97. Jolliffe, I.T. *Principal Component Analysis*; Springer: Berlin/Heidelberg, Germany, 2002. [CrossRef]
98. Davidon, W. *Variable Metric Method for Minimization*; Technical Report; Argonne National Laboratory (ANL): Lemont, IL, USA, 1959. [CrossRef]
99. Niu, D.; Wang, K.; Sun, L.; Wu, J.; Xu, X. Short-term photovoltaic power generation forecasting based on random forest feature selection and CEEMD: A case study. *Appl. Soft Comput. J.* **2020**, *93*, 106389. [CrossRef]
100. Breiman, L. Random Forests. *Mach. Learn.* **2001**, *45*, 5–32. [CrossRef]
101. Behzadian, M.; Otaghsara, S.K.; Yazdani, M.; Ignatius, J. A state-of-the-art survey of TOPSIS applications. *Expert Syst. Appl.* **2012**, *39*, 13051–13069. [CrossRef]
102. Yeh, J.R.; Shieh, J.S.; Huang, N.E. Complementary ensemble empirical mode decomposition: A novel noise enhanced data analysis method. *Adv. Adapt. Data Anal.* **2011**, *2*, 135–156. [CrossRef]
103. Li, H.; Tan, Q. A BP neural network based on improved particle swarm optimization and its application in reliability forecasting. *Res. J. Appl. Sci. Eng. Technol.* **2013**, *6*, 1246–1251. [CrossRef]
104. Jiao, B.; Lian, Z.; Gu, X. A dynamic inertia weight particle swarm optimization algorithm. *Chaos Solitons Fractals* **2008**, *37*, 698–705. [CrossRef]
105. Quan, D.M.; Ogliari, E.; Grimaccia, F.; Leva, S.; Mussetta, M. Hybrid model for hourly forecast of photovoltaic and wind power. *IEEE Int. Conf. Fuzzy Syst.* **2013**. [CrossRef]
106. Grimaccia, F.; Mussetta, M.; Zich, R.E. Genetical swarm optimization: Self-adaptive hybrid evolutionary algorithm for electromagnetics. *IEEE Trans. Antennas Propag.* **2007**, *55*, 781–785. [CrossRef]
107. Koprinska, I.; Wu, D.; Wang, Z. Convolutional Neural Networks for Energy Time Series Forecasting. In Proceedings of the International Joint Conference on Neural Networks (IJCNN), Rio de Janeiro, Brazil, 8–13 July 2018.

108. Torres, J.F.; Troncoso, A.; Koprinska, I.; Wang, Z.; Martínez-Álvarez, F. Big data solar power forecasting based on deep learning and multiple data sources. *Expert Syst.* **2019**, *36*, e12394. [CrossRef]
109. Zhang, T.; Lv, C.; Ma, F.; Zhao, K.; Wang, H.; O'Hare, G.M. A photovoltaic power forecasting model based on dendritic neuron networks with the aid of wavelet transform. *Neurocomputing* **2020**, *397*, 438–446. [CrossRef]
110. Jiang, T.; Wang, D.; Ji, J.; Todo, Y.; Gao, S. Single dendritic neuron with nonlinear computation capacity: A case study on XOR problem. In Proceedings of the 2015 IEEE International Conference on Progress in Informatics and Computing, PIC 2015, Nanjing, China, 18–20 December 2016; pp. 20–24. [CrossRef]
111. Wolff, B.; Lorenz, E.; Kramer, O. Statistical learning for short-term photovoltaic power predictions. *Stud. Comput. Intell.* **2016**, *645*, 31–45. [CrossRef]
112. Siddiqui, M.U.; Abido, M. Parameter estimation for five- and seven-parameter photovoltaic electrical models using evolutionary algorithms. *Appl. Soft Comput. J.* **2013**, *13*, 4608–4621. [CrossRef]
113. Opara, K.R.; Arabas, J. Differential Evolution: A survey of theoretical analyses. *Swarm Evol. Comput.* **2019**, *44*, 546–558. [CrossRef]
114. Glover, F.; Laguna, M. *Tabu Search*; Springer: New York, NY, USA, 1997. [CrossRef]
115. Liang, Y.; Wang, X.; Zhao, H.; Han, T.; Wei, Z.; Li, Y. A covariance matrix adaptation evolution strategy variant and its engineering application. *Appl. Soft Comput. J.* **2019**, *83*, 105680. [CrossRef]
116. Long, W.; Wu, T.; Jiao, J.; Tang, M.; Xu, M. Refraction-learning-based whale optimization algorithm for high-dimensional problems and parameter estimation of PV model. *Eng. Appl. Artif. Intell.* **2020**, *89*, 103457. [CrossRef]
117. Zhang, Y.; Ma, M.; Jin, Z. Backtracking search algorithm with competitive learning for identification of unknown parameters of photovoltaic systems. *Expert Syst. Appl.* **2020**, *160*, 113750. [CrossRef]
118. Saxena, A.; Sharma, A.; Shekhawat, S. Parameter extraction of solar cell using intelligent grey wolf optimizer. *Evol. Intell.* **2020**. [CrossRef]
119. Mirjalili, S.; Mirjalili, S.M.; Lewis, A. Grey Wolf Optimizer. *Adv. Eng. Softw.* **2014**, *69*, 46–61. [CrossRef]
120. Rezk, H.; Arfaoui, J.; Gomaa, M.R. Optimal parameter estimation of solar pv panel based on hybrid particle swarm and grey wolf optimization algorithms. *Int. J. Interact. Multimed. Artif. Intell.* **2021**, *6*, 145–155. [CrossRef]
121. Li, M.; Li, C.; Huang, Z.; Huang, J.; Wang, G.; Liu, P.X. Spiral-based chaotic chicken swarm optimization algorithm for parameters identification of photovoltaic models. *Soft Comput.* **2021**, *25*, 12875–12898. [CrossRef]
122. Meng, X.; Liu, Y.; Gao, X.; Zhang, H. A New Bio-inspired Algorithm: Chicken Swarm Optimization. *Lect. Notes Comput. Sci.* **2014**, *8794*, 86–94. [CrossRef]
123. Yang, X.; Gong, W. Opposition-based JAYA with population reduction for parameter estimation of photovoltaic solar cells and modules. *Appl. Soft Comput.* **2021**, *104*, 107218. [CrossRef]
124. Rao, R. Jaya: A simple and new optimization algorithm for solving constrained and unconstrained optimization problems. *Int. J. Ind. Eng.* **2016**, *7*, 19–34. [CrossRef]
125. Sattar, M.A.E.; Sumaiti, A.A.; Ali, H.; Diab, A.A. Marine predators algorithm for parameters estimation of photovoltaic modules considering various weather conditions. *Neural Comput. Appl.* **2021**, *33*, 11799–11819. [CrossRef]
126. Faramarzi, A.; Heidarinejad, M.; Mirjalili, S.; Gandomi, A.H. Marine Predators Algorithm: A nature-inspired metaheuristic. *Expert Syst. Appl.* **2020**, *152*, 113377. [CrossRef]
127. Yang, S.; Max, N.; Xie, S.; Li, L.; Zhao, T. Photovoltaic cell model parameter optimization using micro-charge field effect P systems. *Eng. Appl. Artif. Intell.* **2021**, *104*, 104374. [CrossRef]
128. Nian, B.; Fu, Z.; Wang, L.; Cao, X. Automatic detection of defects in solar modules: Image processing in detecting. In Proceedings of the 6th International Conference on Wireless Communications, Networking and Mobile Computing, WiCOM 2010, Chengdu, China, 23–25 September 2010. [CrossRef]
129. Anwar, S.A.; Abdullah, M.Z. Micro-crack detection of multicrystalline solar cells featuring shape analysis and support vector machines. In Proceedings of the 2012 IEEE International Conference on Control System, Computing and Engineering, ICCSCE 2012, Penang, Malaysia, 23–25 November 2012; pp. 143–148. [CrossRef]
130. Deitsch, S.; Buerhop-Lutz, C.; Sovetkin, E.; Steland, A.; Maier, A.; Gallwitz, F.; Riess, C. Segmentation of Photovoltaic Module Cells in Uncalibrated Electroluminescence Images. *Mach. Vis. Appl.* **2018**, *32*, 84. [CrossRef]
131. Karimi, A.M.; Fada, J.S.; Liu, J.; Braid, J.L.; Koyuturk, M.; French, R.H. Feature Extraction, Supervised and Unsupervised Machine Learning Classification of PV Cell Electroluminescence Images. In Proceedings of the IEEE 7th World Conference on Photovoltaic Energy Conversion, WCPEC 2018—A Joint Conference of 45th IEEE PVSC, 28th PVSEC and 34th EU PVSEC, 2018, Waikoloa, HI, USA, 10–15 June 2018; pp. 418–424. [CrossRef]
132. Bartler, A.; Mauch, L.; Yang, B.; Reuter, M.; Stoicescu, L. Automated detection of solar cell defects with deep learning. In Proceedings of the 26th European Signal Processing Conference (EUSIPCO), Rome, Italy, 3–7 September 2018; pp. 2035–2039. [CrossRef]
133. Karimi, A.M.; Fada, J.S.; Hossain, M.A.; Yang, S.; Peshek, T.J.; Braid, J.L.; French, R.H. Automated Pipeline for Photovoltaic Module Electroluminescence Image Processing and Degradation Feature Classification. *IEEE J. Photovoltaics* **2019**, *9*, 1324–1335. [CrossRef]
134. Ying, Z.; Li, M.; Tong, W.; Haiyong, C. Automatic Detection of Photovoltaic Module Cells using Multi-Channel Convolutional Neural Network. In Proceedings of the 2018 Chinese Automation Congress, CAC 2018, Xi'an, China, 30 November–2 December 2018; pp. 3571–3576. [CrossRef]

135. Akram, M.W.; Li, G.; Jin, Y.; Chen, X.; Zhu, C.; Zhao, X.; Khaliq, A.; Faheem, M.; Ahmad, A. CNN based automatic detection of photovoltaic cell defects in electroluminescence images. *Energy* **2019**, *189*, 116319. [CrossRef]
136. Balzategui, J.; Eciolaza, L.; Arana-Arexolaleiba, N. Defect detection on Polycrystalline solar cells using Electroluminescence and Fully Convolutional Neural Networks. In Proceedings of the 2020 IEEE/SICE International Symposium on System Integration, SII 2020, Honolulu, HI, USA, 12–15 January 2020; pp. 949–953. [CrossRef]
137. Mathias, N.; Shaikh, F.; Thakur, C.; Shetty, S.; Dumane, P.; Chavan, D.S. Detection of Micro-Cracks in Electroluminescence Images of Photovoltaic Modules. *SSRN Electron. J.* **2020**. [CrossRef]
138. Su, B.; Chen, H.; Chen, P.; Bian, G.; Liu, K.; Liu, W. Deep Learning-Based Solar-Cell Manufacturing Defect Detection with Complementary Attention Network. *IEEE Trans. Ind. Inform.* **2021**, *17*, 4084–4095. [CrossRef]
139. Girshick, R. Fast R-CNN. In Proceedings of the 2015 IEEE International Conference on Computer Vision (ICCV), Santiago, Chile, 13–16 December 2015; pp. 1440–1448. [CrossRef]
140. Demirci, M.Y.; Beşli, N.; Gümüşçü, A. Efficient deep feature extraction and classification for identifying defective photovoltaic module cells in Electroluminescence images. *Expert Syst. Appl.* **2021**, *175*, 114810. [CrossRef]
141. Abadi, M.; Agarwal, A.; Barham, P.; Brevdo, E.; Chen, Z.; Citro, C.; Corrado, G.S.; Davis, A.; Dean, J.; Devin, M.; et al. TensorFlow: Large-Scale Machine Learning on Heterogeneous Systems. 2015. Available online: tensorflow.org (accessed on 10 September 2022).

batteries

MDPI

Review

A Comparative Review on Energy Storage Systems and Their Application in Deregulated Systems

Mitul Ranjan Chakraborty [1], Subhojit Dawn [2,*], Pradip Kumar Saha [3], Jayanta Bhusan Basu [1] and Taha Selim Ustun [4,*]

1 Department of Electrical Engineering, Siliguri Institute of Technology, Siliguri 734009, West Bengal, India
2 Department of Electrical & Electronics Engineering, Velagapudi Ramakrishna Siddhartha Engineering College, Vijayawada 520007, Andhra Pradesh, India
3 Department of Electrical Engineering, Jalpaiguri Government Engineering College, Jalpaiguri 735102, West Bengal, India
4 Fukushima Renewable Energy Institute, AIST (FREA), Koriyama 963-0298, Japan
* Correspondence: subhojit.dawn@gmail.com (S.D.); selim.ustun@aist.go.jp (T.S.U.)

Abstract: Electrical energy is critical to the advancement of both social and economic growth. Because of its importance, the electricity industry has historically been controlled and operated by governmental entities. The power market is being deregulated, and it has been modified throughout time. Both regulated and deregulated electricity markets have benefits and pitfalls in terms of energy costs, efficiency, and environmental repercussions. In regulated markets, policy-based strategies are often used to deal with the costs of fossil fuel resources and increase the feasibility of renewable energy sources. Renewables may be incorporated into deregulated markets by a mix of regulatory and market-based approaches, as described in this paper, to increase the systems economic stability. As the demand for energy has increased substantially in recent decades, particularly in developing nations, the quantity of greenhouse gas emissions has increased fast, as have fuel prices, which are the primary motivators for programmers to use renewable energy sources more effectively. Despite its obvious benefits, renewable energy has considerable drawbacks, such as irregularity in generation, because most renewable energy supplies are climate-dependent, demanding complex design, planning, and control optimization approaches. Several optimization solutions have been used in the renewable-integrated deregulated power system. Energy storage technology has risen in relevance as the usage of renewable energy has expanded, since these devices may absorb electricity generated by renewables during off-peak demand hours and feed it back into the grid during peak demand hours. Using renewable energy and storing it for future use instead of expanding fossil fuel power can assist in reducing greenhouse gas emissions. There is a desire to maximize the societal benefit of a deregulated system by better using existing power system capacity through the implementation of an energy storage system (ESS). As a result, good ESS device placement offers innovative control capabilities in steady-state power flow regulation as well as dynamic stability management. This paper examines numerous elements of renewable integrated deregulated power systems and gives a comprehensive overview of the most current research breakthroughs in this field. The main objectives of the reviews are the maximization of system profit, maximization of social welfare and minimization of system generation cost and loss by optimal placement of energy storage devices and renewable energy systems. This study will be very helpful for the power production companies who want to build new renewable-based power plant by sighted the present status of renewable energy sources along with the details of several EES systems. The incorporation of storage devices in the renewable-incorporated deregulated system will provide maximum social benefit by supplying additional power to the thermal power plant with minimum cost.

Keywords: regulated system; deregulated system; energy storage devices; modern power system; profit; compressed air energy storage

Citation: Chakraborty, M.R.; Dawn, S.; Saha, P.K.; Basu, J.B.; Ustun, T.S. A Comparative Review on Energy Storage Systems and Their Application in Deregulated Systems. *Batteries* **2022**, *8*, 124. https://doi.org/10.3390/batteries8090124

Academic Editors: Luis Hernández-Callejo, Jesús Armando Aguilar Jiménez and Carlos Meza Benavides

Received: 10 August 2022
Accepted: 7 September 2022
Published: 10 September 2022

1. Introduction

Electrical energy storage (EES) systems have demonstrated unique skills in coping with several important aspects of electricity, for instance, hourly changes in demand and pricing [1]. Firstly, EES saves power costs by storing electricity obtained during off-peak hours when its price goes down, for use at peak hours, rather than electricity purchased then at higher costs [2]. Second, in order to increase power supply stability, EES systems assist users when electricity network disruptions occur as a result of natural catastrophes, for example. Thirdly, it preserves and enhances power quality, frequency, and voltage [2]. Electric vehicles with batteries are the most potential off-grid method for replacing the conventional sources with renewable energy [3]. Smart grid relates to power grid updates. Smart grid technology integration makes the grid more adaptable and responsive, with the potential to provide real-time feedback by sharing data among electricity producers and consumers to provide a more sustainable and efficient power supply. EES is one of the key elements in developing a smart grid [4].

1.1. Role of EES

Two properties of electricity cause challenges with its usage while also creating market demand for EES. To begin with, power is consumed at the same moment it is created. To fulfill the shifting demand, the appropriate amount of power must constantly be given. The second trait is that power plants are often positioned distant from where electricity is utilized [5].

1. Because power lines are constantly required, if a line fails (due to congestion or any other reason), the provision of energy is halted; also, because lines are always required, delivering power to mobile applications is problematic.
2. Depending on the locations and amounts of power supply and demand, a large amount of power flow may be focused onto a single transmission line, causing congestion.

1.1.1. Optimization: High Generation Cost during Peak Hours

Generation costs vary across periods. Power suppliers should supplement base-load power plants with less cost-effective but more adaptable kinds of production, such as oil- and gas-fired generators, during peak hours. Costly methods of generating can be shut down during off-peak hours. This surplus can be held in EES and used to lower generating costs. In contrast, EES can reduce energy cost for customers because it can store energy purchased at cheap off-peak prices and use it during prime times in place of costly power. During off-peak time, users can recharge batteries and may also sell to utilities or to other users during peak time [5–7].

1. From the standpoint of utilities, there is a significant opportunity to lower total generating costs by storing electricity during the off-peak hours and reintroducing it into the power system during hours of maximum demand.
2. During peak periods of higher-than-average energy use, power suppliers must supplement the conventional base-load power facilities with less costly but more flexible sources of production, such as oil- and gas-turbine generators.
3. Conversely, from the perspective of customers, EES can cut down the financial burden since it can store electricity purchased at cheap rates during off-peak and use it during peak hours, which would have been costlier if purchased during peak hours.

1.1.2. Continuous and Flexible Supply: Need of the Hour

The main issue for utilities is delivering a consistent and adaptable power supply for consumers, which is a critical quality of energy. If the proper amount of electricity is not accessible when consumers need it, power quality will suffer and service will be disrupted in the worst-case situation. To meet fluctuating power consumption, sufficient amounts of energy should be produced on a regular basis, based on an accurate assessment of demand variations [1,2]. Power generators require two extra functions in addition to the basic generating function. Firstly, producing facilities must have a "kilowatt function" that permits them to generate enough power (kW) as per requirement. Secondly, generating facilities must feature a frequency control mechanism that adjusts the output to fit minute-by-minute variations. To take care of the fluctuating power consumption, adequate amounts of energy should be generated and be available, based on an accurate estimate of demand fluctuations. Such issues are intended to be addressed by EES. When produced electricity is in low supply, pumped hydro has been routinely employed to deliver a huge amount of power [7].

1.1.3. Distance between Generation and Consumer: A Deciding Factor

Consumers' premises are usually located far from power-producing facilities, increasing the likelihood of a power failure [8]. Natural calamities and causes due to human factors trigger system failures that interrupt power supply and have the potential to affect broad areas [9]. When power failures occur, EES will support consumers by continuing to provide electricity. Semiconductor and LCD manufacturing are two industries where voltage sag for just a few milliseconds has an impact on product quality, employ EES [10].

1.1.4. Power Grid Congestion: A Point of Concern

The power flow in transmission networks is determined by the demand and supply. Power congestion can occur during the process of balancing supply and demand [11]. Utility companies strive to avoid future bottlenecks by moving generating output or establishing additional transmission connections. EES, when installed at appropriate places such as substations at the extremities of heavily loaded lines, can help to reduce congestion [12]. This method also helps utilities delay or cease power network reinforcement.

1.1.5. Transmission by Cable: Point of Difficulty

Because power transmission usually entails the use of cables, supplying power to mobile applications and isolated areas is difficult. EES technology, with its mobility and charging functionality, can be beneficial to address this issue. It may be difficult to charge an EV in remote places without access to a power grid, but EES can aid in the creation of a green transportation system that does not rely on traditional IC engines [13].

1.2. Emerging Needs for EES

Two key market needs for EES as a critical and evolving technology are: (i) the use of more renewable energy and lower consumption of conventional fuel and (ii) a future smart grid [1] (shown in Figure 1).

1.2.1. A Step towards Greener Earth: More Renewable Energy, Less Fossil Fuel

On Grid Areas

The variability in the output of renewable sources makes it challenging to regulate the frequency of the system, and if the frequency deviation is too great, system performance may suffer [14].

Thermal generators are not operated at full capacity but rather with a positive and negative output margin (i.e., output increases and decreases) that is utilized to change frequency. If EES can reduce output variation, thermal generator margins can be decreased, and they can run more efficiently [6].

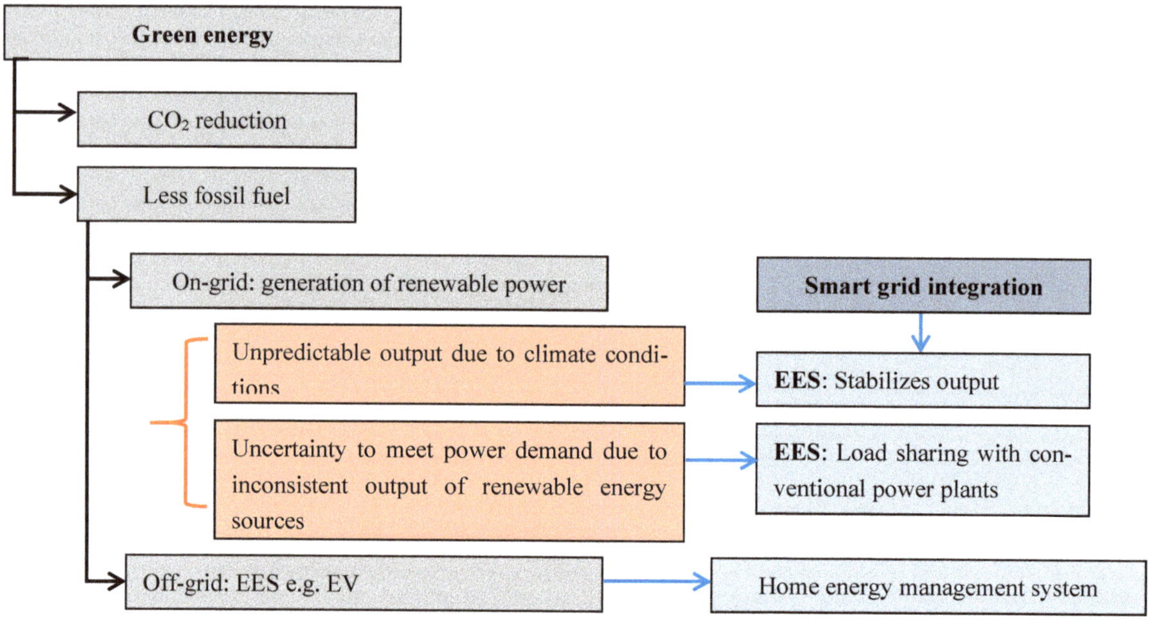

Figure 1. EES: Sustainable option towards greener earth.

Off-Grid Areas

Fossil energy should be replaced with non-fossil energy. This not only eliminates the prohibitive initial costs but also provides a good clean alternative. In particular, low-carbon power produced primarily from renewables ought to take the place of fossil fuels.

1.2.2. Smart Grid

Existing equipment may be energized with EES and be included into the smart grid. By employing a home energy management system to track their actual usage in real time, residential consumers will take an active role in changing their energy consumption patterns [15]. EVs are projected to be a new source of electricity as well as a potential storage medium in a smart grid that uses a portable, distributed energy resource as a load-shifting function, allowing utilities to continue delivering power even as electricity costs rise.

2. Types of Electrical Energy Storage System (EES)

EES systems classified on the basis of the kind of energy consumed are: (i) mechanical, (ii) electrochemical, (iii) chemical, (iv) electrical, (v) thermal and (vi) superconducting magnetic categories [8,12,16]. The classification of EES based on type of energy consumed is shown in Figure 2.

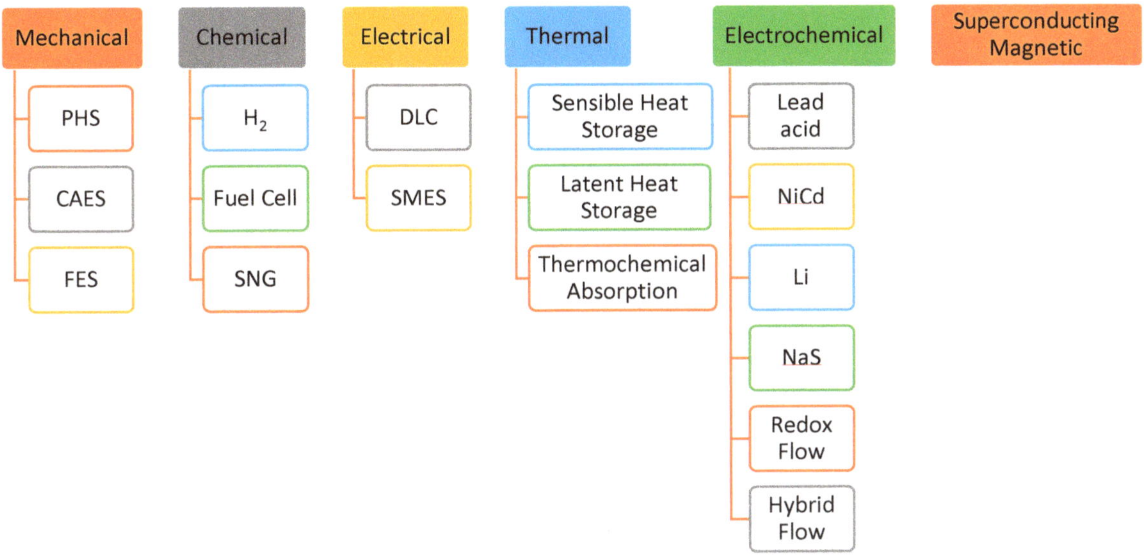

Figure 2. Classification of EES based on type of energy consumed.

3. Working of Electrical Energy Storage System (EES)

3.1. Mechanical Storage Systems

Mechanical energy storage devices store received energy by utilizing kinetic or gravitational forces. These systems are useful in real-world applications due to quality materials, advanced computer control systems, and imaginative design [17]. Mechanical energy storage operates in complicated systems that employ heat, water, or air in conjunction with compressors, turbines, and other machinery.

3.1.1. Pumped Hydro Storage (PHS)

Pumped hydro storage power plants provide for more than 95% of the world's current electrical storage capacity [18]. In pumped hydro storage systems, two water reservoirs at different heights are utilized to pump water during off-peak hours (charging), and as needed, water flows downstream from the top pool to the lower reservoir, driving a turbine that produces electricity (discharging). The efficiency of the PHS plant ranges from 70% to 85% [19]. The main benefits of this system are long life and almost unlimited cycle stability, while its drawbacks are its topography and heavy land use. The world's largest PHS plants have installed capacity of 3003 MW and 2400 MW (as of December 2021), respectively.

3.1.2. Compressed Air Energy Storage (CAES)

CAES has been used in a range of industrial applications since the eighteenth century. Electricity is used to compress air and store it in a subsurface construction or an aboveground system of containers or lines. Subsurface storage options include tunnels, aquifers, and abandoned mines. Diabatic technology is well proven; the plants are highly reliable and can operate without external power [20] (shown in Figure 3). CAES has a large capacity, but it has drawbacks such as low round-trip performance (less than 50%) and geographical constraints.

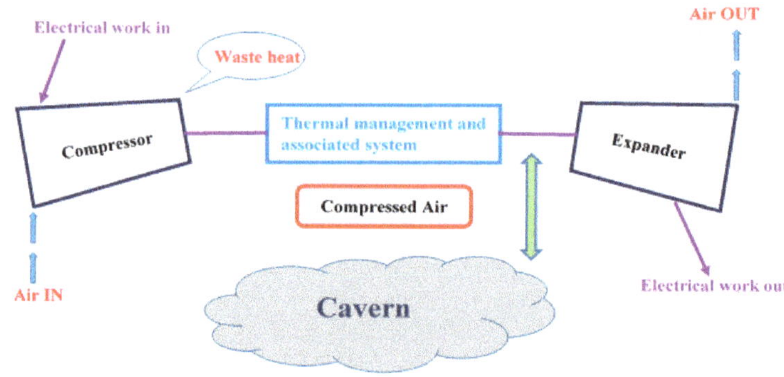

Figure 3. Compressed air energy storage system schematic.

3.1.3. Flywheel Energy Storage (FES)

In flywheel energy storage, kinetic energy is stored in an accelerated rotor which is a massive rotating cylinder. Electricity is supplied to the flywheel using a transmission mechanism and with rise in the speed, amount of stored energy increases [17]. Flywheels are commonly utilized for power quality in industrial and other applications. Flywheels have advantages of exceptional cycle stability and long life, low maintenance, greater power density and the use of environmentally friendly materials. However, it has demerits such as high self-discharge and poor current efficiency [21]. Efforts are focused on improving the management of flywheels as power storage devices for usage in cars and industries for long operation hours (shown in Figure 4).

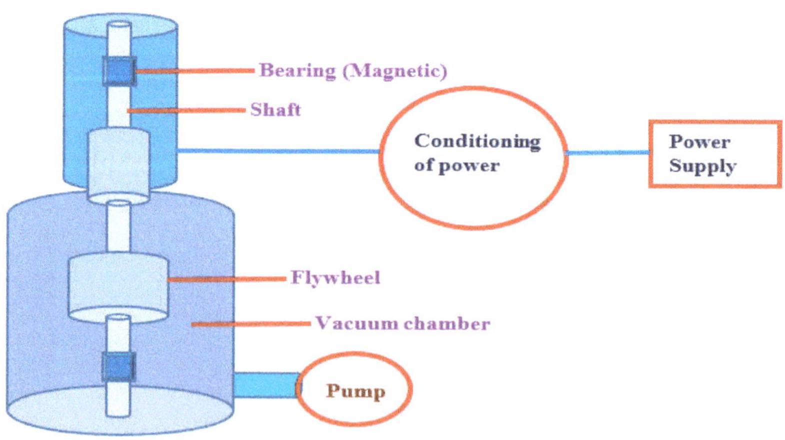

Figure 4. Flywheel energy storage system schematic.

3.2. Electrochemical Storage Systems

Electrochemical energy storage devices have the ability to make a major contribution to the deployment of sustainable energy. Electrochemical energy storage is based on systems with high energy density (batteries) or power density (electrochemical capacitors). High energy and high power densities in the same material are increasingly required in current and near-future applications [17,22]. These are categorized in two types: secondary batteries and flow batteries. The secondary batteries have again classified into following types: lead–acid, NiCd/NiMH, Li-ion, metal–air, sodium–sulfur and sodium–nickel chloride [22].

3.2.1. Secondary Batteries

A secondary battery, or charge accumulator, is a cell or set of cells with reversible cell processes. This implies that the original chemical conditions inside the cell can be restored by allowing current to flow into it, i.e., charging from outside [22].

Lead–Acid Battery (LA)

Lead–acid batteries are the most widely used form of battery in the world, dating back to roughly 1890. Service life is typically 6–15 years, with a service life of 1500 cycles at a % depth of discharge and a cycle efficiency of 80–90% [22–24]. The downsides are lower energy density and the use of lead, a dangerous element that is prohibited or restricted in some locations. Advantages include a good cost/performance ratio, simple recyclability, and a simple charging method. The current focus of lead–acid battery development is to improve their efficiency for micro-hybrid electric vehicles.

Nickel–Cadmium and Nickel–Metal Hydride Battery (NiCd, NiMH)

Before the commercial launch of nickel–metal hydride (NiMH) batteries in 1995, nickel–cadmium (NiCd) batteries had been in use since around 1915. NiMH batteries contain all of the advantages of NiCd batteries, such as greater power density, marginally better energy density, and a larger number of cycles, with the exception of a 10-fold lower maximum nominal capacity. They are far more robust and secure than lithium-ion batteries. However, due to the toxicity of cadmium, they have been limited for consumer use since 2006. NiMH batteries are currently about the same cost as Li-ion battery packs [22].

Lithium-Ion Battery (Li-Ion)

Lithium-ion batteries have been the most important form of storage in portable and mobile applications since about the year 2000. With a cell voltage of only 1.2 Volts, one lithium-ion cell may substitute three NiCd or NiMH batteries [22]. The most significant impediment is the high cost of the unique packaging and incorporated overload protection circuits. Safety is a serious problem in lithium-ion battery technology. Most metal oxide electrodes are thermally unstable and can melt at high temperatures. Lithium-ion batteries feature a monitoring device that prevents overcharging and discharging to lessen this risk [22]. A voltage regulation circuit is often provided to monitor and avoid voltage changes in each individual cell. Lithium-ion battery technology is constantly improving, with plenty of possibilities for advancement. The evolution of cathode materials is being studied [22–26]. The construction of typical Li-ion battery module is depicted in Figure 5.

Metal–Air Battery

A metal–air electrochemical cell's anode is made of pure metal, while the cathode is connected to an infinite supply of air. In the electrochemical process, only oxygen from the air is used. Because of its greater specific energy excluding oxygen (theoretically 11.14 kWh/kg), the lithium air battery is the most enticing of the several metal–air battery chemical couples [22]. Due to lithium's high reactivity to air and humidity, it can catch fire, creating a serious safety risk. Only a zinc–air battery with a theoretical specific energy of 1.35 kWh/kg (without oxygen) is theoretically practical at the moment. It is difficult to design rechargeable zinc–air cells since zinc precipitation from the water-based electrolyte must be properly handled. Although a viable, electrically rechargeable metal–air system could offer low material costs and high specific energy, none has yet attained marketability [22–26].

Sodium–Sulphur Battery (NaS)

In sodium–sulfur batteries, a solid beta-alumina ceramic electrolyte isolates the active constituents (molten sulfur at the anode and molten sodium at the cathode). NaS batteries have a discharge time of 6.0 to 7.2 h and a standard life cycle of around 4500. They are both effective and quick to respond (round-trip efficiency based on AC is around 75%) [23]. Over 200 places in Japan have tested the NaS battery technology, largely for peak shaving.

Many countries employ NaS batteries as well. Although the lack of a heat source is a significant drawback, with correctly sized insulation, the heat developed in the battery may be managed in frequent use by its own reaction heat. These batteries are suited for high-frequency cycling applications [27,28]. The construction of typical NaS battery module is depicts in Figure 6.

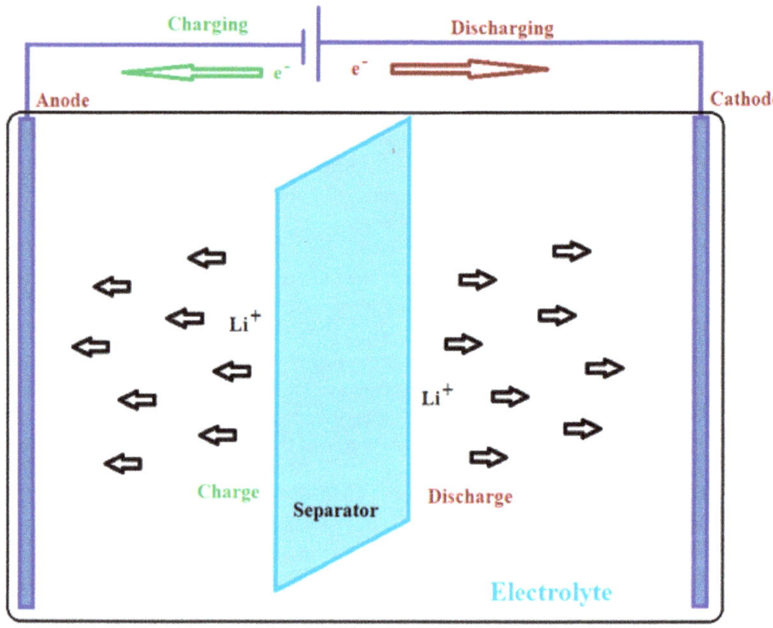

Figure 5. Typical Li-ion battery module.

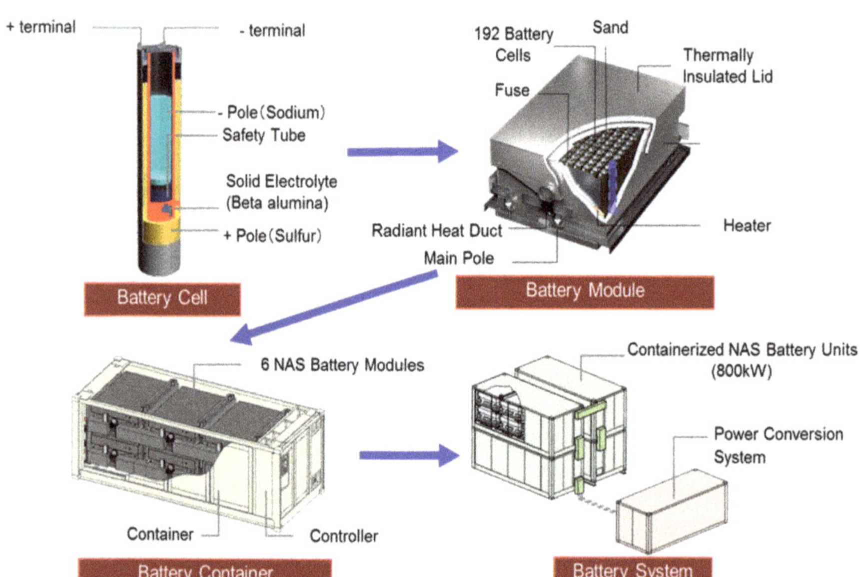

Figure 6. NaS battery system.

Sodium–Nickel Chloride Battery (NaNiCl)

The sodium–nickel chloride (NaNiCl) battery, also known as the ZEBRA battery is a high-temperature (HT) battery that, like the NaS battery, has been available on the market since approximately 1995 [24]. NaNiCl batteries outperform NaS batteries in terms of safety and cell voltage, and they can withstand limited overload and discharge. These batteries have been employed effectively in a variety of electric vehicle designs, and they are a viable alternative for fleet applications. Upgraded variants of the ZEBRA battery with greater power density values for hybrid electric vehicles, as well as high-energy versions for conserving renewable power for load-leveling and industrial purposes, are presently being developed.

3.2.2. Flow Batteries

NASA invented flow batteries in the early 1970s as an EES for long-term space flights [25]. They have the potential to store energy for hours or days and have a power of many megawatts. Flow batteries are of two types: redox flow batteries and hybrid flow batteries.

Redox Flow Battery (RFB)

The electrolytes present at the negative and positive electrodes of a redox flow battery are anolyte and catholyte. During discharge, electrodes are continually provided with dissolved active masses from the tanks; once converted, the product is returned to the tank. During the charge exchange, a current flows between the electrodes, which may be used by a battery-powered device. Redox flow batteries are being studied for use in electric vehicles; however, electrolyte energy density has proved too low thus far. An RFB may potentially be "refilled" in minutes by draining out the emptied electrolyte and replacing it with recharged electrolyte. In RFBs today, many redox couples, such as a Fe-Ti system or a poly S-Br system, have been investigated and tested (shown in Figure 7) [27,28].

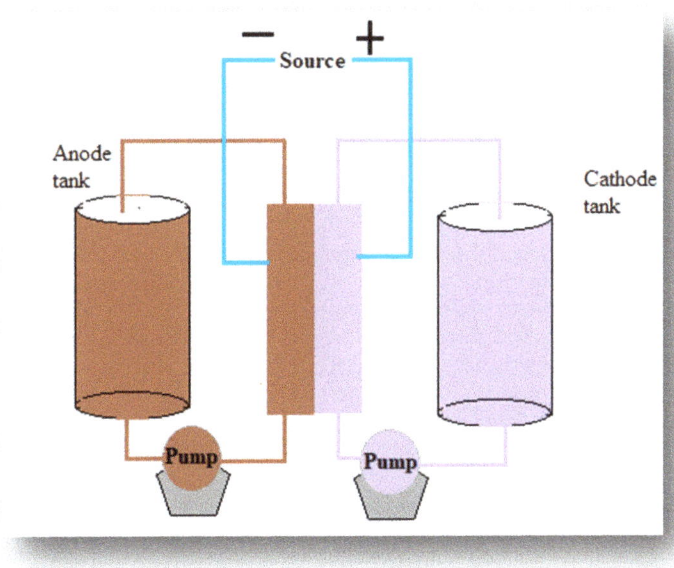

Figure 7. Schematic of redox flow battery.

Hybrid Flow Battery (HFB)

One active mass in a hybrid flow battery (HFB) is kept within the electrochemical cell, while the other is kept externally. The benefits of classic secondary batteries and

RFBs are combined in HFBs. HFBs include the Zn-Ce and Zn-Br systems. The anolyte is a Zn2+ ion-acid solution, and the electrodes are primarily carbon-plastic composites. Exxon pioneered the Zn-Br hybrid flow battery in the early 1970s, and it is now being commercialized by a variety of companies. In addition, 5 kW/20 kWh community energy storage devices are also being developed [22,28].

3.3. *Chemical Energy Storage*

A chemical energy storage system is the only idea that allows for the long-term storage of significant amounts of energy, up to TWh, even as periodic accumulation. SNG and hydrogen may be used in a range of industries, including commuting, movement, heating, and the chemical industry. They have lesser overall efficiency than PHS and Li-ion storage technologies, but are more cost efficient and effective than ordinary batteries [26].

3.3.1. Hydrogen (H_2)

An electrolyzer is a type of electrochemical converter that splits water into hydrogen and oxygen using electricity. It is an endothermic reaction, which indicates that heat is required throughout the process. Hydrogen may be stored under pressure in gas bottles or tanks for nearly indefinite periods of time. Electrolysis releases oxygen into the environment rather than retaining it, and oxygen from the air is utilized to create electricity [26].

3.3.2. Synthetic Natural Gas (SNG)

Methane (synthetic natural gas or SNG) may be synthesized to store energy. SNG can be stored in pressure tanks, underground, or fed directly into the gas infrastructure. To prevent energy losses, CO_2 and H_2 transport to the methanation plant should be avoided. The fundamental drawback of SNG is its low efficiency as a result of conversion losses in electrolysis, methanation, storage, transport, and power production [27]. The overall AC-AC efficiency of 35% is significantly lower than that of hydrogen [13].

3.4. *Electrical Storage Systems*

The classifications of EES are as follows:

3.4.1. Double-Layer Capacitors (DLC)

DLCs, also known as super-capacitors, are a 60-year-old electrochemical double-layer capacitor (DLC) technology. The extremely high capacitance values, on the order of thousands of farads, and the capability to charge and discharge very fast due to extremely low inner resistance are the two important properties. This technology offers a lot of space for advancement because it might result in substantially greater capacitance and energy density than standard capacitors, permitting for more compact designs. Durability, dependability, no maintenance, prolonged lifetime, and functioning across a wide temperature range are further benefits. With the exception of the chemical used in capacitors, which deteriorate in 5–6 years regardless of the number of cycles, the lifetime surpasses one million cycles without degradation. The efficiency is often more than 90%, with discharge times varying from seconds to hours. DLCs are not suitable for long-term energy storage due to their high self-discharge rate, low energy density, and hefty investment needs [28]. As a UPS, a DLC is excellent for bridging small power disruptions. The electric automobile might be used in a unique way, as a buffer system for acceleration and regenerative braking [4].

3.4.2. Superconducting Magnetic Energy Storage (SMES)

SMES devices store magnetic energy in a magnetic field that is generated by a superconducting coil held less than its critical temperature. A temperature of around 4 °K was required at the early age but now materials with higher critical temperatures (about 100 °K) have been developed and are now accessible. Particle detectors for high-energy scientific experiments and nuclear fusion use large SMES systems with more than 10 MW of power [29]. The main benefits of SMES are high overall round-trip efficiency (85–90%), the extremely high power output and the extremely fast reaction time: the required power is practically instantly accessible [30]. The energy can be stored basically as long as the cooling system is running, but longer storage times are restricted by the refrigeration system's energy demand.

3.5. Thermal Storage Systems

Thermal storage systems capture heat from a wide range of sources and preserve it in an insulated storage for later use in industrial and residential applications. Thermal storage systems are used to act as an intermediary between thermal energy demand and supply, making them crucial for the integration of renewable energy sources [31].

There are three forms of thermal storage: sensible heat storage, latent heat storage and thermochemical adsorption and absorption storage [17]. A storage medium can be a liquid or a solid. Thermal energy can only be stored by varying the temperature of the storage medium. A storage system's capacity is determined by the specific heat capacity and mass of the medium used. For latent heat storage, phase change materials (PCMs) are utilized as storage media. Organic (paraffins) and inorganic PCMs (salt hydrates) are also viable options for such storage systems. Latent heat is the energy transmitted during a phase transition, e.g., ice melting [17]. It is also referred to as "hidden" heat since there is no temperature difference during energy transmission. The most well-known latent heat—or cold—storage method is the ice cooler, which uses ice in an insulated container or chamber to keep food cool on hot days. The solid–liquid phase shift is used in the majority of PCMs currently in operation, such as molten salts as a thermal storage device for concentrated solar power (CSP) plants [32–41].

3.6. Superconducting Magnetic Energy Storage

A superconducting magnetic energy storage system (SMES) is a tools that stores electricity from the electrical grid within the magnetic field of a coil contained of superconducting wire with very little energy loss. The SMES systems are categorized into three groups: power supply, control systems and contingency systems [16].

4. Review: A Journey towards the Future with Guidance from the Past

A detailed literature review was conducted on deregulated power systems with the integration of renewable energy sources and energy storage devices. The main objectives of the reviews are the maximization of system profit, maximization of social welfare, and minimization of system generation cost and loss by optimal placement of energy storage devices.

K.C. Divya's study [42] focuses on the incorporation of non-conventional energy sources into the power grid and the usage of energy storage devices for profit maximization. The role of electric hybrid car battery storage systems has been considered. This article proposed that energy storage using battery will play an important role in the sustainable and cost-effective functioning of smart electric grids integrated with renewable energy. There is no single storage system that can fulfill all of the criteria for an ideal EES. Various storage systems are compared by Chen in terms of technological specifications and characteristics, applications, and implementation status [43]. Among the developed technologies, CAES is beneficial in terms of the lowest capital cost. R. Banos looked at some of the major difficulties of renewable energy sources in this research [44], such as generation discontinuity, which is an environment-dependent and continuous development in optimization techniques utilizing computing resources. The current state of the art computational optimization methods is reviewed in this paper, providing a comprehensive picture of the most recent research developments in this subject. Heuristic techniques, Pareto-based multi-objective optimization, and parallel processing have all been discovered to be interesting study areas in the realm of renewable and sustainable energy. In the article [45], Aqeel Ahmed Bazmi discusses the importance of modeling and optimization in the power and supply sectors, as well as the future prospects of optimization modeling as a tool for sustainable energy systems. Modeling and optimization have been found to be effective and valuable methods for solution development in the power and supply sector, particularly for policymakers establishing policies based on extensive assessments of competing technologies and large quantities of scenario studies.

Zhimin Wang [46] developed a unique methodology for energy management in home area using EESs to facilitate energy storage, with the goal of providing wholesale energy at reduced cost and supporting LV distribution networks for network investment reduction. The aim is to create the optimum possible DRs-to-energy-price and network-congestion balance feasible, hence improving customer and network operator advantages. The authors of this work [46] suggest a novel dispatch approach for consumers and DNOs to share ownership of residential energy storage batteries. Ref. [47] discusses various applications of EES technologies in power systems, with a focus on their collaboration with renewable energy sources. The function of ESSs in intelligent micro power grids is also highlighted, as the stochastic nature of renewable energy sources might have an impact on power quality. Each type's applicability in power systems is examined and compared to others. An energy storage system's technological and physical features are also examined in depth. Yanine and Sauma's [48] research focuses on supervisory management of micro-generation systems when connected to the grid and when energy storage is not involved. The goal is to increase energy efficiency, thriftiness, and sustainability. Suggestions have been made that future advancements in smart micro-grid operation should be increasingly focused on recognizing that SHES can be intelligent. Mwasilu [49] conducted a complete evaluation and appraisal of the most recent research and advancements in electric vehicles (EVs) interaction with smart grid, depicting the future electric power system model. The smart V2G system's viability is also addressed. The interactions of electric vehicles with the smart grid as a future energy system model are thoroughly examined in this work.

Zhang [50] presented a two-stage EES-based optimum wind power dispatch system with risk analysis to increase financial advantages through day-ahead operations. Through simulations, the suggested strategy demonstrated promising outcomes in terms of improving financial benefits and risk-reduction capability. Muruganantham, Gnanadass, and Padhy's research [51] demonstrates the several obstacles that DN suffers while adopting RES. This research investigates the significance of energy storage in distributed networks and how to manage the demand. This research provides a high-level overview of the DN's evolution and issues. This provides a quick overview of distribution power flow algorithms, electricity pricing systems and the simultaneous working of DGs and DN. Huang, Xu and Courcoubetis [52] conducted an investigation on three joint market mechanisms to analyze EES investment and operation for locational marginal pricing. The numerical analysis brings out the significance of building integrated storage investment and working mechanisms, while market regulation/schemes focusing simply on EES are unable to produce socially optimal solutions. Das and Bass [53] presented an overview of optimal ESS deployment, size, and operation in power networks in their study. Flywheel energy storage (FES) should also be considered in several distribution network situations. There are many different types of ESSs, each with its own set of benefits and drawbacks. The best ESS for you will be determined by the projected performance improvements, features, and application types. Researchers have already devised various meta-heuristic methodologies for optimization, but there is always room for improvement. Thopil, Bansal, Zhang, and Sharma observed in their research [54] that the abundance of coal-powered generation is not practical, mostly because renewable energy is not yet ready to be the dominant source of energy. Adopting a hybrid and bidirectional energy paradigm, in which customers remain connected to the grid while being fueled by renewable energy sources via small- and medium-scale distributed generators that may be put within the consumer's premises, is suggested as a realistic alternative.

Hirsch (a) defines a microgrid and (b) gives a multidisciplinary portrayal of today's microgrid drivers, practical applications, problems, and future possibilities in the review paper [55]. Proper planning and understanding is needed well in advance to find the most suitable architecture to integrate various distributed energy resources. Various factors including regulations, legal issues, quality of power and financial benefits, etc. will play major roles in deciding the sustainability of microgrids in the long run. Howlader's work [56] on independent ESS to minimize profit uncertainty for retailers in the ISO Market highlighted the problem of financial burden of hour ahead considering load mismatch. This has also concentrated on lowering the cost of IESS installation. This study demonstrates a novel energy market model where IESS is used to compensate for power adjustments. Furthermore, these IESS may be utilized to compensate for predicting errors and solve a variety of other problems. Kong and Jung's research [57] study presents a way for estimating the amount of ESS when there is inadequate data for future PV and WT providers. The predicted ESS size differs from the optimal size with the least amount of error. For future RES suppliers to enhance their profitability, the suggested approach employing polynomial regression is utilized to predict the ESS magnitude. Akbari-Dibavar [58] explored the suitable energy managing techniques of a net-zero emission MPGS incorporating RERs, hydrogen energy systems, and storage units in a deregulate scenario. The robust optimization technique was used to analyze the impacts of wind power uncertainty in order to provide an acceptable level of resilience for the system. Solar and wind power are employed for clean energy generation due to the sustainability characteristic of the micro power grid system (MPGS). Ahmad, Zhang, and Yan [59] provide unique insights into a critical and systematic review of renewable energy and power projection models used as an energy planning tool. The approaches are assessed in terms of prediction applicability, spatial and temporal forecasting accuracy, and relevance to policy and planning objectives. The study's findings help in the recognition of prediction methodologies and allow users to choose the best methods for meeting their intended aims and forecasting criteria. Forecasting capabilities are improving, and some countries are coming closer to developing fully automated smart grids.

Liu, Hu, Kimber and Wang's research gives a complete categorization and assessment of ESS electric grid applications [60]. The most recent optimization and control approaches for each application category were examined. In addition, a cost–benefit analysis for three categories of investors as well as a detailed comparison of market policies regarding ESS involvement in various wholesale markets has been performed. Given the vast variety of improvements in energy storage technologies, the energy storage technologies were critically analyzed in depth and then classified, and comparative studies were conducted to understand the features, limits, and benefits of each energy storage system. Tan, Ramachandaramurthy, Solanki, and Raveendran proposed alternative energy storage system frameworks based on their application [61]. This evaluation included several HESS combinations in which multiple ESS types were blended to provide a better form of energy storage. Mcllwaine, Morrow, Al Kez, and Best [62] undertook a rigorous study of EES and quality of power at the distribution level. The research combined with a Pugh analysis emphasized worldwide trends in power markets with increased renewable energy penetration. The investigation's findings suggest that further study is needed to classify, quantify, and evaluate the installing of bulk energy storage, during distribution.

When RE penetration is low, the electrical market functions efficiently; however, when RE penetration is high, the market is frequently disrupted. Divya Asija threw light on the advancement of renewable energy generation, the inclusion of renewables into the current unregulated power sector, the composition of present power market, main obstacles with RE integration in deregulated power markets, and driving factors [63]. A research study investigated the involvement of a composite energy system comprising wind energy and CAES in the electricity market from the standpoint of a private owner [64]. Due to the high level of unpredictability linked with market values, wind power levels, and regulatory inputs, the problem was modelled using distributionally robust optimization (DRO). The ideal outputs indicate DRO's performance in terms of higher realized earnings and less conservative results. Another study looks at the prospects, problems, and technologies of EVs in a V2G linking system in depth [65]. M.A. Hannan's study demonstrates the benefits of both the EV owners and the power system, as well as relevant suggestions for the future research areas to address existing research gaps and challenges. Dhillon, Kumar, and Singal [66] conducted a detailed analysis of the fundamentals of wind energy, PSP, Wind–PSP System and their present state, applications, and issues with operation in a deregulated market, as well as optimization strategies employed in the advance planning of Wind–PSP System. The researchers proposed optimization strategies such as EA-based, GA with LVQ, HIDSS, and NSGA-II to identify the best feasible solution of complicated computational problems with instabilities for Wind–PSP operation. Global market participants may create a new electricity market architecture in order to reap the benefits of long-term agreements with stakeholders.

Wind energy system modeling is a goal oriented problem that can be solved utilizing advanced computer methodologies. Many algorithms only engage with a sub-model and do not capture the entire picture. The research by Chinmoy, Iniyan, and Goic [67] has focused on essential cost modeling for wind energy projects as well as market associated risk and its mitigation issues. A thorough research on the use of approaches in power balancing in microgrids with renewable generators by Komala, Kumar, and Cherukuri categorized the methods into distinct categories depending on their principle of operation, infrastructure required, and component of the microgrid [68]. The different methodologies, as well as their mathematical models and virtues and drawbacks in application to power balancing in microgrids, have been evaluated. During a literature review, it was discovered that optimal usage of all forms of sustainable energy resources is critical to achieving sustainable energy development (SED). The key problem for SEH modelling is determining the best design/sizing and operating strategy for system components depending on the unpredictability of renewable sources, demand, energy market spot prices, and so on. Lasemi, Arabkoohsar, Hajizadeh, and Mohammadi-ivatloo discovered that uncertainty modelling based on RO and scenario-based stochastic optimization are the most common for SEH modelling [69]. Due to worst-case scenario analysis, a robust method would provide the greatest answer for risk-averse decision makers, whereas a probabilistic approach would provide the optimal answer for risk-neutral decision makers.

Singh and Parida [70] conducted an extensive study on the betterment of the integration of flexible demand as demand response, demand-side management (DSM), and grid proficiency. The evaluation of important data revealed that effective DG allocation will be good for the environment as well as economically favorable for utilities and customers. When DGs are incorporated into the system, the passive distribution or sub-transmission network becomes active, resulting in various technical and economic challenges. Khare, Nema, and Baredar [71] conducted a detailed evaluation of many facets of HRES, focusing on pre-feasibility analysis, optimal size, modeling, control features, and reliability issues. The use of evolutionary techniques and game theory in hybrid renewable energy systems has also been emphasized. Another study looked at current global PHES capacity, technological progress, and hybrid systems (wind-hydro, solar pv-hydro, and wind-pv-hydro) and offered the best options. According to Rehman and Al Hadhrami's research, PHES is the ideal technology for tiny autonomous island grids and huge energy storage, with PHES's efficiency fluctuating in practice between 70% and 80%, with some estimating up to 87% [72]. PHES sizes vary from 1000–1500 MW to 2000–3000 MW across the world. Photovoltaic-based pumped storage systems have only been used on a small scale (few homes only).

The purpose of this study is to provide a complete analysis of current improvements in the ADS's (Active Distribution Systems) operation from the perspective of operational time-hierarchy. In contrast to earlier review publications, prospective applications of ADS devices are evaluated in terms of operating time periods. This study by Ghadi and Ghavidel covers real-world system operations in which network components are initially planned for the stated period ahead, and then their operational status deviations from reference points are updated throughout three time intervals encompassing static, dynamic, and transient periods [73]. There is always a need for DN organizations to investigate current facilities and management systems and then provide some unique practical solutions in the related areas. A critical analysis conducted by Banshwar and Sharma [74] examined the prospects of RES in energy and Ancillary Services (AS) markets and concluded that changes in market designs and norms are still needed in the existing electrical market to integrate energy, AS, and variable energy sources. In another work by Kim and Suharto, storage methods and additional assessments of similar technologies conducted by other scholars were examined [75]. The work has explained the solution techniques to address different difficulties using a case study and also reviewed the assessment parameters.

Tables 1–3 display the summary of reports for considered objective functions, applied system details, and used optimization techniques for the considered pieces of literature. Ghadi and Rajabi's [76] insightful work on the transformation of traditional passive DNs into ADSs, as well as the study based on grid operational features engaged in deregulated electricity market at the distribution level, has provided a new perspective. This study underlines the need to optimize current facility capacity through creative management strategies and practical solutions. Saboori and Hemmati [77] evaluated the challenges of optimal bus position, power rating and energy capacity estimation in distribution networks to improve the functioning of the optimal planning process. While analyzing, energy storage systems and models, as well as their applications and related objective functions, network modelling, solution methodologies and problem uncertainty management, were all taken into account. Zhou and Li's work provides an insight of the design and functional modules of smart HEMS [78], which is critical for a more secure and environmentally friendly energy supply for smart grids. For the purposes of analysis, various non-traditional sources have been considered.

Carreiro and Jorge underline the importance of energy management system aggregators in the Smart Grid framework, particularly in conjunction with demand response programs and technologies that include end-user participation in the provision of ancillary services [79]. They suggest that establishing algorithms, technological benchmarks, and low-cost systems requires deliberate collaboration among academics, industry, and regulators. Modern power management evaluates the performance of various green energy sources against several criteria rather than focusing on a single factor—consumption [80]. This study by Bhowmik and Ray examines the diverse work on separate techniques, integrated approaches, multi-criteria decision-making methodologies, and so on for the green energy planning and scheduling challenge. This study not only confirms that energy management tactics are superior to previous ways, but it also assists scholars and policymakers in implementing the processes. Sundararagavan, Sandhya's research [81] examines the prices of several energy storage systems and identifies the critical aspects that influence their economic feasibility. Rong-Gang Cong [82] identifies several important factors affecting the expansion of renewable energy generation in this article based on a review of current research. Following extensive research, a novel optimization model is developed to optimize future renewable energy generation through the best capacity planning, while taking into account various constraints such as economic, technological, and others. In paper [83], Helder Lopes extensively analyzed several energy storage devices with varying attributes and degrees of maturity. Power rating, discharge duration, energy density in terms of weight and volume, power density, effectiveness, time and cycle durability, and availability have all been compared. Aggarwal, Sanjeev Kumar [84] provides an overview of several price-forecasting approaches used in deregulated systems, as well as an analysis of important difficulties. Lixin Tang [85] presented a policy for a deregulated method to decrease CO_2 emissions in generator scheduling for thermal power stations in his study. The proposal called for a new penalty component depending on emissions. The scheduling maximizes generation profitability based on income gained from sales, cost of generating, and the emissions penalty. Enrique B. CEDEO [86] examines the numerous relationships between the various sections of the deregulated power industry, proposing an integrated model for increasing generation and transmission capacity. The purpose of this methodology is to evaluate and find the best macroeconomic indicative investment ideas.

Table 1. Summary of reports for considered objective function in the literature.

Paper ID	Type		Objective Function			
	Regulated	Deregulated	Profit Max.	Loss Min.	Gen. Cost Min	Social Welfare Max.
[42]	√		√			
[43]						√
[44]	√		√			
[45]	√	√				
[46]	√	√	√			√
[47]	√	√				
[48]	√	√				
[49]	√	√		√		
[50]		√	√	√		
[51]		√	√			
[52]	√	√	√			√
[53]	√	√	√			
[54]		√	√	√		
[56]		√	√	√		
[57]		√	√			
[58]		√	√			
[59]		√				
[60]		√				
[63]		√				
[65]		√	√			
[66]		√				
[67]		√	√			
[68]		√				√
[69]		√				
[70]		√				
[72]	√	√				
[73]		√				
[74]	√	√				
[75]		√				
[76]		√				
[78]		√				
[80]	√					
[81]		√	√			

In paper [87], Pavlos S. Georgilakis proposes a genetic algorithm (GA) solution to the price-based unit commitment problem, which is used by each producing business to maximize its profit in a deregulated market by optimizing its generation schedule. Luo Xing's [88] provides a comprehensive comparison of the most cutting-edge energy storage methods. The study helps to alleviate the problem of selecting acceptable EES technology for a given application and deciding where they would be best integrated into a power generation and distribution system. In his work, Moein Parastegari [89] develops an optimization model for the energy market that includes auxiliary services. The model is used to jointly operate wind farms (WF), pump-storage units (PSU), photovoltaic (PV), and energy storage devices (ESD). The model takes into account WPG, energy and reserve prices, and PV generation unpredictability. A. Zahedi [90] investigated the potential benefits of grid-connected renewable energy-distributed generating in this review paper (RE-DG). It also looked at the factors that are driving the rising use of RE-DG, the technical challenges that come with high RE-DG penetration, and the effect of RE-DG connection points on system voltage. Piyasak Poonpun provided a study on the life-cycle cost of several grid-connected electric energy storage systems in the paper [91]. The results are given as a cost per kilowatt hour of stored and released power. Das [92] how energy storage can curtail risk factors in a competitive power system. In this study, Stephen Frank [93] examines numerous optimization algorithms that have been utilized to achieve optimal power flow (OPF), with an emphasis on their benefits, downsides, and computational aspects. It begins with an overview and then delves into the deterministic optimization methodologies utilized on OPF.

Table 2. Summary of reports for considered system details along with energy storage and renewable energy sources.

Paper ID	Renewable Energy Sources					Energy Storage				
	Wind	Solar	Hydro	Others	Generalized	PSH	Battery	CAES	Others	Generalized
[42]				EDV			√			
[43]						√	√	√		
[44]	√	√	√	Bio, Geo, Hybrid		√	√			
[45]	√	√	√	Bio, Geo				√		
[46]						√				
[47]						√		√	flywheel storage, electrochemical storage	
[48]	√	√								
[49]	√	√	√						EV	
[50]	√						√			
[51]	√	√	√				√		EV	
[52]	√	√	√			√	√	√		
[53]						√	√	√		
[54]	√									
[56]	√	√	√				√			
[57]	√	√					√			
[58]					√				Fuel cell, Hydrogen energy storage	
[59]	√	√		Geothermal						
[60]						√	√	√	√	
[63]					√	√	√	√	Flywheel, thermal	
[65]	√							√		
[66]							√			
[67]	√					√				
[68]	√									
[69]					√	√	√	√	√	√
[70]					√			√		

Table 2. *Cont.*

Paper ID	Renewable Energy Sources					Energy Storage				
	Wind	Solar	Hydro	Others	Generalized	PSH	Battery	CAES	Others	Generalized
[72]	√	√								√
[73]						√				
[74]					√					√
[75]					√					√
[76]						√	√			
[78]						√		√		
[80]					√		√			√
[81]					√					√

Table 3. Summary of reports for used optimization techniques in the literature.

Paper ID	Optimization Techniques					
	PSO	ABC	BAT	GA	Heuristic	Others
[44]	√	√		√		Lagrangian relaxation, quadratic programming and Nelder–Mead Simplex search; heuristic optimization methods, especially genetic algorithms and particle swarm optimization; Pareto-based multi-objective optimization
[45]				√	√	simplex method, dynamic programming, Lagrangian relaxation, sequential quadratic programming, Newton's method and reduced gradient method
[50]					√	LMP
[51]				√		
[53]		√		√		
[57]			√			
[58]						adjustable robust optimization
[65]						robust optimization
[67]				√		
[70]						Monte Carlo
[72]	√			√		Game theory
[81]	√			√		MOCPSO

Ramesh Kumar Selvaraju [94] investigated the efficacy of a deregulated electricity system combined with various energy storage technologies in this study. For determining the LFC controller gain values in a deregulated environment, the Artificial Cooperative Search technique, a new two-population-based optimization strategy, is devised. In paper [95], Patil examines the impact of wind energy system on a deregulated energy market from different perspectives. Bus sensitivity factor and locational marginal pricing have been given special attention. Different optimization algorithms have been investigated and slime mold algorithm has been implemented for the first time in this field. In another work [96], same author examines a hybrid system with energy storage and studies profit maximization in deregulated energy market with imbalance cost improvement. It also covers value at risk and cumulative value at risk factors. In paper [97], Ustun examined integration of EV storage with local solar generation to maximize renewable energy capture without overburdening local distribution network. Driving patterns and solar generation profile are studied along with local load profile to actively control EV batteries to maximize local renewable energy capture,

5. Facts and Analysis of Renewable Energy: A Glimpse

A more significant change in the generating mix is hidden by the total power generation's comparatively high resilience. In particular, generation from renewable sources (wind, solar, biofuels, and geothermal energy, etc.) saw its greatest ever growth despite the decline in overall power consumption. Strong gains in the generation of wind and solar energy were the main drivers of this expansion [98].

The proportion of renewable energy in the world's generation has increased at its quickest rate ever. Around 60% of the increase in worldwide power output over the previous five years has come from renewable sources, with wind and solar power being major among them (shown in Figure 8) [98,99].

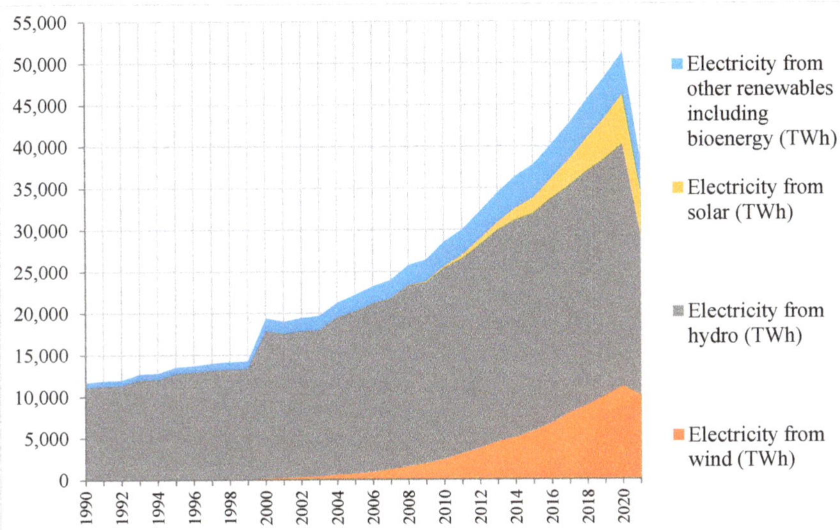

Figure 8. Transition of renewable energy generation around the world.

An emerging market economy is a developing country's economy that is getting increasingly involved in global markets as it expands. A developing economy is one with a low human development index, low growth, low per capita income, and a preference for agriculture-based activities over industrialization and entrepreneurship. In other terms, a developing economy is also known as a developing country or a less developed economy. With increased infrastructure expenditure in Europe, China, and the United States, investments in power networks are anticipated to increase by 10% in future after dropping for the fourth straight year in 2020 due to the COVID-19 epidemic. As part of the effort to attain carbon-free power generation, measures to build more robust and digital grids are being incorporated with ambitious growth and recovery plans.

However, in the Net Zero Emissions by 2050 Scenario, the level of grid investment triples by 2030, particularly for smart grids and digital investments, which should make up around 40% of all investments in this decade (shown in Table 4 and Figure 9) [98,99].

Table 4. Investment spending in electricity networks by region, 2016–2021 in USD billion.

Region	2016	2017	2018	2019	2020	2021
USA	63.1	65	66	71	75.8	77.1
China	86.9	83.7	83.2	76.6	70.7	82.6
Emerging market and developing economies	93.9	88.2	81.1	63.5	53.5	60
Europe	50	48.7	49.5	48.5	51.8	56.7
Rest of the world	17	17.5	15.9	12.6	10.8	12.6

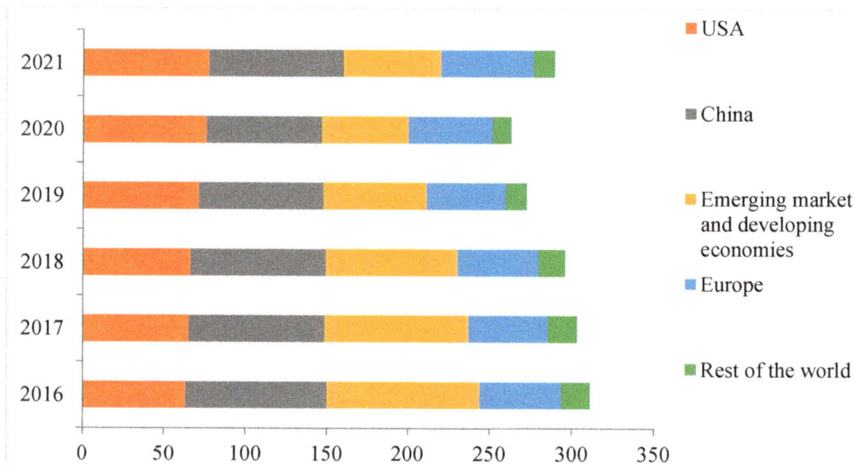

Figure 9. Investment spending in electricity networks by region.

The maximum net generating capacity of power plants and other facilities that employ renewable energy sources to create electricity is used to measure the capability of renewable power generation. The data shows the installed and connected capacity at the end of the calendar year for the majority of nations and technologies (shown in Figures 10 and 11) [99–101].

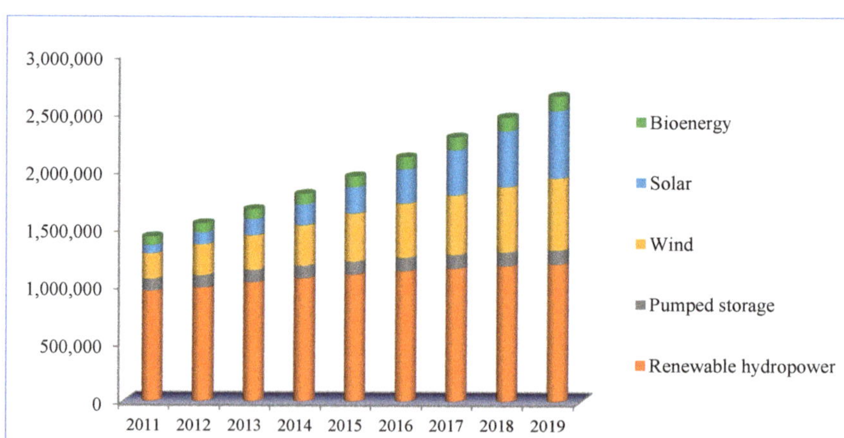

Figure 10. Worldwide renewable electricity capacity (MW) statistics.

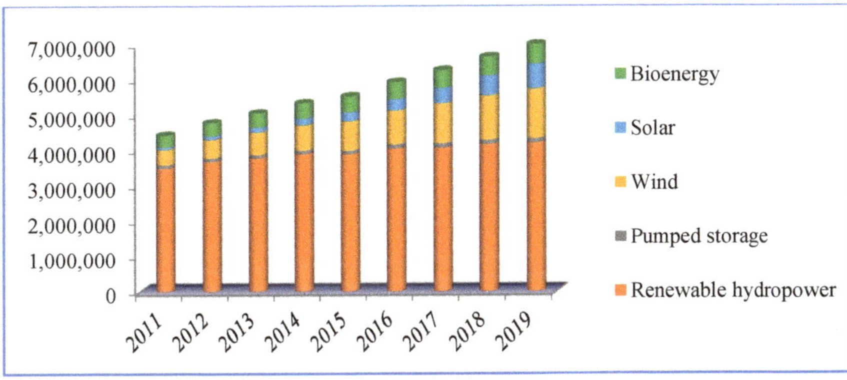

Figure 11. Worldwide renewable electricity generation (MW) statistics.

Comprehensive Energy Storage Roadmap (India)

India has set a target of 40% non-fossil power production in the energy mix by 2030 and is dedicated to lowering GHG emission intensity by 33 to 35% from the level in 2005. In order to achieve this, the percentage of renewable energy (RE) must be scaled up above and above the current goal of 175 GW by 2022. In the upcoming years, grid operators will face a challenge in ensuring grid reliability and the supply of 24 × 7 quality power due to the increased penetration of renewable energy sources and electric vehicles (EV). This will open the door for the deployment of energy storage systems for grid support [102,103]. This will enable utilities to understand the economic opportunities of such systems at various levels of RE and EV penetrations, as well as their impact on grid reliability (shown in Figure 12) [104].

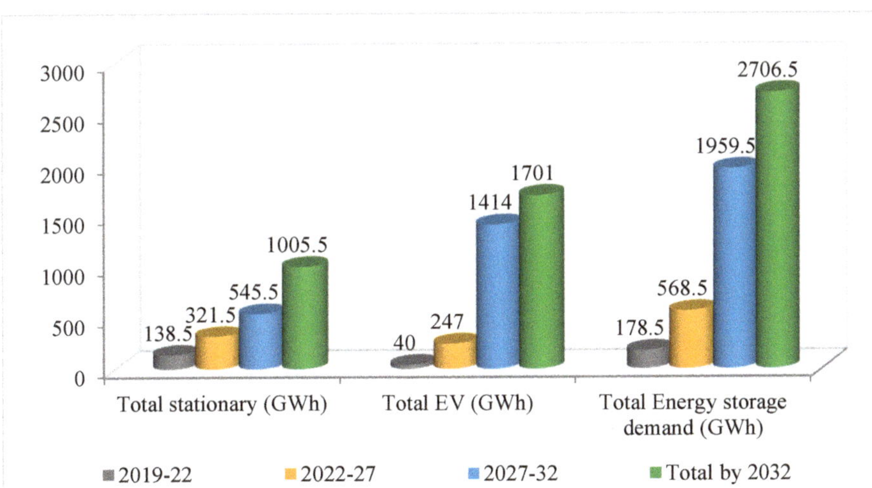

Figure 12. Comprehensive energy storage roadmap of India.

From this section, it is observed that the use of renewable energy system is not an option, it is essential [105]. Due to the discontinuous availability of renewable energy sources, energy storage system is essential for any renewable integrated power system [106,107]. This is especially true for off-grid systems that are more vulnerable to system deviations [108–118]. It may come in different forms, such as hydrogen storage [119], EV battery applications [120] or together with other novel devices such as smart inverters [121–124] In this scenario, this paper provides the clear idea about the different types of energy storage system with the constructions and applications [125].

6. Comparative Study of EES Systems

The comparative study of different types of EEs systems are depicted in Table 5 [126–129] and Table 6 [130–132]. The efficiency, discharge time, cost, and environmental impacts of EES systems are considered for this study.

Table 5. Comparison of EES Systems in terms of efficiency, discharge time and cost [126–129].

System	Max. Power Rating (MW)	Efficiency (%)	Discharge Time	Cost/KW (USD)	Cost/KWh (USD)	Energy Density (Wh/L)
PHS	3000	70–85	4 h–16 h	600–2000	5–100	0.2–2
CAES	1000	40–70	2 h–30 h	400–800	2–50	2–6
FES	20	70–95	sec–mins	250–350	1000–5000	20–80
Lead-acid	100	80–90	1 min–8 h	300–600	200–400	50–80
NiCd/NiMH	40		sec–hours	500–1500	800–1500	60–150
Li-ion	100	85–95	1 min–8 h	1200–4000	600–2500	200–400
Metal-air	0.01	50	secs–day	100–250	10–60	500–10,000
Sodium-sulfur	0.05–8	75–90	sec–hours	1000–3000	300–500	150–250
RFB/HFB	100	60–85	hours	700–2500	150–1000	20–70
H2	100	25–45	min–week		10	600
Fuel Cell	50	60–80	secs–day	10,000		500–3000
SMES	10 MW	95	millisec–secs	200–300	1000–10,000	0.2–2.5
Thermal	150	80–90	hours	200–300	30–60	70–210

Table 6. Comparison of EES Systems in terms environmental impact [130–132].

System	Life Time/Cycles	Environmental Impact	Description of Impact
PHS	30–60 years	-ve	Cutting trees and landscapes for reservoirs
CAES	20–40 years	-ve	Remains from fossil fuel
FES	20,000–100,000	Negligible	
Lead-acid	6–40 years	-ve	Toxic residues
NiCd/NiMH	10–20 years	-ve	Toxic residues
Li-ion	1000–10,000	-ve	Toxic residues
Metal-air	100–300	Very small	Slight residues
Sodium-sulphur	10–15 years	-ve	Toxic residues
RFB/HFB	12,000–14,000	-ve	Toxic residues
H2	5–30 years	Yes	Emission of hydrogen in atmosphere can create disturb in distribution of methane and ozone; thereby causing imbalance.
Fuel Cell	5–15 years	-ve	Remains from fossil fuel
SMES	20 years	-ve	High magnetic field
Thermal	30 years	Small	Releasing charge into atmosphere

7. Conclusions

Presently, while the entire world is concerned about the future of the planet in terms of reducing the carbon footprint and making it a greener one, the electricity industry is focusing on more efficient and sustainable power supply, judicious consumption of energy and CO_2 reduction. While doing so, main areas of research are identified as anticipated growth of renewable generation, design of renewable technology for better performance, design and implementation of smart grids, integration of RNE into smart grid for better energy demand management by optimization techniques.

The followings are important in the present scenario of the electrical system:

- Energy storage systems will play a pivotal role for managing contingency situations apart from acting as integrated part of smart grid.
- The modest and scattered EES market is likely to be large when the smart grid and microgrids are implemented.
- The market for EES systems, particularly small and distributed ones, is growing and will grow in tandem with the renewable energy sector.
- Technical challenges, and also cost and compatibility/sustainability, are potentially critical topics for future initiatives.
- There is scope to work on optimization, power quality and safety issues.
- Upon comparison of different optimization techniques, it has been found that meta-heuristic algorithms outperformed heuristic and linear optimization techniques with the considered objective functions.

Considering the future and investors' interest, it is obvious that the maximization of system profit and minimization of system generation cost and loss will help in increasing societal benefit. This study examines numerous aspects of renewable integrated deregulated power systems and provides an in-depth appraisal of the most recent research advances in this sector. In this context, this study will be helpful in understanding, analyzing and applying the EES technologies for a better tomorrow.

Author Contributions: Conceptualization, M.R.C. and S.D.; methodology, M.R.C., S.D. and J.B.B.; software, M.R.C.; validation, P.K.S. and T.S.U.; formal analysis, S.D.; investigation, M.R.C. and J.B.B.; resources, M.R.C. and S.D.; data curation, P.K.S. and T.S.U.; writing—original draft preparation, M.R.C. and J.B.B.; writing—review and editing, S.D., P.K.S. and T.S.U.; visualization, S.D.; supervision, P.K.S. and T.S.U.; project administration, T.S.U.; funding acquisition, T.S.U. All authors have read and agreed to the published version of the manuscript.

Funding: This research received no external funding.

Institutional Review Board Statement: Not applicable.

Informed Consent Statement: Not applicable.

Data Availability Statement: Not applicable.

Conflicts of Interest: The authors declare no conflict of interest.

References

1. Barik, A.K.; Das, D.; Latif, A.; Hussain, S.; Ustun, T. Optimal Voltage–Frequency Regulation in Distributed Sustainable Energy-Based Hybrid Microgrids with Integrated Resource Planning. *Energies* **2021**, *14*, 2735. [CrossRef]
2. International Renewable Energy Agency. Electricity Storage and Renewables: Costs and Markets to 2030. 2017. Available online: https://www.irena.org/-/media/Files/IRENA/Agency/Publication/2017/Oct/IRENA_Electricity_Storage_Costs_2017_Summary.pdf (accessed on 6 June 2022).
3. Hussain, S.M.S.; Aftab, M.A.; Ali, I.; Ustun, T.S. IEC 61850 based energy management system using plug-in electric vehicles and distributed generators during emergencies. *Int. J. Electr. Power Energy Syst.* **2020**, *119*, 105873. [CrossRef]
4. Nadeem, F.; Hussain, S.S.; Tiwari, P.K.; Goswami, A.K.; Ustun, T.S. Review of Smart and Innovative Energy Storage Systems. In Proceedings of the 2019 International Conference on Vision Towards Emerging Trends in Communication and Networking (ViTECoN), Vellore, India, 30–31 March 2019; pp. 1–6.
5. Buchana, P.; Ustun, T.S. The role of microgrids & renewable energy in addressing Sub-Saharan Africa's current and future energy needs. In Proceedings of the IREC2015 The Sixth International Renewable Energy Congress, Sousse, Tunisia, 24–26 March 2015; pp. 1–6.
6. Niyigena, D.; Habineza, C.; Ustun, T.S. Computer-based smart energy management system for rural health centers. In Proceedings of the 2015 3rd International Renewable and Sustainable Energy Conference (IRSEC), Marrakech, Morocco, 10–13 December 2015; pp. 1–5.

7. International Renewable Energy Agency. From Baseload to Peak: Renewables Provide a Reliable Solution. 2015. Available online: https://www.irena.org/-/media/Files/IRENA/Agency/Publication/2015/IRENA_Baseload_to_Peak_2015.pdf (accessed on 18 June 2022).
8. Ustun, T.S.; Nakamura, Y.; Hashimoto, J.; Otani, K. Performance Analysis of PV Panels Based on Different Technologies after Two Years of Outdoor Exposure in Fukushima, Japan. *Renew. Energy* **2019**, *136*, 159–178. [CrossRef]
9. Waseem, M.; Manshadi, S.D. Electricity grid resilience amid various natural disasters: Challenges and solutions. *Electr. J.* **2020**, *33*, 106864. [CrossRef]
10. Available online: https://clouglobal.com/the-roles-of-electrical-energy-storage/ (accessed on 18 June 2022).
11. Stawska, A.; Romero, N.; de Weerdt, M.; Verzijlbergh, R. Demand response: For congestion management or for grid balancing? *Energy Policy* **2020**, *148*, 111920. [CrossRef]
12. Ustun, T.S.; Hussain, S.M.S.; Syed, M.H.; Dambrauskas, P. IEC-61850-Based Communication for Integrated EV Management in Power Systems with Renewable Penetration. *Energies* **2021**, *14*, 2493. [CrossRef]
13. Brenna, M.; Foiadelli, F.; Leone, C.; Longo, M. Electric Vehicles Charging Technology Review and Optimal Size Estimation. *J. Electr. Eng. Technol.* **2020**, *15*, 2539–2552. [CrossRef]
14. Impram, S.; Nese, S.V.; Oral, B. Challenges of renewable energy penetration on power system flexibility: A survey. *Energy Strat. Rev.* **2020**, *31*, 100539. [CrossRef]
15. Ustun, T.S.; Hussain, S.M.S. Standardized communication model for home energy management system. *IEEE Access* **2020**, *8*, 180067–180075. [CrossRef]
16. European Energy Research Alliance. Energy Storage. Available online: https://www.eera-energystorage.eu/about/sub-programmes.html (accessed on 23 August 2022).
17. Available online: https://energystorage.org/why-energy-storage/technologies/ (accessed on 15 June 2022).
18. Available online: https://www.hydropower.org/factsheets/pumped-storage (accessed on 15 June 2022).
19. Ruiz, R.A.; de Vilder, L.; Prasasti, E.; Aouad, M.; De Luca, A.; Geisseler, B.; Terheiden, K.; Scanu, S.; Miccoli, A.; Roeber, V.; et al. Low-head pumped hydro storage: A review on civil structure designs, legal and environmental aspects to make its realization feasible in seawater. *Renew. Sustain. Energy Rev.* **2022**, *160*, 112281. [CrossRef]
20. Available online: https://www.ctc-n.org/technologies/compressed-air-energy-storage-caes (accessed on 16 June 2022).
21. Pullen, K.R.; Dhand, A. Mechanical and electrical flywheel hybrid technology to store energy in vehicles. In *Alternative Fuels and Advanced Vehicle Technologies for Improved Environmental Performance*; Folkson, R., Ed.; Woodhead Publishing: Sawston, UK, 2014; pp. 476–504. ISBN 9780857095220. [CrossRef]
22. Keshan, H.; Thornburg, J.; Ustun, T.S. Comparison of lead-acid and lithium ion batteries for stationary storage in off-grid energy systems. In Proceedings of the 4th IET Clean Energy and Technology Conference (CEAT 2016), Kuala Lumpur, Malaysia, 14–15 November 2016.
23. Rodrigues, E.; Osório, G.; Godina, R.; Bizuayehu, A.; Lujano-Rojas, J.; Matias, J.; Catalão, J. Modelling and sizing of NaS (sodium sulfur) battery energy storage system for extending wind power performance in Crete Island. *Energy* **2015**, *90*, 1606–1617. [CrossRef]
24. Dustmann, C.-H. Advances in ZEBRA batteries. Journal of Power Sources. *J. Power Sources* **2004**, *127*, 85–92. [CrossRef]
25. Ravikumar, M.K.; Rathod, S.; Jaiswal, N.; Patil, S.; Shukla, A. The renaissance in redox flow batteries. *J. Solid State Electrochem.* **2016**, *21*, 2467–2488. [CrossRef]
26. Hirscher, M.; Yartys, V.A.; Baricco, M.; von Colbe, J.B.; Blanchard, D.; Bowman, R.C., Jr.; Broom, D.P.; Buckley, C.E.; Chang, F.; Chen, P.; et al. Materials for hydrogen-based energy storage—Past, recent progress and future outlook. *J. Alloys Compd.* **2020**, *827*, 153548. [CrossRef]
27. Götz, M.; Lefebvre, J.; Mörs, F.; McDaniel Koch, A.; Graf, F.; Bajohr, S.; Reimert, R.; Kolb, T. Renewable Power-to-Gas: A technological and economic review. *Renew. Energy* **2016**, *85*, 1371–1390. [CrossRef]
28. Cavanagh, K.; Ward, J.K.; Behrens, S.; Bhatt, A.I.; Ratnam, E.L.; Oliver, E.; Hayward, J. *Electrical Energy Storage: Technology Overview and Applications*; CSIRO: Canberra, Australia, 2015; EP154168.
29. Sagadevan, S.; Marlinda, A.; Chowdhury, Z.Z.; Wahab, Y.B.A.; Hamizi, N.A.; Shahid, M.; Mohammad, F.; Podder, J.; Johan, M.R.F. Chapter two—Fundamental electrochemical energy storage systems. In *Advances in Supercapacitor and Supercapattery*; Elsevier: Amsterdam, The Netherlands, 2021; pp. 27–43. [CrossRef]
30. Available online: https://www.sciencedirect.com/topics/engineering/secondary-battery (accessed on 15 June 2022).
31. Wade, N.S.; Taylor, P.C. *VDE—ETG Energy Storage Task Force: Energy Storage in Power Supply Systems with a High Share of Renewable Energy Sources Significance—State of the Art—Need for Action*; VDE: Frankfurt am Main, Germany, December 2008.
32. Perrin, M.; Malbranche, P.; Mattera, F.; Simonin, L.; Sauer, D.U.; Lailler, P.; Joessen, A.; Willer, B.; Dahlen, M.; Ruddell, A.; et al. Evaluation and perspectives of storage technologies for PV electricity. In Proceedings of the 3rd World Conference on Photovoltaic Energy Conversion, Osaka, Japan, 11–18 May 2003; Volume 3, pp. 2194–2197.
33. Espinar, B.; Mayer, D. *The Role of Energy Storage for Mini-Grid Stabilization*; Report IEA-PVPS T11-0X:2011; MINES Paris-Tech/ARMINES: Paris, France, 2011.
34. Atwater, T.B.; Dobley, A. Metal/Air batteries. In *Lindens Handbook of Batteries*; Mcgraw-Hill Professional: New York, NY, USA, 2011; ISBN 978-0-07-162421-X.
35. Worth, B. *Metal/Air*; INVESTIRE: Rome, Italy, 2002.

36. Nadeem, F.; Hussain, S.M.S.; Tiwari, P.K.; Goswami, A.K.; Ustun, T.S. Comparative Review of Energy Storage Systems, Their Roles, and Impacts on Future Power Systems. *IEEE Access* **2019**, *7*, 4555–4585. [CrossRef]
37. Patel, P. Available online: http://spectrum.ieee.org/energy/the-smarter-grid/batteries-that-go-with-the-flow (accessed on 2 July 2022).
38. Sterner, M. Bioenergy and Renewable Power Methane in Integrated 100% Renewable Energy Systems—Limiting Global Warming by Transforming Energy Systems. Ph.D. Thesis, University Kassel, Kassel, Germany, July 2009.
39. International Energy Agency. Prospects for Large Scale Energy Storage in Decarbonised Grids. Report. 2009. Available online: https://www.iea.org/reports/prospects-for-large-scale-energy-storage-in-decarbonised-power-grids (accessed on 12 July 2022).
40. Schossig, P. Thermal Energy Storage. In Proceedings of the 3rd International Renewable Energy Storage Conference, Berlin, Germany, 24–25 November 2008.
41. Fairley, P. Available online: http://spectrum.ieee.org/energy/environment/largest-solar-thermal-storage-plant-to-start-up (accessed on 15 July 2022).
42. Divya, K.C.; Østergaard, J. Battery energy storage technology for power systems—An overview. *Electr. Power Syst. Res.* **2009**, *79*, 511–520. [CrossRef]
43. Chen, H.; Cong, T.N.; Yang, W.; Tan, C.; Li, Y.; Ding, Y. Progress in electrical energy storage system: A critical review. *Prog. Nat. Sci.* **2009**, *19*, 291–312. [CrossRef]
44. Baños, R.; Manzano-Agugliaro, F.; Montoya, F.; Gil, C.; Alcayde, A.; Gómez, J. Optimization methods applied to renewable and sustainable energy: A review. *Renew. Sustain. Energy Rev.* **2011**, *15*, 1753–1766. [CrossRef]
45. Bazmi, A.A.; Zahedi, G. Sustainable energy systems: Role of optimization modeling techniques in power generation and supply—A review. *Renew. Sustain. Energy Rev.* **2011**, *15*, 3480–3500. [CrossRef]
46. Wang, Z.; Gu, C.; Li, F.; Bale, P.; Sun, H. Active Demand Response Using Shared Energy Storage for Household Energy Management. *IEEE Trans. Smart Grid* **2013**, *4*, 1888–1897. [CrossRef]
47. Koohi-Kamali, S.; Tyagi, V.; Rahim, N.; Panwar, N.; Mokhlis, H. Emergence of energy storage technologies as the solution for reliable operation of smart power systems: A review. *Renew. Sustain. Energy Rev.* **2013**, *25*, 135–165. [CrossRef]
48. Yanine, F.F.; Sauma, E.E. Review of grid-tie micro-generation systems without energy storage: Towards a new approach to sustainable hybrid energy systems linked to energy efficiency. *Renew. Sustain. Energy Rev.* **2013**, *26*, 60–95. [CrossRef]
49. Mwasilu, F.; Justo, J.J.; Kim, E.-K.; Do, T.D.; Jung, J.-W. Electric vehicles and smart grid interaction: A review on vehicle to grid and renewable energy sources integration. *Renew. Sustain. Energy Rev.* **2014**, *34*, 501–516. [CrossRef]
50. Zhang, Z.; Zhang, Y.; Lee, W.-J. Energy storage based optimal dispatch scheme for financial improvement and fluctuation mitigation on wind power generation. In Proceedings of the 2017 IEEE Industry Applications Society Annual Meeting, Cincinnati, OH, USA, 1–5 October 2017; pp. 1–7. [CrossRef]
51. Muruganantham, B.; Gnanadass, R.; Padhy, N.P. Challenges with renewable energy sources and storage in practical distribution systems. *Renew. Sustain. Energy Rev.* **2017**, *73*, 125–134. [CrossRef]
52. Huang, Q.; Xu, Y.; Courcoubetis, C. Market Mechanisms for Energy Storage Planning and Operation in a Power Network. In Proceedings of the 2018 Annual American Control Conference (ACC), Milwaukee, WI, USA, 27–29 June 2018; pp. 3869–3874. [CrossRef]
53. Das, C.K.; Bass, O.; Kothapalli, G.; Mahmoud, T.S.; Habibi, D. Overview of energy storage systems in distribution networks: Placement, sizing, operation, and power quality. *Renew. Sustain. Energy Rev.* **2018**, *91*, 1205–1230. [CrossRef]
54. Thopil, M.; Bansal, R.; Zhang, L.; Sharma, G. A review of grid connected distributed generation using renewable energy sources in South Africa. *Energy Strat. Rev.* **2018**, *21*, 88–97. [CrossRef]
55. Hirsch, A.; Parag, Y.; Guerrero, J. Microgrids: A review of technologies, key drivers, and outstanding issues. *Renew. Sustain. Energy Rev.* **2018**, *90*, 402–411. [CrossRef]
56. Howlader, H.O.R.; Akter, H.; Saber, A.Y.; Mandal, P.; Krishna, N.; Senjyu, T. Independent Energy Storage Systems can Minimize Uncertainty of Profit for Retailers in ISO Market. In Proceedings of the 2019 IEEE International Conference on Electro Information Technology (EIT), Brookings, SD, USA, 20–22 May 2019; pp. 584–589. [CrossRef]
57. Kong, J.; Kim, S.T.; O Kang, B.; Jung, J. Determining the size of energy storage system to maximize the economic profit for photovoltaic and wind turbine generators in South Korea. *Renew. Sustain. Energy Rev.* **2019**, *116*, 109467. [CrossRef]
58. Akbari-Dibavar, A.; Daneshvar, M.; Mohammadi-Ivatloo, B.; Zare, K.; Anvari-Moghaddam, A. Optimal Robust Energy Management of Microgrid with Fuel Cells, Hydrogen Energy Storage Units and Responsive Loads. In Proceedings of the 2020 International Conference on Smart Energy Systems and Technologies (SEST), Istanbul, Turkey, 7–9 September 2020; pp. 1–6. [CrossRef]
59. Ahmad, T.; Zhang, H.; Yan, B. A review on renewable energy and electricity requirement forecasting models for smart grid and buildings. *Sustain. Cities Soc.* **2020**, *55*, 102052. [CrossRef]
60. Liu, J.; Hu, C.; Kimber, A.; Wang, Z. Uses, Cost-Benefit Analysis, and Markets of Energy Storage Systems for Electric Grid Applications. *J. Energy Storage* **2020**, *32*, 101731. [CrossRef]
61. Tan, K.M.; Babu, T.S.; Ramachandaramurthy, V.K.; Kasinathan, P.; Solanki, S.G.; Raveendran, S.K. Empowering smart grid: A comprehensive review of energy storage technology and application with renewable energy integration. *J. Energy Storage* **2021**, *39*, 102591. [CrossRef]

62. McIlwaine, N.; Foley, A.M.; Morrow, D.J.; Al Kez, D.; Zhang, C.; Lu, X.; Best, R.J. A state-of-the-art techno-economic review of distributed and embedded energy storage for energy systems. *Energy* **2021**, *229*, 120461. [CrossRef]
63. Asija, D.; Viral, R. Chapter 13—Renewable energy integration in modern deregulated power system: Challenges, driving forces, and lessons for future road map. In *Advances in Smart Grid Power System*; Elsevier Academic Press: Cambridge, MA, USA, 2021. [CrossRef]
64. Aldaadi, M.; Al-Ismail, F.; Al-Awami, A.T.; Muqbel, A. A Coordinated Bidding Model for Wind Plant and Compressed Air Energy Storage Systems in the Energy and Ancillary Service Markets Using a Distributionally Robust Optimization Approach. *IEEE Access* **2021**, *9*, 148599–148610. [CrossRef]
65. Hannan, M.; Mollik, M.; Al-Shetwi, A.Q.; Rahman, S.; Mansor, M.; Begum, R.; Muttaqi, K.; Dong, Z. Vehicle to grid connected technologies and charging strategies: Operation, control, issues and recommendations. *J. Clean. Prod.* **2022**, *339*, 130587. [CrossRef]
66. Dhillon, J.; Kumar, A.; Singal, S. Optimization methods applied for Wind–PSP operation and scheduling under deregulated market: A review. *Renew. Sustain. Energy Rev.* **2014**, *30*, 682–700. [CrossRef]
67. Chinmoy, L.; Iniyan, S.; Goic, R. Modeling wind power investments, policies and social benefits for deregulated electricity market—A review. *Appl. Energy* **2019**, *242*, 364–377. [CrossRef]
68. Komala, K.; Kumar, K.P.; Cherukuri, S.H.C. Storage and non-Storage Methods of Power balancing to counter Uncertainty in Hybrid Microgrids—A review. *J. Energy Storage* **2021**, *36*, 102348. [CrossRef]
69. Lasemi, M.A.; Arabkoohsar, A.; Hajizadeh, A.; Mohammadi-Ivatloo, B. A comprehensive review on optimization challenges of smart energy hubs under uncertainty factors. *Renew. Sustain. Energy Rev.* **2022**, *160*, 112320. [CrossRef]
70. Singh, A.K.; Parida, S.K. A review on distributed generation allocation and planning in deregulated electricity market. *Renew. Sustain. Energy Rev.* **2018**, *82*, 4132–4141. [CrossRef]
71. Khare, V.; Nema, S.; Baredar, P. Solar–wind hybrid renewable energy system: A review. *Renew. Sustain. Energy Rev.* **2016**, *58*, 23–33. [CrossRef]
72. Rehman, S.; Al-Hadhrami, L.M.; Alam, M.M. Pumped hydro energy storage system: A technological review. *Renew. Sustain. Energy Rev.* **2015**, *44*, 586–598. [CrossRef]
73. Ghadi, M.J.; Ghavidel, S.; Rajabi, A.; Azizivahed, A.; Li, L.; Zhang, J. A review on economic and technical operation of active distribution systems. *Renew. Sustain. Energy Rev.* **2019**, *104*, 38–53. [CrossRef]
74. Banshwar, A.; Sharma, N.K.; Sood, Y.R.; Shrivastava, R. Renewable energy sources as a new participant in ancillary service markets. *Energy Strat. Rev.* **2017**, *18*, 106–120. [CrossRef]
75. Kim, J.; Suharto, Y.; Daim, T.U. Evaluation of Electrical Energy Storage (EES) technologies for renewable energy: A case from the US Pacific Northwest. *J. Energy Storage* **2017**, *11*, 25–54. [CrossRef]
76. Ghadi, M.J.; Rajabi, A.; Ghavidel, S.; Azizivahed, A.; Li, L.; Zhang, J. From active distribution systems to decentralized microgrids: A review on regulations and planning approaches based on operational factors. *Appl. Energy* **2019**, *253*, 113543. [CrossRef]
77. Saboori, H.; Hemmati, R.; Ghiasi, S.M.S.; Dehghan, S. Energy storage planning in electric power distribution networks—A state-of-the-art review. *Renew. Sustain. Energy Rev.* **2017**, *79*, 1108–1121. [CrossRef]
78. Zhou, B.; Li, W.; Chan, K.W.; Cao, Y.; Kuang, Y.; Liu, X.; Wang, X. Smart home energy management systems: Concept, configurations, and scheduling strategies. *Renew. Sustain. Energy Rev.* **2016**, *61*, 30–40. [CrossRef]
79. Carreiro, A.M.; Jorge, H.M.; Antunes, C.H. Energy management systems aggregators: A literature survey. *Renew. Sustain. Energy Rev.* **2017**, *73*, 1160–1172. [CrossRef]
80. Bhowmik, C.; Bhowmik, S.; Ray, A.; Pandey, K.M. Optimal green energy planning for sustainable development: A review. *Renew. Sustain. Energy Rev.* **2017**, *71*, 796–813. [CrossRef]
81. Sundararagavan, S.; Baker, E. Evaluating energy storage technologies for wind power integration. *Sol. Energy* **2012**, *86*, 2707–2717. [CrossRef]
82. Cong, R.-G. An optimization model for renewable energy generation and its application in China: A perspective of maximum utilization. *Renew. Sustain. Energy Rev.* **2013**, *17*, 94–103. [CrossRef]
83. Ferreira, H.L.; Garde, R.; Fulli, G.; Kling, W.; Lopes, J.P. Characterisation of electrical energy storage technologies. *Energy* **2013**, *53*, 288–298. [CrossRef]
84. Aggarwal, S.K.; Saini, L.M.; Kumar, A. Electricity price forecasting in deregulated markets: A review and evaluation. *Int. J. Electr. Power Energy Syst.* **2009**, *31*, 13–22. [CrossRef]
85. Tang, L.; Che, P. Generation Scheduling Under a CO_2 Emission Reduction Policy in the Deregulated Market. *IEEE Trans. Eng. Manag.* **2013**, *60*, 386–397. [CrossRef]
86. Cedeño, E.B.; Arora, S. Integrated transmission and generation planning model in a deregulated environment. *Front. Energy* **2013**, *7*, 182–190. [CrossRef]
87. Georgilakis, P.S. Genetic algorithm model for profit maximization of generating companies in deregulated electricity markets. *Appl. Artif. Intell. Int. J.* **2009**, *23*, 538–552. [CrossRef]
88. Luo, X.; Wang, J.; Dooner, M.; Clarke, J. Overview of current development in electrical energy storage technologies and the application potential in power system operation. *Appl. Energy* **2015**, *137*, 511–536. [CrossRef]
89. Parastegari, M.; Hooshmand, R.-A.; Khodabakhshian, A.; Zare, A.-H. Joint operation of wind farm, photovoltaic, pump-storage and energy storage devices in energy and reserve markets. *Int. J. Electr. Power Energy Syst.* **2015**, *64*, 275–284. [CrossRef]

90. Zahedi, A. A review of drivers, benefits, and challenges in integrating renewable energy sources into electricity grid. *Renew. Sustain. Energy Rev.* **2011**, *15*, 4775–4779. [CrossRef]
91. Poonpun, P.; Jewell, W.T. Analysis of the Cost per Kilowatt Hour to Store Electricity. *IEEE Trans. Energy Convers.* **2008**, *23*, 529–534. [CrossRef]
92. Das, A.; Dawn, S.; Gope, S.; Ustun, T.S. A Risk Curtailment Strategy for Solar PV-Battery Integrated Competitive Power System. *Electronics* **2022**, *11*, 1251. [CrossRef]
93. Frank, S.; Steponavice, I.; Rebennack, S. Optimal power flow: A bibliographic survey I—Formulations and deterministic methods. *Energy Syst.* **2012**, *3*, 221–258. [CrossRef]
94. Selvaraju, R.K.; Somaskandan, G. Impact of energy storage units on load frequency control of deregulated power systems. *Energy* **2016**, *97*, 214–228. [CrossRef]
95. Patil, G.S.; Mulla, A.; Dawn, S.; Ustun, T.S. Profit Maximization with Imbalance Cost Improvement by Solar PV-Battery Hybrid System in Deregulated Power Market. *Energies* **2022**, *15*, 5290. [CrossRef]
96. Patil, G.S.; Mulla, A.; Ustun, T.S. Impact of Wind Farm Integration on LMP in Deregulated Energy Markets. *Sustainability* **2022**, *14*, 4354. [CrossRef]
97. Ustun, T.S.; Hussain, S.M.S.; Kikusato, H. IEC 61850-Based Communication Modeling of EV Charge-Discharge Management for Maximum PV Generation. *IEEE Access* **2019**, *7*, 4219–4231. [CrossRef]
98. Available online: https://www.bp.com/en/global/corporate/energy-economics/statistical-review-of-world-energy/power-by-fuel.html (accessed on 8 July 2022).
99. Available online: https://www.iea.org/data-and-statistics/charts/investment-spending-in-electricity-networks-by-region-2016-2021 (accessed on 10 July 2022).
100. Available online: https://www.irena.org/publications/2021/March/Renewable-Capacity-Statistics-2021 (accessed on 10 July 2021).
101. Available online: https://ourworldindata.org/renewable-energy (accessed on 15 July 2022).
102. Dawn, S.; Gope, S.; Das, S.S.; Ustun, T.S. Social Welfare Maximization of Competitive Congested Power Market Considering Wind Farm and Pumped Hydroelectric Storage System. *Electronics* **2021**, *10*, 2611. [CrossRef]
103. Singh, N.K.; Koley, C.; Gope, S.; Dawn, S.; Ustun, T.S. An Economic Risk Analysis in Wind and Pumped Hydro Energy Storage Integrated Power System Using Meta-Heuristic Algorithm. *Sustainability* **2021**, *13*, 13542. [CrossRef]
104. Available online: https://indiasmartgrid.org/reports/ISGF_Report_Energy_Storage_System_RoadmapforIndia_2019to2032_11July2019_Draft.pdf (accessed on 7 July 2022).
105. Yadav, A.K.; Malik, H.; Hussain SM, S.; Ustun, T.S. Case Study of Grid-Connected Photovoltaic Power System Installed at Monthly Optimum Tilt Angles for Different Climatic Zones in India. *IEEE Access* **2021**, *9*, 60077–60088. [CrossRef]
106. Hubble, A.H.; Ustun, T.S. Scaling renewable energy based microgrids in underserved communities: Latin America, South Asia, and Sub-Saharan Africa. In Proceedings of the 2016 IEEE PES PowerAfrica, Livingstone, Zambia, 28 June–2 July 2016; pp. 134–138.
107. Hubble, A.H.; Ustun, T.S. Composition, placement, and economics of rural microgrids for ensuring sustainable development. Sustain. *Energy Grids Netw.* **2018**, *13*, 1–18.
108. Latif, A.; Hussain, S.M.S.; Das, D.C.; Ustun, T.S. Optimization of Two-Stage IPD-(1+I) Controllers for Frequency Regulation of Sustainable Energy Based Hybrid Microgrid Network. *Electronics* **2021**, *10*, 919. [CrossRef]
109. Kumar, K.K.P.; Soren, N.; Latif, A.; Das, D.C.; Hussain, S.M.S.; Al-Durra, A.; Ustun, T.S. Day-Ahead DSM-Integrated Hybrid-Power-Management-Incorporated CEED of Solar Thermal/Wind/Wave/BESS System Using HFPSO. *Sustainability* **2022**, *14*, 1169. [CrossRef]
110. Latif, A.; Hussain, S.M.S.; Das, D.C.; Ustun, T.S. Design and Implementation of Maiden Dual-Level Controller for Ameliorating Frequency Control in a Hybrid Microgrid. *Energies* **2021**, *14*, 2418. [CrossRef]
111. Ranjan, S.; Das, D.C.; Sinha, N.; Latif, A.; Hussain, S.M.S.; Ustun, T.S. Voltage stability assessment of isolated hybrid dish-stirling solar-thermal-diesel microgrid with STATCOM using mine blast algorithm. *Electr. Power Syst. Res.* **2021**, *196*, 107239. [CrossRef]
112. Farooq, Z.; Rahman, A.; Hussain, S.M.S.; Ustun, T.S. Power Generation Control of Renewable Energy Based Hybrid Deregulated Power System. *Energies* **2022**, *15*, 517. [CrossRef]
113. Ulutas, A.; Altas, I.H.; Onen, A.; Ustun, T.S. Neuro-Fuzzy-Based Model Predictive Energy Management for Grid Connected Microgrids. *Electronics* **2020**, *9*, 900. [CrossRef]
114. Abdolrasol, M.G.M.; Hannan, M.A.; Hussain, S.M.S.; Ustun, T.S.; Sarker, M.R.; Ker, P.J. Energy Management Scheduling for Microgrids in the Virtual Power Plant System Using Artificial Neural Networks. *Energies* **2021**, *14*, 6507. [CrossRef]
115. Hussain, I.; Das, D.C.; Sinha, N.; Latif, A.; Hussain, S.M.S.; Ustun, T.S. Performance Assessment of an Islanded Hybrid Power System with Different Storage Combinations Using an FPA-Tuned Two-Degree-of-Freedom (2DOF) Controller. *Energies* **2020**, *13*, 5610. [CrossRef]
116. Chauhan, A.; Upadhyay, S.; Khan, M.T.; Hussain, S.M.S.; Ustun, T.S. Performance Investigation of a Solar Photovoltaic/Diesel Generator Based Hybrid System with Cycle Charging Strategy Using BBO Algorithm. *Sustainability* **2021**, *13*, 8048. [CrossRef]
117. Latif, A.; Hussain, S.M.S.; Das, D.C.; Ustun, T.S. Optimum Synthesis of a BOA Optimized Novel Dual-Stage PI − (1 + ID) Controller for Frequency Response of a Microgrid. *Energies* **2020**, *13*, 3446. [CrossRef]
118. Nadeem, F.; Aftab, M.A.; Hussain, S.M.S.; Ali, I.; Tiwari, P.K.; Goswami, A.K.; Ustun, T.S. Virtual Power Plant Management in Smart Grids with XMPP Based IEC 61850 Communication. *Energies* **2019**, *12*, 2398. [CrossRef]

119. Singh, S.; Chauhan, P.; Aftab, M.A.; Ali, I.; Hussain, S.M.S.; Ustun, T.S. Cost Optimization of a Stand-Alone Hybrid Energy System with Fuel Cell and PV. *Energies* **2020**, *13*, 1295. [CrossRef]
120. Nsonga, P.; Ustun, T.S. Integration of communication standards in Electrical Vehicle Ad-Hoc Networks for smartgrid support. In Proceedings of the 2016 IEEE International Conference on Emerging Technologies and Innovative Business Practices for the Transformation of Societies (EmergiTech), Balaclava, Mauritius, 3–6 August 2016; pp. 106–111.
121. Hashimoto, J.; Ustun, T.S.; Otani, K. Smart inverter functionality testing for battery energy storage systems. *Smart Grid Renew. Energy* **2017**, *8*, 337–350. [CrossRef]
122. Ustun, T.S.; Aoto, Y.; Hashimoto, J.; Otani, K. Optimal PV-INV Capacity Ratio for Residential Smart Inverters Operating under Different Control Modes. *IEEE Access* **2020**, *8*, 116078–116089. [CrossRef]
123. Ustun, T.S.; Hashimoto, J.; Otani, K. Impact of Smart Inverters on Feeder Hosting Capacity of Distribution Networks. *IEEE Access* **2019**, *7*, 163526–163536. [CrossRef]
124. Ustun, T.S.; Sugahara, S.; Suzuki, M.; Hashimoto, J.; Otani, K. Power Hardware in-the-Loop Testing to Analyze Fault Behavior of Smart Inverters in Distribution Networks. *Sustainability* **2020**, *12*, 9365. [CrossRef]
125. Available online: https://www.cecp-eu.in/uploads/documents/events/Webinar_on_Energy_Storage_ISGF.pdf (accessed on 15 July 2022).
126. Moore, J.; Shabani, B. A Critical Study of Stationary Energy Storage Polices in Australia in an International Context: The Role of Hydrogen and Battery Technologies. *Energies* **2016**, *9*, 674. [CrossRef]
127. Felsmann, B.; Mezősi, A.; Szabó, L. Market versus Bureaucracy—Price Regulation in the Electricity Retail Sector. In Proceedings of the Importance of Kornai's Research Today, Budapest, Hungary, 21–22 February 2018.
128. Tawalbeh, M.; Murtaza, S.Z.; Al-Othman, A.; Alami, A.H.; Singh, K.; Olabi, A.G. Ammonia: A versatile candidate for the use in energy storage systems. *Renew. Energy* **2022**, *194*, 955–977. [CrossRef]
129. Kucur, G.; Tur, M.R.; Bayindir, R.; Shahinzadeh, H.; Gharehpetian, G.B. A Review of Emerging Cutting-Edge Energy Storage Technologies for Smart Grids Purposes. In Proceedings of the 2022 9th Iranian Conference on Renewable Energy & Distributed Generation, Mashhad, Iran, 23–24 February 2022. [CrossRef]
130. Asri, L.I.M.; Ariffin, W.N.S.F.W.; Zain, A.S.M.; Nordin, J.; Saad, N.S.C. Comparative Study of Energy Storage Systems (ESSs). *J. Phys. Conf. Ser.* **2021**, *1962*, 012035. [CrossRef]
131. Hossain, E.; Faruque, H.M.R.; Sunny, M.S.H.; Mohammad, N.; Nawar, N. A Comprehensive Review on Energy Storage Systems: Types, Comparison, Current Scenario, Applications, Barriers, and Potential Solutions, Policies, and Future Prospects. *Energies* **2020**, *13*, 3651. [CrossRef]
132. David, B.R.; Spencer, S.; Miller, J.; Almahmoud, S.; Jouhara, H. Comparative environmental life cycle assessment of conventional energy storage system and innovative thermal energy storage system. *Int. J. Thermofluids* **2021**, *12*, 100116. [CrossRef]

applied sciences

MDPI

Article

Optimization of PV and Battery Energy Storage Size in Grid-Connected Microgrid

Selahattin Garip [1] and Saban Ozdemir [2,*]

1 Department of Smart Grids, Graduate School of Natural and Applied Sciences, University of Gazi, Ankara 06500, Turkey
2 Department of Electrical and Electronic Engineering, Faculty of Technology, University of Gazi, Ankara 06500, Turkey
* Correspondence: sabanozdemir@gazi.edu.tr

Abstract: This paper proposes a new method to determine the optimal size of a photovoltaic (PV) and battery energy storage system (BESS) in a grid-connected microgrid (MG). Energy cost minimization is selected as an objective function. Optimum BESS and PV size are determined via a novel energy management method and particle swarm optimization (PSO) algorithm to obtain minimum total cost. The MG was designed to use its own energy as much as possible, which is produced from renewable energy resources. Since it is a grid-connected system, it can demand energy from the grid within the determined limit with penalty. It differs from the studies in the literature in terms of optimizing both parameters such as PV and BESS size, being a grid-connected self-contained MG structure and controlling the grid energy by an energy management algorithm and optimizing the parameter via PSO with an energy management system (EMS). Results are compared for different PV and BESS. Moreover, effectiveness of the novel energy management method with PSO is compared with the genetic algorithm, which is the one of the well-known optimization algorithms. The results show that the proposed algorithm can achieve optimum PV and BESS size with minimum cost by using the new energy management method with the PSO algorithm.

Keywords: energy management; energy storage; microgrid; particle swarm optimization; photovoltaic systems

Citation: Garip, S.; Ozdemir, S. Optimization of PV and Battery Energy Storage Size in Grid-Connected Microgrid. *Appl. Sci.* **2022**, *12*, 8247. https://doi.org/10.3390/app12168247

Academic Editors: Luis Hernández-Callejo, Jesús Armando Aguilar Jiménez and Carlos Meza Benavides

Received: 28 June 2022
Accepted: 16 August 2022
Published: 18 August 2022

1. Introduction

Today, fossil fuels such as coal, oil and natural gas are the main sources of electrical energy generation. However, these fuels cause greenhouse gas emissions and environmental pollution. In addition, while the world energy demand is increasing year by year, fossil fuels' reserves are limited and are about to deplete. Nevertheless, new restrictions are performed by environmental policies to reduce greenhouse gases emissions [1]. The Paris agreement, which was signed by 192 countries plus the European Union, is a promising example to deal with climate change. Countries that signed the agreement are planning to reduce their greenhouse gases emissions [2]. Renewable energy resources (RESs) such as photovoltaic and wind energy systems are environmentally friendly and good alternatives to fossil fuel since they do not cause any harmful gas emissions.

The number of grid-connected RES installations has been increasing year by year. Along with many advantages, these systems have some disadvantages such as intermittency that can cause scheduling, frequency, and voltage regulation problems on the grid [3,4]. Conventional generation systems with fossil fuels have slower responses to regulate frequency deviation in the short term [5]. With the increase in the number and total capacity of the RES installation, these problems and risk on power system stability have become more severe. Installing larger RES systems may overcome this problem [4]. However, it results in high investment cost. Battery energy storage systems (BESS) show up as an effective solution for this problem [3]. A BESS can be advantageous to maintain the balance between supply and demand with its fast dynamic response characteristics compared to conventional generators or other types of energy storage systems [6]. Particularly modern distribution networks are attracting attention for the solution of nanogrid (NG) and microgrid (MG) challenges. Hereby, BESSs are considered as a significant element of modern MGs and smart grids [7].

The MG is a concept that enables effective integration of distributed generation (DG) resources [8]. It is a controllable small network that combines RESs, conventional sources and loads in both grid-connected (on-grid) and island mode (off-grid) [9]. Figure 1 demonstrates a typical MG with these two operation modes [10]. Since DGs' power output characteristics are different from conventional generation systems, the MG should handle power quality problems by itself such as unpredicted fluctuation, robustness of reactive power support, resilience and a reliable system. The BESS is a good choice for maintaining resiliency and reliability with fast and adaptable characteristics. BESS can store the remaining power for later use, thus compensating for unexpected power outages and fluctuations in the RES. Although BESSs and PVs have great advantages in the MG system, they also have some disadvantages. Size and cost are gaining importance as high capacity causes increases in cost and size, while low capacity may not be enough to prevent unexpected power problems and may not meet load demand. Consequently, BESS size must be carefully calculated to determine the optimum size for a given system [8]. Moreover, research has shown that BESSs that are optimally sized for the current loads provide the best performance [11]. Thus, system designers need to find the optimal BESS size according to the specific system to obtain an efficient, reliable, and economical MG system [9].

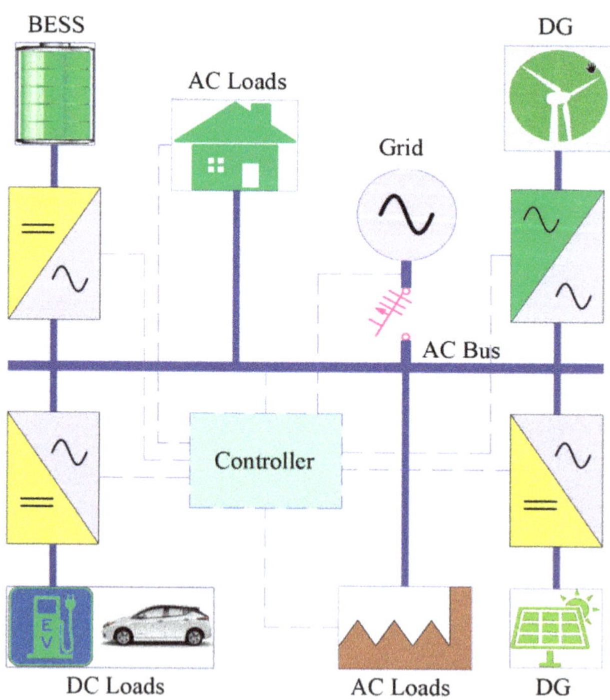

Figure 1. The structure of an AC microgrid.

In the literature, generally one parameter is kept constant and the other parameter is optimized in PV and BESS optimization studies. In most PV and BESS systems, the PV size is kept constant and the BESS size is optimized. A similar approach is used for structures with an energy management system (EMS), and most of them are proposed for island mode operation. BESS sizing is performed according to the system parameters with various methods. Some of the methods can be performed identically to any sized system [10]. Mathematical-based optimization methods are also used for sizing problems. Dynamic programming (DP) and linear programming (LP) are examples of mathematical methods [10]. DP is used in [12], but it is difficult to apply to large-scale systems [13]. LP optimization is chosen as a simpler method in [13], and it is implemented for a small energy storage system (ESS) in [14]. However, it has some problems when it is applied to large scale systems. As a result, LP and DP are not good tools for complex systems [10]. As a remedy, different optimization techniques that are named as probabilistic methods (PMs) have been developed. The Markov chain decision method (MCDM) is one of them and is used for battery sizing optimization due to its simple structure. Energy storage devices are scheduled optimally with an MCDM in [15]. However, probabilistic methods are effective when the number of optimized criteria is less (generally one) [10]. These methods are not suitable for optimizing the two parameters together in interaction with the energy management system in the structure that is the subject of this study.

Since RES output is uncertain, metaheuristic approaches are suggested in many applications. Metaheuristic methods give more accurate results on large and nonlinear optimization problems [16,17]. The Genetic Algorithm (GA) is used for cost reduction and optimization of the energy storage system in a hybrid energy system in [18]. The bottleneck of GA is that its results are not conclusive [13]. The Bat Algorithm (BA) is used to find optimum BESS size for a grid-connected low-voltage MG in [19]. The Grey Wolf Optimization (GWO) algorithm is chosen for optimum BESS sizing and decreasing fuel usage, and GWO performance is compared with BA and PSO in [20]. The Artificial Bee Colony algorithm (ABC) is used to calculation of optimal battery size and operation for revenue increasing in a hybrid power system [21]. The Grasshopper Optimization Algorithm (GOA) is another method used for optimal battery, PV, wind, and diesel sizing in a microgrid [22]. Particle Swarm Optimization (PSO) has simplicity and ease of use among other metaheuristic optimization algorithms, yet it can present a high convergence rate [8]. Its robustness of convergence comes from being less dependent on setting initial points among other methods. The PSO algorithm also needs less parameters than other metaheuristic algorithms. In addition to these, it needs lower data storage [8]. The PSO-based frequency control method for an off-grid microgrid is implemented to evaluate optimum BESS size and reduction in cost [23]. The PSO algorithm is used to find optimum battery size and minimum cost for a grid-connected residential system that currently has an available PV system [24]. Similarly, PSO is selected for battery capacity optimization and effective battery installation for an island mode microgrid in [25]. PSO is used for optimal sizing of wind, PV and tidal as a primary and battery as an auxiliary source considering the reliability index [25]. PSO is also proposed to determine optimal BESS with load shedding [5]. The objective of this paper is to enhance frequency control by load shedding, and thus, operation cost reduces. The cost optimization of a PV and BESS system in the grid-connected MG using PSO is proposed in [26]. However, this study does not use an energy management system.

In this paper, optimum energy storage and PV size considering cost minimization is determined based on the novel energy management method, and the PSO algorithm is proposed for a grid-connected microgrid. In past studies, various algorithms were used for different systems for optimization. According to the literature study, although the PSO algorithm is a common and well-known algorithm, it has not been used as an optimization algorithm for both PV and BESS sizing. In the majority of studies, one of the parameters is kept constant (mostly PV size), and the remaining parameters (mostly battery size) are optimized. In a limited number of studies, the PSO algorithm is used to determine optimal size of the PV system and BESS but only for island mode systems. Most of the remaining studies have not used cost minimization as an objective function or energy management system. A limited number of studies used cost minimization as an objective function or energy management system but with different optimization algorithms [1,8,9,11–13,15,16,18–27]. This paper presents cost minimization as an objective function by finding both optimum PV and BESS sizes and proposes a new optimal energy management method for a grid-connected MG. It is applied to a grid-connected microgrid that consists of a PV system with battery storage. MG is allowed to import energy from the grid with penalty. Thus, by allowing a limited amount of energy to be taken from the grid, it provides a more optimum structure by minimizing the effects of possible instantaneous high power demands. This paper focuses on determining the optimum PV and BESS sizes when the MG supplies energy as much as possible to its loads. The purpose is to create self-sufficient MG with limited grid support by considering cost minimization and defining optimum BESS and PV sizes. Studies are carried out for two different scenarios. In addition, the proposed energy management system with a PSO-based method is compared with GA, which is a well-known optimization algorithms. The results show that the proposed algorithm can achieve optimum PV and BESS size with minimum cost by using the new energy management method with a PSO algorithm. The proposed energy management method provides more flexibility to system designers for various system constraints. This can be accomplished by its configurable parameters. Thus, the new energy management method with PSO can be applied for various systems.

2. System Configuration and Modelling

The grid connected microgrid structure used in this paper is shown in Figure 2. It consists of the BESS, PV, AC bus, grid and load. It is connected to the grid via the AC bus. The battery and PV are connected to the DC bus via DC/DC converters that charge the battery from the PV throughout the DC bus. The DC bus is connected to the AC bus through the DC/AC inverter. The energy management system tracks load demand, available PV power and battery energy level, and it controls charge/discharge status of the battery and decides whether to demand energy from the grid.

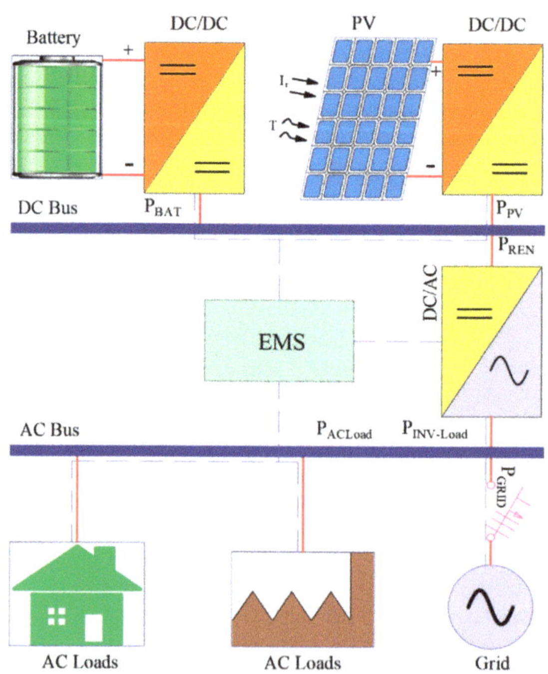

Figure 2. The architecture of the microgrid system.

Solar PV module performance is affected by irradiation, pollution, aging, shading and ambient temperature. Since the aim of this study is not maximum power point tracking design, the effects of these parameters will not be examined, and the net output power of the system will be used directly from the previously obtained data. Two different data sets were used to create different scenarios: case 1 and case 2. The data sets used in this study, which show the relationship between the power produced by the PV system and the load demand, are given in Figure 3a,b. The PV data in Figure 3a are obtained by taking the daily average of the annual data of the International Energy Agency (IEA) future prospect. To test the proposed system for another scenario, the PV and load data in Figure 3b are scaled from data published by the Belgian electricity system operator.

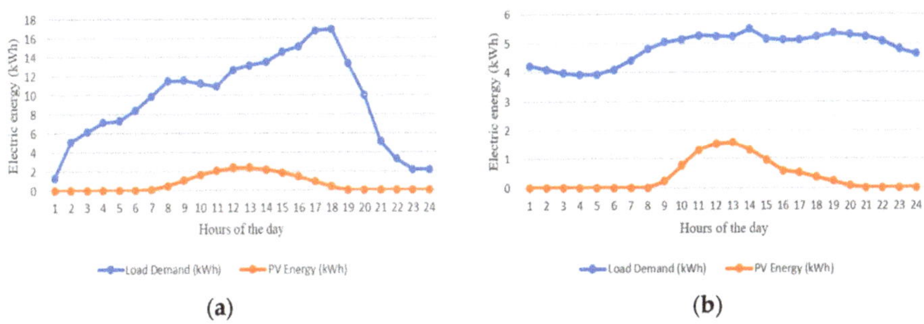

Figure 3. Average load demand and single PV module output for: (**a**) case 1; (**b**) case 2.

The converter efficiency that affects the amount of energy from the PV to the load is given as 95%. Thus, total PV output to load is:

$$P_{pv}(t) = P_{pv}gen(t) \times \eta_{conv} \tag{1}$$

where $P_{pv}gen(t)$ is generated power from the PV modules during time interval t, η_{conv} is the converter efficiency, and $P_{pv}(t)$ is transferred power from the PV to load during time interval t.

The power generation capabilities of PV modules deteriorate from year to year due to aging. Thus, the economic life of a PV is considered as 25 years in this study, and PV modules are considered, as they will not be replaced during system cost calculation. The cost and other parameters are listed in Table 1.

Table 1. PV cost parameters.

Parameter	Value	Unit
Efficiency (n_{PV})	95	%
Capital, Operation and Maintenance Cost	97	USD/Unit

As is known, the minimum and maximum of battery state of charge (B_{soc}) should be defined to prevent shortening the battery life. B_{soc} can be given as:

$$B_{soc}(t) = [E_{bat}(t)/E_{bat,rated}(t)] \times 100\% \tag{2}$$

where $E_{bat}(t)$ is battery energy level and $E_{bat,rated}(t)$ is rated energy capacity [28]. Overcharging and deep discharging of the battery should be prevented, as it will reduce its lifespan and cycle life. Thus, the following limits are defined:

$$E_{bat,\min}(t) \le E_{bat}(t) \le E_{bat,\max}(t) \tag{3}$$

where $E_{bat,\min}(t)$ is minimum energy limit, and it is defined as 0.48 kWh. $E_{bat,\max}(t)$ is a single battery module's maximum energy limit and it is defined as 2.4 kWh.

Battery charging and discharging action defined as below, respectively [28]:

$$E_{bat}(t)[E_{PV}(t) - E_{Load}(t)/\eta_{inv}] \times \eta_{Bch} \quad (4)$$

$$E_{bat}(t)[E_{Load}(t)/\eta_{inv} - E_{PV}(t)] \times \eta_{Bdch} \quad (5)$$

where $E_{PV}(t)$ is the generated energy, and $E_{Load}(t)$ is the load demand during time interval t. η_{Bch}, η_{Bdch} and η_{inv} are battery charging, discharging and inverter efficiencies, respectively, which are defined as 95%.

The capacity of battery modules will also decrease over time. In this work, battery module life is taken as 8 years. Battery modules are replaced three times during system cost calculation. Accordingly, BESS cost and other parameters are given in Table 2.

Table 2. BESS cost parameters.

Parameter	Value	Unit
Efficiency (n_{BESS})	95	%
Capital, Operation and Maintenance and Replacement Cost	493	USD/Unit

Changes in the efficiency of system elements can cause errors. In addition, since the battery capacity and PV must be a certain level as an integer (selected unit has a certain value), it will cause some errors. They can be minimized by reducing the PV unit power and battery unit capacity values. However, using a small PV module and batteries with small capacities may not be both practical and economical. More precisely, this is another optimization problem.

3. Proposed Algorithm

The proposed algorithm will be given in sequence as the objective function, energy management strategy for grid-connected and island modes and the proposed PSO algorithm. First, the PSO algorithm generates random PV and BESS sizes. The proposed energy management algorithm, which also will be explained later, uses these sizing values and generates PV and BESS power output according to the inputs and constraints.

3.1. Objective Function

In this study, the energy cost is chosen as an objective function. The goal is to obtain minimum total energy cost for the MG without compromising defined constraints; thus, the optimum PV and BESS size can be found.

The energy cost (EC) is calculated as:

$$EC = (PV_{Total,energy} \times PV_{\cos t}) + (BESS_{Total,energy} \times BESS_{\cos t}) \quad (6)$$

Here, $PV_{Total,energy}$ and $BESS_{Total,energy}$ are total output energy of PV and BESS, respectively. They are generated from the energy management algorithm in a defined time span. $PV_{\cos t}$ and $BESS_{\cos t}$ are the cost of PV and BESS, which include the capital, replacement, operation and maintenance costs.

3.2. Energy Management Strategy

The management of the power flow is an important process for optimizing the system components and the efficient operation of the system. The proposed energy management strategy can be divided into two parts as island mode and grid-connected mode operation. Figure 4 shows the flowchart of the proposed energy management strategy.

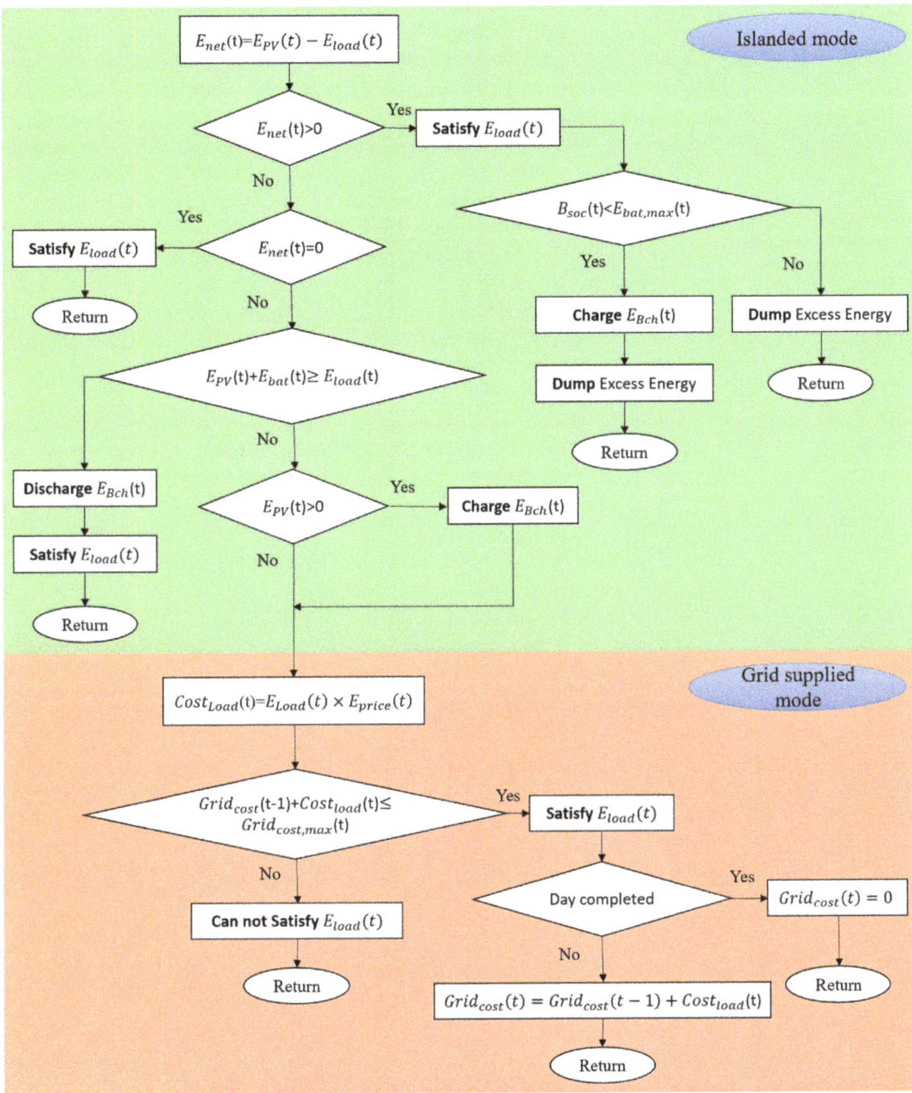

Figure 4. Proposed energy management flowchart.

In the island mode, the MG operates without grid support. The load demand can be satisfied by PV generation and/or BESS available capacity. There is always a balance between the available PV power, BESS capacity and load. Net energy ($E_{net}(t) = E_{PV}(t) - E_{Load}(t)$) is followed, and it is decided that the battery is charged if $E_{net}(t) > 0$ and $B_{SOC}(t) < E_{bat,max}(t)$. Batteries are charged with $E_{Bch}(t)$ until their $E_{bat,max}(t)$ level. If BESS reaches its charge limit and there is still available power in the PV system, this remaining power cannot be used or sold to the grid due to the island mode operation. If $E_{net}(t) = 0$, then there is no excess energy, and thus, load demand is equal to PV generation. If there is not enough PV generation to satisfy load demand ($E_{net}(t) < 0$), EMS controls $Bsoc(t)$ level at that time. If BESS has available energy, batteries can be discharged until their $E_{bat,min}(t)$ level. Finally, if both $E_{net}(t) < 0$ and $B_{SOC}(t) < E_{Load}(t)$, but there is still some available PV power generation (that is, $0 < E_{PV}(t) < E_{Load}(t)$), then batteries are charged by PV power.

For the grid-connected mode operation, the MG operates with grid support. In this study, the aim is to find optimum PV and BESS size for mostly self-sufficient MG in a yearly period. The grid energy is only used for supplying load demand if there is not enough energy in the PV and batteries. In this study, the grid is not used for charging batteries. It is assumed that it is costly to obtain energy from the grid. Thus, there is a grid cost limitation. The MG can be partially or fully supplied from the grid only for a limited time when there is either no or not enough energy in the BESS and PV. The EMS tracks the current energy level of the system components, the status of the PV system and BESS, and if there is not enough energy to be able to supply the load demand ($E_{PV}(t) + B_{SOC}(t) < E_{Load}(t)$), then the MG can obtain energy from the grid with penalty. There is a flag that holds the record of obtained energy from the grid. If energy from the grid exceeds a previously defined limit value, the flag increases. Hence, the PSO algorithm, which is explained below, decides to increase the PV or BESS module capacity to minimize the dependency of grid connectivity by considering the total installation cost.

3.3. PSO Algorithm

The PSO algorithm presents a model of flight patterns of birds and their social behavior for the optimization model, which was proposed by J. Kennedy and R. Eberhart in 1995 [29,30]. Its ties artificial life to the behavior of animal groups, such as bird flocking, fish schooling and swarming theory [30]. The simple explanation of the PSO model can be explained as follows. Each single bird is pointed in the Cartesian Coordinate System (CCS). Their initial location and velocity are assigned randomly. Then, the algorithm is executed with "the nearest proximity velocity match rule"; thus, every bird has the same speed as their closest neighbor. Since iteration maintains in the same direction, all the points will have the same velocity. Because of the simplicity of this structure and not exactly the same as in real situations, a random variable is added to the speed point. In each iteration, aside from meeting "the nearest proximity velocity match", each speed will be added with a random variable that provides convergence to the real case. In this model, every bird can find their maximum points. These can only be local maximum points. After every bird meets, in other words, birds finish their movement on the coordinate system, all the maximum points will be found. The highest value of these maximum points is the global maximum point [31,32].

In this study, PSO is used to find the minimum points, meaning minimum cost. Particles represent PV and BESS module counts (or sizes), and they are initialized randomly in the CCS. n_{Pop} is the swarm size, and it is defined as 50, which means 50 particles. The maximum iteration, *MaxIt*, count is set to 100. The inertia coefficient is set to 1. There are two acceleration coefficients, and both of them are selected as 2.5. Each particle's velocity is zero at the beginning. The objective function, which is explained in the previous section, is called in every iteration to calculate the particle's total cost. Each particle's cost is compared with each other's and the best cost, which is the minimum, is saved as the global best cost. In every iteration, PSO generates random PV and BESS sizes, and their costs are compared with the global best. The lowest value is saved as the new global best. At the end of all iterations' location, which means PV and BESS sizes, of the global best cost is the optimal point [16,33]. Each PV and BESS has a position, and these positions have a velocity. The velocity of the kth particle is:

$$v^j_{k,new} = wv^j_{k,old} + c_1r_1(x^j_{k,pbest} - x^j_k) + c_2r_2 \tag{7}$$

where $v^j_{k,new}$ refers to the recent velocity of the kth particle at jth iteration, the w refers to the inertia weight, $v^j_{k,old}$ refers to previous velocity of the kth particle at the jth iteration, c_1 and c_2 are the acceleration constants, and r_1 and r_2 pair are randomly determined numbers between 0 and 1. The position of the kth particle is renewed as below [16]:

$$x^j_{k,new} = x^{j-1}_{k,old} + v^j_{k,new} \tag{8}$$

where $x^{j-1}_{k,old}$ is the previous position of the *k*th particle from the past iteration [16]. The position *x* is the size of PV and BESS, and in this study, their minimum value is Var_{Min} = 1 and maximum value is Var_{Max} = 50. Figure 5 shows the flowchart of the applied PSO algorithm.

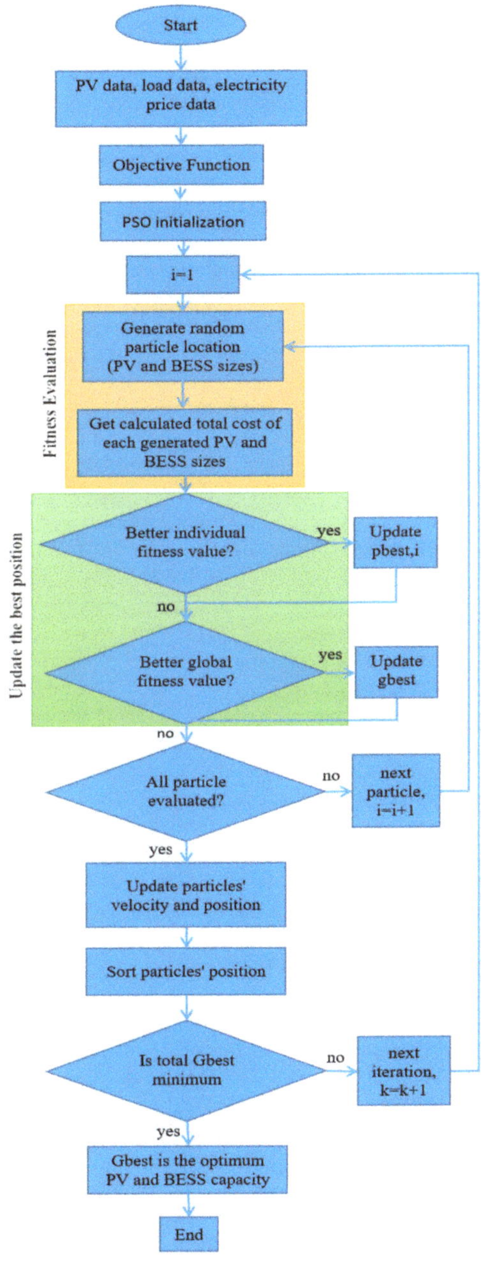

Figure 5. Proposed PSO algorithm flowchart.

4. Results and Discussion

In this paper, optimal sizing of the PV and BESS for MG, which can be operated in island mode and grid-tied mode, is carried out with two different data sets. The data are yearly average of a single PV output and the load. Yearly PV generation data have been taken from the IEA database. Then, they are degraded to 24 h by calculating average values for every hour. Yearly load data have also been taken from the IEA database, but some arbitrary changes have been applied on that data to create the desired test system. Then, they have been degraded to 24 h by calculating average values for every hour for case 1. In the second scenario, the data published by the Belgian electricity system operator is used by scaling. For each case, island and grid-connected mode operations are performed at the same time. The optimization algorithm computes the optimum PV and BESS size with regard to optimization parameters and the total cost of the system for case 1 and case 2. The total system cost includes cost of energy, battery and PV module cost, installation cost, battery degradation, and battery and PV lifetime/replacement cost.

First, the PSO algorithm generates a random PV and battery module size between 1 and 50. The first module (or particle) count is equal to n_{pop}; thus, it is 50. That is, 50 parameters are spread out to different locations randomly at the beginning. Within each iteration, this spreading continues with different velocities depending on c_1, c_2, w, and w_{damp} (damping ratio) values. These are the values that affect the speed and accuracy of the parameter reaching the optimum point. They can be close to the optimum point at the end, but they may take a long time to reach the optimum point due to their slowness. Contrarily, they can find the optimum point fast, but accuracy may not be guaranteed.

The PSO parameters are chosen to obtain faster and more accurate results. The population size is set to 50, the maximum number of iterations is set to 200, c_1 and c_2 are set to 2.5, and w_{damp} is set to 0.99. The number of battery and PV modules is limited from 1 to 50. Then, the novel energy management algorithm calculates total PV and BESS power outputs and how much energy is needed from the grid to supply loads. Here, providing uninterrupted power to the loads is the main concern. For this purpose, the energy management algorithm can decide to demand energy from the grid. However, it should be a limited time and level that is defined by the grid total cost parameter at the system design stage. If the MG loads cannot be supplied by any source, there will be a large increase in the total cost. This effect is controlled by another parameter such as the penalty parameter, and thus, the cost increases. The algorithm selects an optimum level of the PV system and BESS capacity to supply the load with the energy required in a day. After the energy management algorithm is calculated for daily total average PV and BESS energy output, total energy cost can be found. The calculated energy cost is compared by the PSO algorithm for every particle, which equals n_{pop} = 50, along with iterations. The best particle cost over 50 particles (n_{pop} count) is found, and this is called the "particle best cost". This is saved for the next iterations. The particle best cost can be updated with a new value at the next iteration by a particle that holds lower cost. Thus, after all iterations are completed, the best updated "particle best cost" value will be the "global best". This shows the calculated optimum PV and BESS size with minimum cost with defined constraints. The best particle costs between each of the 50 particles inside an iteration and every best cost throughout the iterations can be seen in Figure 6a,b for case 1 and case 2.

At the first iterations, the PSO algorithm generates lower PV and BESS module counts, which means that the PV and the BESS particles are far from the optimum point. (Cost can be seen on the second y-axis in Figure 6a,b. The y1-axis and y2-axis scales are different). Since loads are supplied mostly from the grid, the cost is increasing. After defined the maximum grid cost is exceeded, the total system cost increases faster due to the penalty factor, and the system can obtain supply mostly from renewable energy resources (because increasing the rate of renewable energy use decreases the system total cost). After 200 iterations, calculations were performed by the energy management algorithm. It was found that the optimum PV and BESS module counts were 47 and 28, respectively, and the total cost was USD 40.972 for case 1. Similarly, it was found that the optimum PV and BESS module counts were 24 and 28, respectively, and the total cost was USD 24.186 for case 2.

To prove the results obtained from the proposed method, the total cost variation of the system according to different PV and battery sizes and penalty factor are given in

Figure 7a,b, respectively. At the first point in Figure 7a, there are 45 PV modules and five battery modules. The total system cost at this first point is USD 49.993. In the first area (depicted in Figure 7a), while the number of PV modules is decreasing, the number of battery modules is increasing. It should be considered that the energy cost penalty highly affects the total system cost. The total system cost is increasing slowly until the sixth calculation point. Then, when the number of PV modules is too low, the system cost increases rapidly. At the eighth calculation point, there are five PV modules and 40 battery modules, and this is the highest cost in the figure, which is USD 127.040. At this point, there are not enough PV modules to supply the loads, and there are not enough PV modules to charge this amount of battery modules. Thus, the loads are supplied from the grid for a longer period. At this longer period, as an option, the cost can be increased excessively or the maximum limit can be set in order to prevent taking more energy from the grid. In this study, the maximum level of energy cost that can be taken from the grid has been determined. After reaching the maximum allowable grid supply limit cost, the energy management algorithm cuts off the electricity. Eventually, the total cost will be high in this situation. In the second area, while the number of PV modules has increased, the number of battery modules is low. In this case, the total cost is decreasing because there will be more PV modules to generate energy to supply the loads in the daytime. PV modules can also charge batteries when the number of PV and battery modules are closer to the optimum point. Thus, BESS can supply the loads at night when there is no PV energy. In the third area, both the number of PV and battery modules are increased. The total cost decreases, but at the 18th point, it increases again due to the increased number of battery modules. At the fourth calculation area, both the numbers of PV and battery modules are decreased, and the total cost also starts to decrease. Finally, the number of PV and battery modules reaches the optimum point, such that the total cost is at the lowest value at the 20th point. There are 47 PV modules and 28 battery modules. Total system cost is USD 40.972 at the 20th point. The same study was carried out for case 2. It can be seen from Figure 7b that the cost of the system for 24 PV modules and 28 batteries is USD 24.186. This means that the loads of the MG can be fully supplied by PV and BESS in the daytime, and they can be supplied by BESS most of the night. Therefore, MG can be supplied mostly by its RES, and its dependency to the grid is low. However, the total system cost rises as the number of battery and PV modules continues to decrease because the system needs to import more energy from the grid, which increases the grid supply cost. Another reason is that as the number of both PV and battery modules continues to decrease, the longer the blackout durations occur and thus the penalty cost increases.

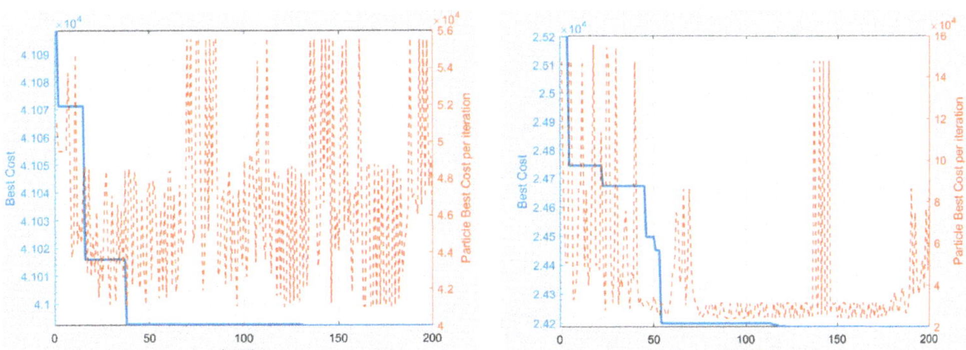

(a) **(b)**

Figure 6. Particle best cost per iteration and best costs through all iterations of the proposed energy management method with PSO: (**a**) case 1; (**b**) case 2.

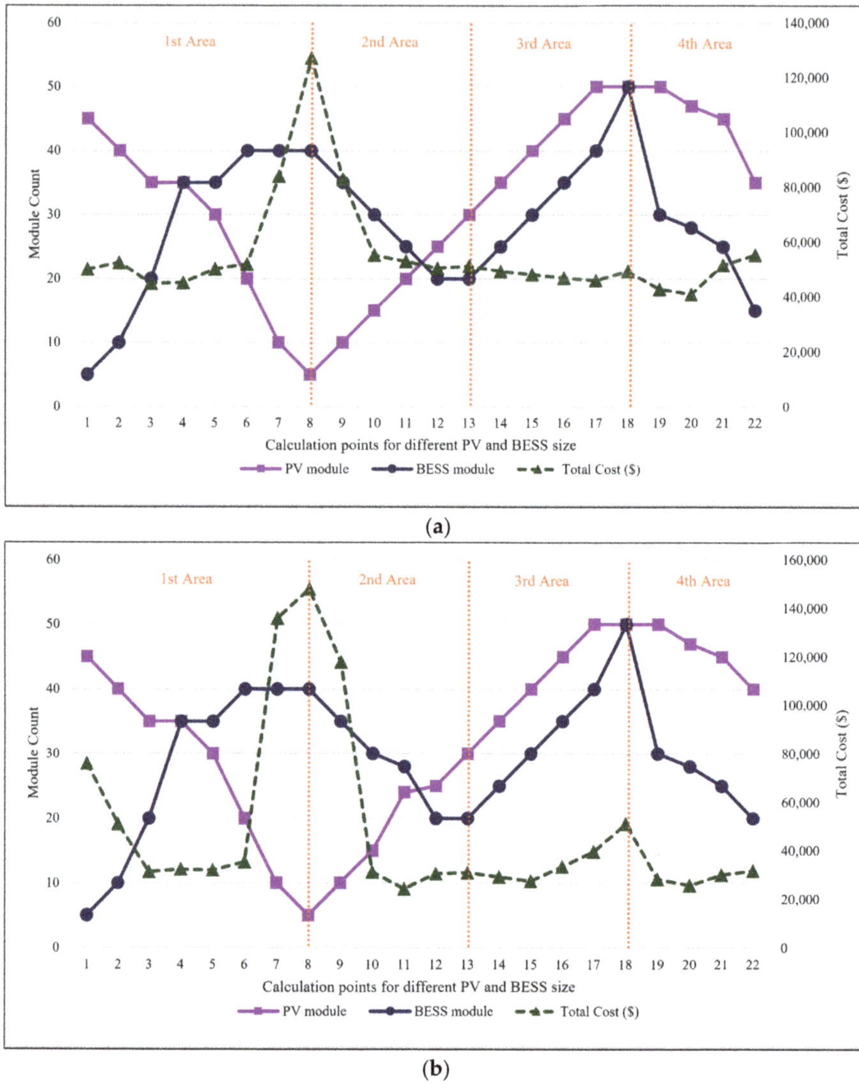

Figure 7. System total cost for various PV and BESS module combinations: (**a**) case 1; (**b**) case 2.

As can be seen, this process is not simple, such as defining the number of PV and battery modules regarding the known changing load demand. PV generation changes according to irradiation and weather conditions. There is an allowed grid supply limit that is defined at the system design stage. Therefore, the energy management algorithm should decide when to charge and discharge the batteries, and when to obtain energy from the grid by considering cost. Eventually, the results show that the proposed optimization algorithm correctly determines the optimum PV and BESS size within the defined constraints. The proposed energy management system with the PSO algorithm has some advantages and superiorities. It also needs only a few initial parameters. In addition, it can be used with different algorithms. Furthermore, the constraints and parameters used in the energy management strategy are also configurable such that they can be easily adapted for different systems. The flexible nature of the proposed approach is its most important strength.

In order to test the performance of the PSO-based method with the energy management algorithm, its performance is compared with GA, which is the one of the well-known optimization algorithms. The obtained results with the GA are shown in Figure 8a,b for case 1 and case 2, respectively. The parameters used in the GA algorithm are as follows: the population size is set to 50, the maximum number of iterations is set to 200, crossover rate = 1, mutation rate = 0.04, and the number of battery and PV module is limited from 1 to 50. As can be seen in Figure 8a,b, the novel energy management method with the PSO algorithm gives better performance than the novel energy management method with the GA algorithm. In addition, the energy management method with GA found that the optimum PV and BESS modules count as 47 and 28, respectively, and the total cost is USD 40.972 at the 192nd iteration for case 1 and the PV and BESS modules count as 24 and 28, respectively, and the total cost is USD 24.186 at the 187th iteration for case 2. It takes more time to find the global point than the proposed algorithm. Furthermore, this comparison is proven that the proposed novel energy management system can also work with other algorithms.

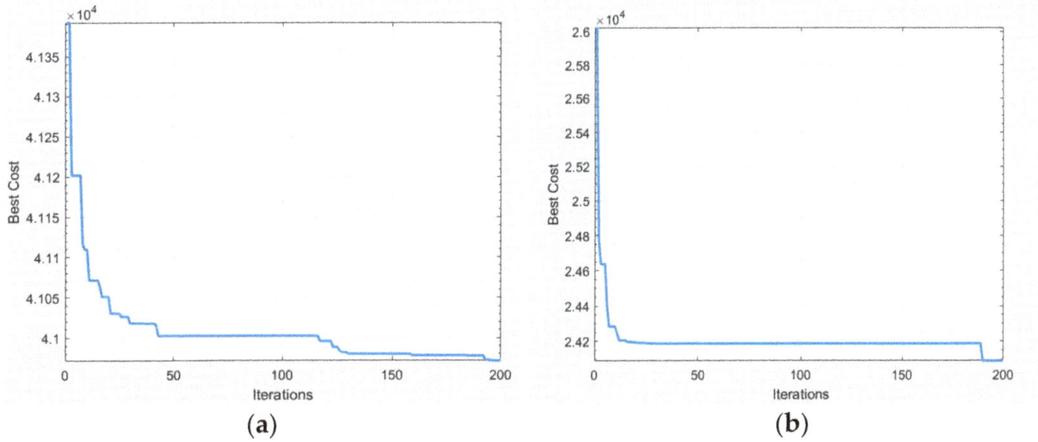

Figure 8. Obtained best cost results obtained from the proposed energy management method with GA: (**a**) case1; (**b**) case2.

Comparisons between existing studies and the proposed study are given in Table 3. A close examination of Table 3 provides an idea of the difference between the proposed system and the other algorithms. In past studies, various algorithms have been used for different systems for optimization. This study differs from other studies in the following aspects. As can be seen from the table, some of the studies in the literature do not use the PSO algorithm for both PV and BESS sizing. In most of the studies, one of the PV and BESS parameters was kept constant, and the other parameter was optimized. Although the PSO optimization algorithm has been proposed for both PV and BESS in a limited number of studies, they have only been used for island mode systems. Most of the remaining studies did not use cost minimization as an objective function or energy management system. A limited number of studies have used cost minimization as an objective function or energy management system, but with different optimization algorithms.

Table 3. Comparison of existing system and proposed system.

Ref.	Optimization Algorithm(s)	Number of Opt. Criteria	System Size	Type of RES	Operation Mode	EMS	Objective Function	Purpose of the ESS
[11]	MILP	Less	Large	PV, BESS	Grid connected	Yes	Minimization of the total annual cost (including both energy and battery degradation-based costs)	Energy sustainability
[12]	DP	Less	Small	PV, BESS	Grid connected	No	Determination of the optimal ESS charging and discharging trajectory with minimum operational cost	Energy sustainability
[13]	LP	Less	Small	PV, Wind, BESS	Grid connected	No	Minimizing the operational costs of the MG and BESS sizing optimization	Peak shaving
[15]	MCDM	Should be less for better effectiveness	Small	BESS (storage plant)	Grid connected	Yes	Calculating the storage power references that maximize the financial gains	Energy sustainability, energy arbitrage
[18]	GA	More	Large	PV, Wind, Diesel generator, BESS	Off-grid	No	Minimization of the system cost and loss of power system probability	Energy sustainability
[19]	Improved BAT	More	Large	Fuel Cell, Micro Turbine, PV, Wind, BESS	Grid connected	Yes	Minimizing the operation cost	MG operation management studies with regard to operation, maintenance and financial points
[20]	GWO	More	Large	Fuel Cell, Micro Turbine, PV, Wind, BESS	Grid connected	No	Minimizing the operation dispatch costs	Energy sustainability, energy arbitrage
[21]	ABC	Average	Average	Wind, Hydro, BESS	Grid connected	No	Maximization of the revenue	Energy sustainability, energy arbitrage
[22]	GOA	More	Small	PV, Wind, Diesel generator, BESS	Off-grid	Yes	Minimization of the DPSP and COE	Energy sustainability
[23]	PSO	Less	Small	PV, BESS	Off-grid	No	Minimization of total BESS cost	Frequency control of the stand-alone microgrid

Table 3. *Cont.*

Ref.	Optimization Algorithm(s)	Number of Opt. Criteria	System Size	Type of RES	Operation Mode	EMS	Objective Function	Purpose of the ESS
[24]	PSO	Less	Small	PV, BESS	Off-grid	No	Minimization of battery capacity to prevent destabilization of system	Energy sustainability
[25]	PSO	More	Small	Wind, PV, Tidal, BESS	Off-grid	Yes	Minimize the annualized cost of the generation system with the constraint having reliability index	Energy sustainability, reliability improvement
[26]	Modified PSO	Less	Large	PV, BESS	Grid connected	No	Minimize the size and site of installation of the PV system	Energy sustainability, power loss min. and voltage profile enhancement of the radial distribution network
[27]	PSO	Less	Large	PV, BESS	Grid connected	No	Maximize the cost profitability of the system	Energy sustainability
Current paper	PSO	More	Small (can be applicable to large systems)	PV, BESS	Off-grid and Grid Connected	Yes	Minimize the cost of the system by finding optimum BESS and PV size	Energy sustainability for self-sufficient system, and it can control grid connection by EMS when it is needed

5. Conclusions

This study presents a PSO-based algorithm with a new energy management strategy to find the optimum PV and BESS size for a grid-connected MG. The MG can operate in island mode and, if necessary, in grid-connected mode with some limitations. The MG structure is designed in such a way that it can demand energy from the grid when there is not enough energy in the PV system and BESS. However, the amount of demanded energy is limited by the system authorities. The aim is to find an optimum PV and BESS size by considering the defined energy cost. This allows the microgrid to be supported from the grid in critical situations, although supplying loads from the RES has priority, regardless of whether the system will demand energy from the grid and/or the amount of energy to be demanded from the grid can be configured with the proposed energy management method. Therefore, the energy management algorithm can be reconfigured and used for various systems and different constraints. To validate the proposed approach, various calculations are carried out for different PV and BESS sizes. Furthermore, to prove the effectiveness of the new energy management method with PSO, it has been compared with GA. Results show that the PSO-based algorithm with the energy management strategy can determine the optimum PV and BESS size, with the minimum cost defining the system constraints. Consequently, PV and battery sizes have been optimized together with the proposed PSO algorithm and novel energy management system. The effectiveness of the system is also explained by comparing the results with different algorithms.

Author Contributions: Conceptualization, S.G. and S.O.; methodology, S.G. and S.O.; software, S.G. and S.O.; validation, S.G. and S.O.; formal analysis, S.G. and S.O.; investigation, S.G. and S.O.; resources, S.G. and S.O.; data curation, S.G. and S.O.; writing—original draft preparation, S.G. and S.O.; writing—review and editing, S.G. and S.O.; visualization, S.G. and S.O.; supervision, S.O. All authors have read and agreed to the published version of the manuscript.

Funding: This research received no external funding.

Institutional Review Board Statement: Not applicable.

Informed Consent Statement: Not applicable.

Data Availability Statement: The dataset associated with this paper can be found in the online version, at https://www.elia.be/en/grid-data?csrt=17010711133344377898 (accessed on 30 July 2022).

Conflicts of Interest: The authors declare no conflict of interest.

Nomenclature

ABC	Artificial Bee Colony
BA	Bat Algorithm
BESS	Battery energy storage system
$BESS_{Total,energy}$	Output energy of BESS
B_{soc}	Battery state of charge
CCS	Cartesian Coordinate System
COE	Cost of energy
c_1, c_2	Acceleration constants
DPSP	Deficiency of power supply probability
DG	Distributed generation
DP	Dynamic programming
E_{bat}	Battery energy level
$E_{bat,max}$	Battery module's maximum energy limit
$E_{bat,min}$	Battery minimum energy limit
$E_{bat,rated}$	Battery rated energy capacity
EC	Energy cost
E_{Load}	Load demand energy
EMS	Energy management system
E_{net}	Net energy

E_{PV}	PV generated energy
ESS	Energy storage system
GA	Genetic Algorithm
GOA	Grasshopper Optimization Algorithm
GW	Grey Wolf Optimization
IEA	International Energy Agency
LP	Linear programming
$MaxIt$	Maximum iteration count
MCDM	Markov chain decision method
MG	Microgrid
NG	Nanogrid
n_{pop}	Particle population count
P_{BAT}	Battery power
P_{GRID}	Grid power
$P_{INV\text{-}Load}$	Inverter power
PM	Probabilistic methods
P_{PV}	PV power
P_{pvgen}	Generated power from PV
P_{REN}	Renewable power
PSO	Particle Swarm Optimization
PV	Photovoltaic
$PV_{Total,energy}$	Output energy of PV
RES	Renewable energy sources
SOC	State of charge
V	Velocity
Var_{Max}	Maximum value of the size of PV and BESS
Var_{Min}	Minimum value of the size of PV and BESS
w	Inertia weight
w_{damp}	Damping ratio
η_{Bch}	Battery charging efficiency
η_{Bdch}	Battery discharging efficiency
η_{conv}	Converter efficiency
η_{BES}	Battery energy storage efficiency
η_{inv}	Inverter efficiency

References

1. Akram, U.; Khalid, M.; Shafiq, S. Optimal sizing of a wind/solar/battery hybrid grid-connected microgrid system. *IET Renew. Power Gener.* **2018**, *12*, 72–80. [CrossRef]
2. Höhne, N.; Gidden, M.J.; den Elzen, M.; Hans, F.; Fyson, C.; Geiges, A.; Jeffery, M.L.; Gonzales-Zuñiga, S.; Mooldijk, S.; Hare, W.; et al. Wave of net zero emission targets opens window to meeting the Paris Agreement. *Nat. Clim. Change* **2021**, *11*, 820–822. [CrossRef]
3. Kelly, J.J.; Leahy, P.G. Sizing battery energy storage systems: Using multi-objective optimization to overcome the investment scale problem of annual worth. *IEEE Trans. Sustain. Energy* **2020**, *11*, 2305–2314. [CrossRef]
4. Zhang, T.; Gooi, H.B.; Chen, S.; Goh, T. Cost-effectiveness studies of the BESSs participating in frequency regulation. In Proceedings of the IEEE Innovative Smart Grid Technologies—Asia (ISGT ASIA), Bangkok, Thailand, 3–6 November 2015. [CrossRef]
5. Chen, S.; Zhang, T.; Gooi, H.B.R.; Masiello, D.; Katzenstein, W. Penetration rate and effectiveness studies of aggregated BESS for frequency regulation. *IEEE Trans. Smart Grid* **2016**, *7*, 167–177. [CrossRef]
6. Aghamohammadi, M.R.; Abdolahinia, H. A new approach for optimal sizing of battery energy storage system for primary frequency control of islanded microgrid. *Electr. Power Energy Syst.* **2014**, *54*, 325–333. [CrossRef]
7. Farrokhabadi, M.; König, S.; Cañizares, C.A.; Bhattacharya, K.; Leibfried, T. Battery energy storage system models for microgrid stability analysis and dynamic simulation. *IEEE Trans. Power Syst.* **2018**, *33*, 2301–2312. [CrossRef]
8. Kerdphol, T.; Qudaih, Y.; Mitani, Y. Battery energy storage system size optimization in microgrid using particle swarm optimization. In Proceedings of the IEEE PES Innovative Smart Grid Technologies, Istanbul, Turkey, 12–15 October 2014. [CrossRef]
9. El-Bidairia, K.S.; Nguyena, H.D.; Jayasinghea, S.; Mahmoudb, T.S. A hybrid energy management and battery size optimization for standalone microgrids: A case study for Flinders Island, Australia. *Energy Convers. Manag.* **2018**, *175*, 192–212. [CrossRef]
10. Yuqing, Y.; Stephen, B.; Chris, M.; Merlinde, K. Battery energy storage system size determination in renewable energy systems: A review. *Renew. Sustain. Energy Rev.* **2018**, *91*, 109–125. [CrossRef]

11. Mulleriyawage, U.G.K.; Shen, W.X. Optimally sizing of battery energy storage capacity by operational optimization of residential PV-Battery systems: An Australian household case study. *Renew. Energy* **2020**, *160*, 852–864. [CrossRef]
12. Choi, Y.; Kim, H. Optimal scheduling of energy storage system for self-sustainable base station operation considering battery wear-out cost. In Proceedings of the Eighth International Conference on Ubiquitous and Future Networks (ICUFN), Vienna, Austria, 5–8 July 2016. [CrossRef]
13. Moghimi, M.; Garmabdari, R.; Stegen, S.; Lu, J. Battery energy storage cost and capacity optimization for university research center. In Proceedings of the IEEE/IAS 54th Industrial and Commercial Power Systems Technical Conference (I&CPS), Niagara Falls, ON, Canada, 7–10 May 2018. [CrossRef]
14. Fedjaev, J.; Amamra, S.; Francois, B. Linear programming based optimization tool for day ahead energy management of a lithium-ion battery for an industrial microgrid. In Proceedings of the IEEE International Power Electronics and Motion Control Conference (PEMC), Varna, Bulgaria, 25–28 September 2016. [CrossRef]
15. Grillo, S.; Pievatolo, A.; Tironi, E. Optimal storage scheduling using markov decision processes. *IEEE Trans. Sustain. Energy* **2016**, *7*, 755–764. [CrossRef]
16. Regis, N.; Muriithi, C.M.; Ngoo, L. Optimal battery sizing of a grid-connected residential photovoltaic system for cost minimization using pso algorithm. *Eng. Technol. Appl. Sci. Res.* **2019**, *9*, 4905–4911. [CrossRef]
17. Beskirli, A.; Temurtas, H.; Ozdemir, D. Determination with Linear Form of Turkey's Energy Demand Forecasting by the Tree Seed Algorithm and the Modified Tree Seed Algorithm. *Adv. Electr. Comput. Eng.* **2020**, *20*, 27–34. [CrossRef]
18. Suresh, M.; Meenakumari, R. An improved genetic algorithm-based optimal sizing of solar photovoltaic/wind turbine generator/diesel generator/battery connected hybrid energy systems for standalone applications. *Int. J. Ambient Energy* **2021**, *42*, 1136–1143. [CrossRef]
19. Bahmani-Firouzi, B.; Azizipanah-Abarghooee, R. Optimal sizing of battery energy storage for micro-grid operation management using a new improved bat algorithm. *Electr. Power Energy Syst.* **2014**, *56*, 42–54. [CrossRef]
20. Nimma, K.S.; Al-Falahi, M.D.A.; Nguyen, H.D.; Jayasinghe, S.D.G.; Mahmoud, T.S.; Negnevitsky, M. Grey Wolf optimization-based optimum energy-management and battery-sizing method for grid-connected microgrids. *Energies* **2018**, *11*, 847. [CrossRef]
21. Paliwal, N.K.; Singh, A.K.; Singh, N.K.; Kumar, P. Optimal sizing and operation of battery storage for economic operation of hybrid power system using artificial bee colony algorithm. *Int. Trans. Electr. Energy Syst.* **2019**, *29*, e2685. [CrossRef]
22. LawanBukar, A.; WeiTan, C.; YiewLau, K. Optimal sizing of an autonomous photovoltaic/wind/battery/diesel generator microgrid using grasshopper optimization algorithm. *Solar Energy* **2019**, *188*, 685–696. [CrossRef]
23. Kerdphol, T.; Fuji, K.; Mitani, Y.; Watanabe, M.; Qudaih, Y. Optimization of a battery energy storage system using particle swarm optimization for stand-alone microgrids. *Int. J. Electr. Power Energy Syst.* **2016**, *81*, 32–39. [CrossRef]
24. Sintianingrum, A.; Khairudin; Hakim, L. Optimization of microgrid battery capacity using pso with considering islanding operation. *J. Eng. Sci. Res.* **2020**, *2*, 1–4. [CrossRef]
25. Bashir, M.; Sadeh, J. Size optimization of new hybrid stand-alone renewable energy system considering a reliability index. In Proceedings of the 11th International Conference on Environment and Electrical Engineering, Venice, Italy, 18–25 May 2012. [CrossRef]
26. Kefale, H.A.; Getie, E.M.; Eshetie, K.G. Optimal Design of Grid-Connected Solar Photovoltaic System Using Selective Particle Swarm Optimization. *Int. J. Photoenergy* **2021**, *2021*, 6632859. [CrossRef]
27. Koskela, J.; Rautiainen, A.; Järventausta, P. Using electrical energy storage in residential buildings—Sizing of battery and photovoltaic panels based on electricity cost optimization. *Appl. Energy* **2019**, *239*, 1175–1189. [CrossRef]
28. Hassan, A.; Al-Abdeli, Y.M.; Masek, M.; Bass, O. Optimal sizing and energy scheduling of grid-supplemented solar PV systems with battery storage: Sensitivity of reliability and financial constraints. *Energy* **2022**, *238*, 121780. [CrossRef]
29. Beltran, A.; Hughes, C.; Das, S. Improved maximum power point tracking of partially shaded pv arrays using particle swarm optimization with zone initialization. In Proceedings of the IEEE 46th Photovoltaic Specialists Conference (PVSC), Chicago, IL, USA, 16–21 June 2019. [CrossRef]
30. Kennedy, J.; Eberhart, R. Particle swarm optimization. In Proceedings of the International Conference on Neural Networks, Perth, WA, Australia, 27 November–1 December 1995. [CrossRef]
31. Wang, D.; Tan, D.; Liu, L. Particle swarm optimization algorithm: An overview. *Soft Comput.* **2018**, *22*, 387–408. [CrossRef]
32. Ignat-balaci, A.; Szilagyi, E.; Petreus, D. Day-Ahead Scheduling, Simulation, and Real-Time Control of an Islanded Microgrid. *Adv. Electr. Comput. Eng.* **2021**, *21*, 89–98. [CrossRef]
33. Nempu, P.B.; Sabhahit, J.N. Stochastic Algorithms for Controller Optimization of Grid Tied Hybrid AC/DC Microgrid with Multiple Renewable Sources. *Adv. Electr. Comput. Eng.* **2019**, *19*, 53–60. [CrossRef]

applied sciences

MDPI

Article

A Resonant Ring Topology Approach to Power Line Communication Systems within Photovoltaic Plants

José Ignacio Morales-Aragonés [1], Matthew St. Michael Williams [2], Víctor Alonso Gómez [1], Sara Gallardo-Saavedra [3], Alberto Redondo-Plaza [3], Diego Fernández-Martínez [3], Francisco José Sánchez-Pacheco [4], Juan Gabriel Fajardo Cuadro [5] and Luis Hernández-Callejo [6,*]

1 Departamento de Física, Universidad de Valladolid, 42004 Soria, Spain
2 Climate Studies Group Mona (CSGM), Department of Physics, University of the West Indies, Kingston 07, Jamaica
3 Departamento Ingeniería Agrícola y Forestal, Universidad de Valladolid, 42004 Soria, Spain
4 Departamento de Tecnología Electrónica, Universidad de Málaga, 29010 Málaga, Spain
5 Departamento de Ingeniería Mecánica, Universidad Tecnológica de Bolivar, Cartagena de Indias 130001, Colombia
6 ADIRE-ITAP, Departamento Ingeniería Agrícola y Forestal, Universidad de Valladolid, 42004 Soria, Spain
* Correspondence: luis.hernandez.callejo@uva.es; Tel.: +34-975129418

Abstract: Within this study, single-cable propagation facilitated by PV strings' wiring characteristics is considered for an adapted design of PLC electronics. We propose to close the communications signal path, resulting in a ring topology where a resonance condition could be implemented. A PLC topology using the resulting circular closed-loop path of a PV series string as its physical communication support is designed and leveraged for practical use. When the path length or the number of transceivers is changed, the resonance properties that come with the circular path as the physical support are affected but are shown to be preserved with the application of automatic adjustable tuning. This automatic tuning guarantees that the resonance improves propagation parameters and reverts the system to its optimal values at the chosen carrier frequency.

Keywords: loop antenna; power line communication; resonance; single-wire transmission; tuning

Citation: Morales-Aragonés, J.I.; Williams, M.S.M.; Gómez, V.A.; Gallardo-Saavedra, S.; Redondo-Plaza, A.; Fernández-Martínez, D.; Sánchez-Pacheco, F.J.; Cuadro, J.G.F.; Hernández-Callejo, L. A Resonant Ring Topology Approach to Power Line Communication Systems within Photovoltaic Plants. *Appl. Sci.* **2022**, *12*, 7973. https://doi.org/10.3390/app12167973

Academic Editor: Gaetano Zizzo

Received: 18 July 2022
Accepted: 8 August 2022
Published: 9 August 2022

1. Introduction

Monitoring and maintenance tasks within solar plants are becoming the focus of many research efforts, and a growing number of sensors and measuring devices are proposed to be installed along different points of the solar module associations, even within each solar module [1]. In this scenario, communications between solar modules and centralization points (combiner boxes) become an essential subject for research. Considering the year-on-year growing amount of solar modules within modern solar plants, the cost of the communications elements (module transceivers, wiring, etc.) becomes a key point in constructing optimal designs for future practical communication implementations. Regarding the cabling, an ideal solution would be a power line communication (PLC) system in order to use the same power cables already installed as a communications physical support, bypassing the need for extra communications wiring.

Some authors have developed works where PLC is used as an automatic transfer switch (ATS); ATS selects the electrical connection circuit in the solar plant [2]. In [3], PLC is used to send the data from the photovoltaic plant through the AC power line. Ref. [4] employs a PLC-based system to avoid the islanding of the PV solar plant; in this work, the authors propose the control of the connection and disconnection devices. Ref. [5] employs a PLC-based communication and control system to control cascade inverters.

Most of the literature regarding PLC over the DC power lines within a photovoltaic (PV) plant makes use of circuitry and several components designed initially for the standard physical support of traditional PLC systems [6,7], namely power wires with two conductors that run parallel to each other. Transmission in these kinds of lines is performed by means of TEM or quasi-TEM modes, and the traditional transmission lines theory is applicable, with of course some limitations, such as the nonuniformity of the distributed parameters. Ref. [8] presents an intelligent PV module monitoring scheme based on a parallel resonant coupling unit, which uses the DC bus as the communication channel and modulates the monitoring data into a 200 KHz carrier for communication. However, the topology of the power cables in each string of PV plants is essentially different since there is a single cable connecting the PV modules in series that runs away from other sections of the line, at least in parts of it. This part of the power line can be modelled as a single cable line, where the propagation mode cannot be TEM and where the classic transmission line theory is not applicable. This difference suggests that the optimal electronics, carrier frequencies and general strategy could be different than the ones intended for traditional PLC systems.

This single-wire topology is imposed by the inherent characteristics of the power wiring within each PV string. As such, in this work, we propose an adapted strategy with electronics for implementing PLC communications over this cabling topology of a single PV string, where we start considering this essential difference from classic PLC standards, since the two-wire physical support common for classic PLC applications is absent in this configuration.

One of the first studies focused within the area of single-cable transmission was performed by G. Goubau [9] in 1950, based on a previous study of A. Sommerfeld from 1899 [10]. These studies showed the possibility of radiofrequency propagation via a surface wave guided by a single conductor, in conjunction with the formulation of a TM propagation mode for the signal. By 2008, Corridor Systems, Inc. (Santa Rosa, United States)registered a patent application [11] for a single-cable transmission line proposed for carrier frequencies from 50 MHz to 20 GHz. Later studies have been published adding updated formulations of the single-cable transmission phenomena [12] and complementing the theory with the calculation of ohmic currents' distributions, losses and characteristic impedance or even creating a classic two-wire-line equivalent model for the single-cable propagation [13]. One of the most interesting conclusions of these works is the possibility of a very-low-loss propagation over a single cable. Therefore, this study evaluates the performance and accompanying possible benefit of the propagation mode of the PV plant single-cable segment and the need for compensatory electronics to sustain the innate benefits of the behaviour of the mode in the event where the cable is altered.

Our proposal for the PV string PLC communications system is to close the single-cable signal path by means of a capacitor connected between the combiner box terminals, in such a way that the line becomes a ring-shaped path for the communications signal, susceptible to being pushed to a loop-antenna-like resonance, improving the signal levels along the line. This work is structured as follows: Section 2 shows the theoretical frame and materials, Section 3 presents the results and Section 4 shows the conclusions and future work.

2. Theoretical Frame and Materials

This section presents the theoretical framework on which the presented work is based as well as the materials used in it.

2.1. Theoretical Frame

The previous documents mentioned regarding single-cable transmission in Section 1 considered frequencies of transmission above 50 MHz (higher than the ones we worked with in this first demonstration stage for our PV–PLC system), whose current distribution is expected to be flowing superficially or in a thin layer close to the surface of the cable (skin effect), with propagation described to be carried out by a surface wave whose phase velocity is c (the speed of light). Our direct measurements of the phase velocity over a line of single PV cable at a frequency of 20 MHz showed speeds around 0.6 c, closer to what is observed in typical coaxial cables [14], even when the propagation mode is different in our single-cable medium (TM) than the one in coaxial cables (TEM). The lower frequency used in our single-cable support suggests that the current distribution flows in a thicker layer inside the conductor, and the influence of the material lowers the phase velocity with respect to the previously mentioned Sommerfeld surface wave. As a result, the single-cable analogy applied to the PV plant cable cannot allow us to assume the same propagation mode described in Section 1, and since our aim with this document is not to establish a theoretical background but to develop a practical approach for implementing a working PLC system specific for PV plants, we assumed the measured phase velocity of 0.6 c and tried to start from a simple model to characterize our system.

The influence of the topology of the physical wiring in the previous literature on PLC systems for PV strings is not usually discussed, probably because the low-frequency carriers proposed do not show limitations related to higher frequencies, such as interferences or resonance issues. Some previous works use an open line topology [7,15,16] and some others propose a current loop topology [17]. The frequencies are normally in the order of 100 kHz or less, that is to say, considering the assumed phase velocity (0.6 c) wavelengths are around 1800 m, much bigger than the usual length of the string. With this work, we try to explore the possibility of setting the carriers of higher frequencies (from 1 MHz on), analysing the effects of the increased frequency and offering a practical solution, which inevitably leads to considering the physical topology of the wiring.

In a single PV string, the power wires starting from a combiner box return to the same box after connecting the modules in series. In essence, it is easy to establish a circular communications path by means of a capacitor connected between the two cables arriving at the combiner box and to bypass capacitors connected through each PV module. This allows for a cable loop configuration to be a communications physical support for a single string. The choice of a closed-loop configuration for the signal path has several advantages for communications purposes. This topology sets all the points along the circular line at the same level regarding attenuation, so differences in reception levels are reduced to a minimum and it allows for a resonance condition to be established for the signal along the loop, since there is a natural reactive impedance associated with a cable loop depending on the relationship between the wavelength and the physical length of the loop, which can be compensated for by some added lumped reactive impedance, leaving only a low-resistance path for the signal.

Adding the ring topology proposed to the aforementioned assumption about phase velocity = 0.6 c, a simple propagation model can be considered. Our starting point is the schematic in Figure 1.

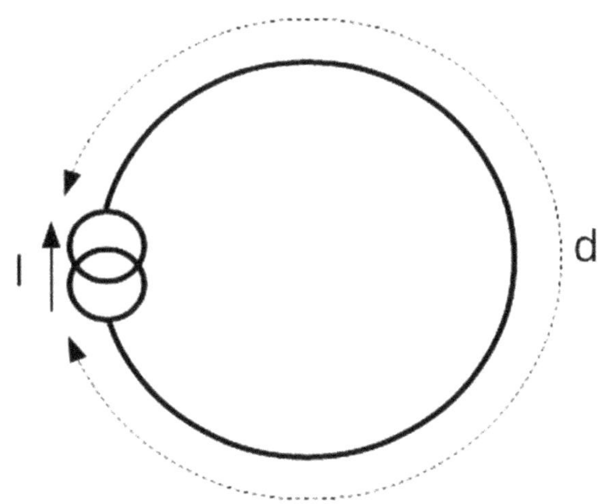

Figure 1. Schematic of the communications loop (signal path) excited by a current generator (I), with a perimeter d.

The cable ring of perimeter d represents a simplified sketch of the signal path, including the PV cable line and bypass capacitors (with enough capacitance for presenting a very low impedance at our working frequency), where the I/O impedance of the transceivers located along the ring is neglected for a first approximation. It is excited by a current generator (which in practice could be the secondary binding of a transformer). A general expression of the current wave along this loop is:

$$I(x) = I\left(A\,e^{-jkx} + B\,e^{jkx}\right) \tag{1}$$

Representing the sum of two waves moving in opposite directions along the ring, where I is the current amplitude with its time dependency:

$$I = I_0\,e^{-jwt} \tag{2}$$

Here, x is a coordinate indicating the distance along the loop measured clockwise from the generator position and k is the wavevector depending on the frequency (f) and phase velocity (0.6 c) and A and B are constants (which could be complex) to be determined.

The boundary conditions imposed are related to the continuity of the current at the generator thus:

$$I(0) = I\,(A+B) = I(d) = I\left(A\,e^{-jkd} + B\,e^{jkd}\right) = I \tag{3}$$

leading to

$$A + B = 1\,;\; A\,e^{-jkd} + B\,e^{jkd} = 1 \tag{4}$$

In general, the solutions of this system of equations lead to (complex) values of A and B different from zero and thus to a stationary wave pattern for the current distribution. These solutions are not desirable to implement a communications system because stationary waves give rise to maximum and minimum current amplitude points along the loop, which means different reception levels since a transceiver could be located at any point in the ring. For the case where

$$\text{kd} = \pi(1+2n) \tag{5}$$

where n is an integer value, this system of equations is incompatible and there is no solution, representing a destructive interference between clockwise and anticlockwise waves. The optimal solutions for our purpose are those where A = 0 or B = 0, which leads to a travelling wave solution of the form:

$$I(x) = I\,e^{-jkx} \text{ or } I(x) = I\,e^{jkx} \tag{6}$$

not presenting maximum or minimum current amplitude points. To obtain these kinds of solutions, it is mandatory that:

$$kd = 2\,\pi\,n \tag{7}$$

that is to say, the perimeter of the loop is an integer multiple of the wavelength.

However, even when the above condition is satisfied, there are stationary wave solutions (A and B $\neq$ zero) satisfying only the extra condition A + B = 1. In order to determine the actual spatial current distribution over a typical PV cable loop, we built a setup comprising 20 m of cable from our real test PV plant (Figure 6, show later) and measured and assessed the string as one single conductor line travelling in a straight fashion from the positive lead of one signal generator to the end of the PV array arrangement where it then turned around, coming back as a loop at a distance of one metre apart from the initiated point at the positive lead, finishing in the negative lead of the signal generator. The topology of the circuit is thus a loop-like one, but the shape is elongated from a circular one. The generator was adjusted for a sinusoidal signal of 9.142 MHz (determined as the signal corresponding to a wavelength equal to the loop length, fulfilling the above condition) and

10 volts of amplitude. The validation of the system in short wire installations is interesting, since for longer installation lengths resonance can be achieved at the same frequency by means of higher-order modes, where more than one wavelength is present in the line. Essentially, if it works for our 20 m loop, it will work also in loops which are 40 m, 60 m, etc., and the fine adjusting for intermediate lengths can be easily achieved by the tuning circuit designed (see Section 2.2).

The current measurements were made in a low-invasive fashion at different points along the loop, sensing the AC magnetic field associated by means of a toroidal ferrite core surrounding the cable, with a 30-turn enamelled cable wiring. This way, an induced AC voltage proportional to the AC current in the line can be measured with an oscilloscope across the terminals of the wiring. The high input impedance of the oscilloscope guarantees that a very low current will flow through the wiring, so a negligible inductance was added to the loop during measurements. Figure 2 shows the results of these measurements taken every metre along the loop, and it shows a clear stationary wave pattern. These results agree with the literature about the current distribution on loop antennas, the theory of which is close to the one here considered.

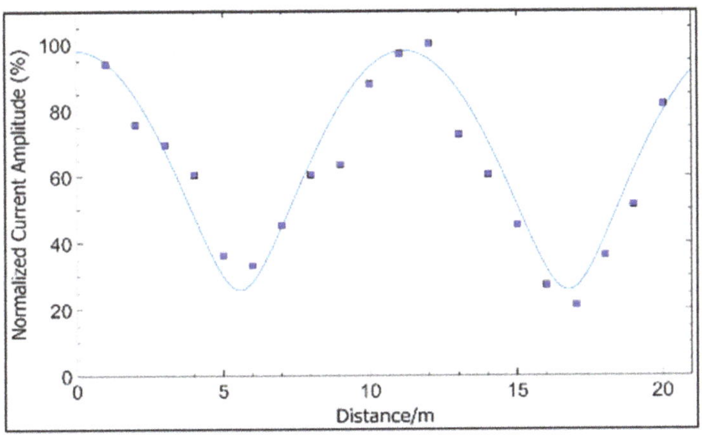

Figure 2. Current amplitude distribution along the loop (normalised (%); measurement points and fitting to a stationary wave envelope function).

The difference between the maximum (100%) and minimum (21%) current amplitude levels from the measurements represents 13.55 dB, which can be managed easily by an automatic gain control (AGC) within receptors. There are two possibilities for fully exploiting the advantages of a resonant loop for communications avoiding significant differences in the signal levels along the loop: (I) Find a simple way to excite only the travelling wave mode over a one-wavelength ring (optimal). (II) Work at carrier frequencies low enough in a fashion that the loop length is a fraction of ¼ of the wavelength or less. This is the slower option, but it still leads to carrier frequencies over 2 MHz (assuming a propagation speed of 0.6 c and a loop length of 20 m), significantly higher than the average previously reported.

Regarding solution I, some authors have faced the theory of travelling wave solutions for loop antennas [18], compared with standing wave solutions [19], and even have determined the necessary conditions for exciting the travelling wave modes [11]. All this work, however, is focused on antenna design and consequently looks for the best radiation properties. For our application, radiation losses are not desired, since they mean power losses and could affect the electromagnetic compatibility of the system. Fortunately, the drastic reduction in the loop area due to the particular shape of our setup will prevent the system from great radiation losses, since radiation integrals obtain contributions of opposite current elements much closer to each other [20]. The travelling wave or quasitravelling wave mode of propagation for our system could be achieved by fulfilling some of the conditions exposed in [18,21].

However, these conditions are not easily applicable in a practical environment, so the insertion in series of a nonreciprocal device is proposed. These kinds of devices will attenuate the waves travelling in one direction, leaving unaffected ones travelling in the opposite direction, leading to a travelling wave propagation mode. Some of the possible practical devices that could accomplish this task are magnetic-circulator-based isolators or unity gain amplifiers referenced to Earth in such a way that waves arriving towards the output will find very low impedances to Earth, being attenuated, but the ones arriving towards the input will find very high impedances and will progress to the output almost unaffected. In this work, we focus on resonance control, and isolator insertion is left for future research.

Regarding solution II, if we work at wavelengths four times the length of the loop or bigger, the phase changes along the loop in $\pi/4$ radians maximum. For a loop length of ¼ wavelength, the general solution for its current wave is:

$$A = \frac{1+J}{2},\ B = \frac{1-J}{2},\ I(x) = I\left(Ae^{-jkx} + Be^{jkx}\right) \tag{8}$$

where kx goes from 0 to $\pi/2$ along the loop length, so the maximum change in the current wave amplitude is:

$$\frac{I_{max}}{I_{min}} = \sqrt{2},\ \frac{I_{max}}{I_{min}} = 1.5\ \text{dB} \tag{9}$$

This value is small enough to avoid AGC systems in receptors and simplifies the electronics, allowing carrier frequencies over 2 MHz as explained before. This carrier (depending on the modulation system) can lead to baud rates around 200 kbps, which are higher than the previous baud rates reported in the literature for PLC within PV plants, so even when this mode of propagation represents the lower baud rate for our model, it still supposedly should provide an improvement on the speeds previously reported.

In addition, the resonance condition can still be achieved for this case because even when the impedance variations related to auto interference are almost not present, there is an inductance associated with the line and reactive impedances representing the transceivers (capacitive or inductive depending on the coupling chosen) that must be compensated for with a lumped element in order to push the loop to resonance. For the higher-frequency (lower-wavelength) option, as a starting point, we can consider a model of signal propagation along a closed loop similar to the ones previously mentioned regarding loop antennas, which show series resonances (impedance close to zero) at frequencies whose wavelengths are integer fractions of the loop lengths [20]. In this way, a constructive interference gives rise to a spatial resonance, and a maximum in the signal amplitude is observed. Since one key goal of a communications system is the integrity of the signal, we must satisfy the above condition in our cable loop in order to work at an optimal point with the best SNR possible.

For a chosen carrier frequency, a loop length is fixed to fulfil the above condition; however, the length of the loop is an imposed parameter depending on the physical dimensions of the installation, and therefore it is necessary to find a way to adjust the natural loop resonance to match the frequency of the carrier. The insertion of a coil in series with the cable loop has the effect of increasing the electrical length seen by the signal (length expressed as a wavelength multiple), that is to say, the loop will resonate at lower frequencies. The opposite of this behaviour is seen with the insertion of a capacitor. As such, the insertion of a reactive component in series with the loop could be used to perform the matching between the carrier and loop resonance frequencies. This effect is shown in Figure 3 from our measurements over a typical PV string cable loop 20 m long with a vector network analyser (VNA) connected to the loop, where the resonance condition is recognized by the minimum in the modulus of the S11 scattering parameter (maximum power sent to the loop) and a sudden change in its phase. Figure 3 shows curves around the first resonance frequency (9.192 MHz without compensation) corresponding to the connection in series of a coil, three values of capacitors and the raw cable.

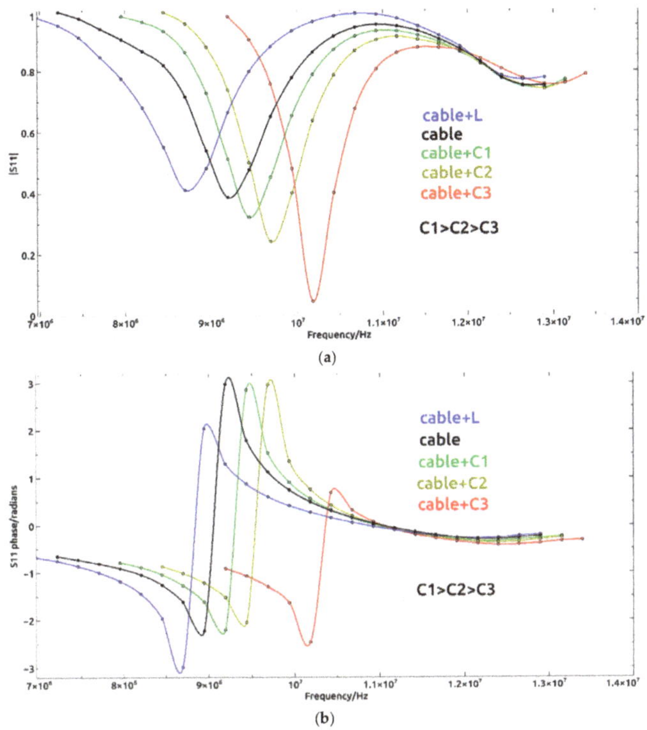

Figure 3. The shift of the loop resonance frequency with the addition of a reactance in series: the modulus (**a**) and phase (**b**) of the scattering parameter-1 measured with a VNA connected to the raw cable loop and with the addition in series of a coil and three increasing values of capacitors.

These measurements have been used to determine more precisely the phase velocity in our PV cable. Since the first cable resonance (9.192 MHz) corresponds to one wavelength in the 20 m long cable:

$$\lambda = \frac{V_f}{f} \Rightarrow V_f = 20mxf = 20 * 9192000 = 0.613c \tag{10}$$

2.2. Materials

Even though a preinstallation length compensation is feasible (installing a fixed reactive series component), employing automatic adjustable tuning would more so be convenient in confining the transmission around an optimal point, in order to account for the inevitable occurrence of small variations in the loop (for example, the addition of one more PV module to a string, which would increase the loop physical length). For this purpose, a simple tuning circuit was designed which was able to show either capacitive or inductive impedance by means of a control voltage. The basic variable component used was a varicap diode whose capacity could be adjusted depending on the inverse voltage applied. The circuit is shown in Figure 4a, and it is composed of a toroidal transformer with the primary (L2) connected to the cable loop and the secondary (L1) connected to the variable capacitor (C) (voltage controlled). The actual implementation circuit for the voltage-controlled capacitor is shown in Figure 4b.

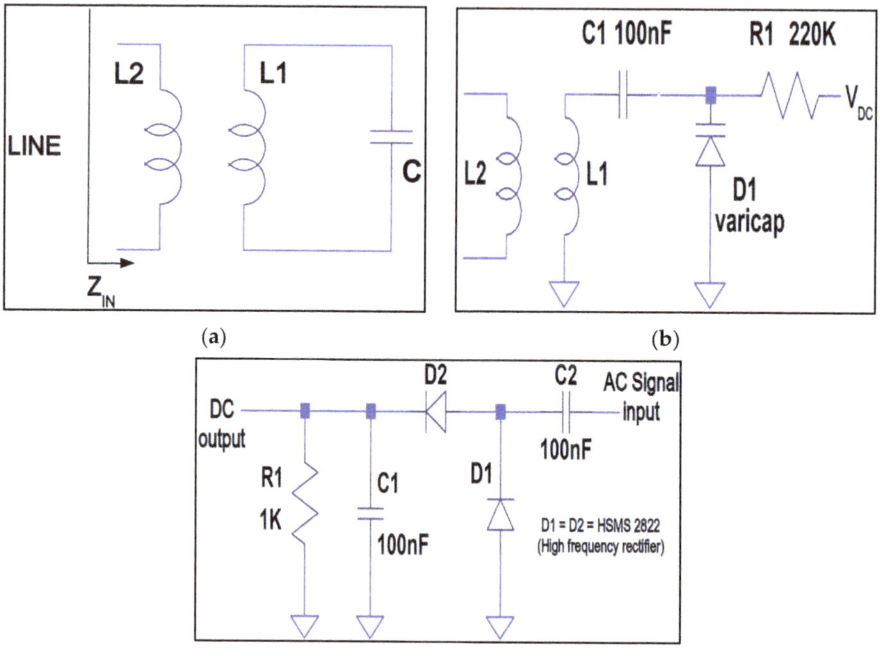

Figure 4. Schematic of tuning circuit in (**a**) its generic representation and (**b**) actual implementation, and (**c**) schematic of voltage doubler detection circuit.

The network analysis of the circuit in Figure 4a leads to an impedance seen from the primary port:

$$Z = jwL_2(1 - w2\ L_1C\ (1 - K2)) \div (1 - w2L_1C) \quad (11)$$

where the mutual inductance coefficient in the transformer is expressed by:

$$M = K\sqrt{L_1L_2} \quad (12)$$

and K is the coupling coefficient, showing a parallel resonance at

$$w_1 = 1 \div \sqrt{L_1C} \quad (13)$$

and a series resonance at

$$w_2 = w_1 \div \sqrt{1 - K2} \quad (14)$$

This impedance has two inductive regions, $(w < w_1)$ and $(w > w_2)$, and one capacitive region: $(w_1 < w < w_2)$. In the centre of the capacitive region with a good coupling in the transformer $(1 - K2 << 1)$, we have $(1 >> w2\ L_1C\ (1 - K2)$ and $(1 << w2L_1C)$, so

$$Z \approx 1 \div \left(jwC\frac{L_1}{L_2}\right) \quad (15)$$

and our circuit works as a capacitor multiplier by the factor $\frac{L_1}{L_2}$, which can be very useful for loops requiring high-capacitive compensation with values higher than the maximum achieved by the varicap diode, for example, for loops with high inductance in series (the inductive coupling of the transceivers).

For loops with low series inductance (the capacitive coupling of transceivers), the usual length of a typical PV string and carrier frequencies over 5 MHz, our circuit is better used in the region around the higher frequency of resonance w_2, where the impedance is close to zero, and for frequencies lower than w_2 the impedance is capacitive, and for frequencies higher than w_2 the impedance is inductive. Around this region and $(1 \ll w2L_1C)$, the impedance can be expressed as:

$$Z = jwL_2(1 - K2) - jL_2 \div (wL_1C) \tag{16}$$

equivalent to a series LC circuit with

$$L_{equ} = L_2(1 - K2) \tag{17}$$

and

$$C_{equ} = \frac{L_1}{L_2}C \tag{18}$$

resonant at w_2. Since the capacitor C can be dynamically adjusted, connecting this circuit in series with the cable loop, we can add series inductance, capacitance or none of them depending on what is required by the loop, and the dynamic adjusting range can be selected with the value of L_2 since it is a common factor in the expression of Z around w_2. The possibility of the capacitance control varying a biasing inverse voltage on the varicap diode allows us to implement a microcontroller-based tuning system that will also require some kind of detection of the resonance condition in the loop.

The carrier amplitude through the primary coil in the circuit in Figure 4a will show a maximum if the loop is pushed to resonance, so if this signal is used as an input for a voltage doubler detection circuit (Figure 4c), we will obtain at the output a DC voltage proportional to the amplitude of the carrier in the loop and the resonance can be detected as a maximum in the DC voltage output. Finally, automatic resonance adjusting can be implemented by sampling the detector DC output with an analogue-to-digital converter (ADC) integrated in the microcontroller (MCU), where the convenient firmware will search for a maximum outputting a DC voltage sweep towards the tuning circuit.

Figure 5 shows the real implementation of the circuits proposed, with all the components labelled. Figure 5a shows the front side of the board and Figure 5b shows the back side.

(a) (b)

Figure 5. Picture of the electronics developed for loop automatic tuning: (**a**) front side; (**b**) back side.

The PV plant used is shown in Figure 6. This plant is located on the facilities of the Duques de Soria campus of the University of Valladolid. The campus is in the city of Soria, Spain. The modules used in this article for the communication line are those on the upper row. Each module has the characteristics shown in Table 1.

Figure 6. PV plant on Duques de Soria campus of the University of Valladolid.

Table 1. Technical characteristics of the PV module.

Model	Cells	Power/W	V_{OC}/V	I_{SC}/A	V_{MPP}/V	I_{MPP}/A
Eoply	72 cells	175	44.35	5.45	36.26	4.83

3. Results

Figure 7 shows the action of the compensation described, where a 20-m-long loop was kept in resonance at 9.5 MHz initially with the corresponding control voltage of 5 V. Then, the length of the loop was shortened to 15.10 m, and with the resonance control electronics off, the resonance frequency was displaced to 11.1 MHz. If we then switched on the electronics, the control voltage was readjusted by the MCU to get the resonance back to 9.5 MHz. All the electronics described above could work to keep the resonance condition on any of the two modes described in Section 2. The S21 modulus was measured with a VNA connected in series with the loop, the tuning circuit and a 50-ohm load. Maximums show the frequencies of minimal attenuation of the propagation along the (loop)-(tuning circuit) chain. The black trace corresponds to the 20 m cable with a control voltage of 5 V in the tuning circuit, where the maximum is over the chosen carrier frequency (9.5 MHz). The blue solid trace corresponds to a variation in the loop length from 20.00 m to 15.10 m with no change in control voltage, and the new maximum is over 11.1 MHz and the carrier (9.5 MHz) is attenuated from |S21| = 0.228 to 0.100. The blue dashed trace corresponds to the new loop length (15.10 m) after the tuning circuit action for compensation, leading to a control voltage of 0.9 V, and the maximum is back over the carrier at 9.5 MHz.

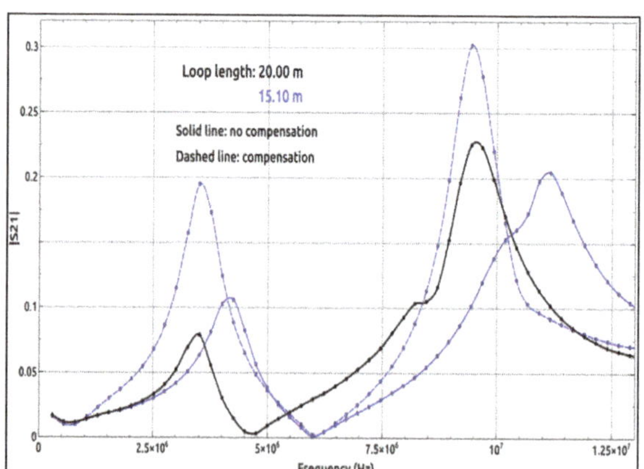

Figure 7. Effect of microcontroller action to retune the loop, in response to variations in cable distance.

During the current distribution measurements in the one-wavelength loop presented before (Figure 2), with an exciting signal of 10 volts in amplitude, the absolute voltages measured across the current sensor were between 70 and 330 mV, which, considering that the sensor sensibility is 10 Volts/Amp, indicates averaged current amplitude values in the order of 200 mA. This suggests that with much lower exciting voltages, communication with reasonable SNRs is possible due to loop resonance, making simpler and low-cost transceivers conceivable.

As a validation of the current levels present in the short-loop configuration (¼ wavelength), measurements were made over an experimental setup composed of a cable loop 20 m long with ten capacitors and ten small toroidal transformers distributed evenly in series simulating the bypass capacitors and the transceiver inductively coupled and installed within each PV module in a real installation. The toroidal transformers had a transformation ratio of 30:1 with the secondary connected to the loop, and their inductances were adjusted to form with the bypass capacitors LC resonators at the carrier frequency, in such a way that the transceivers were LC resonators at the same frequency as the loop resonance. The control electronics were also placed in series with the loop. A carrier frequency of 1 MHz was chosen, fulfilling the condition of a loop shorter than ¼ wavelength, and was injected in the primary of one of the transformers with an amplitude of 5 volts, simulating a transmission from one of the transceivers towards the others. The voltage amplitude measurements at the primaries of the ten transformers receiving the signal are shown in Table 2.

Table 2. Voltage amplitude measurements of the ¼ wavelength configuration.

Receiver	Voltage Amplitude (V)
1	32.3
2	32.1
3	32.3
4	33.5
5	34.6
6	34.0
7	33.5
8	32.7
9	32.5
10	32.8

4. Conclusions and Future Works

An analysis of a ring topology as physical support for a PLC system specially intended for PV strings has been presented, proposing that pushing this loop to resonance optimises the reception levels along the cable. Previous works on this subject had proposed resonant circuits for coupling to the line; we added here to this feature the possibility of working with the whole loop under resonance, improving even more the signal levels along the line. In addition, the conditions needed for the levels to be reasonably equalised has been derived. Control electronics have been designed to keep the communications signal path on resonance, making the system flexible and able to self-adapt to different specific installations or changes within the same setup. As this check and adaptation to resonance is a task that does not need to be accomplished continuously (it would be enough to execute it once per day), the MCU that controls all the communications in a final application could be the same, allowing a lower-cost system.

The insertion of transmission (TX) and reception (RX) circuits along the loop shows reactive impedances (inductive or capacitive coupled to the line) or even small resistances, consequently modifying the resonance condition on the loop; nonetheless, our tests have shown that the addition of the compensation reactance supplied by the circuit described before is able to return the loop to the optimal working point of resonance. The approach here presented is useful as a starting point for further research that could determine the precise influence of lumped impedances along the loop in the communications performance, but for moderate reactances in the TX/RX circuits (enough for signal injection and recovery), the compensation circuit is enough, with only small deviations compared to the behaviour of the raw cable loop.

Two options regarding the relationship between the loop length and the wavelength have been presented: a loop with one wavelength perimeter and a loop with ¼ wavelength perimeter or less. The measurement levels in the loop have shown the possibility of working with relatively low signal levels without compromising the SNR in both options, allowing the design of cheaper and simpler transceivers. The main line of research from our results must be to explore and test for future work the ways proposed to excite the travelling wave mode in the one wavelength loop, which in essence could suppose a great leap forward for equalising the signal levels working at carrier frequencies over 5 MHz, leading to baud rates close to or even over 1 Mbps (depending on the spectral efficiency of the modulation scheme chosen), keeping all the advantages of working in resonance.

Author Contributions: Conceptualisation, J.I.M.-A. and L.H.-C.; methodology, J.I.M.-A., V.A.G., M.S.M.W. and L.H.-C.; validation, J.I.M.-A., V.A.G., M.S.M.W., J.G.F.C. and L.H.-C.; formal analysis, J.I.M.-A. and M.S.M.W.; investigation, J.I.M.-A., V.A.G., M.S.M.W., A.R.-P., D.F.-M., F.J.S.-P. and L.H.-C.; writing—original draft preparation, J.I.M.-A., M.S.M.W., S.G.-S. and L.H.-C.; writing—review and editing, J.I.M.-A., J.G.F.C., S.G.-S. and M.S.M.W.; supervision, L.H.-C.; project administration, L.H.-C.; funding acquisition, J.I.M.-A. and L.H.-C. All authors have read and agreed to the published version of the manuscript.

Funding: This study was supported by the Universidad of Valladolid with the predoctoral contracts of 2020 cofunded by Santander Bank. This study was supported by the Universidad of Valladolid with ERASMUS+ KA-107.

Institutional Review Board Statement: Not applicable.

Informed Consent Statement: Not applicable.

Data Availability Statement: Not applicable.

Conflicts of Interest: The authors declare no conflict of interest.

References

1. Morales-Aragonés, J.I.; Gallardo-Saavedra, S.; Alonso-Gómez, V.; Sánchez-Pacheco, F.J.; González, M.A.; Martínez, O.; Muñoz-García, M.A.; Alonso-García, M.D.C.; Hernández-Callejo, L. Low-Cost Electronics for Online I-V Tracing at Photovoltaic Module Level: Development of Two Strategies and Comparison between Them. *Electronics* **2021**, *10*, 671. [CrossRef]
2. Kurniawan, A.; Taqwa, A.; Bow, Y. PLC Application as an Automatic Transfer Switch for on-grid PV System; Case Study Jakabaring Solar Power Plant Palembang. *J. Phys. Conf. Ser.* **2019**, *1167*, 012026. [CrossRef]
3. Kabalci, E.; Kabalci, Y. A measurement and power line communication system design for renewable smart grids. *Meas. Sci. Rev.* **2013**, *13*, 248–252. [CrossRef]
4. Ropp, M.E.; Aaker, K.; Haigh, J.; Sabbah, N. Using power line carrier communications to prevent islanding. In Proceedings of the Name of the Conference Record of the Twenty-Eighth IEEE Photovoltaic Specialists Conference-2000 (Cat. No. 00CH37036), Anchorage, AK, USA, 15–22 September 2000; pp. 1675–1678. [CrossRef]
5. Evans, D.; Cox, R. Powerline communications strategy enabling fully decentralized control of AC-stacked PV inverters. In Proceedings of the 2017 IEEE Energy Conversion Congress and Exposition (ECCE), Cincinnati, OH, USA, 1–5 October 2017; pp. 2277–2284. [CrossRef]
6. Han, J.; Choi, C.S.; Park, W.K.; Lee, I.; Kim, S.H. PLC-based photovoltaic system management for smart home energy management system. *IEEE Trans. Consum. Electron.* **2014**, *60*, 184–189. [CrossRef]
7. Jonke, P.; Eder, C.; Stockl, J.; Schwark, M. Development of a module integrated photovoltaic monitoring system. In Proceedings of the IECON 2013-39th Annual Conference of the IEEE Industrial Electronics Society, Vienna, Austria, 10–13 November 2013; pp. 8080–8084.
8. Mao, W.; Zhang, X.; Cao, R.; Wang, F.; Zhao, T.; Xu, L. A research on power line communication based on parallel resonant coupling technology in pv module monitoring. *IEEE Trans. Ind. Electron.* **2018**, *65*, 2653–2662. [CrossRef]
9. Goubau, G. Surface Waves and Their Application to Transmission Lines. *J. Appl. Phys.* **2004**, *21*, 1119. [CrossRef]
10. Sommerfeld, A. Ueber die Fortpflanzung elektrodynamischer Wellen längs eines Drahtes. *Ann. Phys.* **1899**, *303*, 233–290. [CrossRef]
11. US8497749B2-Single Conductor Surface Wave Transmission Line System for Terminating E Field Lines at Points Along the Single Conductor-Google Patents. Available online: https://patents.google.com/patent/US8497749B2/en (accessed on 28 July 2022).
12. Roberts, T.E. Theory of the Single-Wire Transmission Line. *J. Appl. Phys.* **2004**, *24*, 57. [CrossRef]
13. Vaughn, B.; Peroulis, D. An updated applied formulation for the Goubau transmission line. *J. Appl. Phys.* **2019**, *126*, 194902. [CrossRef]

14. Santos, A.C.F.; Santos, W.S.; Aguiar, C.E. Electromagnetic wave velocities: An experimental approach. *Eur. J. Phys.* **2013**, *34*, 591–597. [CrossRef]
15. Sanchez-Pacheco, F.J.; Sotorrio-Ruiz, P.J.; Heredia-Larrubia, J.R.; Perez-Hidalgo, F.; de Cardona, M.S. PLC-Based PV Plants Smart Monitoring System: Field Measurements and Uncertainty Estimation. *IEEE Trans. Instrum. Meas.* **2014**, *63*, 2215–2222. [CrossRef]
16. Han, J.; Lee, I.; Kim, S.H. User-friendly monitoring system for residential PV system based on low-cost power line communication. *IEEE Trans. Consum. Electron.* **2015**, *61*, 175–180. [CrossRef]
17. Ochiai, H.; Ikegami, H. PPLC-PV: A pulse power line communication for series-connected PV monitoring. In Proceedings of the 2016 IEEE International Conference on Smart Grid Communications, SmartGridComm 2016, Sydney, Australia, 6–9 November 2016; pp. 338–344.
18. Chen, K.M.; King, R.W.P. A loop antenna coupled to a four-wire line and its possible use as an element in a circularly polarized end-fire array. *Proc. IEE Part C Monogr.* **1962**, *109*, 55. [CrossRef]
19. Adekola, S.A. On the excitation of a circular loop antenna by travelling-and standing-wave current distributions. *Int. J. Electron.* **1983**, *54*, 705–732. [CrossRef]
20. Boswell, A. Loop antennas in the 3-30 MHz band. In Proceedings of the 8th International Conference on High-Frequency Radio Systems and Techniques, Guildford, UK, 10–13 July 2000; pp. 33–36.
21. Li, R.; Bushyager, N.A.; Laskar, J.; Tentzeris, M.M. Determination of reactance loading for circularly polarized circular loop antennas with a uniform traveling-wave current distribution. *IEEE Trans. Antennas Propag.* **2005**, *53*, 3920–3929. [CrossRef]

sustainability

MDPI

Article

Renewable Heat Policy in China: Development, Achievement, and Effectiveness

Chengcheng Xiong * and Mohd Sayuti Hassan

Center for Global Sustainability Studies (CGSS), Universiti Sains Malaysia, Gelugor 11800, Penang, Malaysia; sayuti@usm.my
* Correspondence: xiongchengcheng@student.usm.my; Tel.: +60-4653-888

Abstract: Heat is the largest energy end-use sector, accounting for half of the global final energy consumption and more than 40% of energy-related CO_2 emissions. China produces more than one-quarter of global heat. Policy interventions are of great necessity to overcome the economic and non-economic barriers the sector encounters. The purpose of this study is to explore the evolution history of China's renewable heat policies over the last 20 years and to assess the effectiveness of the current policy system. The evolution of the policies is strongly linked to China's socio-economic background and is driven by various factors at each stage. A policy intensity index model is formulated to further dive into the dynamic characteristics of renewable heating. The results indicate that regulation-based instruments are always preferred, with varying degrees of lag for the other three types of instrument. Since the inception of the clean heating program in 2017, the intensity of renewable heating has increased dramatically, revealing that renewable heating has received increasing policy attention and is gradually becoming a key pillar in the context of climate change targets.

Keywords: renewable heating; policy; instrument

Citation: Xiong, C.; Hassan, M.S. Renewable Heat Policy in China: Development, Achievement, and Effectiveness. *Sustainability* **2022**, *14*, 9297. https://doi.org/10.3390/su14159297

Academic Editors: Luis Hernández-Callejo, Jesús Armando Aguilar Jiménez and Carlos Meza Benavides

Received: 15 June 2022
Accepted: 25 July 2022
Published: 29 July 2022

1. Introduction

China has committed to peaking carbon emissions by 2030 and achieving carbon neutrality by 2060 ("30/60 targets"). Reaching the 30/60 targets demands a dramatic scaling up of clean energy. Heating is the largest energy end-use sector, accounting for half of the global final energy consumption and more than 40% (13.1 Gt) of worldwide energy-related CO_2 emissions in 2020 [1]. More than one-quarter of global heat is produced and consumed in China, where the heat sector remains heavily reliant on fossil energy [2]. Renewable energy plays a critical role in decarbonizing and providing a greener energy supply option [3]. However, for a long time, renewable energy sources of heating (renewable heating) have been neglected in favor of a focus on renewable electricity, with less than 10% of energy supply coming from renewable energy sources.

The deployment of renewable energy in the heating sector should be accelerated to meet 30/60 targets, which will also bring additional benefits, including decreasing air pollution emissions and enhancing energy security [4]. However, there are still many obstacles to overcome, such as high capital costs, low prices of fossil fuels, and subsidies for fossil fuels. Policy intervention is needed to overcome the economic and non-economic barriers faced by the sector. In the past 20 years, China has already formulated some policies to support the deployment of renewable energy in the heating sector; however, there have been few systematic reviews of the existing renewable heating policies, and the key factors in designing the policies are still not well understood.

Besides, the majority of literature on renewable heating focus on specific renewable modes, e.g., geothermal, biomass, wind power heating etc. [5–7], analysis of the effectiveness of renewable heating systems [8–11], and evaluation of the impact of policies on energy use, climate change, and technology penetration [12–14]. These efforts have assisted in gaining a better understanding of the renewable heating system; however, the policy itself is rarely studied, especially when it comes to the implementation level. In China, policy implementation is crucial due to its special administrative structure [15]. Zhang et al. evaluated the energy saving and environmental policies in China, and proposed that the effectiveness of these policies are largely determined by the implementation level factors, including the instruments applied and the intensity of the policy itself [16,17]. However, few studies have evaluated renewable heat policies from these perspectives.

To fill the aforementioned research gaps, this study covers three aspects. Firstly, the development and framework of renewable heat policies formulated in the last 20 years are reviewed to construct a broad picture of the policy system. Secondly, the policies are clustered according to the instruments applied to bolster a quantitative analysis and obtain an in-depth insight into the dynamic development process. Finally, suggestions to establish a more efficient renewable heat policy system are provided. The results of this study will contribute to a comprehensive understanding of China's renewable heat policy and will shed light on policy design and implementation in the future.

2. Methods

In this study, a systematic review of renewable heat policies in China, spanning from 2000 to 2021, is conducted to understand the framework of China's renewable heat policy and to determine the key drivers in policy making. The review approach is developed by referencing several studies, including Zhu Bei et al. [18], Liu Junxia et al. [15], and Chul Kim [19].

To quantify the effectiveness of existing policies, a policy intensity index (PII) model is employed. A policy intensity index is developed based on the game theory of public policy to quantify the impact of policies. The index was first developed by Libecap [20] in 1978, and then widely used to quantify policy effectiveness. The structure of the PII in this paper is established based on previous studies concerning China's energy saving and clean heating policies [16–18,21], and specific modifications are made to precisely reflect the characteristics of renewable heat policy. The modification contains 2 aspects. Firstly, the score scope is narrowed down to 3 degrees. Secondly, the policies are classified into 3 categories instead of 5 categories, to simplify the evaluation.

Table 1 shows a breakdown of the index indicators, which includes 2 primary aspects: issuance level and target level. The policy's authority level is measured by the level of issuance. The issuance level aims to measure the authority degree of the policy through 3 sub-indicators: document type, issuer level, and coverage of the policy. The target level includes 2 indicators, the refinement and the duration of the targets, with the purpose of assessing the stringency degree of the targets in the policy. Each sub-indicator is endowed with the same score rules, from 1 to 3.

In terms of data processing, we gathered policy documents from official government websites, such as the official websites of the state council (SC), the National Development and Reform Commission (NDRC), the Ministry of Ecology and Environment (MEE), and relevant policy databases. A total of 146 relevant policies were collected. The keywords for text analysis included "policy," "regulation," "renewable heating," "geothermal," "biomass," "wind power heating," "clean heating," and "solar heating". Screen criteria was identified to ensure that the policy type was in the form of law, regulation, opinion, measure, or notice. The collected policies were standardized for further quantitative analysis by issue time, issuer, the number of issuers, target type, coverage, etc. The timeline ranged from 2000 to 2021, with the consideration that few policies were released before 2000.

Table 1. The breakdown of PPI indicators.

Indicator	Sub-Indicator	Score Rules
1. Issuance level	1.1. Document type (P1)	➢ Law: 3 ➢ Strategy, guidance opinion, plan, and action plan: 2 ➢ The announcement, notification, and implementation measures: 1
	1.2. Coverage (P2)	➢ National level: 3 ➢ Provincial level: 2 ➢ City level: 1
	1.3. Issuer level (P3)	➢ State council: 3 ➢ Ministries (more than 3): 2 ➢ Less than 2 ministries or provincial governments: 1
2. Target level	2.1. Refinement of targets (P4)	➢ Specific quantitative targets: 3 ➢ Only overall targets: 2 ➢ Only qualitative targets: 1
	2.2. Duration of targets(P5)	➢ More than 5 years: 3 ➢ 3–5 years: 2 ➢ One year or less: 1

After each indicator was scored, the policy intensity for a policy was calculated by multiplying all the sub-indicators. The model is shown in the following formula.

$$P_i = \sum_j^N \left(P1_{j,i} \times P2_{j,i} \times P3_{j,i} \times P4_{j,i} \times P5_{j,i}\right),\ i = [2000, 2021] \quad (1)$$

P_i represents the policy intensity of year i, while j represents the ordinal number of the policy in year i. $P1_{j,i}$ to $P5_{j,i}$ represent the score of the sub-indicator in a specific year of policy j.

3. Development of China's Renewable Heat Policy

3.1. Architecture of Renewable Heat Policies in China

China's current renewable heat policy system has been formed from scratch after more than 20 years of efforts. The architecture of the system is illustrated in Figure 1, to elucidate the inner relationship of the policies. Generally, policies can be classified into several levels. The first level is law, which lays out a fundamental legal framework for the development of renewable heating. The second level is medium- to long-term strategies or plans. The third level focuses on short-term comprehensive plans, including conventional plans and unconventional plans. The fourth level is sub-plans and special policies. The fifth level refers to local plans and policies.

China introduced the Renewable Energy Law in 2005, and since then renewable energy has become a preferential area for energy development. Renewable law lays a legal basis for renewable heating development. To implement the renewable heating law, central and local governments have formulated a series of policies.

Under the renewable law, medium- to long-term strategies are made. "The Medium- and Long-Term Renewable Development Plan" [22] and the relevant climate and environment strategy mainly addressed the principles and priority tasks for making short-term national renewable energy plans in China.

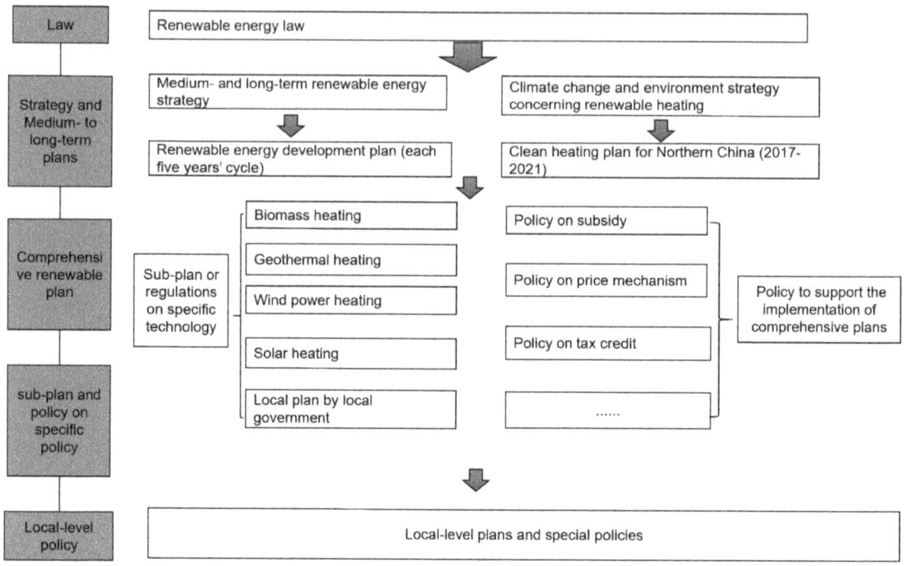

Figure 1. The structure of the renewable policy in China.

Based on strategy and high-level guidance, central governments usually make conventional plans to guide the next five years of development. The first renewable energy development plan was issued in 2012, and China has since began to promulgate a renewable development plan for each five-year period. These plans generally include mandatory goals that should be achieved by the end of the fifth year. Alongside conventional plans, some unconventional plans were also promulgated to address important and urgent issues. In 2017, the Clean Heating Plan for Northern China (2017–2021) (clean heating plan) was released jointly by nine ministries to facilitate the transformation of clean heating, and to control air pollution caused by coal-fired boilers [23]. The implementation of the clean heating plan spanned three years until the end of 2021 and covered "2 + 26" cities in Northern China.

The targets in these comprehensive plans are usually broken down into more specific targets and presented in sub-plans with specific measures. China has issued special plans for geothermal energy heating, biomass energy heating, renewable electricity heating, solar water, and heating policies [24–28]. Additionally, it is of significance to engage in the implementation of comprehensive plans and supporting policies, such as subsidies and tax credits, to show sufficient support and attention.

Local governments commonly release renewable heat policies according to local conditions under the umbrella of national policies. For example, Inner Mongolia issued the "Notice on wind power for heating to support the utilization of wind power for heating in winter seasons" [29]. Furthermore, local governments are also encouraged to formulate local standards to enhance the implementation of renewable policies and regulations. As of 2015, more than 28 provinces had already issued compulsory regulations requiring new buildings to install solar heating systems [30].

3.2. *The Development of Renewable Heating in China*

When the cost of solar water heating systems dropped significantly in the 2000s, renewable heating started to receive more attention from policymakers [30]. Since then, progress in policies has been accelerated to push forward the deployment of renewable energy in the heating sector. The evolution of renewable heat policies has been closely associated with the socio-economic context of the last 20 years [24]. The development of renewable heating policies can be divided into four stages, as shown in Table 2. The key drivers and the role of renewable heating are different in different stages.

Table 2. The four stages of renewable heating development.

Projection Period	**Stage 1**	**Stage 2**	**Stage 3**	**Stage 4**
	2000–2013	**2013–2016**	**2017–2020**	**2021–Present**
Key driver	Increasing demand for consumption	Air pollution issues	Clean heating program to achieve higher air quality	New climate change goals and the need for the energy transition
Key policy	Renewable law, solar heating incentives, and compulsory standards	12th and 13th five-year plans for renewable heating	Plan for clean heating in Northern China (2017–2021)	Notice on implementing renewable heat according to local conditions
Policy goals	No specific goals for renewable heating	Quantitative goals: substitution of coal achieving 100 million by the end of 2015 and 400 million m^2 for solar heating 580 million m^2	Quantitative goals: one billion square meters of geothermal, 2.1 billion square meters of biomass, and 50 million square meters of solar energy	Quantitative goals: the scale of geothermal heating, biomass heating, biomass fuels, solar thermal utilization, and other non-electric utilization reached 60 million tons of standard coal by the end of 2025
Role of renewable heating at the policy level	A supplement to energy use especially in household	A supplement when coal is replaced	One of the main sources of clean heating supply	Imperative solution for climate change goals

- Stage 1: driven by household demand, from 2000 to 2013

Since 2000, China has undergone rapid urbanization, along with a rapid increase in the application of domestic solar water systems. Strong demand from the consumption side drove the significant growth of the installation of solar water heating systems, which achieved 323.1 million m^2 by the end of 2012 [4]. Incentive measures, i.e., subsidies, tax credit, and compulsory standards greatly promoted the deployment of solar heating. The solar heating market expanded rapidly with strong responses, since solar heating was integrated into the energy saving livelihood program (China started to implement the energy saving livelihood program in 2012, by providing subsidies for energy-saving products, including solar water products). This required direct delivery of subsidies for households that were equipped with solar heating systems. However, the year 2014 witnessed the beginning of a decline in the growth rate due to the expiration of the program. The market progress of solar heating is described in Figure 2.

In this stage, solar heating dominated the renewable heating market, with other renewable heating technologies accounting for little market share. This stage is marked by an initial stage of renewable heating, with household demand as the key driving factor.

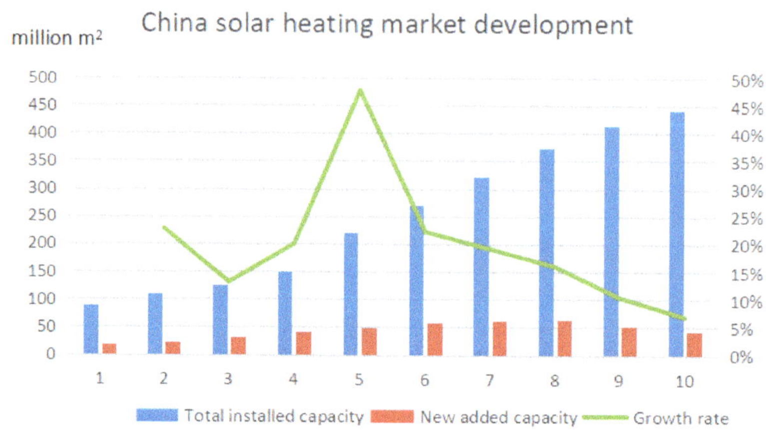

Figure 2. The development of the solar heating market in China.

- Stage 2: driven by air pollution issues, from 2013 to 2017

In 2013, extreme haze weather hit northern areas of China with a high frequency. The increasing environmental awareness of the public and concerns over air pollution instigated heating reform focusing on clean energy transition [31]. In 2013, the state council (SC) issued the Air Pollution Prevention and Control Action Plan [32]. Following this, regulations on the substitution of coal for residential heating became a vital part of national air pollution control and environmental supervision. Moreover, renewable heat policies were introduced accordingly in this stage, focusing on specific technologies, such as biomass and geothermal energy. In 2013, a special policy titled "Guidance opinions on promoting utilization of geothermal energy" [24] was released, clarifying the main objectives, key tasks, and measures required to foster the geothermal heating industry. It is also worth noting that the first renewable heating target for the whole country was confirmed at this stage. In the 12th Five-Year Renewable Energy Development Plan [33], a quantitative indicator for renewable heating, namely the amount of substitution of fossil energy, was established. Subsequently, in the 13th Five-Year Renewable Energy Development Plan, the targets were further enhanced, from 10 billion tce by 2015 to 15 billion tce by 2020 [34].

At this stage, policies on renewable heating began to be more vigorously implemented, and specific quantitative targets for renewable heating development were determined for the first time. The primary policy measures in this stage were aimed at providing resolutions to air pollution issues, and renewable heating was deemed as a supplementary solution to replacing coal.

- Stage 3: driven by demand of higher air quality

In 2017, 10 ministries and commissions jointly promulgated the Plan for Clean Heating in Northern China (2017–2021) [23], which formally defined the energy sources of clean heating, including geothermal energy, biomass, solar energy, natural gas, electricity, industrial waste heat, clean coal combustion, nuclear energy, etc. In the policy, an ambitious target was proposed, striving to achieve a 50% clean heating rate by 2019 and 70% by 2021. Furthermore, renewable heating was for the first time identified as the main source of clean heating energy rather than playing a supplementary role in the previous stage. Specific goals were given for each renewable heating technology, i.e., one billion square meters for geothermal energy, 2.1 billion square meters for biomass, and 50 million square meters for solar energy by the end of 2021.

To ensure the implementation of the plan, regulations on subsidies and price mechanisms were also released at this stage. In 2017, the Ministry of Finance provided subsidies to 43 pilot cities, with an annual subsidy of one billion RMB for municipalities, 700 million RMB for provincial capitals, as well as 500 million RMB for other cities [35]. Following this, the National Development and Reform Commission (NDRC) issued a policy titled "Opinions on Clean Heating Price in North China" to overcome the issue of price distortion, clarifying that renewable energy should be given high priority in heating energy supply and linkage of electric heating, wind, and solar power generation curtailment should be established [36]. Thereafter, the 13th Five-Year Development Plan of Geothermal Energy [37] specified the goals and strengthened the measures, in which specific geothermal modes, shallow geothermal systems, and hot rock geothermal systems were introduced to solve the problem of winter heating where applicable. Meanwhile, in the 13th Five-Year Development Plan of Biomass Energy [26], biomass heating, especially biomass co-generation, was highlighted to supplement the energy supply with regard to the implementation of "coal to gas" or "coal to electricity" policies.

At this stage, renewable heating was formally integrated into the clean heating system as a main source of energy supply, rather than a supplement. Applying more diverse policy instruments could contribute to ensuring the implementation of the clean heating plan.

- Stage 4: driven by ambitious climate change goals, 2021–

In 2021, a comprehensive policy titled "Notice on advancing the development of renewable heating according to local conditions" was launched [38], consisting of six aspects: (1) release special plans for renewable heating, including specific renewable heating goals aligned with the requirements of climate change goals; (2) promote various types of renewable heating technologies according to local conditions; (3) continue promoting pilot demonstration work and major project construction; (4) guarantee policy support of renewable heat; (5) strengthen R&D support for key technical equipment; and (6) improve the government management system for renewable heat. The policy symbolized the significant achievements of clean heating policies in China and initiated a new era of renewable heating development. The 30/60 targets served as the dominant driver for this stage. Subsequently, a series of policies concerning renewable heating were released. The most recent policy is "Several opinions on promoting the development and utilization of geothermal energy" [25], jointly issued by eight ministries, which highlights the importance of geothermal utilization.

In summary, the year 2021 witnessed the transformation of renewable heat policy with the release of a comprehensive renewable heating program. This indicates that renewable heating in China has been formally integrated into the energy transition system as a key measure to address climate change.

4. Evaluation of Renewable Heat Policy

In China, policies have varying impacts according to the authority level of the issuer. Regulations issued by the state council generally have the highest authority, followed by regulations or policies jointly issued by multiple ministries. Policies released by the signal ministry or province (autonomous region, municipality) have the lowest authority. Furthermore, the efficacy of a policy is also greatly related to the document type, i.e., law has the highest efficacy, followed by strategy.

A policy instrument provides the link between policy formulation and policy implementation. The instrument used in a policy has a significant impact on how well it is implemented [39]. Policies can be divided into four categories using the instrument applied: regulation-based policy, fiscal instrument-based policy, price mechanism-based policy, and financial instrument-based policy. Regulation-based instruments include targets, planning, compulsory standards, etc. Fiscal instruments commonly include subsidies, grants, tax credits, etc. Price-based instruments include heat-trading mechanisms, carbon markets, etc. Typical financial instruments are bonds, loans, and direct equity investments [4].

In this section, renewable heat policies are evaluated from two perspectives. First, we focus on the instrument in the policies to find out how the policy is implemented. Second, PII is used to further understand the intensity of the existing policies and their efficacy.

4.1. The Instruments of Renewable Heat Policies

In this paper, the cumulative number is used to quantify the instrument applied in policies. To be specific, policies are classified according to the type of instrument and a specific instrument that appeared in a specific year was counted in the subsequent years. Policies that use several instruments are counted into primary instrument types to avoid double counting. The cumulative numbers of the four types of policies according to instruments applied from 2000 to 2021 are illustrated in Figure 3.

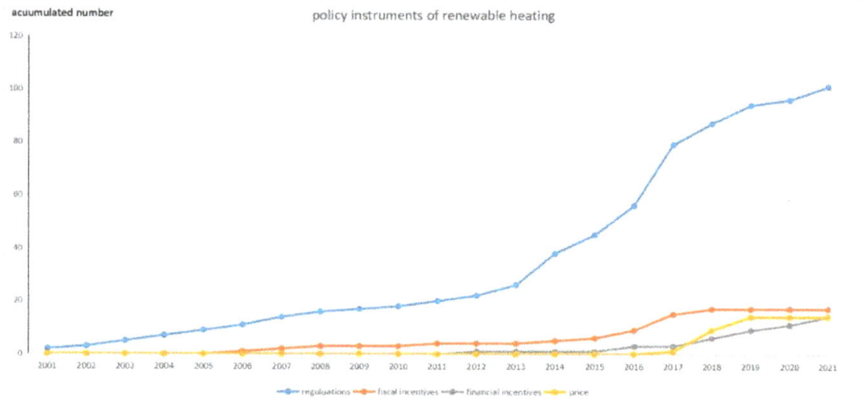

Figure 3. Development of policies according to instruments.

Overall, it demonstrated an upward tendency of all four types of renewable policies since stage 2, particularly for regulation-based policies, which increased from 20 to more than 100 over seven years. Comparatively, it can be seen that the other three types all showed some degree of delay. The amount of fiscal incentive policies has grown moderately since 2015, while the price and financial instrument-based policies began to increase steadily only after 2017.

In terms of the amount of the four types of policies, regulation-based policies are largest with a share of approximately 70% by 2021. Fiscal incentives began to increase in 2017, leading to a share of 11% by the end of 2021, with the remaining two types at less than 20%. Regulation-based policies are preferred, with an amount much higher than the other three instruments, showing a structured imbalance in the application of different instruments.

4.2. Evaluation of Policy Intensity

To further illustrate the dynamic changes in renewable heat policies over the last 20 years, the policy intensity index from 2012–2021 was gained using the model introduced in Section 2. The results are shown in a bubbling figure (Figure 4), where the size of the bubble represents the intensity of the policies of a specific year.

It can be seen that the intensity of the four types of policies showed different patterns, among which regulation-based policies showed the strongest intensity of all four stages. The intensity of regulation-based policies reached the highest level in the year 2017, which can be attributed to a series of policies issued by high-level authorities, including the SC. Fiscal relevant policies followed after and financial instrument-based policies had the lowest intensity. The years following 2017 witnessed an increase in the intensity of the four types of policies, indicating the increasing political will for renewable heating from the policy maker.

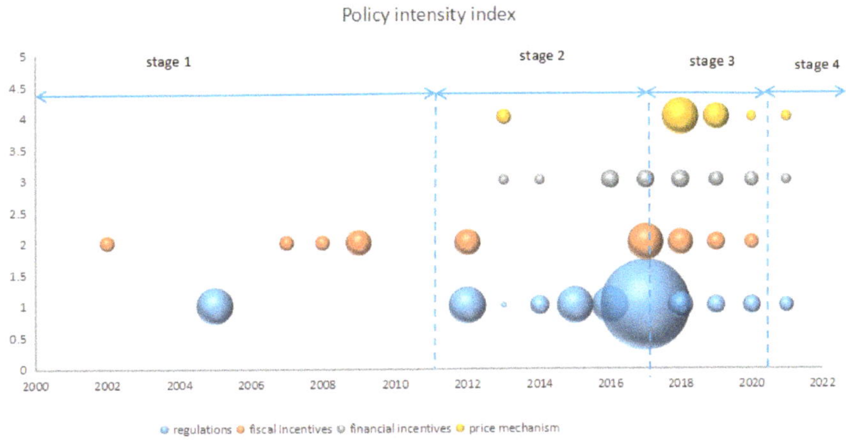

Figure 4. Policy intensity index of the four types of policies.

For the four types of policy instruments, the following can be found if Figures 3 and 4 are combined:

Regulation-based policies dominated in the policy system, which were both largest in amount and strongest in stringency. The high-frequency application of regulation-based instruments in renewable heat policies reflects that the renewable heating market is still primarily driven by governments and is still at an initial stage, where mandatory regulations are greatly needed. However, it is not recommended that regulation-based policies are applied long term, due to the lack of flexibility and cost effectiveness.

Fiscal incentives have been implemented since 2017 when the clean heating program was initiated, marking a major step. However, the incentive options are limited, and subsidy is the most common approach. Fiscal instruments can play a great role in the initial stages by supplementing the capital cost, but put great pressure on the government. More diverse instruments can be deployed when providing fiscal support. One example is the renewable heating incentive (RHI) from the United Kingdom, which is similar to the tariff mechanism in the generation market. The tariff mechanism can be more result-orientated and flexible compared to subsidy.

Market-based policies, including price mechanism and financial incentives, are least used. Market-based policies are seen as the most cost-efficient way to foster a renewable market by internalizing the environmental and economic cost. The application of market-based policies should be greatly accelerated.

5. Conclusions and Suggestions

5.1. Conclusions

China's renewable heat policies have advanced quickly in the last two decades. Numerous policies have been implemented. However, few studies have examined the evolution of these policies. We used the PII model to present the dynamic changes in the renewable heat policies and three conclusions were summarized as follows.

Firstly, we systematically reviewed and illustrated the overall development of renewable heat policies from 2000 to 2021. China has established a renewable heat policy system, which consists of five levels of policy, each with different power and authority. The top levels are law, strategies, and plans, which provide macro guidance for long-term development. The middle levels are comprehensive policies and regulations in the form of five-year plans, special plans, etc., to further specify targets and measures. The policies at the bottom levels are commonly designed for guiding implementation, including supporting policies with incentives, special plans, or regulations, targeting specific renewable heating technology models and local policies.

Secondly, the 20 years were divided into four stages. In stage 1, the main driver was the increasing heating demand from households, when the market for solar water heating soared quickly. In this stage, regulation-based policies, such as compulsory standards, pushed the expansion of the market. Furthermore, incentives also played an important role. The second stage began in 2013. In this stage, issues with air pollution pushed the development of renewable heating, with more political attention paid to renewable heating to supplement the coal-to-gas program in fighting against air pollution. In 2017, China initiated its clean heating plan, when renewable heating began to formally integrate into the clean heating system. The fourth stage was driven by new climate change goals. In this stage, the development of renewable heating became a certainty. The renewable heating development of the past 20 years has a strong link with the socio-economic context, showing various features in different stages. Thirdly, to further dive into the renewable heat policy, we classified the existing renewable heat policies into four categories according to the instruments applied. The results showed that the structure of the four instruments were in disproportion, with regulation-based instruments the most preferred. While the government began to apply other instruments to overcome price distortion and other issues, the impact is still limited. The number of policies with all four instruments increased dramatically since stage 3, when China began to implement its clean heating program. To enhance the implementation of the 30/60 targets, the application of market-based policies are greatly needed.

5.2. Policy Suggestions

To further accelerate the deployment of renewable energy in the heating sector, the following suggestions are proposed:

Firstly, multiple barriers in the renewable heating sector call for a range of policy instruments, often in combination. Regulation-based instruments, which generally lack flexibility and cost-effectiveness, should be supplemented by other policies, especially market-orientated policies. The application of regulation instruments can be strengthened by increasing the application of other types of policy instruments. Instrument diversification is required to strike a balance between cost, adaptability, and efficacy.

Secondly, the incentive system for renewable heating needs to be strengthened. The externality of renewable heating can be partially compensated by incentives. Instead of relying solely on subsidies, the government can use a variety of incentives, such as tax breaks, tariffs for renewable heating sources, etc., to alleviate the financial load. Additionally, innovative measures, such as result-based incentives and loan guarantees, can also be applied when applicable.

Thirdly, market-orientated instruments should be further enhanced in policies. Carbon market and green finance instruments have great potential in further enhancing the development of the renewable heating market. Financial instruments should be more embodied in future policies, such as green loans and green bonds.

Finally, policy coordination should be further strengthened. Coordination can improve the efficiency of resource allocation, as well as avoid unnecessary conflict in implementation.

The present work provides a first and comprehensive review of China's renewable heating policy at the national level. Further research regarding the role of the policy's executor, the local government, would be of great help in understanding the effectiveness of renewable heating. In addition, a considerable amount of work is needed to evaluate the impact of renewable heating on energy transition and carbon emission, to examine the evolution of household heating patterns, and to explore the possible demand-side management option using an experimental approach.

Author Contributions: Writing—original draft preparation, C.X.; supervision, M.S.H. All authors have read and agreed to the published version of the manuscript.

Funding: This research received no external funding.

Institutional Review Board Statement: Not applicable.

Informed Consent Statement: Not applicable.

Data Availability Statement: Not applicable.

Conflicts of Interest: The authors declare no conflict of interest.

Abbreviations

Renewable heating	Renewable energy source for heating
IEA	International Energy Agency
SC	State Council
MEE	Ministry of Ecological Environment
MOF	Ministry of Finance
MOHURD	Ministry of Housing and Urban-Rural Development
NDRC	National Development and Reform Commission
NEA	National Energy Administration
MLR	Ministry of Land and Resource
PII	policy intensity index

References

1. International Energy Agency. *Renewables 2021: Analysis and Forecast to 2026*; International Energy Agency: Paris, France, 2021.
2. National Energy Adiminstration (NEA). *Explanation of Guidance Opinion on Renewable Heating*; National Energy Adiminstration (NEA): Beijing, China, 2019.
3. International Energy Agency. *Renewable Heat Policies: Delivering Clean Heat Solutions for the Energy Transition*; International Energy Agency: Paris, France, 2018.
4. Hu, R.Q. *Market and Policy Research on China Renewable Energy Heat*; China Environment Publishing House: Beijing, China, 2018.
5. Wang, Y.; Liu, Y.; Dou, J.; Li, M.; Zeng, M. Geothermal energy in China: Status, challenges, and policy recommendations. *Util. Policy* **2020**, *64*, 101020. [CrossRef]
6. Xu, J.; Yuan, Z. *An Overview of the Biomass Energy Policy in China*; BE Sustainable: Florence, Italy, 2015. Available online: https://www.besustainablemagazine.com/cms2/overview-of-biomass-energy-policy-in-china/#:~:text=Basic%20framework%20of%20China%E2%80%99s%20biomass%20energy%20development%20policy,and%20rapid%20development%20of%20the%20biomass%20energy%20industry (accessed on 24 July 2022).
7. Wang, W.; Dong, Z. Economic benefits assessment of urban wind power central heating demonstration project considering the quantification of environmental benefits: A case from northern China. *Energy* **2021**, *225*, 120246. [CrossRef]
8. Connor, P.M.; Xie, L.; Lowes, R.; Britton, J.; Richardson, T. The development of renewable heating policy in the United Kingdom. *Renew. Energy* **2015**, *75*, 733–744. [CrossRef]
9. Li, B.; Sun, Y.; Zheng, W.; Zhang, H.; Jurasz, J.; Du, T.; Wang, Y. Evaluating the role of clean heating technologies in rural areas in improving the air quality. *Appl. Energy* **2021**, *289*, 116693. [CrossRef]
10. Ma, S.; Guo, S.; Zheng, D.; Chang, S.; Zhang, X. Roadmap towards clean and low carbon heating to 2035: A provincial analysis in northern China. *Energy* **2021**, *225*, 120164. [CrossRef]
11. Pitelis, A.; Vasilakos, N.; Chalvatzis, K. Fostering innovation in renewable energy technologies: Choice of policy instruments and effectiveness. *Renew. Energy* **2020**, *151*, 1163–1172. [CrossRef]
12. Lim, S.; Huh, S.; Shin, J.; Lee, J.; Lee, Y. Enhancing public acceptance of renewable heat obligation policies in South Korea: Consumer preferences and policy implications. *Energy Econ.* **2019**, *81*, 1167–1177. [CrossRef]
13. Edling, L.; Danks, C. Supporting actors: The role of state policy and private programs in advancing local and renewable heating technology. *Energy Policy* **2021**, *153*, 112266. [CrossRef]
14. Kranzl, L.; Hummel, M.; Müller, A.; Steinbach, J. Renewable heating: Perspectives and the impact of policy instruments. *Energy Policy* **2013**, *59*, 44–58. [CrossRef]
15. Liu, J. China's renewable energy law and policy: A critical review. *Renew. Sustain. Energy Rev.* **2019**, *99*, 212–219. [CrossRef]
16. Zhang, X.; Zhang, P.; Xiu, J. Are energy-saving and emission reduction policy measures effective for industrial structure restructuring and upgrading? *China Popul. Resour. Environ.* **2018**, *28*, 123–133.
17. Zhang, X.; Zhang, P.; Zhang, Z.G.; Li, J. Impact of environmental regulations on industrial structure upgrading: An empirical study on Beijing-Tianjin-Hebei region in China. *J. Clean. Prod.* **2019**, *238*, 117848. [CrossRef]
18. Zhu, B.; Liu, C.; Wei, C. Urban residential heating policy in China: A review. *Energy Build.* **2021**, *253*, 111547. [CrossRef]
19. Kim, C. A review of the deployment programs, impact, and barriers of renewable energy policies in Korea. *Renew. Sustain. Energy Rev.* **2021**, *144*, 110870. [CrossRef]
20. Libecap, G.D. Economic Variables and the Development of the Law: The Case of Western Mineral Right. *J. Econ. Hist.* **1978**, *2*, 338–362. [CrossRef]
21. Liu, C.; Zhu, B.; Ni, J.; Wei, C. Residential coal-switch policy in China: Development, achievement, and challenge. *Energy Policy* **2021**, *151*, 112165. [CrossRef]

22. The National Development and Reform Commission. *The Medium and Long Term Renewable Development Plan*; The National Development and Reform Commission: Beijing, China, 2007.
23. National Development and Reform Commission; National Energy Administration. *Plan for Clean Heating in North China (2017–2021)*; National Development and Reform Commission; National Energy Administration: Beijing, China, 2017.
24. National Energy Administration; Ministry of Finance; National Land Resource Administration; Minstry of Houshold and Urban-Rural Consutruction. *Guidance Opinion on Promoting the Utilization of Geothermal Energy*; National Development and Reform Commission; National Energy Administration: Beijing, China, 2013.
25. National Energy Administration. *Several Opinions on Promoting the Development and Utilization of Geothermal Energy*; National Energy Administration: Beijing, China, 2021.
26. National Energy Administration. *13th Five Years' Development Plan of Biomass Energy*; National Energy Administration: Beijing, China, 2016.
27. National Energy Administration. *Notice on Enhancing Electricity Trading Mechanism and Expanding Wind Power Heating*; National Energy Administration: Beijing, China, 2019.
28. National Energy Administration. *Notice on Enhancing Wind Power Heating in Northern Areas*; National Energy Administration: Beijing, China, 2013.
29. National Energy Administration. *Mongolia Notice on Wind Power for Heating to Support the Utilization of Wind Power for Heating in Winter Seasons*; National Energy Administration: Beijing, China, 2019.
30. Asia Development Bank. *Solar District Heating in the People's Republic of China: Status and Development Potential*; Asia Development Bank: Metro Manila, Philippines, 2019.
31. Zhang, Y.; Zheng, W.; Fang, H.; Xia, J. Clean heating in Northern China: Regional investigations and roadmap studies for urban area towards 2050. *J. Clean. Prod.* **2022**, *334*, 130233. [CrossRef]
32. State Council. *Air Pollution Prevention and Control Action Plan*; State Council: Beijing, China, 2013.
33. National Energy Administration. *12th Five Years' Renewable Energy Development Plan*; National Energy Administration: Beijing, China, 2012.
34. National Energy Administration. *13th Five Years' Renewable Energy Development Plan*; National Energy Administration: Beijing, China, 2016.
35. Ministry of Finance; National Energy Administration; Ministry of House and Urban-Rural Construction; Ministry of Environment Protection. *Notice on Supporting Clean Heating Pilot of Northern Area with Central Goverment Subsidy*; National Energy Administration: Beijing, China, 2017.
36. National Development and Reform Commision. *Opinions on Clean Heating Price in North China*; National Development and Reform Commision: Beijing, China, 2017.
37. National Development And Reform Commission; National Energy Administration; National Land Resource Administration. *13th Five Years'Development Plan of Geothermal Energy*; National Development and Reform Commission; National Energy Administration: Beijing, China, 2017.
38. National Energy Administration. *Notice on Advancing the Development of Renewable Heating According to Local Conditions*; National Energy Administration: Beijing, China, 2021.
39. Ali, M. *Sustainability Assessment: Context of Resource and Environmental Policy*; Academic Publisher: Cambridge, MA, USA, 2013; pp. 99–106.

electronics

MDPI

Article

Prediction of Power Output from a Crystalline Silicon Photovoltaic Module with Repaired Cell-in-Hotspots

Koo Lee [1,2], Sungbae Cho [1,3], Junsin Yi [2,*] and Hyosik Chang [1,*]

1 Graduate School of Energy Science & Technology, Chungnam National University, Daejoen 34134, Korea; engine29@naver.com (K.L.); sbcho@sksel.com (S.C.)
2 College of Information and Communication Engineering, Sungkyunwan University, Suwon 16419, Korea
3 SK Solar Energy Co., Ltd., 112 Jangwookjin-ro, Yeondong-myeon, Sejong-City 34013, Korea
* Correspondence: junsin@skku.edu (J.Y.); hschang@cnu.ac.kr (H.C.); Tel.: +82-31-290-7139 (J.Y.); +82-42-821-8607 (H.C.)

Abstract: Recycling of problematic photovoltaic modules as raw materials requires considerable energy. The technology to restore cells in hotspot modules at a relatively low cost is more economical than replacing them with new modules. Moreover, a technology that restores power by replacing a cell-in-hotspot of a photovoltaic module with a new cell rather than replacing the whole module is useful for operating power plants. In particular, power plants that receive government subsidies have to use certified modules of specific models; the modules cannot be replaced with other modules. Before putting resources into module restoration, predicting the power of a module to be restored by replacing a cracked cell with a new cell is essential. Therefore, in this study, the module output amount after restoration was calculated using the previously proposed relative power loss analysis method and the recently proposed cell-to-module factor analysis method. In addition, the long-term degradation coefficient of the initial cell and the loss due to the electrical mismatch between the initial and new cell were considered. The output of the initial cell was estimated by inversely calculating the cell-to-module factor. The differences between the power prediction value and the actual experimental result were 1.12% and 3.20% for samples 190 A and 190 B, respectively. When the initial rating power and tolerance of the module were corrected, the differences decreased to 0.10% and 2.01%, respectively. The positive mismatch, which restores cells with a higher power, has no loss due to the reverse current; thus, the efficiency of the modules is proportional to the average efficiency of each cell. In this experiment, the electrical mismatches were only 0.37% and 0.34%. This study confirmed that even if a replacement cell has a higher power (<20%) than the existing cell, the power loss is not significantly affected, and heat generation of the existing normal cell is not observed. Hence, it was concluded that when some cells are damaged in a crystalline solar cell, the module could be restored by replacing only those cells instead of disposing of the entire module. However, for commercialization of the proposed method, a long-term reliability test of the module repaired using this method must be performed to confirm the results. Following this, recycling cells instead of recycling modules will be an economical and eco-friendly alternative.

Keywords: cell-in-hotspot; cell replacement; module repair; restoration technology; module recovery; power prediction; electrical mismatch; CTM factor analysis

Citation: Lee, K.; Cho, S.; Yi, J.; Chang, H. Prediction of Power Output from a Crystalline Silicon Photovoltaic Module with Repaired Cell-in-Hotspots. *Electronics* **2022**, *11*, 2307. https://doi.org/10.3390/electronics11152307

Academic Editors: Luis Hernández-Callejo, Jesús Armando Aguilar Jiménez and Carlos Meza Benavides

Received: 23 June 2022
Accepted: 22 July 2022
Published: 24 July 2022

1. Introduction

Renewable energy, including photovoltaic power generation, has steadily increased globally through [1,2] continuous cost-cutting efforts based on eco-friendly elements and low maintenance costs [3,4], despite the high costs and relatively low economy in the early stages of its implementation [5]. Owing to economic security and increased supply of renewable energy [6,7], the achievement of grid parity has recently accelerated [8,9], with a certain percentage of fossil fuel usage steadily being replaced by renewable energy usage, and hence, the total use of renewable energy has increased over the past decade [10]. Expanding the solar energy supply may reduce carbon dioxide emissions and achieve a healthy mix of energy sources to overcome the climate crisis [11,12]. However, the increasing demand for solar energy may cause shortages of the resources used in the advanced production of solar modules [13–15]. In particular, owing to the scarcity of resources such as silver, indium, and bismuth, target material consumptions of 2, 0.38, and 1.8 mg/Wp [16,17], respectively, have been proposed; thus, a significant reduction in material consumption is required to expand renewable energy supply [18].

The large-scale installation of photovoltaic modules results in the problems of economic use of resources during production and processing of waste modules after use [19,20]. By 2050, 80 million tons of accumulated photovoltaic modules are expected to reach their service life worldwide, with 10 million tons in the US alone [21]. With the rapid increase in the installation of photovoltaic modules in countries such as China, the collection and recycling of end-of-life photovoltaic modules is becoming an important task, and various methods of building efficient recycling systems are being investigated [22]. According to previous studies, the predicted accumulated waste that will be generated from 2020 to 2080 in existing solar power plants varies in proportion to solar installations, and is expected to peak at 130,000 tons in 2051 and 141,297 tons in 2054 [23]. Currently, the life of a photovoltaic module is approximately 20–30 years; therefore, the life of photovoltaic modules installed in the early 2000s will expire on a large scale, and the disposal of waste modules will increase rapidly. Photovoltaic modules consist of expensive materials, such as aluminum, silver, copper, tin, and silicon wafers. In addition, they can be used as highly attractive recycled materials in terms of the environmental charges imposed when filling landfills. In the recycling process of a general photovoltaic module, research has primarily focused on recycling by collecting silicon wafers and refabricating them into optimized silicon solar cells [24,25], pyrolyzing organic materials such as ethylene vinyl acetate (EVA) [26], and removing organic materials such as glass and ribbon metals [27,28]. As a research example on recovering the performance of photovoltaic modules, a technology for recovering the insulation resistance of aging modules by injecting coatings based on polyurethane, epoxy, silicone, and synthetic rubber of crystalline photovoltaic modules was introduced [29,30]. However, recycling or reuse technology generally involves removal of frames, junction boxes, or cables, etc., from crystalline photovoltaic modules, followed by thermal or chemical decomposition of the laminated module to collect glass, silicon, metal, and polymer [31,32].

This recycling technology is not currently widely used because it is expensive and the return on investment (ROI) is less than approximately −0.25 as of 2022 [33]. In addition, the recycling method, which involves collecting the raw materials separately, is not applicable to the recovery of damaged modules in an operating power plant because the failure of a part of the module results in the crushing of other usable parts. Accordingly, this paper proposes a technology to recover photovoltaic modules at the same or a higher level of the initial power value by replacing cells at a safety risk, such as power loss and hotspots, owing to damage to some cells of an aged silicon photovoltaic module. Most commercial solar power plants receive subsidies from the government. In this case, only certified modules of a particular model should be used during the generation period. If the module fails, it cannot be replaced by another model. Moreover, owing to the rapid improvement in cell efficiency every year [34,35], the module model continues to change. In commercial power plants, restoring the output of a module by cell replacement is very useful. Technological advancements in the restoration of the module result in a power deviation between the initial and new cell [36,37]. Therefore, when replacing a cell with a new cell having a higher power, the possibility of an electrical mismatch loss occurring should be considered, and the long-term power degradation of the initial cell should be confirmed. Hence, the purpose of the experiment was to determine the extent to which the output improvement of the new cell is reflected in the output of the module to be restored. Previous studies have shown that the prediction of power mainly includes power degradation in modules with hotspots or how much power decreases as a result of EL in modules with potential-induced degradation (PID) [38,39]. However, the purpose of this study was to predict the improvement to power through replacement of damaged cells in a module, which has not been attempted before. The results of this study suggest that the energy and environmental costs of recycling modules can be significantly reduced by reusing waste modules in more diverse states.

2. Experiments

2.1. Methods and Procedures

The overall experiment was conducted in the following order: module power output and defect verification, calculation of grade of originally applied cells, grade verification of replacement cells, power prediction, module power recovery, comparison of predicted power output and experimental results, and application of correction values. First, the defects and power output of the module to be recovered were checked via electroluminescence (EL) measurement and a sun-simulator. EL measurements are used to identify internal defects that cannot be visually identified using EL in solar cells. Table 1 provides the nomenclature for the electrical characteristics of the module.

Table 1. Nomenclature for the electrical characteristics of the module.

I_{sc}	Short-circuit current	I_{mp}	Current at the maximum power output
V_{oc}	Open-circuit voltage	V_{mp}	Voltage at the maximum power output
P_{max}	Maximum power output	FF	Filling coefficient factor

The current corresponding to the cell I_{sc} and the voltage at the same level as the module V_{oc} were applied for the measurement. EL images of the module were captured in several parts of a darkroom, recollected, and displayed on a screen. The EL equipment manufactured by MC Science in Korea was used for the measurements. The simulator measures the module's I_{sc}, V_{oc}, P_{max}, etc., under the standard test condition (STC) at 25 °C, 1 Sun (1000 W/m^2), and air mass 1.5, and corrects the actual temperature to output the calculated value to the screen. The equipment used in this study was a Spire-Nissinbo Sun Simulator. The equipment was calibrated for proper use in the certification test of the photovoltaic module by receiving the AAA in three evaluation items: uniformity, stability, and spectrum. Measurements of power output from equipment are displayed in various ways, i.e., 1–4 digits after the decimal point; however, in this study, the third digit after the decimal point was rounded to two digits to maintain consistency. The CTM (cell to module) factor calculation method was applied to the power analysis of the cells used at the time of manufacturing the target samples and the review of the cells to be replaced [40]. The grade of the applied cell was inversely calculated based on the initial power output of the module disclosed on the Internet by the manufacturer. The module power after cell replacement was predicted after checking the grade of the cell to be replaced.

The CTM coefficient k-factor calculation method was used to analyze the power of the original cell of the target samples and review the replacement cell. Manufacturing modules from cells, models, and formulas for classifying the CTM coefficient k-factor, which affects efficiency or power, and analyzing loss or acquisition mechanisms have been presented in previous research [41,42]. If the dimension data and rated power of a module released by the module manufacturer are the initial power outputs of the module, the module efficiency is calculated to be 13.6%. Because the module power output is calculated from the sum of the CTM coefficient k factor and the initial solar cell power in the module power output calculation model, the power output of the module can be calculated using Equations (1) and (2) [41,43]. The factors i and m in Equations (1) and (2) are variables of the routinely used pie function, and refer to the extension of the CTM factor. The CTM k-factor consisted of 15 types: k_1 (module margin), k_2 (cell spacing), k_3 (cover reflection), k_4 (cover absorption), k_5 (cover/encapsulant reflection), k_6 (encapsulant absorption), k_7 (interconnection shading), k_8 (cell/encapsulant coupling), k_9 (finger coupling), k_{10} (interconnector coupling), k_{11} (cover coupling), k_{12} (cell interconnection), k_{13} (string interconnection), k_{14} (electrical mismatch), and k_{15} (junction box and cabling). The meaning of $I = 3{-}m$ in the $\prod$-function of Equation (1) means CTM k-factor from k_3 to k_{15}. Then, the sum of the cell power outputs from $j = 1{-}n$ from the $\sum$-function is the number n of cells applied to the module.

$$P_{module} = \prod_{i=3}^{m} k_i \cdot \sum_{i=1}^{n} P_{cell.j} \tag{1}$$

$$CTM_{power} = \prod_{i=3}^{m} k_i \tag{2}$$

In terms of module efficiency, factors affecting the entire area of a gap module between modules are important; however, when a module is produced from a cell, a design margin (k_1) to ensure an electrical insulation distance and a loss factor (k_2) owing to the cell interval are not related to a power change. The module efficiency can be expressed by Equations (3) and (4) [41].

$$\eta_{module} = \frac{P_{module}}{E_{STC} \cdot (A_{module} + A_{cell\ spacing} + A_{cells})} \tag{3}$$

$$\eta_{module} = \overline{\eta_c} \cdot (k_1 + k_2 - 1) \cdot \prod_{i=3}^{m} k_i \tag{4}$$

Therefore, according to this model, the module efficiency is proportional to the average efficiency of the cell rather than being dominantly affected by the lowest efficiency. The average efficiency of the cell was calculated by considering the electrical mismatch loss (k_{14}) of the cell to predict the power output of the module to be restored. For the loss caused by the electrical mismatch of cells, studies were published prior to research on the CTM factor, and the widely known definition of RPL is expressed as the difference between the maximum power (P_{mpc}) of n individual cells connected in series to form a cell string or module. RPL can be expressed as Equation (5) from the difference between the sum of the maximum power of all cells and the maximum power of the module.

$$RPL = \frac{\sum_{i=1}^{n} \cdot P_{mpci} - P_{module}}{\sum_{i=1}^{n} \cdot P_{mpci}} \tag{5}$$

In theory, when individual cells operate completely independently, the maximum power output is denoted as P'_{max}, and when the average cell power output value in a group is P_{max}, the calculation of RPL_B (relative power loss of a module using Bucciarelli's equation) is as shown in Equation (6).

$$RPL_B = \frac{P'_{max} - P_{module}}{n \cdot \overline{I_{mp}} \overline{V_{mp}}} \tag{6}$$

The power output after cell replacement and the state inside the module were also confirmed using the EL and Sun simulators. The cell replacement process is discussed in the next section. After cell replacement, the gain factor (power increment of the replacement cell), loss factor (long-term degradation, electrical mismatch), and unidentified tolerance parts of the module track the experimental results and apply the same to the two samples, correct the power predictions, and finally compare them with the results.

2.2. Experiments

Figure 1 presents an EL image of a 6-inch 54-cell 3BB polycrystalline silicon solar module, where the hose power degrades owing to cell damage. Figure 1a shows the first sample of 190-Wp grade, referred to as 190 A for convenience, and its appearance. Figure 1b–d depict EL images of 190 A, the second sample of the 190-Wp class (190 B), and 190 B, respectively. As shown in images (a) and (c), a weak yellowish appearance, which was not severe, was observed. In addition, approximately six to nine dark areas were observed in the EL image (Figure 1b) and approximately six dark areas were observed at 190 B (Figure 1d). In the green-marked cell of (d), the dark area in the cell occurred because of poor soldering between the mutual connector and the busbar.

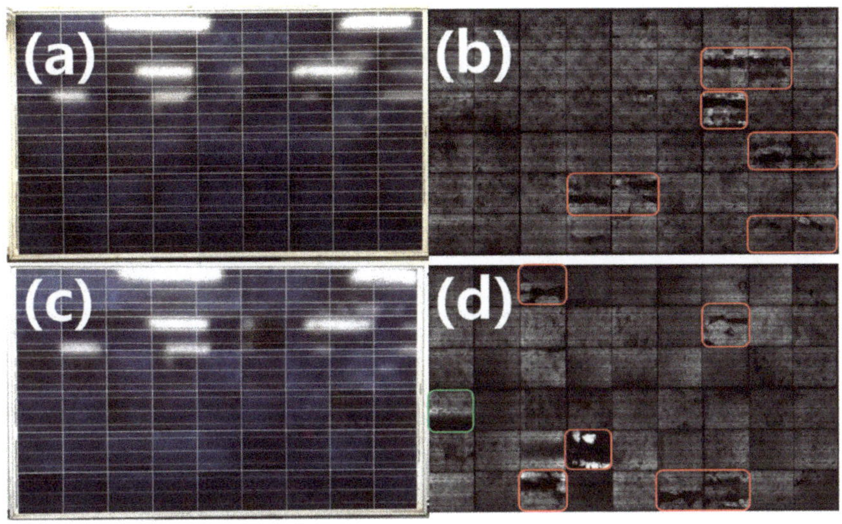

Figure 1. Module appearance and EL images with degraded power output due to cell breakage. (**a**) is the appearance of 190 A, (**b**) is the EL image of 190 A, (**c**) is the appearance of B, and finally (**d**) is the EL image of 190 B.

The modules used in this study included samples collected from commercially operated power plants; however, the current–voltage (I–V) data at the time of manufacture were unknown. Therefore, the electrical characteristics of the model disclosed by the manufacturer were assumed as the initial electrical performance.

Table 2 lists the initial electrical specifications of samples 190 A and 190 B and the electrical data of the failed samples after a certain period of operation. As confirmed in the EL image, the FF was severely degraded by the damaged cells in the middle of the string. For 190 A and 190 B, the power decreased by −21.69% and −26.47%, respectively.

Table 2. Electrical data of the modules in the initial stage and after use.

Sample			P_{max} (Wp)	I_{sc} (A)	V_{oc} (V)	I_{mp} (A)	V_{mp} (V)	*FF*	Tolerance
190 A	54 cells	initial	190.00	7.89	33.00	7.31	26.00	0.73	±3%
		failed	148.80	8.16	32.77	5.16	28.84	0.56	
190 B	54 cells	initial	190.00	7.89	33.00	7.31	26.00	0.73	±3%
		failed	139.70	7.95	32.67	5.67	24.66	0.54	

Figure 2 displays the I–V and voltage–power (V–P) curves of modules 190 A and 190 B. The I–V curves appear step-shaped, while the V–P curves have two or more multi-peaks, which is a typical form caused by the decrease in Isc due to the cracking of a specific cell in a cell string [44].

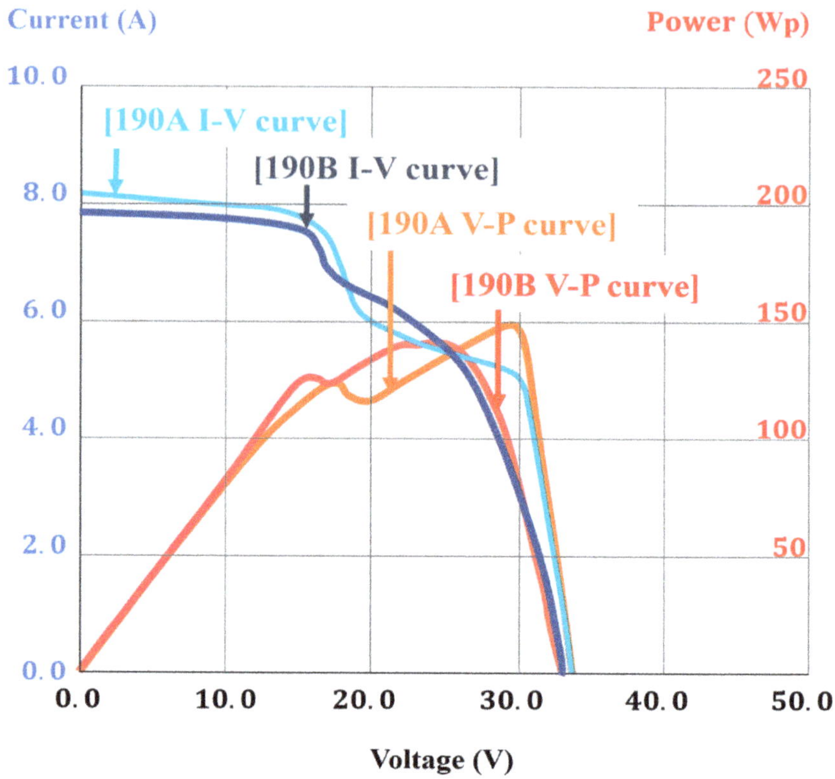

Figure 2. I–V and V–P curves of modules with degraded power output owing to cell breakage.

Figure 3 shows the process of removing the broken cells of 190 A and 190 B cells and replacing them with new cells. (a) First, the module is placed on a hot plate to heat the sun-side and soften the EVA, then, the back sheet is removed from the edge. (b) When the back sheet is completely peeled off, (c) the tape attached to fix the cell-string gap was removed. If it is a material such as EVA, it does not require removal; however, for a tape using polyethylene terephthalate (PET) as a basic material, a gap is formed between the tape and cell owing to the loss of adhesion.

When cleaning the back sheet removal surface or the cell removal area using ethanol or isopropyl alcohol (IPA), the permeated organic solvent may cause solvothermal swelling in the lamination process, or gas may accumulate to cause swelling [32].

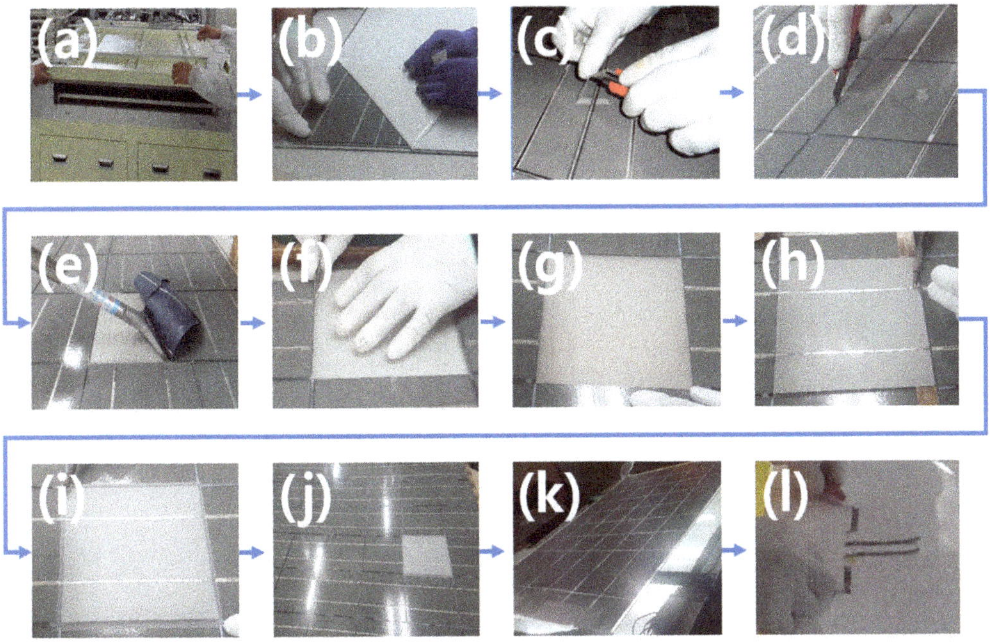

Figure 3. Module recovery process by replacing broken cells. (**d**) Next, the EVA along the boundary of the cell to be removed was cut, and the tab of the cell to be removed was cut by 15 mm or more for electrical connection during recovery. (**e**) Subsequently, the broken cell was removed using a scraper with a blade. (**f**) The remaining EVA was trimmed to the interface of the adjacent cells, and the removed surface was washed. After cell removal, the module was removed from the hot plate and cooled to room temperature. (**g**) EVA was placed between the glass and the new cell to connect it to the module. Here, the size of the EVA is important because it should be perfectly connected to the first EVA of the existing module without leaving a bubble after lamination of the module. Thus, the EVA should be cut accurately with an error of less than 1 mm. If it is larger than the removal surface, stress is applied to the replacement cell, which can cause the cell edge to crack during the lamination process. (**h**) Subsequently, while electrically connecting the new cell and the existing adjacent cell through re-soldering, insulation was applied to prevent the first EVA from melting in the heat. (**i**) The second EVA was slightly thicker than the original size. (**j**) A margin of less than 5 mm should be given, and if it is more than that, the overlapping part of the edge EVA of the cell replaced after lamination is exposed, causing a repair mark. (**k**) Finally, the EVA and new back sheet covering the entire module were laid up. (**l**) The electrical connection was checked, and the lamination process was completed.

3. Result and Discussion

3.1. Power Output Analysis of Initial Cells Applied to Each Sample and Specification of Replacement Cells

The initial CTM of the cells analyzed above was approximately 1.78% based on power, with the median value of 0.9%, as suggested in the optimized module process published in a previous study, and 2.72% of the CTM value of a general photovoltaic module [41]. The calculated power output of the individual cells was approximately 3.6 Wp, which is approximately 14.8% in terms of cell efficiency. Table 3 lists the initial power output of the cell applied to the module using Equations (1) and (2).

Table 3. Initial applied cell grade analysis of sample modules (190 A, 190 B) to be recovered.

CTM Factor (k)	K $_{conventional}$ (%)	CTM $_{power}$ Ratio	Initial CTM $_{power}$ of 190 A, 190 B
Module efficiency (STC)/Power	18.31	98.23%	190.0
k_{15} (junction box and cabling)	−0.05	−0.23%	−0.45
k_{14} (electrical mismatch)	−0.04	−0.19%	−0.36
k_{13} (string interconnection)	−0.03	−0.14%	−0.27
k_{12} (cell interconnection)	−0.037	−0.17%	−0.33
k_{11} (cover coupling)	0.28	1.30%	2.51
k_{10} (interconnector coupling)	0.09	0.42%	0.81
k_9 (finger coupling)	0.17	0.79%	1.52
k_8 (cell/encapsulant coupling)	0.16	0.74%	1.43
k_7 (interconnection shading)	−0.44	−2.04%	−3.94
k_6 (encapsulant absorption)	−0.03	−0.14%	−0.27
k_5 (cover/encapsulant reflection)	−0.01	−0.02%	−0.05
k_4 (cover absorption)	−0.14	−0.65%	−1.26
k_3 (cover reflection)	−0.31	−1.44%	−2.78
Cell efficiency (STC)/Power	21.58	100.00%	193.45

As cells of the same grade were already discontinued, a module was repaired using the 3-bus bar cell, which had the lowest power among the cells currently in use. The P_{max} and cell efficiency of the cell used in the initial manufacture of the module are listed in Table 2. Assuming that the FFs of the module and cell were the same, the cell V_{mp} and V_{oc} were calculated by considering the number of cells from the module V_{mp} and V_{oc}, and I_{sc} and I_{mp} were determined using P_{max} and FF of the cell.

The electrical characteristics of the initial and replacement cells used to restore the modules are presented in Table 4. The tolerance of the initial cell follows that of the module specification sheet.

Table 4. Electrical data of initial cell and replacement cell.

Item	Eff. Cell	P_{max} (Wp)	I_{sc} (A)	V_{oc} (V)	I_{mp} (A)	V_{mp} (V)	*FF*	Tolerance
Initial cell	14.80	3.58	8.07	0.61	7.32	0.49	0.73	±3%
Replacement cell	17.60	4.28	8.62	0.63	8.39	0.51	0.78	±3%

3.2. *Predicting the Power Output of the Restore Module When Applying A Replacement Cell*

The following are the considerations for predicting the power of a module to be recovered when a new cell is installed: the first element is the deviation between the actual power output of the initial module and rated power output. This part was expected to be within the initial tolerance range, and after module recovery, the results and discussion were verified. Next, the power increase of the replacement cell should be added and the value of the field-aged power degradation rate from the initial power of the existing cell should be deducted. Moreover, the loss from the electrical mismatch between the cells should also be considered. The increase in the power of the replacement cell can be easily calculated using Equations (2) and (4). The next part to be considered is the loss caused by the electrical mismatch. A recent study reported that the result of power loss from the electrical mismatch of cells within a module was difficult to determine; however, when the direct parallel configuration of modules was different, the relative power loss (RPL) of the array due to electrical mismatch was 1.3–2.6% [45]. In previous studies, the power loss caused by the electrical mismatch of cells was reported to be approximately 0.009–0.19% [46]; thus, it is already reflected as −0.19% in the CTM factor; therefore, it should be applied conservatively. In the prediction of the power output, the final part to be considered is the loss from power degradation owing to the field aging factor of the existing cell. In general, the rate of power output degradation guaranteed by a module manufacturer is 0.7%/year, which is a guaranteed limit design considering the power degradation caused by the failure of some modules in PV power plants. Referring to the results reported in a previous study, the actual power output degradation rate of more than 80% for crystalline PV modules in PV power plants that have been operated for more than 10 years is approximately 0.27%/year on average [47]. This figure is significantly lower than the limit guaranteed by manufacturers. In this study, we applied this figure to calculate the power output prediction. Table 5 lists the power output predictions for the recovered modules.

Table 5. Power output prediction for recovered modules.

CTM Factor (k)	CTM Power Ratio	190 A (10 New Cells)	190 B (6 New Cells)
Module efficiency (STC)/Power	98.23%	196.40	193.50
Long term degradation of used cell	−0.27% × (% of remaining cell)	−0.34(−0.17%)	−0.47(−0.24%)
k_{15} (junction box and cabling)	−0.23%	−0.46	−0.46
k_{14} (electrical mismatch)	−0.19%	−0.37	−0.37
k_{13} (string interconnection)	−0.14%	−0.28	−0.28
k_{12} (cell interconnection)	−0.17%	−0.34	−0.34
k_{11} (cover coupling)	1.30%	2.60	2.56
k_{10} (interconnector coupling)	0.42%	0.84	0.82
k_9 (finger coupling)	0.79%	1.58	1.56
k_8 (cell/encapsulant coupling)	0.74%	1.49	1.46
k_7 (interconnection shading)	−2.04%	−4.08	−4.03
k_6 (encapsulant absorption)	−0.14%	−0.28	−0.28
k_5 (cover/encapsulant reflection)	−0.02%	−0.05	−0.05
k_4 (cover absorption)	−0.65%	−1.30	−1.28
k_3 (cover reflection)	−1.44%	−2.88	−2.84
Cell power (STC, + power gain)	100.00%	200.32	197.52

For 190 A, 10 broken cells were replaced; thus, (10 × 4.28 Wp) + (44 × 3.58 Wp) = 42.8 + 157.52 = 200.32 Wp is the total power output value of the cell. In 190 B, six cells were replaced: (6 × 4.28 Wp) + (48 × 3.58 Wp) = 25.68 + 171.84 = 197.52 Wp. The results are presented in Table 4. Through the calculation, the predicted power output values of 190 A and 190 B were calculated as 196.40 Wp and 193.50 Wp, respectively. The CTM factor k_i values ranging from k_1 to k_{15}, and k_3 to k_{15} are shown in the table; however, k_1 and k_2 values are not shown in the table nor described here. The CTM factor k_1 is the module margin, which is approximately −2.03% in a typical module, and k_2 is the cell spacing, which is also generally −0.53%. This value is a design factor for the module area and depends on module dimensions. However, the module margin or cell interval for insulation distance affects only the area efficiency of the module and does not produce power by itself; therefore, the calculation of CTM power was excluded from previous research.

3.3. Results of Power Recovery by Cell Replacement of 190 A and 190 B Samples

Figure 4 shows a comparison of the EL images of the modules before and after repair.

In Figure 4c,d, the relatively bright cells are the newly replaced cells. In Figure 4a, when replacing cells of the 190 A sample, one more cell was replaced by damaging adjacent cells while removing the cells from the hot plate, and as the cell replacement operation was repeated, the same mistake was not repeated. Some small cracks not shown in Figure 4a are observed in Figure 4b, which are defects occurring during manual cell removal. However, the result shown in Figure 4d is not much different from that in Figure 4c because cell replacement has become familiar and cell removal progressed much more easily. For an easy recovery process, care should be taken to prevent additional cell cracks when collecting and reinstalling the modules to be repaired.

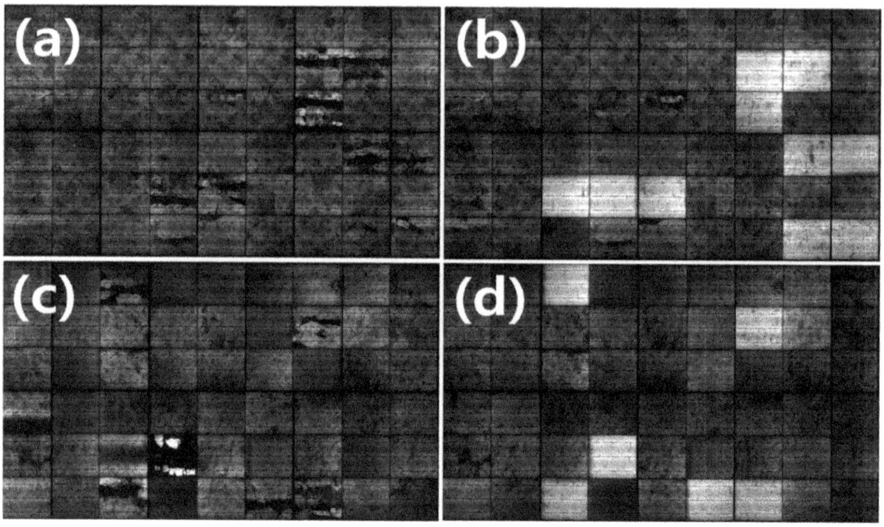

Figure 4. EL images in modules before and after recovery. (190 A, 190 B). (**a**) Among a total of 54 cells from 190 A, 10 cells with a severe crack degree were removed and (**b**) replaced with a new cell to recover. (**c**) Sample 190 B exhibited severe power degradation in approximately six cells, and hot spots due to pore soldering also occurred in the busbar–interconnector connection. (**d**) However, both the power and FF were recovered after cell replacement and pore soldering repair.

Table 6 lists the electrical characteristics of the module before and after recovery. The module power increased by approximately 4.50% to 198.60 Wp from the rated power for 190 A, and by approximately 5.10% to 199.70 Wp for 190 B. We verified that, considering the loss of electrical mismatch between the existing cell and the new cell, the higher power of individual cells had a greater effect on the power of the module.

Table 6. Electrical data of modules before and after recovery.

Item	Replacement		*Pmax* (Wp)	I_{sc} (A)	V_{oc} (V)	I_{mp} (A)	V_{mp} (V)	*FF*	Initial Comparison
190 A	10 cells	before	148.80	8.16	32.77	5.16	28.84	0.56	−21.69%
		recovery	198.60	8.11	32.95	7.54	26.35	0.74	+4.53%
190 B	6 cells	before	139.70	7.95	32.67	5.67	24.67	0.54	−26.47%
		recovery	199.70	7.99	32.89	7.50	26.64	0.76	+5.11%

As mentioned in Section 3.2, when the difference in cell mismatch is not large, the loss due to mismatch is insignificant in the range 0.10–0.19%, and most (>80%) of crystalline photovoltaic modules are only approximately 0.27%/year on average. Therefore, it matches well with the result that predicted that the gain factor would have a greater impact on the final power output of the module than the loss factor [46,47].

Figures 5 and 6 show the I–V curves before and after power recovery for 190 A and 190 B, respectively. The results in Figures 5 and 6 show that I_{sc} and V_{oc} do not change significantly before and after module restoration and that the V–P curve is deformed by cell breakage, the FF is recovered, and the power of the module is restored.

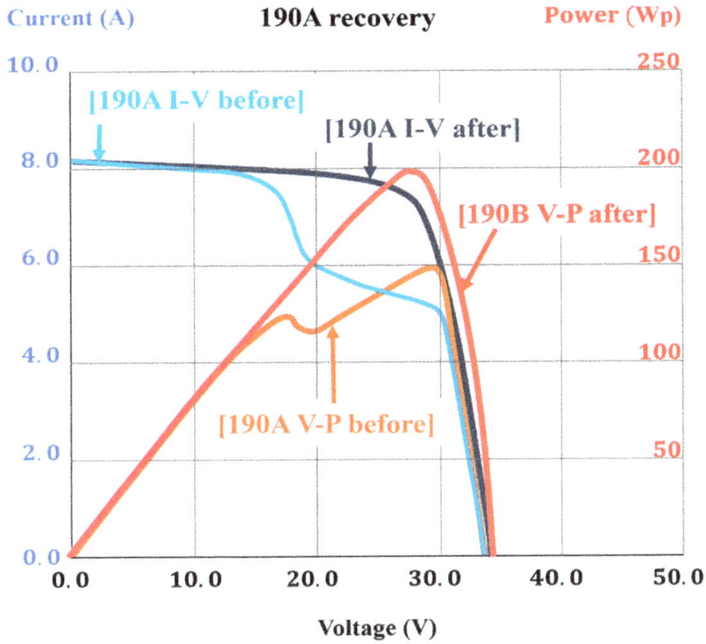

Figure 5. I–V and V–P curves before and after power recovery for sample 190 A.

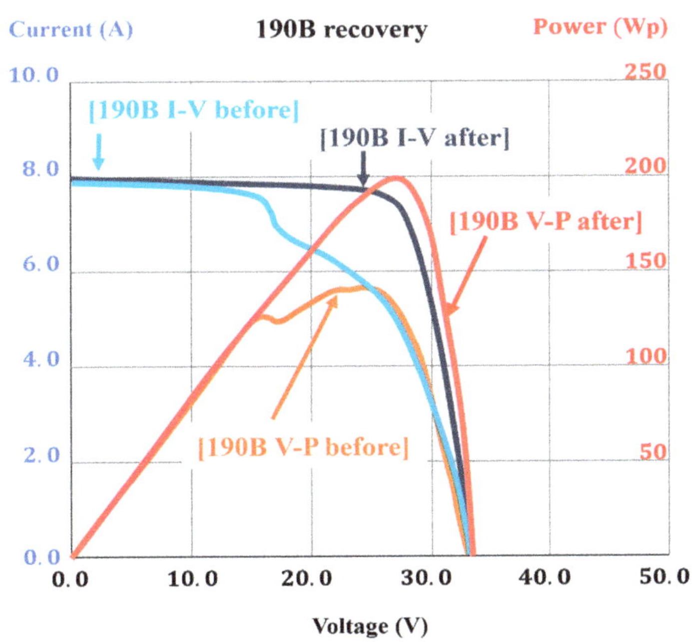

Figure 6. I–V and V–P curves before and after power recovery for sample 190 B.

As shown in Figures 5 and 6, both the cell-in-hotspot-specific stepped I–V and multipeak-shaped V–P curves are recovered.

Figures 7 and 8 show a brief circuit diagram of module 190 A before and after recovery, respectively. In the figures, I_{ph} represents the solar irradiance and I_{pv} represents the power output current. D_1, D_2, and D_3 denote bypass diodes #1–#3, respectively, and R_s denotes the series resistance.

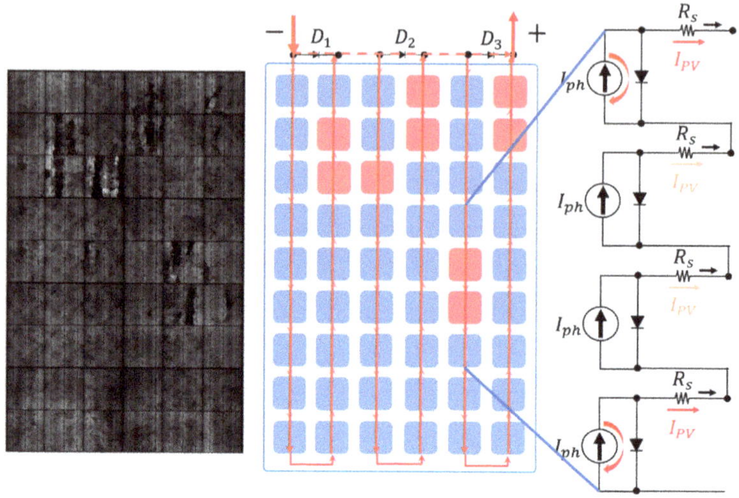

Figure 7. Sub-circuit diagram of 190 A before recovery.

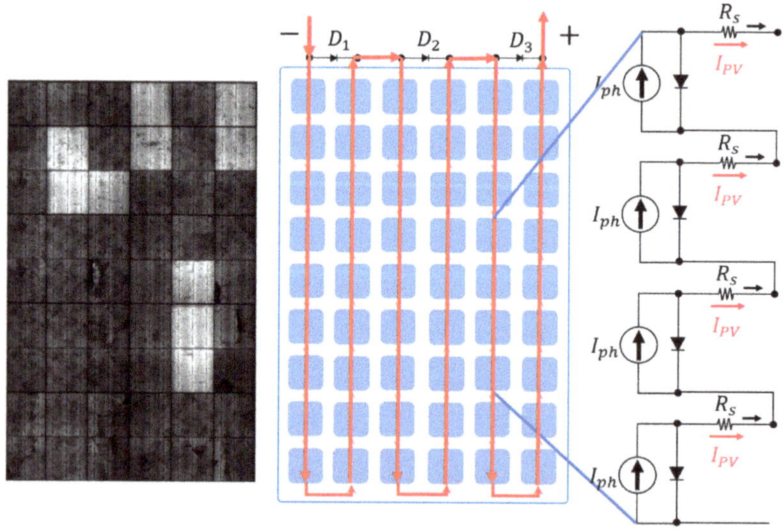

Figure 8. Sub-circuit diagram of module 190 A after recovery.

In the EL image of Figure 7, nine cells were cracked, resulting in resistance loss. In this case, the ratio of shaded (or inactive) areas causing hot spots in the cell increased proportionally with the range of inactive areas between 20% and 50%. If the resistance becomes excessively large over a greater range or if the bypass diode is short-circuited [48], it causes 100% power loss to the entire connected string [49,50]. A part looks relatively brighter around the interconnector immediately next to the dark area of the damaged cell, and the current is concentrated on a part of a cell with relatively low resistance owing to cracks; thus, power loss occurs in the shaded and connected cells.

Figure 8 shows the EL of the module whose power was recovered after the cell replacement of the 190 A sample and its diagram. The picture for 190 B is repeated, so I omit it.

3.4. Comparative Analysis of Power Recovery Results and Predicted Values

Table 7 shows the difference between the predicted power output value obtained using the CTM analysis before module recovery and the value measured after cell replacement.

Table 7. Comparison of predicted and experimental values.

Item	Before Recovery	Predicted Value	Experimental Value	Difference	Tolerance
190 A	148.80	196.40	198.60	+1.12%	±3%
190 B	139.70	193.50	199.70	+3.20%	±3%

Even when applying the power deviation when manufacturing a module, both cases exhibited a positive deviation; therefore, the loss, such as electrical mismatch, in the CTM factor was considered conservative among the possible ranges. The CTM power analysis results at 190 A are shown in Figure 9.

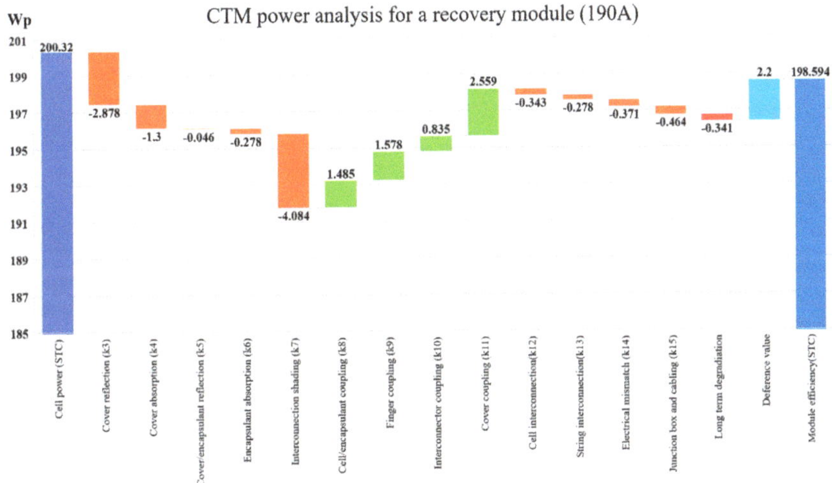

Figure 9. CTM power analysis for a recovered module (190 A).

The sum of the power of the initial and replacement cells was defined as 200.32 Wp using the values calculated in Equation (1) and Table 4, and when CTM factors were applied, the predicted value of 198.60 Wp was determined. Here, if 2.20 Wp, i.e., the difference from the experimental results, was reflected, it was analyzed, as shown in Figure 9. The difference between the predicted and experimental result for 190 A was 1.12%, which fell within 3% of the power output tolerance value of the initial module. The analysis result of sample 190 B indicated that the error was larger. Figure 10 shows the CTM power analysis of the recovered module (190 B).

Sample 190 B was of the same grade as 190 A, and because there were fewer replacement cells (six), the power acquisition from the replacement cell was smaller than that at 190 A; therefore, the total power output of the cell was calculated as 197.52 Wp. In addition, the numbers of remaining cells in 190 A and 190 B were 44 and 48 cells, respectively; thus, the long-term degradation was then calculated to be −0.47 Wp, which is greater than −0.34 Wp for 190 A. The experimental value was 199.70 Wp, i.e., 6.20 Wp higher than the predicted value of 193.50 Wp. This is approximately 3.20% higher than the predicted value of 3% or more, which is the power output tolerance value of the initial module.

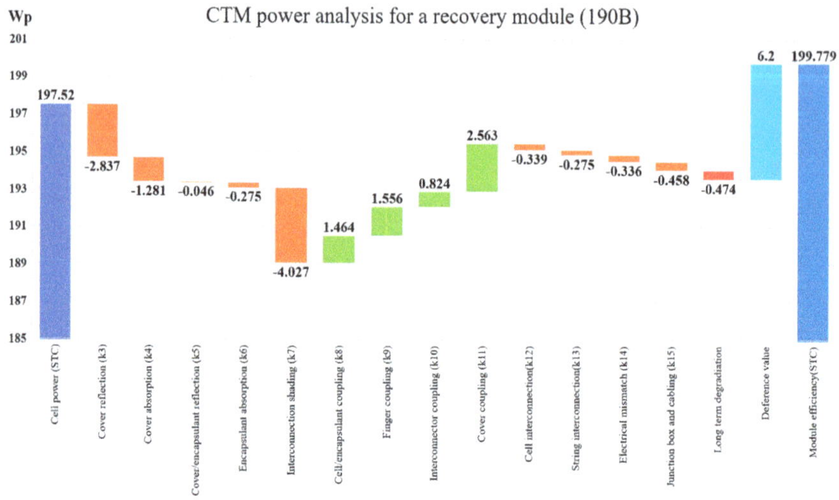

Figure 10. CTM power analysis for a recovered module (190 B).

3.5. Analysis of Prediction Error and Correction of Prediction Value Reflecting Initial Tolerance

The error begins with the sum of the cell power output values. The final power value was 199.70 Wp, and the sum of the calculated cell power values was 197.52 Wp, which began with a difference of 2.18 Wp even if the CTM was assumed to be "0." The value 2.18 Wp was 1.15% of the initial rated power value of 190 Wp, which was within the allowable tolerance range of the module. Therefore, assuming that the initial use cells of 190 A and 190 B were the same, sample 190 B corrected the experimental deviation of 2.18 Wp. Those of 190 A were calculated by adjusting the number of cells to calculate the correction value of 2.00 Wp. Accordingly, the predicted power output values of 190 A and 190 B could be recalculated as listed in Table 8. The initial power output prediction value of sample 190 A was 196.4 Wp. For the power correction value of 1.998 Wp within the tolerance shown in the experimental result, the correction prediction value was 198.4 Wp. Additionally, the error decreased to 0.10% with the final experiment result of 198.6 Wp. When the initial power output prediction value of 193.5 Wp was corrected for 190 B, the corrected prediction value was 195.68 Wp, which was approximately 2.13% lower than the experimental result for 199.7 Wp.

Table 8. Analysis of predicted and experimental values.

Item	Predicted Value	Tolerance Calibration	Correction	Experimental Value	Difference
190 A	196.40	2.00	198.40	198.60	+0.10%
190 B	193.50	2.18	195.68	199.70	+2.01%

When the tolerance value calculated above was added to the initial rated power, the initial power of the module was approximately 192.45 Wp. Based on this, the power before and after module recovery owing to cell damage and the recovery trend of the *FF* are shown in Figure 11.

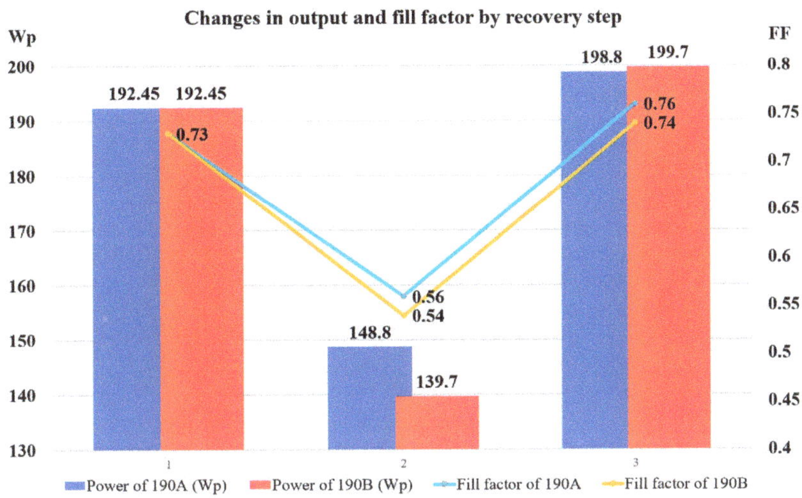

Figure 11. Result of module recovery via cell replacement (power and *FF*).

Table 9 summarizes the initial, failed (before recovery), and recovered (after recovery) values of the power degradation module owing to cell cracking.

Table 9. Electrical data deviation of initial, faulty, and recovered module.

Item			P_{max} (Wp)	I_{sc} (A)	V_{oc} (V)	I_{mp} (A)	V_{mp} (V)	*FF*	Tolerance
190 A	54 cells	initial	190.00	7.89	33.00	7.31	26.00	0.73	±3%
		failed	148.80	8.16	32.77	5.16	28.84	0.56	–
		recovered	198.60	8.11	32.95	7.54	26.35	0.74	
	Rate of decline (initial)		+4.53%	+3.55%	−0.16%	+3.13%	+1.36%	+1.92%	
190 B		failed	139.70	7.95	32.67	5.67	24.67	0.54	–
		recovered	199.70	7.99	32.89	7.50	26.64	0.76	–
	Rate of decline (initial)		+5.11%	+1.28%	−0.33%	+2.57%	+2.42%	+4.11%	

The characteristic of the recovery of the cell in the hotspot module by cell replacement is that the V_{oc} value hardly changes step-by-step but decreases within the error range by step. The largest negative mismatch factor in the phase of the power drop to the cell in the hotspot was I_{mp}, exhibiting a 29.43% decrease at 190 A compared with the initial value, which had the greatest impact on the power decrease of −21.69%. Even in sample 190 B, I_{mp} degradation caused a −22.48% decrease in the cell in the hot spot stage, and a power degradation of −26.47% was also the largest factor. For a positive mismatch with a high power, the I_{sc} and I_{mp} values both increased, and the V_{mp} value decreased step-by-step at 190 A; thus, the factor that most affected the positive mismatch was the increase in I_{sc} and I_{mp}; the increase in I_{mp}, in particular, was the largest factor. Figure 12 shows the EL images of samples (**a**) 190 A and (**b**) 190 B recovered by cell replacement, and (**c**) IR images measuring whether the module generated heat by installing them again in the power plant.

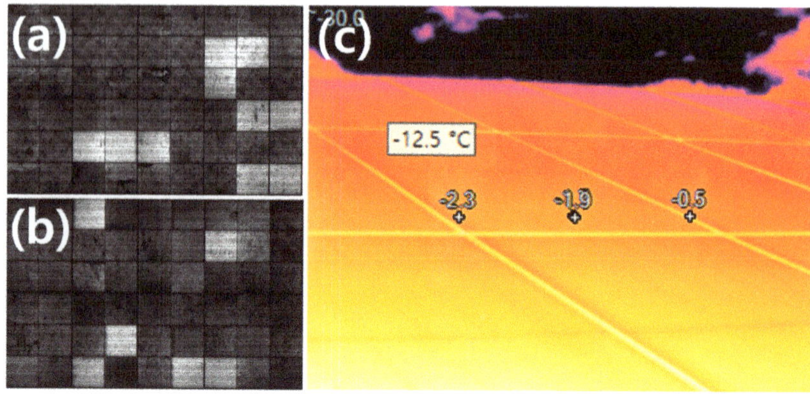

Figure 12. Images of the recovered module (190 A, 190 B). (**a**) is an EL image of the repaired 190A module, and (**b**) is an EL image of the repaired 190B module. (**c**) is an IR image of 190A and 190B re-installed at the plant.

A difference in brightness was observed between the replaced and existing cells in the EL images shown in Figure 12a,b, but in Figure 12c, no significant heat generation was observed in the IR image at the installation site. The IR camera used to measure cell heat generation was a Ti400 FLUKE equipment.

Thus, we confirmed the restoration potential of modules that are underpowered by cells in hotspots in commercial power plants. When some cells are damaged in a crystalline PV module, the module can be restored by replacing those cells instead of discarding the entire module. Assuming that this method restores more than 180 sheets per day on a 200-Wp module basis, the cost of restoring the module is approximately 0.17 $/Wp. This is slightly more than half of the recent crystalline module price of 0.30 $/Wp. However, for commercial use, a long-term reliability test of a module repaired using this method must be performed to confirm the results. Accordingly, reuse of modules instead of recycling will be an economical and eco-friendly alternative.

4. Conclusions

In this study, the power loss caused by the damage of a cell in a module was determined through EL images and I–V and V–P curves of the module, and research was conducted to recover only the damaged cells to be equal to or higher than the initial power of the module. The recovery of modules is important in the electrical serial–parallel design and application of existing structures in PV power plants. Therefore, the grade of the cell applied at the time of module production was calculated using the CTM factor analysis method and applied considering the dimensions and tolerance of the specification sheet of the module presented by the manufacturer. To predict the power of the recovered module, the power degradation factor from the aging factor of the module, not in the existing CTM formula, and the mismatch loss of the cell were checked again and recalculated. The results of the power output prediction calculated using the formula and the power output of the recovered module measured as the experimental result had an error of 1.12% in sample 190 A and 3.20% in sample 190 B. This was determined to be an error, assuming that the rated power output was the initial power output, because the accurate power output of the initial module was unknown. As a result of calibrating the power of approximately 2 Wp by feeding back the initial tolerance from the recovered module power output, the revised prediction was calculated as 198.40 Wp in 190 A and 195.68 Wp in 190 B, and the experimental results indicated error rates of 0.10% and 2.01%, respectively. This study confirmed that even when a replacement cell applied to the recovered module had an average power output of approximately 19.60% (4.28 Wp) higher than that of the existing cell, and I_{sc} had an average value of approximately 8.98% higher (8.62 A), the loss of electrical mismatch did not significantly affect the power, and heat generation of existing normal cells was not observed. In addition, even for modules operated for a long time (>10 years), the power reduction rate is significantly smaller than the 0.70%/year suggested by the module manufacturers. Even if a degradation of approximately 2.40% over 10 years was applied, there was no significant error in the power prediction. As the life of a PV module increases, the recovery technology for discontinued modules becomes a very important economic factor for PV power plants with a considerable remaining operating period. Module recovery technology through cell replacement is useful as an economical reuse and high-value-added technology to prevent power degradation in an operating power plant. A technology to recover a module function by selectively replacing only the necessary cells and recovering a module function, even when it has expired commercially, would be significantly more economical than decomposing and collecting it as raw material. We believe that in future studies, work should continue to verify the effect of electrical mismatch in a wider range of cells on modules as well as the long-term reliability to predict the lifetime of restored modules.

Author Contributions: Conceptualization, formal analysis, writing—original draft preparation, K.L.; investigation, data curation, S.C.; project administration, J.Y.; supervision, project administration, H.C. All authors have read and agreed to the published version of the manuscript.

Funding: This study received no external funding.

Acknowledgments: This research was supported by grants from the New and Renewable Energy Technology Development Program of the Korea Institute of Energy Technology Evaluation and Planning (KETEP), funded by the Korean Ministry of Trade, Industry, and Energy (MOTIE) (Project No. 20203030010060 and 20213030010400).

Conflicts of Interest: The authors declare no conflict of interest.

References

1. Weschenfelder, F.; de Novaes Pires Leite, G.; Araújo da Costa, A.C.; de Castro Vilela, O.; Ribeiro, C.M.; Villa Ochoa, A.A.; Araújo, A.M. A Review on the Complementarity between Grid-Connected Solar and Wind Power Systems. *J. Clean. Prod.* **2020**, *257*, 120617. [CrossRef]
2. Liu, J.; Chen, X.; Cao, S.; Yang, H. Overview on Hybrid Solar Photovoltaic-Electrical Energy Storage Technologies for Power Supply to Buildings. *Energy Convers. Manag.* **2019**, *187*, 103–121. [CrossRef]
3. Mostafa, M.H.; Abdel Aleem, S.H.E.; Ali, S.G.; Ali, Z.M.; Abdelaziz, A.Y. Techno-Economic Assessment of Energy Storage Systems Using Annualized Life Cycle Cost of Storage (LCCOS) and Levelized Cost of Energy (LCOE) Metrics. *J. Energy Storage* **2020**, *29*, 101345. [CrossRef]
4. Zhou, Y.; Cao, S.; Hensen, J.L.M. An Energy Paradigm Transition Framework from Negative towards Positive District Energy Sharing Networks—Battery Cycling Aging, Advanced Battery Management Strategies, Flexible Vehicles-to-Buildings Interactions, Uncertainty and Sensitivity Analysis. *Appl. Energy* **2021**, *288*, 116606. [CrossRef]
5. Poompavai, T.; Kowsalya, M. Control and Energy Management Strategies Applied for Solar Photovoltaic and Wind Energy Fed Water Pumping System: A Review. *Renew. Sustain. Energy Rev.* **2019**, *107*, 108–122. [CrossRef]
6. Mehedintu, A.; Sterpu, M.; Soava, G. Estimation and Forecasts for the Share of Renewable Energy Consumption in Final Energy Consumption by 2020 in the European Union. *Sustainability* **2018**, *10*, 1515. [CrossRef]
7. Kamran, M.; Fazal, M.R.; Mudassar, M.; Ahmed, S.R.; Adnan, M.; Abid, I.; Randhawa, F.J.S.; Shams, H. Solar Photovoltaic Grid Parity: A Review of Issues and Challenges and Status of Different PV Markets. *Int. J. Renew. Energy Res.* **2019**, *9*, 244–260. [CrossRef]
8. Tu, Q.; Mo, J.; Betz, R.; Cui, L.; Fan, Y.; Liu, Y. Achieving Grid Parity of Solar PV Power in China- The Role of Tradable Green Certificate. *Energy Policy* **2020**, *144*, 111681. [CrossRef]
9. Zhang, M.; Zhang, Q. Grid Parity Analysis of Distributed Photovoltaic Power Generation in China. *Energy* **2020**, *206*, 118165. [CrossRef]
10. Samper, M.; Coria, G.; Facchini, M. Grid Parity Analysis of Distributed PV Generation Considering Tariff Policies in Argentina. *Energy Policy* **2021**, *157*, 112519. [CrossRef]
11. Chen, Y.; Wang, Z.; Zhong, Z. CO_2 Emissions, Economic Growth, Renewable and Non-Renewable Energy Production and Foreign Trade in China. *Renew. Energy* **2019**, *131*, 208–216. [CrossRef]
12. Alola, A.A.; Bekun, F.V.; Sarkodie, S.A. Dynamic Impact of Trade Policy, Economic Growth, Fertility Rate, Renewable and Non-Renewable Energy Consumption on Ecological Footprint in Europe. *Sci. Total Environ.* **2019**, *685*, 702–709. [CrossRef] [PubMed]

13. Li, W.; Adachi, T. Evaluation of Long-Term Silver Supply Shortage for c-Si PV under Different Technological Scenarios. *Nat. Resour. Model.* **2019**, *32*, 1–27. [CrossRef]
14. Watari, T.; Nansai, K.; Nakajima, K. Review of Critical Metal Dynamics to 2050 for 48 Elements. *Resour. Conserv. Recycl.* **2020**, *155*, 104669. [CrossRef]
15. Kabeel, A.E.; Sathyamurthy, R.; El-Agouz, S.A.; Muthu manokar, A.; El-Said, E.M.S. Experimental Studies on Inclined PV Panel Solar Still with Cover Cooling and PCM. *J. Therm. Anal. Calorim.* **2019**, *138*, 3987–3995. [CrossRef]
16. Lee, J.S.; Ahn, Y.S.; Kang, G.H.; Ahn, S.H.; Wang, J.P. Development of New Device and Process to Recover Valuable Materials from Spent Solar Module. *Key Eng. Mater.* **2018**, *780*, 48–56. [CrossRef]
17. Zheng, J.; Ge, P.; Bi, W.; Zhao, Y.; Wang, C. Effect of Capillary Adhesion on Fracture of Photovoltaic Silicon Wafers during Diamond Wire Slicing. *Sol. Energy* **2022**, *238*, 105–113. [CrossRef]
18. Kumar, A.; Melkote, S.N. Diamond Wire Sawing of Solar Silicon Wafers: A Sustainable Manufacturing Alternative to Loose Abrasive Slurry Sawing. *Procedia Manuf.* **2018**, *21*, 549–566. [CrossRef]
19. Farrell, C.C.; Osman, A.I.; Doherty, R.; Saad, M.; Zhang, X.; Murphy, A.; Harrison, J.; Vennard, A.S.M.; Kumaravel, V.; Al-Muhtaseb, A.H.; et al. Technical Challenges and Opportunities in Realising a Circular Economy for Waste Photovoltaic Modules. *Renew. Sustain. Energy Rev.* **2020**, *128*, 109911. [CrossRef]
20. Mahmoudi, S.; Huda, N.; Alavi, Z.; Islam, M.T.; Behnia, M. End-of-Life Photovoltaic Modules: A Systematic Quantitative Literature Review. *Resour. Conserv. Recycl.* **2019**, *146*, 1–16. [CrossRef]
21. Walzberg, J.; Carpenter, A.; Heath, G.A. Integrating Sociotechnical Factors to Assess Efficacy of PV Recycling and Reuse Interventions. Available online: https://assets.researchsquare.com/files/rs-151153/v1_covered.pdf?c=1637595475 (accessed on 5 February 2021).
22. Yu, H.; Tong, X. Producer vs. Local Government: The Locational Strategy for End-of-Life Photovoltaic Modules Recycling in Zhejiang Province. *Resour. Conserv. Recycl.* **2021**, *169*, 105484. [CrossRef]
23. Kim, H.; Park, H. PV Waste Management at the Crossroads of Circular Economy and Energy Transition: The Case of South Korea. *Sustainability* **2018**, *10*, 3565. [CrossRef]
24. Heath, G.A.; Silverman, T.J.; Kempe, M.; Deceglie, M.; Ravikumar, D.; Remo, T.; Cui, H.; Sinha, P.; Libby, C.; Shaw, S.; et al. Research and Development Priorities for Silicon Photovoltaic Module Recycling to Support a Circular Economy. *Nat. Energy* **2020**, *5*, 502–510. [CrossRef]
25. Sapra, G.; Chaudhary, V.; Kumar, P.; Sharma, P.; Saini, A. Recovery of Silica Nanoparticles from Waste PV Modules. *Mater. Today Proc.* **2019**, *45*, 3863–3868. [CrossRef]
26. Farrell, C.; Osman, A.I.; Zhang, X.; Murphy, A.; Doherty, R.; Morgan, K.; Rooney, D.W.; Harrison, J.; Coulter, R.; Shen, D. Assessment of the Energy Recovery Potential of Waste Photovoltaic (PV) Modules. *Sci. Rep.* **2019**, *9*, 1–13. [CrossRef]
27. Tao, M.; Fthenakis, V.; Ebin, B.; Steenari, B.M.; Butler, E.; Sinha, P.; Corkish, R.; Wambach, K.; Simon, E.S. Major Challenges and Opportunities in Silicon Solar Module Recycling. *Prog. Photovoltaics Res. Appl.* **2020**, *28*, 1077–1088. [CrossRef]
28. Lee, J.S.; Ahn, Y.S.; Kang, G.H.; Wang, J.P. Recovery of Pb-Sn Alloy and Copper from Photovoltaic Ribbon in Spent Solar Module. *Appl. Surf. Sci.* **2017**, *415*, 137–142. [CrossRef]
29. Voronko, Y.; Eder, G.C.; Breitwieser, C.; Mühleisen, W.; Neumaier, L.; Feldbacher, S.; Oreski, G.; Lenck, N. Repair Options for PV Modules with Cracked Backsheets. *Energy Sci. Eng.* **2021**, *9*, 1583–1595. [CrossRef]
30. Beaucarne, G.; Eder, G.; Jadot, E.; Voronko, Y.; Mühleisen, W. Repair and Preventive Maintenance of Photovoltaic Modules with Degrading Backsheets Using Flowable Silicone Sealant. *Prog. Photovoltaics Res. Appl.* **2021**, 1–9. [CrossRef]
31. Azeumo, M.F.; Conte, G.; Ippolito, N.M.; Medici, F.; Piga, L.; Santilli, S. Photovoltaic Module Recycling, a Physical and a Chemical Recovery Process. *Sol. Energy Mater. Sol. Cells* **2019**, *193*, 314–319. [CrossRef]
32. Xu, X.; Lai, D.; Wang, G.; Wang, Y. Nondestructive Silicon Wafer Recovery by a Novel Method of Solvothermal Swelling Coupled with Thermal Decomposition. *Chem. Eng. J.* **2021**, *418*, 129457. [CrossRef]
33. Zhang, L.; Chang, S.; Wang, Q.; Zhou, D. Is Subsidy Needed for Waste PV Modules Recycling in China? A System Dynamics Simulation. *Sustain. Prod. Consum.* **2022**, *31*, 152–164. [CrossRef]
34. Nayak, P.K.; Mahesh, S.; Snaith, H.J.; Cahen, D. Photovoltaic Solar Cell Technologies: Analysing the State of the Art. *Nat. Rev. Mater.* **2019**, *4*, 269–285. [CrossRef]
35. Andreani, L.C.; Bozzola, A.; Kowalczewski, P.; Liscidini, M.; Redorici, L. Silicon Solar Cells: Toward the Efficiency Limits. *Adv. Phys. X* **2019**, *4*, 1548305. [CrossRef]
36. Green, M.A.; Dunlop, E.D.; Hohl-Ebinger, J.; Yoshita, M.; Kopidakis, N.; Hao, X. Solar Cell Efficiency Tables (Version 58). *Prog. Photovoltaics Res. Appl.* **2021**, *29*, 657–667. [CrossRef]
37. Sinke, W.C. Development of Photovoltaic Technologies for Global Impact. *Renew. Energy* **2019**, *138*, 911–914. [CrossRef]
38. Goudelis, G.; Lazaridis, P.I.; Dhimish, M. A Review of Models for Photovoltaic Crack and Hotspot Prediction. *Energies* **2022**, *15*, 4303. [CrossRef]
39. Dhimish, M.; Tyrrell, A.M. Power loss and hotspot analysis for photovoltaic modules affected by potential induced degradation. *npj Mater. Degrad.* **2022**, *6*, 11. [CrossRef]
40. Hanifi, H.; Pfau, C.; Turek, M.; Schneider, J. A Practical Optical and Electrical Model to Estimate the Power Losses and Quantification of Different Heat Sources in Silicon Based PV Modules. *Renew. Energy* **2018**, *127*, 602–612. [CrossRef]

41. Yousuf, H.; Zahid, M.A.; Khokhar, M.Q.; Park, J.; Ju, M.; Lim, D.; Kim, Y.; Cho, E.C.; Yi, J. Cell-to-Module Simulation Analysis for Optimizing the Efficiency and Power of the Photovoltaic Module. *Energies* **2022**, *15*, 1176. [CrossRef]
42. Ritou, A.; Voarino, P.; Raccurt, O. Does Micro-Scaling of CPV Modules Improve Efficiency? A Cell-to-Module Performance Analysis. *Sol. Energy* **2018**, *173*, 789–803. [CrossRef]
43. Hanifi, H.; Pander, M.; Zeller, U.; Ilse, K.; Dassler, D.; Mirza, M.; Bahattab, M.A.; Jaeckel, B.; Hagendorf, C.; Ebert, M.; et al. Loss Analysis and Optimization of PV Module Components and Design to Achieve Higher Energy Yield and Longer Service Life in Desert Regions. *Appl. Energy* **2020**, *280*, 116028. [CrossRef]
44. Gnoli, L.; Riente, F.; Ottavi, M.; Vacca, M. A Memristor-Based Sensing and Repair System for Photovoltaic Modules. *Microelectron. Reliab.* **2021**, *117*, 114026. [CrossRef]
45. Nnamchi, S.N.; Nnamchi, O.A.; Nwaigwe, K.N.; Jagun, Z.O.; Ezenwankwo, J.U. Effect of Technological Mismatch on Photovoltaic Array: Analysis of Relative Power Loss. *J. Renew. Energy Environ.* **2021**, *8*, 77–89.
46. Forniés, E.; Naranjo, F.; Mazo, M.; Ruiz, F. The Influence of Mismatch of Solar Cells on Relative Power Loss of Photovoltaic Modules. *Sol. Energy* **2013**, *97*, 39–47. [CrossRef]
47. Pascual, J.; Martinez-Moreno, F.; García, M.; Marcos, J.; Marroyo, L.; Lorenzo, E. Long-Term Degradation Rate of Crystalline Silicon PV Modules at Commercial PV Plants: An 82-MWp Assessment over 10 Years. *Prog. Photovoltaics Res. Appl.* **2021**, *29*, 1294–1302. [CrossRef]
48. Niazi, K.A.K.; Akhtar, W.; Khan, H.A.; Yang, Y.; Athar, S. Hotspot Diagnosis for Solar Photovoltaic Modules Using a Naive Bayes Classifier. *Sol. Energy* **2019**, *190*, 34–43. [CrossRef]
49. Lee, C.G.; Shin, W.G.; Lim, J.R.; Kang, G.H.; Ju, Y.C.; Hwang, H.M.; Chang, H.S.; Ko, S.W. Analysis of Electrical and Thermal Characteristics of PV Array under Mismatching Conditions Caused by Partial Shading and Short Circuit Failure of Bypass Diodes. *Energy* **2021**, *218*, 119480. [CrossRef]
50. Teo, J.C.; Tan, R.H.; Mok, V.H.; Ramachandaramurthy, V.K.; Tan, C. Impact of bypass diode forward voltage on maximum power of a photovoltaic system under partial shading conditions. *Energy* **2020**, *191*, 116491. [CrossRef]

electronics

Article

A Modified Modulation Strategy for an Active Neutral-Point-Clamped Five-Level Converter in a 1500 V PV System

Guodong Chen [1] and Jiatao Yang [2,*]

1 Technology Center, Shanghai Electric Power Transmission & Distribution Group, Shanghai 200042, China; chengd@shanghai-electric.com
2 Key Laboratory of Control of Power Transmission and Conversion, Ministry of Education, Shanghai Jiao Tong University, Shanghai 200240, China
* Correspondence: sjtu-sd-zzc@sjtu.edu.cn

Abstract: With the development of 1500 V photovoltaic (PV) systems in recent decades, multilevel inverters such as the five-level inverter have gained much attention for their higher equivalent output frequency and low semiconductor devices' voltage stress. Among five-level inverters, the active neutral-point-clamped five-level (ANPC-5L) inverter is very competitive due to its simple structure and control methods. However, with its conventional commutation strategy, the topology of the ANPC five-level converter has the security risk of overvoltage in the power device when switching to dead time under special conditions, which affects the reliability and safety of the switch state switching process. In this paper, this issue is analyzed in detail and a modified commutation strategy is proposed. Meanwhile, a novel soft start-up method adopted to an ANPC-5L inverter is also proposed. A prototype is also set up to analyze the issue of traditional switching commutation strategies and to verify the effectiveness of the proposed commutation strategy and the soft start-up method.

Keywords: ANPC-5L converter; reliable switching state; modulation; PV grid-tied inverter

Citation: Chen, G.; Yang, J. A Modified Modulation Strategy for an Active Neutral-Point-Clamped Five-Level Converter in a 1500 V PV System. *Electronics* **2022**, *11*, 2289. https://doi.org/10.3390/electronics11152289

Academic Editors: Luis Hernández-Callejo, Jesús Armando Aguilar Jiménez and Carlos Meza Benavides

Received: 18 June 2022
Accepted: 19 July 2022
Published: 22 July 2022

1. Introduction

Photovoltaic generation has been paid more attention recently because of the shortage of fossil fuels and the increasingly serious levels of environmental pollution, which play an important role in PV systems [1]. Compared with previous 1000 V systems, the 1500 V system reduces the number of cables and PV plants, and decreases the line cost and conduction loss [2,3]. Moreover, it provides more voltage range which is used to ensure maximum power point (MPPT) availability by controlling front-end circuits or adjusting the grid-connected voltage [4,5].

Nowadays, multilevel inverters such as the five-level inverter have gained much attention for their high equivalent switching frequency and low voltage stress, which are benefits for increasing the inverter's power density [6,7]. The neutral-point-clamped (NPC) inverter, flying capacitor (FC) inverter and T-type inverter are traditional three-level inverters which have been widely employed in industrial application. The NPC inverter is generally adopted in centralized PV grid-tied inverters because of its simple operation principle and high power level capability [8,9], which are different from the demands of PV string inverters. When used in low bus voltage applications, the T-type inverter is suitable on account that it can reach a higher work frequency, higher conversion efficiency, and higher power density [10,11]. The unbalance of neutral-point voltage is the main issue in multilevel inverters, except in the FC inverter [12]. However, one more floating capacitor is added in each phase, resulting in a larger volume and poorer power density, and its control scheme is more complex.

Many efforts have been made on topologies for photovoltaic multilevel inverters. Five-level topology reduces both the voltage stress of semiconductor devices and the volume of filter inductance compared with three-level topology due to its better harmonic performance, which may lead to loss reduction and system cost reduction. NPC-5L is the usual topology used for five-level topology [13], in addition to other topologies such as cascaded H-bridge five-level (CHB-5L) and FC five-level (FC-5L). Problems such as relatively large switching losses, unbalanced voltages of capacitors, and poor stability have promoted research into five-level topologies [14]. Other different topologies of multilevel inverters have also been adopted in industrial application: the stacked multi-cell (SMC) [15], the H-bridge NPC (H-NPC) [16], the neutral-point piloted (NPP) [17], and the modular multilevel converter (MMC).

The ANPC-5L converter, as shown in Figure 1, has been paid more and more attention since it was proposed [18,19], and is the combination of two types of inverters. One is the FC three-level inverter, the other is the active NPC three-level inverter. The advantages of this inverter consist of low switching losses and the convenience of capacitor voltage balance [20,21]. Switches which are connected in series in this topology switch at fundamental frequency, while the others switch at carrier frequency. Meanwhile, the switching cost of this topology is low because the stress of the switches is $V_{DC}/4$, while V_{DC} is the voltage of DC-link. Moreover, if different switching states are chosen appropriately, the voltage of floating capacitors is easy to balance.

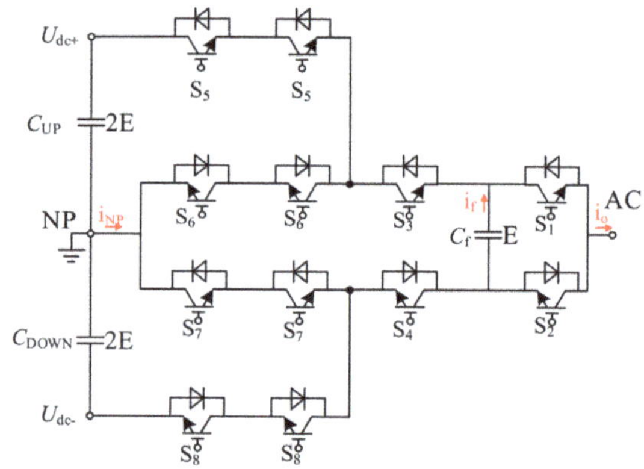

Figure 1. Topology of the ANPC-5L inverter.

Researchers have carried out a lot of work on ANPC-5L modulation technology, flying capacitor voltage control, neutral-point voltage control, and other issues [22,23]. The modulation strategy of the ANPC-5L inverter is simple and reliable most of the time. However, less attention has been paid to the voltage stress of switching devices in ANPC-5L converters, and the voltage stress of switching devices is very important for the safe and reliable operation of inverters. Document [24] analyzes the operation state of inverters based on space vector pulse-width modulation, including 125 vector combinations, and limits the stress of the switching devices by using the safe switching state switching process. Meanwhile, under the conventional modulation scheme, the analyzed inverter has the security risk of overvoltage in the power device when switching to dead time at the zero-crossing point of voltage when the output current is inductive, which affects the commutation safety [25].

In this article, this issue is analyzed in detail and a modified modulation strategy is proposed. In comparison with other modulations, this method provides several free degrees which are used to ensure the elimination of the voltage stress of power devices by choosing favorable circuit states and controlling current commutations. Additionally, it can realize the flying capacitor voltage balance in several carrier wave periods. The implementation of the proposed strategy in digital systems is rather simple. Meanwhile, a novel soft start-up method adopted to the ANPC-5L inverter is also proposed. Experimental results prove that the proposed strategy is valid.

The rest of the paper is organized as follows. Section 2 presents the traditional switching states. Section 3 analyzes the power device's overvoltage issue in detail. Section 4 proposes the modified modulation strategy to solve the potential safety hazards. Section 5 proposes a control method for soft start-up. Section 6 illustrates an inverter prototype for analysis verification. The conclusions are given in Section 7.

2. Traditional Modulation Strategy

As shown in Figure 1, the ANPC five-level inverter has eight power switches S1–S8, a floating capacitor C_F, a upper capacitor C_{UP}, and a lower capacitor C_{DOWN}. For the ANPC-5L inverter, as shown in Table 1, the conventional modulation scheme uses eight basic switching states to produce five voltage levels. There are redundant states at the +E and −E levels (E is 1/4 of bus voltage V_{DC}) which affect the charging or discharging states of the floating capacitors. The balance of the floating capacitor voltages can be realized by choosing appropriate switching states.

Table 1. Traditional switching states of the ANPC-5L inverter.

State	S1	S2	S3	S4	S5	S6	S7	S8	Vo	Vcf	
										Io > 0	Io < 0
V1	0	1	0	1	0	1	0	1	−2E	-	-
V2-1	0	1	1	0	0	1	0	1	−1E	C	D
V3	1	0	0	1	0	1	0	1	−1E	D	C
V4-1	1	0	1	0	0	1	0	1	−0	-	-
V5-1	0	1	0	1	1	0	1	0	+0	-	-
V6	0	1	1	0	1	0	1	0	+1E	D	C
V7-1	1	0	0	1	1	0	1	0	+1E	C	D
V8	1	0	1	0	1	0	1	0	+2E	-	-

The harmonics of the phase disposition (PD) contain a few carrier harmonics because of the different phases of the four carrier waves, as well as DC components, fundamental components, and carrier sidebands. However, some other modulation schemes, such as alternative phase opposition disposition (APOD), phase opposition disposition (POD), and two kinds of phase-shift carriers (PSC), have no carrier harmonics. In terms of single-phase inverters, the harmonic performance of the above-mentioned modulation is just the same due to the signal–energy conservation law. As far as the three-phase system is concerned, the harmonics are quite different. When the carrier waves of the three-phase system are synchronous, the carrier harmonics of the adjacent phases will exactly coincide, which represents that this harmonic will not appear in line voltage. However, carrier sidebands have no similar features. The harmonic performance of PD is the best. The next is APOD and PSC. The worst is POD. The characteristics of line voltage spectrums are far more diverse in the cause of common mode voltage. Eventually, considering the harmonic performance, the PD-PWM method is preferable among various methods. Under the traditional PD modulation of the ANPC-5L inverter, there are four cascaded carrier waves. As shown in Figure 2, comparison with the first carrier wave makes the output voltage change between +2E and +E. Switching between these states changes only two switch devices, and the switching processes are safe. Similarly, the switching processes in comparison with the other carrier waves are also reliable.

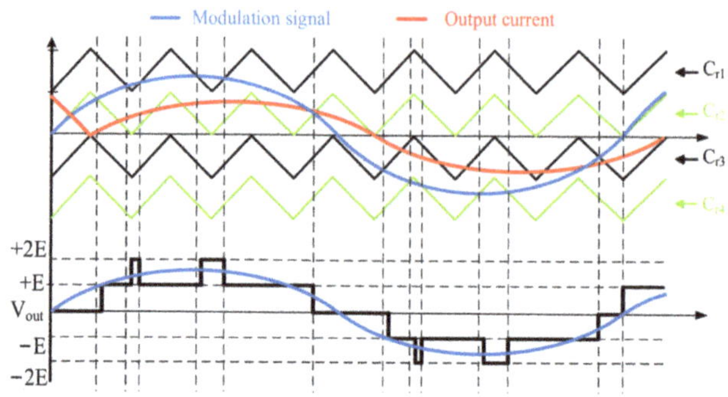

Figure 2. Traditional modulation strategy of the ANPC-5L inverter.

3. Overvoltage Issue of Traditional Modulation

However, there are potential overvoltage issues in the conventional modulation scheme. As shown in Figure 3, according to the counter mode of up–down or down–up, the output voltage will change from +E to −0 or from +0 to −E. Unlike the former switching process, switching between these two states changes six switch devices. Although the dead time of each pair of devices' switch exists, there will be safety problems under certain circumstances.

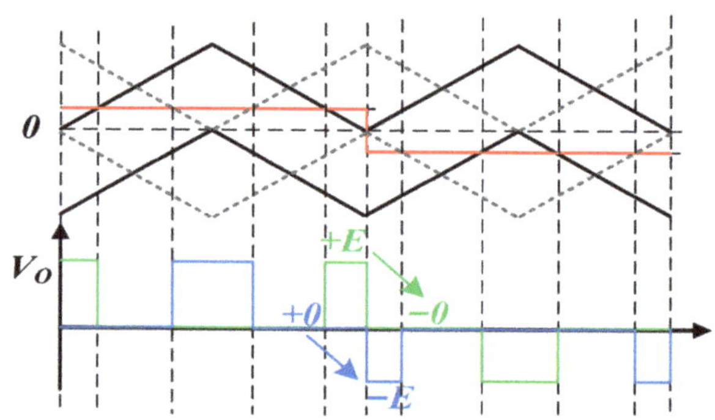

Figure 3. Switching process at zero crossing point.

As shown in Figures 4 and 5, taking the change from +0 to −E as an example, the circuit changes from state V5-1 to V2-1. Due to the existence of the dead zone, all the devices are turned off and there will be an intermediate state (V_{DANGER} (S1-S8:01000000)), which will cause the overvoltage issue.

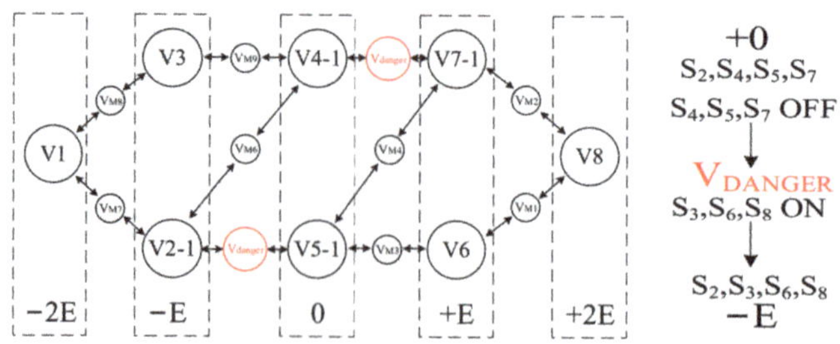

Figure 4. Traditional switching process of the ANPC-5L inverter.

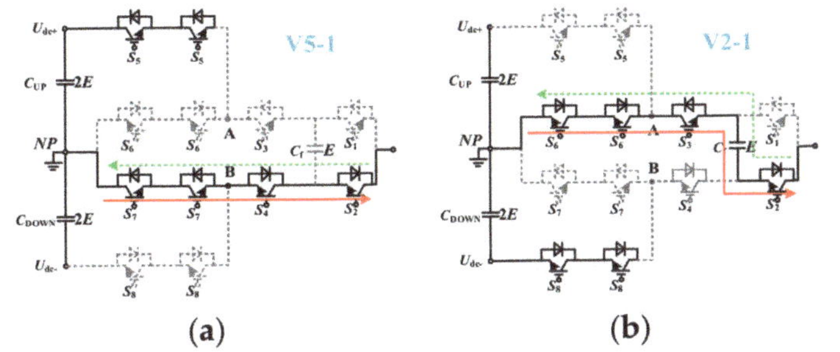

Figure 5. Switching states V5-1 and V2-1 at zero crossing point. (**a**) Switching states V5-1. (**b**) Switching states V2-1.

A detailed analysis is presented as below. In the state of V5-1, the initial states of the parasitic capacitance of the MOSFETs are shown in Figure 6; S3 has a potential difference of E while S6 has 2E. During the dead time of S5, assuming the output current I_O is greater than zero, S3 and S4 are all off. The continuous current path is shown in Figure 6, and the final states of the parasitic capacitance decide the safety of the commutation process.

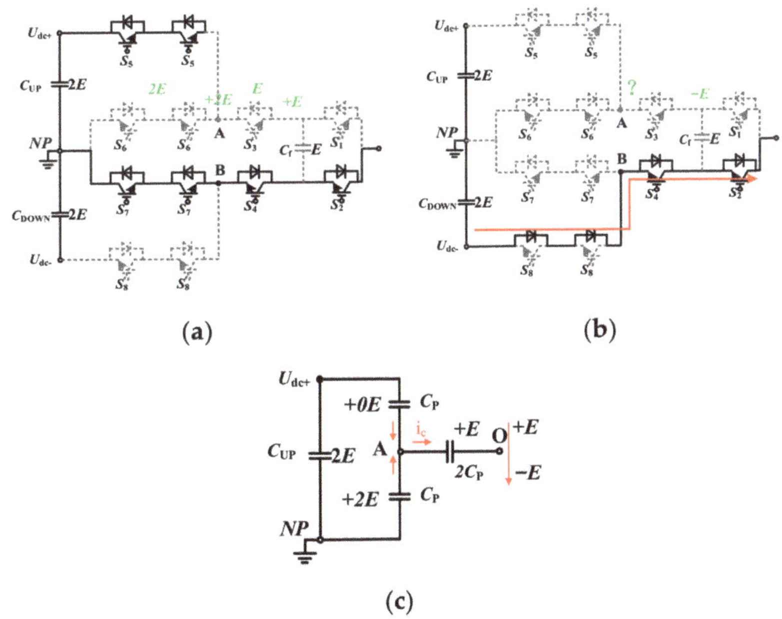

Figure 6. (**a**) Initial potential difference of V5-1. (**b**) Potential difference of V_{DANGER}. (**c**) Equivalent circuit of the charging process from V5-1 to V_{DANGER}.

During the dead time, S3, S5 and S6 are all off, closed switches equivalent to parasitic capacitances. Therefore, as shown in Figure 6, the equivalent circuit of the switching process is equal to the charging of parasitic capacitances.

Because of the charge–balance principle, increased charge on S3 should be equal to the summation of increased charge on S5 and reduced charge on S6. Moreover, the sum of the voltages of the S5 and S6 constants is equal to 2E, and the increased voltage on S5 is equal to the reduced voltage on S6. According to the capacitance definition:

$$dQ = C \cdot dU \tag{1}$$

Equation (2) can be obtained:

$$C_p \cdot dU_{S5} + C_p \cdot dU_{S6} = C_p \cdot dU_{S6} + C_p \cdot dU_{S6} = 2C_p \cdot dU_{S3} \tag{2}$$

The voltage of endpoint O changes from +E to −E by focusing on the steady state of −E, and according to Kirchhoff's Voltage Law (KVL) Equation (3) can be obtained:

$$2E - dU_{S6} = E + dU_{S3} + (-E) \tag{3}$$

Combining the above two formulas, increased voltage on S3 can be obtained:

$$dU_{S3} = E \tag{4}$$

Therefore, as shown in Figure 7, the final voltage of S3 will be 2E. The voltage stress will be higher than the voltage the device can withstand, which may cause overvoltage breakdown and influence the normal operation of the circuit.

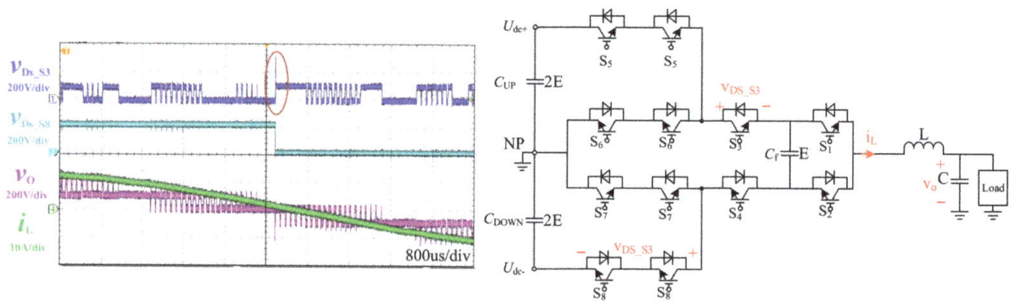

Figure 7. The overvoltage stress issue. (**a**) Experimental waveforms. (**b**) Experimental circuit.

Before the switching process of S5, there are three possible states for S3 and S4. As shown in Figure 8, when S4 is on and S3 is off, or S3 and S4 are both off, assuming the output current IO is greater than zero, the conduction path and the charging process are analyzed as above.

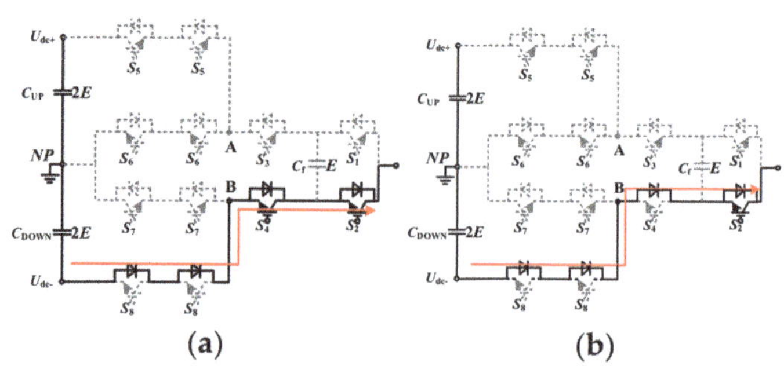

Figure 8. Conduction path of S3 and S4 equals to 0.1 or 0.0. (**a**) Positive current. (**b**) Negative current.

However, when S4 is off and S3 is on, as shown in Figure 9, the switching process is different. If the I_O is greater than zero, turning off S5 causes the conduction of S6's reverse diode, the voltage of point O changes from +E to −E and the initial voltage on S8 is 2E. The process seems similar to the above situation, but the direction of S4 decides that the charging current I_C reduces its voltage to zero; then, the reverse diode will conduct and there will be no overvoltage risk. However, the output voltage goes through 0, +E, −E. The ideal process is 0 to −E, although the commutation is not perfect. Therefore, a modified switching method is proposed to achieve the best commutation.

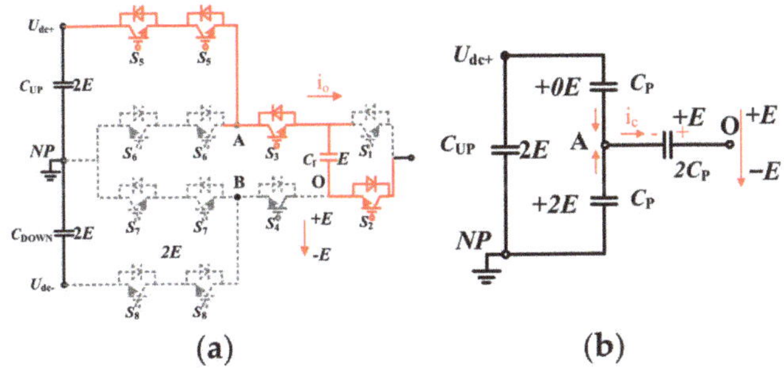

Figure 9. (**a**) Conduction path and potential difference when S3, S4 equals to 1.0 (**b**) Equivalent circuit of the charging process.

4. Proposed Modified Modulation Strategy

The conventional modulation scheme only considers eight simple states. After analyzing the switching states and overvoltage issues, eight other states (V2-2, V2-3, V4-2, V4-3, V5-2, V5-2, V7-2, V7-3) are obtained, as shown in Table 2, which can be included in the state cutover.

Table 2. The ANPC-5L converter's switching states and influence on the voltage of the flying capacitors.

State	S_{x1}–S_{x8}	V_{cf}		V_{ox}	Level
		$I_o > 0$	$I_o < 0$		
V_1	01010101	–	–	$-V_{dc}/2$	−2
$V_{2\text{-}1}$	01100101				
$V_{2\text{-}2}$	01100100	C [a]	D	$-V_{dc}/4$	−1
$V_{2\text{-}3}$	01110100				
V_3	10010101	D	C	$-V_{dc}/4$	−1
$V_{4\text{-}1}$	10100101				
$V_{4\text{-}2}$	10100100	–	–	0	0
$V_{4\text{-}3}$	10110100				
$V_{5\text{-}1}$	01011010				
$V_{5\text{-}2}$	01010010	–	–	0	0
$V_{5\text{-}3}$	01110010				
V_6	01101010	D	C	$V_{dc}/4$	1
$V_{7\text{-}1}$	10011010				
$V_{7\text{-}2}$	10010010	C	D	$V_{dc}/4$	1
$V_{7\text{-}3}$	10110010				
V_8	10101010	–	–	$V_{dc}/2$	2

[a] C: charging; D: discharging.

By combining additional circuit states, a modified modulation strategy and complete state machine are proposed, as shown in Figure 10. Under the proposed modulation scheme, a safe switching process within a power circle can be achieved and there will be no overvoltage stress, as shown in Figure 11.

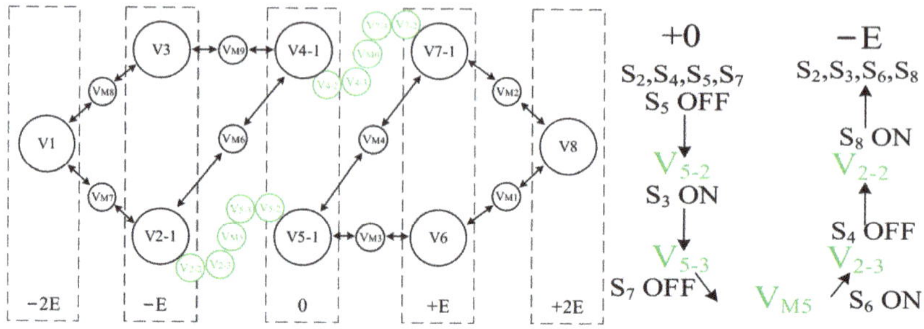

Figure 10. Complete state machine for the ANPC-5L process.

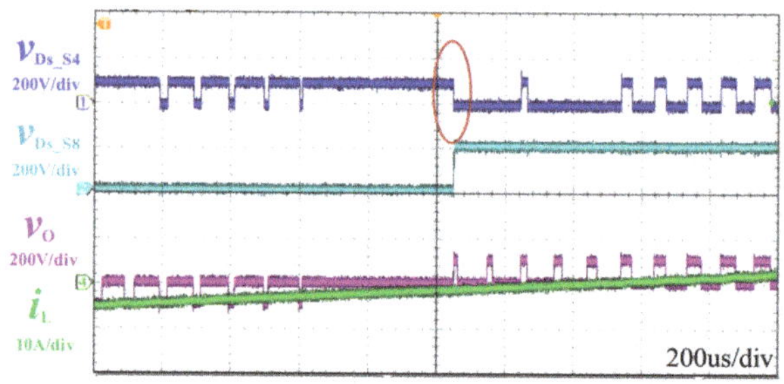

Figure 11. Reliable switching process of proposed modified modulation strategy.

The presented modulation strategy for the ANPC-5L converter is deeply researched in this section. Consisting of modulation signal, output current, driving signal and the voltage of the flying capacitors, Figure 4 shows the schematic diagram. With the scheme that is proposed, the switches, which are series-connected in the ANPC-5L converter, switch at fundamental frequency while the others switch at carrier frequency, which results in low switching losses. In the meantime, low switching and conduction loss can be achieved because the stress of all devices can be restricted to Vdc/4.

A. Flying capacitor voltage balance

The voltage of floating capacitors can be affected by different circuit states. One is E and the other is −E (V6, V7-1 and V2-1, V3), which can cause the capacitor to charge or discharge. To go a step further, we can use +E and −E levels within a sinusoidal voltage wave. If circuit states are selected appropriately, the balance of the voltage of FC will be achieved during the whole cycle.

Figure 12 shows that with a positive output current in V6, the flying capacitor is charging, and with a negative output current in V7-1 the flying capacitor is in a discharging state. With the increasing carrier frequency in each fundamental period, the balance of FC voltage will be better controlled in several carrier periods. Because the carrier wave period is much shorter than the fundamental period, the capacity value of the flying capacitor in the ANPC-5L inverter can be significantly reduced for a definite maximal permissible voltage range, compared with those inverters controlling a fundamental period. Ultimately, with the reduced volume of the flying capacitor, the power density of the inverter increases significantly.

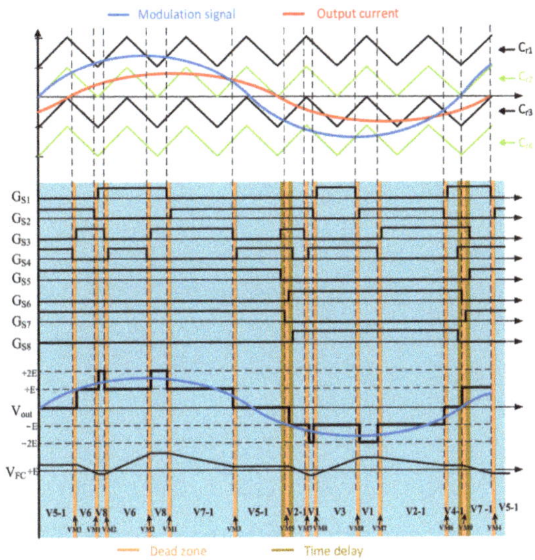

Figure 12. The waveform of the floating capacitor voltage.

B. Commutation

The inverter should always be in the state of safety switch during the commutation process, so the problems of shoot-through and high voltage stress during state cutover should be avoided. To distinguish positive current flow from negative current flow, the circuit can be separated into two parts to comprehend the state commutation.

1. V8 to V6: As shown in Figure 13a, S1 is turned off and after the turn-off delay S2 is turned on. Moreover, with the positive phase current, the state change, current commutation and switching loss occurs at S1 OFF. In contrast, with the negative phase current, the commutation of current and switching loss occurs at S2 ON;
2. V8 to V7-1: As shown in Figure 13b, it is inevitable to turn off S3 and turn on S4. If the phase current is positive, the commutation of current and switching loss occurs at S3 OFF;
3. V6 to V5-1: As shown in Figure 13c, it is inevitable to turn off S3 and turn on S4,and if the phase current is positive, the commutation of current and switching loss occurs at S3 OFF;
4. V5-1 to V7-1: As shown in Figure 13d, it is inevitable to turn off S2 and turn on S1;
5. V5-1 to V2-1: As shown in Figure 13e, it is inevitable to turn off S5, S7, and S4 and turn on S3, S6, and S8 in the cutover from V5-1 to V2-1. If the phase current is negative, the middle state VM5 generates zero voltage level, not −E voltage level.

Moreover, as shown in Figure 12, under the conduction of the state machine, the switching frequency of switch S5 to S8 is the same as modulation frequency, while switching frequency of switch S1 to S4 is the same as carrier frequency. Moreover, the voltage stress of power devices can be limited to Vdc/4. To sum up, shoot-through problems and overvoltage issues will not be caused by the commutation of current and state cutover. Accordingly, the modulation scheme which is proposed in this section is appropriate for the ANPC-5L inverter.

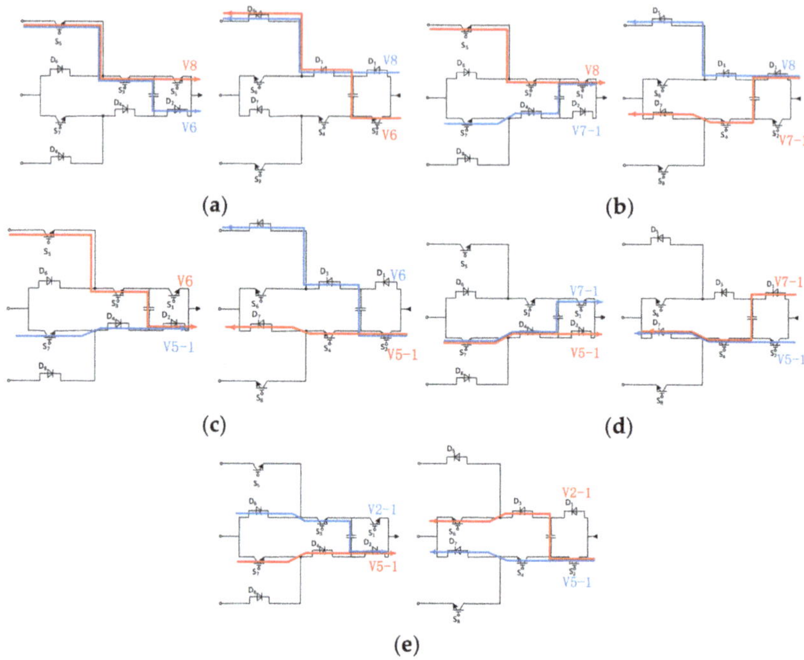

Figure 13. State cutover and current commutation. (**a**) V8 to V6. (**b**) V8 to V7-1. (**c**) V6 to V5-1. (**d**) V5-1 to V7-1. (**e**) V5-1 to V2-1.

5. Proposed Control Method for Soft Start-Up

The overvoltage issue of the ANPC-5L inverter includes two kinds of problems causing high voltage stress. The first kind is caused by the dead zone of the output voltage switching stage of the half-bridge circuit in series, and the second kind is caused by the soft start process. In the traditional inverter's soft start-up scheme, the dynamic change in voltage may lead to the overvoltage of low-voltage devices.

Start-up is an indispensable process in the control of the photovoltaic inverters, especially among power converters with flying capacitors. There will be very large current stresses on DC-bus capacitors, flying capacitors, and voltage stresses on power switches during the buildup of capacitor voltage if the procedure is not well controlled. Connecting current-limiting resistance in series with voltage sources can limit these stresses in conventional ways. In addition, when bus capacitors are pre-charged but flying capacitors are not fully charged, the voltage stress of several switches will increase by using normal working states, as shown in Figure 14.

Motor windings are used as part of a boost circuit to build up the voltage of flying capacitors with a constant pre-charging current for ANPC-5L converters, and the voltage stress of several switches will double. A pre-charge method applied to flying capacitor multilevel inverters is proposed in [26]; however, it requires plenty of AC contactors and even low-voltage DC power supply, which is not suitable for photovoltaic application.

It is clear that further efforts need to be made to reduce the voltage stress of power devices in the process of soft start-up, as well as to produce flying capacitor pre-charging means with less additional auxiliary circuits. The following part presents an analysis and proposes methods to settle these challenges.

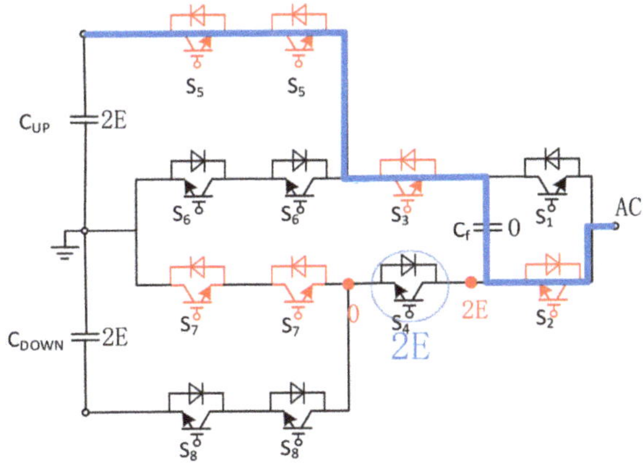

Figure 14. High voltage stress without optimal control.

As shown in Figure 15, there are twelve devices in each phase and switches S3, S4, S5 and S8 are used to connect C_F with C_{UP} and C_{DOWN} in parallel. Then, these capacitors can be charged by the DC-link voltage source in the meantime. Assuming the DC-link voltage source is constant, by controlling the main switches S3, S4, S5 and S8, the voltage of the flying capacitor C_F takes priority over the voltage of C_{UP} and C_{DOWN}, reaching its reference voltage to ensure that the voltage stress of the switches is not higher than $V_{DC}/4$.

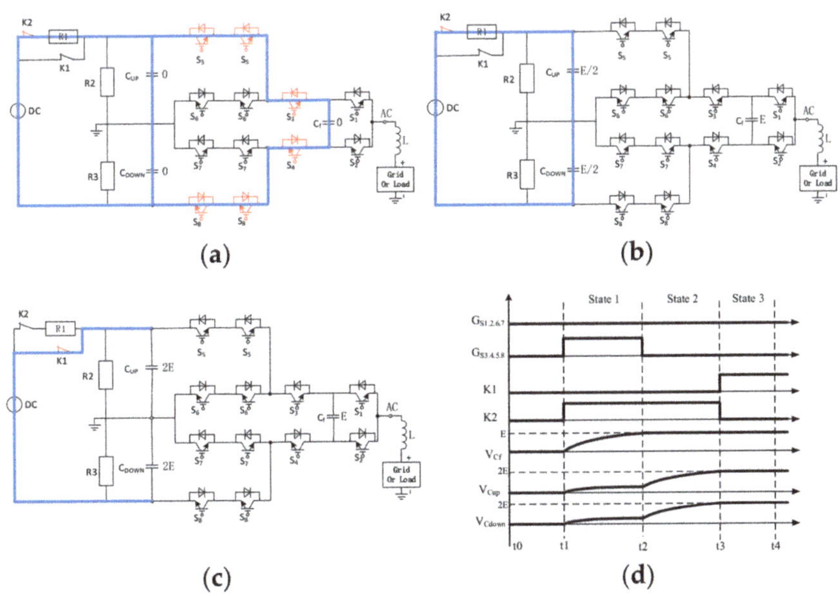

Figure 15. Proposed soft start-up process. (**a**) State 1. (**b**) State 2. (**c**) State 3. (**d**) Sequence diagram.

In a summary, the proposed soft start-up method can be divided into three states, as follows:

- State 1: When the upper bus capacitance C_{UP} and the lower bus capacitance C_{DOWN} are zero, and the flying capacitor C_f is not charged and the contactors K1 and K2 are disconnected, the ANPC five-level single-phase converter is in the initial condition with no energy in the capacitors. As shown in Figure 15a, since the main switches S3, S4, S5 and S8 are turned on and the contactor K2 is connected, the DC-link voltage source charges the upper bus capacitor C_{UP}, the lower bus capacitor C_{DOWN}, and the flying capacitor C_f simultaneously through the current limiting resistance R1. The voltage-divider resistances, R2 and R3, are placed in parallel with each bus capacitor to avoid the influence of the unbalanced characteristics of the upper and lower bus capacitors. In Figure 15d, the voltage of the flying capacitor C_f and bus capacitors increases gradually from t1 to t2;

- State 2: Until the voltage of C_f arrives at E, a quarter of the total bus voltage and power devices S3, S4, S5 and S8 are turned off and the voltage of bus capacitors will be half of flying capacitor voltage. Thus, the voltage stress of power switches is not higher than $V_{DC}/4$ in the entire stage, which meets its request. As shown in Figure 15b, the DC-link capacitors are going to be charged by the DC-link voltage source at the same time with the voltage divider resistances in parallel with each bus capacitor to avoid the influence of the unbalanced characteristics of the bus capacitors;
- State 3: The voltage of C_{UP} and C_{DOWN} increases gradually until they reach their reference value 2E, as shown in Figure 15d from t2 to t3. When C_{UP} and C_{DOWN} are fully charged, the contactor K1 is connected and K2 is disconnected. Then, the ANPC five-level single-phase converter starts up well with enough energy in the capacitors, as shown in Figure 15c and is ready to work.

The proposed method is applicable to the single-phase of the ANPC-5L inverter and pre-charges through the original PV DC voltage source in the photovoltaic application. The power resistance R1 has an impact on the charge current of the whole startup process, but will not influence the final capacitor voltage. While the converter is in H-bridge topology or in three phase topology, it is still sufficiently practical for the tolerance of the voltage stresses. This method uses the DC side power supply to charge the capacitor in the converter, and has the advantages of simple structure, convenient control, fewer additional auxiliary branches, and reliable soft start-up.

6. Experimental Results

To further verify the feasibility of the proposed strategy and theoretical analysis, a low-power prototype was established in the laboratory. Table 3 lists the electrical parameters of the prototype.

Table 3. Electrical parameters of the prototype.

Parameters	Values
Inverter DC-bus voltage	400 V
Output frequency	50 Hz
Converter rating	1 kVA
Switching frequency	10 kHz
Inductance of filter	0.3 mH
Capacitance of DC-link capacitor	2 mF
Capacitance of flying capacitor	660 μF

Figure 16 shows the proposed control scheme for the soft start-up and modulation scheme. V_f is sensed for the balance of the flying capacitor voltage, and i_L is sensed to judge the output current direction. Meanwhile, V_f is sensed for the soft start-up of the inverter. The proposed control scheme can be digitally realized in DSP or CPLD collaborative controllers.

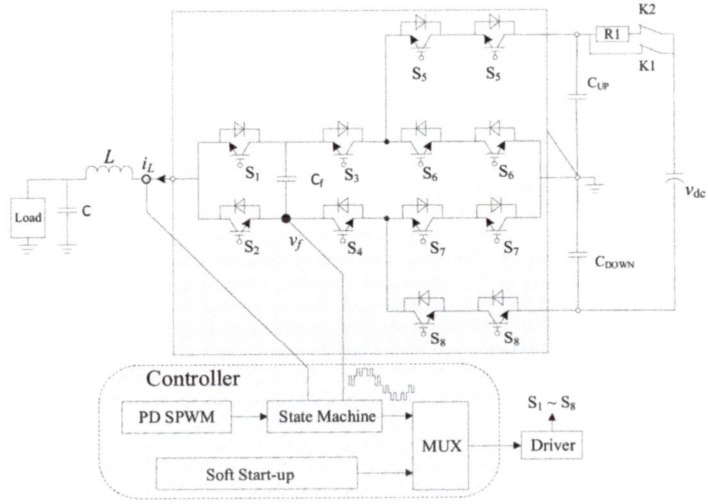

Figure 16. Proposed control scheme for soft start-up and modulation scheme.

The voltage of the DC-bus is 400 V, output voltage is 100 Vac and the switching device is 650 V MOSFET IPW60R190Z. As shown in Figure 17, under the traditional modulation strategy, there exists an overvoltage issue in the dead time at the zero crossing point of the waveform. Under the 400 V DC-bus, the voltage stress is about 300 V due to voltage ringing, and if the device is used under a normal bus of 800 V DC, there will be a risk of breakdown.

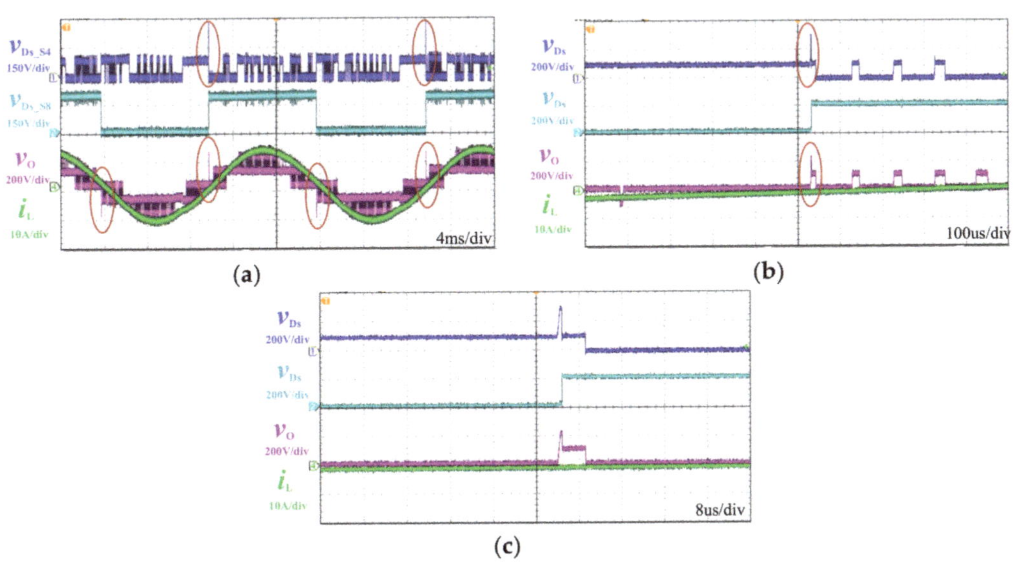

Figure 17. The overvoltage stress issue. (**a**) Under fundamental frequency. (**b**) Under multiple switching cycles. (**c**) Under single switching cycle.

To solve the overvoltage stress issue, the driving signal shown in Figure 18 is applied. As shown in Figure 18, there will be no overvoltage stress or output voltage level jump during the same switching process.

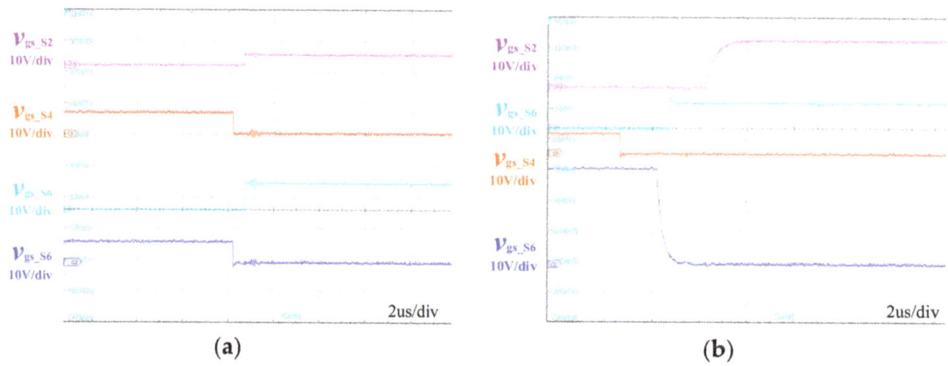

Figure 18. (**a**) Traditional and (**b**) modified modulation strategy.

As shown in Figure 19, under the proposed modified modulation strategy, there is no overvoltage issue in the dead time at zero crossing point of the waveform.

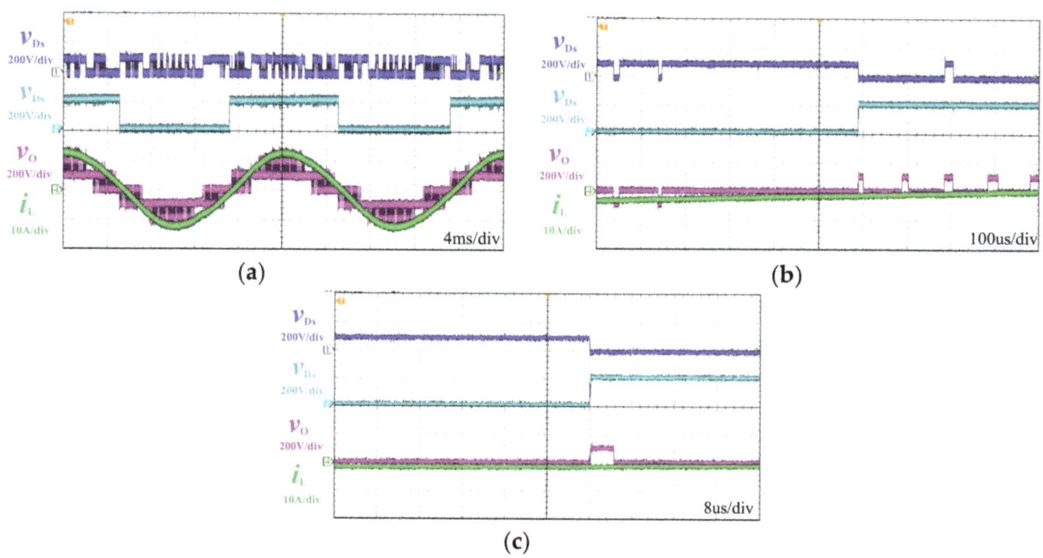

Figure 19. The waveform under the proposed modified modulation strategy. (**a**) Under fundamental frequency. (**b**) Under multiple switching cycles. (**c**) Under single switching cycle.

Figure 20 shows the experimental waveforms of Vc_up, Vc_down and VFc during the soft start-up of the inverter under different charging resistances. As discussed in Section 5, the flying capacitor and DC-link capacitor are charged by the DC-link voltage source in the meantime until the flying capacitor reaches its reference voltage Vdc/4 and the DC-link capacitor is charged to the bus voltage continually by controlling several main switches.

The experimental results have demonstrated that, by using the proposed method, the switching frequency of S5–S8 is the same as the fundamental frequency while the switching frequency of the other switches will switch at carrier frequency. Moreover, the stress of power devices can be no more than Vdc/4. By controlling the main switches S3, S4, S5 and S8, the flying capacitor and bus capacitor are fully charged without the overvoltage problem of the switches.

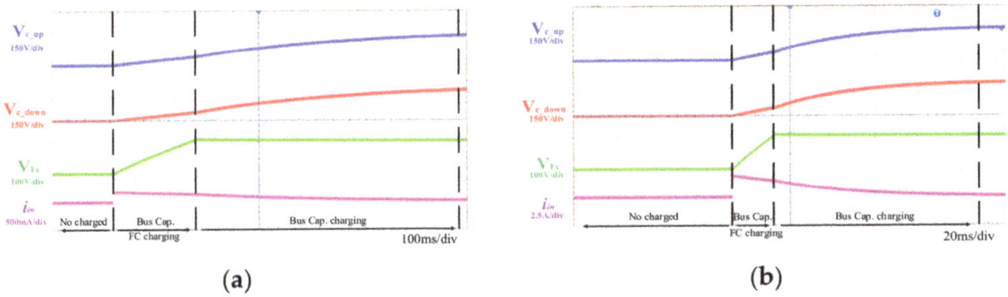

Figure 20. Soft start-up under different charging resistance. (**a**) R1 = 500 Ohms; (**b**) R1 = 50 Ohms.

7. Conclusions

For the ANPC-5L inverter, the traditional modulation strategy has the security risk of the overvoltage of the power device in the switching dead time when the output current is inductive, which affects the commutation safety and leads to an overvoltage issue. In this paper, the overvoltage mechanism is deduced through the voltage stress analysis of different switching states. Meanwhile, a modified modulation strategy is proposed to solve this issue. In comparison with other modulations, this method provides several free degrees which are used to ensure the elimination of the voltage stress of power devices by choosing favorable circuit states and controlling current commutations. Additionally, it can realize the flying capacitor voltage balance in several carrier wave periods. The implementation of the proposed strategy in the digital system is rather simple. Meanwhile, a novel soft start-up method adapted to the ANPC-5L inverter is also proposed. In addition, an experimental prototype is also built to verify the issue of traditional modulation strategy and the validity and feasibility of the proposed modulation strategy.

Author Contributions: Conceptualization, G.C.; methodology, J.Y.; software, J.Y.; validation, G.C. and J.Y.; formal analysis, G.C.; investigation, G.C.; resources, G.C.; data curation, G.C.; writing—original draft preparation, G.C.; writing—review and editing, J.Y.; visualization, J.Y.; supervision, G.C.; project administration, G.C.; funding acquisition, G.C. All authors have read and agreed to the published version of the manuscript.

Funding: This work was supported by Shanghai Outstanding Academic/Technical Leaders Plan (20XD1430700).

Conflicts of Interest: The authors declare no conflict of interest.

References

1. Pires, V.F.; Cordeiro, A.; Foito, D.; Pires, A.J.; Hao, C.; Martins, J.F.; Castro, R. Compensation of Unbalanced Low-Voltage Grids Using a Photovoltaic Generation System with a Dual Four-Leg, Two-Level Inverter. *Electronics* **2022**, *11*, 320. [CrossRef]
2. Gkoutioudi, E.; Bakas, P.; Marinopoulos, A. Comparison of PV systems with maximum DC voltage 1000 V and 1500 V. In Proceedings of the Photovoltaic Specialists Conference (PVSC), Tampa Bay, FL, USA, 16–21 June 2013; pp. 2873–2878.
3. Čorba, Z.; Popadić, B.; Katić, V.; Dumnić, B.; Milićević, D. Future of high power PV plants—1500V inverters. In Proceedings of the 19th International Symposium on Power Electronics (Ee), Novi Sad, Serbia, 19–21 October 2017; pp. 1–5.
4. Stevanovic, B.; Serban, E.; Cobreces, S.; Alou, P.; Ordonez, M.; Vasic, M. DC/DC Stage Contribution to Bus Voltage in 1000 V and 1500 V Grid-Connected Solar Inverters. *IEEE J. Emerg. Sel. Top. Power Electron.* **2022**, 1. [CrossRef]
5. Serban, E.; Ordonez, M.; Pondiche, C. DC-bus voltage range extension in 1500 V photovoltaic inverters. *IEEE J. Emerg. Sel. Top. Power Electron.* **2015**, *3*, 901–917. [CrossRef]
6. Prathaban, A.V.; Dhandapani, K.; Abubakar, A.I.S. Compact Thirteen-Level Inverter for PV Applications. *Energies* **2022**, *15*, 2808. [CrossRef]
7. Gupta, K.K.; Ranjan, A.; Bhatnagar, P.; Sahu, L.K.; Jain, S. Multilevel Inverter Topologies with Reduced Device Count: A Review. *IEEE Trans. Power Electron.* **2016**, *31*, 135–151. [CrossRef]
8. Campanhol, L.B.G.; da Silva, S.A.O.; de Oliveira, A.A.; Bacon, V.D. Dynamic Performance Improvement of a Grid-Tied PV System Using a Feed-Forward Control Loop Acting on the NPC Inverter Currents. *IEEE Trans. Ind. Electron.* **2017**, *64*, 2092–2101. [CrossRef]

9. Busquets-Monge, S. A Simple Virtual-Vector-Based PWM Formulation for Multilevel Three-Phase Neutral-Point-Clamped DC–AC Converters including the Overmodulation Region. *Electronics* **2022**, *11*, 641. [CrossRef]
10. Gonzalez, R.; Gubía, E.; Lopez, J.; Marroyo, L. Transformerless Single-Phase Multilevel-Based Photovoltaic Inverter. *IEEE Trans. Ind. Electron.* **2008**, *55*, 2694–2702. [CrossRef]
11. Schweizer, M.; Kolar, J.W. Design and Implementation of a Highly Efficient Three-Level T-Type Converter for Low-Voltage Applications. *IEEE Trans. Power Electron.* **2013**, *28*, 899–907. [CrossRef]
12. Gateau, G.; Fadel, M.; Maussion, P.; Bensaid, R.; Meynard, T. Multicell converters: Active control and observation of flying-capacitor voltages. *IEEE Trans. Ind. Electron.* **2002**, *49*, 998–1008. [CrossRef]
13. Choi, U.-M.; Ryu, T. Comparative Evaluation of Efficiency and Reliability of Single-Phase Five-Level NPC Inverters for Photovoltaic Systems. *IEEE Access* **2021**, *9*, 120638–120651. [CrossRef]
14. Fazel, S.; Bernet, S.; Krug, D.; Jalili, K. Design and comparison of 4-kV neutral-point-clamped, flying-capacitor, and series-connected H-bridge multilevel converters. *IEEE Trans. Ind. Appl.* **2007**, *43*, 1032–1040. [CrossRef]
15. Ghias, A.M.Y.M.; Pou, J.; Acuna, P.; Ceballos, S.; Heidari, A.; Agelidis, V.G.; Merabet, A. Elimination of Low-Frequency Ripples and Regulation of Neutral-Point Voltage in Stacked Multicell Converters. *IEEE Trans. Power Electron.* **2017**, *32*, 164–175. [CrossRef]
16. Gutierrez, E.; Kouro, S.; Rojas, C.A.; Aguirre, M. Predictive control of an h-npc converter for single-phase rooftop photovoltaic systems. In Proceedings of the Energy Conversion Congress and Exposition (ECCE), Montreal, QC, Canada, 20–24 September 2015; pp. 3295–3302.
17. Tian, K.; Wu, B.; Narimani, M.; Xu, D.; Cheng, Z.; Zargari, N.R. A Capacitor Voltage-Balancing Method for Nested Neutral Point Clamped (NNPC) Inverter. *IEEE Trans. Power Electron.* **2016**, *31*, 2575–2583. [CrossRef]
18. Barbosa, P.; Steimer, P.; Steinke, J.; Winkelnkemper, M.; Celanovic, N. Active-neutral-point-clamped (ANPC) multilevel converter technology. In Proceedings of the European Conference on Power Electronics and Applications, Dresden, Germany, 11–14 September 2005; p. 10. [CrossRef]
19. Guo, W.; Li, R.; Yang, J.; Cao, Y. Circulating Current Suppression Scheme for Interleaved Active Neutral Point Clamped Nine-Level Inverter. *IEEE J. Emerg. Sel. Top. Power Electron.* **2022**, 1. [CrossRef]
20. Lee, S.S.; Yang, Y.; Siwakoti, Y.P. A Novel Single-Stage Five-Level Common-Ground-Boost-Type Active Neutral-Point-Clamped (5L-CGBT-ANPC) Inverter. *IEEE Trans. Power Electron.* **2021**, *36*, 6192–6196. [CrossRef]
21. Sadigh, A.K.; Dargahi, V.; Naderi, R.; Corzine, K.A. Active Voltage Balancing and Thermal Performance Analysis of Dual Flying-Capacitor Active Neutral-Point-Clamped (DFC-ANPC) Inverters. *IEEE Trans. Ind. Appl.* **2021**, *57*, 637–649. [CrossRef]
22. Burguete, E.; Lopez, J.; Zabaleta, M. A New Five-Level Active Neutral-Point-Clamped Converter with Reduced Overvoltages. *IEEE Trans. Ind. Electron.* **2016**, *63*, 7175–7183. [CrossRef]
23. Liu, Z.; Wang, Y.; Tan, G.; Li, H.; Zhang, Y. A novel svpwm algorithm for five-level active neutral-point-clamped converter. *IEEE Trans. Power Electron.* **2016**, *31*, 3859–3866. [CrossRef]
24. Kieferndorf, F.; Basler, M.; Serpa, L.A.; Fabian, J.-H.; Coccia, A.; Scheuer, G.A. ANPC-5L technology applied to medium voltage variable speed drives applications. In Proceedings of the SPEEDAM 2010, Pisa, Italy, 14–16 June 2010; pp. 1718–1725.
25. Yang, J.; Yang, S.; Li, R. A novel and reliable modulation strategy for active neutral-point clamped five-level converter. In Proceedings of the IEEE 3rd International Future Energy Electronics Conference and ECCE Asia (IFEEC 2017-ECCE Asia), Kaohsiung, Taiwan, 3–7 June 2017; pp. 1162–1167.
26. Hong, F.; Liu, M.; Ji, B.; Wang, C. A capacitor voltage buildup method for flying capacitor multilevel Inverters. *Proc. CSEE* **2012**, *32*, 17–23.

energies

MDPI

Article

Impact Analysis of a Battery Energy Storage System Connected in Parallel to a Wind Farm

Nicolau K. L. Dantas [1,*], Amanda C. M. Souza [1,2], Andrea S. M. Vasconcelos [1,3], Washington de A. S. Junior [1], Guilherme Rissi [4], Celso Dall'Orto [5], Alexandre M. A. Maciel [3], José F. C. Castro [2], Yang Liu [4] and Pedro Rosas [2]

1 Institute of Technology Edson Mororó Moura (ITEMM), Recife 51020-280, PE, Brazil; amanda.monteiro@itemm.org.br (A.C.M.S.); andrea.sarmento@itemm.org.br (A.S.M.V.); washington.araujo@itemm.org.br (W.d.A.S.J.)
2 Department of Electrical Engineering and Power Systems, Federal University of Pernambuco (UFPE), Recife 50670-901, PE, Brazil; filho.castro@ufpe.br (J.F.C.C.); pedro.a.rosas@gmail.com (P.R.)
3 Department of Computer Engineering, University of Pernambuco (UPE), Recife 50100-010, PE, Brazil; alexandre.maciel@upe.br
4 CPFL Energy, Campinas 13087-397, SP, Brazil; grissi@cpfl.com.br (G.R.); liuyang1@cpfl.com.br (Y.L.)
5 PSR—Energy Consulting and Analytics, Botafogo 22250-040, RJ, Brazil; celso@psr-inc.com
* Correspondence: nicolau.dantas@itemm.org.br

Abstract: Increasing wind generation insertion levels on electrical grids through power converters may cause instabilities in the AC grid due to the intermittent wind nature. Integrating a Battery Electric Energy Storage System (BESS) in wind generation can smooth the power injection at the Common Coupling Point (PCC), contributing to the power system voltage and frequency stability. In this article, it is proposed to analyze the operation of a lithium-ion battery technology based 1 MW/1.29 MWh BESS connected in parallel with wind generation with a capacity of 50.4 MW. The main characteristics investigated are power smoothing and power factor correction. Experimental results show that BESS contributes to smoothing the active power and correcting the power factor of wind generation, improving the quality of electrical energy at the PCC.

Keywords: storage system; batteries; power smoothing; power factor correction; wind generation; power converter; stability; electrical power quality

Citation: Dantas, N.K.L.; Souza, A.C.M.; Vasconcelos, A.S.M.; Junior, W.d.A.S.; Rissi, G.; Dall'Orto, C.; Maciel, A.M.A.; Castro, J.F.C.; Liu, Y.; Rosas, P. Impact Analysis of a Battery Energy Storage System Connected in Parallel to a Wind Farm. *Energies* **2022**, *15*, 4586. https://doi.org/10.3390/en15134586

Academic Editors: Luis Hernández-Callejo, Jesús Armando Aguilar Jiménez and Carlos Meza Benavides

Received: 10 May 2022
Accepted: 16 June 2022
Published: 23 June 2022

1. Introduction

The growing concern about the environment and the depletion of fossil fuels has given rise to a new scenario to meet the energy needs of society: renewable sources [1,2]. Among the various renewable sources of electricity, wind generation has been presented as the most interesting and the fastest growing in the world [2,3]. According to the Global Wind Energy Council (GWEC) 2022 report, the wind industry had its second-best year in 2021, with nearly 94 GW of capacity added globally. Total global wind power capacity is now up to 837 GW, helping the world avoid more than 1.2 billion tons of CO_2 annually—the equivalent of South America's annual carbon emissions.

The growth of wind energy in the world energy matrix is due to its advantages, such as: it does not emit greenhouse gases; it takes little time to build wind farms; it diversifies the electricity matrix; it is independent of the variation in fuel prices; it is easy to expand the capacity of wind farms; it provides new markets etc. [2]. However, due to the highly uncertain and variable nature of the wind, wind energy can present undesirable characteristics in its generation and impact the Electric Power Systems (EPSs). With an increasing share of EPSs, the uncertainty of wind energy and its power fluctuation will affect the ability of grid operators to balance generation and demand. Furthermore, the significant penetration of wind generation in the grid can harm the Power Quality, the dynamics, and the system reliability [2,4].

According to [2,5], the main concerns regarding the connection of wind generation in electrical systems are related to the impact on the stability and Power Quality of the grid, the ability to compensate for active power fluctuations, and the impact on grid voltage, both short-term and long-term. To minimize some of these problems, additional flexible resources must be used to manage the variability and uncertainty of wind generation. Battery Energy Storage Systems (BESS) can be used to reduce power fluctuations, as well as provide ancillary services (voltage and frequency regulation), manage energy during disturbances (short circuits), and increase network reliability.

There is a wide range of possible BESS applications in the electrical sector, such as power generation, transmission, and distribution, and direct applications to the final consumer. However, it can be said that the attractiveness of each specific solution depends on the characteristics and needs of the applications. For example, BESS can be used for short-term power smoothing in wind farms in a generation. Using BESS, it is possible to mitigate the adverse effects of the power fluctuation in wind generation and, consequently, improve the Electrical Power Quality (EPQ) and the power grid operation. As for the final consumer, BESS can be used to perform arbitrage, charging the batteries during lower-priced hours, and discharging during higher-priced hours (i.e., during peak hours).

The BESS can be idle for a significant fraction of time, depending on the application. Thus, to make this solution more viable, it is possible to merge different applications. For example, BESS systems installed in more robust systems, such as the Brazilian National Interconnected System (SIN), can implement other services such as arbitrage, operating reserve, frequency control, voltage control, and black start. Thus, the merging of applications increases the use of BESS over time.

Energy storage systems have been widely simulated to reduce power fluctuations from wind generation with different control strategies [6–9], and [10] proposed using a storage system integrated into a wind system to reduce high-frequency fluctuations in the generated power, using filters to separate the operation of the inverter controllers and the storage system in the frequency domain. In [11], it is seen that to provide constant supply from a 39.6 MW wind farm, a 2 MWh capacity energy storage system was used to maintain consistent production for one hour and up to 103 MWh to provide consistent output for one day.

In [12], it is seen that the intermittent operation of renewable energy sources, faults occurrences, or PCC disturbances can cause voltage or frequency deviations, resulting in instability problems, which can become severe in weak power networks. In [12], a storage system was used to regulate voltage and frequency in microgrids.

There are many different papers which evaluate the power smoothing of wind generation through BESS. Among the recent papers, there are relevant talks about the problem of the power fluctuation in many different contexts and settings, such as the power smoothing in the context of transmission or distribution, different technologies of BESS, different types of power control for BESS, complementary applications for power smoothing, different settings of connection of BESS, and different ways to evaluate and present the results [2]. However, considering these different contexts and settings there are still topics to be explored and improved. Several papers do not verify the effects of power smoothing in wind generation, as well as do not use numerical indicators to evaluate the performance of the power smoothing techniques. Thus, the main contribution of this article is to analyze the effects of power smoothing in wind generation in a case study at the Campo dos Ventos Wind Complex located in João Câmara, Rio Grande do Norte (RN)—Brazil. The numerical indicator of Maximum Variation Power (MVP) is also used to evaluate the performance of the power smoothing techniques. The MVP indicator quantifies the largest power of wind generation within a predefined time interval.

Furthermore, this paper addresses, in a complementary way, the application of power smoothing, the analysis of the storage system operating in the power factor correction mode, and its impact on Power Quality for the Campo dos Ventos Wind Complex. Experimental results are performed to validate the performance of BESS.

2. Project Description

The purpose of the Research and Development R&D Project PA3026, entitled "Impact Analysis of a Battery Energy Storage System Connected in Parallel to a Wind Farm", is to study energy storage applications from different qualitative and quantitative perspectives. This project is formed by the group of institutions CPFL Energy (Light and Power Company of Sao Paulo State), Institute of Technology Edson Mororó Moura (ITEMM), Federal University of Pernambuco (UFPE), and PSR—Energy Consulting and Analytics.

Brazil still presents a relatively immature environment for the development of energy storage technologies. Faced with regulatory and even non-regulatory gaps, the R&D Project PA3026 seeks to resolve existing uncertainties about the applicability and effectiveness of services and assist recognition through adequate remuneration for these storage systems.

From this perspective, the project proposes to investigate the operationalization of several actions applied to a real wind farm. Among the main functions destined for the storage system identified in the project are produced power smoothing and power factor correction. These two proposals are tested with the operation of a BESS composed, among other components, by a set of Lithium Iron Phosphate (LFP) batteries, which is a lithium-ion battery technology with a capacity of 1 MW/1.29 MWh integrated into an electric power substation of a wind farm. The choice of this technology was due to the benefits that the LFP battery presents, such as: (1) high energy density (about 1932 W/L) [13], (2) high conversion efficiency (90~95%) [14], (3) low self-discharge rate, and (4) fast response time [13]. It should be noted that the service life (>2000 cycles) still needs to be improved and there are potential fire hazards [13,14].

According to [13], flow batteries are safe as they are non-flammable and have a long cycle life (2000 to 20,000 cycles) and do not depend on the depth of discharge. However, the energy density is low, occupies a large amount of land, and the conversion efficiency is low (65–85%) [15]. Hydrogen batteries have the highest specific energy (500–3000 Wh/L) compared to other storage systems and have a high cycle life (about 20,000 cycles). Although hydrogen batteries have a long-life cycle, they have a high initial cost [13]. The lead-acid battery is safe and reliable, but its energy density is low and its cycle times (300–3000 cycles) are limited [16].

The wind farm choice considered the capacity and arrangement of machines with different technologies. The Campo dos Ventos Wind Complex, located at João Câmara—Rio Grande do Norte (RN)—Brazil, a synchronous generator with a full converter and Double Powered Induction Generators (DFIG), was chosen. This farm has twenty-four turbines, each with 2.1 MW rated power, totaling 50.4 MW. Therefore, the 1 MW/1.29 MWh capacity BESS can be analyzed in terms of its real impact on the proposed objectives. The turbines and BESS are connected through a SCADA system. The simplified single-line diagram for the Campo dos Ventos Wind Complex is shown in Figure 1. The installation of BESS at the Campo dos Ventos Electric Substation (ES) is illustrated in Figure 2.

The main objectives of using the storage system are to smooth the wind production through instantaneous power injection and instantaneous power consumption, counterbalancing its instantaneous output and, consequently, removing variations introduced by the wind intermittence acting on each other the wind turbine.

Considering other storage system application options, it is expected to use the remaining storage capacity for reactive power compensation, firstly, to improve the power factor and, secondly, to improve the voltage with the consequent reactive power control.

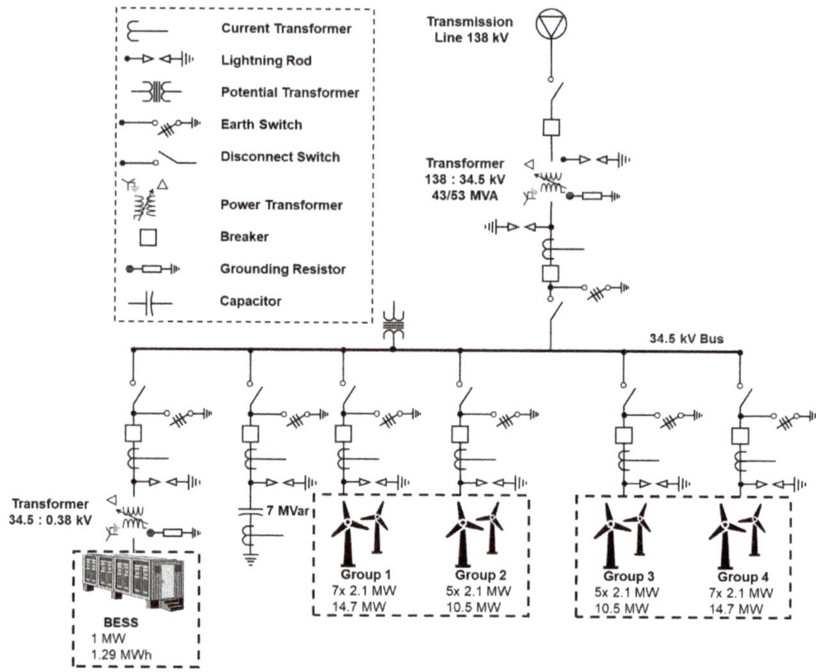

Figure 1. Single-line diagram of the Campo dos Ventos Wind Complex.

Figure 2. Installation of BESS in the Campo dos Ventos Substation.

3. Battery Energy Storage System (BESS)

3.1. General View

In general, the main components of BESS are batteries, battery management system (BMS), energy management system (EMS), power conversion system (PCS), fire detection and suppression system, heating system, ventilation, air conditioning (HVAC), Uninterruptible Power Supply (UPS), container, transformer (if voltage increase is required), cables (primary and secondary) and other auxiliary systems. Figure 3 shows the BESS and some of these components. The system installed at the Campo dos Ventos Wind Complex has two 500 kW PCSs that convert DC energy into AC or vice versa and is connected to the DC side batteries and connected in parallel to the AC side wind generation bus. EMS controls PCSs and communicates with BMS and SCADA and contains all system applications.

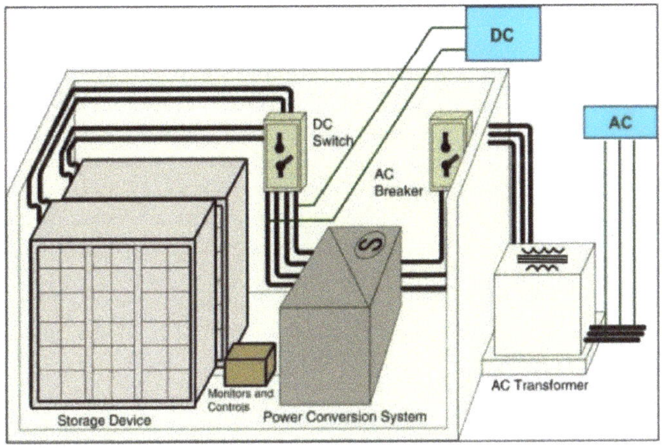

Figure 3. Schematic of BESS and its components [17]. Reprinted from [17] with permission (License 1431469) from U.S. Department of Energy Office of Scientific and Technical Information [17,18].

3.2. Batteries

The battery is one of the main components of the BESS. It is in the battery that energy is stored. There are various battery types integrated into generation, transmission, distribution, and end-consumer worldwide. Each type may use the system differently. Battery selection is mainly focused on the BESS application (e.g., applications that demand more power and energy) and cost-effectiveness ratio due to the high proportion of the battery's monetary value to the total cost of the project (about 50–60%) [19].

LFP batteries were used in the BESS. The LFP battery pack has an output voltage of 51.2 V and a capacity of 180 Ah. Fourteen battery packs are grouped in series in a cluster, totaling an output voltage of 716.8 V and a capacity of 180 Ah. The BESS contains 10 clusters in parallel, leading to an output voltage of 716.8 V and a capacity of 1800 Ah.

3.3. BMS

The BMS plays a vital role in the BESS. The integration of the BESS with the wind generation bus means that multiple batteries are connected in parallel, improving safety and reliability.

The BMS is designed to provide safe operations by monitoring the voltages, currents, and temperature of the cells in the batteries. In addition, the BMS offers the following functions [19–21]:

- Protection for battery cells;
- Evaluation of battery cell recharge and health status;
- Energy balancing between battery cells (including battery charging and discharging patterns).

3.4. EMS

The EMS is responsible for continuously relevant BESS data acquisition and storage, such as voltage, frequency, active and reactive power, power factor, battery cell voltage, etc. This data acquisition and storage can be either in local or remote forms. Additionally, the EMS receives the control setpoints to allow changes in the BESS operating modes and subsystems. The BESS applications of this project are power smoothing, frequency control, voltage control, and power factor correction. Lastly, it contains manual control for active and reactive power injection and absorption.

EMS controls BESS to regulate battery recharges and discharges to achieve optimal efficiency generation requirements. All battery cells are individually monitored to ensure any deviation in performance is detected and corrected before problem occurrences. The EMS can be viewed remotely as needed and communicate with CPFL's local SCADA.

3.5. PCSs

The current project BESS contains two bidirectional PCSs to perform DC/AC and AC/DC conversions. In addition, it controls voltage and frequency, ensuring that the electricity output meets desired connection requirements.

The BESS uses two bidirectional 500 kW PCSs, connected on the DC side to the power bank battery. The PCSs in question contain anti-islanding protection, in which the inverter detects problems in the electrical grid, such as a power outage, and switches off to interrupt the supply. This protection is needed because, after electrical grid problem occurrences, it is assumed that workers will be dispatched to deal with the issue; therefore, it is necessary that the electric power lines are entirely safe and electric current free.

Another functionality of PCSs is stability for under and over voltage/frequency ranges, in which the inverter does not trip if the anomaly duration exceeds a specific period. This function is an essential feature to improve grid stability.

The PCSs' operating modes are:

- P-Q Control mode is when a reference voltage and a constant frequency are supplied by another source (usually the electrical grid). The inverter can change the active and reactive power.
- V-F Control mode (Autonomous Mode): V-F control mode occurs when, regardless of the varying inverter power, the amplitude and frequency of the output voltage are constant. The inverter with V-F control can provide voltage and frequency support to the microgrid during island operation. The inverter acts as a voltage source. The current amplitude and the power factor (PF) will be determined by the sum of the generation (if any) and the consumption load.

The PCSs' operating modes are in four quadrants, as illustrated in Figure 4, both in on-grid and off-grid modes, which means that active power and reactive power can be in four characteristics:

- Consumes active power plus inductive reactive power;
- Consumes active power plus capacitive reactive power;
- Provides active power plus inductive reactive power;
- Provides active power plus capacitive reactive power.

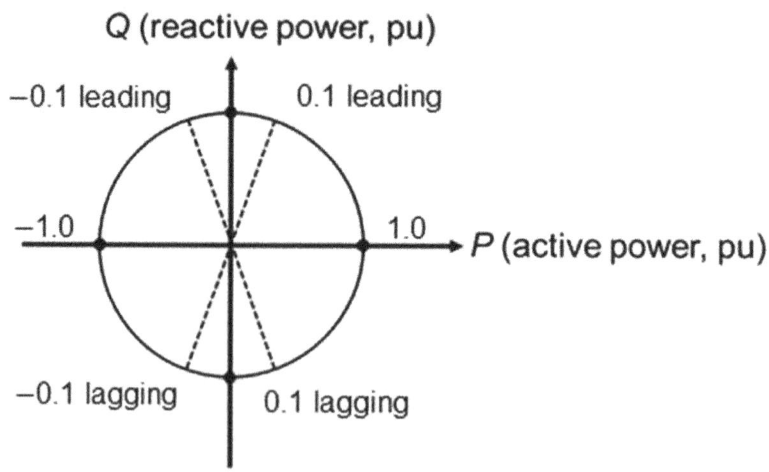

Figure 4. Four-quadrant operation of PCSs.

Control System

The PCS installed on the BESS operates in grid-connected and islanded mode. Active and reactive power control (P-Q Control) is used in grid-connected mode, while constant voltage and frequency control (V-F Control) is employed in islanded mode. These two control strategies are based on [22] and detailed below.

(a) P-Q Control

When the PCS is connected to the electrical grid, it operates in P-Q Control mode. Active and reactive power based on instantaneous active and reactive power theory are shown in Equation (1) [22].

$$\begin{cases} p = 1.5(v_d i_d + v_q i_q) \\ q = 1.5(v_q i_d - v_d i_q) \end{cases} \quad (1)$$

When the q component of the voltage is zero and assuming that the voltage vector is in the d-axis direction, Equation (1) can be represented by:

$$\begin{cases} p = 1.5 v_d i_d \\ q = -1.5 v_d i_q \end{cases} \quad (2)$$

The reference current can be calculated by:

$$\begin{cases} i_{dref} = \frac{P_{ref}}{1.5 v_d} \\ i_{qref} = -\frac{Q_{ref}}{1.5 v_d} \end{cases} \quad (3)$$

where P_{ref} represents active power and Q_{ref} represents reactive power, the expected output values. Figure 5 illustrates the simplified P-Q Control block diagram.

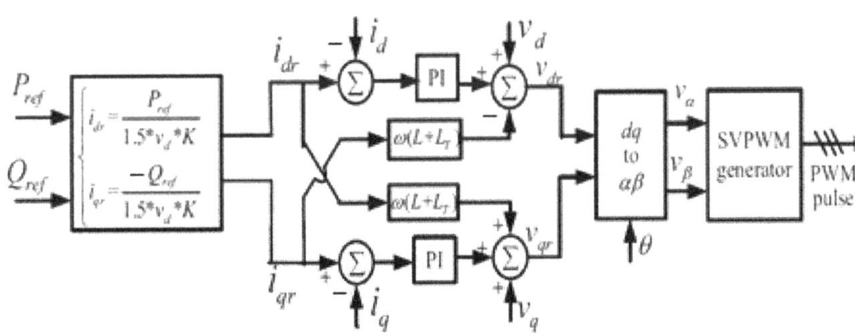

Figure 5. Block diagram of the P-Q Control structure [22]. Reprinted from [22] with permission (License 978-1-4799-7720-8/14) from U.S. Department of Energy Office of Scientific and Technical Information [22].

(b) V-F Control

In the islanded mode operation of PCS, it is controlled as the main power source to supply constant voltage and frequency (V-F Control). To obtain constant V-F Control, a closed loop voltage control structure was adopted. The closed loop voltage control equation using a PI controller is described below [22]:

$$\begin{cases} v_{dr} = k_p \left(1 + \frac{1}{T_i s}\right) * \left(v_{dref} - v_d\right) \\ v_{qr} = k_p \left(1 + \frac{1}{T_i s}\right) * \left(v_{qref} - v_q\right) \end{cases} \quad (4)$$

where the proportional gain is represented by k_p and the integral time constant of the voltage loop controller is represented by T_i, v_d and v_q are the voltages after the coordinate transform, from abc to dq, where these are the components of the d and q axes, respectively.

While v_{dref} and v_{qref} are the voltages after the coordinate transform, also components of the *d* and *q* axes, but these represent the component of the reference voltage (v_{aref}, v_{bref} and v_{cref}). Figure 6 illustrates the block diagram of the V-F Control which is equivalent to Equation (4).

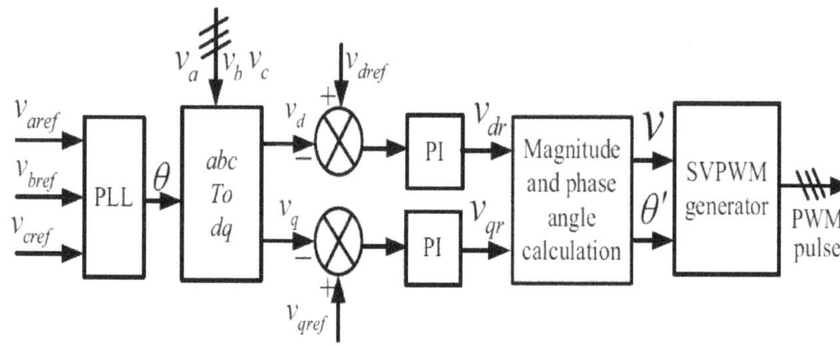

Figure 6. Block diagram of the V-F Control structure [22]. Reprinted from [22] with permission (License 978-1-4799-7720-8/14) from U.S. Department of Energy Office of Scientific and Technical Information [22].

3.6. Fire Detection and Suppression System

The fire detection and suppression system are a set of components used to ensure the safety of the place where it is installed, as well as that of people who transit internally and in the vicinity of the BESS. For this, the system is configured to have quick and efficient responses, ensuring agility to contain/extinguish the fire, as well as ensuring the evacuation of individuals.

The minimum elements that the fire detection and suppression system must include are:

- Control panel or alarm center: equipment responsible for interconnecting all the elements of the system; in other words, it receives and sends warning signals and activation of fire protection devices.
- Sensors/detectors: precision devices that evaluate the conditions of the place where it is installed. The main ones are:
    - Smoke detector;
    - Temperature detector;
    - H2 and H2S detector if the technology is lead.
- Audible alarm: device responsible for emitting audible signals when fire is detected.
- Emergency lights: it is a type of visual alarm, which helps individuals to find the exit from the place.
- Signage: the presence of signs, stickers, and other visual alarms is required, which inform and assist in directing the emergency exit(s).

3.7. HVAC

The HVAC system regards the basic functions of the climatization system, allowing the environment to be in the right conditions for safe and efficient operation. Among the components that make up the HVAC are:

- Heating (H): function of keeping the place at the correct system operating temperature for regions that have low temperature days. Furthermore, it is used to maintain the relative humidity of the air;
- Ventilation (V): used for the renewal of oxygen and air circulation, avoiding the concentration of undesirable gases, as well as removing and/or reducing odors and impurities from the place;

- Air conditioning (AC): used to artificially cool the place, controlling the temperature, and preventing it from becoming high. In addition, this equipment usually has filters, which carry out the removal of impurities and contaminants from the air.

These systems are essential for the proper functioning of the BESS, since by controlling the temperature, leaving it close to the most efficient temperature of the components (25 °C), it increases the productivity of the system. Furthermore, it promotes oxygenation of the place, air filtration, and reduction of air pollutants and the proliferation of fungi/mold.

3.8. *UPS*

The UPS is a secondary power system, which provides emergency power to the load when the primary supply is interrupted. As opposed to generators, the UPS operates very quickly, avoiding interruptions in the power supply.

In general terms, the UPS is made up of converters and batteries, which may or may not have a bypass switch. With respect to BESS, its load is not all the components of the system, but those that must always be kept in operation, allowing that in case of failure some action can be taken.

4. Results and Discussion

In this topic, the power smoothing and power factor correction functions are analyzed, as well as the approach of results and discussions of the data obtained using the two operating modes.

4.1. *Power Smoothing Application*

The results related to power smoothing are analyzed mathematically through the Maximum Power Variation (MVP) indicator. The active power smoothing technique performance of wind turbines is numerically evaluated. The MVP indicator corresponds to the maximum power variation in the wind generation rated power within an established time interval. Energy companies and system operators widely use this indicator from different European countries to restrict wind generation fluctuations, limiting the MVP to 10% in 1 min and 10 min intervals [23,24].

In the present article, a 5 min time interval (MVP5) is used, in which the calculation is performed from the difference between the maximum and the minimum power curve values in the specified interval and, according to that, the rated power of the generation to which the BESS that is connected is obtained. For example, considering a 50 MW wind generation rated power and a power curve whose difference between the highest and lowest value is 10 MW during the 5 min interval, the MVP is 20%. Thus, the lower the MVP value, the better the power smoothing quality. In an ideal case (i.e., a constant curve), the MVP would be 0% for any evaluated interval.

The BESS operating principle is performed through the EMS control system for the power smoothing function. The generation active power on the bus where the BESS is connected is verified. When the EMS verifies the 500 kW variation in a 60 s window, the system acts, absorbing or supplying active power, depending on the current generation status, that is, increasing or decreasing. Figure 7 illustrates a BESS operation based on the power wind generation variation.

In Figure 7, when the wind generation (blue curve) decreases in an interval of 60 s, the BESS supplies active power (orange curve). When wind generation increases, the BESS absorbs active power. The negative sign of active power means that the BESS is providing power and the positive sign of power implies that BESS is absorbing power.

The BESS performance operating in power smoothing mode connected in parallel to a group of wind turbines with 50.4 MW rated active power is illustrated in Figures 8 and 9. These figures show the generation curve behavior and its smoothing. Data from different days are shown.

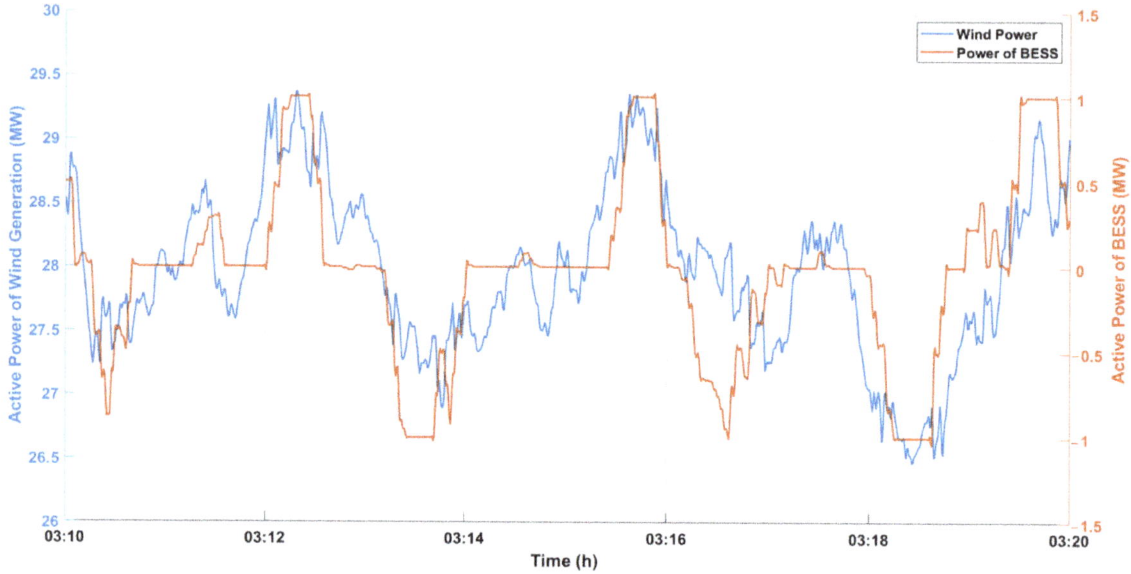

Figure 7. Real-time BESS operation based on power generation variation. The measurement was carried out on 28 July 2021.

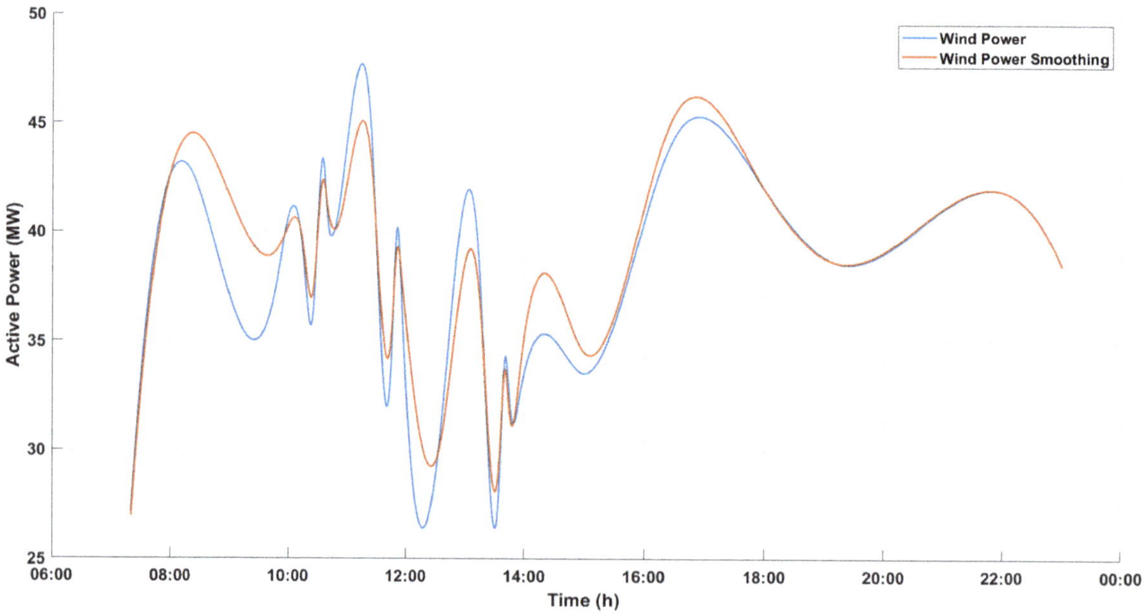

Figure 8. Active power of smoothed wind generation. Real-time measurement performed on 31 August 2021.

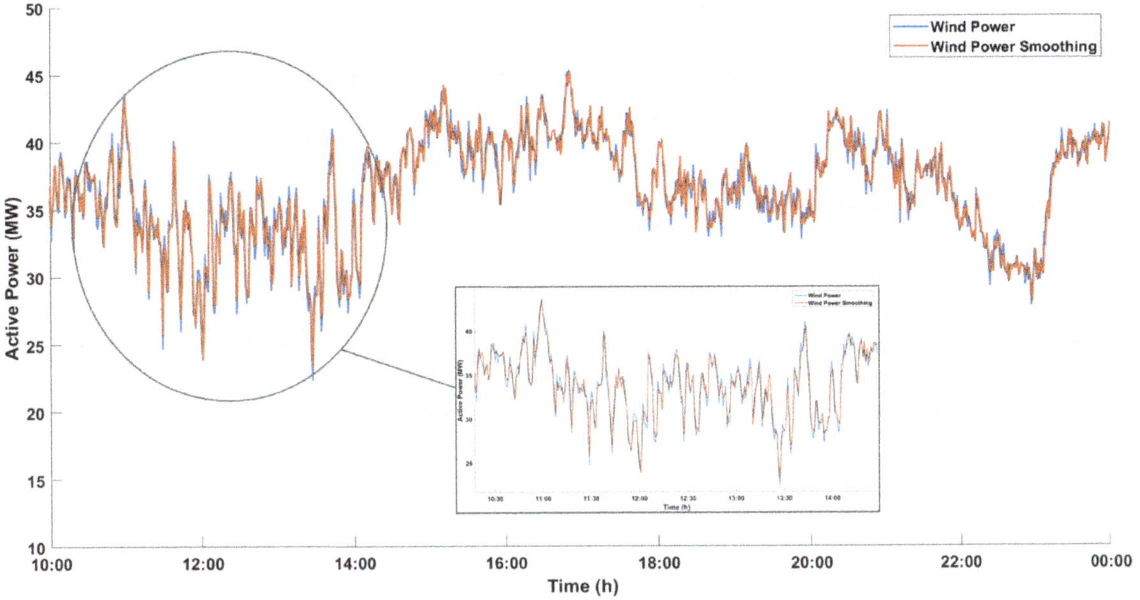

Figure 9. Active Power of Smoothed Wind Generation. Real-time measurement performed on 6 September 2021.

The blue curve indicates the power of the generation without BESS usage, while the smoothed curve is shown in orange. It is observed that the smoothed curves have smaller peaks and valleys and more attenuated curves.

The MVP index is used to evaluate wind generation power smoothing numerically. This analysis has better precision to evaluate power smoothing performed by the BESS in the present wind farm. Thus, in percentage terms, the smoothing effect results with the most significant gains points are presented in Table 1 (measurement in Figure 8) and Table 2 (measurement in Figure 9).

It can be seen from Table 1 that the MVP5 index, with the application of BESS, showed a considerable improvement in different time intervals. The best-obtained result for the 31 August 2021 day was an approximate 3.97% power fluctuation reduction, from 11:20 to 11:25 and 13:10 to 13:15. On the 6 September 2021 day (Table 2), the best result was a 3.97% reduction from 12:25 to 12:30. Different time intervals can be seen in Figures 8 and 9.

Table 1. Main indicators from the 31 August 2021 day.

Case	Time Interval	Maximum Power Value (MW)	Minimum Power Value (MW)	Wind Generation Rated Power (MW)	% Value
Without BESS operation	07:35 to 07:40	39.29	32.00	50.4	14.46%
	09:45 to 09:50	43.71	37.62	50.4	12.08%
	10:20 to 10:25	42.66	35.72	50.4	13.77%
	11:20 to 11:25	43.86	35.99	50.4	15.62%
	11:40 to 11:45	42.35	34.19	50.4	16.19%
	11:45 to 11:50	43.62	34.45	50.4	18.19%
	12:30 to 12:35	33.72	26.67	50.4	13.99%
	13:10 to 13:15	38.37	22.70	50.4	31.09%
	13:30 to 13:35	40.03	28.84	50.4	22.20%
	13:40 to 13:45	34.25	25.39	50.4	17.58%
	17:45 to 17:50	46.05	41.87	50.4	8.29%

Table 1. *Cont.*

Case	Time Interval	Maximum Power Value (MW)	Minimum Power Value (MW)	Wind Generation Rated Power (MW)	% Value
With BESS operation	07:35 to 07:40	38.29	32.83	50.4	10.83%
	09:45 to 09:50	43.52	38.62	50.4	9.72%
	10:20 to 10:25	42.13	36.72	50.4	10.73%
	11:20 to 11:25	42.86	36.99	50.4	11.65%
	11:40 to 11:45	41.53	35.19	50.4	12.58%
	11:45 to 11:50	42.89	35.45	50.4	14.76%
	12:30 to 12:35	32.72	26.76	50.4	11.83%
	13:10 to 13:15	37.37	23.70	50.4	27.12%
	13:30 to 13:35	39.53	29.84	50.4	19.23%
	13:40 to 13:45	33.72	26.39	50.4	14.54%
	17:45 to 17:50	45.69	42.56	50.4	6.21%

Table 2. Main indicators from the 6 September 2021 day.

Case	Time Interval	Maximum Power Value (MW)	Minimum Power Value (MW)	Wind Generation Rated Power (MW)	% Value
Without BESS operation	12:00 to 12:05	34.29	22	50.4	24.38%
	12:05 to 12:10	39.43	26.87	50.4	24.92%
	12:25 to 12:30	38.09	26.73	50.4	22.54%
	13:25 to 13:30	31.22	21.76	50.4	18.77%
	13:30 to 13:35	35.22	25.84	50.4	18.61%
	13:55 to 14:00	36.81	25.64	50.4	22.16%
	14:00 to 14:05	37.23	26.87	50.4	20.56%
	15:30 to 15:35	42.31	36.86	50.4	10.81%
	16:15 to 16:20	44.44	36.84	50.4	15.08%
	19:25 to 19:30	37.72	32.07	50.4	11.21%
	20:40 to 20:45	41.95	35.86	50.4	12.08%
With BESS operation	12:00 to 12:05	33.29	23.00	50.4	20.42%
	12:05 to 12:10	39.43	27.87	50.4	22.94%
	12:25 to 12:30	37.09	27.73	50.4	18.57%
	13:25 to 13:30	30.46	22.76	50.4	15.28%
	13:30 to 13:35	34.22	26.40	50.4	15.52%
	13:55 to 14:00	35.81	26.20	50.4	19.07%
	14:00 to 14:05	36.51	27.87	50.4	17.14%
	15:30 to 15:35	41.64	37.86	50.4	7.50%
	16:15 to 16:20	43.71	37.84	50.4	11.65%
	19:25 to 19:30	36.72	32.72	50.4	7.94%
	20:40 to 20:45	40.95	36.75	50.4	8.33%

4.2. *Power Factor Correction Application*

The power factor is an energy utilization index whose adequate control in wind generation is significant, not only from an electrical energy point of view but also because it is monitored, in the case of Brazil, by the National Electric System Operator, and the power generator may incur fines. In this case, the BESS compensates for the excess reactive power, bringing the power factor within the regulatory limit (currently, in Brazil, the limit power factor is 0.95 in the PCC between wind generation and the transmission grid).

The EMS checks the power factor information generated by the wind turbines in the bus connected to the storage system. A power factor reduction (less than 1.00) activates BESS to operate with capacitive or inductive characteristics, depending on the wind generation power factor behavior (inductive or capacitive).

Figure 10 illustrates the BESS operation behaving with capacitive characteristics when the power factor measured at the bus is less than one (1.00). Therefore, according to the EMS programming, the system acts by injecting reactive power, trying to correct the power factor to the unit value (1.00). In Figure 10, the left scale refers to the system's reactive power (blue legend), and the right scale refers to the power factor value (orange legend) measured on the bus that connects the BESS to the group of wind turbines. It should be noted that the negative sign for reactive power means that the BESS is operating in capacitive mode and the positive sign in inductive mode.

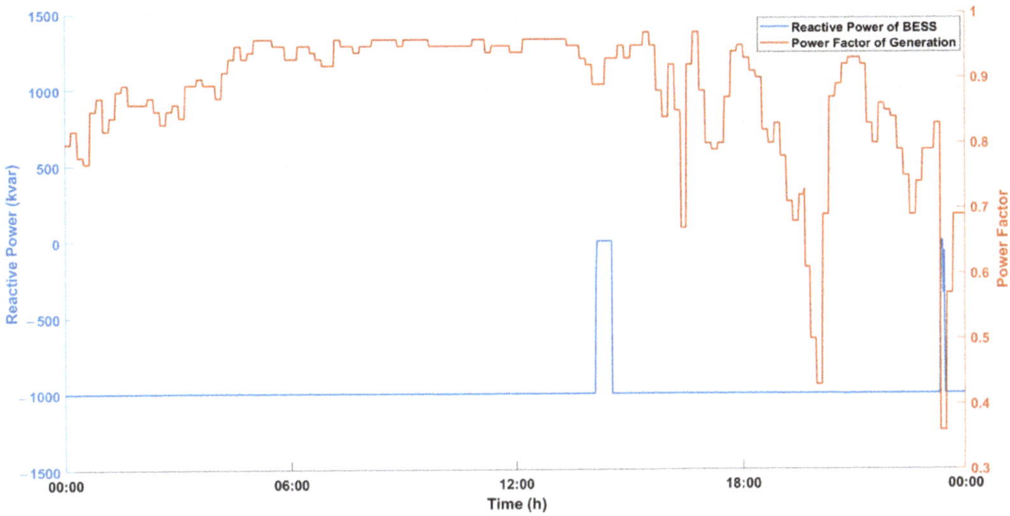

Figure 10. Real-time BESS operation in the power factor correction function. The measurement was carried out on 25 March 2021.

Figure 11 illustrates the BESS operation on 23 March 2021, where it is seen that the system remains without acting while the power factor is unity (1.00). The BESS works by compensating reactive power when there is a drop in the power factor, aiming to establish the unit value. Figure 12 illustrates the operation of BESS on 19 April 2021, where it is seen that the system operates in both capacitive and inductive modes.

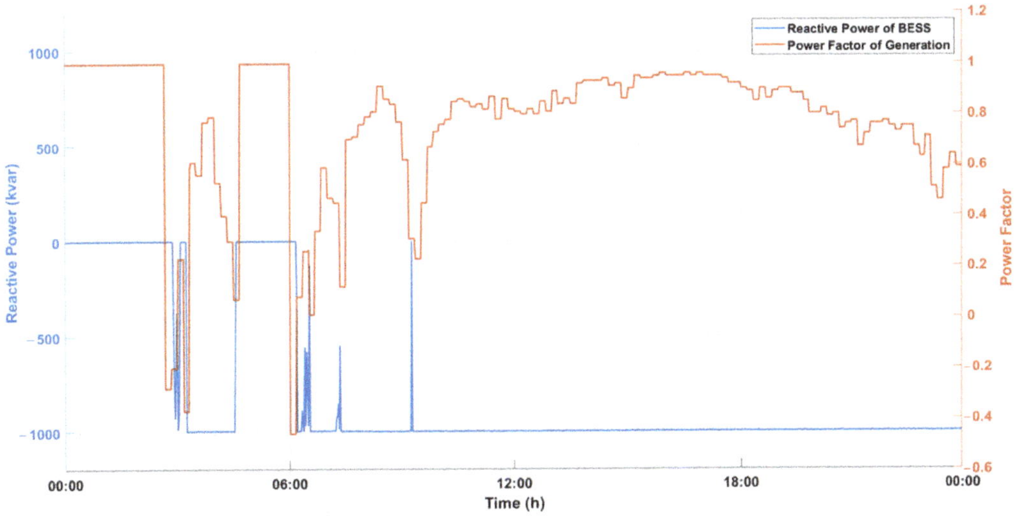

Figure 11. Real-time BESS operation in the power factor correction function. The measurement was carried out on 23 March 2021.

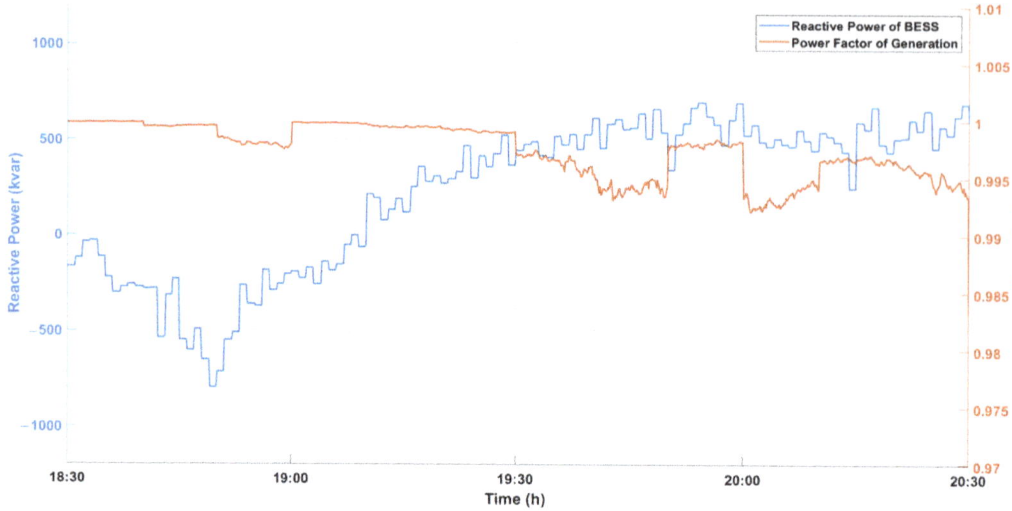

Figure 12. Real-time BESS operation in the power factor correction function. The measurement was carried out on 19 April 2021.

From the 19 April 2021 measurement data (Figure 12), it is possible to graphically characterize the resulting power factor behavior, considering the BESS performance, as illustrated in Figure 13. It can be seen from Figure 13 that the BESS corrects the power factor at different instances of time, aiming at the unit value (1.00), and thus helps to prevent the power factor from falling below 0.95.

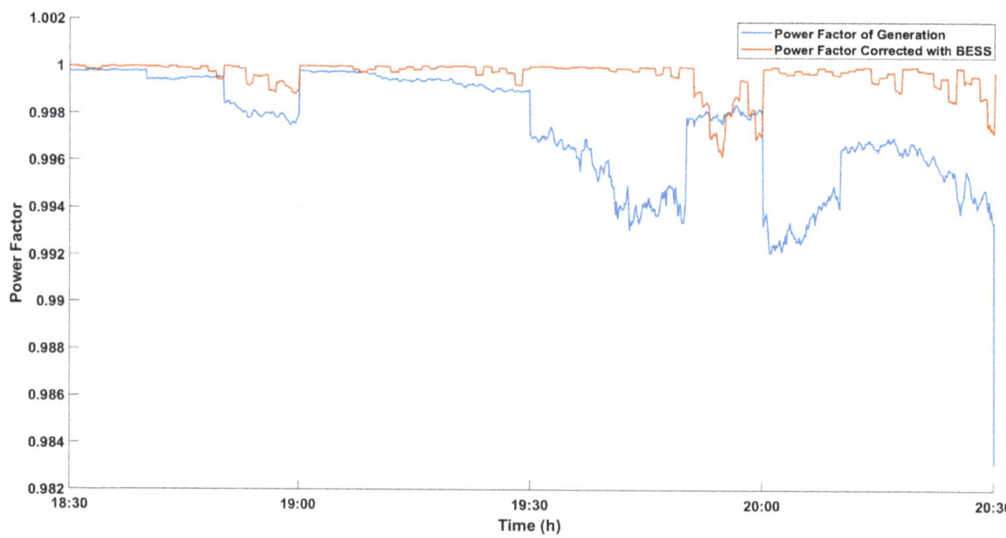

Figure 13. Power factor resulting from the application of BESS. The measurement was carried out on 19 April 2021.

5. Conclusions

From the measurement analysis, it is observed that the power smoothing function implemented in the EMS does not present problems regarding the operating logic and operating time and shows satisfactory results. It was seen that the BESS manages to smooth the wind generation power with gains of up to 3.97% (measurements recorded in Tables 1 and 2) according to the MPV5 indicator. It is noteworthy that this result is considered satisfactory since the BESS rated power is 1 MW, and it is connected to a 50.4 MW wind generation.

From the measurements, it is observed that the BESS usage in power smoothing mode contributes to reducing power fluctuations at the point that connects the power output of the wind farm and the transmission line, generating improvements in the wind farm energy quality.

The power factor correction function performance analysis implemented in the BESS EMS shows that this function does not present problems regarding its operating logic and operating time. It was seen that BESS acted by correcting the power factor whenever necessary, reducing losses to the wind farm.

It should be noted that the constant growth of wind generation should amplify the effect of power fluctuation in transmission, distribution, and microgrid systems. Thus, wind generation should increasingly impact the operation and energy quality of electrical systems. The use of a BESS operating in active power smoothing mode represents a way to circumvent this problem and enable the use of intermittent renewable energy sources.

Author Contributions: Conceptualization, N.K.L.D., A.C.M.S. and A.S.M.V.; methodology, N.K.L.D., A.C.M.S. and A.S.M.V.; software, N.K.L.D.; validation, A.S.M.V., W.d.A.S.J., G.R., C.D. and P.R.; formal analysis, N.K.L.D., A.C.M.S. and A.S.M.V.; investigation, N.K.L.D., A.C.M.S., A.M.A.M. and A.S.M.V.; resources, G.R.; data curation, N.K.L.D. and A.S.M.V.; writing—original draft preparation, N.K.L.D., A.C.M.S., A.S.M.V. and W.d.A.S.J.; writing—review and editing, N.K.L.D., A.S.M.V., P.R., A.M.A.M., J.F.C.C., Y.L. and C.D.; visualization, P.R. and G.R.; supervision, A.S.M.V., W.d.A.S.J., A.M.A.M., J.F.C.C., Y.L. and P.R.; project administration, G.R. All authors have read and agreed to the published version of the manuscript.

Funding: This research was funded by the Program of R&D of the National Electricity Regulatory Agency (ANEEL) and CPFL Energy. This work is related to the Project "PA3026—Insertion of Storage System in Multiple Configurations to Support Wind Generation".

Acknowledgments: This research was carried out by the R&D Program of the National Electric Energy Agency (ANEEL) and CPFL Energia. This work is related to the Project "PA3026 – Insertion of Storage System in Multiple Configurations to Support Wind Generation". The authors thank the R&D Program of ANEEL and CPFL Energia for all the incentives to this Research and Development.

Conflicts of Interest: The authors declare no conflict of interest.

References

1. Yaramasu, V.; Wu, B.; Sen, P.C.; Kouro, S.; Narimani, M. High-power wind energy conversion systems: State-of-the-art and emerging technologies. *Proc. IEEE* **2015**, *103*, 740–788. [CrossRef]
2. Carvalho, W.C. Fuzzy Logic and Energy Storage System for Wind Power Smoothing. Master's Thesis, São Carlos School of Engineering, University of São Paulo, São Paulo, Brazil, 2019; 160p.
3. Nasiri, M.; Milimonfared, J.; Fathi, S. Modeling, analysis, and comparison of TSR and OTC methods for MPPT and power smoothing in permanent magnet synchronous generator-based wind turbines. *Energy Convers. Manag.* **2014**, *86*, 892–900. [CrossRef]
4. Ammar, M.; Joos, G. A Short-Term Energy Storage System for Voltage Quality Improvement in Distributed Wind Power. *IEEE Trans. Energy Convers.* **2014**, *29*, 997–1007. [CrossRef]
5. Abbey, C.; Joos, G. Supercapacitor Energy Storage for Wind Energy Applications. *IEEE Trans. Ind. Appl.* **2007**, *43*, 769–776. [CrossRef]
6. Zhu, J.-H.; Pan, W.-X.; Li, X. Energy storage scheduling design on friendly grid wind power. *Sustain. Energy Technol. Assess.* **2018**, *25*, 111–118. [CrossRef]
7. Korpaas, M.; Holen, A.T.; Hildrum, R. Operation and sizing of energy storage for wind power plants in a market system. *Int. J. Electr. Power Energy Syst.* **2003**, *25*, 599–606. [CrossRef]
8. Teleke, S.; Baran, M.E.; Huang, A.Q.; Bhattacharya, S.; Anderson, L. Control Strategies for Battery Energy Storage for Wind Farm Dispatching. *IEEE Trans. Energy Convers.* **2009**, *24*, 725–732. [CrossRef]
9. Lei, M.; Yang, Z.; Wang, Y.; Xu, H.; Meng, L.; Vasquez, J.C.; Guerrero, J.M. An MPC-Based ESS Control Method for PV Power Smoothing Applications. *IEEE Trans. Power Electron.* **2017**, *33*, 2136–2144. [CrossRef]

10. Xu, G.; Xu, L.; Morrow, D.J.; Chen, N. Coordinated DC Voltage Control of Wind Turbine with Embedded Energy Storage System. *IEEE Trans. Energy Convers.* **2012**, *27*, 1036–1045. [CrossRef]
11. Lu, M.-S.; Chang, C.-L.; Lee, W.-J.; Wang, L. Combining the Wind Power Generation System with Energy Storage Equipment. *IEEE Trans. Ind. Appl.* **2009**, *45*, 2109–2115.
12. Zhao, H.; Hong, M.; Lin, W.; Loparo, K.A. Voltage and Frequency Regulation of Microgrid with Battery Energy Storage Systems. *IEEE Trans. Smart Grid* **2018**, *10*, 414–424. [CrossRef]
13. Hannan, M.; Wali, S.; Ker, P.; Rahman, M.A.; Mansor, M.; Ramachandaramurthy, V.; Muttaqi, K.; Mahlia, T.; Dong, Z. Battery energy-storage system: A review of technologies, optimization objectives, constraints, approaches, and outstanding issues. *J. Energy Storage* **2021**, *42*, 103023. [CrossRef]
14. Piątek, J.; Afyon, S.; Budnyak, T.M.; Budnyk, S.; Sipponen, M.H.; Slabon, A. Sustainable Li-Ion Batteries: Chemistry and Recycling. *Adv. Energy Mater.* **2020**, *11*, 2003456. [CrossRef]
15. Zhang, H.; Sun, C. Cost-effective iron-based aqueous redox flow batteries for large-scale energy storage application: A review. *J. Power Sources* **2021**, *493*, 229445. [CrossRef]
16. May, G.J.; Davidson, A.; Monahov, B. Lead batteries for utility energy storage: A review. *J. Energy Storage* **2018**, *15*, 145–157. [CrossRef]
17. S. N. Laboratories, U. S. Department of Energy, Office of Scientific, and Tehnical Information. DOE/EPRI Electricity Storage Handbook in Collaboration with NRECA. United States. National Nuclear Security Administration: Washington, DC, USA, 2015. Available online: https://books.google.com.br/books?id=XmzhnQAACAAJ (accessed on 20 April 2022).
18. Costa, T.S.; de Fátima Rosolem, M.; Silva, J.L.D.S.; Villalva, M.G. An Overview of Electrochemical Batteries for ESS Applied to PV Systems Connected to the Grid. In *Proceedings of the 2021 14th IEEE International Conference on Industry Applications (INDUSCON), São Paulo, Brazil, 15–17 August 2021*; IEEE: New York, NY, USA, 2021; pp. 1392–1399.
19. Chatrung, N. Battery energy storage system (BESS) and development of grid-scale BESS in egat. In *Proceedings of the 2019 IEEE PES GTD Grand International Conference and Exposition Asia (GTD Asia), Bangkok, Thailand, 19–23 March 2019*; IEEE: New York, NY, USA, 2019; pp. 589–593.
20. Chen, A.; Sen, P.K. Advancement in battery technology: A state-of-the-art review. In *Proceedings of the 2016 IEEE Industry Applications Society Annual Meeting, Portland, OR, USA, 2–6 October 2016*; IEEE: New York, NY, USA, 2016; pp. 1–10.
21. Lawder, M.T.; Suthar, B.; Northrop, P.W.C.; DE, S.; Hoff, C.M.; Leitermann, O.; Crow, M.L.; Santhanagopalan, S.; Subramanian, V.R. Battery Energy Storage System (BESS) and Battery Management System (BMS) for Grid-Scale Applications. *Proc. IEEE* **2014**, *102*, 1014–1030. [CrossRef]
22. Tianwen, Z.; Laijun, C.; Shengwei, M. Control strategy and application of power converter system in battery energy storage system. In Proceedings of the IEEE PES Innovative Smart Grid Technologies, Europe, Istanbul, Turkey, 12–15 October 2014; IEEE: New York, NY, USA, 2014; pp. 1–6.
23. Jiang, Q.; Hong, H. Wavelet-Based Capacity Configuration and Coordinated Control of Hybrid Energy Storage System for Smoothing Out Wind Power Fluctuations. *IEEE Trans. Power Syst.* **2012**, *28*, 1363–1372. [CrossRef]
24. Jannati, M.; Hosseinian, S.H.; Vahidi, B.; Li, G.-J. A survey on energy storage resources configurations in order to propose an optimum configuration for smoothing fluctuations of future large wind power plants. *Renew. Sustain. Energy Rev.* **2014**, *29*, 158–172. [CrossRef]

applied sciences

MDPI

Article

Evaluating and Analyzing the Degradation of a Battery Energy Storage System Based on Frequency Regulation Strategies

Chen-Han Wu, Jia-Zhang Jhan, Chih-Han Ko and Cheng-Chien Kuo *

Department of Electrical Engineering, National Taiwan University of Science and Technology, Taipei 10607, Taiwan; d10907002@mail.ntust.edu.tw (C.-H.W.); d11007004@mail.ntust.edu.tw (J.-Z.J.); wing0990242@gmail.com (C.-H.K.)
* Correspondence: cckuo@mail.ntust.edu.tw

Abstract: The capacity aging of lithium-ion energy storage systems is inevitable under long-term use. It has been found in the literature that the aging performance is closely related to battery usage and the current aging state. It follows that different frequency regulation services, C-rates, and maintaining levels of SOC during operation will produce different battery aging rates. In the simulations, the researchers used real frequency data to generate SOC curves based on the Taiwan frequency regulation services under different C-rates and different battery SOC target levels. Then, the aging formula of lithium iron batteries ($LiFePO_4$ battery, LFP battery) and the proposed improved rainflow counting algorithm were used. The capacity aging situation and economy under different usage scenarios were analyzed. The simulation results showed that using a high C-rate and a low SOC level had a higher net profit, and the income of dReg was more than that of sReg. The SOC of BESS has an important impact on the life cycle. Keeping the SOC at a lower level will help prolong the life cycle and increase the net income. In dReg0.5, maintaining the SOC at 30% would yield 8.5% more lifetimes than 50%, 20.6% more lifetimes than 70%.

Keywords: battery energy storage system (BESS); frequency regulation service; battery degradation; automatic frequency control

Citation: Wu, C.-H.; Jhan, J.-Z.; Ko, C.-H.; Kuo, C.-C. Evaluating and Analyzing the Degradation of a Battery Energy Storage System Based on Frequency Regulation Strategies. *Appl. Sci.* **2022**, *12*, 6111. https://doi.org/10.3390/app12126111

Academic Editors: Luis Hernández-Callejo, Jesús Armando Aguilar Jiménez and Carlos Meza Benavides

Received: 19 May 2022
Accepted: 13 June 2022
Published: 16 June 2022

1. Introduction

In Taiwan, frequency regulation services can be roughly divided into two categories: dynamic regulation (dReg) and static regulation (sReg). There are currently three different modes on the power trading platform, dReg0.5, dReg0.25, and sReg. Applying the three different modes mentioned, there are three different outputs under the same grid frequency. Under long-term operation, the three different modes will obtain completely different state of charge (SOC) battery curves. In the study of [1], the authors said that the higher the SOC, the faster the aging of the battery, which is consistent with the experimental results of [2].Therefore, the aging behavior of the battery energy storage system(BESS) has a great relationship with the SOC curves. It follows that the aging speed of the batteries following the different frequency regulation modes will have completely different aging behavior.

The global energy storage ancillary services market has grown rapidly in recent years, with few large-scale battery storage systems operating and reaching the end of their life cycle so far. Thus, the effect of different ancillary services and control methods on the aging of the BESS is still unknown. Market managers and market participants need a more accurate economic analysis to formulate better market structures or make better bidding strategies [3]. Researchers in [4] evaluated the optimal battery capacity configuration for the frequency regulation market in Germany but not the different frequency regulation services. In [5], the frequency support technoeconomic analysis of energy storage working in conjunction with wind power plants but not pure energy storage plants was evaluated. For this reason, this study proposes to evaluate the aging situation of BESS which is used with different ancillary services and control methods. In this way, this research can help market managers formulate reasonable bidding prices and can help market participants make the best investment decisions.

The aging speed of a battery affects the profit of the system operators due to differences in battery life. The purpose of this paper is to establish a battery aging model based on the SOC curves simulated by different frequency modulation modes and the ratio of different rated capacity to total battery energy, find out its aging characteristics, and evaluate battery aging in each situation.

2. BESS and Frequency Regulation Service

2.1. BESS

BESS includes batteries, battery management systems (BMS), power conditioning systems (PCS), and energy management systems (EMS). Applications of lithium-ion batteries are very common nowadays due to their several advantages, such as high energy and power densities and longer lifetime than that of other technologies [6]. Specifically, the lithium ternary battery ($LiNiMnCoO_2$ battery, NMC battery) and the lithium iron phosphate battery ($LiFePO_4$ battery, LFP battery) are in the mainstream of MW-level battery energy storage systems.

The main applications of BESS can be roughly classified into five categories: large-scale energy services, ancillary services, services for transmission and distribution, client services, and renewable energy integration. Specifically, renewable energy smoothing, frequency regulation services, voltage regulation services, peak-shaving, and load-shifting are some of the most common applications in modern life [7,8]. This study will focus on frequency regulation services, which is going to be discussed in detail in the following section.

2.2. Frequency Regulation Service

Presently, the frequency regulation services that the Taiwan Power Company (TPC) promotes include two different types, namely, dReg and sReg; each service has its corresponding efficiency offer price [9]. The efficiency offer prices corresponding to the frequency regulation are shown in Table 1.

Table 1. Efficiency level and the price of each mode.

Efficiency Level	Frequency Regulation Mode	Efficiency Offer Price ($\frac{NTD}{MWh}\cdot$Wh)
1	dReg0.25	350
2	dReg0.5, sReg	275

2.2.1. Dynamic Regulation (dReg)

dReg can dynamically follow the grid frequency and actively provide corresponding power every second based on the current grid frequency to help maintain the stability of the power system frequency. Such a service is required to respond within a second.

Depending on the corresponding system frequency range and operating power, it can be divided into two types: dReg0.25 and dReg0.5. To be more specific, the operating frequency range of dReg0.25 is from 59.75 to 60.25 Hz, while that of dReg0.25 is from 59.5 to 60.5 Hz. Figure 1 shows the response curve of dReg, and Figure 2 shows its corresponding parameters [9]. The blue area in Figure 1 is the deadband, where the BESS can charge and discharge to maintain the SOC.

2.2.2. Static Regulation (sReg)

sReg does not need to respond as fast as dReg but is required to output 100% of its rated power within ten seconds when the frequency drops to a certain value. The output should return to zero once the frequency increases and reaches a certain frequency. Furthermore, the system is not allowed to start the charging operation until the frequency is over 60 Hz. Figures 3 and 4 show the detailed specifications of sReg [9].

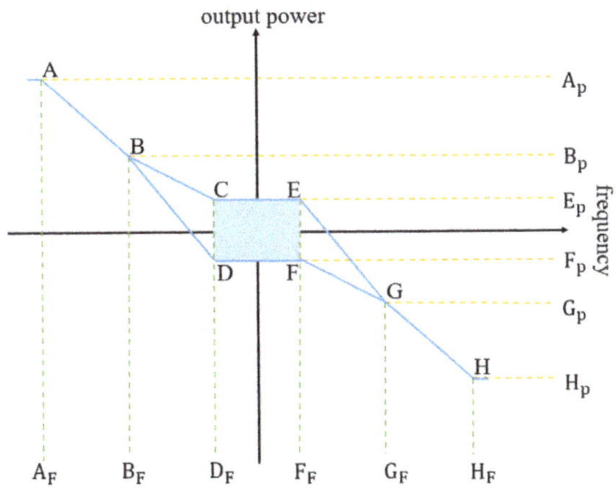

Figure 1. The response curve of the dReg mode.

	Freq.	Symbol	Power	Symbol		Freq.	Symbol	Power	Symbol
dReg0.25	59.75 Hz	A_F	100%	A_P	dReg0.5	59.50 Hz	A_F	100%	A_P
	59.86 Hz	B_F	52%	B_P		59.75 Hz	B_F	48%	B_P
	59.98 Hz	D_F	9 - −9%	E_P/F_P		59.98 Hz	D_F	9 - −9%	E_P/F_P
	60.02 Hz	F	−9 - 9%	F_P/E_P		60.02 Hz	F	−9 - 9%	F_P/E_P
	60.14 Hz	G_F	−52%	G_P		60.25 Hz	G_F	−48%	G_P
	60.25 Hz	H_F	−100%	H_P		60.50 Hz	H_F	−100%	H_P

Figure 2. The response table of the dReg mode.

	Freq.	Symbol	Power	Symbol
sReg	59.88 Hz	C_F	100%	B_P
	59.98 Hz	D_F	0%	-
	60.00 Hz	-	0 − −9%	E_P
	60.25 Hz	F_F	0 − −100%	F_P

Figure 3. The response table of the sReg mode.

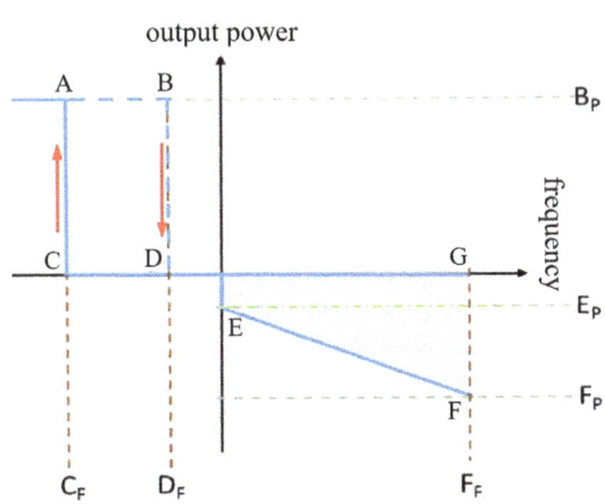

Figure 4. The response curve of the sReg mode.

3. Aging of Lithium-Ion Batteries

3.1. Aging Mechanisms of Lithium-Ion Batteries

The aging mechanisms of lithium-ion batteries are very complicated. With use, various aging mechanisms lead to the loss of active materials and the increase of internal resistance. Tracing back to the sources, the interaction between the electrolyte, the anode, and the cathode, as well as the degradation of the electrolyte itself, are where the aging phenomenon mainly comes from.

Since the aging behavior of batteries has a lot to do with the current degradation status, even if the operating conditions remain the same throughout, they will be completely different at every moment. Based on some research, it can be classified into two categories according to the features of use: calendar aging and cycling aging [10–12], and the aging formula of the lithium iron phosphate battery (LFP battery) quoted in this article can refer to Equations (1) and (2) [13].

3.2. Calendar Aging

Calendar aging refers to the degradation behaviors of capacity and power capability with saving under the same SOC level for a long period. Moreover, calendar aging is closely related to environmental conditions such as storage temperature; that is, the speed of aging varies according to the conditions mentioned above.

According to the study in [13], under 25 degrees Celsius, the capacity fade can be described in Equation (1).

$$C_{f\_cal} = 0.1723 \cdot e^{0.007388 \cdot SOC} \cdot t^{0.8} \tag{1}$$

where C_{f_cal} represents the percentage of capacity fade, SOC refers to the storage SOC level, and t represents the storage period (month).

Figure 5 shows the calendar aging curve drawn by Equation (1). From this figure, it can be seen that the curve is nonlinear and tends to be flat as the storage time increases. In other words, the higher the storage SOC level is, the faster the aging rate will be.

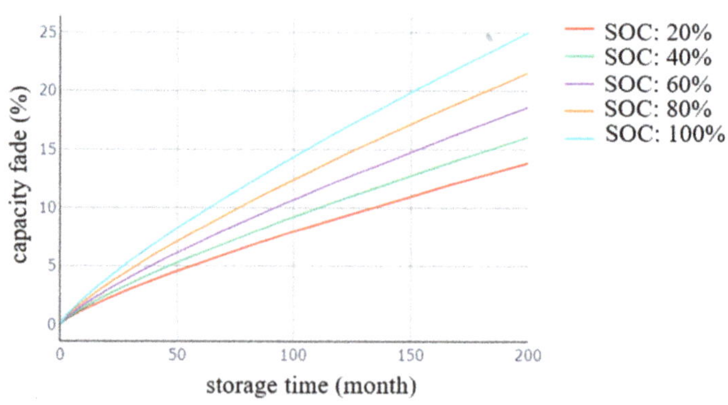

Figure 5. Calendar aging curves.

3.3. Cycling Aging

Cyclic aging refers to the aging caused by the charging and discharging cycle of the battery, and its aging performance has a great relationship with cycle depth (CD). Cycle depth indicates the change in charge state during a cycle and is related to the amount of charge gained or lost during the charge and discharge process.

Based on the research in [13], the relationship between capacity fade and the operation conditions, including cycle depth, average SOC, and the number of cycles (NC), under the conditions of 25 °C storage temperature, is shown in Equation (2):

$$C_{f\_cyc} = 0.021 \cdot e^{-0.01943 \cdot SOC} \cdot cd^{0.7162} \cdot nc^{0.5} \quad (2)$$

where C_{f_cyc} represents the percentage of capacity fade caused by cycling aging, SOC represents the average SOC level during the cycle, and cd and nc represent the cycle depth and the number of cycles, respectively. Figure 6 shows the calendar aging curve drawn by Equation (2).

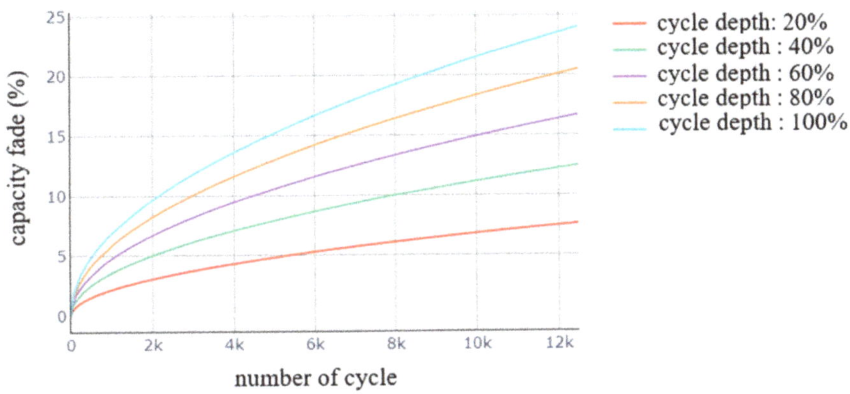

Figure 6. Cycling aging curves.

Compared with calendar aging, the rising slope of cycle aging curve is steeper when the number of cycles is low but tends to be flat when the number of cycles increases. In some cases, the batteries with a deeper cycle depth will have a faster aging rate than the ones with a shallower cycle depth. Sometimes, the difference can be several-fold.

4. Aging Model Establishment

4.1. Introduction of the Aging Model

Since the aging of a battery is a nonlinear problem, it is difficult to apply the traditional mathematical programming method. This research proposes the use of the rain flow algorithm, which has been widely used in the technical and economic analysis of BESS [2,14,15] to solve this nonlinear problem. It extracts the aging features from SOC curves. For details, please refer to Section 4.2.

The aging model of lithium-ion batteries in this study can be roughly divided into two parts: the extraction of aging features and the superposition of aging quantity.

In cycle aging, the feature values of each cycle, such as cycle depth, cycle average SOC, and cycle times, were extracted and recorded as cycle aging events. Meanwhile, in calendar aging, the continuous SOC and its duration were recorded as calendar aging events.

The behavior of battery aging has a lot to do with the current aging state of the batteries; thus, even if the operating conditions remain the same, different aging results will be obtained corresponding to the state of aging. Therefore, it is vital to arrange the aging event sequence in the correct order.

After the extraction, the aging features should be quantified and superpositioned. Seeing that the result from each aging event can not be superpositioned directly because the aging curves are not linear, this study proposed a method named the mapping superposition method to solve the problem.

4.2. The Rainflow Cycle Counting Method

In this study, the four-point algorithm [16] of the rainflow counting method [17] was applied to extract the cycle depth and cycle times.

In the beginning, four consecutive points in the series must be identified and specified as A_1, A_2, A_3, and A_4.Then, calculate the distances S_1, S_2, and S_3 between every two adjacent points. The schematic is shown in Figure 7.

$$S_1 = |A_1 - A_2| \tag{3}$$

$$S_2 = |A_2 - A_3| \tag{4}$$

$$S_3 = |A_3 - A_4| \tag{5}$$

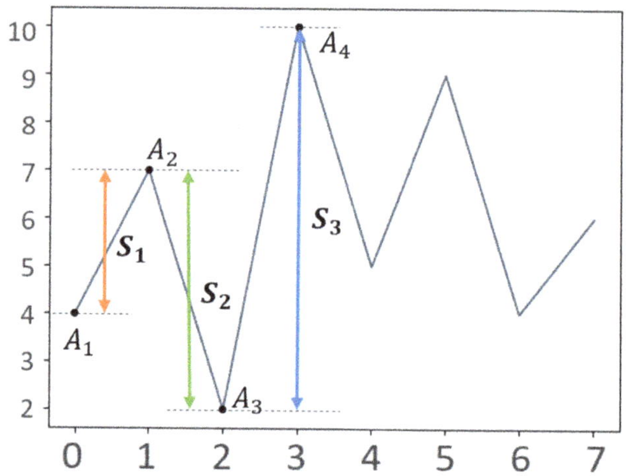

Figure 7. A schematic diagram for the four-point algorithm.

If it meets the condition:

$$S_2 \leq S_1 \text{ and } S_2 \leq S_3 \tag{6}$$

then A_2 and A_3 will be extracted from the series.

Then, the following four adjacent points will be repositioned, operating again until there are no more adjacent four points in the series that meet the conditions. Finally, the remaining sequence will be copied and attached to the end of itself, and then repeated. Examples are described in Appendix A.

4.3. Aging Features' Extraction

Before the calculation, aging features must be extracted. Figure 8 shows the schematic diagram of the extraction architecture.

According to the concept of the rainflow counting method mentioned in [18], the continuous SOC curve is processed to remove unnecessary values, leaving only the relative peaks and relative valleys in the series. Therefore, the original SOC curve is filtered twice, and then the rainflow counting algorithm is used. The schematic is shown in Figure 9.

The original input SOC data has a resolution of 0.5%. Through the first filtering, we can separate the continuous parts of the SOC value from the original SOC sequence and record it as calendar aging events, including the placing time and storage SOC. Since the values between the relative maximums and relative minimums are not meaningful during the rainflow cycle counting method, these values should be removed from the sequence first in the following second filtering to generate a new one with only the relative maximums and minimums. In the final stage, cycling aging features such as cycle depth, the number of cycles, and the average SOC of the cycle will be identified through the rainflow counting algorithm and recorded as cycling aging events in sequence.

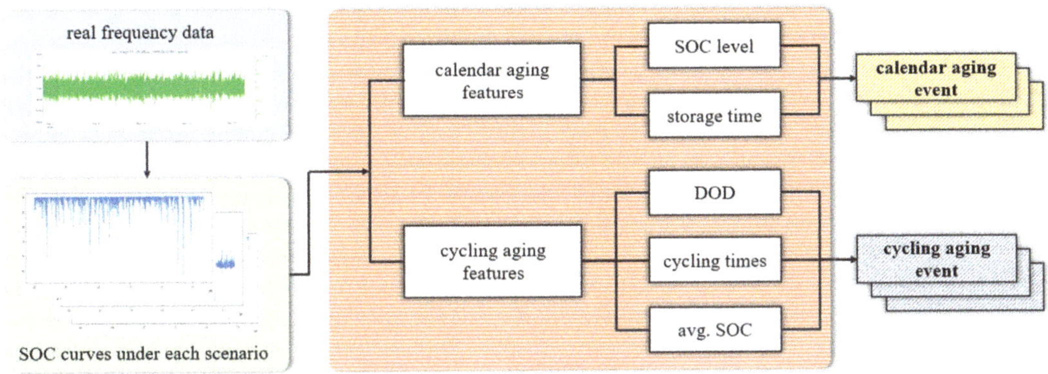

Figure 8. A schematic diagram for aging feature extraction.

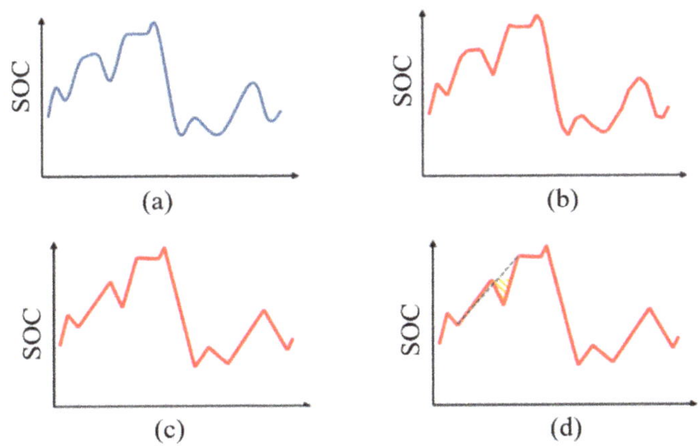

Figure 9. A schematic diagram of the applied rainflow counting method of SOC. (**a**) Raw SOC curve. (**b**) Curve After filtering continuous values. (**c**) Only the relative peaks and valleys are left, and the values between the peaks and valleys are removed. (**d**) Start rainflow cycle counting method.

4.4. Superposition

Because the aging behavior of batteries is nonlinear and is affected by more than one variable, the capacity decay caused by aging cannot be superposed directly. Such as the example shown in Figure 10, we can regard the aging event that "maintains the SOC-level at 50% for three months" as three identical continuous aging events that "maintains the SOC-level at 50% for one month". If we apply the direct superposition method on the latter, which means to superpose the capacity decay by directly adding them together, we will obtain a result that has a considerable disparity from the result of the former.

Seeing the incompleteness of the direct superposition method, another method, named the mapping superposition method, is proposed in this study, as shown in Figure 11. In other words, before the superposition, the equivalent quantity of aging features should be calculated according to its former state of the amount capacity fade.

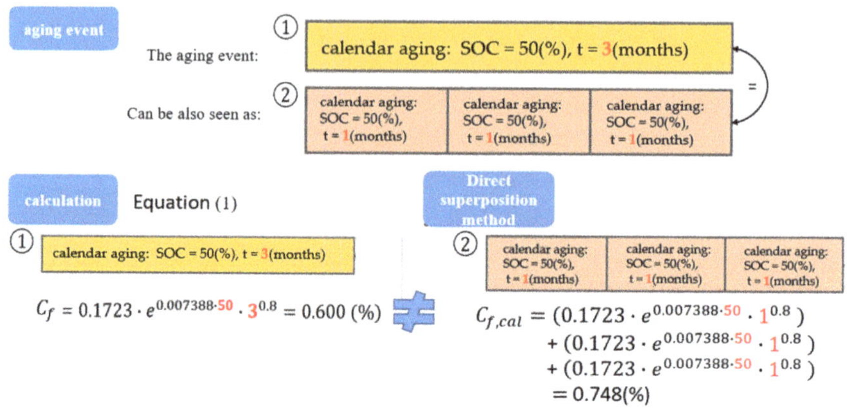

Figure 10. A schematic diagram for direct superposition.

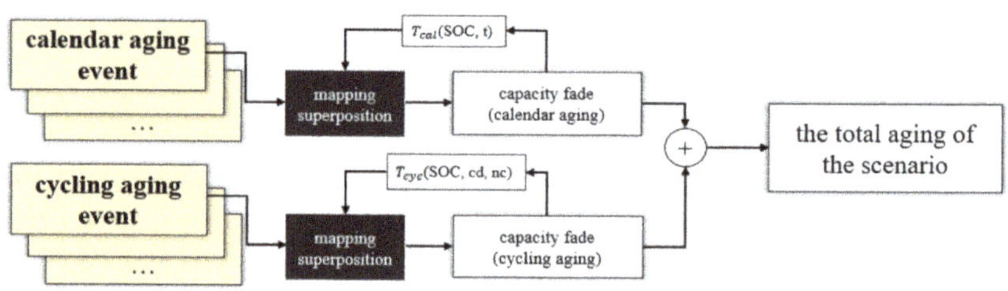

Figure 11. A schematic diagram for superposition.

In calendar aging, the former decay quantity must be "mapped" to the curve corresponding to the SOC level of the aging event to be superposed, so as to obtain the equivalent placing time. The equivalent placing time plus the placing time of this aging event is the equivalent total placing time, which can be used to calculate the aging quantity further.

The relevant calculation formula can be derived from Equation (1):

$$t_{eq} = 9.00831 \cdot {C_{f,cal}}^{1.25} \cdot e^{-0.009235 \cdot SOC} \tag{7}$$

$$C'_{f,cal} = 0.1723 \cdot e^{0.007388 \cdot SOC} \cdot (t_{eq} + \Delta t)^{0.8} \tag{8}$$

In Equation (7), t_{eq} represents the equivalent placing time, $C_{f,cal}$ represents the former calendar aging amount, and SOC represents the storage SOC level of this aging event. According to this formula, the equivalent placing time t_{eq} under the conditions of the aging event can be derived. Meanwhile, in Equation (8), $C'_{f,cal}$ and Δt represent the final calendar aging capacity fade after superposing and the placing time in the aging event, respectively.

On the other hand, in the case of cycling aging, the former aging amount should be "mapped" to the curve corresponding to the conditions including cycle depth and the average SOC of the aging event to be superposed, so that the equivalent number of cycles can be derived. The equivalent number of cycles plus the number of cycles of this aging event is the equivalent total number of cycles, which can be used to calculate the aging amount further.

$$nc_{eq} = 2267.573696 \cdot {C_{f,cyc}}^{2} \cdot e^{0.03886 \cdot SOC} \cdot cd^{-1.4324} \tag{9}$$

$$C'_{f,cyc} = 0.021 \cdot e^{-0.01943 \cdot SOC} \cdot cd^{0.7162} \cdot (nc_{eq} + \Delta nc)^{0.5} \tag{10}$$

Similarly, in Equation (9), nc_{eq}, $C_{f,cyc}$, cd, and SOC represent the equivalent number of cycles, the former cycling aging amount, the cycle depth, and the average SOC of this aging event, respectively. In Equation (10), $C'_{f,cyc}$ stands for the capacity fade after superposing, while Δnc refers to the number of cycles of the aging event.

Applying the mapping superposition method, a result closer to the theoretical values than that when using the direct superposition method can be obtained, as shown in Figure 12.

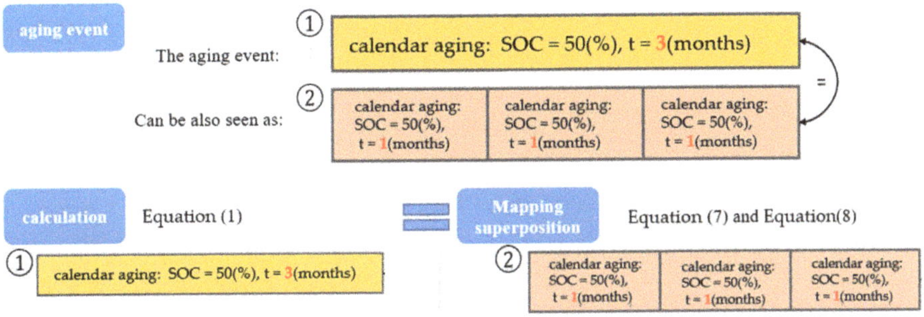

Figure 12. A schematic diagram for mapping superposition.

5. Result of Simulation

5.1. Simulation Scenario

Under the same frequency, different frequency regulation modes will have their required output so that their corresponding SOC curves can be generated. Then, the different aging features can be extracted and recorded from each by implementing the aging model.

Based on the aging results of dReg0.25, dReg0.5, and sReg under different BESS system parameters, including the rated power and the storage capacity, this study will iterate them repeatedly in three months until the capacity fade reaches 20%, i.e., the end-of-life (EOL) criteria. This paper uses SOC curves under nine different scenarios of different modes and different BESS system parameters to simulate the aging results, then discuss the performances of each. Table 2 shows each scenarios.The real frequency data of Taiwan in three different seasons, December 2019, March 2020, and May 2020, were used as input data for every scenario. The frequency distribution and features of the SOC curves are described in Appendix B.

Table 2. Specification of simulation scenarios.

Scenario	Mode	Spec. of BESS	SOC Level Target
1	dReg0.5	5 MW/6.25 MWh	70%
2	dReg0.5	5 MW/6.25 MWh	50%
3	dReg0.5	5 MW/6.25 MWh	30%
4	dReg0.5	5 MW/3.125 MWh	50%
5	dReg0.25	5 MW/6.25 MWh	50%
6	dReg0.25	5 MW/3.125 MWh	50%
7	sReg	5 MW/6.25 MWh	90%
8	sReg	5 MW/12.5 MWh	90%
9	sReg	5 MW/6.25 MWh	70%

5.2. Result and Discussion of Each Scenario

This section compares the battery aging results under each scenario, discussing the relationship between the capacity fade behavior and the system conditions, including the frequency regulation service mode, the rated power, and the storage capacity. Table 3 shows the detailed information of the comparisons.

Table 3. Overview of comparison between scenarios.

	Battery Life Comparison	Scenarios for Comparison
A	dReg0.5 with different SOC level targets	1,2,3
B	dReg under different C-rates	2,4(dReg0.5) 5,6(dReg0.25)
C	dReg0.5 and dReg0.25under the same spec.	2,5
D	sReg under different C-rates	7,8
E	sReg under different SOC level targets	7,9

A. dReg0.5 with the same rated capacity and battery energy but a different SOC level target:

Scenario 1, 2, and 3 were all in dReg0.5 mode with a 5 MW/6.25 MWh BESS. The only difference was the SOC level targets, with 70%, 50%, and 30%, respectively. As shown in Figure 13, Scenario 1, the one that had the highest SOC level target had the shortest lifetime, as described in Section 3.2.

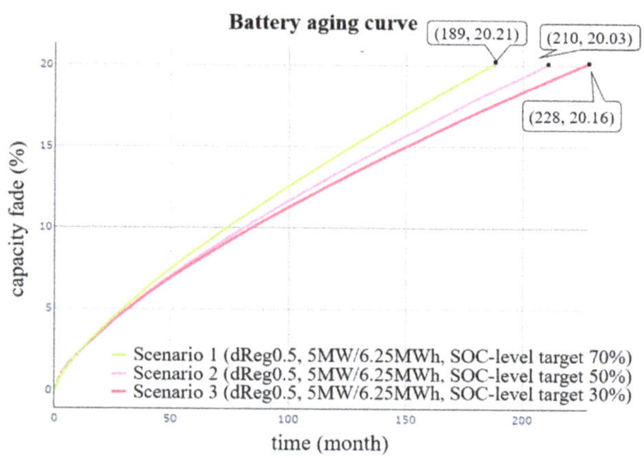

Figure 13. The result under different SOC levels (dReg0.5).

Calendar aging was the main source of capacity fade of the three scenarios, as shown in Figure 14. Since the SOC level target of Scenario 1 was the highest, its capacity fade caused by calendar aging was the highest among the three. Calendar aging was continuously accumulating, so the higher the calendar aging rate, the shorter the life it had. This was probably the reason why Scenario 1 reached its end-of-life the earliest. Maintaining the SOC at 30% yielded 8.5% more lifetimes than 50%, 20.6% more lifetimes than 70%.

B. dReg with the same rated capacity but a different battery energy:

Both Scenarios 2 and 4 used the dReg0.5 mode with the same rated power of 5 MW but had different storage energies of 6.25 and 3.125 MWh, respectively, with the results shown in Figure 15. Similarly, both Scenarios 5 and 6 used the dReg0.25 mode but also differed in the storage energy.

As shown in Figure 16, although most of the aging in the four different situations came from calendar aging, the fade caused by cycling aging in Scenario 4 was significantly higher than that in Scenario 2, and in Scenario 6 it was also higher than that in Scenario 5. Since the SOC level target was 50% for each scenario in this comparison, theoretically, the rate of calendar aging should not have been significantly different. Therefore, it was estimated that with the same SOC level target, the cycle times and depth of the scenario with a higher C-rate would be more than that of the one with a lower C-rate, and thus, it would reach EOL faster.

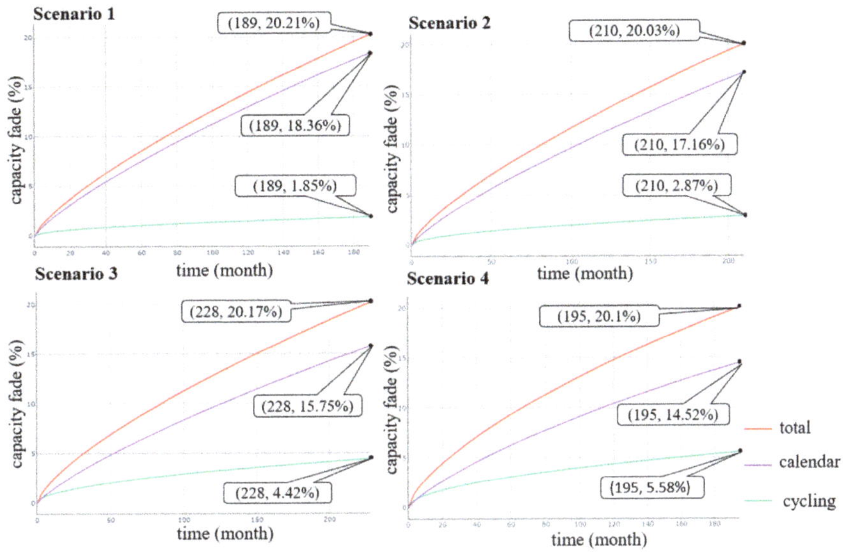

Figure 14. The aging ratio of Scenarios 1 to 4.

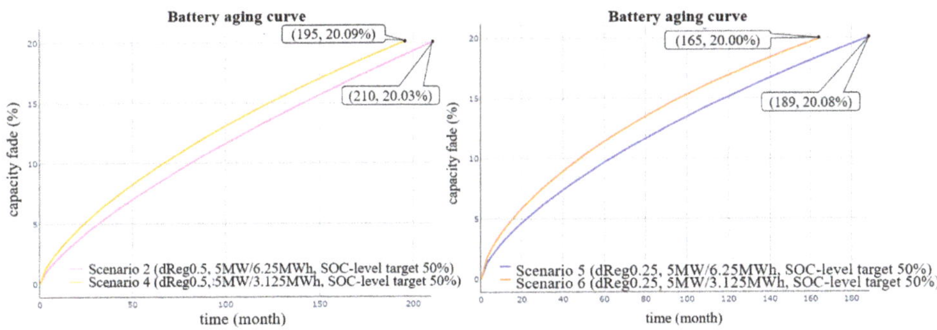

Figure 15. The result under different C-rates using dReg0.5 and dReg0.25.

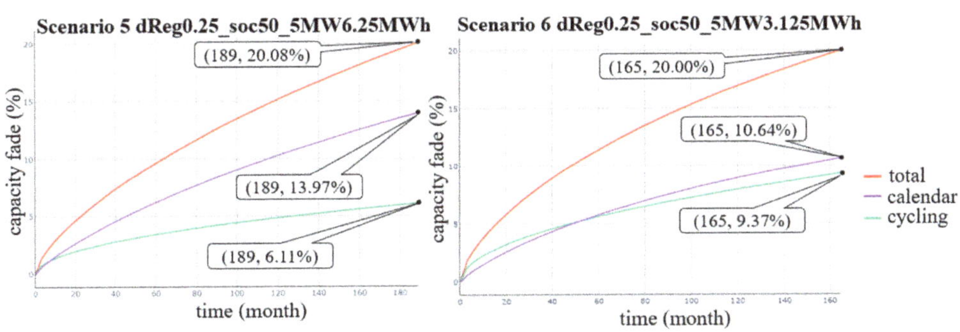

Figure 16. The aging ratio of Scenarios 5 and 6.

C. dReg0.5 and dReg0.25 with the same rated capacity and battery energy:

Scenario 2 and Scenario 5 were dReg0.5 and dReg0.25 at 5 MW/6.25 MWh, respectively, with the results shown in Figure 17. In this case, it can be estimated that for the energy storage system with the same rated capacity and the same total battery energy, the battery lifetime of dReg0.5 mode would be longer than that of dReg0.25.

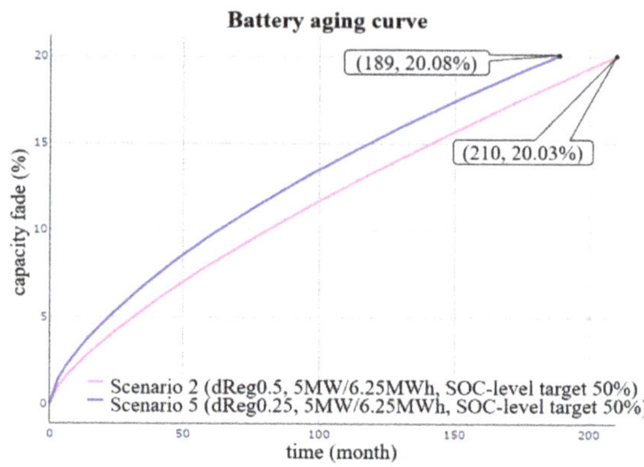

Figure 17. The result under dReg0.5 and dReg0.25.

Compared with dReg0.5, dReg0.25 often needs to carry out more output outside the deadband, so the cycle depth of dReg0.25 is higher than that of dReg0.5 under the same frequency conditions. Therefore, the capacity fade rate caused by cycling aging of dReg0.25 is much faster than that of dReg0.5, which indirectly leads to dReg0.25 reaching the end of battery life faster.

D. sReg with the same rated capacity and SOC level target but different battery energy:

As shown in Figure 18, Scenario 8 with a lower C-rate did lived longer than with a higher C-rate (Scenario 7), but the difference was almost negligible. Considering the cost of installation, the total battery energy of Scenario 8 was twice that of Scenario 7, which may not be economical.

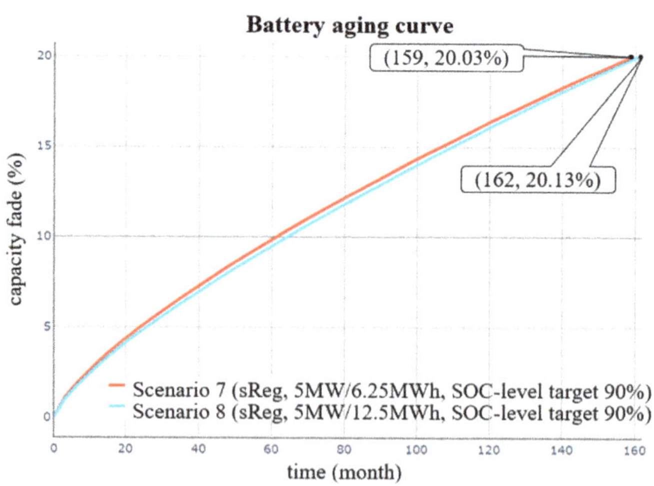

Figure 18. The result under different C-rates (sReg).

As shown in Figure 19, in the scenarios of the sReg mode with the SOC level target at 90%, the proportion of calendar aging was significantly higher than that of cycle aging, especially in the case of Scenario 8, which almost dominated the battery life. Outputting the same power, the SOC variation of the scenario with a lower C-rate will be lower than that of the one with a higher C-rate, so in the long run, the overall capacity fade caused by cycling aging will indeed be lower. However, as the proportion of cycling aging in these scenarios was not high, the effect of reduced cycle aging on the overall battery life was not significant.

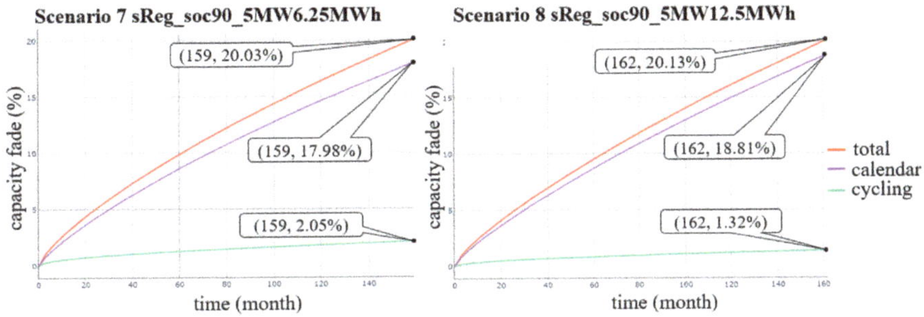

Figure 19. The aging ratio of Scenarios 7 and 8.

E. sReg with the same rated capacity and battery energy but different SOC level target:

As shown in Figure 20, Scenario 7, which had a SOC level target of 90%, reached its end-of-life after only 159 months, while Scenario 9, with a 70% SOC level target, reached EOL through 180 months of use. In conclusion, it could be preliminarily supposed that those with a lower SOC level target had a longer battery life.

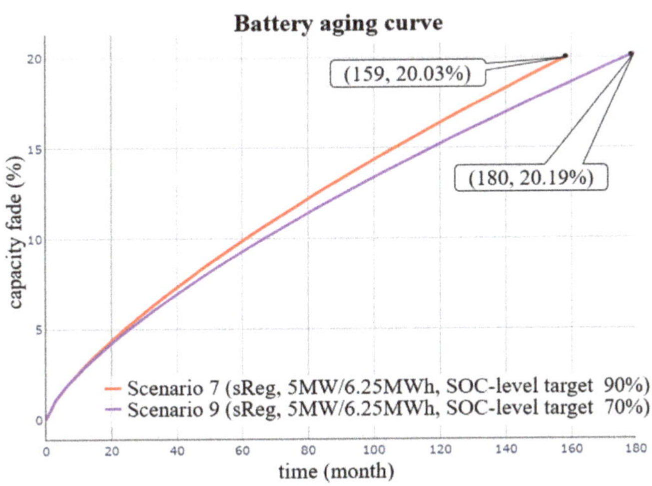

Figure 20. The result under different SOC levels (sReg).

As shown in Figure 21, in Scenario 7, because the SOC level target was higher than that of Scenario 9, the calendar aging rate was higher, which made it reach the end-of-life faster. Simply by lowering the SOC level target, the battery life of Scenario 9 would be about 20 months longer than that of Scenario 7. Therefore, if the SOC level target could be as low as possible within the allowable range, the battery could have a longer lifetime.

Table 4 shows the battery life of all the scenarios. From (Scenario 1, 2, 3), (Scenario 7, 9) it was found that lowering the SOC level significantly improved battery life. From (Scenario 3, 4), (Scenario 5, 6), and (Scenario 7, 8) it was found that the battery life with a lower capacity was shorter.

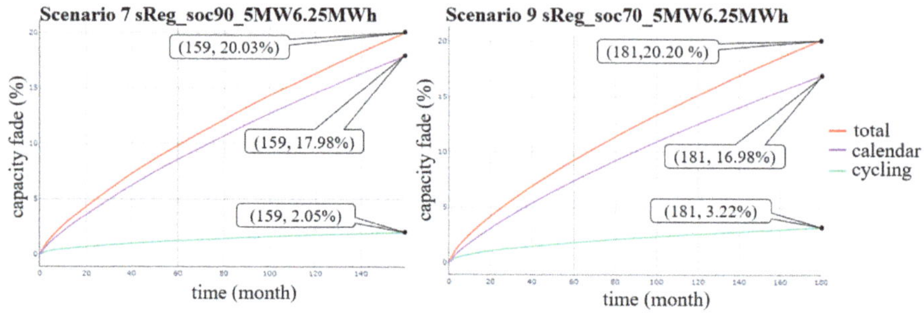

Figure 21. The aging ratio of Scenarios 7 and 9.

Table 4. Battery life overview for each scenario.

Scenario	Mode	Spec. (P_r/E_r)	SOC Level	m_{EOL}(Month)
1	dReg0.5	5 MW/6.25 MWh	70%	189
2	dReg0.5	5 MW/6.25 MWh	50%	210
3	dReg0.5	5 MW/6.25 MWh	30%	228
4	dReg0.5	5 MW/3.125 MWh	50%	195
5	dReg0.25	5 MW/6.25 MWh	50%	189
6	dReg0.25	5 MW/3.125 MWh	50%	165
7	sReg	5 MW/6.25 MWh	90%	159
8	sReg	5 MW/12.5 MWh	90%	162
9	sReg	5 MW/6.25 MWh	70%	180

5.3. Profit Estimation and Comparison

Table 5 is a rough estimate of the system construction costs based on the statistics of the data from various manufacturers [19]. C_t^{SB}, C_t^{SBOS}, and C_t^{PE} represent the cost of the storage block (SB), storage-balance of system (SBOS), and power equipment (PE), while P_r and E_r stand for the rated power and battery energy, and finally, C_t^{BESS} refers to the cost of the entire BESS.

Table 5. Establishment cost estimation of each scenario.

Scenario	P_r (MW)	E_r (MWh)	C_t^{SB} (Thousand NTD)	C_t^{SBOS} (Thousand NTD)	C_t^{PE} (Thousand NTD)	C_t^{BESS} (Thousand NTD)
1	5	6.25	31,850	7350	11,900	51,100
2	5	6.25	31,850	7350	11,900	51,100
3	5	6.25	31,850	7350	11,900	51,100
4	5	3.125	15,925	3675	11,900	31,500
5	5	6.25	31,850	7350	11,900	51,100
6	5	3.125	15,925	3675	11,900	31,500
7	5	6.25	31,850	7350	11,900	51,100
8	5	12.5	63,700	14,700	11,900	90,300
9	5	6.25	31,850	7350	11,900	51,100

This study calculated the profit earned until EOL in each situation, and the related formulas are shown in Equations (11) and (12). To simplify the calculation, there were some simple assumptions in this paper:

$$R_t^{EOL} = \sum_{m=1}^{m_{EOL}} \sum_{d=1}^{d_m} \sum_{h=1}^{24} \left(p_{cap}(m,d,\ h) + p_{eff}(m,d,\ h) \right) \times q(m,d,\ h) \tag{11}$$

$$R_{net}^{EOL} = R_t^{EOL} - C_t^{ESS} \quad (12)$$

$$q(m,d,\ h) = 1\,\{m,d,\ h\,|1\ \leq\ m\ \leq\ m_{EOL},\ 1\ \leq\ d\ \leq\ d_m,\ 1\ \leq\ h\ \leq\ 24\} \quad (13)$$

where R_t^{EOL} and R_{net}^{EOL} represent the total revenue and net profit estimated until the battery's EOL; m_{EOL} represents the number of months until the battery reaches its EOL. p_{cap}, p_{eff}, and q represent the capacity offer price described in Table 1, the efficiency offer price varying with the market bidding price every day, and the quality factor of operation, respectively.

A. dReg0.5 with the same rated capacity and battery energy but different SOC level target:

As shown in Figure 22, Scenario 3 had the highest net profit among the three. The BESS specifications of the three Scenarios were the same so that the construction cost was also the same. Scenario 3 had the longest battery life and could thus earn the most profit.

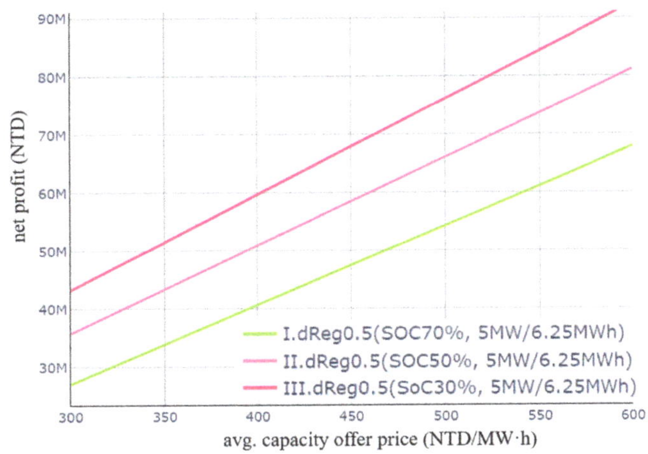

Figure 22. Net profit under different SOC levels (dReg0.5).

B. dReg with the same rated capacity but different battery energy:

As shown in Figure 23, either under the dReg0.5 or dReg0.25 mode, those with a higher C-rate had a higher net profit in the simulation results of this study. Although the battery life of those with a higher C-rate was shorter, the construction cost of their BESS was lower. Therefore, the net profit was still higher than that of those with a lower C-rate. However, these results are only the conclusion from the simulated data in this study; thus, the real results may vary due to the actual construction cost of actual sites and many other factors.

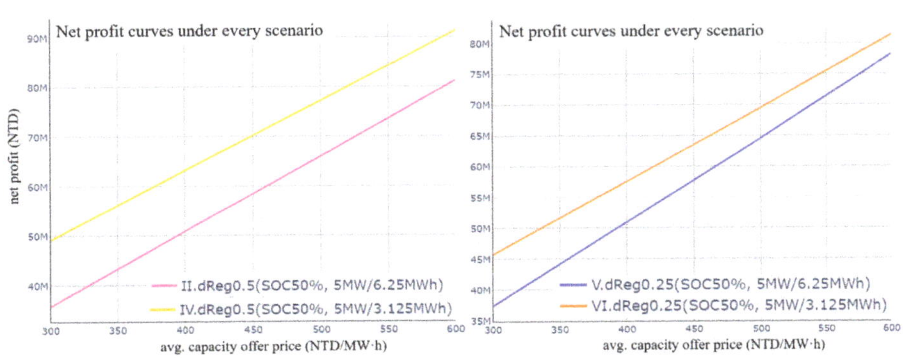

Figure 23. Net profit under different C-rates using dReg0.5 and dReg0.25.

C. dReg0.5 and dReg0.25 with the same rated capacity and battery energy:

As shown in Figure 24, the net profit difference between Scenario 2 and Scenario 5 was small. The two scenarios had the same specifications of BESS, so the construction cost was similar, but dReg0.5 (Scenario 2) had a slightly longer life. dReg0.5 and dReg0.25 have different efficiency offer prices, which is one of the main factors that determine the profits; thus, in this case, the one that had a longer battery life did not necessarily give more net profit.

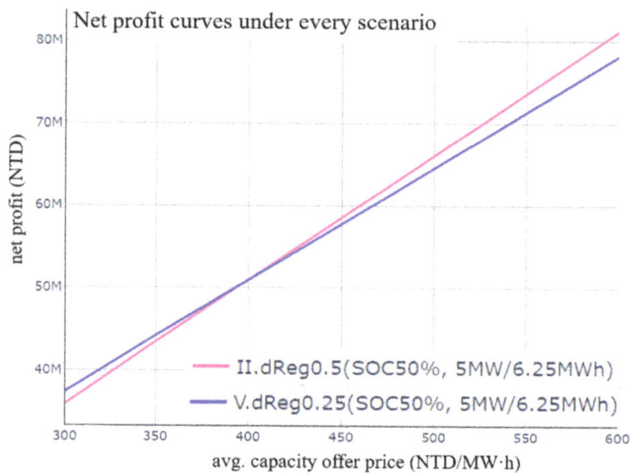

Figure 24. Net profit under dReg0.5 and dReg0.25.

D. sReg with the same rated capacity and SOC level target but different battery energy:

As shown in Figure 25, the battery life in Scenario 8 was slightly longer than that in Scenario 7, but the difference was almost ignorable, so the total revenue should have been about the same. Since the construction cost of Scenario 8 was significantly higher than that of Scenario 7, the former had a dramatically lower net revenue than the latter.

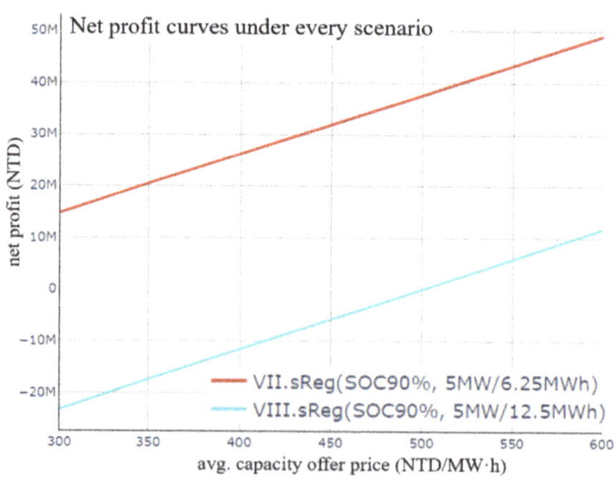

Figure 25. Net profit under different C-rates (sReg).

E. sReg with the same rated capacity and battery energy but different SOC level target:

Figure 26 shows that a lower SOC level target led to a higher net profit because of its longer battery life; thus, Scenario 9 had a higher net profit than Scenario 7 did.

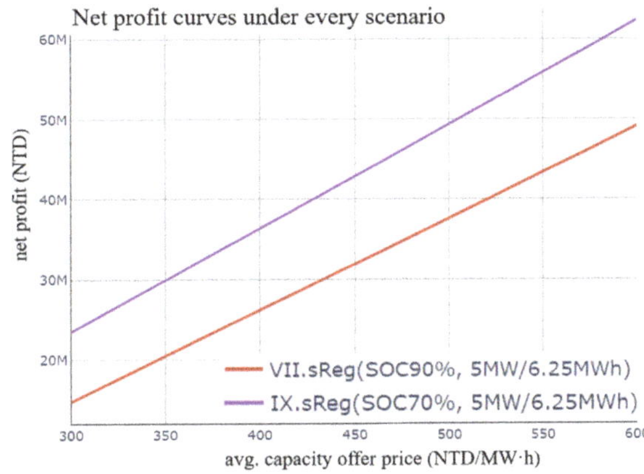

Figure 26. Net profit under different SOC levels (sReg).

For the results of this study, considering the operational requirements of dReg0.5 and dReg0.25 modes, it was not necessary to maintain SOC at a high level. On the contrary, a high SOC level target was necessary for the sReg mode to cope with the situation of long-time full-power output.

Figure 27 shows the net profit curves of all scenarios. The net profits of the scenarios under dReg0.5 were close to those under dReg0.25, but they were all generally higher than that under sReg.

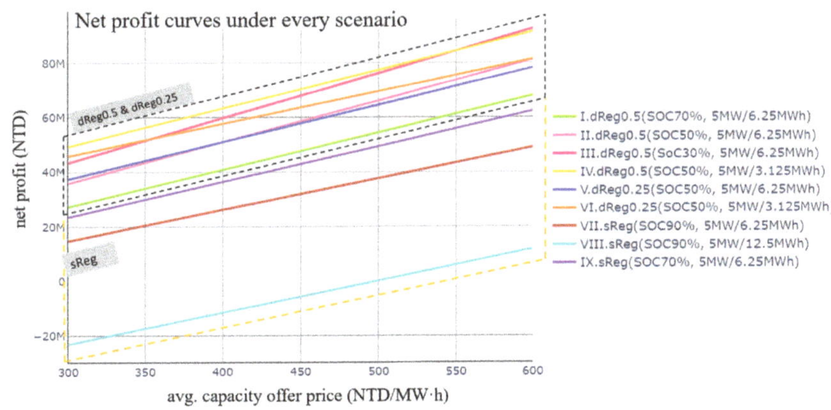

Figure 27. Net profit overview under each scenario.

From this simulation, it was estimated that for either dReg0.5 or dReg0.25 modes, the net profits of those with a higher C-rate would indeed be higher than that of those with a lower C-rate, but the effect did not seem to be that significant. Under the sReg mode, the reduction of C-rate did not bring much benefit to the battery life but led to a significant decrease in net profit due to the increased construction cost. Finally, SOC-level-targeting had a significant impact on overall net profits regardless of what the frequency regulation mode was, so evaluating to which level to set the SOC target is critical to enhancing the profits.

6. Discussion

From (Scenario 1, 2, 3), (Scenario 7, 9) it was be found that a low SOC level gave a higher net profit. Therefore, market participants can increase revenue by reducing the SOC level. From (Scenario 3, 4), (Scenario 5, 6), and (Scenario 7, 8), it was found that no matter how the price changed, the income of a high-C-rate battery was always relatively high and its construction cost was relatively low. It could be inferred that this market is more suitable for high-C-rate batteries. The net profit in Scenario 8 was below zero most of the time, which meant that if the capacity selection is wrong, it may lead to loss of money in the sReg market.

7. Conclusions

This research has shown that there are differences in battery life under different frequency regulation. The following can be observed for each type of frequency regulation:

For systems under dReg0.5, those with a lower SOC level target had a longer battery life and, thus, a higher net profit.

Then, comparing the systems in dReg0.5 and dReg0.25 modes, although the battery life of the system with a lower C-rate was longer, when considering the construction cost, it did not mean that the net profit would be more than that of the system with a higher C-rate.

Finally, for the sReg, since most of the aging comes from calendar aging, the SOC level maintained for a long time seriously affected its battery life. Furthermore, the lower the SOC level, the longer the life. In addition, the effect of a low-C-rate battery on increasing the lifetime was not significant, and the cost was relatively high, so the high C-rate scenario had higher net profit. For a BESS with the same rated power and the same total energy capacity, due to sReg usually being used in response to possible emergencies which needs BESS to maintain a higher SOC level than that in the other two services, it results in a shorter battery life and a lower net income for the batteries operating with sReg.

The limitation of the proposed method is that even for batteries of the same chemical material, batteries manufactured by different battery manufacturers will still have different aging properties, and the parameters of the model need to be adjusted according to different batteries.

Future research will focus on evaluating the optimal capacity configuration of BESS in various ancillary services. Because the BESS of more than hundreds of MW will be composed of dozens of PCS with batteries of a specific capacity, the overall C-rate can be very fine; any number between 0.25–1C is possible, and it is not limited to a specific C-rate. It depends on how many PCS and how many batteries are used. The method proposed in this study can iterate the most suitable configuration according to various regions and various auxiliary services.

Author Contributions: Conceptualization, C.-C.K.; Data curation, C.-H.K.; Funding acquisition, C.-C.K.; Investigation, C.-H.W. and J.-Z.J.; Project administration, C.-H.W. and C.-C.K.; Resources, C.-H.W.; Software, C.-H.K.; Supervision, C.-C.K.; Validation, J.-Z.J.; Visualization, C.-H.K.; Writing—original draft, C.-H.K.; Writing—review & editing, J.-Z.J. All authors have read and agreed to the published version of the manuscript.

Funding: The support of this research by the Ministry of Science and Technology of the Republic of China under Grant No. MOST 111-2622-8-011-006-TE1 is gratefully acknowledged.

Institutional Review Board Statement: Not applicable.

Informed Consent Statement: Not applicable.

Data Availability Statement: All data are provided in this manuscript.

Conflicts of Interest: The authors declare no conflict of interest. The funders had no role in the design of the study; in the collection, analyses, or interpretation of data; in the writing of the manuscript, or in the decision to publish the results.

Appendix A

Figures A1 and A2 show the iteration procedure of the rainflow cycle counting algorithm.

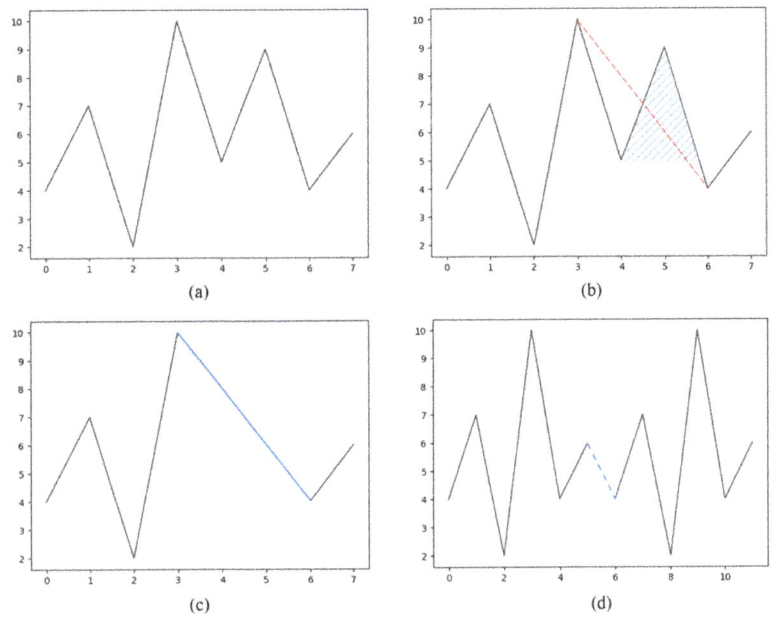

Figure A1. Example of extracting a single cycle using the rainflow counting algorithm. (**a**) It shows a line graph of the original sequence [4, 7, 2, 10, 5, 9, 4, 6] (**b**) Using rainflow counting method extract cycle: 5-9, the depth of this cycle is 4, the average is 7. (**c**) It shows the sequence after the extraction (**d**) Concatenate the original sequence with other sequences.

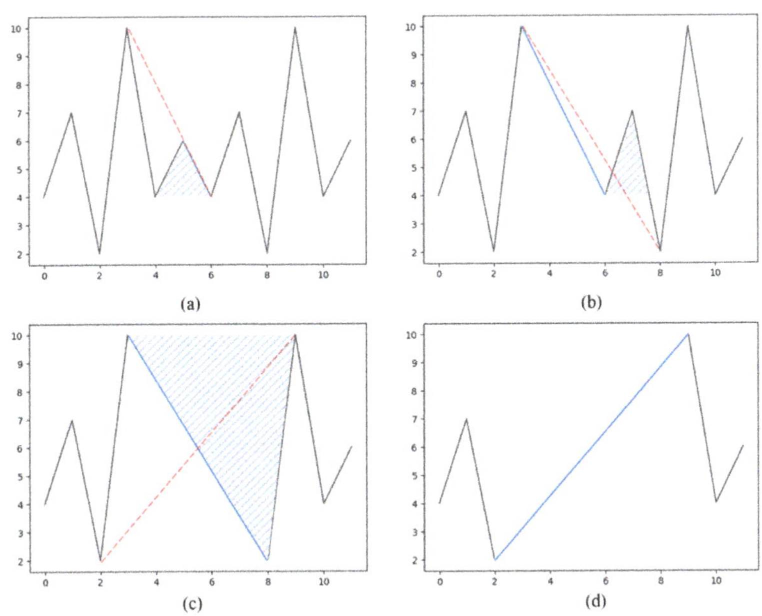

Figure A2. Example of extracting several cycles using the rainflow counting algorithm. (**a**) Extract the cycle of the blue triangle and use the red dotted line to fill the interval, the cycle depth is 2, the average is 5. (**b**)Using the same method to extract cycle [4,7], depth is 3, average is 5.5. (**c**) Use the same method to extract cycle [2,10], depth is 8, average is 6. (**d**) Sequence after Extraction.

Appendix B

	start	end	number of data	probability distribution
1	-	59.5	2	0.0%
2	59.5	59.75	97	0.004%
3	59.75	59.98	931,064	34.762%
4	59.98	60.02	704,761	26.313%
5	60.02	60.25	1,041,642	38.89%
6	60.25	60.5	834	0.031%
7	60.5	-	0	0.0%

Figure A3. Frequency distribution of December 2019.

	start	end	number of data	probability distribution
1	-	59.5	0	0.0%
2	59.5	59.75	141	0.005%
3	59.75	59.98	866,907	33.445%
4	59.98	60.02	743,237	28.674%
5	60.02	60.25	931,352	37.861%
6	60.25	60.5	363	0.014%
7	60.5	-	0	0.0%

Figure A4. Frequency distribution of March 2020.

	start	end	number of data	probability distribution
1	-	59.5	0	0.0%
2	59.5	59.75	4	0.0%
3	59.75	59.98	830,103	32.026%
4	59.98	60.02	791,953	30.554%
5	60.02	60.25	969,838	37.417%
6	60.25	60.5	102	0.004%
7	60.5	-	0	0.0%

Figure A5. Frequency distribution of May 2020.

Features of SOC Curves

The aging features extracted from SOC curves could be roughly classified into two types, one about cycle depth, and another about storage SOC, which closely relate to cycling aging and calendar aging, respectively. Figures A6–A8 show the SOC curves for the simulations of Scenarios 2, 5, and 7 based on the three-month frequency, as described in Section 5.2, and the distribution charts of the two main feature types mentioned above.

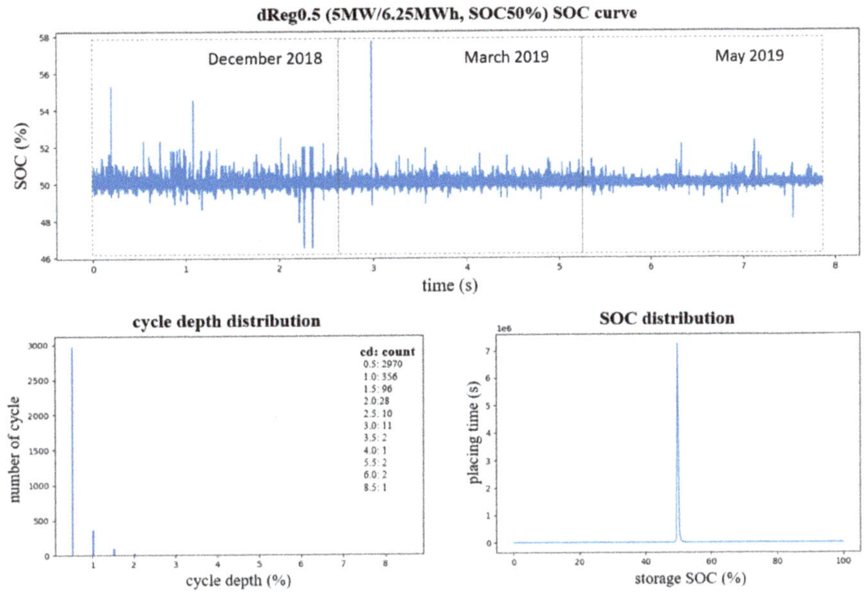

Figure A6. SOC curve and aging features of Scenario 2.

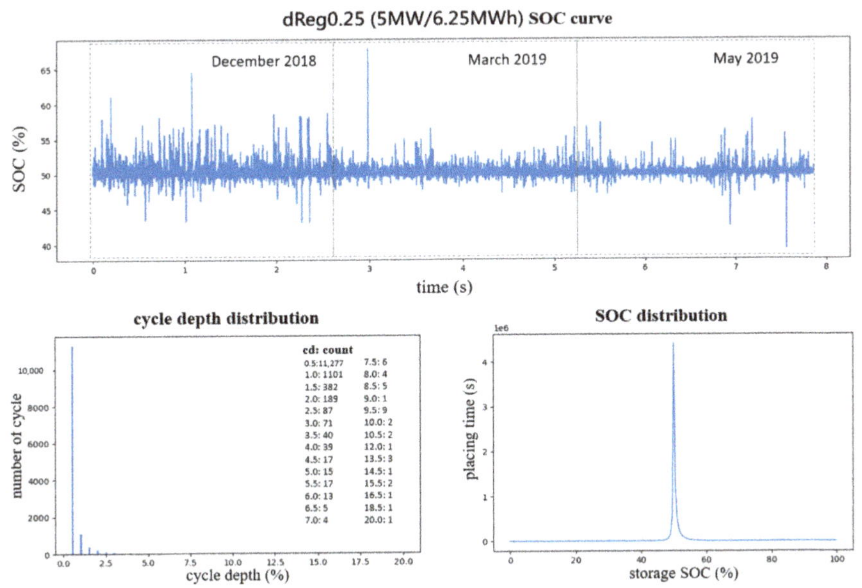

Figure A7. SOC curve and aging features of Scenario 5.

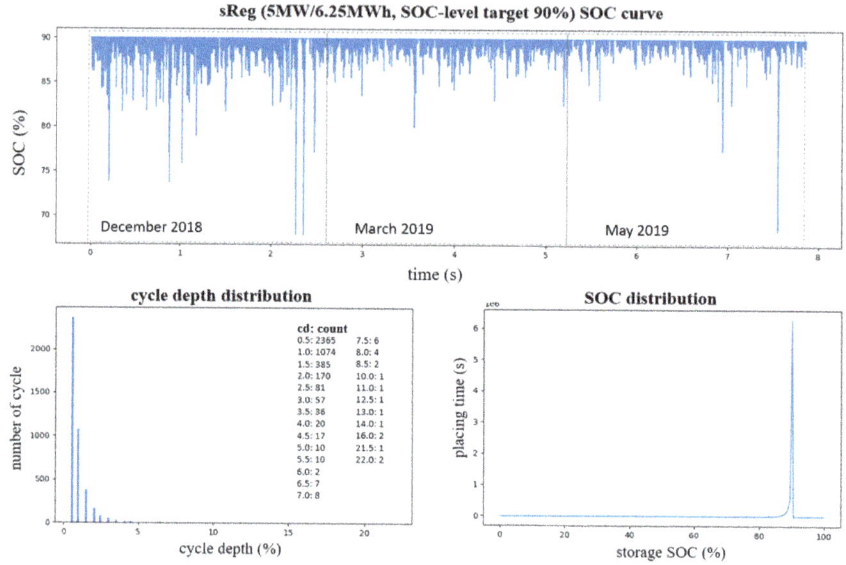

Figure A8. SOC curve and aging features of Scenario 7.

References

1. Smith, K.; Saxon, A.; Keyser, M.; Lundstrom, B.; Cao, Z.; Roc, A. Life prediction model for grid-connected Li-ion battery energy storage system. In Proceedings of the 2017 American Control Conference (ACC), Seattle, WA, USA, 24–26 May 2017. [CrossRef]
2. Xu, B.; Oudalov, A.; Ulbig, A.; Andersson, G.; Kirschen, D.S. Modeling of Lithium-Ion Battery Degradation for Cell Life Assessment. *IEEE Trans. Smart Grid* **2016**, *9*, 1131–1140. [CrossRef]
3. Padmanabhan, N.; Ahmed, M.; Bhattacharya, K. Battery Energy Storage Systems in Energy and Reserve Markets. *IEEE Trans. Power Syst.* **2019**, *35*, 215–226. [CrossRef]
4. Engels, J.; Claessens, B.; Deconinck, G. Techno-economic analysis and optimal control of battery storage for frequency con-trol services, applied to the German market. *Appl. Energy* **2019**, *242*, 1036–1049. [CrossRef]
5. Beltran, H.; Harrison, S.; Egea-Àlvarez, A.; Xu, L. Techno-Economic Assessment of Energy Storage Technologies for Inertia Response and Frequency Support from Wind Farms. *Energies* **2020**, *13*, 3421. [CrossRef]
6. Diouf, B.; Pode, R. Potential of lithium-ion batteries in renewable energy. *Renew. Energy* **2015**, *76*, 375–380. [CrossRef]
7. Hidalgo-Leon, R.; Siguenza, D.; Sanchez, C.; Leon, J.; Jacome-Ruiz, P.; Wu, J.; Ortiz, D. A survey of battery energy storage system (BESS), applications and environmental impacts in power systems. In Proceedings of the 2017 IEEE Second Ecuador Technical Chapters Meeting (ETCM), Salinas, Ecuador, 16–20 October 2017; pp. 1–6. [CrossRef]
8. Oudalov, A.; Buehler, T.; Chartouni, D. Utility Scale Applications of Energy Storage. In Proceedings of the 2008 IEEE Energy 2030 Conference, Atlanta, GA, USA, 17–18 November 2008; pp. 1–7. [CrossRef]
9. Taiwan Electric Power Co., Ltd. Power Dispatching Office, 10 August 2021. Available online: https://etp.taipower.com.tw/web/download/6.%E6%97%A5%E5%89%8D%E8%BC%94%E5%8A%A9%E6%9C%8D%E5%8B%99%E5%B8%82%E5%A0%B4%E4%B9%8B%E4%BA%A4%E6%98%93%E5%95%86%E5%93%81%E9%A0%85%E7%9B%AE%E8%A6%8F%E6%A0%BC.pdf (accessed on 15 May 2022).
10. Keil, P.; Schuster, S.F.; Wilhelm, J.; Travi, J.; Hauser, A.; Karl, R.C.; Jossen, A. Calendar Aging of Lithium-Ion Batteries. *J. Electrochem. Soc.* **2016**, *163*, A1872–A1880. [CrossRef]
11. Kassem, M.; Bernard, J.; Revel, R.; Pelissier, S.; Duclaud, F.; Delacourt, C. Calendar aging of a graphite/$LiFePO_4$ cell. *J. Power Sources* **2012**, *208*, 296–305. [CrossRef]
12. Sarre, G.; Blanchard, P.; Broussely, M. Aging of lithium-ion batteries. *J. Power Sources* **2004**, *127*, 65–71. [CrossRef]
13. Stroe, D.; Knap, V.; Swierczynski, M.; Stroe, A.; Teodorescu, R. Operation of a Grid-Connected Lithium-Ion Battery Energy Storage System for Primary Frequency Regulation: A Battery Lifetime Perspective. *IEEE Trans. Ind. Appl.* **2016**, *53*, 430–438. [CrossRef]
14. Kryonidis, G.C.; Nousdilis, A.I.; Pippi, K.D.; Papadopoulos, T.A. Impact of Power Smoothing Techniques on the Long-Term Performance of Battery Energy Storage Systems. In Proceedings of the 2021 56th International Universities Power Engineering Conference (UPEC), Middlesbrough, UK, 31 August 2021–3 September 2021; pp. 1–6. [CrossRef]
15. Fioriti, D.; Pellegrino, L.; Lutzemberger, G.; Micolano, E.; Poli, D. Optimal sizing of residential battery systems with multi-year dynamics and a novel rainflow-based model of storage degradation: An extensive Italian case study. *Electr. Power Syst. Res.* **2021**, *203*, 107675. [CrossRef]

16. Amzallag, C.; Gerey, J.; Robert, J.; Bahuaud, J. Standardization of the rainflow counting method for fatigue analysis. *Int. J. Fatigue* **1994**, *16*, 287–293. [CrossRef]
17. McInnes, C.; Meehan, P. Equivalence of four-point and three-point rainflow cycle counting algorithms. *Int. J. Fatigue* **2008**, *30*, 547–559. [CrossRef]
18. Stroe, D.-I.; Stan, A.-I.; Diosi, R.; Teodorescu, R.; Andreasen, S.J. Short term energy storage for grid support in wind power applications. In Proceedings of the 2012 13th International Conference on Optimization of Electrical and Electronic Equipment (OPTIM), Brasov, Romania, 24–26 May 2012; pp. 1012–1021.
19. Mongird, K.; Viswanathan, V.; Alam, J.; Vartanian, C.; Sprenkle, V.; Baxter, R. 2020 Grid Energy Storage Technology Cost and Performance Assessment. Available online: https://www.pnnl.gov/sites/default/files/media/file/Final%20-%20ESGC%20Cost%20Performance%20Report%2012-11-2020.pdf (accessed on 15 May 2022).

sustainability

MDPI

Article

Temperature Regulation Model and Experimental Study of Compressed Air Energy Storage Cavern Heat Exchange System

Peng Li [1,2,3], Zongguang Chen [3], Xuezhi Zhou [1,2,*], Haisheng Chen [1,2] and Zhi Wang [4,*]

1 Institute of Engineering Thermophysics, Chinese Academy of Sciences, Beijing 100190, China; 02489@msdi.cn (P.L.); chen_hs@iet.cn (H.C.)
2 National Energy Large Scale Physical Energy Storage Technologies R&D, Center of Bijie High-Tech Industrial Development Zone, Bijie 404004, China
3 Zhongnan Engineering Corporation Limited of Power China, Changsha 410014, China; 02211@msdi.cn
4 School of Mechanics and Safety Engineering, Zhengzhou University, Zhengzhou 450001, China
* Correspondence: zhouxuezhi@iet.cn (X.Z.); wangzhi@zzu.edu.cn (Z.W.)

Abstract: The first hard rock shallow-lined underground CAES cavern in China has been excavated to conduct a thermodynamic process and heat exchange system for practice. The thermodynamic equations for the solid and air region are compiled into the fluent two-dimensional axisymmetric model through user-defined functions. The temperature regulation model and experimental study results show that the charging time determines the air temperature and fluctuates dramatically under different charging flow rates. The average air temperature increases with increasing charging flow and decreasing charging time, fluctuating between 62.5 °C and −40.4 °C during the charging and discharging processes. The temperature would reach above 40 °C within the first 40 min of the initial pressurization stage, and the humidity decreases rapidly within a short time. The use of the heat exchange system can effectively control the cavern temperature within a small range (20–40 °C). The temperature rises and regularly falls with the control system's switch. An inverse relationship between the temperature and humidity and water vapor can be seen in the first hour of the initial discharging. The maximum noise is 92 and 87 decibels in the deflation process.

Keywords: compressed air energy storage; heat exchange system; thermodynamic response; high pressure; charging process; temperature regulation

Citation: Li, P.; Chen, Z.; Zhou, X.; Chen, H.; Wang, Z. Temperature Regulation Model and Experimental Study of Compressed Air Energy Storage Cavern Heat Exchange System. *Sustainability* **2022**, *14*, 6788. https://doi.org/10.3390/su14116788

Academic Editors: Luis Hernández-Callejo, Jesús Armando Aguilar Jiménez and Carlos Meza Benavides

Received: 10 May 2022
Accepted: 30 May 2022
Published: 1 June 2022

1. Introduction

With the gradual development of global carbon emission reduction actions, vigorously developing renewable energy has become an inevitable choice in the new situation. Renewable energy has the advantage of being clean and pollution-free. It has many defects such as instability and difficulty in storage which urgently need corresponding energy storage technology innovation to match. Compressed air energy storage (CAES) is one of the most promising large-scale energy storage technologies. Compared with pumped hydroelectric storage (PHS), CAES is not limited by water source and is a better choice for efficient storage and utilization of clean energy [1].

Today, two existing commercial CAES plants are in operation: a 290 MW unit built in Huntorf, Germany, in 1978, and a 110 MW unit built in McIntosh, AL, USA, in 1991 [2]; the monitoring data of their successful operation bring some valuable validation data for the research related to compressed air energy storage caverns [3]. The research and development progress on energy storage technologies in China has also developed more rapidly [4]. The grid connection of the Feicheng salt cavern advanced CAES plant was realized in 2021 [5]. Other caverns, such as salt caverns [6], abandoned mine caverns [7], underground aquifers [8], and artificial rock-lined caverns [9], can also be used as gas storage design alternatives. Moreover, compared with natural reservoir caverns, artificial caverns with lining, which are more flexible in site selection and more adaptable to the design of large-scale energy storage, are one of the preferred options for achieving energy storage in the future.

Kushnir et al. [10] derived an analytical solution for the temperature and pressure variation of the air in the cavern in adiabatic mode. Then, the theoretical solution for the thermodynamics of the cavern in the heat transfer model was derived based on the air mass and energy conservation equations considering the heat transfer at the cavern wall. It significantly affects the air temperature and pressure variation compared to the adiabatic model [11]. Many scholars also cite this calculation model, and the calculation results are compared with the test data of the Huntorf power station [12]. Kim et al. [13] applied TOUGH-FLAC to study the thermodynamic and mechanical response of lined caverns, and Zhou et al. [14] calculated the heat–flow–solid (THM) coupling process of lined caverns based on COMSOL. Many thermodynamic simulations of CAES caverns show that the temperature field inside the air storage caverns is unevenly distributed and may form extremely high temperatures locally, which poses a significant threat to the lining and surrounding rocks [15]. At the same time, the gas inside the cavern may produce significant temperature fluctuations during the cyclic gas filling and discharging process of the air storage caverns. Under the coupling effect of cyclic temperature and stress, the chambers are prone to thermal stress disasters and safety risks in long-term operation [16]. One of the significant problems of CAES systems is the air temperature rise or fall during the compression or expansion operation, resulting in low efficiency. Some works of literature describe enhancing heat transfer by implementing thermal management measures [17]. Others use numerical and experimental methods to characterize fluid flow patterns and heat transfer behavior at the local level [18–20]. GOUDA proposed a 3D CFD model to simulate the air compression process to achieve near-isothermal operation [21]. It is essential to carry out a thermodynamic simulation of the cavern chamber filling and discharging process and to intervene manually in the possible extreme temperature conditions to realize piezo gas storage power generation [22].

In order to gain insight into the thermodynamic and mechanical response of the operation process of CAES caverns, many countries internationally have carried out experimental cavern tests. Ishihata et al. [23] tested the sealability of a deeply buried underground gas storage reservoir with a test air pressure of 0.9 MPa. However, the test results showed severe cracking of the sealing layer. Swedish scholars GEISSBÜ conducted AA-CAES demonstration plant gas storage adiabatic mode thermal storage test. Due to concrete plug leakage, the air pressure only reached 7 bar, and the thermodynamic response was consistent with the simulation results [24]. An underground lined rock cavern for small-scale pressure gas storage tests as a storage reservoir was tested by Kim. At 100 m underground burial depth, the radius of the cylindrical tunnel designed for gas storage was 2.5 m, and the maximum gas storage pressure was 5 MPa [13]. It is a tremendous challenge for a compressed air energy storage plant to determine whether the test can be conducted for high internal pressure in an underground storage cavern without guaranteeing leakage.

Taking the exploration tunnel of Pingjiang Pumped Storage Plants in Hunan Province, China's first underground gas storage test cavern with a shallowly buried lining of hard rock has been reconstructed to realize the gas storage test with a high internal pressure of 10 MPa. In this paper, we would like to develop a temperature field analysis model for a model underground high-pressure air storage cavern, analyze the temperature fluctuation law of the gas filling and discharging process, and design a heat transfer system in the cavern. Based on thermodynamic, heat transfer, and numerical heat transfer methods, air charging and discharging and heat transfer performance tests in the cavern will be conducted.

2. CAES Cavern Design

The test cavern established in this study is located in the exploration tunnel (PD4) of the underground powerhouse of the Pingjiang Pumped Storage Power Station in China. The design of the cavern is shown in Figure 1. The buried depth of the testing cavern was about 110 m. The length, the inner diameter, the volume, and the inner surface area were 5.0 m, 2.9 m, 28.8 m^3, and 50.6 m^2, respectively. A concrete lining of 0.5 m was set in the testing cavern with a fiber-reinforced plastic (FRP) sealing layer on the surface of the lining. A plug was set at the inlet end of the test chamber to bear the thrust of high-pressure compressed air (maximum design pressure was 10.0 MPa). The test system includes a vehicle-mounted air compressor pressurization system, a charging and discharging pipeline system, cavern gas storage, sealing, and measurement system. The surrounding rock in the flat exploration cave was mainly granite and granite gneiss with a mean value of elastic modulus, deformation modulus, and compressive strength of 63.62 GPa, 35.59 GPa, and 78~130 MPa, respectively. The location of the rock mass was of good quality. The fundamental physical parameters such as density, specific heat, and thermal conductivity of solid materials are shown in Table 1. It was assumed that the changes in solid physical parameters within the calculation temperature range are small and have little effect on the results. The air compressor was designed with a heat storage device to cool down gradually, and the outlet temperature would be cooled down to 30 °C.

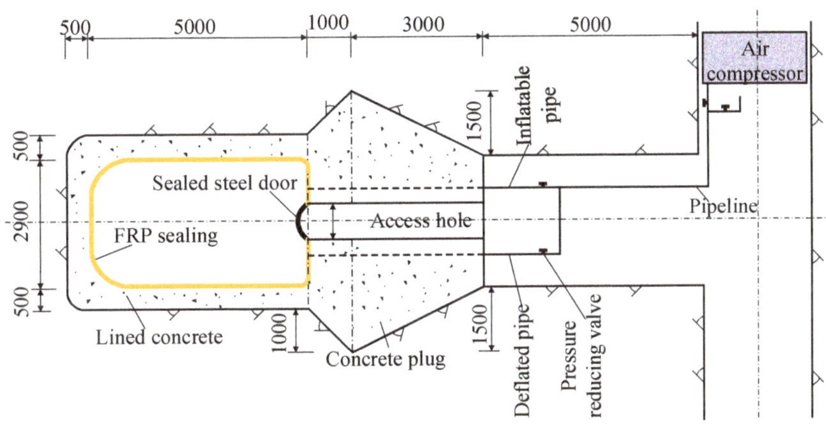

Figure 1. Schematic diagram of the Pingjiang CAES cavern. (Unit: mm).

Table 1. Physical parameters of solid materials.

Material	Young's Modulus (GPa)	Poisson's Ratio	Density (kg/m^3)	Specific Heat (J/kg/K)	Coefficient of Thermal Conductivity (W/m/K)
FRP	2.9	0.3	1800	1260	0.52
Concrete	28	0.167	2500	920	1.74
Rock	18	0.2	2800	920	3.49
Steel door	200	0.3	8030	502.48	16.27

3. Numerical Models of Thermodynamic Processes

3.1. Modeling

The simplified axisymmetric numerical calculation physical model was established according to the CAES cavern structure diagram shown in Figure 2. Considering the low thermal conductivity and small thermal diffusivity of the surrounding rock and concrete layer and the limited influence of the air temperature change in the cavern, the surrounding rock, and concrete areas were simplified to a certain extent. The calculation results had little affection. Because of the small calculation area, the calculation speed would be accelerated, and the calculation cost would be saved. The calculated thickness of the surrounding rock and the cylindrical concrete layer was 1000 mm and 480 mm, respectively.

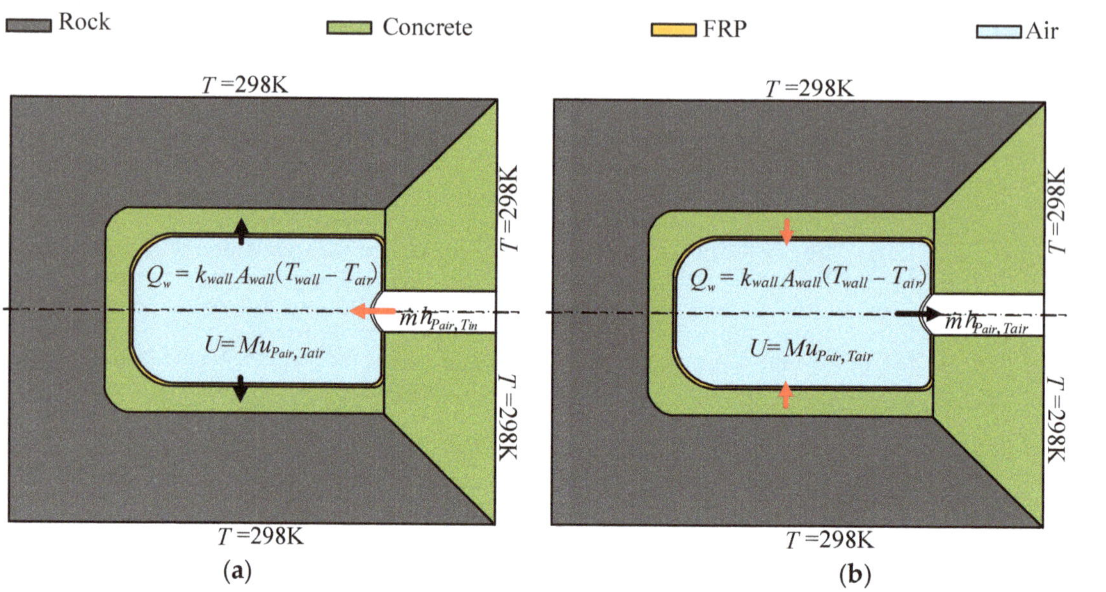

Figure 2. Axisymmetric simplified thermodynamic calculation model: (**a**) charging process; (**b**) discharging process.

The charging–maintaining–discharging of the underground CAES cavern was a complex thermodynamic process coupled with multiple physical phenomena, as shown in Figure 2. The process includes: air compression (charging process)/expansion (discharging process) in the cavern, convection heat transfer between high-pressure air and the wall of the glass fiber-reinforced plastic cylinder, heat conduction inside the solid area such as the glass fiber-reinforced plastic cylinder, concrete layer and surrounding rock, the area absorbs or releases heat due to its heat capacity, as well as convection heat exchange with the outside air in the FRP door area. Considering that the air pressure and temperature fields in the cavern were approximately uniform, a simplified thermodynamic equilibrium equation could be adopted to obtain the internal air's average temperature and pressure. The solid region heat conduction model was coupled to form a complete thermodynamic model of the air storage cavern.

In order to study the thermodynamic process of the gas and solid in the cavern during different charging and discharging processes, numerical simulations were carried out for different charging and discharging flow rates (1000 Nm^3/h, 500 Nm^3/h). The initial temperature field was set to 298 K. Before charging, the pressure of the cavern was 0.1 MPa, and the maximum air pressure was about 10.0 MPa. The gas storage time was two hours, and then the gas was released at (1000 Nm^3/h, 500 Nm^3/h) until the pressure in the cave was close to 0.1 MPa.

According to the conservation of energy, the change of the total internal energy of the air in the cavern is equal to the total enthalpy of the charged/discharged air and the coupled heat transfer between the cylinder wall and the air:

$$\frac{\partial(Mu)}{\partial t} = k_{wall}A_{wall}(T_{wall} - T_{air}) + \frac{\partial m}{\partial t}h_{T_{air},P_{air}} \quad (1)$$

$$\frac{\partial(Mu)}{\partial t} = k_{wall}A_{wall}(T_{wall} - T_{air}) + \frac{\partial m}{\partial t}h_{T_{air},P_{air}} \quad (2)$$

where u is the unit internal energy of air, M is the total volume of air at the current moment, $h_{T_{air},P_{air}}$ is the unit enthalpy of air, $\partial m/\partial t$ is the air flow rate, which is negative when discharged, k_{wall} is the convective heat transfer coefficient between the wall of the cavern and the air, T_{air} is the air temperature, T_{wall} is the average temperature of the cavern surface, Q_{wall} is the coupling heat exchange between the cavern surface and the air, and $Q_{wall} = k_{wall}A_{wall}(T_{wall} - T_{air})$.

From Equation (1), the unit internal energy change can be obtained as:

$$\Delta u = \frac{k_{wall}A_{wall}(T_{wall} - T_{air}) + \Delta m \cdot h_{T_{air},P_{air}}}{M + \Delta m} \tag{3}$$

High pressure air temperature increment can be obtained as:

$$\Delta T = \left.\frac{\Delta u}{c_v}\right|_{P_{air}} \tag{4}$$

The unit internal energy, temperature, and density of the gas at the moment $t + \Delta t$ are:

$$\begin{aligned} u_{(t+\Delta t)} &= u_t + \Delta u \\ T_{(t+\Delta t)} &= T_t + \Delta T \\ \rho_{(t+\Delta t)} &= \rho_{t=0} + \frac{\dot{m} \cdot t}{V} \end{aligned} \tag{5}$$

The pressure value of the compressed air can be calculated by the Peng–Robinson equation:

$$\begin{aligned} a &= 0.45723553\frac{R^2T_c^2}{P_c} \\ b &= 0.07779607\frac{RT_c}{P_c} \\ \kappa &= 0.37464 + 1.54226\omega - 0.26993\omega^2 \\ \alpha &= \left(1 + \kappa\left(1 - \sqrt{T/T_c}\right)\right)^2 \end{aligned} \tag{6}$$

$$\Delta P = \frac{RT}{(v_m - b)} - \frac{\alpha a}{v_m^2 + 2bv_m - b^2} \tag{7}$$

where R is the universal gas constant and $R = 8314.47\left(\text{J} \cdot \text{kmol}^{-1} \cdot \text{K}^{-1}\right)$, T_c is the air temperature in the critical state and $T_c = 132.5306$ K, P_c is the air pressure in the critical state and $P_c = 3.79$ MPa, ω is the eccentricity factor and $\omega = 0.0335$, v_m is the unit molar volume of air and $v_m = MM/\rho$ (m^3/kmol), MM is the air molar mass and $MM = 28.95$ (kg/kmol).

In solid areas, such as glass fiber-reinforced plastic cylinders, concrete lining, surrounding rock, and steel doors, the temperature change can be obtained by solving the two-dimensional axisymmetric unsteady heat conduction energy equation:

$$\rho_s c\frac{\partial T}{\partial t} = \frac{1}{r}\frac{\partial}{\partial r}\left(\lambda_s r\frac{\partial T}{\partial r}\right) + \frac{\partial}{\partial x}\left(\lambda_s r\frac{\partial T}{\partial x}\right) + \dot{Q} \tag{8}$$

where λ_s is the thermal conductivity, ρ_s is the density, $\dot{Q}$ is the internal heat source, which is taken as 0 in this project, the r direction is the radial, and the x-direction is the axial direction.

The boundary condition settings include thermal boundary conditions for the solid region, where the outer surfaces of the surrounding rock and concrete can be set as isothermal wall boundary conditions:

$$T = T_\infty \tag{9}$$

where T_∞ = 298 K is the ambient temperature.

The outer surface of the steel door can be set as the convective heat transfer boundary condition:

$$q'' = k_\infty(T - T_\infty) \tag{10}$$

where k_∞ is the convective heat transfer coefficient between the steel door and the air in the access hole, which can be taken as a constant, $k_\infty = 8\ \text{W/m}^2\text{K}$. The inner surface of the FRP cylinder is the convection heat transfer boundary condition of the inner surface of the FRP cylinder and can be described as:

$$q'' = k_{wall}(T_{wall} - T_{air}) \tag{11}$$

where k_{wall} is the average convective heat transfer coefficient between the wall and the compressed air. According to the formula for the heat transfer coefficient of gas-filled convection in a closed storage tank proposed by Heath et al. [25] and Bourgeois et al. [26], the heat exchange system can be calculated by the following formula:

$$N_u = u(d/D)^{0.5} R_e^{0.67} + 0.104 R_a^{0.352} \tag{12}$$

where N_u is the Nusselt number, the first term on the right is the convective heat transfer effect, which is related to the size of the inlet pipe, and the second term is the natural convection heat transfer effect, which is related to the Rayleigh number $R_a = G_r P_r$. The outgassing process only needs to consider the effects of natural convection.

The relationship between the convective heat transfer coefficient and N_u can be described as:

$$k = \frac{N_u \cdot \lambda_{air}}{D} u(d/D)^{0.5} \tag{13}$$

The above equations can be compiled into the fluent two-dimensional axisymmetric solid thermal conductivity model through a user-defined function (UDF), and the values can be updated after the calculation of each time step to realize the calculation of the temperature field in the entire cavern area.

3.2. Simulated Result

The variation law of air average temperature, inner cavern surface average temperature, and air pressure over time in the whole process obtained by numerical calculation is shown in Figure 3. It can be seen that the whole process is divided into the charging stage, the pressure-holding stage, the discharging stage, and the stop stage. The temperature and pressure changes in each stage have apparent characteristics.

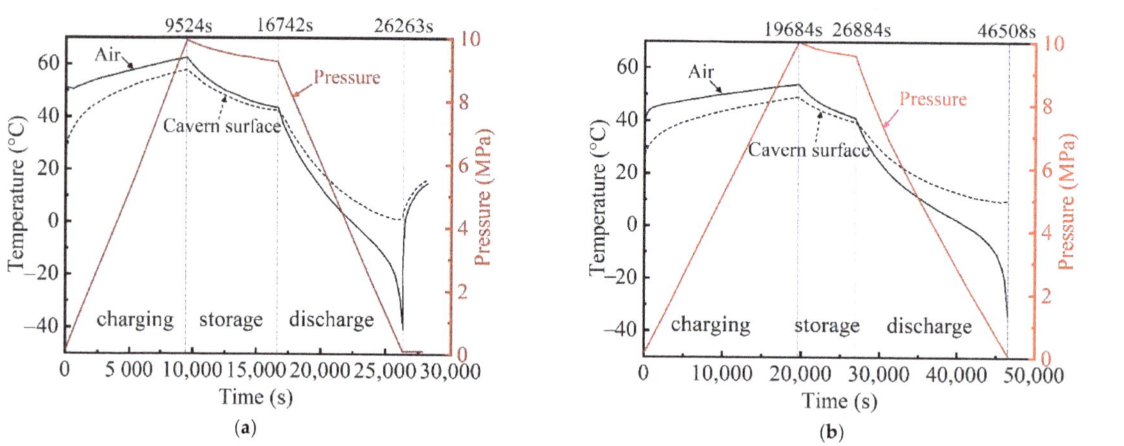

Figure 3. Temperature and pressure variation: (**a**) flow rate of 1000 Nm^3/h; (**b**) flow rate of 500 Nm^3/h.

The average air pressure rises steadily during the charging process. For the two charging and discharging processes with a flow rate of 1000 Nm^3/h and 500 Nm^3/h, the time used to charge the air with 10 MPa is 9542 s and 19,684 s, respectively. The air and cavern temperature increases rapidly at the beginning of the charging, especially in the first 100~200 s. Due to the low air pressure and small total mass in the storage cavern, the compression effect is powerful, and the air temperature rises rapidly from 25 °C to about 50 °C. After that, the air temperature gradually slowed down due to the relative weakening of the compression effect and the combined effect of the high-temperature air on the increase of heat dissipation on the cavern surface. At the end of the charging, the average air temperature reached 62.5 °C and 52.4 °C, and the average cavern surface temperature reached 58.0 °C and 49.4 °C, respectively. In the pressure-holding stage, the inlet pipe is neither filled with air nor released and does not perform work on the air.

At this time, since the air with a higher temperature continues to dissipate heat to the wall with a lower temperature, the average temperature of the air begins to drop continuously, and the inner cavern surface passes the heat through. The heat is transferred to the lower temperature solid area, and the average temperature of the inner cavern surface also continues to drop and gradually approaches the air temperature. After the holding stage, the average air temperature is 43.0 °C and 40.3 °C, respectively, and the average wall temperature is 41.9 °C and 39.5 °C. As the average air temperature dropped, the air pressure in the cave also dropped slightly. During the discharging stage, the average temperature and pressure of the air in the cave dropped rapidly.

Moreover, the air temperature drops rapidly below the average temperature of the inner cavern surface. The cavern surface begins to provide heat for the air, but the heat transfer from the cavern surface is limited. At the end of the discharging stage, the air pressure drops to 0.1 MPa (due to the pressure gradient on the outlet pipe under actual conditions, the pressure in the cavern should be slightly higher than the atmospheric pressure). According to the equation of state, a decrease in air pressure results in a decrease in temperature, and the rate of temperature decrease is proportional to the rate of decrease in air pressure. The average air temperature drops to −40.4 °C and −20.9 °C, and the average temperature of the inner cavern decreases to 1.9 °C and 8.5 °C, respectively. The hot cylinder cavern and surrounding rock heat the air at the stop stage. Both air temperature and pressure increase slightly. The slower the discharging rate, the more heat the compressed air gets from the cavern surface and surrounding rock, and the lower the rate of air temperature drop. It can be seen from the simulation of the air storage process that the temperature in the cavern during the charging and discharging process fluctuates wildly, which brings security risks to the safety and stability of the cavern. It is necessary to control the heat transfer of the air temperature in the cavern.

4. Heat Exchange System Design

Figure 4 shows the schematic diagram of thermodynamics in the air storage cavern with a heat exchange circulation system. With the influx of external air during the charging process, the internal energy gradually increases with air temperature. Part of the heat is transferred to the FRP cylinder through the convection heat exchange between the cavern surface and the air, and then it dissipates through the FRP cylinder and the concrete. Another part of the heat is transferred to the circulating cooling water in the tube. The heated circulating cooling water loses heat through the circulating cooling tower. During the discharging process, the internal energy and air temperature gradually decrease with the release of the air. Due to the temperature difference, the cavern surface and the hot water in the tube transfer heat to the air. In order to maintain the stable air temperature in the cavern, the outside (cavern surface or heat exchange tube) needs to transfer a certain amount of heat to the air or take away a certain amount of heat.

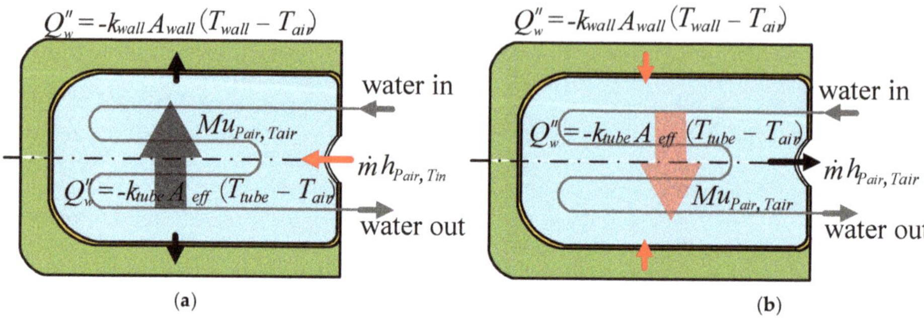

Figure 4. Thermodynamics in the air storage cavern with a heat exchange circulation system: (**a**) charging process; (**b**) discharging process.

The energy conservation equation for the whole thermodynamic system can be expressed as:

$$\frac{\partial(Mu)}{\partial t} = k_{tube} A_{tube} \eta (T_{water} - T_{air}) + k_{wall}(T_{wall} - T_{air}) + \frac{\partial m}{\partial t} h_{T_{air}, P_{air}} \tag{14}$$

where k_{tube} is the convective heat transfer coefficient of the heat exchange tube wall, and T_{water} is the temperature of the water in the heat exchange tube.

Therefore, the heat transferred from the hot water in the heat exchange tube to the air can be obtained:

$$k_{tube} A_{tube} \eta (T_{water} - T_{air}) = \frac{\partial m}{\partial t} (u_{T_{air}, P_{air}} - h_{T_{air}, P_{air}}) + M \frac{\partial u}{\partial t} - k_{wall}(T_{wall} - T_{air}) \tag{15}$$

where $k_{wall}(T_{wall} - T_{air})$ is the heat transfer from the cavern surface to the air, $\partial(Mu)/\partial t$ is the total internal energy change of the air in the tube, $u \cdot \partial m/\partial t$ is the total mass change of air, $(M \cdot \partial u/\partial t)$ is the internal energy change per unit of air, $(h_{T_{air}, P_{air}} \cdot \partial m/\partial t)$ is the energy of the outlet air.

Assuming the air temperature control in the cylinder is stable at 25 °C, which means the temperature change is equal to 0 and the temperature difference between the cavern surface and the air is also kept at 0, the heat transferred from the cavern surface to the air is close to zero:

$$\frac{\partial u}{\partial t} = c_v \frac{\partial T}{\partial t} = 0 \tag{16}$$

$$T_{wall} - T_{air} = 0 \tag{17}$$

In order to keep the air temperature stable, the heat exchange power that the heat exchange tube needs to provide should satisfy the following equation:

$$k_{tube} A_{tube} \eta (T_{water} - T_{air}) = \frac{\partial m}{\partial t} (u_{T_{air}, P_{air}} - h_{T_{air}, P_{air}}) \tag{18}$$

The additional heat required from the heat exchanger piping to maintain a constant air temperature of 25 °C at different pressures is essentially the same. The heating powers for different flow rates are between 30.4 kW~30.7 kW (1000 Nm^3/h) and 15.2 kW~15.3 kW (500 Nm^3/h), respectively.

Typical finned heat exchanger tube dimensions are shown in Figure 5.

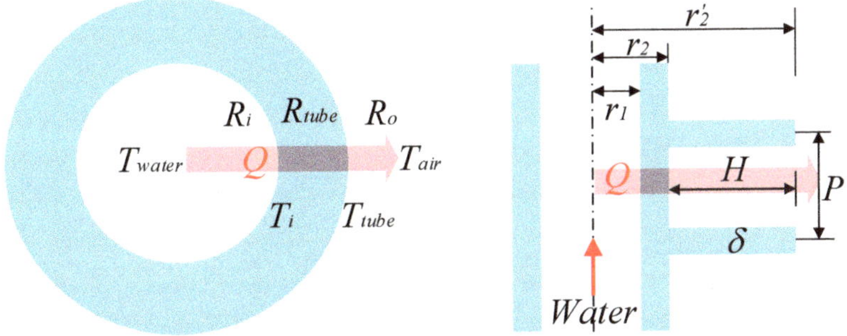

Figure 5. Schematic diagram of finned heat exchange tube structure.

The heat exchange power can be calculated as:

$$Q = \frac{T_{water} - T_{air}}{R_i + R_{tube} + R_o} \quad (19)$$

where Q is the heat exchange power, R_i is the thermal resistance of the water in the tube and $R_i = (1/k_{water})\cdot(r_o/r_i)$, k_{water} is the convective heat transfer coefficient between water and the inner wall of the heat exchange tube and $k = (N_u \cdot D_i)/\lambda_{water}$, R_{tube} is the thermal resistance of the tube, R_o is the external thermal resistance.

According to the Gnielinski equation, the turbulent convective heat transfer Nussle number, N_u, can be calculated as [27]:

$$N_u = \frac{(f/8)(R_e - 1000)P_{r_f}}{1 + 12.7\sqrt{f/8}\left(P_{r_f}^{2/3} - 1\right)}\left[1 + \left(\frac{d}{l}\right)^{2/3}\right]\left(\frac{P_{r_f}}{P_{r_w}}\right)^{0.11} \quad (20)$$

$$\left(\frac{P_{r_f}}{P_{r_w}} = 0.05 \sim 20\right) \quad (21)$$

where the friction factor f can be calculated the turbulent flow resistance coefficient and $f = (1.82lgR_e - 1.64)^{-2}$.

The fluid resistance in a straight tube section can be calculated as:

$$\Delta P = f \cdot \frac{l}{D_i} \cdot \frac{\rho v^2}{2} \quad (22)$$

The thermal resistance of the tube be calculated as:

$$R_{tube} = \frac{ln(r_2/r_1)}{2\pi\lambda_{tube}l} \quad (23)$$

where λ_{tube} is thermal conductivity of the tube.

The outer surface of the tube can be calculated according to natural convection, and the convection intensity can be characterized by the Grashof number G_r:

$$G_r = \frac{g\beta\Delta t D_o^3}{v^2} = \frac{g\beta|T_{tube} - T_{air}|D_o^3}{v^2} \quad (24)$$

$$\beta = -\frac{1}{\rho}\left(\frac{\partial\rho}{\partial T}\right)_p \approx \frac{1}{T_{air}} \quad (25)$$

where ρ is the density (kg/m^3), v is the viscosity (kg/ms), g is gravity acceleration (m/s^2), β is the thermal expansion coefficient, T_{air} is the temperature of the air (K), and T_{tube} is the temperature of the tube (K).

The Nussle number, N_{u_n}, and the heat transfer coefficient, k_n, can be calculated as:

$$N_{u_n} = C(G_r \cdot P_r)^n \quad (26)$$

$$k_n = \frac{N_{u_n} \cdot \lambda}{D} \quad (27)$$

where C and n are constants, when $10^4 < G_r < 5.76 \times 10^8$, $C = 0.48$ and $n = 0.25$; when $5.76 \times 10^8 < G_r < 4.65 \times 10^9$, $C = 0.0445$ and $n = 0.37$; when $G_r > 4.65 \times 10^9$, $C = 0.1$ and $n = 0.333$.

The fin efficiency η_f can be obtained through Figure 6 [28]:

$$\eta_f = \frac{th(mH')}{mH'} \quad (28)$$

$$H' = H_f \cdot \left(1 + 0.35 ln \frac{r_o + H_f}{r_o}\right) \quad (29)$$

where $H' = H + \delta/2$, $A_L = H'\delta$, $\frac{r_2}{r_1} = \frac{r_1 + H'}{r_1}$, $mH = (H')^{3/2}[h/(\lambda A_L)]^{1/2}$.

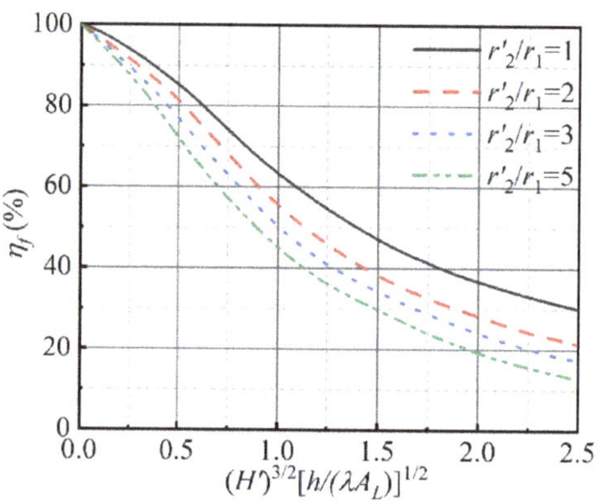

Figure 6. Efficiency curve of Annular Rectangular Rib.

The external thermal resistance R_o can be obtained as:

$$R_o = \left(k_n \cdot \frac{\eta_f A_f + A_o}{A_o}\right)^{-1} \quad (30)$$

where A_o is the outside surface area of tube and $A_o = \pi \cdot D_o \cdot l$, A_f is the fin area and $A_f = 0.5\pi \cdot \left[\left(D_o + 2H_f\right)^2 - D_i^2\right] \cdot \frac{1}{P}$, D_o is the outside diameter of the tube, D_i is the inside diameter of the tube, l is the tube length, P is the fin spacing, and H_f is the fin height.

The heat exchange heat is all carried out by the circulating cooling water, and the internal energy added by the circulating water at the inlet and outlet is equal to the heat exchange power:

$$Q = c_w \dot{m}_w (T_{w,i} - T_{w,o}) = c_w \dot{m}_w \Delta T_w \quad (31)$$

The heat exchange area can be maintained by increasing the number of tubes. The water circuit can be divided into N processes in parallel to control the flow speed and reduce the flow resistance to less than 50 kPa.

The flow of each waterway can be obtained as:

$$\dot{m}_w = \frac{\dot{m}_{w,t}}{N} \quad (32)$$

where $\dot{m}_{w,t}$ is the water flow rate.

Two types of finned tubes with fin height H = 12 mm, fin thickness δ = 0.5 mm, fin pitch P = 6 mm, and tube thickness $(r_2 - r_1)$ = 8 mm are selected (pipe outer diameter is 50 mm and 40 mm, respectively) to calculate the required heat exchange area and waterway loss. The two-way parallel connection method is adopted, and the air temperature, water temperature difference, and water flow rate are 25 °C, 4 °C, and 6.6 t/h, respectively. The calculation results are shown in Figure 7.

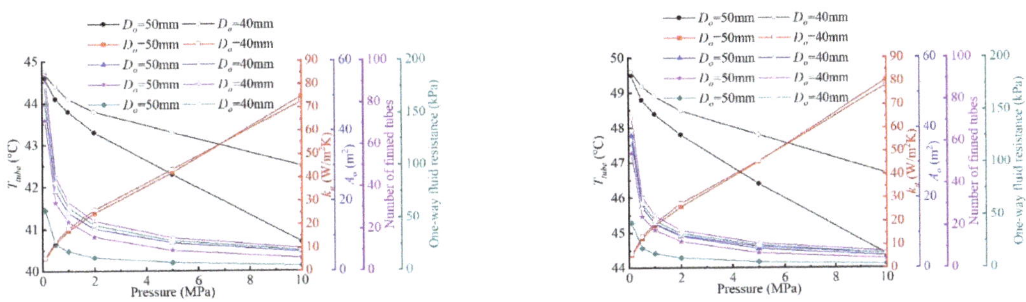

(a) (b)

Figure 7. Influence law of heat exchange system parameters: (**a**) T_{water} = 45 °C; (**b**) T_{water} = 50 °C.

It can be seen from Figure 7 that the tube temperature decreases significantly with the increase of pressure. The higher the circulating water temperature, the higher the tube's temperature. The heat transfer coefficient also increases as the pressure increases, and the higher the water temperature, the greater the heat transfer coefficient. The outer surface area of the tube, the number of ribs, and the fluid resistance decrease rapidly and then slowly with the increase of pressure.

Two schemes have been calculated here to ensure the efficient operation of the heat exchange system. Based on a two-way parallel finned tube system with a single tube length of 4 m, the number of loop tubes needs to be 10 and 13, respectively, for a tube diameter of 50 mm and 40 mm. To provide circulating hot water through an electric heater, it should meet the following requirements: the electric heater must meet the heating power of not less than 47 kW (considering the heat dissipation loss outside the cave), and the heating temperature should not be lower than 50 °C. The water volume of the pump is not less than 10.0 t h^{-1}, and the water head should be equal to the flow resistance of the tube outside the cavern plus 50 kPa.

According to the design and calculation results of the heat exchange system, combined with the internal dimensions of the Pingjiang CAES cavern and the processing limitations of the finned tubes, the final selected finned tubes had an outer diameter of 51 mm, an inner diameter of 40 mm, a fin height of 11 mm, a fin thickness of 1 mm, and a pitch of 5 mm. The structure of the heat exchange system is shown in Figure 8.

The thermodynamic and heat transfer calculations were carried out for the air temperature change in the cavern during the charging–maintaining–discharging of the heat exchange system. The calculation results are shown in Figure 9. Since the heat exchange system could use the circulating water at room temperature to take away a large amount of compression heat generated by air compression during the charging stage. Except that the large amount of heat generated by the severe compression in the early stage could not be quickly discharged, the air temperature in the cave was significantly reduced and stabilized at 35 ± 3 °C (1000 Nm^3/h) and 30 ± 3 °C (500 Nm^3/h). At the same time, due to the decrease in air temperature, the time of the compression stage was prolonged, and the amount of air stored in the cave increased. In the discharging stage, the heated circulating cooling water could continue to provide heat for the low-temperature air to expand and cool. The heat could not be replenished in time due to the violent expansion in the later stage of discharging. The air temperature in the cavern was maintained at about 30 °C

(1000 Nm^3/h) and 25 °C (500 Nm^3/h). A stable temperature was maintained, and air temperatures below 0 °C could be avoided. From the calculation results of the complete charging and discharging cycle, the air temperature was maintained between 20–40 °C (1000 Nm^3/h) and 20–40 °C (500 Nm^3/h). In particular, when the pressure was greater than 1.6 MPa, the temperature range was controlled between 25−38 °C (1000 Nm^3/h) and 25–31 °C (500 Nm^3/h). The heat exchange system can sufficiently suppress the temperature fluctuation during the charging and discharging process.

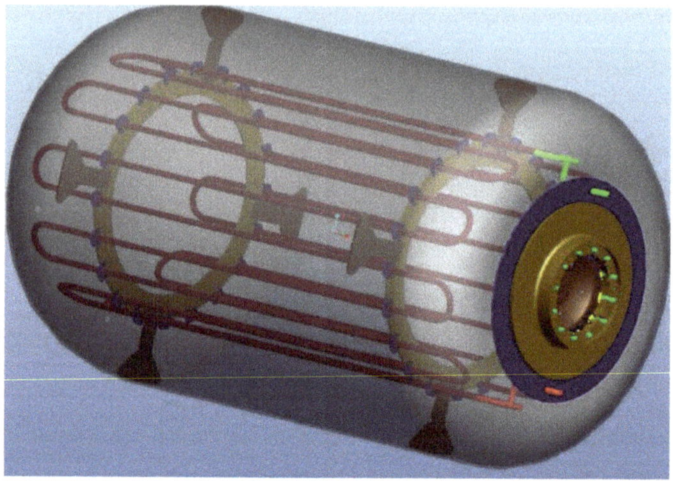

Figure 8. The structure of the heat exchange system in Pingjiang CAES cavern.

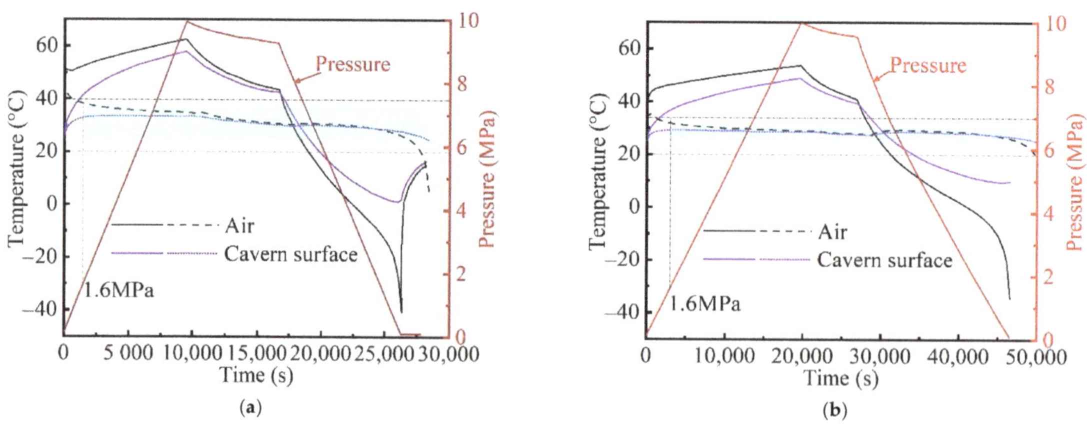

Figure 9. The temperature controlling effect of the heat exchange system: (**a**) flow rate of 1000 Nm^3/h; (**b**) flow rate of 500 Nm^3/h.

5. Experimental Testing of CAES Cavern Heat Exchange System

5.1. Installation of Heat Exchange System

The tube of the heat exchange system was processed following Figure 8. Then it was transported to the on-site cavern for welding and connection one by one (Figure 10). In the welding process, an electric welding cloth was arranged on the entire wall surface to avoid damage to the glass fiber-reinforced plastic by high-temperature welding slag and flame. Due to the heavy weight of the steel tube, a tendon plate should be placed between the outriggers of the temperature control tube and the FRP at different contact positions (left, right, and bottom of the cavern) to protect the FRP from being crushed. Since the temperature-controlled tube was subjected to an external pressure of 10 MPa, if there were a defect in the welding seam, the water in the pipe would be pressed out to form a high-pressure water hammer flow, which was quite dangerous. Therefore, all welds were subjected to penetration non-destructive testing to ensure that the test was 100% qualified. If the weld was defective, repair welding must be carried out.

(a)

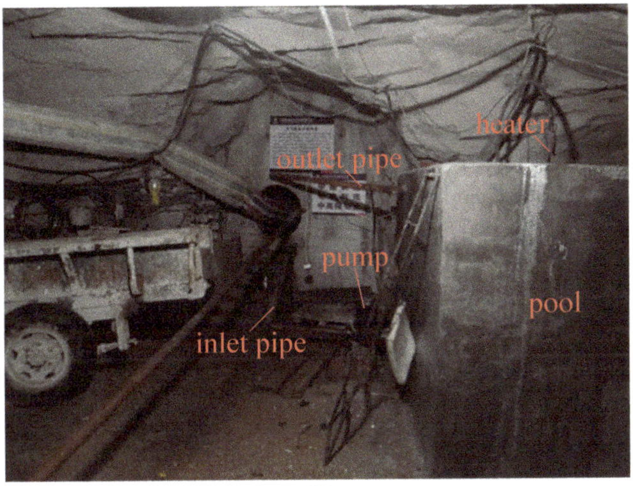

(b)

Figure 10. In situ heat exchange system: (**a**) inside the cavern; (**b**) outside the cavern.

A cube temperature-controlled pool with a side length of 2 m was built outside the CAES cavern. The pool was composed of an upper water outlet pipe, a lower water inlet pipe, a heater, and a vertical water pump. The high-temperature water heated by the heater would be circulated into the compressed air chamber to increase the air temperature to avoid freezing in the cavern. The heating tube of the heater was 1.9 m long, only the lower part of 1.7 m could generate heat, and it would be under the water during operation. The pool's water level was stable at 1.75 m~1.85 m.

5.2. Heat Exchange Test Results

In order to test the operational performance of the heat exchange system, it was designed to turn on the temperature control when the temperature reaches above 40 °C in the charging process. The heating water cycle would be turned on when the temperature drops below 10 °C, and the heater would be turned off, but the water cycle system would be maintained during the discharging process.

The barometer was welded to the charging and discharging pipes next to the test cavern to collect the air pressure. Two sets of thermocouples were used to measure the temperature at 10 positions at different depths (0 m, 1 m, 2 m, 3 m, 4 m, 5 m) in the wall and the middle of the cavern. The hygrometer sensor probe was mounted on the steel door of the plenum inside the hole.

Figure 11 shows the results of the temperature control test during the two CAES conditions. In the initial pressurization stage (when the pressure is less than 1 MPa), regardless of the speed of the pressurization rate, the temperature reaches above 40 °C, and the humidity decreases rapidly within a short time (40 min).

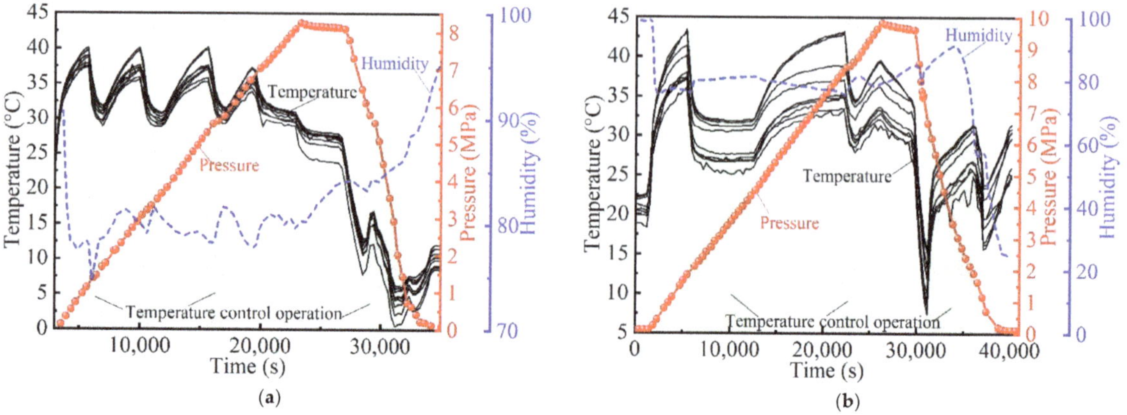

Figure 11. Experimental results of heat exchange system: (**a**) Maximum charging pressure of 8 MPa; (**b**) Maximum charging pressure of 10 MPa.

When the temperature reached above 40 °C, the temperature control system was turned on according to the test plan. There was no significant change in the air pressure and the pressurization rate, but the temperature dropped rapidly, and the humidity increased slightly. During the pressurization stage, the opening temperature of the heat exchange system decreases, and the closing temperature of the heat exchange system increases, showing a reasonable regularity. The air pressure decreased very little during the pressure stabilization stage. Moreover, the temperature decrease was mainly caused by the heat transfer of FRP to the outside. The humidity also increased at this time. Temperature and humidity were inversely related. When the discharging process began, the temperature decreased rapidly, and the humidity increased. After the temperature control system was turned on, the temperature drop could be controlled and even have a recovery trend. However, there was no change in the trend of pressure changes. There was still an inverse relationship between temperature and humidity. There would be noise during the deflation process, with the maximum noise being 92 and 87 decibels, respectively, and water vapor could be seen in the first hour of the initial discharging.

6. Conclusions

In this study, a simulation of the temperature variation law of the underground CAES cavern in the whole cycle of charging–high pressure air storage–discharging was carried out based on thermodynamics and numerical heat transfer methods. Moreover, a pilot cavern was excavated to conduct a thermodynamic process and heat exchange system for practice. Based on the obtained results, the following conclusions can be drawn:

(1) According to the conservation of energy, the thermodynamic equations for the solid and air region were compiled into the fluent two-dimensional axisymmetric model through a user-defined function (UDF), and the values were updated after the calculation of each time step to realizing the calculation of the temperature field in the entire cavern area. The average air pressure rises steadily during the charging process, and the air temperature rises rapidly from room temperature to about 50 °C in the very first moment. At the end of the charging, the average air temperature reached 62.5 °C. At the end of the discharging stage, the average air temperature drops to −40.4 °C, showing a wild fluctuation.

(2) A two-way parallel finned tube heat exchange system of the Pingjiang CAES cavern was designed to provide circulating hot/cool water. The tube temperature decreases significantly with the increase of pressure. Two designing schemes were calculated to ensure the efficient operation of the heat exchange system. A stable temperature was maintained, and air temperatures below zero degrees could be avoided. The air

temperature was maintained between 25 °C and 38 °C when the pressure was greater than 1.6 MPa. The heat exchange system can sufficiently suppress the temperature fluctuation during the charging and discharging process.

(3) According to the experimental testing results of the CAES cavern heat exchange system, it was designed to turn on the temperature control when the temperature reaches above 40 °C in the charging process and below 10 °C in the discharging process. There was no significant change in the air pressure and the pressurization rate, but the temperature dropped rapidly, which means that the heat exchange system can control the temperature within a small range (20–40 °C) without affecting the air charging efficiency. Temperature and humidity were inversely related, and water vapor could be seen in the first hour of the initial discharging. There would be noise during the deflation process, and the maximum noise was 92 and 87 decibels, respectively.

Author Contributions: Conceptualization, P.L. and X.Z.; methodology, P.L.; investigation, P.L., Z.C., H.C. and Z.W.; writing—original draft preparation, P.L., X.Z. and Z.W.; writing—review and editing, X.Z. and Z.W. All authors have read and agreed to the published version of the manuscript.

Funding: This research was funded by the Huxiang Young Talent Project, grant number 2020RC3090, the R&D project of Zhongnan Engineering Corporation Limited of Power China, grant number SQ-KJYF-2021-0137-03 and grant number YFHT-1944, the R&D project of Department of Water Resources of Hunan Province, grant number 2-2021-04426.

Institutional Review Board Statement: Not applicable.

Informed Consent Statement: Not applicable.

Data Availability Statement: Not applicable.

Conflicts of Interest: The authors declare no conflict of interest.

References

1. Mozayeni, H.; Wang, X.; Negnevitsky, M. Dynamic analysis of a low-temperature Adiabatic Compressed Air Energy Storage system. *J. Clean. Prod.* **2020**, *276*, 124323. [CrossRef]
2. Klaas, A.K.; Beck, H.P. A MILP Model for Revenue Optimization of a Compressed Air Energy Storage Plant with Electrolysis. *Energies* **2021**, *14*, 6803. [CrossRef]
3. Jafarizadeh, H.; Soltani, M.; Nathwani, J. Assessment of the Huntorf compressed air energy storage plant performance under enhanced modifications. *Energy Convers. Manag.* **2020**, *209*, 112662. [CrossRef]
4. Zhang, N.; Yang, C.; Shi, X.; Wang, T.; Yin, H.; Daemen, J.J.K. Analysis of mechanical and permeability properties of mudstone interlayers around a strategic petroleum reserve cavern in bedded rock salt. *Int. J. Rock Mech. Min.* **2018**, *112*, 1–10. [CrossRef]
5. Chen, H.; Li, H.; Ma, W.; Xu, Y.; Wang, Z.; Chen, M. Research progress of energy storage technology in China in 2021. *Energy Storage Sci. Technol.* **2022**, *11*, 1052–1076.
6. Yang, C.; Wang, T.; Qu, D.; Ma, H.; Daemen, J.J.K. Feasibility analysis of using horizontal caverns for underground gas storage: A case study of Yunying salt district. *J. Nat. Gas Sci. Eng.* **2016**, *36*, 252–266. [CrossRef]
7. Fan, J.; Wei, L.; Jiang, D.; Chen, J.; Tiedeu, W.N.; Chen, J.; Deaman, J.J.K. Thermodynamic and applicability analysis of a hybrid CAES system using abandoned coal mine in China. *Energy* **2018**, *157*, 31–44. [CrossRef]
8. Li, Y.; Li, Y.; Liu, Y. Numerical study on the impacts of layered heterogeneity on the underground process in compressed air energy storage in aquifers. *J. Energy Storage* **2022**, *46*, 103837. [CrossRef]
9. Ma, Y.; Rao, Q.; Huang, D.; Yi, W.; He, Y. A new theoretical model of local air-leakage seepage field for the compressed air energy storage lined cavern. *J. Energy Storage* **2022**, *49*, 104160. [CrossRef]
10. Kushnir, R.; Dayan, A.; Ullmann, A. Temperature and pressure variations within compressed air energy storage caverns. *Int. J. Heat Mass Transf.* **2012**, *55*, 5616–5630. [CrossRef]
11. Kushnir, R.; Ullmann, A.; Dayan, A. Thermodynamic models for the temperature and pressure variations within adiabatic caverns of compressed air energy storage plants. *ASME J. Energy Res. Technol.* **2012**, *134*, 14005659. [CrossRef]
12. Zhou, S.W.; Xia, C.C.; Du, S.G.; Zhang, P.Y.; Yu, Z. An analytical solution for mechanical responses induced by temperature and air pressure in alined rock cavern for underground compressed air energy storage. *Rock Mech. Rock Eng.* **2015**, *48*, 749–770. [CrossRef]
13. Kim, H.M.; Rutqvist, J.; Ryu, D.W.; Choi, B.H.; Sunwoo, C.; Song, W.K. Exploring the concept of compressed air energy storage (CAES) in lined rock caverns at shallow depth: A modeling study of air tightness and energy balance. *Appl. Energy* **2012**, *92*, 653–667. [CrossRef]

14. Zhou, S.W.; Xia, C.C.; Zhao, H.B.; Mei, S.H.; Zhou, Y. Numerical simulation for the coupled thermo-mechanical performance of a lined rock cavern for underground compressed air energy storage. *J. Geophys. Eng.* **2017**, *14*, 1382–1398. [CrossRef]
15. Chen, X.; Wang, J. Stability analysis for compressed air energy storage cavern with initial excavation damage zone in an abandoned mining tunnel. *J. Energy Storage* **2022**, *45*, 103725. [CrossRef]
16. Wan, F.; Jiang, Z.; Tang, D. The influence of CAES reservoir design parameters on thermodynamic properties. *Energy Storage Sci. Technol.* **2021**, *10*, 370–378.
17. Zhou, Q.; Du, D.; Lu, C.; He, Q.; Liu, W. A review of thermal energy storage in compressed air energy storage system. *Energy* **2019**, *188*, 115993. [CrossRef]
18. Patil, V.C.; Acharya, P.; Ro, P.I. Experimental investigation of heat transfer in liquid piston compressor. *Appl. Therm. Eng.* **2019**, *146*, 169–179. [CrossRef]
19. Li, C.; Wang, H.; He, X.; Zhang, Y. Experimental and thermodynamic investigation on isothermal performance of large-scaled liquid piston. *Energy* **2022**, *249*, 123731. [CrossRef]
20. Neu, T.; Subrenat, A. Experimental investigation of internal air flow during slow piston compression into isothermal compressed air energy storage. *J. Energy Storage* **2021**, *38*, 102532. [CrossRef]
21. Gouda, E.M.; Benaouicha, M.; Neu, T.; Fan, Y.; Luo, L. Flow and heat transfer characteristics of air compression in a liquid piston for compressed air energy storage. *Energy* **2022**, *254*, 124305. [CrossRef]
22. Jiang, Z.; Li, P.; Tang, D.; Zhao, H.; Li, Y. Experimental and Numerical Investigations of Small-Scale Lined Rock Cavern at Shallow Depth for Compressed Air Energy Storage. *Rock Mech. Rock Eng.* **2020**, *53*, 2671–2683. [CrossRef]
23. Ishihata, T. Underground compressed air storage facility for CAES-G/T power plant utilizing an airtight lining. *Int. Soc. Rock Mech.* **1997**, *5*, 17–21.
24. Geissbühler, L.; Becattini, V.; Zanganeh, G.; Zavattoni, S.; Barbato, M.; Haselbacher, A.; Steinfeld, A. Pilot-scale demonstration of advanced adiabatic compressed air energy storage, part 1: Plant description and tests with sensible thermal-energy storage. *J. Energy Storage* **2018**, *17*, 129–139. [CrossRef]
25. Heath, M.; Woodfield, P.L.; Hall, W.; Monde, M. An experimental investigation of convection heat transfer during filling of a composite-fibre pressure vessel at low Reynolds number. *Exp. Therm. Fluid Sci.* **2014**, *54*, 151–157. [CrossRef]
26. Bourgeois, T.; Ammouri, F.; Baraldi, D.; Moretto, P. The temperature evolution in compressed gas filling processes: A review. *Int. J. Hydrogen Energy* **2018**, *43*, 2268–2292. [CrossRef]
27. Sánchez-Escalona, A.A.; Camaraza-Medina, Y.; Retirado-Mediaceja, Y.; Góngora-Leyva, E. Application of the 'Nusselt-Equation Simulated Evolution Method' in Forced Convective Heat Transfer Modeling. *Int. J. Des. Nat. Ecodyn.* **2021**, *16*, 21–32. [CrossRef]
28. Pitts, D.R. *Schaum's Outline of Theory and Problems of Heat Transfer*, 2nd ed.; McGraw-Hill Book Company: New York, NY, USA; Toronto, Canada; London, UK, 1990.

MDPI
St. Alban-Anlage 66
4052 Basel
Switzerland
www.mdpi.com

MDPI Books Editorial Office
E-mail: books@mdpi.com
www.mdpi.com/books

www.ingramcontent.com/pod-product-compliance
Lightning Source LLC
Chambersburg PA
CBHW061445220926
51850CB00004B/20
* 9 7 8 3 0 3 6 5 9 5 5 2 8 *